11/15/82
D0145840

The Physical Universe

A SERIES OF BOOKS IN ASTRONOMY

Editors

Donald E. Osterbrock

Joseph S. Miller

THE PHYSICAL UNIVERSE

An Introduction to Astronomy

FRANK H. SHU

Professor of Astronomy
University of California, Berkeley

University Science Books
Mill Valley, California

University Science Books
20 Edgehill Road
Mill Valley, CA 94941

Library of Congress Catalog Card Number: 81-51271

ISBN 0-935702-05-9

Printed in the United States of America
10 9 8 7 6 5 4 3 2

To my Parents

Preface

The book market is currently flooded with introductory astronomy texts; do we really need another one? My personal answer is, obviously, yes, and I would like to share my reasons for writing yet another elementary astronomy textbook.

This text grew from a set of lecture notes (twenty lectures of 90 minutes each) which I had gradually accumulated and refined during some years of teaching one-semester and one-quarter courses in introductory astronomy at Stonybrook and Berkeley. The book is also suitable for a full-year course. I was encouraged by colleagues and students to turn my notes into a book, because they offered features not found in other textbooks. Foremost, I believe, I have tried not only to describe and catalogue known facts, but also to organize and explain them on the basis of a few fundamental principles. For example, students are told what red giants and white dwarfs are, *and* why stars become red giants or white dwarfs. Similarly, the concepts of "black hole" and "spacetime curvature" are explicitly explained in terms of the *geometric* interpretation of modern gravitation theory. Most importantly, I have tried to emphasize the deep connections between the microscopic world of elementary particles, atoms, and molecules, and the macroscopic world of humans, stars, galaxies, and the universe. In this way, I hope not to shortchange the student into thinking that astronomy represents a large mass of disjoint facts about a bewildering variety of exotic objects. I strongly believe that "gee-whiz astronomy" offers only cheap thrills, whereas the real beauty of astronomy—nay, of science—lies in what Sciama has called *The Unity of the Universe*. If my book succeeds in conveying even a slight appreciation of this unity and of why scientists are drawn to do science, it will have fulfilled my major purpose.

I also had a technical reason for writing this book. I feel that too many introductory texts give an unbalanced treatment of modern astronomy. We happen to know the most details about the solar system, but the solar system does not offer the only important lessons, either scientific or philosphical, to the general reader. This book tries to give a more equal division between basic principles, stars, galaxies and cosmology, and the solar system and life. Moreover, I have not tried to write a book for the lowest common denominator. This book will challenge the minds of even the best undergraduate students. To do less, I feel, would be a failure to carry out my responsibilities as an educator. Nevertheless, it cannot be ignored that elementary astronomy classes attract a spectrum of people with widely divergent backgrounds. Astronomy *does* offer cultural value for the curious business administration major, the life-science major, the future congressman, the bright freshman, the jaded senior physics major, the premed student, and the continuing-education student. I have tried to write a book, therefore, which can be used at two levels: with and without assignment of the Problems. Nonscience majors can read this book for its descriptive discussions, skipping the Problems. Science majors may find it helpful to do the Problems to obtain a *quantitative* feel for the topics covered in the text. To work out the Problems requires only a good background in high-school mathematics and physics (i.e., little or no calculus). The reader who skips the Problems, however, will not be spared from *conceptual* reasoning. Such conceptual reasoning constitutes the backbone of science and much of everyday problem-solving, and to be spared from it is no long-term favor.

Astronomy is, of course, one of the most rapidly developing of the sciences, and many incompletely investigated topics are discussed in this book. I make no apologies for including speculative viewpoints; I only hope that I have clearly labeled speculation as such in writing this book. The frontiers of knowledge always contain treacherous ground, and I would prefer to explore this ground boldly with the reader *and* to be ultimately wrong, rather than be cautiously right at the expense of omitting provocative new ideas which have much scientific promise. There is no onus in being wrong in science as long as the conjectures or conclusions are not based on sloppy reasoning. Some of the most productive paths have come from wrong turns into uncharted territory. There is a famous dictum on this point: what is science is not certain; what is certain is not science. The reader will lose nothing in being told a few things which turn out ultimately to be wrong.

A few comments about units: cgs units are used for all calculations except scaling arguments. The final results may be reexpressed in astronomical units—e.g., solar masses—to obtain more manageable numbers. The use of magnitudes and parsecs is restricted to the Problems (especially those in Chapter 9). In my opinion, the magnitude scale for apparent and absolute brightness hinders, rather than helps, the teaching of astronomy to nonscience majors. For the general beginner, this arbitrary convention is neither needed nor wanted; for the specialized student, working out a few problems provides the best way to become acquainted with the concept. The parsec has the virtue of stressing an observational technique for measuring interstellar distances. However, the light-year is an equally convenient unit of length, and it has the advantage of emphasizing spacetime relationships. Most people find it easier to grasp large differences in time than large differences in length; so they can immediately appreciate that stars are much much smaller than the mean separations between stars if they are told that stars are typically a few light-seconds across, but are typically separated by a few light-years.

Then, there is the thorny problem of proper attribution in a semipopular work like this. Astronomy is a very broad subject, and I make no pretense to know the history of the development of each of the branches. Nevertheless, I consider it worthwhile and only fair to attribute the particularly important contributions in science to their proper discoverers. No one would dream of writing an introduction to English literature without mentioning that the author of *Hamlet* was Shakespeare. In writing this book, I have tried to follow this rule of thumb: if one or a few names stand out as having made the seminal investigations, I have attributed the work to them as founders. Still, my perspective of such history may often be inaccurate; I would appreciate it if knowledgable readers would call errors to my attention.

Many colleagues and friends read at least parts of my manuscript or used it in their courses, and offered constructive criticisms. I wish especially to thank Dick Bond, Judy Cohen, Martin Cohen, John Gaustad, Peter Goldreich, Don Goldsmith, Tom Jones, Joe Miller, and Don Osterbrock. Any errors which remain undoubtedly were not present in the early versions that they saw. Too many people to mention individually kindly gave me permission to reproduce photographs and other materials in their possession. I also greatly appreciate the patience and efforts of my publisher, Bruce Armbruster, my manuscript editor, Aidan Kelly, and my production editor, Dick Palmer, who showed me why true professionals in the art of bookmaking must have a passion for their work. Last, but not least, I wish to acknowledge the help and sustenance of Helen Shu, who not only taught me all the biology I know, but also put up with all the travails of being an author's spouse.

Frank H. Shu

Berkeley, California
December 1, 1981

Note to the Instructor

The Physical Universe was written to be used at several different levels. For an introduction to astronomy for liberal-arts majors, I suggest skipping over the problems set apart by rules and most of the equations displayed in the text. The qualitative meaning of the equations is always spelled out in the surrounding text. Using the book in this way will present the reader with a continuous narrative of the great ideas and discoveries that have helped to shape the modern scientific perspective on humanity's place in the universe. The survey emphasizes explanations, not only descriptions, and it is largely self-contained, in that the book develops internally the important concepts needed for a basic understanding of the fundamental connections between different fields of the scientific endeavor. Students with a good high-school background in science and mathematics should have no difficulty with the equations that appear in the text, and they could benefit from doing an appropriate selection of the problems. The problems occur where they help to amplify points made in the text; they vary greatly in difficulty, both mathematically (from arithmetic to calculus) and conceptually (from single-step verifications to open-ended thought questions requiring a literature search). Some instructors might therefore elect to use *The Physical Universe* for a junior- or senior-level introduction to astrophysics.

Contents

PART I

Basic Principles

The astronomical clock on the south side of the Town Hall in Prague, Czechoslovakia, was constructed in the fifteenth century and depicts the motion of the Sun and Moon about the Earth. Notice the twelve signs of the Zodiac and that the day is divided into twenty-four hours. It was the attempt to understand the motions of celestial bodies that ultimately led to the birth of modern science. (*Observatory and Planetarium Prague, courtesy of A. Rükl.*)

CHAPTER ONE

The Birth of Science

To most people astronomy means stars; stars mean constellations; and constellations mean astrology. In fact, each step of this association contains misconceptions. Modern astronomers deal with more than just stars; they think of stars in terms of more than just constellations; and they use the constellations differently from astrologers. Nevertheless, astronomy and astrology do have the same historical roots, in the geometric patterns formed by the stars in the night sky. Let us begin, therefore, our journey of exploration of the universe with the constellations.

The Constellations as Navigational Aids

In prehistorical times nomadic peoples found that knowing the constellations helped them with directions. The most familiar constellation in the Northern Hemisphere is, of course, the Big Dipper. The Big Dipper is actually part of an astronomical constellation called *Ursa Major*, the Big Bear. Figure 1.1 shows how to use the "pointer stars" of the Big Dipper to find the pole star, Polaris. Polaris indicates the direction north, and it will continue to do so for another thousand years.

It has been speculated that early nomads developed stories to help them remember the various constellations and their relative positions in the sky. These stories pass as entertaining myths today, but in earlier times were

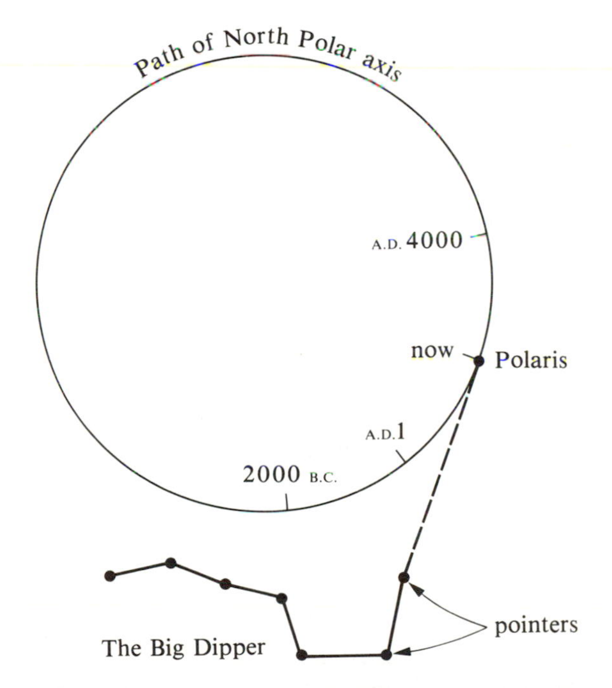

Figure 1.1. The Big Dipper and Polaris. To find the pole star, Polaris, go along the pointers to a distance roughly equal to four times their separation. Polaris lies very nearly in the present direction of the axis of the North Pole of the Earth. However, because of the tidal forces exerted on the Earth by the Sun and the Moon, the spin axis of the Earth makes a slow precession. Every 26,000 years, the polar axis describes a circular path in the sky. Thus, in 2000 B.C. or A.D. 1 the North Pole did not point so nearly at Polaris as now, and neither will it one or two thousand years from now.

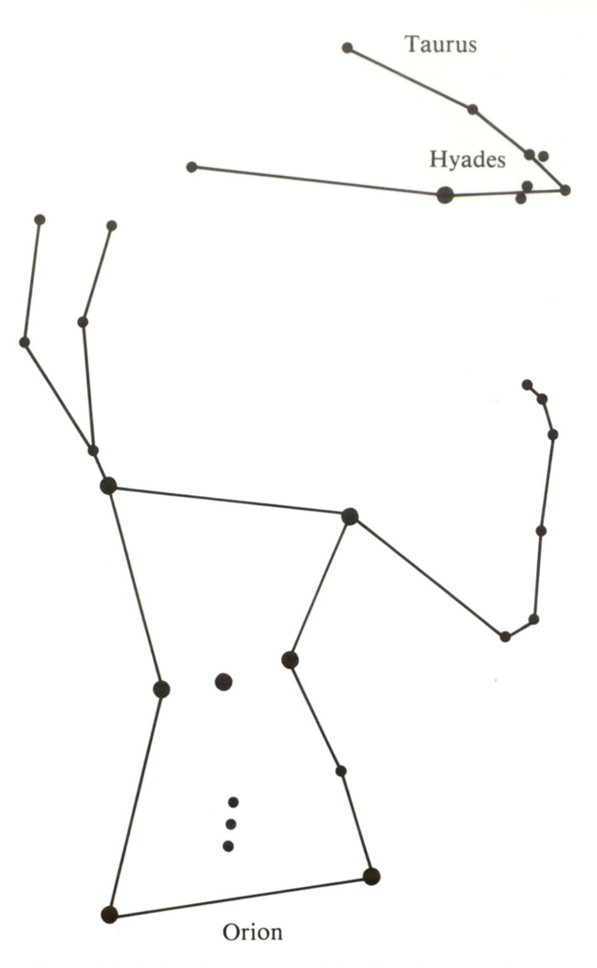

Figure 1.2. Orion, Taurus, and the Hyades star cluster.

probably very important mnemonic devices. Consider, for example, the biblical account of how Samson slew a thousand Philistines with the jawbone of an ass. Later he drinks water out of a hollow place called Lehi. A silly story, you say?

Now, *Lehi* in Hebrew means jawbone, and water in Greek is *hyades*. If we look up in the winter night sky, we find the group of stars known as the Hyades located in the jawbone of Taurus the Bull. Nearby is the mighty warrior Orion (Figure 1.2). Thus, the historian of science Giorgio de Santillana gives the following interpretation for the biblical account: the hero of the original story must be Orion, and that rigmarole about the jawbone of an ass and drinking water out of the hollow of Lehi is simply a mnemonic device for finding the relative positions of the constellations Orion and Taurus and the group of stars called the Hyades.

In other cultures we find the same story but different characters. Thus, the Polynesians—who were excellent navigators—have a story about Maui the Creator using Orion as a net to snare the sunbird. Having caught the sunbird, he proceeds to beat it up with the jawbone of his grandmother!

The Constellations as Timekeeping Aids

When nomads turned to farming, the constellations became equally useful for telling time, especially for keeping track of the seasons. The diurnal rotation of the whole sky forms the unit of time that we know as the day. People noticed that, apart from the daily rotation, the stars appear not to move with respect to one another; consequently, the stars were considered "fixed." Today we understand that the diurnal rotation of the entire sky results simply from the spin of the Earth, and that the relative positions of the stars are fixed simply because of their immense distances from us and from each other. During very long periods of time, say, tens of thousands of years, stars do move detectably with respect to one another, even according to naked-eye observations. Thus, the constellations will not always have their present forms. This fact was not known, of course, to peoples whose recorded history only spanned a few thousand years.

During shorter periods of time, say, a year, people did notice that certain objects did not keep the same positions relative to the "fixed" stars. Rather, they wandered to and fro in a narrow band of the sky, named the Zodiac (Figure 1.3). *Planetes* the Greeks called these wanderers; we now say "planets." Foremost among the "planets" was the Sun; and in ancient times, six other wanderers were known: the Moon, Mercury, Venus, Mars, Jupiter, and Saturn. Today, of course, we no longer think of the Sun and the Moon as planets; instead, we think of the Earth as one of the wanderers, and the Moon as its companion. Today, we explain the to-and-fro wanderings about the Zodiac in terms of the planets' orbits (including that of the Earth) about the Sun, which are confined more or less to a single plane, the "ecliptic" (Figure 1.4). However, for describing the "how" of the planetary motions and not the "why," the ancient description is just as good; in fact it is better for the purposes of an observer on Earth who is trying to visualize the celestial events.

The early Greeks were sophisticated geometers who knew the size, shape, and rotation of the Earth. They possessed a reasonable calculation of the size and distance to the Moon. They also proposed a theoretical method which could have yielded, in principle, the distance of the Sun. Unfortunately, their observational measurements were not precise enough for the last task; so the distance they calculated was a serious underestimate. Despite this underestimate, they still deduced

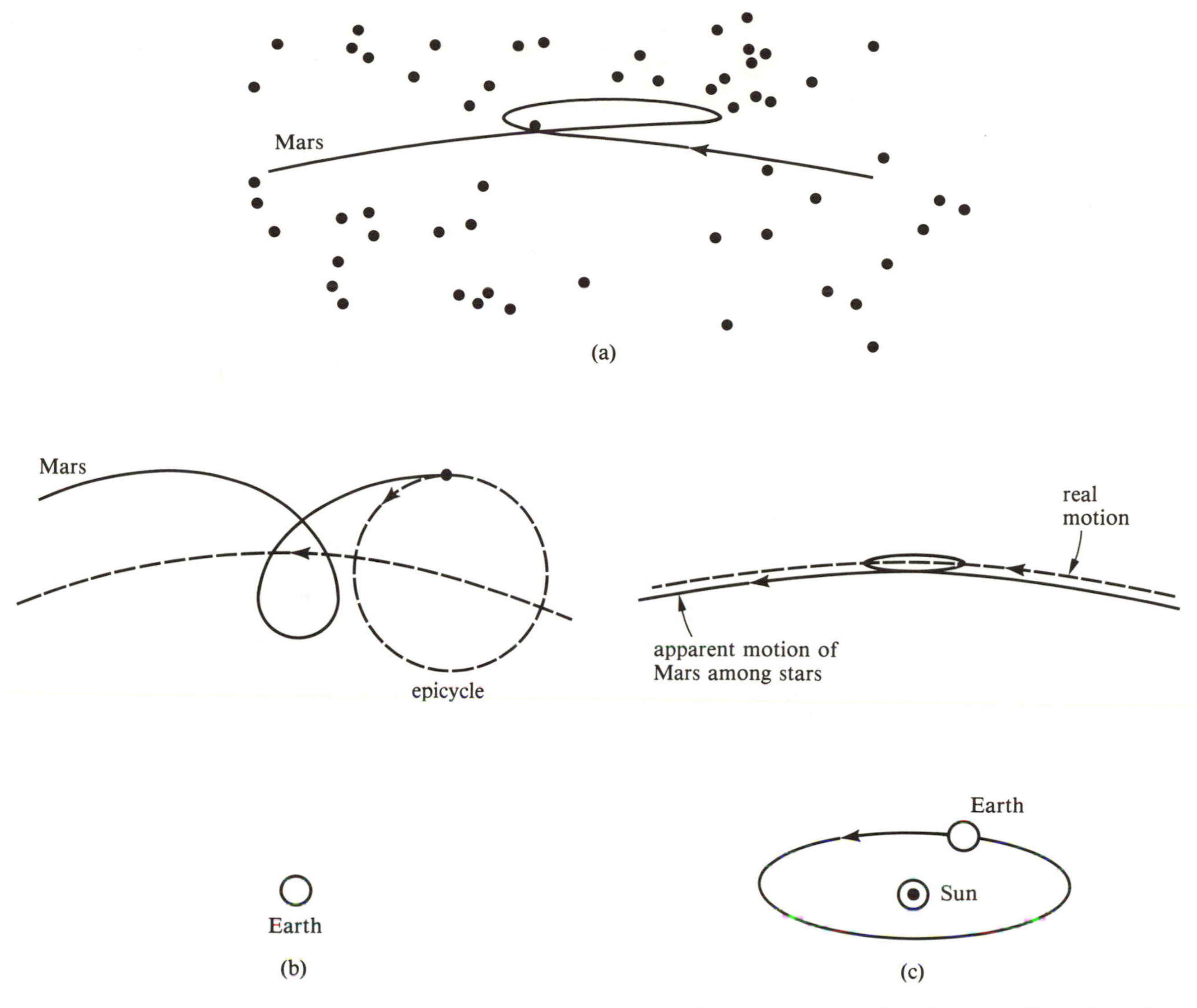

Figure 1.3. The apparent motion of the planet Mars. In an Earth-centered system. Ptolemy required a complicated theory of epicycles to explain these motions. In a Sun-centered system, Copernicus was able to explain the same motions much more simply: namely, at certain points in the motion of the Earth about the Sun, the Earth would seemingly catch up with a planet's projected orbit, and then that planet would appear to go backward.

that the Sun is substantially larger than the Earth. Because of the apparent dominance of the Sun, some of the Greeks speculated correctly that the Earth revolved about the Sun, rather than the other way around. These ideals fell into disfavor when Aristotle argued, on seemingly common-sense grounds, that we could not inhabit a moving and rotating Earth without being more aware of it. For example, Aristotle argued that motion of the Earth should cause foreground stars to be displaced annually with respect to the background (an effect now called **parallax**), whereas no parallax could be detected for any star by the Greeks. We now understand this null result to arise from the very great distances to even the nearest stars (see Problem 3.2).

Problem 1.1. This problem and the next one retrace the early arguments of the Greeks about the sizes of the Earth, Moon, and Sun, and the distances between them.

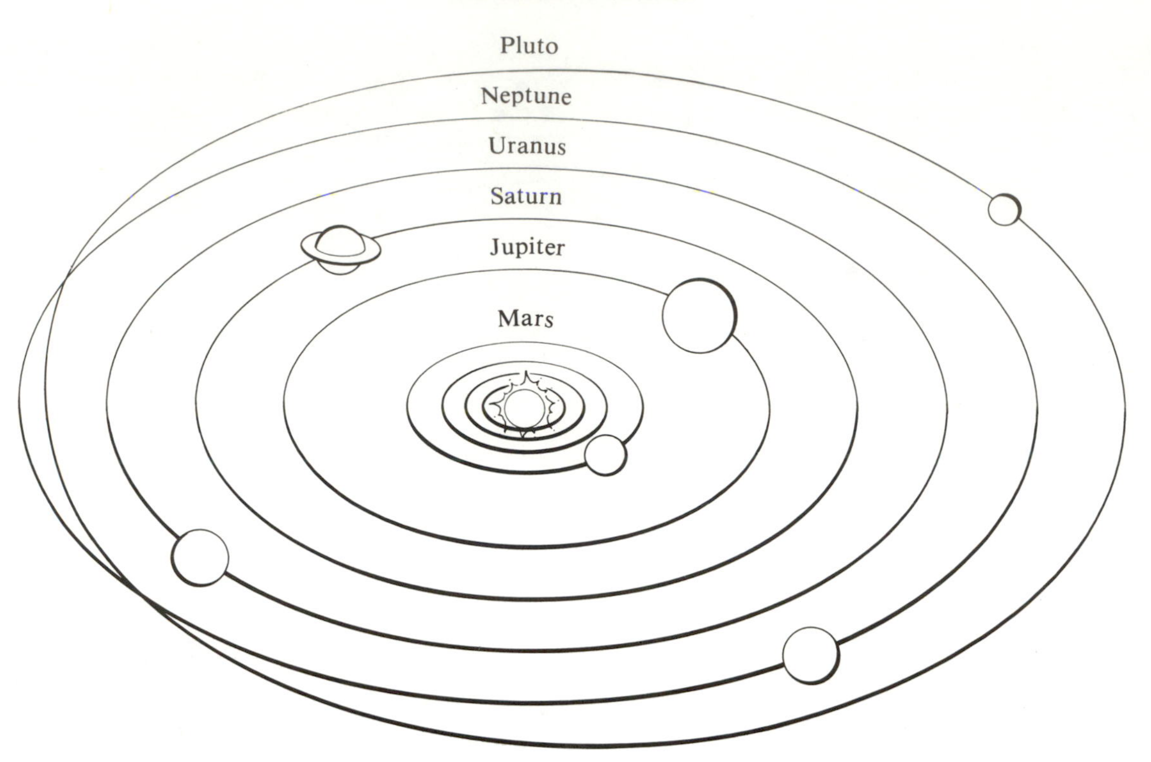

Figure 1.4. The planetary orbits about the central Sun. The plane of the Earth's orbit defines the ecliptic. Except for Mercury and Pluto, all the planetary orbits are nearly circular and lie within several degrees of the plane of the ecliptic. For sake of clarity, the orbits of the three innermost planets, Mercury, Venus, and Earth, are not labeled.

We begin with the size of the Earth. From the shape of the shadow cast by the Earth on the Moon during a lunar eclipse, it can be inferred that the Earth is a sphere. Moreover, the method described in Problem 1.2 shows that the Sun's distance is many times greater than the diameter of the Earth. Erastothenes assumed that the Sun was far enough away that the rays from the Sun are virtually parallel when they strike the Earth (see figure

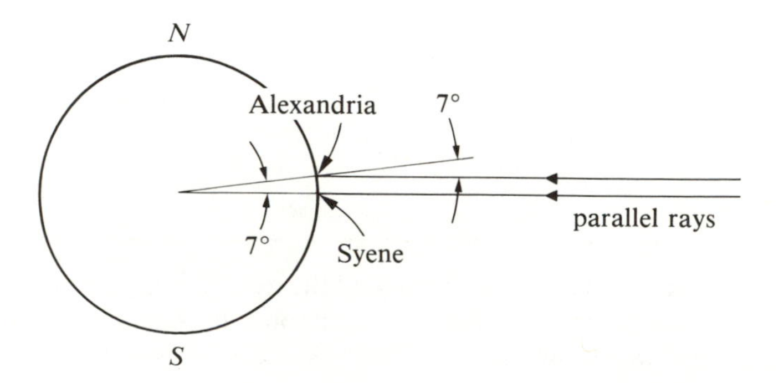

here). Erastothenes then observed that, at noon on the first day of summer, sunlight struck the bottom of a deep well in Syene, Egypt. In other words, the Sun at that time was directly overhead. At the same time in Alexandria, however, the Sun's rays made an angle of about 7° to the vertical. Erastothenes concluded that Syene and Alexandria must be separated by a fraction $7°/360° \cong 1/50$ of a great circle around the Earth. Assume that this distance of 1/50 of the circumference of the Earth can be paced off to be 800 km. Consult Appendix B for the relation between the radius and circumference of a circle, and calculate the radius of the Earth. Compare your answer with the value given by Appendix A.

Reconsider now the observation of the shadow cast by the Earth during a lunar eclipse. Assume still that the rays from the Sun make parallel lines, and draw diagrams to show how a comparison of the curvature of the Earth's shadow on the Moon with the curvature of the Moon's edge allows one to infer the relative sizes of the Earth and Moon. This deduction is a slight modification of the method used by Aristarchos to find that the Moon's diameter is about a third of that of the Earth. The modern value is 0.27. With Erastothenes's value for the radius of the Earth, calculate the diameter D_M of the Moon. Given that the Moon subtends an angular diameter at Earth of about half a degree of arc, calculate the distance r_M of the Moon. *Hint*: If θ_M is the angular diameter of the Moon expressed in radians, and if $\theta_M \ll 1$, show that the formula $\theta_M = D_M/r_M$ holds to good approximation.

Problem 1.2. Aristarchos suggested an ingenious method for measuring the relative distances of the Moon and the Sun. Because the angular sizes of the Moon and the Sun do not change appreciably with time, it can be deduced that they maintain nearly constant distances from the Earth. (The orbits are circular.) From the figure here show how to deduce the ratio of the Moon's distance r_M to the Sun's distance r_S as

$$r_M/r_S = \cos\theta,$$

where 2θ is the total angle subtended at the Earth by the Moon's positions between first and third quarters of the Moon's phases. Unfortunately, the angle θ turns out to be too close to 90° to be practical as a way to tell that the value of $\cos\theta$ is not zero (i.e., that the Sun is not infinitely far way compared to the Moon). Modern measurements using radar reflections show that $r_M/r_S = 2.6 \times 10^{-3}$. Thus, the Sun is about 390 times further away than the Moon. On the other hand, solar-eclipse observations demonstrate that the Sun and the Moon have about the same angular sizes. Argue that this implies the Sun is about 390 times larger than the Moon. Given that the Moon is only 0.27 the size of the Earth, show that the Sun is more than a hundred times larger than the Earth. Thus, unless the mean density of the Sun is much less than that of the Earth, the Sun is also likely to be much more massive than the Earth; and it then becomes more plausible to suppose that such a regal body is the true center of the solar system rather than the Earth. If the Earth can revolve about the Sun, then there can be no great philosophical objection to its spinning about its axis to account for the apparent diurnal rotation of the sky.

Given the value of the size D_M and the distance r_M of the Moon calculated in Problem 1.1, compute now the diameter D_S and distance r_S of the Sun. Convert your answer to light-seconds, and discuss how long it typically takes to bounce radar waves back and forth between objects in the solar system (e.g., between Earth and Venus). Discuss how geometry might be used to deduce the radii of the orbits of Venus and Earth about the Sun. This is the best method to obtain r_S.

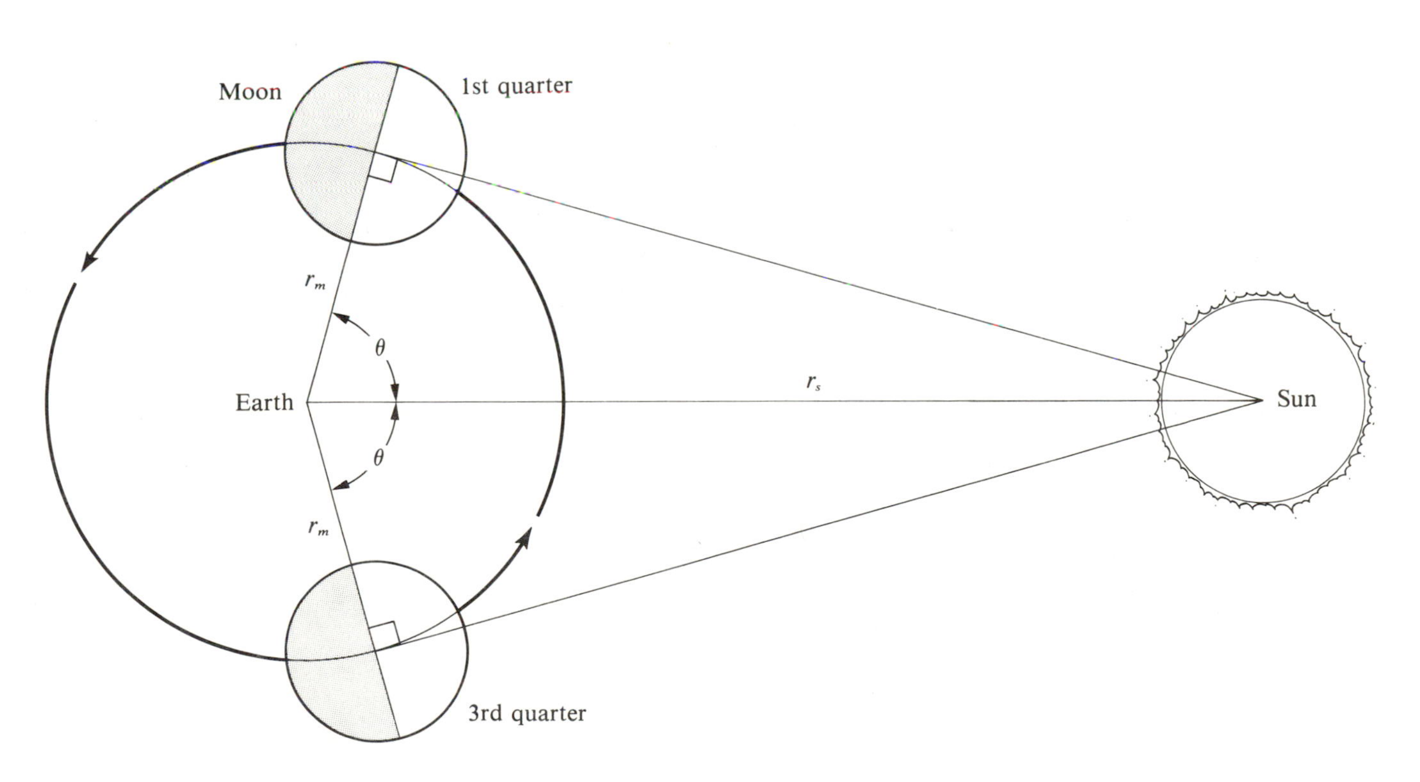

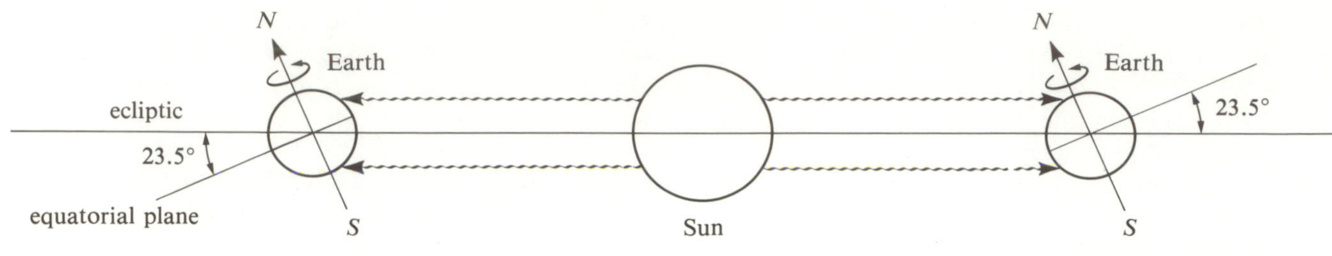

Figure 1.5. The reason for the seasons. Because the equatorial plane of the Earth is inclined by 23.5° with respect to the plane of the ecliptic, the Sun's rays strike the ground more perpendicularly at one point of the Earth's orbit than at the opposite point half a year later. At any one time, however, if it is winter in the Northern Hemisphere, it is summer in the Southern Hemisphere. This geometric result is the same whether we think of the Earth moving around the Sun or the Sun around the Earth. (Note: radii in this drawing are not to scale.)

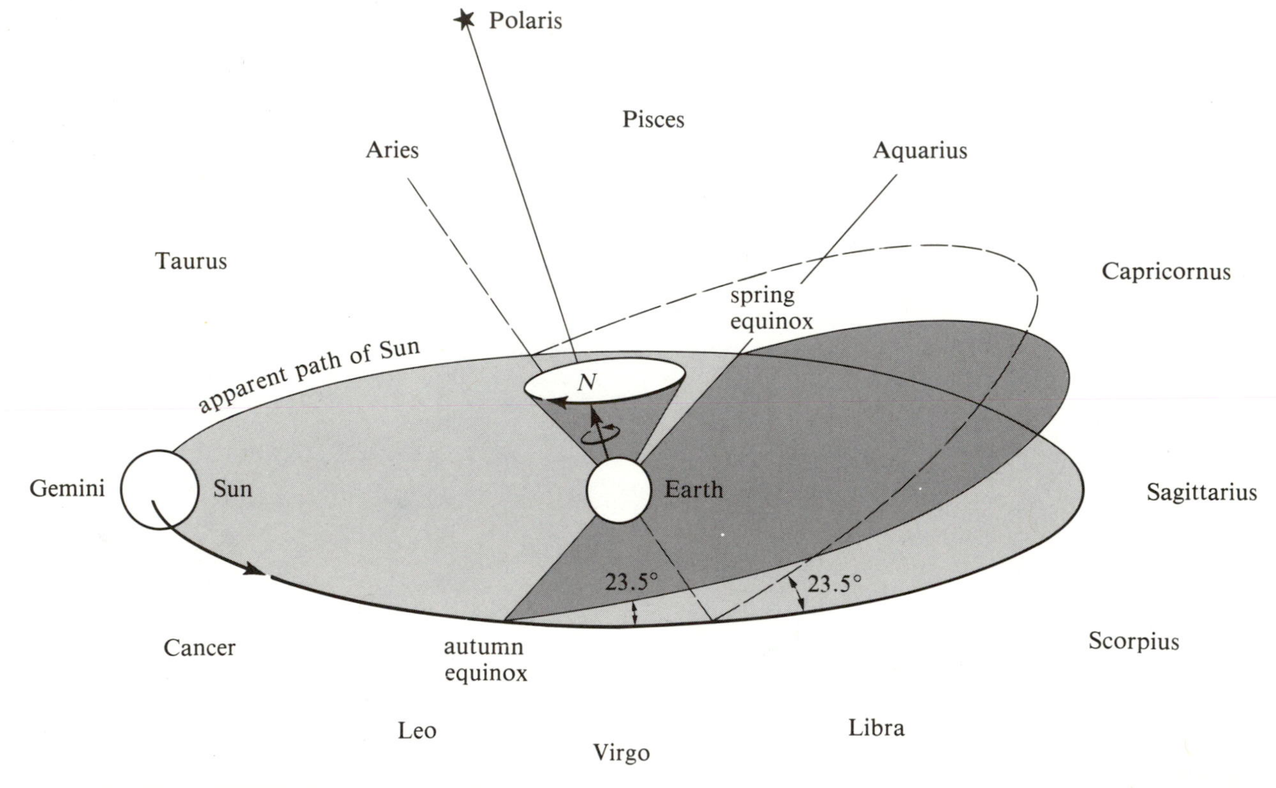

Figure 1.6. The seasons and the signs of the Zodiac. An Earth-bound observer seems to see the Sun move around the Earth once a year in the plane of the ecliptic (lightly shaded oval). At the present epoch, the North Pole of the Earth points toward the star Polaris, and the extension of the equatorial plane of the Earth is shown as the dark semi-oval. The two points of intersection of the Sun's apparent path with the equatorial plane are called the spring and autumn equinoxes. The equatorial plane in 2000 B.C. corresponded to the dashed semi-oval.

The seasons arise because the spin axis of the Earth is tilted with respect to the plane of the ecliptic, which is the apparent path followed by the Sun during the course of one year. Consequently, the Sun's rays at noon fall more vertically during the summer months than during the winter months (Figure 1.5). Spring and fall come when the Sun is at the point of its apparent path where the circle of the ecliptic crosses the circle of the Earth's equator. These intersection points are called the spring and autumn equinoxes (Figure 1.6).

When the Sun is not in the way, stars can be seen in the background all along the ecliptic. There are twelve prominent constellations near the plane of the ecliptic, and these correspond to the twelve signs of the Zodiac. In 2000 B.C., when the Babylonians set up the system of timekeeping, the spring equinox lay in the direction of the constellation of Aries. That is, spring came on March 21, when the Sun entered the "house of Aries," which signaled the beginning of the planting season. However, tidal torques cause a slow precession of the Earth's axis of rotation; so the spring equinox moves backward through the signs of the Zodiac, at about one sign per two thousand years (Figure 1.6). Thus, 2,000 years after the Babylonians had found that the spring equinox lay in Aries, the spring equinox had moved into Pisces. This event coincided approximately with the birth of Christ, and may be why one early symbol of Christianity was the fish. Two thousand years later, the spring equinox is beginning to move into Aquarius (officially in A.D. 2600). This is why our age is sometimes called the coming of the "Age of Aquarius."

The Rise of Astrology

No one knows the exact reasons for the rise of astrology, but it may have been something like the following. To the ancients, it was obvious that the Sun, and to a lesser extent, the Moon, influences events on Earth: witness night and day, the seasons, tides, etc. The Sun and the Moon were like gods. Why not, then the other planets? From very early on, seven wanderers were known: Sun, Moon, Mercury, Venus, Mars, Jupiter, and Saturn. To honor the planetary gods, the Babylonian priests devised the seven-day week and gave the days the names of the planetary gods. In English, the roots of Sunday, Monday, and Saturday can easily be traced to Sun, Moon, and Saturn. In French, the roots of Mercredi, Vendredi, Mardi, and Jeudi can equally easily be traced to Mercury, Venus, Mars, and Jupiter.

Horoscopes were later based on the hypothesis that the positions of the planets in the Zodiac could influence the course of human events just as the positions of the Sun and the Moon affect the seasons and the tides. Especially important was the position of the Sun at the time of one's birth; thus, if one was born in 2000 B.C. between March 21 and April 19, the Sun was in Aries; between April 20 and May 20, the Sun was in Taurus, etc. Even today, one is given a Zodiac sign on the basis of the Babylonian system. The problem is, of course, that today these signs are 4,000 years out of date! For example, the coming of spring, March 21, no longer occurs when the Sun arrives at Aries but at Aquarius. If astrologers were up to date, they should tell people who are Aries that they are really Aquarius, and people who are Aquarius that they are really Sagittarius, etc.

The Rise of Astronomy

In the beginning there was little difference between astronomy and astrology. Indeed, some famous astronomers of the past earned their keep by casting horoscopes for kings and queens. A major divergence of astronomy and astrology came with the work of Nicholas Copernicus in the sixteenth century. Copernicus discovered that he could explain the seemingly complicated motions of the planets in terms of a stationary Sun about which revolved Mercury, Venus, Earth, Mars, Jupiter, and Saturn. The Moon would still have to go around the Earth, but the looping motions of the other planets (which is what observers actually see) became simple to explain if the Earth itself moved about the Sun. (as Figure 1.3 shows, when the Earth's orbital motion carries it past one of the outer, slower planets, that appears to go backwards!) Copernicus's ideas forced a major change in the prevailing philosophy of a human-centered universe, and the Copernican Revolution is justly considered one of the major turning points of science.

Free thinkers like Galileo and Kepler were quick to adopt the Copernican system, and they provided many of the early pieces of astronomical evidence in support of it. Later Galileo was forced by the Church to recant his espousal of a moving Earth, an episode made familiar by Bertolt Brecht's play on the subject. Popular myth has Galileo secretly defiant, muttering the famous phrase "And yet it moves." In any case, the Copernican viewpoint received its ultimate triumph in the seventeenth century, with the work of Isaac Newton. Newton showed that the laws of planetary motion described by Kepler in a Sun-centered system (see Chapter 18) could be derived mathematically from Newton's formulation of the laws of mechanics and gravitation.

Newton also demonstrated how his theory of gravitation could explain the tides raised on Earth by the Sun or Moon. The basic idea was that the Sun or Moon pulled hardest on the side of the oceans facing toward it. less hard on the center of the Earth, and least on the side

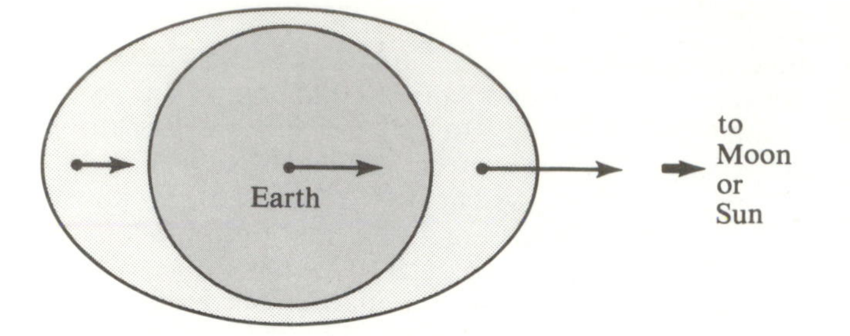

Figure 1.7. The tides. Tides in the oceans arise because the gravity of the Moon or the Sun is greatest on the side facing it and least on the side opposite it. The difference between these two forces is responsible for producing the characteristic two-sided bulge.

of the oceans facing away from it (Figure 1.7). In this way, the oceans would tend to bulge out in *two* directions: on one side because the water is pulled away from the Earth, on the other side because the Earth is pulled away from the water. The difference in force between the two sides of the Earth is called the tidal force. (We shall discuss such forces in other contexts in this book.)

In this manner did astronomy and astrology part ways. With the growing maturity of modern science, astronomy attempted to give *mechanistic* explanations for natural phenomena—seasons, tides, planetary motions—on the basis of laws formulated and tested in laboratories. Astrology continued to attribute *mystical* influences on terrestrial affairs to the planets. Mystical beliefs have a long cultural history, and they die hard. Kepler and Newton were mystics in many ways, and much of the modern world still clings to atavistic concepts.

Modern Astronomy

Today we know that Mercury, Venus, Earth, Mars, Jupiter, and Saturn are only six of the nine planets that go around the Sun; the other three are, of course, Uranus, Neptune, and Pluto. However, as far as we can tell, planets and their satellites are not the major constituents of the universe. Stars are. (Perhaps.) The Sun is a star—the closest one to Earth—and the Sun contains about 99.9 percent of the mass of our solar system. However, the Sun is only one star of myriads that belong to our Galaxy.* Our Galaxy contains more than 10^{11} (one hundred billion) stars! (For a review of the exponent notation, see Box 1.1.) And our Galaxy is only one of myriads in the observable universe. The observable universe contains about 10^{10} (ten billion) galaxies. Thus, in the observable universe there are about 10^{10} galaxies × 10^{11} stars/galaxy = 10^{21} stars: an astronomical number which has no simple English equivalent.

Since numbers like 10^{21} are enormous, almost beyond intuitive comprehension, we cannot hope to study each individual astronomical object. Even if we could see them with our telescopes—and in fact we cannot see more than a small fraction of them—we would not gain much insight from an object-by-object study. A far more practical and informative goal is to look for patterns of behavior among similar groups of objects. This book is organized in terms of the few theoretical concepts which underlie the structure and evolution of the astronomical objects under study. First, we will emphasize the deep connections between how matter is organized on large scales (**macroscopic** behavior) and on small scales (**microscopic** behavior). Second, we will find that there are two main recurring threads in the organizational fabric of macroscopic objects like stars, galaxies, and the universe. These two threads are the **law of universal gravitation** and the **second law of thermodynamics** (Chapters 3 and 4).

Another implication follows from the enormous number, 10^{21}, of stars in the observable universe. Since our life-sustaining Sun is only one star among an enormous multitude, it seems extremely unlikely that we could be either alone in the universe or the most intelligent species in the universe. This realization motivates the discussion on the chances for extraterrestial life and intelligence that we shall pursue in the last part of this book. Again we shall find the second law of thermodynamics to play an integral role in our discussions.

* When Galaxy is capitalized, it refers to our own galaxy.

BOX 1.1
Exponent Notation

$$10^n = \underbrace{1000\cdots000.}_{n \text{ zeroes}}$$

$$10^{-n} = 1/10^n = .\underbrace{000\cdots0001}_{n-1 \text{ zeroes}}$$

multiplication: $10^n \times 10^m = 10^{n+m}$. Also $10^n/10^m = 10^n \times 10^{-m} = 10^{n-m}$.

Important prefixes in exponent notation:
- nano = 10^{-9} (one-billionth)
- micro = 10^{-6} (one-millionth)
- milli = 10^{-3} (one-thousandth)
- centi = 10^{-2} (one-hundredth)
- kilo = 10^3 (one thousand)
- mega = 10^6 (one million)
- giga = 10^9 (one billion)

Example: 1 kilometer = 10^3 meter = $10^3 \times 10^2$ centimeter = 10^5 cm.

Rough Scales of the Astronomical Universe

Given that gravitation and the second law of thermodynamics underlie much of natural phenomena, you will appreciate that an honest understanding of modern astronomy requires a healthy dose of physics. The next three chapters of this book provide the physics needed to understand, on a *qualitative* level, the explanations of the important phenomena. The first step, however, toward a unified appreciation of astronomy is to obtain a physical feel for the scale of the phenomena that we shall be dealing with in this book.

The most fundamental measurements that we can make of an object are its size, its mass, and its age (or duration between one event and another). The units of length, mass, and time that we shall adopt are centimeter (abbreviated cm), gram (abbreviated gm or g), and second (abbreviated sec or s); these are called **cgs units** (Box 1.2). Although it is not immediately obvious, it is nevertheless true that all physical quantities can be expressed as various combinations of powers of cm, gm, and sec. A familiar example is the cgs unit of energy: erg = gm cm^2 sec^{-2}. A less-familiar example is the cgs unit of electric charge: esu = gm$^{1/2}$ cm$^{3/2}$ sec^{-1}. At first sight, temperature seems to be an exception. However, in any fundamental discussion, the temperature (degrees Kelvin or K) always enters in the combination "Boltzmann's constant × temperature," which has the units of energy. Some of the more important physical constants in cgs units are given in Appendix A.

Let us now consider the scales of some astronomical objects. Listed in Table 1.1 are the very rough ("**order of magnitude**," i.e., to the nearest power of ten) sizes, masses, and ages of the Sun, the Galaxy, and the observable universe. For comparison, the same quantities are listed for a small child.

BOX 1.2
The cgs Units

unit of length = centimeter = cm (1 inch = 2.54 cm)
unit of mass = gram = gm or g (1 pound = 454 gm)
unit of time = second = sec or s
(1 year = 3.16×10^7 sec)
unit of force = dyne = gm cm sec^{-2}
(weight of 1 pound on Earth = 4.45×10^5 dyne)
unit of energy = erg = gm cm^2 sec^{-2}
(potential energy of 1 pound at height of 1 foot on Earth = 1.36×10^7 erg)
unit of power or luminosity = erg sec^{-1}
(100 watt = 10^9 erg/sec)

Table 1.1. **Rough scales.**

Object	*Size*	*Mass*	*Age*[a]
child	10^2 cm	10^4 gm	10^8 sec
Sun	10^{11} cm	10^{33} gm	10^{17} sec
Galaxy	10^{23} cm	10^{45} gm	$\gtrsim 10^{17}$ sec
observable universe	10^{28} cm	10^{55} gm	$\gtrsim 10^{17}$ sec

[a] Notice that although the Sun, the Galaxy, and the universe differ appreciably in size and mass, they all have ages comparable in order of magnitude.

The numbers quoted in the last three rows of Table 1.1 are the hard-won efforts of several generations of astronomers and physicists. They result from some truly heroic struggles, and it is a shame that we cannot devote more space in this book to the *history* of astronomy and physics. The principal goal of this book is to develop some appreciation of how numbers like those in Table 1.1 are derived. An important component of such appreciation is understanding the philosophical implications of the results, which can be easy to overlook if we concentrate too much on the specific numerical values.

For example, Table 1.1 gives the age of the universe as somewhat greater than 10^{17} sec. To be more precise, astronomers now believe that the time since creation is between 10 and 20 billion years.* To debate, as experts do, whether the precise age since creation is closer to 10 billion years or to 20 billion years is philosophically and culturally less important than to grasp the following points.

(1) By any human standards, the universe is very old. Nevertheless, the present universe is not infinitely old. There *was* a beginning to time, and that beginning took place on the order of 10^{10} years ago.

(2) By some great stroke of luck, you and I live in an era when it has become possible to measure something as fundamentally interesting as the age of the universe to an accuracy of a factor of two or so. This fortune is well worth some leisurely reflection.

There is another comment we should make about Table 1.1. In quoting a mass and a size, Table 1.1 carefully notes that these estimates apply to the *observable* universe. This distinction between the *universe* and the *observable universe* is important because astronomers have not yet settled the question of whether the universe is *open* or *closed* (Chapter 15). If the universe is open, spacetime (and the number of stars in the universe) is infinite. In other words, although time had a beginning, it has no end, and the amount of space and the number of stars in it are actually infinite. (However, only a finite part of either are observable, even in principle, by us at

* One year = approximately 3×10^7 sec; see Appendix A for more accurate conversion formulae for common astronomical constants.

any finite time after the creation event.) Have you ever contemplated what the truly infinite means? The cosmologist G. F. R. Ellis has, and he makes the following interesting observation. If the number of stars (and planets) is infinite, then all physically possible historical events are realizable somewhere and sometime in the universe. Thus, there may be somewhere and sometime another Earth with a history exactly like our own, except that the South wins the Civil War instead of the North. Elsewhere and othertime, there may even be a reasonable fascimile of Tolkien's Middle Earth! Now, it is highly improbable that such alternative worlds are within our "event horizon"; so we may never be able to communicate with them. Nevertheless, it is fun to speculate on the possibility of such worlds given the truly infinite.

The alternative possibility—that the universe is closed, so that spacetime and the number of stars in it are finite—is almost equally bizarre according to normal conceptions. This possibility leads to the conclusion that the volume of space is finite, yet it does not possess an edge or boundary!

Contents of the Universe

The primary inhabitants of the universe are stars, of which the Sun is only one example (*Color Plate 1*). Stars exhibit a great variety of properties. Besides suns of all colors, there are the tremendously distended **red giants** and the mysteriously tiny **white dwarfs**. The Sun itself is quite an average star, neither very massive nor very light, neither very large nor very small. Why, then, does the Sun appear so bright, but the stars so faint? Because the stars are much further away than the Sun. The Sun is "only" eight light-minutes away, but the nearest star, proxima Centauri, is about four light-years distant. A light-year is the distance traveled by light in one year, and equals 9.47×10^{17} cm. To appreciate how enormous a distance is a few light-years, notice that human travel to the Moon, "only" one light-*second* away, represents the supreme technological achievement of civilization on Earth. Will travel to the stars ever be within our grasp?

The rich variety of stars live in our Galaxy mostly singly or in pairs (Figure 1.8). Sometimes, however, hundreds or thousands of stars can be found in loose groups called **open clusters** (*Color Plate 3*). The Pleiades is a fairly young open cluster, and the **reflection nebulosity** which surrounds the famous "seven sisters" of this cluster attests to the fact that these stars must only recently have been born out of the surrounding gas and dust (*Color Plate 4*). The oldest stars in our Galaxy are found in tighter groups called **globular clusters** (*Color Plate 5*). Rich globular clusters may contain more than a million members.

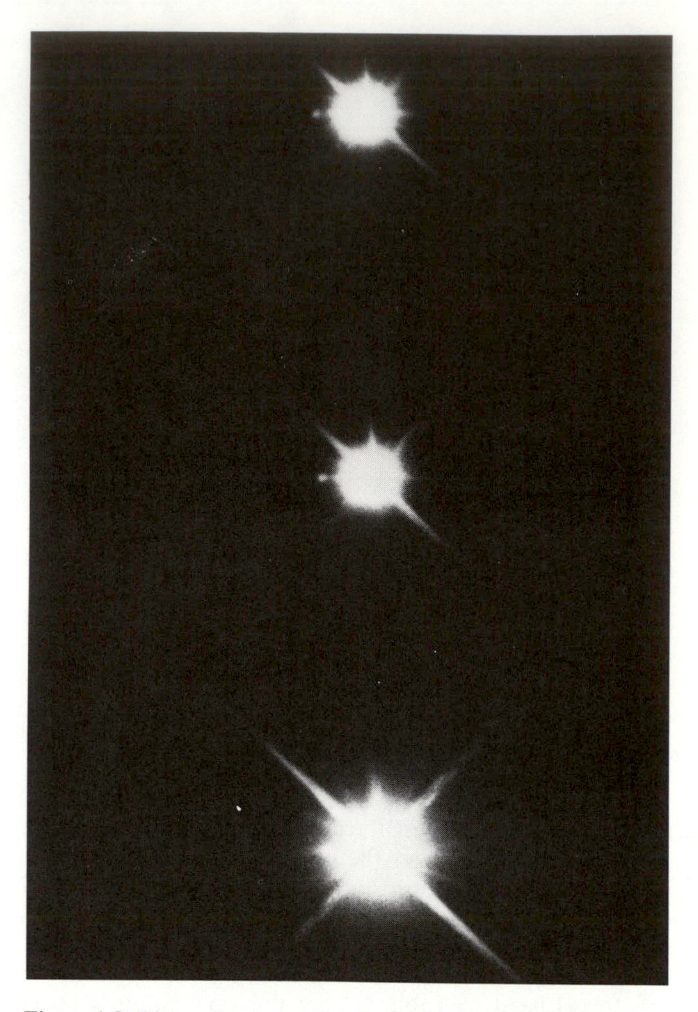

Figure 1.8. These three views of the bright star Sirius show it to be a member of a binary system. The companion of Sirius is a white dwarf. (Lick Observatory photograph.)

The space between the stars is not completely empty. The diffuse matter between the stars is called the **interstellar medium**, but by terrestrial standards it is virtually a perfect vacuum. Clouds of **gas** and **dust**, as well as energetic **cosmic-ray particles** gyrating wildly in **magnetic fields**, reside in the space between stars. Giant fluorescent gas clouds (called **HII regions**) lit up by nearby hot young stars, like the Orion nebula found in the "sword" of Orion (*Color Plate 6*), constitute some of the most beautiful objects in astronomy. Other objects like the Crab nebula (*Color Plate 7*) shine with an eerie light (**synchrotron radiation**) and are now known to be the **remnants** expelled by stars which died in titantic **supernova** explosions. The cores left behind in such cataclysms, **neutron stars** and **black holes**, represent some of the most intriguing denizens of the astronomical kingdom. Other stars die less violently, ejecting the outer shell as a **planetary nebula** (*Color Plate 8*) which

Figure 1.9. A cluster of galaxies in Hercules. (Palomar Observatory, California Institute of Technology.)

exposes the hot central core destinied apparently to become a white dwarf.

Stars and the material between them are almost always found in gigantic stellar systems called **galaxies**. Our own galaxy, the Milky Way System, happens to be one of the two largest systems in the **Local Group** of two dozen or so galaxies. The other is the Andromeda galaxy (*Color Plate 9*); it stretches more than *one hundred thousand light-years* from one end to the other, and it is located about *two million* light-years distant from us. If the kingdom of the stars is vast, the realm of the galaxies is truly gigantic.

Apart from small groups like our own Local Group, galaxies are also found in great **clusters**, containing thousands of members (*Figure 1.9*). Most cluster galaxies are roundish **ellipticals** rather than the more common **spirals** found in the general field (*Color Plates 10–13*). Rich clusters have been found at distances exceeding *three billion* light-years from us. Obviously, the observable universe is a tremendously large place! And modern cosmology tells us that it is expanding.

Clearly, the universe contains many interesting objects, not the least of which are the Earth and its inhabitants (*Color Plate 14*). Of all the objects intensively studied in the solar system (*Color Plates 15–21*), the Earth is the only one known to harbor life. How did we get here? How do we fit into the unfolding drama of the universe? These issues and others are the natural legacy of Copernicus's first inquiries, and they form the subject matter of this book. To survey the entire universe and to develop the important themes fully, our pace must necessarily be fast. Let us start!

CHAPTER TWO

Classical Mechanics, Light, and Astronomical Telescopes

Modern physical science began with classical mechanics and astronomy. The earliest practitioners, such as Galileo and Newton, made important contributions both as physicists and as astronomers. Hence, classical mechanics, the nature of light, and astronomical telescopes make fitting starting points for our study.

Classical Mechanics

In classical mechanics the most fundamental property of matter is **mass**. The fundamental frame of reference for describing the dynamics of matter is an **inertial frame**, any frame which is at rest or, at most, moves at a constant velocity relative to the "fixed" stars. Since the stars are not truly "fixed," we will later need to reconsider this definition; but for now we implicitly assume that henceforth all physical laws are to be stated for such inertial frames.

Galileo discovered the first law of mechanics, and Descartes gave it a general formulation. (According to Stillman Drake, Galileo's criticisms of the Aristotlean viewpoint led him to shy away from stating general principles.) The first law, or the **principle of inertia**, describes the innate resistance of matter to a change in its state of motion:

> A body in motion tends to remain in motion. Unless the body is acted upon by external forces, its **momentum**—the product of its mass and **velocity**—remains constant.

When external forces are present and act on material bodies, the momenta of the bodies no longer remain constant. To describe this situation, Newton formulated the second law of mechanics:

> The time-rate of change of a body's momentum is equal to the applied **force**.

Usually the mass of a body does not change; so the time-rate of change of its momentum is simply its mass times its time-rate of change of velocity. The latter is called **acceleration**, and like velocity, it has both a magnitude and a direction (i.e., it is a **vector**). If we denote the mass of a particle by m, the applied force by $\boldsymbol{F}$ (a vector), and the acceleration by $\boldsymbol{a}$ (another vector), Newton's second law takes the familiar form

$$\boldsymbol{F} = m\boldsymbol{a}, \qquad (2.1)$$

BOX 2.1
The Laws of Mechanics

First law: When $\boldsymbol{F} = 0$, $\boldsymbol{p} = m\boldsymbol{v} =$ constant.

Second law: When $\boldsymbol{F} \neq 0$, $\dfrac{d\boldsymbol{p}}{dt} = \boldsymbol{F}$; usually, $\boldsymbol{F} = m\boldsymbol{a}$.

which is probably the most famous equation in all of physics. A concise summary of the above in mathematical language is given in Box 2.1.

The physical content of $\boldsymbol{F} = m\boldsymbol{a}$ is the following. If we know that forces are acting on a body, we may use $\boldsymbol{F} = m\boldsymbol{a}$ to calculate the acceleration $\boldsymbol{a}$ that is induced by the action of the force $\boldsymbol{F}$. In this way, we may infer the changes of the state of motion produced by the known forces. Conversely, if we observe a body to undergo acceleration, we may infer that forces must be present to produce this change in the body's state of motion.

It is important to notice that acceleration is present even if the velocity maintains a constant magnitude (i.e., speed) but changes direction. The simplest example is circular motion with radius r at a constant speed $v = |\boldsymbol{v}|$. In this important case, the magnitude of the associated acceleration is (see Problem 2.1 for the mathematical derivation)

$$a = v^2/r. \tag{2.2}$$

The direction of $\boldsymbol{a}$ is inward, toward the center of the circle.

Problem 2.1. Consider the polar coordinates (r, θ) of a particle moving in a circle at constant speed v. Let the change in angular position which occurs in time Δt be $\Delta\theta$. The speed, given by $v = r\,\Delta\theta/\Delta t$, is unchanging, but the velocity $\boldsymbol{v}$ does change by an amount $\Delta\boldsymbol{v} = \boldsymbol{v}' - \boldsymbol{v}$ in time Δt because of the change in the direction of motion. Use the diagram below and the small-angle formulae of Appendix B to derive the relation, $a = |\Delta v/\Delta t| = v\,\Delta\theta/\Delta t = v^2/r$, with the direction of $\boldsymbol{a}$ being inward, toward the center of the circle.

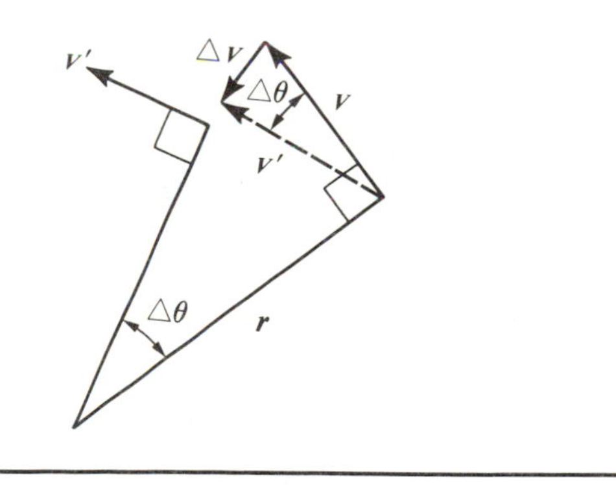

The ancients regarded circular motion as perfect and therefore to be the "natural" state of motion which required no further explanation. However, from the preceding discussion, we see that circular motion entails acceleration (the expression v^2/r is called the *centripetal acceleration*), and therefore requires the application of force to maintain it, even at constant speed. In particular, the (nearly) circular motion of the Earth about the Sun requires the application of force; in this case, as Newton explained, the **gravitational force**.

Figure 2.1. Galileo's experiment of dropping different weights from the leaning tower of Pisa. Historians of science doubt that such experiments ever took place, but there is no question that Galileo performed "thought experiments" that showed bodies of different weights, released from rest and freed from friction, should fall at the same rate. Otherwise, a stone cut in two in mid-air would begin to fall at a different rate, but how could it matter whether the cut is real or imaginary?

We shall describe in more detail in Chapter 3 Newton's theory of gravitation. Here we need only remark that Newton found that mass itself gives rise to a universal force of **attraction** between all forms of matter. Physicists today like to speak of mass as acting as a **source** for a gravitational **force field**. The concept of a force field dispenses with the need for an actual particle whose reaction to the source of gravity is to be observed or to be calculated. The field surrounding a source (any massive body) is a set of (vector) values which are attached to every location in space at each instant in time; *if* a hypothetical particle *were* to be placed at any given point at a specified instant of time, it would feel a force whose magnitude and direction can be deduced from the value of the gravitational field associated with that position and time. To be more precise, it is conventional to define the gravitational field $\boldsymbol{g}$ such that $\boldsymbol{F}$, the force felt by a particle of mass m, equals $m\boldsymbol{g}$. Since $\boldsymbol{F} = m\boldsymbol{a}$, we see that $\boldsymbol{g}$ is numerically equal to the acceleration $\boldsymbol{a}$ produced by the action of the gravitational force. Galileo discovered that, in the absence of air drag, all bodies fall toward the Earth with the same acceleration independent of their physical properties (Figure 2.1), and this experimental fact is incorporated

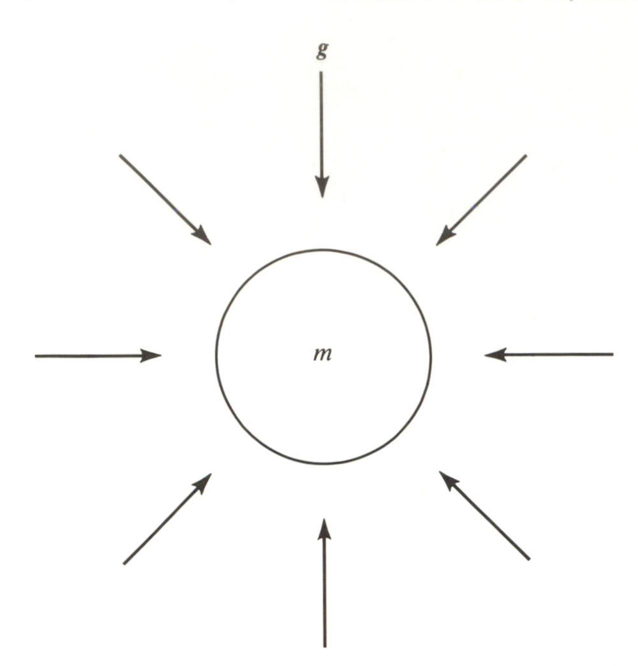

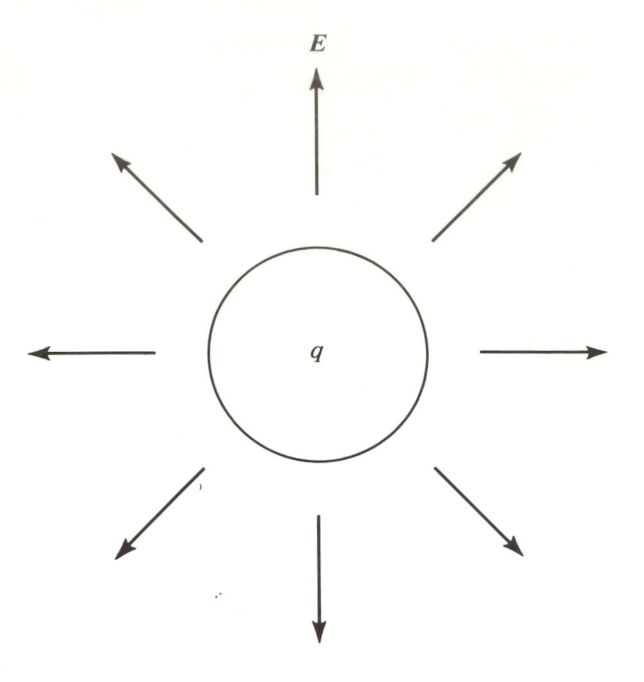

Figure 2.2. The gravitational field $\boldsymbol{g}$ associated with a spherical mass m points radially inward toward the center of attraction. The electric field $\boldsymbol{E}$ associated with a spherical positive charge q points radially outward. The electric *force* experienced by another charge is repulsive (outward) if that charge is also positive, but it is attractive (inward) if that charge is negative. The electric field $\boldsymbol{E}$ associated with a spherical negative charge would point radially inward.

in Newton's formulation of the theory of gravity (Chapter 3). This fact is also the departure point of Einstein's theory of **general relativity**, which is an improvement on Newton's theory of gravitation (Chapter 15).

The Nature of Light

It was discovered later that matter often has other attributes besides mass. For example, matter can also be charged electrically. Moreover, just as mass acts as a source for a gravitational force field, so does electric charge act as a source of an electromagnetic force field. The **electric field** associated with a stationary charge is entirely analogous to the gravitational field associated with a (stationary) mass (Figure 2.2), but Faraday discovered moving charges can also produce a **magnetic field**. Electric and magnetic fields can both accelerate charges; moreover, when charges are *accelerated* by any means whatsoever, they produce a time-varying electromagnetic phenomenon called an **electromagnetic wave**. It was Maxwell who first recognized that **light** is an electromagnetic wave. We now know that electromagnetic waves exist not only as visual light, but also as radio waves, infrared rays, ultraviolet light, X-rays, and gamma rays (Figure 2.3).

The classical picture of light as developed by Maxwell is shown in Figure 2.4. In this figure, $\boldsymbol{E}$ denotes the electric field and $\boldsymbol{B}$ denotes the magnetic field. The length of the corresponding arrows gives the magnitude of $\boldsymbol{E}$ and $\boldsymbol{B}$, and the tail-to-head pointing gives the direction. Notice that both $\boldsymbol{E}$ and $\boldsymbol{B}$ oscillate in position at each instant of time, and in time at each fixed position. This oscillation in space and time can be described as simply the same waveform in space being propagated in the direction perpendicular to both $\boldsymbol{E}$ and $\boldsymbol{B}$ (which are themselves mutually perpendicular) at a speed c, the speed of light. At each instant in time, the waveform in Figure 2.4 repeats itself after a distance λ, which is called the wavelength. Thus, the simple electromagnetic wave in Figure 2.4 has four fundamental properties:

(a) the speed of propagation, c (in a vacuum),

(b) the direction of propagation,

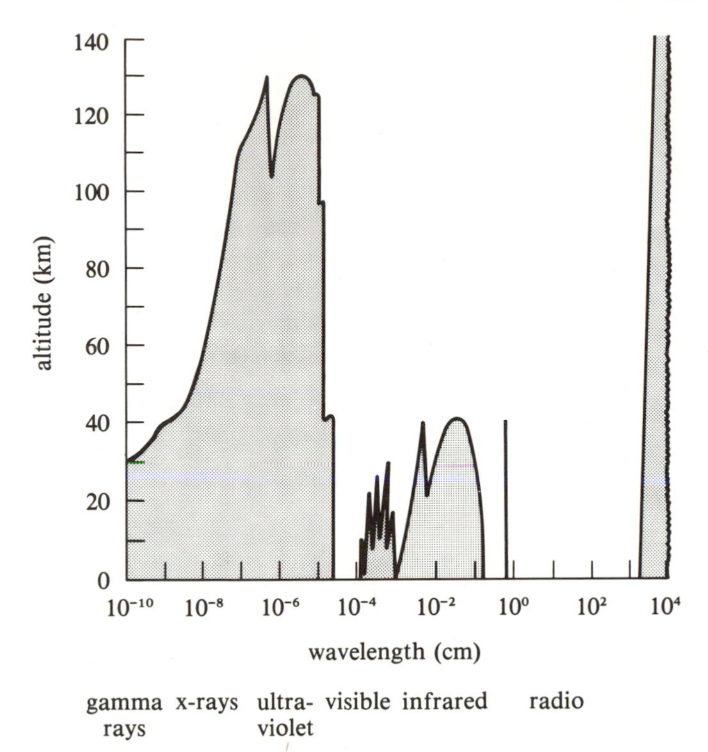

Figure 2.3. The electromagnetic spectrum and its penetration through the Earth's atmosphere. Radio waves have the longest wavelengths; gamma rays, the shortest. Only radio waves, optical light, and the shortest gamma rays have no difficulty penetrating the atmosphere to reach sea level. A few "windows" between the absorption bands of water vapor and carbon monoxide exist in the infrared, but all ultraviolet astronomy and X-ray astronomy has to be carried out above the Earth's atmosphere.

(c) the wavelength, λ,

(d) and the polarization direction (the direction that $\boldsymbol{E}$ points; notice that the direction of $\boldsymbol{B}$ is known to be perpendicular to both $\boldsymbol{E}$ and the direction of propagation).

In general, the light from a source will contain a mixture of waves with different directions of propagation, different wavelengths, and different polarizations. However, one of the great discoveries of Maxwell was that, in a vacuum, all these waves have the same speed of propagation:

$$c = 3.00 \times 10^{10} \text{ cm/sec.}$$

A true appreciation of this important result came only after Einstein's development of the theory of special relativity (Chapter 3).

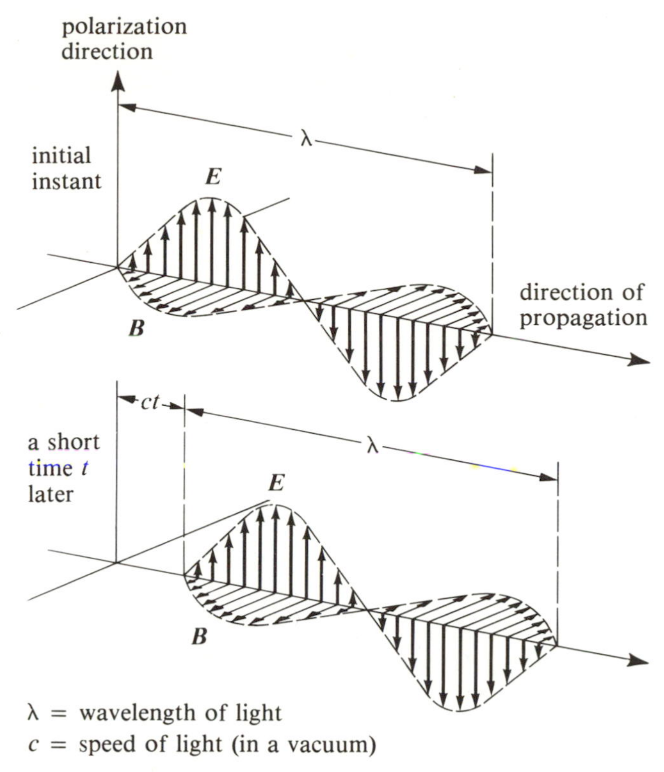

Figure 2.4. The classical picture of light.

The Energy Density and Energy Flux of a Plane Wave

There is a fifth property of light that we have left out in the above discussion, and that is, of course, its intensity. The intensity of the wave depicted in Figure 2.4 should, in some sense, become larger as the magnitudes of $\boldsymbol{E}$ and $\boldsymbol{B}$ (the lengths of the arrows $\boldsymbol{E}$ and $\boldsymbol{B}$) become larger. One possible measure of the intensity of the wave is the **energy** contained per unit volume, which in Maxwell's theory works out to be

$$\text{energy density} = \frac{1}{8\pi}(\boldsymbol{E} \cdot \boldsymbol{E} + \boldsymbol{B} \cdot \boldsymbol{B}) \qquad (2.3)$$

where the dot product $\boldsymbol{A} \cdot \boldsymbol{B}$ between two vectors $\boldsymbol{A}$ and $\boldsymbol{B}$ is calculated according to the rule depicted in Figure 2.5. The dot product of a vector with itself is often written as the square of that symbol without the boldface; thus, $\boldsymbol{A} \cdot \boldsymbol{A} = A^2$, where $A = |\boldsymbol{A}|$ is the magnitude of $\boldsymbol{A}$. Thus, equation 2.3 might also be expressed as

$$\text{energy density} = (E^2 + B^2)/8\pi.$$

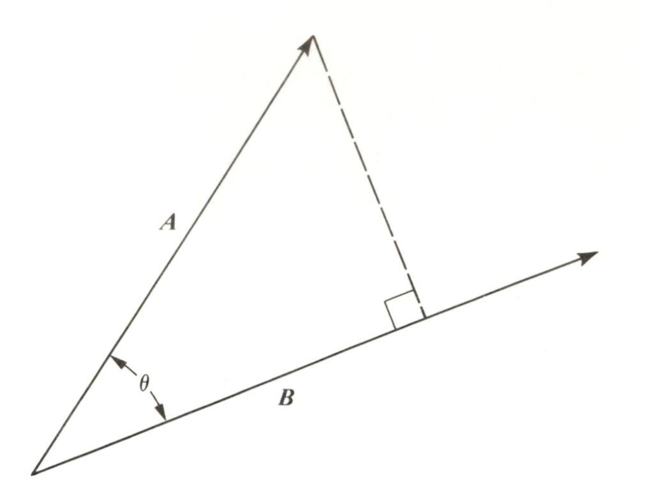

Figure 2.5. The dot product of $\boldsymbol{A}$ and $\boldsymbol{B}$ is given by the formula $\boldsymbol{A} \cdot \boldsymbol{B} = AB\cos(\theta)$, where θ is the angle between $\boldsymbol{A}$ and $\boldsymbol{B}$. Convince yourself that this has the geometric interpretation of projecting the length A onto the direction of B and multiplying the result. Show also that it does not matter whether we project A onto B (as shown above) or B onto A; i.e., that the dot product is commutative: $\boldsymbol{A} \cdot \boldsymbol{B} = \boldsymbol{B} \cdot \boldsymbol{A}$. Notice further that the result $\boldsymbol{A} \cdot \boldsymbol{B}$ may be negative if θ is an oblique angle (greater than 90°), and that $\boldsymbol{A} \cdot \boldsymbol{B} = 0$ if $\boldsymbol{A}$ and $\boldsymbol{B}$ are perpendicular to one another.

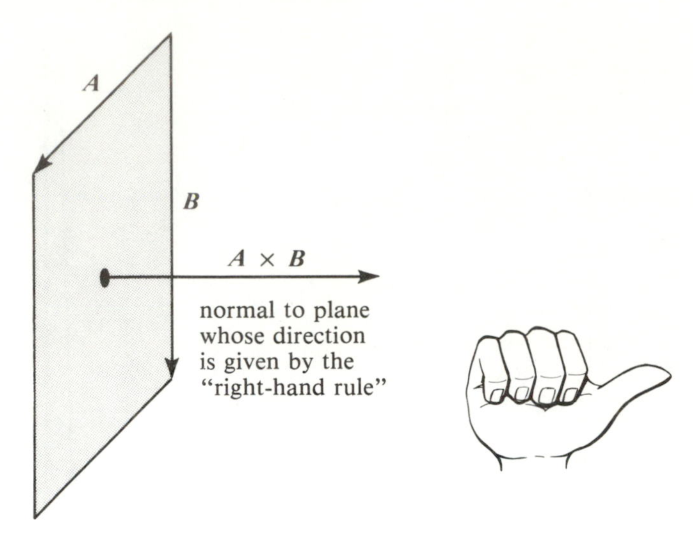

Figure 2.6. The vector cross product of $\boldsymbol{A}$ and $\boldsymbol{B}$, $\boldsymbol{A} \times \boldsymbol{B}$, has magnitude $AB\sin(\theta)$, where θ is the angle between $\boldsymbol{A}$ and $\boldsymbol{B}$ (see Figure 2.5), and direction given by the right-hand rule illustrated above. Convince yourself that the magnitude of $\boldsymbol{A} \times \boldsymbol{B}$ is given by the area formed by the parallelogram associated with $\boldsymbol{A}$ and $\boldsymbol{B}$ (shaded region). Notice that $\boldsymbol{A} \times \boldsymbol{B} = 0$ if $\boldsymbol{A}$ and $\boldsymbol{B}$ are parallel, and is at a maximum if $\boldsymbol{A}$ and $\boldsymbol{B}$ are perpendicular. Notice also that $\boldsymbol{B} \times \boldsymbol{A} = -\boldsymbol{A} \times \boldsymbol{B}$; i.e., the vector cross product does not commute. (In fact, it anticommutes.) As an example of the right-hand rule, verify that the direction of wave propagation in Figure 2.4 is in the direction of the Poynting vector given by equation 2.4.

A better measure of the intensity of the wave is the energy flux $\boldsymbol{f}$ (energy crossing in some direction per unit area per unit time). The energy flux of the wave depicted in Figure 2.4 is given by Poynting's formula:

$$\boldsymbol{f} = \frac{c}{4\pi}(\boldsymbol{E} \times \boldsymbol{B}), \tag{2.4}$$

where the vector cross product $\boldsymbol{A} \times \boldsymbol{B}$ between two vectors is calculated according to the "right-hand rule" shown in Figure 2.6. Since $\boldsymbol{E}$ and $\boldsymbol{B}$ are mutually perpendicular, the vector cross product $\boldsymbol{E} \times \boldsymbol{B}$ has magnitude EB, and the energy flux $\boldsymbol{f}$ has magnitude $f = cEB/4\pi$. Moreover, since E and B are equal for light propagating in a vacuum, we see that f is given by c times the energy density,

$$\text{energy flux} = c \times \text{energy density}, \tag{2.5}$$

an entirely reasonable result, since light always travels at the speed c in a vacuum.

Occasionally, it is convenient to express vectors in terms of their components along three mutually perpendicular axes (Figure 2.7). In the component notation where $\boldsymbol{A} = (A_x, A_y, A_z)$, the dot product of two vectors can be calculated as $\boldsymbol{A} \cdot \boldsymbol{B} = A_xB_x + A_yB_y + A_zB_z$, and the cross product of two vectors can be calculated as $\boldsymbol{A} \times \boldsymbol{B} = (A_yB_z - A_zB_y, A_zB_x - A_xB_z, A_xB_y - A_yB_x)$.

The Response of Electric Charges to Light

For most astronomical objects, the only information that we can obtain comes from the light they emit. Thus, astronomers have become experts in the detection and analysis of the light emitted at all wavelengths (radio, infrared, optical, ultraviolet, X-rays, and gamma rays) by very faint objects. All the known ways of detecting light are based on the single principle that light is an electromagnetic wave (consisting, as we now know, of many particles called **photons**; see Chapter 3). Thus, not only is light always generated by the acceleration of charged particles, but, once it is generated, its presence is always detected by the acceleration it produces in other charged particles.

Usually such charged particles are **electrons** because electrons have the highest charge-to-mass ratio of all particles and are, therefore, easiest to accelerate. To see this, let us note that the force $\boldsymbol{F}$ exerted on a charge q moving at velocity $\boldsymbol{v}$ in an electric field $\boldsymbol{E}$ and a magnetic field $\boldsymbol{B}$ is given by Lorentz's formula,

$$\boldsymbol{F} = q\left(\boldsymbol{E} + \frac{\boldsymbol{v}}{c} \times \boldsymbol{B}\right). \tag{2.6}$$

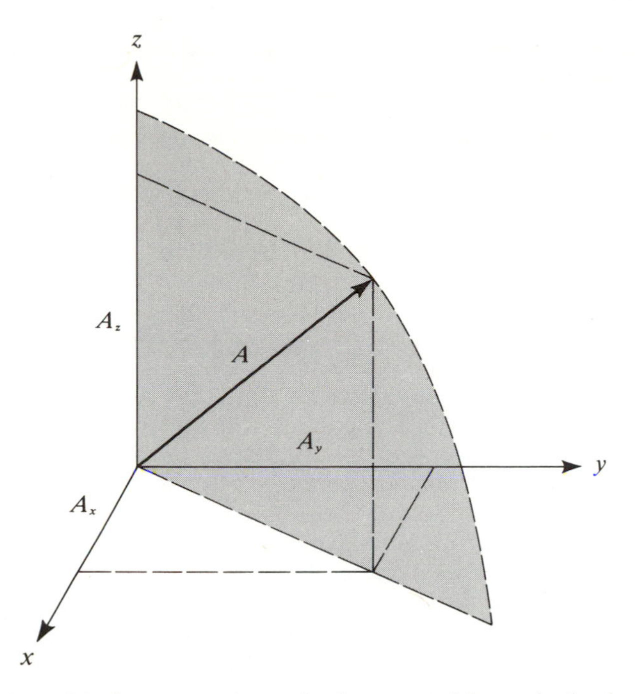

Figure 2.7. Any vector $\boldsymbol{A}$ may be decomposed by projecting its length onto three mutually perpendicular directions. The results of those projections, A_x, A_y, A_z, are called the x, y, z components of the vector $\boldsymbol{A}$, and we may sometimes write the triplet (A_x, A_y, A_z) to mean the vector $\boldsymbol{A}$ itself. Notice, however, that the concept of the vector $\boldsymbol{A}$ exists independently of any particular component decomposition (i.e., independent of the coordinate representation). Nevertheless, in practical calculations, it is convenient to use a decomposition into components because we may then just ideal with ordinary numbers for each component.

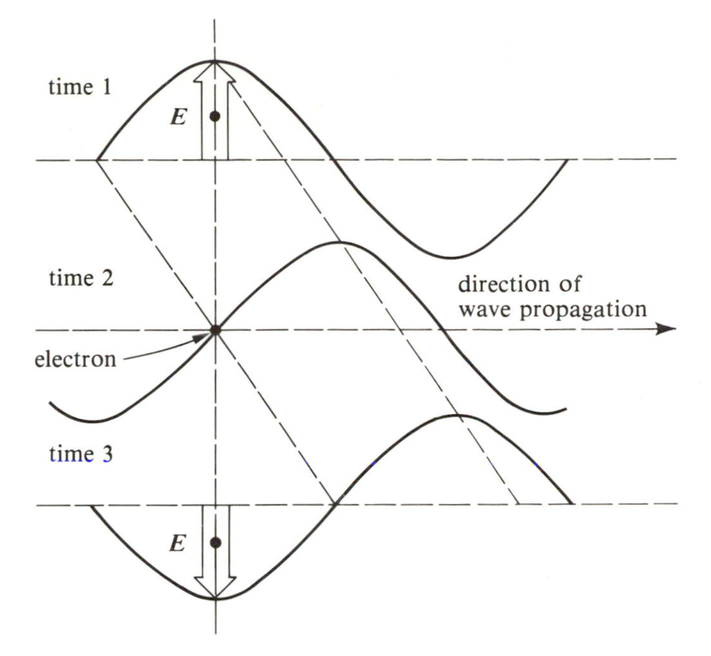

Figure 2.8. The response of an electron to a passing electromagnetic wave. The electric and magnetic vectors, $\boldsymbol{E}$ and $\boldsymbol{B}$, are equal in an electromagnetic wave; consequently, the electric force on a charge, $q\boldsymbol{E}$, is generally much larger than the magnetic force, $q(\boldsymbol{v}/c) \times \boldsymbol{B}$, as long as the velocities are much less than the speed of light. Thus, a free electron (shown above) displaces up and down "in phase" with the corresponding $\boldsymbol{E}$ vector, whereas a bound electron will usually displace up and down "out of phase" with the corresponding $\boldsymbol{E}$ vector (assuming a "natural period of oscillation" shorter than that of visible light).

From Newton's law, $\boldsymbol{F} = m\boldsymbol{a}$, we see that the acceleration $\boldsymbol{a}$ of charges in given electric and magnetic fields is proportional to the ratio q/m, which is highest for electrons. Hence, the interaction of light with matter, whether in the human eye, in the chloroplasts of green plants, or in photographic film, occurs because light interacts with the electrons in matter, either those bound to atoms or those free to move about (like the conduction electrons in metals).

When an electromagnetic wave passes by an electron, the electron tries to oscillate sympathetically ("in tune") with the passing oscillatory wave (Figure 2.8). These sympathetic vibrations underlie such familiar phenomena as the reflection and refraction of light, but a detailed explanation would take us too far afield. Suffice it to say that the electronic oscillations generate secondary sources of waves which either reinforce or interfere destructively with the original wave (Figure 2.9). As a result, the original wave appears to be refracted into a different direction, or to be reflected, or both.

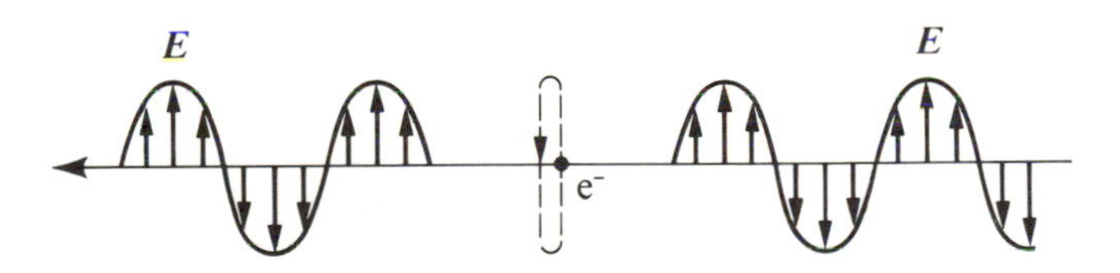

Figure 2.9. The generation of an electromagnetic wave by an oscillating electron. As the electron oscillates up and down, the electric field—which in a stationary electron points radially toward the electron—tries to follow the oscillatory motion. The time-varying components of the electric vector far from the electron alternate up and down in space because they "remember" the electron's position as it was one light-travel time back. At a fixed point in space the electric vector oscillates up and down in time with a period equal to the oscillation of the electron. The diagram is highly simplified in that: (a) we have not drawn in the oscillatory *magnetic* field induced by the electron's motion; (b) the electromagnetic fields *near* the electron (within several wavelengths) have a more complicated structure than a simple outgoing electromagnetic wave; and (c) the oscillating electron actually radiates in all directions except along its oscillation length.

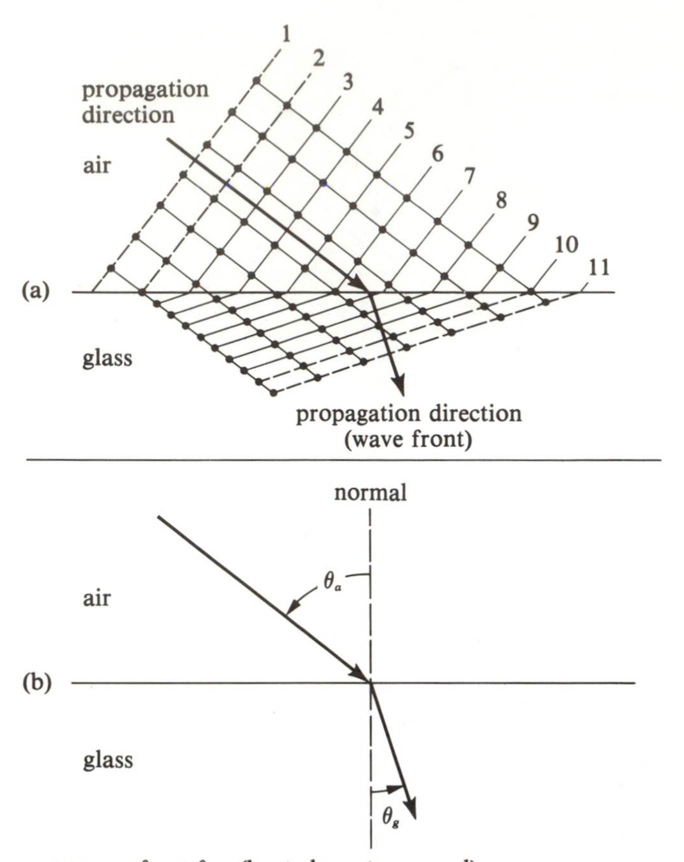

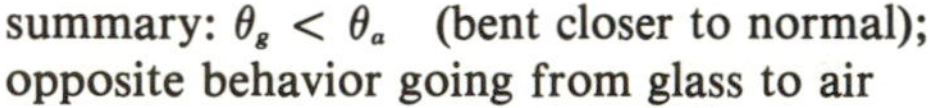

Figure 2.10. Refraction.

Astronomical Telescopes

Astronomical telescopes can be constructed which use either **refraction** or **reflection** to gather light over a large collecting surface and focus it to a much smaller area. This property—gathering more light and focusing it, and *not magnification*—is the primary purpose of most astronomical telescopes.

Refraction

To understand the principle of refracting telescopes, we must first understand refraction on a macroscopic level. (The word **macroscopic** in this book means on a scale which is large compared to atoms; it is to be contrasted with the word **microscopic**.) We have already mentioned that the speed of light in a vacuum, c, is a very large number, 3×10^{10} cm/sec. In fact, c is the fastest possible speed at which individual photons (or, indeed, any physical *entity*) can travel in any medium at all. There is a different kind of speed (the "phase velocity") associated with the progress of wavecrests (the positions where photons are most likely to be found). The phase velocity also equals c in a vacuum, but it may exceed c in exceptional materials. Usually, however, the sympathetic oscillations of the electrons in the atoms and molecules of a nonconducting medium impede the average progress of the photons, and the speed of wavecrests, v, is also less than c, its value in a vacuum.

Now, air is more like a vacuum than is glass; so it is easy to believe that the speed of light in air, v_a, is greater than the speed of light in glass, v_g. Consider what this means when a whole collection of photons are incident as a plane wave on an air-glass interface at an oblique angle (Figure 2.10). The photons (the dots) are imagined to come from a very distant source so that they arrive in waves whose fronts are connected by the dashed lines. The initial direction of the propagation of the photons and the wavefront is indicated by the heavy top arrow. As the photons enter the glass interior, they are effectively slowed down; so they travel less distance in a given interval of time than their companions who have still not entered the glass. The net effect is to "turn" the direction of the wavefront defined by the collection of photons so that the propagation of the wavefront in the glass is in a direction closer to the normal. (The **normal** is the direction perpendicular to the air-glass interface.) Notice that this "turning" is strictly a property of the entire collection of photons, i.e., a property of the *wave*; no individual photon can be said to have suddenly "turned" in direction upon entering the glass. The quantitative relation between the incident direction, θ_a, and the emergent direction, θ_g, is given by Snell's law (Problem 2.2).

Problem 2.2. Use the diagram below to consider how far the wavefront travels in interval Δt. Let v_a and v_g be the speeds of the *wavefront*; show

$$\frac{v_a \Delta t}{\cos(\pi/2 - \theta_a)} = \frac{v_g \Delta t}{\cos(\pi/2 - \theta_g)}.$$

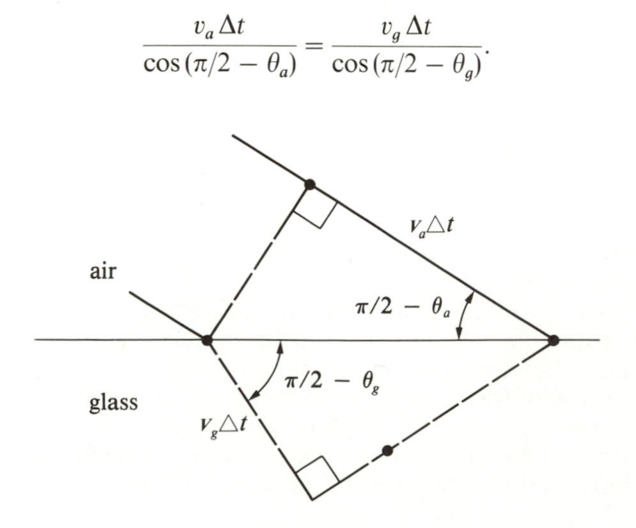

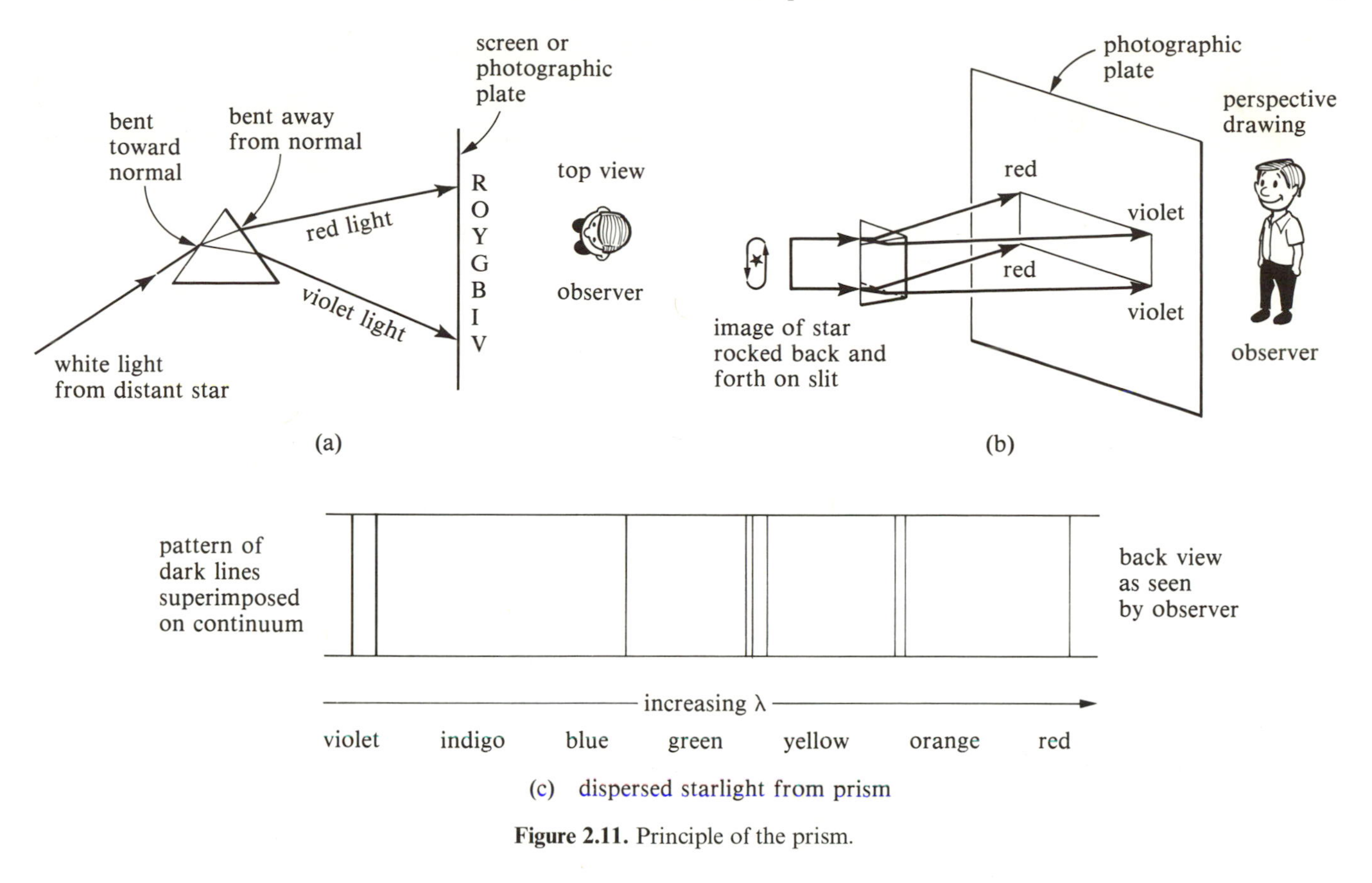

Figure 2.11. Principle of the prism.

Convert the above equation into Snell's law: $n_a \sin\theta_a = n_g \sin\theta_g$, where the indices of refraction of air and glass are given, respectively, by $n_a = c/v_a$ and $n_g = c/v_g$, with $n_g > n_a > 1$.

In Figure 2.10a we have drawn the air-glass interface as if it were an infinite flat plane. This is an adequate approximation if the edge-to-edge dimension of the interface is very large compared to the wavelength of light; otherwise there are complicated diffraction effects, which we shall discuss later. In the diffractionless approximation, we can summarize the reasoning by Figure 2.10b, where we draw only the directions of the propagation of the wavefronts as being **rays** of light. Using this ray-tracing technique (also called "geometric optics"), we are now prepared to explain some familiar and important phenomena.

The Principle of the Prism

In Figure 2.10 we assumed implicitly a unique speed of light v_a in air and another unique speed v_g in glass, independent of wavelength λ. Only in a vacuum does light of different wavelengths have a unique speed c. In matter, light of different wavelengths travels at different speeds because the electrons bound to atoms and molecules have different abilities to respond to rapidly varying and slowly varying electromagnetic waves (refer back to Figure 2.8). In particular, the "natural period of oscillation" of such electrons corresponds usually to the ultraviolet (Chapter 3); thus violet light will generally interact more with matter than blue light, and blue light more than red. This means that refraction in glass will turn violet light more than blue light, and blue light more than red light. As is shown in Figure 2.11a, this fact explains how a prism breaks up white light into its component colors.

Astronomers use this "dispersive" property of prisms to analyze the wavelength composition of starlight ("spectral analysis"). Figure 2.11b shows a simplified setup of how this is done. In order to use as much as possible of the photographic plate (the most common and inexpensive photon-detection device), the image of a star is rocked back and forth mechanically along the length of an entrance slit. The light which enters the slit is then dispersed by a prism. (In high-accuracy work, a diffraction grating is used instead of a prism.) The dispersed light is then allowed to fall onto a photographic plate. From the point of view of the observer drawn in

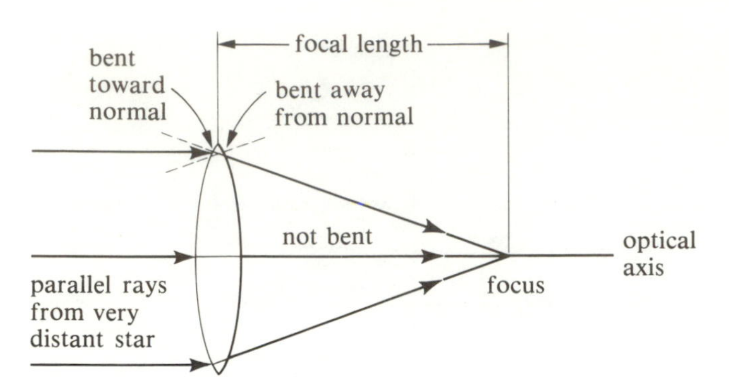

Figure 2.12. Principle of the lens.

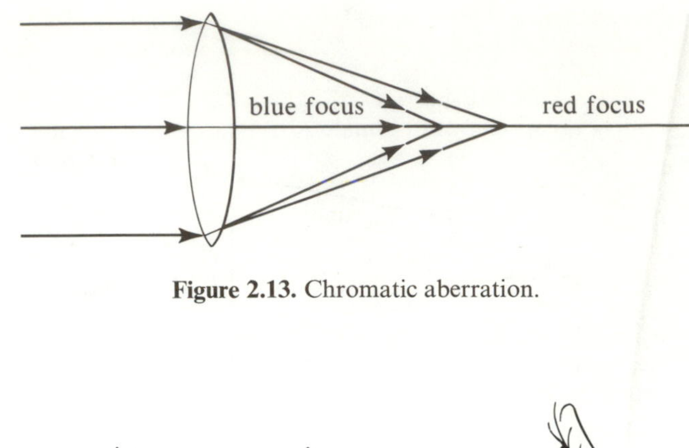

Figure 2.13. Chromatic aberration.

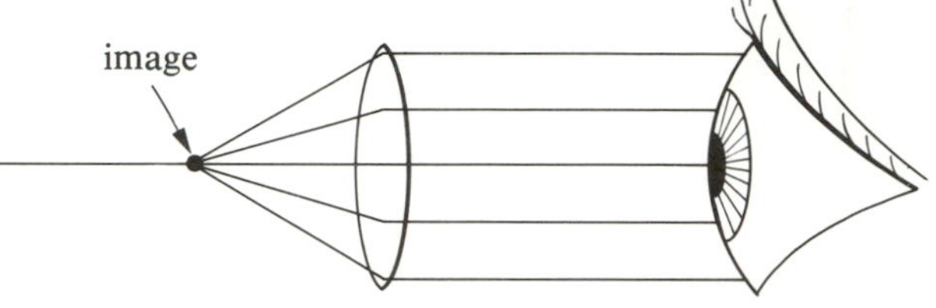

Figure 2.14. The principle of the eyepiece.

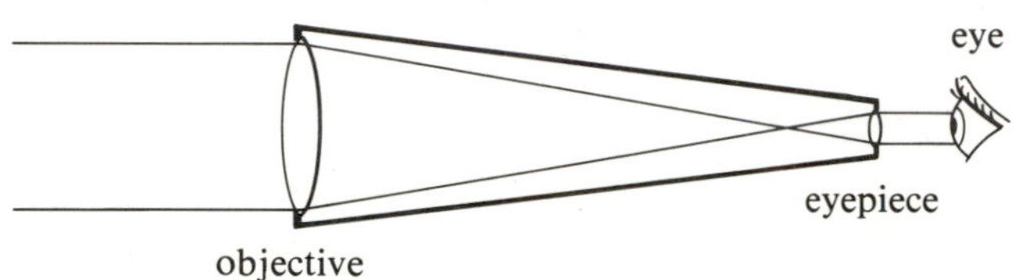

Figure 2.15. A simple refracting telescope.

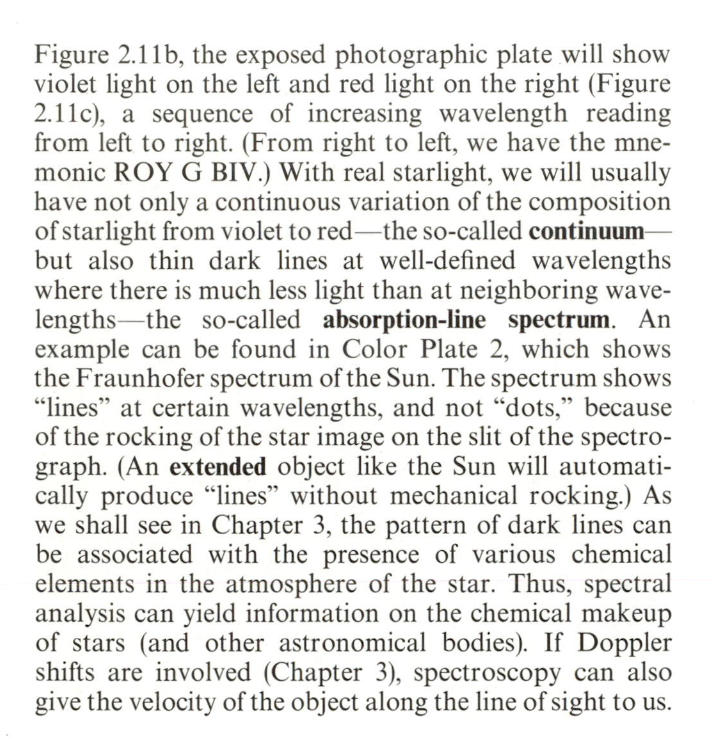

Figure 2.11b, the exposed photographic plate will show violet light on the left and red light on the right (Figure 2.11c), a sequence of increasing wavelength reading from left to right. (From right to left, we have the mnemonic ROY G BIV.) With real starlight, we will usually have not only a continuous variation of the composition of starlight from violet to red—the so-called **continuum**—but also thin dark lines at well-defined wavelengths where there is much less light than at neighboring wavelengths—the so-called **absorption-line spectrum**. An example can be found in Color Plate 2, which shows the Fraunhofer spectrum of the Sun. The spectrum shows "lines" at certain wavelengths, and not "dots," because of the rocking of the star image on the slit of the spectrograph. (An **extended** object like the Sun will automatically produce "lines" without mechanical rocking.) As we shall see in Chapter 3, the pattern of dark lines can be associated with the presence of various chemical elements in the atmosphere of the star. Thus, spectral analysis can yield information on the chemical makeup of stars (and other astronomical bodies). If Doppler shifts are involved (Chapter 3), spectroscopy can also give the velocity of the object along the line of sight to us.

The Principle of Lenses and the Refracting Telescope

The main optical element in a refracting telescope is the objective lens, which collects light over the entire area of the aperture and forms an image of the observed object at the focus (Figure 2.12). The principle of the lens is easily understood from Figure 2.11, since light bends within the lens in the same way that it does in the prism. Indeed, the rim of the lens is shaped like a prism.

To discuss the focusing ability of a lens, imagine a source like a star, which is essentially a single point of light. Since the star is practically at infinity, the rays from the star arrive at Earth nearly on parallel lines. A perfect lens would bring these parallel rays to a single point at the focus. This point is, of course, the correct image for our idealized star. In practice, however, starlight is made up of light of different colors. Since blue light is bent more by a glass lens than red light, chromatic aberration results, as shown in Figure 2.13. A photographic plate placed at the blue focus will show a blurred red image, and vice versa at the red focus.

Let us ignore this chromatic aberration (or assume that steps can be taken to compensate for it) in order to discuss light-gathering power. Clearly, a lens can be used in "reverse" of the above description; i.e., if an object (or the image of a real object) were placed at the focus of a lens, the rays from it, in passing through the lens, would leave in parallel lines (Figure 2.14). A human eyeball placed behind this lens would perceive the object (or image) as if it were at infinity, which turns out to be the most relaxing condition for visual observations. Thus, the combination of a large objective lens plus an eyepiece (Figure 2.15) serves as an astronomical telescope

Figure 2.16. In 1921, Albert Einstein posed with the staff of Yerkes Observatory in front of the 40-inch refractor, the largest in the world. (Yerkes Observatory photograph, University of Chicago, Williams Bay, Wisconsin.)

(a refractor), which enables an astronomer to gather starlight over a much larger area than his or her own eyeball could. Electronic devices ("image intensifiers"), with or without photographic film, allow the astronomer to "observe" to much fainter limits than is possible with naked-eye observations, even on telescopes with very large apertures (Figure 2.16).

Different apparent sizes (magnifications) of the final image—for example, the apparent separation of a double star—result from using different eyepieces (Problem 2.3). However, the light-gathering power is determined by the square of the diameter of the objective, not by the eyepiece. For an extended object, say, the Moon, spreading the same amount of light over a larger area (by high magnification) decreases the apparent surface brightness of the observed object, but a net gain may still result for the clarity of detailed features if the object is sufficiently bright. Thus, high magnification may be important for solar-system observations, but it is generally not needed for other types of astronomical investigations.

Problem 2.3. To discuss magnification, imagine a telescope pointing directly toward a star on its optical axis (see diagram). Imagine another star lying at an angle θ_o with respect to the first. The rays from both stars will pass through the center of the objective lens on a straight line. These rays will be brought into focus with rays which pass through other parts of the objective at a distance behind the objective equal to its focal length, f_o. If the eyepiece is placed its focal length, f_e, further back, the double-star image would in turn pass through the eyepiece and be collimated into two parallel beams at an angle θ_e to each other. An astronomer behind the eyepiece will perceive an inverted image of two stars separated by an angle θ_e, whereas the actual double star is only separated by the angle θ_o. Thus, the angular separation has been *magnified* by the ratio θ_e/θ_o. To calculate this magnification in terms of the properties of the telescope, derive the relationship

$$f_o \tan \theta_o = f_e \tan \theta_e.$$

The angles θ_o and θ_e are usually quite small; therefore use the small-angle approximations of Appendix B to show that the magnification θ_e/θ_o is given by the ratio of the two focal lengths f_o/f_e. A typical amateur's telescope may have a one-meter objective focal length and a 40-millimeter eyepiece. What would its magnification be? Does this magnification apply only to the observation of double stars?

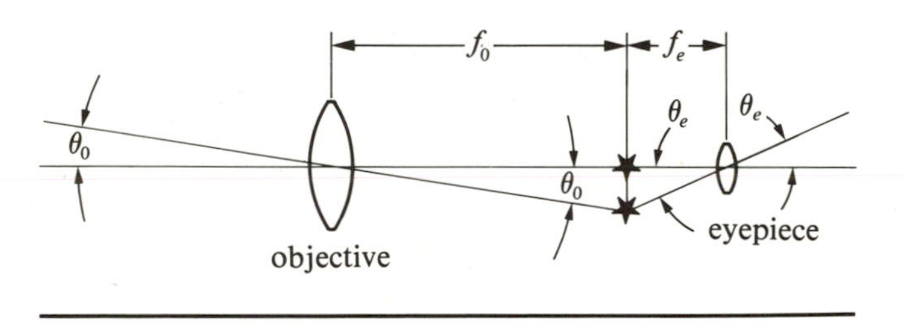

Reflection

In Figure 2.10 the rays which fell on an air-glass interface were drawn as if they would all be refracted. In practice, there will also always be some reflection, as well as some absorption. If a thin layer of metal is deposited onto the glass surface and is polished, the reflectance will be increased, and we will have made a mirror. This everyday fact—metals are *shiny*—is a result of the microscopic fact that metals contain many free electrons which are not bound to the atoms in the lattice of the metal. (That is why metals are good electrical conductors.) These free electrons respond quite readily to the presence of electromagnetic waves, and they allow almost no radiation of sufficiently long wavelength to pass through the

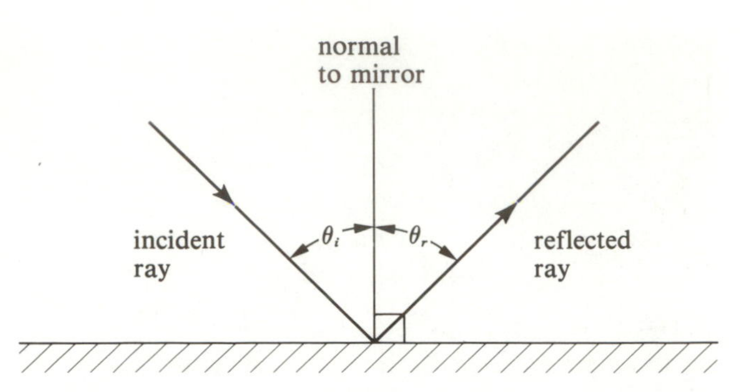

Figure 2.17. The law of reflection. The angle of reflectance is equal to the angle of incidence: $\theta_r = \theta_i$.

metal. Thus, a mirror will reflect almost all the optical light which falls on it.

The geometric properties of a flat mirror are easily deduced by knowing only some simple wave properties of light. Light will reflect from a flat surface in such a way that the angle of reflectance (relative to the normal) is equal to the angle of incidence (Figure 2.17). What fraction of the light is reflected or absorbed or transmitted depends on the properties of the mirror (for example, the kind of metal used: aluminium, silver, or gold) and the wavelength of the incident light (radio, optical, or X-rays), but the incident and reflection angles are always symmetric, because the wave propagation before and after reflection takes place in the same medium (Problem 2.4).

Problem 2.4. Consider the wavefront AB which moves to $A'B'$ in time Δt. In another interval Δt, B' will have moved to B'', and A' will have moved (by reflection) to A''. Since the rays $A'A''$ and $B'B''$ are moving in the same medium (say, air), the distances $A'A''$ and $B'B''$ are equal. Show now that the angles $B'B''A'$ and $A''A'B''$ are equal.

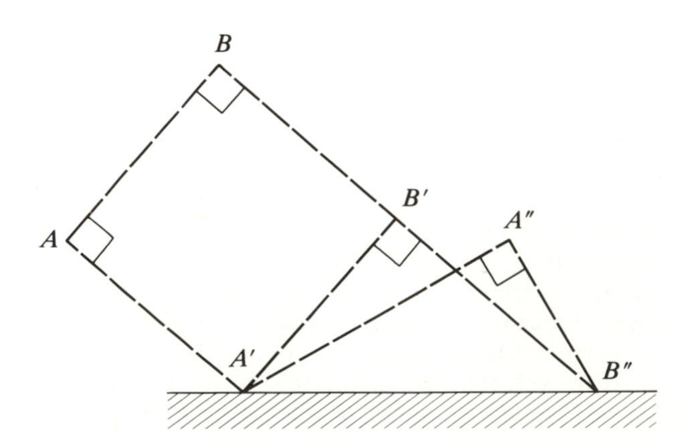

Show also that these angles are, respectively, the complements of the angles of reflectance and incidence, thereby proving the law of reflection.

The Principle of Parabolic Mirrors and the Reflecting Telescope

The high reflectance of metallic surfaces allows reflecting telescopes for use with waves of any length from radio to ultraviolet to be constructed on the same general principle (Figure 2.18). To provide good imaging, the surface of the reflecting element (a mirror or a dish) should be paraboloidal (a parabola of revolution), and the surface should not contain any irregularities larger than a fraction of the wavelength of the light that needs to be focused. Since visual light is much shorter than radio waves, optical telescopes are made by depositing a metal film on accurately figured glass; whereas the much coarser technique of directly shaping a metal dish (which may even have small holes in it to make it lighter!) can be used to construct a radio telescope.

Why does the mirror or dish have to be paraboloidal? To understand this, we need first to review the properties of ellipses. You may remember how to draw an ellipse (Figure 2.19) from high-school geometry. You take a length of string, tie it into a loop, place the loop over two tacks or nails (which become the two foci of the ellipse), take a pencil, and push the point tight against the string. The locus traced out by the pencil tip, keeping the string taut, is an ellipse. Consider what this means for the geometry of an ellipse. Since the string is taut, the lines joining the foci and the pencil tip are straight lines. Moreover, the direction that the pencil has to be pushed against the string to keep it taut and the direction that the pencil tip can move are mutually perpendicular, i.e., the former direction is the *normal* to the ellipse. It should also be obvious that the normal bisects the angle made by the two lengths of string from the foci.

Imagine now making a mirror in the shape of an ellipse, and putting a point source of light at one of the foci (Figure 2.20). All the light which leaves this source in the plane of the page will strike the ellipse somewhere and reflect off it, with the angle of reflectance equal to the angle of incidence. Given the properties of the ellipse, however (see Figure 2.19), this means that each ray will be brought to the other focus! Hence, if we want to collect *all* the light which leaves the source—including those rays which travel out of the plane of the page—at a single point at the other focus, we need only make our mirror an ellipse of revolution. Such an ellipsoid (a prolate spheroid) would represent the ultimate mirror in terms of light-gathering power for the assumed geometry. Of course, if one had funds to construct only part of the ellipsoidal mirror (Figure 2.21), one would be able

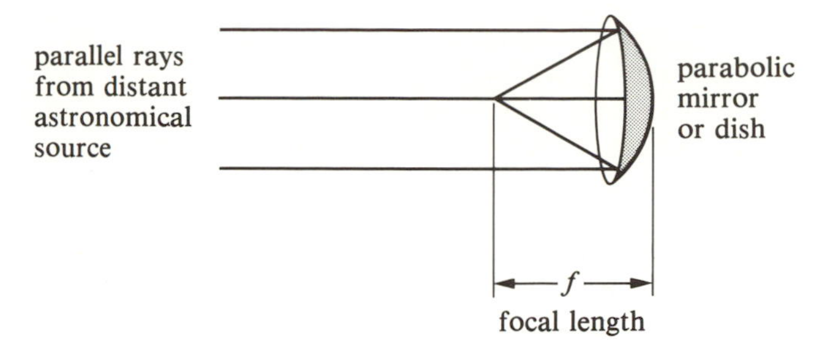

Figure 2.18. The principle behind optical and radio telescopes.

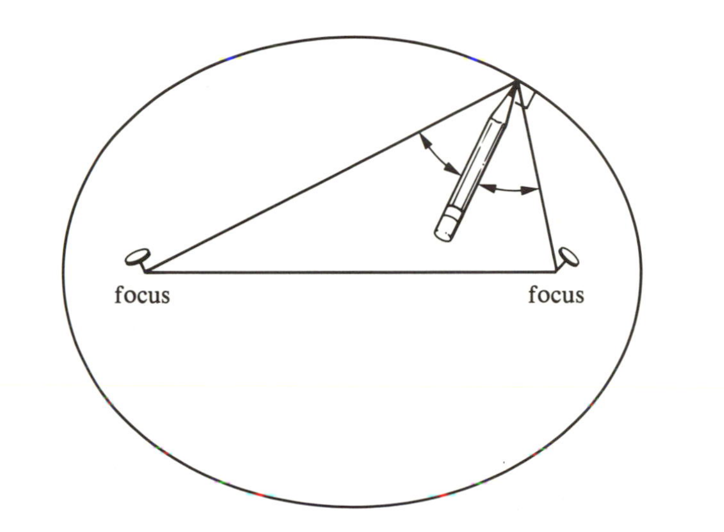

Figure 2.19. How to draw an ellipse. A long length of string produces a rounder ellipse than a short length.

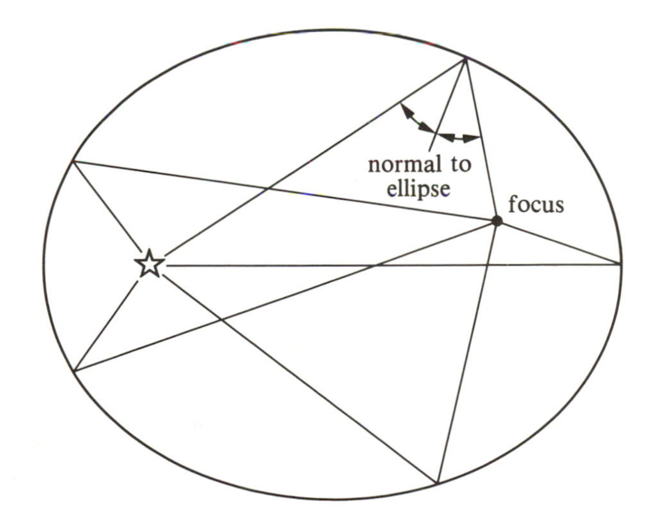

Figure 2.20. An elliptical mirror brings the light from a point source at one focus to the other focus.

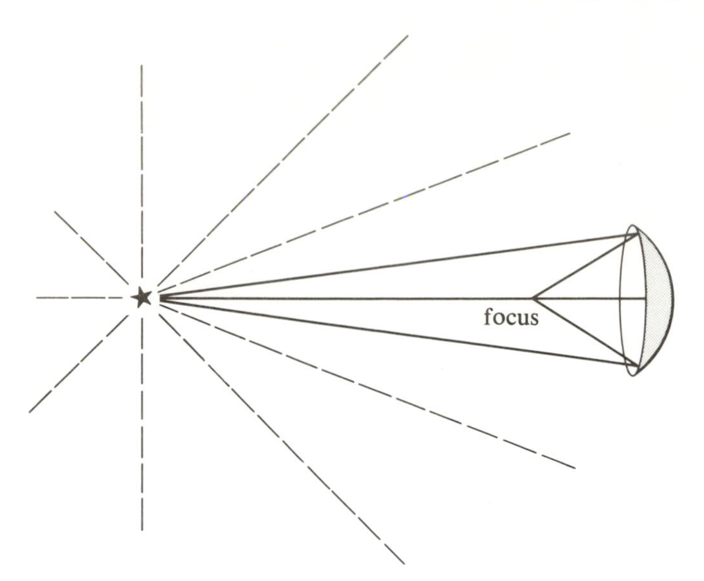

Figure 2.21. Part of a complete ellipsoidal mirror will capture only part of the light which leaves a point source.

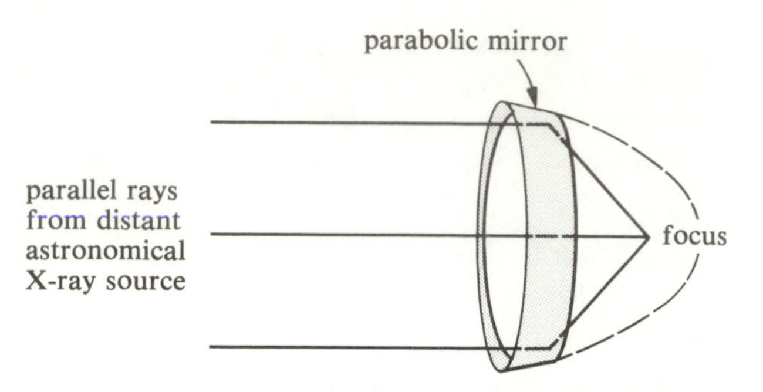

Figure 2.22. An example of a X-ray telescope which works by grazing incidence reflections. Notice that the focus is brought to the back of the mirror arrangement.

Figure 2.23. The Einstein Observatory X-ray telescope has a nested set of surfaces with an effective collecting area of 300 cm^2 that brings X-rays to a focus by grazing-incidence reflections. (Courtesy of Perkin-Elmer.)

to collect only the corresponding fraction of all the rays which left the light source.

Imagine now moving the point source of light to infinity. If the two foci of an ellipse become infinitely separated, the ellipse becomes a parabola. Thus, to bring the light rays from a very distant point source like a star to a single focus, the objective of a reflecting telescope should be a paraboloidal mirror (Figure 2.18). This provides the answer to our original question. To collect as much light as possible, clearly, the aperture of the paraboloid should be made as large as available funds will allow.

X-rays are much more penetrating than electromagnetic radiation of longer wavelengths. X-rays which are incident almost perpendicularly on a metallic surface will be mostly absorbed rather than reflected. This problem is less severe at grazing incidences (especially if one uses gold instead of aluminum); so X-ray telescopes are built to bring the rays to a focus by grazing-incidence reflections (see Figure 2.22). This configuration results in much less collecting area for a given amount of metallic surface; and since the telescope has to be put above the Earth's atmosphere, X-ray astronomy is a much more expensive proposition per experiment than optical or radio astronomy. That is the price one pays to be an X-ray astronomer (Figure 2.23).

There are two big advantages of reflectors over refractors even for use confined to optical wavelengths. First, there is no chromatic aberration. Second, reflectors can be built much larger, because the primary mirror can be supported mechanically all along its back, whereas the lens must be supported only along its edges. Other practical considerations—such as that a mirror can have some bubbles in the glass without affecting its performance—means that all large modern telescopes are constructed as reflectors rather than as refractors.

The one disadvantage of a reflector is, of course, that its prime focus is *in front* of the mirror. Thus, for an astronomer or for instruments to observe at the prime focus, an appreciable fraction of the incoming light may be blocked (Figure 2.24a). This problem can be partially solved in several ways. The first solution was invented by Newton. In the Newtonian arrangement (Figure 2.24b), a secondary mirror is used to bring the light to a focus at the side of the telescope tube. This arrangement tends to make the system lopsided; so a more satisfactory solution was devised by Cassegrain. In the Cassegrain arrangement (Figure 2.24c), a secondary mirror is used to bring the light to a focus behind the telescope tube through a hole in the primary mirror. None of the above

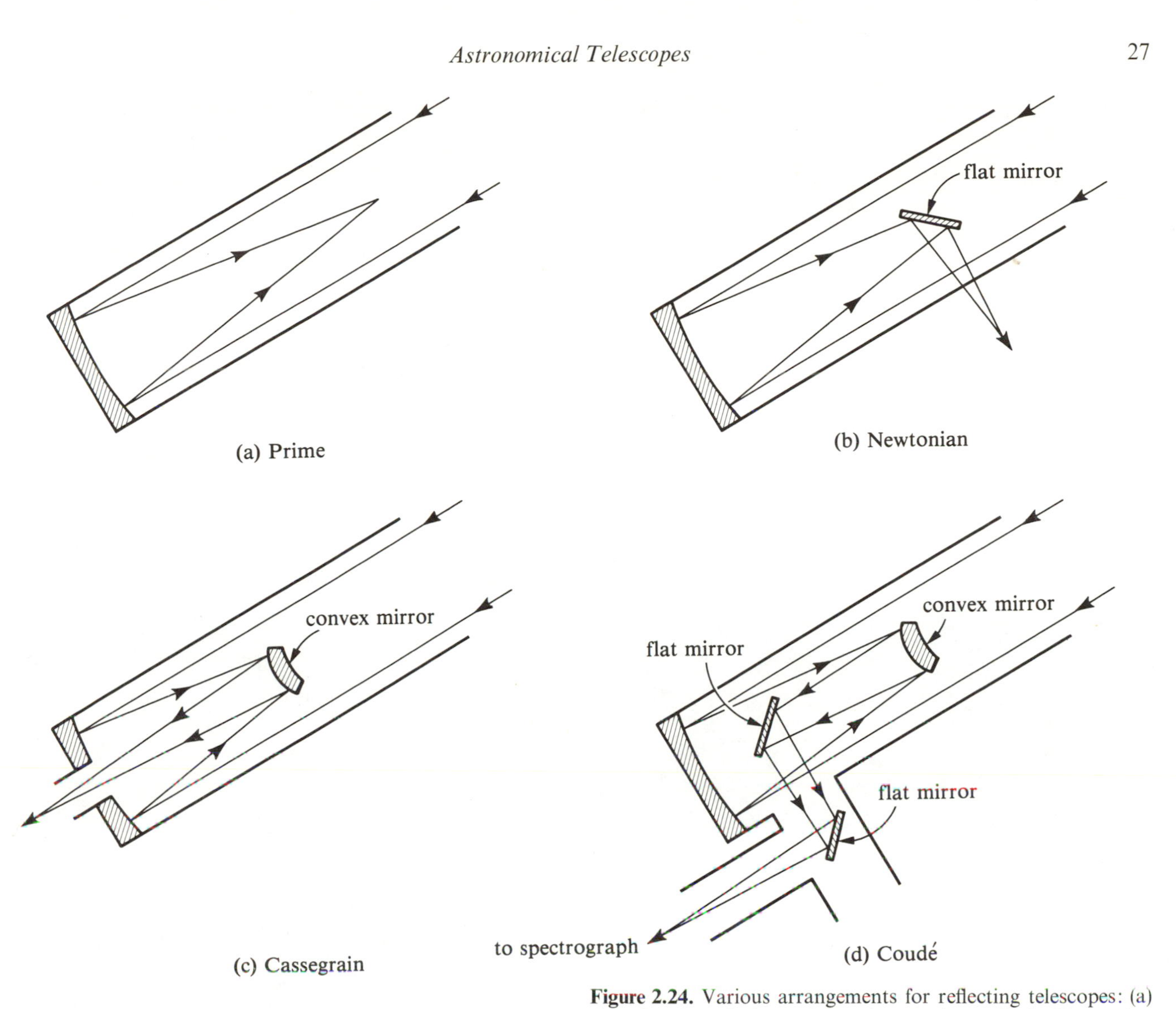

Figure 2.24. Various arrangements for reflecting telescopes: (a) the prime focus in front of the primary mirror; (b) the Newtonian focus at the side of the telescope; (c) the Cassegrain focus at the back of the telescope; (d) the Coudé focus underneath the telescope.

arrangements are satisfactory for making very precise spectroscopic observations, because the bulky and heavy spectrographs would strain on the telescope tube if they were suspended in front, to the side, or in back of the primary mirror. For such observations, the Coudé arrangement was devised, whereby a series of secondary mirrors bring the light down the polar axis of the telescope drive to a fixed focus in a room below the telescope (Figure 2.24d). In the Coudé room, large pieces of equipment can be mounted in a stationary configuration.

Angular Resolution

The ability to distinguish fine details in the image is an important advantage of a telescope compared to a "light bucket," which merely collects light over a large area without attempting to bring it to an accurate focus. This ability is called **angular resolution**. Figure 2.25 illustrates the concept of angular resolution in its simplest context. Imagine a telescope pointing directly toward a star on its optical axis. Imagine another star lying at an angle θ with respect to the first. A perfect focus would result in two point images slightly displaced from one another. Such a perfect focus is, however, never attainable in practice, for the following reasons.

Ground-based optical telescopes must contend with the blurring produced by turbulence in the Earth's atmosphere. The varying amount of refraction in a turbulent atmosphere makes the focused optical image dance around ("seeing") on the photographic plate. If

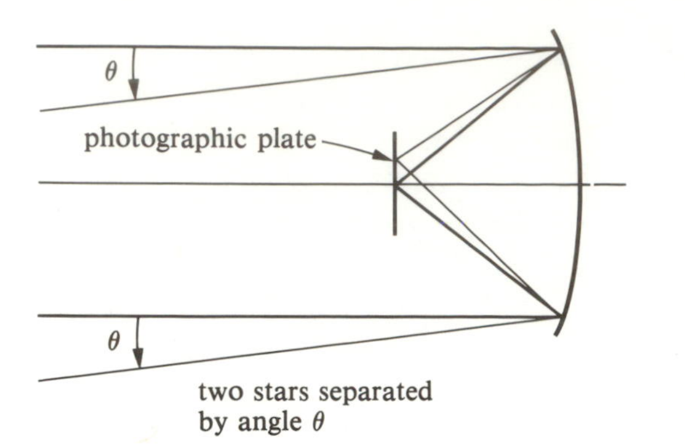

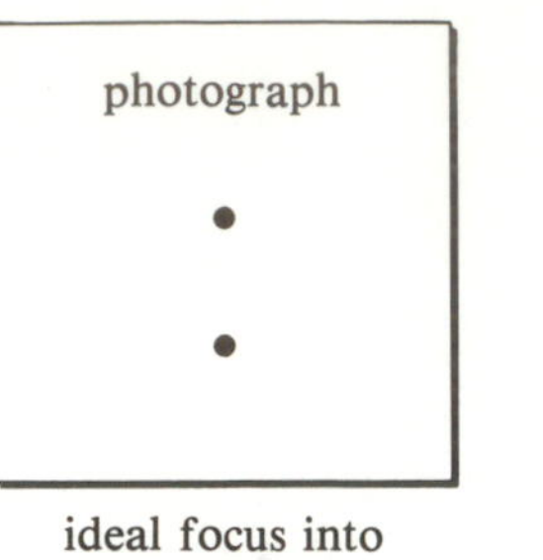

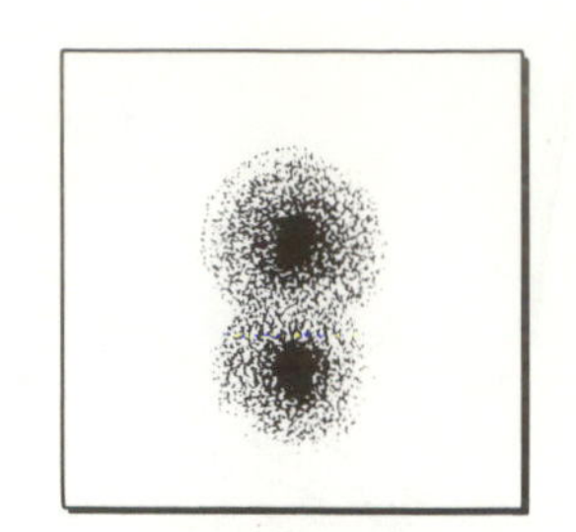

Figure 2.25. The concept of angular resolution applied to a double star.

Figure 2.26. Because of diffraction, the shadow cast by a razor blade is not perfectly sharp. (Courtesy of John Wilder, Berkeley Astronomy Department Photo Lab.)

Figure 2.27. The 100-meter radiotelescope at Bad Muenstereifel-Effelsberg, Germany. (Max-Planck Institute für Radioastronomie at Bonn.)

the source is sufficiently bright, astronomers can use a series of short exposures to correct partly for this effect by means of a technique invented by Antoine Labeyrie called "speckle interferometry." For faint objects, this technique is impractical, and atmospheric turbulence limits the seeing image of a point source to about one second of arc, even at the best sites. (One second of arc equals 1/3600 of a degree of arc, and corresponds roughly to the angle subtended by a dime at a distance of about two miles.) In other words, if two stars are separated by less than one second of arc, the observer cannot tell from the blurred images that more than one point of light is actually present.

Radio waves do not suffer from this atmospheric blurring. The angular resolution of modern radio telescopes is limited by another effect: the diffraction produced by the wave nature of light. As we noted earlier, the ray-tracing method to determine the focusing of electromagnetic waves assumes that the wavelength is negligibly small compared with the aperture of the telescope. On angular scales comparable to the ratio of the wavelength λ of light to the diameter D of the telescope's collecting surface, however, diffraction effects produce an unavoidable blurring even of point sources. The reason is that the secondary waves produced by the interaction of light and matter at the edges of the objective interfere constructively and destructively to give a pattern of light and dark fringes which, on small angular scales, is quite different from that predicted by ray tracing. Figure 2.26 shows that the shadow cast by a razor blade is not perfectly sharp. Diffraction of light

Figure 2.28. The 200-inch (5-meter) telescope on Palomar mountain has a large enough prime-focus area to accommodate an observer plus some equipment. (Palomer Observatory, California Institute of Technology.)

allows one to see (slightly) around corners, as diffraction of sound allows one to hear around corners. For a circular aperture, the blurring produced by diffraction limits the angular resolution to an amount given by

$$\theta_{\text{diffraction limit}} = 1.22\lambda/D.$$

Thus, the world's largest fully steerable single-dish telescope (the 100-meter dish in West Germany; Figure 2.27) cannot resolve two radio sources closer than about 9 minutes of arc when it operates at a wavelength of 21 cm.

Problem 2.5. What is the diffraction limit in seconds of arc of a 5-meter telescope (Figure 2.28) at a wavelength of 5000 Å? (1 Å $= 10^{-8}$ cm.) Compare your answer to the seeing limit.

Hence, the diffraction-limited resolution of single-dish radio telescopes (even the Arecibo dish—Figure 2.29—of

Figure 2.29. The 300-meter Arecibo radiotelescope sits in a natural bowl in Puerto Rico and uses the rotation of the Earth to help turn the aim of its beam. (Arecibo Observatory, N.A.I.C., Cornell University.)

Figure 2.30. The Space Shuttle will launch a 2.4-meter telescope into orbit around the Earth, which will greatly improve the focus and clarity of observed astronomical objects. (NASA, Marshall Space Flight Center, artist's rendition.)

Figure 2.31. A vast array of 1,000 radiotelescopes, each of 100-meter diameter, has been comtemplated in Project Cyclops as a means to search for extraterrestrial intelligence. (NASA, Ames Research Center.)

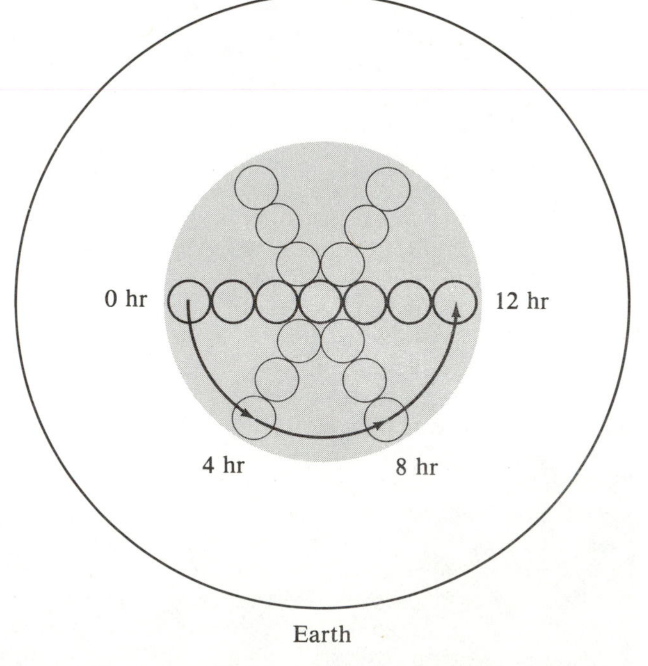

Figure 2.32. The principle of aperture synthesis illustrated by an imaginary line of radio telescopes at the North Pole. During one twelve-hour period, the rotation of the Earth causes this line to sweep out a full circle. Thus, this configuration of telescopes mimics the angular resolution of a filled aperture with a diameter equal to the baseline (shaded area).

Figure 2.33. The very large array of radiotelescopes in Soccoro, New Mexico, has movable elements arranged in a Y. (The National Radio Astronomy Observatory VLA program is operated by Associated Universities, Inc., under contract with the National Science Foundation.)

300-m diameter, which is not fully steerable) is typically much worse than the seeing-limited angular resolution of large optical telescopes.

There are two direct remedies available to improve the angular resolution of astronomical telescopes. For optical work, a dramatic improvement could be made by getting above the Earth's atmosphere (above the "seeing"). This is one principal motivation for the Space Telescope project (Figure 2.30). Even the space telescope, however, will not reach its diffraction-limited potential because of mechanical jitter associated with trying to point the telescope accurately.

For radio work, the computer recombination of the signals from more than one radio telescope can increase the effective baseline D of the observations, and thus lower the ratio of λ/D which governs the diffraction limit. A brute-force technique would be merely to build a huge array of telescopes like that proposed for project Cyclops (Figure 2.31). A more elegant solution, devised by Sir Martin Ryle of England, uses the rotation of the Earth. In the simplest case, imagine a line of telescopes placed at the North pole (Figure 2.32). During a 12-hour period, the rotation of the Earth would swing the line of radio telescopes through a full circle, and thus allow them to mimic the resolving power (but not the collecting area) of a vast single-dish with the corresponding aperture diameter. Such powerful full-synthesis instruments have been built in England and in Holland. The most powerful such synthesis instrument is the very large array (VLA) in New Mexico, whose movable telescopes are so arranged that they need not rely on the rotation of the Earth to obtain (nearly) complete aperture coverage (Figure 2.33).

Less than full synthesis is possible with simultaneous observations by steerable radio telescopes located far from one another. Such very long baseline interferometric (VLBI) work has provided the most detailed information so far on the structure of certain astronomical radio sources (Chapter 13). Multi-mirror telescopes have also been built for ground-based optical work (Figure 2.34), but the motivation here is to increase the collecting area rather than to improve on the angular resolution. Large

Figure 2.34. The Multiple Mirror Telescope (MMT) consists of six 72-inch mirrors, having a combined collecting area equal to one 176-inch telescope. (The MMT is a joint facility of the Smithonian Institution and the University of Arizona.)

optical interferometers are impractical because it is difficult to keep track of the phase (where the peaks and troughs lie) of optical light and maintain alignment of the mirrors to a fraction of a wavelength.

Astronomical Instruments and Measurements

The popular image of the astronomer observing by eye at the end of a telescope (Figure 2.15) is badly outdated. The human eye is a relatively poor photon detection and storage instrument. All manner of electronic equipment exists today to help the astronomer measure the brightness of astronomical objects (photometry), calculate the wavelength distribution and spectral features from such bodies (spectrophotometry and spectroscopy), assess the amount of polarization in this light (polarimetry), and fix the position of the object in the sky (positional astronomy); see Box 2.2. The one indispensable tool of the modern astronomer is the electronic computer, which can be used to run many of these instruments more-or-less automatically, guide the telescope, reduce data, and help construct theories which interpret the observational results. For better or worse, modern astronomy, like the rest of modern life, becomes more complex with every passing day. It is a tribute to the human brain and the scientific method that the important themes contained in the rapidly growing body of astronomical knowledge can still be sorted out in terms of relatively simple organizational ideas. These themes and ideas are the frame of this book, but the foundations are the basic astronomical measurements listed in Box 2.2.

BOX 2.2
Some Fundamental Astronomical Measurements

(1) Parallax. Apparent angular shift of nearby objects against the positions of distant objects, due to the changing perspective as the Earth revolves around the Sun. This shift yields the distance to the nearby object by triangulation, since the Earth-Sun distance (the **astronomical unit**) is known.

(2) Proper motion. Real angular shift of nearby objects against the positions of distant objects due to the motion of the nearby object perpendicular to our line of sight. This shift yields the component of velocity perpendicular to the line of sight if the distance to the object is known.

(3) Doppler shift. Shift in wavelength of a spectral line compared to its position for a laboratory source. Such shifts are usually associated with motion along the line of sight. (See Chapter 3 for appropriate formulae.)

(4) Apparent brightness. The total brightness of an astronomical body as it appears relative to some standard. If the distance to the object is known, a measurement of its apparent brightness allows us to calculate its intrinsic brightness, i.e., its luminosity.

(5) Color. The difference in apparent brightness when measured through two filters that pass light of two different wavelength bands (for example, a blue filter and a yellow filter). This technique provides a quantitative measure of the wavelength distribution of the continuum light, and can be used to estimate the surface temperatures of stars (Chapter 9).

CHAPTER THREE

The Great Laws of Microscopic Physics

Physics and mathematics are the basic theoretical tools which astronomers use to interpret the data collected by their telescopes. They are used to construct models of specific astronomical objects and theories of the basic physical mechanisms at work. This does not imply that we can proceed deductively from the known laws of physics and derive all of astronomy. For one thing, physics is not yet complete; for another, astronomical systems are too complex. Astronomers need to rely on observations to keep their theoretical speculations from going too far astray. Nevertheless, we need a general appreciation of the *architecture* of physics to understand why it is so important to astronomy. This is the purpose of Chapters 3 and 4.

Figure 3.1 is an abbreviated blueprint of the architecture of physics. From this figure we see that **mechanics** is the single unifying theory which underpins all physics, and, therefore, presumably all natural phenomena. The two basic components of twentieth-century mechanics are **quantum mechanics** and **special relativity**, and they both grew from (and eventually supplanted) **classical mechanics**.

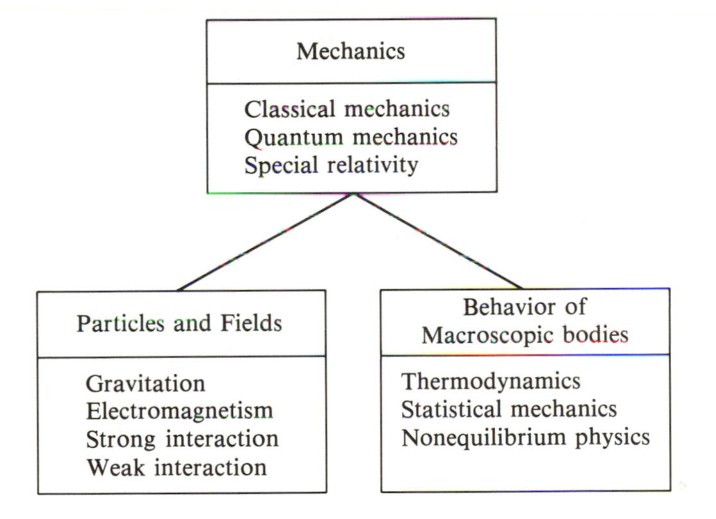

Figure 3.1. Architecture of physics.

Mechanics

In Chapter 2 we saw that at the center of classical mechanics lay Newton's second law:

time-rate of change of a particle's momentum = applied force.

Thus, given the applied force, Newton's second law allows us to calculate the change in the state of motion of a particle, and, by superposition, the state of motion of any collection of particles. It is therefore one of the most fundamental discoveries of modern science that all forces in the physical world originate in only *four* basic forms:

(a) the gravitational force,

(b) the electromagnetic force,

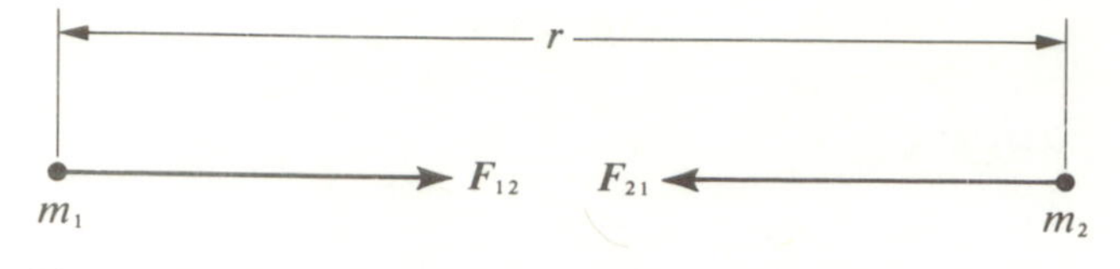

Figure 3.2. The gravitational force $\boldsymbol{F}_{12}$ exerted on mass m_1 by mass m_2 at a distance of r is attractive and directed along the line of centers.

(c) the strong nuclear force,

(d) the weak nuclear force.

A basic theoretical program begun by Einstein and still pursued actively today is the development of a "unified field theory," which would explain all the above forces in terms of a single "superforce." A major development of the 1970s was the apparently successful unification of the electromagnetic and weak forces by Weinberg and Salam. The single "superforce" remains to be found, although a name, "supergravity," has already been proposed for one of the more promising candidates.

The Universal Law of Gravitation

The first, and perhaps most important, general force law to be discovered was Newton's recognition that matter itself is a source of force for influencing the motion of other matter. In particular, Newton proposed that the property of **mass** gives rise to a universal force of **attraction**. For point masses, this force is directed along the line of centers and varies as the product of the two attracting masses divided by the square of the separation distance. The proportionality constant G is called the universal gravitational constant, and its value in cgs units is

$$G = 6.67 \times 10^{-8}\ \mathrm{gm}^{-1}\ \mathrm{cm}^3\ \mathrm{sec}^{-2}.$$

In formula form, we have that the force exerted on mass m_1 by mass m_2 has magnitude

$$F_{12} = Gm_1m_2/r^2, \tag{3.1}$$

and is directed from 1 to 2 (Figure 3.2). The force exerted on mass 2 by mass 1, $\boldsymbol{F}_{21}$, is identical in magnitude as $\boldsymbol{F}_{12}$, but oppositely directed; i.e., $\boldsymbol{F}_{21}$ is directed from 2 to 1. Occasionally, people will write formula (3.1) with a minus sign on the right-hand side to remind themselves that the gravitational force is always attractive; however, such reminders can also cause confusion unless we write the forces in their full vector-notation forms. In any case, when the forces are treated as vectors, the force due to finite-sized bodies can be deduced by considering them to be a superposition of many point masses and by

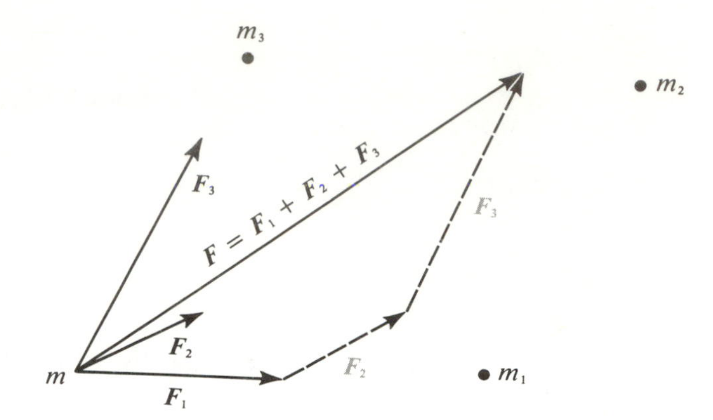

Figure 3.3. The resultant gravitational force $\boldsymbol{F}$ exerted on mass m by masses m_1, m_2, and m_3 is the vectorial sum of the individual forces $\boldsymbol{F}_1$, $\boldsymbol{F}_2$, and $\boldsymbol{F}_3$ formed by "lining up" these three vectors "head-to-tail." This case is easily generalized to an arbitrary number of attracting mass points.

adding according to the force law for point masses (Figure 3.3). In particular, for two spherical bodies, Newton found that the force of attraction was the same as if all the mass of each body were concentrated at a single point at its center. Newton delayed publication of his important research concerning the motion of the Moon until he had proved this result for spherical bodies. The result does not hold for nonspherical bodies.

Gauss's Formulation of the Law of Gravitation

We see from the above discussion that Newton's original formulation of his law of gravitation is so cumbersome for nonpoint bodies that even a genius like himself had difficulties making practical calculations. A much more useful formulation was devised by Gauss; it is mathematically equivalent to Newton's law, but greatly simplifies practical calculations (especially if one knows vector calculus). We shall briefly discuss Gauss's law, not only for later applications, but also because it makes more concrete the usefulness of the concept of a gravitational field $\boldsymbol{g}$ (see Chapter 2).

Suppose we have a distribution of matter characterized by a mass density ρ which can be an arbitrary function of position $\boldsymbol{x}$ and time t. Let $\boldsymbol{g}$, which is generally also a function of $\boldsymbol{x}$ and t, be the gravitational field which arises from the entire mass distribution. We begin by constructing an *arbitrary* volume V which encloses part or all or none of the mass (Figure 3.4). Let A be the surface area of the volume V, and let $g_\perp$ be the component of the gravitational field $\boldsymbol{g}$ which is directed along the *inward* direction perpendicular to any point on the surface A. (We reverse the usual convention here to avoid

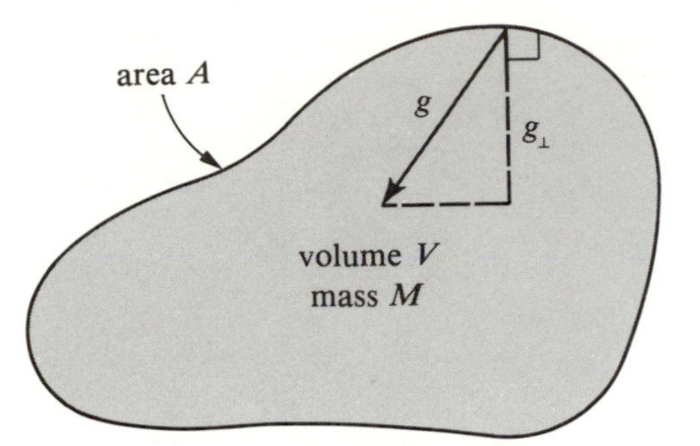

Figure 3.4. Gauss's law states $\langle g_\perp \rangle A = 4\pi GM$.

dealing with too many negative signs.) Let the average value of $g_\perp$ on the surface A be called $\langle g_\perp \rangle$. Gauss's law states that the average value $\langle g_\perp \rangle$ multiplied by the area A equals $4\pi G$ times the mass M contained in volume V:

$$\langle g_\perp \rangle A = 4\pi GM. \tag{3.1'}$$

Equations (3.1) and (3.1′) are completely identical in mathematical content, but the far greater power of equation (3.1′) is that it allows us to choose the mass distribution $\rho(\mathbf{x}, t)$ and the size and shape of the volume V completely arbitrarily.

To demonstrate explicitly the power of Gauss's law, consider the problem which stymied Newton for so long (Figure 3.5). Enclose the spherical distribution of mass in a concentric sphere with radius r greater than the radius R of the mass distribution. By symmetry, $g_\perp$ everywhere on our larger sphere must have the same value, and this value must equal both the average value $\langle g_\perp \rangle$ and the magnitude g of vector $\mathbf{g}$. Therefore, Gauss's law yields

$$g = 4\pi GM/A = GM/r^2,$$

since the area of a sphere of radius r is $4\pi r^2$ (Appendix B). In the above, M is the entire mass of our spherical body, since r is greater than R; therefore, Gauss's law yields that the gravitational attraction of a spherical body is the same as if all the mass were concentrated at a single point at its center: $g = GM/r^2$ with $\mathbf{g}$ pointing radially inwards (Q.E.D.).

Conservation of Energy

From the laws of mechanics it is often possible to derive certain **conservation principles**, i.e., the fact that certain quantities do not change in time although everything else may. One of the most important conservation principles is that of the conservation of energy. For example, from Newton's second law of motion and Newton's law of gravitation for two isolated point masses

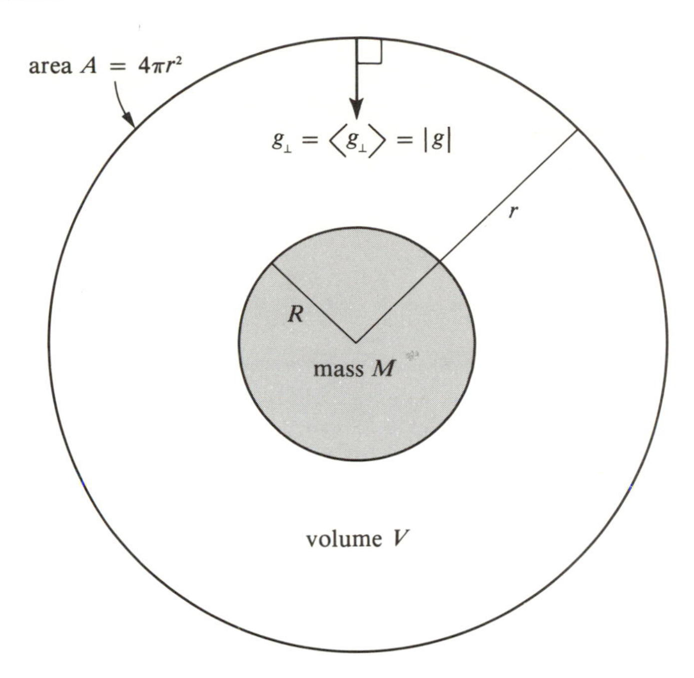

Figure 3.5. Gauss's proof of Newton's theorem about spherical bodies: $\langle g_\perp \rangle A = 4\pi GM \Rightarrow g = GM/r^2$ with $\mathbf{g}$ pointing radially inward.

(or spherical bodies), it is possible to prove that the total energy—which is the sum of the kinetic energies and the potential energy associated with gravitation—is a conserved quantity. In formula form we have

$$\text{total energy} = E = \frac{1}{2} m_1 v_1{}^2 + \frac{1}{2} m_2 v_2{}^2 + \left(-\frac{G m_1 m_2}{r} \right) = \text{constant}, \tag{3.2}$$

where v_1 and v_2 are the instantaneous speeds of masses m_1 and m_2 and r is the separation distance between the two bodies. Now, without using calculus, you may think it miraculous that statement (3.2) follows deductively from $\mathbf{F} = m\mathbf{a}$ plus Newton's theory of gravity, but with calculus, the proof is not difficult (Problem 3.1). You may complain that this makes the concept of conservation of energy very mathematical. It *is* mathematical. Energy is not something that you can touch, smell, or count. Kinetic energy, $mv^2/2$, can be said to be the energy associated with the motion of a material body, but that statement does not make the concept of kinetic energy any less mathematical. Similarly, gravitational energy $(-Gm_1m_2/r)$ can be said to be the potential for doing work by gravitational forces, but that statement does not make the intrinsic concept less mathematical either.

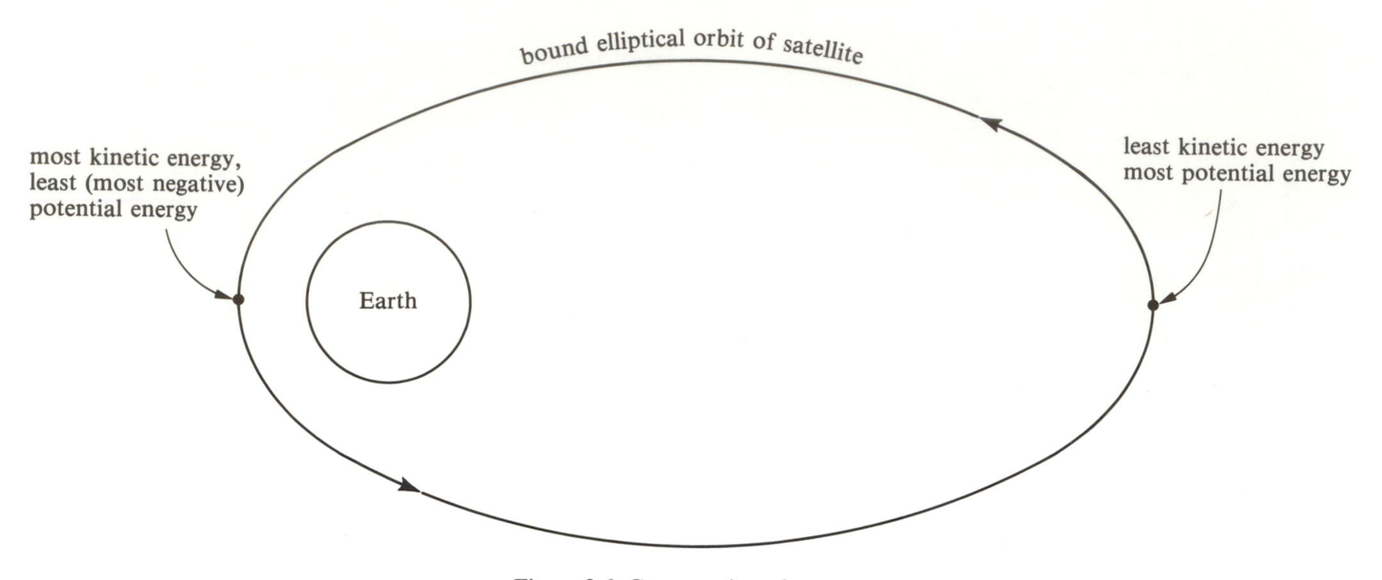

Figure 3.6. Conservation of energy.

Problem 3.1. This problem is for those who know calculus. To prove equation (3.2), show that Newton's laws imply, for two isolated masses m_1 and m_2 which feel only each other's mutual gravitation:

$$m_1 \frac{d\boldsymbol{v}_1}{dt} = \frac{Gm_1m_2}{r^2}\left[\frac{1}{r}(\boldsymbol{r}_2 - \boldsymbol{r}_1)\right],$$

$$m_2 \frac{d\boldsymbol{v}_2}{dt} = \frac{Gm_1m_2}{r^2}\left[\frac{1}{r}(\boldsymbol{r}_1 - \boldsymbol{r}_2)\right],$$

where $\boldsymbol{r}_1$ and $\boldsymbol{r}_2$ are the position vectors of masses 1 and 2, respectively; r is the magnitude of their difference; and the quantities in the square brackets yield the direction of the gravitational force (draw your own diagram). The velocities $\boldsymbol{v}_1$ and $\boldsymbol{v}_2$ are given by the derivatives of $\boldsymbol{r}_1$ and $\boldsymbol{r}_2$:

$$\boldsymbol{v}_1 = \frac{d\boldsymbol{r}_1}{dt}, \qquad v_2 = \frac{d\boldsymbol{r}_2}{dt},$$

Take the dot products of the first set of equations with the second set, and add the results to obtain

$$m_1\boldsymbol{v}_1 \cdot \frac{d\boldsymbol{v}_1}{dt} + m_2\boldsymbol{v}_2 \cdot \frac{d\boldsymbol{v}_2}{dt} = \frac{Gm_1m_2}{r^2}\frac{1}{r}\left[(\boldsymbol{r}_2 - \boldsymbol{r}_1) \cdot \frac{d\boldsymbol{r}_1}{dt} + (\boldsymbol{r}_1 - \boldsymbol{r}_2) \cdot \frac{d\boldsymbol{r}_2}{dt}\right].$$

With $v_1{}^2 = \boldsymbol{v}_1 \cdot \boldsymbol{v}_1$, $v_2{}^2 = \boldsymbol{v}_2 \cdot \boldsymbol{v}_2$, and $r^2 = (\boldsymbol{r}_2 - \boldsymbol{r}_1) \cdot (\boldsymbol{r}_2 - \boldsymbol{r}_1)$, derive the following identities:

$$\frac{d}{dt}(v_1{}^2) = 2\boldsymbol{v}_1 \cdot \frac{d\boldsymbol{v}_1}{dt},$$

$$\frac{d}{dt}(v_2{}^2) = 2\boldsymbol{v}_2 \cdot \frac{d\boldsymbol{v}_2}{dt},$$

$$\frac{d}{dt}(r^2) = 2(\boldsymbol{r}_2 - \boldsymbol{r}_1) \cdot \frac{d}{dt}(\boldsymbol{r}_2 - \boldsymbol{r}_1).$$

Show, therefore, that we may write the previous equation as

$$\frac{d}{dt}\left(\frac{1}{2}m_1v_1{}^2 + \frac{1}{2}m_2v_2{}^2 - \frac{Gm_1m_2}{r}\right) = 0,$$

which implies the conservation of total energy, equation (3.2).

The important thing to realize is that as the two bodies move, their kinetic energies (which are always positive) and their gravitational potential energy (which is always negative) may be converted back and forth into one another, but the sum total remains constant in time. As an example, Figure 3.6 shows the bound elliptical orbit of an Earth satellite. Artificial satellites are much less massive than the Earth; the kinetic energy of the Earth that results from the attraction of artificial satellites is completely negligible. Thus, in applying equation (3.2) to this example, we need only worry about the kinetic energy and potential energy of the satellite. The potential energy of the satellite is most negative when it is closest to the Earth; therefore, at this point (perigee), the kinetic energy (and the speed) of the satellite must be the greatest to keep the algebraic sum of kinetic and potential energies constant. In contrast, the potential energy of the satellite is least negative when it is furtherst from the Earth; therefore, at this point (apogee), the kinetic energy (and the speed) of the satellite must be the least. In the space age, these facts have become familiar knowledge. We see here that they are simple consequences of the principle of conservation of energy.

In the above and elsewhere, we have used a word, "bound," which we would now like to define more precisely. Consider the general case of equation (3.2) again, and assume that we refer to a frame which is at rest with respect to the center of mass of the system. For the two bodies to escape each other's gravitational clutch, i.e., for the separation r to become infinite, the total energy must be *positive*. To see this, notice that at $r = \infty$, the gravitational energy, $(-Gm_1m_2/r)$, vanishes, and E becomes a sum of squares, which must be positive. Conversely, if E is not positive, the two bodies cannot separate infinitely. Thus, when $E < 0$, we say that the system is **bound**. To be more precise, we should draw the dividing line between unbound orbits (when $E > 0$) and bound orbits ($E < 0$) at $E = 0$, when the two bodies can barely separate to infinity, where their relative velocity is zero. For the gravitational two-body problem, Newton found that the shapes of unbound orbits are hyperbolas; bound orbits are ellipses; and the transition $E = 0$ cases are parabolas (Figure 3.7).

Bound states are always a possibility with attractive forces. In particular, more than two bodies may be involved in a bound state. The planets in the solar system are bound to the Sun; the stars in the Milky Way are bound to the Galaxy; and galaxies may be bound to clusters of galaxies. An interesting question is whether the universe is bound to itself. The binding agent in other examples need not be gravitation; it could be electric forces, as with the electrons in an atom, or it could be nuclear forces, as with the protons and neutrons in the nucleus of an atom.

The difference in energies between the hypothetical state where all bodies of the system are infinitely separated from one another and the actual bound state is referred to as the **binding energy** of the system. Defined in this way, the binding energy is always a *positive* quantity. For example, two gravitating bodies have a binding energy $0 - E = -E$, where E as given by equation (3.2) is *negative* for elliptical orbits. The minus sign introduced by our convention for defining binding energy as the *difference* between two states avoids the inconvenience of discussing negative values of E.

We shall find the principle of conservation of energy to be useful in discussing all such cases. Clearly, this principle is a very powerful one; it is justifiably considered one of the *great* laws of physics. Its greatness comes from its generality; indeed, it sometimes appears in very subtle forms. Two examples are shown in Figure 3.8. High-school students are taught the "law of the lever" and the "$1/r^2$ law of radiation." However, these are not really separate laws; they are just two more applications of the general principle of conservation of energy. To derive the "law of the lever," notice that the top and bottom states of Figure 3.8(a) do not involve any kinetic energy. They involve the same potential energy, since weight 1 in the top state is twice as high as weight 2 in the bottom state. Clearly, then, it takes no extra energy to transform the top state into the bottom state—but this is precisely the "law of the lever" (Q.E.D.). To derive the "$1/r^2$ law of radiation" (Figure 3.8b), notice that this law requires a steady source of radiation, say, a star. Assume there is no absorption. Then the same amount of energy per unit time (**luminosity**, L) must cross successive spheres of larger and larger radii centered on the

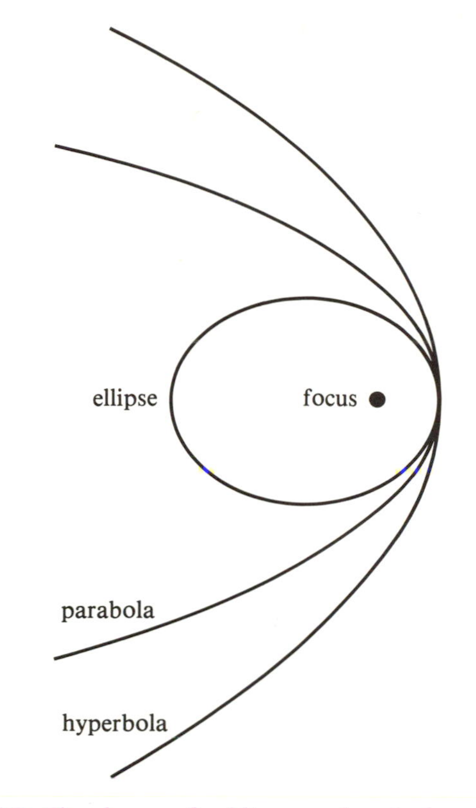

Figure 3.7. The shapes of orbits around a single gravitating point of force. Bound orbits ($E < 0$) are ellipses with the source of force at one focus. The transition cases ($E = 0$) are parabolas. Unbound orbits ($E > 0$) are hyperbolas. In all these cases, the motion is confined to a plane, and the radius vector sweeps out "equal areas in equal times," a law found empirically by Kepler for the elliptical orbits of planets and later shown by Newton to be a general consequence of the principle of conservation of angular momentum (see Figure 3.21).

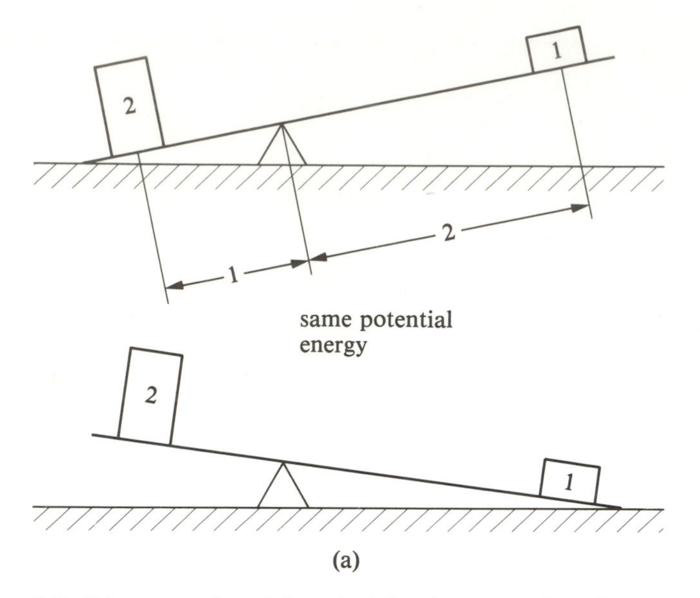

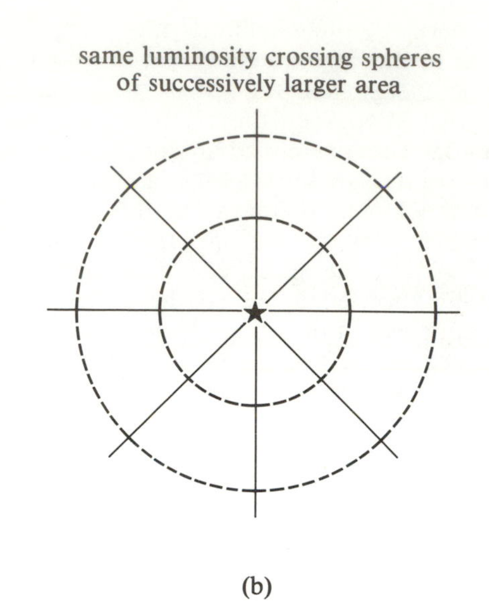

Figure 3.8. Other examples of the principle of conservation of energy. (a) The law of the lever. (b) The $1/r^2$ law of radiation.

source. The energy per unit area per unit time (**energy flux**, f) measured at a distance r from the source equals the luminosity divided by the surface area of a sphere of radius r (see Appendix B):

$$f = L/4\pi r^2. \tag{3.3}$$

In other words, for given luminosity L, the apparent brightness f must fall off as the inverse square of the distance r from the source—but this is precisely the statement of the "$1/r^2$ law of radiation" (Q.E.D.).

Problem 3.2. Does the light source have to have spherical symmetry for the $1/r^2$ law of radiation to apply? *Hint*: Consider the situation of a cone of light, i.e., the light directed into any solid angle.

Problem 3.3. This problem reconstructs the method by which Christian Huygens first estimated the distances to the stars. Suppose you view the Sun through a pinhole in a dark room. Suppose the angular diameter of the pinhole to be 600,000 times smaller than the angular diameter of the Sun. The image in the pinhole appears to you to be equally bright as a star which you observed the previous night and which you suspect to be a sun similar to our own. Assume that the images of stars are circular disks, and argue from the $1/r^2$ law of light that the star is 600,000 times further away than the Sun. If the Sun-Earth distance is 1.5×10^{13} cm, how far away is the star? Convert your answer to light-years.

The Electric Force

Just as mass acts as a source of a gravitational force field, so does electric charge act as a source of an electric force field. For point charges, Coulomb found the electric force to be directed along the line of centers and to vary as the product of the two charges divided by the square of the separation distance. In cgs units the proportionality constant is unity; i.e., it is "absorbed" into the cgs definition of a unit of charge. In formula form, Coulomb's law states that the force exerted on charge q_1 by charge q_2 has magnitude

$$F_{12} = q_1 q_2/r^2. \tag{3.4}$$

(See Figure 3.9.)

Ordinary matter contains both positively charged particles (protons, in the nuclei of atoms) and negatively charged particles (electrons, in the shells surrounding the nuclei of atoms). Electrons turn out always to have the same mass, $m_e = 9.11 \times 10^{-28}$ gm, and the same

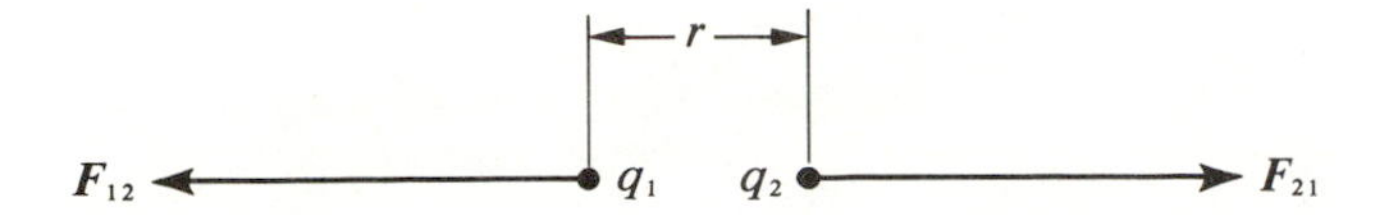

Figure 3.9. The electric force $\boldsymbol{F}_{12}$ exerted on charge q_1 by charge q_2 at a distance r is directed along the line of centers. This force is repulsive if the two charges have the same sign, and is attractive if the two charges have opposite sign. The electric force $\boldsymbol{F}_{21}$ exerted on charge q_2 by charge q_1 is equal in magnitude to $\boldsymbol{F}_{12}$ but oppositely directed. The magnitude of $\boldsymbol{F}_{12}$ is given by q_1q_2/r^2. The resultant electric force due to many point charges can be obtained by vector superposition just as for gravitation (Figure 3.3).

charge, $-e = -4.80 \times 10^{-10}$ esu (Appendix A). Indeed, in the quantum-mechanical theory of electrons, it is physically impossible to distinguish one electron from another. Protons are much heavier (about 2,000 times) than electrons, but their charge is identically opposite that of the electron, $+e$. Surely, this is not a mere coincidence, but must arise from an integral characteristic of matter (Chapter 16).

Relative Strengths of Electric and Gravitational Forces

Since electricity and gravity are both $1/r^2$ forces, it is easy to compare their relative strengths, say, for two electrons (Figure 3.10):

$$\frac{F_{\text{electric}}(2\text{ electrons})}{F_{\text{gravity}}(2\text{ electrons})} = \frac{ee/r^2}{Gm_em_e/r^2} = \frac{e^2}{Gm_e^2}$$
$$= \text{pure number} = 4.17 \times 10^{42}.$$

Problem 3.4. (a) Calculate the ratio of the electric force to the gravitational force between two protons (See Appendix A.) (b) Formulate Gauss's law for the electrostatic force law.

For bare charges, the electric force is *much, much* stronger than the gravitational force. For ordinary situations ("ordinary" by terrestrial standards), electromagnetic forces dominate the scene. Electromagnetic forces explain all of chemistry and biology, not to mention electronics. The only reason gravity can play a role at all is that electric charges come with both positive and negative signs, whereas mass comes in only one sign: positive. Thus, if matter is electrically neutral in bulk (containing equal numbers of electrons and protons, on the average), the electric forces cancel in bulk, and gravitational forces will dominate in bodies of sufficient

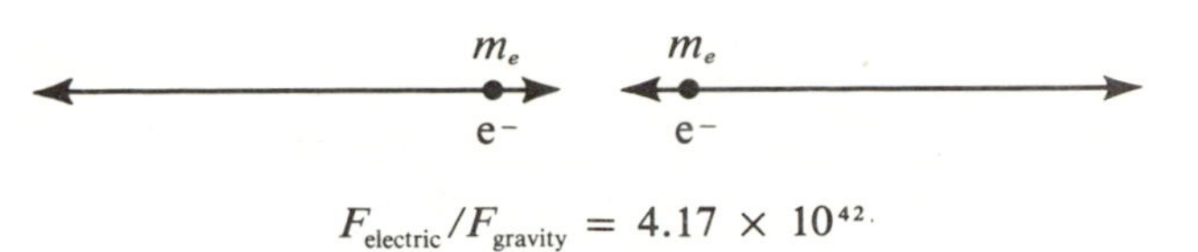

Figure 3.10. The repulsive electric force is much stronger than the attractive gravitational force between two electrons.

mass. *Sufficient* mass is needed for the following reason. Even if electrons and protons come in equal numbers, they are usually not exactly spatially coincident in ordinary matter. Thus, electromagnetic forces do not cancel out perfectly even in perfectly neutral matter. The leftover forces of electromagnetic origin in such cases account for many of the chemical and biological properties of terrestrial matter. However, the strength of the leftover forces decreases with distance at a faster rate than $1/r^2$; consequently, gravity in bodies of *sufficient* mass can always win out over such "short-range" forces.

Electromagnetism

In Chapter 2 we have already discussed the fact that stationary charges produce only electric fields, but moving charges can also produce magnetic fields. Accelerated charges emit electromagnetic radiation, one type of which is visible light. The unification of electricity, magnetism, and light was effected in the nineteenth century, largely by the efforts of Faraday and Maxwell (Box 3.1). However, the inner beauty of this unification was not fully appreciated until Einstein's development of Relativity Theory. We shall discuss special relativity

BOX 3.1

The Laws of Electromagnetism

(1) Electric charges act as sources for generating electric fields. In turn, electric fields exert forces that accelerate electric charges.
(2) Moving electric charges constitute electric currents. Electric currents act as sources for generating magnetic fields. In turn, magnetic fields exert forces that deflect moving electric charges.
(3) Time-varying electric fields can induce magnetic fields; similarly, time-varying magnetic fields can induce electric fields. Light consists of time-varying electric and magnetic fields that propagate as a wave with a constant speed in a vacuum.
(4) Light interacts with matter by accelerating charged particles. In turn, accelerated charged particles, whatever the cause of the acceleration, emit electromagnetic radiation.

in more detail shortly, but here we wish to make only two remarks.

First, Einstein adopted the basic philosophy that the laws of physics should apply with equal validity to moving and stationary observers—a "relativity" principle traceable back to Galileo. Further, he postulated that the notion of absolute motion is a fiction; the only thing that counts is relative motion. Thus, what is a stationary charge for one observer is a moving charge for an observer who moves relative to the first. But the first observer notices only electric fields (stationary charge), whereas the second notices both electric and magnetic fields (moving charge). This fact can be reconciled with the relativity principle only if motion transforms electric fields into magnetic fields, and vice versa. Indeed, Maxwell's equations for electromagnetism contained the correct transformation properties implicitly; it took Einstein's genius to recognize their physical consequences explicitly. The unification of electricity and magnetism involves, therefore, this "relativity" principle: no absolute distinction exists between electric and magnetic fields; what we call one or the other depends on our state of motion.

Second, although Coulomb's law, (equation 3.4) for stationary charges looks very similar to Newton's law (equation 3.1) for gravitation, the behavior of moving charges in Maxwell's theory is very different from the behavior of moving masses in Newton's theory. In particular, accelerated charges produce *electromagnetic radiation* in Maxwell's theory; accelerated masses do *not* produce *gravitational radiation* in Newton's theory. The crucial difference is the following. In Maxwell's theory, when one charge moves relative to another, the electromagnetic information about this change of configuration is passed from one charge to another "only" at the speed of light. In Newton's theory, when one mass moves relative to another, the gravitational information is transmitted instantaneously. This "action at a distance" was a distasteful feature even to Newton himself. Einstein felt these inadequacies of Newton's theory even more deeply, and his theory of **general relativity** was invented to provide a more satisfactory basis for gravitation. In particular, general relativity *does* predict that gravitational radiation is generated when masses are accelerated, a prediction which has yet to be verified experimentally, more than sixty years after the invention of general relativity. The experimental detection of such gravitational waves is currently being actively pursued by several groups throughout the world (Figure 3.11). The task is difficult (and, so far, controversial) because gravity is intrinsically such a weak force (compared, say, to electromagnetism). Only astronomical bodies possess enough mass to provide appreciable amounts of gravitational radiation, and the strongest sources (having the best combination of mass and acceleration) are far away. Thus, they are very difficult to detect even if we know where to look ("feel" would be a more apt word, since gravitational radiation is to be detected by its oscillatory tugging of matter). We shall have more to say on gravitational radiation and general relativity in Chapters 7, 10, and 15.

Figure 3.11. Joseph Weber, who invented, analyzed, and developed gravitational antennas. (Courtesy of Professor Joseph Weber.)

Nuclear Forces

In ordinary matter, the strong and weak interactions have only a short range, extending over about 10^{-13} and 10^{-15} cm, respectively. These forces are central to the structure and dynamics of atomic nuclei (which have characteristic dimensions of about 10^{-13} cm), but they do not manifest themselves in ordinary life. They intrinsically involve quantum-mechanical descriptions; hence, we shall delay discussing their properties until we come to thermonuclear reactions inside stars (Chapter 6).

Quantum Mechanics

Quantum mechanics grew from the difficulties that arose when classical mechanics and electrodynamics were

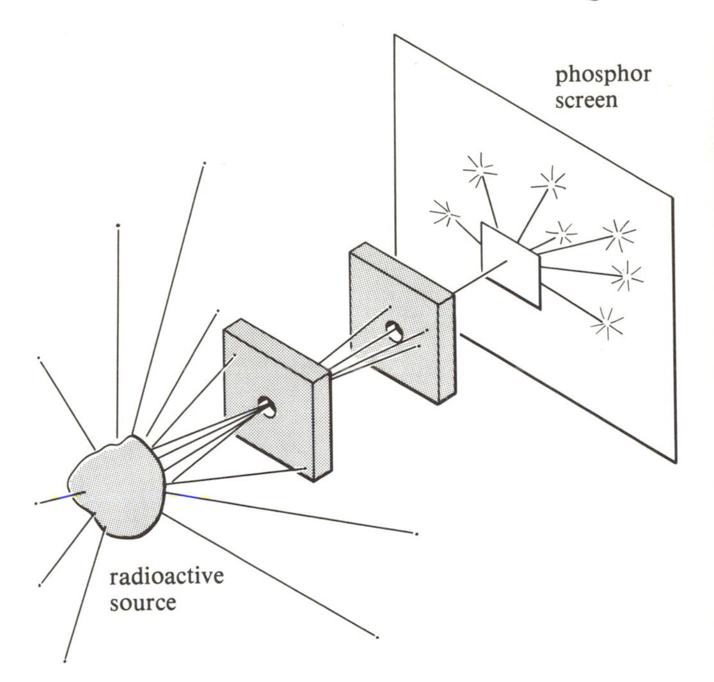

Figure 3.12. Rutherford's scattering experiment. A radioactive source emits fast nuclei of helium atoms (alpha particles). The alpha particles are collimated into a narrow beam defined by small holes in two thick walls of lead. The narrow beam strikes a thin foil of gold. The gold atoms deflect part of the beam onto a phosphor screen, which glows when struck by the alpha particles. The many large-angle scatterings produced by the gold foil led Rutherford to infer that the massive, positively charged part of the gold atom must reside in a small dense nucleus.

Figure 3.13. An aerial photograph of the Stanford linear accelerator, which runs under Highway 280. (Stanford Linear Accelerator Center.)

applied to models of microscopic bodies like atoms. Rutherford showed in a series of scattering experiments (Figure 3.12) that most of the mass and positive charge of an atom had to lie in a tiny nucleus. These experiments were the forerunners of all modern accelerator experiments, which aim at a better understanding of the fundamental structure of matter by studying what happens when very energetic particles are smashed against one another (Figures 3.13 and 3.14). Physicists were led by Rutherford's momentous discovery to view the atom as a bound state of negatively charged electrons circling the positively charged nucleus, much like a miniature solar system with electric forces replacing gravitational forces as the binding agent. However, in this classical picture the centripetal acceleration (see equation 2.2) of an electron caused by the attraction of the nucleus should lead the electron to radiate electromagnetic waves. This loss of energy must lead to orbital decay; i.e., the electron must dip closer to the nucleus and become more *bound*, so that the sum of the negative orbital energy E and the radiated energy adds to a constant value, as required by the conservation of total energy. In other words, this

Figure 3.14. Photograph inside the tunnel enclosing the proton synchrotron of Fermilab in Batavia, Illinois. (Fermi National Accelerator Laboratory.)

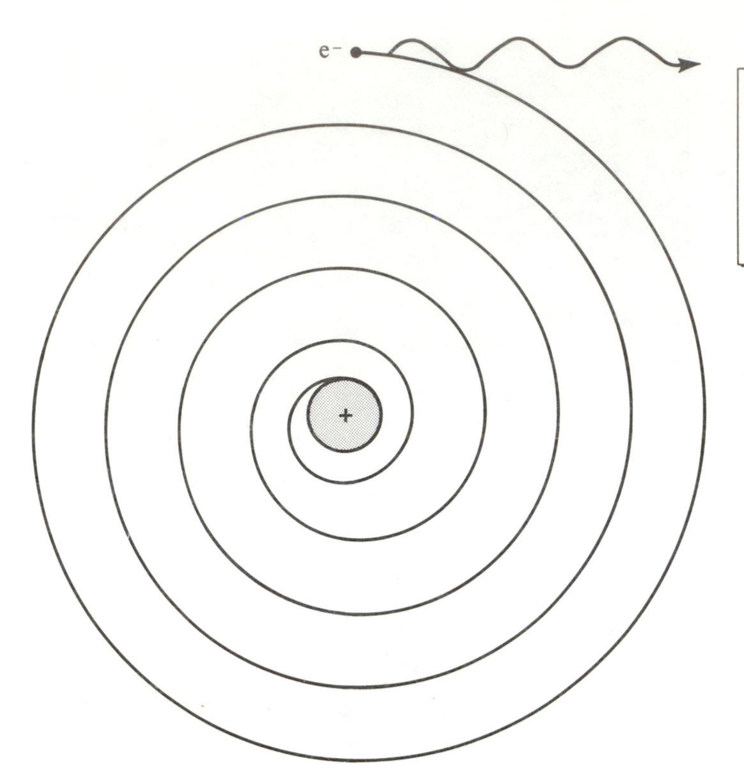

Figure 3.15. Orbit decay of an electron in an atom according to the classical theory. This classical model is in direct contradiction with the observed stability of atoms and the discrete nature of their radiation.

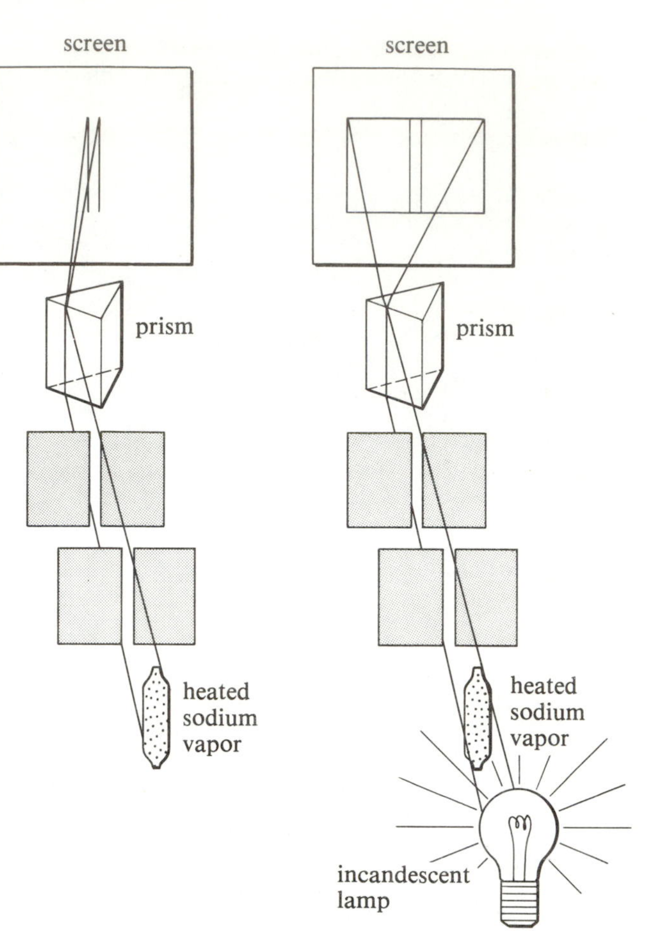

Figure 3.16. When the light from a heated sodium vapor is collimated and dispersed, two bright lines (the Sodium D lines) appear in emission on the screen. If a bright source of continuum radiation is placed in back of the sodium vapor, two dark absorption lines appear at the same wavelengths against the bright background of the continuum. This is an example of Kirchoff's law of radiation for gases.

classical model of the atom required the electron to spiral fairly quickly into the nucleus of the atom (Figure 3.15). Yet atoms obviously are stable, and when they emit radiation, they do not emit continuously at all wavelengths, as would result from a continuous spiraling of the electron into the nucleus. Instead, it was observed that when atoms are kept far apart from one another (in a gaseous state; see Figure 3.16), they radiate only at certain specific wavelengths which are always the same for the same type of atom. The experimental facts, therefore, were in direct contradiction with the theoretical classical model of the atom.

Gradually, during the following two decades, physicists came to realize that the laws of classical mechanics were only approximately true, and broke down when applied to microscopic bodies like atoms. To describe the physics of microscopic objects, a new underpinning was needed. That underpinning is quantum mechanics.

In classical mechanics, matter behaves like particles, and light behaves like waves. Quantum mechanics endows both matter and light with intrinsic properties of *both* waves and particles. The **wave-particle duality** of matter and light is the single most important concept introduced

BOX 3.2

Important New Concepts Introduced by Quantum Mechanics.

(1) Wave-particle duality, which leads to
  (a) quantized energy levels for bound states,
  (b) Heisenberg's uncertainty principle for free states.

(2) Pauli exclusion principle, which, coupled with quantized values for the angular momentum (including spin), leads to
  (a) an explanation of the periodic table of the chemical elements,
  (b) fundamentally different statistical behavior of fermions and bosons.

by quantum mechanics, and it manifests itself in the existence of a new fundamental constant of nature:

$$\text{Planck's constant} = h = 6.63 \times 10^{-27}\ \text{gm cm}^2\ \text{sec}^{-1}.$$

Any formula which has h in it contains quantum-mechanical considerations and cannot be derived from classical mechanics. We now summarize some of the important implications for matter and radiation. (See also Box 3.2.)

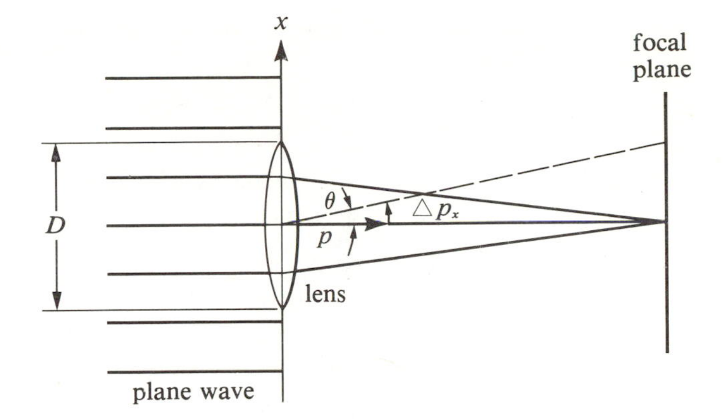

Figure 3.17. The diffraction limit according to quantum mechanics. In the classical ray-tracing technique, the parallel rays of light from a plane wave which enter a lens would be brought to a perfect focus in the focal plane as is shown by the convergence of the solid lines. The wave theory of light shows that a perfect focus is not possible because of the diffraction effects associated with the finite aperture of the lens. From a quantum point of view, a photon could have entered the lens anywhere along the diameter of the lens. This uncertainty in x-position is associated with an uncertainty Δp_x in the photon's x-component of momentum. Consequently, a photon, which in the absence of this uncertainty would have been brought to the optical axis of the focal plane, may now be deflected through an angle θ. For $\Delta p_x \ll p$, the original momentum of the photon, θ is approximately given by $\Delta p_x/p$.

The Quantum-Mechanical Behavior of Light

In 1905, Einstein showed that the photoelectric effect could be understood only if light came in discrete packets (photons) of energy E related to the wavelength of the light by the formula

$$E = \frac{hc}{\lambda}. \tag{3.5}$$

Since the energy E carried by light was known to be c times its momentum p, de Broglie later proposed that each photon carried momentum

$$p = \frac{h}{\lambda}. \tag{3.6}$$

The relationship to Maxwell's concept of light (see Figure 2.4) is that the crest of the wave (where $\mathbf{E}$ and $\mathbf{B}$ are largest) corresponds to the maximum probability for being the location of photons with the properties given by equations (3.5) and (3.6). Notice that these two equations state that shorter-wavelength photons carry more energy and more momentum than longer-wavelength photons. The qualitative implications of these two equations are even more important than their quantitative results. *Nothing epitomizes the wave-particle duality of light better than these two equations*; they lie at the heart of later developments by Bohr, Heisenberg, and Schrödinger. It is therefore ironic (some physicists would say tragic) that Einstein in his advanced years wished to repudiate the probabilistic interpretation of modern quantum mechanics with the rejoinder: "God does not throw dice."

The probabilistic implications of quantum mechanics are best summarized by the set of rules known as **Heisenberg's uncertainty principle**. The most easily grasped of these rules is the following. According to Heisenberg, the position and momentum of a free particle in any direction cannot be *both* measured to arbitrary accuracy. In fact, the product of the *uncertainties* in the measurements of the position and the momentum must typically exceed Planck's constant h. (A more accurate estimate replaces h by $\hbar/2 = h/4\pi$ for the minimum uncertainty product.) In formula form we may write this rule as

$$(\Delta x)(\Delta p_x) > h, \tag{3.7}$$

where we have written the subscript x on p_x to remind ourselves that only the component of the vector momentum in the same direction as the position measurement x shares in the uncertainty. The physical meaning of equation (3.7) is the following. We may choose to measure either the position or the momentum of a particle with very high accuracy, but because of *unavoidable* quantum-mechanical effects, the accuracy of measuring the other variable must necessarily be very poor. For example, if Δx is very small, Δp_x must be at least as large as $h/\Delta x$. Since h is a very small quantity, such fundamental uncertainties are not very apparent in ordinary experience; but in dealing with atomic and subatomic phenomena, they are paramount. On microscopic scales, *nature is intrinisically "fuzzy."*

A macroscopic example will help to strike home the lesson that wave-particle duality lies at the heart of Heisenberg's uncertainty principle. Consider again the diffraction limit of a lens, as shown in Figure 3.17. In Chapter 2, we noted that the wave theory of light predicts that a lens of diameter D cannot focus a parallel beam of light with wavelength λ to within an angle better than the diffraction limit:

$$\theta = 1.22(\lambda/D).$$

How do we interpret this result in terms of a particle theory of light? Let x be the direction associated with the diameter of the lens in the plane of the page. The incident (infinite) plane-wave of photons is known to contain *exactly* zero x-component of momentum (i.e., $\Delta p_x = 0$), but this is compatible with the uncertainty principle, since in a plane wave we have no idea where any photon is in the x-direction (i.e., $\Delta x = \infty$). Consider now the photons which are focused by the lens. Such photons are known to have passed somewhere within one diameter of the lens center ($\Delta x = D$). According to Heisenberg's uncertainty principle, upon passing through the lens, each photon is given an unpredictable kick

$$\Delta p_x > h/\Delta x = h/D$$

in the x-direction beyond that required for a perfect focus. This results in a random angular deflection of amount given by the ratio of Δp_x to the original momentum of the photon:

$$\theta = \frac{\Delta p_x}{p} > \frac{h/D}{h/\lambda} = \frac{\lambda}{D}.$$

The more precise wave analysis shows that the coefficient of the inequality is 1.22 instead of unity for this case. In order of magnitude, however, we see that diffraction-limited angular resolution arises as a result of the uncertainty principle, and is, therefore, unavoidable (assuming quantum mechanics to be correct). Of the two ways of looking at things, Heisenberg's uncertainty principle is much more general because it applies to matter as well as to radiation. (Thus, electron microscopes also have a "diffraction limit" to their best resolution.)

The Quantum–Mechanical Behavior of Matter

In a monumental Ph.D. thesis of extreme brevity, de Broglie proposed that the formula $p = h/\lambda$ held also for particles of matter. Indeed, this formula is normally used in a reverse sense; namely, that associated with every particle of momentum p is a (probability) wave with a wavelength λ given by

$$\lambda = h/p.$$

Let us apply this concept to the structure of the hydrogen atom. The hydrogen atom has a single proton as its nucleus, and a single (bound) orbital electron. For simplicity, visualize the hydrogen atom as a stationary proton with an electron which circles it. Following the work of Bohr, de Broglie proposed that the size of the electron's circular orbit is not arbitrary; instead, the circumference must be an integral number times the wavelength (3.5) associated with the motion of the electron (Figure 3.18):

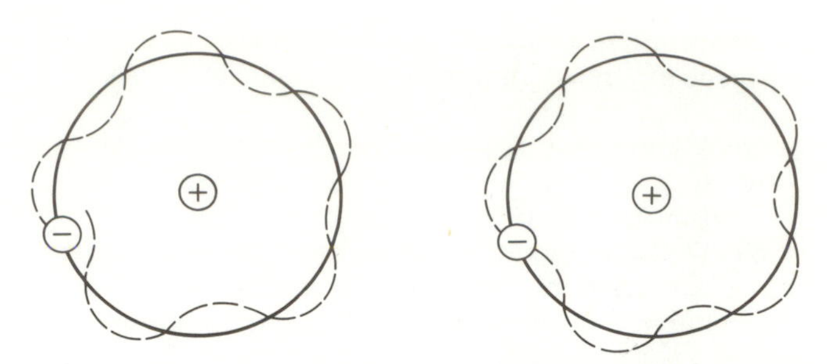

Figure 3.18. The Bohr-de Broglie interpretation of quantized spacings for circular orbits. If the number of wavelengths $\lambda_e = h/p_e$ in the circumference of the orbit of the electron is not equal to an integer n (left diagram), then that orbit is not allowed. The only allowed circular orbits have an integral number (like 5 in the right diagram) of wavelengths in the circumference.

$$2\pi r = (\text{integer}) \cdot \lambda_e,$$
$$\text{where } \lambda_e = h/p_e = h/m_e v,$$

and where v is the speed of the electron's orbital motion. The latter can be derived from $F = ma$, with $F = e^2/r^2$ (see equation 3.4) and with $a = v^2/r$ (see equation 2.2), to yield the allowable values of the radius r:

$$r = n^2(\hbar^2/m_e e^2),$$

where n is an integer called the **principal quantum number**, $\hbar = h/2\pi$ and is called ***h*-bar**, and the combination $(\hbar^2/m_e e^2)$ is called the **Bohr radius**.

Problem 3.5. Verify the algebra which leads to the above result and calculate the numerical value of the Bohr radius.

The smallest value allowed for r is obtained by taking $n = 1$; this corresponds with the *ground* electronic state of atomic hydrogen. The next smallest value is obtained by taking $n = 2$; this corresponds to the *first* excited state of atomic hydrogen. Notice that for circular orbits, the first excited state has a radius which is $2^2 = 4$ times larger than the radius of the ground state. Because of nature's intrinsic fuzziness, one should not speak of electron trajectories but of electron (probability) "shells"; nevertheless, 1 times the Bohr radius, 4 times the Bohr radius, etc., are good estimates for the size of the hydrogen atom when it is in its ground state, first excited state, etc. Even in its ground state, therefore, the hydrogen atom is about 10^5 times bigger in size than the 10^{-13} cm of its nucleus. *Ordinary atoms are mostly empty space* between a small dense nucleus and a point-like electron which has the highest probability of being found in discrete shells about the nucleus.

The above result has profound implications for the nature of the world that we live in. First, atoms are stable because electrons cannot gradually spiral in toward the nucleus, as in the classical theory. Electrons must reside in discrete shells; in particular, the electron in the hydrogen atom cannot come closer to the nucleus than one Bohr radius. Second, even though atoms are mostly empty space, we do not plunge into the center of the Earth; our bodies' matter does not "slip" around the atoms of the Earth under the latter's gravitational pull. When the electronic shells in the atoms of our feet (or shoes) push against the electronic shells of the atoms that make up the ground, huge electric forces (compared to gravitational forces) are set up that prevent further interpenetration. In this manner does the Earth feel like like solid ground, despite the fact that it is mostly empty space!

Spectrum of the Hydrogen Atom

Let us now combine the above concepts in a study of the radiation emitted and absorbed by atoms. Let us start with the hydrogen atom. Assuming circular orbits, we have already shown that only the values

$$r = n^2(\hbar^2/m_e e^2),$$

with n equal to an integer, are allowable for the radius r. If we continue to ignore the motion of the proton (which is much less than that of the electron), we can see that the allowable values for the orbital energy (kinetic energy plus electric potential energy),

$$E = m_e v^2/2 + (-e^2/r),$$

must also be quantized (have only discrete values). These values are given by

$$E = -n^{-2}(m_e e^4/2\hbar^2). \tag{3.8}$$

Problem 3.6. Refer to the algebra of Problem 3.5 and derive equation (3.8) for the energy levels of atomic hydrogen. The combination $(m_e e^4/2\hbar^2)$ has the units of energy; what is its numerical value?

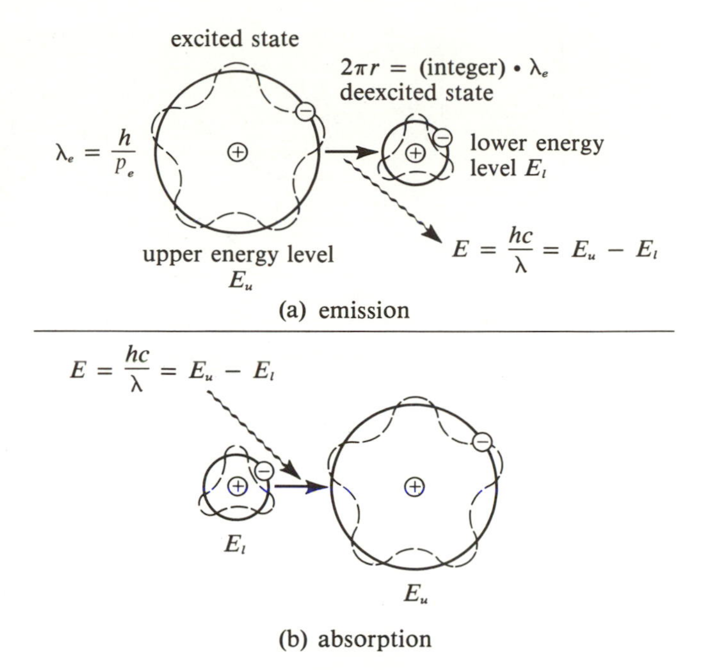

Figure 3.19. Spectral lines of atoms.

Notice that E is *negative* for these *bound* states, and that E approaches zero as $n \to \infty$ (ionization limit). In the ionization limit, the electron moves so far from the nucleus that it becomes detached from it. An ionized hydrogen atom consists, therefore, of a free proton and a free electron.

Consider now a hydrogen atom in an excited state (say, $n' = 5$), and suppose the electron to jump spontaneously to a less excited state (say, $n = 3$) as shown in Figure 3.19a. The orbital energy $E_{n'}$ of the former state is algebraically greater (less negative) than the orbital energy E_n of the latter state. The energy difference $E_{n'} - E_n$ is carried off by the spontaneous **emission** of a photon, whose energy E is related to its wavelength λ by Einstein's formula (3.5). To satisfy the conservation of energy, E must equal $E_{n'} - E_n$; i.e., the wavelength of the photon must be given by

$$\lambda = \frac{hc}{E_{n'} - E_n}.$$

Since $E_{n'}$ and E_n have definite fixed values which depend only on the atom under consideration, the resulting wavelength λ for the emitted photon also has a definite fixed value. The reverse process is also possible (Figure 3.19b): where a photon of wavelength λ given by the above formula is *absorbed* by a hydrogen atom in state n, causing the atom to jump up to the more excited state n'. The detection of a spectral line at this wavelength, either in

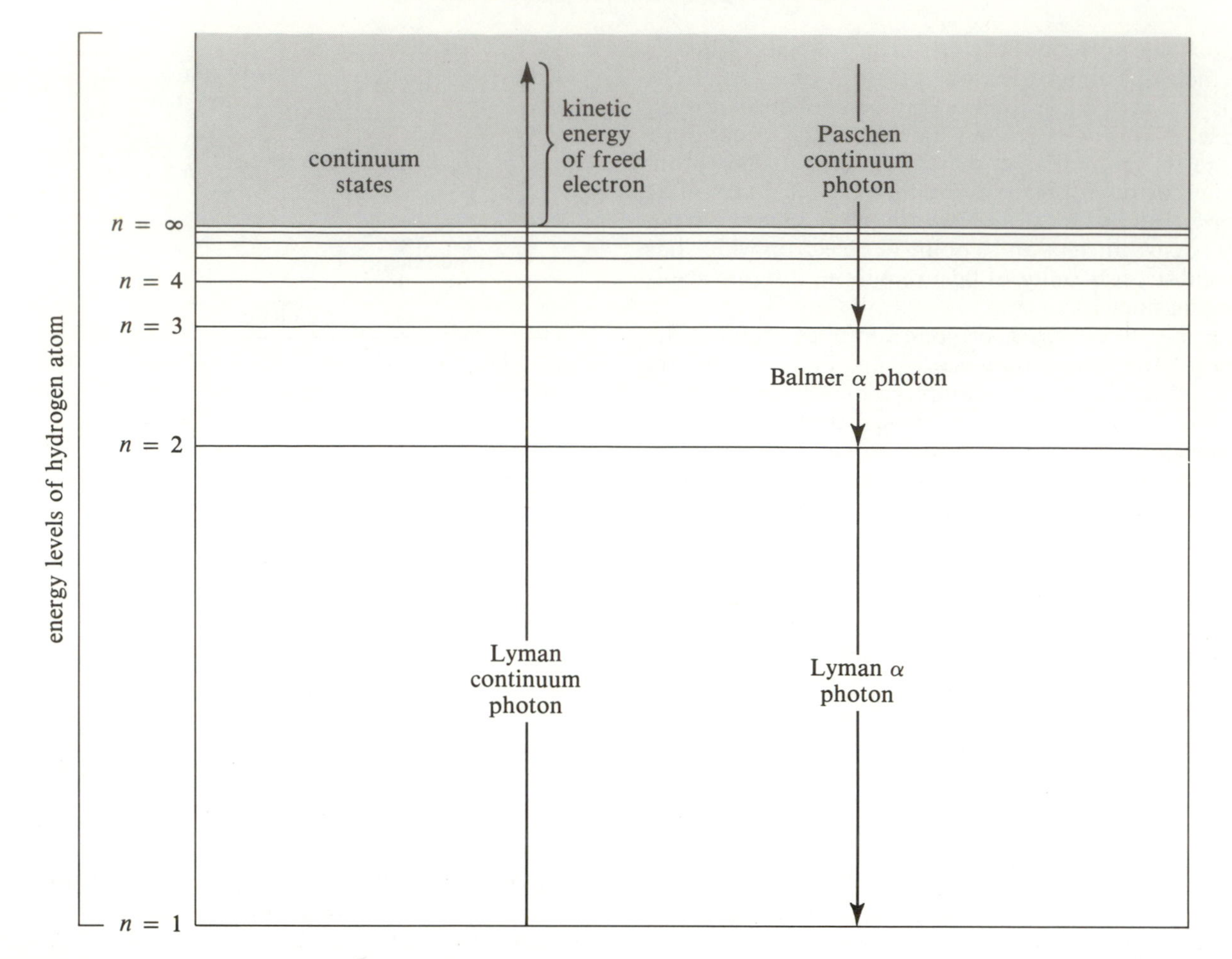

Figure 3.20. Radiative transitions to and from the continuum. Bound states of hydrogen are represented by the horizontal lines labeled $n = 1, n = 2, \ldots, n = \infty$, with the height from the ground state ($n = 1$) proportional to the energy difference of the states. Above the last bound state, $n = \infty$, exists a continuum of free states (ionized hydrogen). An electron can be freed into the continuum from the ground state by the absorption of a Lyman continuum photon. Conversely, a free electron may be captured, say, into the level $n = 3$ with the emission of a Paschen continuum photon. The electron in $n = 3$ may then cascade to the ground level by the subsequent emission, say, of a Balmer α photon and a Lyman α photon.

emission or in absorption, would constitute spectroscopic evidence for the presence of hydrogen.

Problem 3.7. For atomic hydrogen show that equation (3.8) results in the expression

$$\lambda = 4\pi\left(\frac{\hbar c}{e^2}\right)\left(\frac{n^2\hbar^2}{m_e e^2}\right)\left(\frac{n'^2}{n'^2 - n^2}\right),$$

where n' and n are, respectively, the principal quantum numbers associated with the upper and lower levels. The combination $\hbar c/e^2$ is known as the inverse of the "fine-structure constant" and has an approximate value of 137. The combination $n^2\hbar^2/m_e e^2$ is the size of the hydrogen atom in the lower state. Thus, if the last term is of order unity, the hydrogen atom typically emits and absorbs radiation with wavelength roughly 10^3 times its own size. (This fact leads to an important approximate method for calculating radiation probabilities, but it lies outside the scope of this book.) Let us calculate here merely the numerical value of the wavelength of the $n' = 2$ to $n = 1$ transition (Lyman alpha emission). Are the $n' > 1$ to $n = 1$ transitions (Lyman series) in the infrared, optical, or ultraviolet? Calculate the wavelength of the $n' = 3$ to $n = 2$ transition (Balmer alpha, also called Hα, in emission). Are the $n' > 2$ to $n = 2$ transitions (Balmer series) in the infrared, optical, or ultraviolet? If $n' = 3$ to $n = 2$ is Hα emission, and $n' = 4$ to $n = 2$ is Hβ emission, what do you think $n' = 5$ to $n = 2$ is called? *Hint*: Consult the Greek alphabet listed in Appendix C.

In addition to the discrete line transitions mentioned above, a hydrogen atom is also capable of **scattering** radiation and making transitions to and from the **continuum**. For example, consider a hydrogen atom in its ground state, $n = 1$. It takes energy $E_\infty - E_1$, corresponding to a photon of wavelength 912 Å, to cause a transition to the ionization limit, $n = \infty$. Photons with wavelength longer than 912 Å (the Lyman limit) can be absorbed only if they correspond to one of the discrete wavelengths of the Lyman series (refer to Problem 3.7). If they do not correspond to one of these discrete wavelengths, they can at best be scattered by the hydrogen atom; i.e., their direction of propagation would be changed, but their energy and wavelength would be unaffected. This scattering by atoms or molecules is, of course, the *microscopic* cause of the phenomena of refraction and reflection discussed in Chapter 2.

Photons with a continuum of wavelengths *shorter* than 912 Å may **photoionize** the hydrogen atom, with the excess photon energy above the energy of the Lyman limit contributing to the kinetic energy of the freed electron (Figure 3.20). Similar ionization phenomena prevent extraterrestrial X-rays from penetrating the Earth's atmosphere (refer back to Figure 2.3). In reverse, a free electron may be captured by a free proton with the emission of a continuum photon. If the electron is captured in an excited state of hydrogen, it may be subsequently **cascade** to the ground level by the sequential emission of a number of line photons. Such **recombination radiation** plays an important role in the fluorescence seen in Color Plates 6 and 8.

Angular Momentum

In principle, the spectra of atoms other than hydrogen could be explained in the same way as the spectrum of atomic hydrogen. In practice, some complications arise from the presence of more electrons. The most important of these complications have to do with quantum-mechanical considerations of the electron's angular momenta.

The concept of angular momentum has a classical origin like that of the concept of linear momentum and energy. For a single particle, the angular momentum $\boldsymbol{j}$ relative to any point (the origin of the coordinate system, say) is the vector cross product of the radius vector $\boldsymbol{r}$ from that point to the particle and the linear momentum $\boldsymbol{p}$ of the particle:

$$\boldsymbol{j} = \boldsymbol{r} \times \boldsymbol{p}, \tag{3.9}$$

where the vector cross product can be visualized geometrically with the aid of Figure 3.21. The motion of any one of the planets about the Sun can be reduced to a problem of a body orbiting a center of force. In this case $\boldsymbol{j}$ is a conserved quantity. Figure 3.21 explains how the conservation of $\boldsymbol{j}$ implies Kepler's "equal areas rule" and the fact that the planet's orbit is confined to a plane (Chapter 18). Kepler's laws are actually only approximate, because the planets interact with each other as well as with the Sun. However, the angular momentum of the entire solar system is still conserved as long as we take the vector sum of all the individual angular momenta. In fact, for any *isolated* collection of particles, the total angular momentum is a conserved quantity, both in magnitude and in direction (Figure 3.22).

For solid body rotation about an axis of symmetry, it is possible to write the total angular momentum $\boldsymbol{J}$ of a body in a form reminiscent of the formula $\boldsymbol{p} = m\boldsymbol{v}$ for linear momentum. This formula is

$$\boldsymbol{J} = I\boldsymbol{\omega}, \tag{3.10}$$

where I is called the **moment of inertia** (units = gm cm^2) and $\boldsymbol{\omega}$ is the angular velocity (units = radian/sec). The moment of inertia is a measure of the resistance of a rotating body to having its state of rotation changed, just as mass is a measure of the resistance of a moving body to having its state of translation changed. A macroscopic example of the conservation of angular momentum is provided by a spinning skater. When she pulls in her outstretched arms and decreases her moment of inertia I, a noticeable increase in her rate of spin $\boldsymbol{\omega}$ results because of the tendency to conserve the product $\boldsymbol{J} = I\boldsymbol{\omega}$ (Figure 3.23).

Angular momentum is clearly a useful and important concept in classical mechanics. It is even more useful and more important in quantum mechanics. Applied to the structure of atoms, quantum-mechanical angular momentum comes in two varieties:

(a) orbital angular momentum of an electron about the nucleus, which for high quantum numbers can be thought of in terms of the classical analogue, equation (3.9);

(b) spin angular momentum of an electron, which has only one value and therefore corresponds only vaguely to the classical analogue, equation (3.10).

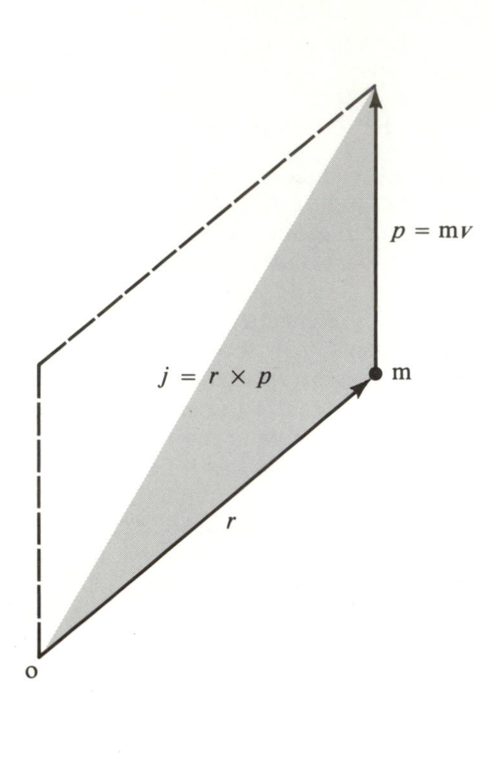

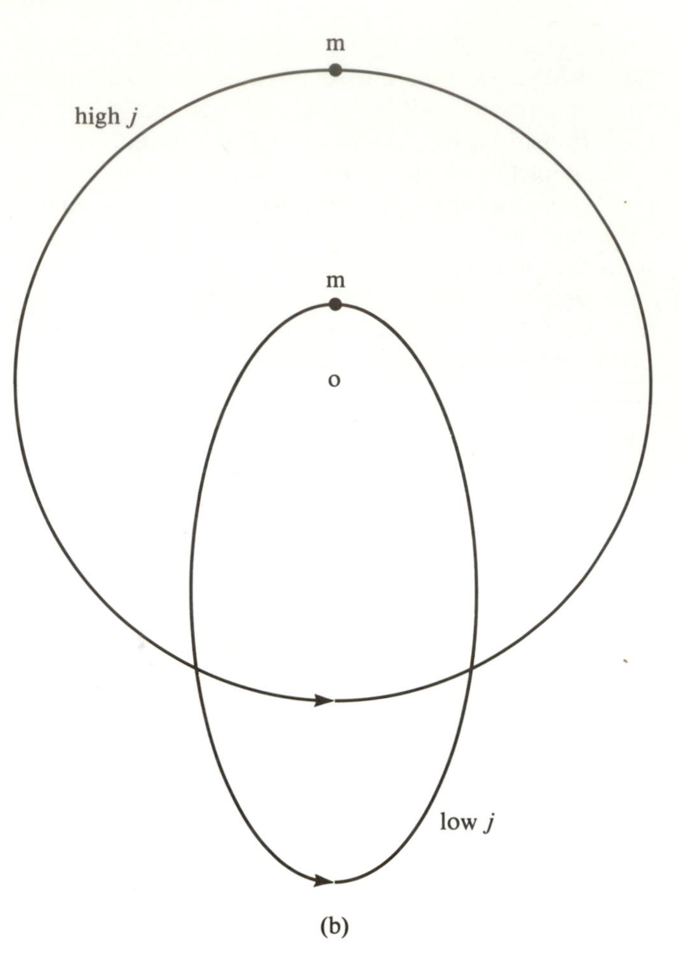

Figure 3.21. (a) Angular momentum $\boldsymbol{j}$ of a point mass m with respect to the point O (the origin usually) is given by the vector cross product of the radius vector $\boldsymbol{r}$ from O to m and the linear momentum $\boldsymbol{p}$ of the particle. The direction of $\boldsymbol{j}$ is given by the right-hand rule, and its magnitude is equal to the area of the parallelogram associated with $\boldsymbol{r}$ and $\boldsymbol{p}$ (refer back to Figure 2.6). Kepler's second law for the motion of the planets about the Sun (see Chapter 17) states that equal areas are swept out in equal times by the radius vector $\boldsymbol{r}$ from the Sun to the planet (O to m in the above diagram). But the area swept out per unit time is proportional to the shaded triangle in the above diagram, which in turn is just half of the magnitude of $\boldsymbol{j}$. Thus, we see that Kepler's second law is equivalent to the conservation of angular momentum for the planet. Moreover, since the direction of $\boldsymbol{j}$ is also conserved, we see that the planetary orbit is confined to a plane. (b) For gravitational binding to a center of force O of given energy E, orbits with low values of angular momentum j are more elongated (containing less area) and dip closer to O than orbits with high j.

Orbital Angular Momentum

Quantum angular momentum can be thought of as vectors with funny projection properties. The orbital angular momentum of an electron in an atom can take

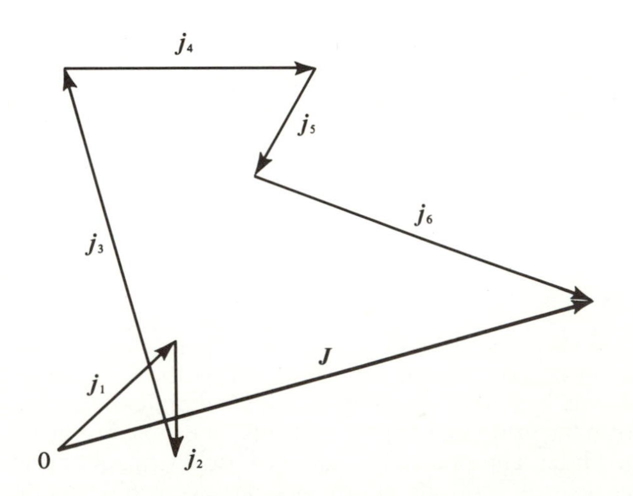

Figure 3.22. The total angular momentum $\boldsymbol{J}$ of any collection of point particles can be obtained by adding vectorially ("head to tail") their individual angular momenta, $\boldsymbol{j}_1, \boldsymbol{j}_2, \boldsymbol{j}_3$, etc.

Figure 3.23. The spin-up of an ice-skater. When the twirling skater has her arms out, her moment of inertia I is large and her rate of spin ω is small for a given angular momentum $J = I\omega$. When she pulls her arms in, her moment of inertia I decreases, and the conservation of angular momentum implies that her rate of spin ω increases.

on only integral multiples of $\hbar$ for its magnitude. Call this integer l. If the principal quantum number is n, l can take on the values 0, 1, 2, up to $n - 1$. Moreover, the orbital angular momentum can only have $2l + 1$ quantized directions such that the component m_l of the vector $\boldsymbol{l}$ along any fixed direction takes on only the values $-l$ to $+l$ in increments of 1. Thus, if $n = 1$ (innermost radial shell), l can equal only 0 and m_l can only equal 0. If $n = 2$ (next shell out), l can equal either 0 or 1; if $l = 0$, m_l can again only take on the single value 0, whereas if $l = 1$, m_l can take on three possible values, $-1, 0, +1$.

Problem 3.8. List the possible combinations of l and m_l if $n = 3$.

Spin Angular Momentum

Elementary particles also possess a spin angular momentum of magnitude, $s\hbar$, where s is either an integer (0, 1, 2, ...) or a half-integer (1/2, 3/2, 5/2, ...). Unlike the orbital quantum number l, however, s has a definite value which depends only on the type of particle and not on its quantum state (this is true only for elementary particles, not for compound particles). The component m_s of the vector $\boldsymbol{s}$ along any fixed direction can generally take on $2s + 1$ possible values, from $-s$ to $+s$ in increments of 1. Thus, the electron, which has spin 1/2, has only two possible spin orientations: $m_s = -1/2$ or $+1/2$ (antiparallel or parallel). The photon has spin 1; so one might think that it should have three possible spin orientations: $m_s = -1, 0, +1$ (antiparallel, perpendicular, or parallel). In fact, because the photon travels at the speed of light, c, only two of these spin states can manifest themselves: $m_s = -1$ or $+1$ (antiparallel or parallel to the direction of propagation). The two independent spin states of the photon correspond to the classical fact that light has two independent senses of polarizations. (In Figure 2.4 we took these to be the two states of *linear* polarization, but we could as well take them to be two states of *circular* polarization in which the radiation field would carry [spin] angular momentum.)

Quantum Statistics

Particles with half-integer spins are called **fermions** (after Enrico Fermi); particles with integer spins are called **bosons** (after Satyendra Bose). Fermions include the electron, the proton, and the neutron, which constitute the vast bulk of the observable material universe. There is a fundamental difference between fermions and bosons because of the quantum-mechanical rule known as **Pauli's exclusion principle**. The origin of this rule—like that of many others in quantum mechanics—is highly mathematical, having to do with the symmetry and antisymmetry of allowable wavefunctions. This mathematical nature of modern quantum mechanics makes it very difficult to get an intuitive grasp of the subject. What is important for us is the result for fermions:

> No two identical fermions can occupy the exact same quantum state.

The "quantum state" for the bound electron of an atom means the set of numbers n, l, m_l, and m_s. We may characterize this result colloquially by the statement that electrons behave antisocially: *they don't like to be any place where there's already another electron.*

Photons, being bosons, do not have to obey the Pauli exclusion principle. As a consequence, as many photons as one likes can occupy, in principle, the same quantum state. Thus, the wavelike aspect of each photon—see equation (3.6)—has a chance to reinforce the wavelike aspects of other similar photons and to produce a *macroscopic* effect. Indeed, it is only where very many photons are superimposed coherently that we have Maxwell's classical description of electromagnetic radiation (see Figure 2.4). Conversely, Pauli's exclusion principle *does* forbid a similar superposition of electrons or, more generally, of any of the fermions that make up ordinary matter. This is why the wavelike aspects of such particles—see equation (3.5)—were so late to be discovered. The intrinsic wave properties of ordinary matter cannot be manifested macroscopically. They can enter via the particles, so to speak, only one at a time. Nevertheless, because the physical world *is* built up with these particles, one at a time, the subtle but overwhelmingly important implications of the quantum basis of matter touches every aspect of our daily existence. Let us begin with a discussion of the chemical properties of matter.

The Periodic Table of the Elements

With the information supplied above, we are in a position to elucidate the electronic shell structure of multi-electron atoms. The basic principles are spelled out in Box 3.3.

As an important application of the ideas in Box 3.3, let us now consider the periodic table of the chemical elements (Figure 3.24, also Appendix D).

Hydrogen (H), containing one proton in its nucleus and one orbital electron is the simplest atom; it is also the first entry in Mendeleev's periodic table. In its ground state, the electron occupies $n = 1$, $l = 0$, $m_l = 0$, and either $m_s = +1/2$ or $m_s = -1/2$.

BOX 3.3

Spherical approximation of electronic shell structure of multi-electron atoms

(a) In the lowest order of approximation, consider each electron in a multi-electron atom to be stationed in a spherically symmetric electric field provided by the attraction of the nucleus and the average repulsion of the rest of the electrons. Inner electrons "shield" outer electrons from feeling the full positive charge of the nucleus; thus the binding force does not generally vary as $1/r^2$. In this approximation, the orbital angular momentum of each electron is conserved, but its energy will depend on the quantum number l in addition to the principal quantum number n.

(b) Low values of the quantum number l associated with the orbital angular momentum allow the electron to sample regions which are closer to the nucleus for a given value of n. Thus, low values of l tend to correspond to lower energies than high values of l for the same n. Occasionally, the l consideration can overcome the n consideration. For example, the $n = 4$, $l = 0$ orbitals are lower in energy than the $n = 3$, $l = 2$ orbitals.

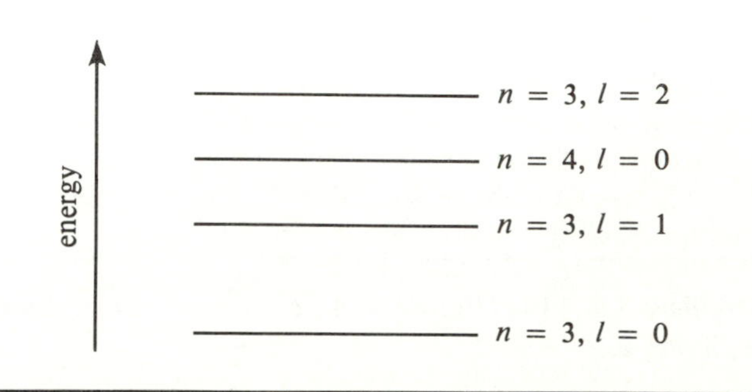

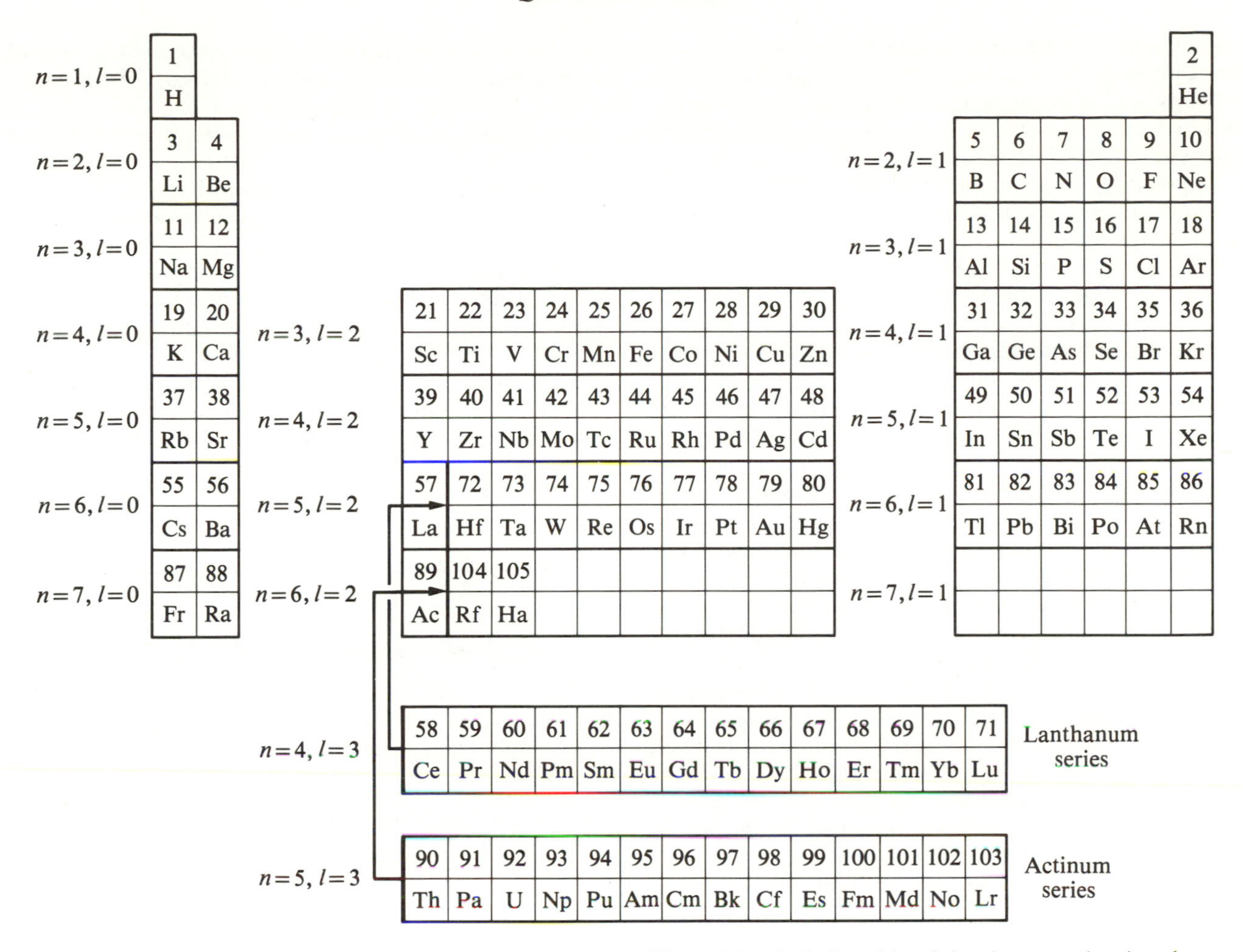

Figure 3.24. Periodic table of the elements, showing the separation into shells and subshells (adapted from B. H. Mahan, *University Chemistry*, Addison-Wesley, 1965).

Helium (He), containing two protons (plus neutrons) in its nucleus and two orbital electrons, is the next entry. In helium's ground state, both electrons occupy $n = 1$, $l = 0$, and $m_l = 0$. To satisfy Pauli's exclusion principle, therefore, one electron has to have $m_s = +1/2$; the other, $m_s = -1/2$.

Lithium (Li), containing three protons in its nucleus and three orbital electrons, is the next entry. The first two electrons go into the lowest shell exactly as helium, but then the $n = 1$ shell becomes filled. No more than two electrons (with opposite spins) can go into any fixed set of n, l, and m_l (here, $n = 1$, $l = 0$, and $m_l = 0$). Thus, the third electron of lithium must go into the $n = 2$ shell. If lithium is in its ground state, this $n = 2$ electron has $l = 0$, $m_l = 0$, and $m_s =$ either $+\frac{1}{2}$ or $-\frac{1}{2}$. Since the outermost shell of lithium has one electron (the "valence" electron) and since the valence electrons determine, to a large extent, the chemical properties of the element, lithium belongs to the same family as hydrogen, which also has just one valence electron. Elements with an excess or deficit of one valence electron (i.e., atoms with unpaired electrons) are especially reactive. More generally, elements like to combine to form compounds (i.e., atoms sharing electrons to form *molecules*) with each atom acquiring enough outer electrons to possess filled shells. Such configurations are lower in energy than the nonfilled states. Thus, helium, which possesses a filled $n = 1$ shell in its natural state, is chemically inert; it is a "noble gas." Lithium, possessing an extra outer electron, would like to donate this electron to some other atom. This property makes lithium a "metal."

Beryllium (Be), containing four protons in its nucleus and four orbital electrons, is the next entry in the periodic table. The first two electrons go into the $n = 1$ shell, just as in helium. The next two go into $n = 2$, $l = 0$, $m_l = 0$ (with opposite spins). Thus, beryllium contains two outer electrons just like helium; however, the $n = 2$ shell is not filled yet. It can take 6 more electrons (two each in $l = 1$ with $m_l = -1, 0,$ or $+1$). Thus, beryllium is not in the same family as helium, but forms a family of its own. Beryllium likes to donate two electrons; hence, beryllium is also a metal.

Six elements from beryllium is the next element in the helium family of "noble gases," neon (Ne). Neon has a complete $n = 2$ shell. The element just before neon is fluorine (F). Fluorine misses just one electron to have a filled $n = 2$ shell. Thus, chemically, fluorine would like to add an electron. The element before fluorine, oxygen (O), likes to add two electrons, for example, by combining with two hydrogen atoms to form a water **molecule** (H_2O). Roughly speaking, therefore, the elements on the left side of the periodic table are electron donors; the ones on the right (except for the noble gases) are electron acceptors. Carbon (C), with four valence electrons, is in the middle of this range, and it is a particularly versatile element in forming a rich variety of compounds. It is, of course, the basis of life on Earth.

Now, it might be thought that the next row down from neon, which starts filling the $n = 3$ shell, would have eighteen entries (refer back to Problem 3.8). Instead, this row only has eight entries: sodium (Na) to argon (Ar). When argon is reached, the $l = 0$ and $l = 1$ orbitals of the $n = 3$ shell are filled. According to Box 3.3, the $n = 3$, $l = 2$ orbitals are higher in energy than the $n = 4$, $l = 0$ orbitals. Thus, *before* the $l = 2$ orbitals of $n = 3$ are filled, a new row of chemical elements starts, with potassium (K) and calcium (Ca) adding the two electrons corresponding to the $n = 4$, $l = 0$ orbitals. *Then*, the ten electrons which can go into $n = 3$ and $l = 2$ are added—scandium (Sc) through zinc (Zn)—before we pick up up the six electrons of $n = 4$ and $l = 1$, gallium (Ga) through krypton (Kr). With the outermost $n = 4$ and $l = 1$ subshell filled, krypton behaves as argon does with its outermost $n = 3$ and $l = 1$ subshell filled, namely, like a noble gas. With this and similar complications, the entire periodic table can be explained on the basis of known quantum-mechanical principles.

An important issue can be raised before we leave the topic of the chemical elements. If an atomic nucleus existed containing a million protons, the rules we outlined above would be perfectly content to organize its million orbital electrons into shells, subshells, etc. *But no such element exists.* Why is that? In Chapter 6, we shall see that the fault lies not in the outer electronic structure of the atom, but in the nucleus.

Atomic Spectroscopy

In the previous section, we have used the approximation that an electron in a multi-electron atom feels only the radial attraction of the nucleus and the *spherically symmetric* part of the repulsion due to the other electrons. This approximation suffices for describing the ground state of an atom and for interpreting the *main features* of the periodic table of the elements, but the transitions to and from the excited states of an atom require an understanding of some additional complications.

These complications are associated with the angular momentum of the electrons. For a neutral atom, only the *valence electrons* generally need be considered, because the electrons in the inner *filled* subshells are usually in the lowest energy states consistent with Pauli's exclusion principle and have, in net, a spherically symmetric distribution. For the valence electrons, the important effects turn out to be those outlined in Box 3.4.

BOX 3.4

Atomic Spectroscopy and LS Coupling

(a) Electrons in the inner filled subshells have, in net, a spherically symmetric distribution. The valence electrons do not; consequently, their vector angular momenta, $\boldsymbol{l}$ and $\boldsymbol{s}$, are not individually conserved. In LS coupling, the vector *sums*, $\boldsymbol{L}$ and $\boldsymbol{S}$, of the orbital and spin angular momenta of all the valence electrons are assumed to be separately conserved.

(b) Quantum states with aligned spin angular momenta (large values of S) tend to have electrons which avoid each other (like Pauli's exclusion principle). This reduces the positive potential energy associated with the mutual repulsion of electrons. Thus, states with large values of S tend to have *lower* energies than states with small values of S.

(c) Quantum states with aligned orbital angular momenta (large values of L) correspond classically to electrons which revolve in the same sense about the nucleus. This again reduces the interaction between the electrons. Thus, states with large values of L tend to have *lower* energies than states with small values of L.

(d) The orientation of the spins of the electrons relative to their orbital angular momenta (characterized by the magnitude J of the total angular momentum $\boldsymbol{J} = \boldsymbol{L} + \boldsymbol{S}$) leads to "fine structure." A spinning electron acts like a bar magnet, and can interact with the magnetic field associated with the motion of the orbiting electrons. The result is that spin-orbit states which are aligned (large values of J) tend to have *higher* energies than misaligned spin-orbit states (small values of J). A similar phenomenon involving the interaction of the spin of the electron with the spin of the nucleus is responsible for the "hyperfine structure" that gives rise to the 21-cm line of atomic hydrogen (Chapter 11).

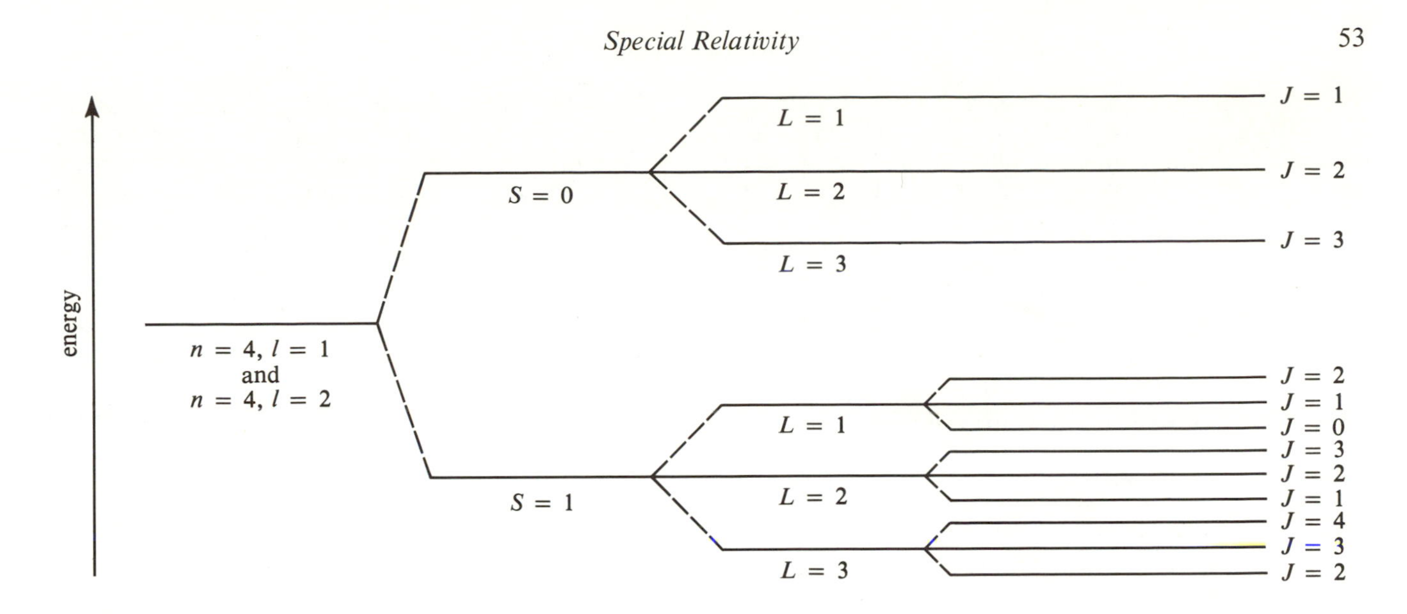

The "energy level diagram" here shows a case with two valence electrons with $n = 4$, $l = 1$, and $n = 4$, $l = 2$. The "final" possible energy levels labelled by the different combinations of S, L, and J can be further "split" into $2J + 1$ levels (corresponding to the different possible values of m_J) in the presence of an external magnetic field (Zeeman effect). Clearly, transitions between these states and a comparably complicated set of other states can lead to a very rich and complex pattern of spectral lines! An atom like iron, with a relatively high cosmic abundance and with many electrons, can produce hundreds of lines in the visible portion of the spectrum alone.

Problem 3.9. Let each right-hand level of the diagram in Box 3.4 be split into $2J + 1$ "degenerate" states by the Zeeman effect (application of an external magnetic field). Verify that 60 states in total would result.

Under normal circumstances, not all possible transitions are equally likely. Quantum mechanics provides certain "selection rules" which predict the more commonly observed transitions between different energy levels. These "allowed" transitions generally represent energy that one photon can carry off easily (in the emission case) and still conserve the total angular momentum of the system. Other transitions are mathematically possible, but are considered "forbidden" because the chances are good *under terrestrial conditions* that before the atom can radiate by such a "forbidden' transition, a collision with another atom or molecule will deexcite the atom collisionally. Collisions are much more rare in the interstellar medium because atoms are spaced much further apart on the average there; so "forbidden" transitions play an important role in the radiation produced from gas clouds in the spaces between and surrounding stars.

This concludes our discussion of quantum mechanics. We have obtained a taste of how quantum mechanics can provide a coherent account of the chemical properties of the elements and the spectroscopic properties of atoms. From atoms are made molecules, and from molecules is made the macroscopic world with which we are familiar. The extraordinary success of quantum mechanics in these fields provided the crowning touch to the efforts of the atomic physicists of the 1920s. To achieve this beautiful reconciliation between the mechanics of the atom and the mechanics of the macroscopic world, Bohr supplied the philosophical principles, and Schrödinger and Heisenberg supplied the mathematical equations. And the world has not been the same since then.

Special Relativity

Besides quantum mechanics, classical mechanics received in the twentieth century another major modification. This modification was special relativity, as formulated by Albert Einstein. Its basis is just two ideas: one old, one new (Box 3.5). The old idea is the **relativity principle**.

BOX 3.5

The Basic Postulates of Special Relativity

(a) The relativity principle: the laws of physics are the same for all inertial observers.

(b) The constancy of the speed of light: the speed of light, $c = 3.00 \times 10^{10}$ cm/sec, is the same for all inertial observers, independent of their velocity or motion relative to the source of the light.

We have already discussed, in connection with electromagnetism, the concept that the laws of physics (not necessarily the *appearance* of phenomena) are the same for all inertial observers. They do not depend on the observer's velocity with respect to the "fixed stars." The new and astounding idea is that the speed of light in a vacuum, c, is the same for all inertial observers. In other words, the speed of light equals 3.00×10^{10} cm/sec independent of how fast the observer moves relative to the source of the light!

Consider what this second idea means. I shine a flashlight and observe the photons to rush away from me at 3.00×10^{10} cm/sec. You move along the path of the outwardly rushing photons at half the speed of light relative to me. I ask you how fast the photons are rushing past you, fully expecting to hear "1.50×10^{10} cm/sec." Imagine my surprise when you tell me "3.00×10^{10} cm/sec"! Imagine the world's surprise when Michelson and Morley told them "3.00×10^{10} cm/sec"!

How can this answer be right? Yet, right it is, checked time and time again in many different experiments. How do we reconcile this answer with our common-sense notion of space and time? It took the genius of Einstein to see the way:

> Space and time cannot be thought of as separate entities. They are intrinsically linked, and it is only a combination of the two which will appear the same to two observers who move relative to one another.

Indeed, speaking colloquially, we may say that the great achievement of Einstein's special relativity theory is precisely this unification of space and time. This unification has several well-known astounding implications, which we shall proceed to develop in a series of problems now, using no mathematics more complicated than high-school algebra and geometry. It is a fallacy that Einstein's theory of special relativity is difficult to understand. Indeed, David Hilbert, the famous mathematician, was fond of saying that every schoolboy in Göttingen could derive special relativity. And so they can—after Einstein has shown the way, and we accept the two basic postulates of special relativity (Box 3.5).

Before we begin our discussion, I should remark that the preferred modern derivation of the results of special relativity proceeds by a different route, one that makes explicit use of the four-dimensional tranformation properties of spacetime. However, the concise derivations possible with such a formalism can be appreciated only after one has already developed some intuitive feeling for relativity theory, and I believe it is best developed by retracing the historical route discussed below.

Time Dilation

The first result we wish to derive is the concept of **time dilation**. This concept can be illustrated forcefully by the so-called twin paradox (see Figure 3.25):

> Once there were a woman and a man who were twins. The woman left on a rocket ship for a trip to Alpha Centauri, traveling at 86.6 percent of the speed of light. The man stayed at home. The man aged five years before the woman arrived at Alpha Centauri (4.34 light-years from the solar system). The woman aged only half as much, two and a half years.

How can this be? The answer is that the man who stayed "stationary" thinks that time is "dilated" for the woman who travels relative to the "fixed" stars: all physical processes seem to slow down for her, including her biological clock. According to the woman, however, everything is perfectly natural; the pace of life and everything else proceeds on the rocket trip just as it did at home on Earth. She brought along a clock just to make sure; to make our reasoning simple, we shall suppose this clock to be constructed as follows. It is a "light" clock in which a ray of light bounces back and forth between two flat mirrors separated by a distance d. Each time the light strikes one of the mirrors, the clock makes a ticking sound; so, by multiplying the number N of ticks by the distance d and dividing by the speed of light $c = 3.00 \times 10^{10}$ cm/s, the woman can keep track of the passage of time, t_0. The man hears (through a radio link) the same number of ticks, N, but for him the light ray must make a zigzag path in order to bounce between two *moving* mirrors. Thus, the total path length traveled by the light ray to make N ticks appears longer to the man than the path length Nd according to the woman. Since the speed of light is the *same* for both of them, the time elapsed for the man, t, seems longer than the time elapsed for the woman, t_0. A simple geometrical construction (Problem 3.10) yields the quantitative relation:

$$t = \gamma t_0, \tag{3.11}$$

where

$$\gamma \equiv (1 - v^2/c^2)^{-1/2} > 1$$

and v is the speed of the woman. Notice, especially, that the woman thinks that the clock works normally, just like at home!

Problem 3.10. Show from Figure 3.25 and Pythagoras's theorem (Appendix B) that each diagonal path traveled by the ray of light as seen by the man is given by $(d^2 + s^2)^{1/2}$, where s is the distance traversed (perpendicular to the mirror arrangement) per tick. Clearly, if t is the total elapsed time for the man, $s = vt/N$. Set now the total path length $N(d^2 + s^2)^{1/2}$ equal to ct, and solve for t. Note that Nd/c is t_0 and derive equation (3.11). Verify that if $v/c = \sqrt{3}/2 = 86.6$ percent, t is twice t_0. How much time dilation is there if you fly in a jet plane at 3×10^4 cm/sec?

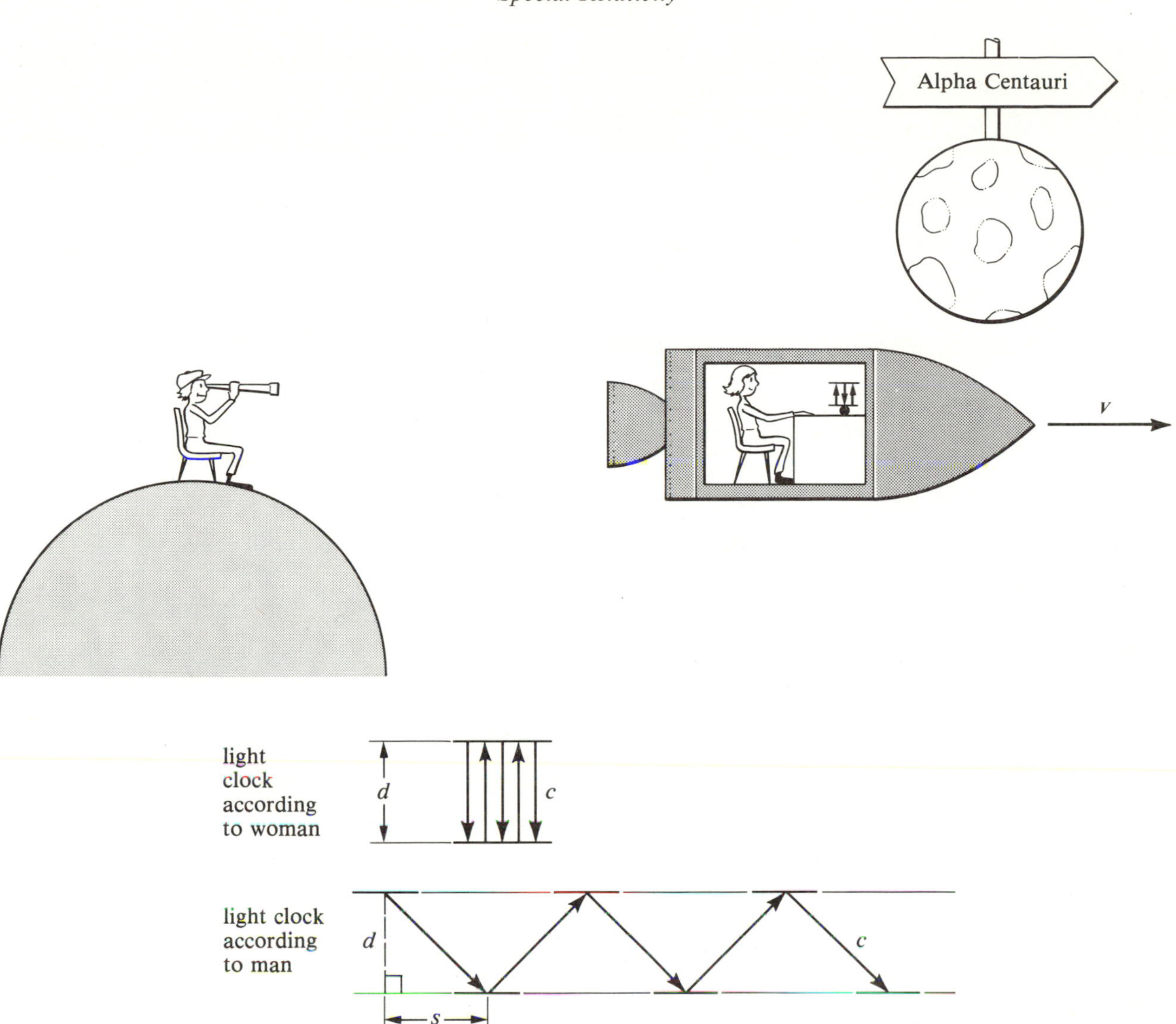

Figure 3.25. A rocket trip to Alpha Centauri at relativistic speeds provides an illustration of time dilation and the so-called "twin paradox."

Now, where is the paradox in this? Well, according to the principle of relativity, shouldn't the woman feel justified in thinking that it is she who is at rest and that it is the man (and the rest of the universe) who is moving? Thus, shouldn't she think that she ages more quickly than her brother because it is he who is moving? And so she does! However, there is still no paradox, because while she is moving there is no way they can get together to compare wrinkles. To compare themselves, she has to decelerate to a stop, turn around, accelerate back, decelerate, and come to a stop at the Earth. While decelerating and accelerating (with respect to the universe), she is no longer in an inertial frame, and we would have to use a more complicated argument to figure out what is going on. But the net result would be that after she arrives back on Earth, she would indeed be younger than her twin brother! If this is confusing, just accept the result (3.11) and note that the woman's position is not identical to the man's, because it is she who is moving with respect to the "fixed" stars, not he.

Lorentz Contraction

Now, the persistent reader may argue that there is still something wrong. Isn't Alpha Centauri more than four light years away? How could the woman think she

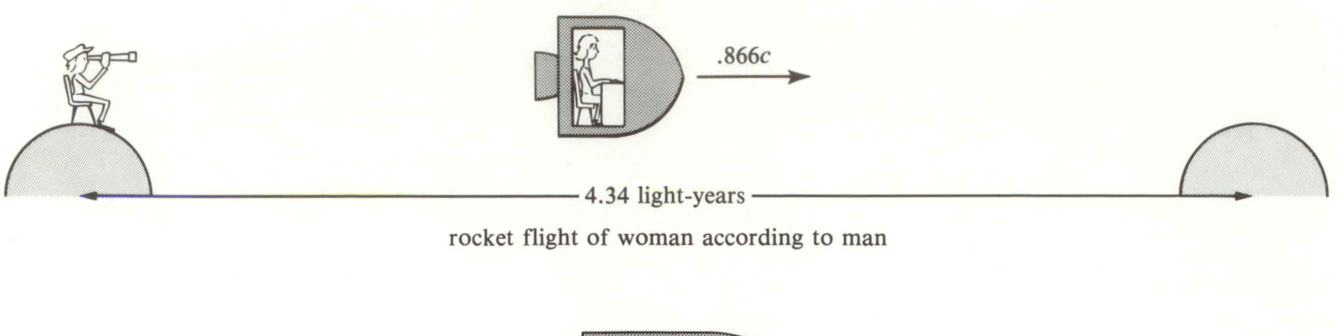

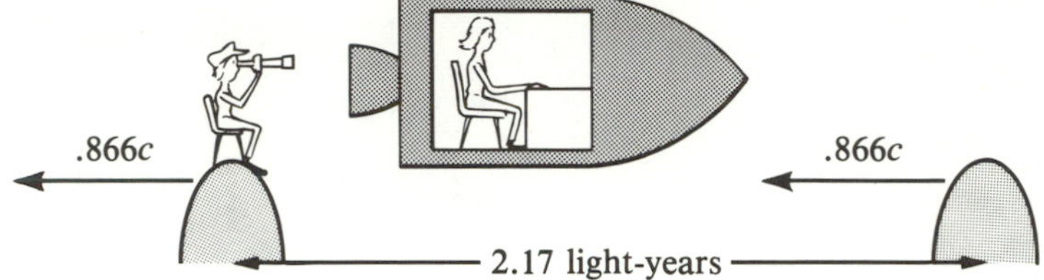

Figure 3.26. Two points of view of travel between solar system and Alpha Centauri at 86.6 percent of the speed of light. In the top diagram, the man sees a foreshortened rocket and sees the woman taking five years to traverse 4.34 light years at $0.866c$. In the bottom diagram, the woman sees everything suffering the Lorentz contraction. In particular, she sees the solar system and Alpha Centauri separated a distance of 2.17 light-years, which take 2.5 years to pass by her at $0.866c$.

could get there in two and a half years if she moves relative to Alpha Centauri at only 86.6 percent of the speed of light, or $0.866c$? The answer is that for the man, Alpha Centauri is indeed more than four light-years away, and that is why he thinks she takes five years to reach it traveling at $0.866c$. But for the woman, she thinks that she is at rest and that it is the solar system and Alpha Centauri which are whizzing by her at $0.866c$. And because of the phenomenon called Lorentz contraction (Figure 3.26), what is 4.34 light-years for the man (along the direction of motion) is only half that distance for the woman. That is, the woman would claim that Alpha Centauri and the Sun are spaced only 2.17 light-years apart, and that's why it takes them only two and a half years to whiz by her traveling at $0.886c$. The woman and the man are both right. According to Einstein, any observer has the right to consider himself or herself to be at rest.

A more general statement of the Lorentz contraction phenomenon is the following.

> If the length of a measuring rod in its frame of rest is L_0, and if it moves along its length at a speed v with respect to some observer, that observer thinks the length of the rod is shorter than L_0. In fact, the observer thinks the rod has contracted to a length L given by
>
> $$L = L_0/\gamma. \tag{3.12}$$

Notice that lengths perpendicular to the direction of relative motion are not affected by the Lorentz contraction. That's why both the man and the woman saw the same distance d to separate the clock's mirrors in Figure 3.25.

Problem 3.11. Retrace the physical arguments to satisfy yourself that equation (3.12) necessarily follows from equation (3.11) if no paradox is to result.

The reader might object that there may be something peculiar to our use of light to measure time and space intervals. Maybe other kinds of clocks and rulers would not exhibit the transformation properties associated with special relativity. In fact, physicists have good reason to believe that all *valid* formulations of the laws of physics must conform to the requirements of special relativity. More to the point, physicists *know* that Maxwell's laws of electromagnetism (and their quantum generalization) do conform with the requirements of special relativity; therefore, *any* clock or ruler that is constructed using the principles of electromagnetism will automatically show the effects of time dilation and Lorentz contraction. This explains our confidence in claiming that even *biological* clocks will be affected (according to an observer in relative motion) despite the fact that no one knows in detail how biological clocks work. It suffices to know only that biochemists have good evidence that all biological functions are ultimately based on the principles of electromagnetism (Chapter 19).

Problem 3.12. A more direct and laborious proof of the Lorentz contraction formula is the following. (a) The diagram shows that it is possible to measure the length of a rod by measuring the round-trip light-travel time across the rod and back. For a nonmoving rod of length L_0, a photon leaving from one end at time 0 strikes the other end at time t_0', reflects, and reaches the original end at time t_0. Clearly, $L_0 = ct_0'$ and $L_0 = c(t_0 - t_0')$ describes the two phases of this simple experiment. Adding the two expressions, we obtain the expected result: $2L_0 = ct_0$. (b) Consider now this same experiment performed on a moving rod (or equivalently, observed by a moving observer). At time 0 the photon leaves one end and strikes the other end at time t'. In the interim, the other end has moved a distance vt' further away; thus, if the length of the moving rod is L, this phase of the experiment satisfies the relation, $L + vt' = ct'$. After reflecting off the other end at time t', the photon returns to the original end at time t. In the interim the original rod has moved closer by a distance $v(t - t')$; thus, this second phase of the experiment satisfies the relation, $L - v(t - t') = c(t - t')$. Eliminate t' from these two expressions to obtain the result, $2L = (c^2 - v^2)t/c$. Use equation (3.11) together with $2L_0 = ct_0$ to derive equation (3.12).

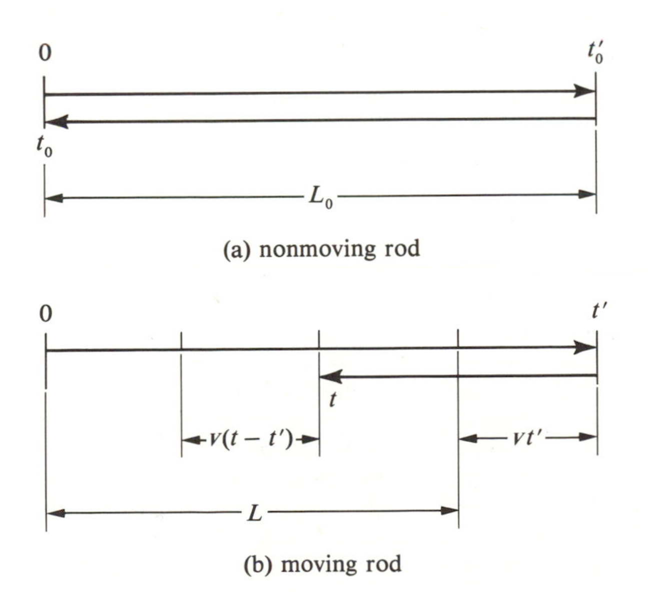

(a) nonmoving rod

(b) moving rod

Relativistic Doppler Shift

Having derived these basic formulae, we are now prepared to discuss the **relativistic Doppler shift**. In the section on atomic spectra, we saw that certain atoms (and for that matter, molecules) will emit and absorb light of very definite wavelengths. These wavelengths were calculated for an observer at rest with respect to the source; consequently, they should really be labeled as λ_0, the rest wavelengths. The Doppler shift arises because motion of the source (the atoms) cause the observed wavelength λ to shift away from the rest value λ_0.

In particular, motion along the line of sight away from the observer causes **redshifts** (increases of the wavelengths), and motion along the line of sight toward the observer causes **blueshifts** (decreases of the wavelengths).*

Relativistic effects of objects which (seemingly) move rapidly *along* the line of sight are frequently encountered in astronomical objects. If v is the component of velocity along the line of sight (here, v is positive for recession and negative for approach), the ratio of the observed wavelength λ to the rest wavelength λ_0 is given by

$$\frac{\lambda}{\lambda_0} = \left(\frac{1 + v/c}{1 - v/c}\right)^{1/2}. \tag{3.13}$$

It is conventional to write $\lambda/\lambda_0 = 1 + z$, where $z = (\lambda - \lambda_0)/\lambda_0$ and is called the "redshift" if it is positive, "blueshift" if it is negative. For speeds small compared to the speed of light, z is given by the approximate (nonrelativistic) formula:

$$z = v/c. \tag{3.14}$$

Equation (3.14) can be obtained from equation (3.13) by the so-called "binomial expansion," but it is more general than that derivation would seem to indicate. Equation (3.14) holds as long as z is small compared to unity and as long as the *square* of the velocity (including both parallel *and* perpendicular components) is small compared to the *square* of the speed of light.

* Relativistic effects make a time-dilation contribution to the classical formula in such a way that redshifts occur even if the source moves perpendicularly to the observer. This transverse Doppler effect is usually negligible in astronomical objects (a possible exception may be an object called SS433), and we shall ignore it in most of this book.

Problem 3.13. The diagrams here show a light source emitting a wavetrain, say, of Hα radiation, with N wave crests between the observer and the source. For a nonmoving source (a), the wavelength is λ_0, and N times λ_0 is equal to c times the light-travel time t_0 from the source to the observer: $N\lambda_0 = ct_0$. A moving source (b), in the time t it takes to emit N wave crests, will have moved an additional distance vt away. The head of the wavetrain in this time travels a distance ct to the observer. Thus, the N wave crests are at this instant stretched out over the distance $ct + vt = (c + v)t$; so $N\lambda = (c + v)t$. Dividing this expression by $N\lambda_0 = ct_0$, we obtain $\lambda/\lambda_0 = (1 + v/c)(t/t_0)$. The time t elapsed for the second observer is related to the time t_0 during which the nonmoving source emits N wave crests by the time-dilation formula (3.11). (This is the relativistic part of the effect; the nonrelativistic derivation sets $t = t_0$.) Derive now the relativistic Doppler formula (3.13). The redshift z of a certain quasar is observed to be 3.5. Set $1 + z = \lambda/\lambda_0$ and use the relativistic Doppler-shift formula to estimate v/c for the quasar if its redshift is assumed to be associated with recessional velocity along the line of sight.

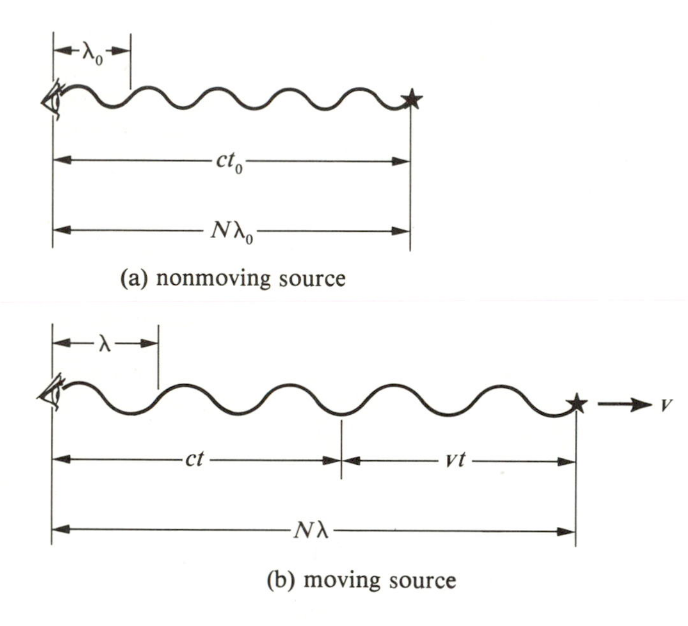

(a) nonmoving source

(b) moving source

A practical question now arises. Suppose an astronomer sees a spectral line at 5200 Å coming from some astronomical object. How does the astronomer know whether this line is due to magnesium being redshifted slightly, or whether it is due to, say, Lyman α (with rest wavelength equal to 1216 Å) being redshifted greatly? With only one line, the astronomer cannot know for sure. A small redshifted magnesium line may be more probable than a huge redshifted hydrogen line, but the astronomical universe presents many surprises. To obtain a reliable line identification, *more than one line is needed.* For example, the hydrogen lines (Lyman α, Lyman β, Hα, Hβ, etc.) have definite wavelength ratios relative to one another, and this *pattern of ratios* is unaffected by a common redshift factor $(1 + z)$. It was precisely this fact which allowed Maarten Schmidt to identify in 1963 that the quasar 3C273 had a fairly large redshift. The reverberations from that important discovery (S. Chandrasekhar, then editor of the *Astrophysical Journal*, had the press run overtime in order to include the announcement of Schmidt's discovery) still echoes strongly in astronomy (Chapter 13).

Relativistic Increase of Mass

Let us return, however, to our discussion of special relativity. The next important result that we wish to derive is the formula for the **relativistic increase of mass** (Problem 3.14). The result is:

If a mass has the rest value m_0 when it is stationary, it has the increased value m,

$$m = \gamma m_0 \tag{3.15}$$

where

$$\gamma = (1 - v^2/c^2)^{-1/2},$$

when it moves at speed v.

This result has the important implication that the speed of material particles cannot reach, much less exceed, the speed of light. To see this, remember that the original form of Newton's second law of motion (Box 2.1) reads $d\mathbf{p}/dt = \mathbf{F}$ where $\mathbf{p} = m\mathbf{v}$. When relativistic effects are important (i.e., when v^2 becomes comparable to c^2), equation (3.15) states that m is a function of v and will change if v changes. In this case $\mathbf{F} = d\mathbf{p}/dt \neq m\mathbf{a}$, where $\mathbf{a} = d\mathbf{v}/dt$. Thus, when force continues to be applied, a material particle will generally be pushed closer and closer to the speed of light. But it cannot reach it, because as v approaches c, the inertia m of the particle increases according to equation (3.15) without bound. Thus, $\mathbf{p}$ would increase without bound, and the application of either infinite force for a finite time or finite force for an infinite time would be required, according to $d\mathbf{p}/dt = \mathbf{F}$, to increase $\mathbf{p}$ or v any more. Since neither infinite force nor infinite time is ever available in practice, it is not possible to accelerate any particle with finite rest mass m_0 up to the speed of light. Conversely, to exist, anything with *zero* rest mass (like photons and gravitons) must move at speed c.

Problem 3.14. The diagram here shows the elastic collision of two balls which are identical when both are at rest. The collision is observed by two different observers, A and A', with observer A' arranged to be moving with the x-velocity of the 2 ball, $v_{2x} = -v_{1x}'$, with respect to observer A. The perpendicular distance y traveled by ball 2 is the same for both A and A'; however, the time t elapsed to make this traversal for observer A is longer (by the description of A) than the time t_0 elapsed for the moving observer A'. The ratio of t to t_0 is given by the time-dilation formula:

$$t/t_0 = (1 - v_{2x}^2/c^2)^{-1/2}.$$

By this reasoning, derive the velocity transformation formula for $v_{2y} = y/t$ and $v_{2y}' = y/t_0$:

$$v_{2y}' = v_{2y}(1 - v_{2x}^2/c^2)^{-1/2}.$$

But from the obvious symmetry of the left-hand and right-hand diagrams, we may conclude $v_{2y}' = v_{1y}$. Thus, derive

$$v_{1y} = v_{2y}(1 - v_{2x}^2/c^2)^{-1/2}$$

a relation between the y-velocities of the two balls in the left-hand diagram. The fact that v_{1y} has to be greater than v_{2y} to produce the scattering as shown suggests that ball 2 has greater mass than ball 1 by virtue of its extra x-motion; the combination of a larger mass and a smaller y-velocity produces a y-momentum for ball 2 identical to that of ball 1. Since we expect, therefore, m to be a function of speed, let us write the conservation of y-momentum according to observer A as

$$m(v_{1y})v_{1y} = m(|\mathbf{v}_2|)v_{2y}.$$

Show that the preceding equation implies

$$m(|\mathbf{v}_2|) = m(v_{1y})(1 - v_{2x}^2/c^2)^{-1/2}.$$

Consider the limit $v_{1y} \to 0$ and $v_{2y} \to 0$ where the two balls do not scatter at all. In this limit we have one stationary ball with mass $m(v_{1y}) = m_0$ and another ball which is identical except that it moves only in the x direction: $|\mathbf{v}_2| = v_{2x} \equiv v$. Call $m(v) = m$, and derive equation (3.15).

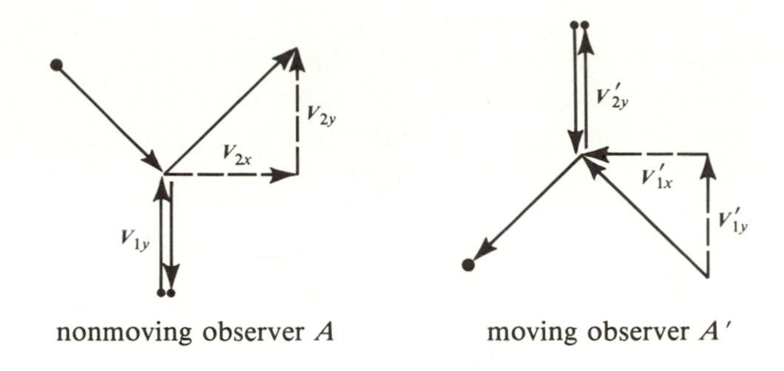

(The motion of A' with respect to A is $v_{2x} = -v_{1x}'$ in the x-direction.)

Equivalence of Mass and Energy

Our final derivation is Einstein's famous demonstration of the **equivalence of mass and energy**. This equivalence results from the famous formula

$$E = mc^2, \tag{3.16}$$

a relationship which has justifiably become as famous as $\mathbf{F} = m\mathbf{a}$ to people in all walks of life. The motivation for the identification of equation (3.16) is considered in Problem (3.15). For our discussion here, let us accept equation (3.16) and show that this acceptance implies a possibility for converting energy into mass and vice versa.

Problem 3.15. Consider a single moving ball whose rest mass is m_0. According to equations (3.15) and (3.16), we should identify the energy E of the freely moving ball as

$$E = m_0c^2(1 - v^2/c^2)^{-1/2}.$$

To see that this identification makes sense in the nonrelativistic limit, note that the binomial expansion of $(1 - x)^{-1/2}$ for small x is $1 + x/2 + \cdots$. In this way, show, for $v^2 \ll c^2$, that

$$E = m_0c^2 + \tfrac{1}{2}m_0v^2 + \cdots.$$

The first term on the right-hand side, m_0c^2, is called the rest energy and is a constant. An additive constant does not change any of the classical considerations of total energy. The next term on the right-hand side, $m_0v^2/2$, we recognize as the classical expression for the kinetic energy of a particle. For a free particle, there is no other form of energy; so the above expression gives the correct answer in the nonrelativistic limit. This provides the original motivation for the identification $E = mc^2$ for a freely moving particle. This discussion also suggests that the relativistic generalization for the kinetic energy is $E - m_0c^2 = (m - m_0)c^2$.

Problem 3.16. Suppose the rest mass of 1 gm of matter could be annihilated to give pure energy. How many 100-watt bulbs could this amount of energy keep burning for one year? The most exothermic (energy-releasing) chemical reaction per unit mass occurs when two ordinary hydrogen *atoms* combine electronically to form a molecule, H_2. Such a reaction produces 4.5 eV of energy (1 eV $= 1.6 \times 10^{-12}$ erg). On the other hand, two ordinary hydrogen nuclei which combine to form a deuterium nucleus, ^2H (see Chapter 6), release 1.4 MeV of energy (1 MeV $= 10^6$ eV). How much more energetic roughly are nuclear reactions than chemical reactions?

To see the conversion of energy into mass, consider the inelastic (sticky) collision of two idealized "putty" balls (Figure 3.27). Before colliding each ball has speed v and mass m related to its rest mass m_0 by equation (3.15). After the collision, the two balls stick and are brought to rest. What is each ball's rest mass m_0' after collision? Well, energy is conserved. According to equation (3.16), the energy of each ball before collision is mc^2; after collision, $m_0'c^2$. The conservation of energy requires

$$mc^2 + mc^2 = m_0'c^2 + m_0'c^2,$$

i.e.,

$$m_0' = m = \gamma m_0. \tag{3.17}$$

The left-hand part of equation (3.17), $m_0' = m$, merely states the conservation of mass: namely, the mass after collision, m_0', is equal to the mass before collision, m. Thus, if mass (multiplied by c^2) is considered a form of energy, the old idea of separate conservations of mass and energy is supplanted in relativity by the *single unified principle of the conservation of mass-energy*. The right-hand part of equation (3.17) contains the desired result: the mass m before collision is greater, because of motion, than the *rest* mass m_0 before collision. Thus, the final result,

$$m_0' = \gamma m_0,$$

implies that the *rest mass after the collision, m_0', is greater than the rest mass, m_0, before the collision*! Evidently, there was conversion of kinetic energy into rest

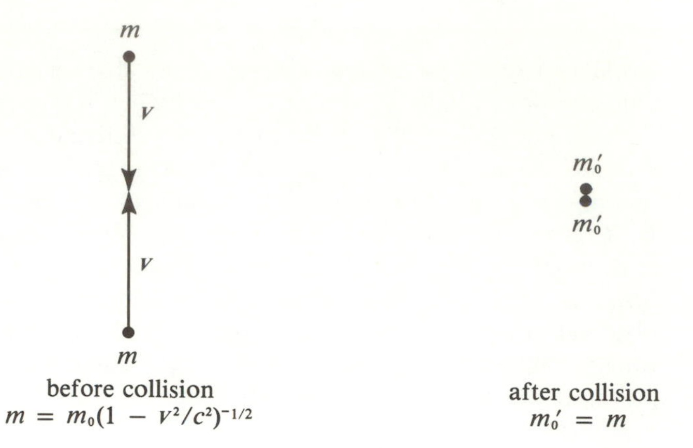

Figure 3.27. The conversion of kinetic energy into rest mass in the inelastic collision of two idealized putty balls.

mass. If it is possible in this fashion to convert energy into rest mass, it is easy to believe that the conversion of all or some fraction of the rest mass into energy is also possible. And so it is. And therein lies the power of the Sun (Chapter 5) and the hope of humanity.

Relativistic Quantum Mechanics

Now, it could be argued that *real* putty balls, as distinguished from our idealized ones of the last example, would not gain *rest* mass in an in elastic collision. Real putty balls are made of many, many molecules, and in an inelastic collision of real putty balls, the kinetic energy of bulk motion would be converted into heat, the random thermal energy of many many molecules (Chapter 4). Thus, it might be argued that only the apparent rest mass of the putty balls has been increased, because the individual molecules have been set into more agitated *microscopic* motions. The real rest mass of the putty balls would not have been increased, because if all the molecules were brought to rest (cooled), we would have the same number of molecules as before the collision, and the same rest mass.

This argument, while superficially correct, would miss the central point. The fact that real matter is constituted of atoms and molecules is the concern of quantum mechanics, not of special relativity *per se*. Within the logical structure of special relativity, it is perfectly permissible, and perfectly *correct*, to imagine *idealized* putty balls where inelastic collisions would create, not heat, but more putty. This discussion therefore leads naturally to the important issue: what is the *real* situation in inelastic collisions where *both* special relativity and quantum mechanics apply?

The marriage between special relativity and quantum mechanics into a single framework, **relativistic quantum mechanics**, is today's unifying theory which underlies the modern description of the physical world. The major new feature of relativistic quantum mechanics is, indeed, that it allows the *creation* and *destruction* of particles in accordance with Einstein's prescription $E = mc^2$. Not all particles imaginable, however, can be created and destroyed. In particular, the creation of matter has a minimum energy requirement (a "threshold"), because it is not possible to create particles of finite rest mass which are lighter than electrons. Moreover, even at the energy range available to modern accelerators (refer back to Figures 3.13 and 3.14), it is possible to produce rest mass from pure energy only in the form of matter-antimatter pairs. Indeed, the concept of antimatter originated with Dirac's marriage of quantum mechanics and special relativity in his theory of the electron (whose antimatter counterpart is the positron). We leave a more detailed discussion of such topics in **particle physics** for chapters 6 and 16. Here, let us merely complete the train of thought which began this subsection. According to relativistic quantum mechanics, if we were to throw putty balls at *relativistic* speeds at one another, we could create actual rest mass from pure energy. However, at the energies per particle now available to us in terrestrial laboratories, we could create matter only by creating an equal amount of antimatter. Thus, inelastic collisions *could*, in principle, give us more putty (and not just heat), but we would also create some antiputty! How then did the universe manage to create (apparently) unequal amounts of matter and antimatter? This deep unanswered question lies at the very frontier of modern physics and cosmology; we shall examine it in Chapter 16.

Concluding Philosphical Remarks

Some people have the mistaken notion that science proceeds by the accumulation of more and more facts. I hope, this chapter has helped to dispel some of this impression. We have seen that a truly fundamental idea in science *reduces* the number of facts that have to be remembered. It accomplishes this simplification by unifying what seemed previously to be disjoint pieces of information. Thus, Newton's work is great because it unified the mechanics of Heaven and Earth. Maxwell's work is great because it unified electricity and magnetism and light. Bohr's work is great because it led to a unified account of the chemical elements and a unified view of macroscopic and microscopic phenomena. And Einstein's work is great because it unified the concepts of space and time, and of mass and energy. Whatever future discoveries remain to be made, one thing is certain: they are just as likely to astound and delight us with the beauty of further simplification and unification.

CHAPTER FOUR

The Great Laws of Macroscopic Physics

In Chapter 3 we discussed the basic laws of physics which govern the dynamics of a small number of particles and their associated force fields. When we deal with a large number of particles, it is possible, *in principle*, to calculate their dynamics by superimposing all the various interactions. However, *in practice*, such a procedure is computationally beyond our capabilities. Moreover, such a detailed investigation would make it very easy for us to lose sight of the "forest for the trees." In dealing with macroscopic bodies that contain many many atoms and molecules, therefore, scientists have historically resorted to techniques different from those described in Chapter 3. (Refer, especially, to Figure 3.1.)

Thermodynamics

The first technique is empirical and is based on careful observations of controlled experiments. Scientists and engineers found that the results of *all* such experiments could be summarized in four general rules of great power and scope. These fundamental rules are known today as the *zeroth*, *first*, *second*, and *third* law of thermodynamics (Box 4.1).

Of these four rules, the most important by far is the second law of thermodynamics. Indeed, the second law of thermodynamics is the supreme jewel of macroscopic physics. Together with the law of universal gravitation, the second law of thermodynamics provides the central theme which governs the drama of the universe. Because of this crucial role, it behooves us to understand the second law better. Let us try to achieve this better understanding by restating the second law in a number of different ways, and by giving a few examples.

BOX 4.1
The Laws of Thermodynamics

Zeroth law: Heat always diffuses from hot to cold, with the temperature becoming uniform when thermodynamic equilibrium is reached.

First law: Heat is a form of energy. When this is taken into account, energy is always conserved.

Second law: Some forms of transformation of one kind of energy to another do not occur in natural processes. The allowed transformations in a closed system are always characterized by a nondecreasing entropy (introduction of more macroscopic disorder). In open systems where the entropy may be kept constant, the allowed transformations are always characterized by a decrease in the amount of (free) energy available to do useful work.

Third law: There is an absolute zero to temperature (0 K) where all forms of matter become perfectly ordered (usually perfect crystals).

Alternative Statements of the Second Law of Thermodynamics

Processes which involve macroscopic bodies are not usually time-reversible. Natural processes tend to proceed in the direction so that order (structure) on the macroscopic scale tends to be replaced by disorder. Physicists say that the total *entropy* in the universe always increases (or, at best, remains constant); it never decreases. Entropy is a precisely definable *negative* measure of the complexity of macroscopic structures; the less information needed to specify the architecture of something, the higher is the entropy associated with that object. Thus, the second law of thermodynamics states that time proceeds in the direction in which things wear out. To construct a great cathedral requires a lot more information than to create a rubble of rock and stone, and natural processes always proceed to reduce cathedrals to rubble.

Stated thus colloquially, the second law of thermodynamics sounds obvious and trivial. It is not, for two reasons.

First, it is amazing, but nevertheless true, that such vague concepts as order-disorder, informational complexity, etc., can be defined in a mathematically precise way. Moreover,

> the precise formulation of the entropy S is such that the product of the temperature T and the change ΔS in the entropy during an incremental process is equal to the amount of heat added to the body: heat added = $T\,\Delta S$.

Second, the fundamental irreversibility of most natural processes on the macroscopic level (the concern of all ecologists) seems, at first sight, to be at odds with the fundamental time-reversibility of the microscopic laws of physics. How is this apparent paradox resolved?

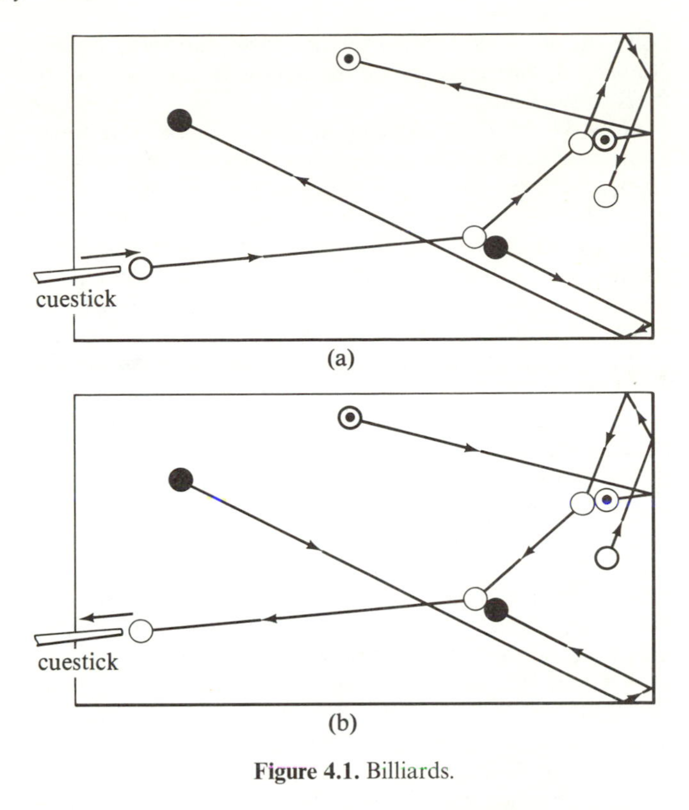

Figure 4.1. Billiards.

The Statistical Basis of Thermodynamics

To fix ideas, let us consider a concrete example. The game of billiards (not pool!) is played on a four-sided table with three balls: your cue ball (white without dot), your opponent's cue ball (white with dot), and a red ball. In the simplest version, the object is to score points by striking the other two balls with your cue ball. Figure 4.1(a) shows a simple shot. You stroke your cue ball, which strikes the red ball and then the opponent's cue ball.

Since the game seems to involve only a few objects—three balls and some walls (elastic cushions)—we should be able to analyze the dynamics of the billiard shot on the basis of classical mechanics. For example, we could check to see if energy is conserved. This seems an easy task, since there is apparently only kinetic energy of motion in the problem and no complications with potential energies (assuming the game is played on a flat table). A cursory check of each collision would show that the total kinetic energy is indeed conserved after each collision, with the two balls carrying off the same total energy as that possessed by one ball before the collision. Precise measurements made over a longer period of time would show a discrepancy, however. For example, if you wait long enough, all three balls eventually come to rest, and the total kinetic energy is then obviously zero. Does this mean that energy is not conserved after all? No. The oversight is regarding the three balls as if they were the only three objects in the system. The actual balls are not elementary entities, but are made of many molecules—and so is the billiard table and the air in the room. On striking each other, the balls do not make a perfectly elastic impact; some energy goes into making the individual molecules in the ball jiggle around. The jiggling of the ball's molecules constitutes an increase in the **heat content** of the ball, which comes at the expense of the total kinetic energies of the colliding balls. Similarly, as the balls roll, they set the molecules in the table and in the air of the room to jiggling. This phenomenon manifests itself on the macroscopic scale as **friction**. If the heat energy generated by frictional conversion of ordered motion and the heat energy transferred by thermal conduction is taken into account, we would find that the total energy is indeed conserved, precisely as demanded by the first law of thermodynamics.

Now, where does the second law of thermodynamics come into play in this little story? On the microscopic level, all the processes described above—the collision of the balls and the jiggling of the molecules—are perfectly time-reversible. That is, if we were to take a movie of the billiard shot and run it backward, the movie would make perfect microscopic sense. All the microscopic laws of physics would be obeyed in each elementary encounter of one molecule with another. Yet, the movie run backward in this fashion is clearly ridiculous in a macroscopic sense (Figure 4.1b). The billiard shot in Figure 4.1(a) is made countlessly every day all across the nation; the billiard shot in Figure 4.1(b) has never been made in the history of the world. Consider what making the latter shot constitutes. Remember we are running an actual shot backward in time. The shot begins with slightly hot billiard balls at rest on a slightly hot table in a slightly hot room. Gradually and spontaneously, the jiggling molecules in the air, in the table, and in the room decide to organize themselves and start to move the balls across the table. Your cue ball (white without dot) smashes into your opponent's cue ball (white with dot) with the jiggling of the atoms arranged precisely to bring your opponent's cue ball to rest and to send your cue ball careening off faster than before! This scenario is repeated once more, with the red ball being brought to rest and your cue ball now acquiring a very fast speed. Finally, your cue stick is brought into action as you deftly jerk the cue stick backwards and bring your cue ball precisely to rest—cooling it down in the process! A beautiful shot, a miraculous shot, one that I can confidently predict that you have never made or seen made. It is not, strictly speaking, an *impossible* shot; after all, it is "merely" the time-reverse of a very ordinary actual billiard shot. Yet the chances of that many molecules coming together precisely and cooperating to bring off such a shot are so vastly improbable that we might as well say "never" in practice. This *statistical* improbability is what reconciles the strong adjectives "always" and "never" in the usual statements of the second law of thermodynamics with the fundamental time-reversibility of physics at a microscopic level.

On this statistical level, we see that the second law of thermodynamics states that the organized complexity on large scales tends to be replaced by complexity on smaller scales until there is statistical uniformity on the macroscopic level and unresolvable structures on the microscopic level. Thus, in our actual billiard shot (Figure 4.1a), there is first the organized rushing around of your cue ball, then localized hot spots as it collides with the other balls, then diffusion of the heat to cause more and more jiggling of individual molecules. The information needed to specify the motion of all the molecules in the initial state (one moving cue ball) has not really disappeared; it has merely flowed into the many seemingly random jiggling of countless molecules. This *effectively* irretrievable loss of information on the macroscopic scale makes it, in practice, impossible to recover the initial state from the final state without additional input from the outside world. And according to the second law of thermodynamics, such additional input would always increase the entropy of the outside world in such a way that, on the whole, the universe "runs down."

Statistical Mechanics

From the above discussion, it should be clear that there exists a second way to treat the *equilibrium* behavior of macroscopic bodies (refer to Figure 3.1 for the overall blueprint of the architecture of physics). This second way is called **statistical mechanics**, and it involves a statistical treatment of the mechanics of very large numbers of particles and photons. With large enough numbers of particles, the method of statistical mechanics recovers the results of thermodynamics, and, in addition, gives more detailed information about specific systems.

Statistical mechanics, as formulated by Boltzmann and Gibbs and later extended into the quantum regime, has as its basis the following concepts. Suppose we have an *isolated* system of N identical particles (for example, $N = 10^{25}$ nitrogen molecules in a room). A specification of the **microstate** would give the position and momentum of every single particle (within an uncertainty and indistinguishability compatible with quantum mechanics if we are dealing with quantum statistical mechanics). Such a description contains far more detail than we need for an adequate specification of the system's **macrostate**. For the latter purpose, it suffices to give a *statistical* description of how many particles are in a certain portion of physical space with momenta between a certain interval. That is, we are interested in the *distribution* of particles in small bins of physical space and momentum space (Figure 4.2). For example, a macrostate might consist of 1,000 particles in the first bin, 2,500 in the second, 6,200 in the third, etc. The important point is that in specifying the *macrostate*, we are not interested in whether a certain particle has this position and that momentum, only in there being 1,000 particles in the first bin, 2,500 in the second, 6,200 in the third, etc. This means that each *microstate* corresponds to a unique *macrostate*, but the converse is *not* true. A given macrostate will generally correspond to many many microstates, with the total number determined by how many permutations of the detailed microscopic configurations will lead to the same macroscopic distribution. The connection with thermodynamics is made on the basis of the concepts in Box 4.2.

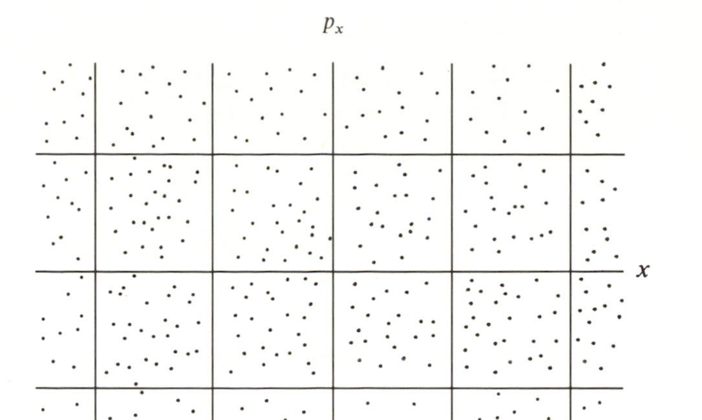

Figure 4.2. The distribution of particles (dots) in terms of a particle's x-coordinate and its x-component of momentum. A microscopic description requires a specification of the exact location in x and p_x for every particle (within the limitations of quantum mechanics). A macroscopic description requires only the specification of how many dots are in each bin. In particular, the set of numbers of dots in all the bins constitutes a *macrostate*.

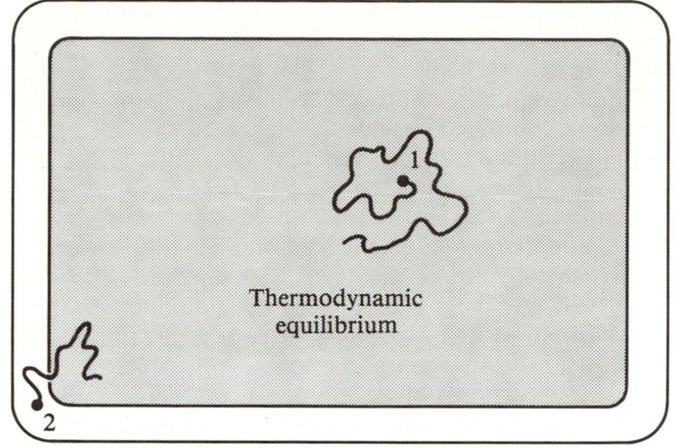

Figure 4.3. Schematic representation of the second law of thermodynamics and the approach to thermodynamic equilibrium. The bordered area represents the set of all microstates consistent with the constraints of the system. The shaded area corresponds to the subset of microstates which belong to the family representing thermodynamic equilibrium. System 1, which is initially in thermodynamic equilibrium, tends to stay in thermodynamic equilibrium. System 2, which is not in thermodynamic equilibrium, tends to wander into thermodynamic equilibrium.

BOX 4.2

Basic Postulates of Statistical Mechanics

(a) In an ensemble of randomly prepared systems of N particles, all possible microstates compatible with the constraints on an isolated system are equally likely. Let W represent the number of different microstates which correspond to the same macrostate. The state of thermodynamic equilibrium is the macrostate which has the *maximum* value of W consistent with the constraints placed on the system (fixed total energy and spatial volume, usually). Moreover, when N is large, the number of microstates W corresponding to the equilibrium macrostate constitutes the vast majority of all possible microstates. Thus, in the absence of all other information, a given system is likely to be in a microstate corresponding to thermodynamic equilibrium.

(b) The entropy S of a macrostate is equal to Boltzmann's constant k times the natural logarithm of W:

$$S = k\ln(W).$$

Since the natural logarithm is a monotonic function, S attains its maximum value when W does. Thus, of all possible macrostates, the one corresponding to thermodynamic equilibrium has the *maximum* entropy (for an isolated or closed system).

Often we *know* that a *given* system is not in thermodynamic equilibrium. How do the considerations in Box 4.2 help us to see that this system will head toward thermodynamic equilibrium? That is, how do these considerations imply the second law of thermodynamics, the tendency for the entropy to increase in time until it acquires its maximum possible value at thermodynamic equilibrium? Figure 4.3 gives the schematic answer. A given *microstate* of an entire system is represented by a single point in this diagram. The set of all possible microstates compatible with the constraints is represented by the set of all points in the bordered area. (Postulate [a] gives all points equal weight.) The subset of microstates which correspond to the equilibrium macrostate is given by the shaded area; the latter occupies almost all of the total bordered area. Now, consider a *given* system in some initial microstate. As a result of motions and interactions of particles within the system, the microstate of the system will change in time; i.e., the system point in this diagram will move from its initial point. If the system was initially in thermodynamic equilibrium (case 1), its subsequent evolution in time is very unlikely to carry the system point out of the vast shaded region. In other words, a system in thermodynamic equilibrium tends to remain in thermodynamic equilibrium. Stated in terms of the entropy, once a system has achieved its maximum value of the entropy, the entropy will remain constant (if the system is isolated). On the other hand, if the system was initially *not* in thermodynamic equilibrium (case 2), then its subsequent evolution in time is very likely to carry it closer to the shaded region. In other words, a system out of thermodynamic equilibrium tends to head toward thermodynamic equilibrium. Stated in terms of the entropy, a system with a less than maximal value of the entropy will tend to increase its entropy in time. Eventually, *if we wait long enough*, the system point 2 will wander into the shaded region, i.e., the system will reach thermodynamic equilibrium and acquire its maximum value of the entropy. The mathematical development of these ideas constitutes the subjects of equilibrium and nonequilibrium statistical mechanics.

For the purposes of this book, we shall need only the qualitative implications of statistical mechanics for the bulk behavior of matter, and we shall summarize the quantitative relations for the bulk behavior of a collection of photons under conditions of thermodynamic equilibrium. First, however, we make a few general observations, as follows.

(a) Thermodynamic concepts apply only to large numbers of particles and photons. The laws of thermodynamics are intrinsically statistical in nature. Thus, there is no such thing as the temperature of a single atom.

(b) In thermodynamic equilibrium, a collection of particles and photons will usually have a distribution of energies. It is statistically very unlikely that one particle could hog all the system's energy for itself. In fact, for simple systems, the number of particles at high energies will decrease exponentially with increasing energy, a fact known as Boltzmann's law.

(c) The most interesting phenomena are, of course, exhibited by systems in which thermodynamic equilibrium does *not* prevail—yet. This book deals with the nonequilibrium phenomena of basic interest to astronomy. The key to understanding the time evolution of such systems is to realize that, *in some fundamental sense*, they must be trying to attain thermodynamic equilibrium. It is the obstacles set in their path that provide the plot for the drama of the universe. One of the most fundamental is the complication presented by gravity. Whether gravity in the long run can prevent the thermodynamic death of the universe remains to be seen.

Thermodynamic Behavior of Matter

We stated the basic postulate of statistical mechanics as the tendency for systems of fixed total energy to seek out the macroscopic state of *greatest entropy*. If we do not deal with isolated systems, i.e., if we deal with systems where external inputs of energy may occur, an equivalent statement of the second law of thermodynamics is that systems of fixed entropy tend to seek out the macroscopic state of *least energy*. For material systems these rules provide a tension between wanting more freedom in physical space and more freedom in momentum space. Thus, material systems follow the general rule stated in Box 4.3.

BOX 4.3

The Thermodynamic Behavior of Matter

At relatively low temperatures, a material system prefers more binding energy. At relatively high temperatures, a material system prefers more spatial freedom and more unbound particles. (Here the term "relatively" refers to whether the thermal energy is large or not compared to the typical binding energy associated with the attractive forces at play. Remember, more binding energy means the system is more bound—has greater *negative* energy.)

Everyday experience corresponds to this rule. For example, if we raise the temperature (at fixed pressure) from low values to high values, we have the sequence of events:

> ice → liquid water → water vapor → dissociation of water molecules into hydrogen atoms and oxygen atoms → ionization of atoms to form a "plasma" of electrons and protons and oxygen ions.

We shall see that similar considerations apply to thermonuclear reactions inside stars and in the first three minutes of the creation of the universe.

The Properties of a Perfect Gas

Our example above leads us to a discussion of the three states of ordinary matter: solid, liquid, and gas. The key to the understanding of these states is that ordinary matter is composed of electrically neutral material, usually molecules. Electrically neutral matter cannot, of course, transmit long-range forces (except for gravitation, which becomes important only for bodies of astronomical size). Thus, when molecules are on average separated by distances that are very large in comparison with their intrinsic sizes (Figure 4.4), they will behave like independently and freely moving particles. This state characterizes a perfect gas.

Consider the properties of a perfect gas in local thermodynamic equilibrium at a temperature T, and suppose that the effects of quantum mechanics and special relativity are negligible. Let $\boldsymbol{v}$ be the velocity excess of a gas particle compared to the mean velocity $\boldsymbol{u}$ of all particles in its spatial neighborhood. The velocity $\boldsymbol{v}$ is called the *random velocity*, and the velocity $\boldsymbol{u}$ is called the *bulk velocity* or *fluid velocity*. For a classical perfect gas, Maxwell and Boltzmann used the methods of statistical mechanics to show that the number of gas particles of

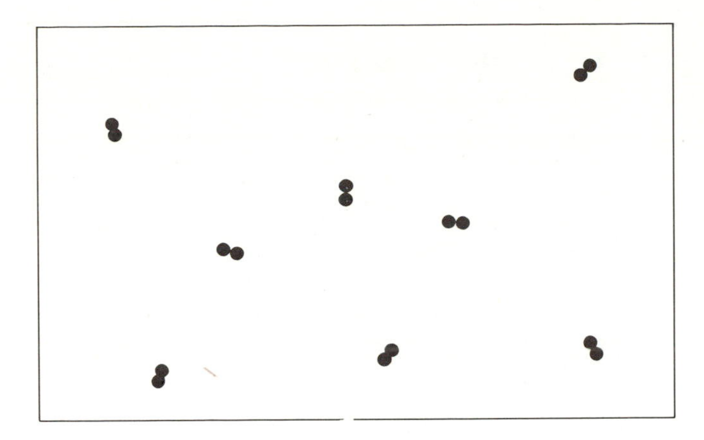

Figure 4.4. A gas has considerable empty space between its constituent molecules. As an example, the spacing between nitrogen molecules N_2 in the air in this room is about 10 times larger on average than the typical size of the electronic shells surrounding these diatomic molecules. As a consequence, the weak molecular forces in the gas play little role in its bulk macroscopic properties.

mass m with random speeds in a small range between v and $v + dv$ is proportional to the expression

$$\exp\left(-\frac{mv^2}{2kT}\right)4\pi v^2\, dv. \tag{4.1}$$

In the above, k is Boltzmann's constant, 1.38×10^{-16} erg/K (Appendix A), and $4\pi v^2\, dv$ is the "volume" of a spherical shell in "velocity space" of "radius" v and "thickness" dv.* Notice that the quantity $mv^2/2$ represents the random kinetic energy E of a particle with random velocity v. Thus, the Maxwell-Boltzmann distribution (4.1) is proportional to the factor $e^{-E/kT}$ that is characteristic of thermal distributions of matter. Numerically, the expression (4.1) means the following. If we count the number of molecules per unit volume of velocity space, there will be $e^{-1} = 0.3679$ fewer, statistically, molecules of energy $E = kT$ than molecules of energy $E = 0$; e^{-1} fewer yet molecules of energy $E = 2kT$; etc. If a batch of molecules were prepared which did not satisfy the expression (4.1), elastic collisions between the molecules (and other less-efficient processes) would usually quickly restore the thermal distribution (4.1).

Even the above statistical description contains more information than we need for many macroscopic purposes. For example, we may be interested only in the net force per unit area exerted on the container wall by a gas confined inside; we do not want a blow-by-blow account of how each molecule strikes the wall and rebounds. This force per unit area is called the **pressure** P of the gas. For a classical perfect gas which obeys the distribution of random velocities (4.1), it is possible to relate the pressure P to the number per unit volume of gas particles, **number density** n, and their temperature T. This relation is called the **perfect gas law**:

$$P = nkT. \tag{4.2}$$

Similarly, it is possible to show that the kinetic energy of random motions per unit volume, **internal energy density** $\mathcal{E}$, is given by the expression

$$\mathcal{E} = \tfrac{3}{2}nkT. \tag{4.3}$$

The mnemonic for remembering this result is that each gas particle has, on average, an energy of $kT/2$ for each translational degrees of freedom; there are three translational degrees of freedom (x, y, and z directions); and there are n particles per unit volume; therefore $\mathcal{E} = (kT/2)(3)(n)$. Equations (4.2) and (4.3) plus the laws of thermodynamics suffice to give a complete description of the local state of a perfect gas.

* The notation $\exp(x)$ is an alternative way to write e^x.

Problem 4.1. To derive equation (4.2), let us proceed in two steps. The pressure of any gas is the force per unit area that it exerts on a real or imaginary wall. This force per unit area equals the momentum transferred per unit time per unit area from the gas to the wall:

momentum transferred across area A in time t =
[total number of particles] ·
[average momentum per particle in x-direction] =
[(number density) · volume] ·
[average x-momentum per particle] =
$[n(Av_xt)] \cdot [p_x] = nAv_xtp_x$.

Verify the algebra given above and convince yourself that there are no factors of 2 missing from the final formula. (For example, rebound off a real wall would double the momentum transfer per particle, but only half the particles would be moving toward the wall.) To get the pressure P = momentum transferred per unit time per unit area, divide by t and A. In this manner, show $P = n\langle v_xp_x\rangle$, where we have put angular brackets around the product v_xp_x to indicate that it is really the combination which should be averaged, not the quantities individually. For an isotropic distribution of particles (no preferred direction), $\langle v_xp_x\rangle = \langle v_yp_y\rangle = \langle v_zp_z\rangle$. Show now that

$$P = \tfrac{1}{3}n\langle vp\rangle.$$

where $vp = \mathbf{v} \cdot \mathbf{p} = v_xp_x + v_yp_y + v_zp_z$. Written in the above form, we see explicitly that the definition of P does not depend on a special choice of axis.

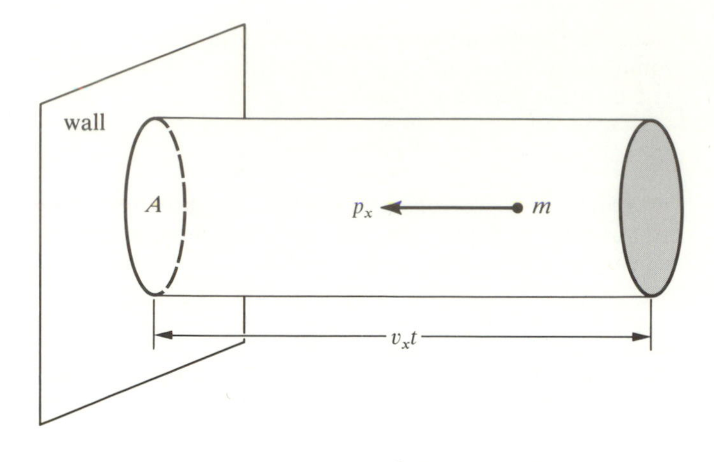

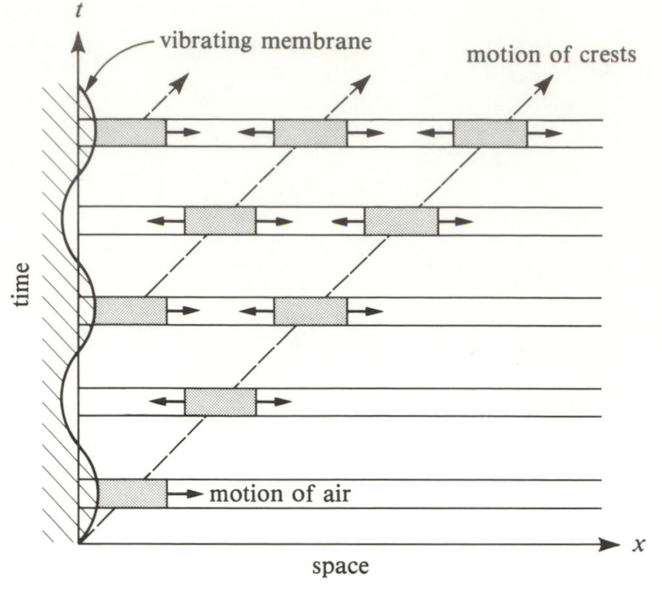

Figure 4.5. The generation and propagation of acoustic waves by a vibrating membrane placed in still air. The separation in space at any given time between consecutive compressed regions (= wave crests) is the wavelength λ; the separation in time at any given spatial location between consecutive wave crests (or wave troughs) is the wave period τ. For a small amplitude wave, the wavelength, period, and speed satisfy the relation: $\lambda = c_s\tau$, where c_s, the acoustic speed, depends only on the temperature and mean molecular mass of the gas through which the acoustic disturbance propagates. Light satisfies a similar relation, $\lambda = c\tau$, but it differs from an acoustic wave in that (a) it is a *transverse* wave, and not a *longitudinal* one, and (b) it, unlike sound, can propagate even across a vacuum.

Problem 4.2. Let us now apply the general result derived in Problem 4.1 to calculate the pressure of a classical perfect gas. In classical mechanics, the momentum is given by $p = mv$, with m equal to the rest mass. But $\langle mv^2\rangle$ is just twice the average kinetic energy due to thermal motions. The temperature T in the statistical mechanical derivation of equation (4.1) is defined so that for a perfect gas, $\langle mv^2/2\rangle = 3kT/2$. Show now that this leads to the perfect gas law, $P = nkT$.

Problem 4.3 (for those who know calculus). The average $\langle mv^2\rangle$ associated with the distribution given by expression (4.1) is by definition

$$\langle mv^2\rangle = \frac{\int_0^\infty mv^2 \exp(-mv^2/2kT)4\pi v^2\,dv}{\int_0^\infty \exp(-mv^2/2kT)4\pi v^2\,dv}.$$

Transform variables by defining $x = mv^2/2kT$, and show that the above equation becomes

$$\langle mv^2\rangle = 2kT\frac{\int_0^\infty x^{3/2}e^{-x}\,dx}{\int_0^\infty x^{1/2}e^{-x}\,dx}.$$

Integrate the top integral once by parts to show that it equals 3/2 times the bottom integral. In this manner, derive the result $\langle mv^2\rangle = 3kT$ quoted in Problem 4.2.

Acoustic Waves and Shock Waves

The property of compressibility allows a gas to transmit signals known commonly as sound. Sound constitutes only one example of a general class of macroscopic oscillations, or waves, that can propagate in material media. We will later discuss several other types of waves, but acoustic radiation remains perhaps the most important of all such material disturbances. Sound waves arise when a local pressure excess or deficit causes the neighboring air to compress or decompress in response to this excess or deficit. In a dynamical situation, there will usually be some overshoot, and the overcompressed or overdecompressed region will reexpand or recontract. The resulting oscillations of pressure (and density and temperature) yield the characteristic signature of an acoustic wave.

To see in greater detail the typical generation and propagation of an acoustic wave, let us consider the example depicted in Figure 4.5. The sinusoidal path indicated by the hatched curve in Figure 4.5 indicates schematically the position in spacetime of a vibrating membrane (for example, in a loudspeaker or a human voice box). Each rightward excursion of the membrane compresses the air immediately to its right; each leftward excursion decompresses the same air. The compressed (dark) portions of the air subsequently reexpand into the decompressed (light) portions, with the usual dynamical overshoot. The air molecules on average just oscillate back and forth in an attempt to follow the motion of the vibrating membrane. However, the *pattern* of compressed regions marches steadily to the right, as denoted by the diagonal dashed lines in our spacetime diagram. This propagation of the compressed regions (the crests of the sound wave) and the rarefied regions (the troughs of the sound wave) occurs with the acoustic speed c_s, where c_s^2 for a perfect monatomic gas is given by

$$c_s^2 = \frac{5P}{3\rho} = \frac{5kT}{3m},$$

with P, ρ, T, and m being, respectively, the pressure, mass density, temperature, and mean molecular mass of the gas. (The perfect gas law, equation 4.2, can be written in the form $P = \rho kT/m$ with $\rho = mn$.)

Problem 4.4 (for those who know calculus). To derive the formula $c_s^2 = 5P/3\rho$ without using partial differential equations, we need to resort to a fairly sophisticated physical argument. Consider an observer who moves with the wave in Figure 4.5; i.e., consider an observer who travels at the same speed c_s as the wave crests. Let $u\ (\simeq c_s)$ be the fluid velocity of the gas relative to the observer. For a small amplitude wave, this observer sees a steady flow of gas moving past, and the conservation of mass and momentum requires the associated fluxes to be constant with respect to such an observer. (Otherwise, mass or momentum would continuously build up or drain from some local region.) The mass flux obviously has the form of mass density times the fluid velocity; i.e., mass flux $= \rho u$. The momentum flux has two contributions: the momentum flux P which arises because of the random motions of the gas particles (see Problem 4.1); and the momentum flux ρu^2 which arises because the mean momentum density ρu is translated at the mean velocity u; i.e., total momentum flux $= P + \rho u^2$. Thus, mass and momentum conservation with respect to an observer who travels with the wave takes the forms

$$\rho u = \text{constant},$$
$$P + \rho u^2 = \text{constant}.$$

The term ρu^2 is often referred to as the "ram pressure"; however, it differs from the kinetic pressure P in not being exerted equally in all directions (i.e., ρu^2 is not an "isotropic" pressure).

To get rid of the constants in our above conservation relations, consider the spatial variation of the material properties at neighboring points in the wave. Show by taking differentials that

$$\rho\, du + u\, d\rho = 0,$$
$$dP + 2\rho u\, du + u^2\, d\rho = 0.$$

Eliminate du and derive $dP = u^2\, d\rho$. For a small amplitude wave, the total fluid velocity u is almost entirely due to the motion of our observer at the speed c_s; therefore, show

$$c_s^2 = \frac{dP}{d\rho},$$

where the derivative is evaluated as we follow the motion of the gas from one point in the wave to another. If we ignore the effects of electromagnetic radiation, viscosity, or thermal conduction, this motion takes place with each parcel of gas not exchanging any heat with the surrounding gas ($T\,ds = 0$, where s is the entropy per unit mass). Thus, the change, $d(3kT/2m)$, in internal energy per unit mass must equal the rate of doing work on the gas by the pressure, $-Pd(\rho^{-1})$, where ρ^{-1} is the volume per unit mass. With $kT/m = P/\rho$ for a perfect gas, show that $dP/d\rho = 5P/3\rho$, which yields the desired expression, $c_s^2 = 5P/3\rho = 5kT/3m$.

More generally, the square of the adiabatic speed of sound is given by $c_s^2 = (\partial P/\partial\rho)_s$, where the last derivative means the rate of change of P when ρ is varied holding the specific entropy s constant (no addition of heat). Terrestrial air is composed primarily of diatomic molecules (N_2 and O_2), for which the internal energy per unit mass contains an additional contribution of kT/m for two extra degrees of rotational freedom. If there is no addition of heat, the change in internal energy per unit mass equals the rate of doing work per unit mass: $d(5kT/2m) = -Pd(\rho^{-1})$. With $kT/m = P/\rho$, show now that $dP/d\rho = 7P/5\rho$, where by $dP/d\rho$ we really mean $(\partial P/\partial\rho)_s$. Because 7/5 is less than 5/3, we may say that terrestrial air is less springy than a monatomic gas, because some of the work of compression goes into making molecules rotate faster rather than increasing the random translational motions. With $c_s^2 = 7kT/5m$, calculate the numerical value for the speed of sound in air ($m = 29m_p$) at a temperature of 290 K. How does c_s compare with the speed of light, c?

Figure 4.5 shows that the wavelength λ, period τ, and acoustic speed c_s, of a sinusoidal sound wave satisfies the relation $\lambda = c_s\tau$. This equation is often written as a *dispersion relationship* which gives the radian frequency, $\omega = 2\pi/\tau$, in terms of the wavenumber, $k = 2\pi/\lambda$. Show that linear acoustic waves satisfy the dispersion relation $\omega^2 = k^2 c_s^2$.

In a self-gravitating homogeneous medium of density ρ, the propagation of acoustic waves is modified by the gravitational attraction of a compressed region for its own center. This attraction reduces the natural frequency which the increased pressure would otherwise like to reexpand the compressed regions. The modification enters the dispersion relationship in the form $\omega^2 = k^2 c_s^2 - 4\pi G\rho$, as Sir James Jeans was the first to show. Since negative values of ω^2 (imaginary wave frequencies) imply exponentially growing disturbances, argue that disturbances of sufficiently long wavelengths (sufficiently small values of k) in a self-gravitating homogeneous medium are unstable with respect to gravitational collapse. Derive the critical wavelength to be $\lambda_J = (\pi/G\rho)^{1/2} c_s$. Compute the numerical value of λ_J for air at standard temperature and pressure, and compare this value with the thickness of the Earth's atmosphere. Is the atmosphere of the Earth in any danger from suffering Jeans's instability? Is the effect of self-gravity important for terrestrial acoustic waves of frequency equal to, say, 2π times 440 cycles per second? *Hint*: Compare $4\pi G\rho$ to ω^2.

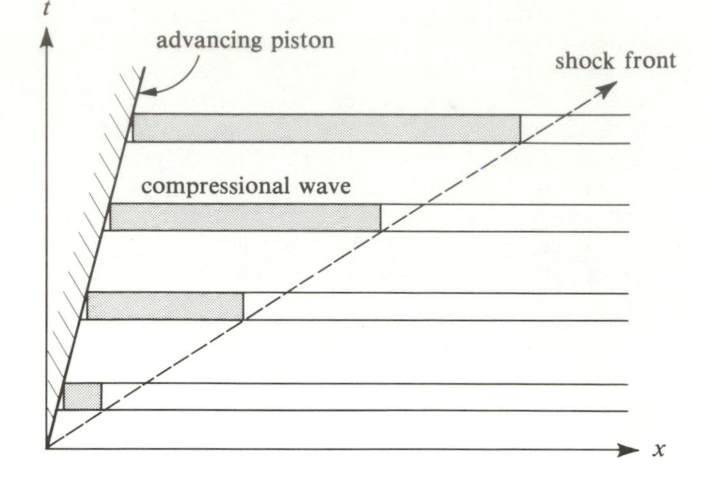

Figure 4.6. The generation and propagation of a shock wave by a piston which advances steadily into a long column of still air (light region). The compressed region (dark piece of air column) moves with the piston.

In our above example, we implicitly considered only the behavior of "linear" acoustic waves, which are generated by small displacements of the membrane from its equilibrium position. Let us now discuss the behavior of "nonlinear" acoustic waves which are generated by large displacements of a moving obstacle. Consider, first, the motion of a piston into a long cylinder of trapped air (Figure 4.6). Because the air cannot flow past the piston, air must continuously pile up in a high-pressure region and be forced to move to keep ahead of the piston. A detailed analysis shows that the front of this high-pressure region propagates into the undisturbed air ahead of the piston at a speed which exceeds the speed of sound c_s associated with small amplitude ("linear") waves. The undisturbed air, therefore, receives no signal that a piston is coming until the front reaches it all of a sudden. The air is then quickly overtaken by the compressional wave and is "shocked" from a low (undisturbed) value of the pressure (and density and temperature) to a high value as the gas passes through the front. The length scale within which this transition occurs is comparable to the distance which a molecule can freely fly before colliding with another molecule. This mean free path is usually very small in comparison with the interesting macroscopic length scales of the problem. Shock waves are, therefore, highly nonlinear compressional waves where the compression of the gas occurs almost discontinuously in one jump.

What happens if air can flow around the moving obstacle? Let us consider an airplane flying in the Earth's atmosphere (Figure 4.7). In particular, let us first describe the airflow from the point of view of the pilot. The air well ahead of the airplane rushes toward the pilot at the airspeed of the plane. The air on one streamline must decelerate to a full stop at a point on the nose (the "stagnation point"), and from there slowly creep around the plane's surface eventually to leave from the vicinity of the tail. How the air is brought to rest at the nose of the airplane depends on whether the airplane is flying supersonically or subsonically. If the aircraft is flying subsonically, the air ahead of the oncoming plane can be forewarned of the obstacle presented by the plane and be brought smoothly to rest at its nose. However, if the plane is flying supersonically, the oncoming air must suddenly decelerate in a bow shock from supersonic relative speeds to subsonic relative speeds (at least, in the component of velocity perpendicular to the shock front). For an aircraft with a blunt nose, the bow shock stands off a finite distance in front of the nose. An observer on the ground would describe the airplane as pushing a postshock region of high pressure ahead of it. The extension of this region of high pressure causes the observer to hear a "sonic boom" as the shock front at large distances from the airplane passes over the ground.

Shock waves can, of course, be created by mechanisms other than moving solid bodies. The explosion of a bomb, for example, is accompanied by the production of a strong shockwave. Generally speaking, strong shock waves are created whenever energetic events lead to large *relative motions* or pressure discrepancies between neighboring parcels of gas. The observable universe contains many violent events, and it should not surprise us, there-

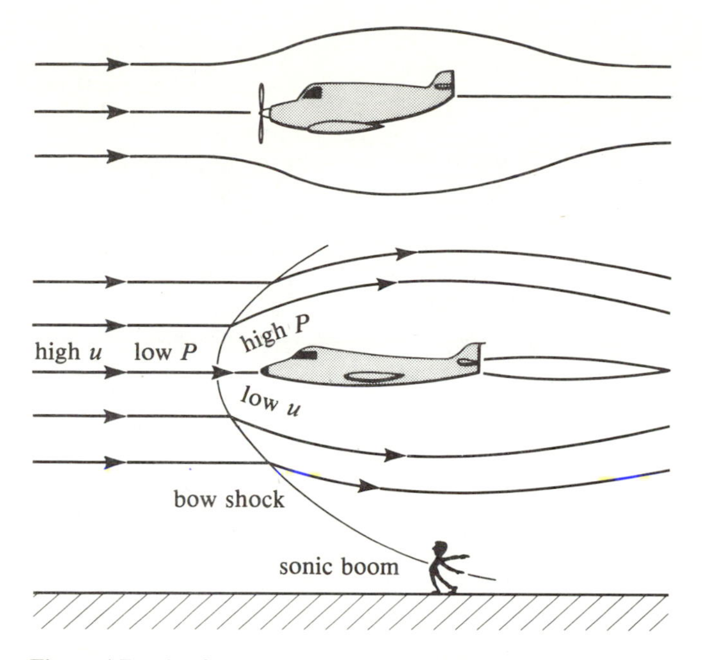

Figure 4.7. The flow of air around an aircraft has a different pattern for a plane which flies subsonically (top) and one which flies supersonically (bottom). A plane which flies faster than the speed of sound is preceded by a bow shock, which brings upstream gas of high relative speed u and low pressure P to a condition of low relative speed u and high pressure P that enables the postshock gas to flow (more or less) smoothly around the aircraft. The sudden transition from low pressure to high pressure as the bow shock sweeps by an observer on the ground produces the sonic boom associated with low-flying jets.

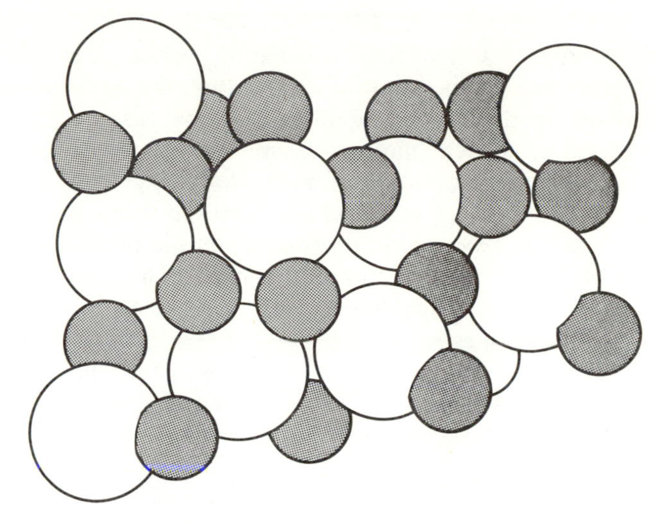

Figure 4.8. The molecules of a liquid are packed so closely together that the electronic shells practically touch. The strong repulsive forces between the electronic shells accounts for the near incompressibility of ordinary liquids. In a molecule like water, however, the electric charge is not distributed perfectly uniformly about the nuclei. As a result, the hydrogen atoms (dark blobs) contain a slight excess of positive charge, and the oxygen atoms (light blobs), a slight excess of negative charge. As a consequency, the water molecules attempt to orient their hydrogen atoms closer to their neighbor's oxygen atoms. This orientation produces the attractive binding force which holds liquid water molecules together against the disruptive influence of the repulsive force and random thermal motions.

fore, to find that shockwaves play an important role in many different astrophysical contexts.

Real Gases, Liquids, and Solids

In our discussion of the thermodynamics and mechanics of gases, we have concentrated on the properties of perfect or **ideal** gases. No real gas is, of course, perfect. Who is? Ordinary gases are made from molecules which do exert mild forces on each other. In Chapter 3, we mentioned that because the positively charged constituents (protons) and negatively charged constituents (electrons) of molecules are not spatially coincident, even neutral molecules can exert short-range forces on each other. Van der Waals pointed out that incompletely canceled electric forces have an *attractive-repulsive duality*. For example, the water molecule is formed by a polar covalent bond. The bonding electrons are shared unequally by the nuclei; so the two hydrogen atoms are asymmetric relative to the oxygen atom (Figure 4.8). As a consequence, the molecule contains a slight excess of positive charge near the hydrogen atoms and a compensating excess of negative charge near the oxygen atom. If water molecules are pushed close to one another, the molecules will try to orient their hydrogen ends toward the oxygen ends of neighboring molecules. Clearly, this orientation produces a net attraction between water molecules. If this attraction is sufficiently strong, compressed water vapor may make a phase transition to turn into *liquid* water. This phase transition generally releases some energy ("latent heat") associated with the attractive binding of liquid water molecules for one another.

The *repulsive* part of the short-range force arises in the following manner and has the following consequence for the properties of liquid water. The attractive force in liquid water brings the molecules as close as possible without causing the electronic shells of neighboring molecules to overlap. We may say that the molecules, in some sense, "touch," but in liquid water, retain sufficient freedom that they can easily slip around one another. If one tried to push the molecules even closer together, the overlapping of electronic shells would produce very strong repulsive forces that prevent further compression. Thus, liquid water maintains nearly a constant density over a wide range of pressures. These two properties—ability to flow easily while being nearly incompressible—distinguish a liquid from other states of matter.

Problem 4.5. Consider a mixture of perfect gases all at the same temperature T. Show that the Maxwell-Boltzmann distribution predicts that the gases which correspond to light molecules, small values of m, will have a greater proportion of molecules at high random velocities than the gases which correspond to heavy molecules, large values of m. The distributions of energies, $E = mv^2/2$, will be the same (equipartion of energy), but at each value of E, light molecules possess greater speeds v than heavy ones.

The greater thermal agitation of light molecules in comparison with heavy ones is a general property of matter in all states, not only the gaseous state. On this basis, argue why light molecules like hydrogen might be expected to have a lower solidification temperature than heavy ones like calcium silicate (a familiar component of terrestrial rocks) or iron. However, molecular weight cannot be the whole story, since helium remains a liquid to lower temperatures than molecular hydrogen. Why? *Hint*: Which species has the stronger molecular forces, judging from their chemical reactivities?

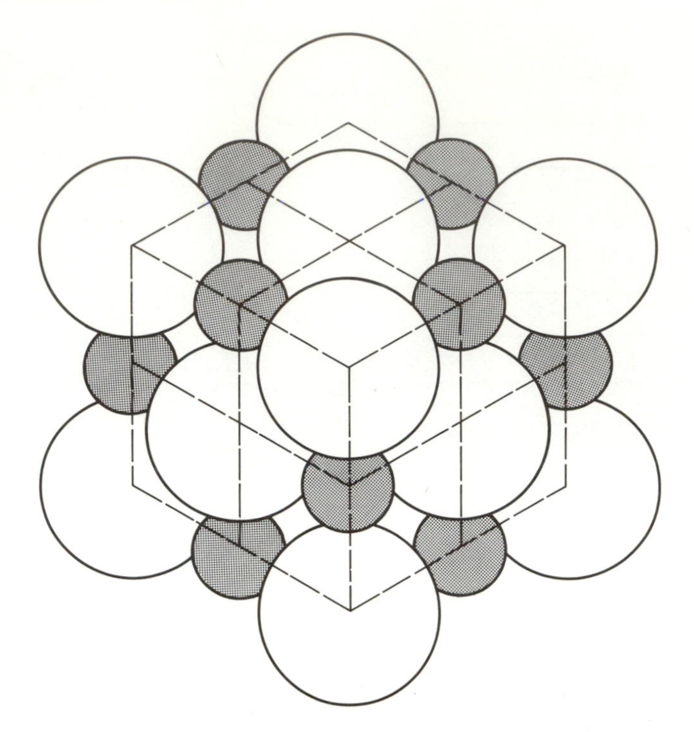

Figure 4.9. The molecules or ions of a solid are well-ordered on a lattice array. As a consequence, solids are quite rigid typically and do not flow easily. Instead, when stressed, solids tend to fracture or break into smaller crystalline pieces. The crystalline structure shown here is that of sodium chloride, table salt. The dark spheres are sodium ions (positively charged), and the light spheres are chlorine ions (negatively charged). Solid water, ice, forms a more complicated crystalline structure which makes use of the ability of hydrogen atoms to bond to *two* oxygen atoms at the same time. Such hydrogen bonds also exist, but are less permanent, in liquid water (see Figure 4.8). The open (hexagonal) crystalline structure of ice makes the solid phase less dense than the liquid phase.

What happens if we cool liquid water, say, at constant pressure? As the temperature drops, the thermal agitation of the individual molecules becomes less and less. Their tendency to twist and turn is reduced, and at some value of the temperature, the attractive short-range forces becomes strong enough in comparison with the thermal agitation to cause almost all the molecules to become perfectly oriented. In other words, the molecules are locked into place relative to one another—as it turns out, on a crystal lattice (see Figure 4.9)—and the matter would *solidify*. As long as the temperature remains above absolute zero, there will remain some thermal agitation (at $T = 0$ K, there is a minimum level of agitation dictated by quantum-mechanical principles), but the displacements of the molecules will statistically be small in comparison with their average separations. This property—the rigidity of the lattice structure—is the most important property of the solid state.

To finish our thermodynamic description of the three states of matter, let us consider the phase diagram of water. In Figure 4.10 we have plotted schematically the regions in the temperature-pressure plane where water can be found in one or more of its ordinary states under thermodynamic equilibrium. As we can see, at relatively high temperatures and low pressures, water exists in the gaseous or vapor state. The gaseous state of water is separated from the liquid and solid states by two curves. These two curves define the maximum pressure (the saturation vapor pressure) that can exist for water vapor (say, in the air) at a given temperature, without the precipitation of the water vapor in either liquid or solid form (rain or snow). If we have a container of saturated water vapor, and we were to decrease its volume at a constant temperature, the compression would not produce a higher internal pressure. Instead, part of the low-density vapor would be converted into higher-density liquid or solid form (depending on the maintained temperature), which occupies less volume per unit mass. In the temperature-pressure plane, therefore, the system would remain at the appropriate point on the phase curve until all the vapor has been converted into liquid or solid. Past that stage, any attempt to decrease the volume further will lead to very large resisting internal pressures.

Notice that a similar curve separates the liquid and solid phases of water. Water, however, differs from most substances in that the pressure needed to solidify the liquid *increases* with decreasing temperature. Alter-

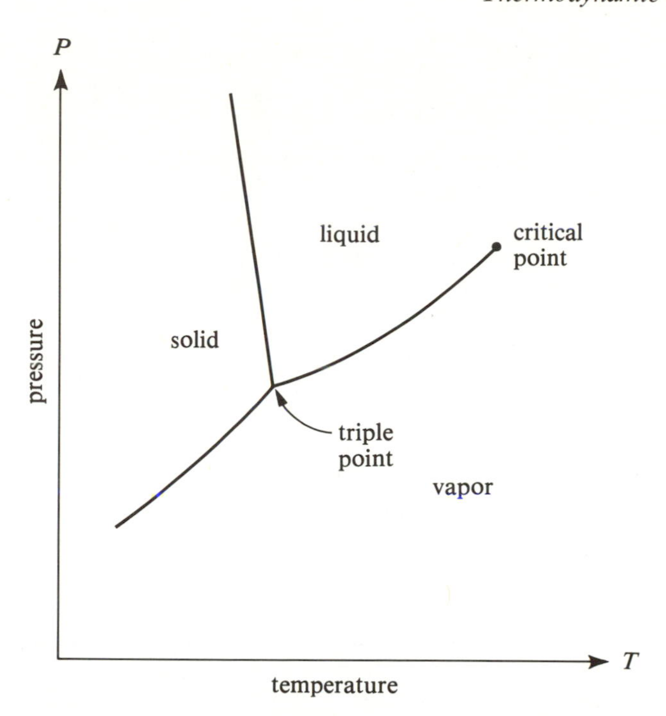

Figure 4.10. Schematic phase diagram for water. The triple point represents the only temperature and pressure where all three phases of water: vapor, liquid water, and ice, can simultaneously exist in equilibrium. Beyond the black dot, the critical point, the distinction between vapor and liquid vanishes.

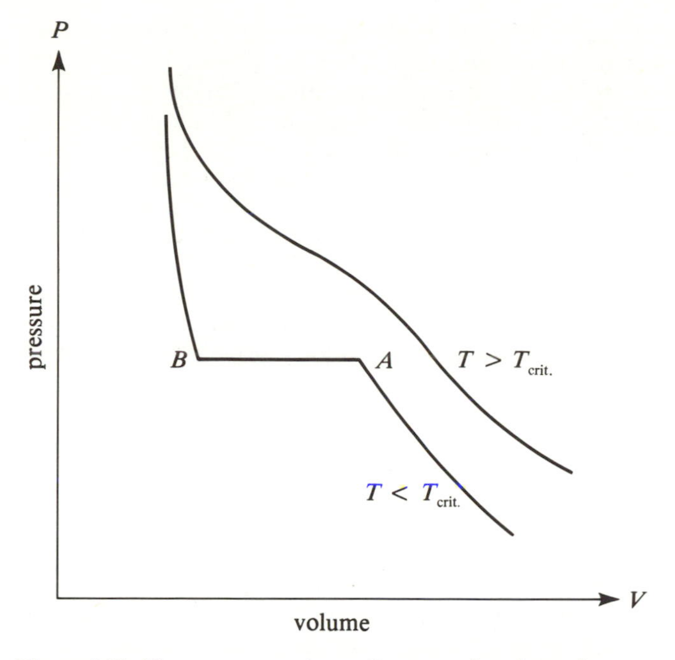

Figure 4.11. The pressure-volume diagram of an imperfect gas. When a gas is compressed at a constant temperature less than the critical value T_{crit}, there comes a point A when it begins to make a phase transition to a liquid phase. The system stays at the same pressure upon further compression until all the gas has been converted to the liquid phase at point B. Further attempts to compress the liquid is resisted by a very rapid increase of the pressure. If the same system is compressed at a constant temperature greater than the critical value T_{crit}, no phase transition is observed. The easy compressibility of early states gives way gradually and smoothly to essential incompressibility of later states.

natively, we may say that the temperature needed to melt the solid *decreases* with increasing pressure (which explains why glaciers slide since their base liquefies under the weight of the ice mass). These properties are equivalent to saying that at a given pressure and temperature, ice is less dense than water: ice cubes float! This property of water—the solid phase being buoyant in the liquid phase—turns out to be lucky for fish, since otherwise lakes would start freezing in winter from the bottom (instead of from the top) and would eventually freeze solid.

The point where all three phase curves intersect in Figure 4.10 is called the triple point of water. It represents the only pressure and temperature where all three phases of water can be simultaneously in thermodynamic equilibrium. Notice also the critical point labeled with a solid dot in Figure 4.10. For temperatures higher than the critical temperature, 647 K, the thermal agitation of the molecules is so great that there is no true distinction between the gaseous and liquid states. If the temperature is held above 647 K, and the pressure is slowly increased for a container of water, the easy compressibility of earlier states is smoothly followed by the difficult compressibility of later states, but there is never a definite point at which we can say we have a transition from gas to liquid. At high temperatures, the mild attractive forces of water are too weak to play a significant role (Figure 4.11).

Before we leave this topic, we should warn the reader that we have simplified matters considerably. In actuality, solid water is known to possess at least *eight* different crystalline states which are observable at very high pressures. At low temperatures and pressures, such as those encountered in interplanetary or interstellar space, ice may form as an *amorphous solid* rather than as a regular hexagonal array. In amorphous solids, individual crystals exist, but their sizes are typically so small that the aggregate takes on a powdery or other nondescript shape.

Water is also an unusual molecule in that it has a relatively large separation of positive and negative charge distributions. This allows liquid water to self-ionize slightly: 2 H_2O molecules react to form H_3O^+ and OH^-, each with a concentration of one part in 10^7. (Chemists say that liquid water has a *p*H of 7 in their scale of acids and bases.) The relatively large charge separation of water makes it a versatile molecule, both chemically and spectroscopically. Liquid water is an excellent solvent, and water vapor is an extremely potent absorber of infrared radiation. Both facts are crucial to the maintenance of life on Earth (Chapters 17–20).

Ordinary ice is an example, albeit a somewhat unusual example, of a molecular crystal. Ice is held together by the so-called **hydrogen bonds**, elucidated by Linus Pauling, that allow a hydrogen atom to bond to two atoms instead of only one. Hydrogen bonds also play an important role in the structure of DNA, the central molecule of biological life that we shall discuss in Chapter 19. Another type of solid is the ionic crystal, of which sodium chloride (table salt) is probably the best-known example. The sodium-chloride lattice has an especially simple structure. In the sodium-chloride crystal, positively charged sodium ions (sodium atoms which have lost one valence electron) alternate with negatively charged chlorine ions (chlorine atoms which have gained one valence electron) on a cubic array. At room temperature and pressure, solid sodium chloride is preferred over the gaseous state because the former contains more binding energy (Problem 4.6). Metals form crystals in which the valence electrons are free to roam the entire lattice. Such substances are, therefore, good conductors of electricity and heat. Semiconductors have bound valence electrons, but the binding is weak, and this property is exploited by the modern electronics industry. Other crystals—for example, diamond—form a continuous network of chemical bonds; diamond is, in effect, one single molecule.

Problem 4.6. As a pedagogical exercise, let us imagine a hypothetical one-dimensional sodium-chloride crystal consisting of alternating sodium and chlorine ions with the ionic spacing L. If a sodium ion were partnered only by a lone chlorine ion in the gaseous state, the Coulomb energy (negative of the binding energy) would be $-e^2/L$, since the charge of the sodium ion is $+e$ and that of the chlorine ion, $-e$. Let us show now that our sodium ion contains more binding energy by belonging to a crystal lattice. It has two immediate chlorine ions as neighbors; this contributes $-2e^2/L$ to the Coulomb energy. It has two sodium ions as next neighbors at a distance $2L$ each; this contributes $+2e^2/(2L)$ to the Coulomb energy. Continue in this manner to show that the total Coulomb energy of our sodium ion is given by the sum (infinite sum if we have an infinite one-dimensional lattice),

$$-\frac{2e^2}{L}\left(1-\frac{1}{2}+\frac{1}{3}-\frac{1}{4}+\cdots\right).$$

Show by summing in pairs (which physically corresponds to the shielding provided by charges of opposite signs) that the quantity inside the parenthesis exceeds 1/2. To perform the sum exactly, notice that the function $\ln(1+x)$ has the Taylor series expansion,

$$x - x^2/2 + x^3/3 - x^4/4 + \cdots,$$

which is convergent for $|x| < 1$. We obviously should not use the expansion for $x = -1$, because $\ln(0)$ is undefined (the reason for the finite radius of convergence), but we can hope that setting $x = +1$ will be all right. Take this simple approach, and show that the sum $1 - 1/2 + 1/3 - 1/4 + \cdots$ can be evaluated as $\ln(2) = 0.693\ldots$. Can you devise a physical or mathematical argument to justify our trick? In this manner, show that the binding energy per ion of our hypothetical lattice equals $1.386e^2/L$. For the actual three-dimensional sodium-chloride crystal, a **Madelung sum** yields $1.75e^2/L$. In other words, the ionic solid has 75 percent more binding energy than the gaseous state. This is a typical result in the following sense. In *any* medium, gaseous, liquid, or solid, where electric charges of opposite sign exist but the overall medium is neutral, the Coulomb energy per particle will roughly be given by nearest neighbor effects because of the shielding of distant charges.

Box 4.4 summarizes what we need to know here about the three ordinary states of matter. At very high temperatures, matter can acquire a fourth state: the **plasma** state, where molecules are dissociated and atoms are ionized. Plasmas do not occur naturally on Earth (except in flames and lightning discharges), but in fact they constitute the dominant form of visible matter in the universe. As we shall see in the next chapter, the matter in the Sun (and in most stars) is in the plasma state.

BOX 4.4

The Three Ordinary States of Matter

Solids contain molecules or ions which are more-or-less rigidly locked in place in a regular crystal lattice. This structure gives solids their cohesive strength and rigidity, and explains why they are the most bound of the three ordinary states of matter.

Liquids contain molecules which are also tightly packed together by attractive molecular forces, but the individual molecules can still move and twist relative to one another. As a result, liquids are quite incompressible, but are able to flow.

Gases contain a lot of empty space between the constituent molecules. This accounts for their easy compressibility and ability to flow. Gases, unlike liquids, are compressible fluids.

Macroscopic Quantum Phenomena

So far we have described states of matter where quantum-mechanical considerations enter only on the microscopic level, in the structure of the molecular constituents. Given the molecules and the forces they exert on each other, we could use purely classical considerations to deduce what would happen when we have such molecules in bulk. There are situations, however, where quantum considerations can affect dramatically even the bulk macroscopic state. These situations inevitably involve low temperatures or high densities. If the density is very high, as in a white dwarf or a neutron star, the temperature need not be very low (Chapter 7).

To understand why this state of affairs arises, we must explicitly recognize that all real matter is composed of elementary particles (such as electrons) which are fundamentally indistinguishable from one another (see Chapter 6). The discrete units that these elementary particles combine to form—say, helium-4 atoms composed of two neutrons, two protons, and two electrons—are also fundamentally the same. The indistinguishability of helium-4 atoms changes, for example, how we should count the number of microstates that correspond to the same macrostate depicted in Figure 4.2. In classical statistical mechanics, if helium-4 atom A is found in bin 1 and helium-4 atom B in bin 2, we would count that microstate as different from having helium-4 atom A in bin 2 and helium-4 atom B in bin 1. But in quantum statistical mechanics (i.e., in the real world), there is no such thing as helium-4 atom A and helium-4 atom B, only helium-4 atoms. Thus, the "two" situations described previously correspond to only one microstate: a helium-4 atom in bin 1 and a helium-4 atom in bin 2. Obviously, quantum-mechanical considerations greatly reduce the total number W of microstates that classical statistical mechanics abscribes to a given macrostate, and this might require correspondingly large changes in the macroscopic properties of matter.

In actual practice, large effects are observed only at low temperatures or high densities ("high occupation numbers" in Figure 4.2, where some of the bins are densely occupied by particles). For low occupation numbers, quantum considerations reduce the W deduced by classical statistical mechanics by nearly the same (enormous) factor for all macrostates of the system. Consequently, for low occupation numbers (ordinary phenomena), the classical methods yield probabilities ($\propto W$) for different macrostates which remain the *same* relative to one another, and therefore yield the *correct* thermodynamic results for transformations of one such macrostate to another. In particular, since $S = k \ln (W)$, the *difference* in entropies of two macrostates of the same physical system is not affected by using the incorrect counting procedure. This *difference* in entropies, ΔS, when multiplied by the temperature T, equals the amount of heat needed to transform one thermodynamic state of equilibrium to another.

Problem 4.7. Let W_1 be the number of microstates which corresponds to macrostate 1, and W_2, the number to macrostate 2. Show that the difference in entropies of macrostate 2 and macrostate 1 depends only on the ratio W_2/W_1, not on their individual values.

Notice that the point where S equals zero is left ambiguous in classical statistical mechanics. In quantum statistical mechanics,

$$S = k \ln (W) = 0 \quad \text{when} \quad W = 1,$$

i.e., when there is only one microstate possible for the given macrostate. This state is one of perfect macroscopic order, and it is reached at a temperature of absolute zero, $T = 0\,\text{K}$. The requirement $S = 0$ at $T = 0$ is part of the usual formulation of the "third law of thermodynamics" (see Box 4.1). Another part of the third law conjectures that it is not possible to reach absolute zero in a finite number of thermodynamic steps, but this issue is not important for any discussion in this book.

What constitutes the state of perfect order for real matter? In systems made of molecules with appreciable attractive forces, the state of perfect order is a perfect crystal, where each molecule is located on a geometrically precise lattice and executes the minimal vibration about the equilibrium point consistent with Heisenberg's uncertainty principle. Are there any substances with such weak molecular attraction that they can remain gaseous or liquid all the way down to absolute zero? If there are, what would their properties be at low temperatures?

Before we address the first question, let us consider the second question, in terms of a **perfect gas**, that is, a system of particles which exert *zero* forces on each other. The quantum statistics (see Chapter 3) of such a gas depends on whether it is composed of fermions (particles with half-integer spin) or bosons (particles with whole integer spin). Fermions must obey the Pauli exclusion principle; bosons need not. Let $dn_f(\mathbf{p})$ or $dn_b(\mathbf{p})$ denote the number density of fermions or bosons, respectively, with spin s and with momenta between $\mathbf{p}$ and $\mathbf{p} + d\mathbf{p}$. If these indistinguishable fermions or bosons are in thermodynamic equilibrium at temperature T, then the number densities $dn_f(\mathbf{p})$ and $dn_b(\mathbf{p})$ satisfy the perfect gas distributions:

$$dn_f(\mathbf{p}) = \frac{(2s+1)}{\exp[(E - C)/kT] + 1} (h^{-3} 4\pi p^2\, dp), \quad (4.4)$$

$$dn_b(\boldsymbol{p}) = \frac{(2s+1)}{\exp[(E-C)/kT]-1}(h^{-3}4\pi p^2\,dp), \quad (4.5)$$

where $E = mc^2$ is the energy of a particle of relativistic mass $m = m_0(1 - v^2/c^2)^{-1/2}$, and C is a quantity which depends on the number density and temperature. In equations (4.4) and (4.5), the factor $2s + 1$ enters into the counting of states because there are $2s + 1$ independent spin orientations for a material particle of spin s. The factor h^{-3} enters because Heisenberg's uncertainty principle essentially defines a minimum cell of size h^3 for the product of spatial and momentum volumes in Figure 4.2. (For a more precise explanation, see Problem 4.8.) In equations (4.4) and (4.5), $\{\exp[(E-C)/kT]+1\}^{-1}$ and $\{\exp[(E-C)/kT]-1\}^{-1}$ are the "occupation numbers" associated with Fermi-Dirac and Bose-Einstein statistics that we referred to earlier in our discussion. They play the same role for quantum statistical mechanics that the Boltzmann factor, $\exp(-E/kT)$, plays for classical statistical mechanics. In particular, notice that the occupation number for fermions is of this form: 1 over a number larger than 1. Thus, the occupation number for fermions is always less than 1, a direct result of Pauli's exclusion principle, which forbids more than one fermion from occupying a single quantum state. (For free particles, a quantum state is a "volume" h^3 in the six-dimensional space of ordinary space and momentum space.)

The presence of the $+1$ or -1 in the denominators of equations (4.4) and (4.5) is insignificant at high temperatures or low densities (see Problem 4.8); however, they may make a big difference at low temperatures or high densities. To deduce what happens at low temperatures is somewhat difficult because of the complicated mathematics in equations (4.4) and (4.5). In Chapter 7, we shall take an approximate but physically more revealing route to deduce the thermodynamic behavior of a perfect Fermi-Dirac gas at low temperatures or high densities. For the properties of a material Bose-Einstein gas at low temperatures or high densities, see Problem 4.9.

Problem 4.8 (for those good at calculus). To calculate the number of quantum states associated with the translational degrees of freedom of a particle, imagine enclosing it inside a cubic box of sides L oriented in the x, y, and z directions. Quantum mechanics requires the magnitude of the x-momentum $|p_x| = h/\lambda_x$ to have values such that the side L in the x-direction has a half or whole integer number of wavelengths,

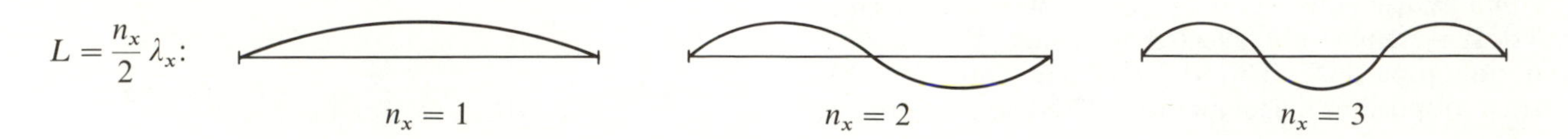

In other words, $|p_x| = hn_x/2L$ with $n_x = 1, 2, 3, \ldots$ Similar considerations hold in the y and z directions. Therefore, the allowable values for the square of the momentum is given by

$$p^2 = p_x{}^2 + p_y{}^2 + p_z{}^2 = \frac{h^2}{4L^2}(n_x{}^2 + n_y{}^2 + n_z{}^2).$$

Define $n^2 = n_x{}^2 + n_y{}^2 + n_z{}^2$, and write the above as $p = hn/2L$, where n need not be an integer, although n^2 is, of course. Each distinct triplet of positive integers (n_x, n_y, n_z) corresponds to a single translational quantum state. (A complete quantum state includes also a specification of one of the $2s + 1$ spin orientations.) Consider the three-dimensional lattice of positive integers with spacing equal to 1. The quantity n is geometrically the length of the radius vector from the origin $(0,0,0)$ to (n_x, n_y, n_z). When n_x, n_y, or n_z change, so will n and p. If

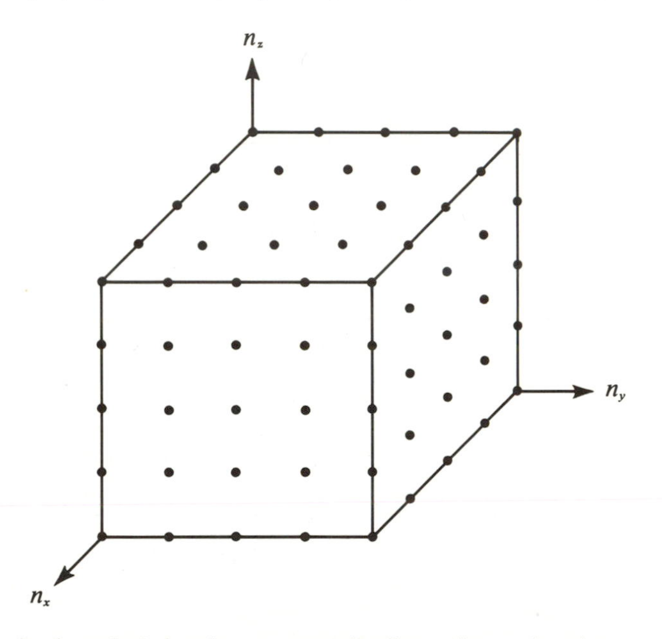

the length L is of macroscopic dimensions, the change of p will be almost continuous, and we may use the approximation of differential calculus. The complete spherical shell $4\pi p^2\,dp$ corresponds to a spherical shell only in the first octant of the space (n_x, n_y, n_z), since the latter numbers must be positive. How many triplets of integers are contained in the latter shell? *Hint*: Notice that each lattice point lies at the vertex of eight unit cubes, and that each unit cube has eight vertices. Except for negligible edge effects, argue that the number of vertices (or distinct translational quantum states) equals the volume $(\pi/2)n^2\,dn$ of the shell in the first octant. Thus, show that the number of translational quantum states $(\pi/2)n^2\,dn$ in momentum volume $(4\pi p^2\,dp)$ and spatial volume L^3 equals $h^{-3}L^3 4\pi p^2\,dp$. Per unit spatial volume, show that the latter quantity yields the expression adopted in the parentheses of equations (4.4) and (4.5).

For nonrelativistic motions, $E = m_0c^2 + m_0v^2/2$ and $p = m_0v$ (see Problem 3.15). At high temperatures or low densities, the quantity $(m_0c^2 - C)/kT$ turns out to be a large positive number. Show that the two distributions (4.4) and (4.5) both become the Maxwell-Boltzmann distribution in this limit.

Problem 4.9. In the limit $T \to 0$, bosons can all occupy the lowest energy state available to them, and this is, in fact, predicted by equation (4.5). This property is called "Bose condensation," but it differs from the ordinary condensation of a gas into a liquid in that Bose condensation takes place in momentum space, not in physical space. A gas of bosons at low temperatures placed in a large container can be expected to possess virtually no random motions. This would mean that neighboring streams of such bosons could have very different bulk velocities (shear), yet would exhibit no friction, because there would be no migration of bosons from one stream to the other to average out the difference in momenta.

Helium-4 is a boson of spin 0; its two neutrons, protons, and electrons all have oppositely directed spins of 1/2 each. Helium-4 atoms have a very slight attraction for each other, and this attraction unfortunately prevents helium-4 from staying gaseous at very low temperatures. At atmospheric pressure, helium-4 liquefies below 4.3 K, and it should remain a liquid all the way to absolute zero. However, below 2.2 K, helium-4 makes a phase transition to become two liquids. One of the liquid phases exhibits superfluid properties, the ability to flow without any apparent friction. The other liquid is an ordinary fluid. Why do you suppose superfluidity is readily observed in helium-4, but not in helium-3? *Hint*: Helium-3 has one neutron, two protons, and two electrons. Unless helium-3 forms pairs, is it a fermion or a boson?

Thermodynamic Behavior of Radiation

In contrast to matter, the statistical mechanics of a radiation field in thermodynamic equilibrium at temperature T is extremely simple and unvaried. This simplicity results for two reasons. First, photons have negligible interactions with each other; therefore, they can be treated as a perfect Bose-Einstein gas. Second, photons all fly at the speed of light; consequently, they possess zero rest mass. They are thereby easily created and destroyed in electromagnetic interactions with matter. As a further result, the quantity C in equation (4.5) is zero for photons.

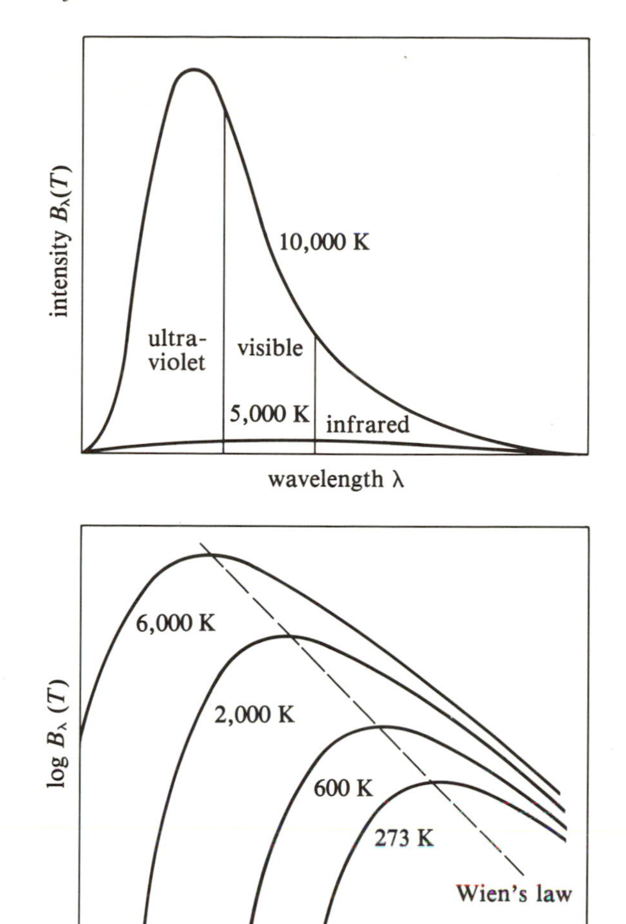

Figure 4.12. Thermal radiation.

It was Max Planck who first derived the law for the intensity of the radiation emitted at different wavelengths λ by any sufficiently opaque body of temperature T (Figure 4.12). This derivation initiated quantum mechanics, and Planck's constant h made its first appearance in the theory of thermal radiation (which is a more appropriate name than the more common "black-body radiation"). Thus, Planck's statistical-mechanical derivation of the law of thermal radiation is one of the greatest triumphs of theoretical physics.

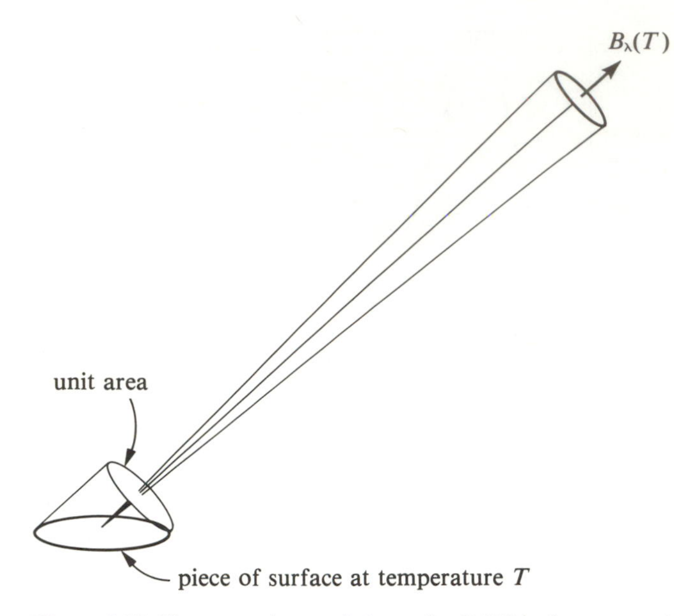

Figure 4.13. The monochromatic intensity $B_\lambda(T)$ is the power at wavelength λ radiated per unit wavelength per unit solid angle into a small cone from a piece of the surface of an opaque body with uniform temperature T, with the piece having unit area in projection perpendicular to the axis of the cone.

Our derivation will follow a different route than Planck originally followed. First, we notice that $E = hc/\lambda$ and $p = h/\lambda$ for a photon. Next, we notice that the factor $2s + 1$ in equation (4.5) should be replaced by the factor 2, because even though the photon has a spin of 1, only two of the spin orientations can manifest themselves physically (see Chapter 3). Finally, rather than use the quantity $dn_b(\mathbf{p})$, astronomers prefer to consider the energy radiated per second per unit wavelength per unit solid angle, toward the direction $\mathbf{p}$, from a surface which has a unit area in perpendicular projection to the direction of emission (Figure 4.13). This mouthful is given a name by astronomers (so that they don't have to repeat the definition too often); it is called the **monochromatic specific intensity**. For thermal radiation, the monochromatic specific intensity $B_\lambda(T)$ can be shown from equation (4.5) to be given by Planck's law,

$$B_\lambda(T) = \frac{2hc^2}{\lambda^5}\frac{1}{e^{hc/\lambda kT} - 1}. \tag{4.6}$$

Notice that equation (4.6) does *not* depend on any property of the body which radiates it except for its temperature T. *Any* body—black, brown, or white—which is at a *uniform* temperature T and in thermodynamic equilibrium with its own radiation field (basically requiring it to be sufficiently opaque) will emit photons with a spectral distribution of wavelengths given by Planck's law. Moreover, the rate of emission does not depend on the direction of emission, and the radiation comes out with no net sense of polarization. This universal character of "blackbody radiation" makes Planck's law truly one of the great laws of macroscopic physics. To summarize, thermal radiation from an opaque body of uniform temperature T:

(a) has a certain spectral composition (distribution with wavelength) characterized only by the temperature of the source;
(b) has a certain specific intensity;
(c) is isotropic (the same emission in all directions);
(d) is completely unpolarized.

Problem 4.10 (for those who know elementary calculus). To derive equation (4.6) from equation (4.5), show that

$$B_\lambda(T)\,d\lambda = -(cE/4\pi)\,dn_b(\mathbf{p}),$$

where the $-$ sign is needed because increasing p corresponds to decreasing λ; i.e., $p = h/\lambda$ implies $dp = -(h/\lambda^2)\,d\lambda$.

Problem 4.11. For fixed temperature T, define the dimensionless wavelength x by $x = \lambda kT/hc$. Show now that the dimensionless monochromatic specific intensity can be written

$$\frac{h^4c^3}{2(kT)^5}B_\lambda(T) = \frac{1}{x^5(e^{1/x} - 1)}.$$

Plot the left-hand side versus the right-hand side for a range in x from 0.00 to 3.00. For what value of x does the right-hand side have a maximum value? What does this imply for the corresponding numerical value of the product λT? Compare the shape of your curve with Figure 4.3.

It does not matter what the emitting body is made of: gold, lead, human flesh, or garbage; all thermal bodies emit the same if they have the same temperature T and if they are sufficiently opaque. Hence astronomers cannot tell the chemical constitution of a star by examining the star's continuum radiation, which comes basically from relatively opaque layers. Instead, astronomers must look for clues in the line spectrum, formed in the more diffuse upper layers of the star's atmosphere.

Planck's law, equation (4.6), implies two other important results: Wien's displacement law and the Stefan-Boltzmann law.

(a) *Wien's displacement law* (Problem 4.11): for thermal radiation, if the temperature T increases, there are more photons with shorter wavelengths (higher energies). In particular, the wavelength of maximum emission, λ_{max}, shifts with changing temperature according to the law

$$\lambda_{max} T = 0.29 \text{ cm K}. \tag{4.7}$$

(b) *Stefan-Boltzmann law* (Problem 4.13): for thermal emission from the surface of a hot body of temperature T, the rate of emission per second per cm^2 (i.e., the energy flux f leaving the surface) is proportional to the fourth power of the temperature T. In formula form, the flux of energy from the surface of a thermal radiator is given by

$$f = \sigma T^4, \tag{4.8}$$

where σ is the Stefan-Boltzmann constant (Appendix A).

The Stefan-Boltzmann law implies that the emission of thermal radiation from the surface of an opaque body rises very quickly with increasing temperature. A body which is twice as hot will radiate *sixteen* times faster.

Problem 4.12. How much power would a light bulb give off if it has a filament of radiating area 1 cm^2 which is heated up to 2,000 K? Convert your answer to watts. At what wavelength is the radiation peaked? Is this in the infrared, visible, or ultraviolet? Why does an ordinary light bulb feel hot? Why is tungsten used as the filament? *Hint*: At atmospheric pressure, most substances vaporize before reaching 2,000 K.

Problem 4.13. To derive equation (4.8) from equation (4.6) requires a good command of calculus. Let us first derive another formula that we shall need later in this book. From Problem 4.10, we recognize the energy density $\mathscr{E}_{rad}$ of the radiation field as given by the integral of $(4\pi/c)B_\lambda(T)$ over all wavelengths. If we use the integration variable $y = 1/x$, where x is defined by Problem 4.11, show that $\mathscr{E}_{rad}$ is given by

$$\mathscr{E}_{rad} = \frac{4\pi}{c}\int_0^\infty B_\lambda(T)\,d\lambda = \frac{8\pi(kT)^4}{(hc)^3}\int_0^\infty \frac{y^3\,dy}{(e^y - 1)}.$$

The last integral is a pure number equal to $\pi^4/15$. To prove this requires quite a bit of work, which we shall outline below. However, before embarking on this mathematical exercise, let us note that we have already deduced an important result, namely, $\mathscr{E}_{rad} \propto T^4$. This result was, in fact, first derived by thermodynamical arguments, but deriving the value of the proportionality constant turned out to require the use of (quantum) statistical mechanics. To calculate it exactly, write $(e^y - 1)^{-1}$ as $e^{-y}(1 - e^{-y})^{-1}$. Since $e^{-y} < 1$ for all positive y, we may expand by the usual formula to obtain $(1 - e^{-y})^{-1}$ as $1 + e^{-y} + e^{-2y} + e^{-3y} + \cdots$. Use the summation notation to show now

$$\int_0^\infty \frac{y^3\,dy}{e^y - 1} = \sum_{n=1}^\infty \int_0^\infty y^3 e^{-ny}\,dy.$$

Transform variables one more time, $z = ny$, to show that the integral on the right-hand side equals $3!/n^4 = 6/n^4$, where we have made use of a standard formula for the gamma (or factorial) function. Thus, show that

$$\mathscr{E}_{rad} = aT^4,$$

where the radiation constant $a = 48\pi\zeta(4)k^4/(hc)^3$, with $\zeta(4)$ equal to the Riemann zeta function of argument 4:

$$\zeta(4) = \sum_{n=1}^\infty \frac{1}{n^4}.$$

A standard (but difficult!) exercise in complex variable theory proves $\zeta(4) = \pi^4/90$. Assume this to be true (or check it to your satisfaction on your electronic calculator), and compare your value of a with that given in Appendix A. From the definition of $B_\lambda(T)$, use spherical polar coordinates to show that the outward energy flux is given by

$$f = 2\pi\int_0^{\pi/2}\cos\theta\sin\theta\,d\theta\int_0^\infty B_\lambda(T)\,d\lambda = \frac{c}{4}\mathscr{E}_{rad}.$$

In this manner, derive equation (4.8) with $\sigma = ca/4$, also given in Appendix A.

Problem 4.14. Radiation can also exert a pressure by transferring the momentum of the photons to the matter. Let us calculate the pressure P_{rad} associated with a black-body radiation field. Use Problem 4.1 together with $v = c$ and $E = cp$ (appropriate for photons) to deduce

$$P_{rad} = \tfrac{1}{3}n_{rad}\langle E\rangle.$$

The product of the number density n_{rad} of photons and the average energy $\langle E\rangle$ of a photon in the distribution is, by definition, equal to $\mathscr{E}_{rad}$, the energy density. Therefore, show that for a black-body radiation field,

$$P_{rad} = \tfrac{1}{3}aT^4.$$

The factor 1/3 between the pressure and energy density of a relativistic gas (here, a photon "gas") is to be contrasted with the factor 2/3 for a nonrelativistic gas (cf. equations 4.2 and 4.3).

An Example

You may have found the preceding discussion of the second law of thermodynamics and the properties of thermal radiation a little esoteric; so perhaps you dismissed the topics as having very little relevance for everyday life. On the contrary, although they may be taken for granted, the second law of thermodynamics and thermal radiation are the *central* facts of everyday existence.

Consider, for example, the so-called "energy crises." How can there be an energy crisis if energy is conserved? Why don't we just use the same energy over and over again, recycling it like used soda-pop bottles? The answer, of course, is that the second law of thermodynamics tells us there is a *quality* to energy just as there is a quantity. The energy contained in the chemical bonds of gasoline is of higher quality than the low-grade heat which comes from the exhaust of a car. And according to the second law of thermodynamics, the latter cannot be converted to the former without degrading some other natural resource. In a crucial sense, therefore, the "energy crisis" is a misnomer; it should be called the "entropy crisis."

To pursue this topic further, it has often been said that the Sun is our ultimate source of energy for sustaining life on Earth. Is it? In fact, the Earth gains no *net energy* from sunshine. If it did, the Earth would continuously heat up, and soon it would be hotter than Hell. For every erg of sunshine which falls on the Earth, the Earth returns one erg of infrared radiation back to space. (In fact, because of the slow leakage of residual heat from a hot interior [Chapter 17], the Earth actually sends back slightly more than it receives.) The Earth has long since reached a balance between heating by sunshine and cooling by infrared radiation (Problem 4.15).

Problem 4.15. Show that the total luminous power P intercepted by the Earth is

$$P = \pi R_\oplus{}^2 (L_\odot/4\pi r^2),$$

where $\pi R_\oplus{}^2$ is the cross-sectional area of the Earth, $L_\odot$ is the luminosity of the Sun, and r is the distance of the Earth from the Sun. Take into account the rotation of the Earth by calculating the value $\bar{f}$ of the *average* energy flux incident on the Earth, where this average is performed by spreading P over the entire *surface* area $4\pi R_\oplus{}^2$ of the Earth:

$$\bar{f} = P/4\pi R_\oplus{}^2 = L_\odot/16\pi r^2.$$

(Notice that $\bar{f}$ differs by a factor of 4 from the **solar constant**, defined as the energy per second which falls *perpendicularly* on an element of unit area outside the Earth's atmosphere.) Ignore the role of the Earth's atmosphere, and assume that only 61 percent of $\bar{f}$ is absorbed by the Earth, the other 39 percent being reflected from cloudtops, the ground, and the oceans. The former flux of light heats the surface to an average temperature T, and the Earth's surface reradiates the energy effectively as a black body at the rate σT^4 per unit area. Compute T numerically from the formula $\sigma T^4 = 0.61\bar{f}$. Is your answer reasonable? *Hint*: There should be a slight discrepancy between T and the observed average surface temperature of about 290 K. This discrepancy is due to the "greenhouse effect" of the carbon dioxide (CO_2) and water vapor (H_2O) in the Earth's atmosphere (Chapter 17). The greenhouse effect keeps the average surface temperature above the freezing point of water, 273 K.

The crucial role played by the Sun for life on Earth, then, is that it gives us high-grade energy in the form of sunshine, and we return low-grade energy in the form of infrared radiation. In the process of transforming the former into the latter, a minute fraction of the free energy is stored temporarily in plants, which in turn transfer it (with inevitable degradation) to animals and to humans. Occasionally, a human expends this free energy in the effort of producing a symphony. Again, entropy—not energy—lies at the root of the issue (Chapters 19 and 20).

Philosophical Comment

It is a truism that science spawns technology. It is less well-known, but nevertheless true, that technology advances science. Nowhere does the latter statement ring more forcefully than for thermodynamics. The steam engine was developed by an inventor who knew nothing about thermodynamics; it is debatable whether thermodynamics, the ultimate concern of all ecologists, would have been born without the invention of the steam engine. Humanity owes a far greater debt to technology than it is fashionable to admit in today's troubled world.

PART II

The Stars

Omega Centauri (NGC5139), a globular cluster observable from the Southern Hemisphere. (The Cerro Tololo Inter-American Observatory.)

CHAPTER FIVE

The Sun as a Star

We ended Chapter 4 with an assertion that it is the high *quality* of sunshine which maintains life on Earth. However, the *quantity* of sunshine is also important for maintaining the comfortable average temperature of about 290 K on the *surface* of the Earth. (The *effective* temperature of the infrared reradiation is about 246 K [refer back to Problem 4.15]). A happy combination, then, of quality and quantity of sunshine is what makes Earth a thriving garden rather than either a blazing desert or a frigid wasteland.

Exactly how high-grade is sunshine, and how much of it is produced by the Sun? To answer these questions, let us start with directly measurable quantities and deduce the luminosity and effective temperature of the Sun. From direct measurements, the **solar constant**—the energy which falls per unit time per unit area perpendicularly outside the Earth's atmosphere—is known to be

$$f = 1.36 \times 10^6 \text{ erg sec}^{-1} \text{ cm}^{-2}.$$

From a variety of different measurements (e.g., see Problem 1.2), the Earth is known to be located approximately at a distance from the Sun of

$$r = 1 \text{ AU} = 1.50 \times 10^{13} \text{ cm}.$$

In other words, light takes 500 seconds or eight and a third minutes to get from the Sun to the Earth. Given the above values for f and r, it is easy to deduce, from the principle of the conservation of energy (the "$1/r^2$ law of radiation"), that the luminosity $L_\odot$ of the Sun must be

$$L_\odot = 3.90 \times 10^{33} \text{ erg sec}^{-1}.$$

The above value for $L_\odot$ represents an enormous total power (Problem 5.2). The Earth manages to intercept less than one billionth of this power, and even this minuscule fraction is mostly wasted by immediate reradiation in the infrared back to space (see Problem 4.15).

Problem 5.1. Show that the total amount of energy per unit time crossing an imaginary sphere centered on the Sun with radius r is equal to $f \cdot 4\pi r^2$. From the principle of the conservation of energy, argue that this quantity must equal the luminosity of the Sun:

$$L_\odot = f \cdot 4\pi r^2.$$

Calculate the numerical value of $L_\odot$ given $f = 1.36 \times 10^6$ erg sec^{-1} cm^{-2} and $r = 1.50 \times 10^{13}$ cm.

Problem 5.2. Assume that a 100-watt bulb occupies about 30 cm^2 of area, and cover the surface of the Earth with such 100-watt bulbs. Clearly, this experiment is totally beyond human capabilities at the present time, but if it could be carried out, it would make the Earth quite bright—by our standards. How would the total power requirement compare with the luminosity of the Sun?

What effective temperature characterizes the radiation emitted by the Sun? To answer this question, we need to know the linear size of the Sun. The angular diameter of the Sun as seen on Earth is about 32 minutes of arc. Since the Earth-Sun distance is known to be 1 AU = 1.50×10^{13} cm, simple trigonometry (Problem 5.3) yields the radius of the Sun as

$$R_\odot = 6.96 \times 10^{10} \text{ cm},$$

more than one hundred times the radius of the Earth. If the Sun emitted its luminosity like a blackbody with effective temperature T_e, we would have the relation

$$L_\odot = \text{emision per unit area of blackbody}) \times (\text{surface area}) = \sigma T_e^4 \cdot 4\pi R_\odot^2.$$

Although the Sun is not a perfect blackbody, as we shall soon see, we can nevertheless use the above formula to define the *effective temperature* of the Sun. *The effective temperature of any star is the surface temperature that the star would have if it were a perfect blackbody radiating its given luminosity.* Thus, the effective temperature of the Sun is

$$T_e = (L_\odot/\sigma 4\pi R_\odot^2)^{1/4} = 5{,}800 \text{ K},$$

which is about 20 times hotter in absolute terms than it is here on the surface of the Earth. It is this temperature difference between the *quality* of sunshine and the *quality* of "earthshine" which maintains life on Earth.

Problem 5.3. Given that $L_\odot = 3.90 \times 10^{33}$ erg sec^{-1}, $r = 1.50 \times 10^{13}$ cm, and that the angular diameter of the Sun is $\theta = 32$ minutes of arc, deduce the linear size of the Sun and its effective temperature. In other words, verify the numbers given above (to two significant figures).

The Atmosphere of the Sun

We may check our derived value $T_e = 5{,}800$ K by comparing the observed spectral-energy distribution of sunlight (distribution of emission at different wavelengths) with the spectral-energy distribution of a true blackbody of temperature 5,800 K (which we can compute using

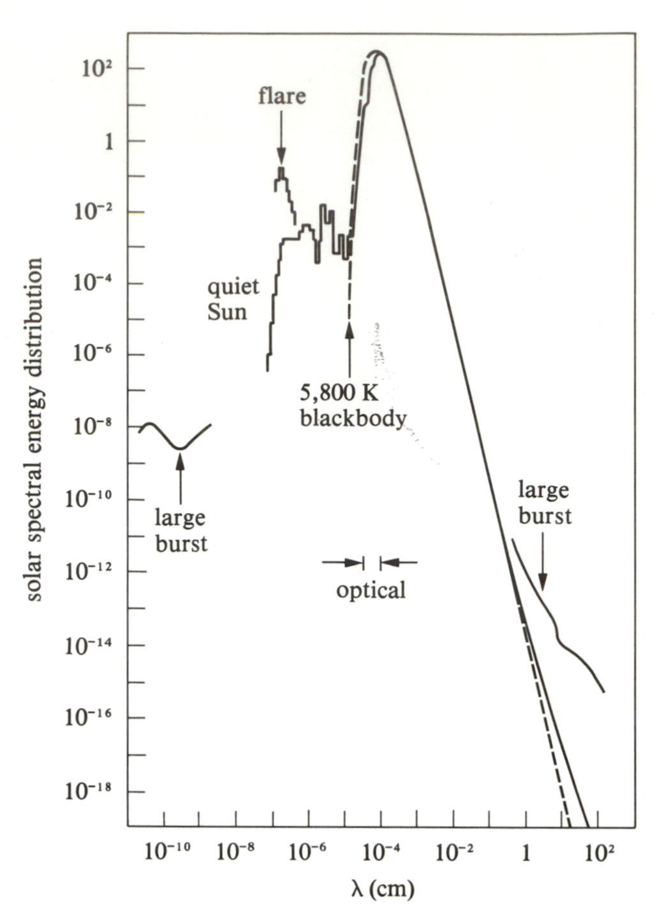

Figure 5.1. The solar spectral energy distribution.

Planck's law, equation 4.6). This comparison is shown in Figure 5.1.

As you can see, the agreement is reasonably good from the ultraviolet to the infrared. Discrepancies arise in the X-ray, far ultraviolet, and radio portions of the electromagnetic spectrum, and these discrepancies are especially noticeable during flare and burst activity in the Sun. Even gamma rays are detectable in a large burst on the Sun. However, the total energy contained at these wavelengths is small, especially in the quiet Sun. Near the peak of the Sun's emission (in the yellow part of the visible spectrum; that's why the Sun looks yellow), we can reasonably claim that a 5,800 K blackbody gives a fair representation of most of the output of the Sun.

Even at optical wavelengths, however, Figure 5.1 shows detectable departures in the spectral-energy distribution of the continuum radiation. Moreover, at higher dispersions (Color Plate 2) dark absorption lines appear in the solar spectrum. These clues suggest that the Sun is *not* at a uniform temperature, and that its surface (or atmospheric) layers are not in perfect thermodynamic equilibrium with its radiation field. (Otherwise, the emission *would* satisfy Planck's law.) The temperature

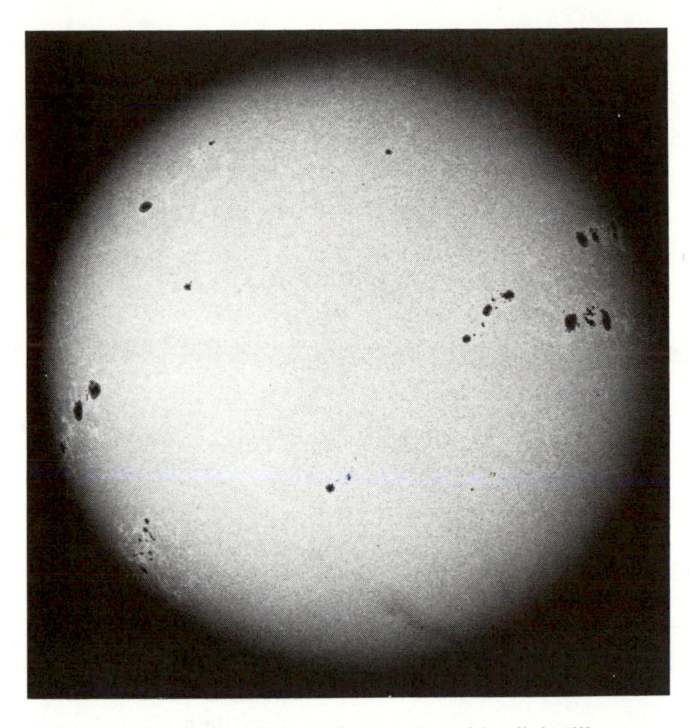

Figure 5.2. A photograph of the Sun in white light illustrates the phenomenon of limb darkening and the fact that the Sun often possesses dark spots. (Palomar Observatory, California Institute of Technology.)

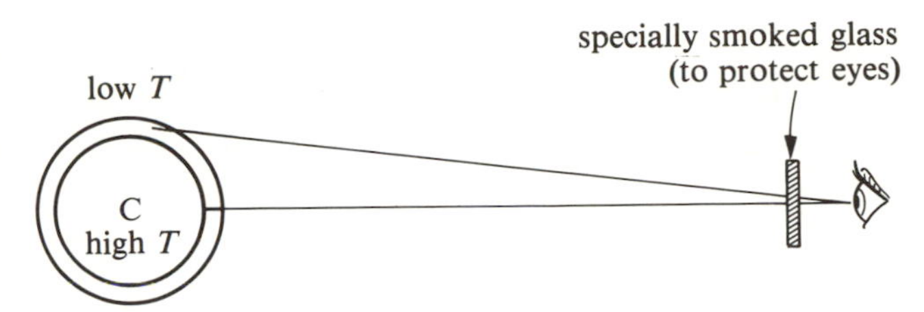

Figure 5.3. The explanation for limb darkening in the Sun. For the same optical path lengths traversed through the outer layers of the Sun (the short solid lines), the eye sees to deeper and hotter layers along a line of sight aimed through the Sun's center than along one aimed through the Sun's limb.

5,800 K must evidently characterize only part of the surface layers.

If the Sun is not at a uniform temperature, is it hotter or cooler in its interior than at its surface? Well, the energy is flowing *out* of the Sun. Since the zeroth law of thermodynamics states that heat flows from hot to cold (Box 4.1), we may infer that the Sun is hotter on the inside. This inference can be checked observationally by the phenomenon known as **limb darkening**. Figure 5.2 shows a picture of the Sun in white light. We see from this picture that the edge of the Sun (its "limb" in astronomical nomenclature) appears darker than its center. The explanation is given in Figure 5.3, which shows two lines of sight for an observer, one directed toward the limb of the Sun, the other toward its center. For the same (optical) path lengths traversed through the outer layers of the Sun (the short solid lines), the eye sees closer to the point C (the geometric center of the Sun) along a line of sight which passes through the Sun's center than along one through the Sun's limb. Since the deeper layers are hotter, and therefore intrinsically brighter, the limb seems dark relative to the center. This is the origin of "limb darkening."

The evidence of absorption lines can be interpreted to give the same conclusion. Let us *assume* the conclusion that the atmosphere of the Sun is characterized by cooler rarefied regions which overlie hotter dense regions (Figure 5.4a), and let us then show that dark lines can be naturally expected. Thermal photons flow from the deeper layers with a continuous distribution of wavelengths. Some photons with the proper wavelengths will be absorbed by atoms situated in the upper layers (Figure 5.4b). The excited atom may then deexcite in two different ways (Figure 5.4c).

(1) It may deexcite collisionally with the excess energy carried off as extra kinetic energy of the two colliding partners. Since the temperature of the gas in the upper layers is lower than the temperature of the radiation field emerging from the lower layers, statistically there are more radiative excitations followed by collisional deexcitations than there are collisional excitations followed by radiative deexcitations. Thus, there is more absorption than there is emission, and the net effect is the removal of some line photons from the underlying continuum radiation field. This mechanism for producing an intensity at the wavelength of the line which is darker than the intensity at the wavelengths of the neighboring continuum is called "true absorption."

(2) The radiatively excited atom may also deexcite radiatively, generally with the emission of a line photon traveling in a different direction than the original photon. This process of "resonant scattering" does not constitute "true absorption," but it can also lead to the appearance of a dark line. No net effect would arise if every microscopic process were statistically balanced by its reverse process; indeed, the statistical condition of "detailed balance" is equivalent to the condition of true thermodynamic equilibrium. However, in a stellar atmosphere, the presence of a boundary prevents "detailed balance" from ever being achieved for the process of resonant scattering. There exists a basic asymmetry at the surface of a star: a hot, bright star lies to one side; cold, dark space lies to the other. Because of this asymmetry, there is a greater statistical tendency for photons originally traveling along our line of sight (assuming we are situated outside of the star!) to be scattered out of our line of sight than there is for photons not originally

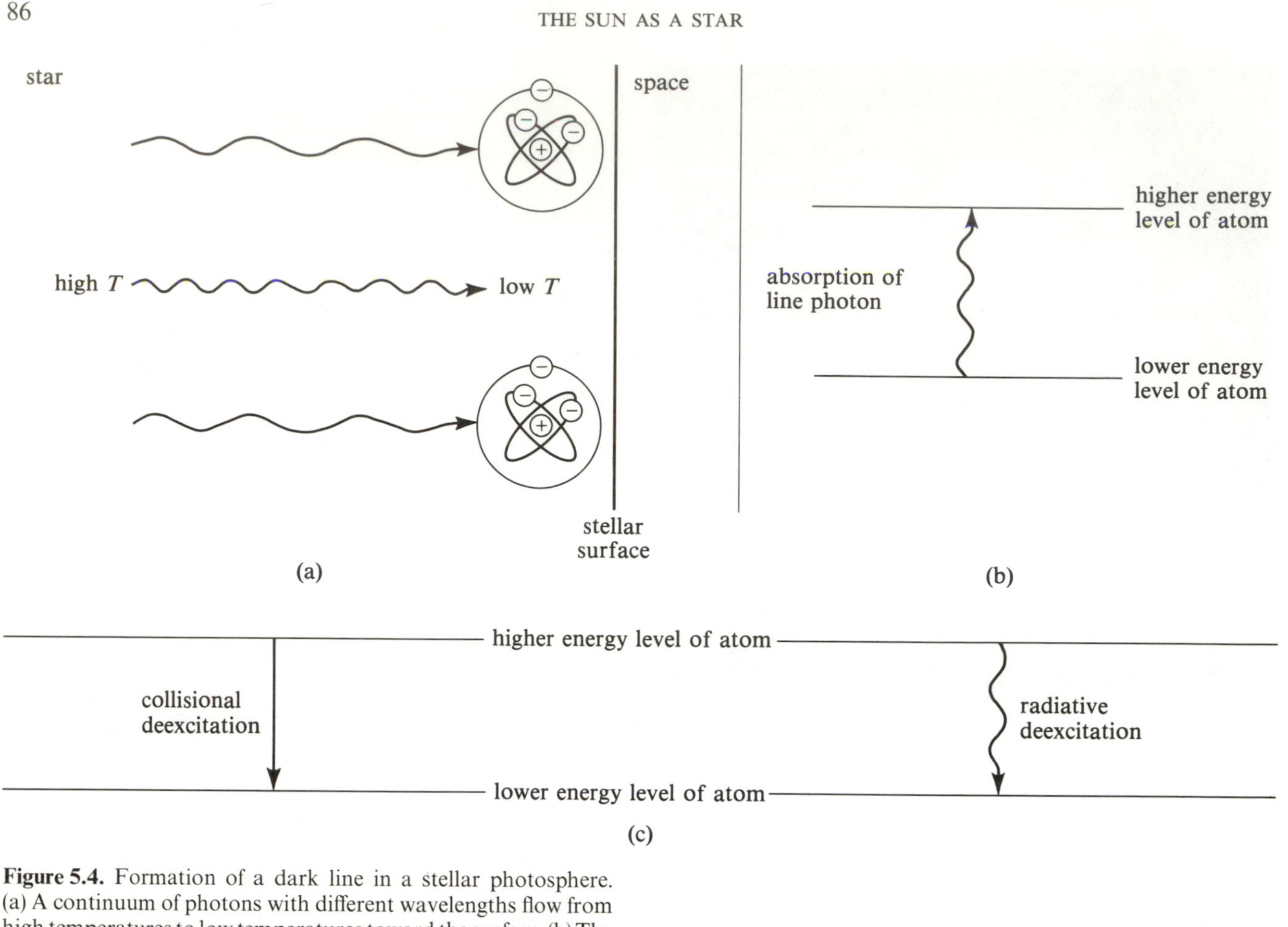

Figure 5.4. Formation of a dark line in a stellar photosphere. (a) A continuum of photons with different wavelengths flow from high temperatures to low temperatures toward the surface. (b) The absorption of a line photon causes an atom to enter an excited state. (c) The excited atom subsequently deexcites either collisionally (true absorption) or radiatively, by the emission of a photon in a different direction (scattering).

in our line of sight to be scattered into our line of sight. (For example, there are no inward-going photons from cold, dark space to scatter into our line of sight, but there are lots of outward-going photons to scatter out of our line of sight into cold, dark space in different directions from us.) The net loss of line photons from our line of sight compared to the neighboring continuum photons leads to a dark "scattering" line. In general, dark lines are produced by a combination of "true absorption" and "pure scattering," but for simplicity we shall henceforth refer to all such dark lines as "absorption lines." The quantitative study of the formation of the continuum radiation and the pattern of absorption lines in the outer layers of a star is part of the discipline called **stellar atmospheres**. We shall comment in greater detail on this subject in Chapter 9.

The layers that can be analyzed in this fashion only constitute the outer 0.1 percent of the radius of the Sun. These superficial layers, where optical photons can fly to us more-or-less directly, are known collectively as the **photosphere** of the Sun. What are conditions like in the deep interior of the Sun?

The Interior of the Sun

To discover the conditions in the deep interior of the Sun, astronomers have to resort to theory, for there are no direct observations of the Sun's interior. (Solar neutrinos constitute an important exception, but we will discuss them in Chapter 6.) This theory is complex, but only in detailed application. Its main qualitative ideas are quite simple.

To start, we must first obtain one additional datum on the Sun, namely, its mass. The best way to calculate the mass of any celestial object is from the orbital characteristics of a nearby orbiting body (if any). For the Sun, we know of many orbiting companions. In particular, the Earth orbits the Sun approximately in a circle with radius $r = 1.50 \times 10^{13}$ cm and period $P = 1$ year $= 3.16 \times 10^7$ sec. Hence, the speed v of the Earth's motion about the Sun is

$$v = \frac{\text{circumference}}{\text{period}} = \frac{2\pi r}{P} = 2.98 \times 10^6 \text{ cm/sec},$$

which is about 67 thousand miles per hour! If we ignore the motion of the Sun, we may set $\boldsymbol{F} = m\boldsymbol{a}$ in the form where $\boldsymbol{a}$ is the centripetal acceleration (magnitude $= v^2/r$) and $\boldsymbol{F}$ is the gravitational force (magnitude $= GM_\odot m/r^2$, with $m =$ mass of the Earth). Solving the resultant expression for the mass of the Sun, $M_\odot$, we obtain

$$M_\odot = \frac{rv^2}{G} = 2.0 \times 10^{33} \text{ gm}. \tag{5.1}$$

Problem 5.4. Verify the algebra and the numerical value of equation (5.1). Also prove that the total energy $E = mv^2/2 + (-GM_\odot m/r)$ of the Earth's orbital motion is equal to one-half of its potential energy $(-GM_\odot m/r)$.

Thus, the Sun is by far the most massive body in the solar system. It is about 300,000 times more massive than the Earth.

Previously, we noted that the Sun is about 100 times larger than the Earth. Hence, the mean density of the Sun is not very different from the mean density of the Earth. The mean density of the Sun is

$$\text{mean density} = \frac{\text{mass}}{\text{volume}} = \frac{M_\odot}{4\pi R_\odot{}^3/3} = 1.4 \text{ gm/cm}^3,$$

which is to be compared with the average density of the Earth, 5.5 gm/cm^3. The density of liquid water is, of course, 1 gm/cm^3. Does this mean that the interior of the Sun consists of material intermediate between that of water, which is a liquid, and that of the Earth, which is mostly a solid? (See Box 4.4 for the properties of gases, liquids, and solids).

To answer this question, let us estimate whether the conditions of pressure and temperature inside the Sun would allow atoms to retain their electronic shells. To estimate the pressure, let us start with the observation that the Sun is not shrinking or expanding rapidly. For the Sun to remain mechanically static, all forces must be very nearly in balance (the condition of **hydrostatic equilibrium**). Apply this condition to any two vertical levels in the Sun. Since pressure is a force per unit area, the upward push at the bottom level must balance the downward push at the top level and the downward pull of gravity on the mass in between. Choose now the upper level to be at the surface of the Sun. The pressure at the surface must be nearly zero if the outside is nearly a vacuum, and this allows us to deduce the familiar result that the pressure at an arbitrary bottom level must balance the weight of a column of unit cross-sectional area of matter above it (Figure 5.5). In particular, the outward push of pressure at the center must balance the inward pull per unit area of the entire Sun. The great mass of the Sun implies that the pressure needed at the center to support the overlying layers against its own weight must be enormous. In such high-pressure regions, neutral atoms and molecules would be pushed so hard against one another that their electronic shells could not possible survive intact (Problems 5.5 and 5.9). Thus, the matter in the interior of the Sun must be composed almost entirely of bare nuclei and free electrons, a state of matter that physicists call a **plasma**. Although such a mixture of ions and electrons is electrically neutral on average, the sizes of the particles (nuclei $\sim 10^{-13}$ cm, electrons = literally points, as far as we can tell) are much smaller than atoms or molecules. Consequently, even at the relatively high mean density of the interior of the Sun, there exists plenty of room between the individual particles (Figure 5.6). In other words, the plasma in the Sun behaves as a *perfect gas*.

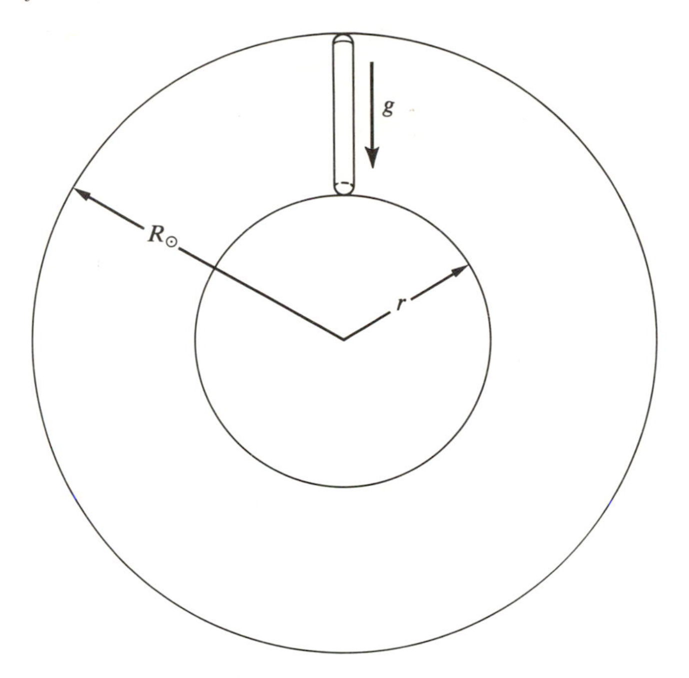

Figure 5.5. Hydrostatic equilibrium involves the balance of gravitational force and pressure force at each level in the star. To be precise, in hydrostatic equilibrium, the pressure at radius r equals the weight of the vertical column of matter of unit area above it. When the forces are balanced in this way at all values of r, no net force acts on any element of gas, and the star remains mechanically steady (hydrostatic).

Figure 5.6. The solar plasma contains the bare nuclei of atoms (open circles) plus their stripped orbital electrons (black dots). The mean separation between particles is much larger than their individual sizes, and the medium is electrically neutral as a whole. The electric energy associated with nearest neighbors is small compared to the thermal energy. Thus the solar plasma can be considered to be a perfect gas.

Problem 5.5. Consider packing hydrogen atoms in the ground state in a cubic array with each side of the cube equal to a Bohr diameter, $2r_1$, where r_1 is the radius of the first Bohr orbit calculated in Problem 3.5. Argue that the mean density of matter in this configuration equals $(m_p + m_e)/(2r_1)^3$, and calculate the numerical value of this quantity. How does this value compare with the mean density of the Sun, 1.4 gm/cm^3?

Assume (as we shall soon calculate in Problem 5.9), that the central pressure of the Sun equals 2.1×10^{17} dyne/cm^2. If hydrogen exists as neutral atoms at the center of the Sun, this is the force per unit area exerted on each electronic shell. The electric force which binds the electron to the proton in a hydrogen atom equals on average $e^2/r_1{}^2$ for a hydrogen atom in the ground electronic state. Divide this by the area $(2r_1)^2$ of one face of our hypothetical cube to find the force per unit area, $e^2/4r_1{}^4$, associated with the electric binding. Calculate this quantity numerically, and compare its value with the central pressure of the Sun. Do you think electronic shells of hydrogen atoms can remain intact in the conditions at the center of the Sun? Thus, the coincidence found in the first part of this problem must be meaningless for the Sun. In Chapter 17, however, we shall find that the calculation is not an empty exercise for Jupiter.

Problem 5.6. In Chapter 3, we mentioned that Newton proved an important theorem for spherical bodies, namely, that the gravitational force outside a spherical distribution of matter acted if all the mass were concentrated at a single point at the center. Newton also proved another important result for spherical bodies: if you are *inside* a body of mass density $\rho(r)$, then you only feel the attraction of the mass $M(r)$ interior to a sphere of your radius r. The attraction of the mass exterior to such a sphere cancels to zero. *Prove this theorem. Hint*: Consider a spherical shell of radius $r' > r$ and uniform thickness $\Delta r' \ll r'$. Construct a small solid angle $\Delta\omega$ as shown in the diagram here, and show that the masses intercepted by the two oppositely directed cones of solid angle $\Delta\omega$ are given, respectively, by $\rho(r')\Delta r' d_1{}^2 \Delta\omega \cos\theta$ and $\rho(r')\Delta r' d_2{}^2 \Delta\omega \cos\theta$, where θ is the angle between the radii $\boldsymbol{r}'$ to the two mass elements and their vector displacements $\boldsymbol{d}_1$ and $\boldsymbol{d}_2$ from $\boldsymbol{r}$. Show now that the gravitational fields produced at $\boldsymbol{r}$ by the two mass elements have equal magnitude, $G\rho(r')\Delta r' \Delta\omega \cos\theta$, and are oppositely directed. Deduce, therefore, that the gravitational attraction of these two mass elements cancels. Since the entire mass distribution outside the sphere of radius r can be considered to consist of the sum of all such pairs of mass elements, infer that only the mass $M(r)$ contributes to the net gravitational force acting on an observer situated at r. (Q.E.D.)

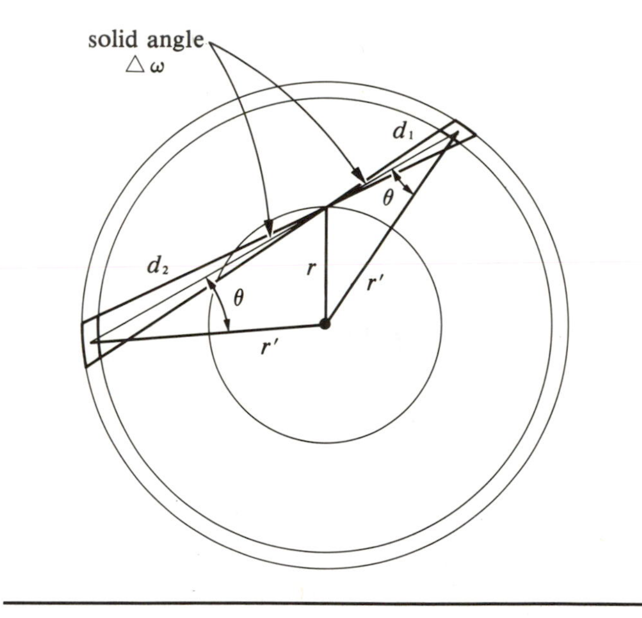

Problem 5.7. Give a far simpler proof of the above result by using Gauss's law.

Problem 5.8. Show that the gravitational force per unit volume acting on matter of density $\rho(r)$ at radius r has magnitude $GM(r)\rho(r)/r^2$. Show that the weight of a column of unit area above radius r equals, therefore, the sum (integral) of $GM(r)\rho(r)\Delta r/r^2$ from r to the surface R. How might you roughly estimate this sum (integral)?

Problem 5.9. The pressure at the center of the Sun equals the weight per unit area of the material on top. Let g be the average gravitational field felt by such a column of material of mass per unit area μ. In order of magnitude g must roughly equal $GM_\odot/R_\odot^2$, and μ must roughly equal $M_\odot/R_\odot^2$. Thus we have the order-of-magnitude estimate for the central pressure, P_c, of

$$P_c = \mu g \sim (M_\odot/R_\odot^2)(GM_\odot/R_\odot^2) = GM_\odot^2/R_\odot^4.$$

A more accurate calculation yields a numerical coefficient of about 19, i.e.,

$$P_c = 19 \cdot \frac{GM_\odot^2}{R_\odot^4} = 2.1 \times 10^{17}\,\text{gm cm}^{-1}\,\text{sec}^{-2}.$$

Verify this numerical value. How does this pressure compare with the Earth's atmospheric pressure at sea level? *Hint*: the air mass per unit area above sea level is $\mu = 1.03 \times 10^3$ gm/cm^2, and the gravitational acceleration at the surface of the Earth is $g = 9.80 \times 10^2$ cm/sec^2. The atmospheric pressure at sea level is called one **atmosphere**.

The pressure of a perfect gas is associated with the random thermal motions of the particles making up the gas. In particular, if the gas has a number density n of a certain kind of particles and a temperature T, the **perfect gas law** states that these particles contribute to the total pressure an amount given by equation (4.2):

$$P = nkT. \tag{5.2}$$

The central mass density ρ_c in the Sun is about 110 times its average value 1.4 gm cm^{-3}, i.e., $\rho_c = 150$ gm cm^{-3}. The mean particle mass m_c at the center (averaged over ions and electrons) is somewhat less than the mass m_p of the proton: $m_c = 1.5 \times 10^{-24}$ gm. Thus, the central number density $n_c = \rho_c/m_c = 1.0 \times 10^{26}$ cm^{-3}. With the pressure $P_c = 2.1 \times 10^{17}$ erg cm^{-3} deduced from Problem 5.9, equation (5.2) implies that the Sun's central temperature must equal 15 million degrees Kelvin!

$$T_c = 1.5 \times 10^7\ \text{K}.$$

Problem 5.10. Verify the numerical values for T_c given in the text.

At the densities and temperatures characteristic of the interior of the Sun, the matter is quite opaque. (That is why we can only see into the Sun a little bit.) Consequently, we can expect the matter and the radiation field to be very nearly in thermodynamic equilibrium with each other locally. The thermal radiation field associated with a central temperature of 15 million degrees contains mostly X-rays, as a quick calculation with Wien's displacement law, equation (4.7), would easily show. X-rays interact quite strongly with matter, and the X-rays in the interior of the Sun fly only about half a centimeter on the average before they crash into a material particle and are either absorbed (and subsequently reemitted) or scattered into a completely different direction. (This means that even Superman with his X-ray vision would not be able to see beyond the tip of his nose if he flew into the interior of the Sun!) Thus, the conditions inside the Sun—and inside other stars—is a scalding inferno with a hurly-burly of particles and photons rushing about madly, smashing violently into each other, and generally seeming to make little progress getting anywhere in net.

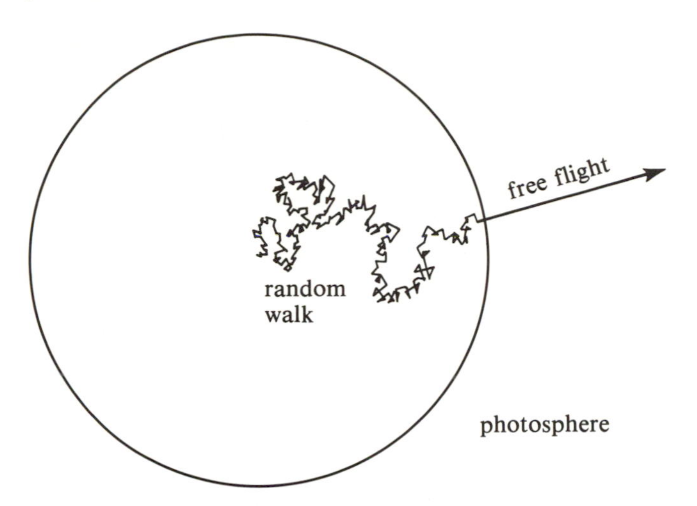

Figure 5.7. Radiative transfer in a star. In the stellar interior, photons suffer many interactions with matter and only slowly diffuse outward by a tortuous series of "random walks" which eventually bring them to the photosphere. In the process, one X-ray photon is degraded into many optical photons. In the photosphere, the photons encounter relatively little resistance from matter, and they can freely stream from the stellar surface in a straight line.

Radiative Transfer in the Sun

It was the astrophysicist Eddington who first appreciated that there is a net slow effect in the seemingly crazy dance performed by the photons inside stars like the Sun. The dance of the photons is nowadays called a "random walk" and is to be distinguished from free flight. By such a random walk, a photon is able gradually to zigzag from the center of the Sun and eventually make its way to the surface (Figure 5.7). In executing this tortuous path, the photon suffers many interactions with matter and is transformed each time. On the average, then, the photons are degraded from being X-ray photons characteristic of the central temperature of 1.5×10^7 K to optical photons characteristic of the effective temperature of 5,800 K at the surface. This "degradation" of high-quality X-ray photons to relatively "low-quality" optical photons, which results from the diffusion process from center to surface, is only to be expected from the second law of thermodynamics. To obtain an idea of how tortuous a path is taken to get from the center to the surface, Problem 5.11 asks you to estimate that the typical diffusion time to "walk" out of the Sun is 30,000 years. This is to be contrasted with the just over two seconds required if a photon could only "fly" out in a straight line! The very slow outward progress of the collection of photons is what regulates the luminosity of the Sun to be a steady 3.90×10^{33} erg/sec (Problem 5.12).

Problem 5.11. The opacity of the Sun at an average density of 1.4 gm/cm^3 and an average temperature of 4.5×10^6 K implies a "mean free path" l for a photon of about 0.5 cm before it interacts with matter. Consider the expected x-position after $N + 1$ interactions, $\langle x_{N+1} \rangle$. If the photon starts off at the origin (the center of the Sun), it has equal probability of going left or right, and after $N + 1$ such random steps, the average x-position will still be zero:

$$\langle x_{N+1} \rangle = \langle x_N \rangle = \cdots = \langle x_1 \rangle = 0.$$

(Just as many photons would have gone to the left as to the right statistically.) Consider, now, the expected *square* of the x-position after $N + 1$ random steps, $\langle {x_{N+1}}^2 \rangle$ which is a measure of the displacement irrespective of sign. This quantity is related to its previous value after N interactions by the probabilistic notion that the photon has equal chance of going left or right, i.e.,

$$\langle {x_{N+1}}^2 \rangle = \tfrac{1}{2}\langle (x_N - l)^2 \rangle + \tfrac{1}{2}\langle (x_N + l)^2 \rangle = \langle {x_N}^2 \rangle + l^2,$$

where we have made use of the fact that $\langle x_N \rangle = 0$. Verify the algebra leading to the above result, and show by induction that

$$\langle {x_N}^2 \rangle = N l^2,$$

which is the fundamental formula of random-walk theory. Thus, in one dimension, to cover the root-mean-square distance $\langle {x_N}^2 \rangle^{1/2}$ in random steps of size l requires $N = \langle {x_N}^2 \rangle / l^2$ steps. In three dimensions, to cover the distance $R_\odot$ in random steps of size l requires, therefore, $N = 3{R_\odot}^2/l^2$ steps. The total photon flight time to random walk a distance $R_\odot$ is therefore $t = Nl/c = 3{R_\odot}^2/lc$. Compute the numerical value of t assuming $l = 0.5$ cm and compare this value with the hypothetical free-flight time $R_\odot/c$ if only the photon could fly in a straight line from the center out to the surface.

Problem 5.12. Let us now estimate what the slow leakage of photons from the interior of the Sun calculated in Problem 5.11 implies for the luminosity of the Sun. The energy density of thermal radiation at a temperature T is proportional to the fourth power of T, with the proportionality constant = radiation constant, a, whose numerical value is given in Appendix A. Argue that the leakage rate $L_\odot$ is given by

$$L_\odot = \frac{\text{(volume)} \times \text{(radiation energy per unit volume)}}{\text{random walk time to cover distance } R_\odot}$$

$$= \frac{(4\pi {R_\odot}^3/3)(aT^4)}{3{R_\odot}^2/lc},$$

where we have used the results of Problem 5.11 for the denominator. Calculate the numerical value of $L_\odot$, assuming an average temperature of $T = 4.5 \times 10^6$ K, and compare this theoretical value with the observed value $L_\odot = 3.90 \times 10^{33}$ erg/sec.

The Source of Energy of the Sun

As the photons slowly leak out of the Sun, new photons are created to take their place, because the matter is still hot. Problem 5.13 asks you to estimate that the thermal-energy content of the hot plasma is about 10^3 times greater than the instantaneous energy content of the thermal radiation field. Since the leakage time for the photons is 3×10^4 years, it would take the Sun $10^3 \times (3 \times 10^4 \text{ yr}) = 3 \times 10^7$ years to radiate away the thermal energy it has now in the plasma and the radiation field.

Problem 5.13. Let us calculate what the ratio of plasma-energy density to radiation-energy density is at the center of the Sun. The plasma-energy density is $(3/2)n_c kT = 3P_c/2$, whereas the radiation-energy density is $a{T_c}^4$. With $P_c = 2.1 \times 10^{17}$ and $T_c = 1.5 \times 10^7$ in cgs units, what is the desired ratio? A rough value of 10^3 holds for the ratio throughout the interior of the Sun.

Now, life has existed on Earth for more than three billion years (Chapter 19), and during that interval, at least, the Sun must have been shining more or less stably with a luminosity close to its present value. We now see that the store of thermal energy present in the Sun can withstand such a steady drain for only one percent of 3 billion years; so there must be some other, much larger supply of energy which explains the Sun's power.

One source of energy discussed seriously by the physicists Kelvin and Helmholtz during the nineteenth century is gravitational energy. The basic idea is fairly simple and is based on the **virial theorem**. The Sun is a statistically steady object, and one can prove a result for it like that found in Problem 5.4 for the statistically steady orbit of the Earth about the Sun: the total energy E of the Sun is numerically equal to half its gravitational potential energy. In the Sun, it is thermal energy of the plasma rather than kinetic energy of orbital motion that produces the resistance to gravitational contraction. Nevertheless, the total energy E, which is the sum of the thermal energy and the gravitational potential energy, is equal, by the virial theorem, to half the gravitational potential energy:

$$\text{thermal energy} + \text{grav. pot. energy} = E = \tfrac{1}{2}\,\text{grav. pot. energy}. \quad (5.3)$$

The above equation can be solved for "thermal energy" to obtain

$$\text{thermal energy} = -\tfrac{1}{2}\,\text{grav. pot. energy}. \quad (5.4)$$

The gravitational potential energy is always negative for a self-gravitating body, and would become more negative if a body of fixed mass were to contract. In such a contraction, formula (5.3) shows that only half the energy released gravitationally would remain in the star; the other half would presumably be released as radiation. Moreover, formula (5.4) shows that the thermal energy would *increase* if the gravitational potential energy were to become more negative. This increase is understandable, since classical gases usually get hotter when compressed; however, this result has the following, very important consequence:

As a ball of self-gravitating classical gas loses energy by radiation and is forced to contract, if it has no compensating sources of energy, it gets hotter, not colder!

Contrast this behavior with that of a cooling ember. The ember is hotter than its surroundings; so it radiates into its surroundings. As it radiates, it cools, and eventually it is able to come into thermodynamic equilibrium with its surroundings (achieve the same temperature). Not so for a classical self-gravitating gas (Figure 5.8). As such a body radiates into a cooler surrounding (the universe), the self-gravitating gaseous body becomes hotter and hotter, and therefore increases the disparity between its temperature and that of its surroundings. The cooperation of gravity and the second law of thermodynamics in this case has produced a situation which appears to violate intuitive concepts, in particular, the zeroth law of thermodynamics (Box 4.1). However, this violation is as yet only apparent, since the zeroth law only requires heat to flow from hot (the Sun) to cold (the universe) and for the temperature to become uniform when the system has *reached* global thermodynamic equilibrium. For reasons to be discussed shortly, the Sun has not *yet* reached thermodynamic equilibrium with the universe; therefore, we can postpone our concern that the thermodynamics of self-gravitating bodies appears paradoxical.

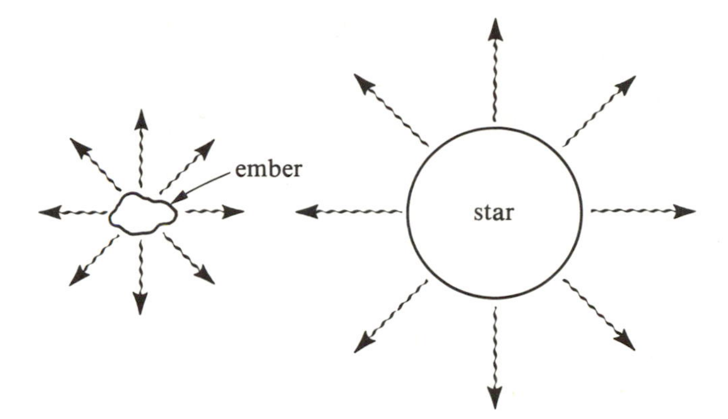

Figure 5.8. A self-gravitating star differs from a cooling ember in one very important respect. A hot ember will radiate into its cooler surroundings, cool down, and arrive eventually at the same temperature as its surroundings. A hot star, without any source of energy other than its internal heat and self-gravitation, will also radiate into its surroundings. But instead of cooling down, it will contract and heat up! Thus, a star made of a perfect classical gas tends to evolve toward states with a greater and greater temperature disparity with the rest of the universe.

The obstacle preventing the Sun from reaching thermodynamic equilibrium is, at present, not gravitational contraction, but thermonuclear reactions. To see that gravitational contraction cannot have provided the long-lasting energy supply required by the fossil evidence, we need only enquire how long gravitational contraction could have maintained the present luminosity of the Sun. The answer turns out to be 3×10^7 years (Problem 5.14), the same time-scale as for the Sun's thermal store of energy. This answer is no coincidence, because equation (5.3) implies that only half the released gravitational energy is available for radiation, and this half is numerically equal to the present thermal-energy store (by equation 5.4), whose usefulness we have already calculated to be 3×10^7 years. Indeed, the virial theorem shows that only the combination of thermal energy and gravitational energy produces a net source for radiation in a quasi-statically contracting self-gravitating body.

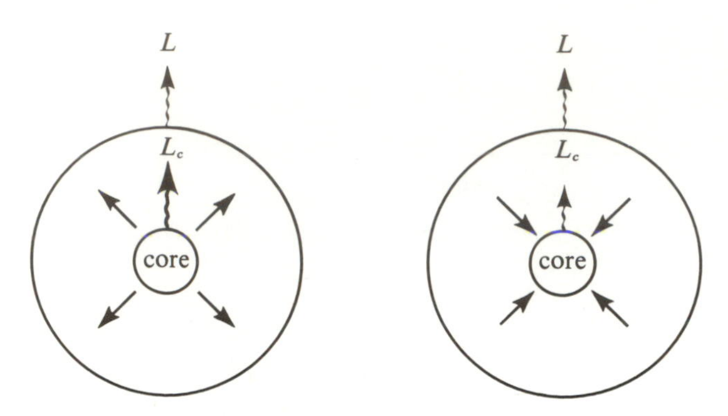

Figure 5.9. The Sun's safety valve for thermonuclear fusion. If the power L_c produced in the core exceeds the luminosity L escaping from the surface, the net input of energy into the Sun would cause it to expand slightly (outward straight arrows). This would lower the central pressure and the central temperature, lowering the rate of energy generation until $L_c = L$. Conversely, if the power L_c should be less than L, the net loss of energy would cause the Sun to contract slightly (inward straight arrows). This would increase the central pressure and the central temperature, raising the rate of energy generation until $L_c = L$.

Problem 5.14. The gravitational potential energy W of a self-gravitating sphere of mass M and radius R depends on the detailed distribution of mass within the sphere, but it is generally of order of magnitude $-GM^2/R$. For the Sun, $W_\odot = -2 \cdot GM_\odot{}^2/R_\odot$. What is the time-scale $t = -(1/2)W_\odot/L_\odot$ over which gravitational contraction could have supplied the power radiated by the Sun at its present rate?

Thus, nineteenth-century physics had encountered a dilemma about the Sun's ultimate source of energy. The physicists were inclined to believe that Kelvin-Helmholtz contraction was the answer, but the geologists pointed to the mounting geological and fossil evidence which indicated that a much longer-lasting source of energy was required. In this argument, the physicists have proven to be wrong, and the geologists right. We now know that the basic energy source of the Sun is thermonuclear fusion, the process by which heavy chemical elements are built up from light ones via nuclear reactions which take place at high temperatures. We shall discuss the nature of thermonuclear reactions in more detail in Chapter 6. For now, we are content to make two comments.

First, at the central temperatures of about 15 million degrees Kelvin in the Sun, only hydrogen reactions can proceed near the core. The requirement that these reactions provide a stable "burning" lifetime of greater than 3 billion years then implies that the Sun must be made mostly of hydrogen.* Thus, the Sun's energy source supplies the first important clue that the predominant form of matter in the universe is the simplest form of all possible elements, hydrogen. And therein lies a hint to the creation of the universe.

Second, although thermonuclear energy is a very powerful source, it does not have infinite capacity to offset radiation from the surface of a star. But gravity is relentless, and the second law of thermodynamics is inexorable. Thus, sooner or later, *every* star—including the Sun—must confront the paradox posed by its getting hotter and hotter while losing more and more energy to a cold and dark universe. This confrontation and its ultimate resolution provides the central plot to the life stories of the stars. It is a theme which repeats itself with delightful variations in almost all major branches of astronomy.

The Stability of the Sun

You may have heard that the H-bomb works on the same general principle as the thermonuclear reactions which power the Sun. Why then does the Sun "burn" stably for billions of years? Why does it not explode like an H-bomb?

The answer is that the Sun has a built-in safety valve. It "burns" by "controlled thermonuclear fusion," a condition which laboratory physicists still cannot reproduce here on Earth despite its potential boon for the "energy crisis." How does the Sun's safety valve work? (See Figure 5.9.)

Imagine a perturbation in the Sun which causes it to produce energy in the core by thermonuclear reactions faster than this energy leaves the surface as the luminosity of the Sun. This extra net input of energy would cause the Sun to expand slightly, just the opposite to Kelvin-Helmholtz contraction, when a net loss of energy would cause the Sun to contract slightly. The expansion would tend to lower the temperature in the core, reducing the rate of energy generation by thermonuclear reactions, and thereby curing the original overproduction of nuclear energy. Conversely, if an initial perturbation caused an underproduction of nuclear energy, the Sun would suffer a net loss of energy and would contract slightly. This contraction would raise the temperature in the core, raising the rate of energy generation by thermonuclear reactions, and thereby again curing the original problem. The facts that the Sun's matter acts like a perfect gas and that the hot solar plasma is confined by self-gravity give the Sun a built-in stabilizer against runaway thermonuclear reactions. In this way does the Sun avoid the

* "Fuse" would be a better choice of wording than "burn," since the latter conjures up chemical reactions, not nuclear ones. However, the conventional astronomical usage is "burn," and we shall stick with it, using quotation marks whenever there might be some confusion.

Figure 5.10. The Poloidal Divertor Experiment (PDX) at Princeton is a major ongoing research project to study the feasibility of harnessing fusion power on Earth. The basic idea behind magnetic-confinement fusion is to use strong magnetic fields to deflect electrically charged particles in a hot plasma and prevent them from striking and eventually destroying the walls of the reactor vessel. Some energetic particles inevitably reach the walls, and they sputter loose heavy ions, which are thus introduced as impurities into the plasma. The PDX is designed to eliminate most of these impurities, which would otherwise radiate efficiently in combination with the electrons of the plasma, and prevent the plasma from reaching the high temperatures (about 10^8 K) needed for thermonuclear ignition under laboratory conditions. (Courtesy of Princeton Plasma Physics Laboratory.)

rapid expansion of gases that characterizes a bomb explosion.

Confinement of a hot plasma under laboratory conditions is much more difficult. To initiate thermonuclear reactions even in the more readily "burned" heavy isotopes of hydrogen (see Chapter 6) requires temperatures in excess of a million degrees. All material walls would melt well before ignition is reached. The goal of much of modern **plasma physics** is to find other confinement mechanisms. Self-gravity is impractical without astronomical masses, because gravity is such a weak force intrinsically. The two most promising proposals involve **magnetic confinement** and **inertial confinement** (Figures 5.10 and 5.11). The former aims at true "controlled

Figure 5.11. One inertial confinement scheme uses the power of many lasers to implode a tiny fusion pellet. This photograph shows six of the twenty lasers of the Shiva system which can deliver 30 trillion watts of optical light in one billionth of a second to a target the size of a grain of sand. (Lawrence Livermore Laboratory photograph)

thermonuclear fusion," but formidable problems remain because the plasma escapes from within its "magnetic walls." The latter hopes to achieve ignition by bombarding a small pellet of nuclear "fuel" with laser beams or particle beams until it implodes. The implosion is followed by an explosion, if nuclear ignition is achieved; so the energy released by inertial-confinement schemes is not so much "controlled" thermonuclear fusion as it is a miniature H-bomb. (The initial implosion of a *real* H-bomb is triggered by the focused explosion of a A-bomb, from which comes much of the radioactivity of such weapons.) In any case, nuclear fusion and solar energy probably offer the best chances for the long-range solution of the "energy crisis," and it is interesting to note that these *practical* possibilities are all rooted in the *scientific* enterprise of understanding better the inner workings of the Sun and the stars.

Summary of the Principles of Stellar Structure

At this point we are poised at the brink of a theoretical understanding of why the Sun is as big as it is. *The present radius of the Sun has just the right value to maintain a central temperature which provides a rate of nuclear-energy generation that exactly balances the leakage rate due to the "random walk" of the photons from center to surface.* This conclusion closes the logical gap in our entire deductive chain of reasoning, where we had to rely on observations to give us the mass, luminosity, and radius of the Sun. Now we see that both the luminosity (Problem 5.12) and the radius (previous subsection) can be theoretically deduced, in principle, from detailed considerations of the Sun's structure, providing we are given the mass and the chemical composition of the Sun. A summary of the important concepts is given in Box 5.1. As we shall see more generally in Chapter 8, it is precisely these two quantities, mass and (initial) chemical composition, which determine the structure and evolution of any (isolated) star.

BOX 5.1
Principles of Stellar Structure

(a) *Hydrostatic Equilibrium.* For a star to be mechanically in equilibrium, the pressure at every level must equal the weight of a column of material of unit cross-sectional area on top.

(b) *Energy Transfer.* Photons in the interior of the star carry heat outward by random walking from regions of high temperature to regions of low temperature. At the photosphere of the star, the photons make a transition from "walking" to "flying." If the luminosity required to be carried out is too large for radiative transfer to handle stably, convection results.

(c) *Energy Generation.* For a star to be thermally in equilibrium, the energy carried outwards by radiative diffusion and/or convection past any spherical surface must be balanced by an equal release of nuclear energy in the interior of that sphere. If the nuclear source is inadequate, gravitational contraction must result. In stars like the Sun, the size of the star has adjusted itself so that exactly as much energy is lost from the surface as is released by nuclear reactions in the interior.

The Convection Zone of the Sun

Figure 5.12 shows schematically the structure of the Sun from core to photosphere as currently understood from detailed modelling. The solar core is defined to be that region (containing about 10 percent of the mass of the Sun) where the temperatures are high enough to yield appreciable rates of hydrogen fusion into helium (Chapter 6). The solar photosphere is defined to be the outer layers where the photons stop "walking" (where it's "optically thick") and begin "flying" (where it's "optically thin"). In the subphotospheric layers lies the solar **convection zone.** It arises for the following reason.

The Sun generates nuclear energy in its core at a rate numerically equal to $L_\odot = 3.90 \times 10^{33}$ erg/sec. This luminosity is ultimately radiated from the photosphere according to the formula

$$L = 4\pi R^2 \sigma T_e^{\,4}, \tag{5.5}$$

where L is $L_\odot$ and R is $R_\odot$ for the Sun. We have already seen that the effective temperature T_e has to equal 5,800 K in order to radiate the luminosity $L_\odot$ from a sphere of radius $R_\odot$. At the photospheric temperatures and pressures characteristic of the Sun, the valence electrons of the metallic elements are easily detached; i.e., the metals are partially ionized. However, at the relatively low photospheric temperatures of the Sun, hydrogen atoms prefer more binding energy and remain neutral. In fact, hydrogen, possessing only one valence electron in its neutral atomic state, has a weak tendency to want to add one of the free electrons released by the metals to complete its $n = 1$ shell. (Chapter 4 led us to think of hydrogen as belonging chemically to the same family as the alkali metals—lithium, sodium, etc.—but it is almost equally valid to put hydrogen in the same chemical family as the halogens, fluorine, chlorine, etc.) Thus, in the Sun's photosphere—and, indeed, in the photospheres of all cool stars—hydrogen exists partially in the form of the H^- *negative ion* (proton plus *two* orbital electrons). The importance of this remark is that H^- is a considerable source of continuum opacity, as first suggested by Wildt and computed quantitatively by Chandrasekhar. Because of this, in the Sun's outer layers—as in that of all cool stars—there is so much resistance to the free flow of outward radiation that radiative transport alone cannot carry the requisite amount (required for net energy balance) stably. As a consequence, convection develops in the photospheric and subphotospheric layers. In the convection zone (Figure 5.13) bubbles of gas which happen to be slightly hotter than their surroundings are buoyant and rise, whereas bubbles which happen to be slightly colder than their surroundings sink. Statistically, there is no net transport of *mass* by the rising and sinking bubbles, but there is a net transport of *heat* (convective transport).

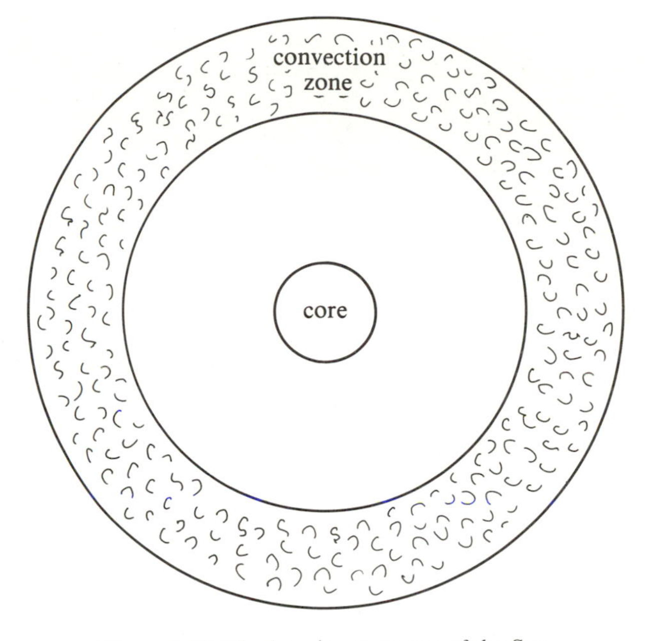

Figure 5.12. The interior structure of the Sun.

The "granular" appearance of the Sun in white light (Figure 5.13) provides observational evidence for such convective motions in the outer layers of the Sun. Similar outer convection zones are believed to exist for all stars with relatively cool photospheres. The outer convection zone of the Sun is believed to extend about one-third of the way into the center. Very cool stars may be completely convective.

Problem 5.15. Consider a parcel of gas embedded in the interior of the Sun. This parcel tends to maintain the same pressure as its surroundings. If at first it is accidentally hotter than its surroundings and therefore less dense, it is buoyant and will rise. As it rises it will cool, even if it loses no heat to its surroundings (i.e., even if its entropy per unit mass remains fixed), because it will expand to adapt to the new pressure surroundings. To carry heat radiatively outward, the new surroundings are also cooler than the original surroundings. If the surroundings' temperature has decreased faster than the temperature of the expanding parcel, the parcel is still buoyant and will continue to rise. At a given pressure, the temperature of all substances is a monotonic function of the entropy per unit mass. Show by the above argument that the criterion for the onset of convection, derived first by Karl Schwarzschild, amounts to the statement that convection will arise if the entropy per unit mass of

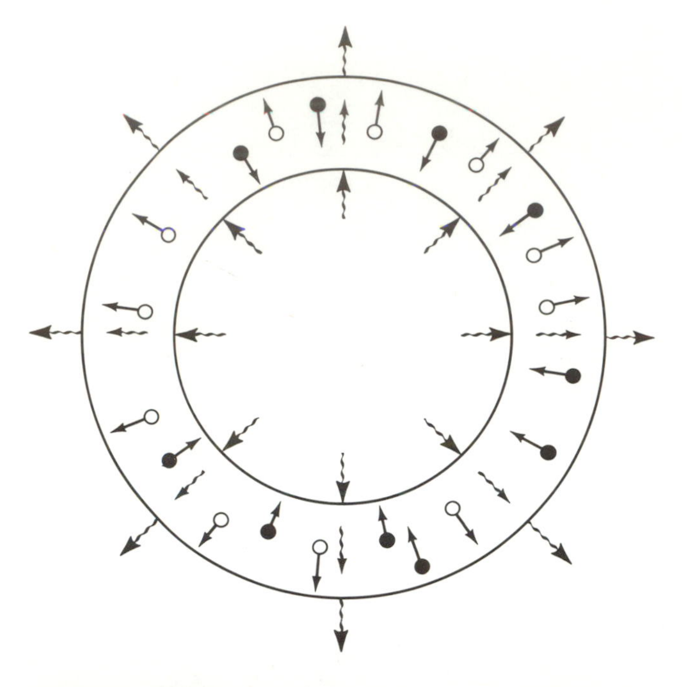

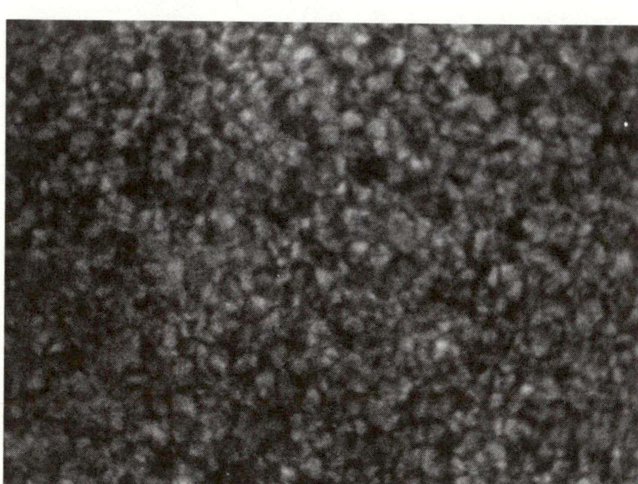

Figure 5.13. The outer convection zone of the Sun. (a) Light energy (wavy lines) enters the base of the convection zone which the high-opacity regions of the convection zone find difficult to carry outward without producing conditions that induce convection. In the convection zone bubbles which are slightly colder than their surroundings sink (filled circles) while those slightly hotter than their surroundings rise (unfilled circles). The net effect is to produce a convective heat flux which helps to carry out the requisite amount of energy per second. Convection is inefficient in the photosphere of the Sun (top of the convection zone), and the power generated in the core is ultimately carried away almost entirely by radiation from the surface. A small amount of energy is carried outward by mechanical waves in the material medium. (b) A highly magnified portion of the Sun's surface, showing the granulations produced by the underlying convection cells. (Palomar Observatory, California Institute of Technology.)

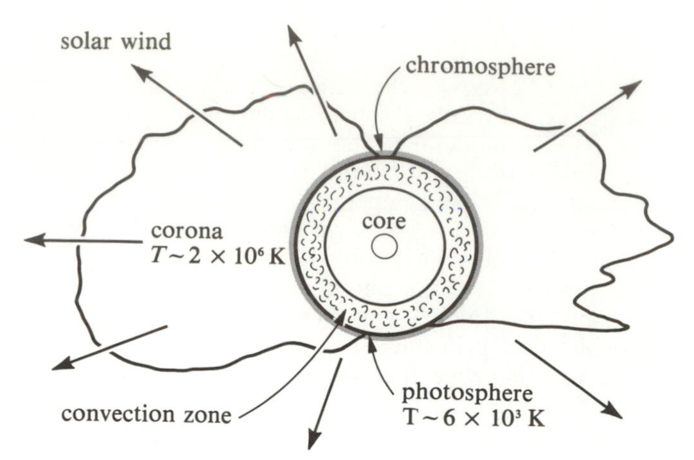

Figure 5.14. Outer layers of the Sun.

the general surroundings decreases in the upward direction. Derive in words the same criterion by considering a parcel which started accidentally colder than its surroundings. Why does high opacity or high luminosity cause convection. *Hint*: Do they produce a steeper rate of decrease of the temperature outward? What does that do to the distribution of entropy per unit mass?

The Chromosphere and Corona of the Sun

So far, we have discussed the Sun as if it has a perfectly sharp outer edge, coinciding approximately with the location of its photosphere. The total thickness of the photosphere, where the photons make a transition from "walking" to "flying," occupies, as we have already remarked, only 0.1 percent of the radius of the Sun; so the Sun does look as if it has a sharp edge (Figure 5.2). This explains the conventional definition of the photosphere of the Sun as its "surface." This description is entirely appropriate for the radiation field, but it oversimplifies the actual situation for the matter distribution. There is gas belonging to the Sun which lies above the photosphere, in the chromosphere and the corona. Indeed, the chromosphere itself is a transition zone between the properties of the gas in the photosphere and those of the gas in the corona (Figure 5.14). The gas temperature as we work our way outward from the photosphere initially decreases, but later it begins to climb dramatically, until temperatures around 2 million degrees Kelvin are reached in the corona. This temperature is much higher than the photospheric temperature of

Figure 5.15. During a total eclipse of the Sun by the Moon, it becomes possible to see the Sun's tenuous corona. (Lick Observatory photograph.)

around 5,800 K; however, the light output from the photosphere vastly dominates that from the corona, because the corona gas is so rarefied that it has little emissive (and absorptive) power. The gas in the chromosphere and the corona is "optically thin." For this reason, it is not easy to see the corona in optical light unless the light from the photosphere is blocked, either naturally by an eclipse of the Sun by the Moon (astronomers are very fortunate that the apparent angular sizes of the Sun and the Moon are very nearly equal), or artificially by a coronagraph (Figure 5.15).

Several interesting observational consequences result from the dramatic increase of temperature in the chromosphere and the corona. First, when the light from the solar photosphere is blocked by a solar eclipse, we see light arising from the chromosphere wherein hot gas lies above cool gas. This situation gives rise to emission lines instead of absorption lines (contrast with Figure 5.4). This is the origin of the "flash spectrum" associated with emission lines in the chromosphere observed during solar eclipses. It was such an observation which first revealed the presence of an element (through its spectral lines) that was previously unknown on Earth. That element is helium (named after Helios, the Sun). Helium is quite rare on Earth, but it is now known to be the second most abundant element (after hydrogen) in the universe. Helium is also the second most simple chemical element. This is a second important clue to the creation of the universe.

Second, the 2 million degree temperature of the solar corona means that this gas can emit X-rays. This explains the presence of X-rays in the solar spectrum displayed in Figure 5.1. X-ray photographs of the Sun look quite different from the usual optical pictures (Figure 5.16; contrast with Figures 5.2 and 5.15). Much information about the structure and dynamics of the solar corona is being gleaned currently from satellite X-ray observations of the solar corona.

What heats the solar corona to such high temperatures? Surely, it cannot be the photospheric photons,

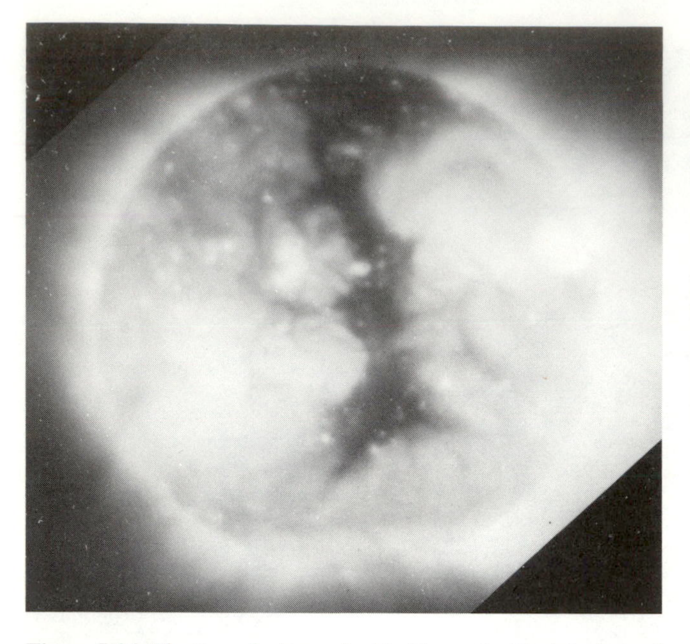

Figure 5.16. The Sun photographed in X-rays by Skylab showed coronal holes and looked altogether quite different from its appearance in visual light. (NASA, courtesy of L. Golub and the Harvard-Smithonian Center for Astrophysics.)

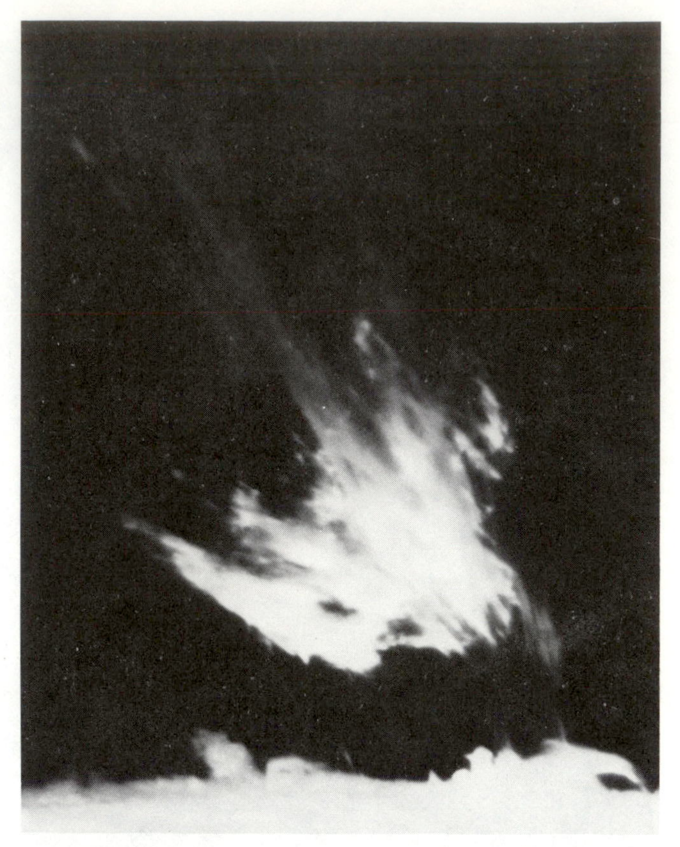

Figure 5.17. A prominence on the Sun 205,000 miles high is dramatically captured in the light (K line) of singly-ionized calcium. (Palomar Observatory, California Institute of Technology.)

since (a) they pass right through the corona without much interaction with the matter, and (b) they have an energy distribution characteristic of a 5,800 K blackbody and are therefore totally unable (by the second law of thermodynamics) to heat any gas up to 2×10^6 K. Some astronomers believe that solar corona is heated by the acoustic waves ("noise") and other kinds of mechanical waves which are generated in the solar convection zone lying below the solar photosphere. These waves propagate upward into the more rarified corona where they steepen into shock waves and dissipate their energy into heat. This heats the corona, which has very little heat capacity, to such high temperatures that the gravitational field of the Sun finds it hard to hold onto the gas in the outer parts of the solar corona. As Eugene Parker first deduced theoretically, these portions blow continuously from the Sun to form a "solar wind" which pervades the solar system (Figure 5.14). It is now possible to collect the particles in this wind by satellites, and thus to verify that the outer parts of the Sun, at least, are made primarily of hydrogen.

Magnetic Activity in the Sun

In addition to the solar wind, the upper reaches of the solar atmosphere also manifest other violent activity in the form of flares, loops, arches, etc. (Figure 5.17). This activity may well provide a substantial input of energy and particles into the expanding solar corona. The active regions seem intimately associated with the presence of magnetic fields in the surface layers of the Sun. Magnetic field activity can be studied qualitatively by observing sunspots, or more quantitatively, as Horace Babcock has done, by measuring the Zeeman splitting in atomic lines.

What is the origin of the solar magnetic field? To appreciate the subtlety of this question, let us first enquire how a bar magnet works (Figure 5.18a). In Chapter 3, we asserted that magnetic fields are produced by moving charges, i.e., by electric currents. A clue to the origin of the magnetic field in a bar magnet comes from realizing that the structure of this field is identical to that produced by an electrical current circulating through a coil of wire (Figure 5.18b). Ordinary matter also contains myriads of similar electrical currents in the form of electrons orbiting about atoms. A microscopic magnetic field is also associated with the spin of an electron. However, in ordinary matter, the many tiny microscopic

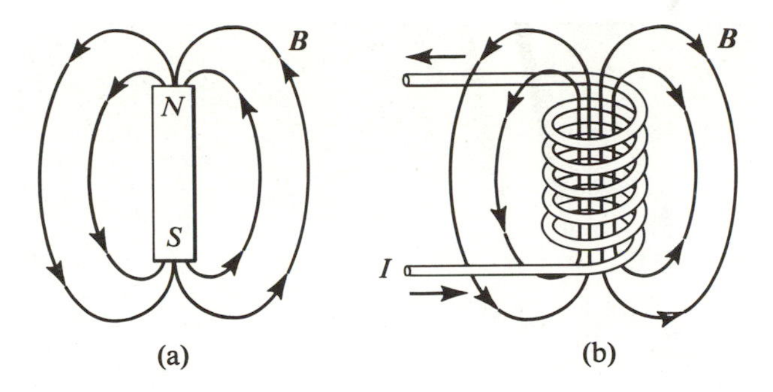

Figure 5.18. Comparison of the magnetic fields of a bar magnet and that produced by an electric current circulating through the coil of a wire. Notice especially that the magnetic field lines never really end on an object. Magnetic field lines could end on or originate from an object only if particles called magnetic monopoles existed. Such particles have not been found in nature.

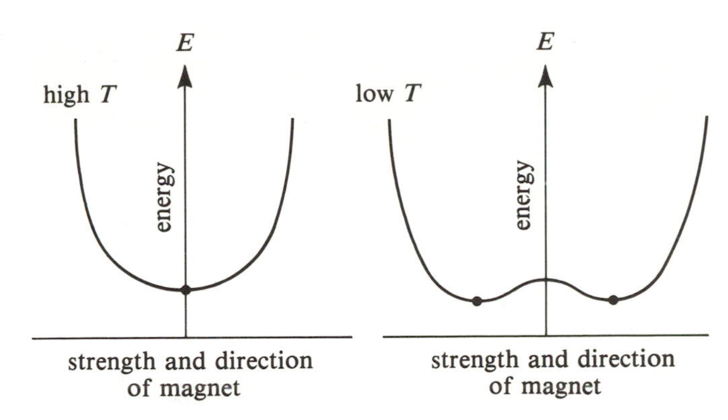

Figure 5.19. Schematic energy versus magnetization curve for a bar of ferromagnetic material. At high temperatures, the state of minimum total energy corresponds to zero magnetization. At low temperatures the state of zero magnetization is actually a local maximum of total energy. There are two minima of total energy which correspond to spontaneous magnetization with the north pole of the magnetic dipole at the left or right end of the bar. The actual bar magnet will settle on one of these two states (denoted by dots); which one it chooses is largely a matter of accident. Notice that the phenomenon of spontaneous magnetization is another example of our general rule for the thermodynamic behavior of matter. At high temperatures, the state of least energy (for fixed entropy) corresponds to more freedom for the atomic magnets to twist and turn. At low temperatures, ferromagnetic materials prefer to gain more magnetic binding energy by giving up some freedom for the atomic magnets to twist and turn.

atomic magnets are usually randomly oriented relative to one another, so the net *macroscopic* magnetization is zero. In **paramagnetic** materials, it is possible to align the atomic magnets by sticking a bar of the material between coils carrying an electric current. This configuration can dramatically reinforce the magnetic field of the imposed current, and this constitutes the principle of an electromagnet. An even more interesting phenomenon can occur in ferromagnetic materials: spontaneous magnetization at sufficiently low temperatures (Figure 5.19). At high temperatures, the atomic magnets jiggle too violently to point collectively in any one direction. The state of least macroscopic energy corresponds to zero net magnetization. At low temperatures, however, ferromagnetic materials actually find it energetically favorable to align many or even all of the atomic magnets. The actual direction of the magnetization—whether the north pole of the magnet ends up on the left end of the bar or the right end—is not predetermined, if one does not place the bar in an external magnetic field. Left to itself, the bar has a perfectly symmetric choice between the two valleys in Figure 5.19. However, the resolution must be one valley or the other; the system cannot stably occupy the middle ground, which is the top of a hill. If the bar magnet is produced by gradually cooling a length of ferromagnetic material, the selection of the final direction occurs accidentally: a small region of one predominant sense of magnetization grows and influences larger and larger neighboring domains to follow its example. The result is a *permanent* magnet with a definite sense of polarity which can be destroyed only by heating up the magnet to a temperature above a critical value, the Curie point. The Curie point of ferromagnetic materials is typically 800 K.

Could the Sun be a permanent magnet in the above sense? No! The Sun is everywhere much too hot to be able to maintain any macroscopic orientation of many microscopic magnets. The existence of magnetic fields in the Sun requires the presence of *macroscopic electric currents*. What is the origin of these currents? One thing we know for sure is that they are not a remnant from a primeval era. The Sun's interior, although an excellent electrical conductor, is sufficiently dissipative that primeval currents and fields would have largely disappeared during the lifetime of the Sun. Moreover, observationally, it is known that the large-scale structure of the magnetic field in the Sun reverses directions every eleven years! What was the north magnetic pole becomes the south magnetic pole, and vice versa.

The eleven-year cycle of the Sun has many other accompanying effects, some of which will be discussed in Chapter 17. For now, we wish only to remark that it is correlated with flare activity and the number of sunspots observed at different latitudes on the Sun. The latter effect is particularly interesting and was discovered in 1904 by E. W. Maunder. In Figure 5.20, we display Maunder's "butterfly diagram" where the latitude of sunspots is plotted as a function of time. At the beginning of each cycle, sunspots form near $\pm 30^{\circ}$ from the equator. Later, more and more spots form closer and closer to

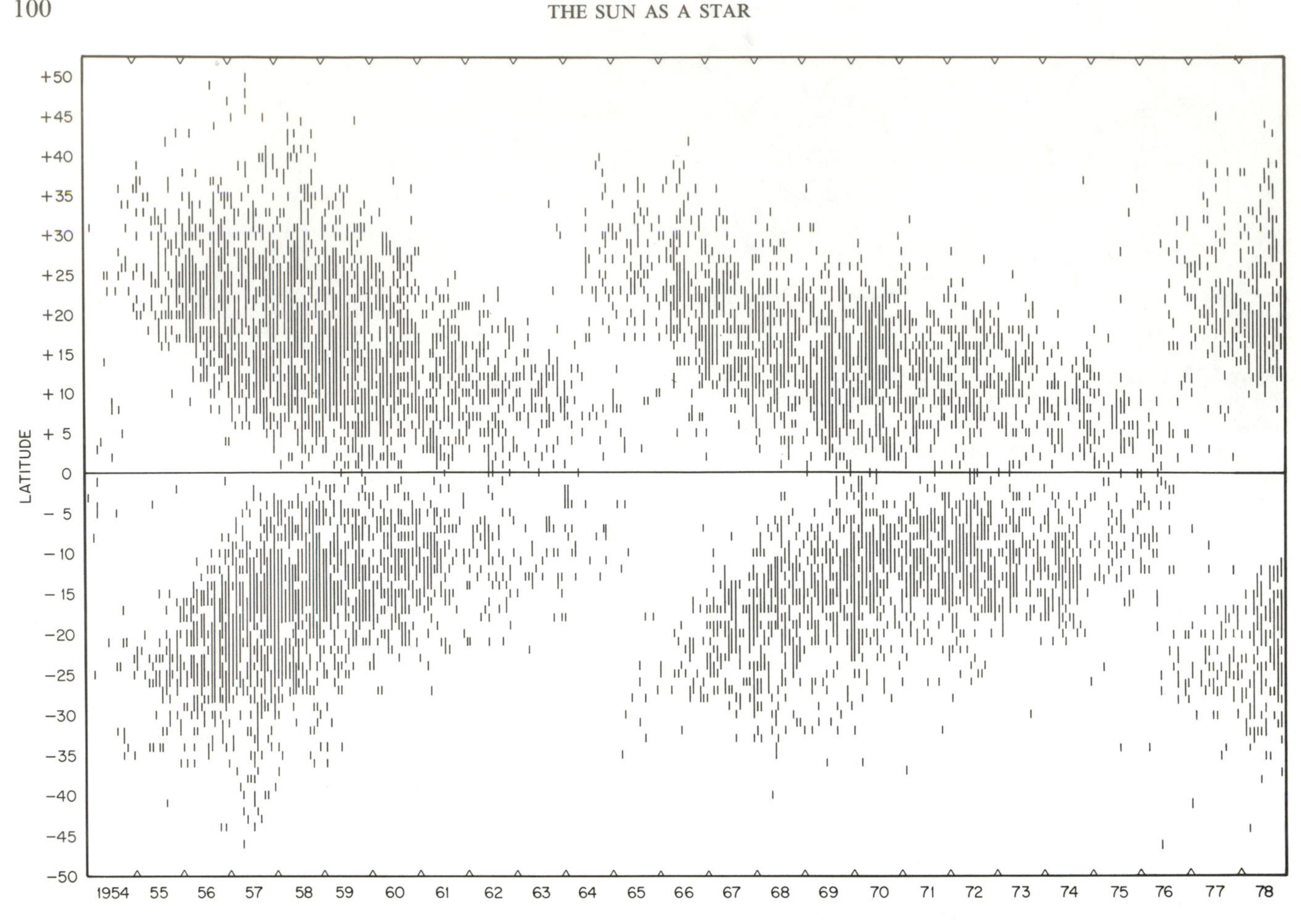

Figure 5.20. Maunder's butterfly diagram is so named because the solar cycle produces a distribution of sunspots whose latitude plotted against time looks like a butterfly. (Courtesy of Robert Howard, Mt. Wilson and Las Campanas Observatories.)

the equator. Near the end of the cycle there are very few spots, and all of these exist close to the equator. Then begins a new cycle of spots appearing near $\pm 30^\circ$. (The pattern of data points in this diagram resembles the wings of a butterfly, which gives the diagram its name.)

Sunspots are known to be regions of intense magnetic field; so the eleven-year sunspot cycle must be related to the mechanism which generates the magnetic field in the Sun. The generally accepted idea is dynamo action and was advanced by T. G. Cowling and E. N. Parker. This dynamo action requires a complex interaction between (differential) solar rotation and solar convection. Simplified dynamo equations have been solved numerically by Robert Leighton, and they yield solutions which somewhat resemble the chaotic field structures that are deduced actually to exist on the Sun (Figure 5.21). This topic, and that of the acceleration of fast particles in solar flares, provides much of the impetus in current research on the physics of the Sun. However, the scientific issues are too technically complex to be discussed in greater detail here.

The Relationship of the Sun to Other Stars and to Us

Because the Sun is so close to us and so central to our continued existence on Earth, it is easy to forget that it is just one star among countless myriads. Thus, a novice (or a congressman) might understandably ask, if it is so important to study the Sun, why should we waste time and money on other esoteric astronomical pursuits? There are two possible answers to this question. The first is the aethestic answer: no great civilization can afford to think that the pursuit of pure knowledge for

Figure 5.21. The magnetic-field structure of the Sun, as deduced from ground-based observations, is usually quite chaotic. The average strength of the magnetic field on the surface of the Sun is comparable to that on the surface of the Earth. However, in localized spots, the solar magnetic field can rise to thousands of times higher than the average value.

its own sake is a waste of time or effort. Science has a beauty and uplifting spirit which rivals any of the other cultural attainments of humanity. This aethestic response arose in a recent congressional hearing. When asked how particle physics contributes to the defense of our country, Robert Wilson replied that it makes the country worth defending. The second possible reply is the practical answer: the history of science has shown time and time again the folly of directing all one's efforts to gain a narrow specific goal. The lunar missions are often cited as a counterexample, but going to the Moon was an enormous *engineering* feat. It did not require any new scientific breakthroughs (except possibly for computer technology, which benefitted from the lunar program, but was proceeding along independently of it anyway); the scientific breakthroughs required for the lunar missions were achieved in the seventeenth, eighteenth, and nineteenth century by scientists who were motivated to seek pure knowledge for its own sake.

A different example is provided by the ongoing thermonuclear fusion programs, humanity's best hope for a long-term solution to the "energy crisis." These programs are relatively well-developed today because of the earlier curiosity of astronomers and physicists about the basic energy source of the Sun (Chapter 6). Moreover, magnetic schemes for the confinement of plasmas are rooted in the scientific attempt to understand the trapping of charged particles in the "bottle" of the Earth's magnetic field (Chapter 17). Various instabilities that thwart practical fusion goals occur also in the plasma and magnetic fields of the Sun's upper atmosphere, and control of the cooling of laboratory plasmas by heavy-ion impurities borrows from astronomers' knowledge of similar processes in interstellar space (Chapter 11). Finally, inertial confinement schemes share several basic mechanisms with studies of the supernova phenomenon in high-mass stars (Chapter 8).

It *is* important, culturally and practically, to understand how the Sun works. Apart from the solar neutrino problem (Chapter 6), astronomers can justifiably claim that they have achieved a good understanding of the deep interior of the Sun. Conditions there are mechanically static and close to *local* thermodynamic equilibrium. The theory under such conditions is relatively simple. Nevertheless, this understanding could not have been achieved without the prior study of such "esoteric" subjects as fluid mechanics, thermodynamics, radiative transfer, and nuclear physics. Even more to the point, a detailed understanding of the Sun's interior was not obtained until astronomers had arrived at a *general* understanding of stellar structure and evolution (Chapter 8). This required the "esoteric" study of many other stars.

Astronomers' understanding of the outer regions of the Sun—especially the chromosphere and corona, and their interaction with the solar rotation and convection zone—is considerably less complete. Very complex processes (e.g., dynamo action) are involved, and pure reasoning alone cannot be expected to provide a deductive theory for these regions. Yet it is precisely these regions which have the most direct impact on terrestrial affairs. Historical precedent leads us to expect that a detailed understanding of complex phenomena requires a study of many examples. Thus, it is interesting that recent ultraviolet and X-ray observations carried out in space satellites have shown that a cool atmosphere (implying an outer convection zone) plus stellar rotation is necessary for significant chromospheric activity but not for coronal activity. The development of ultraviolet and X-ray astronomy, however, was primarily spurred by the desire to understand "esoteric" objects like the interstellar medium, binary X-ray sources, and active galactic nuclei. The point is that what is "esoteric" in the 1960s may become "very practical" in the 1980s, and what is a "waste of taxpayers' money" in the twentieth century may become "vital to the national (or global) interest" in the twenty-first. There is no way to foretell future developments in science; that is one of the features which makes science so fascinating.

CHAPTER SIX

Nuclear Energy and Synthesis of the Elements

In Chapter 5 we stated that the energy source which now maintains the Sun's luminosity is nuclear fusion. Fusion is the process by which heavy chemical elements are built up ("synthesized") from light ones, in the process usually liberating some energy. For example, Hans Bethe found that the most important nuclear reactions in stars like the Sun are fusion reactions which produce, in net, one helium nucleus from combining four hydrogen nuclei:

$$4H \rightarrow He.$$

Now, the mass of the hydrogen nucleus is $1m_p$, where m_p is the mass of the proton, whereas the mass of the helium nucleus is $3.97m_p$. Thus, in converting four hydrogen nuclei into one helium nucleus, a deficit $m = 0.03m_p$ of mass results. This change in mass must liberate an amount of energy E corresponding to Einstein's famous formula:

$$E = mc^2 = 0.03m_pc^2.$$

Since $0.03m_p$ is 0.7 percent of the original mass $4m_p$, astronomers like to say that hydrogen fusion reactions have a 0.7 percent efficiency.

Given the above efficiency, and given that the Sun is composed mostly of hydrogen, it is easy to compute that if the entire store of the Sun's energy were available to spend, the Sun could continue to shine with its present luminosity for about 7×10^{10} years (70 billion years). In fact, as we shall see in Chapter 8, the Sun's structure should change drastically after it has used up only about 13 percent of its total store of hydrogen. Thus, the Sun can remain as it is for only about 9×10^9 years (9 billion years). Since the Sun is believed to be about 4.6 billion years old (Chapter 18), we may say that the Sun is now in its middle age.

Problem 6.1. Assume that the Sun was initially composed 70 percent by mass of hydrogen. How many total hydrogen nuclei, N, were there originally in the Sun if its total mass is 2.0×10^{33} gm? What is the total nuclear energy supply $NE/4$, where $E = 0.03m_pc^2$, if all the hydrogen could be fused into helium? As Iben has remarked, core-hydrogen exhaustion occurs in the Sun after 10 percent of the total supply has been used, but a period of hydrogen burning in a thick shell extends the stable luminous phase of the Sun. What is the lifetime of this phase if 13 percent of the total supply were available to radiate at the luminosity $L_\odot$? Convert your answer from seconds to years.

Having summarized the importance of nuclear fusion reactions for the Sun (and the stars), we may now turn to a more detailed examination of how such reactions work. To understand this subject more than superficially, we must look at how the physical structure of matter is understood today.

Matter and the Four Forces

It is currently believed that all ordinary forms of matter are made up of only two fundamental kinds of elementary particles: **leptons** and **quarks** (Box 6.1), which are both fermions (spin-1/2 particles). These fermions can possess four different kinds of charge: strong (or "color"), electric, weak, and gravitational (or "mass"). Associated with these attributes are four forces which arise from the exchange of four different kinds of bosons (Figure 6.1). The exchange of bosons with odd-integer spin (e.g., spin = 1 in units of $\hbar$) gives rise to forces of repulsion between *like* particles (e.g., two electrons); even-integer spin (e.g., spin = 2) to forces of attraction. The supreme goal of present-day particle physics is to describe the properties of the elementary fermions (leptons and quarks) and their four forces in terms of a single superforce. The electromagnetic and weak interactions were apparently combined successfully in 1971 by the efforts of Weinberg, Salam, Ward, and 't Hooft. In 1974, Georgi and Glashow presented an attractive candidate for the grand unification of three of the four fundamental forces: strong, electromagnetic, and weak. These three suffice to explain all of present-day particle physics where gravity plays little role (see, however, Chapter 16 for a possible superunification of all four forces).

There are, apparently, six different kinds of leptons, which come in three pairs:

(a) electron and its neutrino,

(b) muon and its neutrino,

(c) tau lepton and its neutrino.

There are, apparently, also six different kinds ("**flavors**") of quarks, which come in three pairs:

(a) up and down quark (u and d),

(b) strange and charmed quark (s and c),

(c) top and bottom quark (t and b).

The theories under consideration which hope to unify the strong, weak, and electromagnetic forces envisage the leptons and quarks to be paired into three "generations":

(a) electron and its neutrino, up and down quark,

(b) muon and its neutrino, strange and charmed quark,

(c) tau lepton and its neutrino, top and bottom quark.

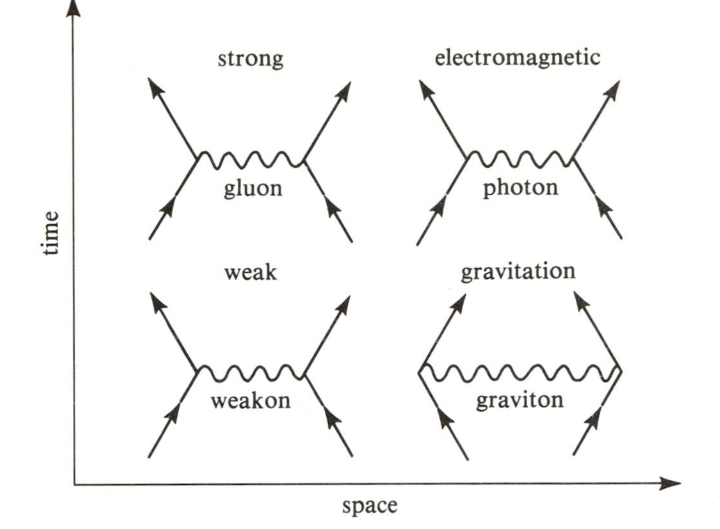

Figure 6.1. The four forces in relativistic quantum mechanics are mediated by the exchange of integer-spin particles (bosons). Of the four forces, only gravity gives rise to attractive forces between *like* particles (same type of color charge, electric charge, weak charge, or mass). This difference arises because the graviton is spin 2, whereas the gluon, photon, and weakon are spin 1. The spacetime diagram for gravitation may be interpreted as follows. Two masses originally separating from one another exchange a virtual graviton which attracts the two masses together again. The graviton is considered "virtual" because it does not exist long enough to become observable. The spacetime diagrams for the other forces have similar interpretations, except that repulsion replaces attraction.

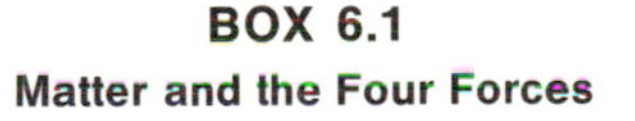

BOX 6.1

Matter and the Four Forces

All ordinary forms of matter are made of two kinds of elementary particles:

(a) leptons, which include the electron, that participate in weak and electromagnetic (if charged) interactions;

(b) quarks, which, taken three at a time, make up the proton and the neutron, which participate in strong, weak, and electromagnetic interactions.

In addition, leptons and quarks are affected by gravitation, the only force that universally affects all forms of matter-energy. On the scale of atomic nuclei, the forces can be ranked in increasing strength as: gravitation, weak interaction, electromagnetism, and strong interaction. However, this ranking depends on the energy available for the interaction; at superhigh energies, it is thought that all four forces have the same strength.

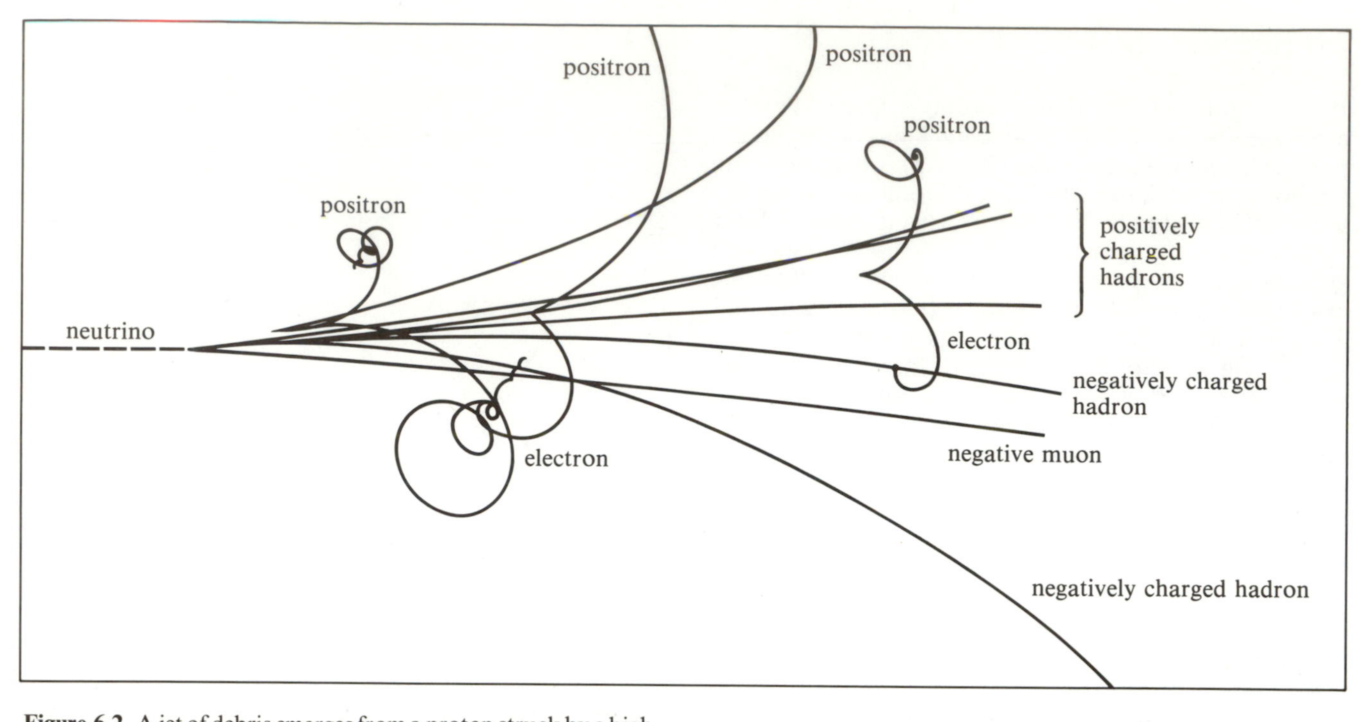

Figure 6.2. A jet of debris emerges from a proton struck by a high-energy neutrino. The event is captured photographically via trails of bubbles left as charged particles pass through a chamber of superheated hydrogen. The neutrino which triggered the event leaves no track because it is electrically neutral. Relative to the original line through the centers of mass of the neutrino and the proton, the scattered muon and jet of hadrons (particles which feel the strong force) emerge at large angles. Such wide-angle scatterings are reminiscent of the Rutherford scattering experiment (see Figure 3.9) and suggests that the proton itself is made of small hard subunits, presumably quarks. (From Jacob and Landhoff, *Scientific American*, March 1980, p. 67.)

For our purposes here, only the first "generation" is important. In Chapter 16, when we describe the conjectures concerning the first 10^{-4} sec of creation, we will return to a more general discussion of particle physics.

Protons and Neutrons

The nuclei of ordinary atoms are made of protons and neutrons. The proton and neutron are the simplest and most important examples of a class of particles called **baryons** (the word "baryon" derives from Greek, meaning the "heavy one"). Murray Gell-Mann and George Zweig proposed in 1963 that the proton and neutron were not truly elementary particles, but were made of three quarks. At the time of their proposal, it had long been known that the proton and neutron acted as if they had finite size (about 10^{-13} cm), but only in the 1970s did high-energy scattering experiments provide indisputable evidence that the proton and neutron were made of small hard subunits (Figure 6.2). Presumably these constituent particles are the quarks of the theory of Gell-Mann and Zweig.

Quarks were originally invented to explain, not the internal structure of the proton and the neutron, but the diversity of the particles which feel the strong nuclear force. In particular, the proton was envisaged to consist of two up quarks (of opposite spins) and one down quark; the neutron, of two down quarks (of opposite spins) and one up quark. Since the quarks each have spin 1/2, both the proton and the neutron would have net spin 1/2 and would be fermions (Table 6.1). In addition to flavor and spin, the quarks have other attributes, two of which are electric charge and **baryon number**. The up quark has an electric charge $= +2/3$ (in units of $e = 4.80 \times 10^{-10}$ esu) and a baryon number $= +1/3$; the down quark has an electric charge $= -1/3$ and a baryon number $= +1/3$ also. Thus, the proton has an electric charge $= +2/3 + 2/3 - 1/3 = +1$, and a baryon number $= +1/3 + 1/3 +$

Table 6.1. **The proton, the neutron, and their constituent quarks.**

Particle	*Spin*	*Electric charge*	*Baryon number*
Up quark (u)	1/2	+2/3	+1/3
Down quark (d)	1/2	−1/3	+1/3
Proton (uud)	1/2	+1	+1
Neutron (udd)	1/2	0	+1

$1/3 = +1$. The neutron has an electric charge $= +2/3 - 1/3 - 1/3 = 0$, and a baryon number $= +1/3 + 1/3 + 1/3 = +1$. To remind ourselves of these important results, let us write the symbols p for the proton, and n for the neutron, with the left-hand *sub*script giving the electric charge, and the left-hand *super*script giving the baryon number:

$$\text{proton} = {}^{1}_{1}\text{p}, \ \text{neutron} = {}^{1}_{0}\text{n}. \tag{6.1}$$

In more recent developments of the quark theory, quarks are also endowed with another property: "color." Each quark may take on the hues red, blue, and green, but these have nothing to do with ordinary colors. Rather, the quark colors represent a kind of charge that is associated with the strong nuclear force. Indeed, it is the exchange of eight kinds of gluons containing color and anticolor that provides the attractive force which binds quarks of different colors together. The detailed theory of this internal force, called "quantum chromodynamics," is not needed for this book. In any case, rigorous calculation in quantum chromodynamics is very difficult using current techniques, and physicists have developed semiclassical models ("strings", "bags", etc.) to describe aspects of the strong force which "glues" quarks together. We shall see that similar models are helpful for seeing how protons and neutrons are bound together inside the nuclei of atoms. For now, let us merely remark that the three colors, red, blue, and green, are represented in equal proportions in protons and neutrons. Thus, these baryons are "white" or colorless.

Electrons and Neutrinos

The shells of atoms are populated by electrons. Unlike the proton or neutron, the electron appears in all experiments conducted so far to have no internal structure. As nearly as we can tell, the electron is a pointlike and truly elementary particle. The electron has an electric charge $= -1$ and a baryon number $= 0$ (it is not a baryon). In the notation introduced previously, therefore, we may write an electron as

$$\text{electron} = {}^{0}_{-1}\text{e}. \tag{6.2}$$

The electron is also colorless. Indeed, all the particles we will discuss in this book (apart from quarks and gluons) are colorless; so we will not need to keep track of "color." Most natural phenomena are colorblind, and this explains why we are using subscripts and superscripts to keep track only of electric charge and baryon number.

Associated with the electron is a neutrino, denoted by the symbol ν_e. The electron neutrino has an electric charge $= 0$ and a baryon number $= 0$. The electron and its neutrino both have spin 1/2 (they are fermions like the proton and neutron); but, unlike the proton and neutron, the electron and its neutrino have very little rest mass,* and so they are called leptons (Greek for "light ones"). In this book, we shall have little occasion to discuss other types of neutrinos; so we shall write ν without the subscript e to denote the electron neutrino. (This notation may turn out to be best in any case, if, as has been suggested, a neutrino oscillates between the various electron, muon, and tau forms.) Thus, in our subscript and superscript notation for electric charge and baryon number, we may write

$$\text{neutrino} = {}^{0}_{0}\nu. \tag{6.3}$$

Particles and Antiparticles

For each particle, relativistic quantum mechanics allows an antiparticle, which is denoted with the same symbol but with an overbar. Thus, the antiproton consists of two antiup quarks and one antidown quark (composed of the three anticolors: antired, antiblue, and antigreen). The antiproton therefore has an electric charge $= -1$ and a baryon number $= -1$ (it is an antibaryon):

$$\text{antiproton} = {}^{-1}_{-1}\bar{p}.$$

The proton and antiproton have identical masses, but otherwise their properties are exact opposites.

The opposite of an electron, the antielectron, is called a *positron*. It is denoted by e^+ to distinguish it from an electron e^-. In more detail,

$$\text{positron} = \text{antielectron} = {}^{0}_{1}\bar{e}.$$

Historically, it was Dirac's theory of the electron which first led to the prediction of antiparticles, in particular, the prediction of the positron. Dirac arrived at his

* It used to be thought that the neutrino had zero rest mass, but recent experiments have cast doubt upon this assumption.

epochal conclusion by his attempt to marry the theory of special relativity, as developed by Einstein, and the theory of wave mechanics, as developed by Schrödinger, into a single theory—relativistic quantum mechanics—for discussing the dynamics of the electron.

We have already discussed in Chapter 3 the central role of Heisenberg's uncertainty principle in quantum mechanics. Applied to the simplest case of a free particle, this principle leads to equation (3.7),

$$\Delta x \cdot \Delta p_x \geq h, \tag{6.4}$$

for the intrinsic uncertainties in position and momenta of a particle that we can measure. A completely analogous expression applies to the uncertainties in energy and time at which the particle possessed this energy:

$$\Delta t \cdot \Delta E \geq h. \tag{6.5}$$

In the language of quantum mechanics, space (x) and momentum (p_x) are said to be **complementary variables**, and the same is true of time (t) and energy (E).

In relativity theory, however, it is space and time which have a close relationship (Chapter 3). To incorporate all the symmetries required in a relativistic theory of quantum mechanics, therefore, the basic equations should in some sense possess certain symmetries with respect to the pairs (space, time), (space, momentum), (time, energy). This, in turn, requires starting expressions relating momentum and energy to have a closely symmetric form. Even for a free particle, however, the classical expression relating (kinetic) energy E and momentum p are manifestly not symmetric,

$$E = p^2/2m,$$

since E enters only with the first power and p enters with the second. Schrödinger's wave mechanics begins by giving a quantum-mechanical interpretation to the above equation. Since his starting point thereby lacks the symmetry required by relativity theory, it is not surprising that Schrödinger's equation constitutes a basis for quantum mechanics only in the nonrelativistic regime.

To make the correct relativistic generalization, we must start with the relativistic analogue of $E = p^2/2m$. For example, we might start with Einstein's famous equation $E = mc^2$, where E is now the total energy of a free particle and m is the relativistic mass $m_0(1 - v^2/c^2)^{-1/2}$ with m_0 being the rest mass and v being the speed. For quantum mechanics, however, we want a relation between the energy E and the momentum $p = mv$. Problem 6.2 asks you to show that the required expression can be written

$$E^2 - p^2c^2 = m_0{}^2c^4, \tag{6.6}$$

which is manifestly symmetric, since both E and p now enter as squares, with the right-hand side equal to a constant (a Lorentz invariant, in the language of special relativity).

Problem 6.2. Eliminate the variables m and v from the expressions $E = mc^2$ and $p = mv$, with $m = m_0(1 - v^2/c^2)^{-1/2}$ to obtain equation (6.6). Notice that for $m_0 = 0$, we recover the relation, $E = pc$, valid for photons.

The quantum-mechanical interpretation of equation (6.6), found first also by Schrödinger, does lead to a formalism which treats space and time on an equal footing. (The resulting equation is now called the Klein-Gordon equation, and it turns out to apply to particles of spin zero.) However, the solution for the allowed energy states, obtained by taking the square root of equation (6.6) to solve for E, appeared objectionable. Whenever one takes the square root of a positive number, one obtains two answers, which are equal but opposite in sign. In our case, $E = \pm(m_0{}^2c^4 + p^2c^2)^{1/2}$. What could be meant by negative energy states? Dirac had the insight to believe that *both* signs for E may correspond to physical reality: the positive sign corresponding to the solution for an electron; the negative sign, when interpreted properly, to a positron.

However, the equation $E = (m_0{}^2c^4 + p^2c^2)^{1/2}$ for an electron does not treat E and p symmetrically, since E enters in the first power, p in the second, which was the problem with the classical equation. To rectify this shortcoming, Dirac took a radical step and assumed that it must be possible to write the square-root expression as an entity linear in the rest energy m_0c^2 and the *vector* momentum $\boldsymbol{p}$,

$$(m_0{}^2c^4 + p^2c^2)^{1/2} = \boldsymbol{a} \cdot \boldsymbol{p}c + bm_0c^2, \tag{6.7}$$

where $\boldsymbol{a}$ and b are, respectively, vector and scalar numbers. However, it is easy to show that no set of four ordinary numbers a_x, a_y, a_z, and b can satisfy equation (6.7). Problem 6.3 asks you to show that a_x, a_y, a_z, and b have to be *matrices* which anticommute with each other but not with themselves (i.e., $a_xa_y = -a_ya_x$, but $a_xa_x \neq -a_xa_x$). Similar matrices were used previously by Pauli to describe electrons behaving as if they had spin 1/2. In other words, Dirac found that, in order to satisfy the symmetry requirements of *both* quantum mechanics and special relativity, electrons not only had to have antiparticles (positrons), but also had to have spin angular momentum of the amount $h/2$! A full discussion of this beautiful point would require much more advanced mathematics than is appropriate for this book. Let us be content here that experiments have amply confirmed Dirac's basic deductions on these two points.

However, we should remark that Dirac's original interpretations of positrons as "holes" left in an infinite prevasive sea of electrons that possess *negative* energy states is no longer accepted as the most natural viewpoint. It is far more desirable to view positrons as genuine particles equal to electrons in every respect (including having positive mass-energy).

Problem 6.3. (for those familiar with matrix algebra). Since equation (6.7) might have to be interpreted as a matrix (or *operator*) equation, we should be careful about the order of multiplication. Write the square of the magnitude of $\boldsymbol{p}$ as $p^2 = p_x p_x + p_y p_y + p_z p_z$ and the dot product of $\boldsymbol{a}$ and $\boldsymbol{p}$ as $\boldsymbol{a} \cdot \boldsymbol{p} = a_x p_x + a_y p_y + a_z p_z$. Then square both sides of equation (6.7) by multiplying each side by itself. Set the resulting coefficients of the various products of $p_x c$, $p_y c$, $p_z c$, and $m_0 c^2$ equal. In this manner, derive the requirements,

$$a_x a_x = a_y a_y = a_z a_z = bb = 1,$$

$$a_x a_y + a_y a_x = a_y a_z + a_z a_y = a_z a_x + a_x a_z = a_x b + b a_x$$
$$= a_y b + b a_y = a_z b + b a_z = 0.$$

Show that the above relations cannot be satisfied if a_x, a_y, a_z, and b are treated as ordinary numbers. However, take a_x, a_y, a_z, and b to be the following 4×4 matrices (the number 4 enters because spacetime has four coordinates):

$$a_x = \begin{pmatrix} 0 & 0 & 0 & 1 \\ 0 & 0 & 1 & 0 \\ 0 & 1 & 0 & 0 \\ 1 & 0 & 0 & 0 \end{pmatrix}, \quad a_y = \begin{pmatrix} 0 & 0 & 0 & -i \\ 0 & 0 & i & 0 \\ 0 & -i & 0 & 0 \\ i & 0 & 0 & 0 \end{pmatrix},$$

$$a_z = \begin{pmatrix} 0 & 0 & 1 & 0 \\ 0 & 0 & 0 & -1 \\ 1 & 0 & 0 & 0 \\ 0 & -1 & 0 & 0 \end{pmatrix}, \quad b = \begin{pmatrix} 1 & 0 & 0 & 0 \\ 0 & 1 & 0 & 0 \\ 0 & 0 & -1 & 0 \\ 0 & 0 & 0 & -1 \end{pmatrix},$$

where $i = (-1)^{1/2}$. Now, show $a_x a_x = 1$, etc., and $a_x a_y + a_y a_x = 0$, etc., where 1 and 0 in the latter equations are interpreted as the unit and null matrices:

$$1 = \begin{pmatrix} 1 & 0 & 0 & 0 \\ 0 & 1 & 0 & 0 \\ 0 & 0 & 1 & 0 \\ 0 & 0 & 0 & 1 \end{pmatrix}, \quad 0 = \begin{pmatrix} 0 & 0 & 0 & 0 \\ 0 & 0 & 0 & 0 \\ 0 & 0 & 0 & 0 \\ 0 & 0 & 0 & 0 \end{pmatrix}.$$

Problem 6.4. Describe the quarks which make up an antineutron. Give the symbol for an antineutron. An antineutrino. What is an antihydrogen atom? What kind of spectral lines does an antihydrogen atom have? *Hint*: Photons come in only one kind, neither matter nor antimatter.

The Quantum-Mechanical Concept of Force

The naive concept of force is of a push or a pull exerted on an object by an adjacent object. This simple concept received a setback from Newton's formulation of gravitation as an "action at a distance" (Chapter 3). In relativistic quantum mechanics, one returns, with some added sophistication, to the older concept of force as a push or a pull exerted by neighboring objects in spacetime. Relativity and quantum mechanics taken together define what one means by "neighboring" objects that can influence one another. The detailed action of quantum forces is closely linked to symmetry properties and conservation principles, ideas whose basis we will pursue in Chapter 16. Here we will merely examine the elementary notion that the forces of nature, the pushes and pulls, are ultimately mediated by the exchange of bosons (see Figure 6.1). On the face of it, this view seems absurd. If two electrons exert an electric force on each other by exchanging photons, why don't we see light fly back and forth between all charged bodies? Why do we have to wait until charges are accelerated before we see photons?

The answer is that the exchanged photons are not *real* photons in a crucial sense. The photons which give rise to the electrostatic force of repulsion between two electrons (or attraction between a proton and an electron) are **virtual** photons. Virtual particles are those ephemeral entities which are created (emitted) and destroyed (absorbed) within a time interval Δt that is too short to allow their detection by any physical means. To be specific, suppose a photon of energy E is created, and that it is destroyed a time interval Δt later, with $\Delta t \leq h/E$. The creation of this photon has violated the principle of the conservation of energy during the brief period of its existence, because with stationary charges there is an energy discrepancy $\Delta E = E$ between the state before the photon appeared and the state after it appeared. However, if this debt (borrowed from the vacuum) is returned very shortly afterward, the product $\Delta E \cdot \Delta t$ will *not* satisfy Heisenberg's uncertainty principle, equation (6.5), and it will be impossible physically to detect the embezzlement. To see another difference between virtual photons and real photons, consider how far a virtual photon can travel in time Δt. Traveling at the speed of light c, the photon can get a distance $c\,\Delta t$ before it is absorbed by a charge. If $\Delta t \leq h/E$, where $E = hc/\lambda$ and λ is the wavelength of the photon (equation 3.6), we see that the distance $c\,\Delta t$ traveled by a virtual photon is always less than one wavelength. (If $h/2$ is used in place of h, we have less than $1/4\pi$ times a wavelength.) A disturbance with less than one wavelength cannot be detected as being "wavy." In classical terms, we would describe the field of virtual photons which transmits the electromagnetic force as the nonwavy (or longitudinal) "Coulomb field" associated with electric charges (Chapter 3).

The above calculation also indicates why the electrostatic force is a long-range force. If the virtual photon has very little energy E (very long wavelength λ), it may get very far from its source before being absorbed by another charge. Thus, charges may influence one another over very long distances r, but the force associated with the influence falls off as $1/r^2$ because of the small energies of the exchanged photons at large separations.

Virtual photons can even be created by a charged particle and absorbed by the same charged particle (Figure 6.3). Historically, such "closed loop diagrams" in spacetime caused great computational difficulties for early workers in quantum electrodynamics. The loop can occupy arbitrarily small intervals of time, so that photons of arbitrarily large energies can formally be considered. This leads to an undue influence of the electron on itself (!), and gives rise to troublesome infinities when one tries to interpret what the mass of the electron is. The difficulty was solved by Feynman, Schwinger, and Tomanaga, who invented a "renormalization procedure" to take care of the troublesome infinities. By now, quantum electrodynamics gives exceedingly accurate results for all calculated processes, and it has become the standard by which other physical theories are judged. In particular, on the basis of Figure 6.3, we may think of an electron as always being surrounded by a cloud of virtual photons. If the electron is violently accelerated by some external means, some of this cloud may be shaken loose and given enough energy to become real photons. These real photons are then seen as light created by the process of accelerating an electron. (Compare with Figure 2.9.)

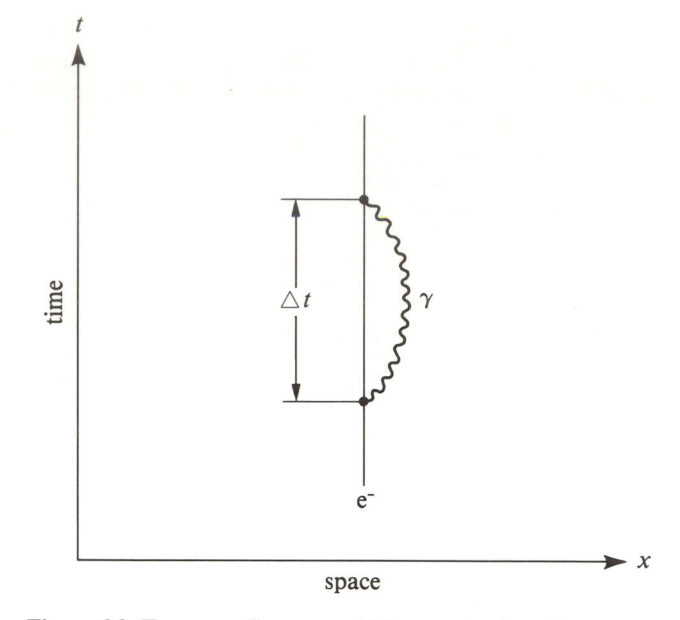

Figure 6.3. Feynman diagrams which contain closed loops, representing in this example the emission and absorption of a virtual photon by the same electron, present a challenge to quantum field theories.

Let us summarize our findings. Since the maximum speed of any particle is c, the maximum distance $c\Delta t$ that a *virtual* boson can travel in time $\Delta t \leq h/E$ is set by the minimum energy that this boson can have. For particles with rest mass m_0, the minimum energy is $E_0 = m_0c^2$ achieved when the momentum $p = h/\lambda$ is zero (see equations 3.5 and 6.6). Thus, the *range* of a force associated with the exchange of bosons of rest mass m_0 is characteristically given by h/m_0c, an expression called the **Compton wavelength**. If the rest mass m_0 of the exchanged boson is zero, as it is for the photon, graviton, or gluon (all of which fly at the speed of light c and satisfy the Einstein relation $E = hc/\lambda$), the range h/m_0c of the corresponding force is infinite. Thus, the electrostatic force between electric charges, the gravitational forces between masses, and the color force between quarks are all believed to be long-range forces.* The electric force between unshielded charges falls off as $1/r^2$ because photons are themselves uncharged. Gravitons do gravitate, but only very mildly; thus, gravity is expected to depart from $1/r^2$ only in intense field situations when gravity can substantially deflect even particles traveling at the speed of light (see Chapters 7 and 15). In contrast, gluons themselves both contain color and interact strongly; thus, the color force between quarks may not grow weaker as the distance between quarks is increased. If this conclusion is correct, many physicists have speculated that it may take infinite energy to separate quarks by an infinite distance. Free quarks may then be impossible to produce.

Nuclear Forces and Nuclear Reactions

The Strong Nuclear Force

The three quarks that make up a proton or a neutron are bound together by the color force, but the exchange of colored gluons cannot occur beyond the bound par-

* If more than one boson has to be exchanged in Δt, the corresponding force may be short-range even if the bosons have zero rest mass. Van der Waals forces are short range for this reason; they can be thought of as involving the simultaneous exchange of two or more photons.

ticle, because the proton or neutron must remain colorless overall. Thus, it is believed that protons and neutrons interact with one another by exchanging bosons other than gluons. The strong force between protons and neutrons is mediated by bosons consisting of a bound (colorless) quark-antiquark pair. These quark-antiquark pairs have rest masses which are in between those of baryons and leptons; so the quark-antiquark pairs are called **mesons** (Greek meaning the "medium ones"). Especially important are the π mesons, which are made of up or down quarks and their antiparticles. The combination of an up quark and an anti-up quark ($u\bar{u}$) or a down quark and an anti-down quark ($d\bar{d}$) yields a neutral π meson, denoted by us as ${}^0_0\pi$. The combination ($u\bar{d}$) yields a positively charged π meson, ${}^0_1\pi$; and the combination ($\bar{u}d$) yields a negatively charged π meson, ${}^{\ 0}_{-1}\pi$. (Shorthand notations for ${}^0_0\pi$, ${}^0_1\pi$, and ${}^{\ 0}_{-1}\pi$ are π^0, π^+, π^-.) In any case, the spins of the quark and antiquark are antiparallel in all the π mesons; so the spin of each π meson is zero. Being even-integer-spin bosons, therefore, the π mesons transmit attractive forces between like baryons. In terms of the strong force, protons and neutrons are similar baryons. (The electric charge which distinguishes the proton from the neutron is sensed only by the electromagnetic force; the difference in constituent quark "flavors" [up or down] is sensed only by the weak force; the slight difference in mass is sensed only by gravity.) Thus, the exchange of π mesons between protons and neutrons (or neutrons and neutrons, or protons and protons) corresponds to attraction. This attractive force is what binds the protons and neutrons together in the nucleus of an atom. Moreover, because the π mesons have nonzero rest mass, the strong nuclear interaction is a *short-range force*.

Problem 6.5. Calculate the range h/m_0c of the attractive part of the nuclear strong force if the exchanged π meson has a rest mass $m_0 = 0.15m_p$, where m_p is the rest mass of the proton. The π meson contains two of the three quarks that make up a proton or an antiproton. Why does it have only 0.15 of the rest mass? *Hint*: Is there a mass deficit associated with the binding of quarks?

Historically, the meson theory for explaining the strong interaction between baryons was invented in 1935 by Yukawa, well before the development of the quark theory outlined in this Chapter. The actual behavior of the strong force in the nucleus of an atom is, however, more complicated than that suggested by the consideration of π mesons alone. For example, the strong nuclear force is known to become repulsive at sufficiently small separation distances, presumably because of the exchange of mesons with odd-integer spin. Such mesons might arise from the binding of quarks and antiquarks with aligned spins (meson spin = 1), or from the union of two π mesons, converting their orbital angular momentum into spin angular momentum (the rho meson with spin 1 is a known "resonance" of this kind). Thus, the strong nuclear force between protons and neutrons (colorless combinations of quarks) has an attractive-repulsive duality that is reminiscent of the van der Waals force between molecules (neutral combinations of electric charges).

The Weak Nuclear Force

In addition to the strong nuclear force, the protons and neutrons can interact via the weak force (the exchange of weakons, sometimes called intermediate vector bosons, as depicted in Figure 6.1). The range of the weak force has been estimated to be less than 10^{-15} cm. This means that the mass of the weakons must be of the order of 100 times that of a proton. It is hoped that they will soon be detected experimentally.

Weak interactions are important for the nuclear reactions that synthesize the heavy elements inside stars, because only the weak force can convert a proton into a neutron (and vice versa). The nuclei of ordinary hydrogen atoms are purely protons, but the nuclei of the heavy elements generally contain as many neutrons as protons. Thus, to fuse hydrogen into heavier elements, the weak force must be invoked to convert about half the initial protons into neutrons, and the strong force must be called upon to bind these protons and neutrons together.

In a *weak* interaction between quarks, it is possible to transform an up quark into a down quark (and vice versa). To conserve electric charge in such a transformation ($+2/3 \leftrightarrow -1/3$), an electron or a positron must be involved, and an antineutrino or a neutrino must also appear in order to conserve lepton number. As an example, a neutron outside the nucleus of an atom will decay via the weak interaction in about 10 min into a proton, an electron, and an antineutrino (Figure 6.4),

$$ {}^1_0\mathrm{n} \rightarrow {}^1_1\mathrm{p} + {}^{\ 0}_{-1}\mathrm{e} + {}^0_0\bar{\mathrm{v}}. \tag{6.8}$$

To keep track of the conservation of electric charge, we merely need to check that the left-hand subscripts add to the same total on both sides of the reaction: thus, $0 = 1 - 1 + 0$. To keep track of baryon conservation, we need to do the same with the left-hand superscripts: thus, $1 = 1 + 0 + 0$. To keep track of lepton conservation, we need only remember that an electron is a lepton, whereas an antineutrino is an antilepton.

The decay of a free neutron by the weak interaction takes about 10^3 sec (10 minutes), a relatively short time by human standards. It is very long compared to the

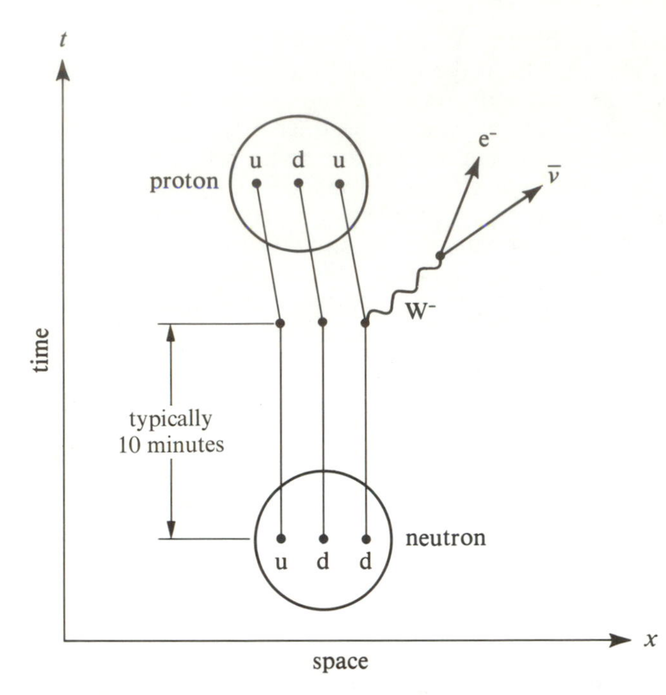

Figure 6.4. The decay of a free neutron in the quark model. A down quark emits a W^- (negatively charged weakon) which decays into an electron and an antineutrino. The down quark in the process is changed into an up quark, which together with the remaining up and down quarks constitute a proton. In the end, it is the energy difference between the neutron and the proton which enables the creation of the electron and the antineutrino.

typical lifetime, 10^{-23} sec (roughly the light-travel time across 10^{-13} cm), of unstable particles which decay by the strong interaction. Atoms which are neutron-rich in their nuclei may be unstable to the conversion of one or more neutrons into protons, with the subsequent emission of an electron and an antineutrino. Atoms which are proton-rich in their nuclei may be unstable to the conversion of one or more protons into neutrons, with either the emission of a positron and a neutrino or the capture of an inner shell electron and the emission of a neutrino. (The emission of a neutrino or an antineutrino in any nuclear reaction is a tip-off that the weak force is involved.) Since electrons (or positrons) emitted this way were originally called beta rays, elements which were radioactive in this sense were said to suffer **beta decay**. The lifetime of a radioactive element against beta decay can vary greatly—from 10^3 sec to 10^{18} sec—depending on quantum selection rules and the energy available to drive the reaction.

Problem 6.6. Other spin-1/2 baryons, besides the proton and the neutron, include the lambda, three sigma, and two xi particles. Of these, the proton has the least rest energy mc^2. Plotted in an energy diagram, the rest energies of these spin-1/2 baryons look like the following:

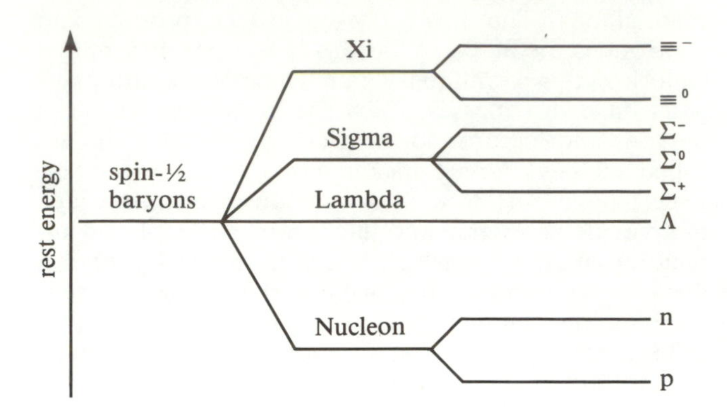

If this reminds you of atomic spectroscopy (compare with Box 3.3), good, it's supposed to. Particle physicists envisage the various particles of different rest energies to arise from a single baryonic state that is "split" by various interactions (like the electromagnetic "splitting" at the end); so the proton is the ground state for spin-1/2 baryons. The other particles, given time, will decay into the proton plus mesons. For example, the lambda can decay into a proton plus a negative pi meson: ${}^1_0\Lambda \rightarrow {}^1_1\text{p} + {}_{-1}^{\;0}\pi$ in a typical time 10^{-10} sec. This is longer than 10^{-23} sec, because the lambda contains a "strange" quark and neither the proton nor the pi meson does. Since the neutron does not contain a strange quark, why is it not even more unstable than the lambda? *Hint*: $m_n - m_p = 0.0014m_p$, but $m_\pi = 0.15m_p$.

Atomic Nuclei

The nucleus of the ordinary hydrogen atom consists of a single proton; hence, when the elemental nature of the particle is to be emphasized, a proton can also be written as ${}^1_1\text{H}$,

$${}^1_1\text{p} = {}^1_1\text{H}.$$

In addition to this most common form, hydrogen also has two **isotopes** (Figure 6.5):

(a) **deuterium** = ${}^2_1\text{H}$, which consists of a bound nuclear state of a proton and a neutron;

(b) **tritium** = ${}^3_1\text{H}$, which consists of a bound nuclear state of one proton and two neutrons.

Deuterium and tritium *atoms*, like those of ordinary hydrogen, contain only one orbital electron to balance the charge of the nucleus. It is this orbital electron (va-

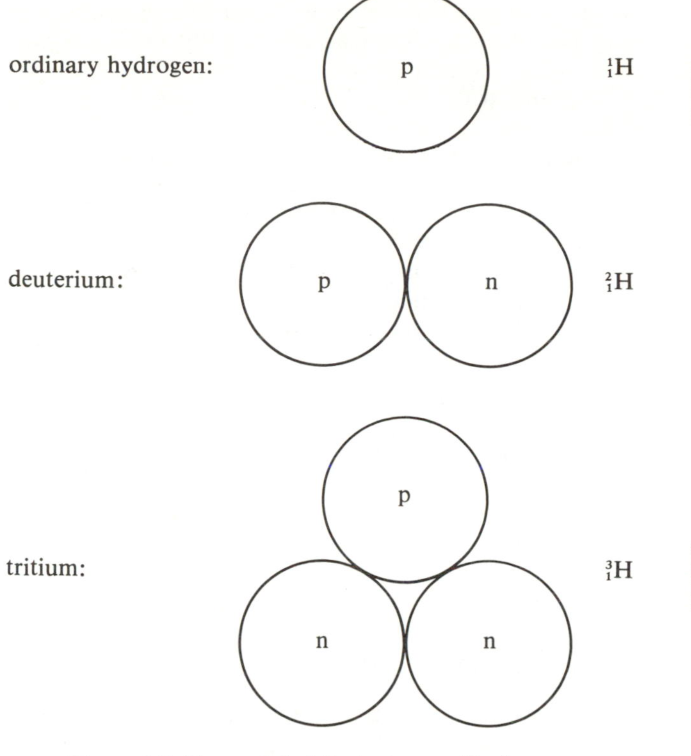

Figure 6.5. The nuclei of the isotopes of hydrogen.

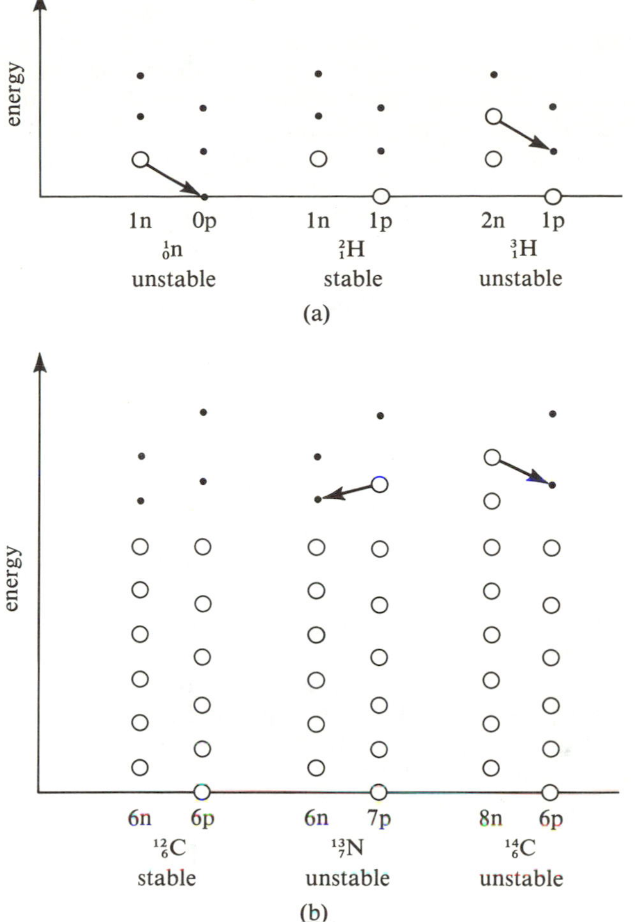

Figure 6.6. Beta radioactivity tends to keep the number of protons and neutrons approximately equal in atomic nuclei of moderate atomic weight. The carbon-12 nucleus in the left columns is stable. The nitrogen-13 nucleus in the middle columns is unstable, and will decay into a stable carbon-13 nucleus. The carbon-14 nucleus in the right columns is unstable, and will decay into a stable nitrogen-14 nucleus. Notice that electric repulsion increases the successive spacings of the proton stack relative to the neutron stack. To keep the two stacks approximately level in very heavy atomic nuclei therefore requires more neutrons than protons.

lence electron) which accounts for the chemical properties. Since all three isotopes of hydrogen have the same number of outer electrons, they all have *identical chemical properties.** The nuclear properties of deuterium and tritium (also called heavy hydrogen) are, however, quite different from those of ordinary hydrogen. In particular, the proton and neutron in the deuterium nucleus are bound together by mesons ("leftover quark forces") in such a way that their total mass is slightly *less* than the sum of the masses of a free proton and a free neutron. This fact is usually expressed by saying that deuterium has **binding energy** relative to the free particles. (Thus, the usual symbol ^A_ZE for an element E of "atomic number Z,"' equal to nuclear electric charge, and "atomic weight A," equal to nuclear baryon number, involves a slight approximation.)

Why does the bound neutron in the deuterium nucleus not decay via the process given in equation (6.8), as a free neutron will? Because of Pauli's exclusion principle (Chapter 3). Protons and neutrons in the nucleus of an atom are arranged in discrete energy levels, much as the electrons in the shells of an atom are. The guiding principle in both cases is Pauli's exclusion principle, which forbids two identical fermions to occupy the same quantum state. Protons and neutrons are fermions (they both have spin 1/2); so they must satisfy Pauli's exclusion principle in their arrangement in the nucleus of an atom. Figure 6.6a shows a schematic diagram of the energy levels of protons and neutrons in the simplest possible atomic nuclei. Since protons can be distinguished from neutrons, they each form an energy stack.

Consider a hypothetical atomic nucleus with only one neutron (on the left). Such a nucleus, a free neutron, is unstable even in its ground energy state, because that

* Subtle physical effects can enter because of the differences in the mass of the nucleus, and such effects can lead to fractionization; but we will ignore such phenomena in this book.

state lies above the ground energy state of a single proton. The neutron will therefore beta-decay, via the process given in equation (6.8), to a proton, the nucleus of an ordinary hydrogen atom. Consider now a deuterium nucleus with one neutron and one proton, each in the bottom level of its own stack (middle). The neutron can no longer decay into a proton, because the proton state, with an energy level lower than that of the neutron, is already occupied. To convert the neutron into a proton at the next available proton level would require energy that the nucleus does not possess. The deuterium nucleus is therefore stable. The same argument shows now why a tritium nucleus (on the right) *is* unstable (with a half-life of 12.5 years) toward the decay

$$^{3}_{1}\mathrm{H} \rightarrow {}^{3}_{2}\mathrm{He} + {}^{\ 0}_{-1}\mathrm{e} + {}^{0}_{0}\bar{\nu}.$$

Extension of the above ideas explains why atomic nuclei like to keep the number of neutrons and protons approximately equal, i.e., $A - Z \simeq Z$ or $A \simeq 2Z$. Figure 6.6b shows the situation schematically for the nuclei of carbon-12, $^{12}_{6}\mathrm{C}$, nitrogen-13, $^{13}_{7}\mathrm{N}$, and carbon-14, $^{14}_{6}\mathrm{C}$. The carbon-12 nucleus (on the left) is stable, because converting either a proton into a neutron or a neutron into a proton would require removing the top particle from one stack and putting it *higher* on the other stack. Imagine now adding a proton to carbon-12 and getting nitrogen-13 (middle). Protons repel one another electrically, in addition to being attracted by the same strong binding force as neutrons. Because the electric force is long-range, but the strong force is short-range, as we add more protons, the spacing between adjacent energy levels in the proton stack grows larger more quickly than the energy level spacing in the neutron stack. The topmost (seventh) proton in $^{13}_{7}\mathrm{N}$ is higher than the topmost (seventh) neutron in $^{13}_{6}\mathrm{C}$. Nitrogen-13 is therefore unstable by beta decay (more accurately, by inverse beta decay) into carbon-13:

$$^{13}_{7}\mathrm{N} \rightarrow {}^{13}_{6}\mathrm{C} + {}^{0}_{1}\bar{\mathrm{e}} + {}^{0}_{0}\nu.$$

Consider now the situation of carbon-14 (on the right). The topmost (eighth) neutron in $^{14}_{6}\mathrm{C}$ is higher than the topmost (seventh) proton in $^{14}_{7}\mathrm{N}$. Carbon-14 is therefore unstable by beta decay into nitrogen-14:

$$^{14}_{6}\mathrm{C} \rightarrow {}^{14}_{7}\mathrm{N} + {}^{\ 0}_{-1}\mathrm{e} + {}^{0}_{0}\bar{\nu}.$$

The half-life of carbon-14 is 5,900 years, and it provides a very useful method of radioactive dating for anthropologists and other scientists (Chapter 18). It should now be clear why quantum effects like to keep the number of protons and neutrons approximately equal in stable atomic nuclei of low to moderate atomic weight. Moreover, given that the spacing between protons grows progressively larger than that between neutrons as the weight of the nucleus grows, we can also understand why very heavy nuclei like lead-206, $^{206}_{82}\mathrm{Pb}$, prefer to have more neutrons than protons, i.e., prefer $A > 2Z$.

To summarize, the bound neutrons and protons of a stable atomic nucleus can no more decay (by the weak interaction) than a bound electron in the shell of an atom can fall to a lower energy level if there is already another electron there. Chapter 3 explained how the resulting electronic shell structures of atoms lead to the physical basis for the periodic table, and therefore to the physical basis for chemistry. The analogy with electronic shell structure is pursued further in so-called "shell models" of atomic nuclei, where one finds "magic nuclei" to be especially stable, much as "noble gases" are especially chemically inert. In nuclear-shell models, one also finds excited states of atomic nuclei to be possible, just as excited states of electrons in atoms are possible. Excited atomic nuclei, however, may deexcite by emitting particles other than photons. In the rest of this book, we will discuss only the ground states of atomic nuclei, unless we explicitly note otherwise.

Thermonuclear reactions

As explained in the preceding, deuterium, in contrast to tritium, is a stable (albeit rare) isotope of hydrogen. Deuterium can be produced from ordinary hydrogen nuclei (protons) by the nuclear reaction

$$^{1}_{1}\mathrm{H} + {}^{1}_{1}\mathrm{H} \rightarrow {}^{2}_{1}\mathrm{H} + {}^{0}_{1}\bar{\mathrm{e}} + {}^{0}_{0}\nu. \qquad (6.9)$$

The appearance of the neutrino on the right-hand side of equation (6.9) shows that the reaction is mediated in part by the weak interaction. Furthermore, the two initial particles on the left-hand side are both positively charged, and thus tend to repel one another electrically (Coulomb repulsion). For the two protons to come close enough together to allow the attractive part of the nuclear strong force to dominate and create a bound nuclear state, they must move toward each other rapidly. (Remember that the nuclear forces are short-range, whereas the electric force is long-range.) The high relative speed necessary to overcome the Coulomb repulsion can be produced artificially in terrestrial accelerators, or it can arise naturally and randomly in stellar interiors, where a high central temperature causes large random thermal motions. The common fusion reactions usually involve charged particles; therefore, they require high temperatures to proceed and are called **thermonuclear reactions**. Fusion reactions involving light particles require only enough thermal energy to overcome the initial Coulomb barrier; once the reaction has been completed, there is a net liberation of energy in accordance with $E = mc^2$. This liberated energy appears as extra kinetic energy of the reaction products. Thus, in reaction (6.9) the deuterium nucleus and the positron leave with kinetic energies much larger than typical for the thermal energies of the reacting protons. In subsequent collisions with other particles in the medium, they tend to share this excess more equitably, leading to a net heating of the surrounding medium. In addition, the positron will eventually annihilate with an existing electron, giving rise to two gamma rays. The gamma rays will interact with the surrounding medium to give additional heating. The neutrino, in contrast, interacts so weakly with the surrounding matter (it is an uncharged lepton) that it usually escapes unimpeded even from the deep interior of a star; so the energy carried off by neutrinos is usually not a source of heat for a star. Except for this (usually minor) correction, the mass deficit of the reaction product (binding energy) becomes the effective heat input into the surrounding medium.

Even at high temperatures, reaction (6.9) occurs very slowly, because the reaction is mediated, in part, by the weak interaction. Therefore, it has a low probability of occurring per collision of two hydrogen nuclei (or as physicists say, the reaction has a small *cross section*). Nevertheless, reaction (6.9) is of primary importance for humans, because it is the first step in a chain of reactions which produces energy in the Sun.

The Proton-Proton Chain

The entire chain of reactions which ultimately produces, in net, one ordinary helium nucleus from hydrogen nuclei in the Sun reads

$$ {}^{1}_{1}\mathrm{H} + {}^{1}_{1}\mathrm{H} \rightarrow {}^{2}_{1}\mathrm{H} + {}^{0}_{1}\bar{e} + {}^{0}_{0}\nu, \tag{6.10} $$

$$ {}^{2}_{1}\mathrm{H} + {}^{1}_{1}\mathrm{H} \rightarrow {}^{3}_{2}\mathrm{He} + {}^{0}_{0}\gamma, \tag{6.11} $$

$$ {}^{3}_{2}\mathrm{He} + {}^{3}_{2}\mathrm{He} \rightarrow {}^{4}_{2}\mathrm{He} + {}^{1}_{1}\mathrm{H} + {}^{1}_{1}\mathrm{H}. \tag{6.12} $$

These reactions, the proton-proton chain, require a few comments.

(1) Reaction (6.10) gives the chain its name; it starts with the reactants, proton + proton.

(2) The symbol ${}^{0}_{0}\gamma$ in reaction (6.11) means a gamma ray, a high-energy photon. It might seem that reaction (6.11) requires nothing other than ${}^{3}_{2}\mathrm{He}$ on the right-hand side, given ${}^{2}_{1}\mathrm{H} + {}^{1}_{1}\mathrm{H}$ on the left-hand side. However, to conserve both energy and linear momentum, two particles cannot generally combine to become one (of a known rest mass); a second product particle is needed. Here the second particle must have zero charge, zero baryon number, and zero lepton number; so it must be a photon. Its presence indicates that both electromagnetism and the strong force are involved.

(3) Reaction (6.12) involves two reactants of charge +2. Why wouldn't it be easier to form ${}^{4}_{2}\mathrm{He}$ by combining ${}^{3}_{2}\mathrm{He}$ with ${}^{1}_{1}\mathrm{H}$, a proton, which is plentiful in stars and has a charge of only +1? Because such a reaction would need the mediation of the weak interaction (Problem 6.7), and so is less likely than reaction (6.12).

Problem 6.7. Fill in the blanks: ${}^{3}_{2}\mathrm{He} + {}^{1}_{1}\mathrm{H} \rightarrow {}^{4}_{2}\mathrm{He} +$ ____ + ____.

---

(4) Clearly, reactions (6.10) and (6.11) must occur twice for each reaction (6.12). Thus, six protons appear on the left-hand sides and two protons and one helium-4 nucleus appear on the right-hand side. In net, four protons (four hydrogen nuclei) have been converted into one helium nucleus. In Problem 6.2 we calculated the net energy input of this reaction by comparing the rest energies of the reactants and the reaction products. We see now that a correction is needed, because the energy released in the form of neutrinos is not utilizable by the Sun; but for the proton-proton chain, the neutrino correction is small.

It is sometimes stated that it is fortunate that the first step of the proton-proton chain involves the weak interaction, because otherwise the Sun would liberate energy so fast it might explode like a bomb. This statement is false. We saw in Chapter 5 that the Sun's luminosity is determined by the rate at which heat leaks out of its interior; the Sun adjusts its *radius* (and thereby its central temperature) to give a steady rate of energy generation just equal to this rate of heat leakage. *If* the first step of the proton-proton chain were easier than the weak interaction actually allows, the Sun would have stopped its contraction from a more diffuse state at a larger radius (and therefore a lower central temperature), so that the energy generation rate would still have equaled the appropriate leakage rate. What would have changed is that we might have lived near a red star instead of a yellow one. A lot of things might then have been different. Biological evolution might have led to eyes most sensitive to the red part of the spectrum instead of the yellow. Biological evolution might have progressed more slowly because the "quality" of sunshine would have been less than that of the real Sun. But we would not have lived near a bomb.

The CNO Cycle

Nevertheless, the above discussion does show that the central temperature in stars like the Sun which run on the proton-proton chain is higher than the charges of the reacting species alone would require. Thus, Bethe was motivated to consider whether reaction schemes "burning" four hydrogens (in net) to get one helium, but involving higher-charge species as **catalysts**, might not be competitive with the proton-proton chain. The most important of these alternative schemes is the CNO cycle:

$$^{12}_{6}C + ^{1}_{1}H \rightarrow ^{13}_{7}N + ^{0}_{0}\gamma, \quad (6.13)$$

$$^{13}_{7}N \rightarrow ^{13}_{6}C + ^{0}_{1}\bar{e} + ^{0}_{0}\nu, \quad (6.14)$$

$$^{13}_{6}C + ^{1}_{1}H \rightarrow ^{14}_{7}N + ^{0}_{0}\gamma, \quad (6.15)$$

$$^{14}_{7}N + ^{1}_{1}H \rightarrow ^{15}_{8}O + ^{0}_{0}\gamma, \quad (6.16)$$

$$^{15}_{8}O \rightarrow ^{15}_{7}N + ^{0}_{1}\bar{e} + ^{0}_{0}\nu, \quad (6.17)$$

$$^{15}_{7}N + ^{1}_{1}H \rightarrow ^{12}_{6}C + ^{4}_{2}He. \quad (6.18)$$

Problem 6.8. Explain the right-hand sides of reactions (6.13) to (6.18).

Two comments are in order here.

(1) Notice that (6.18) recovers the original $^{12}_{6}C$ nucleus; thus, the carbon, nitrogen, and oxygen nuclei (CNO) are unaffected (in net) by the reactions, once these reactions have reached steady state, where each step takes place once to produce (in net) one helium nucleus from four hydrogen nuclei. Thus, the reaction sequence is truly cyclic, and we could have started the cycle anywhere with the addition of a proton to any one of the stable nuclei $^{12}_{6}C$, $^{13}_{6}C$, $^{14}_{7}N$, $^{15}_{7}N$; consequently, the part of the CNO cycle shown above (there are less likely "branches" not shown) is often called the CN cycle.

(2) The beta decays at (6.14) and (6.17) involve nuclei in which the protons outnumber the neutrons; so they require the mediation of the weak interaction. Why don't these steps hold up the cycle as does the first step of the proton-proton chain? The answer is that the beta decays involve a completely assembled nucleus: once $^{13}_{7}N$ or $^{15}_{8}O$ have been created by the previous strong-interaction step, they will decay sooner or later by the weak interaction; one need only wait the requisite 870 and 178 sec (see Figure 6.6). In contrast, in the proton-proton reaction, the weak interaction must work in the short span of time that the two protons are whizzing by each other. This is what makes that reaction so improbable. This is also why the CNO cycle can become the dominant source of hydrogen burning in stars whose central temperatures are just a little hotter than that of the Sun, despite the need for nuclei with larger atomic numbers than are needed for the proton-proton chain.

Temperature Sensitivity of Thermonuclear Reactions

We have seen that the microscopic details of the various thermonuclear processes inside stars can become quite complicated. However, the basic *macroscopic* tendency is clear. From a thermodynamic point of view, the macroscopic tendency is merely another application of the statistical-mechanical tendency for matter that we stated in Box 4.3:

> At relatively low temperatures, a material system prefers more binding energy.

Although a temperature of 15 million K in the center of the Sun sounds awfully high by terrestrial standards, it corresponds to thermal energies which are much smaller than typical nuclear-binding energies. Thus, the matter in the Sun's core would prefer to be in the more-bound nuclear state of helium than in the less-bound form of hydrogen. Statistically, of course, this is true all over the Sun, and not only at the core, but only at the core do the temperatures rise high enough for the reacting charged particles to overcome the repulsive Coulomb forces. In fact, even at the Sun's center, the reacting particles are forced to "tunnel" under the Coulomb barrier (instead of "leaping" over it) to reach the nuclear distances where attractive forces dominate. Higher-energy particles can tunnel more vigorously, but they are rarer in a given thermal distribution. Higher temperatures imply, therefore, greater numbers of particles capable of "effective tunneling." This temperature sensitivity explains why nuclear reactions now proceed only at the core of the Sun.

Problem 6.9. The encounter of two nuclei of charges q_1 and q_2 and masses m_1 and m_2 can be analyzed in terms of a single particle of reduced mass $m = m_1m_2/(m_1 + m_2)$ moving at relative velocity v in the mutually repulsive Coulomb field. Let v be the relative speed at infinite separation. Since the electric potential energy is zero at infinity, argue that the total energy of the reduced mass is $mv^2/2$. Show that according to classical mechanics,

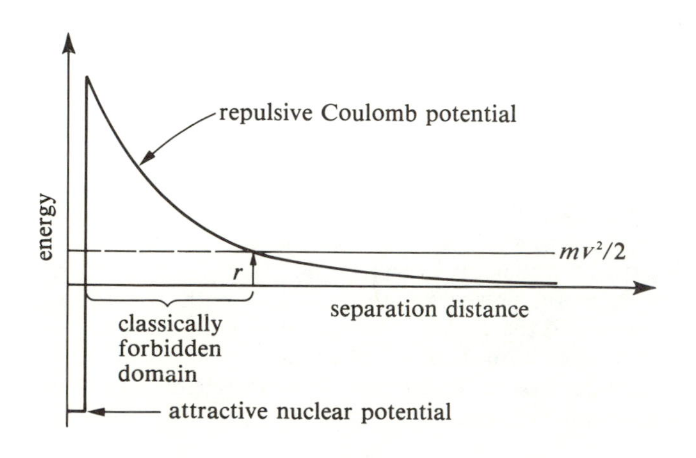

the two masses cannot approach closer than the distance r, where the electric potential energy q_1q_2/r equals $mv^2/2$; i.e., the reduced mass encounters a "barrier" at

$$r = \frac{2q_1q_2}{mv^2}.$$

Quantum mechanics allows the particle to "tunnel" closer than r, but the probability of reaching nuclear separations (assumed to be small compared to r) decreases exponentially with the ratio of r to the de Broglie wavelength $\lambda = h/mv$ associated with the momentum mv of the particle at infinity. In fact, Gamow found that the

$$\text{penetration probability} \propto \exp(-2\pi^2 r/\lambda) \propto \exp(-4\pi^2 q_1q_2/hv).$$

The probability for a nuclear reaction is proportional to the penetration probability multiplied by the nuclear "cross section" once the particle has penetrated to nuclear distances.

From the above formula, we see that the faster the relative speed v, the higher is the penetration probability. However, in conditions of local thermodynamic equilibrium at temperature T, there are few particles with high relative speeds v. For a classical gas, the probability of having the relative speed v is given by the Maxwell-Boltzmann distribution (equation 4.1),

$$\text{probability of relative speed } v \propto \exp(-mv^2/2kT).$$

The net probability of coming within nuclear distances is therefore proportional to the product of two exponentials: one which increases with increasing v, the other which decreases with increasing v. Clearly, the total probability is maximized when the arguments of the two exponentials are comparable with each other (calculus provides a factor of 2), i.e., when

$$v = (4\pi^2 q_1q_2kT/hm)^{1/3}.$$

Compute the corresponding numerical value of r for the proton-proton reaction at $T = 1.5 \times 10^7$ K when v has this "most effective" value for nuclear reactions. How does r compare with 10^{-13} cm?

The rate of nuclear reactions is proportional to the maximum value of the net probability. Show that this maximum value equals

$$\exp[-\tfrac{3}{2}(4\pi^2 q_1q_2/h)^{2/3}(m/kT)^{1/3}].$$

Notice that this last expression is higher for higher values of T and for lower values of the charges and masses. Define and calculate the quantity

$$T_0 = (3/2)^3(4\pi^2 q_1q_2/h)^2(m/k)$$

for hydrogen reactions. Plot the resulting probability factor $\exp[-(T_0/T)^{1/3}]$ as a function of T from $T = 10^5$ K to $T = 10^8$ K. Draw appropriate conclusions concerning the temperature sensitivity of thermonuclear reactions.

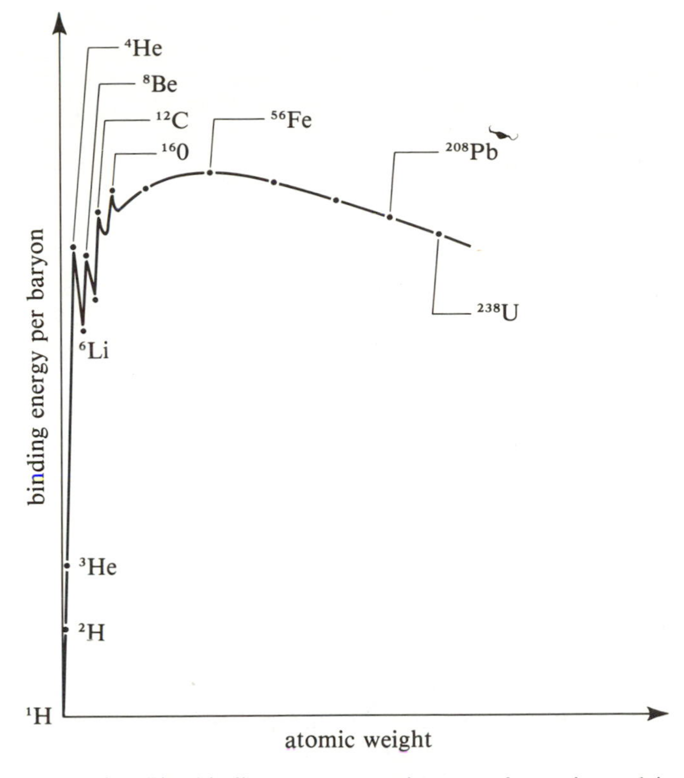

Figure 6.7. The binding energy per baryon of atomic nuclei. Notice the high binding energy per baryon of helium-4 nuclei. Helium-4 has about 7/8 of the binding energy per baryon of iron-56.

Binding Energies of Atomic Nuclei

Now, if it is advantageous for hydrogen to become helium, is it also advantageous for helium to become something heavier, and so forth and so on? Yes and no. Figure 6.7 plots the binding energy per baryon of a nucleus against the atomic weight (the total number of baryons) of that nucleus. From this plot, we see that deuterium ^{2}H has more binding energy per baryon than ordinary hydrogen ^{1}H; that helium-3 has more than deuterium; and that helium-4 has more than helium-3. These facts explain why the reactions of the proton-proton chain are driven in the direction they go in.

However, Figure 6.7 illustrates two competing tendencies. First, it is true that Figure 6.7 shows a *general* tendency for the binding energy per baryon to become greater and greater as we go to heavier and heavier nuclei. However, this tendency reverses past iron-56, ^{56}Fe. Second, even the first general tendency is offset for helium-4 by the fact that the stable nuclei immediately heavier than helium-4—such as lithium-6, ^{6}Li—actually contain *less* binding energy per baryon than helium-4. Let us comment on these features before we address the question of how the elements heavier than helium-4 in the periodic table were produced.

The *general* tendency for the binding energy per baryon to increase until iron and to decrease thereafter can be explained in relatively simple terms. The structure of the nucleus of an atom can be understood roughly as a balance between the attractive and repulsive parts of the strong nuclear force between protons and neutrons, modified by the electric repulsion between the protons. Let us first consider the role of the attractive-repulsive duality of the strong nuclear force. As we mentioned earlier, this duality is shared by the van der Waals force that is responsible, for example, for many of the properties of liquid water. Water makes a phase transition from a vapor to a liquid (see Figure 5.5) when the individual molecules have been pushed close enough for the attractive part of the van der Waals force to bind many, many molecules into a single droplet. Once water has become a liquid, the attractive and repulsive parts of the van der Waals force quickly reach an equilibrium with the internal pressure of the droplet. The resulting equilibrium causes liquid water to maintain a nearly constant density throughout a wide range of physical conditions. The very short-range repulsive part of the van der Waals force rises steeply with decreasing molecular separation, and keeps liquid water very incompressible if one tries to decrease the droplet volume by external means. Left to itself, a liquid drop will become spherical in shape, because the net attractive part of the van der Waals force prefers the least surface area for a given volume (the distinguishing property of a sphere). This geometry maximizes the binding energy, since then the number of particles missing an attractive neighbor on one side is minimized. The macroscopic manifestation of this microscopic effect is called "surface tension." Many of the same considerations apply on a fantastically reduced scale to the nucleus of a complex atom. Indeed, a very successful model for the structure of atomic nuclei invented by Bohr is called the "liquid-drop model."

Consider now the balance between the net attraction of the strong nuclear force and the repulsion of the positively charged protons. As long as the nucleus is not too big, the nuclear force, being short-range, tends to win over the repulsion of the electric force. It is then advantageous to add another baryon, which can be either a proton or a neutron. On the average, half of these will be protons, for the reasons outlined in Figure 6.6. As long as the nuclear force dominates, it is advantageous for light nuclei to add baryons and become heavier. However, they do so by maintaining nearly constant nuclear density; therefore the heavier nuclei also become physically bigger. The increase in nuclear binding energy per baryon therefore saturates with iron-56; its nucleus is the largest in which the constructive short-range strong force has any advantage over the destructive long-range electric force. The further addition of baryons leads, actually, to a loss of binding energy per baryon; i.e., binding energy per baryon reaches a maximum with iron-56 and declines thereafter. This has a number of very important consequences.

First, very heavy nuclei, like uranium, are quite unstable because they contain so many baryons. They occasionally find it advantageous to "spit out" a helium-4 nucleus (also called an "alpha particle") to help alleviate their problems. A helium-4 nucleus can be spit out intact because of its relatively high binding energy per baryon (see Figure 6.7). This mode of radioactive decay is called alpha decay. By a series of alpha and beta decays, in fact, uranium will decay to lead, a process which is very useful for the radioactive dating of terrestrial rocks (Chapter 18). This instability of very heavy nuclei explains why past a certain point they are not found to occur naturally. In fact, uranium, with an atomic number of 92, is the heaviest naturally occurring element found on Earth. All the heavier elements listed in the periodic table are made by humans. If any of these originally existed on Earth, they would long have since decayed to some lighter species.

Second, if a very heavy nucleus has a neutron added to it (which it will absorb without any intermediate electric repulsion), the nucleus may become so unstable that it wants to split into two more-or-less equal pieces, a process called nuclear *fission.* If the fission process itself releases more free neutrons (because heavy nuclei tend to be more neutron-rich than smaller fragments), then obviously one could produce a **chain reaction** if a **critical mass** of the fissionable material is gathered together. Uncontrolled fission chain reactions lead to an A-bomb; controlled fission chain reactions, to a nuclear reactor. The health danger of the radioactive wastes from fission reactors results primarily from the long decay lifetimes of some of these wastes that are toxic if ingested or inhaled. Gamma-ray emission is quite penetrating, but alpha particles are stopped by thick rubber gloves.

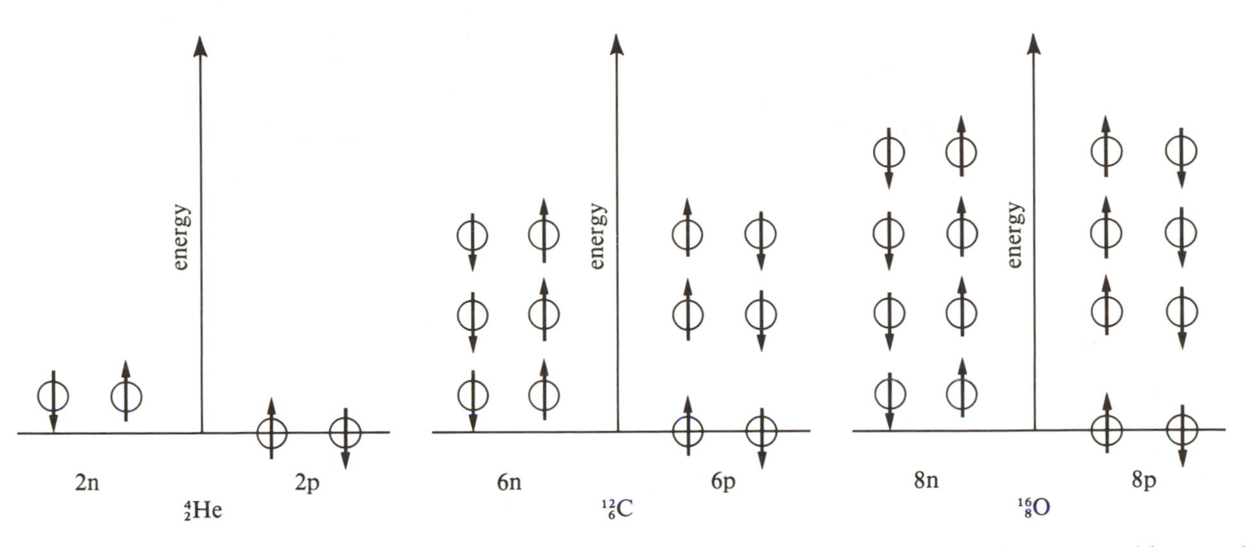

Figure 6.8. Because neutrons and protons with oppositely aligned spins can stack two at a time, atomic nuclei with even atomic weight and even atomic number are especially stable.

Third, the fact that the binding energy per baryon reaches a peak with iron has the following general consequence for the evolution of (high-mass) stars. Once a star has produced iron from the fusion of lighter elements, it is in deep, deep trouble. No further nuclear reactions can *release* energy to offset the steady drain of heat flowing from the stellar surface. Iron is the ultimate slag heap of the universe. No matter how hot a star's interior may get, its ultimate supply of nuclear energy is finite, because it cannot extract nuclear energy from iron. A low-mass star may have even less nuclear energy available, because its interior may never achieve the conditions necessary to "burn" hydrogen all the way to iron. The resolution of these dilemmas for the stars is a topic we leave for Chapters 7 and 8.

We can now see that it is generally advantageous for a star to create heavier and heavier elements. However, we are left with two obstacles. First, having seen how this works to create helium-4 from ordinary hydrogen, we arrive at the impasse noted at the beginning of this subsection: ^{4}He is more bound than its immediate neighbors. The existence of this impasse has a simple explanation.

In the discussion of Figure 6.6, we simplified matters by ignoring the fact that protons and neutrons can actually stack two at a time (with opposed spins) in any energy level (Figure 6.8). This fact implies that nuclei having even numbers of protons and neutrons are more stable than their immediate neighbors. Thus, helium-4 (2 protons and 2 neutrons), beryllium-8 (4 protons and 4 neutrons), carbon-12 (6 protons and 6 neutrons), oxygen-16 (8 protons and 8 neutrons), etc, all form local peaks in the nuclear binding energy diagram of Figure 6.7. Indeed, helium-4 is so stable that two separate helium-4 nuclei contain more binding energy than one beryllium-8 nucleus. How do helium reactions inside stars overcome this barrier to the formation of heavier and heavier elements? Second, how are the elements beyond the iron peak created if thermonuclear fusion stops being advantageous? Let us address these questions one at a time.

The Triple Alpha Reaction

Edwin Salpeter and Fred Hoyle were the astrophysicists who resolved the helium puzzle. Salpeter proposed that two helium-4 nuclei could come together to form a relatively unstable beryllium-8 nucleus:

$$^4_2\text{He} + ^4_2\text{He} \rightarrow ^8_4\text{Be} + ^0_0\gamma.$$

The beryllium-8 nucleus will alpha-decay into two helium-4 nuclei in a typical time of 2.6×10^{-16} sec (Problem 6.9). Salpeter showed that there is a small but nonnegligible fraction of beryllium-8 nuclei when the formation-destruction reactions have come into equilibrium at sufficiently high temperatures and densities. (For example, a helium gas at $T = 10^8$ K and $\rho = 10^5$ gm/cm^3 has about one beryllium-8 nucleus for every 10^9 helium-4 nuclei.) He therefore conjectured that once in a while a third helium-4 nucleus could fuse with the beryllium-8 nucleus to form a carbon-12 nucleus:

$$^8_4\text{Be} + ^4_2\text{He} \rightarrow ^{12}_6\text{C}^* + ^0_0\gamma.$$

Carbon-12 is a stable nucleus and, therefore, could complete the reaction sequence (actually in an excited nuclear state; that's the reason for the asterisk). In net, three helium-4 nuclei (or three alpha particles) have come together to form one carbon nucleus (with the emission of two gamma rays). This net result gives the process its name: the triple alpha reaction (Box 6.2).

Given the two-stage nature of the triple alpha reaction, Hoyle pointed out that the reaction rate would be prohibitively slow at *reasonable* stellar-interior temperatures and densities unless the formation of carbon-12 occurs with higher than normal probability (i.e., higher than normal "cross section"). Hoyle therefore proposed that the reaction had to be a "resonant" one where the formation of carbon-12 occurs in an excited nuclear state, with the energy level located above the ground level of the reactants by an amount very nearly equal to the "most effective energy for tunneling" at temperatures of about 10^8 K (Figure 6.9). At the time of Hoyle's suggestion, such an excited state of carbon-12 was not experimentally known. The subsequent experimental verification that such a state exists was therefore a glorious triumph for theoretical nuclear astrophysics.

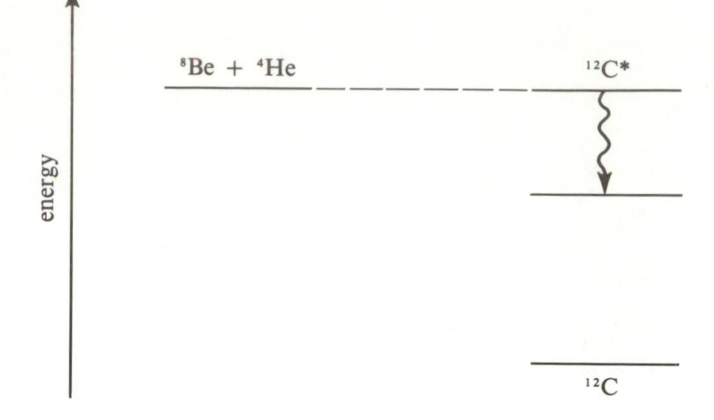

Figure 6.9. Energy-level diagram for the production of carbon-12 from the fusion of beryllium-8 and helium-4. A beryllium-8 nucleus may combine with a helium-4 nucleus to give an excited state of carbon-12. (Nuclear states have "shell structure" similar to the atomic "shell structure" discussed in Chapter 3.) Usually the excited carbon-12 nucleus will spit the helium-4 nucleus back out (alpha decay), but occasionally it will deexcite radiatively via the emission of a gamma-ray, eventually cascading to the ground level of carbon-12.

Problem 6.10. Does the alpha decay of beryllium-8 proceed by the weak interaction or by the strong interaction?

Hoyle has made the following philosphical comment about his discovery. The existence of such an excited nuclear state of carbon-12 results from several small accidents of nuclear physics. No fundamental law of nature requires—as far as we know—the existence of such an excited state with just the right properties to enable the triple alpha reaction to proceed under reasonable stellar conditions. Does this mean that the creation of carbon, which is the basis of life on Earth, would have been impossible except for this accidental coincidence of nuclear physics? In Chapters 7 and 8, we shall argue that the answer to this question is no. Independent of the minor details of microscopic nuclear physics, the general *macroscopic* conditions inside stars of sufficiently high mass will *inevitably* lead to the conditions favorable for producing the synthesis of chemical elements up to the iron peak. The key phrase in this issue is "reasonable stellar conditions." In fact, we have no *a priori* expectations of what constitutes "reasonable stellar conditions" except from *observations* of real stars. Such observations automatically take into account the many quirks of nuclear physics. Had the actual nuclear physics been different, the stars would also have been different.

BOX 6.2
The Triple Alpha Reaction

The triple alpha reaction is a two stage process which allows the burning of helium into heavier elements:

$$ {}^4_2\text{He} + {}^4_2\text{He} \rightarrow {}^8_4\text{Be} + {}^0_0\gamma, $$
$$ {}^8_4\text{Be} + {}^4_2\text{He} \rightarrow {}^{12}_6\text{C}^* + {}^0_0\gamma. $$

The formation of carbon-12 in an excited nuclear state constitutes a "resonant" reaction at stellar temperatures of about 10^8 K.

The General Pattern of Thermonuclear Fusion

Once the leap-frogging of helium to carbon has taken place, the way is clear, in principle, to the synthesis of the rest of the chemical elements in the periodic table up to iron. For example, once carbon has been created from helium, the carbon nuclei may sometimes combine with some of the helium nuclei to form oxygen:

$$ {}^4_2\text{He} + {}^{12}_6\text{C} \rightarrow {}^{16}_8\text{O} + {}^0_0\gamma. $$

Alternatively, the carbon nuclei may fuse with themselves to produce magnesium:

$$ {}^{12}_6\text{C} + {}^{12}_6\text{C} \rightarrow {}^{24}_{12}\text{Mg} + {}^0_0\gamma. $$

Clearly, a great variety of such reactions are possible once enough elements have already been synthesized.

Modern nuclear astrophysics is infamous for the complex network of reactions which have to be considered in realistic situations. For this book, we wish to draw attention only to some main themes, the most important of which is the following.

Heavier elements generally have larger electric charges. To fuse them into yet heavier elements requires overcoming greater Coulomb barriers than for elements with small electric charges; so such reactions will not be initiated until the star acquires higher temperatures. Thus, helium reactions generally require higher temperatures than hydrogen reactions; carbon reactions, higher temperatures than helium reactions, etc. This pattern is a general feature of thermonuclear reactions of charged nuclei inside stars:

> To burn heavier and heavier elements by thermonuclear reactions generally requires higher and higher temperatures.

As we shall see in Chapters 7 and 8, gravity and the second law of thermodynamics conspire to produce rounds of higher and higher temepratures inside stars. And in stars of high-enough mass, these two processes can produce almost arbitrarily high temperatures. Astronomers therefore believe that virtually all the heavy elements in our bodies were made originally deep inside one or more relatively heavy stars, which subsequently spewed out this material in titanic explosions. Eventually this enriched material was incorporated into the formation of the solar system, and into the making of our bodies. Thus, in a fundamental sense, we are all made of star stuff.

The *r* and *s* Processes

There is an exception to the temperature rule if a source of free neutrons is present. Since a neutron is electrically neutral, even a slow one may be absorbed by an atomic nucleus, making the latter heavier by one atomic mass unit (one baryon). The absorption of too many neutrons (and one may be one too many) may make an atomic nucleus unstable against beta decay. Whether the actual nucleus will beta-decay (converting a nuclear neutron into a proton) or capture another neutron before it has a chance to beta-decay is the distinguishing feature of the so-called *s* and *r* processes, as propounded by Burbidge, Burbidge, Fowler, and Hoyle (Figure 6.10).

First, what is the source of free neutrons? A few nuclear reactions between charged nuclei will produce free neutrons as an end product. The simplest example is the fusion of deuterium and tritium to produce helium:

$$^{2}_{1}\mathrm{H} + ^{3}_{1}\mathrm{H} \rightarrow ^{4}_{2}\mathrm{He} + ^{1}_{0}\mathrm{n}. \qquad (6.19)$$

Although the fusion reaction (6.19) is of great practical importance (Box 6.3), it is relatively unimportant in most astrophysical contexts. Tritium does not exist naturally, being radioactive and having a half-life of only 12.5 years; deuterium is cosmically relatively rare, having an abundance of about 2×10^{-5} compared to ordinary hydrogen. Deuterium is, of course, made in the first step of the proton-proton chain inside stars, but essentially all that deuterium is destroyed in subsequent interactions with protons. However, other nuclear reactions can also produce free neutrons. For example, a fraction of all helium-4 and carbon-13 collisions or oxygen-16 and oxygen-16 collisions will release a neutron:

$$^{4}_{2}\mathrm{He} + ^{13}_{6}\mathrm{C} \rightarrow ^{16}_{8}\mathrm{O} + ^{1}_{0}\mathrm{n},$$
$$^{16}_{8}\mathrm{O} + ^{16}_{8}\mathrm{O} \rightarrow ^{31}_{16}\mathrm{S} + ^{1}_{0}\mathrm{n}.$$

The free neutrons produced in these (and other) ways may subsequently be captured by other heavy nuclei to produce elements (or isotopes of elements) which cannot be synthesized by the ordinary charged-nuclei fusion reactions. In particular, neutron capture offers a natural means of forming elements *beyond the iron peak.*

BOX 6.3

Hydrogen Fusion as an Energy Source

Reaction (6.19) is central to the hopes of humanity for fusion as a terrestrial energy source. The charges of the reacting species are small (+1 each), and the liberation of energy per reaction is relatively large. Nevertheless, with present-day confinement techniques (magnetic or inertial), the density of the heavy hydrogen plasma and its confinement time are relatively low. To reach the "breakeven point" where the reactions produce more energy than is put in to get the reactions going under present-day laboratory conditions requires the plasma to be heated up to temperatures in excess of 100 million K. This goal is now within sight, but significant engineering problems remain even if "scientific feasibility" should be demonstrated. A serious difficulty is the following. The liberation of free neutrons in reaction (6.19) constitutes no great problem in itself. However, the neutrons so liberated are free to strike the metal walls of the reactor vessel. This weakens the walls eventually, and also makes them radioactive. Thus the walls must be periodically dismantled and disposed of; the seriousness of this procedure, for both health and economics, varies with the different fusion schemes. For this reason, it would probably be wise to continue the search for a viable fusion scheme along many fronts instead of settling for the one which demonstrates "scientific feasibility" first.

Problem 6.11. If tritium does not occur naturally, where does the tritium come from that fusion reactors (and the first H-bomb) use as fuel? *Hint*: How would you make

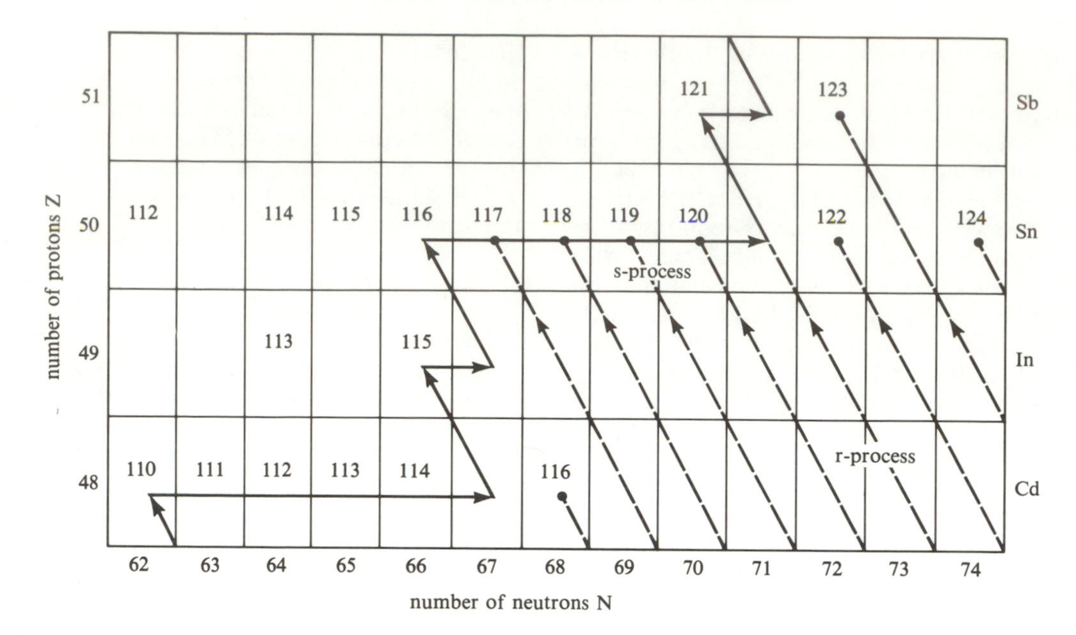

Figure 6.10. The synthesis of elements beyond the iron peak. The r-process and s-process contribute different isotopes of cadmium (Cd), indium (In), tin (Sn), and antimony (Sb). Neutron capture (horizontal line segments) followed by beta-decay (diagonal line segments) of unstable isotopes (not labelled by A = N + Z) characterizes the s-process. The r-process results when heavy bombardment with neutrons make very neutron-rich nuclei that subsequently beta-decay to stable isotopes. For example, tin has 10 stable isotopes, of which A = 116 to 120 lie on the s-process path (but 117 to 120 could also be made from the r-process), while A = 122 and 124 can only be made from r-process. Tin-112, 114, and 115 are believed to be made from a much rarer third process, called the p-process. (Adapted from Clayton, Fowler, Hull, and Zimmerman 1961, *Ann. Phys.*, *12*, 331.)

tritium, say, from deuterium? (Lithium is actually more practical.) Can you now see how a hybrid fission-fusion scheme might work?

In ordinary circumstances inside stars, the rate of production of free neutrons is relatively *slow*; consequently, the element formed by neutron capture will generally have a chance to beta-decay if it is unstable ("radioactive"). The sequence of elements formed this way—by neutron capture followed by beta decay—are called ***s*-process elements** (*s* for "slow"). In the extraordinary circumstances which accompany a supernova event, the rate of production of free neutrons may become very *rapid*; consequently, the element formed by neutron capture will not generally have a chance to beta-decay before it captures another neutron (and another and another). After the period of neutron production, of course, the synthesized very heavy elements will have a chance to decay (by the emission of alpha and beta particles). But the sequence of elements formed this way—called the ***r*-process elements** (*r* for "rapid")—will generally differ isotopically from the sequence formed by the *s*-process. (Some overlap will exist for certain isotopes.) A major part of nuclear astrophysics concerns itself with the reconstruction of the history of such synthesis events in the Galaxy to see if we can understand the observed abundances of the elements in the periodic table (Figure 6.11).

Apart from hydrogen and helium, the elements which lie in the "mainstream" of thermonuclear fusion, i.e., carbon, nitrogen, and oxygen (CNO), are the most abundant in Figure 6.11. This lends credence to the idea that the synthesis of the elements heavier than helium took place mostly inside stars. (The light elements like lithium, beryllium, and boron are believed to have a different origin, either in the "big bang" or in cosmic-ray interactions with the interstellar medium.) We also see the iron-peak elements to be relatively abundant, a natural expectation given the energetic desirability of the formation of such elements in the interiors of massive stars. Finally, we see that radioactive elements like uranium

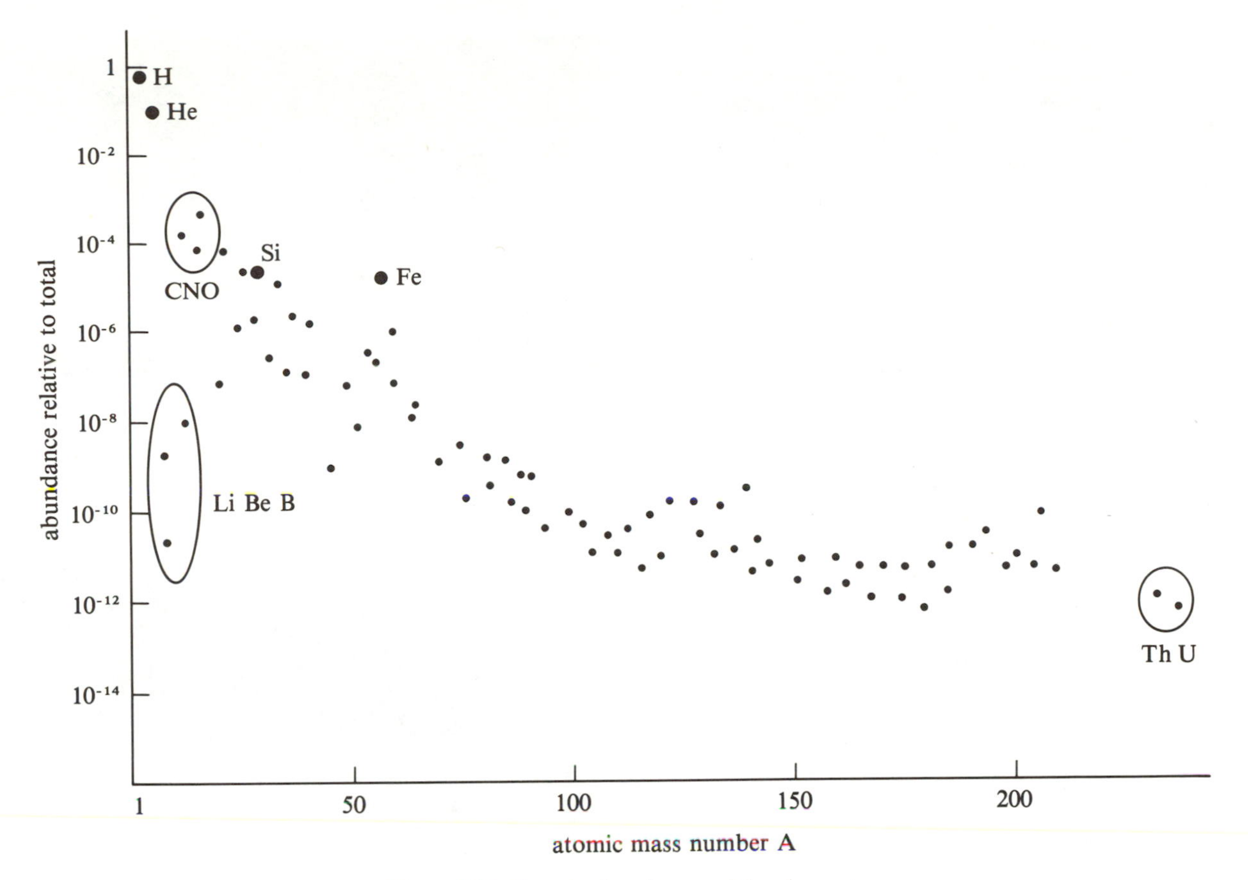

Figure 6.11. Cosmic abundances of the elements.

and thorium, while rare, are not absent in the natural environment. *The presence of such unstable nuclei proves that the synthesis of some of the chemical elements, at least, cannot have taken place infinitely long ago.* (Otherwise, these radioactive elements would long since have disappeared.) This realization allows scientists to "date" the formation of these radioactive elements (Chapter 18).

The Solar-Neutrino Experiment

As we have seen, there is very strong circumstantial evidence that the basic energy source inside ordinary stars is thermonuclear fusion, and that this process would explain how the heavier elements in the periodic table came into existence. Nevertheless, it would be nice to see some more direct evidence of the fusion process inside stars, and in particular inside the Sun. Desire for such evidence motivated proposals by Pontecorvo, by Alvarez, and by Bahcall for a number of solar neutrino experiments, one of which has been carried out by Raymond Davis, Jr.

The idea is the following. Of all the products of thermonuclear fusion reactions, the only one which can reach us directly from the core of the Sun is the neutrino. Only the neutrino flies out in a straight line from the center of the Sun (because it interacts only weakly with matter), and only it can carry direct information about the conditions there. Unfortunately, the very weak interaction of neutrinos with matter also makes them hard to detect experimentally. Only the highest-energy neutrinos can be detected with present-day resources.

Problem 6.12. The opacity κ of most materials for neutrino capture is on the order of 10^{-20} cm^2/gm (cross section of about 10^{-44} cm^2 per nucleus). Calculate the mean free path $l = 1/\rho\kappa$ for neutrino capture at the center of the Sun, where $\rho \sim 10^2$ gm/cm^3. Convert your answer to light-years and compare with the radius of the Sun.

Consider, for example, the neutrinos produced by the first step of the proton-proton chain,

$$^1_1\text{H} + {}^1_1\text{H} \rightarrow {}^2_1\text{H} + {}^0_1\bar{\text{e}} + {}^0_0\nu.$$

Such neutrinos emerge from the reaction with relatively little energy, because they must share the small mass-deficit energy of the reaction with the light positrons.

These low-energy neutrinos, although copiously produced, could be detected only by a vast amount of exotic material (to be discussed later). A slightly better situation holds for a variation of the above reaction, wherein the positron is transposed to the left-hand side to become an electron:

$$ {}^{1}_{1}\mathrm{H} + {}^{\ 0}_{-1}\mathrm{e} + {}^{1}_{1}\mathrm{H} \rightarrow {}^{2}_{1}\mathrm{H} + {}^{0}_{0}\nu. $$

This three-body encounter between a *p*roton, an *e*lectron, and a *p*roton, is called the "pep" reaction, and it produces a somewhat higher energy neutrino than the proton-proton reaction. However, the "pep" reaction occurs about 400 times less frequently than the proton-proton reaction, which involves a two-body encounter.

Davis's experiment was designed to detect the neutrinos produced by a even less likely branch for producing helium. When helium-3 is produced by the second step of the proton-proton chain (equation 6.11), it may combine with helium-4 to form beryllium-7; the beryllium-7 may then combine with a proton to form boron-8; boron-8 is unstable, and will beta-decay into beryllium-8, which in turn alpha-decays into two helium-4 nuclei:

$$ {}^{3}_{2}\mathrm{He} + {}^{4}_{2}\mathrm{He} \rightarrow {}^{7}_{4}\mathrm{Be} + {}^{0}_{0}\gamma, $$
$$ {}^{7}_{4}\mathrm{Be} + {}^{1}_{1}\mathrm{H} \rightarrow {}^{8}_{5}\mathrm{B} + {}^{0}_{0}\gamma, $$
$$ {}^{8}_{5}\mathrm{B} \rightarrow {}^{8}_{4}\mathrm{Be} + {}^{0}_{1}\bar{\mathrm{e}} + {}^{0}_{0}\nu, $$
$$ {}^{8}_{4}\mathrm{Be} \rightarrow {}^{4}_{2}\mathrm{He} + {}^{4}_{2}\mathrm{He}. $$

Here, we need only note that ${}^{8}_{5}\mathrm{B}$ is a highly unstable nucleus; so the neutrino emitted in the third step of the above is quite energetic. These neutrinos interact sufficiently with chlorine-37 to lead to a viable detection scheme. The actual experiment involved putting 100,000 gallons of cleaning fluid (C_2Cl_4) one mile underground in the Homestake Gold Mine (to provide adequate shielding against non-neutrino events) and trying to detect the few dozen capture events that would occur over every few months! The capture of a solar neutrino by a chlorine-37 nucleus in the cleaning fluid would produce a radioactive argon-37 atom, via the reaction

$$ {}^{37}_{17}\mathrm{Cl} + {}^{0}_{0}\nu \rightarrow {}^{37}_{18}\mathrm{Ar} + {}^{\ 0}_{-1}\mathrm{e}. $$

The argon can be flushed out of the cleaning-fluid tank, and the radioactive atoms can be "counted" by the following method. Argon-37 captures one of its inner-shell electrons via the reaction

$$ {}^{37}_{18}\mathrm{Ar} + {}^{\ 0}_{-1}\mathrm{e} \rightarrow {}^{37}_{17}\mathrm{Cl} + {}^{0}_{0}\nu, $$

which leaves the chlorine-37 atom in an excited *electronic* state. This capture does not occur via "action at a distance" (which special relativity forbids), since quantum mechanics allows a small but finite probability for an orbital electron to be found inside the nucleus. In any case, after such a capture, the excited chlorine atom will deexcite by emitting photons that can be detected by standard techniques (see Figure 6.12).

Figure 6.12. Raymond Davis, Jr. (top) and the solar neutrino experiment. A tank containing 100,000 gallons of cleaning fluid is placed in a deep mine and used as a neutrino detector. (Brookhaven National Laboratory.)

Davis's findings are astonishing. At the time I am writing, Davis's results implied a rate of nuclear reactions inside the Sun which is only about a fourth of that required to maintain the present luminosity of the Sun! Does this mean that the main energy source of the Sun is not fusion reactions after all? Probably not, but the alternative explanations are complex.

First, the experiments might be in error. This possibility is remote, given the care with which Davis has carried out his experiments.

Second, perhaps we understand nuclear reactions or neutrino physics less well than we think. It has been proposed that the neutrino may spend time oscillating between three states: the electron neutrino, the muon neutrino, and the tau neutrino. If so, Davis's experiments may be detecting only the fraction of neutrinos (about half of them) that are in the electron-neutrino state. For such oscillations to be possible, the neutrino must have a nonzero rest mass, a deduction which has important implications for cosmology (Chapter 16). But this possibility would explain only half the divegence from the expected value.

Third according to Bahcall, another great uncertainty may exist in nuclear physics. This uncertainty comes from the necessity of extrapolating to very low energies the measured reaction rates for

$$^{3}_{2}\text{He} + ^{4}_{2}\text{He} \rightarrow ^{7}_{4}\text{Be} + ^{0}_{0}\gamma.$$

Reactions closer to the main chain suffer less uncertainty, because they can be tied more closely to the *observed* solar luminosity. For example, the neutrinos released from the proton-proton reaction (6.10) are known with almost absolute certainty as long as nuclear reactions do provide the basic power of the Sun. Thus, it would be important to check on the neutrinos which are produced by the proton-proton reaction and by the "pep" reaction. This is why astrophysicists are anxious to conduct solar-neutrino experiments using indium, gallium, or lithium ($^{125}_{49}$In, $^{71}_{31}$Ga, or $^{7}_{3}$Li) at the detecting medium. Many tons are needed just to capture one neutrino per day! Looking for a needle in a haystack is easier work than hunting for solar neutrinos.

Fourth, perhaps we understand the input physics of stellar structure and evolution calculations less well than we think. This is especially true concerning calculations of stellar opacities; however, improved opacity calculations have generally worsened the solar-neutrino discrepancy, not improved it. Nevertheless, the neutrinos Davis is trying to detect come from a insignificant side branch of the main proton-proton chain that is believed to be responsible for the energy generation in the Sun. This side branch could have a rate very different from the theoretically calculated rate without significantly affecting the observed luminosity of the Sun. Moreover, this side branch is very sensitive to the exact temperature of the center of the Sun, and it is conceivable that present-day solar models have significant remaining inaccuracies concerning opacities, nuclear-reaction cross sections, etc.

Of course, the Sun may temporarily be not generating as much nuclear energy in its core as it is leaking photons from its surface. For example, it has been suggested that the Sun may undergo long-term oscillations wherein it periodically underproduces and overproduces nuclear energy in comparison with its average value. Such cycles may even be related to the cycles of ice ages on Earth. In this view, we happen to live in an era of underproduction. No reasonable mechanism has been suggested, however, that could produce such long-term oscillations. John Eddy has claimed that an examination of the historical records on solar eclipses and planetary transit times across the disk shows that the Sun has been shrinking, but Irwin Shapiro and his colleagues in a reanalysis find no such effect. It has even been suggested that the Sun contains a small black hole at its center (Chapter 7 explains black holes), and that the gravitational swallowing of matter by this black hole accounts for a substantial part of the energy release from the center. The problem with all such suggestions is that, apart from making the Sun very different from the other stars, they have a difficult time explaining what we know from the biological and geological record: that the solar luminosity has been more or less steady during at least the last 3.5 billion years.

We see, therefore, that despite intense study for the past few decades, we still do not completely understand the interior of the Sun. Very precise measurements of possible variations in the solar luminosity from outer space would be helpful. Such measurements would not only have astronomical importance, but would also help scientists evaluate the importance of changes in the Sun for changes in the global climate. Purely scientific goals frequently have important practical side effects.

Speculations about the Future

We have come a long way since the nineteenth century, when scientists thought that the chemical elements were immutable; that is, although you may combine hydrogen atoms with oxygen atoms to form water molecules, the hydrogen and oxygen atoms could be recovered by other chemical or physical means. We now know that the conservation of chemical elements is only an approximate law, not an exact one. Chemical elements are conserved when only electromagnetic forces are involved; they can be transmuted by the introduction of nuclear forces.

Moreover, even the building blocks of the nucleus of the atom—the proton and the neutron—are not inviolate. The weak interaction can transform a neutron into a proton (and vice versa). The constituent quarks of the proton and the neutron maintain their "flavors" ("up" or "down") when only strong forces ("color" forces, which are mediated by gluons) are involved; the quark flavors can be changed by the introduction of weak forces.

We have used these guiding principles—plus the notion of the conservation of baryons and leptons—to outline a theory of nuclear reactions. This theory is perfectly adequate to describe the energy regimes encountered in the interiors of stars and in present-day terrestrial accelerators. However, at higher energies is it possible that baryons and leptons might not be conserved? The trend of modern particle physics answers yes.

We have mentioned that physicists now believe that they have succeeded in combining the theories of electromagnetic and weak interactions; this would mean that, at a deeper level, the electromagnetic force and the weak force are fundamentally intertwined in a single framework. The theories currently being pursued to unify the electromagnetic-weak force with the strong

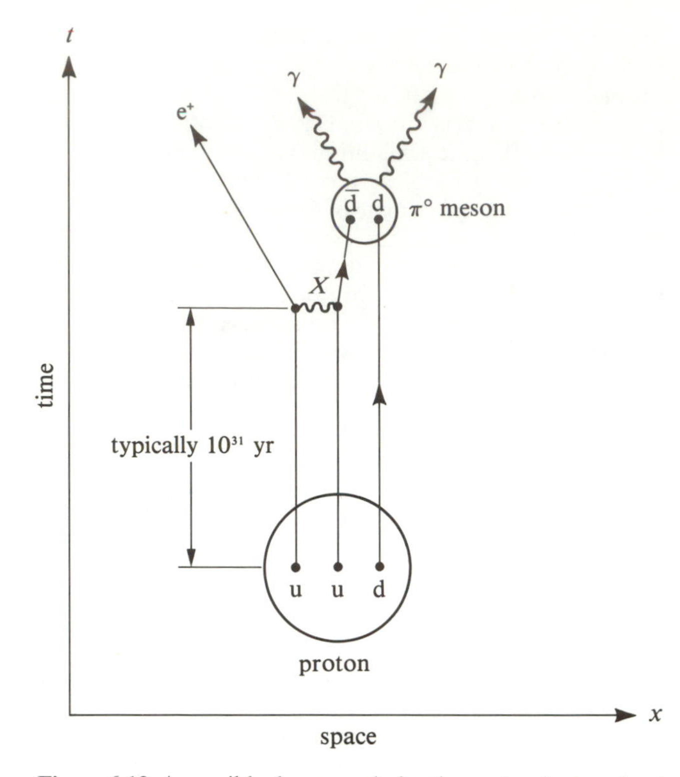

Figure 6.13. A possible decay mode for the proton in grand unified theories. Two up quarks interact via the hyperweak force and are changed into a positron and an antidown quark. The antidown quark pairs with the remaining down quark of the original proton to form a π^0 meson. The π^0 meson has a mean lifetime of about 10^{-16} sec before it decays into two photons.

("color") force seem to require a "hyperweak" force which could change quarks into leptons (and vice versa). This approach may explain why there are six "flavors" of quarks and leptons. Success would mean, of course, that baryon and lepton conservation is only an approximate law, not an exact one, just as chemical-element conservation has become only an approximate law and not an exact one. This nonconservation would become important only at energies per particle which are 10^{15} times larger than the rest energy of the proton, energies well beyond the reach of contemporary accelerators, terrestrial or cosmic. Thus, the nonconservation of baryons and leptons has almost no practical consequences, but it does have the following mind-blowing philosophical consequences. It would mean that the proton is not a truly stable particle, i.e., that the nuclei of atoms are doomed for extinction. Ultimately, if the current trend of ideas are correct, the proton should decay. For example, a proton might decay into a positron and a π^0 meson, with the π^0 decaying in 10^{-16} sec into two photons (Figure 6.13). We need not, however, immediately take out insurance against the disintegration of the material universe, because the mean lifetime of the proton is estimated to be some 10^{31} years. If the universe should last that long, the last survivors will learn how evanescent material things truly are.

If baryons and leptons are ultimately not distinct, what then about the distinction between fermions (half-integer-spin particles) and bosons (integer-spin particles)? Is there not an unpleasant asymmetry between the particles which constitute matter (quarks and leptons, which are fermions) and the particles which transmit forces (bosons)? At higher energies yet, is it possible that fermions might change into bosons (and vice versa)?

Again the trend of modern particle physics answers yes. Some physicists have outlined a super unification of all four forces—strong, weak, electromagnetic, and gravitational—under the banner of one superforce, which has been called "supergravity." Supergravity would allow the conversion of fermions into bosons (and vice versa), if an energy per particle of about 10^{19} times the rest energy of the proton were available.

Problem 6.13. Using dimensional analysis, construct a quantity having the units of mass by multiplying together various powers of the three most fundamental physical constants: G, h, and c. This quantity is called the Planck mass. Calculate the numerical value of the Planck mass, and compare it with the mass of a proton.

Energies *per particle* like 10^{15} and 10^{19} times the rest energy of the proton are practically never encountered in the present universe; however, such energies are not out of the reach of the universe when it was very young. Do we live in an age when we can outline a *scientific* theory of how matter itself came to be created? We will return to such heady topics when we discuss the earliest stages of the history of the universe in Chapter 16.

Problem 6.14. Calculate the kinetic energy $mv^2/2$ of a car of mass $m = 2 \times 10^6$ gm (about two tons) and speed $v = 1.2 \times 10^3$ cm/sec (about 30 mph). Compare this with the rest energy m_pc^2 of the proton. How fast would our car have to move to have a kinetic energy equal to $10^{19}\ m_pc^2$?

CHAPTER SEVEN

The End States of Stars

Having considered the Sun, we wish now to contemplate how stars end in general before we discuss in Chapter 8 how they live. There are two reasons for this apparent digression. First, before we undertake a detailed examination of the life history of the stars, it will help us to know where they are going. An analogy would be that a discussion of biological development could be aided by the knowledge that the end product of aging is death.* Second, neither the theory nor the observation of stars has progressed far enough to allow the recounting of the life history of any star from birth to death in one continuous thread. There are missing segments; hence, it is informative to know that astronomers now believe that all such stories have only four possible endings.

This belief arises in part from an application of thermodynamics to obtain an overview of the evolution of stars. From our study of the Sun in Chapters 5 and 6, we can generalize and state that the two most important properties governing ordinary stars are

(a) Ordinary stars are self-gravitating, and therefore have to be hot inside to sustain the thermal pressure that resists the inward pull of gravity.

(b) On the other hand, space on the outside is dark and cold (a remarkable fact that we take for granted for now). Thus, heat flows continuously from the star to the universe.

Now, as long as a star behaves as a classical gas, there is no true thermodynamic equilibrium possible under the above circumstances. The star will continue to lose heat to the universe. This, as we explained in the last chapter, will eventually force the star to contract and to get hotter and hotter in comparison with the universe. In this never-ending struggle against gravity and the second law of thermodynamics, nuclear-energy sources can provide only *temporary* respites. The nuclear stores of energy, though very large, are not infinite. When they have run out, the star faces an inevitable end to its brilliant career. Death may come convulsively—in a violent explosion—or it may come more quietly, in a lingering slide toward darkness. But violent or lingering, death is as inevitable for the stars as for us.

In the war between self-gravity and thermodynamics, what endings lie ahead for stars? Astronomers believe four are possible (see also Box 7.1):

(a) *Nothing* may be left; if the final explosion is sufficiently violent, all the matter may be dispersed effectively into interstellar space. This would represent an ultimate victory for thermodynamics.

* In biology the time-development of an individual is "aging." The word "evolution" is reserved for the time development of a population (or a species). In astronomy the word "evolution" is (mis)used to describe the aging of individual stars, but this misusage is by now so ingrained that we shall continue to follow it.

BOX 7.1
The Four End States of Stellar Evolution

(a) *Nothing.*
(b) *White dwarf*: typical mass $= 0.7M_\odot$; typical radius $= 10^9$ cm, which is about the size of the Earth.
(c) *Neutron star*: typical mass $= 1.4M_\odot$; typical radius $= 10^6$ cm, which is somewhat larger than the height of Mount Everest.
(d) *Black hole*: typical mass $= M$ greater than 2 or $3M_\odot$ (?); Schwarzschild "radius" $= 2GM/c^2$ (a nonrotating black hole).

(b) A *degenerate dwarf* may be left. The exposed core of an aged star, such an object would be blazing hot to begin with (so degenerate dwarfs are often also called white dwarfs), but it would eventually cool and turn black. This would represent a truce mediated by the quantum behavior of electrons.

(c) A *neutron star* may be left. The imploded core of a highly evolved star, such an object may have a brief fling as a pulsar, but it too would eventually come into thermodynamic (lifeless) equilibrium with the universe. This would represent a truce mediated by the quantum behavior of baryons.

(d) A *black hole* may be left. The Dracula of stellar corpses, it lies in wait hoping to ensnare more matter to share its fate. This would represent an ultimate victory for self-gravity.

Let us now take a closer look at this roster. Since the first item, nothing, requires no further discussion, we begin with degenerate or white dwarfs.

White Dwarf

The first white dwarf to be discovered historically is a companion of Sirius (the "dog star"). White dwarfs typically have masses on the order of one solar mass, and sizes comparable to that of the Earth (which is about 100 times smaller than the Sun). White dwarfs therefore have average densities which are about $100^3 = 10^6$, a million times greater than that of the Sun:

$$\text{mean density of white dwarf} = \frac{\text{mass}}{\text{volume}} = 10^6 \text{ gm/cm}^3.$$

One sugarcube of white-dwarf matter brought to Earth would weigh more than a car! Obviously, a white dwarf is immensely dense and compact. What holds a white dwarf up against its enormous self-gravity? To answer this question, we must return to the quantum-mechanical behavior of particles, and specifically to the quantum statistics of electrons.

Electron-Degeneracy Pressure

In Chapter 3 we remarked that electrons are fermions, and satisfy the quantum-mechanical rule known as Pauli's exclusion principle:

> No two electrons can have exactly the same quantum mechanical state.

For free electrons, knowledge of the quantum-mechanical state is restricted by Heisenberg's uncertainty principle (see also Problem 4.8):

> It is impossible to define the position and momentum of a particle to an accuracy which is better than Planck's constant h for the product of the uncertainties: $(\Delta x)(\Delta p_x) > h$.

Taken together, Pauli's exclusion principle and Heisenberg's uncertainty principle imply that a gas containing free electrons will exhibit pressure even at a temperature of absolute zero. This quantum-mechanical contribution to the pressure of an electron gas is called **electron-degeneracy pressure**. We assume, of course, that the electrons are mixed with a background of ions, so that the overall medium is electrically neutral.

Electron-degeneracy pressure arises for a simple reason. In a gas, pressure results from the random motions of the particles (review Problem 4.1). In a classical gas, such motions arise because the gas has thermal energy (as long as $T \neq 0$). An electron gas has no thermal motions at $T = 0$, but it does have motions caused by the quantum-mechanical effects described above, especially at high densities, when the mean separation Δx between one electron and its nearest neighbor is very small. In this situation, the two electrons must have momenta that differ by at least an amount $h/\Delta x$; otherwise they would violate Pauli's exclusion principle. (In this discussion we will ignore the factor of 2 that arises because electrons may differ in quantum-mechanical states by having antiparallel spins.) If Δx is small (high densities), Δp_x must be correspondingly large; so electrons compressed to high densities would generally have high random speeds $v = p/m_e$ relative to each other, especially because m_e is small. These large random motions can give a degeneracy pressure which is much larger than the thermal pressure. It is the very large electron-degeneracy pressure which supports a white dwarf against its enormous self-gravity.

Problem 7.1. Consider the formula $P = nv_xp_x$ derived in Problem 4.1, and apply it to a degenerate electron gas. Let the number density of electrons be n_e; then $\Delta x = n_e^{-1/3}$ is the average interparticle spacing. Consequently, the mean x-momentum must be of the order of $p_x \sim \Delta p_x \sim h/\Delta x = hn_e^{1/3}$. If these motions are nonrelativistic, the mean x-velocity must be given by $v_x = p_x/m_e$, where m_e is the rest mass of the electron. Show now that the electron degeneracy pressure P_e is roughly given by the expression $h^2n_e^{5/3}/m_e$. The precise formula obtained by considering equation (4.4) reads

$$P_e = 0.0485\,\frac{h^2 n_e^{5/3}}{m_e}.$$

In a white dwarf, we have overall charge neutrality. Thus, if we have n_+ ions per cm^3 of atomic number Z and atomic weight A, and n_e electrons per cm^3, we require

$$Zn_+ = n_e.$$

The mass density ρ is given by $\rho = Am_pn_+ + m_en_e \simeq Am_pn_+$ because electrons are so light. Show therefore that

$$n_e = \frac{Z}{A}\frac{\rho}{m_p}, \qquad P_e = 0.0485\,\frac{h^2}{m_e}\left(\frac{Z}{A}\right)^{5/3}\frac{\rho^{5/3}}{m_p^{5/3}}.$$

In a typical white dwarf, the average density might correspond to $\rho = 10^6$ gm/cm^3, and the average internal temperature might be $T = 10^7$ K. Assume $Z/A = 0.5$, and compare the thermal pressure n_ekT with the electron-degeneracy pressure $0.0485\ h^2n_e^{5/3}/m_e$. Which is dominant? Assume that the ions are fermions, and argue that the ion-degeneracy pressure is given by $0.0485\ h^2n_+^{5/3}/Am_p$. Why is this pressure much smaller than the electron-degeneracy pressure? How about the ion thermal pressure n_+kT? (In most white dwarfs, the ions contain an even number of spin-1/2 baryons, and are therefore really bosons.)

Mass-Radius Relation of White Dwarfs

Under the conditions which prevail inside white dwarfs, the total pressure is virtually independent of the temperature (refer back to Problem 7.1). The primary dependence is on the density. For nonrelativistic motions of the electrons, the pressure varies as the 5/3 power of the density. R. H. Fowler found that this dependence leads to the following interesting relationship between the mass and the radius of a white dwarf:

> If we compare two white dwarfs of different masses, the more massive white dwarf has the smaller radius.

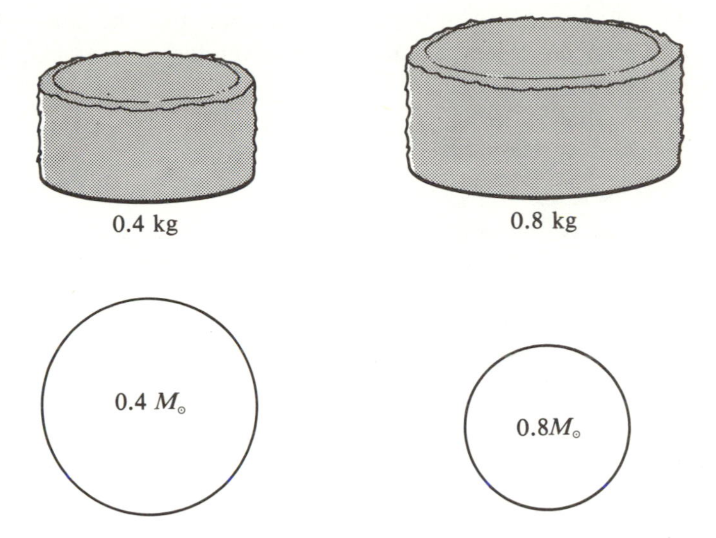

Figure 7.1. The difference between chocolate cakes and white dwarfs. A chocolate cake of 0.8 kilogram which is twice as massive as a chocolate cake of 0.4 kilogram would have twice the volume. A white dwarf of $0.8M_\odot$ which is twice as massive as a white dwarf of $0.4M_\odot$ would have only half the volume.

This result is very different from ordinary experience (Figure 7.1). If we make a chocolate cake twice as massive as another, the more massive cake will occupy twice the volume of the less massive. Not so with white dwarfs. If we make one white dwarf twice as massive as another (within limits), the more massive white dwarf will occupy only *half* the volume of the less massive (Problem 7.2). This strange result underscores once again that objects which are held together by their self-gravity can surprise us with their delightfully unusual behavior.

Problem 7.2. According to Problem 5.9, the central pressure required to support the weight of a column of matter of unit area must roughly equal GM^2/R^4, where M and R are the mass and radius of the object. For a self-gravitating sphere where $P \propto \rho^{5/3}$, an accurate calculation yields

$$P_c = 0.770\,\frac{GM^2}{R^4},$$

and the corresponding central density is given by

$$\rho_c = 5.99 \cdot \bar{\rho} = 1.43\,\frac{M}{R^3}, \quad \text{where} \quad \bar{\rho} = \frac{M}{4\pi R^3/3}.$$

Use the arguments of Problem 7.1 to show that the central electron-degeneracy pressure is given by

$$(P_e)_c = 0.0485 \frac{h^2}{m_e}\left(\frac{Z}{A}\right)^{5/3} \frac{\rho_c^{5/3}}{m_p^{5/3}}.$$

Set the available pressure $(P_e)_c$ equal to the required pressure P_c, and solve for R to obtain the mass-radius relation:

$$R = 0.114 \frac{h^2}{Gm_e m_p^{5/3}}\left(\frac{Z}{A}\right)^{5/3} M^{-1/3}.$$

Notice that this last equation implies that the volume is inversely proportional to the mass. Compute R numerically for $M = 0.5M_\odot$ and $M = 1M_\odot$. Compare these sizes with the size of the Earth.

If we try to make white dwarfs that have masses much larger than a solar mass, an even stranger phenomenon happens. The large gravities associated with a high mass compress such a white dwarf to very high internal densities. The average interparticle spacing Δx would then become very small, and the electrons would gain very high relative momenta, in accordance with $\Delta p_x \sim h/\Delta x$. At some point, the nonrelativistic increase of the velocity $v_x = p_x/m_e$ must saturate, because the velocity v_x must begin to approach the speed of light for very large values of p_x. Under these circumstances, the electron-degeneracy pressure must increase less quickly with increasing density; indeed, the ultrarelativistic electron-degeneracy pressure increases only as the 4/3 power (instead of the 5/3 power) of the density (Problem 7.3).

Problem 7.3. Consider anew the formula $P = nv_x p_x$. Applied to a degenerate electron gas, this formula yields the rough estimate $P_e \sim hv_x n_e^{4/3}$ (see Problem 7.1), where we do not use the nonrelativistic formula $v_x = p_x/m_e$. In the ultrarelativistic regime, v_x approaches its maximum value, c, the speed of light. Thus, $P_e \sim hcn_e^{4/3}$ in the ultrarelativistic regime. A more precise calculation using equation (4.4) yields

$$P_e = 0.123hcn_e^{4/3}.$$

At what value of electron density n_e does the nonrelativistic formula for P_e have the same numerical value as the ultrarelativistic formula? *Hint*: Set $0.0485h^2 n_e^{5/3}/m_e$ equal to $0.123hcn_e^{4/3}$, and solve for n_e. Given this value of n_e, verify that the typical velocity implied by $v_x = p_x/m_e \sim hn_e^{1/3}/m_e$ is indeed relativistic. How does this density compare with that inferred for the center of the $1M_\odot$ white dwarf of Problem 7.2?

Because a relativistic electron gas provides less pressure support at a given density than the nonrelativistic formula would indicate, the radius of high-mass white dwarfs must be smaller than indicated by the mass-radius relation derived in Problem 7.2. Indeed, Chandrasekhar showed that if the mass were big enough, the radius of the white dwarf would formally want to shrink to zero! The limit, radius = 0, is reached at a mass now known as Chandrasekhar's limit (see Problem 7.4):

$$M_{\text{Ch}} = 0.20\left(\frac{Z}{A}\right)^2 \left(\frac{hc}{Gm_p^2}\right)^{3/2} m_p, \tag{7.1}$$

where Z and A are the average atomic number and atomic weight of the ions in the white dwarf, and where the other symbols have their usual meanings. Equation (7.1) is one of the most beautiful and important formulae in all of theoretical astrophysics. You need not memorize it. Just ponder as one of the glories of science that one of the most important properties of such apparently complicated objects as stars can be summarized by a single simple equation containing nothing except some fundamental physical constants whose values were measured in terrestrial laboratories for entirely different reasons!

White dwarfs are an end state of stellar evolution where we can expect all the hydrogen to have been converted to some heavier chemical element. Since the ratio of atomic number to atomic weight, Z/A, for most heavy elements (until we approach and pass the iron group) is 0.5, the numerical value of Chandrasekhar's limit can be calculated (see Problem 7.4) from equation (7.1) to be

$$M_{\text{Ch}} = 1.4M_\odot.$$

In other words, the maximum mass possible for a white dwarf is only 40 percent more than the mass of the Sun. All white dwarfs whose masses have been observationally measured have masses which are less than this maximum value. This result offers one important piece of supporting evidence in favor of Chandrasekhar's theory.

Problem 7.4. Substitute the expression $n_e = Z\rho/Am_p$ derived in Problem 7.1 into the expression for P_e given in Problem 7.3 to obtain

$$P_e = 0.123 \frac{hc}{m_p^{4/3}}\left(\frac{Z}{A}\right)^{4/3} \rho^{4/3}.$$

For a self-gravitating sphere where $P \propto \rho^{4/3}$, hydrostatic equilibrium requires the central pressure and density to have the values

$$P_c = 11.0 \frac{GM^2}{R^4}, \qquad \rho_c = 54.2\bar{\rho}, \quad \text{where} \quad \bar{\rho} = \frac{M}{4\pi R^3/3}.$$

Set the available pressure $(P_e)_c$ with $\rho = \rho_c$ equal to the required pressure P_c; show that R cancels out; and solve for M. In this way, derive equation (7.1). Notice that a white dwarf with the mass of Chandrasekhar's limit contains $0.20(hc/Gm_p{}^2)^{3/2}$ protons or neutrons, where the quantity $hc/Gm_p{}^2$ is a dimensionless number (the inverse of the "gravitational fine-structure constant"). Compute M_{Ch} numerically for $Z/A = 0.5$.

Source of Luminosity of a White Dwarf

If white dwarfs have exhausted their nuclear supply of fuel, why do (some) degenerate dwarfs shine? Because they have nonzero internal temperatures as relics from a more fiery past. Thus, white dwarfs resemble cooling embers, glowing ever more faintly as they radiate and slide to everlasting darkness. Unlike ordinary stars made of gases with classical properties, a white dwarf does not contract appreciably (and become hotter) as it radiates. It is degeneracy pressure and not thermal pressure which holds up a white dwarf against its self-gravity. Thus, a white dwarf can lose energy, cool, and come into true thermodynamic equilibrium with a cold universe. In the process, the electric forces between the ions will eventually dominate the random thermal motions of the ions. As has been studied by Van Horn and Savedoff, the ions will then form a crystal lattice (see Figure 4.9), and the white dwarf can be said to solidify. The electrons will whiz around degenerately through this crystal lattice; consequently, a solid white dwarf resembles a terrestrial metal in many respects. Stars which end as white dwarfs (as our Sun probably will) can rest in peace.

Problem 7.5. The average spacing r between ions is $n_+{}^{-1/3}$, where n_+ is the number density of ions. Thus, the electrostatic interaction energy between ions is of order $(Ze)^2/r = Z^2e^2n_+{}^{1/3}$ per ion. On the other hand, the average thermal energy per ion is kT (actually, $3kT/2$). Calculate the critical temperature for which kT equals $Z^2e^2n_+{}^{1/3}$ for a helium white dwarf of mean density $\rho = Am_pn_+ = 10^6$ gm/cm^3. What is the physical meaning of this critical temperature (in order of magnitude)?

Neutron Stars

Imagine a physical process which tries to compress the core of a star beyond white-dwarf densities. For example, if the core had a mass larger than Chandrasekhar's limit, would the radius really reach zero? A zero volume is possible for the electrons, but would the nuclei of atoms permit such an arbitrarily large compression? The answer is a complicated no and yes. Let us begin by looking at the "no" part of the answer; namely, let us begin by considering neutron stars.

In the imaginary compression pictured above, at some point almost all the free electrons will be forced to combine with the protons in the nuclei of atoms to form a mass of neutrons. This self-gravitating mass of neutrons is a neutron star. For simplicity, let us begin by considering the neutrons to be free particles so that they constitute a perfect gas of fermions. If we further adopt a nonrelativistic description for the hydrostatic equilibrium of this self-gravitating sphere of neutrons, we can show that a neutron star satisfies the following mass-radius relationship (Problem 7.6):

$$R = 0.114 \frac{h^2}{Gm_p{}^{8/3}} M^{-1/3}, \tag{7.2}$$

where R and M are the radius and mass of the neutron star. For a mass $M = 1.4M_\odot$, the above formula yields a radius $R = 1.5 \times 10^6$ cm, about 9.6 miles. Hence, a neutron star is only about the size of a large city!

Problem 7.6. Derive equation (7.2) for the mass-radius relation of a classical neutron star. *Hint*: Follow the derivation of Problem 7.2 for the mass-radius relation of a white dwarf, and show that the desired result follows if one ignores the difference between the mass of a neutron and the mass of a proton. For a given mass $M < M_{\text{Ch}}$, notice the neutron star is smaller than the corresponding white dwarf by a factor of $(Z/A)^{5/3}m_p/m_e$, i.e., by a factor of about 10^3.

Two considerations restrict the validity of the above mass-radius relation, although our order-of-magnitude estimates remain correct. Consider, first, what the mean density of matter must be for a mass $1.4M_\odot = 2.8 \times 10^{33}$ gm, and a radius 1.5×10^6 cm:

$$\text{mean density} = 2 \times 10^{14} \text{ gm/cm}^3.$$

A sugarcube of neutron-star stuff on Earth would weigh as much as all of humanity! This illustrates again how much of humanity is empty space (Figure 7.2).

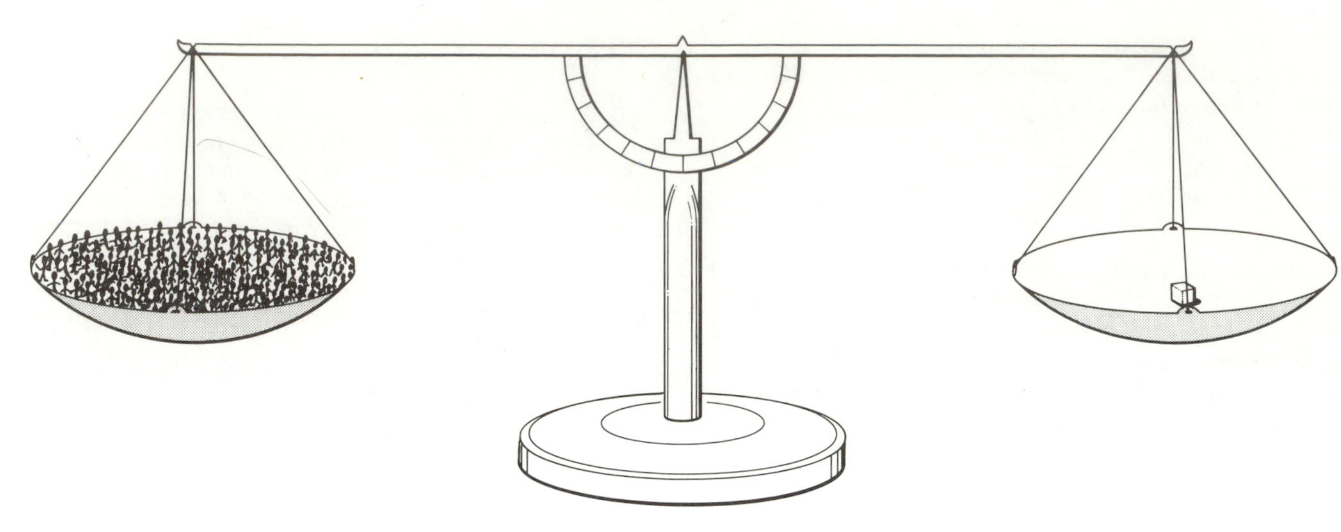

Figure 7.2. A sugarcube of neutron star stuff weighs as much as all of humanity.

Compare, now, this density with the mass density of a single neutron with mass 1.7×10^{-24} gm and "radius" 10^{-13} cm:

$$\frac{\text{mass of neutron}}{\text{volume of neutron}} \sim 4 \times 10^{14} \text{ gm/cm}^3.$$

The rough coincidence between the mean density of our hypothetical neutron star and the mean density of a single neutron implies that the individual neutrons in a neutron star are so tightly packed that they almost "touch." Under these conditions (i.e., nuclear densities) we should not ignore the role of the strong force between the individual neutrons. In some sense, therefore, a neutron star is "just" another atomic nucleus, with two differences. First, the "glue" that holds together a neutron star is self-gravity instead of meson-exchange forces. Second, the "atomic weight" A of a neutron star is 10^{57} or so, instead of the paltry 10^2 or so which characterizes even the "heavy" atomic nuclei we are familiar with on Earth!

Consider, next, the strength of the gravitational field on a neutron star. A neutron star is only somewhat larger than a big mountain (Figure 7.3); yet a neutron star has a mass comparable to the Sun. Even though the mountain has the force of the whole Earth behind its pull, you could climb up and down Mount Everest a hundred thousand times on the energy you would spend to climb one cm on a neutron star! Another measure of the enormous gravity of a neutron star can be obtained by computing the speed required for a rocket to escape its gravitational clutch. Let m be the mass of the rocket; v_e, the escape speed; M, the mass of the neutron star; and R, its radius. For the rocket barely to escape, the sum of

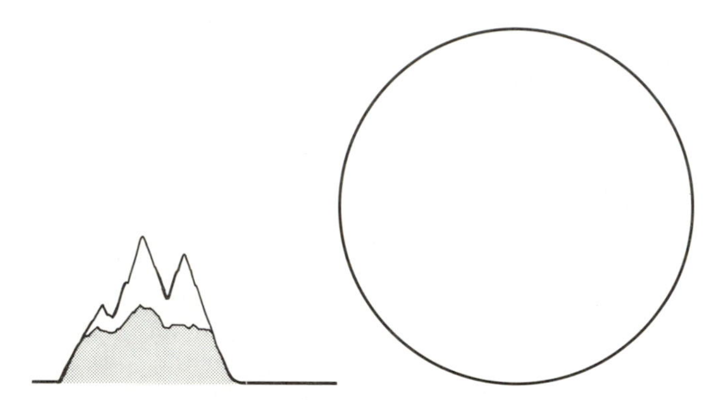

Figure 7.3. A neutron star is only somewhat larger than Mount Everest.

its kinetic energy and potential energy must be zero (refer back to Chapter 3):

$$\frac{1}{2} m v_e^2 + \left(-\frac{GMm}{R}\right) = 0.$$

If we cancel out m and solve for v_e, we obtain

$$v_e = (2GM/R)^{1/2}, \qquad (7.3)$$

which for a neutron star with $M = 1.4 M_\odot$ and $R = 1.5 \times 10^6$ cm yields

$$v_e = 1.6 \times 10^{10} \text{ cm/sec},$$

more than half the speed of light! Clearly, the gravity on the surface of a neutron star is much too large for

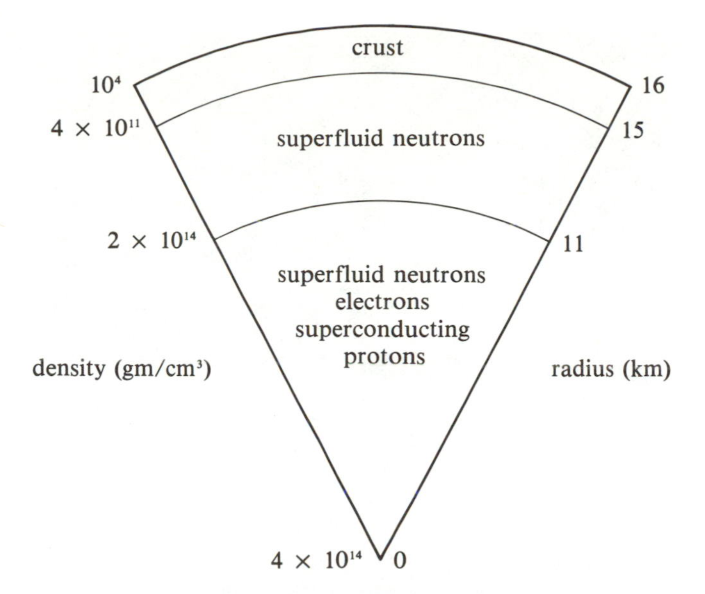

Figure 7.4. A possible model for a neutron star with a (gravitational) mass of $1.3M_\odot$ (after V. R. Pandharipande, D. Pines, and R. A. Smith, *Ap. J.*, **208**, 1976, 550). The solid crust and the superfluid interior may both play a role in the so-called "pulsar glitches." Occasional cracking or slippage of the crust may trigger neutron starquakes. The change in the spin rate may be restricted by the quantized amount of angular momentum contained in the superfluid neutrons that are pinned to normal matter.

any conventional rocket to be able to escape its pull. Indeed, the gravity of a neutron star is so large that it has squashed most of the free electrons into the nuclei of atoms and forced them to combine with the protons to form neutrons. This is the reason, of course, why the "atomic weight" A of the neutron star is so much larger than its "atomic number" Z ($Z = 0$ if the neutron star is electrically neutral).

The nuclear density and the large gravitational field of neutron stars imply that a proper treatment of their structure should take into account nuclear interactions and Einstein's modification of Newton's theory of gravity. Physicists' understanding of the former is incomplete; therefore one of the hopes of neutron-star theorists is that observations of real neutron stars may provide us with important clues concerning the microphysics of nuclear interactions.

It is now thought that neutron stars have solid outer crusts composed of heavy nuclei (like iron) and electrons (Figure 7.4). Interior to this crust, the material consists mostly of neutrons, although there will be some protons and electrons as well. At a sufficiently deep level, the neutron density may become high enough to give rise to a superfluid (see Problem 4.9). Neutrons are fermions, which usually cannot exhibit the phenomenon of "Bose condensation." However, some of the neutrons may form "pairs," as in the electron-pairing mechanism discovered by Bardeen, Cooper, and Schrieffer to explain the phenomenon of superconductivity. Indeed, the deep interior of a neutron star is believed to be not only superfluid but superconducting.

Upper Mass Limit for Neutron Stars

Do neutron stars also have a maximum possible mass analogous to the Chandrasekhar limiting mass appropriate for white dwarfs? The answer is yes, but the quantitative computation of the numerical value of this mass is much less certain than that for white dwarfs, because nuclear physics is still incomplete. Nevertheless, there are good theoretical arguments that, no matter how strongly repulsive nuclear forces become at ultrahigh densities, they cannot resist gravity if the mass of an object becomes large enough, for two reasons. First, special relativity sets a limit on how "stiff" matter can be. The stiffness of matter can be measured by how quickly the pressure rises with increasing density to resist further compression. But the rate of change of pressure with increasing density is related to the speed of sound in the material (see Problem 4.4). Matter cannot be so stiff that the speed of sound exceeds the speed of light, and this sets an upper limit on how strongly repulsive nuclear forces can become at ultrahigh densities. Second, energy is always involved in the generation of repulsive force fields, and in general relativity (Chapter 15), energy acts just like mass as a source of gravitational attraction. Thus, the more matter tries to resist its own gravity, the more does gravity pull on matter. And for large-enough ratios of mass to radius, the pull of gravity becomes irresistable. Thus, Rhoades and Ruffini have used very general theoretical arguments to place a firm upper limit of about three solar masses for the mass of a neutron star. The self-gravity of a more massive aggregation of neutrons would literally crush the neutrons into a state of infinite density. This singular state, which we shall discuss when we come to black holes, provides the "yes" part of our answer to the original question of whether the nuclei of atoms could be compressed to zero volume. One can scientifically debate whether physics itself is meaningful under such singular conditions.

Neutron Stars Observable as Pulsars

Before we discuss black holes, let us ask whether neutron stars exist anywhere in the universe besides astronomers' imaginations. Radio astronomers were the first to discover celestial objects which correspond to the neutron stars hypothesized by Baade and Zwicky and calculated to be possible by Landau and by Oppenheimer and Volkoff. In 1967 Bell (now Burnell) and Hewish

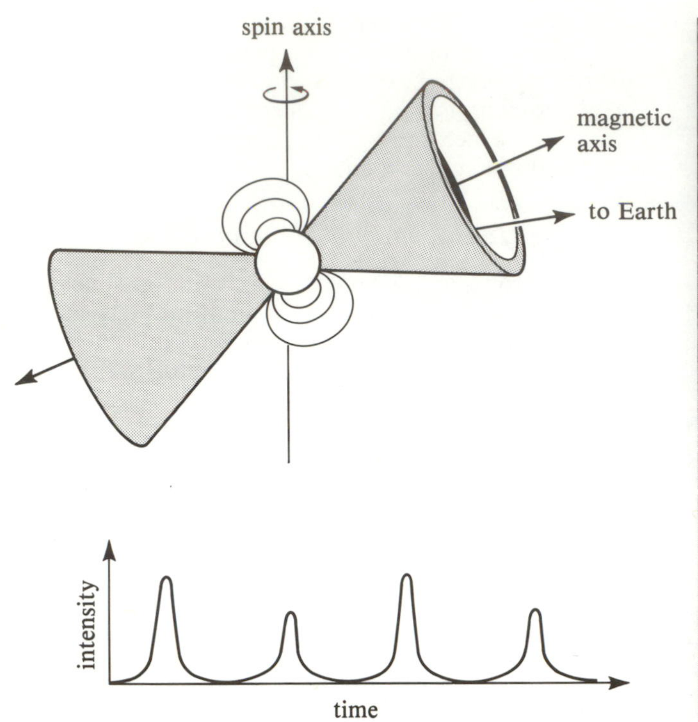

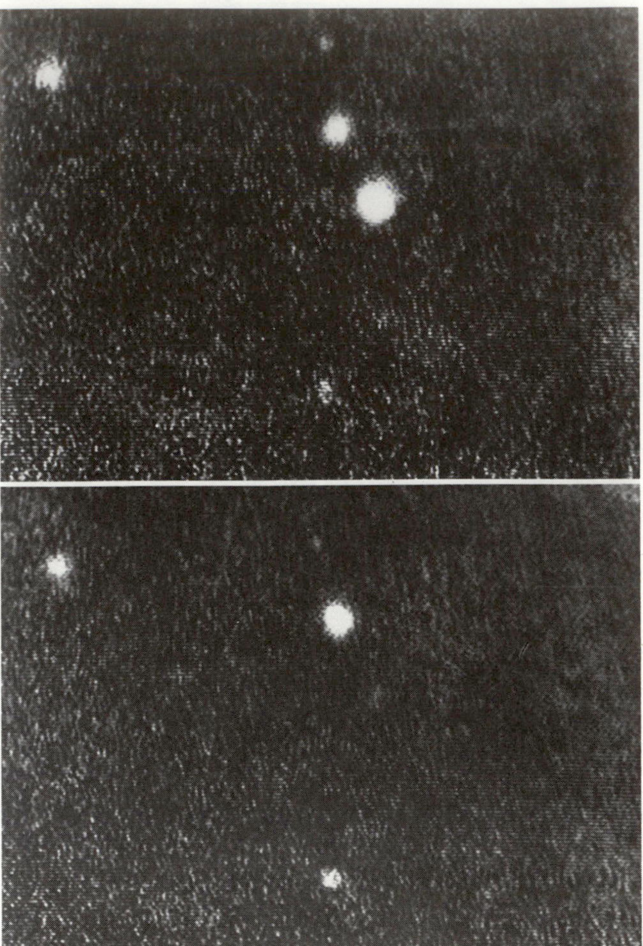

Figure 7.5. a. Lighthouse model of pulsar as spinning, magnetized neutron star. As a "hollow" cone sweeps past the radio observer's line of sight, one side of the hollow cone contributes the larger main radio pulse and the other side, the smaller interpulse. **b.** The emission from the Crab pulsar extends to the optical, allowing television photographs of it to be taken at minimum and maximum light. (Lick Observatory photograph)

discovered regularly pulsing radio sources, ultimately to be called **pulsars** (but initially designated only half-jokingly as "little green men"). Upon further investigation, pulsars were deduced by Gold to be magnetized, spinning neutron stars. Such a star can apparently produce a rotating beam of radiation, which leads to a series of regularly spaced pulses when observed by anyone who happens to lie in the path of the sweeping beam (Figure 7.5). The effect is not unlike that of a lighthouse, except that the sweeping beam may often take the form of a "hollow cone," and that the interception of the two sides of the cone produces a "main pulse" and an "interpulse."

What is the radiation mechanism of pulsars? This is a controversial subject, and it is only now that astrophysicists have gradually begun to converge on a promising solution. An important point was made by Goldreich and Julian, who pointed out in 1969 that the combination of rapid rotation plus strong magnetic fields must, by Maxwell's equations, induce strong electric fields near the surface of the star. These electric fields should force electric charges to flow from the surface of the star. It is now thought that charged particles may flow out of the magnetic polar caps of the neutron star, essentially parallel to the magnetic field lines (Figure 7.6). The acceleration of the charged particles as they try to follow the curved trajectories required by the magnetic field structure will cause them to radiate.* The high energies and densities of the resulting radiation field may lead to the creation of a "pair plasma" in the space surrounding the star, where electrons and positrons are created and annihilated in great profusion. Gamma-ray astronomers hope to detect the line photons resulting from this annihilation to test the validity of these models.

Although astronomers and physicists are still far from having a detailed theory for the processes which

* Sturrock, Ruderman, and Sutherland have been especially active in developing such models, but Michel and Arons, Fawley, and Scharlemann have also made important contributions.

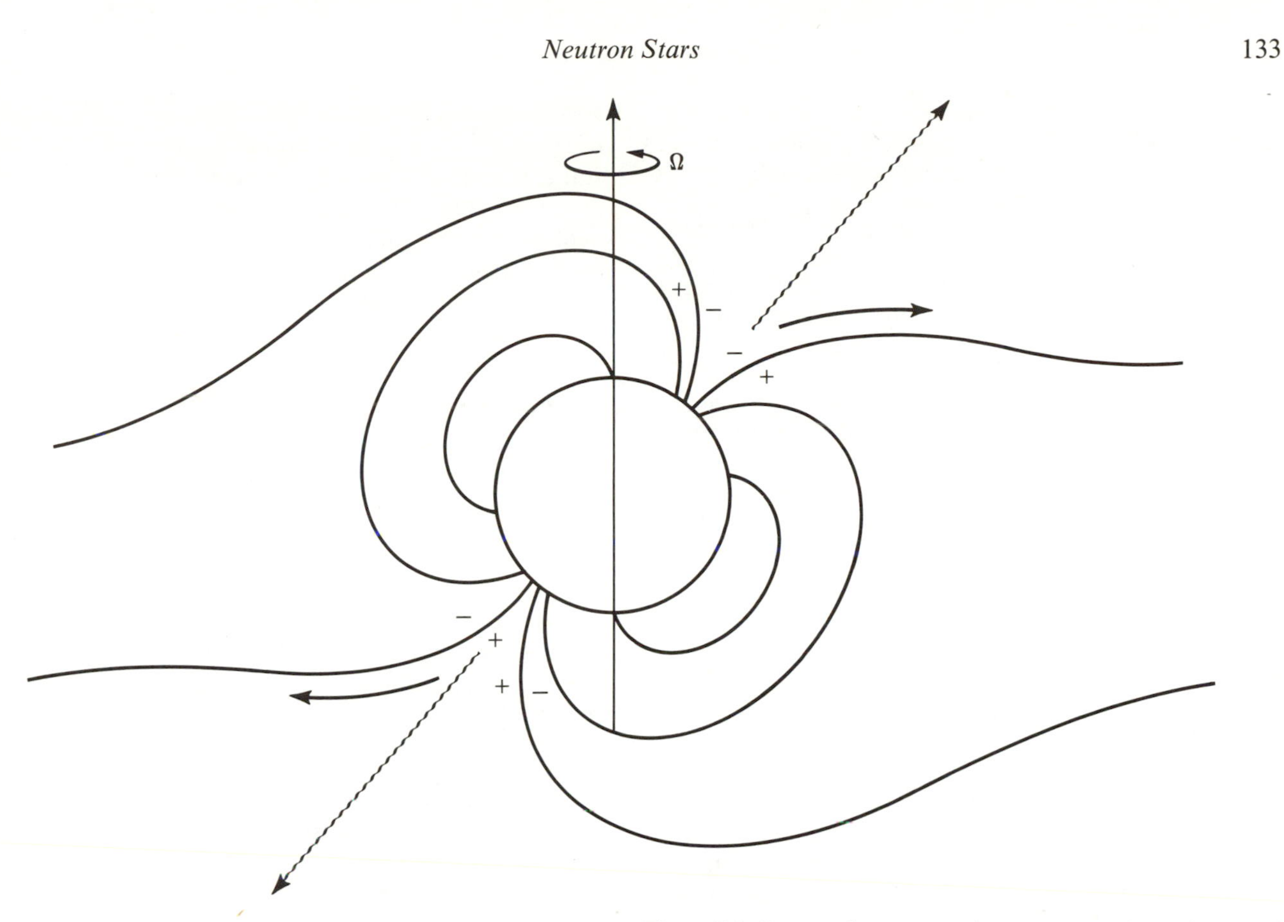

Figure 7.6. Proposed magnetosphere structure for pulsars and radiation mechanism.

produce the observed spectrum and conical pattern of radiation from pulsars, the gross energetics are reasonably well understood, largely because of observations of the Crab pulsar. At the center of the Crab nebula, found in the constellation of Taurus (Color Plate 7), lies the Crab pulsar. The Crab nebula itself is a cloud of gas located at the site of a gigantic stellar explosion observed and recorded by Chinese astronomers in 1054 (see Chapter 11). This association of the Crab nebula with a historical supernova remnant was known well before the discovery of the Crab pulsar. The Crab nebula has been extensively studied by modern astronomers interested in the remnants of supernova explosions. From such studies a puzzle had arisen. The diffuse part of the Crab nebula (not the filaments) shone with an eerie light, identified by Shklovskii to be synchrotron radiation, arising from relativistic electrons gyrating wildly in magnetic fields (Figure 7.7). The total power needed to explain the light from radio waves to X-rays and the acceleration of the nebula is now known to be about 3×10^{38} erg/sec. This is a very large luminosity even by stellar standards. Moreover, the energy output of the

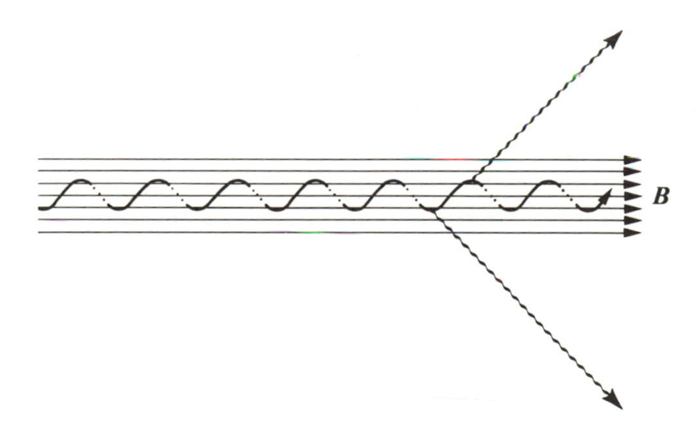

Figure 7.7. Synchrotron radiation. A relativistic electron of charge $-e$, moving at velocity $\boldsymbol{v}$ in a magnetic field $\boldsymbol{B}$, will be deflected by a force equal to $-e(\boldsymbol{v}/c) \times \boldsymbol{B}$ to spiral around the magnetic field lines. The resultant acceleration causes the electron to radiate synchrotron photons which travel primarily in the instantaneous direction of motion of the relativistic electron. This type of radiation, first encountered in terrestrial particle accelerators called "synchrotrons," is produced in many astrophysical contexts, and is also called "nonthermal emission."

Crab nebula is very high-quality to boot. What could be the cosmic accelerator producing this phenomenon?

The discovery of the Crab pulsar provided the answer to this long-standing puzzle. The Crab pulsar's radio pulses were soon observed to be arriving at slightly but steadily increasing intervals. If the "clock" behind the pulses was to be identified with the spin of a neutron star, then the spinning motion must be slowing down. Gold reasoned that the rotational energy of the neutron star must be steadily transferred into other forms, for example, accelerating electrons. Although the theoretical study of these processes remains incomplete, the *observed* slowing-down rate of the Crab pulsar was about right to supply the 3×10^{38} erg/sec necessary to explain the Crab nebula (Problem 7.7). Thus, the discovery of the Crab pulsar answered two astronomical issues:

(a) What is the energy source for the Crab nebula? The Crab pulsar.

(b) Was anything left of the core of the star which exploded as a supernova in 1054? Yes, the Crab pulsar, a neutron star.

Interestingly enough, in 1966, two years before pulsars were discovered observationally, John Archibald Wheeler and Franco Pacini had speculated theoretically that the basic energy source of the Crab nebula may be a spinning neutron star!

Problem 7.7. The moment of inertia of a homogeneous sphere of mass M and radius R is $I = (2/5)MR^2$. What is the value of I for $M = 1.4M_\odot$ and $R = 1.5 \times 10^6$ cm? A more realistic model of a neutron star of "baryon mass" $= 1.4M_\odot$ might have a "radius" $R = 1.1 \times 10^6$ cm and a moment of inertia $I = 9 \times 10^{44}$ gm cm^2 (according to some calculations of Malone, Johnson, and Bethe). If such a body spins with an angular speed ω, it has an angular momentum $J = I\omega$, and rotational energy $E = I\omega^2/2$ (see equation 3.10). Assume the Crab pulsar to have a moment of inertia $I = 9 \times 10^{44}$ gm cm^2. How much rotational energy does the Crab pulsar currently have if it has a spin period $2\pi/\omega = 0.033$ sec? The Crab pulsar is slowing down at a rate which would bring its spin to a halt in about 2,500 years. What would be the average power lost by the pulsar if its present rotational energy were dissipated over 2,500 years? (The instantaneous power is twice as high.) How does this compare with the luminosity of the Crab nebula?

Suppose a body spinning originally at a rate ω_1 with a radius R_1 collapses gravitationally to a radius R_2, conserving its mass M and its angular momentum J. Express the ratios of the new and old spin rates ω_2/ω_1, and the new and old rotational energies E_2/E_1, in terms of the ratio R_1/R_2. By what factor would the core of a star spin faster if it were to collapse from a radius typical of a white dwarf, 10^9 cm, to the dimensions typical of a neutron star, 10^6 cm? By what factor would the rotational energy increase in such a collapse? Where ultimately does this extra energy come from?

The Masses of Neutron Stars

To calculate the mass of any celestial object reliably, astronomers must, as we mentioned in Chapter 5, measure the orbital properties of a nearby companion. With one notable exception (to be discussed in Chapter 10), all the known pulsars are single stars. However, neutron stars are also believed to exist in certain binary-star systems (binary X-ray sources to be discussed in Chapter 10); so it is sometimes possible to measure their masses. Within the considerable errors of such measurements, the indications are consistent with the masses of all newly formed neutron stars being about $1.4M_\odot$. The coincidence between this value and Chandrasekhar's limiting mass for a white dwarf is probably a clue to the process by which neutron stars are produced in the course of the evolution of high-mass stars (Chapter 8).

Black Holes

We saw that the escape speed from the surface of a neutron star was typically a healthy fraction of the speed of light. What radius R would an object of mass M have to have if the escape speed from its surface were equal to the speed of light? If we naively apply formula (7.3), set v_e equal to c, and solve the result for R, we obtain the desired radius—the *Schwarzschild radius*—as the expression

$$R_{\text{Sch}} = \frac{2GM}{c^2}. \tag{7.4}$$

Apart from quantum-mechanical effects, any object whose radius R becomes smaller than its Schwarzschild radius is doomed to collapse to a single point. No known force in nature can resist this collapse to singularity with an infinite density. The numerical value of the Schwarzschild radius for a $1M_\odot$ black hole (if there is a way to form them) is

$$R_{\text{Sch}} = 3 \times 10^5 \text{ cm for } M = 1M_\odot,$$

and it is proportionally larger for larger values of M. Hence, stellar-mass black holes are only a few times smaller than neutron stars of comparable mass.

The expression $2GM/c^2$ which appears on the right-hand side of equation (7.4) was first derived by the nineteenth-century physicist Pierre Laplace. His derivation, which we have reproduced, contains two errors of assumption. The first error is to suppose that the kinetic energy of light is given by $mv^2/2$. It is not. The second error is to suppose that Newton's law of gravitation continues to hold when even light finds it difficult to escape the gravitational clutch of a massive body. It does not. Karl Schwarzschild was the first person to give the correct derivation for the size of an object—later named a **black hole** by John A. Wheeler—which would not let even light escape from its gravitational pull. For this reason, the expression $2GM/c^2$ is called the "Schwarzschild radius." The two errors of Laplace's derivation *happen to cancel* to give the correct-looking formula.

Schwarzschild's analysis shows that a photon which starts with wavelength λ_0 at position r outside a spherical gravitating mass M will have a longer wavelength λ when it emerges at infinity. The relation between λ and λ_0 is given by the **gravitational redshift** formula:

$$\frac{\lambda}{\lambda_0} = \left(1 - \frac{2GM}{c^2 r}\right)^{-1/2}. \tag{7.5}$$

Notice that if $r = 2GM/c^2$, the above formula predicts that the photon will be redshifted to infinite wavelength λ and zero energy hc/λ. In other words, if a body is smaller than its Schwarzschild radius, all photons which tried to leave its surface would be redshifted into nonexistence before they could reach infinity. This is the ultimate meaning of the Schwarzschild radius (7.4).

The correct derivation of equation (7.5) requires the use of Einstein's improvement of Newton's theory of gravitation. Einstein's theory of gravitation is, of course, general relativity. General relativity yields Newtonian gravity as an acceptable approximation when the gravitational field is relatively weak. For example, suppose in equation (7.5) we are at a value of r which is much greater than the Schwarzschild radius, $2GM/c^2$. Then, if we take the reciprocal of equation (7.5) and expand the resulting right-hand side by the binomial theorem for $2GM/c^2r \ll 1$, we obtain the equation

$$\frac{\lambda_0}{\lambda} = 1 - \frac{GM}{c^2 r}.$$

Multiplying by the energy $E = hc/\lambda_0$ of the photon at position r and identifying $m = E/c^2$ as the associated "gravitating mass" of the photon, we obtain

$$\frac{hc}{\lambda} = \frac{hc}{\lambda_0} - \frac{GMm}{r}. \tag{7.6}$$

The physical interpretation of equation (7.6) in terms of the Newtonian principle of conservation of energy is obvious. It states that the kinetic energy of the photon at infinity is equal to the sum of the kinetic and potential energies of the photon at r. The result is non-Newtonian, because we use quantum mechanics to identify hc/λ as the kinetic energy of a photon, and Einstein's equivalence between mass and energy $E = mc^2$ to identify $m = h/c\lambda$ as the "gravitating mass" of a photon.

We should not think, however, that Einstein's theory of gravity represents merely a quantitative modification of Newton's theory. General relativity gives quantitatively and *qualitatively* very different results from Newtonian theory when the escape speeds of matter from any gravitating mass approach the speed of light, as they must by definition near the Schwarzschild radius of a black hole. The most important qualitative distinction is the geometric interpretation of spacetime, r and t.

Gravitational Distortion of Spacetime

Spacetime is very distorted from that appropriate to special relativity in the vicinity of a black hole. Indeed, to be precise, we should not even think of $2GM/c^2$ as the "radius" of the black hole, since there is no operational way for us to measure such a "radius." If we tried to lay a tape measure across the black hole, the tape measure would be swallowed by the black hole. If we try to send light across the black hole and bounce it back with a mirror to time the round trip, the photons would fall into the black hole and never emerge from the other side. All such attempts to measure the "radius" of a black hole are doomed to fail.

Something we can measure is the circumference of the **event horizon** (Figure 7.8). The event horizon is the surface where outwardly traveling photons would *barely* be able to escape to infinity. If we are suitably suspended, we could imagine walking around the spherically symmetric event horizon of a Schwarzschild black hole and measuring the circumference. In this way, we would obtain a distance $4\pi GM/c^2$, where M is the mass obtained by applying Newton's laws to a small body orbiting the black hole at a very large distance. *Defining* the circumference measured this way, $4\pi GM/c^2$, to be $2\pi R_{\text{Sch}}$, we would then recover equation (7.4). In a similar fashion, we could imagine lowering many sheets of graph paper toward the event horizon and counting up how many sheets are needed to cover the latter. The surface area measured in this way would turn out to be $16\pi G^2M^2/c^4$, i.e., $4\pi {R_{\text{Sch}}}^2$ with our definition of R_{Sch}.

The recovery of the usual Euclidean relationship between the circumference of a circle and the surface area of a sphere is a consequence of the angular symmetry of the problem. The *radial* direction suffers distortions from our usual geometric perceptions. To experience this distortion, imagine conducting the following surveying experiment. First, place yourself in a

Figure 7.8. Although it is not possible to measure the "radius" of a black hole, one can imagine measuring its circumference by encircling its event horizon with a tape measure. The circumference obtained this way is equal to $2\pi R_{Sch}$, where $R_{Sch} = 2GM/c^2$, and M is the "gravitational mass" of the black hole.

powerful rocket ship at a great distance from the center of the black hole whose Schwarzschild "radius" is 3 km. Next, fly in a circle about it. What is the circumference? Two π times 30 km, you reply. At this point, you might feel justified to deduce that you are 30 km from the black hole, but don't jump to conclusions. Try flying 10.67 km toward the black hole. Now, fly in a circle around it. What is the circumference? Two π times 20 km, you reply. Have we made a mistake? Shouldn't you be 30 km − 10.67 km = 19.33 km, not 20 km, from the black hole? Well, try flying 21.92 km from your original starting point toward the black hole. Now, fly in a circle around it again. What is the circumference? Two π times 10 km, you reply. Now, you begin to become alarmed because Euclidean geometry predicts you should be 30 km − 21.92 km = 8.08 km, not 10 km, from the black hole. What's going on? Try flying 28.52 km from your original starting point toward the black hole. You protest, won't that take you inside the event horizon of the black hole? If the Schwarzschild "radius" is 3 km, isn't it only 27 km from your original starting point? Don't worry; try flying 28.32 km inwards. See, you're OK; you can still fly in a circle around it. What is the circumference? Two π times 5 km, you reply, badly shaken. Clearly, the strong gravitational field of the black hole must be distorting the usual spatial relationships. Figure 7.9 summarizes our thought-experiment findings in the form of an "embedding diagram."

Because of the spherical symmetry, flight circles of arbitrary inclination around the black hole are equivalent. Let us plot only one angle of inclination. *If* the geometry of space had been Euclidean, we could have made a "polar plot" of the relationship between radial distance and circumference on a flat sheet of paper (Figure 7.9a). For our thought-experiment findings, we actually need a curved sheet of paper (Figure 7.9b). In Figure 7.9b the direction perpendicular to the curved paper has no physical meaning. The two directions on the surface of the curved paper correspond to physical directions *in a single plane*. Remember that the physical third dimension (the direction perpendicular to our physical plane) has been left off our "polar plot." The "curvature" of our two-dimensional plot embedded in a third *plotting* dimension is purely a mathematical device constructed to allow us to visualize geometrically the distortion produced in space by the gravitational field of the black hole. If we had insisted on plotting a plane on a flat sheet of paper, one cm near the center would represent a different distance than one cm further out. The "curvature" of actual three-dimensional space cannot be visualized by us three-dimensional creatures; we need additional (artificial) spatial dimensions to plot three-dimensional space curvature. Differential geometry allows general relativists, however, to calculate symbolically the three-dimensional space curvature that they cannot visualize geometrically. The symmetry of the present problem allows us to throw away one of the actual spatial dimensions and to use the "embedding diagram" of Figure 7.9b. One glance at this embedding diagram suffices to show that you have to go a greater radial distance inward than Euclidean geometry (Figure 7.9a) would predict to measure a given smaller circumference. At great distances from the black hole, the distortions from Euclidean geometry are negligibly small, but the distortions become severe near the event horizon. In particular, the outsider's notion of space ends with the event horizon, because the "curvature" associated with the gravity of the black hole is a curvature of spacetime, and not just a curvature of space alone. Space and time change roles across the event horizon.

To explore the curvature of spacetime near a black hole, let us imagine the following thought experiment. Let us imagine lowering you very carefully by a very strong cable toward the event horizon of a black hole. What do you see?

Well, when you are still far from the event horizon, you can see the constellations much as they usually appear in that part of the universe. As you approach closer to the event horizon, the positions of the stars in

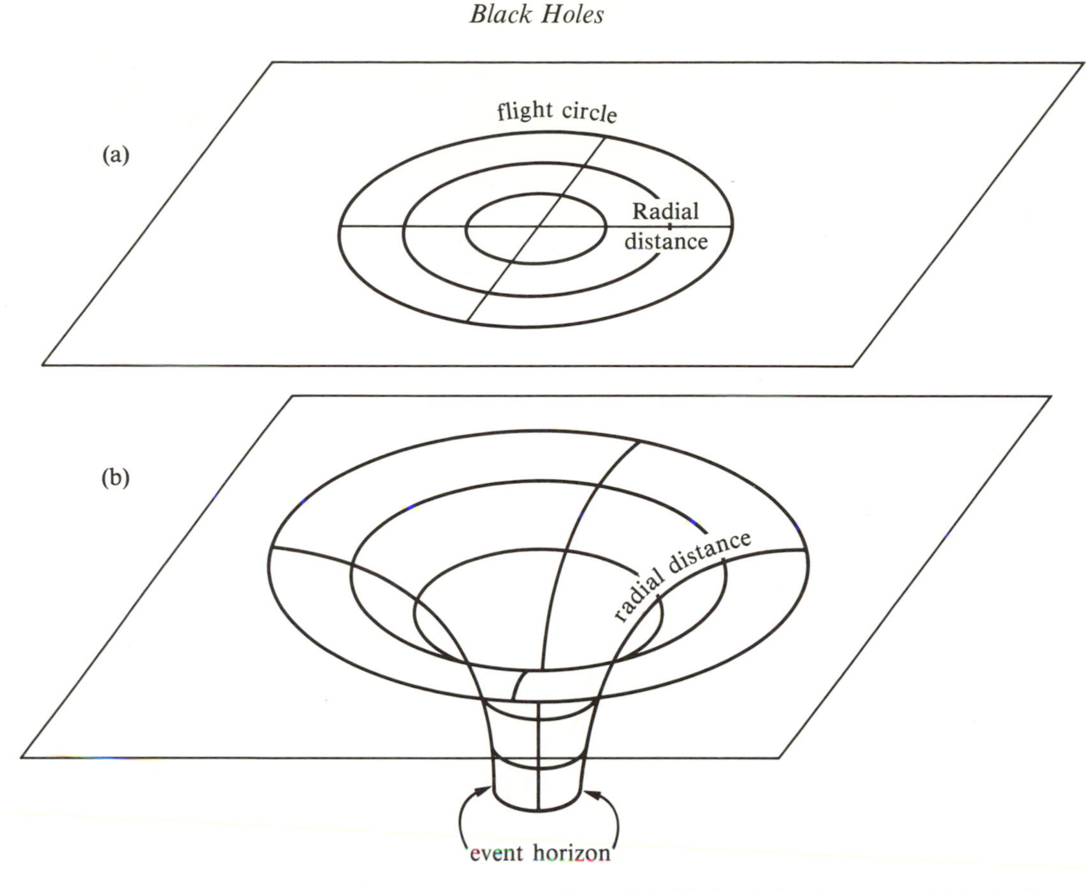

Figure 7.9. Flight circles in a *plane* (a) in Euclidean geometry, (b) in the non-Euclidean geometry near the event horizon of a black hole. In case (b), one has to travel a greater distance inward than in case (a) to have a flight circle of given smaller circumference. The radial direction in both cases is as indicated. At great distances from the event horizon (not drawn), the "curvature" of our embedding diagram becomes negligibly small, and the flight circles of case (b) have nearly the same geometry as case (a).

the constellations shift and distort. Do you begin to get a sinking feeling? When you are lowered to where the circumference is 1.5 times the circumference of the event horizon, you can look perpendicular to the black hole and see the back of your own head! Don't look toward the black hole. You can't see anything in that direction. It's completely dark toward the black hole. Look straight ahead, *perpendicular* to the black hole. You can now see the back of your own head because the photons leaving the back of your own head can orbit around the black hole to strike your eyes directly (Figure 7.10). It's a nice trick, and it's done with gravity, not with mirrors.

Now, I continue to lower you very carefully nearer and nearer to the event horizon. Careful! Watch out! With a sickening snap, the cable breaks, and you fall irretrievably toward the black hole. It's no use blaming me; I used the strongest cable money can buy. The problem is that as you approach the event horizon, no force in nature—not the electromagnetic force, not the strong force, and certainly not the weak force—can prevent you from falling toward the black hole (Problem 7.8). As you fall, it appears to me—stationed comfortably at a safe distance from the black hole—that it takes you formally an infinite time to reach the event horizon. Nevertheless, once you start falling, effective communication ends between us in about a millisecond, because all photon signals you may frantically try to send out to me either are gravitationally redshifted to undetectably long wavelengths by the time they reach me or fall with you into the black hole (Figure 7.11). In a sense, space and time seem to reverse roles as you approach the event horizon of the black hole. When you were far from the black hole,

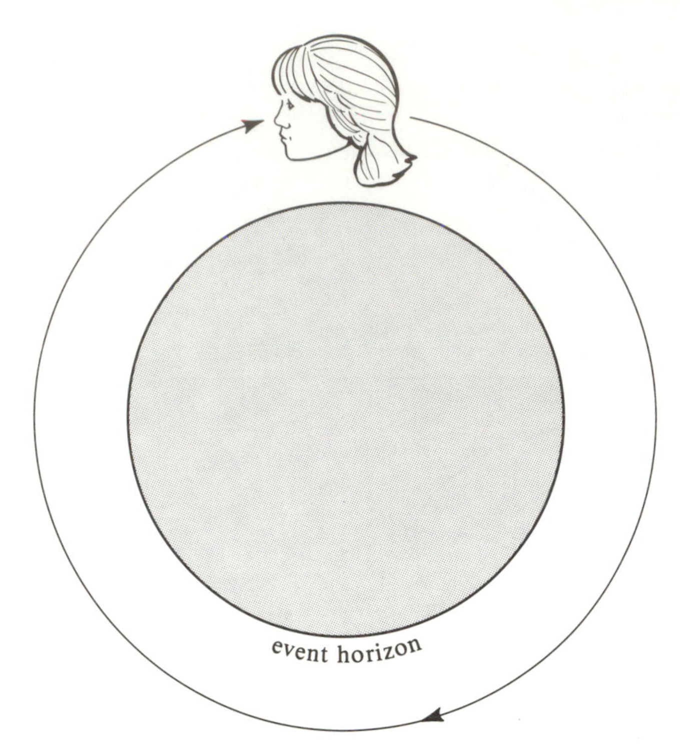

Figure 7.10. When at a circumference equal to 1.5 times the circumference of the event horizon of a black hole, a suitably suspended astronaut can see the back of her own head without the benefit of any mirrors.

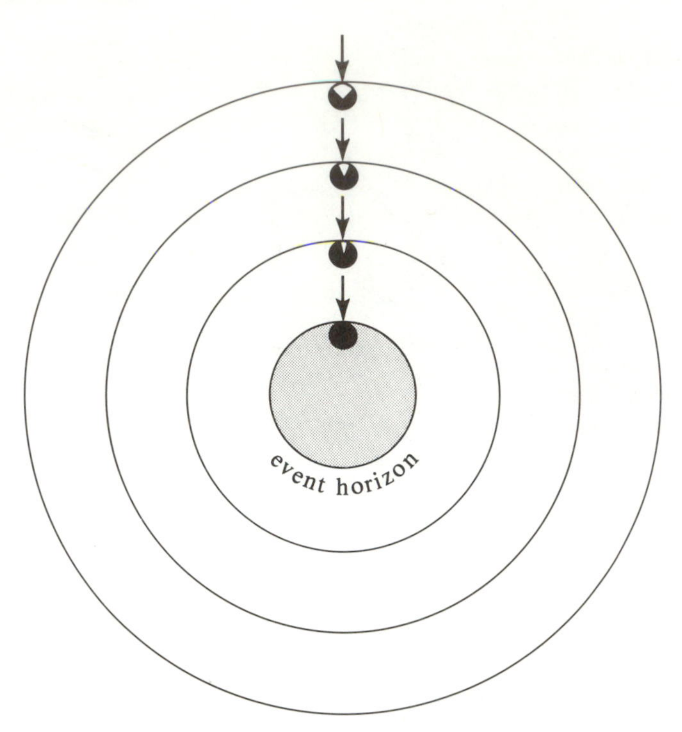

Figure 7.11. Journey to the center of a black hole. As an adventuresome explorer falls toward a black hole, he sends out light signals in all directions. The explorer sees locally the photons race away from him in all directions at the speed of light c (small circles). However, only the photons traveling in the outward direction of the light cone will escape from the gravitational clutch of the black hole to reach eventually any outside observer. The rest of the photons (dark part of small circle) are doomed to fall into the black hole with the explorer. The escape cone becomes smaller and smaller as the explorer approaches the event horizon. In particular, when he reaches it, only photons going exactly radially outward can escape.

by exerting enough effort, I could hold your position steady with respect to the black hole. But there was nothing that I could do to stop the forward march of time; I could not, for example, stop you from aging. As you fall toward the event horizon of the black hole, there is nothing I can do to stop you from falling, but time seems to have stopped moving for you!

Problem 7.8. The acceleration g of a freely falling body near a Schwarzschild black hole is given by the formula

$$g = \frac{GM/r^2}{(1 - 2GM/c^2r)^{1/2}},$$

when at a radius where $2\pi r$ is the circumference. Show that far from the event horizon, the above formula reduces to the usual Newtonian expression. Show also that when $r = R_{\text{sch}}$, the gravitational force becomes infinite, and therefore irresistable. An infinite quantity is not very convenient to work with; hence, it is convenient to define the "surface gravity" which is just the numerator of the preceding formula evaluated at the event horizon, i.e.,

$$\text{“surface gravity”} = GM/R_{\text{Sch}}{}^2 = c^4/4GM.$$

A similar quantity can be defined for rotating black holes, which do not possess complete angular symmetry, but the "surface gravity" also works out to a constant (depending only on the mass M and angular momentum J).

Let Δt_∞ be the time interval elapsed for an observer at infinity. Corresponding to this, the time elapsed for a *fixed* local observer is

$$\Delta t_0 = (1 - 2GM/c^2r)^{1/2}\,\Delta t_\infty.$$

Show that a local observer far from the black hole ages at the same rate as the observer at infinity, and that the former's aging rate stops in comparison with the latter's when the event horizon is approached.

A particle falling with a parabolic velocity (zero binding energy) takes time (observed by the falling particle)

$$\Delta t = \frac{2}{3}\left[\frac{r}{c}\left(\frac{r}{R_{\text{Sch}}}\right)^{1/2} - \frac{R_{\text{Sch}}}{c}\right]$$

to fall from a "radius" r to the event horizon. Calculate Δt numerically if $R_{\text{Sch}} = 3$ km and $r = 10$ km. The observer at infinity thinks that it takes an infinite time for the particle to reach the event horizon.

How do you react as you fall? According to your watch, it takes only a finite time to reach and *cross* the event horizon. Indeed, it takes you only a finite time to reach the very center of the black hole, where a singular catastrophe awaits you. Nothing very peculiar happens as you cross the event horizon, except that all photons which you send outward using your flashlight cannot escape the clutches of the black hole either. Well, actually, one peculiar thing does happen to you as you fall that I forgot to mention before sending you on this journey. As you fall feet first toward the center of the black hole, the gravitational pull on your feet is considerably stronger than the gravitational pull on your head. This "tidal force" soon stretches you out in a long long string. Sorry. As a consolation, if you survive this modern-day torture rack, as you cross the event horizon, you will have entered another "universe" whose properties (apart from mass and rotation) none of us on the "outside" will ever sample.

Thermodynamics of Black Holes

The above discussion of black holes ignores some interesting quantum-mechanical effects discovered only recently. On page 135 we noted that the surface area of a nonrotating black hole is proportional to the square of the mass M of the black hole. Classically, matter-energy can only fall into black holes, never fall out; so the mass of a nonrotating black hole can only increase, never decrease. This result holds also for rotating black holes, providing we replace the word "mass" by the phrase "surface area." Indeed, a theorem about this result devised by Stephen Hawking is known as the *second law of black holes*:

> In any natural process, the surface area of the event horizon of a black hole always increases or, at best, remains constant; it never decreases (in the absence or quantum effects).

The wording of this theorem deliberately imitates that of the second law of thermodynamics, which states that entropy always decreases, never increases. In fact, corresponding to the zeroth, first, second, and third law of thermodynamics (Box 4.1), theorems have been derived by Bardeen, Carter, Hawking, and Penrose called the zeroth, first, second, and third law of black-hole dynamics.

Problem 7.9. Let us use the second law of black-hole dynamics to calculate the maximum efficiency of black-hole processes as the second law of thermodynamics can be used to calculate the maximum efficiency of heat engines. As an example, consider the problem of the maximum energy extractable—say, in the form of gravitational radiation—from the merger of two black holes of equal mass M. Assume a head-on collision so that the initial and final black holes are nonrotating. In this simple situation, increase of the surface area implies

$$(16\pi G^2/c^4)M_{\text{final}}{}^2 > (16\pi G^2/c^4)M^2 + (16\pi G^2/c^4)M^2.$$

The energy released after all the commotion has died is given by $E = M_{\text{initial}}c^2 - M_{\text{final}}c^2$, where $M_{\text{initial}} = 2M$. The efficiency ε is defined to be $E/M_{\text{initial}}c^2$. From these considerations calculate the maximum possible value of ε. Numerical simulations show that the actual amount of gravitational radiation generated by head-on collisions of identical black holes involves on the order of 1 percent efficiency. How does this compare with the theoretical maximum?

The analogy between the second law of black-hole dynamics and the second law of thermodynamics led Jacob Bekenstein in 1972 to propose that the surface area of a black hole's event horizon is proportional to its entropy. This bold suggestion that a black hole is physically a *thermal* body was truly revolutionary, potentially as important for physics as the early thermodynamical discussions of the properties of blackbody radiation. (A black hole is the ultimate blackbody, since it absorbs every photon which falls on it.) For the idea to have merit, however, it must be possible: (a) to define precisely what is meant by the "entropy" of a black hole; and (b) to associate the concept of a temperature with a black hole.

Bekenstein proposed to define the entropy of the black hole on the basis of the theorem, "A black hole has no hair," which was proved by Carter, Hawking, Israel, and Robinson. This theorem has the following meaning. After a black hole is created by gravitational collapse and settles into an equilibrium configuration, an outside observer can sample only three possible properties of the black hole: its mass, its angular momentum (if it rotates), and its electric charge. Usually, the last is zero, because any black hole which contains a net charge will quickly attract an equal number of opposite charges. Thus, the only practical observables of a black hole are its mass and angular momentum; all other details ("hair") about the makeup of the matter of the black hole is lost in the gravitational collapse. This implies a large loss of information, and since information content is the negative of entropy, it should be possible to assign an entropy to the black hole. Adopting the methods of statistical mechanics (Chapter 4), Bekenstein proposed that the entropy of a black hole is proportional to the natural logarithm of the total number of quantum microstates possible for a black hole of given total mass and total angular momentum. This definition indeed makes the entropy proportional to the surface area of the event horizon.

To associate a temperature with a black hole, Bekenstein used the analogy of thermodynamics. The zeroth law of thermodynamics states that the temperature of a body is uniform at thermodynamic equilibrium. The zeroth law of black-hole dynamics states that the "surface gravity" (Problem 7.8) at the event horizon of a black hole is uniform at equilibrium. Should the temperature of the black hole therefore be taken to be proportional to its "surface gravity?" This identification is reinforced by the first laws of thermodynamics and black-hole dynamics. The first law of thermodynamics equates the product of the temperature and a change of the entropy with the amount of heat which is added to a body, heat being a form of energy. The first law of black-hole dynamics equates the product of the "surface gravity" and the change in surface area with the amount of energy added to the black hole. (Actually, the two are only proportional to one another. This proportionality is most easily derived for a nonrotating black hole; see Problem 7.10.) Thus, the first laws of thermodynamics and black-hole dynamics support the notion that temperature is proportional to the surface gravity of a black hole.

Problem 7.10. (for those who know more calculus). Problem 7.8 showed that the "surface gravity" g_s of a nonrotating black hole is inversely proportional to the mass M. Given that the surface area A is proportional to M^2, show that the infinitesimal change dA is proportional to M dM. In this manner, show $g_s dA \propto dM$. On the other hand, the first law of thermodynamics states $T dS = dQ$, where T is the temperature, S the entropy, and Q the heat. Summarize in words how these two statements are related. *Hint*: $E = Mc^2$.

Bekenstein's analysis suggested that the temperature of a black hole with a finite mass is nonzero. This leads to several paradoxes, because the four laws of black-hole physics were originally derived using classical physics plus general relativity, whereas Bekenstein's proposal contained elements from quantum theory. For example, thermal bodies with a finite temperature should emit thermal radiation (in accordance with Planck's law). Yet in the classical approximation, matter-energy could only fall into black holes; nothing could ever emerge from black holes. Hawking realized that this conclusion might not hold quantum-mechanically; therefore, Hawking considered the quantum-mechanical processes of the creation and destruction of particles near the event horizon of a black hole.

In modern relativistic quantum mechanics, the vacuum is not the uninteresting empty state it is in classical mechanics. Instead, modern physicists speak of the vacuum as a place where all manner of "virtual processes" of the creation and destruction of particle-antiparticle pairs occur all the time. The particles and antiparticles are called "virtual" because their acquiring permanent existence would violate the principle of conservation of energy or momentum (Chapter 6). Nevertheless, they can appear transiently as long as they do not exist long enough to be observed. Heisenberg's uncertainty principle—in the form $(\Delta E)(\Delta t) > h$—allows the observation of particle-antiparticle creation of mass-energy $\Delta E = 2mc^2$, providing they exist for a period of time $\Delta t > h/\Delta E$. As long as the creation and destruction of such "virtual" entities follow on the heels of each other within less time (Figure 7.12), we can have no inkling of this "vacuum fluctuation" (Problem 7.11).

Problem 7.11. Calculate $h/\Delta E$ for $2m = 2m_e$, the mass of an electron-positron pair.

In quantum mechanics, the exchange of virtual photons and gravitons across space "explains" Coulomb's force law for charged particles and Newton's force law for gravitating masses (see Figure 6.1). More dramatic successes are encountered in quantum electrodynamics, especially for the electric charge of an electron. The "bare charge" predicted for an electron by quantum electrodynamics is infinite. Because of the continual creation and destruction of electron-positron pairs in the quantum vacuum, however, a real electron will attract virtual positrons toward itself and repel virtual electrons. The resulting cloud of excess positive charge surrounding the real electron cancels most of its bare charge, leaving the net small charge, $-e$, that is measured by experiments carried out at large distances from the electron. Sampled at closer distances where the layers of shielding are partially penetrated, the measured charge would in-

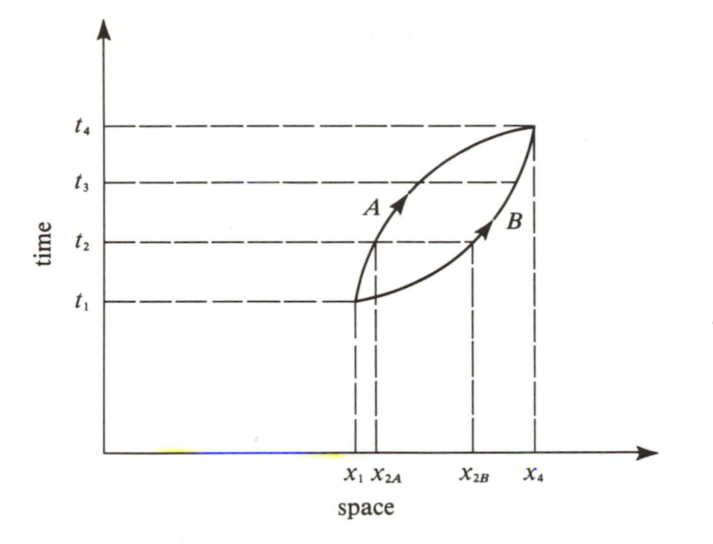

Figure 7.12. The creation and destruction of a particle-antiparticle pair in a vacuum is represented schematically in a Feynman diagram by two curved trajectories in spacetime which begin and end at two vertices (points of intersection). When the entities are created at time t_1 at the location x_1, they may move at different velocities. In the above example, entity A covers less space, $x_{2A} - x_1$, during duration, $t_2 - t_1$, than entity B covers, $x_{2B} - x_1$. Eventually, however, the attractive forces between the two unlike entities bring them together again at the same spacetime point (x_4, t_4), and they annihilate each other. In accordance with Heisenberg's uncertainty principle, the particle and antiparticle do not exist for sufficient distance or time to make their momenta or energy observable; they constitute a "virtual" pair.

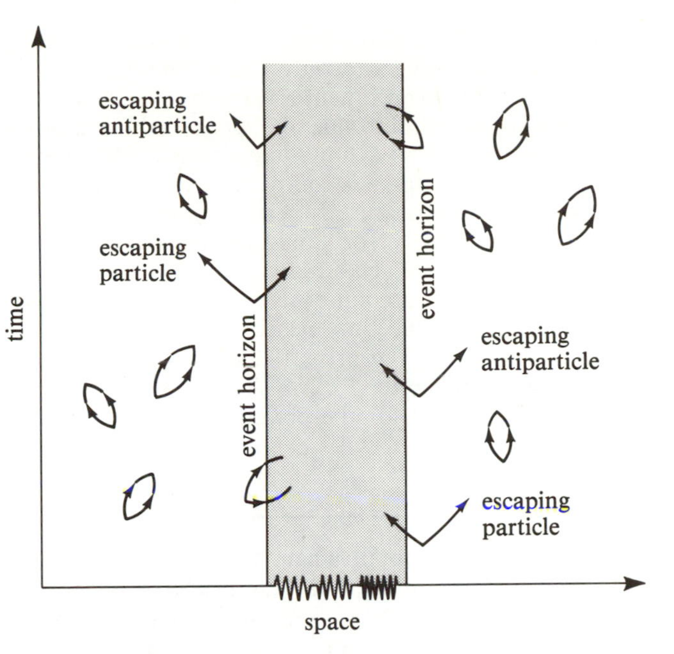

Figure 7.13. Hawking process near the event horizon of a black hole. The creation of particle-antiparticle pairs usually ends with their destruction a short time later. Occasionally, however, the tidal forces of the black hole can cause one of the pair to fall into the black hole, enabling its liberated partner to "tunnel" safely to infinity and escape. Notice the assumption that the black hole is (quasi) stationary implies that the two sides of the event horizon remain fixed in space. Actually, this diagram is even more schematic than usual, because the actual spacetime near a black hole is highly curved. Chapter 15 explains the concept of spacetime curvature in greater detail.

crease in magnitude. Precisely such effect has been detected in the so-called "Lamb shift" of the spectral lines of the hydrogen atom. The implied increase in electromagnetic force above the usual value when electric charges are approached very closely plays an important role in "grand unification theories" based on the idea that electromagnetic, weak, and strong forces should have the same intrinsic strength at high energies (Chapters 6 and 16).

Hawking imagined similar virtual processes to occur near the event horizon of a black hole (Figure 7.13). Usually, destruction follows closely upon the heels of creation, and no detectable change emerges at infinity. However, occasionally, a newly created particle and antiparticle are accelerated differently by the tidal force of the black hole. One of the two falls into the black hole (carrying negative energy), and its partner gains enough energy to materialize and escape to infinity. Quantum mechanics has allowed the particle (or antiparticle) to escape by "tunneling" under an energy barrier that classical mechanics considers impenetrable. The appearance of mass-energy at infinity is compensated for by a decrease of the mass-energy of the black hole, with net conservation of mass-energy. In this way, a black hole begins to "evaporate." To his surprise and delight, Hawking discovered that the statistical effect of many such emissions leads to the emergence of particles with a *thermal* distribution of energies! (Photons cannot be directly created by the Hawking process, but the electromagnetic interaction of the created thermal particles leads to a black-body radiation field if the conditions are optically thick.) Moreover, the temperature T associated with these thermal distributions for a nonrotating black hole is given by the formula

$$kT = \frac{hc^3}{16\pi^2 GM}, \tag{7.7}$$

which is inversely proportional to the mass M, and therefore directly proportional to the "surface gravity," just as Bekenstein had surmised!

Classically, Planck's constant, $h = 6.63 \times 10^{-27}$ erg sec, can be considered to be negligibly small, and equation (7.7) would imply that the temperature to be associated with a black hole is zero. Thus, in the classical approximation, a black hole can only absorb photons, never emit them. However, quantum mechanics assigns a small but nonvanishing value to h; hence, all black holes with finite mass M have finite temperatures. As long as the mass of a black hole is comparable to a stellar mass or larger, this temperature is negligibly small (Problem 7.12), and very little error would be introduced by approximating it to be at absolute zero. In this approximation, black holes are not observable except from their gravitational influence on nearby matter. When we discuss binary X-ray sources (Chapter 10) and active galactic nuclei (Chapter 13), we shall present some observational evidence for the presence of black holes in our universe.

Problem 7.12. To obtain a heuristic derivation of equation (7.7), proceed as follows. The uncertainty principle allows vacuum fluctuations to produce particles and antiparticles of energy ΔE for a time Δt such that $\Delta E \Delta t \sim \hbar/2$. If the maximum distance that a virtual pair can separate is comparable to half the circumference of the event horizon, $c\Delta t/2 \sim \pi 2GM/c^2$, then one of the pair has a reasonable chance of escaping to infinity while its partner falls into the black hole. If the energy ΔE that materializes at infinity is characteristic of a thermal distribution, $\Delta E \sim kT$, show that the temperature T of the black hole is given by equation (7.7). Notice, of course, that we have deliberately "fine-tuned" our argument unjustifiably to obtain the correct numerical coefficient. Use equation (7.7) to compute T when $M = 1\ M_\odot$, and when $M = 10^{15}$ gm, the mass of a mini-black-hole. What is the size R_{Sch} for $M = 10^{15}$ gm? Observationally, there seem to be few, if any, such mini-black-holes in the universe. (See next problem.)

Problem 7.13. It might be thought that if a black hole has temperature and radiates, losing surface area (and therefore entropy), this would violate the second law of thermodynamics. In fact, as Bekenstein has explicitly shown, the entropy lost by the black hole is more than made up by the entropy produced in the thermal radiation. The second law of thermodynamics, stated in its most general form, is thus still satisfied, since the total entropy of the universe, a black hole plus outside, has increased in the evaporation process.

Ignore particulate emission, and suppose for simplicity that a black hole radiates only electromagnetic energy at a rate $L = 4R_{\text{Sch}}{}^2\sigma T^4$, where R_{Sch} is given by equation (7.4), T by equation (7.7), and σ by Appendix A as $(2\pi^5/15)(k^4/h^3c^2)$. Show now that the energy-loss rate can be written as

$$L = \frac{hc^6}{30720\pi^2 G^2 M^2},$$

which demonstrates that a black hole would not radiate if h were zero, and that low-mass black holes emit more quantum radiation than high-mass black holes. Compute L numerically for $M = 10^{15}$ gm. In what part of the spectrum is this radiation mostly emitted?

The time-rate of decrease of the mass-energy content, Mc^2, of an isolated black hole must equal the above expression for L. Argue thereby that Mc^2/L gives roughly the time t required for a black hole of mass M to evaporate completely. Because the evaporation accelerates as M decreases, calculus makes the actual time shorter by a factor of 3. Show now that

$$t = 2560\pi^2(2GM/c^2)^2(M/h).$$

Calculate t for $M = 2 \times 10^{14}$ gm. Convert your answer to billions of years. Draw appropriate astronomical conclusions.

Problem 7.14. Suppose an evaporating black hole reaches a mass m such that its Compton length h/mc, the scale on which it is "fuzzy" (see Chapter 6), becomes comparable to its Schwarzschild radius, $2Gm/c^2$. Show that the argument of Problem 7.12 must break down, because any particle or antiparticle emitted would have $\Delta E = \Delta mc^2$, comparable to the mass-energy contained by the black hole as a whole (no pun intended). Except for factors of order unity, show that the value of m when this happens is given by the Planck mass

$$m_P = (hc/G)^{1/2}.$$

Compute m_P numerically, and compare it with the mass of the proton m_p. How does m_P relate to Problem 6.13?

To discuss black holes of mass m_P or less requires a quantum theory of gravity. Such a theory does not yet exist (see Chapter 16); general relativity is a classical theory of the gravitational field (the structure of spacetime). Hawking's discussion of evaporating black holes is therefore semiclassical, in that it treats particles quantum-mechanically but the gravitational field (spacetime) classically.

Concluding Philosophical Remarks

At the end of Chapter 3, we noted that the truly great achievements always unify large bodies of knowledge previously thought to be disconnected. By this criterion, the accomplishments of Chandrasekhar, Bekenstein, and Hawking can truly be termed great. With the derivation of the limiting mass for white dwarfs, Chandrasekhar demonstrated explicitly the intrinsic unity of the world of elementary particles and the world of stars. His work also opened up the frontiers of research on compact objects. In Chapter 8, we will see the crucial role played by his theory of white dwarfs for the entire story of stellar evolution.

Bekenstein and Hawking's achievement is equally remarkable. We have noted throughout this book that two central themes recur in all of astronomy: gravitation and the second law of thermodynamics. Modern gravitation theory reaches its purest form in the theory of black holes. Classically, it would be correct to say that they are pure gravity. Yet quantum-mechanically, in these amazingly simple objects whose souls are completely expressed at the event horizon, Bekenstein and Hawking have found that gravitation and thermodynamics meld into a single beautiful scheme. And in this union, we discover that black holes are, after all, not truly stable end states of matter. Given enough time, black holes will get hotter and hotter relative to the universe and evaporate away!

Is this failure of the ability of quantum mechanics to rescue stars from the final ravages of the war between gravity and thermodynamics unique to black holes? No. If current ideas are right that protons and neutrons are not fundamentally stable states of matter (Chapter 6), then, given enough time, even the baryons inside white dwarfs and neutron stars will ultimately decay into photons and leptons. The positrons will annihilate with the electrons, and then the universe will be left with nothing but neutrinos and photons, which fly away from each other at the speed of light. Are all material entities then merely temporary expedients in nature's scheme to turn matter into energy? Among the four end states of stars (Box 7.1), is the only one that is truly compatible with the inexorable demands of thermodynamics, the uninteresting state: nothing? A possible escape may yet exist for matter from this terrible sentence. This escape hinges on the crucial issue of whether infinite time is indeed available to all objects in the universe (Chapter 15). In the arena of the ultimate fate of the entire universe, gravity may yet triumph over thermodynamics.

CHAPTER EIGHT

Evolution of the Stars

In Chapter 5, when discussing the structure of the Sun, we mentioned in passing that the two most important characteristic governing the properties of a star were its initial mass and its chemical composition. From observations of the orbital motions of binary stars (Chapter 10), we know today that ordinary stars come in a variety of masses. The lightest have somewhat less than a tenth of a solar mass; the heaviest, somewhat greater than fifty solar masses. We understand this result theoretically in the following way.

Gaseous bodies less than about $0.08M_\odot$ do not possess enough self-gravity to compress their central regions to sufficiently high temperatures to ignite the nuclear fusion of hydrogen; so they cannot shine with starlight produced by nuclear reactions. Such bodies—like Jupiter—are not called stars. (They may generate some light and heat by gravitational contraction. Jupiter generates twice as much energy as it reflects in sunlight. However, gravitational contraction to *planetary* sizes cannot compete seriously with nuclear energy as a major energy source in the universe.)

Bodies greater than about $60M_\odot$, on the other hand, possess so much self-gravity that their interiors are compressed to very high temperatures. Under such conditions, radiation pressure begins to dominate greatly matter pressure. (Radiation pressure is proportional to the fourth power of the temperature; whereas, for given density, matter pressure is only proportional to the first power.) Stars evidently cannot exist *stably* or cannot form in the first place, with such high masses. In any case, very high-mass stars are not found in nature, as was first emphasized by Eddington.*

The initial chemical composition of stars can be deduced by a variety of techniques. The most direct was initiated by Payne (later Payne-Gaposchkin) and consists of analyzing the line spectrum of the radiation which comes from the photosphere of a star. One then assumes that the photosphere is uncontaminated by nuclear processing in the interior, so that the photospheric composition is basically primitive. For most stars, this assumption is a good one, but mixing with enriched matter from the interior seems to take place in some stars. For example, some stars called S stars (Chapter 9) contain the radioactive element technetium in their spectra. The half-lives of all the isotopes of technetium are less than a few million years, quite short by stellar standards. Consequently, technetium at least must have recently been synthesized in the interiors of these stars in question, then brought to the surface by some process or another. Other stars, called R and N stars, show anomalously high abundances of carbon in their spectra. However, stars with peculiar chemical abundances in their photospheres constitute a definite minority of all stars

* An exception may exist in 30 Doradus, a region in the Large Magellanic Cloud, which some astronomers believe is kept ionized by the ultraviolet photons pouring from a single star of a few thousand solar masses.

which have been observed. Henceforth, when we talk about the chemical composition of stars, we shall mean the *photospheric* composition unless otherwise noted.

From spectroscopic studies of normal stars and their surroundings, it has been learned that most stars initially contain about 70 percent hydrogen by weight and about 28 percent helium. The initial mass fraction of heavier elements is small and varies considerably from 2 to 3 percent in stars like the Sun, to 0.1 to 0.01 percent in stars found in globular clusters. Stars which are relatively rich in heavy elements—such as our Sun—are called **population I** stars; stars which are relatively poor in heavy elements—such as globular-cluster stars—are called **population II** stars. It is generally believed that population I stars represent a later generation of stars, formed after some enrichment of the gas clouds between stars had already taken place, and that population II stars represent an earlier generation, from when elements heavier than hydrogen and helium were generally quite rare in the gas clouds which gave birth to the stars. No star with *zero* heavy-element abundance has ever been found; however, some astronomers have theorized that such stars do exist and have dubbed this hypothetical *earliest* generation of stars, **population III**.

The main lesson to be learned from the study of chemical composition of the stars is that stars evidently start their lives predominantly made of hydrogen and helium. These two elements are vastly overrepresented in the universe in comparison with what nuclear statistical equilibrium would favor at typical temperatures in the universe today, since at low temperatures, matter prefers more binding energy, i.e., more heavy elements like iron. This predominance of hydrogen and helium in the present universe suggests that the universe started under conditions of extremely high temperature (Chapter 16). Under such conditions, matter prefers more particles, i.e., more simple elements like hydrogen and helium.

In any case, we learn from the above discussion that the Sun is quite a mediocre star, in terms of both its chemical composition and its mass. It is reasonably rich in heavy elements, but not as rich as extreme population I stars. It also has a lowish mass, but even that is quite normal, since the Galaxy turns out to have many more stars of low mass than stars of high mass. We shall soon see that the age of the Sun is also quite mediocre.

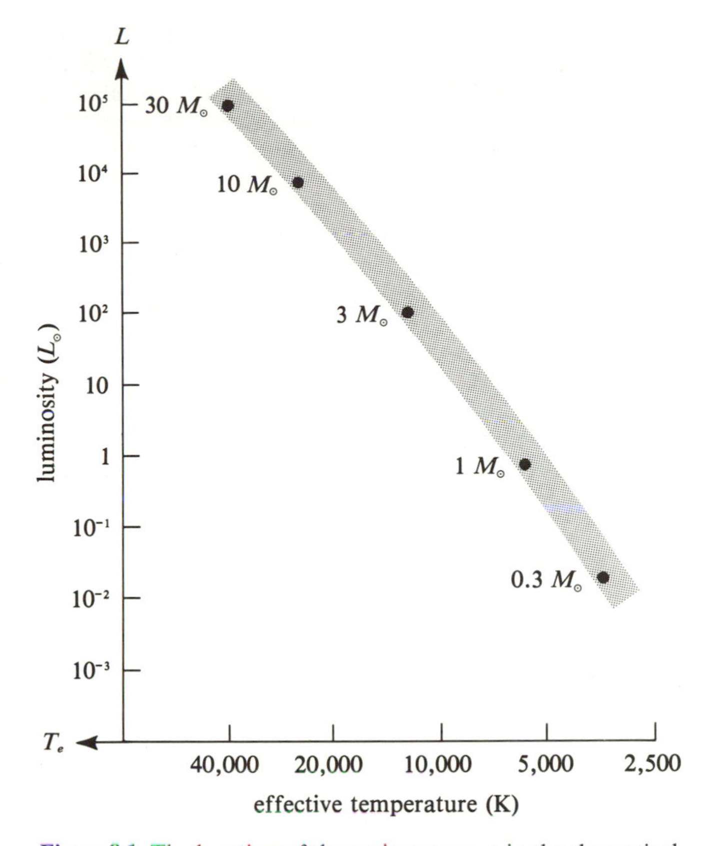

Figure 8.1. The location of the main sequence in the theoretical Hertzsprung-Russell diagram. The points give the theoretical positions of stars of various masses which have just begun their lives as main-sequence stars (the so-called "zero-age main sequence").

Theoretical H-R Diagram

Many properties of stars are conveniently discussed in terms of the location of a star in the theoretical Hertzsprung-Russell (H-R) diagram. The theoretical H-R diagram is a plot of the luminosity L of a star versus its effective temperature T_e.* Most stars when plotted in such a diagram are found to be located on a diagonal band called the **main sequence** (Figure 8.1). Stars on the main sequence are essentially chemically homogeneous, and are burning hydrogen into helium in their cores. In this state, where the stars have the same fractional amounts of hydrogen, helium, etc., everywhere throughout the star (except for the partial nuclear processing in the core), the stars can quietly and steadily shine for the greatest period of their luminous lifetime. Consequently, most easily observed stars for which we can derive L and T_e observationally (Chapter 9) lie on the main sequence.

Properties of Stars on the Main Sequence

In the above we stated that the two important properties which characterize the main sequence are chemical homogeneity and core-hydrogen burning. What, then,

* See Chapter 5 for a review of the definitions of these quantities.

distinguishes stars along the main sequence? I.e., what makes one star lie at one position on the band, and another star at another position? It turns out that different initial amounts of elements heavier than hydrogen and helium make relatively little difference in the gross properties on the main sequence. Similarly, the amount of hydrogen used up in the core, which depends on a star's age, can shift its position inside the band relatively little, as long as the core hydrogen is not *all* used up. The major factor which determines a main-sequence star's position in the H-R diagram is its *mass*. High-mass stars expend energy much faster than low-mass stars. As Eddington first discussed, the rate at which heat leaks out of stars by radiative diffusion yields a luminosity L which depends roughly on the mass M to the fourth power (as an average for the entire range of masses to be found on the main sequence; see Problem 8.1):

$$L \propto M^4.$$

Problem 8.1. In Problem 5.12 we showed that the rate of radiative leakage of photons from the interior of a star was given by

$$L = \frac{(4\pi R^3/3)(aT^4)}{3R^2/lc},$$

where R is the radius of the star, T is its mean interior temperature, a is the radiation constant, c is the speed of light, and l is the mean free path for the "random walk" of a photon. To derive a mass-luminosity relationship, we wish to express T and l in terms of M and R. Moreover, for our purposes here, we are interested only in proportional relationships. It turns out that the opacity in main-sequence stars is such that the mean free path l works out to be

$$l \propto T^{3.5}/\rho^2 \quad \text{for stars with low to medium mass,}$$
$$l \propto 1/\rho \quad \text{for stars with high to very high mass,}$$

where ρ is the mean density. The reason for these dependences is briefly the following. At very high temperatures, all the electrons are stripped from atoms, and the primary source of opacity is the scattering of X-rays from free electrons. This scattering depends only on the number of electrons per unit volume; i.e., the mean free path l in high-mass stars should be proportional to the reciprocal of the mass density ρ (the higher the density, the shorter the mean free path). At lower temperatures, some electrons are still bound to the ions, and these inner-shell electrons contribute most to the X-ray opacity. However, the number of such incompletely ionized contributors goes down if the temperature goes up, and the number rises if the density goes up. The mean free path l is inversely proportional to the number of opacity contributors, and l in low-mass stars works out to be $\propto T^{3.5}/\rho^2$.

Argue that (see Problem 5.9)

$$\rho \propto M/R^3 \quad \text{and} \quad P \propto GM^2/R^4,$$

where P is the total pressure. In stars with low to high mass, P can be taken to be the gas pressure, but in stars with very high mass, radiation pressure dominates. Argue therefore,

$$P \propto \rho T \quad \text{for stars with low to high mass,}$$
$$P \propto T^4 \quad \text{for stars with very high mass.}$$

Use the above formulae now to show the results:

$$L \propto M^{5.5}/R^{0.5} \quad \text{for stars with low to medium mass,}$$
$$L \propto M^3 \quad \text{for stars with high mass,}$$
$$L \propto M \quad \text{for stars with very high mass.}$$

To burn hydrogen in the core at the requisite rate in stars with low to medium mass, $R \propto M$. Thus, we may take $L \propto M^5$ to be representative of them (ignoring the fact that stars with very low mass are completely convective). Since stars with very high mass are extremely rare, argue that $L \propto M^4$ gives a good compromise for the entire range of masses on the main sequence.

The relationship $L \propto M^4$ means, for example, that a $10M_\odot$ main-sequence star radiates roughly 10^4 times more energy per second than a $1M_\odot$ main-sequence star (like the Sun). A $10M_\odot$ star has about ten times the energy reserve of a $1M_\odot$ star, because both have approximately 10 percent of their total hydrogen to use for core-hydrogen burning on the main sequence. But a $10M_\odot$ star spends its energy reserve at 10^4 the rate a $1M_\odot$ star does; thus, a $10M_\odot$ star should have a main-sequence lifetime which is only one-thousandth the main-sequence lifetime of the Sun. In other words, a $10M_\odot$ star only has about 10^7 yr to live as a main-sequence star. More generally, since the energy store $E \propto M$, and $L \propto M^4$, we may say that the lifetime $t_{\text{life}} = E/L$ is proportional to M^{-3}:

$$t_{\text{life}} \propto M^{-3}.$$

As Problem 8.1 shows, this relationship breaks down for both very heavy and very light stars, because L is then no longer accurately proportional to M^4. In particular, the main-sequence lifetimes of the heaviest stars turn out to be virtually independent of M, and are in the neighborhood of a few million years.

Stars less massive than about $2M_\odot$ fuse hydrogen into helium by the proton-proton chain, stars more massive do it by the CNO cycle (refer back to Chapter 6). The balance of energy release by hydrogen burning in the core and energy flow to the surface leads to a radius R on the the sequence which is roughly proportional to the mass:

$$R \propto M.$$

Thus, a $10M_\odot$ star is roughly ten times the size of the Sun. (Actually, it is somewhat smaller, since $R \propto M$ is only accurate on the lower main sequence. On the upper main sequence, $R \propto M^{0.6}$ gives a better approximation.) If we combine $R \propto M$ and $L \propto M^4$ with $L = (4\pi R^2)\sigma T_e{}^4$, we can derive

$$T_e \propto M^{1/2}$$

on the main sequence. Thus, high-mass stars have hotter surfaces and higher luminosities than low-mass stars on the main sequence. This explains why the main sequence falls in a diagonal band on the theoretical H-R diagram. (Please notice that convention decrees effective temperature T_e to be plotted increasing to the *left*.)

Since the Sun has an effective temperature of 5,800 K, we can estimate that a $10M_\odot$ main sequence star would have a surface temperature of about 20,000 K, whereas a $0.5M_\odot$ star would have a surface temperature of only 4,000 K. Thus, the Sun is a yellow star; a $10M_\odot$ main-sequence star would be blue; and a $0.5M_\odot$ main-sequence star would be red. *High-mass main-sequence stars are intrinsically brighter and intrinsically bluer than low-mass main-sequence stars.*

Stars of a given mass have their smallest sizes as normal stars when they are on the main sequence; thus, main-sequence stars are also called *dwarfs*. Main-sequence dwarfs are not to be confused with white dwarfs, which are not normal stars, and which are quite a bit smaller than their main-sequence counterparts of the same mass. White dwarfs of a given mass have the same radii, but they may have different locations in the H-R diagram if they have different effective temperatures. This effect is shown in Figure 8.2; notice, in particular, that as a white dwarf cools to come into thermodynamic equilibrium with the universe, it follows an evolutionary track which is a locus of constant radius in the theoretical H-R diagram. The locus followed by a normal star as it evolves is much more complicated, as we shall now discuss.

What happens when a star uses up its core hydrogen? The details depend on whether we are talking about

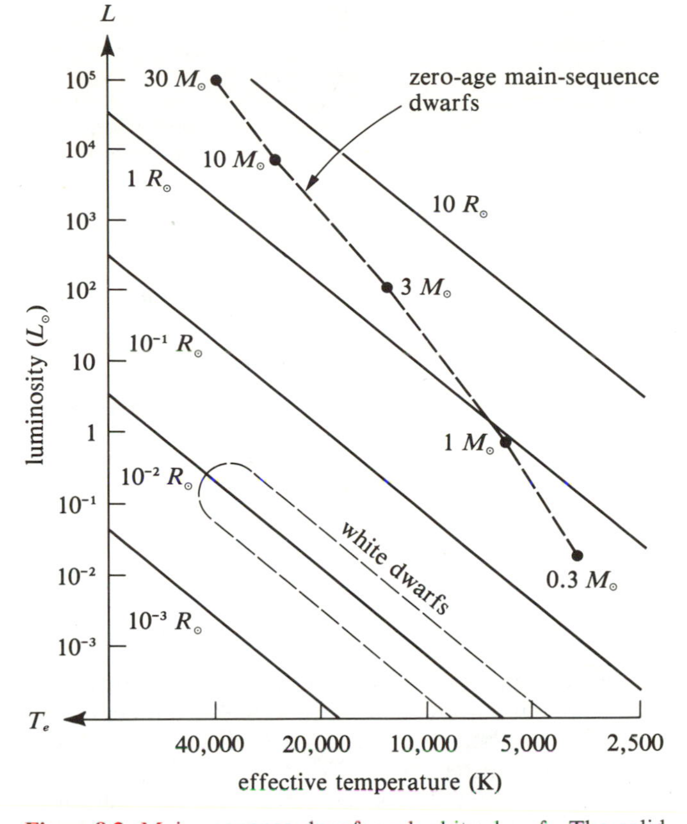

Figure 8.2. Main-sequence dwarfs and white dwarfs. The solid diagonal lines give loci of constant radii. Thus, a $1M_\odot$ zero-age main-sequence star has a radius just slightly less than $1R_\odot$; a $10M_\odot$ zero-age main-sequence star, somewhat less than $10R_\odot$. A typical white dwarf might have radius $10^{-2}R_\odot$, and as it cools, it would slide down the H-R diagram along the appropriate locus of constant radius.

low-mass stars or high-mass stars, but the central issue is similar for both. The central issue is the exhaustion of successive rounds of nuclear fuel as the star tries to combat the relentless pull of self-gravity and the inexorable flow of heat to the outside.

Evolution of Low-Mass Stars

Ascending the Giant Branch

To fix ideas, let us first discuss a low-mass star like our Sun. After hydrogen has been exhausted in the core, heat continues to leak out. Since there is no more nuclear energy generation in the core to make up the deficit, the core must contract gravitationally, much as Kelvin and Helmholtz originally envisaged in the nineteenth century

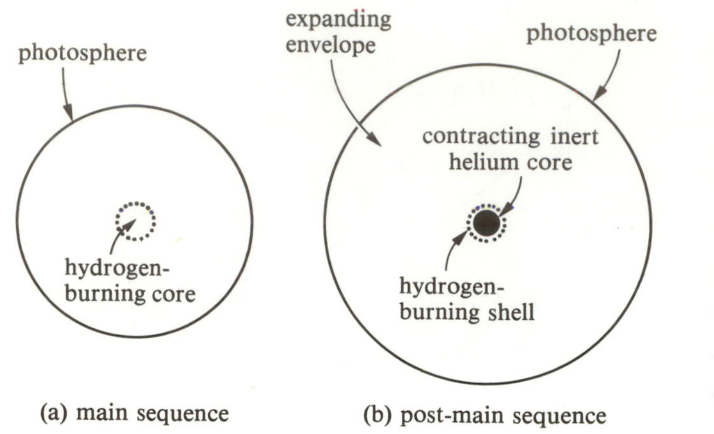

Figure 8.3. The structure of a star (a) on the main sequence and (b) as it begins to leave the main sequence because of core-hydrogen exhaustion.

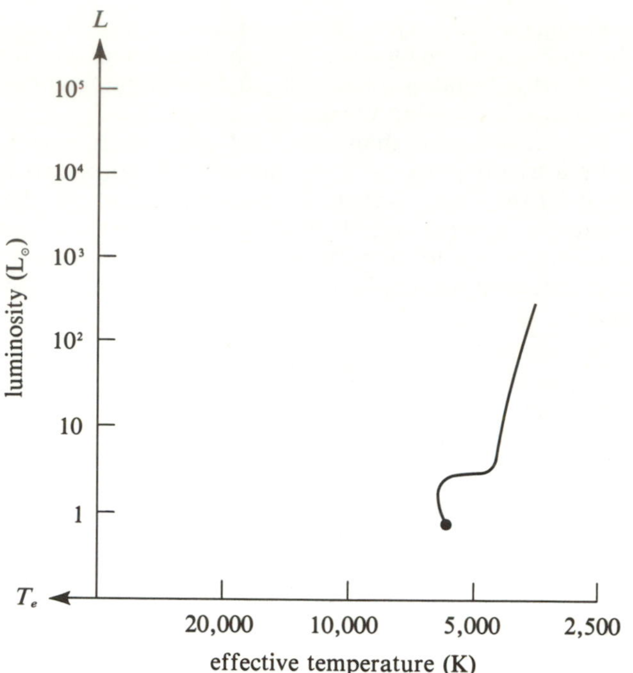

Figure 8.4. Ascent of a low-mass star to the red-giant branch. (Adapted from Icko Iben, *Ann. Rev. Astr. Ap.*, **5**, 1967, 571.)

(Chapter 5). As the core contracts, it heats itself up as well as the layers just above it. At the new higher temperatures, hydrogen can begin to burn in a shell just outside the hydrogen-exhausted core (Figure 8.3). However, the helium core itself still has no nuclear-energy generation, and as it continues to lose heat to the cooler overlying layers, it must continue to contract. This contraction is abetted as the surrounding hydrogen-burning shell drops more and more helium "ash" onto the core. The shrinkage of the core accompanied by the addition of more mass makes the gravity at the border of the core stronger and stronger, so the gravitational field felt by the hydrogen-burning shell becomes stronger and stronger. But the pressure in the shell equals the weight of a column of material of unit area above it (Chapter 5). This pressure must therefore try to increase to counterbalance the increasing gravity of the core. The pressure of the ordinary gas in the shell can be increased, in accordance to the perfect-gas law, either by raising the density or by raising the temperature. In fact, both occur, and both increase the rate of hydrogen burning in the shell.

However, not all the high luminosity generated in the shell finds its way to the surface. As long as the envelope remains **radiative**, the luminosity that it can carry is limited by the photon diffusion rate. The latter is nearly fixed for a star of given mass (Problem 8.1). The difference between the luminosity generated in the shell source and that leaving the surface goes into heating up the intermediate layers, causing them to expand. This expansion increases the total radius R; given a nearly constant value for the surface luminosity L, there must be a decrease of the effective temperature T_e, in accordance with the relation $L = 4\pi R^2 \sigma T_e^{\,4}$. The immediate post-main-sequence

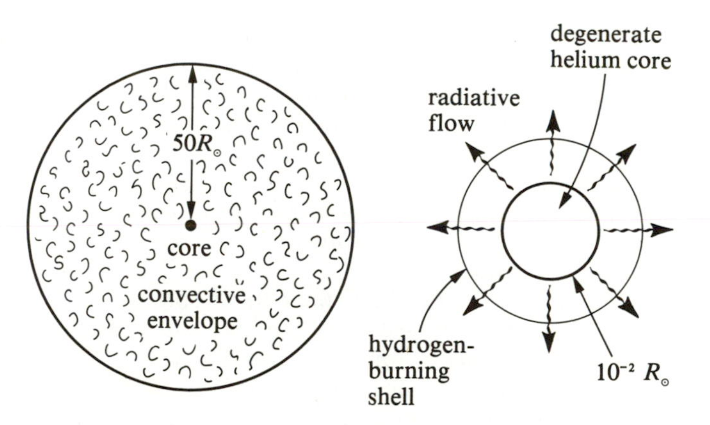

Figure 8.5. The structure of a red giant. The left figure shows the entire star from core to photosphere. The right figure shows an enlarged picture of the region near the core. Notice that the core, which may contain about half the total mass of a low-mass star at this point, occupies only one ten-billionth of the total volume.

evolution of a radiative star therefore moves the star's position more-or-less horizontally to the right in the H-R diagram, turning the dwarf star into a **subgiant**. The cooling and expanding surface layers cause the star to turn *red* in outward appearance.

As the star expands, however, the effective temperature cannot continue to fall to arbitrarily low values. Hayashi and his coworkers pointed out that the ability of the photospheric layers to prevent the free streaming of photons drops rapidly with decreasing temperatures (Problem 8.2). This ease for photon streaming in turn leads to a minimum temperature below which T_e is prevented from falling. The existence of such a temperature barrier forces the evolutionary tracks of low-mass stars in the H-R diagram, sooner or later, to travel almost vertically upward, turning the red subgiant into a **red giant** (Figure 8.4). The accompanying increase in the amount of shell luminosity which makes its way to the surface is too much for radiative diffusion to carry outward stably, and the entire envelope of the red giant becomes **convective** (Figure 8.5).

Problem 8.2. Let us try to understand Hayashi's important result qualitatively. A stellar photosphere is defined to be that level (a) where photons stop "walking" and start "flying," and (b) where the gas temperature T approximately equals the effective temperature T_e defined in terms of the surface luminosity L and radius R by $L = 4\pi R^2 \sigma T_e^4$. Argue that condition (a) implies that the mean free path of the photons, l, must be comparable to the thickness H of the photosphere. Argue that hydrostatic equilibrium requires the pressure P to balance the weight $\rho H g$. Use the perfect-gas law to derive now a relationship between l and T. In a cool stellar atmosphere, l is very sensitive to T and relatively insensitive to ρ. Show that this requires T to have almost a fixed value throughout a wide range of photospheric densities ρ. Thus, show that given R, the luminosity L must therefore accord with this value of T_e because of condition (b). If this required L exceeds the value which can be carried stably by radiative diffusion in the envelope (see Problem 8.1), argue that the envelope will become convective.

Meanwhile, the core continues to contract, and in a low-mass star, the free electrons become so tightly packed that they become degenerate (Chapter 7). If we could artificially peel off the overlying layers of a red giant at this point, the core would be essentially a low-mass (about $0.4M_\odot$) helium white dwarf. The very large border gravities associated with this "white dwarf" cause the hydrogen in the shell source to burn furiously, sending the star quickly up the red-giant branch. At the tip of the red-giant branch (in Figure 8.4), the temperatures in the core rise to about 10^8 K, which is high enough to ignite helium and "burn" it into carbon via the triple-alpha process (Chapter 6).

The Helium Flash and Descent to the Horizontal Branch

The ignition of helium in the core of a low-mass star, however, occurs under degenerate conditions; so it lacks the safety-valve feature which characterizes core-hydrogen burning on the main sequence (Chapter 5). There, a slight increase in the core temperature will lead to a pressure change that decreases the temperature. Here, the pressure increases primarily because of degeneracy effects, not because of thermal motions; so an increase of the core temperature leads to an overproduction of nuclear energy without a compensating pressure increase and a compensating expansion. Hence, an increase of the temperature, once helium has been ignited, tends to lead to a runaway production of nuclear energy:

higher temperatures → more nuclear energy
→ even higher temperatures, etc.

Therefore, helium burning turns on in low-mass stars with a "flash," as was first suggested by Mestel and verified in detailed calculations by Schwarzschild and Härm. So much energy is released in the "flash" that the core's temperature rises enough to remove the degeneracy. Normal thermal pressure then dominates over electron-degeneracy pressure, and the core expands. This expansion lowers the border gravity of the core, which weakens the hydrogen-shell source. Thus, although the star now has two nuclear energy sources—helium burning in the core, and hydrogen burning in a shell—the prodigious shell source is now so weakened that the star actually produces *less* luminosity than before. The lowered total luminosity is too little to keep the star in its distended red-giant state, and the star both shrinks in size and becomes intrinsically dimmer (Figure 8.6).

After the helium flash is completed, the core contains an ordinary (i.e., nondegenerate) helium plasma which is stably fusing helium into carbon. Surrounding this core is a hydrogen-burning shell, whose strength depends on the mass of the overlying envelope. This state of core-helium burning and shell-hydrogen burning is called the **horizontal branch** (Figure 8.7). The exact location of a horizontal-branch star in the theoretical H-R diagram depends not only on its initial mass and chemical composition on the main sequence, but also on the amount of envelope mass lost by the star as it ascended the red-giant branch. This mass loss can be expected because, with the low gravity at its distended surface, a red giant finds it even more difficult to retain coronal gas than the Sun (see the discussion on the solar wind in Chapter 5). Theoretical reasoning alone cannot yet predict quantitatively the amount of mass loss to be expected, but observations of red-giant stars show substantial mass loss. For a group of stars which start on the

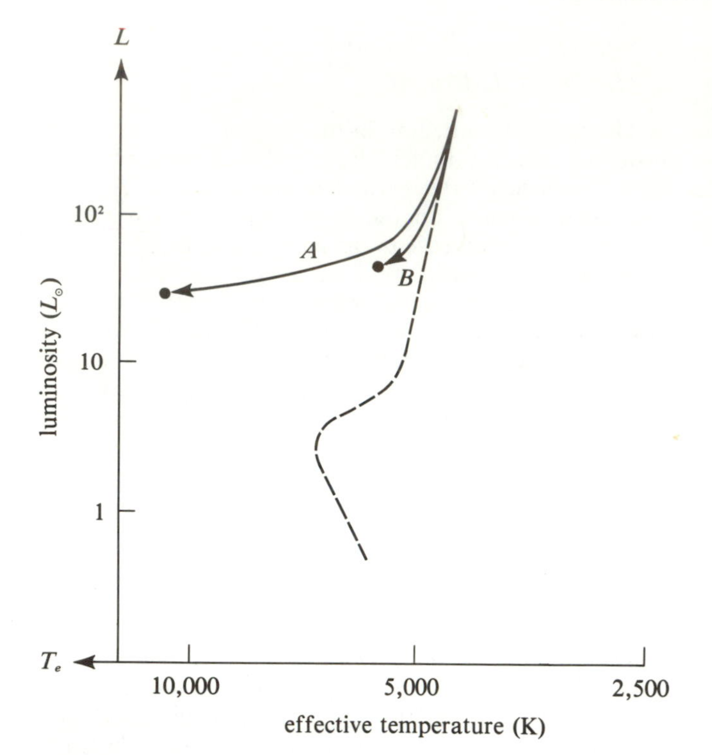

Figure 8.6. Descent of a low-mass star with poor heavy-element abundances (Population II star) from the tip of the red-giant branch to the horizontal branch. Track A corresponds to a star which suffered a relatively large loss of mass during the red-giant phase of stellar evolution. Track B corresponds to a star which suffered relatively little loss of mass. (Adapted from Icko Iben, *Ann. Rev. Astr. Ap.*, **5**, 1967, 571.)

lower main sequence with similar initial masses and chemical compositions, the stars which lose more mass on the red-giant branch end up on the horizontal branch with smaller envelope masses and, therefore, weaker shell sources in addition to the core source. In particular, a horizontal branch star which had lost *all* its envelope mass would be a chemically homogeneous helium star burning helium into carbon at its center. Such a state is often called the "helium main sequence" by analogy with the usual (hydrogen) main sequence. Even a relatively low mass (say, $0.5M_\odot$) helium star appears quite blue because of its moderately large luminosity and moderately small radius. Most horizontal-branch stars, however, have a finite envelope mass on top of such a helium-burning core, and the hydrogen-burning shell associated with the weight of this envelope keeps the envelope relatively distended. Thus, true horizontal-branch stars tend to have slightly higher luminosities than a "helium-main-sequence" star of the same core mass, as well as appreciably lower effective temperatures. If we started with a group of stars, therefore, with similar initial masses and chemical compositions, and if this group suffered varying amounts of envelope-mass loss ascending the red-giant branch, we should expect them to end on the horizontal branch with nearly the same luminosities but with different effective temperatures. In other words, such a group would occupy a *horizontal* locus in the H-R diagram, and this feature gives these stars their name. Faulkner found the extent of this horizontal locus to be most pronounced for stars which start with very low heavy-element abundances; this is entirely consistent with the observational fact that only globular clusters have well-defined "horizontal branches" (Chapter 9).

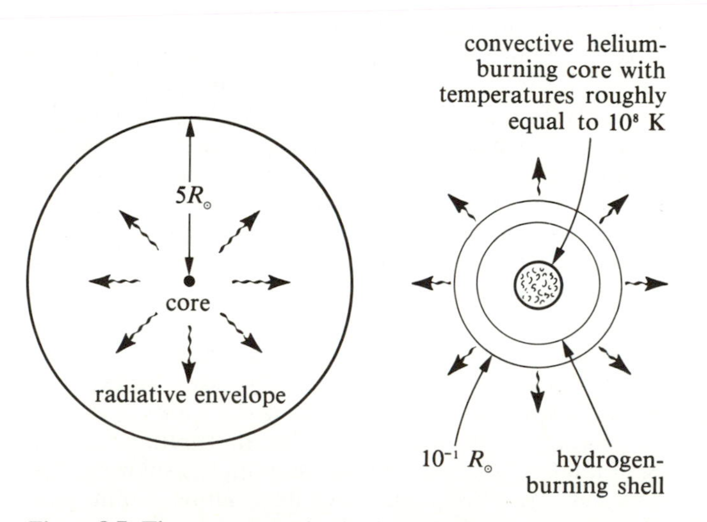

Figure 8.7. The structure of a horizontal-branch star. The left figure shows the entire star from core to photosphere. The right figure shows an enlarged picture of the region near the core.

Problem 8.3. To deduce the approximate location in the H-R diagram of a "helium main-sequence star" of $0.5M_\odot$, let us reconsider Problem 8.1. For a chemically homogeneous star of low mass which has a radiative interior, Problem 8.1 showed that the luminosity L satisfied:

$$L \propto RT^{7.5}/\rho^2,$$

with R equal to the radius, T the average internal temperature, and ρ the average mass density. If ordinary gas pressure dominates, T is related to the pressure P and the number density n by the formula

$$T \propto P/n.$$

The number density n and the mass density are related via $\rho = mn$, where m is the mean mass of all the constituent particles. Let $X =$ the mass fraction of hydrogen $= m_p n_H/\rho$, where n_H is the number density of hydrogen nuclei, and let $Y =$ the mass fraction of helium $= 4m_p n_{He}/\rho$, where n_{He} is the number density of helium nuclei. In the interior of the star, hydrogen and helium are completely ionized; thus, $n = 2n_H + 3n_{He}$, since each hydrogen atom contributes two particles (a hydrogen nucleus and an electron), and each helium atom contributes three particles (a helium nucleus and two electrons). With $Y = 1 - X$ (extreme Population II star which has virtually no heavy elements), show that the mean mass m is given by

$$m = \frac{4}{5X + 3} m_p.$$

Retrace now the arguments of Problem 8.1 to show that

$$T \propto Mm/R, \qquad L \propto m^{7.5}M^{5.5}/R^{0.5} \propto m^7 M^5 T^{0.5}.$$

Suppose a $1M_\odot$ hydrogen-main-sequence star ($X = 0.75$) has a luminosity of $1L_\odot$, a central temperature of 1.5×10^7 K (to be able to burn hydrogen in the core), and a radius of $1R_\odot$. Argue that a $0.5M_\odot$ "helium-main-sequence" star ($X = 0$) with a central temperature of 1.0×10^8 K (to be able to burn helium in the core) has a luminosity of $24L_\odot$ and a radius of $0.17R_\odot$. What is the effective temperature of such a star? How do these estimates compare with Figures 8.6 and 8.7?

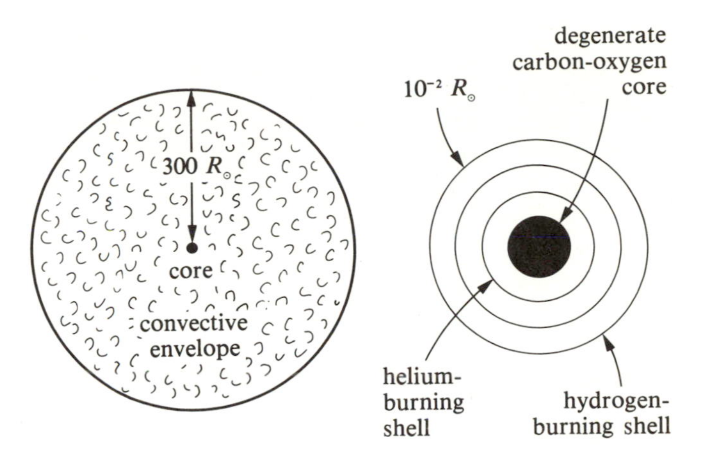

Figure 8.8. The structure of an asymptotic giant. The figure on the left shows the entire star from core to photosphere. The figure on the right shows an enlarged picture of the region near the core.

Ascending the Asymptotic Giant Branch

What happens when helium in the core of a horizontal-branch star is exhausted (by burning into carbon and oxygen)? Well, obviously, the core must contract, which increases the pressure and temperature of the overlying layers. Thus, helium ignites in a shell just outside the core, and hydrogen burns in a shell outside of that. The star is now in a double-shell-burning stage. The mass of the inert carbon-oxygen core continues to increase, and it continues to contract just as the helium core did when the star ascended the red-giant branch for the first time. Indeed, the energy generation in the (two) shell sources must proceed at an ever-increasing pace just as before, and the rapidly increasing luminosity must distend the overlying envelope just as before. Thus, the star must ascend the red-giant branch again, and the double-shell-source phase is also known as the **asymptotic giant branch** (Figure 8.8). Eventually, the shrinkage of the core again causes the free electrons to become degenerate. If the overlying envelope were now to be stripped away (say, by mass loss), the core would be a hot carbon-oxygen white dwarf. Now, however, the mass of the degenerate core is *larger* than before because of the additional "ash" in the core, and the radius of the incipient white dwarf is *smaller* than its helium counterpart at the tip of the red-giant branch.* Thus, the gravity of any overlying shell sources would be correspondingly larger, forcing them to generate higher luminosities yet. Stars at the end of the double-shell-burning phase may become red **supergiants**. At such tremendous rates of expenditures of energy, the star cannot live much longer.

What happens to stars at these very late stages of stellar evolution is theoretically quite uncertain; several complications make detailed computation difficult. One of these is the onset of "thermal relaxation oscillations" (discovered by Schwarzschild and Härm) when the helium-shell source becomes spatially very thin. The origin of this instability is very different from the "helium flash" discussed earlier. There, an initial overproduction of nuclear energy leads to a runaway because of the degeneracy of the nuclear-burning region. Here, an initial overproduction of nuclear energy also leads to a thermal runaway, but for an entirely different reason. Here, the nuclear burning region is nondegenerate, but it is a spatially *thin* shell. Thus, with the input of excess nuclear energy, the layer can and will expand. But the expansion of a *thin* shell does little to relieve the weight of the overlying material; this material is lifted only a little. Thus, the weight hardly changes, and therefore the pressure that the thin shell has to maintain to offset this weight also hardly changes. Meanwhile, the temperature has increased, and if the rate of nuclear-energy generation is

* Remember that a high-mass white dwarf has a smaller radius than a low-mass white dwarf.

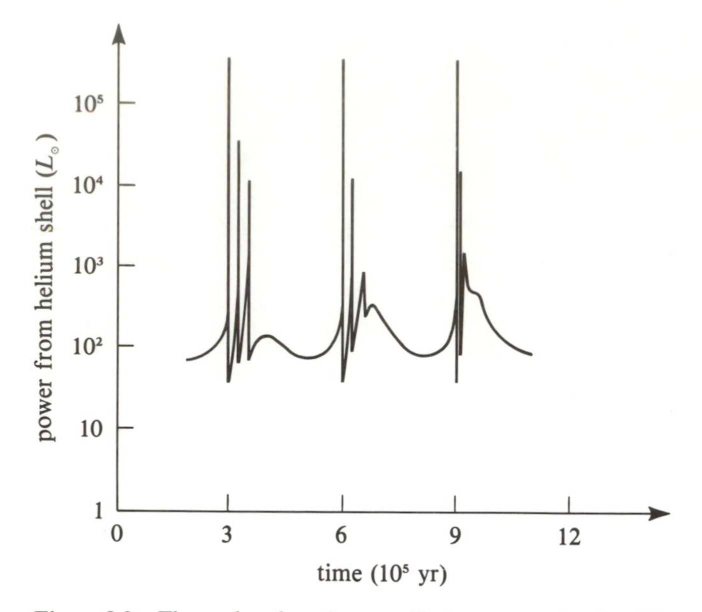

Figure 8.9. Thermal relaxation oscillations associated with helium-shell flashes. (Adapted from M. Schwarzschild and R. Härm, *Ap. J.*, **150**, 1967, 961.)

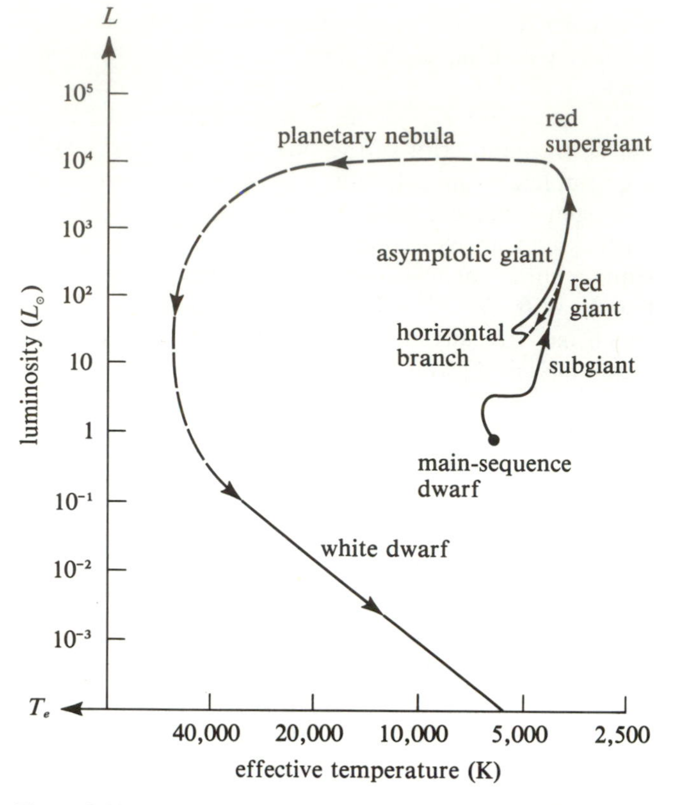

Figure 8.10. The complete evolution of a low-mass star from the main sequence to a white dwarf. The track from the asymptotic giant branch to a white dwarf (via a planetary nebula) is uncertain and is shown as a dashed curve.

sufficiently sensitive to temperature changes—as the triple-alpha reaction is—then it will also increase further before the excess heat has a chance to diffuse away. Thus, a thermal runaway ensues. The runaway is checked only after considerable expansion of the layer and the appearance of convection to carry away the excess heat. But the basic problem remains. After the runaway is checked, and when the star tries to adopt the "natural" double-shell-burning configuration appropriate for this stage of its evolution (i.e., when it tries to "relax" back to the "natural" stage of equilibrium), it finds itself in the same difficulty. Thus, the star undergoes a series of "thermal relaxation oscillations" which consist of one or more sharp pulses of extra energy generation followed by relatively long periods of quiet evolution (Figure 8.9). Each thermal runaway is followed by the development of a convection zone which extends from the helium-burning shell almost to the hydrogen-burning shell.

Some workers, notably Ulrich and Scalo, have suggested that the inner convection zone may actually connect to the outer convective envelope through the hydrogen burning shell (see Figure 8.8). If this happens, the products of helium burning, carbon and oxygen, as well as *s*-process material (Chapter 6), may be brought to the surface of the asymptotic red giant. This speculation may explain the so-called "carbon stars" as well as the peculiar behavior of a star known as FG Sagittae. Herbig and Kraft and their colleagues at the Lick Observatory have found that FG Sagittae has drastically changed in surface temperature and chemical composition (especially for *s*-process elements) just in the course of one decade! To find such profound changes in a normal star occurring right before one's eyes must have provided a unique experience in astronomy.

Planetary Nebulae and White Dwarfs

Another complication is that considerable mass loss may occur on the asymptotic giant branch. Observations suggest that asymptotic branch stars beyond the red-giant tip lose mass very rapidly. Among the many promising mechanisms which have been considered is the suggestion that small specks of dust may form in the cool atmospheres and be driven out subsequently by the radiation pressure of the star. Quantitative calculations, unfortunately, are difficult. Observations indicate that stars that originally had less than about six solar masses seem to lose so much mass during such high-luminosity stages that they become (perhaps periodically) planetary nebulae illuminated by a hot central core (Color Plate 8).

This hot central core is presumably an incipient white dwarf, with mass necessarily below the Chandrasekhar limit $1.4M_{\odot}$. From the central-star stage of planetary nebulae, the exposed core burns out its hydrogen and helium shells, loses its extended envelope, and descends the H-R diagram to enter the region occupied by white dwarfs proper. Figure 8.10 summarizes the complete evolution of a low-mass star from the main sequence to a carbon-oxygen white dwarf. The final approach to a white dwarf from an asymptotic giant star is shown in dashed lines, to emphasize that the theory is incomplete for these late stages of stellar evolution.

Take a good look at Figure 8.10. You may be staring at the future of the Sun. When the Sun swells up to become a red giant for the first time, it will occupy about 30° in the sky. What a sunset that would make! When the Sun swells up for the second time (assuming it does not lose so much mass that it becomes a helium white dwarf), it may well engulf the Earth. Any inhabitants, of course, would long since have been roasted. Be consoled at least that the Sun itself will ultimately be able to rest in peace as a senescent white dwarf.

Problem 8.4. Suppose the time-average luminosity of a $1M_{\odot}$ star after it leaves the main sequence (including the thick shell-burning phase discussed in Problem 6.3) to be $20L_{\odot}$. Suppose further that in ascending the giant branch the helium core increases by about $0.3M_{\odot}$. How long does it take such a star to ascend to the tip of the red-giant branch? Express your answer in billions of years. What ratio of red-giant stars to main-sequence stars is to be expected for $1M_{\odot}$ stars? What would their relative contributions be to the total light of a galaxy?

Evolution of High-Mass Stars

What happens to stars which start on the main sequence with masses greater than six solar masses? The evolution of these high-mass stars differs both quantitatively and qualitatively from the evolution of the low-mass stars just described. The quantitative difference is that high-mass stars evolve much more quickly. Martin Schwarzschild has likened a low-mass star to a poor man who has little and spends thriftily. A high-mass star is like a rich man who has much and spends prodigiously. Carrying the analogy further, we may say that very soon the rich man discovers that he has no further cash reserves. However, unlike the poor man who, finding himself in the same predicament, fades into obscurity, the rich man blows himself apart! The reason for this qualitatively different ending is that a high-mass star has enough mass, even after mass loss, to produce a white-dwarf core with a mass arbitrarily close to Chandrasekhar's limit, $1.4M_{\odot}$. Thus, the border gravities outside the core can grow almost arbitrarily large, and force the surrounding layers to as large a pressure and temperature as is needed to burn elements all the way to the iron peak. Once the iron peak is reached, the star cannot help but come to a catastrophic end. This overview of the evolution of high-mass stars succintly summarizes the central importance of Chandrasekhar's theory of white dwarfs for the story of stellar evolution. Let us now fill in some of the details.

Approach to the Iron Catastrophe

In high-mass stars, core-hydrogen exhaustion, helium-core contraction, and shell-hydrogen ignition occur pretty much as for low-mass stars, with only minor differences. The main difference is that in stars with initially more than $2.25M_{\odot}$, core-helium ignition occurs before the core has contracted very far. Helium ignites in the cores of these higher-mass stars under nondegenerate conditions, and there is no helium flash. During helium-core burning, hydrogen-shell burning continues at about the same rate as before core ignition. During core contraction, the star moves generally to the right (lower T_e) in the H-R diagram. After core ignition, the star moves generally to the left (higher T_e). In high-mass stars, these rightward (core exhaustion) and leftward (core ignition) excursions occur with only a slight systematic increase of the luminosity; hence, the evolutionary tracks of high-mass stars occur virtually horizontally in the H-R diagram (Figure 8.11). In very-high-mass stars, the nuclear evolution in the central regions of the star occurs so quickly that the outer layers have no time to respond to the successive rounds of core exhaustion and core ignition, and there is only a relatively steady drift to the right of the H-R diagram before the star arrives at the pre-supernova state.

In any case, helium-core burning and hydrogen-shell burning is followed by core-helium exhaustion. The carbon-oxygen core then contracts, and ignites a helium-shell source below the hydrogen-shell source, as before. In its turn, the carbon-oxygen core will also ignite. Carbon burning has a very high temperature sensitivity and a relatively large energy release per gram of fuel. Thus, it is important to know whether it is ignited under degenerate conditions or not, and that depends on whether the original mass was lower than eight solar masses or not. In stars of less than $8M_{\odot}$, Arnett suggested that the ignition of the carbon-oxygen core under degenerate conditions might lead to a detonation wave and a supernova explosion. However, there are some observational objections to this suggestion, and several theoretical proposals have been advanced to avoid the dilemma posed by Arnett's suggestion. In stars whose original mass was higher than about $10M_{\odot}$, the carbon will

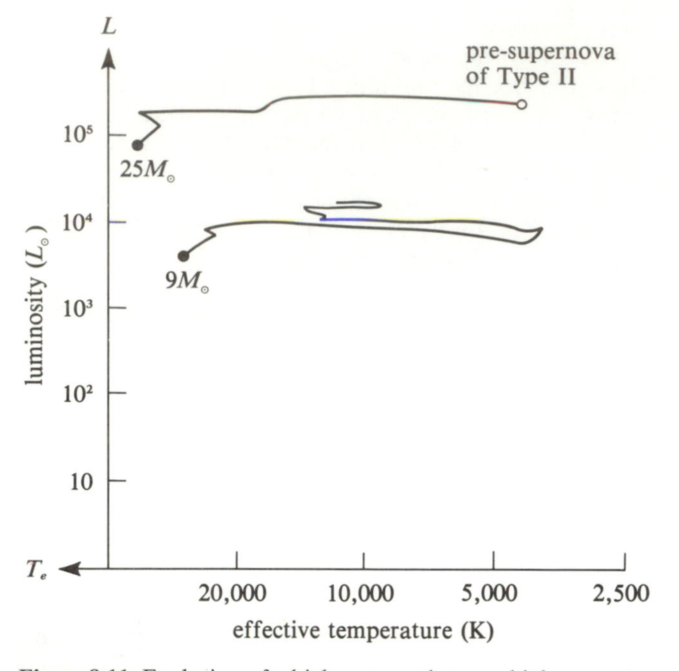

Figure 8.11. Evolution of a high-mass and a very-high-mass star in the Hertzsprung-Russell diagram. (Adapted from Icko Iben and from Weaver and Woosley, *Ninth Texas Symp. Rel. Ap.*, *Ann. N.Y. Acad. Sci.*, **336**, 1980, 335.)

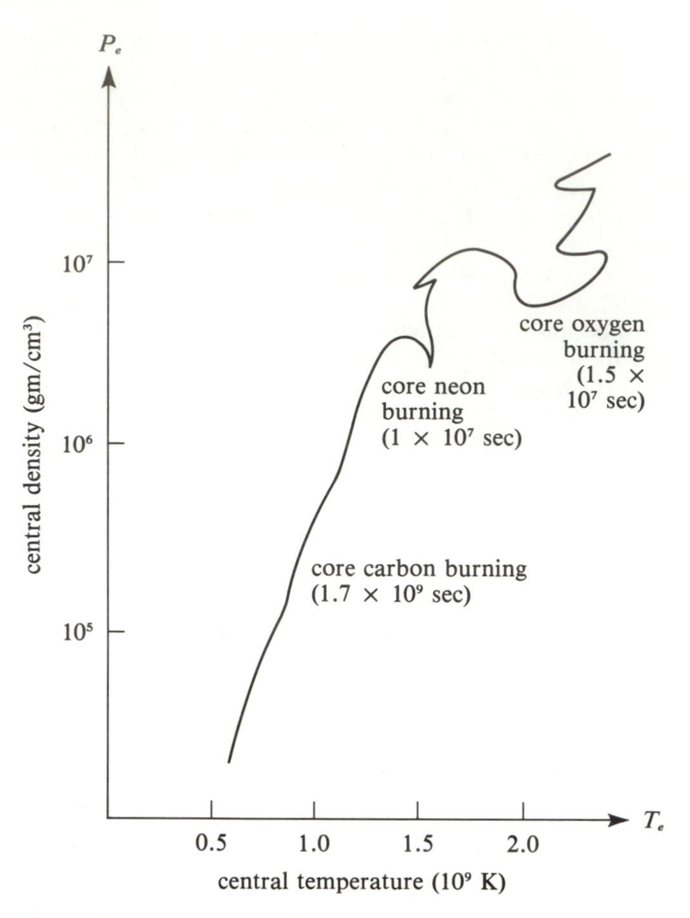

Figure 8.12. Evolutionary history of center of a $25M_\odot$ star in late stages. The central density and temperature increase to higher and higher values on timescales less than a year as the iron catastrophe is approached. (Adapted from Weaver and Woosley, *Ninth Texas Symp. Rel. Ap.*, *Ann. N.Y. Acad. Sci.*, **336**, 1980, 335.)

ignite at several hundred million K, before electron degeneracy becomes important. All stars of sufficiently high mass will therefore evolve by successively using up one round of fuel in the core, undergoing core contraction, achieving core ignition of previous ash into new fuel, etc. (Figure 8.12). In the simplest models (proposed first by Fowler and Hoyle, and calculated by Arnett and by Paczynski), the core will be surrounded by more and more shell sources, much as an onion is surrounded by many layers (Figure 8.13). The cycle of turning ash into new fuel proceeds faster and faster as the now degenerate core approaches closer and closer to the Chandrasekhar limiting mass. Finally, iron is produced in the core, and the story reaches a climax.

Supernova of Type II

Iron, as we explained in Chapter 6, is the ultimate slag heap of the universe, for which no further nuclear extraction of energy is possible. When the core has become iron, it has no recourse except to contract and heat up catastrophically. Under these conditions, copious numbers of neutrinos are released, which worsen the problem. Every channel for the loss of heat causes the core to contract that much more. Finally, when very high temperatures are reached—on the order of several *billion* K—the core reaches the regime where the temperatures can be considered to be high, even by nuclear-physics standards. At this point we arrive at the conditions where the other half of our rule of thumb for the thermodynamic behavior of matter holds (Box 4.3):

> At relatively high temperatures, matter prefers more spatial freedom and more unbound particles.

Following this rule, the iron begins to photodisintegrate into alpha particles (helium nuclei) and neutrons (Problem 8.5), undoing much of the nuclear evolution which preceded this catastrophe. As was first suggested by Hoyle, this undoing robs the core of heat, and the collapse accelerates. In the resulting fiery furnace, the alpha particles themselves photodisintegrate, extracting much of the core's heat to overcome the large binding energy of helium nuclei. The core now suffers catastrophic col-

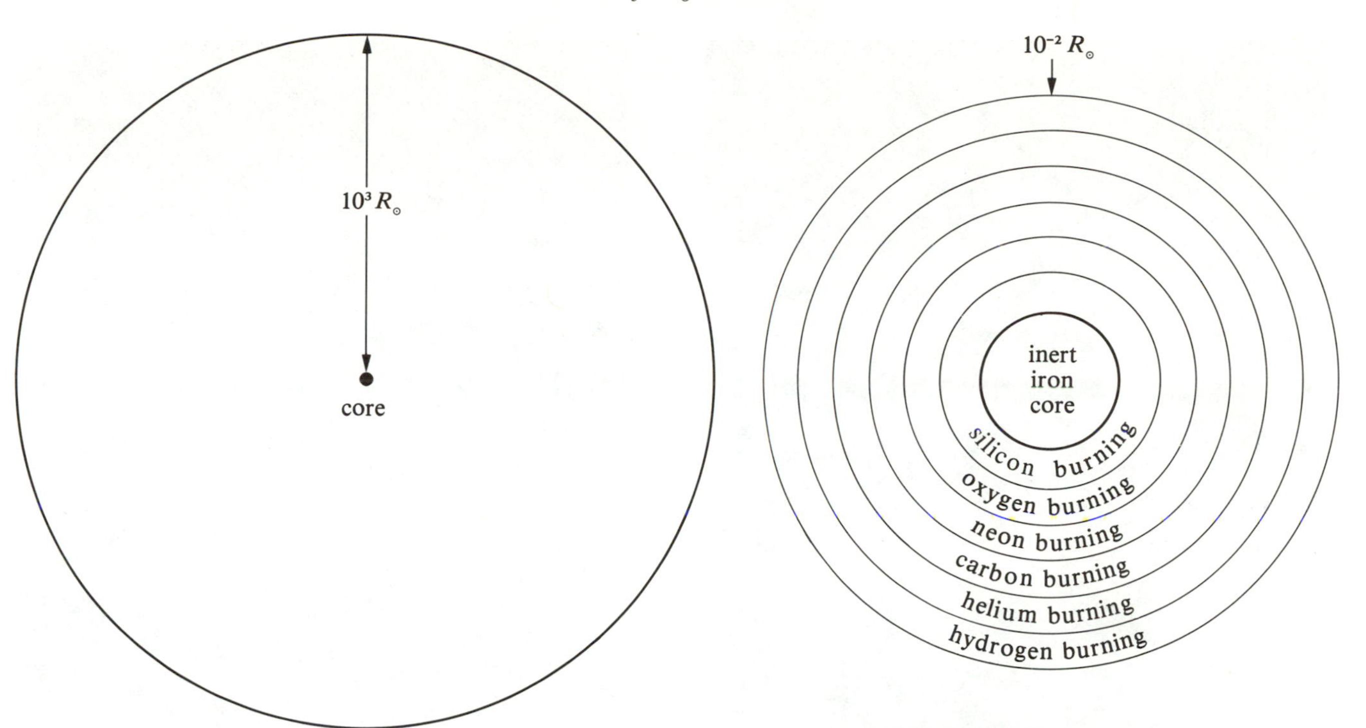

Figure 8.13. The onion-ring structure of a pre-supernova star (a very evolved star of high mass). The diagram on the left shows the dimensions of the entire star, a red supergiant, from core to photosphere. The diagram on the right shows the nuclear-burning regions near the inert iron core.

lapse, and begins to fall freely under its own self-gravity. The rapidly rising density squeezes almost all the free electrons into the protons to form a huge hot mass of neutrons at nearly nuclear densities (Problem 8.6). Whether the core collapse is arrested at nuclear densites and forms a neutron star, or whether it continues to collapse and forms a black hole, is not known. Theoretical calculations initiated by Colgate and continued by Wilson and others hope to resolve these difficult issues at the very frontier of stellar-evolution theory.

Problem 8.5. Fill in the blanks on the following reactions which occur in a bath of energetic photons:

$$^{54}_{26}\text{Fe} \rightarrow 13\ ^{4}_{2}\text{He} + \_\_\_\_,$$
$$^{56}_{26}\text{Fe} \rightarrow 13\ ^{4}_{2}\text{He} + \_\_\_\_,$$
$$^{4}_{2}\text{He} \rightarrow 2\ ^{1}_{1}\text{H} + \_\_\_\_.$$

Speculate on how the particles represented by the blanks might be useful for the *r*-process of nucleosynthesis (Chapter 6). What are the difficulties associated with getting these particles from the imploded core into the envelope?

Problem 8.6. Because of the complications presented by very high temperatures, the formation of neutrons discussed in the text is too difficult for us to discuss here. Let us consider a simple account of the neutronization of *cold* matter. The mass of a free proton m_p is less than the mass of a free neutron m_n by about $2.5m_e$, where m_e is the mass of an electron. In the nuclei of very bound atoms, however, the mass of the proton is about $50m_e$ less than the mass of a neutron. To convert these protons into neutrons requires the combining electrons to have an energy $E = 50m_ec^2$; such electrons clearly need to be relativistic. Calculate the density at which the energy associated with electron degeneracy will equal this energy E. *Hint*: At an electron density n_e, the electrons typically have a momentum $p_e \sim hn_e{}^{1/3}$. If we assume ultrarelativistic formulae, such electrons have an associated energy $E_e = cp_e \sim chn_e{}^{1/3}$. A more accurate

Figure 8.14. A supernova appeared between May 10, 1940, and January 2, 1941, in NGC 4725, a spiral galaxy in Coma Berenices. The earlier photograph on the left shows the galaxy; the later photograph on the right shows that a supernova had exploded during the interim in one of the spiral arms. (Palomar Observatory, California Institute of Technology.)

calculation yields $E_e = ch(3n_e/8\pi)^{1/3}$. Set $E_e = E$, and derive the required electron density $n_e = (8\pi/3)(50m_ec/h)^3$. Compute the numerical value of n_e. Assume an equal number of protons and electrons per unit volume, and estimate the mass density $\rho = 2m_pn_e$.

The fate of the envelope is also uncertain. Further nuclear reactions are likely in the high-temperature environment, but these reactions are sensitive to the history of the previous nucleosynthesis. An "onion-ring" model is usually adopted for the calculations, but Nomoto and Sugimoto have argued that this layered distribution may be somewhat smeared out by the outer-convection zone. We do not have even theoretical answers to the following questions? Will nuclear reactions in the envelope, driven perhaps by a detonation wave, lead to a net explosion of the outer layers? Or will these layers follow the collapse of the core either (a) to form a black hole, or (b) to be ejected in a rebound of the core when it reaches neutron-star densities? Or will the envelope be driven outward by a combination of neutrino-energy deposition and radiation pressure? In any case, the observational picture is clear. Some supernova explosions have managed to leave neutron stars behind, and the mass of such a neutron star, when measurable, has turned out to be tantalizingly close to $1.4M_{\odot}$! The latter is, of course, the Chandrasekhar limit associated with the "white dwarf" core of the pre-supernova star for all models of the late stages of evolution for high-mass stars.

At peak light output, observed supernova explosions can outshine a whole galaxy of stars (Figure 8.14). Supernovae are rare and dramatic occurrences, and astronomers are anxiously awaiting to study one at close range. (Our Galaxy seems overdue for a bright supernova.) The study of the expelled gases from historical supernova explosions in our own Galaxy has revealed that some do show evidence of considerable nuclear processing. These findings reinforce the belief that supernovae do eject substantial amounts of processed material into the interstellar medium. These heavy elements are incorporated evidently in subsequent generations of stars (and planetary systems) which condense from such interstellar gas clouds. Thus, although supernovae seem to be exotic events, they are actually basic to understanding how we and our Earth came to be.

Before leaving the topic of supernova explosions, we should note that at least two different kinds of supernova explosions are now known. The kind we have discussed here is called type II, and is associated with the death of massive stars. Type I is apparently associated with the death of an old population of stars, namely, population II stars. (It is unfortunate that the nomenclature "type II supernovae" turned out to correspond to population I stars, and "type I supernovae" to population II stars.) No consensus has been reached on how type I supernovae arise, but in Chapter 9 we shall see that population II stars generally correspond to low-mass stars. We have thus arrived at an apparent paradox: the story of stellar evolution presented in this Chapter predicts that low-mass stars should end as white dwarfs, not in supernova explosions. A promising way out of this predicament is the suggestion that type I supernovae arise not from the evolution of single stars, but from the evolution of stars in close binary systems (Chapter 10). Thus, we see that, although stellar-evolution theory has been actively studied for more than fifty years, important unsolved problems still remain.

Concluding Philosophical Comment

Zeldovich and Novikov have made the following intriguing philosophical point about the picture of the formation of a neutron star sketched here. They note that stars begin their lives as a mixture mostly of hydrogen nuclei and their stripped electrons. During a massive star's luminous phase, the protons are combined by a variety of complicated reactions into heavier and heavier elements. The nuclear binding energy released this way ultimately provides entertainment and employment for astronomers. In the end, however, the supernova process serves to undo most of this nuclear evolution. In the end, the core forms a mass of neutrons. Now, the final state, neutrons, contains *less* nuclear binding energy than the initial state, protons, and electrons. So where did all the energy come from when the star was shining all those millions of years? Where did the energy come from to produce the sound and the fury which is a supernova explosion? Energy is conserved; who paid the debts at the end? Answer: gravity! The gravitational potential energy of the final neutron star is much greater (negatively; that's the debt) than the gravitational potential energy of the corresponding main-sequence star (Problem 8.7). So, despite all the intervening interesting nuclear physics, ultimately Kelvin and Helmholtz were right after all! The ultimate energy source in the stars which produce the greatest amount of energy is gravity power. This is an important moral worth remembering and savoring.*

Problem 8.7. If M and R are the mass and radius of a neutron star, we might be tempted to estimate its gravitational potential energy as $-GM^2/R$. This would give order-of-magnitude accuracy, but it misrepresents an important general-relativistic effect. This effect is the mass deficit: namely, the mass of a gravitationally bound system of particles is less than the sum of the masses of the free particles. The effect is not unique, of course, to general relativity; it exists for any relativistically correct binding agent, such as nuclear forces. The mass deficit ΔM is related to the *binding energy* ΔE by Einstein's formula $\Delta E = \Delta Mc^2$. Thus, when a mass M of particles is gathered gravitationally from a great distance apart, the gravitational binding energy ΔMc^2 must be released in the form of photons, neutrinos, gravitational radiation (gravitons), kinetic energy of expelled matter, etc. The mass deficit for a neutron star is typically $\Delta M = 0.1M$. Gravitational condensation for a neutron star therefore has about a 10 percent efficiency in converting rest mass into energy. How does this compare with nuclear fusion (where conversion of hydrogen to helium extracts about

* If we regard a neutron star as one gigantic atomic nucleus, we may also say that nuclear processes plus gravity have succeeded in converting many atomic nuclei into one nucleus. Problem 8.7 then shows that the ultimate energy source for the entire output of the star is the relativistic binding energy of the final end state.

7/8 of the total possible from conversion to iron)? Calculate numerically the gravitational binding energy ΔE for a "baryon mass" $M = 1.4M_{\odot}$. How does this compare with the total radiated energy if the progenitor star shined for 6 million years with a luminosity $L = 10^{38}$ erg/sec? How does ΔE compare with the energy 10^{51} erg released in *observable* form (kinetic energy of expelled matter and photons) in a typical supernova explosion of type II? Are neutrinos and gravitational radiation therefore copiously emitted during a supernova outburst? How would *you* go about detecting these latter emissions?

CHAPTER NINE

Star Clusters and the Hertzsprung-Russell Diagram

In Chapter 8 we have seen that the H-R diagram, which plots luminosity versus effective temperature, is a very useful tool for discussing the theoretical evolutionary tracks of stars. To compare the results of the theoretical calculations with observed stars, we need to see how to generate the observational counterpart of the theoretical H-R diagram.

The Observational H-R Diagram

Luminosity

To obtain the luminosity of a star, we need to make two measurements (Box 9.1):

(a) We need the apparent brightness f of a star, which is the total energy received from the star per unit time per unit area at the Earth. (For the Sun, we called this quantity the solar constant.) In practice, the apparent brightness is usually measured only within a range of wavelengths—for example, through colored filters—and corrections ("bolometric corrections") need to be applied to obtain the apparent brightness for all wavelengths (Problem 9.1).

(b) We need the distance r of the star away from us. For nearby stars the distance can be calculated directly by parallax measurements (Box 2.2). Figure 9.1 explains how such measurements work in principle. This direct method works well out to a distance of about three hundred light-years (Problem 9.2). For more distant stars, we have to rely on indirect methods. Such indirect distance measurements constitute one of the fundamental problems of observational astronomy. We will discuss a few of these methods as we proceed in this book.

Knowing the apparent brightness f and the distance r of a star, we may obtain its luminosity from the equation

$$L = f \cdot 4\pi r^2. \tag{9.1}$$

BOX 9.1

Important Astronomical Measurements of Stars

1. To ascertain the luminosity, we need both
   (a) the apparent brightness f, which is usually obtained by UBV photometry with corrections for interstellar reddening and extinction, and
   (b) the distance r to the star.
2. To ascertain the effective temperature, we need either
   (a) the true $B-V$ color of the star, which usually requires a correction for interstellar reddening, or
   (b) the spectral type of the star, which does not require a correction for interstellar reddening, but is more time-consuming to obtain.

Of the above data, the most difficult to secure is generally the distance r to the star. (See Box 9.5.)

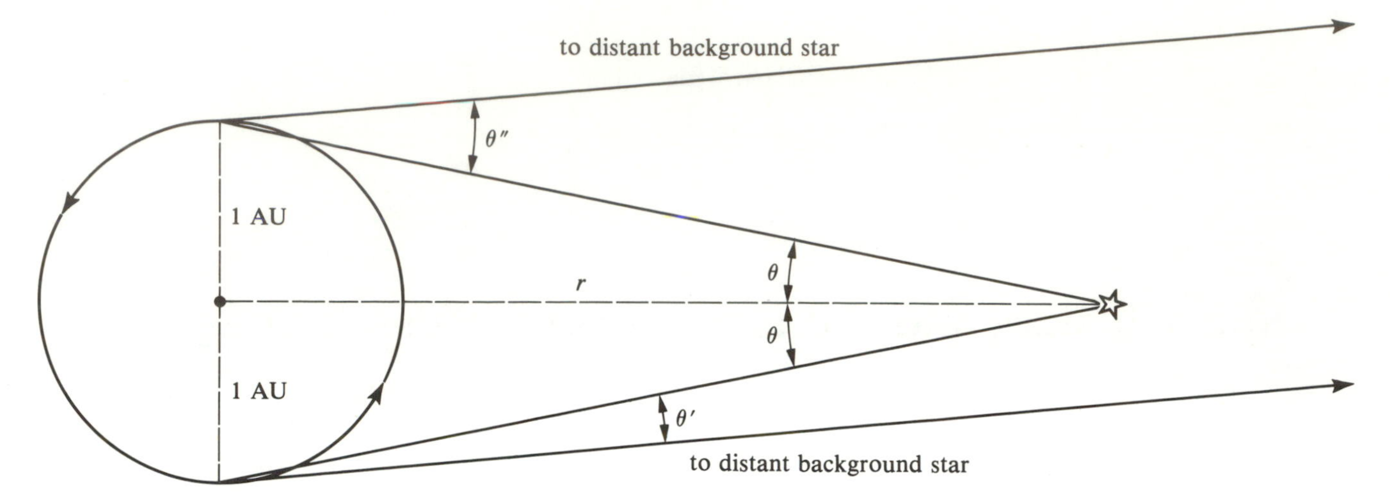

Figure 9.1. Distance measurement by trigonometric parallax. During six months of the Earth's motion about the Sun, a nearby star is observed to shift its apparent angular position with respect to distant background stars. In the above example, it is first observed to lie at an angle θ' to the left of a distant background star; six months later, it is seen at an angle θ'' to the right of the distant background star. Approximating the background star to be infinitely far away, it is easy to calculate that $(\theta' + \theta'') = 2\theta$, where θ is the angle subtended by 1 AU at the distance r of the nearby star. If θ is measured in radians, the small-angle formula yields the distance of the nearby star as $r = 1\ \text{AU}/\theta$. The angle θ is called the parallax of the star.

Problem 9.1. The response of the human eye works on the basis of a geometric progression rather than an arithmetic progression. Thus, if a human subject is asked to select a light bulb which is intermediate in brightness between 25 watts and 100 watts, the human subject will choose $(25 \text{ watts} \times 100 \text{ watts})^{1/2} = 50$ watts rather than $(25 \text{ watts} + 100 \text{ watts})/2 = 67.5$ watts. The early Greeks devised a system of classifying the apparent brightness of stars with first-magnitude stars being the brightest, second-magnitude being the next brightest, etc., with sixth-magnitude stars being those barely discernible to the best of human eyes. It turned out that their sixth-magnitude stars were roughly 100 times fainter than their first-magnitude stars. The modern magnitude classification therefore begins by postulating a difference of five magnitudes to equal exactly a factor of 100 in apparent brightness. In mathematical language, the magnitude scale is logarithmic (because the eye is). If m_1 and m_2 are the apparent magnitudes of two stars with apparent brightness f_1 and f_2, the apparent magnitude system is defined so that

$$m_2 - m_1 = 2.5 \log_{10}(f_1/f_2).$$

Notice that the fainter star has the bigger apparent magnitude. What is the magnitude difference if $f_1 = 10^4 f_2$? Are there stars fainter than sixth-magnitude? Are there zeroth-magnitude stars? Are there any stars with negative magnitudes? (*Hint*: consider the nearest star.)

Problem 9.2. A parsec (abbreviated pc) is defined to be the distance at which a star would have a parallax (see Figure 9.1) of one second of arc. How many light-years equal one pc? The smallest angle which can be reliably measured for parallax purposes is about 0.01 second of arc. Suppose that 0.08 stars/pc^3 are observable near the solar position in the Galaxy. How many stars, in principle, exist which could have their distances measured by the parallax method?

Problem 9.3. The absolute magnitude M of a star is the apparent brightness that star would have if it were placed at a standard distance of 10 pc away. From the $1/r^2$ law of radiation, deduce, therefore, the proper relation between absolute magnitude M and apparent magnitude m:

$$M = m - 5 \log_{10}(r/10 \text{ pc}).$$

The absolute (bolometric) luminosity of the Sun is 4.72. Argue, therefore, that the absolute magnitude M of a star with luminosity L is given by

$$M = 4.72 - 2.5\log_{10}(L/L_{\odot}).$$

What is the luminosity of a star whose distance is 60 pc and whose apparent magnitude is 3.61? Express your answer in solar luminosities.

Effective Temperature

Strictly speaking, astronomers define the effective temperature T_e of a star with luminosity L and radius R by the equation

$$L = (4\pi R^2)(\sigma T_e{}^4). \tag{9.2}$$

However, it is generally not possible to measure directly the radius R of stars; so equation (9.2) is more useful as a way to find R after L and T_e have been ascertained by independent means.

There are essentially two independent means for measuring T_e, both based on the idea that the effective temperature is an indicator of the *quality* of the radiation field emergent from stellar photospheres. The first method, **UBV color photometry**, uses the distribution of *continuum* radiation at different wavelengths. The second method, **spectral classification**, uses the distribution of *absorption lines* in the dispersed spectrum of a star (see Box 9.1).

UBV Photometry and Spectral Classification

The basic idea behind *UBV* photometry ("multicolor broad-band photometry") is to measure the proportions of radiant energy put out by a thermal body at ultraviolet (U), blue (B), and visual (V) wavelengths. (The V filter is centered roughly on the yellow part of the electromagnetic spectrum.) These proportions depend on the surface temperature of the opaque body: the hotter the body, the greater is the proportion of shorter wavelengths (Chapter 4). Schematically, then, if an astronomer measures the apparent brightnesses f_U, f_B, and f_V of a star at ultraviolet, blue, and visual wavelengths, the ratios will depend only on the effective temperature of the star;

$$f_V/f_B = \text{function of } T_e,$$
$$f_B/f_U = \text{function of } T_e,$$

since the inverse-square dependence of the individual fs on the distance r will divide out when we take their ratios. Thus, a measurement of either f_V/f_B (whose logarithm is proportional to $B - V$ as defined by optical astronomers, who like to work in apparent magnitudes) or f_B/f_U ($U - B$) would, in principle, enable one to calculate the effective temperature T_e. In practice, both ratios are usually needed, because interstellar dust (Chapter 11) along the line of sight often reduces f_U, f_B, and f_V from their true values ("extinction"). Moreover, this extinction is "selective," in that there is a bigger effect at shorter wavelengths than at longer wavelengths ("reddening"). Consequently, there is a distortion of the ratios f_V/f_B and f_B/f_U from their true values. Since the amount of distortion differs for the two ratios, however, a measurement of both ratios allows astronomers to remove the distortion, i.e., "to correct for interstellar reddening." This correction procedure is illustrated in Figure 9.2. The amount of reduction of f_V can also be

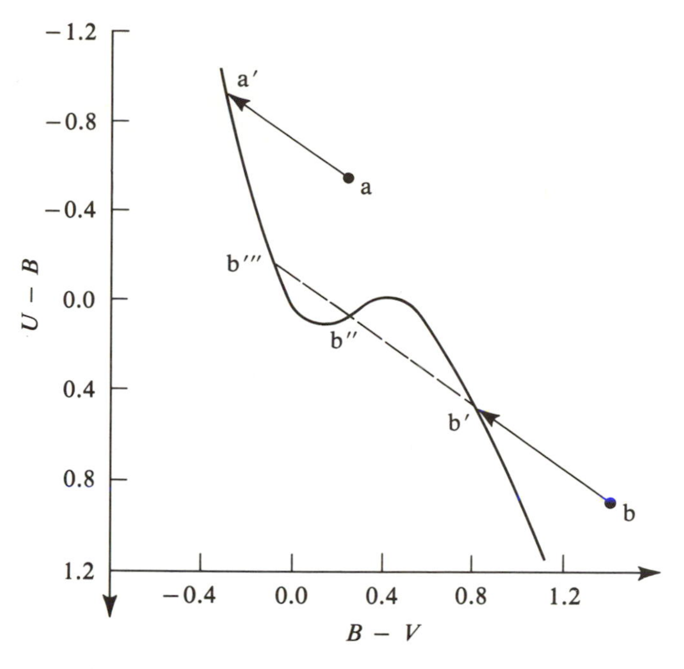

Figure 9.2. Correction for interstellar reddening. In the absence of interstellar reddening, the $B - V$ and $U - B$ colors of stars are found to lie along the wavy curve. (Notice that $U - B$ is conventionally plotted to increase downward.) Suppose star a is found to lie off this curve. We assume it has suffered interstellar reddening, and move it upward and leftward along a diagonal line of known slope until it falls on the unreddened curve at a'. The displacement in $B - V$ from a to a' (0.4 magnitudes here) is called the color excess E_{B-V}. If star b were also measured to lie off the curve, we would also move it backward along the "reddening curve" of known slope; but now there are three possible choices for the unreddened position: b', b'', and b'''. But if star a and star b lie in the same cluster, then the light from them has passed through the same line of sight, and star b should therefore have the same color excess E_{B-V} as star a. We would therefore deduce that point b' gives the true colors of star b.

Table 9.1. **Bolometric corrections.**[a]

$B-V$	Bolometric corrections: Main sequence	Giants	Supergiants
−0.4	4.1	—	—
−0.2	1.9	—	2.9
0.0	0.68	—	0.7
0.2	0.24	—	0.4
0.4	0.07	—	0.2
0.6	0.05	0.05	0.3
0.8	0.17	0.26	0.3
1.0	0.43	0.52	0.5
1.2	0.78	0.76	0.8
1.4	1.2	1.0	1.0
1.6	2.1	1.6	1.4
1.8	—	2.8	2.0
2.0	—	—	2.9
2.2	—	—	4.4

[a] Adapted from C. W. Allen, *Astrophysical Quantities*.

Table 9.2. **Spectral type, color, and effective temperature.**[a]

Spectral type	Main sequence $B-V$	Main sequence T_e (K)	Giants $B-V$	Giants T_e (K)
O5	−0.45	35,000	—	—
B0	−0.31	21,000	—	—
B5	−0.17	13,500	—	—
A0	0.00	9,700	—	—
A5	0.16	8,100	—	—
F0	0.30	7,200	—	—
F5	0.45	6,500	—	—
G0	0.57	6,000	0.65	5,400
G5	0.70	5,400	0.84	4,700
K0	0.84	4,700	1.06	4,100
K5	1.11	4,000	1.40	3,500
M0	1.24	3,300	1.65	2,900
M5	1.61	2,600	—	—

[a] Adapted from C. W., Allen, *Astrophysical Quantities*.

deduced once the "reddening" is known, because ordinary interstellar dust produces a constant ratio of "total to selective extinction." In this manner, the true values of f_U, f_B, and f_V can be obtained. Knowing these true values then allows the astronomer to deduce the apparent brightness of the star over all wavelengths ("apply bolometric corrections"; see Table 9.1). Thus, *UBV* photometry ultimately serves two purposes: (1) to measure T_e (Table 9.2); (2) to measure f.

Problem 9.4. A certain star has a measured V magnitude equal to 13.54 and a measured B magnitude of 14.41. Thus, the measured value of $B - V$ is $14.41 - 13.54 = 0.87$, consistent with a reddish color. However, from the measured U magnitude, the star can be deduced to have a color excess due to interstellar reddening, $E_{B-V} = 0.25$. Normal interstellar dust produces a ratio of visual extinction A_V to color excess equal to 3.0. Thus, we may infer that the star has suffered a visual extinction $A_V = 3.0 \times 0.25 = 0.75$. The measured $B - V = 0.87$ can now be corrected for interstellar reddening to give a true $B - V =$ color index $= 0.87 - E_{B-V} = 0.62$, and the measured apparent magnitude of V can now be corrected to give a true apparent magnitude of $V = m_V = 13.54 - A_V = 12.79$. A star with a color index 0.62 would have an effective temperature from 5,500 K to 5,800 K, depending on whether it was a yellow giant or a yellow main-sequence dwarf. Either would have a bolometric correction of about 0.07; so the bolometric apparent magnitude $m = m_V - 0.07 = 12.72$. Convert the latter value into a value for f in units of erg $\text{sec}^{-1}\ \text{cm}^{-2}$. *Hint*: The Sun, absolute magnitude = 4.72, would have to be placed at what distance (without interstellar extinction) to have an apparent magnitude equal to 12.72? How does this relate to the solar constant?

Problem 9.5. Suppose we had not made the corrections for interstellar reddening and extinction. We would then have assumed from the $B - V$ measurement that we were dealing with a red star. (If we had bothered to check our $U - B$ measurement, we would have noticed a discrepancy, because the measured $U - B$ is inconsistent with this interpretation.) A red star with a color index equal to 0.87 (the uncorrected $B - V$ value) would have an effective temperature of about 4,600 K and a bolometric correction of about 0.3. Thus, we would have miscalculated the bolometric apparent magnitude to be $m = V - 0.3 = 13.2$. By what factor would we have erred in estimating f? In estimating T_e? Comment on the importance of taking into account the (nuisance) effects of interstellar dust.

The basic idea behind spectral classification is that, for a given chemical composition, the pattern of absorption lines which are formed in the photosphere of a star depend on the temperatures and pressures which exist in the photosphere. Indeed, a great variety of stellar spectra can be observed for different stars (Figure 9.3). This great variety arises more from the variations of the

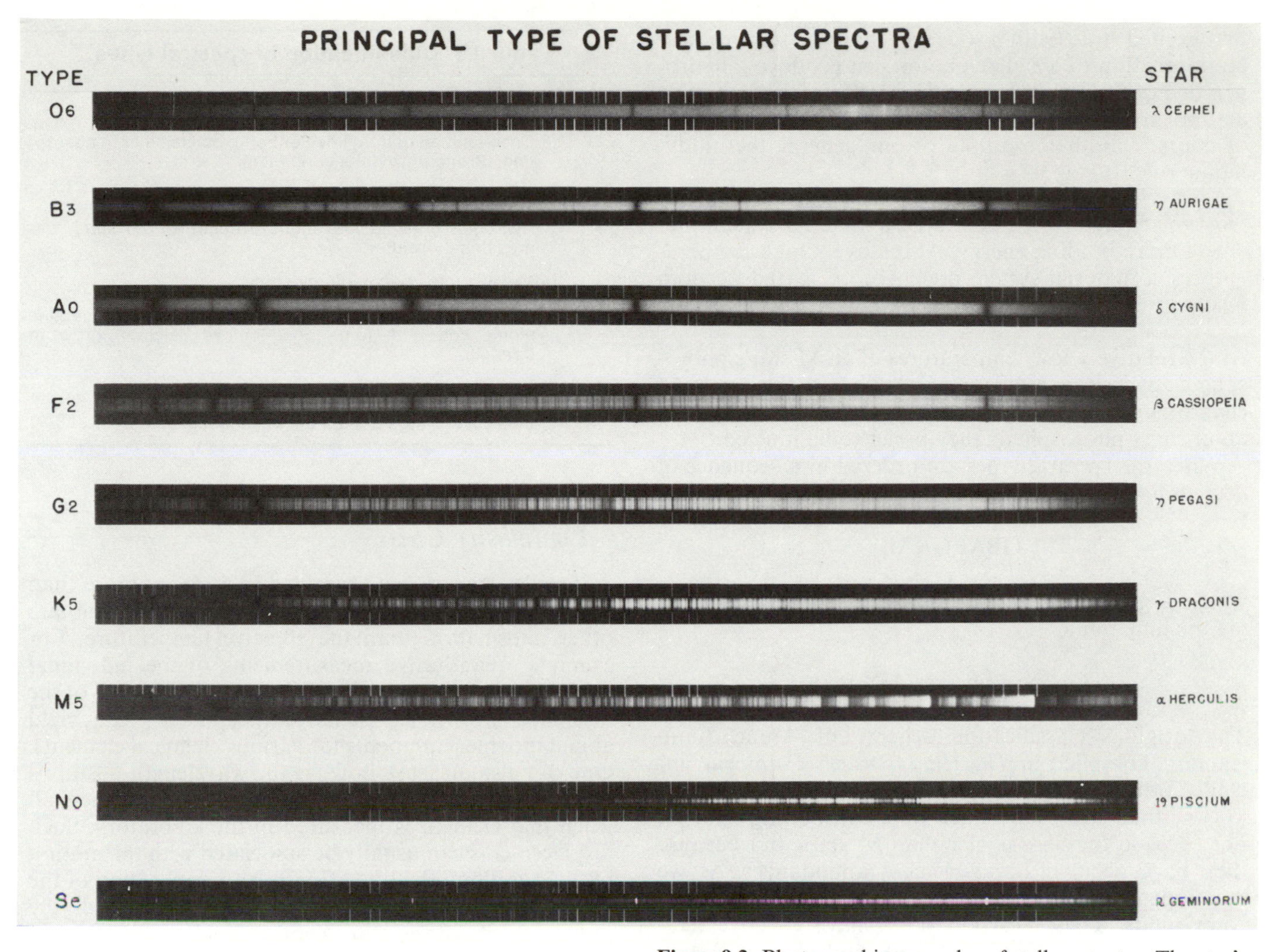

Figure 9.3. Photographic examples of stellar spectra. The star's spectrum is in the middle of each strip; the bright lines above and below each stellar spectrum yield reference wavelengths from a laboratory source. Notice that the hydrogen absorption lines (dark stripes against a bright background) are strongest in the A0 star, whereas the "e" attached to the Se star indicates that it exhibits emission lines (bright stripes against a less bright background in addition to the more common absorption lines. (Palomar Observatory, California Institute of Technology.)

effective temperatures of stars than they do from variations of the chemical compositions or surface pressures. Thus, to a first approximation, the spectral type of a star yields an estimate of its effective temperature:

$$\text{spectral type} = \text{function of } T_e.$$

Historically, spectral types were assigned by Cannon and Pickering alphabetically in an order determined primarily by the decreasing strength of the hydrogen-absorption lines. Thus, A stars were strongest in hydrogen lines; B stars next strongest; etc. It was soon realized that this assignment does not order stars in effective temperature. Through the work of Saha, astronomers learned that O stars have weak hydrogen lines because their photospheres are so hot that hydrogen is almost completely ionized; whereas M stars, which are alphabetically close to O stars, have weak hydrogen lines for a completely different reason. M stars have such cool photospheres that there is little atomic hydrogen in the first excited state (with $n = 2$; see Chapter 3) from where an absorption of a stellar photon can produce a hydrogen line of the Balmer series (Hα, Hβ, etc) that is visible at optical wavelengths. Saha's important discovery is, of course, another example of our general thermodynamic rule (Box 4.3):

> At relatively low temperatures, a material system prefers more binding energy. At relatively high temperatures, a material system prefers more spatial freedom and more unbound particles.

At the relatively low temperatures of an M star's photosphere, hydrogen atoms prefer to be in their ground electronic state. At the relatively high temperatures of an O star's photosphere, they prefer to be ionized.

When the spectral types are ordered in a sequence of *decreasing effective temperatures*, they read

OBAFGKM,

which a generation of astronomy students memorized via the mnemonic

Oh, be a fine girl, kiss me!

The original classification scheme of Annie Jump Cannon, compiled in *The Henry Draper Catalogue*, included other spectral types, such as N, R, and S. It is now known that such stars have different-looking spectra, not because they are cooler than M stars, but because they have peculiar heavy-element abundances. As we mentioned in Chapter 8, R and N stars are rich in carbon compounds. These "carbon stars" have effective temperatures comparable to those of K and M stars. S stars have photospheres with enhanced abundances of *s*-process elements. It is currently speculated that these stars are in the double-shell-burning phase of stellar evolution, in which nuclear processing in the interior has been accompanied by mixing to the stellar surface. In any case, the additional classes R, N, S are often appended in parentheses to the main spectral types:

OBAFGKM (RNS).*

Table 9.3 explains the major characteristics of the prominent spectral lines used to type stellar spectra. Notice that spectral types are independent of interstellar reddening, and that finer subdivisions are introduced by appending the numbers 0 to 9 to each spectral class. Table 9.2 then yields the effective temperature.

* For women or monster-movie fans, appropriate mnemonics might be: Oh, be a fine guy, kiss me! (Right now! Smack!); Overseas broadcast: A flash! Godzilla kills Mothra! (Rodan named successor.)

Table 9.3. **Classification of spectral types.**

Type	*Main characteristics*
O	Once-ionized helium lines either in emission or in absorption. Strong ultraviolet continuum.
B	Neutral helium lines in absorption.
A	Hydrogen lines at maximum strength for A0 stars, decreasing thereafter.
F	Metallic lines become noticeable.
G	Solar-type spectra. Absorption lines of neutral metallic atoms and ions (e.g., once-ionized calcium) grow in strength.
K	Metallic lines dominate. Weak blue continuum.
M	Molecular bands of titanium oxide noticeable.

Luminosity Class

If stellar spectra are examined in terms of more than the pattern of spectral lines, one can sometimes obtain other information than the effective temperature. For example, quantitative measurements of the individual line strengths plus a theory of line formation (part of the discipline known as "stellar atmospheres") can yield abundance measurements for various chemical elements. One can also measure how far the wavelength positions of known lines have been displaced from where they fall when that element is measured in the laboratory. Such displacements can usually be associated with the motion of the star along the line of sight, and application of the Doppler shift-formula, equation (3.14), then yields the radial velocity of the star. Another application is that if the spectral lines show evidence of "Zeeman splitting" (see Box 3.3 and Problem 3.9), one can deduce the

BOX 9.2
Common Uses of Stellar Spectral Analysis

(a) Classification of spectral type.
(b) Classification of luminosity class.
(c) Measurement of photospheric chemical abundances.
(d) Measurement of radial velocity of star from Doppler shift of center of spectral line.
(e) Measurement of stellar rotation from additional broadening of spectral line by the rotational velocity.
(f) Measurement of mass inflow or outflow from asymmetries in the line profile.
(g) Measurement of photospheric magnetic fields by the Zeeman effect.

strength of the surface magnetic fields which exist in the star. Box 9.2 summarizes the common uses of spectral analyses.

The widths of the absorption lines of stars can yield information about the surface pressures. For a given effective temperature, atoms under high pressure will collide much more frequently than atoms in a low-pressure environment. The greater frequency of collisions means that line photons which are emitted and absorbed in radiation processes possess less well-defined energies (partly as a consequence of Heisenberg's uncertainty principle and partly as a result of the perturbing electric fields of nearby atoms). Thus, the absorption lines produced in a high-pressure environment will generally be considerably broader than the lines produced in a low-pressure environment. In addition, for a given temperature, atoms are thermally ionized more easily at low pressures than at high pressures (assuming the pressure is not so high that it pushes electronic shells on top of one another, as in the center of the Sun). These subtle effects provide a way to distinguish between giant stars and dwarf stars of the same effective temperature. Giant stars typically have narrower ("sharper") lines and exhibit slightly higher stages of ionization than dwarf stars of the same effective temperature, because the photospheres of the giant stars are at lower pressures. In other words, giant stars of the same spectral type (the same stages of ionization) as dwarf stars have slightly lower effective temperatures (see Table 9.2).

The physical reason for the lower photospheric pressures in giant stars is simple. Giants are much more distended than dwarfs and, therefore, have much lower surface gravities: $g = GM/R^2$. (The radius difference between giants and dwarfs more than makes up for any possible mass difference.) The lower surface gravity of giants requires less gas pressure to provide hydrostatic support of the photosphere (Box 5.1). Quantitative analysis of the line spectrum of a star can therefore yield information about the star's surface gravity $g = GM/R^2$. More qualitatively, we may say that narrow lines are associated with big stars (large R); broad lines, with little stars (small R). For a given effective temperature T_e, large values of R correspond to large values of the luminosity L in accordance with equation (9.2); so there is a direct correlation between the width of the spectral lines and the **luminosity class** of a star of given spectral type (see Box 9.3). Figure 9.4 outlines the boundaries of these luminosity classes in the Hertzsprung-Russell diagram. Notice that "dwarfs" refer to main-sequence dwarfs and not white dwarfs, which form a separate category from ordinary stars.

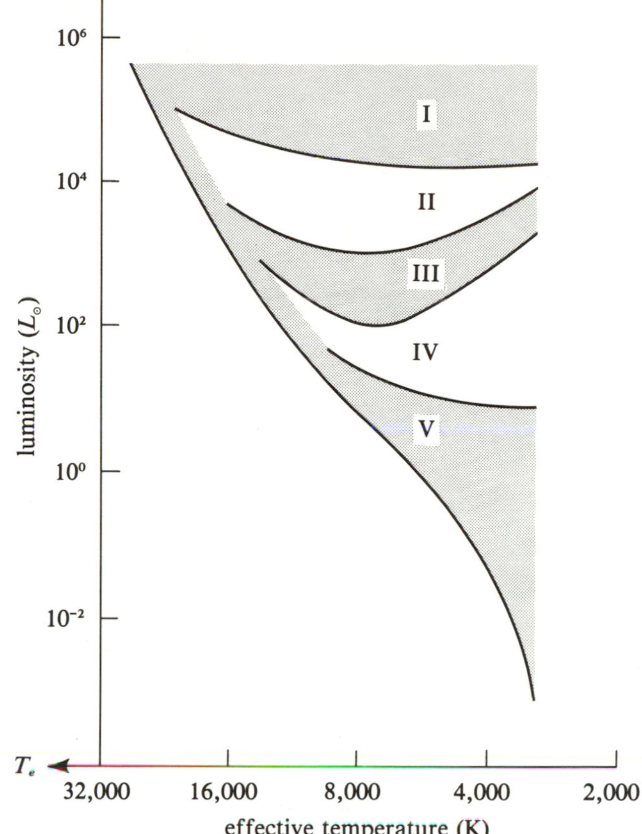

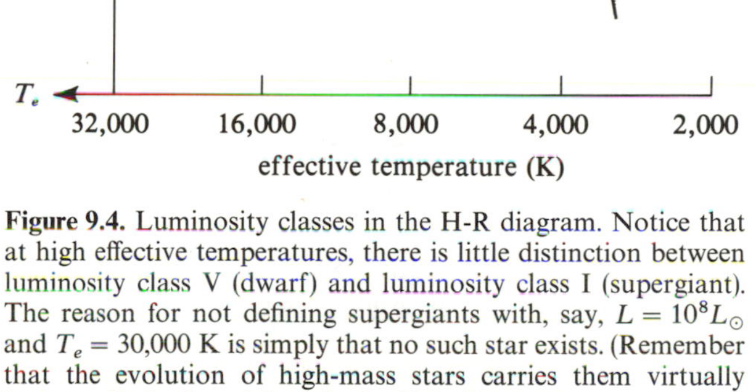

Figure 9.4. Luminosity classes in the H-R diagram. Notice that at high effective temperatures, there is little distinction between luminosity class V (dwarf) and luminosity class I (supergiant). The reason for not defining supergiants with, say, $L = 10^8 L_\odot$ and $T_e = 30{,}000$ K is simply that no such star exists. (Remember that the evolution of high-mass stars carries them virtually *horizontally* back and forth in the H-R diagram.)

BOX 9.3
Stellar Luminosity Class

Luminosity class I: supergiants (often subdivided Ia and Ib)
Luminosity class II: bright giants
Luminosity class III: giants
Luminosity class IV: subgiants
Luminosity class V: dwarfs

In principle the Morgan-Keenan system of a two-parameter specification of a star's spectral type and luminosity class plus a measurement of the star's apparent brightness could yield an indirect assessment of its distance (Problem 9.6). In practice, this method of obtaining distances does not yield high accuracy, because

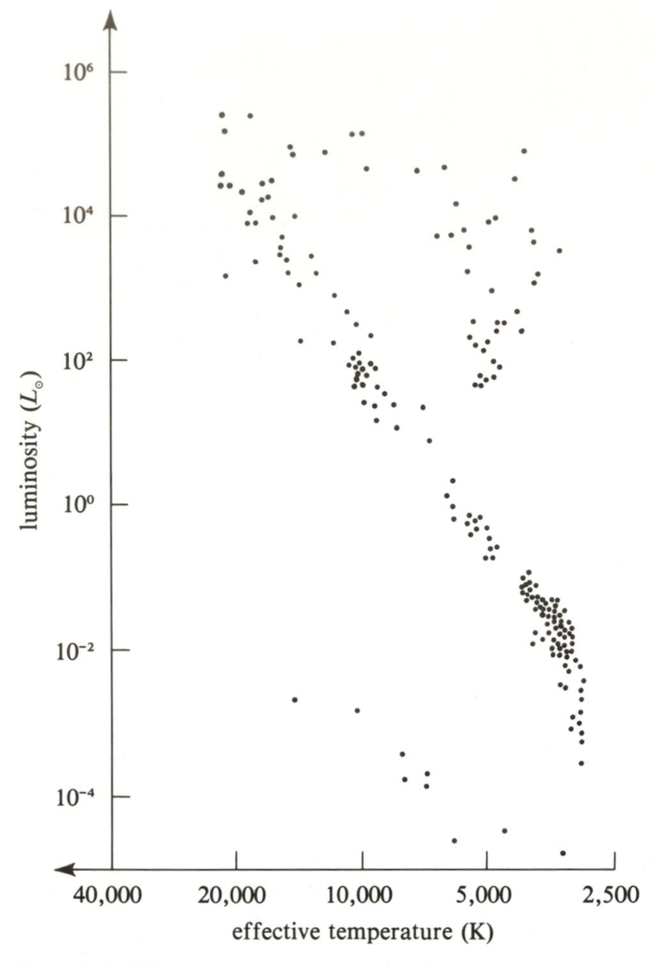

Figure 9.5. The Hertzsprung-Russell diagram of nearby stars. Main-sequence stars and white dwarfs are fairly clearly delineated.

of the relatively large spread of luminosities possible for each luminosity class, especially toward the later spectral types (Figure 9.4).

Problem 9.6. Assume that spectroscopy is carried out for the star discussed in Problem 9.4, and that it can be classified as a G2 V star. Is the spectral type G2 consistent with the deduced color index 0.62? Assume that the star has the same absolute magnitude 4.72 as the Sun. Given that it has a deduced apparent magnitude 12.72, what is the distance r of this star? Why do you think this indirect method for obtaining distances is called "spectroscopic parallax"?

The H-R Diagram of Nearby Stars

Nearby stars can have their distances calculated directly, by the method of "trignometric parallax" (Figure 9.1); so the observational H-R diagram for nearby stars can be constructed by calculating L and T_e for the stars individually. Figure 9.5 shows the resulting H-R diagram for such nearby stars. The outstanding feature of Figure 9.5 is the clear demarcation of the main-sequence band, where most of the stars can be found. We understand this result theoretically as arising from the fact that luminous stars spend most of their active lifetimes as main-sequence stars (Chapter 8). Giants and white dwarfs are also found in Figure 9.5. Spectroscopy of the former show them to have very narrow lines; the latter, very broad lines. Indeed, these stars, which can have their distances directly calculated, constitute a calibration for the luminosity-class scheme.

The interpretation of Figure 9.5 is complicated by the fact that the nearby stars comprise a mixture of stars with different masses, different initial chemical compositions, and different ages. With such a mixed bag, it is difficult to sort out the different effects which would produce scatter in the diagram. Moreover, it is very time-consuming to obtain the distance (and thereby the luminosity) of each star individually. For comparisons with theory, therefore, it would be convenient to have a group of stars which all have the same initial chemical composition, the same age, and (nearly) the same distance, so that they differ from each other only in mass. Such groups are available; they are called star clusters.

The H-R Diagram of Star Clusters

Observationally, astronomers have found the two different kinds of star clusters in the Galaxy listed in Box 9.4. A cluster's angular size is always small; so all its stars can

---

BOX 9.4

Two Kinds of Star Clusters

(a) Open clusters typically contain 10^3 stars, are found in the disk of the Galaxy, and have heavy-element abundances characteristic of Population I stars.

(b) Globular clusters typically contain 10^6 stars, are found in the halo of the Galaxy, and have heavy-element abundances characteristic of Population II stars.

Figure 9.6. (a) The open clusters h and χ Persei, and (b) the globular cluster M3. The angular diameters of h and χ Persei are about half a degree each, and the angular diameter of the bright core of M3 is about 3 minutes of arc. The small angular extent of a star cluster implies that all the stars belonging to the cluster lie at nearly the same distance from us. (Lick Observatory photograph.)

be approximated to lie at the same (unknown) distance from us. Figure 9.6 indicates the angular scales of clusters.

Open Clusters

The H-R diagrams of two open clusters, NGC 2362 and the Praesepe, are plotted in Figure 9.7, with apparent brightness f replacing the intrinsic brightness or luminosity L. The reason for this replacement is that f is directly measurable, whereas L is not. The form of the H-R diagram is not changed by this replacement (except for the labeling of the vertical scale) if the stars in a given cluster lie at the same distance r.

Notice that portions of the main sequence are easily discernible in both star clusters; however, the lowest portions are missing in both clusters (compare with Figure 9.5), because observations can be carried out only to a certain limiting faintness (apparent brightness). The arrows in Figure 9.7 mark off segments of the main sequence in NGC 2362 and Praesepe which contain the same range of effective temperatures, and in which the stars presumably comprise the same range of masses. The fact that the black arrows are higher in Praesepe than those in NGC 2362 must then mean that the Praesepe cluster is closer to us, so that the same stars appear brighter. That is, in the Praesepe cluster we are able to observe further down the main sequence, because the same limiting apparent brightness allows us to observe intrinsically fainter stars in the closer cluster.

To make meaningful comparisons of different open clusters, we must correct for the differences in relative distance. Trumpler was the first to do this by sliding the observational H-R diagram—with apparent brightness replacing luminosity—up or down until the main sequences line up. The amount that we have to slide the diagram up or down tells us how much further or nearer

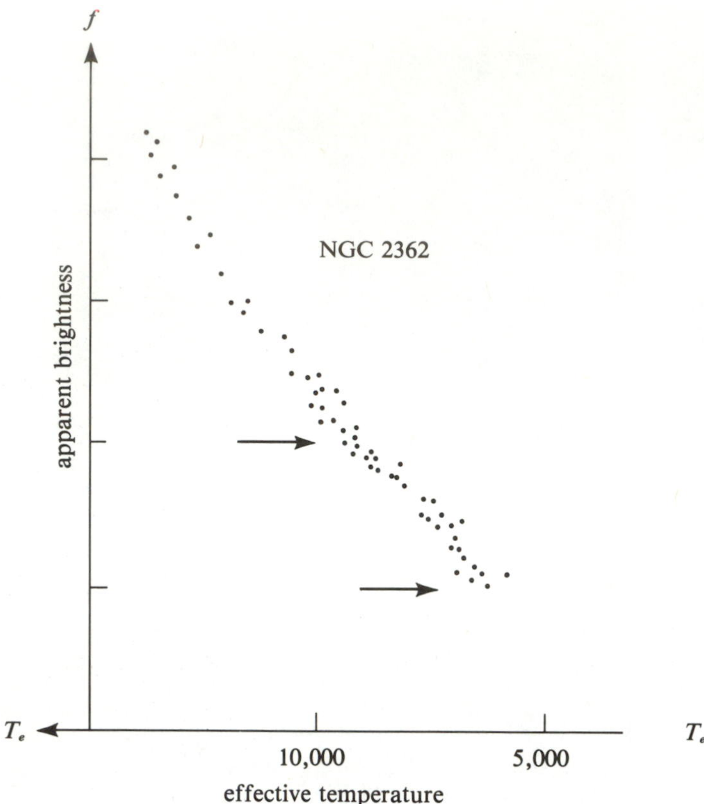

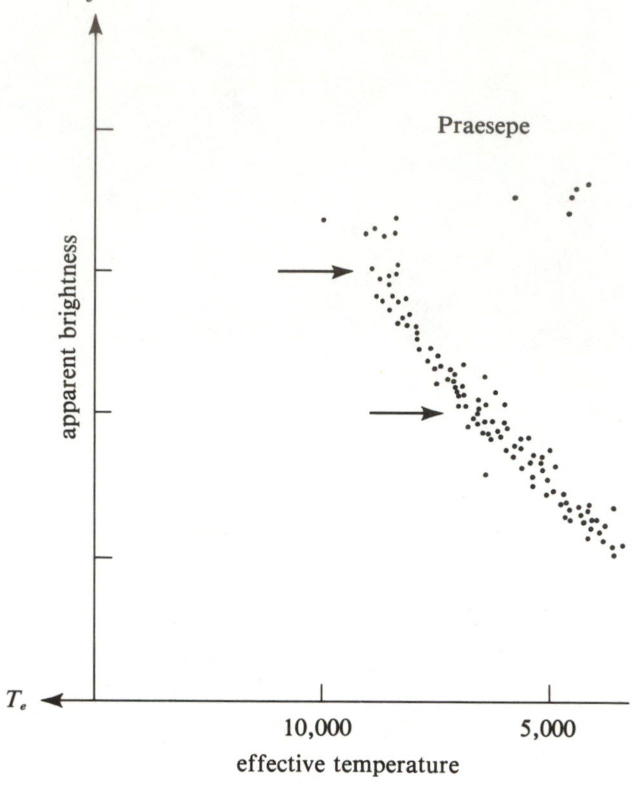

Figure 9.7. The H-R diagrams of the NGC 2362 and the Praesepe open clusters. The segments of main-sequence stars between the two arrows in each diagram refer to intrinsically similar stars; the segment in Praesepe appears displaced upward from the segment in NGC 2362 only because the former cluster is closer than the latter. (Adapted from H. L. Johnson and W. W. Morgan, *Ap. J.*, **117**, 1953, 313, and H. L. Johnson, *Ap. J.*, **116**, 1952, 640.)

the cluster, relative to some standard. Figure 9.8 shows the result for a number of open clusters when the Hyades cluster is taken as the standard. If we can then discover the distance to the Hyades cluster (as we shall soon discuss), we can deduce the distances to all the other open clusters in the diagram from the amount that we had to slide them up to make their main sequences line up with that of the Hyades.

The H-R diagram standardized to the Hyades, however, already tells us much when we interpret it by means of the theory of stellar evolution developed in Chapter 8. Imagine constructing a theoretical star cluster with a variety of stars that have different masses, but the same initial chemical composition, so that, at the initial time, all the stars lie on the "zero-age main sequence" (Figure 9.9; compare with Figure 8.1). After about 10^7 years, all the stars heavier than $10M_\odot$ will have evolved to their end states (i.e., off the diagram), whereas a star just equal to $10M_\odot$ will just be moving off the main sequence (by traversing rapidly to the right). Stars less than $10M_\odot$, on the other hand, will hardly have moved at all. After 10^9 years, all stars heavier than $2M_\odot$ or so will have evolved to their end states (off the diagram, if we do not plot white dwarfs), whereas a star of just $2M_\odot$ will have started to ascend the red-giant branch for the first time. After 10^{10} years, even a $1M_\odot$ star will begin to ascend the red-giant branch. At each such instant of time, of course, stars just lighter and just heavier than the star under discussion will be travelling the same path just behind or just ahead. Thus, successive snapshots of the locations of all the stars in the cluster taken at different instants of time will show an effect like the peeling of a banana. As Stromgren and Kuiper were the first to point out, this is precisely what we see in Figure 9.8 when we go from NGC 2362 to h and χ Persei to Pleiades to Hyades to Praesepe to NGC 188. The clusters so named

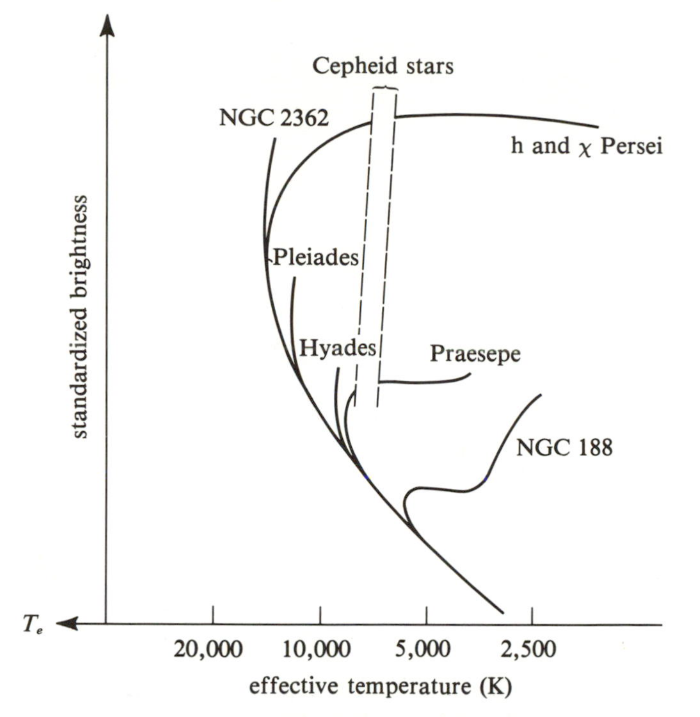

Figure 9.8. Schematic composite H-R diagram for open clusters whose main sequences have been "lined up" with the Hyades. Stars are conventionally not plotted in the strip indicated by dots, because stars whose mean state falls in this part of the diagram ("Cepheid instability strip") invariably prove to be pulsating stars. In addition, a gap (the Hertzsprung gap) usually exists between the turnoff point on the main sequence and the red giants, because stars evolve so quickly through this region that there is little probability of seeing them there. (Adapted from Allan Sandage, *Ap. J.*, **125**, 1957, 435.)

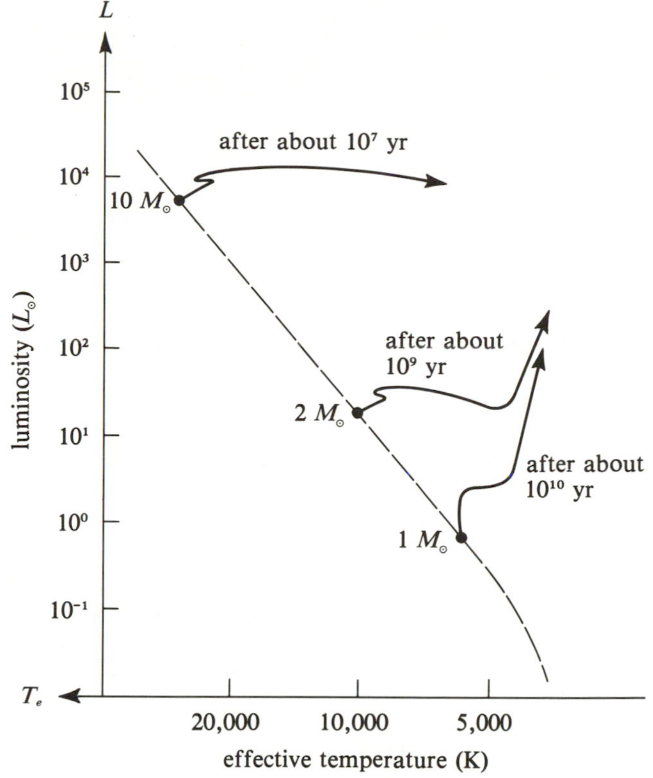

Figure 9.9. The evolution of stars of different masses away from the main sequence. (Adapted from Icko Iben, *Ann. Rev. Astr. Ap.*, **5**, 1967, 571.)

must therefore be arranged in a sequence of increasing age, with NGC 2362 being the youngest and NGC 188 being the oldest open cluster in our diagram. Astronomers believe that NGC 2362 is about 10^7 years old, and that NGC 188 lies between 6 and 10 billion years old.

Distance to the Hyades Cluster

From the above discussion, we see that we would have a fundamental calibration of the distances to the open clusters if we could measure the distance to the Hyades cluster. The Hyades cluster happens to be close enough to us, and happens to be moving with a component of velocity directed away from our line of sight, so that it has a deducible change of angular size with time. These facts allow the distance to the Hyades cluster to be calculated by what is called the moving-cluster method. The geometric ideas are simple, and its essentials are shown in Figure 9.10. Let v be the radial velocity of the cluster (measurable by the Doppler shifts of the stars), t the elapsed time, θ' and θ one-half the angular diameter at the earlier and later times. Under the approximation that the angular size of the cluster is small (so that θ and θ' expressed in radians are much less than unity), the geometric situation depicted in Figure 9.10 results in the formula

$$r = vt\theta/\Delta\theta,$$

where $\Delta\theta = \theta' - \theta$ is the measured change of angular radius during time interval t. The principle used to obtain this distance r to the cluster is analogous to the principle that one uses every day to cross the street. You stand on the curb and observe a car to come at you at some (unknown) speed v. You gauge mentally that the car changes fractionally in angular size $\Delta\theta/\theta$ by 10 percent in an interval $t = 1$ sec. From this data, you intuitively know that you have time $r/v = t\theta/\Delta\theta = 10$ sec to cross the street before the car is upon you. If you know that the car is traveling at the speed limit $v = 25$ mph, you could also compute the actual distance $r = vt\theta/\Delta\theta$ that

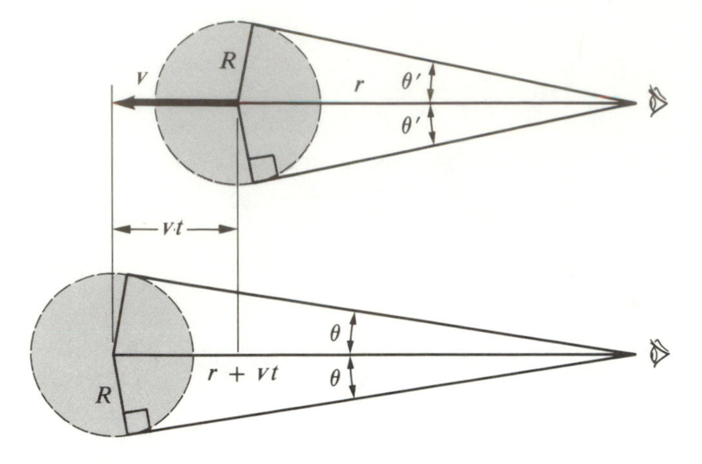

Figure 9.10. The "moving cluster" method for obtaining distance. The Hyades star cluster has a component of velocity v directed away from the observer. At an earlier time the angular size appears to be $2\theta'$. After a time t, the angular size appears to be smaller, 2θ. Because the radius R of the cluster cannot statistically have changed much in the interim available for the observations, the change in angular size must be attributed to the increased distance $r + vt$ of the cluster. Geometry yields the relations: $R = r\sin(\theta')$, and $R = (r + vt)\sin(\theta)$. Eliminating R and solving for r yields

$$r = [vt\sin(\theta)]/[\sin(\theta') - \sin(\theta)].$$

Using the small-angle approximation of Appendix B gives the simplified expression $r = vt\theta/\Delta\theta$, where $\Delta\theta = \theta' - \theta$.

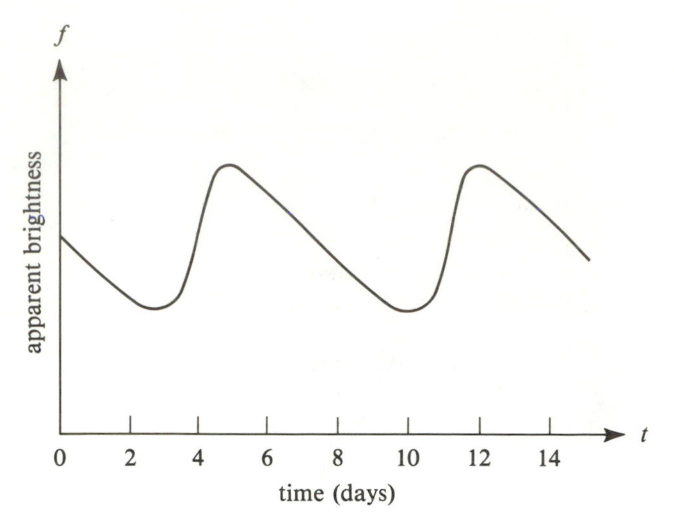

Figure 9.11. The light curve of a Cepheid variable. The period here is about a week (interval between, say, peak light to peak light).

the car is from you, but this latter computation is not essential to survival.* Alternative checks help you to judge the distance to the car. For example, your two eyes yield a "trignometric parallax," your knowledge of the intrinsic sizes of cars can be used as a "standard ruler" to compare with the angular size of the oncoming car, and your ears allow you to compare the "apparent loudness" produced by the car with your estimate of its "intrinsic loudness." Similar checks are available to the astronomer who measures distances to celestial objects.

The moving-cluster method has established the distance to the Hyades cluster to be about 140 light-years, which is quite close to the value proposed by Hodge and Wallerstein on the basis of other evidence. The method of "main-sequence fitting" then yields the distances to all open clusters whose main-sequence stars are sufficiently bright to be observable. In this manner, astronomers can convert the observed H-R diagrams from apparent brightnesses to luminosities. They can then make direct comparisons with the theoretical H-R diagrams calculated in computer models of real stars. This comparison is, by now, quite satisfactory, especially for the absolute locations of the main-sequence stars in the H-R diagram.

What about star clusters that are too far away to allow astronomers to measure the properties of their relatively faint main-sequence stars? There is another way to discover their distances, by the use of Cepheid variable stars.

Cepheid Period-Luminosity Relation

In many star clusters are found stars which vary in total light output in a periodic fashion (Figure 9.11). Such stars are called variable stars, and the most famous periodically varying stars in open clusters are called Cepheids. Cepheids always lie in a narrow band in the H-R diagram (see Figure 9.8), and they are not usually plotted in observational H-R diagrams, precisely because their apparent brightnesses do vary periodically in time.

In a study of Cepheids located in the Small Magellanic Cloud (a satellite galaxy of the Milky Way), Henrietta Leavitt discovered that Cepheids with longer intervals between maximum light have larger time-average luminosities. Shapley proposed that this *period-luminosity relationship* resulted from Cepheids being pulsating stars. Eddington pointed out that stars might become unstable toward pulsations if the opacity in the outer layers satisfied a certain "reverse valve" principle (Problem 9.7). Later workers discovered that this "reverse valve" principle could work in Cepheid variables providing the hydrogen and helium ionization zones lie neither too deep (to feel the effects of pulsation) nor too shallow (to contain too little mass) in the outer envelopes of the stars. This condition results in the "instability strip" being

* Science-fiction fans may compare this argument with Dave Weichart's reasoning in Hoyle's novel *The Black Cloud*.

confined to a fairly narrow region of effective temperatures (Figure 9.8). The weak dependence of ionization on pressure causes the "instability strip" to slope slightly toward lower T_e for larger stars. However, for a weakly changing T_e, the luminosity L of a star is larger for larger values of the radius R. Thus, higher-luminosity Cepheids are intrinsically larger than lower-luminosity Cepheids, and it is only to be expected that it would take the former longer than the latter to pulsate in and out.

Problem 9.7. An internal-combustion engine works because alternating cycles of expanding and contracting gases drives a piston up and down. A Cepheid pulsates for the same reason, but the underlying cause of the alternating cycles of high pressure and low pressure is different. In an internal-combustion engine, the underlying cause is a periodically varying supply of heat generated by the cyclic ignition of injected fuel. In a Cepheid, the underlying cause is a periodically varying "valve" which has gone awry. Plot the radius of a star as a function of time. For a star in mechanical equilibrium, this plot (the dotted lines) is a horizontal line at the position of the equilibrium radius. Imagine initially pushing the star's radius below this equilibrium value and then releasing it to move freely. The subsequent reaction of the star is represented as solid curves. Clearly, the star will begin to oscillate about its equilibrium state. In an ordinary star, these oscillations will be slowly damped, and the radius will eventually return to its equilibrium value (stable case). In a Cepheid, these oscillations will actually increase in amplitude at first, and in each cycle the radius will overshoot its previous value on each side of the equilibrium (unstable case). Eddington termed this situation "overstable," to distinguish it from the ordinary case of "instability," where an initial departure from equilibrium sets up forces which drive the system further from equilibrium in the same direction (such as in the helium "flashes" discussed in Chapter 8). In a Cepheid, an initial departure from equilibrium sets up forces which more than suffice to return the system back to equilibrium. The overreaction of the system causes the system to overshoot both sides of equilibrium until the oscillations become so nonlinear that other considerations enter. Here, we wish only to examine the basic cause of the initial overstability.

The first thing to realize is that the pulsation amplitude is largest in the outermost layers of a star. Thus, the nuclear sources of energy are undisturbed by the pulsation, and the basic instability mechanism must involve the behavior of the outer layers. When the star is maximally compressed (smallest R), the star is hottest. Throughout most of the mass of the outer layers of an ordinary star, higher temperatures imply lower opacities. Consequently, more photons leak out from the overcompressed, high-pressure, regions than in the unperturbed state. The "valve" opens, and relieves some of the overpressure which drives the radius outward. Thus, the star does not expand as much as it would have in the absence of the "valve effect." When the star is maximally

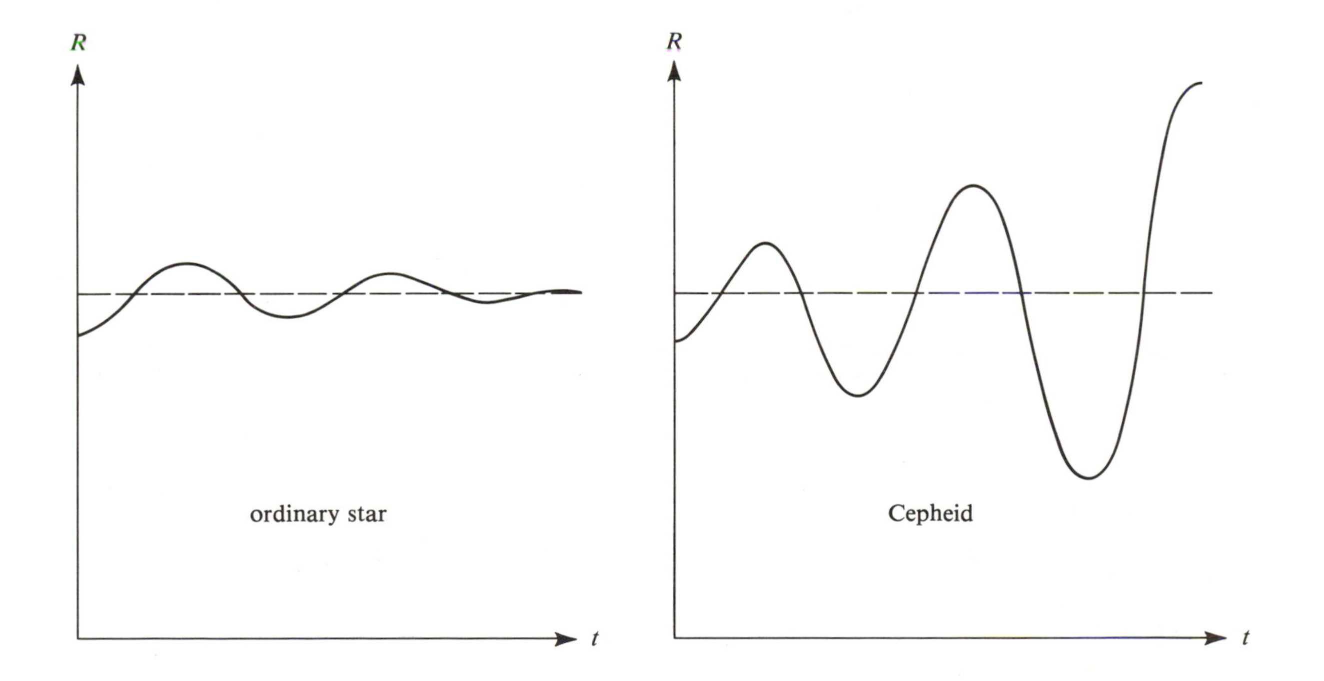

expanded (largest R), the star is coolest. Throughout most of the mass of the outer layers of an ordinary star, lower temperatures imply higher opacities. Consequently, the "valve" shuts, and the trapping of the heat causes the pressure to build up faster in the recontracting star, so that the radius falls less inward than during the previous cycle. Thus the oscillations are gradually suppressed in an ordinary star. Describe what you think happens in a Cepheid. Why should the opacity *increase* for increasing temperatures in regions where hydrogen and helium are partially ionized?

It is now known that a single star of high mass may cross the "Cepheid instability strip" (Figure 9.8) many times as it zigzags across the H-R diagram during its evolutionary lifetime (see Figure 8.11).* Observationally, the period-luminosity relationship for Cepheid variables is, by now, very well-established (Figure 9.12). The discovery of this relationship is one of the most important in optical astronomy, because it opened up the frontiers of extragalactic research (Chapter 13). The period-luminosity relation of Cepheids provides the major method by which astronomers extend the distance scale beyond the nearby open clusters. Box 9.5 gives a summary of the important stellar-distance indicators.

To see how the Cepheid method works, notice from Figure 9.12 that if you find a Cepheid variable and observe its light variations to find its period, you can deduce from the period-luminosity relationship what its time-averaged luminosity should be. Then, an additional measurement of its time-averaged apparent brightness allows you to deduce what its distance is (from the $1/r^2$ law of radiation). Since long-period Cepheids are especially bright stars (supergiants), they can be seen from very far away, even when their main-sequence friends are too faint to observe. In particular, such Cepheids can be used to calculate the distances to the nearest galaxies (Chapter 13).

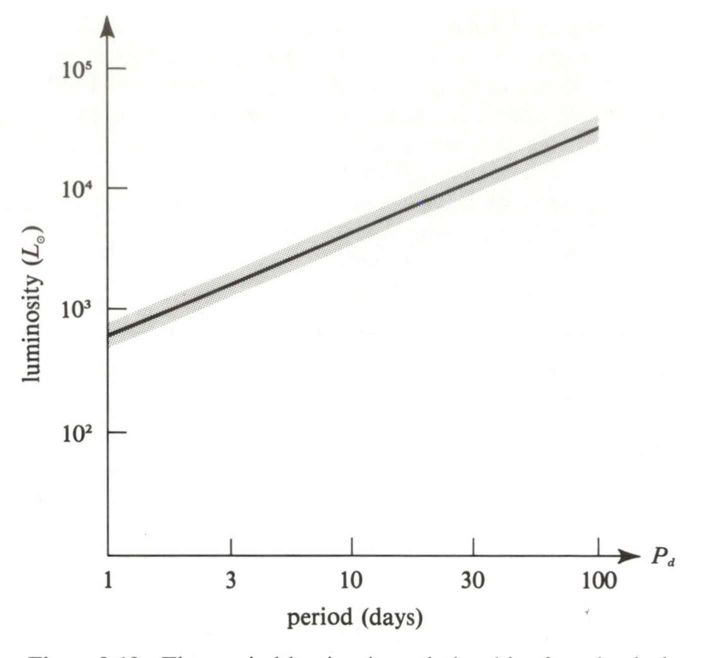

Figure 9.12. The period-luminosity relationship for classical Cepheids. The shaded strip centered on the mean line indicates schematically that there are slight deviations from a one-to-one relation, because of both theoretical reasons and observational uncertainties.

BOX 9.5

Distances to the Stars

Object	*Method*
Nearest stars	Trigonometric parallax.
Hyades open cluster	Moving-cluster method.
Open clusters	Main-sequence fitting to the Hyades.
Classical Cepheids	Period-luminosity relationship calibrated from the Cepheids in open clusters.
Field RR Lyrae stars	Statistical methods to calibrate luminosity.
Globular clusters	RR Lyrae stars as "standard candles."
Type II Cepheids	Period-luminosity relationship calibrated from Cepheids in globular clusters.

Comparison with theory checks both the theory and the internal consistency of the above methods. For nonvariable stars not found in clusters, there is not yet any accurate way to calculate distance if the star is too far away to exhibit trigonometric parallax. For such stars, the only distance indicator is the so-called method of "spectroscopic parallax."

Problem 9.8. Suppose you found a Cepheid with a period of 30 days in the Andromeda galaxy M31, and you measured the Cepheid to have a mean apparent magnitude of 18.7. Verify from the period-luminosity relation given in Figure 9.12 that such a Cepheid should have a mean absolute magnitude of about -5.5. Compute now the distance to the Andromeda galaxy. Convert

* Computer studies of nonlinearly pulsating stars, first caried out by Robert F. Christy, now provide a sensitive test of stellar-evolution theory when detailed comparisons are made between the pulsation characteristics of model stars and those of real Cepheid variables. According to Icko Iben, the agreement is reasonably good, but nagging small discrepancies remain.

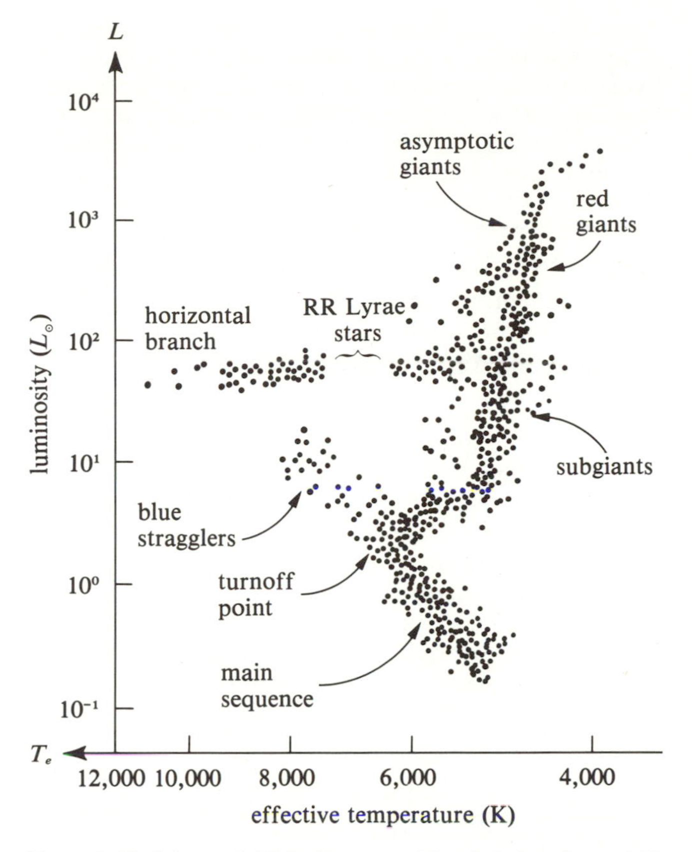

Figure 9.13. Schematic H-R diagram of the globular cluster M3. The distance to the cluster has been derived on the assumption that the luminosity of the RR Lyrae stars is $50L_\odot$. The diagram is somewhat schematic, because the conversion from the $B - V$ and V measurements of Johnson and Sandage to T_e and L is somewhat uncertain for Population II stars. (Adapted from H. L. Johnson and A. R. Sandage, *Ap. J.*, **124**, 1956, 379.)

your answer to light-years. How long ago did the light leave the Cepheid in M31 finally to enter your telescope? If you could fly instantaneously to M31, do you think you could find your Cepheid?

Globular Clusters

The second type of star cluster which exists in the Galaxy is the globular cluster. Globular-cluster stars are Population II stars, and differ from open cluster stars both in heavy-element abundance and in general age. Plotted in Figure 9.13 is the observed H-R diagram of the globular cluster M3. Notice that this diagram differs from the H-R diagram of even a very old open cluster like NGC 188 in several details (compare Figures 9.8 and 9.13). In particular, the long "horizontal branch" in M3—containing a gap of unplotted variable RR Lyrae stars, which we will soon discuss—has no counterpart in NGC 188. This difference is one manifestation of the very low heavy-element abundance in globular clusters like M3. The "turnoff point" in the "peeling of the banana of the main sequence" in globular clusters is also typically quite low. This means that even quite light stars (say, $0.8M_\odot$) in globular clusters have had a chance to evolve significantly, and this fact attests to the great age of the globular-cluster stars. Detailed comparisons of theory and observations yield typical estimates of about 12 billion years for the ages of globular clusters. Since globular-cluster stars are the oldest known objects in our Galaxy, it is likely that our own Galaxy formed something like 12 billion years ago.

Certain details in Figure 9.13 seem to violate the general picture of stellar evolution outlined in Chapter 8. For example, the subgiant branch of M3 contains too broad a swath of stars to accord with theoretical calculations. Some of these stars have turned out to be foreground stars, not really members of the globular cluster. Another cause for the broad swath may be a greater variation of the initial heavy-element abundance in stars of a given cluster than originally thought plausible. Notice also the presence of "blue stragglers" in the H-R diagram of M3. The "blue stragglers" seem to be main-sequence stars actually in the cluster that are more massive than the stars at the "turnoff point." This is, of course, not possible under the theory of single-star stellar evolution outlined in Chapter 8 if all the stars in the globular cluster were born at about the same time. A promising speculation about the "blue stragglers" is that they are (or were) members of close binary systems in which a nearby companion has fairly recently dumped some additional mass onto a previously low-mass star. The addition of mass shifted the latter's position to the main-sequence location appropriate for the new total mass. We shall discuss the interesting freaks which can be produced by such close binary interactions in Chapter 10.

RR Lyrae Variables

As we said, the horizontal branch of a globular cluster contains a class of pulsating stars—RR Lyrae variables—which are closely analogous to Cepheids, except that RR Lyrae stars, being horizontal-branch stars instead of giants or supergiants, are smaller generally than the Cepheids found in open clusters. Thus, RR Lyrae stars have shorter periods than the classical Cepheids. The periods of the former are shorter than one day; the periods of the latter, longer than one day. Moreover, RR Lyrae stars all have about the same time-averaged luminosity (because they lie on the *horizontal* branch); so they can also be used as distance indicators once their

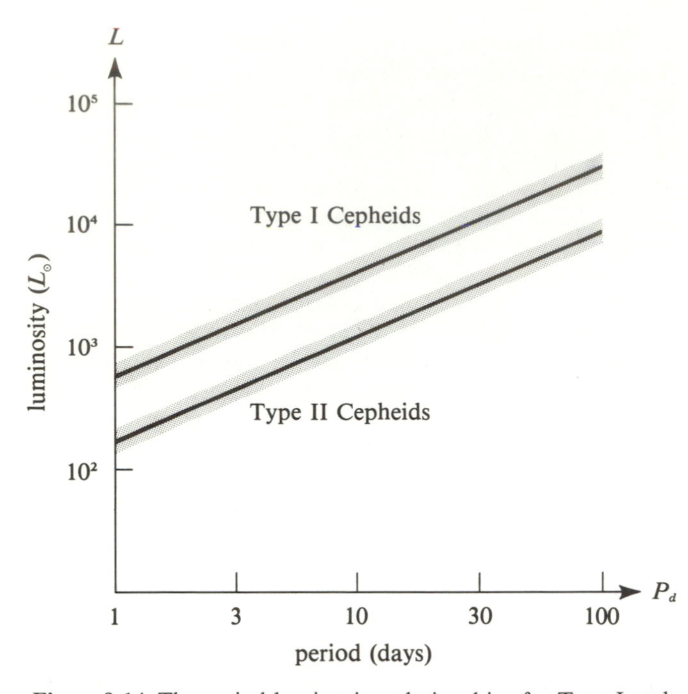

Figure 9.14. The period-luminosity relationships for Type I and Type II Cepheids. Notice that the luminosity of Type II Cepheids is about 4 times less at every period than the luminosity of classical Cepheids.

absolute brightness has been calibrated. Unfortunately, there is no globular cluster close enough to let us measure its distance directly as we did for the Hyades open cluster. The absolute brightness of RR Lyrae stars is actually obtained by means of a series of steps that involve getting the distance statistically to RR Lyrae stars in the general field. Once this has been done, it becomes possible to use RR Lyrae stars as distance indicators. In fact, they are the means by which astronomers discover the distances to the globular clusters themselves.

Population II Cepheids

In addition to the RR Lyrae stars, globular clusters occasionally also contain a rarer class of pulsating stars, called Type II Cepheids or W Virginis stars. These stars are fully analogous to the classical Cepheids ("Type I Cepheids"), but are Population II stars. Population II Cepheids also satisfy a period-luminosity relationship, but it is different from that satisfied by Population I Cepheids (Figure 9.14). For a given period, the mean luminosity of a Population II Cepheid is about four times fainter than the mean luminosity of a Population I Cepheid. It was only after the calibration of the distances to the open clusters by main-sequence fitting and of the distances to the globular clusters by using RR Lyrae stars that astronomers fully realized the distinction between the two types of Cepheids. Before that time, astronomers had confused the two types, and had therefore underestimated the distances to external galaxies.

Problem 9.9. Suppose in Problem 9.8, you had mistakenly thought you were dealing with a Type II Cepheid. By what factor would you have underestimated the distance to the Andromeda galaxy?

Astronomers have since discovered that Type I and Type II Cepheids of the same period can be distinguished by the *shapes* of their light curves. Another distinction lies in the fact that Population I stars and Population II stars are now known to have very different spatial distributions in a galaxy (Chapter 12). This fact, coupled with the differences in general age and heavy-element abundance between Population I and Population II stars, must be clues to the history of the formation and evolution of galaxies.

Dynamics of Star Clusters

Our present knowledge of the formation of stars from diffuse clouds of gas and dust in the interstellar medium (Chapter 11) is consistent with the statement that most stars, and perhaps all stars, were born in clusters. Yet the Sun is not now in a cluster, and neither are the great majority of Population I stars in the Galaxy. How do we reconcile these two statements?

To answer this question requires an understanding of the dynamics of a stellar system like a star cluster. The study of such systems falls within the boundaries of the discipline called "stellar dynamics." Important early work on this problem was carried out by Jeans and by Chandrasekhar, but a refined understanding of the dynamics of star clusters came only with the later work of Spitzer, King, Lynden-Bell, and others. For our purposes here, a simplified account will suffice.

Because the sizes of stars are small compared to their separations, even in a dense cluster of stars, the stars move freely past one another without suffering direct collisions. To a first approximation, a spherical stellar system like a star cluster can then be understood dynamically as a balance between the inward gravitational attraction on each star by the cluster as a whole, and the inertia associated with that star's random motion relative to the cluster center. Each star's orbital motion carries it periodically closer to and further from the cluster

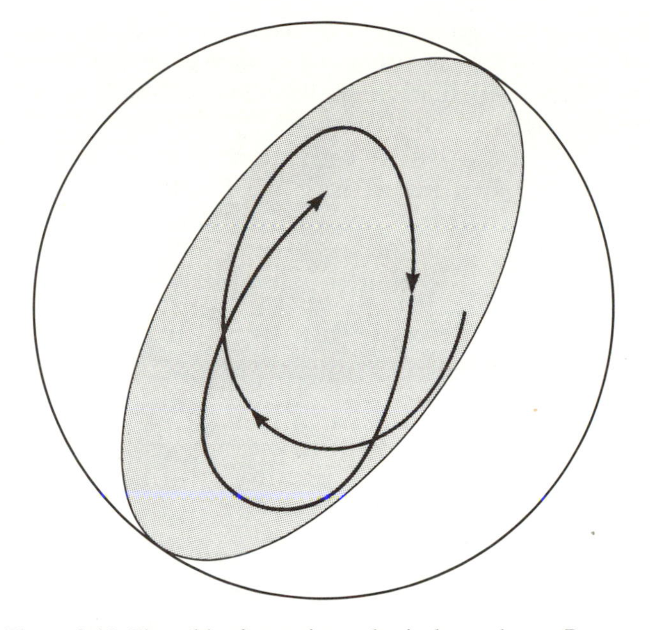

Figure 9.15. The orbit of a star in a spherical star cluster. Because the mass distribution of the star cluster is neither homogeneous nor concentrated all at the center, the orbit of the star is generally not closed. However, because the vector angular momentum of the star is conserved in a spherical star cluster, the orbit takes place entirely in a single plane (shaded ellipse). Encounters with other stars gradually destroy the latter properties of the smooth orbits.

center, but statistically no net effect is caused by the interaction of a star with the gravitational field of the cluster as a whole (Figure 9.15). (We ignore here "violent relaxation" effects that might occur in the early stages of the formation of a star cluster.) There is, however, a slow net effect because of the fact that a star cluster is made of a discrete number of stars. Stars feel slight gravitational tugs as they pass one another randomly. These small random deflections are superimposed on the smooth orbit drawn in Figure 9.15. The resulting tendency is to establish a distribution of stellar velocities which is entirely analogous to the thermodynamic (Maxwell-Boltzmann) distribution of random velocities established by random encounters of particles in a classical gas (Chapter 4 and Problem 6.9). Such distributions in a cluster are characterized by a dispersive speed V which is analogous to the expression $(3kT/m)^{1/2}$ for a classical gas. Most stars have random speeds comparable to V, but a true Maxwell-Boltzmann distribution has a "tail" of particles with speeds v much in excess of V. Random gravitational encounters in the star cluster attempt to establish such a "thermodynamic" distribution, but they can never quite succeed, for the following simple reason:

> Stars which exceed the escape speed from the cluster will be lost from the cluster.

Indeed, application of the "virial theorem" demonstrates that the typical escape speed v_e is only twice the typical dispersive speed V:

$$v_e = 2V.$$

Problem 9.10. The virial theorem for a statistically steady, spherical, self-gravitating star cluster states that the total kinetic energy of N stars with typical random speed V is equal to minus one-half of the total gravitational potential energy. If R is the average separation between any two stars (assumed to be of equal mass m), the gravitational potential energy of the pair is $-Gm^2/R$. There are $N(N-1)/2$ possible pairings of N stars; consequently, the total gravitational potential energy is $-GN(N-1)m^2/2R$. The virial theorem can now be written

$$NmV^2/2 = GN(N-1)m^2/4R.$$

Since R is a typical size for the cluster (the "core radius"), argue that the typical escape speed from the cluster is given by

$$mv_e^2/2 = G[(N-1)m]m/R.$$

From the above two relations, show $v_e^2 = 4V^2$, i.e., $v_e = 2V$. Estimate R numerically if $N = 10^3$, $m = 1M_\odot$, and $V = 1$ km/sec (open cluster); and if $N = 10^6$, $m = 0.5M_\odot$, and $V = 20$ km/sec (globular cluster).

The escape of high-speed stars requires the remaining cluster stars to readjust their random velocity distribution to repopulate the high-speed tail. In the simple case of N equal-mass stars with a cluster-core radius R, this readjustment typically takes time (Problem 9.11)

$$t_{\text{relax}} = \left(\frac{R}{V}\right)\frac{N}{12\ln(N/2)}.$$

The fraction of high-speed stars which exceed the escape speed then works out to give a characteristic evaporation timescale (for the cluster to reduce its number of stars by a factor of $e = 2.718\ldots$),

$$t_{\text{evap}} = 96t_{\text{relax}}.$$

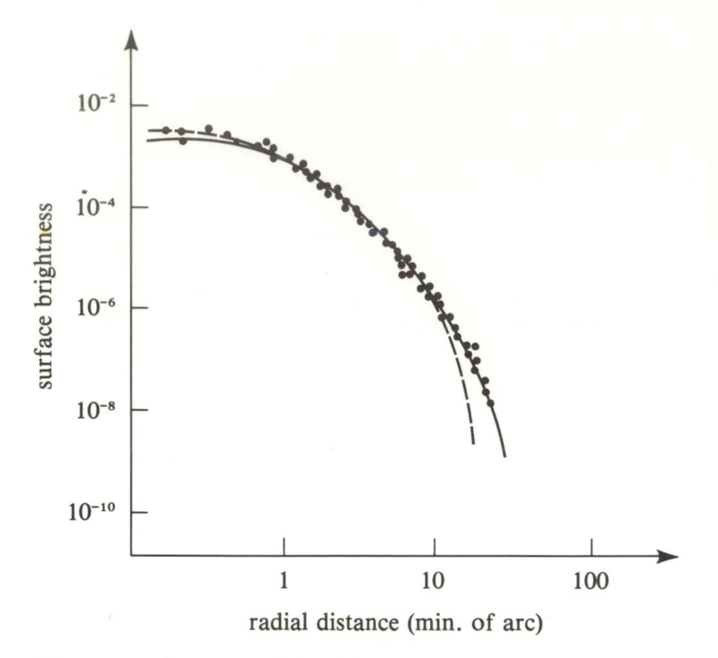

Figure 9.16. Structure of the globular cluster M3. The dots give the observational measurements of the mean surface brightness (some standardized unit) as a function of the radial distance from the center of the cluster. The dashed curve gives the best single-mass "King model" fit for the observational points. Clearly, this simple model fails to follow the outer points because equal-mass stars which populate the core to the desired degree fail to populate the envelope sufficiently. A better overall fit is achieved by the solid curve, which adopts a multi-mass "King model." (Adapted from Da Costa and Freeman, *Astrophysical Journal*, **206**, 1976, 128.)

For a typical open cluster, we may compute t_{evap} to be about 3×10^9 yr; for a typical globular cluster, to be about 8×10^{10} yr (Problem 9.12). Thus, we can understand quantitatively why all except the richest and most compact open clusters will have completely evaporated if their ages are comparable to the age of the Galaxy (8 to 13×10^9 yr). We expect globular clusters also to have undergone some dynamical evolution, but we can also appreciate why they are still around after about ten billion years. Detailed models of the spatial distribution of stars in globular clusters to be expected on the basis of dynamical theory have been constructed and compared with observations. On the whole, the comparisons are pretty favorable, but it has become clear in recent studies that really good fits require a more realistic assumption than a single mass species for the cluster (Figure 9.16).

Problem 9.11. A rough estimate of the relaxation time for stellar encounters proceeds as follows. Imagine every star to have a sphere of influence of cross-sectional area πr^2, such that if any other star were to enter this sphere of influence, it could be said to have suffered a strong encounter. The relaxation time t_{relax} is defined to be the time between successive encounters. In other words, if the number density of stars is n, and if the typical random velocity of a star is V, we want the cylindrical volume swept out by that star's sphere of influence in time t_{relax}, $\pi r^2 V t_{relax}$, to contain one other star. Thus, we

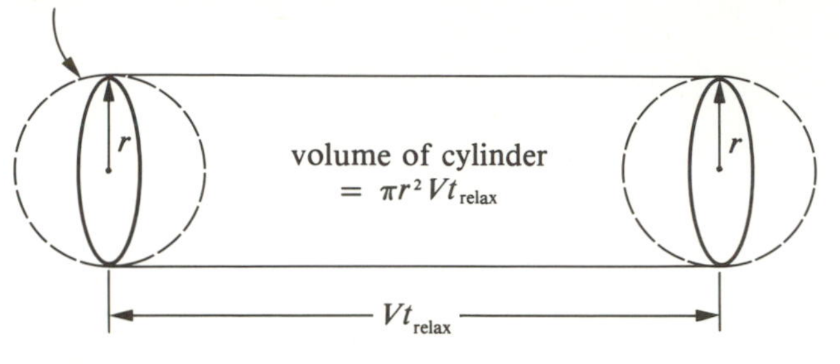

want $n(\pi r^2 V t_{relax}) = 1$, and this condition defines t_{relax} to be

$$t_{relax} = 1/n\pi r^2 V.$$

What should we choose for r here? A sensible choice would seem to be the radius at which the gravitational potential energy of a pair of stars is equal to the typical random kinetic energy per star:

$$Gm^2/r = mV^2/2.$$

This definition certainly suffices to define r for a strong encounter. Substitute the above value of r into our previous formula for t_{relax} and show that our rough estimate becomes

$$t_{relax} = \frac{V^3}{4\pi G^2 m^2 n}.$$

A more careful analysis shows that many small deflections are more efficient for the problem of stellar encounters than one big one, by a factor of about $\ln(2R/r)$, where $2R$ is the cluster-core diameter. This correction reduces t_{relax} from our naive estimate to the expression

$$t_{relax} = \frac{V^3}{4\pi G^2 m^2 n \ln(2R/r)},$$

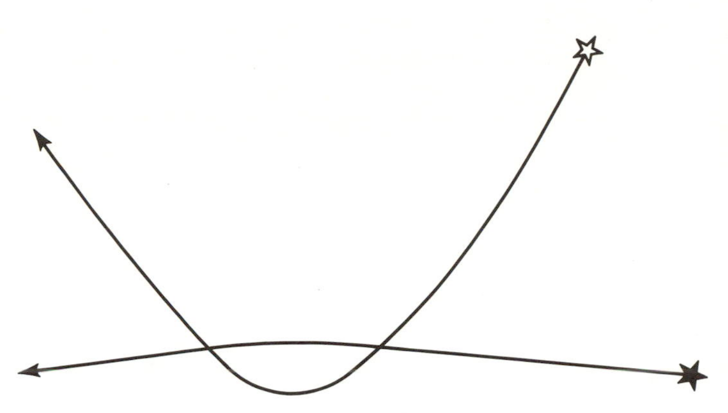

Figure 9.17. In a gravitational encounter between a heavy star (filled star) and a light star (empty star), the light star is deflected greatly, but the heavy star is deflected little. The statistical tendency of many such encounters is to establish "equipartition of kinetic energies" in the random-velocity distributions of light and heavy stars.

which differs from the "reference time" of Spitzer and Härm by a factor of $(1/2)(3/2)^{3/2} = 0.9186$. Define the typical number density of stars by the relation $n(4\pi R^3/3) = N$, approximate $N \gg 1$, and use the results of Problem 9.10 to convert the preceding equation into the form

$$t_{\text{relax}} = \left(\frac{2R}{V}\right) \frac{N}{24 \ln (N/2)}.$$

For what value N does t_{relax} equal the dynamical crossing time $2R/V$? Argue that this result means that orbits in the sense of Figure 9.15 constitute a useful first approximation for most clusters.

Problem 9.12. Use the data given in Problem 9.10 to compute $t_{\text{evap}} = 96t_{\text{relax}}$ for typical open clusters and globular clusters.

When more than one mass species is included in the calculations, the evaporation process is greatly accelerated. The reason is simple. In encounters between light stars and heavy stars, the light stars are deflected greatly, but the heavy stars are deflected little (Figure 9.17). (As an extreme example, consider the case of the Earth and an artificial satellite. The satellite whizzes around, but the Earth hardly moves at all.) In statistical equilibrium, the stars of all masses tend to have distributions characterized by "equipartition of kinetic energy." (This tendency is a general thermodynamic result, and is not restricted in validity to gravitational interactions.) If statistically the quantity $mv^2/2$ has a fixed value, stars with small masses m will tend to have larger speeds v than average, whereas stars with large masses will tend to have smaller speeds than average. The fast, light stars thus find it easier to exceed the escape speed and they will evaporate preferentially, whereas the heavy stars tend to sink to the center of the cluster. The contraction of the cluster core gravitationally, however, releases binding energy, which requires the heavy stars to move faster. The heavy stars share their newly found kinetic energy with the light stars in subsequent encounters, and this accelerates the evaporation of the latter. The overall tendency is to form a "core-envelope" structure reminiscent of the structure of a red giant star (Chapter 8). Here, however, we are speaking of the spatial distribution of a whole cluster of stars, each treated as a mass point. Detailed calculations show that the inclusion of multi-mass species in star-cluster models can cause evaporation to proceed more than ten times faster than our simple estimates supposed. Much evolution should therefore have occurred for the cores of some of the densest globular clusters, and it has been conjectured that "core collapse" may have been initiated in some of these systems to give a very massive (say, $10^3 M_\odot$) black hole at the center. The observational support for such a scenario is, however, rather meager at the present time.

Possible complications for the above theoretical picture include the following effects. The tidal forces of the Galaxy would help to evaporate stars; they cannot realistically prevent core collapse. On the other hand, Heggie has shown that binary stars in the cluster may be very efficient in ejecting stars from the cluster. The interaction of a third star with a loosely bound binary tends to make the latter even looser, but a tightly bound binary tends to throw the intruder out faster than it came in (Figure 9.18). Thus, a possible end state of cluster evolution in such cases is the evaporation of all the stars from the cluster, leaving behind only one very tightly bound binary-star system. The observational situation for globular clusters is unclear, because they now seem to lack close binary systems.

Although the details of the dynamical evolution of star clusters is somewhat complex, the overall trend is clear. It is another example of the interplay between the second law of thermodynamics and the law of universal gravitation. On the macroscopic scale of star clusters, the individual stars can be considered to be microscopically small; so the inexorable operation of the second law of thermodynamics and the relentless pull of self-gravity leads to a continuing flow of energy and stars from the cluster to its environment. Catastrophe befalls the cluster, because no true thermodynamic state of equilibrium exists for a realistic self-gravitating cluster of stars. It remains to be seen whether the final outcome of this "gravothermal catastrophe" is a completely

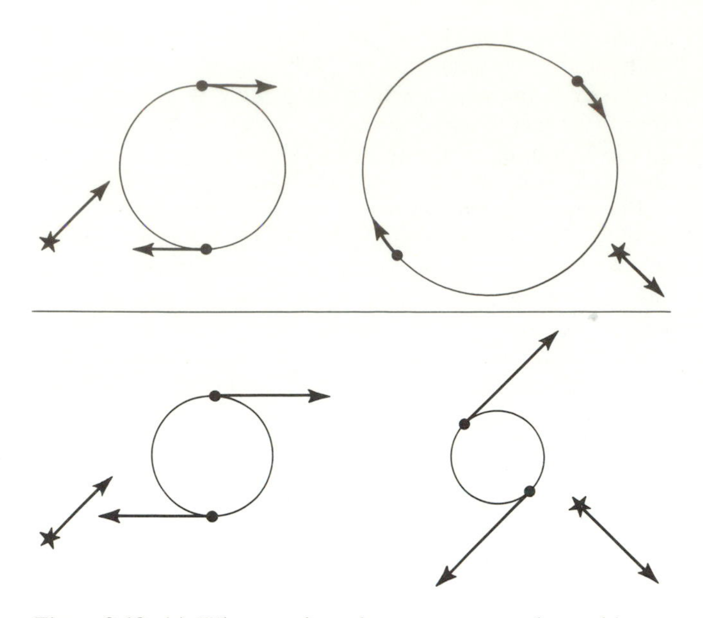

Figure 9.18. (a) When an intruder encounters a loose binary, whose stars orbit at a velocity less than that of the intruder, the intruder tends to give energy to the binary and loosen it even more (or even disrupt it completely). (b) When an intruder encounters a tight binary whose stars orbit at a velocity greater than that of the intruder, the pair tend to give energy to the intruder and become even tighter. Notice the familiar gravitational result that when a binary system loses energy, the orbiting stars speed up! This result implies that there is no true thermodynamic equilibrium possible between the distribution of binaries' internal degrees of freedom and the stars' external (translational) degrees of freedom.

dispersed system of stars, or whether a black hole may be left in some circumstances.

Philosophical Comment

For pedagogical efficiency, we have presented the properties of real stars and star clusters after first discussing the prevailing theory. This presentation may have given the reader the misimpression that observations are subservient to theory in astronomy, and serve only as a check to pure thought. Nothing could be further from the truth. Almost always the seminal observational discoveries were made before the final synthesis of a coherent theory. For example, the Cannon-Pickering spectral classification, the Hertzsprung-Russell diagram, and the Adams-Kohlschütter luminosity criteria were constructed on purely observational bases long before anyone learned how to interpret them theoretically. Natural science is distinguished from pure mathematics by the crucial role played by experiments and observations. This is why progress is slower in science than in mathematics. There are many self-consistent, logical structures, but most of these give a wrong description of the actual universe, not because they are illogical, but simply because nature does not happen to work that way. Without experiments and observations to guide it, the human mind would err and err again in trying to decipher the riddles of the actual universe.

CHAPTER TEN

Binary Stars

So far we have discussed only single stars, which are free to pursue their evolution in isolation from other stars. In the actual Galaxy, however, most stars come in pairs or multiples. It has been estimated that more than half of all the stars in the sky have other stars orbiting them as companions. In a small percentage of these cases, the two stars are actually close enough to have profound effects on each other's life history.

Observational Classification of Binary Stars

Only a very small minority of binary stars can be detected as **visual binaries**, where one can see both components of a well-separated pair orbiting in ellipses about one another. A famous visual binary is Sirius, in which Sirius A is a main-sequence star of spectral type A1, and Sirius B is a white dwarf of spectral type A5, orbiting about one another with a period of 49.9 years (Figure 10.1). Even through a good telescope, it is not easy to see the white dwarf in the glare of its much brighter companion. Indeed, Sirius was first discovered to be a double star, not by the direct detection of Sirius B, but because Sirius A was seen to "wobble" in the sky. That is, Sirius was found to be an **astrometric binary** before it was discovered to be a visual binary. If we take into account the effects of projection, the geometric forms of the orbits of visual binaries accord with the theory of Newton (Figure 10.2):

> The bound orbits of binary stars are ellipses when viewed from either star, and are two ellipses when viewed by an observer at rest with respect to the center of mass of the system. These various ellipses are of the same shape, but not of the same size.

Stars which have small angular separations are not necessarily physically associated. For example, people with good eyes can see that the "second star" from the end of the handle of the Big Dipper is actually double (Figure 10.3). The two stars making up this pair, Mizar

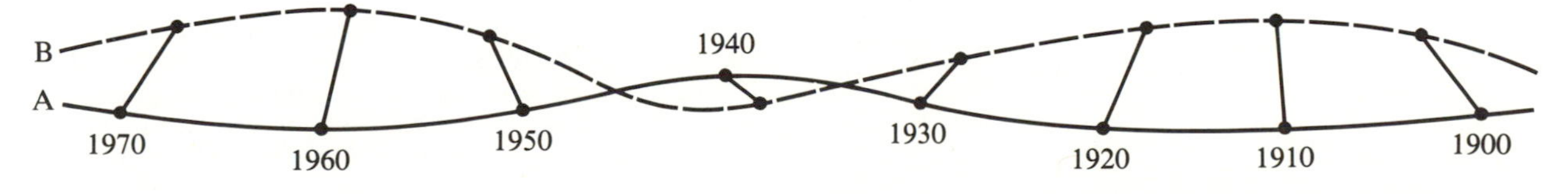

Figure 10.1. Sirius A and B wobble about a straight line which describes the motion of the center of mass through space.

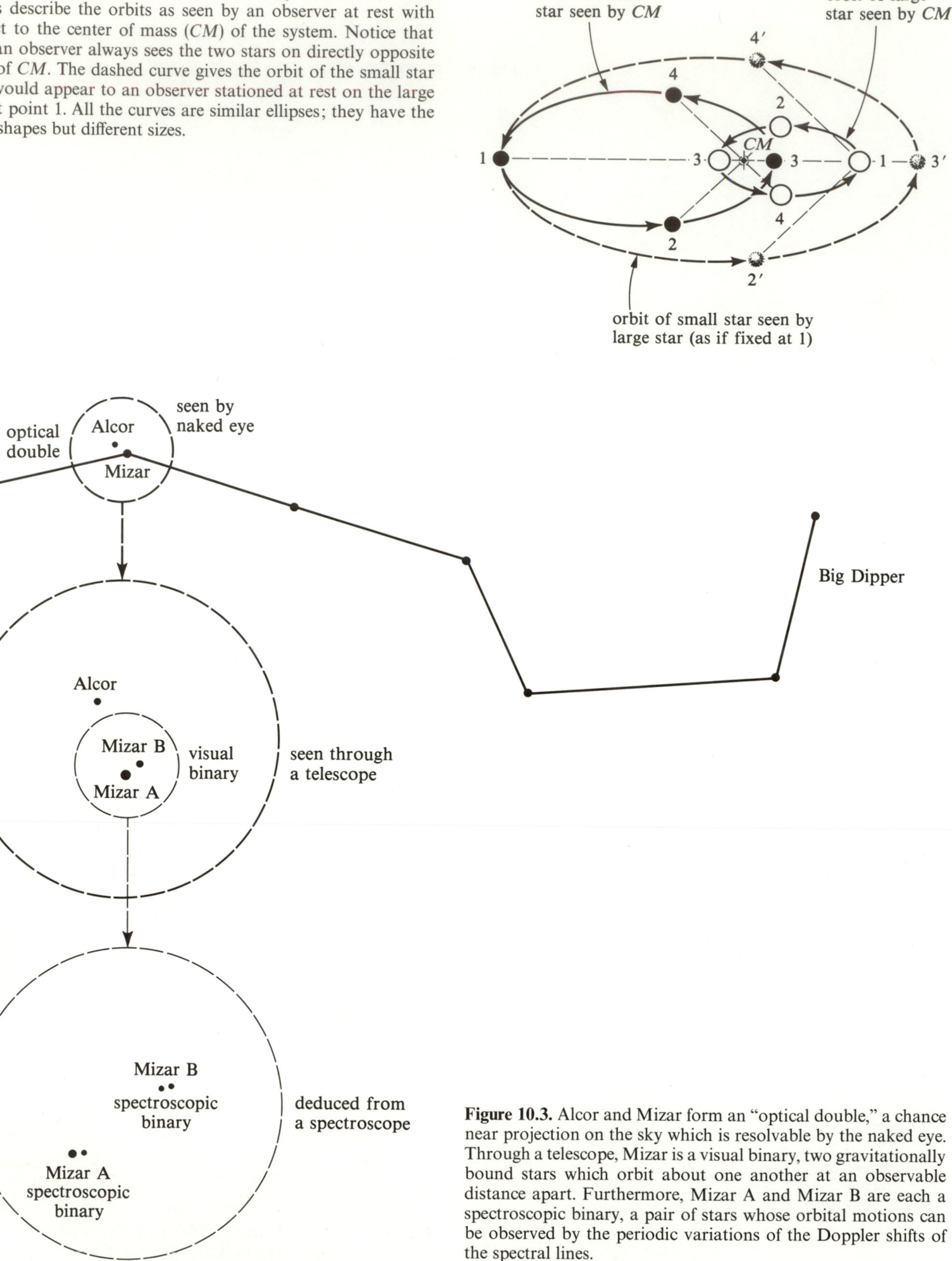

Figure 10.2. The geometric forms of binary orbits. The solid curves describe the orbits as seen by an observer at rest with respect to the center of mass (CM) of the system. Notice that such an observer always sees the two stars on directly opposite sides of CM. The dashed curve gives the orbit of the small star as it would appear to an observer stationed at rest on the large star at point 1. All the curves are similar ellipses; they have the same shapes but different sizes.

Figure 10.3. Alcor and Mizar form an "optical double," a chance near projection on the sky which is resolvable by the naked eye. Through a telescope, Mizar is a visual binary, two gravitationally bound stars which orbit about one another at an observable distance apart. Furthermore, Mizar A and Mizar B are each a spectroscopic binary, a pair of stars whose orbital motions can be observed by the periodic variations of the Doppler shifts of the spectral lines.

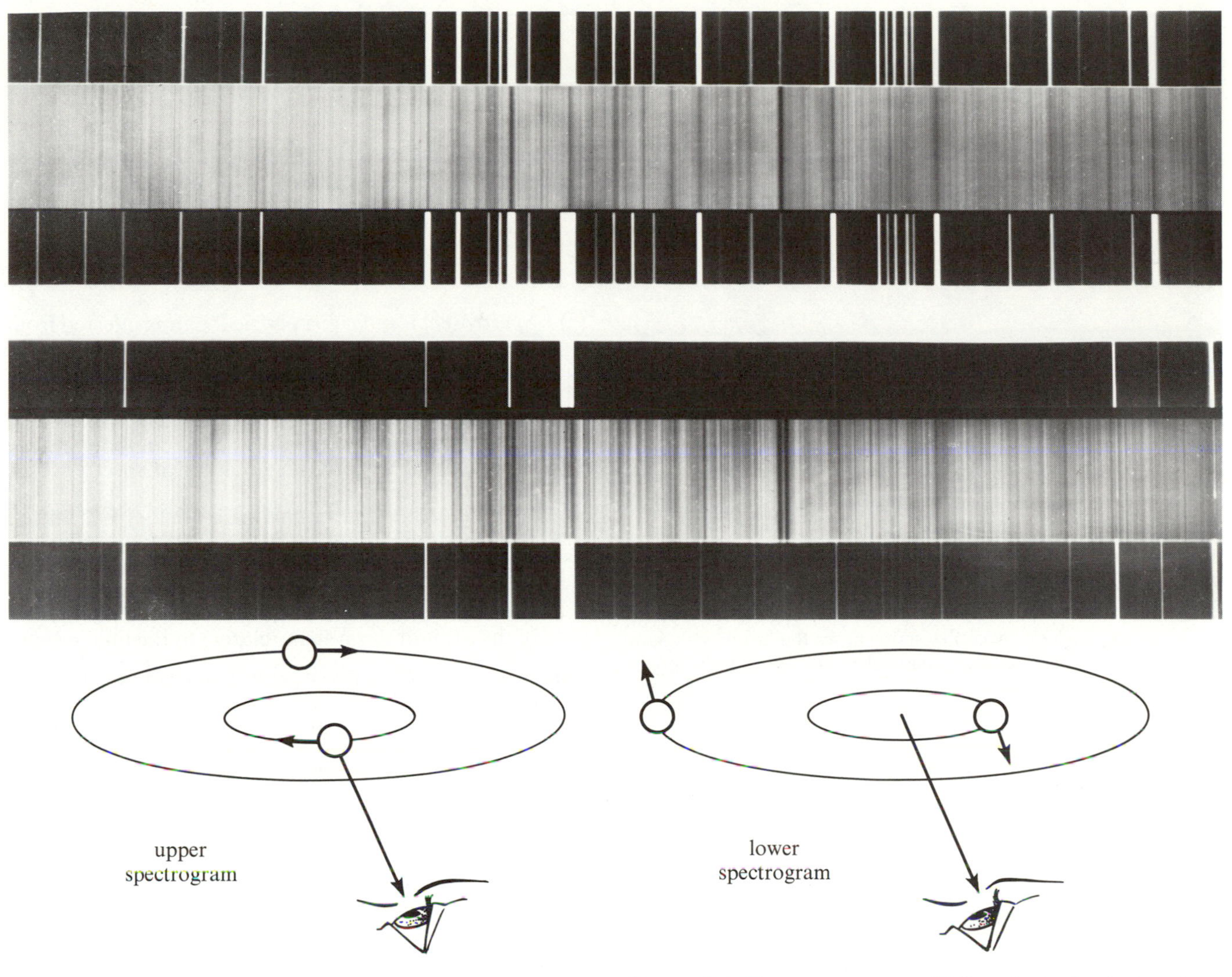

Figure 10.4. Two spectrograms of Mizar A obtained a few days apart in a binary cycle which repeats every 20.5 days. In the upper spectrogram, the absorption lines of the two stars lie on top of one another. At this time the two stars' orbital motions must be directed perpendicular to our line of sight. In the lower spectrogram, the absorption lines are separated by Doppler shifts which correspond to a difference of more than 100 km/sec in relative velocity along the line of sight. At this time, the revolution of the two stars about a common center of mass must have carried the stars to a configuration where one star is moving toward the Earth (blueshift) and the other star is moving away from the Earth (redshift). (Yerkes Observatory photograph.)

and Alcor, happen by chance to lie nearly along the same line of sight; they are merely an "optical double." However, viewed through a telescope, Mizar is a visual binary, whose two components—Mizar A and Mizar B—slowly revolve around one another. To make life more interesting, when astronomers examined the light of each component spectroscopically, they discovered that Mizar A and Mizar B are themselves binary stars! The angular separation of the two stars that make up Mizar A and of those that make up Mizar B is too small to resolve even through a large telescope, but spectra of the system reveal evidence of the binary motion (Figure 10.4). Thus,

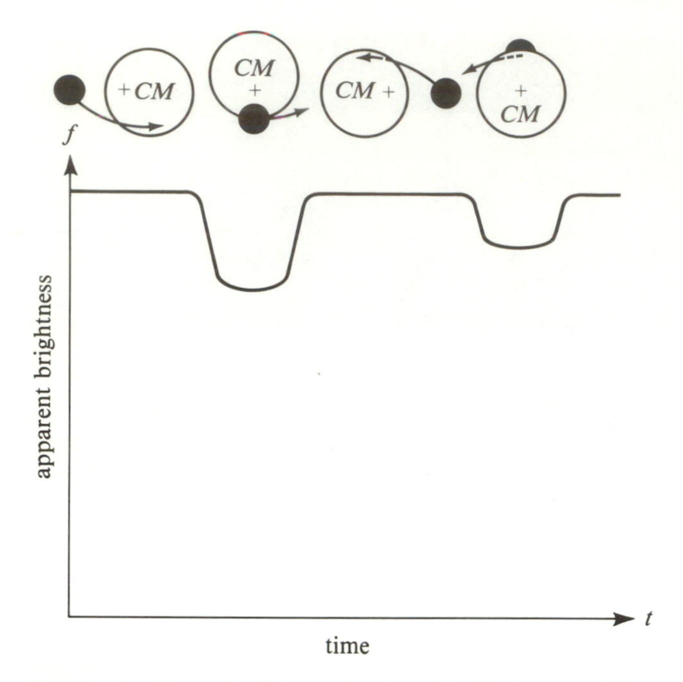

Figure 10.5. Schematic light curve of an eclipsing binary. Deeper eclipses result when the cooler star eclipses the hotter star than when the hotter star eclipses the cooler. The pattern of eclipses repeat once every binary period.

Mizar A and Mizar B are **spectroscopic binaries**, and the entire Mizar system is a quadruple system. Spectroscopic binaries are called **single-lined** if only one star's spectrum is observable; **double-lined**, if both stars' are.

Binary stars whose angular separations are too close to be resolved (and this includes most binaries) may also have their double nature revealed if the two stars eclipse one another. We can detect this only if we happen to lie fairly close to the orbital plane of the system. An **eclipsing binary** is defined observationally to be a system where the apparent brightness varies periodically because of successive eclipses of one star by another (Figure 10.5). A particularly spectacular eclipsing binary is Algol. Every 2.9 days, a primary eclipse drops the apparent brightness of Algol by more than a factor of two. For reasons to be explained later, such deep eclipses imply a state of evolution for the system that seems not to accord well with the theory of single-star evolution developed in Chapter 8.

For double-lined spectroscopic binaries that also eclipse, we can deduce all important properties of the system (except the distance). In particular, for such systems we can directly measure the masses and the radii of the two stars (Problem 10.1). Thus, double-lined spectroscopic binaries that eclipse play an important role in observational astronomy as checks on certain predictions of the theory of stellar structure and evolution. (Visual binaries can also provide masses, but they are usually too wide to have much chance of eclipsing.) Box 10.1 summarizes the observational classification of binaries.

BOX 10.1
Observational Classification of Binaries

(1) *Astrometric binary*: the physical association of two stars is inferred, although only one star is actually observed, because the proper motion of the visible star wobbles in the sky.
(2) *Visual binary*: two stars are seen as separate images that orbit about one another with the passage of time.
(3) *Spectroscopic binary*: a physical pair is inferred from spectroscopic observations which show a periodic variation of the Doppler shift of the spectral lines. If the spectral lines of both stars are observed, the system is a double-lined spectroscopic binary. If the spectral lines of only one component are observed, the system is a single-lined spectroscopic binary.
(4) *Eclipsing binary*: a bound pair of stars is deduced from periodic changes of the total light from the system that can be interpreted in terms of eclipses of one star by the other. An analysis of the light curve can often yield an estimate of the inclination of the orbit of the system relative to the line of sight.

Stellar masses for the two components can be measured in three cases:
(a) visual binary with analyzed orbital shape, if the distance to the system is known;
(b) double-lined spectroscopic and eclipsing binary with analyzed light and velocity curve;
(c) visual binary with analyzed orbital shape, if the radial velocities of the two components are known.

Cases (a) and (b) are much more common than case (c), which requires the two stars to be both close to one another and very close to us.

Problem 10.1. As a simple example, suppose that Doppler measurements show that the radial velocities of two stars have the variations with time shown in Figure **a**. From the exactly sinusoidal nature of the variations, we can deduce that the orbit is a circular one with binary period $P = 11.0$ days. Let the masses of the two stars be denoted by m_1 and m_2 with $m_2 < m_1$. If r equals the (unknown) relative separation, the stars m_1 and m_2 maintain constant distances from the center of mass ("CM," marked with CM in Figure **b**) given, respectively, by $r_1 = [m_2/(m_1 + m_2)]r$ and $r_2 = [m_1/$

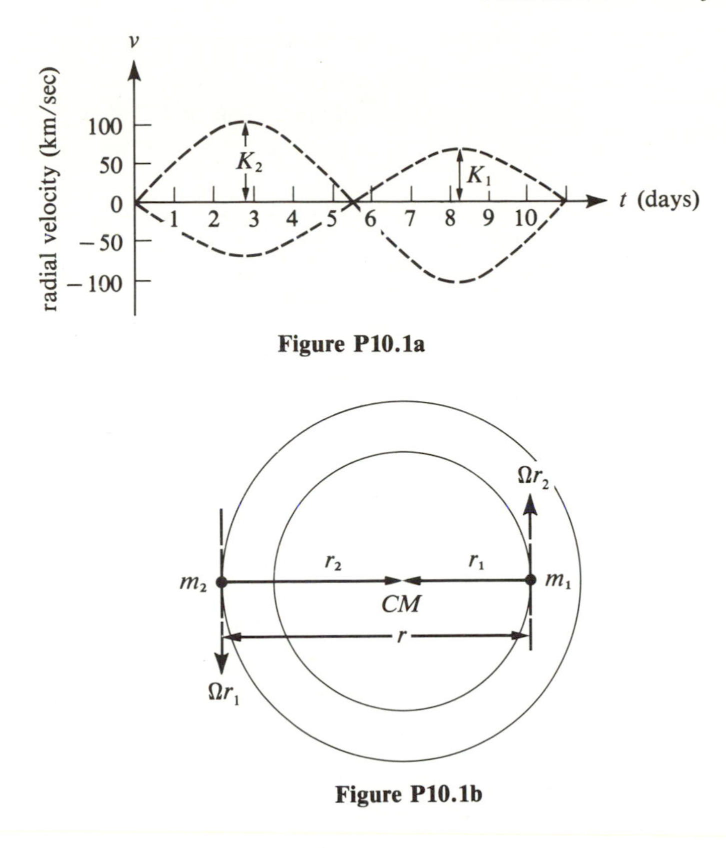

Figure P10.1a

Figure P10.1b

$(m_1 + m_2)]r$. The speeds of stars m_1 and m_2 about the center of mass are also constant and are given, respectively, by Ωr_1 and Ωr_2, where $\Omega = 2\pi/P$ is the angular speed of revolution. Show that Newton's second law, applied to either star, yields the requirement

$$\Omega^2 = G(m_1 + m_2)/r^3.$$

Show also that if the normal to the orbital plane is inclined by an (unknown) angle to our line of sight (so that $i = 90°$ corresponds to edge-on viewing), the maximum Doppler velocities to be observed for stars m_1 and m_2 relative to the center of mass are, respectively, $K_1 = \Omega r_1 \sin(i)$ and $K_2 = \Omega r_2 \sin(i)$. Since $r_1 + r_2 = r$, eliminate r_1 and r_2 to derive the following relations:

$$r = [(K_1 + K_2)/\Omega] \sin(i),$$
$$m_2/m_1 = K_1/K_2,$$
$$(m_1 + m_2) = \Omega^2 r^3/G.$$

Thus, given $K_1 = 75$ km/sec and $K_2 = 100$ km/sec with $P = 2\pi/\Omega = 11$ days, we can find the mass ratio m_2/m_1, but we cannot obtain either the total mass $(m_1 + m_2)$ or the orbital separation r without also knowing the orbital inclination i.

To obtain i, let us suppose that the system also happens to be an eclipsing binary with a light curve as shown in Figure **c**. Argue that the fact that the secondary eclipse occurs exactly half-way between the primary eclipses reinforces our deduction that the orbit is circular and not eccentric. Argue also that the flat bottoms of the eclipses imply that the eclipses must be total and not partial. (*Hint*: Why would partial eclipses show a continuous variation of blocked light?) Since the duration of the eclipses is short compared to the orbital period, show that the radii R_1 and R_2 of the stars must be relatively small compared with their separation r. From the last two facts, argue that the inclination angle i must be close to 90°.

Assume that i does equal 90° and calculate the numerical values of m_1, m_2, and r. Express your answer in $M_\odot$ and $R_\odot$. How much would your answer have changed if i had been, say, 80°?

It is also possible, given the above data, to calculate the radii of the two stars. Consider the eclipse geometry

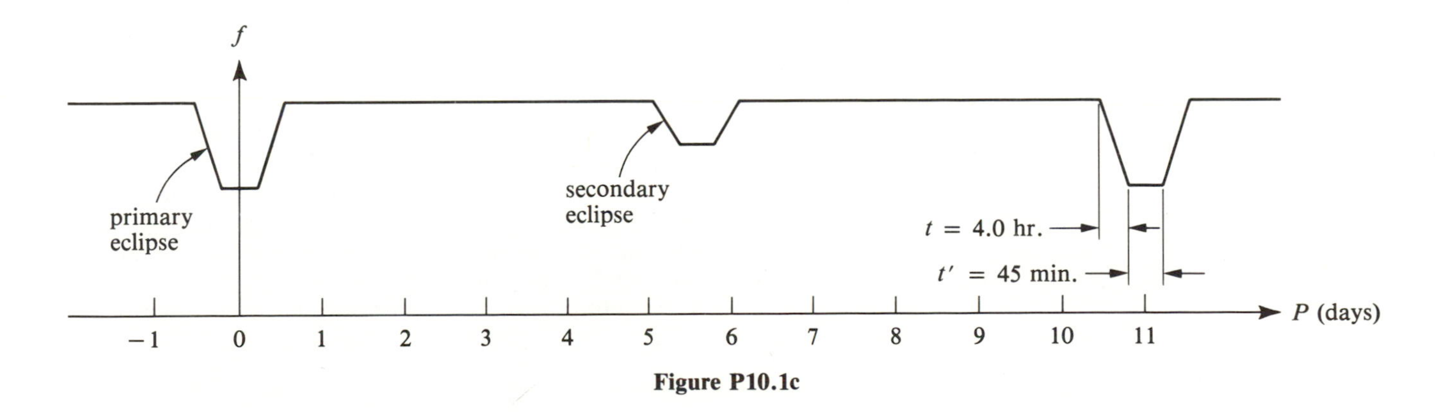

Figure P10.1c

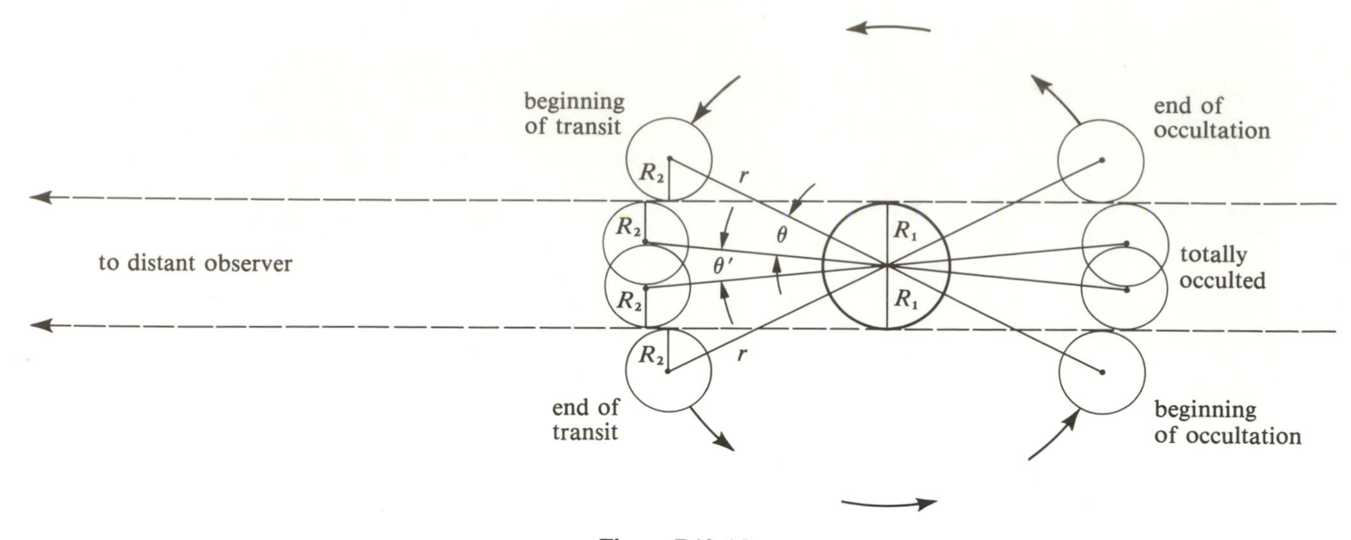

Figure P10.1d

shown in Figure **d**, which is drawn from the point of view of star 1 when we assume $i = 90°$. Argue that the angles θ and θ' are related to the time intervals t and t' shown in Figure C by the ratios:

$$\theta:\theta':2\pi = t:t':P.$$

Since R_1 and R_2 are small here in comparison with r, show that we may approximate θ and θ' by the formulae

$$\theta = 2R_2/r,$$
$$\theta' = (2R_1 - 2R_2)/r.$$

Given that $\theta = 2\pi(t/P)$ and $\theta' = 2\pi(t'/P)$, compute now the radii R_1 and R_2. Express your answer in $R_\odot$, and verify that R_1 and R_2 are indeed small in comparison with r. In actual situations, astronomers have more complicated methods to deduce m_1, m_2, r, i, R_1, and R_2 from the double spectra and the light curve, even when the system has an eccentric orbit and a less favorable inclination to the line of sight.

The Formation of Binary Stars

The proximity of the two stars in most close binary systems, compared to the large distances which typically separate stars, suggests that the two companions were born together. In dense star clusters, Hills and Press and Teukolsky have calculated that one star might capture another to form a bound state in the presence of a third body or of energy dissipation. However, the vast majority of the binaries which exist in our Galaxy must have been born in a bound state.

To investigate the process of binary star formation, Abt and Levy conducted an observational survey of nearby stars. Their findings on the companions of sunlike stars are especially interesting, for they may shed light on the formation of our own solar system. About 57 percent of the nearby sunlike stars turn out to have at least one stellar companion detectable by some means. The data also seem to indicate two different types of double stars. In binaries with orbital periods shorter than about 100 years, the masses of the two stars tend to be roughly comparable. In binaries with orbital periods longer than about 100 years, the masses of the two stars tend to be independent of one another. Abt and Levy interpret the first finding to indicate that the formation of a short-period binary takes place by the fission of a single rotating mass of gas into two nearly equal pieces as the gas mass contracts gravitationally in the process of star formation. They interpret the second finding to indicate that the formation of a long-period binary takes place by the more-or-less independent condensation of two (bound) pieces of gas. Moreover, since it is easy to overlook a distant, low-mass companion, Abt and Levy estimate that perhaps 75 percent of sunlike stars have at least one stellar companion. Of the remaining 25 percent, they speculate that perhaps 80 percent have planets. Our Sun would then be among the 20 percent of sunlike stars which have planets rather than stars as companions. This estimate is, however, based on a quite uncertain extrapolation.

Evolution of the Orbit Because of Tidal Effects

Many binary systems with periods of about ten days or less exhibit circular orbits. Moreover, when the spectra of the component stars are examined carefully for evidence of rotational broadening (see Box 9.2), the stars are often found to have **synchronized spins**. That is, they spin about their rotation axes in such a way that they always present the same face toward their revolving companion. What is the explanation for this behavior? As usual, the detailed answer is somewhat complex, but the general pattern is simple. It again involves the interplay of the second law of thermodynamics and the universal law of gravitation.

If two stars are close enough, they can exert tides on one another that can significantly distort their shapes (Figure 10.6). If either the orbits are not circular or the spins are not synchronized, the tides that are raised will generally cause different parts of the stars to move differently. These differential motions excited by the tidal forces will usually be dissipated into heat, either by a generalized kind of friction or by compressional effects. This heat is ultimately radiated into space. The continuous drain of the energy raised in the tides has to come at the expense of either the orbital energy or the spin energies of the two stars. Thus, the continuous radiation of the excess energy into space must tend to drive the system into the lowest-energy state consistent with the conservation of the total angular momentum. The state which has this property is a circular orbit with the two stars always presenting the same faces toward each other (synchronized spins).

To see that a circular orbit plus synchronized spins do constitute an equilibrium state for the problem, notice that this state has no motions whatsoever if we go into the frame which corotates with the orbit. Except for the effects of stellar evolution, once a binary system has achieved this state, it need not ever change. In the orbit frame, the two stars continue to bulge out toward one another, but the tidal distortion has reached an equilibrium shape which involves no relative motion for any piece of matter. Except for special circumstances where the spin angular momentum of one of the bodies constitutes a large fraction of the total angular momentum of the system, the state of a circular orbit and synchronized spins is a *stable* state of equilibrium (energy a local minimum and not a local maximum). Thus, short-period binaries which contain stellar components whose sizes

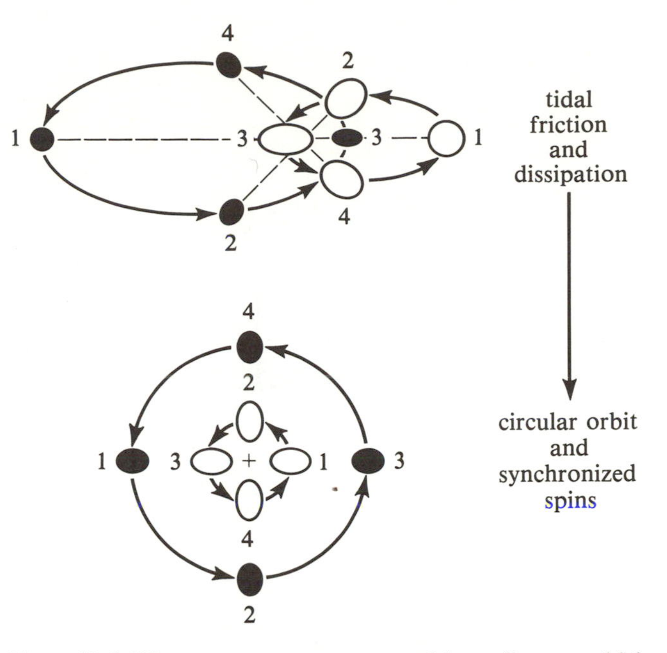

Figure 10.6. When two stars are separated by a distance which is not much greater than their individual sizes, tidal forces can raise bulges in the stars. The tidal interactions lead to a slow change of the orbit shape and the spin rates of the stars. Except in very unusual circumstances, the long-term evolution tends to bring the system to a state of circular orbits and synchronized spins.

are not very small compared with the orbital separations almost invariably have had enough time for tidal effects to have circularized their orbits and synchronized the spins. (If one of the stars is a compact object, like a white dwarf, neutron star, or a black hole, only the spin of the normal star may be synchronized.) This property allows the following physical basis, devised originally by Zdenek Kopal, for classifying close binaries.

Problem 10.2. Do you think the fact that *uniform rotation* (of which circular orbit plus synchronized spins represent only a special case) is the lowest energy state of a mechanical system is true only for gravitating systems? Have you ever played with mobiles? How do you explain the fact that no matter how differently you set the parts of a mobile to spinning initially, the whole system eventually prefers to rotate uniformly? Can you think of any other astronomical body which has a circular orbit and always presents the same face toward its companion? *Hint*: Consider the companion of the Earth. What do you think happened here?

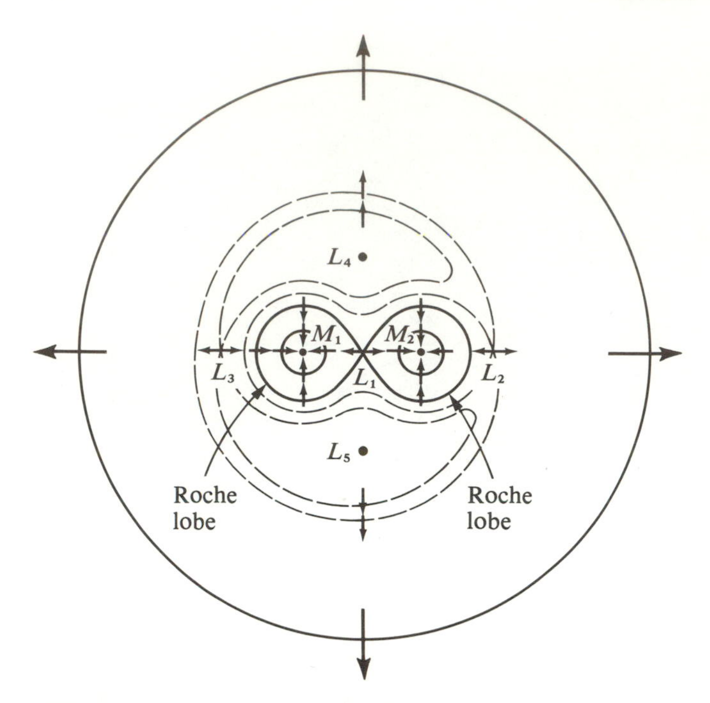

Figure 10.7. The Roche equipotential surfaces plotted in the equatorial plane for two point masses with a mass ratio equal to 2/3. The short arrows indicate the direction of the effective gravitational field in the frame of reference which corotates with the orbital motion. The effective gravity vanishes at the five Lagrangian points L_1, L_2, L_3, L_4, L_5. The first three, L_1, L_2, L_3, lie along the line joining the two mass points; the last two, L_4, L_5, form equilateral triangles with the two mass points, M_1 and M_2. The sideways "figure 8" which passes through the L_1 point contains the two Roche lobes.

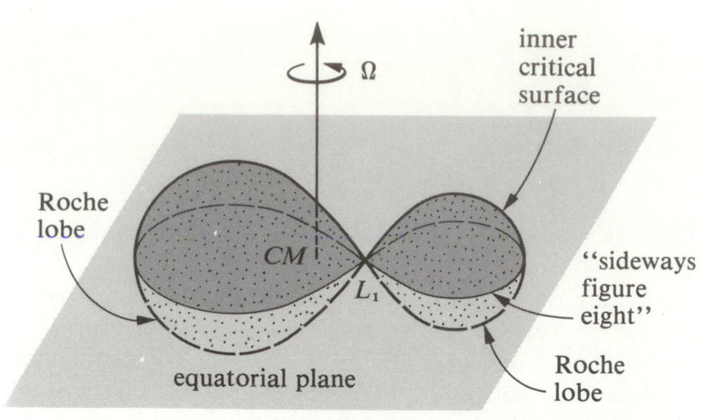

Figure 10.8. A schematic depiction of the three-dimensional structure of the Roche lobes of a binary system with a circular orbit. The rotation of the entire system takes place with angular speed Ω with respect to an axis which passes through the center of mass CM. This axis is perpendicular to the equatorial plane of the system.

Classification of Close Binaries Based on the Roche Model

Consider a binary-star system which has a circular orbit. If we go into a frame of reference which corotates with the circular orbital motion, we can follow Roche and define **equipotential surfaces** whose normals give the direction of the local **effective gravity** (Figure 10.7). The effective gravity takes into account the real gravity and the effect of being in a rotating frame of reference (the "centrifugal force"). Near the center of either star, the presence of the other star and the fact of being in a rotating frame of reference are relatively unimportant compared to the gravitational attraction of the star at hand. Thus, near the center of either star, the effective gravity must essentially point radially inward, and the equipotential surfaces must essentially be spheres that encircle the center of the star. On the other hand, in the equatorial plane far from both stars, the effective gravity must be dominated by the outwardly directed "centrifugal force." Thus, far from both stars, the equipotential surfaces must intersect the equatorial plane in circles that enclose both stellar centers. To summarize, then, the equipotential surfaces when drawn in the equatorial plane are circles which enclose each star separately when near either star, and which enclose both stars jointly when far from both stars. The change in topology of these curves therefore takes place through a sideways "figure eight" as shown in Figure 10.7. The full three-dimensional equipotential whose equatorial section is the sideways "figure eight" is called the "inner critical surface," and it is depicted schematically in Figure 10.8. The two halves of the inner critical surface are called the **Roche lobes**. The two Roche lobes touch at a single point, the L_1 point.

To grasp the significance of the L_1 point, consider Figure 10.7. Imagine a walk along the line joining the centers of the two stars, starting near the center of the star on the the left. As you walk toward the other star, the effective gravity tries to pull you back to the left; the star on the left exerts the greater influence. As you approach the L_1 point, however, the star on the right begins to make itself felt, and the net effective gravity weakens. At the L_1 point itself, you are in "no man's land," because you are pulled equally by both sides and the effective gravity vanishes. When you cross into the Roche lobe of the star on the right, the effective gravity reverses direction, and you feel yourself in the grasp of the star on the right. The significance of the L_1 point, then, is that atoms placed at rest to the immediate left of it "belong" to the star on the left; atoms to the right, to the star on the right.

Equipotentials in general have the following significance. The surfaces of constant pressure in stars that are

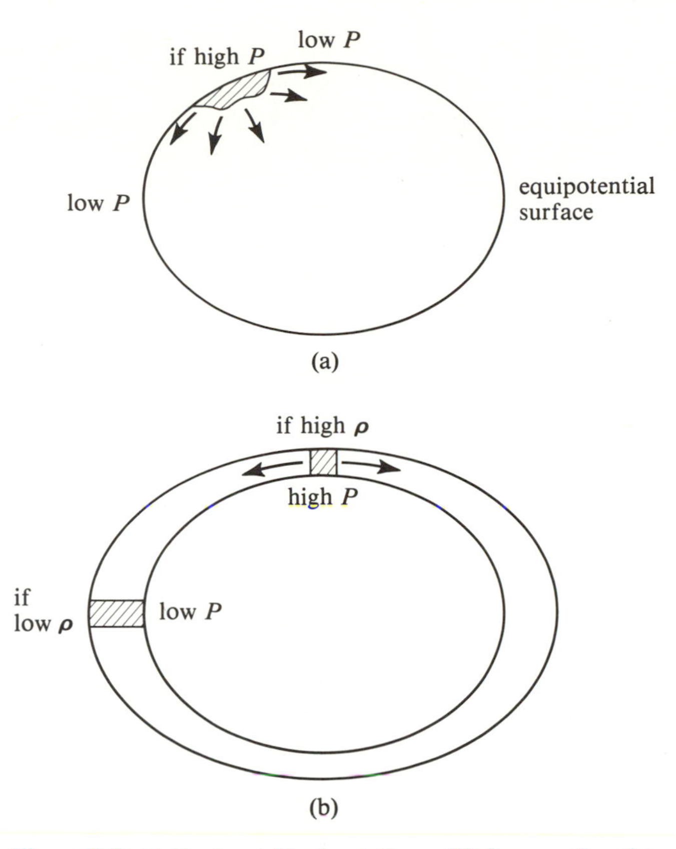

Figure 10.9. (a) Horizontal hydrostatic equilibrium requires the pressure P of a synchronously spinning star to be uniform on an equipotential surface, because an area of high pressure would otherwise be unopposed in its expansion along the surface. (b) Vertical hydrostatic equilibrium requires the density ρ to be uniform, because the weight of a column of high density would otherwise result in a high pressure at its base.

motionless in the orbit frame must coincide with the equipotential surfaces. To see the truth of this statement, suppose the reverse. Suppose the pressure is *not* constant along an equipotential. Because the effective gravity, by definition, has no component *along* an equipotential (the effective gravity is *perpendicular* to equipotentials), there would then be no force to resist the uneven pressure. Thus, the gas in the high-pressure regions of the equipotential surface would expand into the low-pressure regions, and after a short mechanical readjustment, the pressures all along the equipotential would have been equalized. *Horizontal* hydrostatic equilibrium, therefore, requires the pressure to be constant on equipotential surfaces (Figure 10.9).

Similarly, it can be shown that *vertical* hydrostatic equilibrium requires the density to be constant on equipotential surfaces. To see the truth of this statement, suppose that the density is not constant along some equipotential surfaces. Consider, now, the pressure along an equipotential surface beneath the ones in question. For vertical hydrostatic equilibrium, the pressure at each point on the latter surface must equal the weight of a column of matter above it (Box 5.1). But if the density is not uniform on equipotential surfaces above the level in question, the weight will be different at different points. This would make the pressure uneven on the equipotential surface, and the resultant flows set up would proceed to alleviate the difficulty, i.e., to make the density uniform on equipotential surfaces. If the pressure and density are both uniform on equipotential surfaces, the equation of state of the gas will generally make the temperature also uniform on equipotential surfaces.

Now, the photosphere of a star is roughly a surface of constant density; that is, the photosphere is nearly the surface where the gas density has dropped to *zero* relative to its typical interior values. To this level of approximation, we can say that the photosphere of a synchronously spinning star must lie on one of the equipotential surfaces of the Roche model. This conclusion forms the basis for Kopal's classification scheme of close binary stars (see Box 10.2).

BOX 10.2
Classification of Close Binaries

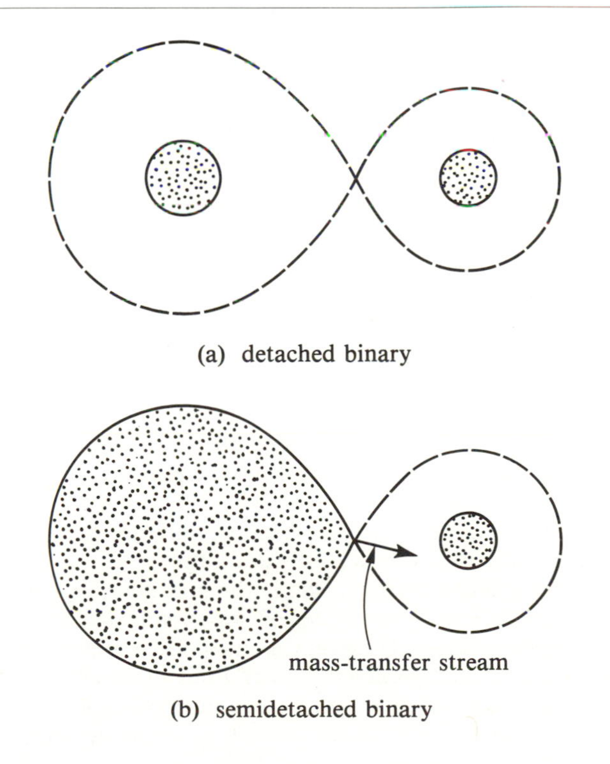

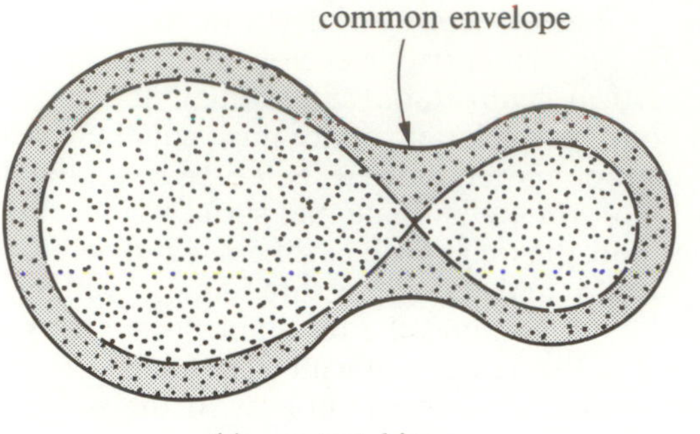

(c) contact binary

(a) *Detached binaries* are systems where both stars' photospheres lie beneath their respective Roche lobes. Such a pair of stars interact with each other significantly only by means of their mutual gravitational attraction. They are important observationally as a source of information about masses and radii, but they will not concern us further in this chapter.
(b) *Semidetached binaries* are systems where one of the photospheres coincides with its Roche lobe, and the other star's photosphere lies beneath its Roche lobe. The lobe-filling star is called the *contact component*; the other star, the *detached component*. Thus, the *system* is semidetached. The central dynamical feature which distinguishes semidetached binaries from detached binaries, as we shall soon see, is mass transfer from the lobe-filling star to its companion, a process that Hoyle quaintly termed "dog eats dog."
(c) *Contact binaries* are systems where both stars' photospheres equal or exceed their Roche lobes. In the latter case, which is the prevalent one, a common envelope lies above the inner critical surface that surrounds both stars and hides them from individual view. The theory of such objects is currently very controversial; and we will comment only briefly on contact binaries in this book.

Mass Transfer in Semidetached Binaries

Gerard Kuiper was the first to realize that objects like what we now call semidetached binaries must undergo mass transfer. Kuiper's analysis used a particle trajectory approach, which was later questioned by Kevin Predergast. In what follows, we shall use the gas-dynamical description devised by Lubow and Shu.

Consider what happens when the more massive member of a detached binary begins to expand in response to core evolution (Chapter 8). Its photosphere moves out and eventually coincides with the Roche lobe of the star. At this point, the system becomes a semidetached binary, and the lobe-filling star must begin to transfer

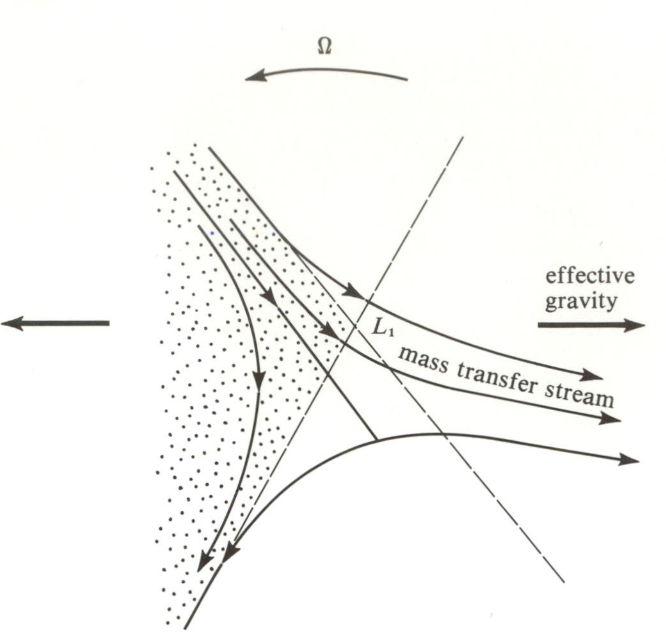

Figure 10.10. The cause of mass transfer in semidetached binaries. On the left side of the inner Lagrangian point L_1, gas that belongs to the lobe-filling star faces a vacuum on the right side. Since the effective gravity vanishes by definition at L_1, the gas near L_1 cannot be restrained from rushing toward the vacuum. The Coriolis force cause by the rotation of the system turns some of the gas back toward the lobe-filling star, but gas high up in the atmosphere ends up on the other side of L_1, where it is then pulled ever more strongly toward the detached companion, resulting in a mass-transfer stream.

mass to its companion. The reason is simple. In the lobe-filling star, gas on one side of L_1 faces a vacuum on the other side (Figure 10.10). This is also true elsewhere at the photosphere, but elsewhere the tendency of the gas to want to expand outward because of the internal pressure is exactly balanced by the inward force of the effective gravity. Thus, the photosphere remains stationary except for a very slow outward expansion produced by core evolution. This balance between pressure forces and gravity is what we have called (vertical) hydrostatic equilibrium. However, hydrostatic equilibrium is not possible at L_1, because the effective gravity vanishes at L_1 by definition. Thus, at L_1—indeed, in a small neigborhood around L_1—nothing prevents the gas of the lobe-filling star from rushing outward toward the vacuum on the side of the other star. Once this gas reaches the inside of the Roche lobe of the other star, the effective gravity will pull the gas ever more strongly toward the detached component. The rotation of the system as a whole (the "Coriolis force") prevents the gas stream from falling directly toward the detached star. Thus, in the corotating frame, we see the gas stream rush off at an angle which lags behind the line joining the centers of the two stars.

(In an inertial frame, we would see the detached star revolve away from the onrushing gas stream.) The lost gas is replaced by the expansion of the underlying layers, and this results in a continuous transfer of mass in a thin stream from the lobe-filling star to the detached companion.

The future evolution of the two stars is therefore quite different from that of normal single stars. But before we discuss these evolutionary aspects, let us first finish our formal classification of close binaries.

Energy Transfer in Contact Binaries

If both stars fill their Roche lobes, either because of evolutionary expansion or because of being born that way, mass transfer will proceed until the pressures are balanced across L_1. With gas of equal pressure on both sides of L_1, there is no dynamical need for further mass transfer, and gas can be supported in the region above the two Roche lobes in a common envelope around both stars. Stars with this property are called contact binaries. If the common envelope of the contact binary is sufficiently thick, it will effectively obscure the intrinsic nature of the two underlying stars. In particular, the common photosphere which lies (nearly) on an equipotential enclosing both stars above the common envelope must exhibit nearly equal temperatures at all points along its surface (Figure 10.11), rather than the differences in effective temperatures that we would expect if the two underlying stars were separate, single stars. As Osaki and Lucy were the first to point out, there must be a mechanism in the common envelope to mix the two separate luminosities emerging from the interior of the two Roche lobes into the joint luminosity of the system that emerges from the common photosphere.

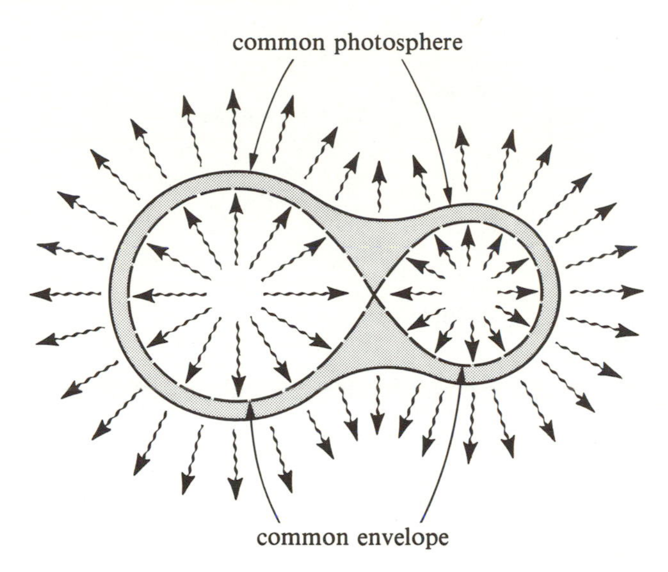

Figure 10.11. The common-envelope model of contact binaries like W Ursae Majoris stars. A common envelope, which encloses the two interiors beneath the Roche lobes, redistributes the uneven flow of heat from the two interiors and results in radiation from a common photosphere that has a nearly uniform effective temperature. In other words, the surface brightness per unit area on the two sides is the same even though the two cores must generate nuclear energy at very different rates.

The most numerous contact binaries are W Ursae Majoris stars, in which two main-sequence stars share a common outer envelope. These stars have periods typically only a fraction of a day (say, eight hours); so they are a favorite of variable-star observers, who can get a complete light curve in one night's observing. Shown in Figure 10.12 is the light curve of W Ursae Majoris itself, the prototype of the systems under discussion. Notice that the eclipses produce a very different-looking light curve from that produced in detached binaries (compare with Figure 10.5) in two important respects:

(a) The maxima of the light curve are rounded,
(b) The minima of the light curve are of nearly equal depths.

Feature (a) implies that the shapes of the two stars must be severely tidally distorted, because two spherical stars would present the same total radiating area, and therefore the same apparent brightness, to the observer as they revolve until entering an eclipse. Feature (b) implies that the effective temperatures of the two stars must be nearly the same even though their masses have a ratio of about 0.4. If the effective temperatures were those appropriate for two separate main-sequence stars, the eclipse when the small, cool star blocks the large, hot star would be deeper than the other eclipse. The deduction that the effective temperatures of the two stars are nearly the same is consistent with the lack of appreciable changes of either color or spectral type as the two stars revolve around each other. These facts are most simply interpreted in terms of the two stars' sharing a common envelope as depicted in Figure 10.11.

Problem 10.3. Draw some typical eclipse geometries for two spherical stars of different sizes. Show that less light reaches the observer when the cool star eclipses the hot star than when the hot star eclipses the cool star. Argue also that the light level remains (approximately) constant when the two stars are not in eclipse. Actually, the light level does not remain constant, because of the "heating" effect of the two sides of the stars which face each other. Reason out why this "heating" effect is often called "reflection." *Hint*: The net effect must be to raise the temperature of the outer layers sufficiently to *re-radiate* all the energy coming from the other star, as in the case of the Earth and the Sun discussed in Chapter 5.

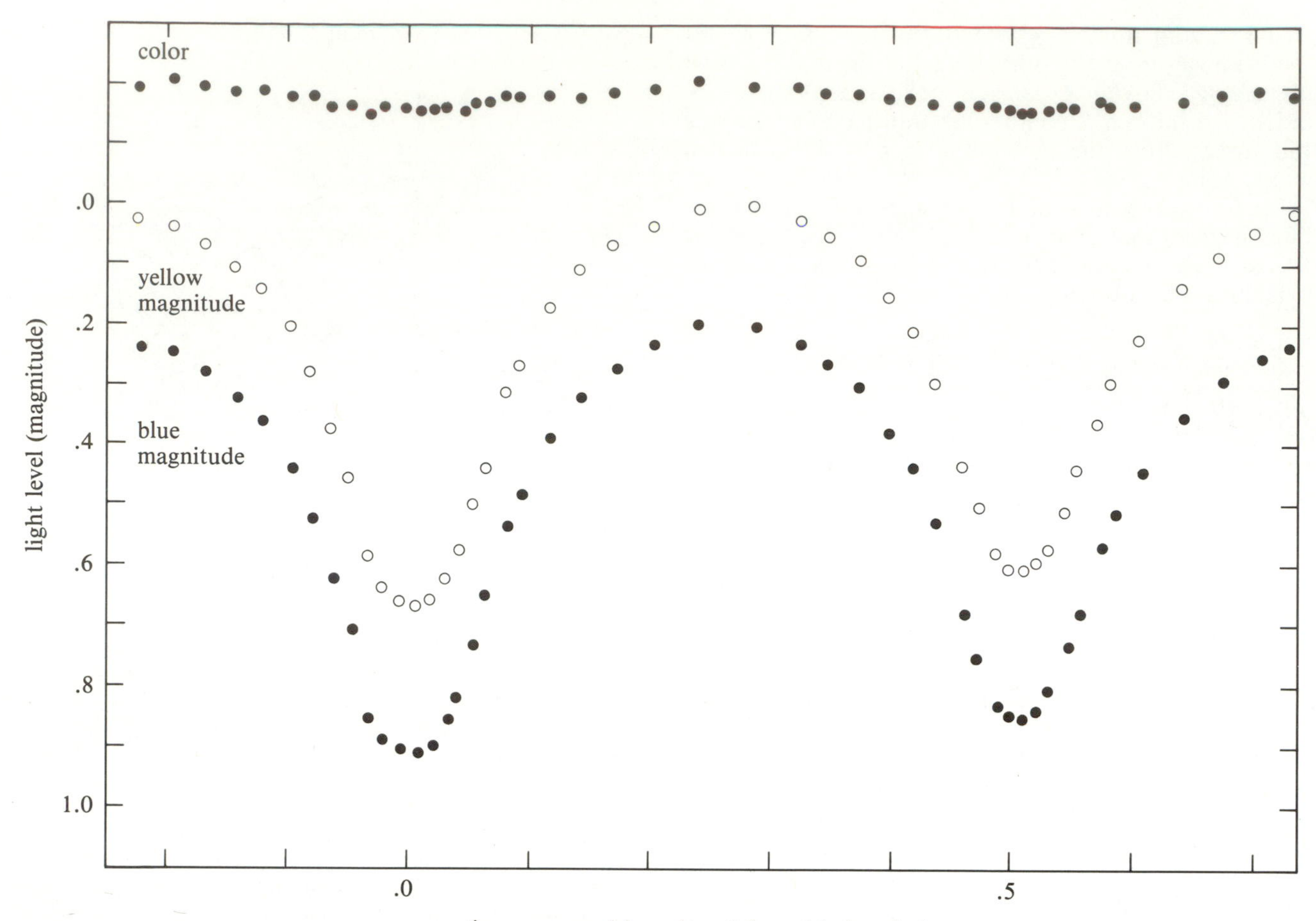

Figure 10.12. The light curve of the contact binary W Ursae Majoris. The curved maxima imply that the shapes of the two stars are very distorted because of their proximity to one another. The nearly equal depths of the two minima imply that the effective temperatures of the two stars must be nearly equal. When combined with the knowledge that the mass ratio of the system is 0.4 and that the two stars are main-sequence objects, these data lead to the conclusion that W Ursae Majoris is a contact binary; indeed, it is the prototype of a class of binary stars. (L. Binnendijk, *A.J.*, **71**, 1966, 340.)

The indirect deduction of the geometric shapes of W Ursae Majoris stars given above has recently received direct verification from detailed spectroscopic studies of such systems. The basic idea, is quite simple (Figure 10.13). The net result is that the profile of a narrow spectral line of a contact binary should be rotationally broadened at elongation to mimic the geometric shape of the body. Detailed analyses of the spectra of several W Ursae Majoris systems do reveal them to have the geometric shapes which are expected for the common-envelope interpretation of these binaries.

The preceding discussion of the properties of the outer layers of contact binaries is, by now, relatively well-established and noncontroversial, but there is no universal agreement on the proper theory for their internal structure. Very different hypotheses have been put forward, whose merits astronomers are currently working hard to assess. The issues are interesting, because if stars are considered the "atoms" of the universe, then contact binaries are the "diatomic molecules." Just as the two atoms of a diatomic molecule may share their outer electrons in a covalent bond, so the two stars of a

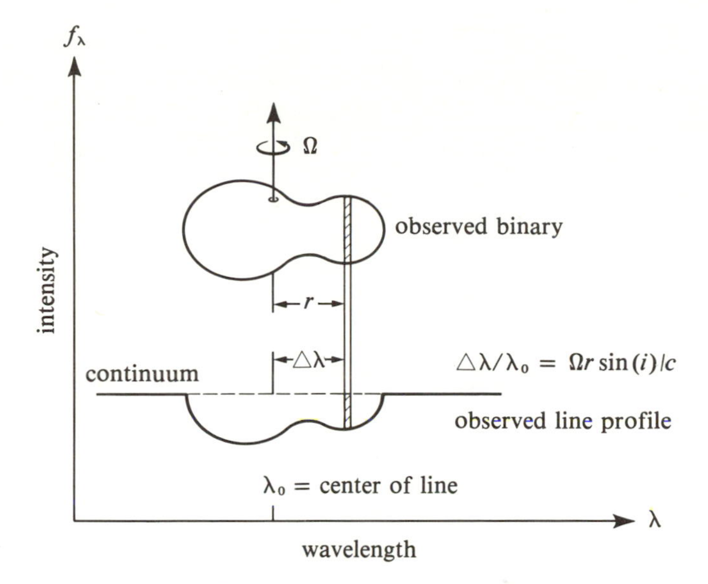

Figure 10.13. The profile of an isolated absorption line in the spectrum of a contact binary should mimic the shape of the observed binary. The reason is that the spectral type of a contact binary is nearly uniform over its entire surface; therefore, each element of photosphere makes a contribution at the Doppler shift appropriate for its line-of-sight velocity and in intensity proportional to its projected surface area. For a uniformly rotating object, elements of the photosphere which lie the same projected distance r from the rotation axis have the same projected line-of-sight velocity, $\Omega r \sin(i)$, where i is the inclination angle of the rotation axis to the line of sight. Thus, the line profile of a spectral line when the system is at elongation looks like the bottom half of the shape of the binary.

contact binary may share their outer material in a common envelope. And just as it would be wrong to think that the properties of a diatomic molecule are simply the combined properties of its two constituent atoms, so is it wrong to think that the properties of a contact binary are simply the combined properties of its two constituent stars.

Evolution of Semidetached Systems

As interesting as contact binaries are, the greatest variety of phenomena associated with close binary stars is exhibited by semidetached systems. Whereas the two components of a contact binary have to be similar ("normal") stars, the detached, mass-receiving, component of a semidetached system may be a normal star, or a white dwarf, or a neutron star, or a black hole. Let us now discuss examples of each of these in turn (see Box 10.3).

BOX 10.3
Semidetached Binaries

The appearance of a semidetached binary in which mass is transferred from a normal lobe-filling star to its companion depends largely on the physical nature of the detached companion.

(1) If the detached component is a normal star, we have Algol-like systems. Most such systems are in the phase of slow mass transfer, where an evolved, less-massive star loses mass to an unevolved, more-massive companion. The latter is often the hotter but smaller star in the system; therefore, if eclipses occur, they are very deep.

(2) If the detached component is a white dwarf, we have cataclysmic variables, novae, or dwarf novae. Between outbursts, the light from such systems is dominated by the accretion disk and a hot spot at its edge which is formed by the impact of the mass transfer stream. The outbursts in novae are believed to be thermonuclear, arising from a sudden, cataclysmic ignition of the hydrogen newly added to an otherwise hydrogen-exhausted white dwarf. Only a small amount of mass is ejected per outburst; so nova outbursts may repeat themselves in the same system after an appropriate passage of time.

(3) If the detached component is a neutron star or black hole, we have X-ray binaries. Binary X-ray sources are believed to divide into two classes: low-mass systems, where mass transfer is effected by the classical mechanism of Roche-lobe overflow; and high-mass systems, where stellar winds may play a more dominant role. Where a neutron star is the detached companion, most of the X-rays arise when the accreting matter strikes the surface of the neutron star. Where a black hole is the detached companion, most of the X-rays arise when matter spiraling through an accretion disk is heated to very high temperatures outside the event horizon of the black hole.

Detached Component = Normal Star: Algols

When discussing eclipsing binaries earlier, we mentioned Algol ("the demon," in Arabic) as an example. Algol, we also noted, is exceptional for the depths of its primary eclipses. We now understand these deep eclipses to arise from a partial blocking of the light from a main-sequence star of spectral type B8 by a subgiant of spectral type K0. In order for the apparent brightness to drop by more than a factor of two, however, the subgiant must be bigger in size than the main-sequence star yet be less luminous (reconsider Problem 10.3). This state of affairs is possible only if the main-sequence star is considerably more massive than the subgiant, a deduction which is in accord with the observational evidence pointing to about $3.7M_\odot$ for the main-sequence star and about $0.8M_\odot$ for the subgiant. But according to the theory of stellar evolution reviewed in Chapter 8, a $3.7M_\odot$ star should have a considerably shorter main-sequence lifetime than a $0.8M_\odot$ star. We naturally assume the two stars were formed at the same time, yet in Algol it is the less-massive star which is the more evolved! This apparent contradiction, an outgrowth of the observational work of Struve, Sahade, and their coworkers, became known as the "Algol paradox."

Problem 10.4. Why can't a less-massive main-sequence star be brighter than a more-massive subgiant?

A hint of an explanation came from the work of Kopal, Wood, and others, who showed that the subgiants in Algol-like systems (called simply "Algols" for short) fill their Roche lobes, and therefore, are probably transferring mass to their detached companions. The subgiant therefore might originally have been the more massive member of the binary system, and this would explain why it evolved first. However, this discussion leaves unexplained why we rarely, if ever, see a more-massive star transferring mass to a less-massive companion in Algols. Crawford found the correct full resolution of the Algol paradox. He proposed that the initial phase of mass transfer from the more-massive component to the less-massive companion is so rapid that we hardly ever have a chance of catching a system in the act. Crawford speculated that the mass-transfer rate slows down only after the originally more-massive star becomes the less-massive star, and vice versa. Only in this latter, slow phase do we have much chance to observe the systems, which he then identified with the Algols.

Full evolutionary studies by Kippenhahn, by Paczynski, by Plavec, and their collaborators, provided the theoretical justification for Crawford's speculation. Figure 10.14 shows the important phases for the accepted theoretical picture. Phase (a) is the starting configuration: a close binary consisting of a more-massive main-sequence star and a less-massive main-sequence star, orbiting in a circle about the center of mass. After core-hydrogen exhaustion, the more-massive component swells up to become a subgiant, while its less-massive companion is still on the main sequence. Shortly thereafter, in phase (b), the subgiant comes into contact with its Roche lobe and begins to spill material toward the other star. If no mass or angular momentum is lost from the system, and if the lobe-filling star maintains synchronous rotation, the orbit size shrinks as mass transfer proceeds from the more-massive star to the less-massive star. This effect, plus the growing "lobe of influence" of the mass-gaining star, causes the Roche lobe to shrink on the mass-losing star. The contracting Roche surface of the lobe-filling star then tends to squeeze matter out of the star like air out of a balloon with a hole in it. The continued transfer of mass causes the Roche lobe of the mass-losing star to shrink even more, and the process runs away in a genuine mass-transfer instability, with the maximum rate limited only by the mass-losing star's ability to adjust its structure to the changing conditions. So much mass is now transferred that the more-massive star quickly becomes the less-massive star. The rapid mass-transfer phase ends only after the mass-accepting star contains about as much mass as the originally more-massive component. In this phase (c), the orbital separa-

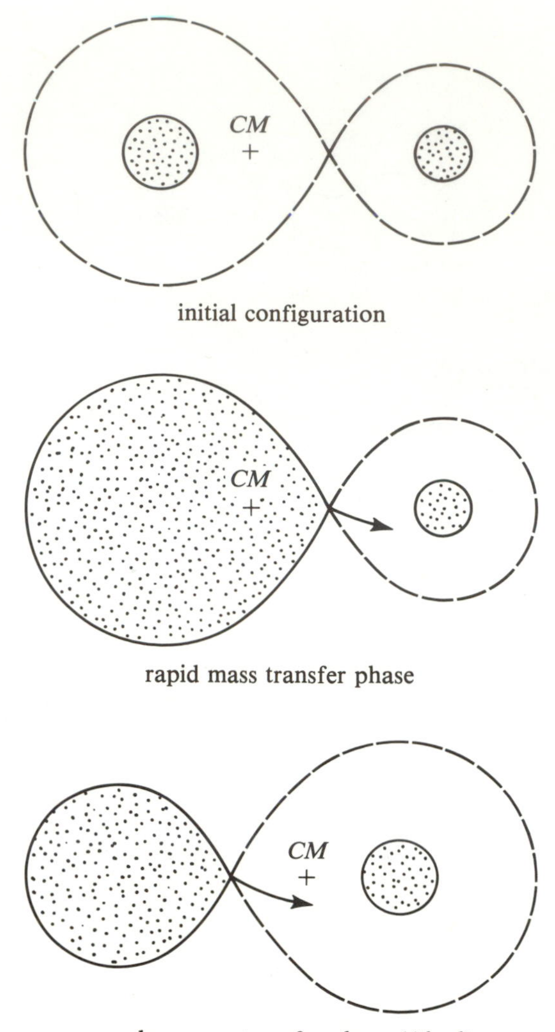

Figure 10.14. Theoretical picture of the evolution which leads to an Algol-type binary system.

tion and Roche lobe of the mass-losing star both expand in size with additional mass transfer. The system now enters a phase of slow mass transfer, in which the rate is dictated by the evolutionary expansion of the subgiant because of changes in its deep interior. This slow phase of mass transfer from the evolved subgiant (now the less-massive star) to the (now more-massive) main-sequence companion is identified with the Algols. Since the slow phase is considerably longer-lived than the rapid phase, most of the mass-transfer binaries with two normal stars that we see are in the slow phase (Algols) rather than in the fast phase (Beta-Lyrae systems, perhaps). Except for some slight remaining discrepancies, this picture provides a satisfactory resolution of the "Algol paradox."

A similar scenario has been invoked to explain the "blue stragglers" found in some globular clusters (Chapter 9). It has been proposed that "blue stragglers" represent the unevolved companion of a close binary system which has been the recent recipient of some additional mass. This speculation would then explain why some main-sequence stars in the cluster are more massive than the stars in the cluster that have already begun to evolve away from the main sequence. Unfortunately, no real evidence of a binary nature for the "blue stragglers" has been forthcoming. More recent proposals to explain "blue stragglers" have involved more complicated schemes which convert a pair of stars into a single one.

Another unresolved issue has arisen in the theory of close binary evolution. Calculations, initiated by Benson, which follow the evolution of the mass-receiving star have shown that a possibly severe complication may develop in the phase of rapid mass transfer. In this phase, the unevolved star receives mass faster than it can adjust to a new state of equilibrium appropriate to its newfound mass. In some cases, this inability to adjust causes the mass-receiving star to expand and come into contact also with its Roche lobe. The system then acquires a *contact configuration* reminiscent of W Ursae Majoris stars (see Figure 10.11). There is much current research underway on what is the proper theory for the structure of the interior of such objects; so we do not yet know what the next step in their evolution is. These difficulties and others show that astronomers have arrived at a substantial but not complete understanding of the problem of close binary evolution.

Accretion Disks

The first star to evolve in Algol-like systems (and perhaps in contact binaries) may eventually turn into a white dwarf, a neutron star, or a black hole. If the binary system is not disrupted completely by this process, the remaining normal companion may then evolve to fill its Roche lobe, initiating mass transfer in the opposite direction from earlier (Figure 10.15). Since the detached component is now a compact object (white dwarf, neutron star, or black hole), the mass transfer stream does not strike the compact object directly. Instead, the stream orbits around the object to strike itself and to produce an orbiting ring of gas. It is generally believed, following the initial suggestion of Prendergast and Burbidge, that such rings will broaden to form a swirling disk of material which gradually settles onto the compact object. Such disks are called *accretion disks*, because accretion is the word used by astronomers to denote any process whereby an object gains mass from its surroundings.

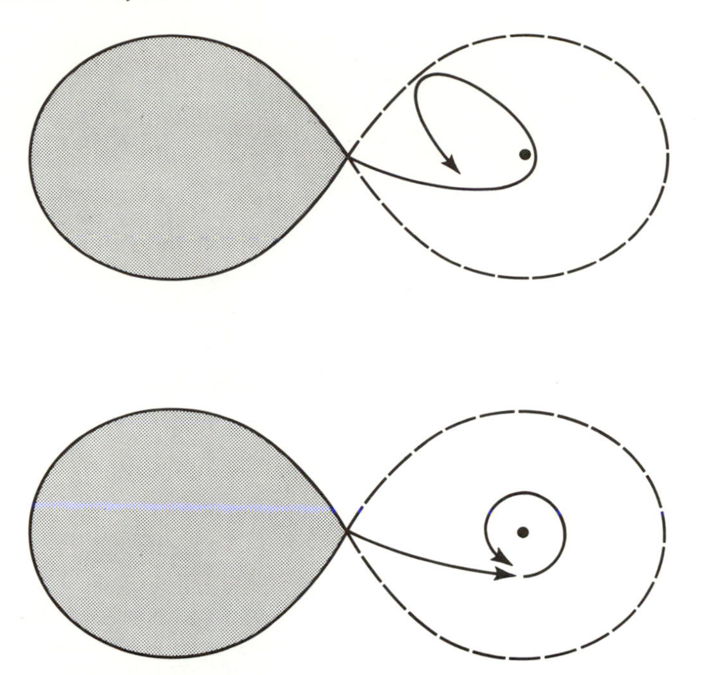

Figure 10.15. Formation of an accretion disk by mass transfer from a lobe-filling star toward a compact companion (white dwarf, neutron star, or black hole). The latter is too small to intercept the mass transfer stream directly; so the stream orbits around the compact star to strike itself. The high-speed impact dissipates the kinetic energy of relative motion but not the angular momentum; so ultimately an accretion disk forms and encircles the compact object. (Adapted from S. H. Lubow and F. H. Shu, *Ap. J.*, **198**, 1975, 383.)

It has been suggested (by Shakura and Sunyaev, and by Lynden-Bell, Pringle, and Rees, among others) that some sort of anomalous friction inside the accretion disk would force the material to spiral slowly inward to settle eventually on the compact object (Figure 10.16). In the process, the matter is heated to very high temperatures, and the inner parts of the accretion disk may emit copious amounts of ultraviolet light or even X-rays (Problem 10.5). What the compact object plus its accretion disk looks like depends on whether the compact object is a white dwarf, a neutron star, or a black hole.

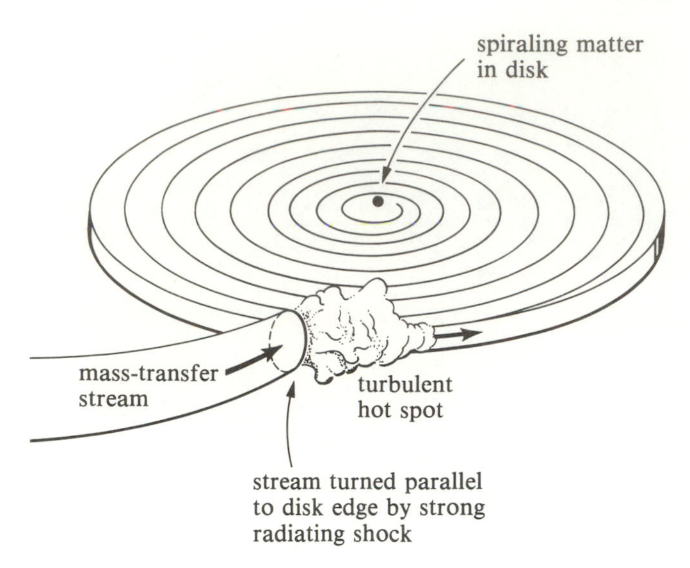

Figure 10.16. Schematic diagram of accretion disk which surrounds the compact component of a mass-transfer binary. Inside the accretion disk some sort of anomalous friction apparently causes the matter to spiral slowly onto the central compact object. The release of orbital energy by frictional dissipation heats up the disk and causes it to radiate. In cataclysmic variables, the radiation released by shock-impact heating at a hot spot is also observable.

Problem 10.5. It might be thought that matter which spirals from radius r_1 to radius r_2 through a rotating accretion disk would gain the total orbital-energy difference associated with those two radii. If the central body has mass M and the spiraling matter has mass m, Problem 5.4 yields this orbital-energy difference as (minus) one-half of the potential energy difference:

$$\Delta E = (1/2)(GMm/r_2 - GMm/r_1).$$

Detailed calculations show that the energy transferred outward because of the difference in frictional rubbing on the inner and outer edges of any annulus increases the local energy release in the outer parts of the accretion disk by a factor of 3. This excess heating is made up by a decrease in the local heating rate of the inner parts of the accretion disk, where most of the total energy release occurs anyway. (The frictional transport of heat thus behaves like Robin hood, robbing from the rich to give to the poor.) Ignore this effect, and assume that the expression for ΔE given above represents a good average value for the entire accretion disk. If mass m is transferred in time t to give a steady accretion rate $\dot{M} = m/t$, argue that the luminosity of the accretion disk between radii r_1 and r_2 is given by

$$L = (\dot{M}/2)(GM/r_2 - GM/r_1).$$

Calculate L numerically for a typical accretion disk in a cataclysmic variable where $M = 1M_\odot$ (a white dwarf), $\dot{M} = 10^{-8}M_\odot\ \text{yr}^{-1}$ (mass transfer from a red dwarf), $r_1 = 10^{11}$ cm, and $r_2 = 10^9$ cm. What fraction of this luminosity comes from energy release between $r = 2 \times 10^9$ cm and $r_2 = 1 \times 10^9$ cm?

Suppose that the upper and lower faces of the accretion disk radiate like blackbodies with a local effective temperature T_e. Argue that the average effective temperature for the entire disk can be calculated from the formula:

$$2(\pi r_1^2 - \pi r_2^2)(\sigma T_e^4) = L.$$

Figure 10.17. The outburst of Nova Cygni as followed in a sequence of photographs by Ben Mayer. Within two days the nova brightened by a factor of about a million. (Courtesy of Ben Mayer, Chairman, ENVEL Corp.)

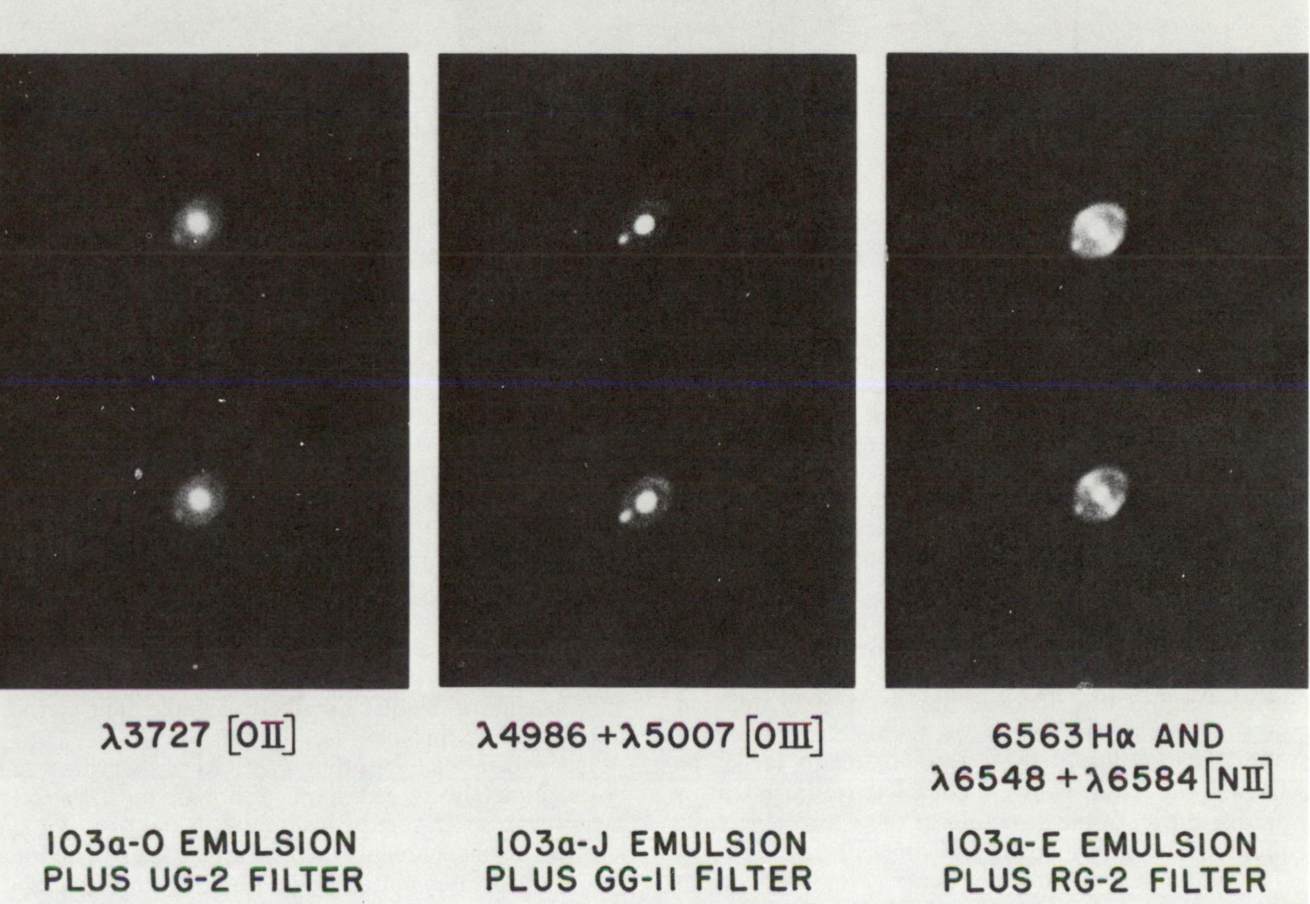

Figure 10.18. Three photographs taken in 1951, 17 years after the outburst of Nova Herculis, show an ejected shell of gas in the light of various emission lines. (Palomar Observatory, California Institute of Technology.)

Compute T_e numerically for the above example. What is the average T_e for the part of the accretion disk between $r = 2 \times 10^9$ cm and $r_2 = 1 \times 10^9$ cm? Where in the electromagnetic spectrum does this accretion disk radiate mostly as a whole? The innermost part? What do these considerations imply for the possible observation of accretion disks in cataclysmic variables?

Detached Component = White Dwarf: Cataclysmic Variables

The observational work of Joy, Walker, and Kraft has shown that all eruptive variables, such as classical novae, recurrent novae, and dwarf novae (collectively called **catacylsmic variables**), are probably interacting binary systems. The light coming from these systems is characterized by long intervals of relative quiet interrupted by sudden outbursts which may increase the apparent brightness of the object by a factor of 10 (in the smallest-amplitude dwarf novae) to more than 10^6 (in the classical novae, Figure 10.17). Eruptions may be separated by only a few weeks (in dwarf novae), and by more than a hundred years (in recurrent novae). It has been proposed that probably all cataclysmic variables have repeated outbursts, but the repetition interval in the classical novae may be so long (say, 10^4 yr) that we have not yet detected it. Ejection of a small amount of matter (typically about $10^{-5} M_\odot$) is observed to accompany novae outbursts (Figure 10.18). Little, if any, ejection of gas occurs from the system during dwarf-novae outbursts.

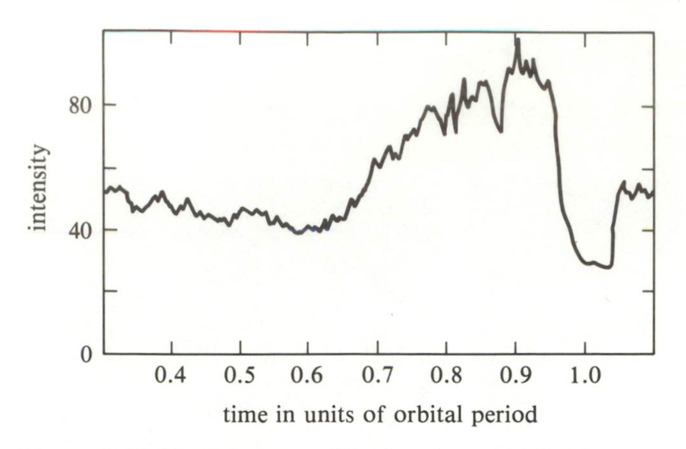

Figure 10.19. The light curve of the dwarf nova, U Geminorum, during a quiescent period between outbursts. The rapid irregular flickering probably arises from a turbulent hot spot formed by the impact of the mass-transfer stream with the edge of the accretion disk. Notice that the flickering disappears when the hot spot is eclipsed. (Adapted from Warner and Nather, *MNRAS*, **152**, 1971, 219.)

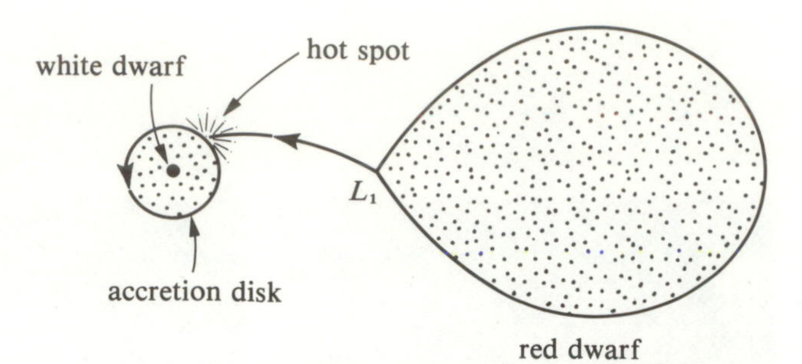

Figure 10.20. The model of dwarf novae proposed by Krzeminski and Smak and by Warner and Nather. Much of the light from the system between outbursts comes from the flickering hot spot. The eclipse of this hot spot by the red dwarf as the system revolves produces the light curve shown in Figure 10.19 for U Geminorum.

These properties of the eruptions of cataclysmic variables (largely deduced from the early observational work of McLaughlin and Payne-Gaposchkin) make it clear that novae and dwarf novae are completely different from supernovae. The former involve a violent but minor outburst which leaves the system intact to repeat its behavior; the latter involve a complete readjustment or even disruption of the structure of the entire object. Be careful not to confuse novae and supernovae.

The outburst cycle in cataclysmic variables does not represent the orbital period of the binary system. The latter are typically much shorter, on the order of six hours or so in dwarf novae, for example. The binary period of these systems can be deduced from the light curve between outbursts. Figure 10.19 shows the eclipse light curve of the dwarf nova, U Geminorum. Notice the asymmetric shape of the light curve plus the irregular flickering which occurs outside of eclipse. From the study of such observational data, Krzeminski and Smak, and Warner and Nather, independently proposed the model of dwarf novae shown in Figure 10.20. The central features of their model are the following. A lobe-filling red dwarf (a near main-sequence star) transfers mass in a steady stream to a closely orbiting companion (a white dwarf). The stream strikes a disk (or ring) of matter which encircles the white dwarf. The impact of the gas stream on the disk creates a "hot spot" which radiates away the excess kinetic energy of the stream. In U Geminorum, the eclipse of this off-center hot spot produces the dip in the light curve shown in Figure 10.19. This interpretation would explain why the light curve is asymmetric which it would not be if, say, the eclipse were merely blocking the light from the white dwarf. The fact that the "flickering" disappears during eclipse then implies that the light coming from the hot spot varies erratically on a wide range of timescales, which Robinson has measured to be as small as a fraction of a second. I think we can understand this erratic behavior most simply in terms of the unstable nature of the post-impact flow, in which large differences in the properties of the stream and the edge of the disk are likely to make their merging together quite turbulent (see Figure 10.16).

The reason for the outburst activity in dwarf novae is not completely agreed-upon, although most astronomers think some sort of instability in the disk, for unknown reasons, suddenly increases the effective frictional forces and causes matter to be dumped onto the white dwarf. In between such active outbursts, the matter flowing from the red dwarf is temporarily stored in the disk (or ring). Some support for this type of picture has come from the analysis by Smak and Paczynski of the sizes of the accretion disk during and between outbursts.

The outbursts of novae eruptions are almost certainly different from those of dwarf novae. Careful analysis of the spectra of the ejected material from Novae Aquilae by Weaver allowed the geometry of the ejected shell to be deduced (Figure 10.21). It turns out to be remarkably similar to the direct photographs taken later of Nova Herculis (Figure 10.18). The observational evidence is compatible with the idea that the basic energy source of the nova process is a thermonuclear explosion on the surface of a white dwarf. Bath and Shaviv have suggested that the violent ejection of matter from the system relies on the action of radiation pressure.

Many people have studied the theoretical problem of the initiation of a thermonuclear explosion on the surface of a white dwarf. The basic idea is that if hydrogen-rich material from a lobe-filling ordinary star settles onto

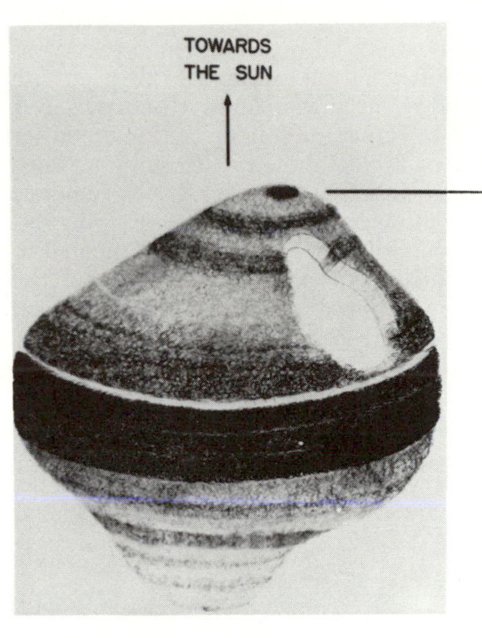

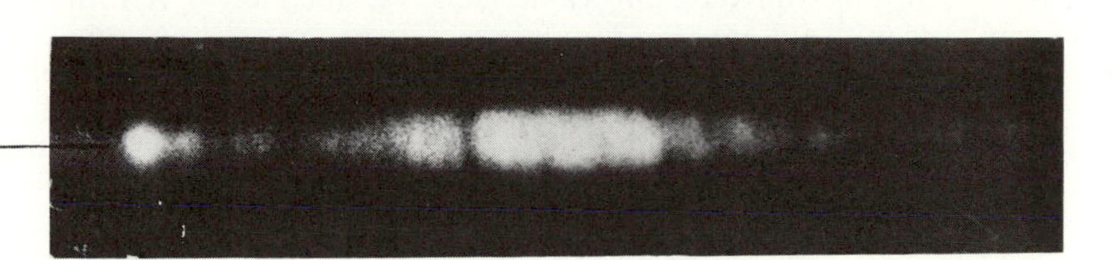

Figure 10.21. From analysis of the shapes of the spectral lines of Nova Aquilae, which exploded in 1918, we can reconstruct an image of its ejected shell. (From H. F. Weaver, *Highlights of Astronomy*, **3**, 1974, 509.)

a hot degenerate dwarf, enough material may accumulate to trigger a violently unstable process of nuclear burning. This situation may arise, for example, if the newly added hydrogen is partially mixed with the ambient matter of a carbon-oxygen white dwarf. Here the increase in the rate of hydrogen burning caused by the CNO cycle (where carbon and oxygen play the role of catalysts; see Chapter 6) may lead to an explosive situation, as has been demonstrated by Starrfield, Sparks, and Truran. A violent flash may consume all the newly added material in a nuclear holocaust. After the nova eruption, the system subsides for a while until a critical new supply of fuel is accumulated from the continuing transfer of mass from its normal companion.

Problem 10.6. Suppose mass transfer at a rate of $10^{-8} M_\odot \text{ yr}^{-1}$ is accumulated during a period of 10^2 yr, and suppose that the hydrogen in this mass is then all burned in a short interval of time. How much nuclear energy E would be released (consult Problem 6.3), and what would be the mean luminosity if most of this energy were to escape in a month? How does this outburst luminosity compare with the accretion luminosity calculated in Problem 10.1? Compute also the total energy released gravitationally in the accretion process. Which source of energy, gravity or nuclear, is the more powerful cataclysmic variables?

Except for scale, the outbursts of novae, recurrent novae, and dwarf novae are quite similar; so it has been disappointing that the most promising mechanisms offered so far to explain the outbursts in novae and dwarf novae have turned out to be so dissimilar. Other incompletely resolved problems about the cataclysmic variables include their evolutionary history. What class of objects are the progenitors of cataclysmic variables? Are their progenitors the W Ursae Majoris contact binaries, which have comparable orbital separations (as has been suggested by Kraft); or are they much wider binaries which have been whittled down by orbit decay through a phase of double-core evolution (as has been suggested by Ostriker and Paczynski). What ultimately becomes of cataclysmic variables? Does the white dwarf retain enough mass from the mass-transfer process to be forced ultimately past the Chandrasekhar limit? Would the subsequent conversion of a white dwarf into a neutron star release enough gravitational binding energy in observable form to account for type I supernovae, as has been speculated by Iben and Whelan? Or does the white dwarf always eject enough mass in nova eruptions to keep safely below the Chandrasekhar limit?

These and other unanswered questions promise to keep astronomers busy investigating cataclysmic variables for some time in the future.

Detached Component = Neutron Star: Binary X-ray Sources

The corona of the Sun was known for a long time from optical observations to be hot enough to emit X-rays, and direct verification of this deduction was obtained from X-ray detectors sent up in sounding rockets by Friedman and his collaborators. Cosmic X-rays from beyond the solar system were first detected by similar techniques in 1963 by Giacconi, Gursky, Paolini, and Rossi. The strongest such X-ray source was an unresolved object lying in the direction of the constellation of Scorpius. Because it was the first X-ray source to be discovered "in" this constellation, it was named Scorpius X-1. Shortly thereafter, Hayakawa and Matsuoka, and Novikov and Zeldovich, suggested that Scorpius X-1 might derive its X-ray emission from accretion of matter by a neutron star or a black hole in a close binary system. Optical spectra taken of Scorpius X-1 by Sandage and collaborators showed the spectrum of Scorpius X-1 to resemble that of an old nova, and this fact seemed to reinforce the idea that this X-ray source was a close binary system. However, until very recently, no optical observations revealed decisive evidence for a binary nature of Scorpius X-1. There the matter rested until 1971, when the first X-ray satellite UHURU provided dramatic and conclusive evidence for the binary nature of two other unresolved X-ray sources: Hercules X-1 and Centaurus X-3. X-rays have now been detected from the coronae of other stars and from pulsars, but as a general

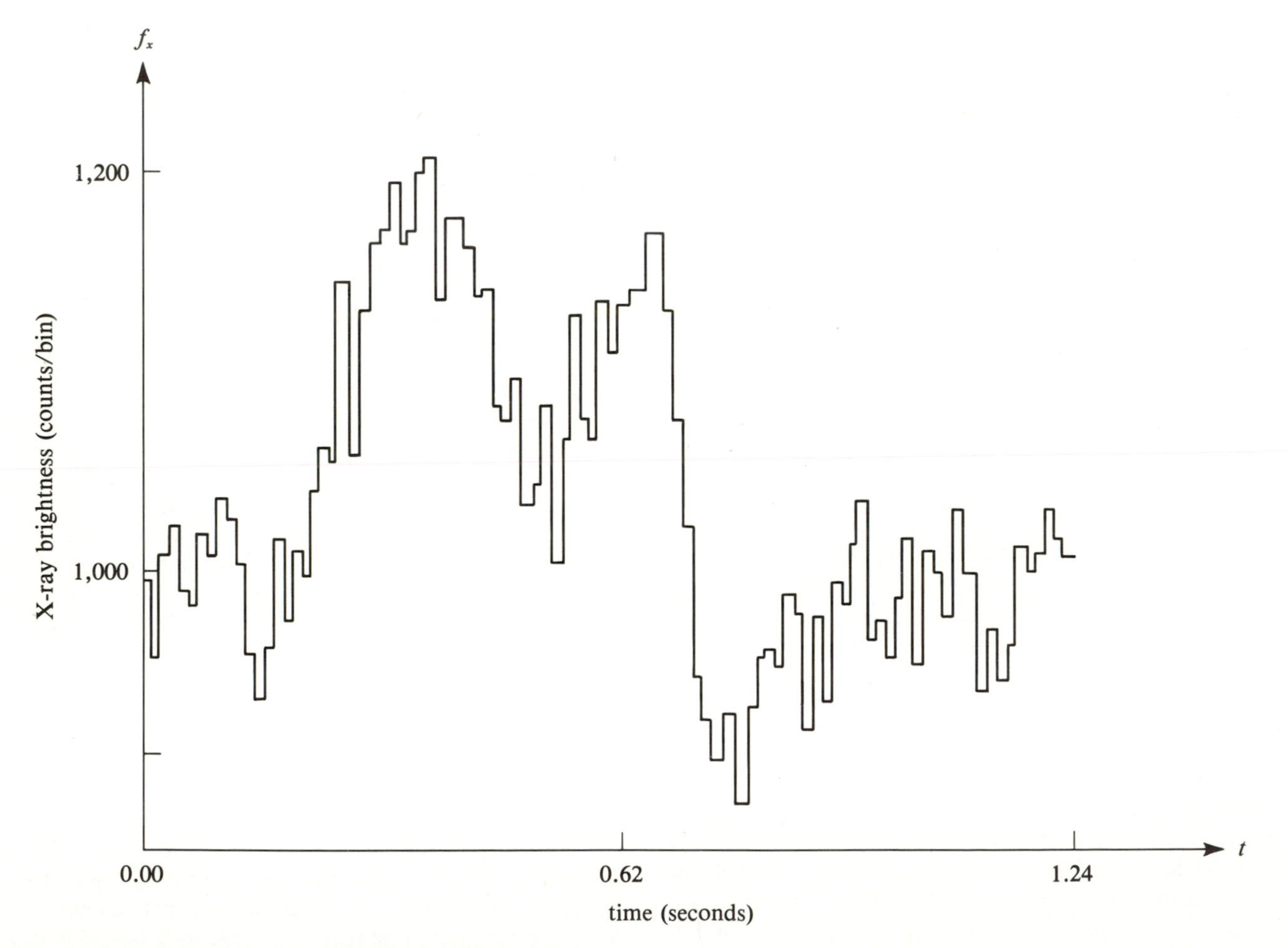

Figure 10.22. An X-ray pulse from Hercules X-1. The pulse shape roughly repeats itself once every 1.2372253 seconds. (Adapted from Doxsey *et al.*, *Ap. J. Lett.*, **182**, 1973, L25.)

rule the strongest point-like sources for X-rays in our Galaxy are all mass-transfer binaries which contain a compact object. In particular, Scorpius X-1 is now known to be a binary X-ray source.

The evidence that Hercules X-1 is binary is especially compelling. The X-rays from this system arrive at the satellite in pulses separated by 1.24 seconds (Figure 10.22). Moreover, this pulse period varies slightly and sinusoidally with another periodicity equal to 1.7 days. In other words, sometimes the periods arrive with separations slightly shorter than 1.24 sec, sometimes with separations slightly longer. The simplest interpretation of these facts is to suppose that the X-ray source is a neutron star which has a spin period of 1.24 seconds and which orbits about a companion with a binary period of 1.7 days. The 1.7-day variation of the 1.24-sec pulses could then be explained in terms of the Doppler shift caused by the orbital motion of a pulsing light source (Problem 10.7).

Problem 10.7. Review the derivation of the Doppler formula given in Problem 3.13 for the shift in wavelength of a monochromatic light source. Now, consider the problem of the observed pulse period P of a light source which emits pulses of (unknown) period P_0 in its own rest frame. Suppose the source to have a speed v and a component of velocity $v_{\|}$ along the line of sight to the observer. Show that the ratio of the observed period P to the rest period P_0 is given by the Doppler formula:

$$P/P_0 = (1 + v_{\|}/c)(1 - v^2/c^2)^{-1/2}.$$

Suppose Hercules X-1 to be in a circular orbit with angular speed Ω and distance r_X from the center of mass of the system, and suppose that this orbit has an inclination i with respect to the observer. Show that the observed period P can then be expected to vary sinusoidally with a repetition interval equal to $2\pi/\Omega$ and with a fractional amplitude proportional to $\Omega r_X \sin(i)/c$.

The identification of 1.7 days as the period of the binary orbit is further strengthened by the observed fact that the X-ray source undergoes eclipses every 1.7 days. In the same direction as Hercules X-1, optical astronomers had long been aware of a star, HZ Herculis, whose light varied periodically. Further work showed that the light curve of HZ Herculis varied with the exact same period, 1.7 days, as the deduced orbital period of Hercules X-1; so it was inferred that HZ Herculis is the optical companion of a neutron star, Hercules X-1.

The physical model which has been developed and generally accepted for the Hercules X-1/HZ Herculis system is the following (Figure 10.23). A normal star, HZ Herculis, and a neutron star, Hercules X-1, orbit about one another with a binary period 1.7 days. HZ Herculis fills its Roche lobe and transfers matter at a rate of about $10^{-9} M_\odot$ per year. This matter drops down the deep gravitational well of the neutron star, releasing about 10^{37} erg/sec at a temperature of about 10^8 K when it strikes the surface of the neutron star. Because the neutron star has a strong magnetic field, the incoming gas may be "funneled" to the polar regions of the neutron star. If the neutron star spins with a period of 1.24 seconds and if the magnetic axis is not aligned with the spin axis, an outside observer will observe the X-rays to emerge from the hot poles with a period equal to the spin period. Part of this pulsing beam of X-rays is intercepted by HZ Herculis. Because the intercepted X-radiation contains much more power than the intrinsic luminosity of HZ Herculis, the side of HZ Herculis which faces Hercules X-1 is strongly heated. As Bahcall and his collaborators first explained, this X-ray heating causes one side of HZ Herculis to be both appreciably hotter and appreciably brighter than the side which is not exposed. This "blistering effect" shows up as a periodic variation of the optical light received from the system, as the orbital motion of the system causes HZ Herculis to point its heated and unheated sides toward us alternately on a 1.7-day cycle (Problem 10.8).

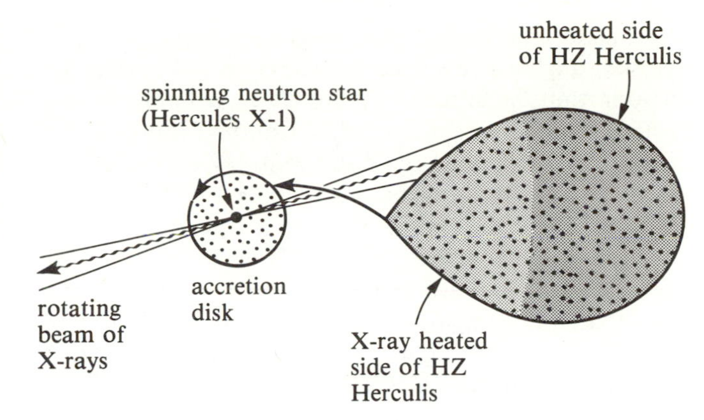

Figure 10.23. The physical model that has been developed for the system Hercules X-1/H Z Herculis.

Problem 10.8. If the model for Hercules X-1 discussed in the text is correct, should maximum optical light be received when the X-ray source is in mid-eclipse or when it is in front of HZ Herculis, half an orbital cycle later? How do you think X-ray astronomers deduce that the X-ray source has a temperature of about 10^8 K? *Hint*: At what wavelength would the maximum emission occur for a thermal body at 10^8 K?

How do you think a luminosity of X-rays $L_X = 10^{37}$ erg/sec roughly is deduced for Hercules X-1? *Hint*: To convert the observed X-ray flux f_X to a luminosity, a distance measurement is needed. Can you think of any method for estimating the distance to HZ Herculis?

Given $L_X = 10^{37}$ erg/sec and an effective temperature $T_e = 10^8$ K for the emission of this X-radiation, what is the area A of the emitting source? How does this area A compare with the surface area of a neutron star? Argue now that this comparison makes plausible the idea of funneling the accreting matter to a small polar region of the neutron star, Hercules X-1.

Suppose that the gravitational binding energy of a mass m which drops to the surface of the neutron star to be 0.1 mc^2 (see Problem 8.7). How does this energy compare with the amount released if the accreted matter were subsequently all converted by nuclear reactions to iron? Let the mass m accreted in time t correspond to a steady rate $m/t = \dot{M}$. Argue that the X-ray luminosity should be roughly $L_X = 0.1\dot{M}c^2$. Given that $L_X = 10^{37}$ erg/sec roughly, calculate the mass transfer rate $\dot{M}$ from HZ Herculis to Hercules X-1.

Astronomers have also detected optical pulsations with a period of 1.24 sec from the Hercules X-1/HZ Herculis system. From these observations and the information available from the X-ray studies, Middleditch and Nelson have deduced a mass for the neutron star, Hercules X-1, which is very close to Chandrasekhar's limit, $1.4M_\odot$. This deduction assumes that HZ Herculis exactly fills its Roche lobe and that the pulsed optical light comes from a "reprocessing" of the pulsed X-rays intercepted by the "pointy nose" of HZ Herculis. Although this method is very ingenious and plausible, it is based on the assumption, difficult to verify, that HZ Herculis exactly fills its Roche lobe; hence we would like to obtain an independent measurement of the masses of the two stars in the Hercules X-1/HZ Herculis system. We would have such an independent measurement if we could find the radial velocity of HZ Herculis spectroscopically, in much the way that the Doppler-shift formula is applied to the X-ray observations to yield the radial velocity of Hercules X-1. Unfortunately, although spectra of HZ Herculis are not difficult to obtain, their interpretation is greatly confused by the uneven heating of this star (so that the center of light is not the center of mass) and by the presence of substantial amount of gas between the two stars (due presumably to the process of mass transfer). Much observational work is therefore still needed to measure the properties of this intriguing system.

Many other binary X-ray sources are also now known. Astronomers believe that such binary X-ray sources

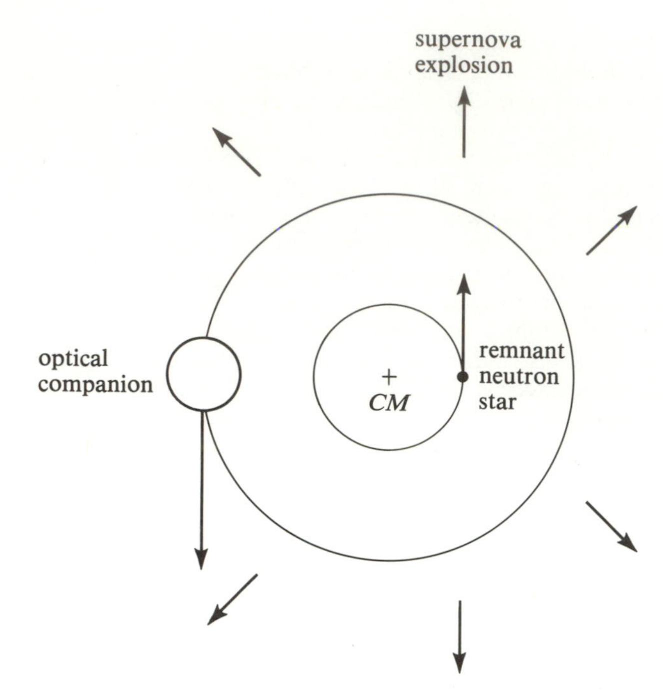

Figure 10.24. The loss of a large fraction of the binding mass by a supernova explosion in a binary system will unbind the two stars. The reduced gravitational attraction of the two stars will not be able to deflect the large relative velocities of the two stars; so they will be flung from the system like stones from a sling.

divide into two classes. The first class contains low-mass systems like Scorpius X-1 and Hercules X-1, where matter is accreted onto a neutron star because of mass transfer from a normal-star companion that overflows its Roche lobe. The second class contains high-mass systems like Centaurus X-3, where matter may be accreted onto a neutron star by the capture of matter blowing away from the normal-star companion (usually an O or B giant or supergiant) in a stellar wind. An accretion disk should always form in the first class; whether it does also in the second class is not generally known.

An interesting problem with the binary X-ray sources is their evolutionary history. Before the discovery of binary X-ray sources, it was generally thought that neutron stars would not be found in binary-star systems. The reasoning went as follows. To form a neutron star requires a supernova explosion, in which a substantial fraction of the mass of the star would be expelled from the system. In a high-mass binary system, the first star to supernova would be the more massive component. Thus, it is likely that more than half of the total mass would be blown away, and the loss of this much binding mass would unbind the system (Figure 10.24). Indeed, the newly formed neutron star would be catapulted out of the system at roughly its pre-explosion orbital velocity, and this would be consistent with observations that many pulsars (Chapter 7) have a high space velocity. (The same mechanism could account for the normal-star companion becoming a "runaway" O or B star.)

After the discovery of binary X-ray sources, it was realized in hindsight that mass transfer in close binaries might prevent the disintegration of the binary system at the instant of the supernova explosion. The basic idea is similar to that proposed for the resolution of the "Algol paradox": the more-massive star would indeed evolve first, but before it completed its evolution (i.e., before it encountered the "iron catastrophe"), it would have transferred enough mass to its companion to have become the less-massive component. The supernova explosion of the evolved but less-massive star may now leave the binary system intact, although the binary orbit would generally become quite noncircular with the loss of any appreciable binding mass. Thus, a remaining problem with theories which hope to disentangle the past history of binary X-ray sources is what mechanisms recircularize the orbit. Tidal damping of some sort (Figure 10.6) is undoubtedly involved, but apart from some general work by Zahn, the details of the process remain somewhat obscure.

It is possible, of course, that the neutron star now found in a binary X-ray system of low total mass was formed when accretion of matter pushed a white dwarf "over the brink" past the Chandrasekhar limit. The conversion of a white dwarf into a neutron star would result in a loss of about 10 percent binding mass even if no mass were expelled in an explosion from the system! (The gravitational binding energy of the neutron star might come out in the form of photons, neutrinos, or gravitons; see Problem 8.7.) According to Einstein's theory of general relativity, a white dwarf of $1.44M_{\odot}$ which became a neutron star would have the gravitational pull of a $1.3M_{\odot}$ object. Thus, if the binary orbit was circular to begin with, it would be only slightly noncircular after the conversion process, and recircularization would then be relatively easy.

What is the final state for the evolution of binary X-ray sources? Again, the details are highly uncertain, but the overall answer is likely to be the following. If synchronous spin of the normal star can be maintained, the normal star will eventually swell up and fill its Roche lobe if it has not already done so. After a while it would begin to dump so much mass so rapidly toward its compact companion, that the X-ray source will be "smothered" by an overload of absorbing material. The normal star may then be completely "cannibalized" by its companion, leaving the system as a single neutron star (or black hole) surrounded perhaps by a relic accretion disk. The matter in this accretion disk may slowly dribble onto the central object, perhaps reactivating the X-ray source for an interval of time. On the other hand, the normal star may not be completely "eaten" before it turns into a white dwarf, neutron star, or black hole. If the death throes of this star do not disrupt the binary system, the final outcome will be two compact objects which gradually spiral in toward one another (because of general-relativistic effects, as we will soon discuss) and become a single compact object. If synchronous spin of the normal star cannot be maintained, as is probable for high-mass binary X-ray sources, where a significant fraction of the total angular momentum resides in the spin of the normal star (refer back to the discussion of orbit evolution), the normal star may swell past its Roche surface without transferring matter to its companion, and may well "swallow" its compact companion whole. If so, a phase of "double-core" evolution would proceed (Figure 10.25). If the system is not interrupted by a titatic explosion in the interim, eventually the two cores shrouded in a gigantic common envelope will spiral toward one another because of friction from the common envelope. The final outcome here will also be a single star.

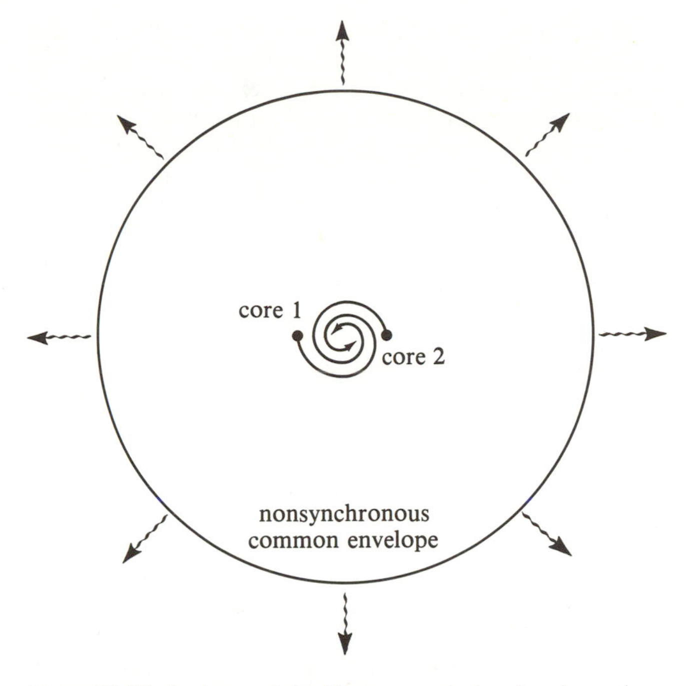

Figure 10.25. A phase of double-core evolution has been theorized by Ostriker and Paczynski if the expanding envelope of an evolving star is unable to maintain synchronism and swallows the compact companion. The envelope will then contain two cores, which will gradually spiral toward one another as a result of frictional drag and orbit decay.

The final states of binary-star evolution in all circumstances are therefore the same as the final states of single-star evolution: nothing left, white dwarf, neutron star, or black hole. It would be very hard to distinguish the end products of binary-star evolution from the end products of single-star evolution, except perhaps that the former probably have a high angular momentum.

Grindlay at Harvard, and Bradt, Clark, Lewin, and their colleagues at MIT, have found X-ray "burst sources" in globular clusters and in the nuclear bulge of our Galaxy that may represent evidence for an evolutionary scenario like one of those outlined. The properties of these "burst sources" are very similar to those of the "classical" binary X-ray sources, but with two differences. First, the "burst" X-ray sources have periods of steady X-ray emission interrupted by sudden flare-ups in increased X-ray activity. It is now believed (largely because of the work of Woosley, Taam, Picklum, and Joss) that the outbursts are thermonuclear in origin. The idea is that a neutron star accretes matter steadily from a disk of surrounding gas. The accompanying release of gravitational binding energy from the accretion process accounts for the steady X-ray source. Every now and then, however, the build-up of a layer of unburnt helium or carbon on the surface of the neutron star undergoes a thermonuclear "flash," releasing an X-ray burst. Although the peak luminosity during the burst is large, the duration of the burst is short (seconds). Thus, the total energy released per burst is roughly a hundred times smaller than the total energy released by the steady source between outbursts (typically hours). This is another reminder that gravity in truly compact objects is potentially a much more powerful source of energy than nuclear interactions.

Problem 10.9. Suppose a mass m drops onto a neutron star. Compare the fraction of mc^2 which is released in gravitational binding energy and in nuclear binding energy (for helium or carbon burning). In this manner, recover the 100:1 ratio discussed in the text.

There is also an exceptional "rapid burster," which is to X-ray burst sources as a machine gun is to muskets. Lamb, Pethick, and Pines, Arons and Lea, and Elsner and Lamb have proposed that the rapid series of bursts in this source results from instability in the accretion flow onto the neutron star as the infalling plasma encounters the magnetic field of the neutron star (Figure 10.26). Alfven had shown many years ago that a magnetic field tends to be "frozen" in a plasma because charged particles tend to spiral around magnetic field lines. That is, if the magnetic field threads the plasma, the plasma tends to carry the field around in its motion, whereas if the magnetic field is external to the plasma, the plasma wants to push the magnetic field out of its way, and the magnetic field tries to exclude the plasma. This tendency

Figure 10.26. The rapid burster and the magnetospheric structure proposed to explain its activity as a series of accretion instabilities.

is, of course, the physical basis for all "magnetic confinement" schemes for controlled thermonuclear fusion on Earth.

Consider now the case of an accreting magnetized neutron star. The infalling plasma tries to fall directly onto the neutron star, but the magnetic field of the latter resists being squashed. A quasi-equilibrium is reached wherein the inrushing matter is arrested in a standing shockwave by the magnetic field. However, the weight of the halted matter continues to press on the magnetic field and to try to force an entry. The magnetic-field configuration resists this entry, but its ability to hold up the weight of a heavy gas is limited by its curvature. Under some conditions, this quasi-equilibrium configuration is unstable; that is, the configuration is "top-heavy" and would like to "turn over," with the heavy plasma falling to the bottom and the magnetic field rising to the top. The infalling matter creates a burst of X-rays when it strikes the surface of the neutron star. In the meantime, the magnetic field, which has buckled outward somewhat upon relief of its oppressive load, begins to be pressed inward again by a new load. When conditions of instability are again reached, a new cycle of unloading begins with an accompanying fresh burst of X-rays. Although the details for this mechanism is elaborate, we see that the basic energy source for this model of the "rapid burster" (devised by Fabian, Lamb, and Pringle) is again gravity power.

The second difference between X-ray burst sources and the classical binary X-ray sources is more intriguing. Despite intensive observational searches, no direct evidence has been found that any X-ray burst source is a binary system. None of them, for example, suffer X-ray eclipses. Milgrom has proposed that this may result from an observational selection effect, wherein only observers whose line of sight is nearly perpendicular to the accretion disk can see an unobscured X-ray source. Given such a viewing geometry, the same observer could not see X-ray eclipses, because the usual mass-transfer dynamics produce an accretion disk which lies in the same plane as the orbital plane (Figure 10.15).* Paczynski has made the alternative suggestion that the X-ray burst sources are actually single neutron stars with a surrounding relic accretion disk. This suggestion is intriguing for stellar evolution, because it would imply that some Population II binary stars have succeeded in evolving close to the ideal end state of a single compact object. It would also imply that accretion disks around a single central object may continue to feed the central object at a slow and steady rate for a relatively long time.

Clearly, the subject of binary X-ray sources is a very young one. Consequently, many theoretical ideas are possible, and many have, in fact, been proposed to explain relatively few observational facts. Mark Twain once remarked that science was wonderful, because it gave such a wholesale return of conjecture for a trifling investment of fact! In citing various theories here, I have presented what seemed to me to be the most promising approaches; but our learning which of them, if any, turn out to be correct will require a larger investment of observational fact.

* There is some evidence accumulated by Boynton and his collaborators that the accretion disk in Hercules X-1/HZ Herculis may "flop" about the orbital plane with a 35-day period.

Interesting Special Examples of Close Binary Stars

Cygnus X-1: A Black Hole?

The first X-ray source to be discovered "in" the constellation of Cygnus has a highly variable and irregular X-ray emission. The irregular variations occur on time-scales as short as tens of milliseconds, which is the light-travel time across about 10^9 cm. This is a direct indication that in Cygnus X-1, we must be dealing with a compact stellar object. Accurate X-ray locations led to more precise radio positions by Braes and Miley, and by Hjellming and Wade. The latter allowed an optical identification of a normal-star companion to Cygnus X-1. Comparison of the interstellar reddening of this companion with that of other stars in the field led to the conclusion that it is a BO supergiant at a distance of at least 8,000 light-years. The spectrum of this BO supergiant has a Doppler shift which varies periodically with a 5.6-day binary orbit. The amplitude of the associated velocity variation, coupled with the estimate that the BO supergiant probably has a mass near $30M_\odot$, implies that the unseen X-ray companion must itself be fairly massive; otherwise the X-ray companion could not exert enough gravitational pull to make the BO supergiant wobble in orbit as much as it does. The minimum mass deduced in this way for Cygnus X-1 (assuming the optimum geometry for producing the observed velocity variations) has been estimated by Bolton to be about $6M_\odot$. Since the X-ray source must be a compact object (a white dwarf, neutron star, or a black hole) and since the minimum mass of the X-ray source of $6M_\odot$ exceeds the maximum possible mass for both a white dwarf and a neutron star, most astronomers believe that Cygnus X-1 must in fact be a black hole. To be sure, other proposals have been put forward to evade this conclusion, such as that the system might be a triple, with the $6M_\odot$ shared by a neutron star and a closely orbiting unseen companion. However, the simplest and most direct conclusion is that Cygnus X-1 is indeed a black hole.

If this conclusion is correct, the X-rays from the system would not come, of course, directly from the black hole itself. Rather, they must arise from the hot gases in the accretion disk before this matter plunges into the event horizon of the black hole. The properties of the matter in such relativistic regions have been studied by many people, especially by Thorne and his colleagues at Caltech. Although the general-relativistic effects have been fairly thoroughly investigated by now, the main uncertainty in such calculations concerns what the "anomalous friction," which is hypothesized to transport angular momentum outward and mass inward in the accretion disk, might actually be. Much important work remains to be carried out for Cygnus X-1, not the least of which is nailing down the tremendously significant conclusion that in Cygnus X-1 we must be dealing with a black hole.

An interesting point can be made concerning the accretion of matter by a black hole. What the outside observer perceives depends little either on the mass of the black hole or on what swirls down the deep gravitational well. For a wide range of conditions, much of the rest-mass energy of the accreted material will emerge in the form of energetic photons near the event horizon of the black hole. Thus, a small black hole—say, with the mass of the Moon—could provide a fanciful solution to the waste-disposal problem and energy crisis faced by humanity. Dump one ton of garbage per day onto such a black hole, and receive enough high-grade X-rays back to satisfy the needs of everyone on Earth for that day. There does remain, of course, the minor problem of how to compress the Moon inside its Schwarzschild radius of 10^{-2} cm!

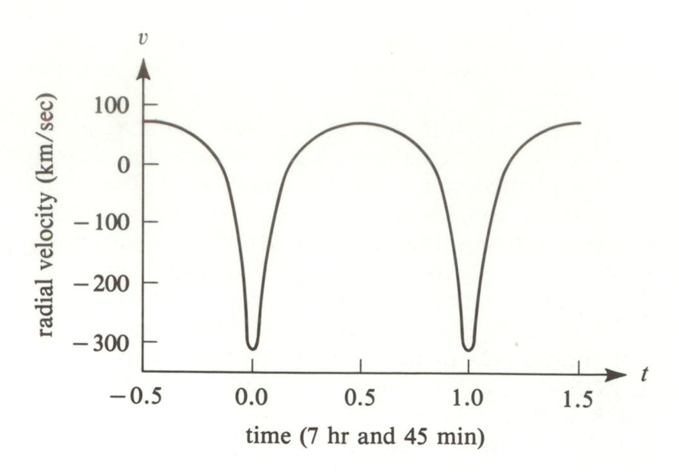

Figure 10.27. The velocity curve of the binary pulsar. The nonsinusoidal behavior of the velocity curve, which has a period of 7.75 hours, indicates that the orbit is quite noncircular (an eccentric ellipse).

The Binary Pulsar: A Confirmation of Einstein's General Relativity?

We have already mentioned that astronomers had generally expected to find pulsars as single stars. It therefore came as a great surprise when Taylor and Hulse announced in 1974 the discovery of a radio pulsar in a binary system. This binary pulsar has a pulse period of about 59 milliseconds which varies by about one part in a thousand every 7.75 hours. Interpreted as Doppler shifts (see Problem 10.7), these variations imply orbital velocities on the order of a thousandth the speed of light. The periodic variations of the resulting velocity curve are highly nonsinusoidal (Figure 10.27), indicating that the orbit is quite eccentric. Because we can keep track of time to an accuracy of 1 part in 10^{12}, we can detect general-relativistic effects in the evolution of the orbit of the binary pulsar within a period of years. We need consider only two such effects here.

The first effect is the rotation of the line of apsides. We have already remarked that a $1/r^2$ force law (Newtonian gravity for spherical bodies) yields an ellipse as the most general geometric form for a bound orbit. The long axis of this ellipse joins the point of closest approach and the point of greatest separation, points which are called "apsides" in astronomy. The Newtonian theory of binary stars treated as mass points predicts that this long axis, or "line of apsides," should remain fixed in direction in an inertial frame. In Einstein's theory, however, the gravitational force between two mass points does not vary exactly as $1/r^2$; consequently, general relativity predicts that the line of apsides of an eccentric orbit should rotate slowly in space. Indeed, Einstein made this prediction originally for the so-called "advance of the perihelion" of Mercury by an amount of 43 seconds of arc per century, a prediction which brilliantly explained a puzzling existing discrepancy between observations and the prediction of Newtonian theory. For the binary pulsar, the proximity of the two orbiting bodies should produce a much larger effect. Happily, the radial velocity curve of the observed pulsar does show the effect of a rapid rotation of the line of apsides (Figure 10.28). Unhappily, this effect could also arise in Newtonian theory if the companion of the Taylor-Hulse pulsar has a nonnegligible size, since tidal bulges or flattening due to rapid rotation could then cause the companion to assume a nonspherical shape, and the gravitational attraction of a nonspherical body also deviates from a $1/r^2$ law. (Indeed, the so-called "apsidal motion" induced by this classical effect is a standard test of theories of stellar structure.) Now, a substantial size for the unseen body is ruled out by the lack of observed eclipses of the pulsar for an inferred inclination of 54°. However, if the companion is a helium star (Chapter 8) or a rapidly rotating white dwarf, its tidal effects would confuse the observa-

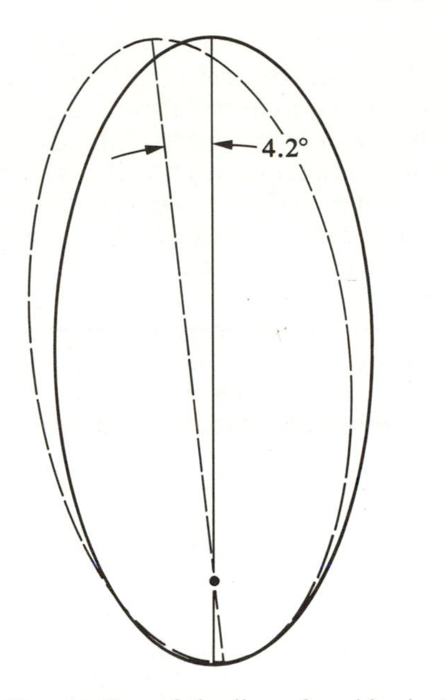

Figure 10.28. The rotation of the line of apsides in the binary pulsar amounts to 4.2 degrees per year. This effect has been interpreted to arise because the gravitational attraction of the companion deviates from a $1/r^2$ force law because of general-relativistic corrections.

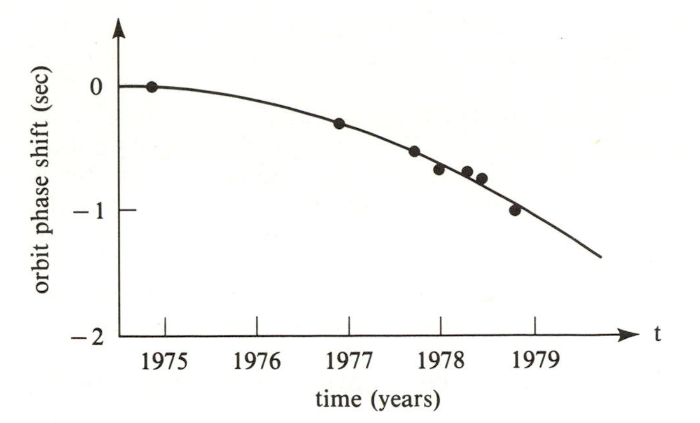

Figure 10.29. The decrease of the orbital period of the binary pulsar. This decrease is consistent with gravitational radiation causing a slow decay of the orbital separation as the two stars spiral slowly toward one another.

tions enough to prevent a decision test of the predictions of general relativity against the observational facts.

The second effect implied by general relativity is a decrease in the orbital period of the system. Newton's theory applied to the bound orbit of two mass points predicts that the binary period should remain absolutely constant. In Einstein's theory, however, two stars which revolve about one another suffer accelerations. Such accelerations generate gravitational radiation (Chapter 3). The radiation of energy and angular momentum by this process must come at the expense of orbital decay. Orbital decay causes the two stars to spiral slowly toward one another, giving rise to systematically shorter and shorter periods. This decrease of binary period has now been observed by Taylor, Fowler, and McCulloch (Figure 10.29). Taken together with the rotation of the line of apsides (and with more subtle effects), this decrease of the binary period constitutes very strong evidence for the basic validity of the general-relativity theory. Now, other explanations could be found for the decrease of the binary period, just as other explanations could be found for the rotation of the line of apsides. However, the power of the general-relativistic explanation is that a *single* mechanism reproduces all the observational results, both qualitatively and quantitatively. Without general relativity, *many* arbitrary assumptions would have to be invoked to reproduce the same observational features. The philosophical principle called "Occam's razor" leads scientists to prefer the simpler and more elegant explanation.

Enough information can be gleaned from the timing observations (assuming general relativity to be correct) to deduce the masses of the two stars in the system. Both masses turn out to be approximately the Chandrasekhar limiting mass of $1.4M_\odot$. This fact, plus the inference that the unseen body is compact (because otherwise it should produce an "apsidal motion" in excess of the general-relativistic value), leads to the conclusion that the binary pulsar is likely to consist of *two* neutron stars. It would be wonderful if the second neutron star were also a pulsar and if its beam swept in our direction, but so far all attempts to find such a second pulsar have failed.

The history of the stellar evolution which would have produced two neutron stars in a single binary system must have made interesting reading. The only plausible reconstruction of events offered so far is the "double-core" hypothesis discussed earlier for the likely evolution of X-ray binaries. In this model, a neutron star would be swallowed in the expansion of its normal companion, and would begin to spiral toward the center of the latter under the resulting frictional forces. If the companion explodes as a supernova before the neutron star reaches the center, most of the expelled mass will lie beyond the binary orbit. Such an explosion may leave the orbit quite eccentric, but it need not unbind the remaining two neutron stars. Presumably, the observed pulsar is the most recently formed neutron star. The future of this unfolding drama will see the completion of the merger process, driven by the inexorable loss of energy and angular momentum by gravitational radiation from the system. The end product is likely to be a single rotating black hole, a highly distorted rip in the fabric of spacetime.

It should not be thought that these processes were uniquely proposed for the evolution of the binary pulsar. After all, Einstein's theory of gravitation is every bit as universal as Newton's. Long before the discovery of the binary pulsar, Kraft, Greenstein, and Mathews had already proposed that gravitational radiation should be an important factor in the evolution of W Ursae Majoris stars, the contact binaries discussed earlier in this chapter. However, these close binaries do not provide the "clean" test of general relativity that the binary pulsar does. In the other binaries, considerations besides gravity enter into the problem. In contrast, two neutron stars come close to the ideal of two purely gravitating masses. The only possible improvement would be two black holes in orbit about one another, but they would produce no electromagnetic emission by which we could observe them!

This situation would improve, of course, if we could detect the gravitational radiation from such close binaries directly instead of by inference from changes in orbital properties. However, despite ingenious ongoing efforts, the construction of gravitational-wave detectors sensitive enough to be useful remains one of the most challenging endeavors in experimental science. It has been estimated, for example, that even the strongest astronomical sources of gravitational radiation would impart less energy to a typical detector than a match placed at a distance of a million miles! This example gives a striking demonstration of the weakness of gravity in phenomena on the laboratory scale. Yet, despite this intrinsic weakness, gravity rules the astronomical universe. This is why any study that improves our understanding of gravity, such as the study of the binary pulsar, is an important enterprise.

SS433

The final object that we wish to discuss in this chapter is a weird system called SS433, the 433rd member of a catalog of peculiar stars compiled by Stephenson and

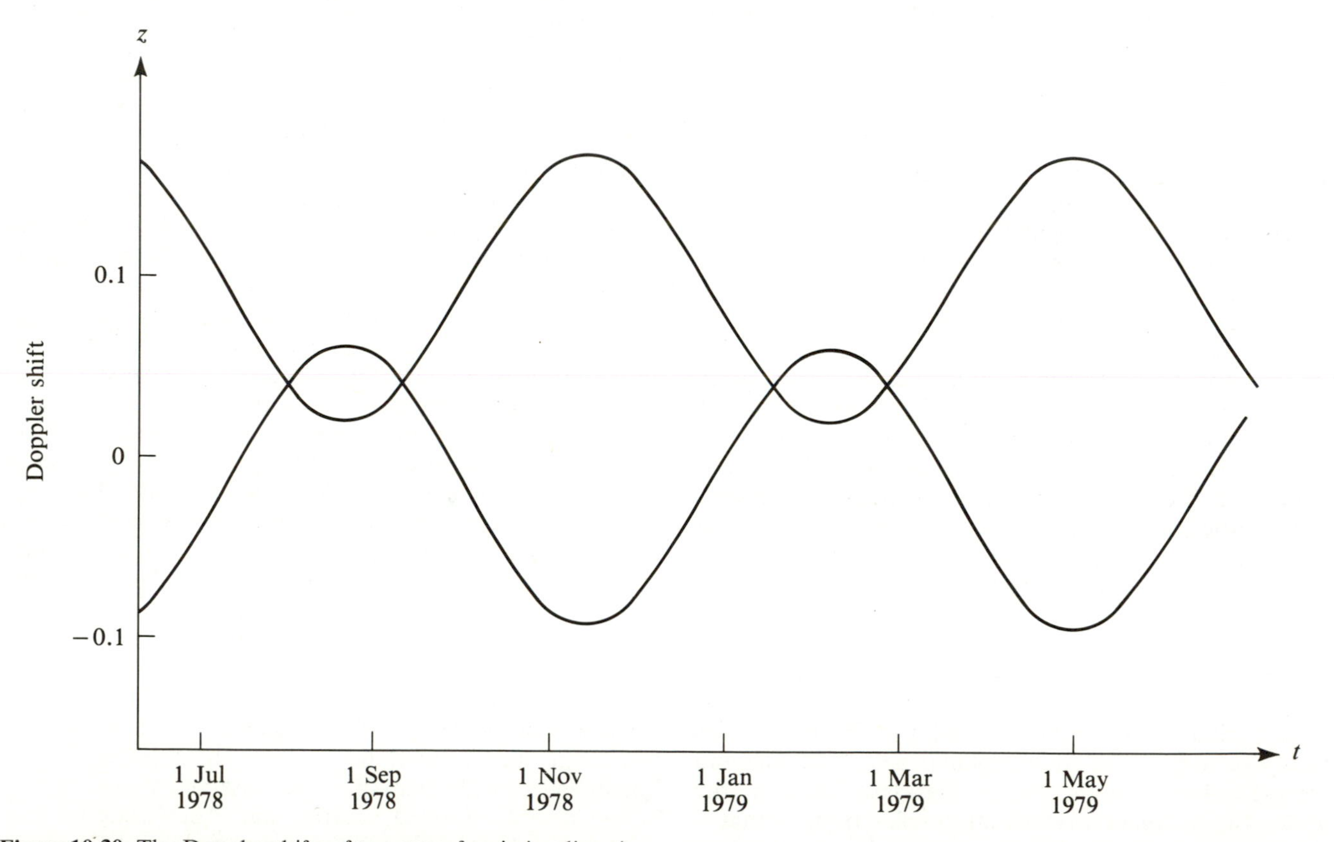

Figure 10.30. The Doppler shifts of two sets of emission lines in the object SS433. The fact that z has magnitudes of 0.1 to 0.2 implies that the corresponding motions must amount to a substantial fraction of the speed of light.

Sanduleak. No compelling model has been offered for this system at the time I write. Found as an optical object, SS433 was subsequently observed to possess variable radio and X-ray emission. There is a hint that SS433 may be associated with a supernova remnant. Early in 1979, Margon and his coworkers obtained high-quality spectra of the system that showed the presence of intense emission lines located at rather puzzling wavelengths. Apart from a set of emission lines at their ordinary locations, there existed another two sets whose ratios of wavelengths revealed them to be associated with hydrogen and helium at enormous Doppler shifts. Moreover, both redshifts and blueshifts occur, and the positions of the two sets of lines shift back and forth in time with a period of about 164 days (Figure 10.30). At maximum separation, the inferred velocity difference along the line of sight is more than one-fourth the speed of light! The very high speeds, combined with the relatively long period, implies that the combination cannot be due to the orbital motion of two ordinary stars.

Problem 10.10. Make the (incorrect) assumption that the period of 164 days and the relative velocity of 0.8×10^{10} cm/sec are those of a circular binary orbit at angular speed Ω between two stars at a distance r apart. Ignore relativistic effects and use the formula

$$\Omega^2 = G(M_1 + M_2)/r^3,$$

where M_1 and M_2 are the masses of the two stars (see Problem 10.1). Assume, therefore, $2\pi/\Omega = 164$ days and $\Omega r = 0.8 \times 10^{10}$ cm/sec, and calculate the total mass implied for the system, $M_1 + M_2$. Can this mass be that of two ordinary stars?

Nevertheless, strong circumstantial evidence exists that SS433 is within our Galaxy, not outside it. In particular, recent spectroscopic studies (by Crampton, Cowley, and Hutchins) of the "nearly stationary" emission and absorption lines which are also present in the system reveal a binary orbital period of about 13 days. Such a period is quite typical of other interacting binaries in the Galaxy. What, then, is the origin of the high Doppler shifts and the 164-day periodicity of the rapidly moving emission lines? The now most-popular model involves two beams of matter ejected relativistically in opposite directions from a compact object, perhaps a neutron star or a black hole (Figure 10.31). A precession of the ejection axis might account for the 164-day periodicity. Milgrom has made the interesting suggestion that the acceleration of the two beams might be due to the action of radiation pressure. It remains to be seen whether SS433 represents a new class of astronomical object, or whether it is an isolated example of a very weird system.

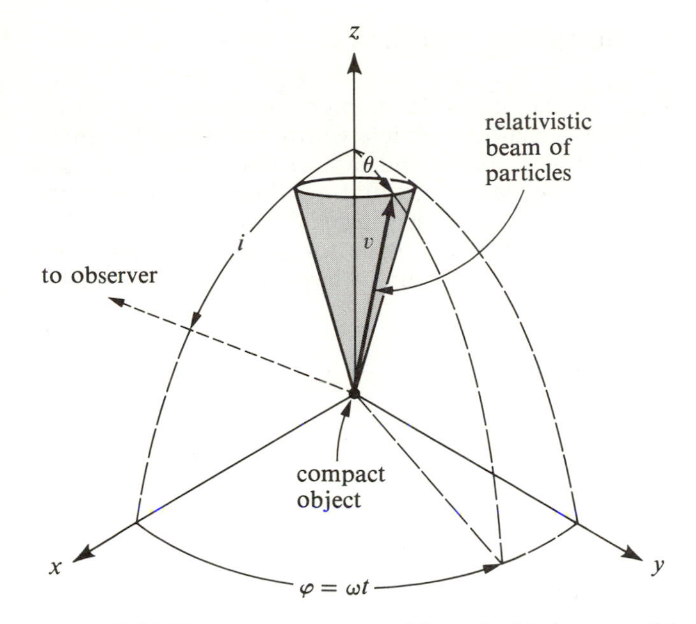

Figure 10.31. The geometry proposed for a double-beam model to explain the large periodic Doppler shifts in SS433. (For clarity, the opposite, lower beam has not been drawn.)

Problem 10.11. Assume that the enormous Doppler shifts of SS433 are due to two relativistic beams traveling at speed v in diametrically opposite directions (see Figure 10.31). The Doppler shift of each beam is given by the relativistic formula (consult Problem 10.7)

$$1 + z = (1 + v_{||}/c)(1 - v^2/c^2)^{-1/2},$$

where $v_{||}$ is the component of velocity parallel to the line of sight. Let the z-axis be the fixed axis about which the two beams precess, and let i be the angle of inclination of our line of sight to this axis. If the geometry is given by Figure 10.31 then show that $v_{||}$ is given by

$$v_{||} = v_z \cos(i) + v_x \sin(i),$$

with $v_z = v\cos(\theta)$ and $v_x = v\sin(\theta)\cos(\varphi)$. Suppose the precession angle φ increases with time as $\varphi = \omega t$, where $2\pi/\omega = 164$ days, and assume $\theta = 17°$, $i = 78°$, with $v/c = \pm 0.27$ for the two beams. Plot the resulting values for z as a function of time, and compare your plot with Figure 10.30.

Concluding Remarks

Pairs of stars in mortal embrace provide a rich variety of phenomena not possible with single stars. The interplay between the second law of thermodynamics and the universal law of gravitation ultimately shatters the fragile peace which exists in such cramped quarters. The resulting conflict can make for brilliant rejuvenations of the end states of stars if they reside in mass-transfer systems. Thus, if burnt-out white dwarfs are supplied with fresh hydrogen fuel, they can find an explosive new lease on life. No wonder past peoples thought such objects represented "new stars" or *novae.* Neutron stars and black holes respond even more vigorously to the cannibalizing of a close companion. The copious release of X-rays when matter falls or spirals into the deep potential wells of such objects testifies once again to the truly awe-inspiring power of gravity.

PART III

Galaxies and Cosmology

The Sombrero galaxy (M104, type Sa/b). (The Cerro Tololo Inter-American Observatory.)

CHAPTER ELEVEN

The Material Between the Stars

Between the stars lies a vast amount of material in the form of gas and dust. The interstellar medium of our Galaxy probably has a mass of several billion $M_\odot$. Thus, the mass of the material between the stars is a non-negligible fraction—a few percent—of the mass of all the visible stars in the Galaxy. Yet the presence of interstellar matter is much less obvious than the presence of stars, since the mass contained in ordinary stars has been compacted by self-gravity into a readily observable dense state. In contrast, the interstellar gas is spread very thinly over the vast distances between the stars; this diffuse gas is much more rarefied than the best so-called "vacuums" produced in terrestrial laboratories. Self-gravity plays a relatively minor role for this gas, and it should not surprise us to learn that even astronomers were slow to realize the existence of interstellar gas and dust in the Galaxy.

The Discovery of Interstellar Dust

More than 200 years ago William Herschel described "holes in the sky," where there were an apparent deficit of stars (Figure 11.1). The most plausible explanation for this deficit was the existence of obscuring material along the line of sight which blocked the light from the background stars. Claims were occasionally made for the definitive discovery of other observational evidence for this absorption, but careful follow-up analyses invariably showed these claims to be based on either incorrect reasoning or faulty data. The conclusive proof for the presence of a general and selective absorption came in 1930, with the epochal work of R. J. Trumpler on the properties of star clusters.

We have already discussed (in Chapter 9) Trumpler's method of lining up the main sequences of open clusters to deduce their relative distances. In fact, when Trumpler first carried out this procedure, he discovered an anomaly. It concerned the intrinsic sizes of open clusters, and led to irrefutable evidence for the presence of fine grains of solid matter in interstellar space, matter that is now called **interstellar dust**.

We may reproduce Trumpler's reasoning as follows. Let D be the real (linear) diameter of an open cluster and r equal its distance. Since D should be independent of the open cluster's distance r, we expect the angular diameter $\theta = D/r$ to decrease with increasing distance. Thus, we expect the square of the angular diameter,

$$\theta^2 = (D/r)^2,$$

to vary on the average as the inverse square of the cluster's distance r from us. The apparent brightness $f = L/4\pi r^2$ of (the main sequence stars of) an open cluster should also vary as the inverse square of its distance from us, because, on the average, the intrinsic brightness (luminosity) L of a typical cluster should be independent of its position relative to the Earth. Thus, if apparent brightnesses of open clusters are plotted against angular

Figure 11.1. The Coalsack nebula in the southern Milky Way gives the impression of a large region in the sky where there is a marked deficit of stars. In fact, the apparent "hole in the sky" is explained by the obscuration of the light of many background stars by an intervening cloud of dark material. (Courtesy of Bart J. Bok.)

diameters squared, we should expect theoretically to see a straight-line relation (Figure 11.2a). The actual result is shown schematically in Figure 11.2b. The interpretation of this result is the following.

First, there is **scatter** about the theoretical line, because there must be intrinsic variations of the luminosities L and diameters D of open clusters about some mean values. (That is, some open clusters are intrinsically more or less luminous than average and some are intrinsically bigger or smaller than average.)

Second, there is a *systematic departure* of the observed points from the theoretical line. This effect might be explained in several ways.

(a) Perhaps far-away clusters look intrinsically bigger than their measured apparent brightnesses f would predict. If this were a real effect, then our position in the universe would be special, since open clusters which are close to the Earth would then be intrinsically small, and those which are far away would be intrinsically large. Such a special location for Earth has been anathema for

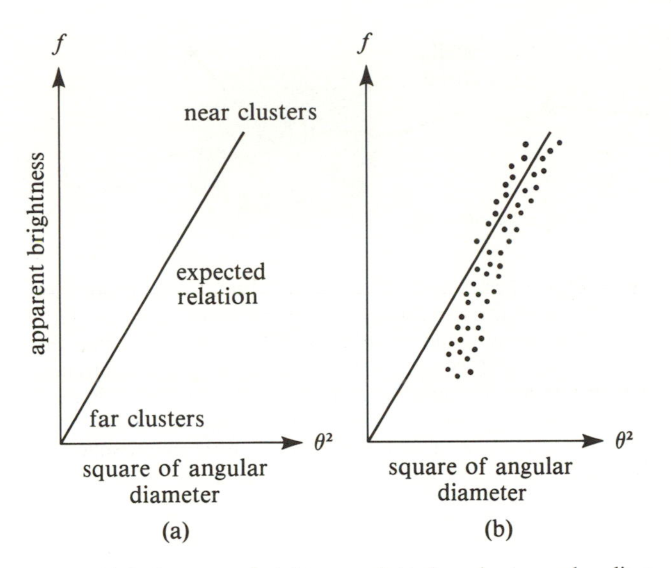

Figure 11.2. Apparent brightness plotted against angular diameter squared for open clusters. (a) The expected relation in the absence of extinction. (b) A schematic representation of the observed relation (the dots are data points) found by Trumpler. The deviation of the dots from the expected relation enabled Trumpler to prove conclusively that there is a general interstellar extinction by dust.

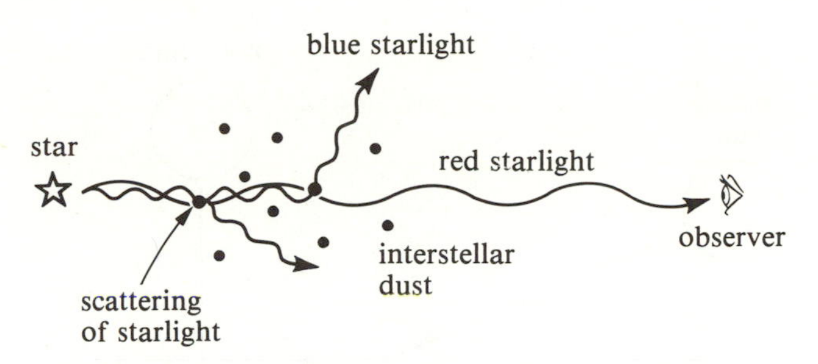

Figure 11.3. The mechanism of interstellar reddening.

astronomers since Copernicus, and this interpretation may therefore be rejected.

(b) Could this effect be one of observational selection? That is, could this effect arise because observers find it easier to see intrinsically big open clusters than intrinsically small ones if they are far away? Trumpler was well aware of such observational biases, and he was able to show convincingly that his data set did not contain errors of this kind.

(c) Perhaps far-away clusters look fainter than their measured squares of angular diameters θ^2 would predict. The Copernican viewpoint prevents us from thinking that this could be a real effect.

(d) The remaining possibility is, then, that far-away clusters have been dimmed by a general obscuration of starlight which increases with increasing distance, as more and more obscuration occurs along the line of sight. This interpretation is the one accepted today, and it was the one given by Trumpler.

Astronomers now know that this obscuring material is in the form of small solid specks, dust grains whose chemical composition may be silicates (like sand) or carbon-containing compounds (like graphite or silicon carbide). The obscuration of starlight is believed to arise from a combination of true absorption and scattering, and this combination goes by the general name of **extinction** of starlight. Most of the dust grains are slightly smaller than the wavelength of visual light, since, apart from a general extinction of starlight, interstellar dust also produces a **reddening** of starlight. Figure 11.3 helps to explain the physical basis of this phenomenon. (See also Box 11.1.)

Interstellar reddening occurs because blue light is scattered out of a beam of starlight directed toward us more than red light is. A similar effect occurs in the Earth's atmosphere, because of the scattering of sunlight by atmospheric molecules. The light from the setting Sun (or rising Sun) has to travel through a greater column of air and undergoes more scattering and more

---

BOX 11.1

The Effects of Interstellar Dust on Starlight

(1) *Interstellar extinction*: the general obscuration of background stars through the absorption and scattering of their visible light by intervening dust grains suspended in space.

(2) *Interstellar reddening*: the progressive reddening of starlight by interstellar dust grains because blue light suffers selectively more extinction than red light.

(3) *Interstellar polarization*: the selective elimination of starlight with polarization vectors aligned parallel to the long axes of dust grains. If the long axes of dust grains are aligned perpendicular to the interstellar magnetic field, the remaining starlight tends to be polarized in a direction parallel to the interstellar magnetic field. (See Figure 11.33.)

The scattering of starlight by interstellar grains also leads to reflection effects, whereby we can see interstellar matter by means of reflected starlight. The most obvious examples are the so-called reflection nebulae; but at low-enough light levels, even the darkest of dark clouds show some reflection of starlight.

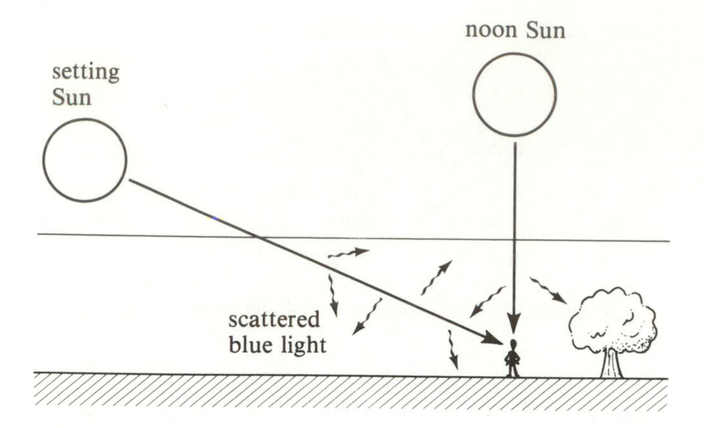

Figure 11.4. Why the setting Sun looks redder than the noonday Sun. A similar phenomenon explains why starlight that has undergone more interstellar extinction also suffers more reddening.

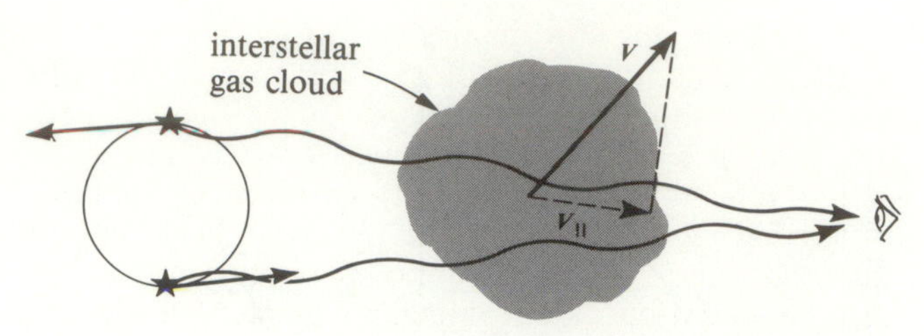

Figure 11.5. The presence of "stationary lines" in the absorption-line spectra of spectroscopic binaries can be explained in terms of an intervening cloud of cold gas. The absorption lines produced by the latter would show a constant Doppler displacement appropriate for the cloud's component of velocity $v_{\|}$ along the line of sight. They would not share in the periodic shifting back and forth of the absorption lines formed in the photospheres of the two orbiting stars.

reddening than light from the noon Sun (Figure 11.4). On a microscopic level (Chapter 3), we understand this result to arise because air molecules interact more strongly with blue light than with red light and, therefore, scatter blue light preferentially out of the sunbeams, leaving mostly reddened sunlight. *Exactly* the same process cannot, however, explain interstellar reddening. Because they are so tiny, individual atoms and molecules are quite inefficient at scattering visual light. For them to produce the observed obscuration of starlight would require totally implausible amounts of interstellar gas. To block light efficiently, the blocking agent must be about the same size as (or larger than) the wavelength of the light to be blocked. For example, a brick blocks visual light quite well, but it hardly hinders long-wavelength radio waves. On the other hand, to redden visual light, the reddening agent must be about the same size as (or smaller than) the wavelength of visual light. Interstellar reddening could not be caused by bricks, because a brick blocks red light as much as blue light. The required combination of blocking efficiency and reddening efficiency implies that interstellar extinction and reddening must be caused by solid dust grains which are slightly smaller than the wavelength of visual light.

The conclusion that interstellar extinction and interstellar reddening are both caused by interstellar dust is, of course, subject to observational check. If the conclusion is valid, there should be a correlation between extinction and reddening; stars which suffer more extinction should also suffer more reddening. Trumpler found that this correlation does exist; and subsequent work has shown the ratio of "total to selective extinction" to be generally constant in the Galaxy (Chapter 9).

The Discovery of Interstellar Gas

Historically, the conclusive evidence for the existence of interstellar gas was presented somewhat earlier than the evidence for interstellar dust. In 1904, Hartmann discovered that a set of absorption lines of once-ionized calcium did not undergo the periodic Doppler shifts of the absorption lines in the spectrum of a spectroscopic binary. In other words, whereas the wavelengths of the absorption lines from the photospheres of the stars in the binary system shifted back and forth as the stars revolved in their orbits, there were also absorption lines whose wavelengths remained stationary. Hartmann concluded that the "stationary lines" arose from absorption produced by a cold interstellar cloud of gas which lay between the binary system and Earth (Figure 11.5). There may be also a fixed displacement of the interstellar absorption line from the position of a similar line produced in a terrestrial laboratory, because the cloud may have a component of velocity $v_{\|}$ along the line of sight. However, the latter velocity does not vary on the timescale which characterizes the orbital motion of the background binary system. The interstellar lines therefore appear "stationary" relative to the changing pattern of lines presented by the spectroscopic binary.

Hartmann's hypothesis of an intervening gas cloud did not gain immediate acceptance, because the absorbing gas might have resided near the star in question rather than a long distance away. This issue of whether the absorbing gas is "circumstellar" or "interstellar" remains difficult to resolve in any one case even today. The "interstellar" interpretation was demonstrated by Plaskett, Struve, Eddington, and Bok, who showed that the ionization stages, or Galactic distribution, or velocities of the "stationary lines" were often incompatible with a "circumstellar" interpretation.

Further insight into the interstellar gas was provided by Beals, Adams, Munch, and Zirin, who found that many stars show multiple interstellar absorption lines, i.e., the same lines, say, of once-ionized calcium at several different Doppler velocities. Moreover, the interstellar lines were relatively narrow in comparison with the absorption lines produced in the photospheres of the stars in question. These facts indicated that the gas producing the interstellar absorption lines must be relatively cold and diffuse, and must come in discrete lumps ("clouds") which move more or less coherently as a unit. This notion, that interstellar space is not empty (a true vacuum) but is filled with moving clouds of gas, remains the central concept underlying most studies of the medium between the stars.

Different Optical Manifestations of Gaseous Nebulae

We now know that the dust and the gas in interstellar space are intimately mixed, and that both generally reside in clouds or complexes of clouds. The ratio of dust mass to gas mass in the Galaxy is about 1 percent. Since the gas itself is only a few percent of the mass of the stars in the Galaxy, interstellar dust constitutes a very minor fraction of the total mass of our Galaxy. Nevertheless, this dust has a great influence on how we see the Galaxy, because it obscures so much starlight in many directions.

Problem 11.1. A rough estimate of the mass fraction of dust in the Galaxy proceeds as follows. Approximate a spherical dust grain of radius R to have a cross-sectional area πR^2 for blocking visual light. (This "shadow" formula breaks down if R is comparable to or smaller than the wavelength of visual light, but it is valid enough for our purposes here.) A photon mean-free-path l is defined to be the length between successive encounters with dust grains (contrast with Problem 9.11). If the number density of dust grains is n, show that

$$l = 1/n\pi R^2.$$

If a beam of photons from a star travels a distance l toward an observer, the beam will suffer an extinction in intensity by a factor of $e = 2.718\ldots$. Extinction observations indicate that at the position of the Sun in the Galaxy, l equals about 3,000 light-years. Given the estimate $R = 10^{-5}$ cm, calculate the value of n in units of cm^{-3}. How many interstellar dust grains would you expect to find in a volume equal to a football stadium? (Assume a typical dimension of 100 meters in all three directions.)

Solid material typically has a mass density of 2 gm/cm^3. Estimate now the mass m of a typical interstellar dust grain. Let V be the volume in which one typically finds one solar mass of stars in the Galaxy. If $V = 300\ (\text{lt-yr})^3$, what is the mass $M = nVm$ of dust grains in this same volume? What, then, is the mass fraction of dust compared to stars at the solar position in the Galaxy? (In actual practice, the average mass fraction is even smaller, because the dust is confined to a layer in the Galaxy which is about three times thinner than the stars.)

The appearance of clouds of gas and dust depends, in part, on the wavelength regime in which they are observed and, in part, on how close they are to neighboring stars. We first explore the different optical appearance of interstellar gas clouds. Such gas clouds are also generally called **gaseous nebulae** (See Box 11.2).

BOX 11.2

Optical Classification of Gaseous Nebulae

(1) *Dark nebulae*: observed by the obscuration of background stars or some other background which is otherwise bright (such as an HII region).
(2) *Reflection nebulae*: observed by the scattered light from embedded stars. The spectrum is the (reflected) absorption-line spectrum of the embedded stars.
(3) HII *regions*: bright ionized regions surrounding newborn hot and bright stars (of spectral type O and B). The spectrum is dominated by emission lines. Thermal radio-continuum emission is found.
(4) *Planetary nebulae*: similar to HII regions, but the exciting object is a very hot evolved star in the throes of death. Planetary nebulae also tend to be denser and more compact than optical HII regions.
(5) *Supernova remnants*: optical emission usually strongest from filaments, whose spectra are dominated by emission lines. In a young supernova remnant like the Crab nebula, an amorphous region which emits continuum light by the synchrotron process may coexist with the filamentary structure. The radio emission from supernova remnants is invariably nonthermal, but the X-ray emission and optical-line emission may arise from thermal processes in a shock-heated gas.

Figure 11.6. The dark globules in this photograph show up particularly well because they lie in front of the extended background of light from the northwest quadrant of the Rosette nebula, a Galactic HII region. (Lick Observatory photograph.)

Dark Nebulae

A gas and dust cloud placed in a rich field of background stars would block most or all of the starlight behind it. We would easily see many stars to the sides of the dust cloud (and a few in front of it); hence such a dark nebulae would manifest itself as one of Herschel's "holes in the sky" (Figure 11.6). Especially interesting among the dark nebulae are the round ones studied by Barnard and by Bok. The regular shapes of the "Bok globules" suggest that these objects are self-gravitating, and led Bok to propose that they are probably sites where new stars are forming. This suggestion is almost certainly correct for the dark clouds in giant complexes, as we will discuss later. Whether isolated Bok globules are also collapsing to form stars is being debated now.

Reflection Nebulae

A gas and dust cloud which surrounds a star or a group of stars can shine by reflected light. This effect was demonstrated observationally by Hubble, and explained theoretically in terms of scattering from dust grains by Russell in 1922 (see Figure 11.7). The reflection nebula

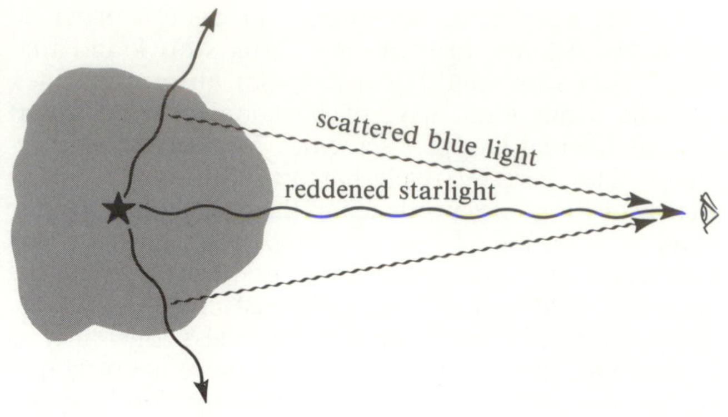

Figure 11.7. Why a reflection nebula looks bluer than its illuminating star. Compare this diagram with Color Plate 4.

tends to look bluer than the direct light from the star because blue light is scattered more than red light by interstellar dust grains (see Color Plate 4). The same reasoning explains why the daytime sky looks bluer than the Sun. When we look at the daytime sky, we are looking at scattered sunlight; so the sky looks blue, although it is illuminated by a yellow sun.

Problem 11.2. If you were to take a spectrum of a reflection nebula, would you see absorption lines, emission lines, or no spectral lines? How would this help to show that the illumination is by reflection from the central star?

Thermal Emission Nebulae: HII *Regions*

Hydrogen atoms in an interstellar gas cloud located near a very hot star—say, a newborn O or B star—will be exposed to the copious outpouring of ultraviolet radiation from such a star. Ultraviolet photons with energies greater than the Lyman limit (see Chapter 3) can ionize the hydrogen atoms. The part of an interstellar cloud where hydrogen has been once-ionized (the maximum possible for hydrogen atoms) is called an **HII region**. The roman numeral II distinguishes once-ionized hydrogen, HII, from neutral atomic hydrogen, HI. Bengt Stromgren showed that the division between HII regions and HI regions will be quite sharp if the surrounding gas cloud is so massive that all the ultraviolet photons from the central O or B star are used up before the HII region can encompass the entire cloud (Figure 11.8). Inside the HII region, the hydrogen plasma is constantly trying to recombine to form neutral hydrogen atoms, but the plasma is kept almost completely ionized by the con-

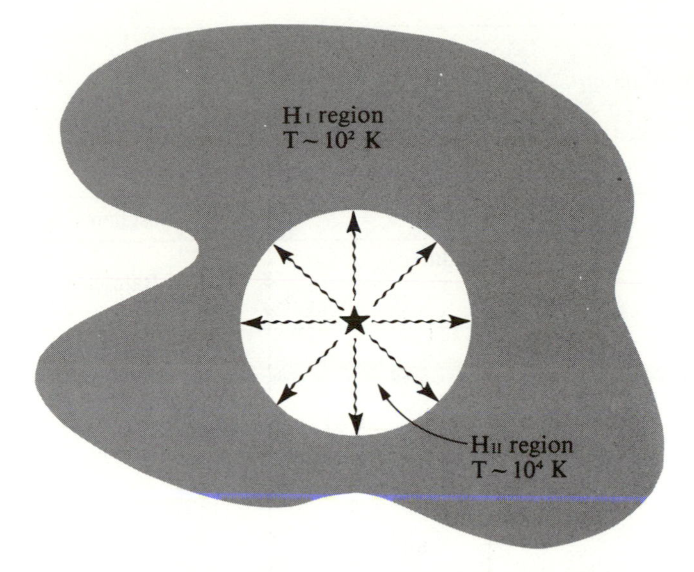

Figure 11.8. A hot and bright star embedded in a cloud of cold atomic hydrogen gas (HI) or cold molecular hydrogen gas (H_2) will ionize a roughly spherical region around itself (HII region). The size of this HII region under equilibrium conditions is given by Stromgren's requirement that the total number of recombinations occurring inside the HII region per unit time equal the total number of ionizing photons emitted per unit time by the central star.

tinued outpouring of ultraviolet photons from the central source. These ultraviolet photons break apart any newly formed hydrogen atoms, and the ions and electrons so formed keep recombining to form new atoms. The part of the HII region where the ultraviolet output of the central star is able to keep a balance between recombination and ionization is called the **Stromgren sphere**.

Problem 11.3. To calculate the size of the Stromgren sphere, idealize the problem by considering a pure hydrogen gas of uniform density which surrounds a single hot star. Let N_* be the number of ultraviolet photons beyond the Lyman limit which leave per unit time from the central star. Assume that each such photon will ultimately ionize one and only one hydrogen atom. Let $\mathscr{R}$ be the number of recombinations of protons and electrons into hydrogen atoms per unit volume per unit time. In a steady state, the total number of recombinations in the Stromgren sphere of radius r must balance the total number of ionizations:

$$\mathscr{R}(4\pi r^3/3) = N_*.$$

Given $\mathscr{R}$ and N_*, this equation would allow us to find r. To obtain $\mathscr{R}$, let us note that recombination at interstellar densities is a two-body process (involving for each recombination one proton and one electron). Thus the number of recombinations per unit volume $\mathscr{R}$ must be proportional to the product of the number densities of protons and electrons, $n_p n_e$. The proportionality factor is denoted by α, and it is called the "recombination coefficient." Thus,

$$\mathscr{R} = \alpha n_p n_e = \alpha n_e^2,$$

where we require $n_p = n_e$ for overall charge neutrality. Show now that the Stromgren radius r is given by the formula

$$r = (3N_*/4\pi\alpha n_e^2)^{1/3}.$$

The "recombination coefficient" α is a function of the temperature of the hydrogen plasma. For temperatures characteristic of Galactic HII regions, α has the approximate value 3×10^{-13} cm³ sec⁻¹. Assume $n_e = 10$ cm^{-3}; compute r when $N_* = 3 \times 10^{49}$ sec^{-1} (O5 V star), $N_* = 4 \times 10^{46}$ sec^{-1} (BO V star), and $N_* = 1 \times 10^{39}$ sec^{-1} (G2 V star). Convert your answer to light-years. What types of main-sequence stars have appreciable HII regions?

Recombination leads to a process whereby a gas cloud surrounding a hot bright star, or a group of such stars, can shine in the visual part of the spectrum. This mechanism, **fluorescence**, works both in planetary nebulae (see Color Plate 8), where a recently ejected circumstellar gas shell is lit up by a hot dying star, and in HII regions (see Color Plate 6), where a neighboring gas cloud is lit up by one or more newborn O or B stars. The modern theory of nebular emission lines originated with Zanstra and Menzel and Baker. To see the essential idea, let us consider an HII region.

We have already stressed that a balanced process of (photo) ionization and (radiative) recombination occurs inside HII regions. The radiative recombination of a proton and an electron will usually form an excited state of neutral hydrogen, which subsequently cascades to the ground state by the emission of additional photons (see Figure 3.20). Because gaseous nebulae are so rarefied and the radiative processes are so efficient, the vast majority of hydrogen atoms are in the ground electronic state. If the nebula contains enough hydrogen, each Lyman-continuum photon from the star will ultimately

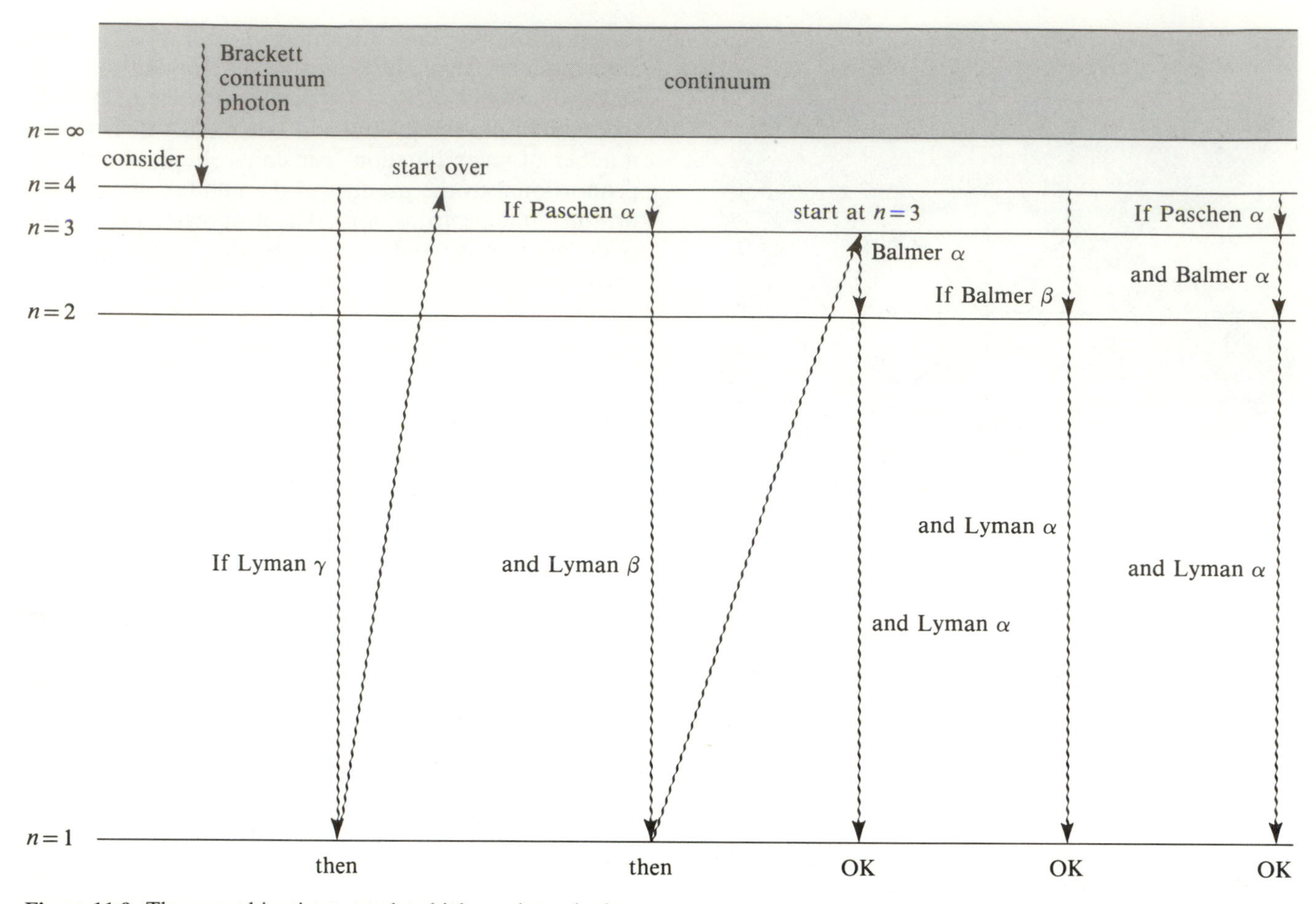

Figure 11.9. The recombination cascade which produces hydrogen emission lines in HII regions and planetary nebulae can take many possible paths. But in a nebula which is optically thick to the Lyman transitions, all such paths ultimately produce one Balmer photon plus one Lyman α photon plus, perhaps, lower-energy continuum and line photons (the paths marked OK).

be degraded into a Lyman α-line photon, plus a Balmer-line photon, plus perhaps lower-energy photons.

To see this, consider a stellar ultraviolet photon which ionizes a hydrogen atom. The ejected electron eventually recombines with another proton, and cascades to the ground state.

As an example, consider the capture of the free electron to the principal quantum number $n = 4$ (Chapter 3). This capture, which releases a "free-bound" continuum photon, can subsequently cause the emission of several line photons. The possibilities are: (a) a Lyman γ photon; or (b) a Paschen α photon plus a Lyman β photon; or (c) a Balmer β photon plus a Lyman α photon; or (d) a Paschen α photon plus a Balmer α photon plus a Lyman α photon (Figure 11.9). In cases (c) and (d), we have the desired creation of a Balmer-line photon and a Lyman α photon (plus perhaps one Paschen-line photon). Since the vast majority of the hydrogen atoms are in the ground electronic state, the Lyman α photon will scatter resonantly from these atoms and will have to "random walk" its way out of the HII region. The Balmer photon is free to escape from the nebula and fly directly toward an optical astronomer. In cases (a) and (b), we have the creation of Lyman-line photons with greater energies than Lyman α, because the final jump to the ground state bypassed the level $n = 2$. Nevertheless, subsequent interactions of the Lyman β and Lyman γ photons will lead to the desired degradation.

Consider the emitted Lyman β photon. It will soon be absorbed by a hydrogen atom in the ground electronic state. The hydrogen atom so excited to $n = 3$ will either (1) jump immediately to the ground level, emitting another Lyman β photon, in which case we start again, or (2) cascade to the ground level, emitting a Balmer α photon plus a Lyman α photon, and we will have the same net result as in case (d). Similarly, we may easily deduce that the emitted Lyman γ photon will result finally in either case (c) or case (d). A single Lyman continuum photon therefore always transforms to an optically observable Balmer-line photon plus a Lyman α photon plus continuum or lower-energy line photons. Although the details of this fluorescence process are complicated, the entire chain of degradation of high-quality photons to low-quality ones is, of course, another example of the second law of thermodynamics in action.

The result that the total number of Balmer-line photons created by the recombination process equals the total number of stellar ultraviolet photons beyond the Lyman limit finds an important application in planetary nebulae. Zanstra pointed out that an observational measurement of the flux of Balmer-line photons then allows an estimate of the stellar ultraviolet flux. The latter is not directly measurable, because of the intervening absorption of the interstellar medium and the Earth's atmosphere. Measuring the visual continuum flux then allows the optical astronomer to calculate the ultraviolet "color" of the central star and, therefore, its approximate surface temperature. The "Zanstra" temperature of the central stars of planetary nebulae calculated thus is typically several tens of thousands of degrees. These very high surface temperatures are consistent with the idea that the central stars of planetary nebulae represent the exposed cores of highly evolved stars (Chapters 8 and 9).

The theory of the recombination-cascade process is by now very well-developed, with important modern contributions made by Brocklehurst, Mathis, O'Dell, Osterbrock, Pottasch, and Seaton. In particular, Miller has shown that long-standing discrepancies between theory and observations resulted from observational error. There is now excellent agreement about the relative intensities of the various recombination lines of hydrogen emitted by HII regions and planetary nebulae. Deviations from the expected theoretical ratios (the "Balmer decrement") in these objects can usually be explained in terms of reddening by interstellar dust.

The balance of ionization and recombination in HII regions leads also to a heating of the hydrogen plasma, which in turn yields additional radiation processes. The heating mechanism works as follows. In general, the central star will have a distribution of ultraviolet photons with different energies beyond the Lyman limit. When a neutral hydrogen atom in the ground electronic state absorbs a photon with energy greater than the Lyman limit, the liberated electron will carry off the excess in the form of kinetic energy (see Figure 3.20). For a hot central star, this excess energy is appreciably greater than the thermal energies of the gas particles in the HII region; so, upon subsequent collisions with such particles, the newly liberated electron will tend to share its kinetic energy more equitably with the other particles. This leads in net to a heating of the HII region. In a steady state, this heat input is balanced by the following heat output.

The thermal distributions of electrons and ions suffer continual collisions. A fraction of such collisions are inelastic and raise an incompletely ionized atom—say, singly or doubly ionized oxygen—to an excited electronic state. Before the atom has a chance to give back its excitation energy in a superelastic collision with another particle, the atom may deexcite radiatively by the emission of a photon. Because of the rarefied state of the gas and the low fractional abundance of oxygen, this line photon may be free to leave the system. The loss of its energy constitutes a net cooling of the region (Figure 11.10). The balance of heating and cooling leads to an equilibrium temperature of about 10^4 K for Galactic HII regions. Such a temperature is two orders of magnitude greater than the roughly 10^2 K characteristic of HI regions which do not have access to the ionizing radiation of hot bright stars (see Figure 11.8).

The cooling radiation from an HII region (or a planetary nebula) is also interesting because it results in emission lines which have not been seen in terrestrial laboratories! We have already mentioned that gaseous nebulae are so rarefied that almost all atoms are in their ground electronic states. Consider now the collisional excitation of atoms from their ground levels (Figure 11.10). At a temperature of about 10^4 K, most of the collisions between thermal particles are too weak to raise atoms to an excited level with more than a small energy difference from the ground state. Atomic hydrogen has no such easily accessible excited states, but atoms such as singly ionized nitrogen, oxygen, sulpher, doubly ionized oxygen, or neon do. The accessible low-energy states of these relatively abundant atoms are **metastable** levels, meaning that the atom has no "permitted" radiative transitions whereby it can quickly emit a photon and drop to a lower level if it is left alone for long enough, say, an hour. Under normal terrestrial conditions, such an atom would be jostled by collisions with other atoms every nanosecond or so; consequently, it would tend to deexcite collisionally before it had a chance to deexcite radiatively. Lowering the pressure drastically in a laboratory would enhance the chances for any *particular* atom to deexcite radiatively, but the total number of radiative deexcitations would in fact go down, because there would be fewer atoms in the laboratory container. In the very extended conditions of interstellar space, however, there are so many atoms that even a low rate of radiative deexcitations can lead to detectable emission of photons. Under rarefied-enough conditions, every upward transition by an inelastic collision is followed by the emission of a "forbidden" photon, and the "forbidden" lines can become as strong as or stronger than the "permitted" lines (Problem 11.4). A famous example is

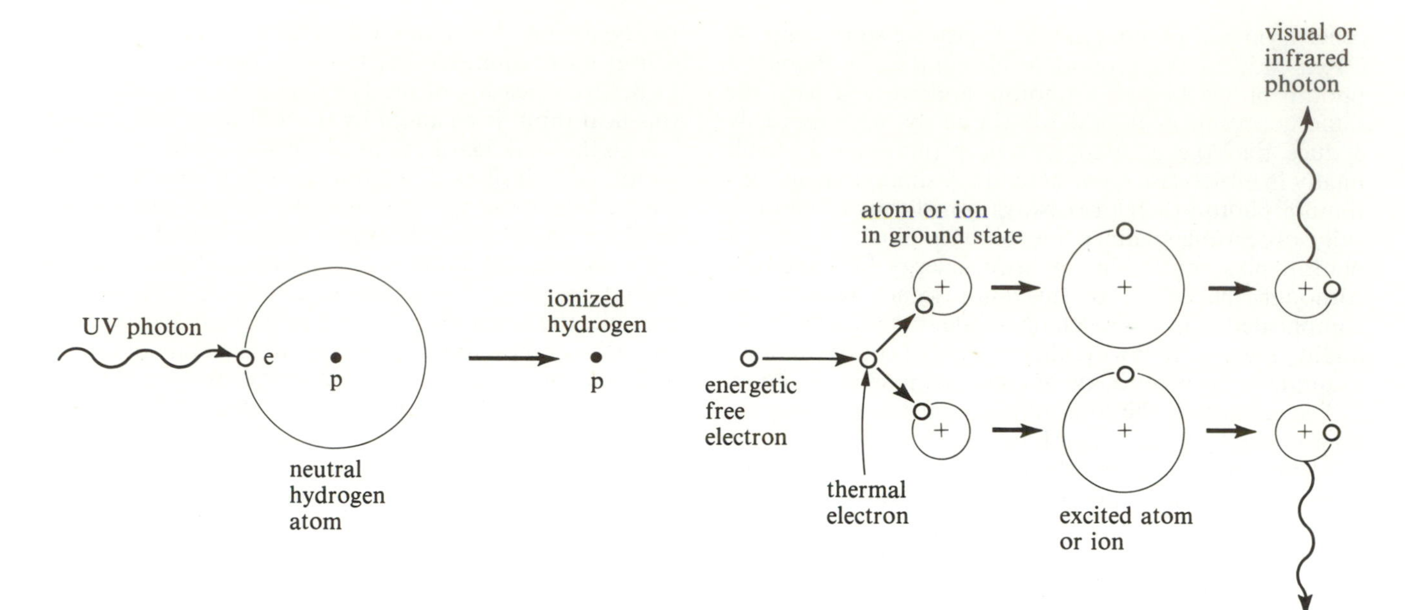

Figure 11.10. The mechanism of heating HII regions and its balance by radiative cooling. An ultraviolet photon energetically ejects an electron from a neutral hydrogen atom (one of the few which have recombined in an HII region). This energetic free electron collides subsequently with other free electrons (or ions or neutrals), and shares its kinetic energy more equitably with them. This constitutes a net source of heat for the HII region, with the ultimate energy reservoir being the ultraviolet photons from the central star. The heat input is balanced by the following radiative output. Thermal electrons collide with ions or atoms in their ground energy levels, and excite them to higher levels. The kinetic energy lost in such inelastic collisions is radiated away when the excited atoms deexcite radiatively. The emitted visual or infrared photon is usually free to escape from the nebula, because of the extreme rarefaction of interstellar gas clouds. The equilibrium temperature which results from a balance of heating and cooling in HII regions is generally somewhat less than 10^4 K.

the pair of green emission lines at wavelengths of 4959 and 5007 Å (Figure 11.11). For a long time, astronomers were puzzled by the ubiquitous appearance of this pair at strengths comparable to those of the hydrogen lines in the spectra of bright nebulae. No identification of the lines could be made in terms of known terrestrial sources. It was suggested that perhaps the lines were due to an unknown element peculiar to interstellar nebulae; "nebulium" was proposed as its name. (This history parallels the naming of "helium," which was first discovered in the flash spectrum of the solar chromosphere; see Chapter 5.) Bowen provided in 1927 the correct explanation for the nebular green lines, namely, that they arose from "forbidden transitions" of doubly ionized oxygen for the reasons given in our discussion above. Today, astronomers distinguish "forbidden" transitions from "permitted" ones by putting square brackets around the responsible atom; so the nebular green lines are denoted by [OIII] 4959 and [OIII] 5007.

Problem 11.4. The number of Balmer-line photons emitted per unit time in an HII region is equal to the total number of hydrogen recombinations. (Each recombination produces one Balmer photon, according to Figure 11.9.) Show now that the luminosity in all the Balmer lines is proportional to

$$L(H) \propto V\alpha(T)n_p n_e,$$

where V is the volume of the HII region, $\alpha(T)$ is the recombination coefficient (see Problem 11.3), n_p is the proton number density, and n_e is the electron density.

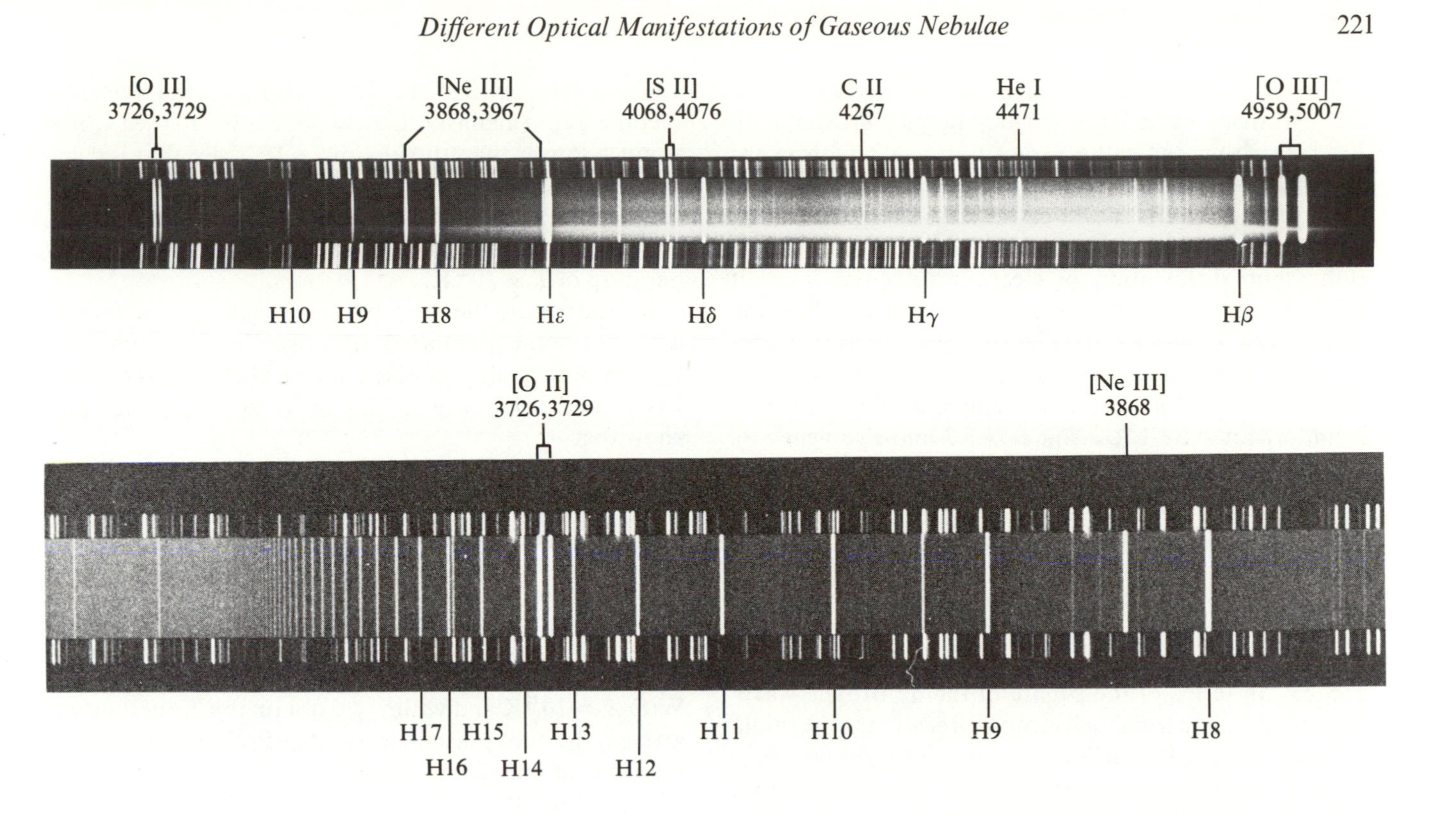

Figure 11.11. Two photographic spectrograms of the Orion nebula, a Galactic HII region whose emission is dominated by bright lines. (a) The top spectrum shows that the forbidden lines of OIII, OII, and NeIII can reach intensities comparable to or larger than the permitted lines of H and HeI. The hydrogen lines in the visible part of the spectrum (the Balmer lines) are denoted by the convention discussed in the text; thus, Hε can also be called H7, because it arises from the transition n = 7 to n = 2. The slight curvature of the lines is caused by the use of a prismatic spectrograph. (Lick Observatory photograph, courtesy of George Herbig.) (b) The bottom spectrum extends to shorter wavelengths (given in Ångströms) than the top one, although there is some overlap, e.g., the lines of [OII] and [NeIII]. Notice the very large number of hydrogen lines that decrease in intensity toward shorter wavelengths (the Balmer decrement) until they end in a bright Balmer continuum to the left. The hydrogen emission arises as a result of a recombination cascade. (Courtesy of Lawrence Aller.)

Consider now collisionally excited transitions between an upper and lower level of an atom (e.g., the "nebular" transitions of OIII lumped together for simplicity), where state 2 is separated from state 1 by energy E_{21}. Let us calculate the number density n_2 of atoms in state 2 if we know the number density n_1 in state 1 (assumed to be the ground level). If the medium remains optically thin to photons of energy E_{21}, upward transitions from 1 to 2 take place via inelastic collisions with thermal electrons at a rate per unit volume of $n_e n_1 \gamma_{12}(T)$, where $\gamma_{12}(T)$ is the collisional excitation coefficient when the electron velocity distribution is characterized by a temperature T. Downward transitions from 2 to 1 can take place either by the spontaneous radiation of photons of energy E_{21} or by superelastic collisions with thermal electrons. The total downward rate per unit volume is given by $n_2 A_{21} + n_e n_2 \gamma_{21}(T)$, where A_{21} is Einstein's coefficient for spontaneous emission and $\gamma_{21}(T)$ is the collisional de-excitation coefficient. If we ignore transitions from higher states (say, level 3) into and from levels 1 and 2 (to keep the discussion simple), show that the steady-state population levels in 1 and 2 satisfy

$$n_2 A_{21} + n_e n_2 \gamma_{21}(T) = n_e n_1 \gamma_{12}(T).$$

Upward collisional transitions have an energy threshold: electrons must have kinetic energies greater than E_{21} before they can cause a transition $1 \to 2$. Downward collisional transitions have no such threshold. Thus, the fraction of electrons which contributes to γ_{12} is smaller than that which contributes to γ_{21}. For a Maxwell-Boltzmann distribution of electron velocities (consult equation 4.1), the fraction is given by the Boltzmann factor $\exp(-E_{21}/kT)$. In addition, the transition rates γ_{12} and γ_{21} will differ by a factor equal to the number of equivalent quantum states of the final level: g_2 for state 2, and g_1 for state 1 (consult Box 3.3 and Problem 3.9). We therefore derive the relationship

$$\gamma_{12}/\gamma_{21} = (g_2/g_1)\exp(-E_{21}/kT),$$

where γ_{21} will have a value given approximately by the product of the electron thermal speed $(kT/m_e)^{1/2}$ and a typical superelastic cross section $\sigma_{21} \sim \hbar^2/m_e kT$. Interpret σ_{21} in terms of the square of the de Broglie wavelength of a thermal electron. (Recall the formula $\lambda_e = h/p_e$ from Chapter 3.) Assume $\gamma_{21}(T)$ can be calculated in this or a more accurate fashion, and show that n_2 can now be obtained in terms of n_1 via the formula

$$n_2 = n_1 \frac{n_e\gamma_{21}(T)}{A_{21} + n_e\gamma_{21}(T)} \frac{g_2}{g_1} e^{-E_{21}/kT}.$$

What condition does n_e have to satisfy in order for Boltzmann's law,

$$n_2 = n_1(g_2/g_1)\exp(-E_{21}/kT),$$

to apply? Interpret your answer physically.

If the forbidden-line atoms occupy the entire volume V of the HII region (or planetary nebula), the luminosity $L(E_{21})$ of photons with energy E_{21} must be proportional to the total number of downward radiative transitions per unit time, $n_2 A_{21} V$, if we assume optically thin conditions. Thus, show

$$L(E_{21}) \propto (n_1 V) \frac{g_2}{g_1} e^{-E_{21}/kT} \left[\frac{n_e A_{21}\gamma_{21}(T)}{A_{21} + n_e\gamma_{21}(T)}\right].$$

For fixed T, sketch schematically the behavior of $L(E_{21})$ and $L(H)$ as functions of n_e in a thought experiment where we compress a fixed mass of material. (V decreases with $n_e V =$ constant and $n_1 V \cong$ constant.) In particular, indicate the qualitative change in behavior of $L(E_{21})$ when n_e increases beyond the critical value A_{21}/γ_{21}. For "forbidden" optical transitions, A_{21} might have a typical value of $A_{21} \sim 10^{-2}$ sec^{-1}, implying a characteristic lifetime $A_{21}{}^{-1}$ against radiative decay of $\sim 10^2$ sec, which is much longer than the lifetime $\sim 10^{-8}$ sec that is characteristic of excited states with permitted transitions. Assume $\gamma_{21} \sim \hbar^2 m_e{}^{-3/2}(kT)^{-1/2}$ and $A_{21} \sim 10^{-2}$ sec^{-1}, and calculate the critical density $n_{cr} = A_{21}/\gamma_{21}$ for a temperature of $T = 10^4$ K.

Consider now the ratio of the strengths of the forbidden lines to the (permitted) hydrogen lines. Assume that the proportionality coefficients in the expressions for $L(E_{21})$ and $L(H)$ have similar order of magnitudes, and show that

$$\frac{L(E_{21})}{L(H)} \sim \frac{n_1 g_2}{n_p g_1} e^{-E_{21}/kT} \left(\frac{\gamma_{21}/\alpha}{1 + n_e/n_{cr}}\right).$$

For a heavy element like oxygen (OIII), $n_1 g_2/n_p g_1$ might typically be $\sim 10^{-3}$, whereas $E_{21} = hc/\lambda_{21}$ with $\lambda_{21} =$ 5000 Å for "forbidden transitions" in the optical regime. With $T = 10^4$ K and with γ_{21}/α obtainable from the data given in the above discussion and in Problem 11.3, graph the ratio of line strengths for various values of n_e from 10^3 cm^{-3} to 10^8 cm^{-3}. For what $n_e = n_w$ do forbidden lines become considerably weaker than the hydrogen lines?

Because the nebular "forbidden" lines are excited primarily by collisions with thermal electrons, the observed line strengths can be used to calculate the densities and temperatures of HII regions as well as their chemical compositions. Sophisticated methods for these purposes have been devised by Aller, Ambartsumian, Osterbrock, Seaton, and others. The actual analyses are too complicated to discuss here; let us merely summarize the principal findings.

The densities of optical HII regions range from 10 to 10^3 particles per cm^3. Lower-density HII regions do exist, but the low emission of "diffuse" HII regions make them difficult to observe. Higher-density HII regions have also been inferred to exist, but the large obscuration by the dust which usually accompanies "compact" HII regions makes them also difficult to observe optically. The temperatures of Galactic HII regions have been inferred to be generally somewhat less than 10^4 K. These values are in reasonably good agreement with refined treatments of the theory of heating and cooling that we discussed earlier. The chemical compositions of Galactic HII regions turn out to be similar to that of Population I stars. This is a welcome finding, inasmuch as all evidence indicates that the most newborn stars are formed out of gas with the same chemical composition as that which now surrounds the O and B stars as HII regions.

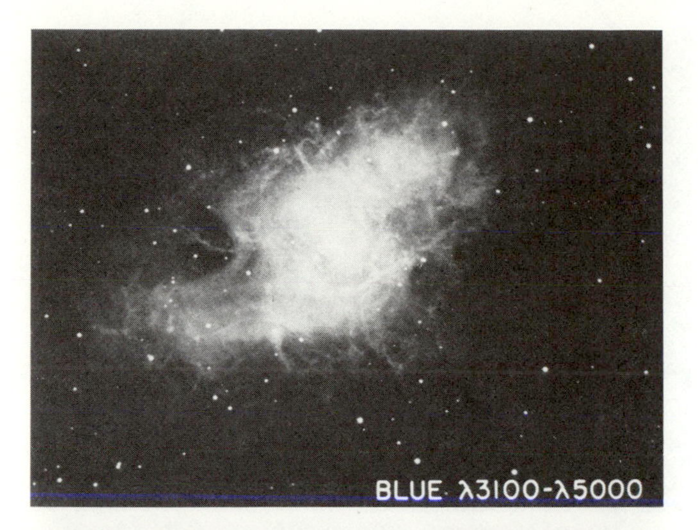

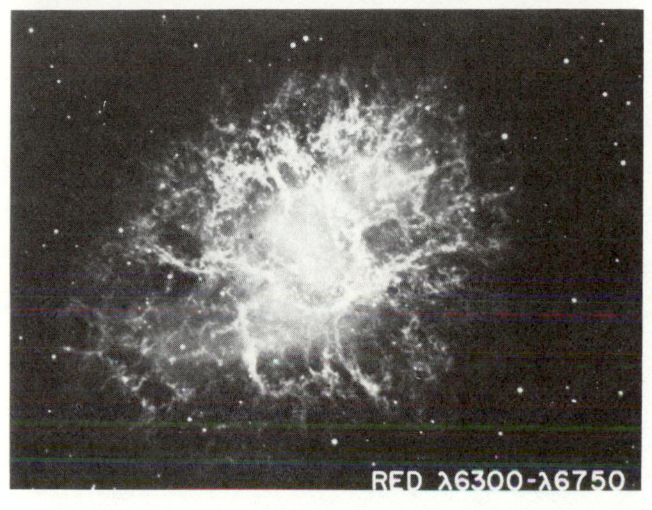

Figure 11.12. (a) Four photographs of the Crab nebula in blue, yellow, red, and infrared light. The range of wavelengths in Ångströms transmitted by the filter is indicated in the lower right-hand corner of each frame. Notice that the filaments show up especially prominently in the red photograph because of the intense emission in the Balmer alpha lines of hydrogen (Hα) from the filaments. (Mt. Wilson and Las Campanas Observatories, Carnegie Institution of Washington.)

Nonthermal Emission Nebulae: Supernova Remnants

In 1844 Lord Rosse discovered the most fascinating emission nebula in the Galaxy. This nebula is located toward the constellation of Taurus, but it has the shape of a crab (Color Plate 7); hence it has become known as the Crab nebula. In Chapter 7 we mentioned that the Crab nebula is known to be a supernova remnant. We will now retrace the research that led to this important conclusion.

The light from the Crab nebula arises from two different sources: there is an amorphous part that is now known to be synchrotron radiation (Chapter 7), and a filamentary part. Spectra of the luminous filaments show them to be dominated by emission lines similar to those found in HII regions; however, the mechanisms of excitation in these filaments are probably quite different from those in HII regions. Examination of the Doppler shifts of the emission lines and the sequential positions (proper motions) of the filaments in photographs taken over a course of decades led Duncan to conclude in 1921

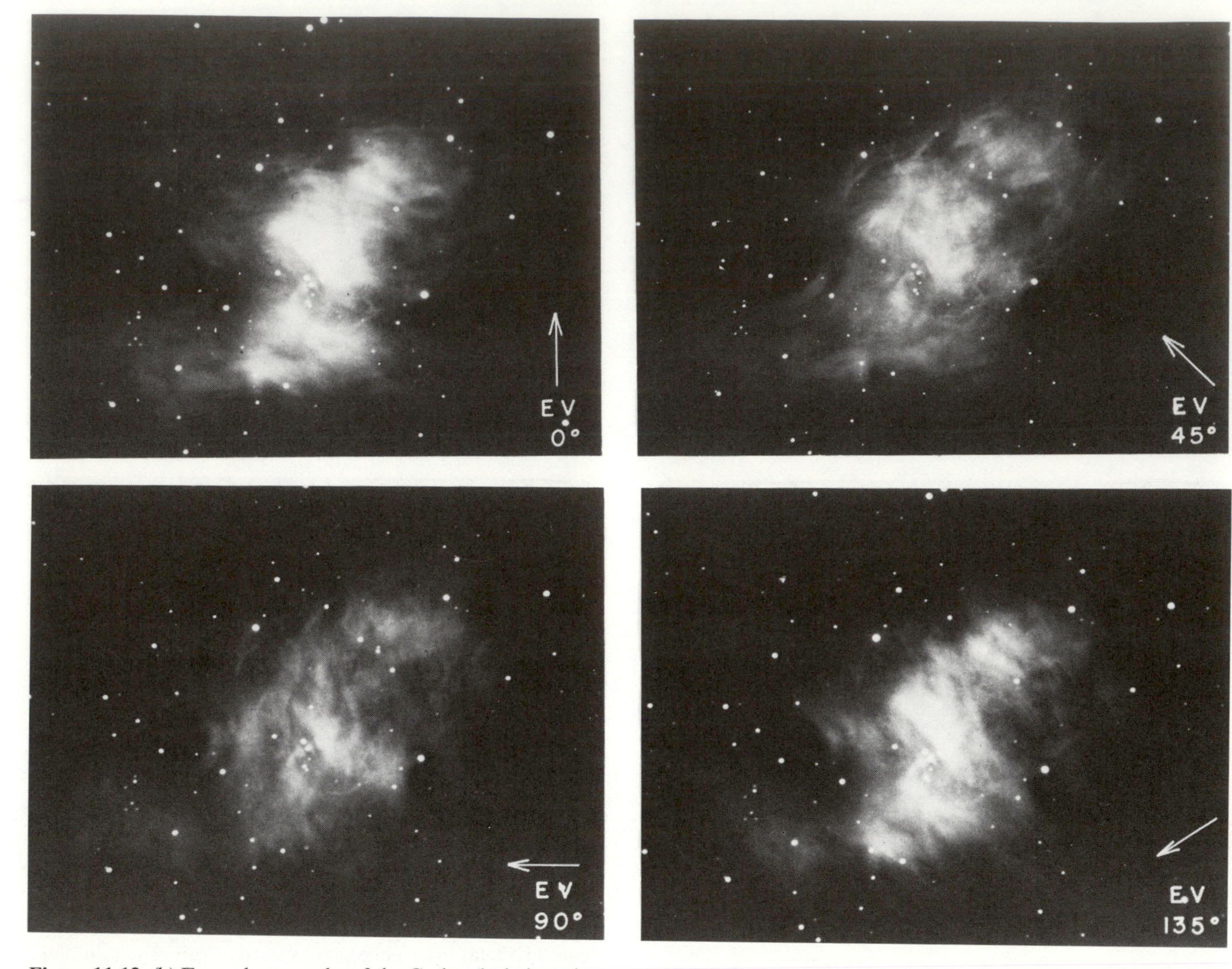

Figure 11.12. (b) Four photographs of the Crab nebula in polarized light. The direction of the polarization vector is indicated by the arrow in the lower right-hand corner of each frame. The very different appearance of the Crab nebula in different polarizations attests to the strong synchrotron emission from the amorphous regions. (Palomar Observatory, California Institute of Technology.)

that the filaments are expanding in all directions from a faint blue object at the center of the nebula. The rate of expansion was consistent with the hypothesis that the central object had exploded nine centuries earlier. (Strictly speaking, the light from the explosion would have reached the Earth about nine centuries earlier.) This conclusion led to a search of the historical records to find out whether a bright "new" star had appeared in the constellation of Taurus during the eleventh century. No such an event had been recorded in the relatively backward Western civilizations, but Japanese and Chinese astronomers had recorded a bright "new" star—visible in daylight—which appeared in Taurus on July 4, 1054, and did not fade from view until April 17, 1056. In 1942 Mayall and Oort published definitive evidence that the Crab nebula constitutes the shreds of a star whose explosion was seen in 1054 as one of the brightest supernovae on record.

Several other filamentary nebulae have now been identified with historical supernovae. Examination of the chemical compositions of the filaments have revealed that they are often enriched in heavy elements. This finding supports the idea that the filaments are basically the ejecta from a highly evolved star in which consider-

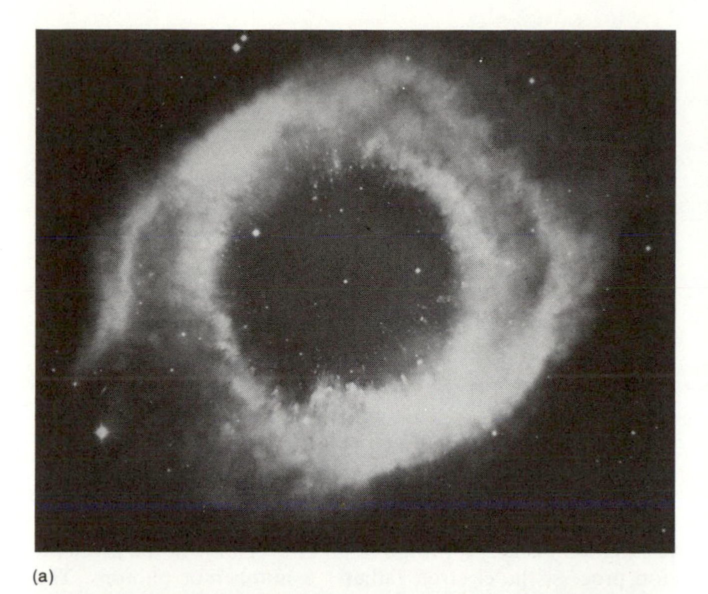

(a)

(b)

(c)

Figure 11.13. Comparison of the optical appearance of some emission nebulae. (a) A planetary nebula, NGC 7293. (b) An H II region, the Rosette nebula. (c) An old supernova remnant, the Cygnus loop or the Veil nebula. Supernova remnants emit nonthermal radio continuum radiation, but the optical emission from the filaments probably arises by thermal processes. The central cavities seen in all three of the objects above probably arose from energetic interactions with the central star(s) that exist (or had existed) in the middle. (Palomar Observatory, California Institute of Technology.)

able nuclear processing has occurred (Chapter 8). It also supports the view that most of the enrichment of the interstellar medium in heavy elements comes from supernova explosions.

As interesting as the luminous filaments in the Crab nebula are, in fact most of the light of the Crab comes from the relatively featureless nebulosity. The amorphous part of the nebula has a continuous spectrum, and the light of this continuum is strongly polarized. Polarization of starlight of several percent could be produced by aligned interstellar dust grains, as we will soon discuss, but the optical polarization in the Crab nebula was much too strong to be explained by such a mechanism (Figure 11.12). Moreover, polarization of the Crab's radio emission was found shortly afterward, and interstellar dust cannot possibly affect radio waves. Alfven and Herlofson, Kipenheuer, and Ginzburg made the correct suggestion that the radio emission arises by the synchrotron process (see Figure 7.7); and Shklovskii boldly extended this explanation to the eerie radiation in the optical continuum. The polarization vector of most of the synchrotron radiation should be perpendicular to the direction of the local magnetic field. Chapter 7 explained that the ultimate source of the radiating high-energy electrons is now believed to be the faint blue object at the center of the Crab nebula's expansion, the Crab pulsar. To emphasize the very different radiation mechanisms at work, astronomers refer to H II regions as *thermal* emission nebulae and to supernova remnants as *nonthermal* emission nebulae (Figure 11.13).

Radio Manifestations of Gaseous Nebulae

We mentioned in passing that supernova remnants possess nonthermal radio continuum emission (synchrotron radiation). Do HII regions also emit radio waves? Yes, both continuum and line radio waves. A thermal origin underlies both. We first consider the origin of the line radiation.

Radio Recombination Lines

Van de Hulst and Kardashev were the first to point out that the same recombination-cascade process which leads to hydrogen emission lines in the optical could result, for the high n states, in hydrogen emission lines in the radio. Consider, for example, the capture of an electron by a proton in a bound state represented by a very high value of the principal quantum number n. As this electron cascades to the ground state, it may pass through the level $n = 109$ by making a radiative transition from $n = 110$ to $n = 109$. The result would be the emission of a hydrogen 109α photon, a photon whose wavelength falls in the radio part of the electromagnetic spectrum (Problem 11.5). Many other radio recombination lines of hydrogen are, of course, also possible; the strongest of these tend to be the "alpha" member of the series, because the probability is highest for the cascade to proceed from level $n + 1$ to level n.

Goldberg pointed out that similar radio recombination lines should be present for atoms other than hydrogen. The radio recombination lines of hydrogen, helium, and carbon have since been observed by many radio astronomers. The observational importance of such lines for the study of Galactic HII regions cannot be exaggerated. Because the emission is in line radiation and not in continuum, a Doppler-shift measurement is possible and will give velocity information. Moreover, because of the long wavelengths of the radio-line emissions, they are unaffected by interstellar dust. This implies that radio recombination lines can be used to explore compact HII regions as well as to probe the Galaxy-wide distribution of HII regions, as Mezger and his coworkers have done, in a manner impossible at optical wavelengths because of obscuring by interstellar dust.

Problem 11.5. Consult Problem 3.7 and calculate the wavelength of the hydrogen 109α transition.

Thermal Radio Continuum Emission

Because the hydrogen plasma of an HII region is hot, it can emit continuum radiation. If the nebula were opaque to this emitted radiation, then only the radiation from the surface of the nebula could freely escape, and we would expect a Planck spectrum for this emergent radiation (Chapter 4). In fact, except at the very longest radio wavelengths, a typical HII region is transparent ("optically thin") to its internally generated thermal continuum radiation; so the observed spectrum has the characteristics of the local microscopic emission process, which in general is non-Planckian.

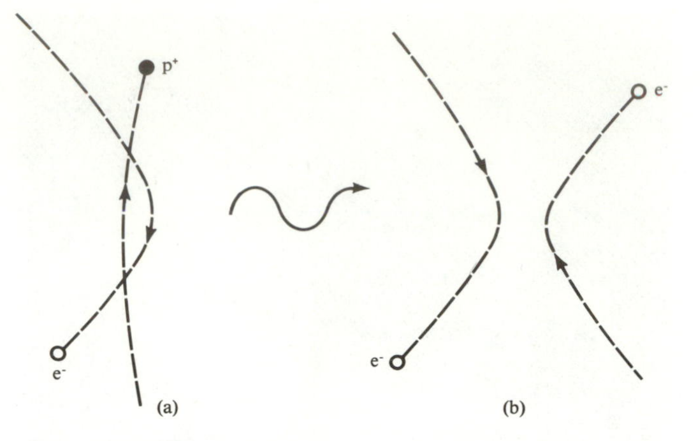

Figure 11.14. The microscopic mechanism responsible for producing thermal radio continuum (free-free emission). (a) A free electron is momentarily deflected by the electric attraction of a passing ion (usually a proton in an HII region). In the acceleration process, the electron radiates a number of photons. The energy carried off by photons of different frequencies (speed of light divided by wavelength) is a pretty flat distribution at radio frequencies, because the encounter time is short compared to the period of vibration of radio waves. (b) When two free electrons deflect one another by their mutual electric repulsion, no net radiation emerges. The oscillatory parts of the electric field associated with accelerating electrons (see Fig. 2.9) exactly cancel when we have two electrons possessing equal but opposite motions.

The detailed microscopic process responsible for generating thermal radio continuum emission in HII regions is called the "free-free" mechanism, and its principles are illustrated in Figure 11.14. Electrons at a temperature of about 10^4 K randomly encounter protons or other ions in the HII region. The Coulomb attraction between the charged particles accelerates the electrons briefly during such encounters, but the electron need not be captured by the proton or ion. (The transition is from one "free' state to another "free" state.) Accelerated electrons produce electromagnetic radiation (Chapter 3); in particular, the "free-free" mechanism in HII regions produces radio continuum emission. The typical duration of the encounter is quite short; so the intensity of thermal radio continuum emission as a function of wavelength is quite flat (Figure 11.15). This spectral

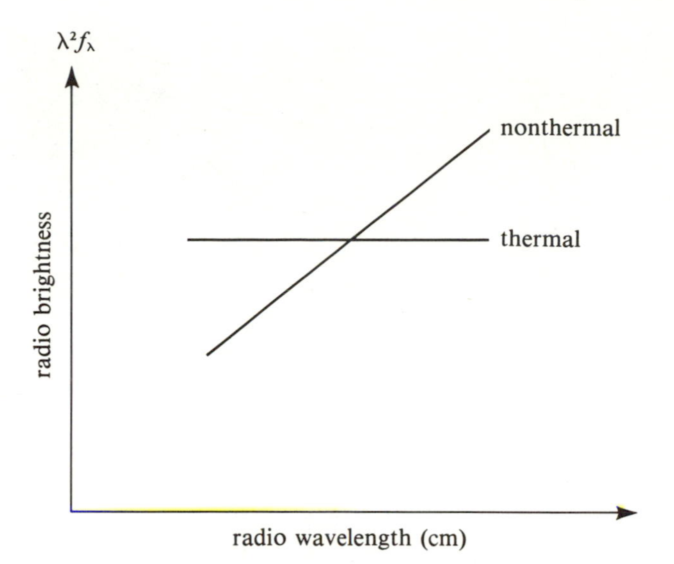

Figure 11.15. Contrast between the spectral distribution of the continuum radio emission from a thermal source and a nonthermal source. The vertical axis for radio brightness is labeled $\lambda^2 f_\lambda$ because this quantity is traditionally given per unit frequency instead of per unit wavelength.

characteristic of thermal radio continuum emission makes it easily distinguishable from nonthermal radio continuum emission, which characteristically increases in intensity with increasing wavelength. Another distinguishing attribute of thermal emission is its lack of intrinsic polarization.

Problem 11.6. Consider the deflection of an electron by a proton if they pass one another at a distance r at closest approach. A large deflection can be considered to have taken place if the Coulomb potential energy at closest approach is comparable to the typical thermal energy per particle in the HII region:

$$e^2/r = kT,$$

where e is the magnitude of the charges of an electron and a proton, k is Boltzmann's constant, and T is the temperature. Let m_e be the mass of the electron, and show that the typical duration of a close encounter between a thermal electron and a proton is given by

$$t = r(kT/m_e)^{-1/2} = m_e{}^{1/2}e^2(kT)^{-3/2}.$$

Compare t with the period of oscillation of electromagnetic radiation of wavelength λ, λ/c. For $T = 10^4$ K, show that t is much shorter than λ/c when λ has a value characteristic of radio waves, say, 10 cm. Therefore, for producing radio waves, the electron was accelerated sharply. This circumstance produces a flat radio spectrum for thermal radio emission. (In Fourier-analysis language, a "delta function" gives a flat Fourier decomposition.)

Historically, the radio detection of HII regions in continuum radiation came before their detection in radio recombination lines. Indeed, the earliest radio survey of Galactic HII regions by Westerhout was carried out in the continuum. Many HII regions in the Galaxy are now known by their number in the Westerhout survey; for example, a famous radio HII region is W49.

21-cm Line Radiation from HI *Regions*

So far we have discussed optical and radio techniques whereby relatively hot ionized gas clouds can be seen in emission. Such HII regions are, however, relatively rare in the Galaxy, because they can be found only near new born O and B stars, which are also relatively rare. The vast bulk of the interstellar medium is in relatively cold gas clouds ($T \sim 10$ to 100 K) where the hydrogen exists in either atomic or molecular form. Such gas clouds can be detected optically either from the general obscuration of starlight produced by their embedded dust grains or from the optical absorption lines of the atoms of heavy elements which are mixed in with the hydrogen. However, optical techniques allowed astronomers to sample only gas clouds which are relatively near to the solar system, and this sampling is quite indirect, since the most abundant species, hydrogen gas, is not directly measured. An observational dilemma therefore seemed to have been reached, because the hydrogen gas in the cold clouds is too cold to have any measurable emission at optical wavelengths.

This dilemma was broken in 1945, when van de Hulst announced that it should be possible to observe atomic hydrogen at a radio wavelength of 21 cm. Van de Hulst's theoretical prediction bore fruit in 1951 with the almost simultaneous detection of the 21-cm line by Ewen and Purcell and by Muller and Oort. The discovery of the 21-cm line is probably the most important achievement in the astronomical study of the interstellar medium. The mechanism that produces this line is illustrated in Figure 11.16.

The lowest orbital-energy state (ground state) of atomic hydrogen actually is subdivided into two levels by a quantum-mechanical effect known as hyperfine structure. A spinning proton and a spinning electron act like two bar magnets. The magnetic polarity of the proton is the same as its spin direction; that of the electron is opposite to its spin direction. In the ground state

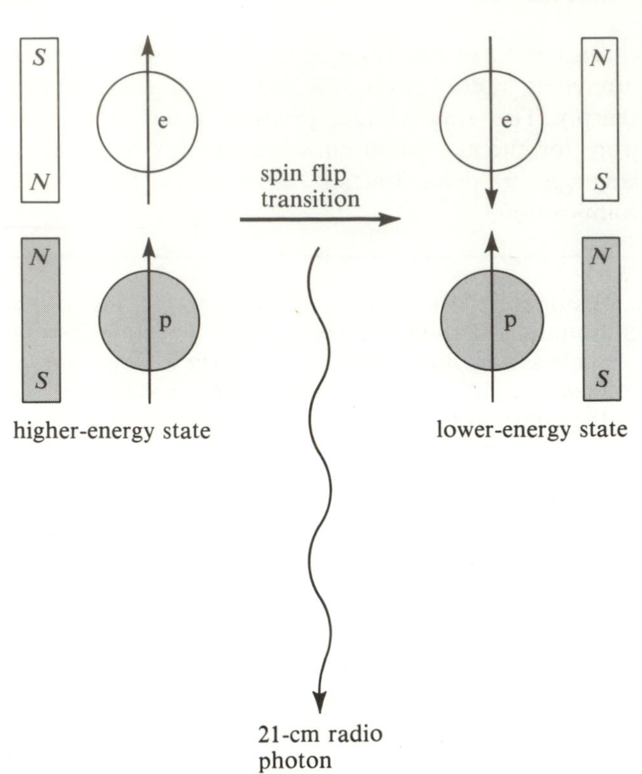

Figure 11.16. The spin-flip transition of atomic hydrogen in its ground electronic state which leads to the emission of a 21-cm radio-line photon.

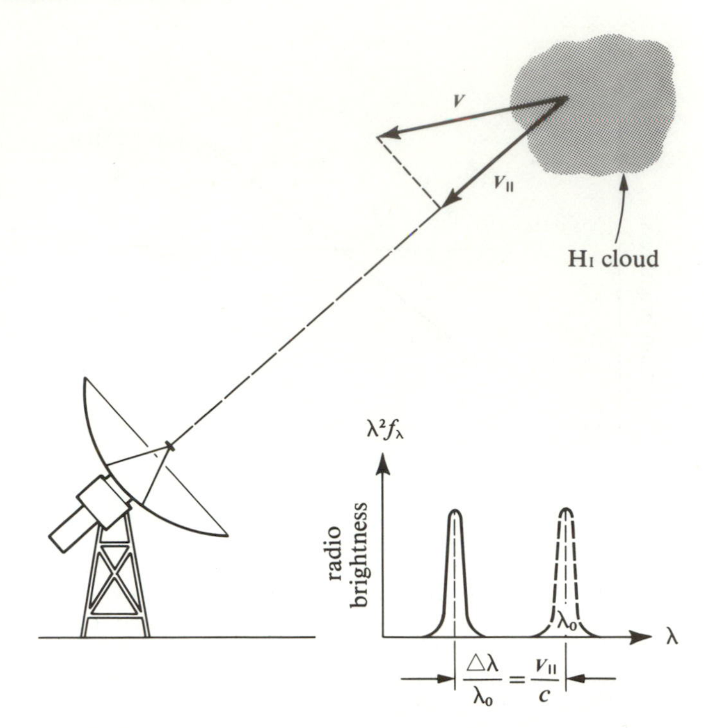

Figure 11.17. Knowledge of the rest wavelength λ_0 of the spin-flip transition of atomic hydrogen allows radio astronomers to measure the velocity component along the line of sight, $v_{||}$, of a hydrogen-gas cloud by applying the (nonrelativistic) Doppler-shift formula.

of the hydrogen atom, the spatial location of the electron is described by a probability distribution which is spherically symmetric relative to the proton. The interaction energy of the bar magnets is, however, different at different points in this electron shell, and the two resulting hyperfine states act as if they were oriented as shown in Figure 11.16. The higher-energy state corresponds to parallel spins (bar magnets with opposite polarities on a single line); the lower-energy state, to antiparallel spins (bar magnets with same polarities on a single line). If you have ever played with bar magnets, you know that the first configuration will spontaneously try to flip to the second configuration. So will a hydrogen atom. A hydrogen atom which finds itself in the parallel state can spontaneously make a "**spin flip transition**" to the lower-energy antiparallel state, with the energy difference being carried off by a radio photon with a wavelength of 21 cm.

The 21-cm spin-flip transition of atomic hydrogen cannot be observed in terrestrial laboratories, because the transition is highly forbidden (as described earlier in this chapter). However, the sum total of rare occurrences over very long paths (typically hundreds of light-years) does allow the astronomical detection of the 21-cm line from interstellar gas clouds. The clouds of gas from which 21-cm radiation can be detected are called HI regions. The radio-astronomical study of the interstellar medium provides the only method by which atomic hydrogen gas can be directly observed, from one end of the Galaxy to the other. The method is doubly important, because a measurement of the Doppler shift of the 21-cm-line radiation also yields the component of velocity of the atomic hydrogen along the line of sight (Figure 11.17).

Problem 11.7. The Einstein coefficient for spontaneous emission of a 21-cm-line photon is $A_{21} = 2.85 \times 10^{-15}$ sec^{-1} (see Problem 11.4). The smallness of A_{21} implies that collisions of hydrogen atoms with each other dominate the hyperfine-level populations. Argue that this implies

$$n_2 = n_1(g_2/g_1)\exp(-E_{21}/kT).$$

For $E_{21} = hc/\lambda_{21}$ and $\lambda_{21} = 21$ cm, show that $E_{21}/kT \ll 1$ for temperatures T comparable to 100 K. Since $g_2/g_1 = 3$ here, derive the approximation $n_2 = 3n_1 = (3/4)n_H$,

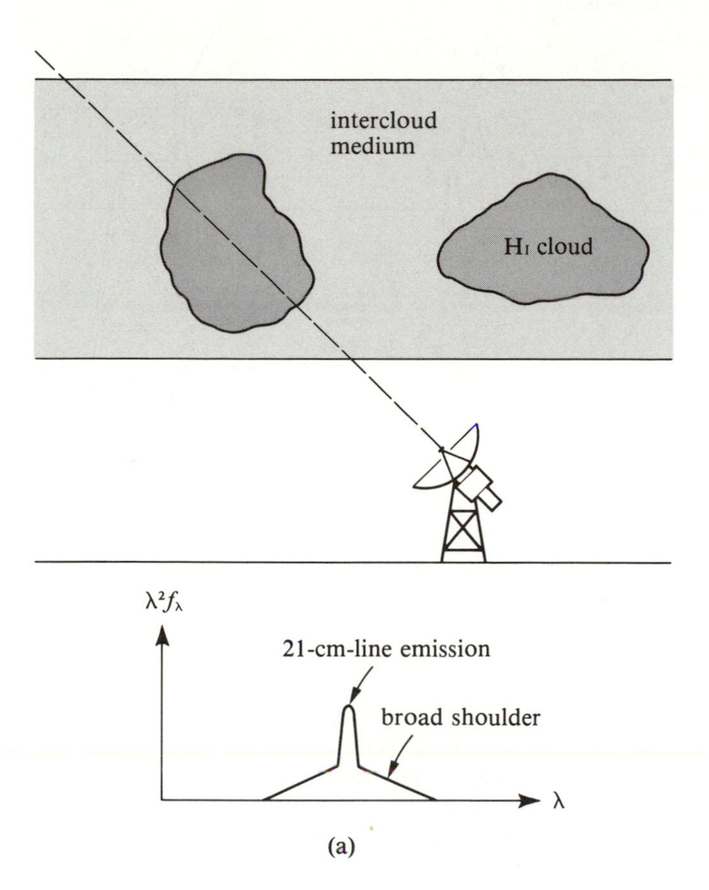

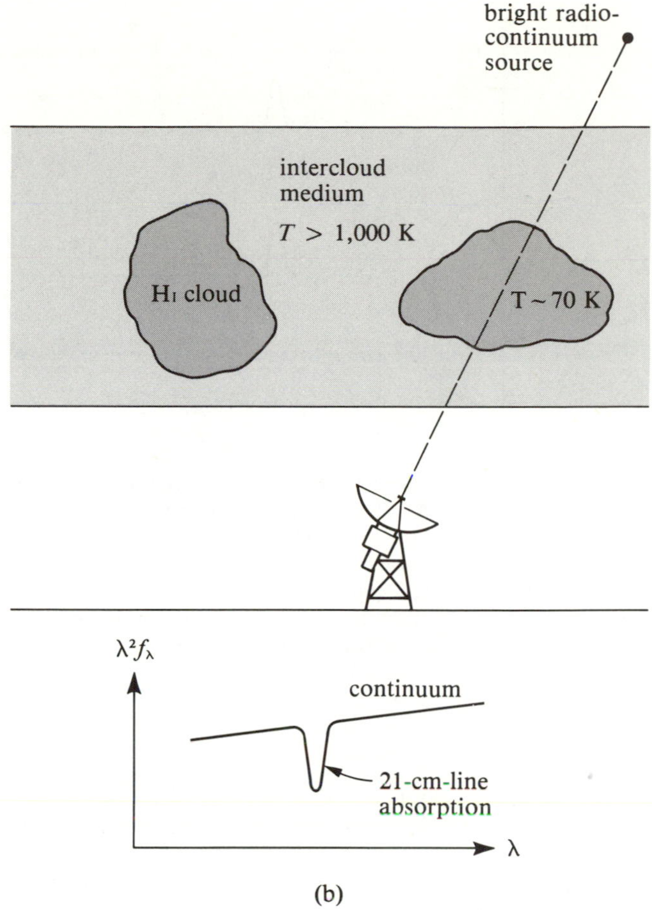

Figure 11.18. Results of emission- and absorption-line measurements of the 21-cm radio transition. (a) When the radio telescope is pointed in a direction which contains no bright background source of radio continuum emission, two components of 21-cm-line emission are generally seen. The narrow component has been abscribed to the emission from cold HI clouds. The broad shoulders have been attributed usually to emission from a warm HI intercloud medium (light gray region). (b) When the radio telescope is pointed toward a bright radio continuum source, a narrow absorption feature can often be seen at a wavelength of 21 cm. This absorption feature can be attributed to an intervening HI cloud of temperature somewhat less than 100 K. Broad shoulders of absorption are not seen, indicating that the gas responsible for this feature in emission is too hot ($T > 1{,}000$ K) to produce absorption.

where n_H is the number density of all hydrogen atoms (most of which are in either one of the two hyperfine ground electronic states). Now show that the total number of 21-cm-line photons emitted per unit time per unit volume is given by $n_2 A_{21} = (3/4) n_H A_{21}$, which depends only on the density n_H and not on the temperature T. If you assume optically thin conditions, how can this fact be used to obtain the distribution of atomic hydrogen gas from measurements of the 21-cm-line emission? A brief explanation in words will do.

The 21-cm line has now been observed both in emission and in absorption against the bright background of strong radio continuum sources (Figure 11.18). The

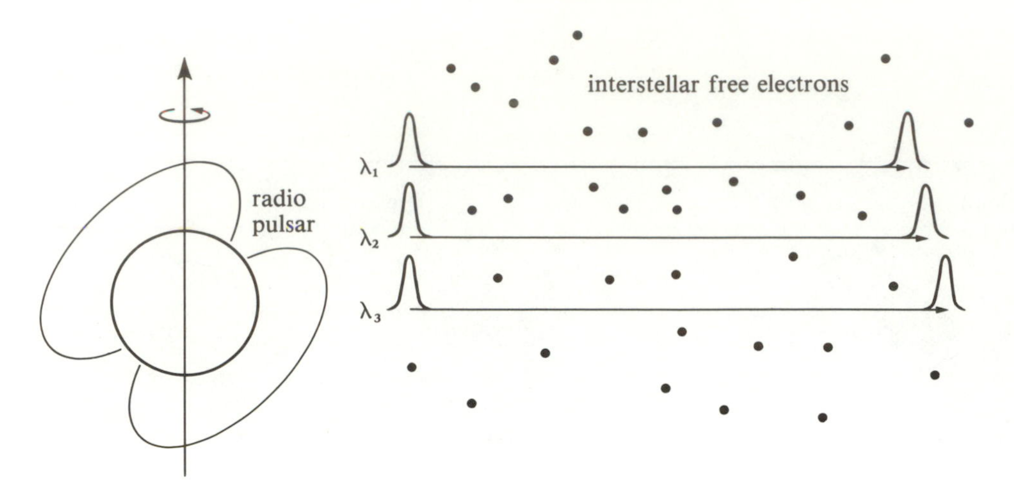

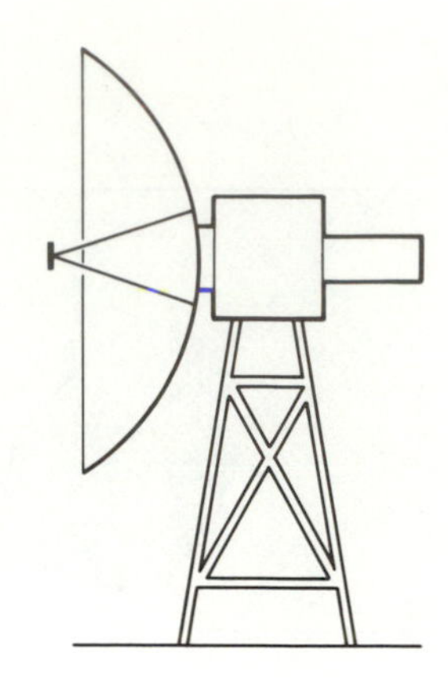

Figure 11.19. Pulsar dispersion. Radio pulses at three different wavelengths, $\lambda_1 > \lambda_2 > \lambda_3$, leave simultaneously from the pulsar. As they travel through the interstellar medium, the radio waves of longer wavelength interact more readily with the *free* electrons which exist in interstellar space than do radio waves of shorter wavelengths. Consequently, the former travel more slowly than the latter, and arrive later at the radio telescope.

emission measurements yield the total amount of atomic hydrogen along the line of sight at each velocity interval, and the absorption measurements provide an estimate of the temperature of the absorbing hydrogen gas. A curious result has arisen from such studies. As was first pointed out by Clark, the 21-cm emission in many directions seems to contain two components. The first component consists of narrow lines at wavelengths which are slightly displaced from the rest wavelength, a feature consistent with the standard picture of many cold and dicrete interstellar gas clouds moving at different velocities in interstellar space. The second component consists of substantially broader lines, which appear as "shoulders" beneath the narrow components (Figure 11.18a). Absorption studies by Radhakrishnan and others show only the narrow components (Figure 11.18b); so the broad-component gas must be too hot to produce measurable 21-cm-line absorption, although it is detectable in emission. The temperature of the absorbing narrow component is about 70 K.

The interstellar gas is also partially ionized on the large scale. Radio-astronomical studies of pulsars revealed that radio pulses at longer wavelengths were arriving slightly later than those at shorter wavelengths. The spread or "dispersion" of arrival times was consistent with an interpretation that radio pulses of all wavelengths start from the pulsar at the same time; but, because free electrons in interstellar space interact with the radio waves, the radio waves of longer wavelengths travel, as a group, at slower speeds than those of shorter wavelengths (Figure 11.19). In particular, the pulsar dispersion measurements implied that the interstellar hydrogen was ionized a few percent *on the average*.

The preceding observations motivated Pikelner, Field, Goldsmith, and Habing, and Spitzer and Scott, to propose that the HI gas in the Galaxy comes in two phases: a cold ($T \sim 10^2$ K) and dense ($n \sim 10$ H atoms/cm^3) phase which is concentrated in clouds, and a warm ($T \sim 10^4$ K), rarefied ($n \sim 10^{-1}$ H atoms/cm^3), and partially ionized phase which constitutes the general medium between the clouds (the "intercloud medium"). The two phases would be in rough pressure equilibrium, which would explain why interstellar clouds maintain their integrity and do not expand and disperse even though they are not gravitationally bound in general (Figure 11.20).

In the original versions of the theory, the two phases had such disparate temperatures because the same heating mechanism (postulated to be the energetic ejection of electrons from hydrogen atoms because of ionization by low-energy cosmic rays) would be balanced by two distinct cooling mechanisms in high- and low-density HI regions. At relatively high densities, cooling by the excitation of fine-structure transitions (Chapter 3) would keep the gas at the relatively low temperature of about 10^2 K. At lower densities, this cooling mechanism gives diminishing returns, and the gas must heat up substantially before cooling again balances heating. When

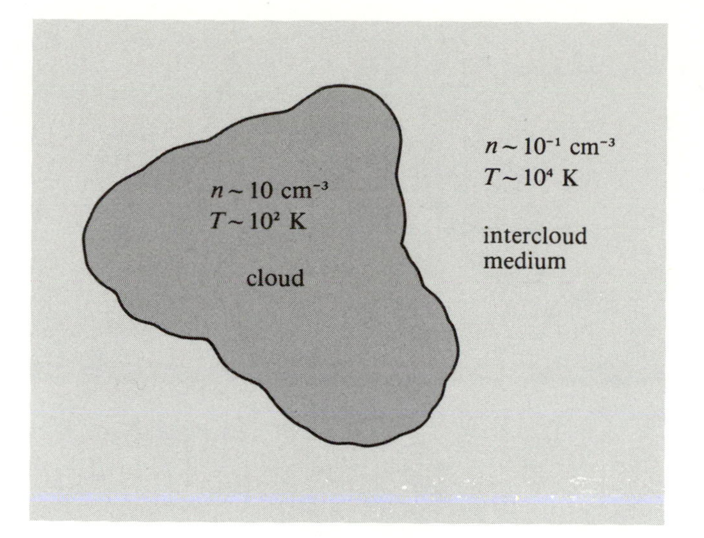

Figure 11.20. The two-phase model of the interstellar medium envisages cold, relatively dense, HI gas confined to clouds in rough pressure equilibrium with a warm, relatively diffuse, HI gas which constitutes the intercloud medium. This simple picture has proven to be too simplified to represent all the observed phenomena which occur in the actual interstellar medium.

the temperature is about 10^4 K, inelastic collisions can excite hydrogen, oxygen, iron, and other abundant elements to higher electronic states, which, upon radiative deexcitation, provide a second source of efficient cooling.

As elegant as this two-phase model of the interstellar medium is, it has not survived the ultimate test: that is, other effects implied by it have not been verified by new observations. Watson and Weisheit were instrumental in pointing out the fundamental problem. The same postulated flux of low-energy cosmic rays (or soft X-rays, in an alternative proposed by Werner and Silk) which keeps the intercloud medium partially ionized (to explain pulsar dispersion) would also lead to certain definite distributions of ionization levels of various atomic and molecular species in dense HI or H_2 clouds. These later predictions have not been substantiated by subsequent observations. In addition to these observational difficulties, there arose the theoretical issue of whether low-energy cosmic rays could ever propagate freely from their place of origin (presumably the site of supernova explosions). The alternative suggestion, that soft X-rays provide the requisite ionization and heating, suffers from the fact that the observed flux of soft X-ray is too low by an order of magnitude to do the desired job. The two-phase model of the interstellar medium has therefore been generally abandoned by workers in the field. The original 21-cm-line studies, however, still stand as an empirical fact, and they remain to be explained on a sound theoretical basis. Perhaps a modified version of the two-phase model will eventually prevail.

Problem 11.8. Why do you suppose that the ultimate test of physical theories concerns phenomena which the original theory was not specifically designed to explain? Why should not an explanation of known facts be considered sufficient for a true scientific theory? In particular, what would you think of a theory which contains as many arbitrary assumptions as there are observed facts? Do you think such a "theory" is anything more than a rewording of the original observed or experimental facts? Given these considerations, why do you think the two-phase model of the interstellar medium deserves to be called an elegant theory, although it is probably ultimately wrong? Discuss the merits of Karl Popper's view that science proceeds by the falsification of theories rather than by their verification.

Radio Lines from Molecular Clouds

At densities of 10^2 particles per cm^3 or greater, and at temperatures of 10 K or so, the hydrogen in interstellar gas clouds no longer tends to remain in atomic form. Instead, hydrogen atoms tend to react chemically with other hydrogen atoms rather quickly to form molecular hydrogen:

$$H + H \rightarrow H_2.$$

McCrea and McNally, and Gould and Salpeter, suggested that this reaction would proceed in interstellar space via the catalytic action of dust grains. The surfaces of the latter provide useful reaction sites to absorb the excess energy and momentum released by the reaction (Figure 11.21). The conversion of atomic hydrogen into molecular hydrogen implies that the 21-cm-line radiation becomes useless for investigating the coldest and densest regions of interstellar space. This proves to be unfortunate, since many interesting processes—in particular, the formation of new stars—take place only in the densest parts of interstellar space.

Problem 11.9. Assume interstellar dust grains to have the properties described in Problem 11.1. If n_H is the number density of hydrogen atoms and n_d is the number density of dust grains, assume that the initial ratio n_d/n_H is given by 10^{-12}. Let T be the temperature of the hydrogen atoms, so that $(kT/m_H)^{1/2}$ is the thermal speed

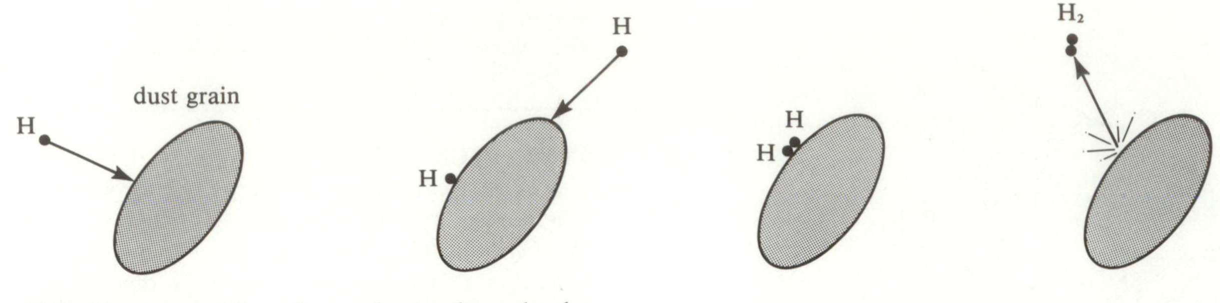

Figure 11.21. The proposed formation mechanism for molecular hydrogen, H_2, in interstellar space. The dust grain serves as a catalyst to drive the energetically favorable chemical reaction, $H + H \rightarrow H_2$. The newly formed H_2 molecule will be able to survive the harsh environment of outer space, where ultraviolet photons tend to dissociate H_2 back into H atoms, only if it is sheltered in the relatively dense environment of dark clouds and molecular clouds.

v_T of hydrogen atoms. If R is the radius of a typical (spherical) interstellar dust grain, show that the number of hydrogen atoms striking any given dust grain per second is given approximately by $n_H v_T \pi R^2$. Thus, show that the total number of hydrogen atoms striking any dust grain at all per second per cm^3 is given by $n_d n_H v_T \pi R^2$. Assume that every hydrogen atom which strikes a dust grain sticks to it long enough to find another hydrogen atom, and to combine with it to form a molecule of H_2, which is then released back to the gas phase. Show now that the timescale for converting hydrogen atoms to hydrogen molecules is given by

$$t = n_H/n_d n_H v_T \pi R^2 = 1/n_d v_T \pi R^2.$$

Calculate the numerical value of t if the initial value of $n_H = 300\ \text{cm}^{-3}$ and $T = 10$ K. Convert your answer to millions of years. What does this imply for the minimum age of molecular hydrogen clouds?

Molecular hydrogen, H_2, turns out to have no radio lines; so, unlike atomic hydrogen, it cannot be observed directly by radio-astronomical techniques. Optical astronomers, however, have detected absorption lines of the molecules CH, CH^+, and CN in interstellar space. Indeed, theoretical analyses by Bates and Spitzer and by Herzberg in the 1950s suggested that a variety of simple molecules should be able to form in the dense parts of of gas clouds that are shielded from the destructive effects of stellar ultraviolet photons. Shklovskii pointed out that several such molecules may be observable by means of the radio-spectral lines which arise when the rotation of the nuclei interacts with the electron orbital angular momenta. In 1963, Weinreb, Barrett, Meeks, and Henry made the first radio-astronomical detection of an interstellar molecule. The molecule they found was OH. The identification of OH was especially secure because four separate lines—hyperfine splittings of the transition described at a wavelength of 18 cm by Shklovskii—of this molecule can be detected at relative strengths which are in good accord with theory.

Shortly after the discovery of OH in radio absorption lines, OH was observed in emission from the vicinity of some HII regions which had previously been identified by Westerhout as sources of intense radio continuum emission. The OH emission in these objects, detected by a Berkeley group headed by Weaver and by a Cambridge group including Weinreb, Meeks, and Barrett, had a very high intensity, peculiar line-ratio strengths, very small line widths, and very high degrees of polarization, and varied on a timescale of days. The emission intensity is so high that, if it arose by thermal processes, the temperature would have to be on the order of 10^{12} K! It was clear that the emission process in such objects was nonthermal, and Perkins, Gold, and Salpeter soon made the accepted suggestion that **maser** action lay behind this intense emission by OH.

Maser is an acronym for *m*icrowave *a*mplification by the *s*timulated *e*mission of *r*adiation, and it is the antecedent of a similar phenomenon in optical light called **laser**. Masers were invented in terrestrial laboratories by Townes and Schawlow. For maser action to take place, there has to be an "inversion" in the ratio of the numbers of molecules in the excited and ground states compared to the ratio expected for local thermodynamic equilibrium (Figure 11.22). In thermodynamic equilibrium,

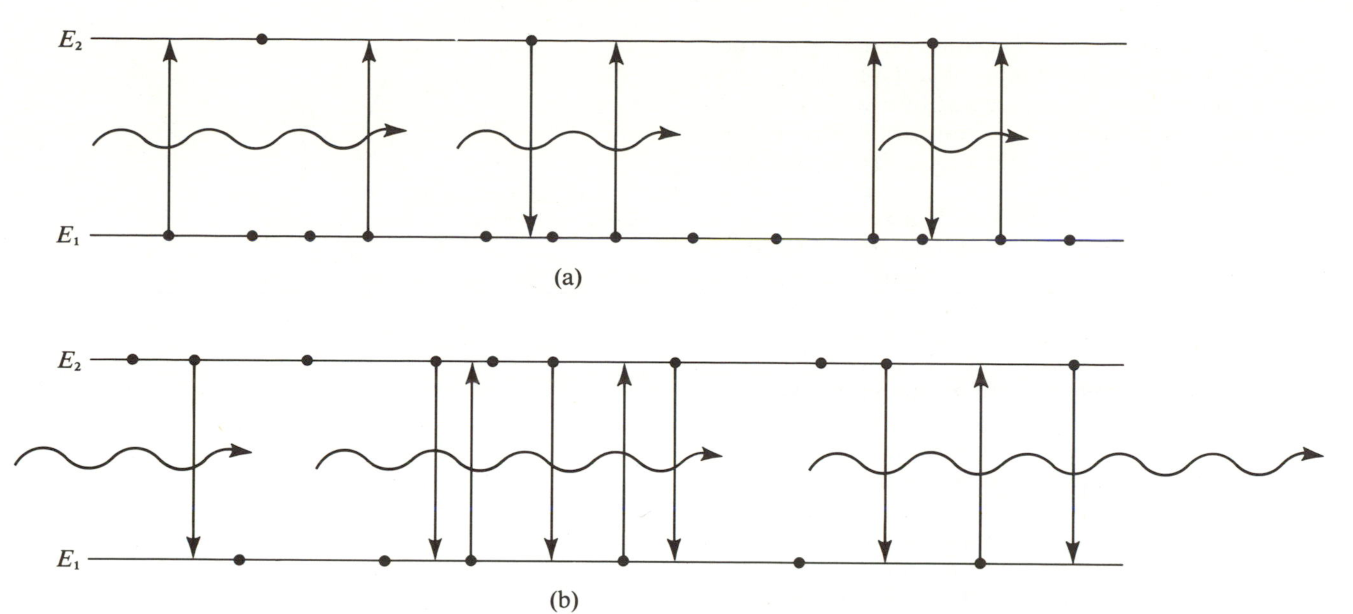

Figure 11.22. The origin of maser action. (a) Under normal circumstances there are more molecules (dots) in the lower-energy state E_1 than there are in the upper state E_2. Thus, if radio or microwave radiation of wavelength corresponding to the line transition between these two states comes along, this radiation will induce more upward transitions (absorptions) than downward transitions (emissions). The net effect is to produce a progressive diminution of the original signal. (b) Under extraordinary circumstances, more molecules may be pumped to the upper state than remain in the lower state; if so, an incident wave of line radiation will be progressively amplified by inducing more downward transitions than upward transitions. This maser emission differs from random thermal emission, in that the molecules emit coherently.

there is a natural (Boltzmann) distribution of molecules at various energy levels of internal degrees of freedom. Most of the molecules would be in the ground state, the next largest number would be in the first excited state, etc. (see Chapter 4). Now, atoms and molecules in the interstellar medium are usually not in local thermodynamic equilibrium, since too many particles are in the ground state rather than the higher-energy states, given the temperature of the region (as measured by the translational motions of the particles). Thus, atoms and molecules in the interstellar medium are usually found with level distributions precisely the opposite of what is needed for laser or maser action. At energy levels corresponding to radio transitions, this discrepancy is small, but it is usually present. For OH in certain astronomical circumstances, however, enough energy is somehow pumped into the system to overpopulate the excited energy levels. Given such an overpopulation of excited levels (which in terrestrial laboratories are effected by radiative or chemical "pumps"), a stray line photon may stimulate an avalanche of coherent radiative deexcitations. This avalanche would then lead to the observed intense maser emission. A continuous maser would result if the molecules continue to be pumped from the ground state to the excited level.

Radio astronomers now know that there are at least two different kinds of OH maser sources. The first kind is associated with the birth of young stars, and was the type of source discovered near compact HII regions in the earliest detections; the second kind is associated with old stars in very late stages of stellar evolution. In both kinds the maser emission arises in very special, dense regions ($n \sim 10^8$ cm^{-3}) within a few light-days of the central star or protostar.

Normal (i.e., "thermal") OH emission has by now been detected in interstellar clouds of less unusual circumstances. Such observations constitute a powerful radio-astronomical tool for investigating the physical conditions of the dense, cold molecular clouds which reside in interstellar space.

By 1968 astronomers had tacitly assumed that interstellar molecules were always composed of two atoms, one being hydrogen, and the other being hydrogen (usually) or one of the more common reactive elements (such as C or O). This naive belief was shattered that year by the announcement by Cheung, Rank, Townes, Thornton, and Welch of the detection of ammonia (NH_3) in the direction of the Galactic center. This startling discovery was followed by the detection of water vapor (H_2O) in several regions of interstellar space by the Berkeley group. By 1981, the list of discovered interstellar molecules had grown to more than fifty, some of them being relatively complex organic compounds. For example, vast quantities of ethyl alcohol (CH_3CH_2OH) have been discovered near the center of our Galaxy; evidently our Galaxy has also discovered that the best place to store liquor is in the middle of the system!

Obviously, at relatively low temperatures, when suitably high densities are reached, interstellar matter prefers to form molecules rather than remain as atoms. This preference provides another example of the action of the second law of thermodynamics at low temperatures. The formation of molecules in interstellar space allows the matter to take advantage of the higher chemical binding energy of larger and larger molecules.

Besides maser emission, interstellar molecules have provided another surprise. The discoverers of formaldehyde (H_2CO), Snyder, Buhl, Zuckerman, and Palmer, found that formaldehyde was always seen in absorption, even when there was no apparent bright background source of radio continuum emission whose radiation the formaldehyde could absorb! It turns out that formaldehyde is absorbing the cosmic background of microwave radiation which pervades the universe (Chapter 16). For formaldehyde to be able to do this, its upper energy level must be "refrigerated" by processes which are formally the opposite of maser action.

Interstellar molecular lines, particularly those of OH and CO (carbon monoxide), are currently used to explore the large-scale distribution and motions of molecular clouds in the Galaxy. Pioneering work of this kind has been carried out by Solomon and Scoville, Gordon and Burton, and Cohen and Thaddeus. Comparison of the strengths of interstellar molecules which are the same chemically but contain different isotopes of the constituent atoms (for example, $^{13}C^{16}O$ and $^{12}C^{18}O$) yields information about the synthesis rates of various elements in different parts of the Galaxy. Work of this kind by Penzias and his coworkers at Bell Laboratories promises to offer interesting confrontations with theories of the history of nucleosynthesis in the universe.

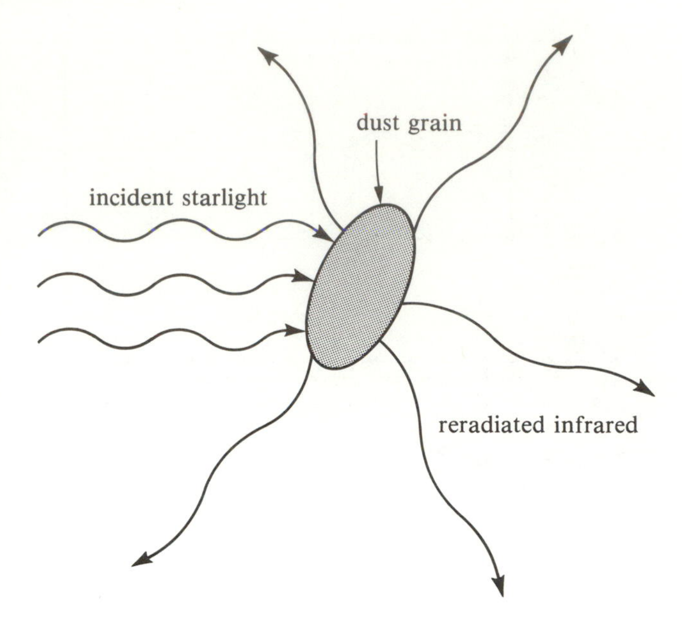

Figure 11.23. Thermal emission from dust grains usually results because the dust grains are exposed to an incident flux of stellar photons. A fraction of such visible or ultraviolet photons are absorbed, heating the grain to a higher temperature. The equilibrium temperature is reached when the thermal reemission by the warm interstellar grain in the infrared balances the absorption of starlight. In dark dust clouds, this equilibrium temperature is on the order of 10 K, but dust near stars may be heated up to several hundreds of degrees. Notice that this process, apart from scale, is entirely analogous to how sunlight keeps the Earth warm.

Infrared Manifestations of Gaseous Nebulae

In principle, the same thermal processes that give rise to optical and radio lines and to radio continuum emission in HII regions could also give rise to infrared lines and infrared continuum emission from the same nebulae. In practice, there are observational complications. The Paschen lines of recombining hydrogen atoms and the fine-structure lines of collisionally excited atoms, such as singly ionized neon, have now indeed been detected by infrared spectroscopy of HII regions. Infrared continuum emission from HII regions could be expected from the free-free process (see Figure 11.14), but the infrared continuum radiation from warm interstellar dust has generally turned out to be more powerful (Figure 11.23).

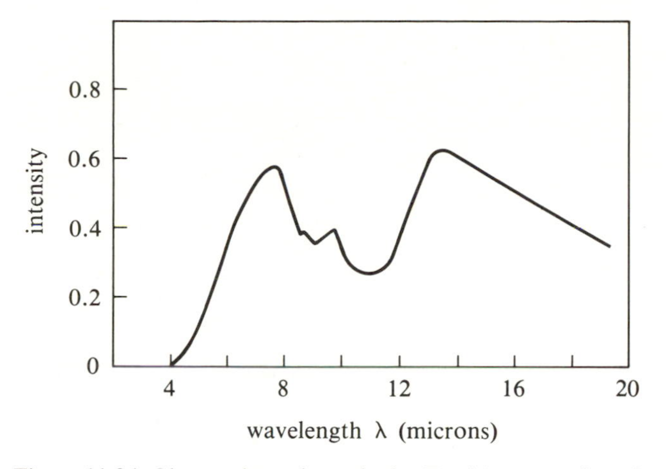

Figure 11.24. Observations through the Earth's atmosphere in the infrared is confined to a few "windows" which fall in between major emission bands of various molecules in the Earth's atmosphere.

The advantages of infrared astronomy are that it can yield information on areas not visible to optical astronomers and on a smaller angular scale than is resolvable by radio astronomers. The disadvantage is that infrared astronomy is relatively difficult to perform, for several reasons.

First, water vapor, carbon dioxide, and ozone in the Earth's atmosphere absorb much of the infrared radiation from outer space. Only a few "windows" of transparency exist at infrared wavelengths (Figure 11.24). Consequently, in infrared astronomy there is a great advantage to observing from the highest possible sites, where there is little water vapor overhead (high mountains, balloons, airplanes, or outer space).

Second, all lukewarm bodies emit copiously in the infrared; so the thermal pollution of the environment provide a "background" of infrared radiation which can overwhelm the signal coming from distant astronomical sources. This background contamination can be partly eliminated by "chopping" (looking on and off the astronomical source, and differencing the two signals to subtract out the background), but statistical fluctuations can limit the accuracy of the measurement. The infrared contribution of the radiation from the astronomer's own equipment can be minimized by cooling it to a very low temperature in liquid nitrogen or liquid helium, but one cannot thus cool the entire telescope, or the Earth's atmosphere, or the astronomer!

Problem 11.10. Use Wien's displacement formula (Chapter 3), and calculate the wavelength of maximum emission from a thermal body of temperature equal to 300 K. Draw appropriate conclusions about the importance of eliminating the background in infrared astronomy.

The very fact that all warm bodies radiate infrared radiation, although a local nuisance, accounts for the existence of many of the infrared sources found in astronomy. For example, an early survey by Leighton, Neugebauer, and their colleagues at Caltech turned up about 20,000 near-infrared sources which were essentially pointlike. A few of these sources turned out to be well-known optical stars, such as Betelgeuse, which is the brilliant red supergiant in the shoulder of Orion (Figure 11.25); but most of them turned out to be stars whose outer atmospheres are too cool for them to radiate much of their total energy in the visual part of the electromagnetic spectrum. Later infrared surveys at longer wavelengths by Walker and Price of the Air Force Cambridge Research Laboratories turned up both pointlike and extended objects. Apart from the nuclei of galaxies, including that of our own Galaxy, most of the extended sources seem to be stars deeply enshrouded in thick clouds of gas and dust. The dust absorbs most of the optical light of the embedded star, making it practically invisible to optical astronomers. In the process, the dust is heated up, and in a steady state the dust reradiates all the absorbed starlight as infrared radiation. That the star is surrounded by a dense cloud or shell of gas and dust suggests either that the gas and dust have been ejected from a star which is old and highly evolved, or that the star has been newly born from the cloud of gas and dust.

An especially intriguing infrared source is the Becklin-Neugebauer object, in the Orion nebula (Figure 11.26). It is a pointlike source which radiates very strongly in the infrared, but hardly at all in the optical. The central source is so heavily obscured by surrounding clouds of dust that even if it radiated intrinsically in visible light, this visible light would be attenuated by a factor of about 10^{30} before it could reach us. If the Sun were similarly buried in a cloud of gas and dust (as in Hoyle's novel *The Black Cloud*), the Earth would suffer perpetual night. Near the Becklin-Neugebauer object is a diffuse source of infrared radiation called the Kleinmann-Low nebula. A major astronomical puzzle is what energy source powers the infrared radiation of the Kleinmann-Low nebula. Both the Becklin-Neugebauer object and the Kleinmann-Low nebula lie within the densest part of a giant molecular cloud which is believed to be slightly behind the Orion HII region. Taken together, the optical, infrared, and radio investigations reveal the Orion complex to be apparently a cauldron of active star formation in our own Galaxy. It is especially exciting to think that the Becklin-Neugebauer object may be a massive star being born before our very eyes.

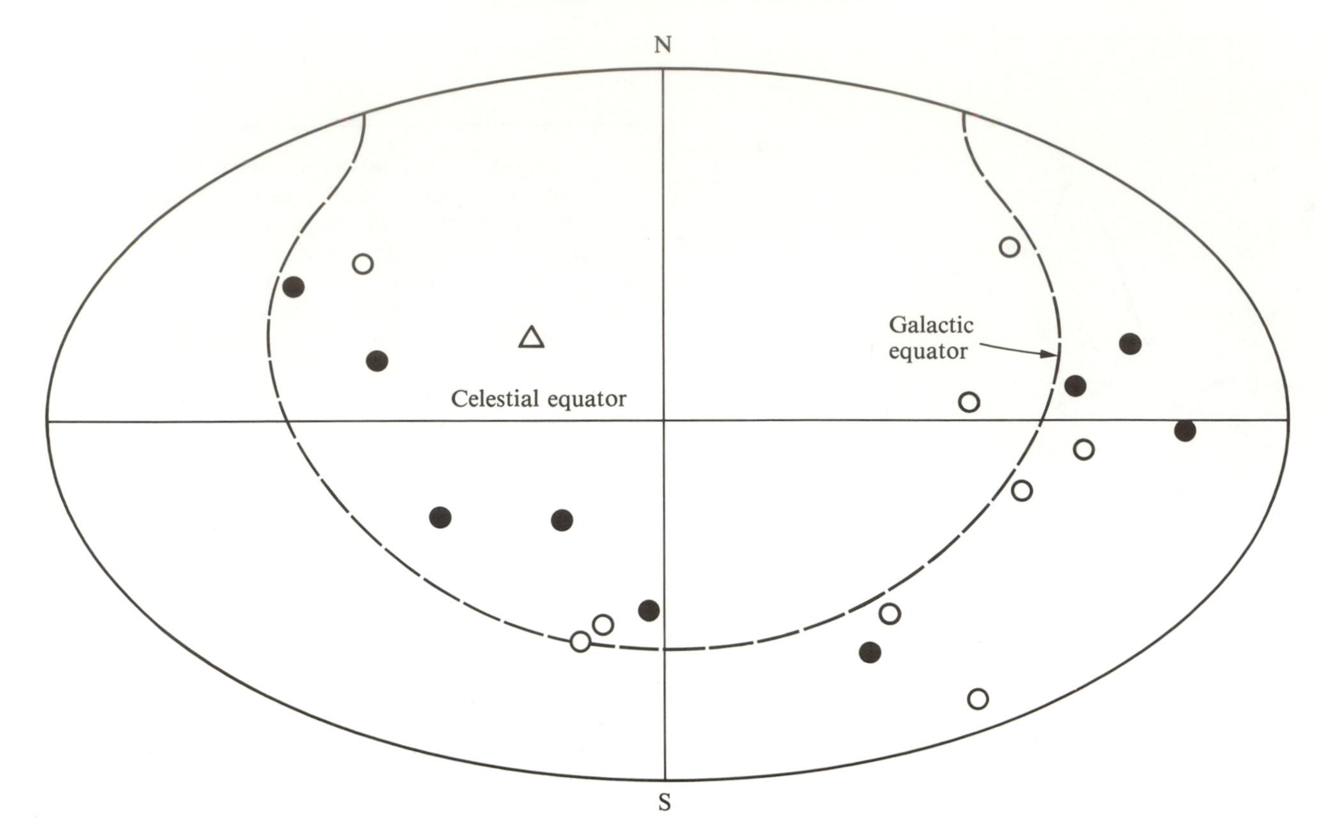

Figure 11.25. The ten brightest near-infrared and optical sources are plotted on the sky in celestial coordinates, where the projection of the Earth's equatorial plane runs horizontally through the middle of the diagram and the north and south spin poles lie at the top and bottom. In this diagram, the projection of the Galactic equator (the Milky Way) forms the dashed band. A comparison of the brightest infrared stars (filled circles) with the brightest optical stars (open circles) reveals only one common member (triangle = Arcturus). (For further details, see Neugebauer and Becklin, *Scientific American*, **228**, April 1973, 28.)

Figure 11.26. The Orion nebula is a Galactic HII region which harbors many newly born stars. The intriguing infrared sources, the Becklin–Neugebauer object and the Kleinmann–Low nebula, are found in the molecular cloud associated with the Orion nebula. (Palomar Observatory, California Institute of Technology.) ⟶

Ultraviolet, X-ray, and Gamma-ray Observations of the Interstellar Medium

In the modern era, astronomical observations are possible not only at radio, infrared, and optical wavelengths, but also at ultraviolet, X-ray, and gamma-ray wavelengths. Morrison has pointed out that we usually think of high-energy radiation as being more penetrating than low-energy radiation; consequently, it seems paradoxical that radio, infrared, and optical astronomy can be carried out on the ground, but ultraviolet, X-ray, and gamma-ray astronomy must be carried out above the Earth's atmosphere. The reason for this was given in Chapter 3, but it bears repetition and expansion.

When absorption occurs over a *continuum* of wavelengths, as in a solid, higher-energy photons are, indeed, more penetrating than lower-energy photons. Solids do possess individual absorption lines and bands, but these are usually so broad that they overlap one another and merge into a continuum. The greater penetration by high-energy photons under these circumstances explains why internal organs and bones in the human body can be photographed in X-rays but not in visible light.

In contrast, the absorption lines and bands of the Earth's *gaseous* atmosphere are well-separated, especially at optical wavelengths. Although the atmosphere is quite opaque at the centers of such lines or bands, low-energy photons between those wavelengths find the Earth's atmosphere to be quite transparent. On the other hand, high-energy photons can ionize the atoms and molecules in the Earth's atmosphere. Photons with a continuum of energies above the ionization limit of any atomic or molecular species can be absorbed, and it is this *continuum* opacity which prevents all but the most energetic of gamma-ray photons from reaching the ground (see Figure 2.3).

Even rocket or satellite observations above the Earth's atmosphere cannot entirely avoid the absorption of high-energy photons from celestial sources. The ionization of hydrogen and heavier elements in the *interstellar medium* is very severe for photons which are not very energetic and, therefore, not very penetrating. It was therefore thought that extreme ultraviolet astronomy—say, observations at wavelengths between 912 Å and 100 Å—would always remain an unfulfilled dream. Bowyer and his group, however, took advantage of the Apollo-Soyuz joint space mission between the US and USSR to demonstrate that extreme ultraviolet observations could, in fact, be made, at least for the nearest celestial sources, for which interstellar absorption would be minimal. Their successful detections of extreme ultraviolet radiation have turned out to be primarily of thermal emission from extremely hot white dwarfs or from the coronae of ordinary stars. Extreme ultraviolet astronomy promises to be a powerful tool in the investigation of these and related objects.

Ultraviolet photons less energetic than the Lyman limit (i.e., with wavelengths longer than 912 Å) can escape absorption by the intervening interstellar atomic hydrogen. Satellite observations at such ultraviolet wavelengths have given us new insights into the nature of the interstellar medium. Particularly revealing has been the study of the ultraviolet absorption lines produced by interstellar molecular hydrogen, H_2, along the line of sight toward bright hot stars. The first detection of molecular hydrogen in interstellar space by this technique was carried out by Carruthers from rocket experiments, and the satellite plus rocket observations help to confirm the general theoretical expectation that H_2 should be the dominant form of gaseous hydrogen in the denser interstellar clouds.

The satellite observations also show, again by means of the analysis of ultraviolet absorption lines, that heavy elements are much less abundant in interstellar clouds than inside Population I stars. This finding is consistent with the suggestion that the heavy elements in the interstellar medium are largely locked up in the form of interstellar grains. There is not general agreement on whether this locking-up took place primarily during the condensation of solid material that had been ejected from stars, or by the subsequent deposition of "icy mantles" on preexisting cores of dust grains in the interstellar medium proper.

Another important result from the satellite ultraviolet observations has been the discovery of an amount larger than expected to five-times ionized oxygen, OVI, in the directions toward many O and B stars. This highly ionized form of oxygen can exist in substantial quantities only if the ambient medium is relatively hot, say, in the neighborhood of several hundred thousand degrees. The existence of interstellar matter with slightly higher temperatures, say, $\sim 10^6$ K, has also been deduced from rocket and satellite observations of a patchy background of soft X-ray emission in the Galaxy. A general diffuse background of both hard X-rays (wavelengths less than about 10 Å) and soft X-rays (wavelengths greater than about 10 Å) has been found and studied. The hard X-ray background, which appeared to be amorphous in the early observations carried out at low angular resolutions, may be largely composed of an enormous number of unresolved discrete sources that lie outside our own Galaxy. Indeed, recent results with the imaging Einstein satellite observatory suggest that quasars that strongly emit X-rays may provide the bulk of this hard X-ray background. In contrast, the soft X-ray background has turned out to be truly amorphous, and appears to be associated with the interstellar medium of our own Galaxy. If soft X-rays originate from beyond the Galaxy, we should see a significant absorption of such X-rays by the gas in the Magellanic Clouds when we look at them

Figure 11.27. A giant HII region, 30 Doradus, which lies in the Large Magellanic Cloud, one of two satellite galaxies of the Milky Way System. For a view of the entire Large Magellanic Cloud, see Color Plate 29. Although the Large Magellanic Cloud contains many fewer stars than the Milky Way System, the former has fractionally much more gas. This may account in part for why its largest HII region, 30 Doradus, is so much larger than any known for the Milky Way System, although other contributing factors may be that the Large Magellanic Cloud has a barred structure, which induces severe compressions of the interstellar gas, and that 30 Doradus may contain a star of a few *thousand* solar masses which provides a powerful source of ultraviolet light. In any case, the large fractional gas contents of the Large and Small Magellanic Clouds would make them potent absorbers of soft X-rays if such radiation came from much beyond our Galaxy. (The Cerro Tololo Inter-American Observatory.)

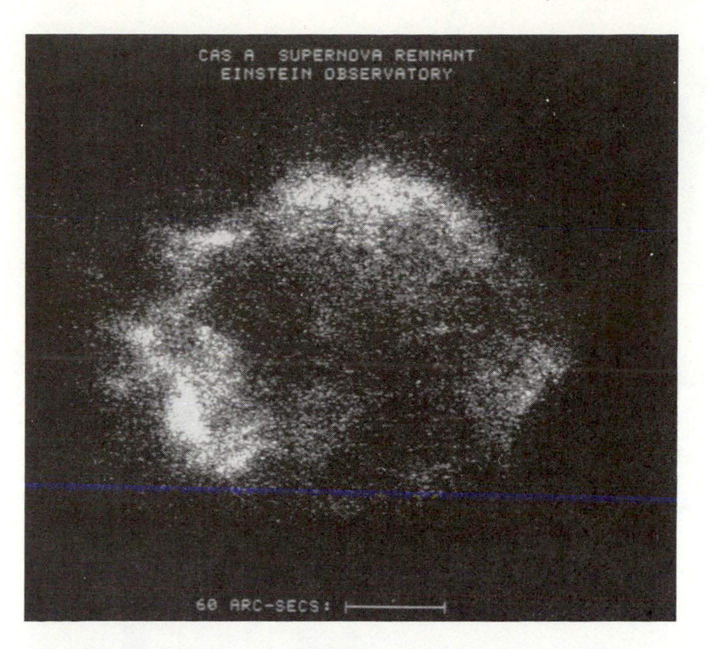

Figure 11.28. In contrast with the Cygnus Loop (see Fig. 11.13c), Cassiopeia A, shown here by its X-ray emission as observed by the Einstein Observatory, is a relatively young supernova remnant. No historical record exists of its explosion, but measurements of the expansion velocity of its faintly visible optical filaments indicate that it cannot be older than about 300 years. (Courtesy of Riccardo Giacconi and the High-Energy Astrophysics Division of the Harvard-Smithonian Center for Astrophysics.)

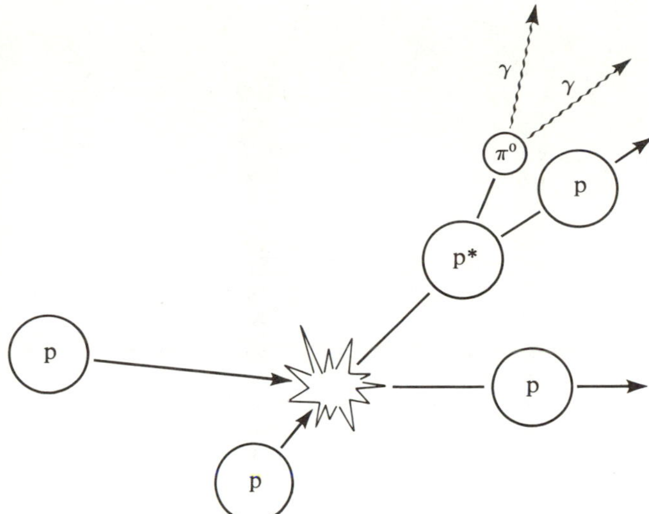

Figure 11.29. The production of gamma rays in the interstellar medium. A high-energy proton (cosmic-ray proton) strikes a thermal proton (nucleus of a hydrogen atom) in the interstellar medium, and produces a proton in an excited nuclear state (p*). The excited proton decays by the emission of a π^0 meson, which further decays to give two very energetic gamma-ray photons.

(Figure 11.27), but we do not. Some, and perhaps all, of the soft X-ray emission may come from supernova remnants like the Cygnus loop and Cassiopeia A, known to be extended sources of X-ray, optical, and nonthermal radio emission (Figure 11.28). Largely from the work of Cox, Kraushaar, and their colleagues at Wisconsin, we now know that the volume which emits soft X-rays contains a hot, patchy, and rarefied gas which extends for at least a few hundred light-years around the solar system.

These observations have led to the revival of an old proposal by Spitzer that "coronal gas" at a temperature of a million degrees or so might be quite common in the Galaxy. In the latest versions of this picture (developed by Cox and Smith and by McKee and Ostriker), this hot, rarefied gas might constitute much of the medium between interstellar clouds. Such a model for the intercloud medium would differ substantially from the "two-phase" theory discussed earlier, in which the intercloud medium was taken to be partially ionized HI at a temperature of about 10^4 K. It remains to be seen how the different results of radio, ultraviolet, and X-ray observations of the general interstellar medium can be reconciled with one another.

Gamma rays also come from the interstellar medium. Since gamma rays are very energetic photons (wavelengths less than 0.1 Å) and since even the hottest regions of interstellar space are too cool to emit gamma rays, these very energetic photons probably must originate by a nonthermal process. One accepted mechanism for producing a significant fraction of the observed diffuse gamma-ray flux involves high-energy interactions between relativistically moving protons and ordinary hydrogen nuclei, which are found in great profusion in the interstellar medium (Figure 11.29). The microscopic process is, briefly, as follows. A proton suffers a high-energy collision with another proton, and is excited to a high nuclear-energy level. Such a proton can decay to a lower level by the emission of a real π° meson (see Chapter 6). Thus, a π° meson is to an excited baryon what a photon is to an excited atom, except that a free π° meson is an unstable particle. The bound quark-antiquark pair ($u\bar{u}$ or $d\bar{d}$) it represents will decay into two gamma rays in about 10^{-16} sec.

Cosmic Rays and the Interstellar Magnetic Field

The crucial ingredient in the mechanism for producing gamma rays in the interstellar medium is the presence

Figure 11.30. Energetic cosmic-ray particles that traverse lexan plastic leave ionization tracks in it, which can be rendered visible by chemical etching. (Courtesy of P. Buford Price.)

of relativistic protons. Since the work of Hess at the beginning of the twentieth century, it has been known that very energetic charged particles exist in outer space. At the highest energies they arrive on Earth from all directions, only somewhat deflected by the Earth's magnetic field. In the early days of research, there was some confusion about how to interpret the ionization tracks produced by these invaders from outer space (Figure 11.30), and they were given the general name cosmic rays. We now know that cosmic rays are very energetic material particles, composed primarily of protons, although the nuclei of elements heavier than hydrogen are also present.

Since the cosmic-ray particles move at speeds close to that of light, they could not possibly be confined to the Galaxy by gravitation. The gravity of even the entire Milky Way is too weak to deflect significantly the motion of any cosmic-ray particle. Does this mean that cosmic rays are not confined to the Galaxy? No, on galactic scales, even the wild motion of an ultrarelativistic charged particle can be deflected by a moderately weak interstellar magnetic field.

Problem 11.11 Consider the circular motion of a cosmic-ray proton of charge e, rest mass m_0, and momentum $\boldsymbol{p} = m\boldsymbol{v}$, where $m = \gamma m_0$, with $\gamma = (1 - v^2/c^2)^{-1/2}$, which takes place in a plane perpendicular to a uniform magnetic field $\boldsymbol{B}$. If r is the radius of the circle and $\omega = v/r$ is the angular speed, show that the time rate of change of the momentum $d\boldsymbol{p}/dt$ has magnitude ωp and is directed toward the center of the circle. What does the sense of $\boldsymbol{v}$ have to be so that the Lorentz force $\boldsymbol{F} = e(\boldsymbol{v}/c) \times \boldsymbol{B}$ is also directed toward the center of the circle? (Consult Chapter 3.) Set $d\boldsymbol{p}/dt = \boldsymbol{F}$, and show

$$\omega = eB/\gamma m_0 c, \qquad r = \gamma m_0 cv/eB.$$

A particle is considered ultrarelativistic if $\gamma \gg 1$. Compute numerically the values of ω and r for a cosmic-ray proton for which $\gamma = 10^5$ and which moves in an interstellar magnetic field of strength $B = 3 \times 10^{-6}$ gauss. (Note esu-gauss = dyne.) How does r compare with the distances between stars?

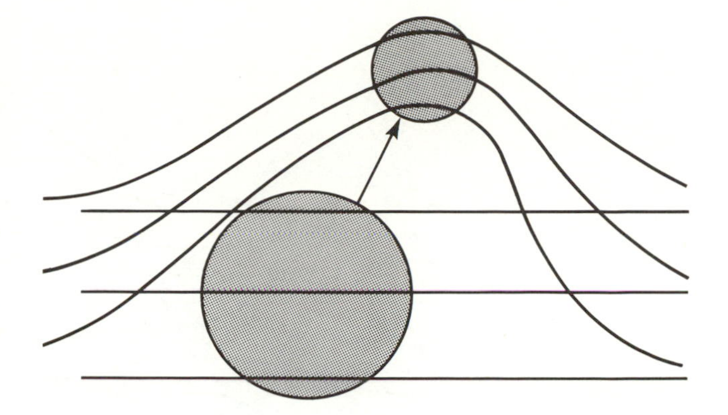

Figure 11.31. In diffuse interstellar clouds, the magnetic field is effectively frozen to the matter. Thus, if the cloud moves elsewhere and contracts, the field lines must move with the cloud and be compressed by it. The magnetic field behaves as if it were rubber bands tied to the gas cloud. Thus, after the cloud has moved and contracted, the stretched and curved field lines exert a force which tries to pull the cloud back to where it came from and to prevent further contraction.

What confines the magnetic field of the interstellar medium to our Galaxy? The weight of the thermal gas. Because the ions and electrons of a partially ionized thermal gas also tend to spiral around the magnetic field lines, the gas and magnetic field are frozen together (Figure 11.31). The thermal gas cannot escape the gravitational attraction of the Galaxy; by this indirect means, interstellar magnetic fields and cosmic rays are confined to our Galaxy (Figure 11.32). Parker pointed out, however, that this confinement scheme is top-heavy and unstable: the thermal gas tends to drain into "valleys," and the cosmic rays and magnetic fields tend to buckle upward in-between to form "mountains" (Figure 11.33). A number of astronomers have speculated that Parker's instability might explain the origin of giant complexes of HI and H_2 clouds which contain on the order of $10^6 M_\odot$ of gas and dust.

Cosmic rays also contain relativistic electrons, and it is the wild spiraling motion of these energetic electrons in interstellar magnetic fields which gives rise to synchrotron radio emission. To produce the characteristic spectrum observed for nonthermal radio emission (see Figure 11.15); the number of cosmic-ray electrons of particle energy E must drop off as a negative power of E, say, as $E^{-\beta}$. This (slow) power-law decrease of the energy distribution differs markedly from the (rapid) exponential decrease characteristic of thermal distributions (Chapter 4), and it is this feature which gives rise to the nomenclature "nonthermal." The power β required to explain the observed nonthermal radio emission by the synchrotron process turns out to agree well with the power law in the energy distribution of cosmic-ray electrons observed outside the Earth's magnetosphere by rocket and satellite experiments.

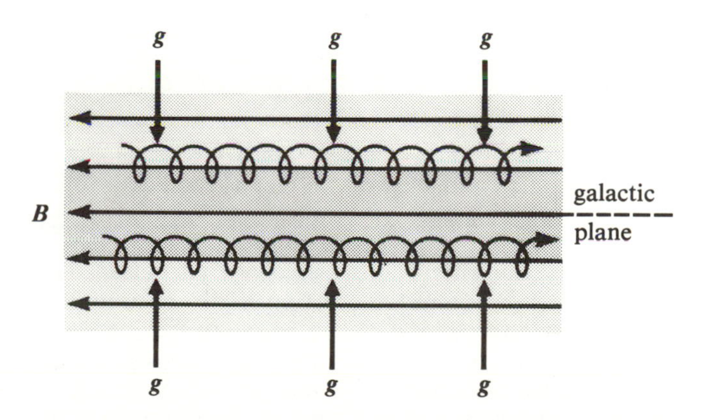

Figure 11.32. Cosmic-ray particles do not escape from the Galaxy, despite their great kinetic energies, because they are forced to spiral around the interstellar magnetic field lines. The magnetic field $\boldsymbol{B}$ of the Galaxy in turn is confined to the Galaxy by being frozen to the thermal gas (shaded regions), which is concentrated toward the central plane of the Galaxy by its own weight. The direction of the local gravitational field $\boldsymbol{g}$ is indicated here by the heavy arrows.

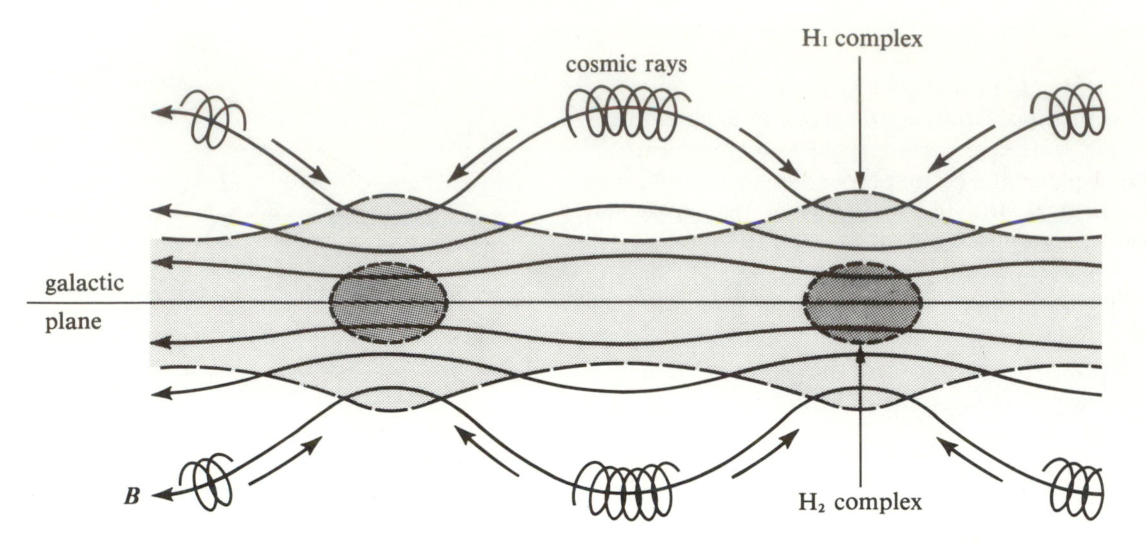

Figure 11.33. The confinement of the cosmic ray particles and the interstellar magnetic field by the mechanism outlined in Figure 11.32 tends to be unstable. The heavy thermal gas would prefer to be lower in the gravitational well of the plane of the Galaxy, and the cosmic rays and the magnetic field would prefer to rise higher. Because the magnetic field is frozen in the thermal gas, the system cannot simply overturn. But the thermal gas can drain downward along the field lines, and the magnetic field and cosmic rays can buckle upward in between. It has been proposed that this is the mechanism by which huge complexes of HI and H_2 gas clouds are formed.

Problem 11.12. Why must experiments to observe cosmic-ray electrons be carried out above the Earth's magnetosphere? *Hint*: The Galactic synchrotron radiation is primarily due to electrons with γ between 10^3 and 10^5. What is the radius of gyration of such electrons according to Problem 11.11 (with m_0 = electron rest mass) in the Earth's magnetic field of strength ~ 1 gauss? How does this compare with the size of the Earth?

What is the origin of cosmic-ray particles? This problem is still one of the great mysteries of twentieth-century astronomy, and it has been studied by many astrophysicists. A very important clue is undoubtedly the chemical composition of the cosmic-ray particles. We have already noted that most of the cosmic rays are protons, which should come as no surprise, given the great predominance of hydrogen nuclei in the universe. Heavy nuclei, however, are also present in an unbiased sample of cosmic rays; and among these heavy nuclei, we find that elements of the iron group are a bit more abundant than they are elsewhere in the universe. This overrepresentation of the iron peak possibly suggests that cosmic rays originate as ejecta from very evolved stars, in particular, from supernovae. Most astronomers were convinced that this identification was correct; unfortunately, there are theoretical problems with the idea that cosmic rays come *directly* from supernova explosions. Lerche, Wentzel, Kulsrud, and others have shown that when cosmic rays try to stream away from a supernova explosion, they excite disturbances in the magnetic-field structure which prevent them from propagating very far from their origin. Although a way may yet be found around this difficulty, the problem has led many workers to abandon a direct origin in supernovae for cosmic rays, and to return to modified versions of Fermi's idea that cosmic rays might be accelerated by various fluid motions in the interstellar medium. The most promising idea in this category involves the boosting of charged particles with ordinary energies to relativistic ones by repeatedly reflecting them from "magnetic mirrors" associated with "collisionless shocks." In this theory, the chemical composition of cosmic rays would be determined by the ionization level of the ambient medium.

The magnetic field in the interstellar medium can also be important for the dynamics of the ordinary gas, as has been stressed by Mestel for several years. Thus, the interstellar magnetic field may provide the major support for the denser pieces of molecular clouds against the pull of their own self-gravity (Figure 11.34). Mouschovias has suggested that the tension from the curved magnetic-field lines may help to prevent the envelope of a collapsing interstellar cloud from falling into the central core which is destined to become a star. In this way, one might explain why star formation is a relatively inefficient process, with only a small fraction of the total mass of a gravitationally unstable gas cloud actually being incorporated inside stars for each star-forming episode.

The interstellar magnetic field can also influence the dynamics of interstellar dust grains, as was pointed out in a particularly important example by Davis and Greenstein. They showed that rotating nonspherical dust grains would tend to orient their long axes perpendicular to the interstellar magnetic field under certain circumstances (Problem 11.13), and would then block starlight with polarization vectors parallel to the grain's long axis more than starlight with polarization vectors perpendicular to the long axis (Figure 11.35). A star's intrinsic light may contain a random mix of polarization

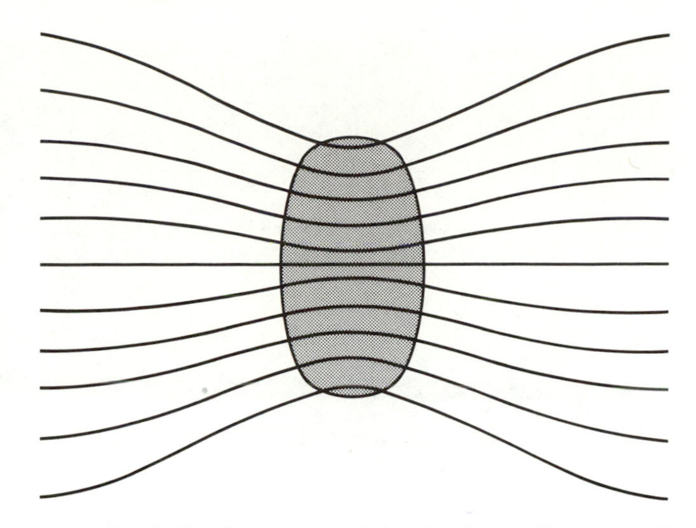

Figure 11.34. The forces exerted by the interstellar magnetic field can help to support a dense interstellar cloud against the inward pull of its self-gravity.

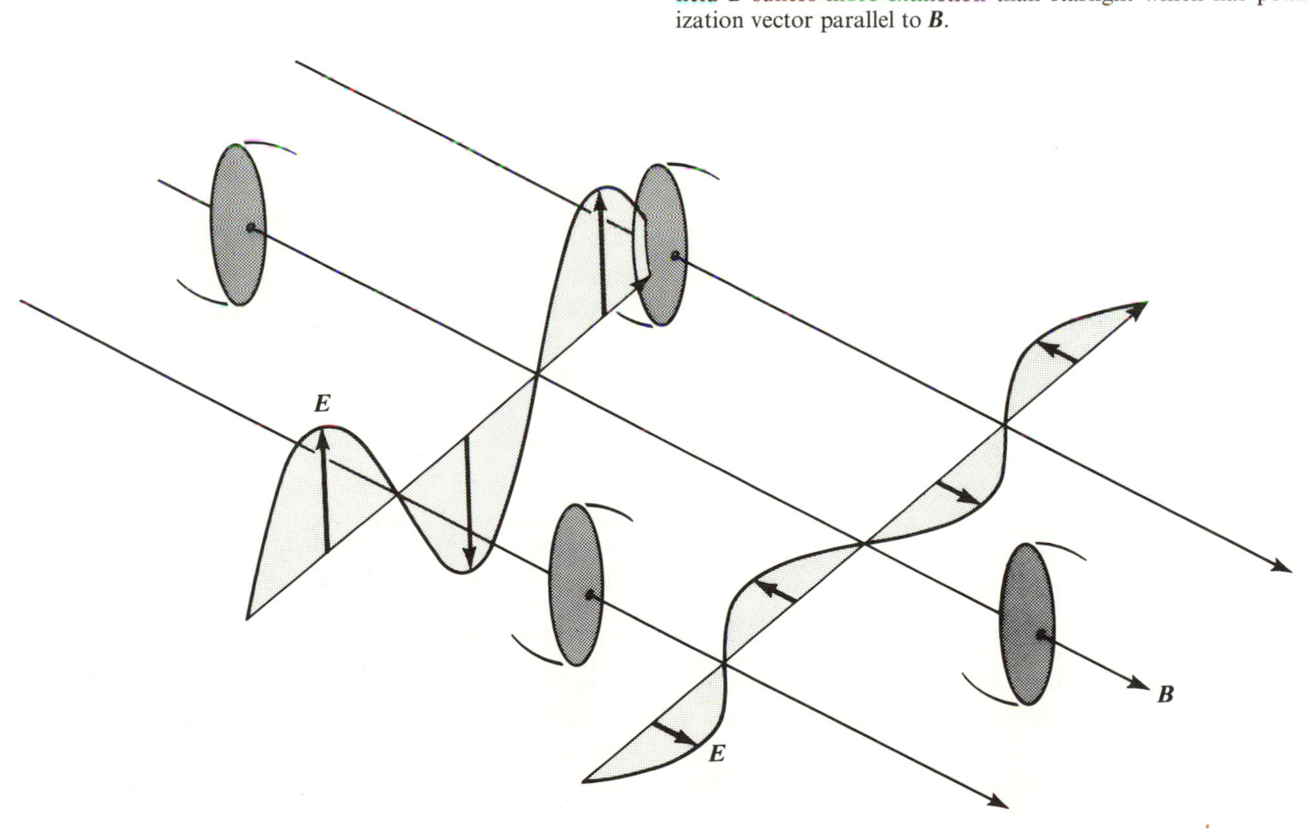

Figure 11.35. The tendency of rotating paramagnetic dust grains to align their long axes perpendicular to the interstellar magnetic field gives rise to interstellar polarization. Starlight which has polarization vector ***E*** perpendicular to the interstellar magnetic field ***B*** suffers more extinction than starlight which has polarization vector parallel to ***B***.

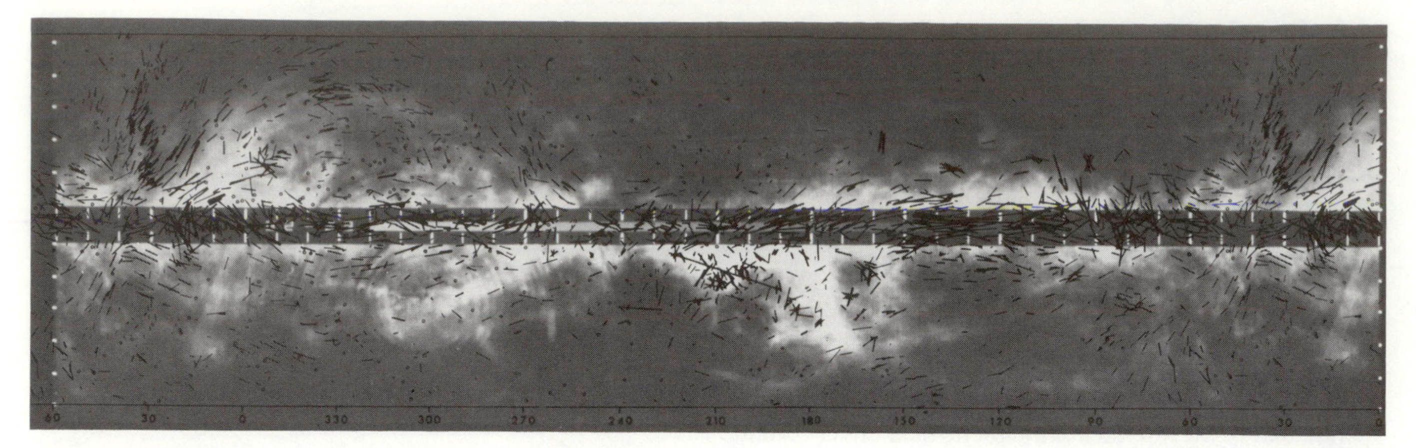

Figure 11.36. The direction of the polarization of starlight (short lines) as measured by Mathewson and Ford are superimposed on a low-contrast print of the hydrogen-gas distribution in Galactic latitude (vertical axis) and longitude (horizontal axis). The polarization direction of the nonthermal radio continuum emission (not shown) is largely perpendicular to the polarization direction of starlight. This is consistent with an interpretation of synchrotron emission for the former, and dust-grain alignment by magnetic fields for the latter. Notice that the hydrogen-gas filaments in the figure lie mostly parallel to the polarization directions of starlight, indicating that the gas concentrations are mostly elongated in the directions *parallel* to the local magnetic field. This probably implies that the gas filaments cannot be strongly self-gravitating. (From Cleary, Heiles, and Haslam, *Astr. Ap. Suppl.*, **36**, 1979, 95.)

states, but the light with polarization vectors perpendicular to the interstellar magnetic field will be scattered or absorbed more than other light. A distant observer of the starlight would, therefore, see more light with polarization parallel to the magnetic field than with polarization perpendicular to it. This effect is called the **polarization of starlight**, and it was discovered observationally in 1949 by Hiltner and Hall.

Problem 11.13. One condition required for the Davis-Greenstein mechanism to work is that the dust should be colder than the gas atoms which hit it and try to make it spin randomly. Without going into the details of the Davis-Greenstein theory (which is rather complicated), can you think of a reason why the mechanism should not work if the gas and dust temperatures are equal? *Hint*: Uniformity of temperature is characteristic of thermodynamic equilibrium. Can order (polarization) arise spontaneously from disorder (no polarization) when the important parameters of a system are (locally) in thermodynamic equilibrium? Can you think of a reason why the dust should generally be cooler than the gas? *Hint*: Suppose both to be heated ultimately by starlight, and that both come into energy balance, so that heating balances cooling. What do you suppose is the radiative cooling efficiency of dust grains compared to that of gas atoms? Can you guess the sense of interstellar polarization that would result if the gas were cooler than the dust?

We can check the prediction of the sense of polarization observationally. On the one hand, the Davis-Greenstein mechanism predicts that starlight should be polarized parallel to the interstellar magnetic field; on the other hand, the synchrotron mechanism predicts that nonthermal radio emission should be polarized perpendicular to the interstellar magnetic field. Thus, in the same parts of the sky, the polarization of starlight should be perpendicular to the polarization of synchrotron emission. It is (Figure 11.36). The check is so satisfactory that astronomers now routinely use the direction

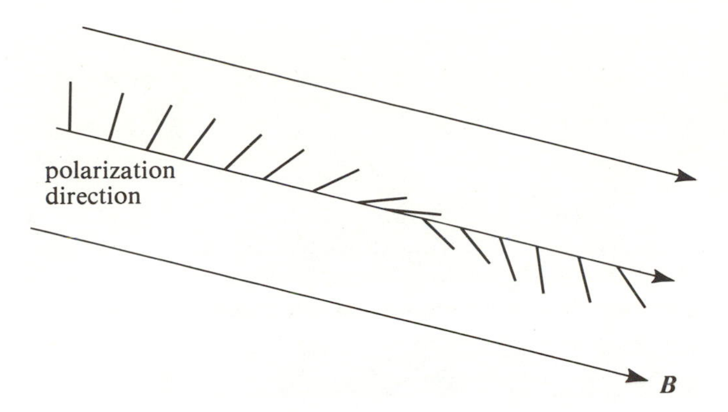

Figure 11.37. Intrinsically polarized radio emission (say, from an extragalactic nonthermal radio source) which travels in the interstellar medium will suffer Faraday rotation. The effect is strongest for radio waves which propagate in the direction parallel to the interstellar magnetic field ***B*** (as shown here).

of interstellar polarization as an indicator of the direction of the interstellar magnetic field.

The strength of the interstellar magnetic field can, in principle, be calculated from measurements of the Zeeman effect (Chapter 3) in radio lines. In practice, this technique is difficult to carry out, but Verschuur has been able to obtain a few positive results. The surveys of Heiles and Troland, however, seem to indicate that the magnetic-field strength in interstellar clouds is generally abnormally low. Interstellar magnetic fields can affect radio radiation in other ways as well. For example, thermal electrons gyrate clockwise as seen by an observer looking down the direction of the interstellar magnetic field. Such gyrating electrons cause radio photons of one spin to propagate slower than radio photons with the opposite spin along the direction of the magnetic field (Figure 11.37). (Photons, being spin 1 particles which travel at the speed of light, have only two possible orientations of their spins: parallel and antiparallel to the direction of their motion.) This difference leads to an effect called "Faraday rotation," whereby the polarization vector of some intrinsically linearly polarized radio continuum source (say, a quasar) will be systematically rotated as the radio waves propagate in the magnetic interstellar medium. The amount of this rotation depends on the wavelengths of the radio radiation, the strength of the component of the magnetic field along the line of sight, and the number of free electrons along the line of sight. Measurements of the "Faraday rotation" at several different wavelengths, coupled with measurements of pulsar dispersion, can therefore yield information about the magnetic-field strengths and number density of free electrons in interstellar space. In particular, the average magnetic-field strength deduced for the interstellar medium is about 3 microgauss.

Interactions Between Stars and the Interstellar Medium

At various points in this chapter, we have dwelt upon how important the action of stars is for the appearance and motions of the interstellar medium, and vice versa. Nowhere is this mutual dependence more dramatic than in the birth and death of stars.

The Death of Stars

The death of stars influences the interstellar medium in several important ways. There is much observational evidence that stars in late stages of stellar evolution shed appreciable amounts of gas via stellar winds. In giants and supergiants of late spectral type, Woolf and Ney found infrared evidence for dust accompanying the mass outflow. The dust grains must have condensed from the heavy elements in the gas, either when this gas was in abnormally cool regions of the atmosphere of the star or when the gas cooled upon expansion in the wind. Gilman, Salpeter, Weymann, and others have proposed that the wind may have been produced, or at least accelerated, by radiation pressure pushing outward on the newly formed dust grains. Frictional drag between the dust and the ambient gas would then pull the latter along.

What kind of dust is formed seems to depend on whether carbon or oxygen is relatively more abundant in the atmosphere of the star. Carbon and oxygen tend to combine with each other to form the very stable gaseous compound carbon monoxide, CO. Once this process is completed, the leftover atoms of the more abundant species can combine with atoms of other elements or with themselves to form other molecules. Some of these molecules—for example, the compounds containing silicon—can apparently aggregate together and condense into small, solid grains, much as soot forms in the cooler parts of a candle flame. In normal M supergiants, where oxygen is the more abundant element, this theory correctly predicts that silicates should be the predominant form of dust. In carbon stars, where carbon is the more abundant element, this theory predicts the predominant species of dust to be silicon carbide or graphite.

Infrared observations show that dust grains can also form in the gas shells from nova explosions and in planetary nebulae. Clayton has speculated that dust grains with peculiar isotopic abundances may form in the ejecta of supernovae. The death of stars may therefore not only generally enrich the interstellar gas in heavy elements, but also provide much of the dusty soot which exists between the stars.

Dying stars may also provide much of the mechanical agitation of the interstellar medium, which has a rather

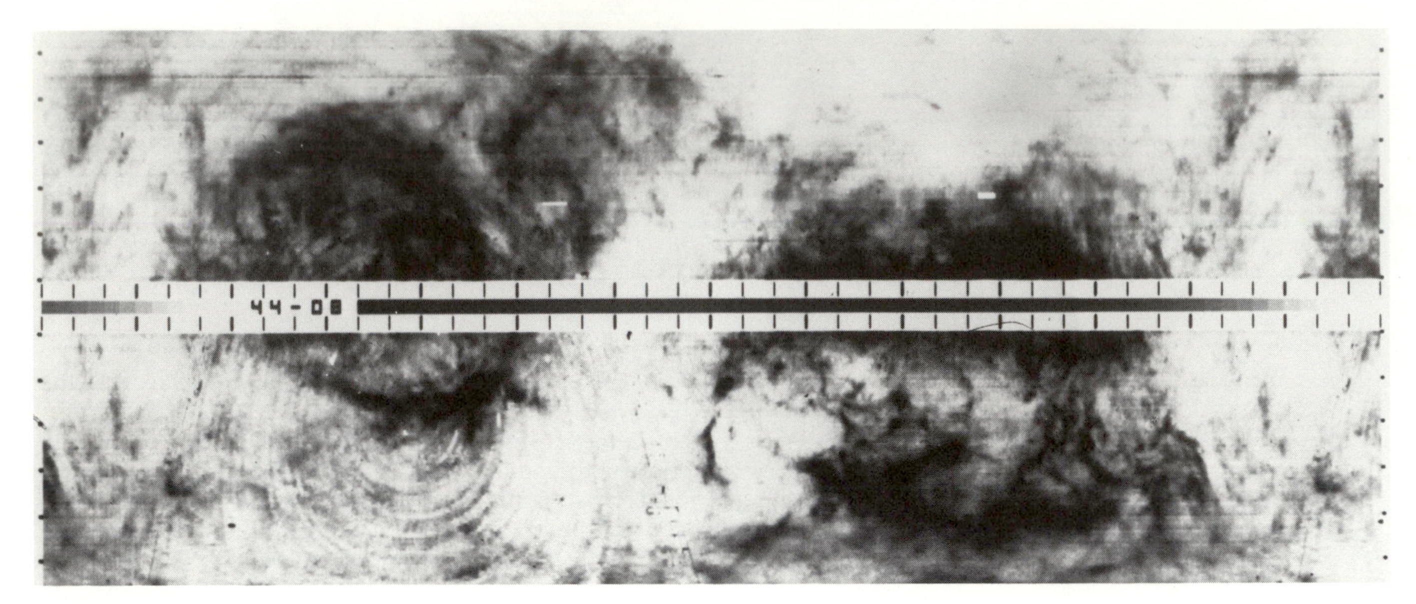

Figure 11.38. The distribution of atomic hydrogen within a narrow velocity range about −8 km/sec at all Galactic longitudes and most latitudes (except for a strip 10° above and below the Galactic equator). Notice the many filaments and occasional large holes in the hydrogen-gas distribution. Both may have been produced by supernovae events. The pattern of concentric ovals in the lower left is an artifact that arises because the observations in the Southern Hemisphere were performed using a radiotelescope that scans with the help of the rotation of the Earth. (From Colomb, Pöppel, and Heiles, *Astr. Ap. Suppl.*, **40**, 1980, 47.)

lumpy appearance (Figure 11.38). The most violent example is provided by supernova explosions. Oort was the first to realize that the expanding ejecta from such explosions would compact the surrounding interstellar gas and push it into a thin shell ahead of the ejecta, much as a snowplow compacts and pushes snow (Figure 11.39). In this way the supernova explosion could impart momentum to the surrounding medium and perhaps account for the random velocities seen in interstellar clouds.

In the snowplow model, much of the kinetic energy of the original explosion would be converted into heat, which is subsequently radiated away in the thin dense shell behind the shock wave that advances into the ambient interstellar medium (Problem 11.11). Thus, once

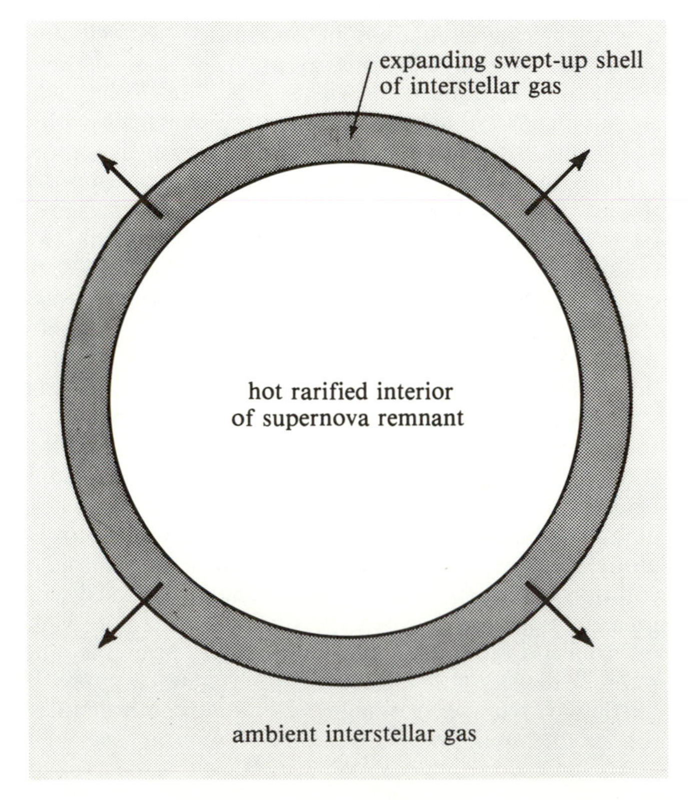

Figure 11.39. In the late stages of the evolution of a supernova remnant, the expanding shell of gas tends to sweep up the surrounding ambient medium and compact into a dense state. Compare this theoretical model with the actual situation in an old supernova remnant like the Cygnus loop (Figure 11.13c). →

a supernova remnant enters the snowplow (or radiative) phase of its evolution, its efficiency for accelerating interstellar matter is not great. Detailed calculations show that the formation of a thin, dense shell because of radiative cooling of the shocked interstellar gas is a sensitive function of the density of the pre-existing intercloud medium and the properties of the embedded interstellar clouds. If the density of the intercloud medium is low, the radiative phase is delayed, and the hot interior of the blast wave produced by the supernova explosion could expand to much greater radii before being dissipated. In particular, Cox and Smith propose that under certain conditions, neighboring supernova remnants may overlap and merge their hot interiors to form a network of "supernova tunnels." McKee and Ostriker propose a variant of this scheme, whereby the overlapping hot interiors merge to form a general intercloud medium of about 10^6 K. Large regions containing low-density "coronal gas" undoubtedly exist in our Galaxy; however, it is not certain what fraction of the volume of interstellar space is occupied by such gas.

Problem 11.14. Suppose a spherical shell identified as a supernova remnant is observed with radius r and with outward expansion speed v. Assume the mass density of the ambient medium to have the uniform value ρ_0; then the supernova remnant must have swept up mass $M = \rho_0 4\pi r^3/3$. Let the original mass M_0 be ejected at speed v_0. If we ignore communication between different parts of the shell (via the thermal pressure of the hot interior), and suppose that each piece of the shell preserves its outward linear momentum as it sweeps up more material initially at rest, we have the snowplow model. Show that the snowplow model implies

$$(M + M_0)v = M_0 v_0.$$

The original kinetic energy E_0 of the ejected material equals $M_0 v_0{}^2/2$. The present kinetic energy E of the shell equals $(M + M_0)v^2/2$. Show that the ratios E/E_0 and v/v_0 are given by

$$E/E_0 = v/v_0 = M_0/(M + M_0).$$

In a typical supernova explosion, $M_0 = 4M_\odot$ worth of matter might be ejected from the central star at speeds $v_0 = 5{,}000$ km/sec. What is the original kinetic energy E_0? (Compare your answer with Problem 8.7.) How much mass M would have to be swept up to bring the supernova shell speed v to a value of 10 km/sec, which is typical of the random velocities of interstellar clouds? In the process, by what factor is E reduced from E_0? What must happen to the lost energy in the snowplow model? (In actual practice, the snowplow model should not be applied to the early evolution of a supernova remnant, in which expansional energy rather than momentum tends to be conserved.)

What is the radius r when $v = 10$ km/sec if $\rho_0 = 2 \times 10^{-24}$ gm/cm^3 (corresponding to about 1 hydrogen atom per cm^3)? Convert your answer to light-years.

For $M \gg M_0$, show that v is inversely proportional to r^3; deduce, therefore, that the remnant slows down as it expands. With the time-rate change of radius proportional to r^{-3}, calculus allows us to show the age τ of the remnant to be four times shorter than if it had expanded to radius r at a constant speed v: $\tau = r/4v$. Compute τ for our hypothetical supernova remnant.

Problem 11.15. Suppose that the coronal gas in the interior of a supernova remnant occupies an average volume $V = 4\pi r^3/3$ and survives for a time τ. Assume supernova explosions to occur randomly in the Galaxy, and let the rate of supernova explosions per unit volume per unit time be R. Show that the fraction f of the volume occupied by the interiors of supernova remnants in a statistical steady state can be expected to be

$$f = Q/(1 + Q),$$

where Q is the dimensionless quantity $RV\tau$. *Hint*: Proceed as follows. Let f_1 be the probability that some randomly chosen point is inside a supernova remnant at some instant t_1, and let f_2 be the same probability at instant t_2, a short time later. Clearly, the probability f_2 equals the probability f_1 that the point was in a remnant at time t_1 multiplied by the probability $[1 - (t_2 - t_1)/\tau]$ that the remnant has survived to time t_2, plus the probability $(1 - f_1)$ that the point was not in a remnant at time t_1 multiplied by the probability $[1 - (t_2 - t_1)/\tau]$ that the remnant has been created at that point during the elapsed interval $(t_2 - t_1)$:

$$f_2 = f_1[1 - (t_2 - t_1)/\tau] + (1 - f_1)RV(t_2 - t_1).$$

In a statistically steady state, f_2 must equal f_1. Denote $f_2 = f_1 = f$, and derive the desired result $f = Q/(1 + Q)$. Estimate R from the observed fact that about one supernova explosion occurs every 50 years in a galactic volume of about 3×10^{12} cubic light-years. What is the corresponding value of Q, and therefore of f, if r and τ are given by Problem 11.14? By what factor does Q need to be increased to make the filling fraction f of coronal gas equal, say, to 0.5? Can you suggest reasons why Q might be increased from the value implied by Problem 11.14?

The Birth of Stars

The Galaxy is believed to be more than 10^{10} years old; yet there exist type O stars in the Galaxy whose main-sequence lifetimes are not more than 3×10^6 years. The formation of massive stars, such as Rigel in the constellation Orion, must therefore be an ongoing process in the Galaxy. This conclusion is reinforced by the finding of radio astronomers that OB stars are almost always surrounded by dense clouds of gas and dust. Evidently these young blue-white stars have not had enough time to wander far from the molecular-cloud cradles of their birth.

The sudden appearance of a brilliant hot star in a cloud of H_2 or HI gas must produce dramatic changes in the system. The dense hydrogen gas near the star will become completely ionized, to form a compact HII region (Figure 11.40). This HII region probably has somewhat higher densities and definitely much higher temperatures than the surrounding ambient gas (H_2 or HI region). Sooner or later, the much greater thermal pressures of the HII region will cause it to expand into the surrounding gas at speeds comparable to the thermal speed of the HII gas. This expansion provides another mechanism whereby interstellar gas can be pushed to speeds of the order of 10 km/sec.

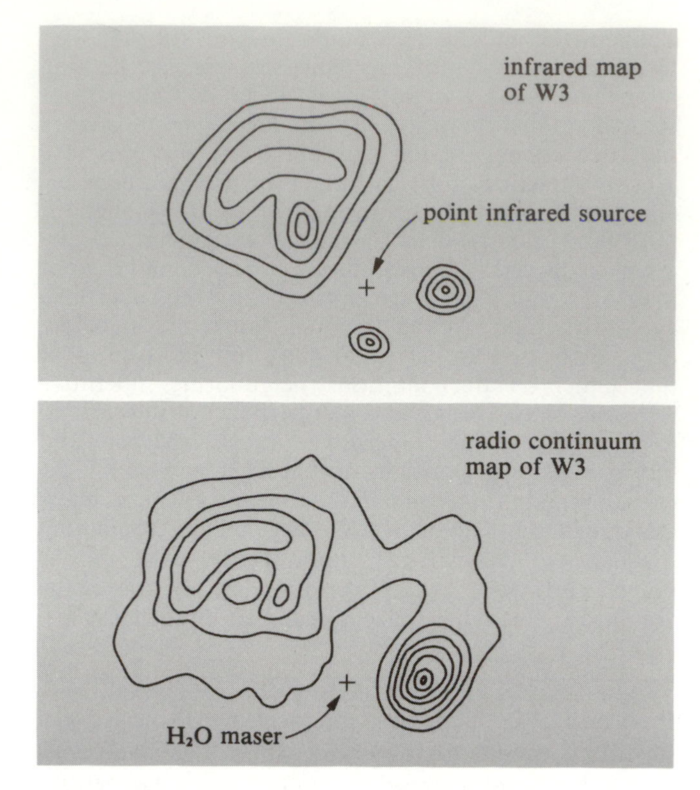

Figure 11.40. Comparison of the infrared and radio-continuum maps of the compact HII region W3 shows that they are very similar. There is also a point infrared source at the same position as a water-vapor (H_2O) maser. All this is evidence that this region, which is completely obscured by dust at optical wavelengths, is an active site of star formation. Sooner or later the high thermal pressures of this compact HII region will cause it to expand into the neighboring space. When the obscuring cloud of dust has been pushed away, the newly formed stars and surrounding ionized gas will probably be an HII region like those familiar from photographs of optical HII regions.

The expansion of HII regions into neighboring areas may also push pieces of the latter to become gravitationally unstable and to collapse to form new OB stars. Thus, once initiated, OB star formation might spread like an infectious disease throughout the whole molecular-cloud complex (a proposal made popular recently by Elmegreen and Lada). This proposal would explain why Blaauw has observed subgroups of OB stars, apparently formed at different times over a 10^7-year interval, in a loose association near molecular complexes. This process may be aided by Blitz's discovery that giant molecular complexes tend to be intrinsically quite clumpy.

Optical and infrared observations have yielded evidence that low-mass stars are also currently being born in giant molecular-cloud complexes. The story begins in the 1940s, with Joy's discovery of T Tauri stars, which have strikingly peculiar properties. Later studies by many astronomers (in particular, by Herbig and Kuhi) have elucidated some of the peculiarities.

(1) T Tauri stars are invariably embedded in dense patches of gas and dust. This fact necessitates observations in the infrared, because the neighboring dust can absorb the visible light of the star and reradiate it in the infrared. The dark clouds where T Tauri stars are found often also contain peculiar bright nebulae called Haro-Herbig objects (Figure 11.41), which were once thought to be contracting protostars; but the work of Strom, Strom, and Grasdalen points to their being gas and dust clouds which have been lit up by a flux of radiation or particles from a nearby T Tauri star.

(2) T Tauri spectra are usually dominated by intense emission lines. This feature is conventionally believed to be an indicator of strong chromospheric activity. (It is not entirely clear why strong chromospheric activity should be preferred in young stars; hence much research is being devoted to deciphering this riddle.)

(3) T Tauri stars show spectroscopic evidence for strong stellar winds. This is also taken to be an indicator of strong chromospheric activity.

(4) The brightnesses of T Tauri stars can change erratically, within a few hours, and we do not yet have any explanation for this.

Figure 11.41. A photograph of T Tauri, the prototype of T Tauri stars. The arcs of reflection nebulosity near the star constitute "Hind's variable nebula." The Herbig-Haro emission nebulosity very near the star shows on this photo only as an asymmetry of the overexposed star image (which has spikes because of diffraction inside the telescope tube). (Lick Observatory photograph, courtesy of George Herbig.)

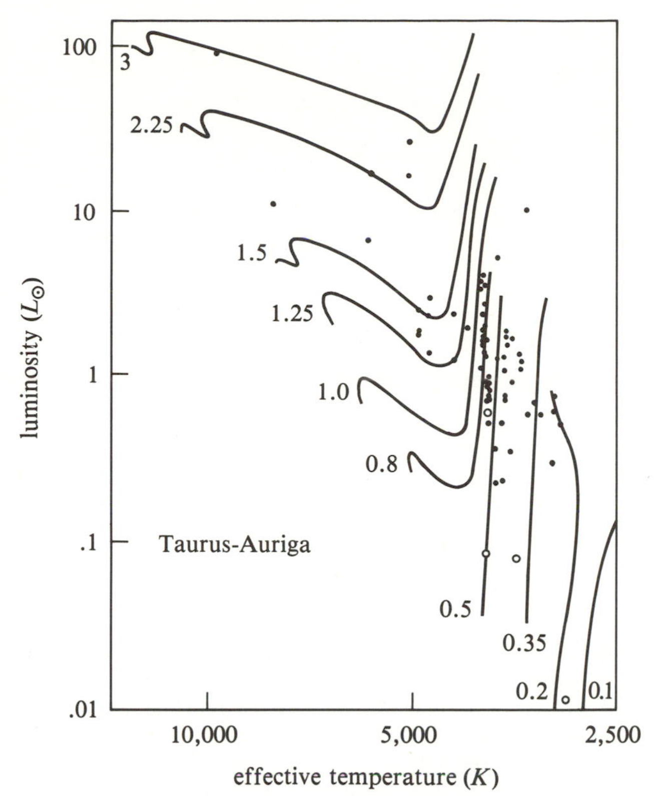

Figure 11.42. The Hertzsprung-Russell diagram of T Tauri stars in the Taurus dark-cloud complex. The curves depict classical tracks for pre-main-sequence stars, which derive their luminosities from quasistatic contraction. (Adapted from Cohen and Kuhi, *Ap. J. Suppl.*, **41**, 1979, 743.)

(5) T Tauri stars have unusually large amounts of lithium in their atmospheres. Since lithium is readily consumed by nuclear reactions in stellar interiors, the high lithium abundance is usually also taken as evidence of surface activity and extreme youth.

The best direct evidence that T Tauri stars are young, however, comes from their locations on the Hertzsprung-Russell diagram. Such diagrams have been constructed for T Tauri stars only relatively recently (by Strom and his coworkers, and by Cohen and Kuhi). Observationally, it is very hard to assign effective temperatures (because of the spectral peculiarities of T Tauri stars) and bolometric brightnesses (because of the need to add the infrared fluxes to the optical fluxes). The Hertzsprung-Russell diagram constructed by Cohen and Kuhi for the T Tauri stars in the Taurus dark-cloud complex is particularly revealing (Figure 11.42). Almost all the T Tauri stars in this region lie substantially above the theoretical zero-age main-sequence curve for Population I stars. This fact is roughly consistent with the hypothesis that all these stars are quite young and have only recently contracted out of the interstellar medium.

The theoretical work leading to this conclusion was begun by Henyey, Hayashi, and Larson (Figure 11.43). Henyey and his coworkers envisaged a gradual quasi-static contraction of a protostar through a sequence of radiative states which grew gradually smaller and smaller. For a star of low to moderate mass, the mass-luminosity-radius relation (see Problem 8.1) ensures that L will rise only slightly as the star contracts to smaller R along a radiative track. Hayashi and his coworkers pointed out, however, that the photospheres of stars cannot reach arbitrarily low values (see Problem 8.2); a "forbidden region" at low effective temperatures is inaccessible to the gas photosphere of a hydrostatic star. The contraction of pre-main-sequence stars from radii larger than a certain size must take place along an almost vertical track. The higher up this vertical segment a star is, the more convective its interior will be. Gaustad, among others, pointed out that the contraction of an interstellar cloud to stellar dimensions could not proceed entirely quasi-statically. Somewhere along the line, the radiative efficiency of the protostar would be so great, and thermal support of the object would be lost so fast, that a stage of dynamical collapse was inevitable. The question therefore naturally arose: where along the

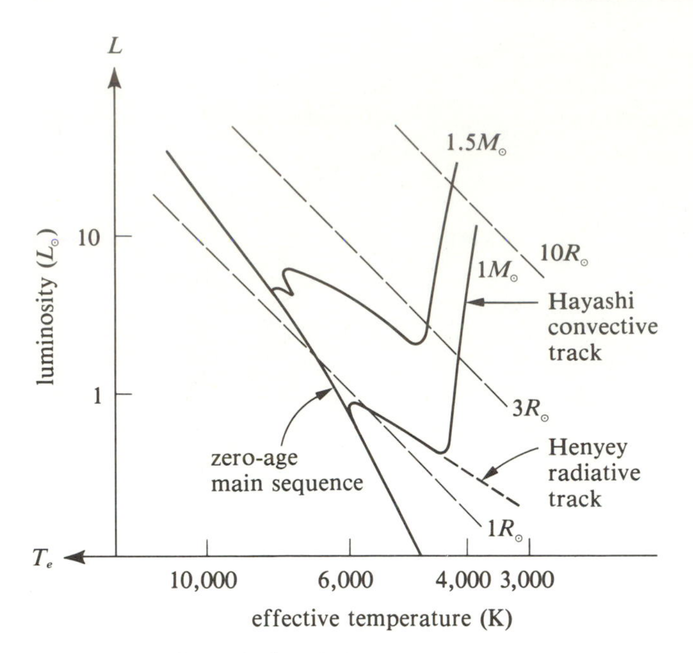

Figure 11.43. Theoretical evolutionary tracks for the pre-main-sequence phase of quasi-static contraction. Tracks are shown for a $1M_\odot$ and a $1.5M_\odot$ star. The dashed diagonal lines give the loci of constant radius. The $1.5M_\odot$ track has an extra bump just before reaching the main-sequence position that the $1M_\odot$ track does not have, because the heavier star must make an extra adjustment to burn hydrogen in equilibrium by the CN cycle (see Chapter 6).

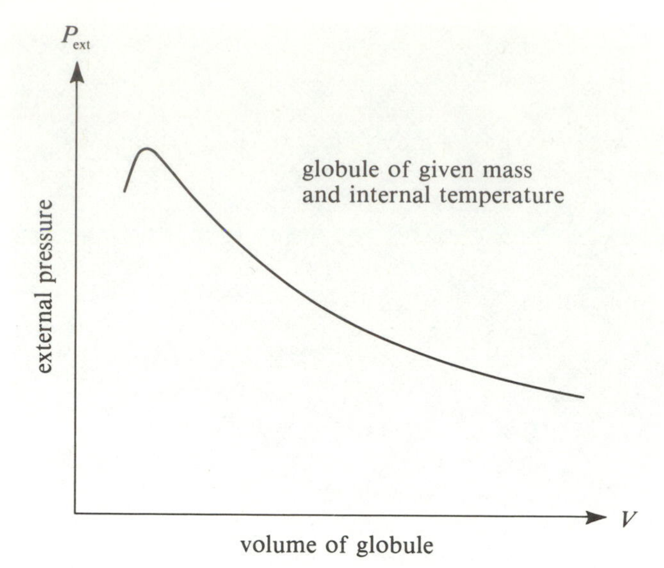

Figure 11.44. The relation between the volume of a self-gravitating isothermal gas cloud and the external pressure exerted on its surface. At small volumes the self-gravity becomes dominant, so that a decrease in the volume does not require an increase of the external pressure.

Hayashi-Henyey track would a star first make its appearance after the phase of dynamical collapse was over? Dynamical calculations carried out by Hayashi and his coworkers, by Larson, and by a number of other groups have given conflicting answers to this question. Some computations yield stars which start their quasi-static stage of contraction very high up the convective track; other computations yield stars which miss the convective phase altogether and start on the radiative portion of the track. The numerical calculations are very difficult to carry out accurately on an electronic computer, because one must follow changes in density by more than twenty orders of magnitude in the conversion of a piece of an interstellar cloud to a star. Recently, calculations have been published which use special techniques to avoid the numerical difficulties of the earlier calculations. The resulting stars of low to moderate mass are found to begin their quasi-static phase of contraction to the main sequence moderately high up on the convective "Hayashi tracks."

The picture of the earliest stages of a star's life that is emerging needs closer scrutiny. Imagine, for simplicity, a spherical globule of mass M whose temperature T remains nearly constant at 10 K as long as the gas and dust remain transparent enough to their own cooling radiation. As this globule is compressed, say, by an unspecified increase of the external pressure P_{ext}, the cloud volume V will decrease. Since the bulk cloud may be expected to behave like a perfect gas when the volume is large, V will at first decrease inversely with increasing P_{ext}. As V decreases, however, we expect self-gravity to become more and more important. Past a certain critical radius, which has been calculated by Bonnor and Ebert (Problem 11.16), the self-gravity becomes so strong that a further decrease in V requires no increase of P_{ext} (Figure 11.44). Past this point, the globule must become dynamically unstable to gravitational collapse.

Problem 11.16. In Chapter 5 we stated the virial theorem for an *isolated* gaseous body which is in hydrostatic equilibrium. This theorem has a form equivalent to $2U + W = 0$, where U and W are the thermal and self-gravitational energies of the gas mass. For a gaseous sphere which is partially confined by an external pressure P_{ext}, the right-hand side of the corresponding equation is not zero, but is equal to the radius r times the product of the surface area $4\pi r^2$ and the external pressure P_{ext}. In other words,

$$2U + W = 4\pi r^3 P_{ext}.$$

For an isothermal nondegenerate gas, $U = 3NkT/2$, where N equals the total number of particles in the cloud. For a total mass M and a mean molecular weight m, $N = M/m$. The self-gravitational energy of the cloud can be expected to be of the form $W = -\alpha GM^2/r$, where α is a pure number of order unity whose exact value depends on how concentrated the center of the globule is. Eliminate r in favor of $V = 4\pi r^3/3$, and show that the virial theorem can be written

$$P_{\text{ext}} = \frac{MkT}{mV} - \frac{\beta GM^2}{V^{4/3}},$$

where we have defined $\beta = (4\pi/3)^{1/3}\alpha/3$. Further define the nondimensional volume $v = V(GMm/kT)^{-3}$ and the nondimensional pressure $p = P_{\text{ext}}[G^3M^2(m/kT)^4]$, and show that the above equation becomes

$$p = \frac{1}{v} - \frac{\beta}{v^{4/3}}.$$

Plot p versus v under the assumption that β = constant = 0.45, and show that p has a maximum value of 1.1 at $v = 0.22$. Thus, show that the critical values of P_{ext} and V are given by $P_{\text{ext}} = 1.1G^{-3}M^{-2}(kT/m)^4$ and $V = 0.22(GMm/kT)^3$.

Problem 11.17. The external pressure on a globule in an HII region might be on the order of $P_{\text{ext}} = 10^{-11}$ dyne/cm^2. What is the critical mass M required to produce gravitational collapse in a globule of temperature $T = 10$ K under these circumstances? What is the radius r of such a globule? Assume the globule to be made of pure molecular hydrogen.

The dynamical collapse proceeds most quickly in the dense central regions of the globule. In many circumstances the dense central regions tend to fall away from the nearly hydrostatic outer regions to form an opaque core. When the core becomes opaque enough that the compressional heat can no longer be radiated away as quickly as it is generated by the collapse, the core's infall is arrested in a strong shock wave. After some transitory phenomena have died down, there is a hydrostatic object in the center of the globule which may be termed a protostar (Figure 11.45). The initial mass of the central protostar may be only $10^{-3}M_\odot$ or so, but it slowly accretes matter as the surrounding gas and dust fall onto it. This accretion phase builds up the protostar to stellar masses on timescales of 10^4 to 10^6 years, depending on the initial conditions which triggered the collapse.

Problem 11.18. The material in the outer regions of the globule begin to fall toward the central star only after the lower layers have dropped out from the upper layers. The information that the "bottom has dropped out" can propagate outward at a speed roughly given by the thermal speed $v_T = (kT/m)^{1/2}$. Show from Problem 11.17 that the radius r of the critical state is given by $r = 0.37GM/v_T{}^2$ and, therefore, that the time $t = r/v_T$ for the whole cloud to have fallen in toward the center is given by $0.37GM/v_T{}^3$. Since M is the mass of the whole globule, the mass accretion rate by the central protostar must be given roughly by $M/t = 3v_T{}^3/G$. Compute the numerical value of the mass-accretion rate for a temperature $T = 10$ K. Convert your answer to $M_\odot$ yr^{-1}, and comment on the time it takes to build up the central mass to stellar masses.

The structure and dynamics of the system during the accretion phase have been computed for the idealized problem of spherical collapse. The central features, starting from inside out, are:

(1) a hydrostatic core, the protostar;
(2) a radiating shock which defines the surface of the core and into which infalling matter flows and is stopped;
(3) a dust-free zone where the outward-traveling photons from the radiating shock have heated the gas to high-enough temperatures to destroy the dust;
(4) an opaque dust shell of infalling matter which transforms the outward-traveling optical photons into diffusing infrared photons;
(5) a dusty "false photosphere," where the infrared photons have been degraded to such long wavelengths that they can fly almost freely out of the surrounding infalling cloud.

During the accretion phase, an optical astronomer's view of the central protostar would be obscured by the overlying dust, and an infrared astronomer would see only the reprocessed radiation from the dusty "false photosphere." For a sufficently hot and bright central protostar, the dust-free region outside the accretion shock may have its hydrogen completely ionized, and may be detectable by radio astronomers as an extremely compact HII region. Exciting prospects lie ahead for the observations of the earliest stages of the formation of a star.

On the theoretical front, Bodenheimer and Black have emphasized that the idealized spherical-collapse

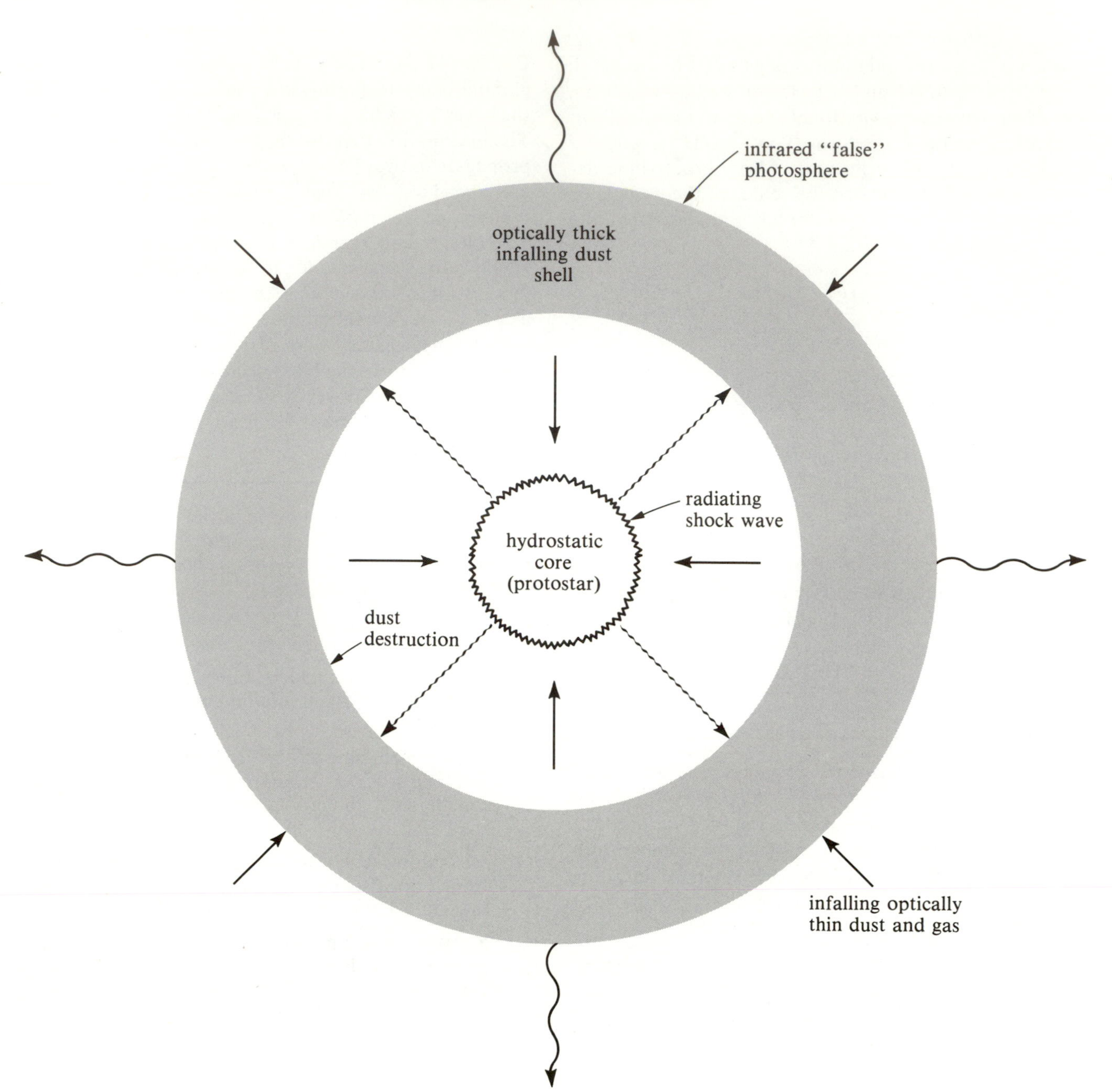

Figure 11.45. The theoretical structure of a low-mass protostar in the phase of accreting matter from a collapsing cloud envelope. The central protostar is hidden by an optically thick shell of infalling dust. (Adapted from Stahler, Shu, and Taam, *Ap. J.*, **241**, 1980, 637.)

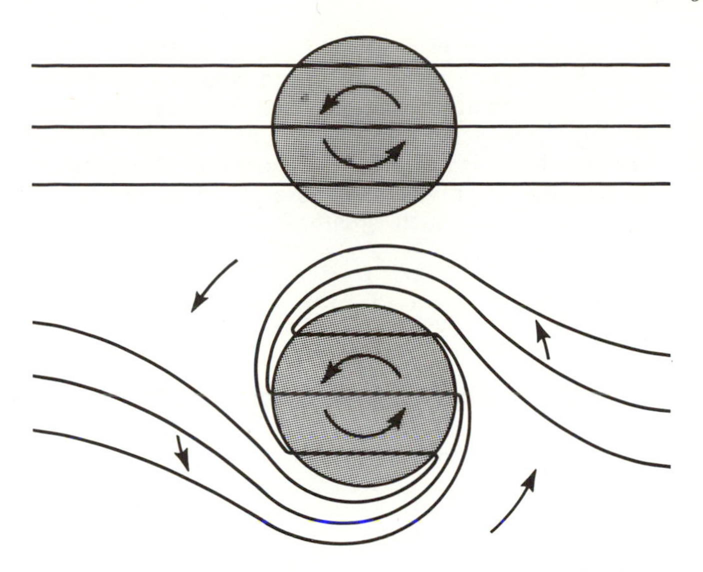

Figure 11.46. Magnetic-field lines which thread a rotating interstellar cloud may help to remove angular momentum from the cloud by transferring it to the surrounding medium. The frozen-in field lines wrap up as the cloud turns, and the resulting curved field lines tend to spin up the surrounding gas. More and more of the ambient gas is set into motion as the rotating disturbance propagates away from the gas cloud as an Alfven wave.

calculations are only a first step. Rotation of the piece of gas cloud which ultimately becomes a star or a pair of stars must play an important dynamical role when the collapsing object has reached a relatively small dimension. Magnetic fields probably play an important role in transporting most of the angular momentum of a contracting interstellar cloud to its surroundings, as was first suggested by Mestel and Spitzer.

The basic idea is relatively straightforward (Figure 11.46). Because of field freezing, the magnetic-field lines which thread both the cloud and the intercloud medium are twisted as the cloud rotates relative to the surrounding medium. The tension in the twisted magnetic-field lines imparts a force to the surrounding gas which makes it try to corotate with the cloud. The setting of the intercloud medium into rotation removes angular momentum from the cloud. In the language of magnetohydrodynamics, the rotating cloud radiates a torsional Alfven wave into the surrounding medium that transports angular momentum away from the cloud. The magnetic field affects, of course, only the charged particles directly; the neutral atoms and molecules are coupled to the magnetic field only by collisional drag between neutrals and ions. Thus, as the cloud contracts, the neutrals are pulled to the center in preference to the charged particles, which feel the resistance of the magnetic field that is tied to the surrounding matter. The slow drift of the neutrals relative to the ions, called "ambipolar diffusion," ultimately leads to a situation in which the combined forces of magnetic fields and gas pressure can no longer resist the inward gravitational pull of the concentrated material. This inner piece of the cloud then undergoes gravitational collapse, presumably much as has been described for the idealized problem without rotation and magnetic fields. Mouschovias believes that the cloud decouples from the magnetic field when enough angular momentum has been transferred to allow the collapsing cloud to form either a binary system or a single star with a surrounding planetary system. The theoretical and observational distinction between these two possibilities remains a challenge for the future.

Problem 11.19. There is a simple argument that a mechanism like ambipolar diffusion must remove the magnetic flux that would otherwise be frozen into the gas which undergoes star formation. Consider a magnetic field of uniform strength B_0 which threads a spherical gas cloud of radius r_0. The total magnetic flux contained by the gas cloud equals the product of the field strength and the cross-sectional area: $\Phi = B_0 \pi r_0^2$. Suppose such a gas cloud were to contract spherically, preserving its value of $\Phi = B\pi r^2$. To form a star from the general interstellar medium, show that an average increase in density by a factor of about 10^{24} is needed. What is the corresponding factor in the decrease of the radius? Given this ratio of r/r_0, what would be the magnetic-field strength B of the final star if the original frozen-in field strength had its general interstellar value, $B_0 = 3 \times 10^{-6}$ gauss? How does the implied value of B compare with Babcock's estimates (via the Zeeman effect) that the largest magnetic fields in normal stars are several thousand gauss?

Concluding Remarks

The study of the interstellar medium, once merely a major nuisance for optical astronomers, has become an important branch of modern astronomy. The subject offers fascinating variety, because, as long as self-gravity is not a prime factor, the regions to be studied extend far through interstellar space. The consequent extreme rarefaction of matter and dilution of the radiation field produce local conditions which are as far removed from thermodynamic equilibrium as anywhere in the universe. This fact provides both the interest and the challenge of the investigation of the interstellar medium. A plethora of competing processes vie for our attention, many of which beguile us with their individual allures. An unfortunate side effect of this multitude of riches is that it becomes hard to tell the "forest from the trees." For our purposes in this book, the most important astronomical theme is the interaction of the interstellar medium with the embedded stars. In particular, the birth, life, and death of stars are essential to keep the material between the stars churning with activity. In turn, the interstellar medium, bound gravitationally to the Galaxy, traps the heavy elements spewed out by dying stars. When pockets of gas and dust become dense and cold enough to become self-gravitating, the enriched material goes into forming a new generation of stars and planets. We see therefore that it is again the (indirect) interplay between the demands of gravity and those of the second law of thermodynamics which breathes life into the material between the stars.

CHAPTER TWELVE

Our Galaxy: The Milky Way System

When the ancient Greeks stared at the summer night sky, they could see a faint band of light stretching from horizon to horizon (Figure 12.1). They likened this band to a river of milk, and the modern word "galaxy" is derived from the Greek word for milk. In particular, our Galaxy—also called *the* Galaxy—is named the Milky Way system. Today we know that the band of light associated with the Milky Way comprises countless numbers of individual stars. Most of the visible stars in our Galaxy lie in a gigantic thin disk. We see this disk edge-on, because our solar system sits nearly in the mid-plane of this disk, more toward the outer rim than toward the center (Figure 12.2). Figure 12.2 also explains why the Milky Way appears more prominent during the summer months for an observer in the Northern Hemisphere than during the winter months.

The honor of discovering the true shape and size of our own Galaxy fell on the astronomer Harlow Shapley. Until Shapley's work, the method used to survey the Galaxy was star counts. The number of stars at each interval of apparent brightness in different directions of the night sky was counted and binned. The distribution of stars with different apparent brightnesses then yielded, in principle, the distribution of stars in space. As the simplest example, suppose that the fractional occurrence of a star with given intrinsic brightness (luminosity L) does not vary with position in the Galaxy. Further, suppose that stars of all intrinsic brightnesses were distributed uniformly throughout the universe. Then a simple argument shows that the number of stars $N(f > f_0)$, with apparent brightness f greater than a certain amount f_0, should be inversely proportional to the 3/2 power of f_0 (Problem 12.1),

$$N(f > f_0) = A f_0^{-3/2}, \tag{12.1}$$

where A is a proportionality constant.

Problem 12.1. A star of luminosity L at a distance r_0 will have an apparent brightness $f_0 = L/4\pi r_0^2$. All stars of the same luminosity but with distances $r < r_0$ will have apparent brightnesses greater than f_0. Suppose that the number density of stars of luminosity L, $n(L)$, does not depend on r (a spatially uniform distribution), and show that the number of stars of intrinsic brightness L which have $f > f_0$ is given by

$$N_L(f > f_0) = n(L) 4\pi r_0^3/3 = \frac{n(L) L^{3/2}}{3(4\pi)^{1/2}} f_0^{-3/2}.$$

The total number of stars with f greater than the limit f_0, independent of the intrinsic brightness L, is given by the sum of the above expression over all possible L. Show that this operation yields expression (12.1), and derive as a byproduct the expression for the proportionality constant A. What would be the qualititative influence of

Figure 12.1. (a) A photomosaic of the Milky Way, from Sagittarius to Cassiopeia. (Palomar Observatory, California Institute of Technology.)

Figure 12.1. (b) A map of the Milky Way hand-drawn by Martin and Tatjana Kesküla under the direction of Knut Lundmark. (Lund Observatory.)

interstellar extinction on the above line of reasoning? Of a nonuniform spatial distribution?

Equation (12.1) implies the following. Suppose one counts the number of stars brighter than a certain f_0 to be 1,000. Then equation (12.1) predicts that there should be eight times more stars, or 8,000 stars, brighter than $f_0/4$. The increase occurs because going to fainter light levels allows us to sample stars which lie at greater distances. In practice, when actual stars counts were carried out, the number of stars were found to increase less quickly with decreasing limiting light level f_0 than the formula (12.1). Since equation (12.1) was derived assuming a uniform distribution of stars, the conclusion was reached that the actual spatial distribution of stars must

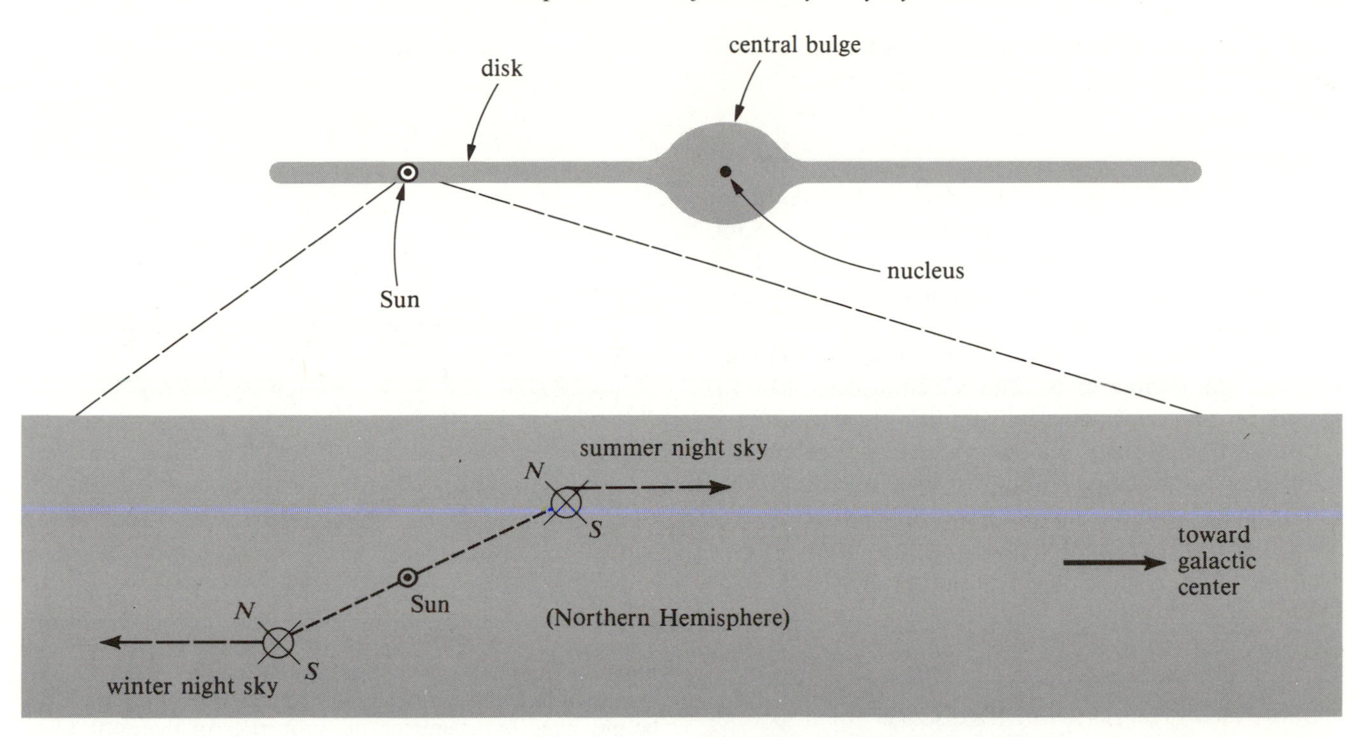

Figure 12.2. A schematic drawing of the Milky Way System viewed edge-on by an observer outside the Galaxy. The bottom, expanded view illustrates why an observer situated in the Northern Hemisphere has a more prominent view of the Milky Way during the summer months than during the winter months. In the summer, such an observer's night sky lies in the direction toward the Galactic center; in the winter, toward the anticenter. The scale of the bottom drawing is greatly distorted. The size of the solar system is actually a hundred million times smaller than the thickness of the Galactic disk.

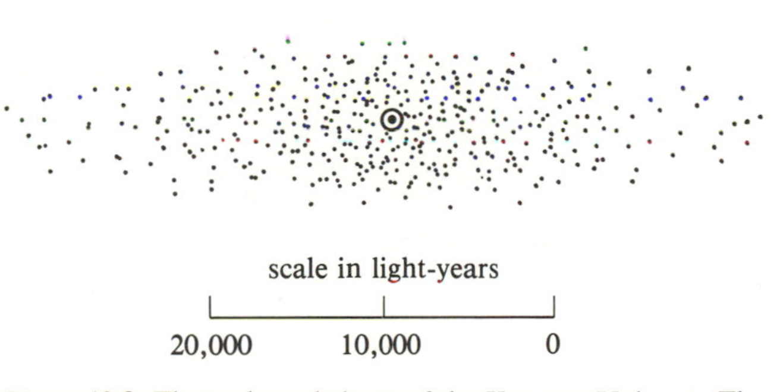

Figure 12.3. The scale and shape of the Kapteyn Universe. The Sun is located nearly at the center of the system, and the thickness is about one-fifth of the total radial extent. We know today that interstellar extinction greatly distorts the true sense of the distribution of the stars around us.

fall off with increasing distance from us, i.e., that we must be near or at the center of the distribution of all stars in the sky. Since the thinning out of the number of stars was obviously quicker in some directions than in others, it was concluded that we must live near the center of a flattened "stratum of fixed stars" in which the thickness of the Milky Way is about one-fifth of the diameter. This work culminated with Kapteyn, who calculated the spatial dimensions of this configuration (the "Kapteyn universe") by measuring the average distances of stars by statistical arguments (Figure 12.3).

The True Shape and Size of the Milky Way System

We now know that Kapteyn and the astronomers who preceeded him were drastically misled in their star counts because of the thick tracts of dust which obscure the central plane of the Milky Way (Figure 12.1). This obscuration causes the stars to "thin out" much more quickly with increasing distance from us than they would in the absence of interstellar dust. It was Shapley in 1917 who mounted the first serious challenge to the Kapteyn model.

Problem 12.2. The work of Kapteyn and his colleague van Rhijn was not useless. Their method of star counts is still used to derive the important quantity $n(L)$ in problem 12.1, now called the "luminosity function." Discuss qualitatively what information is needed besides $N(f > f_0)$ if we are to obtain the function $n(L)$. Discuss also the "Malmquist bias," the selection caused by the fact that we can see intrinsically luminous stars at greater distances than we can see stars which are intrinsically faint. To eliminate the Malmquist bias, we clearly need a "distance-limited" sample of stars rather than a "brightness-limited" sample. Discuss how one might proceed to obtain such a distance-limited sample. For example, how might one make use of proper-motion surveys?

Shortly after Henrietta Levitt's discovery of the period-luminosity relationship for Cepheids in the Small Magellanic Clouds, Shapley had proposed correctly that pulsation lay behind the variability of these stars. Pulsating stars, the RR Lyrae variables, were known to exist in globular clusters, and Shapley proceeded to apply the Cepheid period-luminosity relationship to the RR Lyrae stars to calculate the distances to the globular clusters in which they were found. Shapley did not know of interstellar dust, nor that RR Lyrae stars differ from classical Cepheids (Chapter 9); so he overestimated the distances to the globular clusters. However, modern measurements still use his basic method, and his general conclusions about the shape and size of our Galaxy have stood the test of time (Figure 12.4).

Shapley found that the globular clusters in the Galaxy have a roughly spherical distribution. He identified the center of the spatial distribution of globular clusters with the center of our Galaxy. The center of the distribution of roughly 200 globular clusters is believed today to lie about 30,000 light years from us. The globular cluster stars ("halo population") generally lie well out of the central plane of obscuring dust clouds and, therefore, suffer much less interstellar extinction than the stars in the disk of the Milky Way ("disk population"). The Sun, which is a disk star, is very far removed from the center of the Galaxy. Because of the dust obscuration, we cannot even *see* the Galactic center, much less be anywhere near it. In terms of its philosphical import for humanity, Shapley's dethronement of the Sun from the center of our stellar system has often been compared to Copernicus's dethronement of the Earth from the center of our planetary system.

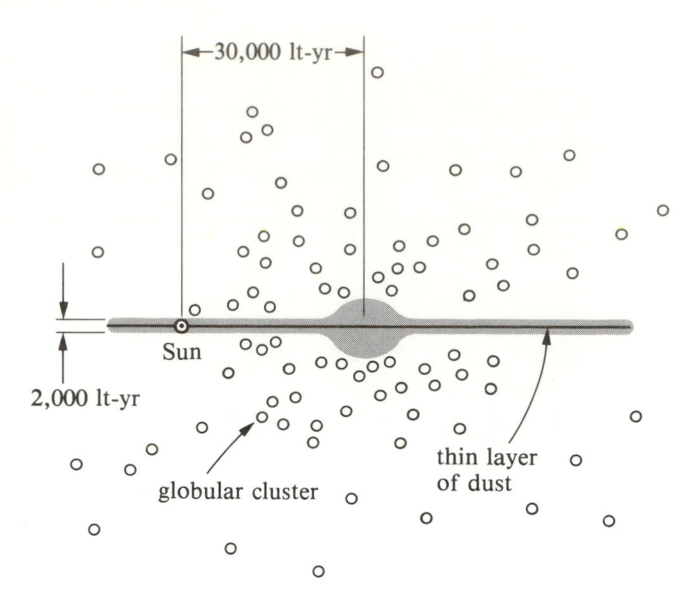

Figure 12.4. A modern view of the structure of our Galaxy. There are three major components: a thin disk of stars, gas, and dust; a central bulge; and a halo of old stars. Globular clusters are found in the roughly spherical halo, and their distances can be measured by examining the properties of their RR Lyrae stars. By this method, Shapley was able to calculate the distance to the center of the roughly spherical distribution of globular clusters. This distance is assumed also to be the distance to the Galactic center, a value now given as roughly 30,000 light years.

Although optical means cannot reach the very center of our Galaxy, astronomers have succeeded in probing the nucleus by radio, infrared, X-ray, and gamma-ray techniques. From such studies the realization has grown that the heart of our Galaxy may be witnessing energetic events which are a mild version of what goes on in the nuclei of much more active galaxies. The extraordinary events that take place in such objects will be discussed in Chapter 13, and we will consider the nucleus of our Galaxy there. Here we will concentrate on the global structure and dynamics.

Stellar Populations

As an outgrowth of Shapley's work on the structure of our Galaxy and Baade's study of the Andromeda galaxy, M31, astronomers came to realize that stars of high heavy-element abundance (Population I) differ from stars of low heavy-element abundance (Population II) in spatial location in the galaxy. The stars found in the disk are mostly Population I stars. The central bulge seems to comprise a mixture of Population I and II stars,

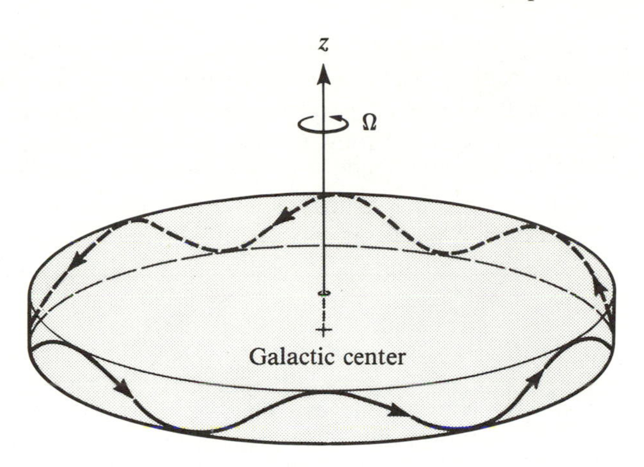

Figure 12.5. The circular speed of the orbit of a disk star is large compared to the random component of its velocity in the vertical (z) direction. As a consequence, the z-excursions of the star from the central plane $z = 0$ are small in comparison with the radius of the nearly circular orbit, and the entire collection of disk stars forms a flattened distribution about the galactic center.

Figure 12.6. The central bulge of the Galaxy consists of stars which do not have a strong sense of mean rotation. These stars swarm about the galactic center (marked with a cross) more or at less at random, like bees around a hive. The balance between the swarming motion and the attraction of gravity lends the central bulge its roughly spherical shape. The exact shape of the central bulge of our Galaxy is not well-known because of the heavy obscuration by dust toward this region.

and the visible halo of the Galaxy appears to be composed virtually exclusively of Population II stars.

Stellar Motions and Galactic Shape

We can understand the different shapes acquired by the various stellar populations in terms of their orbital characteristics in the Galaxy. Our Galaxy, like other galaxies, is a bound stellar system wherein the orbital motions of the individual stars keep them from all being collected in a dense mass at the Galactic center by the gravitational field of the entire system. The disk of the Galaxy contains mostly stars (but also a small mass fraction of gas and dust clouds), which travel in nearly circular orbits about the Galactic center (Figure 12.5). The circular speeds are large compared to the small superimposed random motions; this accounts for the highly flattened shape of the stellar distribution in the disk (Figure 12.4). The central bulge of the Galaxy contains mostly stars with small circular speeds; thus, these stars form a roughly spherical distribution (Figure 12.6). The Galactic halo contains stars of large random velocities, and these stars are less tightly bound to the Galaxy than the stars in the central bulge. It is not known whether there is a smooth transition from the stars in the central bulge to the stars in the visible halo.

Missing-Mass Problem

The thickness of the Galactic disk in the solar neighborhood can be measured for stars such as K giants, which are relatively bright and numerous and, therefore, relatively easy to see even at great distances from the central plane. Combined with a measurement of the random velocities of these stars, the thickness of the disk of K giants allows astronomers to deduce the strength of the vertical component of the local gravitational field (Problem 12.3). From the measured gravitational field, one can deduce the mass of material needed locally to provide the observed gravity. We can then compare the gravitationally required total mass with the available mass observed to exist in the form of stars and gas clouds. When the astronomer Jan Oort first carried out this program, he discovered that the observed mass is only about half the required mass. This discrepancy is only the first of a series which plague similar (dynamical) mass measurements of objects with galactic scale or larger. These discrepancies have become collectively known as the **missing-mass problem**. The missing-mass problem is one of the major unresolved issues of modern astronomy (Box 12.1). The important question is:

> Does there exist in the universe a vast amount of nonluminous matter which remains undetected (and perhaps undetectable) except for its gravitational influence?

BOX 12.1
The Missing Mass Problem.

(1) Oort's analysis of the vertical motions of stars in the solar neighborhood of the Galactic disk shows that twice as much mass must exist gravitationally as can be accounted for by luminous matter.
(2) The rotation curves of spiral galaxies stay flat out to substantial distances from the galactic center, indicating that the total gravitating mass in such systems must be at least a few times larger than that which can be seen in the optically visible parts.
(3) The velocities of galaxies relative to one another in binary systems exceed the binding capabilities of the galaxies unless they have masses which are at least several times larger than is usually inferred from the amount of light they possess.
(4) The random velocities of galaxies in a rich cluster are so large typically that the cluster would be unbound unless the total mass of the cluster is at least ten times larger than the sum of masses contained in the optical parts of the member galaxies.

Problem 12.3. The steady-state distribution of stars in the disk of the Galaxy is analogous to the problem of the hydrostatic equilibrium of the Earth's atmosphere if we replace the air molecules by K giants and the Earth by the Galaxy. Let ρ_K be the mass density of K giants in the central plane $z = 0$, and let v_z be the mean random velocity in the z-direction of such stars. Then the "pressure" associated with the z-random motion of the K giants at $z = 0$ must be given by the formula (refer back to Problem 4.1): $P_K = \rho_K v_z^2$. In a steady state, as many stars must cross each level in z during the upward portion of their orbital oscillation as cross in the downward portion, and the partial pressure P_K must equal the partial weight per unit area of K giants on either side of the plane $z = 0$. Let H be the effective thickness of the distribution of K giants, and $\bar{g}_z$ be the mean value of the z-component of the Galactic gravitational field that they sample; then the weight of a column of K giants with unit cross-sectional area is given by $\rho_K \bar{g}_z H$. Apply the condition of "hydrostatic equilibrium" and show that $\bar{g}_z = v_z^2/H$. Thus, derive $\bar{g}_z$ by requiring only measurements of the velocity dispersion v_z and scale height H of a (tracer) distribution of K giant stars.

For a plane-parallel slab of gravitating matter, use Gauss's law (Chapter 3), with volume V in the form of a rectangular box, to show that the field g_z at a large height from the central plane requires a mass per unit area $\mu = g_z/2\pi G$ to serve as a source. Assume $g_z = 2\bar{g}_z$, and compute the required total surface density of matter if $v_z = 18$ km/s and $H = 900$ lt-yr. How does the required μ compare with the available amount of about 8×10^{-3} gm/cm^2 in the form of observable stars and gas?

The Local Mass-to-Light Ratio

From star counts, one can also estimate the mean surface brightness of the Galactic disk in the solar neighborhood. If we now divide the surface mass density (mass per unit area, μ) by the surface brightness (luminosity per unit area, $\mathscr{L}$), we obtain a quantity that astronomers call the local mass-to-light ratio, $\mu/\mathscr{L}$. In solar units, this local ratio works out to be about 5, i.e.,

$$\mu/\mathscr{L} = 5M_\odot/L_\odot,$$

a value which is typical of the visible parts of disk galaxies in general. We note that the mass-to-light ratio of the Sun is, by definition, 1 in solar units. The fact that $\mu/\mathscr{L} = 5$ in the immediate Galactic vicinity of the Sun therefore has several interesting implications.

First, it means that the mean gravitating mass in the Galaxy has a lower efficiency for producing light per unit mass than the Sun. This conclusion is consistent with the observed fact that most stars in the Galaxy are faint K or M dwarfs. Such stars have low masses and even lower luminosities (Chapters 8 and 9). Of course, we know little about the "missing mass", except that it has even less light per unit mass than K or M dwarfs; so we arrive at the revealing view that most of the light of the Galaxy (except possibly at infrared wavelengths) comes from a disproportionally small number of high-mass blue stars or evolved late-type giants. Most of the mass, even in the disk of the Galaxy, exists in the form of low-mass faint dwarfs (or perhaps unobservable black dwarfs or black holes).

A second important conclusion concerns the production of helium. As Hoyle first noted, matter with such a low mean efficiency for generating luminous energy cannot have produced the present amount of helium in the universe. The Sun, with a mass-to-light ratio of 1, manages to convert only about 10 percent of its hydrogen mass into helium in 10^{10} yr (consult Problem 6.3). Matter with a mass-to-light ratio of 5 will manage by nuclear processes to convert only 2 percent of its hydrogen mass into helium in 10^{10} yr, which is roughtly the age of the Galaxy. Most of the helium produced this way would be locked up inside the stars and not be incorporated in new generations of stars. But even if this helium could get out, it would contribute less than 1/10 of the observed cosmic helium content of about 25 to 30 percent by mass. Of course, nuclear processing in stars might have occurred faster in the past than now, but astronomers believe that the helium produced in stars cannot much exceed 3 times the stellar heavy-element production (about 2 percent by mass). The bulk of the cosmic helium abundance therefore must have a nonstellar origin. Most astronomers now believe it originated during the first few minutes of the creation of the universe (Chapter 16).

Mass of the Halo

The observed mass of all luminous stars in the halo of the Galaxy has been estimated by Maarten Schmidt to be less than a few percent of the mass of the disk. Many astronomers hold the belief, popularized by Ostriker and Peebles, that the halo may also contain a very large and very massive invisible component. Recent estimates place the spatial extent and total mass of the halo at roughly 10 times the mass and extent of the disk. An unsubstantiated idea is that stars of such a massive halo constitute a Population III, an earliest generation of stars which have virtually zero initial heavy-element content. There are plausible but not conclusive arguments which support both sides of the massive halo controversy. This issue, like many involving the question of "missing mass," still awaits definitive resolution.

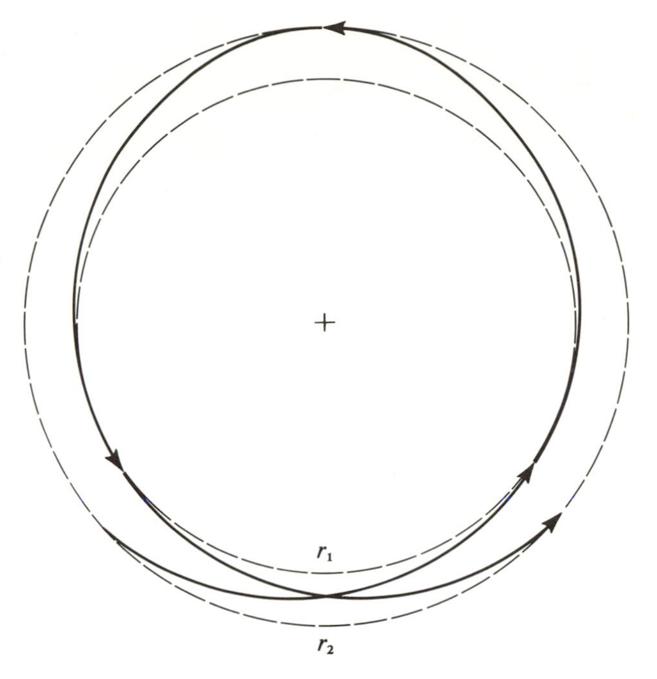

Figure 12.7. The orbit of a star projected into the plane of the disk. For a disk star, the orbit is nearly circular, with the radius confined to lie between an inner value r_1 and an outer value r_2. The time required to return to r_2, having started there, is generally less than the time required to go once around; so the orbit does not close on itself as Newtonian orbits around a mass point do.

Differential Rotation of the Galaxy

The most studied component of the Galaxy is the disk, and the dominant fact of the disk is differential rotation. The stars in the disk travel in nearly circular orbits around the center of the Galaxy (Figure 12.7). So do the interstellar gas clouds; their random motions, which are superimposed on top of the circular motion, are even smaller than that of the disk stars. The rotation of the stars and gas clouds around the Galactic center does not occur uniformly, like the rotation of a solid body. Galactic rotation takes place differentially, with the inner parts requiring less time to travel once around than the outer parts.

It might seem surprising that a feature as dominant as differential rotation would have been discovered only about fifty years ago. The flattening of the Milky Way provided a clue to the rotation of the stellar system, but this clue is indirect and inconclusive; moreover, flattening would take place whether the galactic disk rotates uniformly or differentially. More direct arguments awaited Lindblad and Oort's studies of stellar motions in the local neighborhood of the Sun.

The Local Standard of Rest

Part of the difficulty in measuring the rotation of the disk is that the Sun is a disk star and shares in this rotation. The Sun travels nearly with the mean motion of the disk stars and gas clouds in the solar neighborhood; consequently, this material appears to be nearly at rest relative to the solar system. The **local standard of rest** is defined in astronomy to follow the mean motion of the disk material in the solar neighborhood. The Sun has a slight known motion relative to this local standard of rest; however, the non-circular component of the Sun's total velocity is small enough that, for this book, we can take the solar motion to be synonymous with that of the local standard of rest.

There are a few stars in the solar neighborhood which have a large velocity relative to the local standard of rest. Such "high-velocity stars" exhibit a curious asymmetry. There are far fewer stars with large velocities in one particular direction than in the opposite direction (Figure 12.8). Lindblad provided the first correct explanation for this asymmetry. He suggested that the "high-velocity stars" actually form a distribution which has relatively low rotational velocities and large noncircular velocities (a mild form of the behavior exhibited by halo stars).

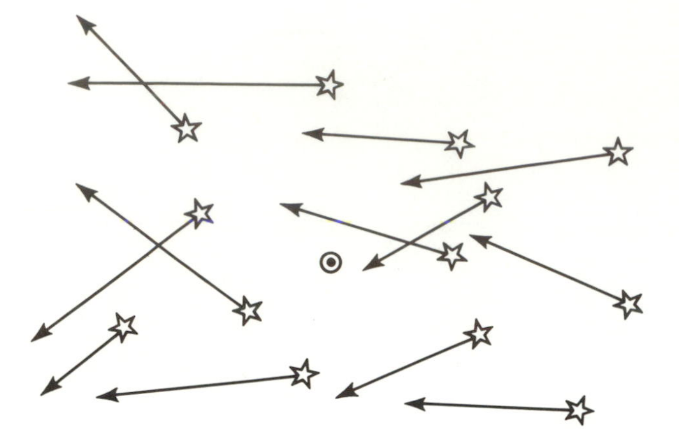

Figure 12.8. The asymmetry of "high-velocity" stars. Stars which move at velocities greater than about 65 km/sec relative to the local standard of rest seem to move exclusively from right to left (in the usual way that such diagrams are plotted); there are no known stars of "high velocity" which move from left to right. The explanation for this asymmetric behavior is given in the text.

The Sun and the other disk stars rotate rapidly past the former population; so the "high-velocity stars" seem to be streaming past us in the opposite direction. We think of ourselves as being at rest, but relative to an inertial observer at the center of the Galaxy, the "high-velocity stars" have smaller mean motions than we.

Local Differential Motions

The first studies of the effects of differential Galactic rotation were necessarily restricted in spatial scale, because the Galaxy is a big place, and optical astronomy allows us to sample only our local neighborhood inside the disk. The motion of a star within several thousand light-years of the Sun can be thought to consist of two parts: (a) a mean rotation which it shares with all other disk stars in its immediate vicinity; and (b) a random motion, superimposed on the mean, which is different for different stars. Surprisingly enough, differential rotation affects a local observer's perception of both the mean motions *and* the random motions. (Box 12.2). Even more surprising is that Lindblad's discovery of the latter preceded Oort's discussion of the former. Here, we shall find it conceptually easier to discuss the mean motions before the random motions.

Mean motions. Oort's analysis of the local effects of differential rotation on the field of mean motions is illustrated in Figure 12.9. On the average, stars interior to the Sun's position in the Galaxy take less time to go once around the Galaxy than stars located at the same radial distance from the Galactic center as the Sun. On the other hand, stars exterior to the Sun's position take more time to go once around. Thus, the Sun is like a car in the middle lane of a circular freeway. Faster cars (stars) pass the Sun on the inside, while the Sun overtakes slower cars (stars) on the outside. Referred to the motion of the Sun (or more accurately, to the mean motion of the stars in the immediate vicinity of the Sun), stars #1, 3, 5, and 7 have no motion along the line of sight to the Sun, whereas stars #2 and 6 have radial velocities of approach, but stars #4 and 8 have radial velocities of recession. The radial velocities of stars like Cepheids are readily measured, together with their distances from the Sun, and from such measurements two results emerge.

First, the expected pattern of mean radial motions outlined in Figure 12.9 is indeed observed. In particular, stars in the center and anticenter directions and in the circular directions have virtually zero mean radial velocity relative to the Sun, whereas maximum radial velocities of approach or recession at a given distance

---

BOX 12.2

Effect of Differential Rotation on the Mean and Random Motion of Stars.

(1) The Galaxy rotates differentially in the sense that stars in the disk which are closer to the Galactic center take less time to go once around than stars which are further. The mean motions of stars near the Sun then behaves like cars on a freeway, with stars outside the solar circle being in a slower lane and stars inside the solar circle being in a faster lane than stars on the solar circle. The pattern of shearing motions implied by this picture is indeed observed when the average differential motions of stars are examined carefully.

(2) Individual disk stars also possess small excesses and deficits of velocities compared with the mean of all stars in their immediate neighborhood. The differential rotation of the Galaxy also affects the pattern of random velocities. In a disk where the period of revolution increases outward, one expects theoretically to find the random velocities of stars in the direction of rotation to be systematically larger in magnitude than the random motions in the direction pointing radially away from the galactic center. The observed pattern of random velocities show exactly this behavior. In combination with a measurement of the shear in the mean motions, a quantitative analysis of the difference of random velocities in the radial and tangential directions allows astronomers to deduce all the important parameters characterizing the local Galactic rotation (the Oort constants, A and B).

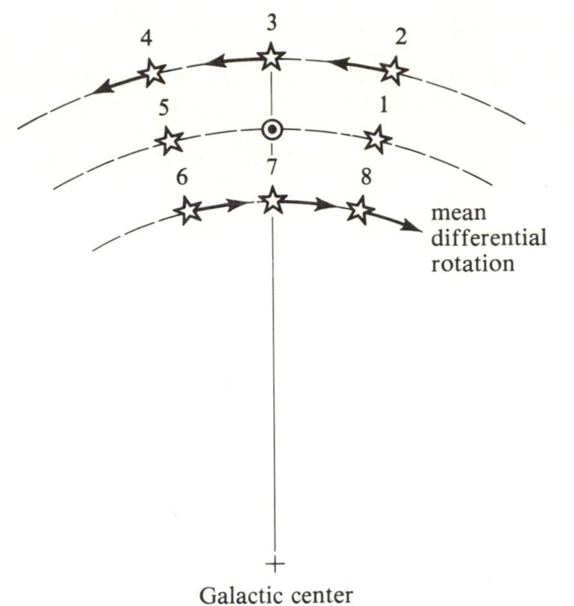

Figure 12.9. The pattern of motions to be expected for the differential mean motions of stars within a few thousand light-years of the Sun.

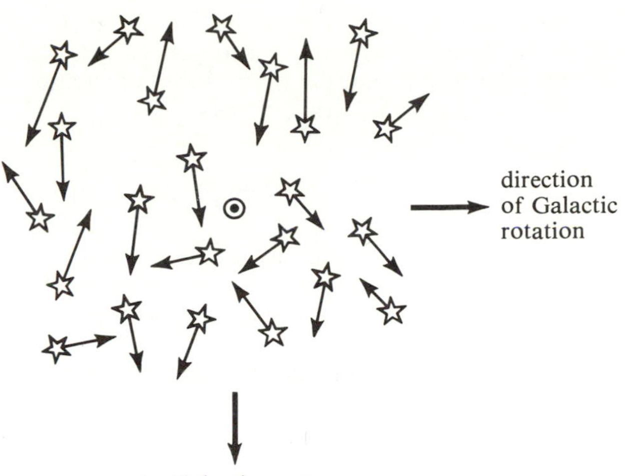

Figure 12.10. The random velocities of stars in the solar vicinity tend to be larger toward the Galactic center than in the direction of Galactic rotation. The unequal velocity dispersions in the two directions give evidence that the Galactic rotation occurs differentially and not uniformly.

are seen in the 45° positions. This pattern of local differential motions lends credence both to Oort's analysis and to Shapley's view that the Sun lies in an off-center position in the Galaxy.

Second, the quantitative measurements allow an assessment of the amount of *shear* that is present in differential Galactic rotation. This shear rate is proportional to the rate of change of the angular speed of rotation Ω with increasing distance r from the Galactic center, and is called Oort's constant A in astronomical circles.

In principle, just as the local shear rate (Oort's constant A) is obtainable from an analysis of the mean *radial velocities*, so should the *local* rotation rate (vorticity, related to Oort's constant B) be obtainable from an analysis of the mean *proper motions*. In fact, a simple thought experiment shows that this is generally quite a difficult task. Imagine that the Galaxy did rotate uniformly. Would it then be possible to measure the *absolute* rotation rate (angular velocity Ω) by examining the angular displacements of the disk stars with respect to one another? Clearly not, for in such a situation, the disk would rotate as a solid body and there would be no average displacements of one star with respect to another! By examining the proper motions of the disk stars relative to each other in the actual Galaxy, we would be able to recover information only about the proper motion which arises because of shear (*differential* rotation). In other words, examining the proper motions in this fashion would recover only Oort's constant A, not Oort's constant B. To obtain Oort's constant B, we would need a frame of reference other than that provided by the disk stars themselves. An inertial frame tied, say, to the quasars (which do not share in the rotation of the Galaxy), or to a platform that removes the dynamical effects of a moving Earth, might serve in principle. In practice, neither method yet yields very accurate measurements for Oort's constant B. This discussion should give an inkling of the subtleties involved in defining a true inertial frame once we give up the notion that the stars are "fixed" (Chapter 2). We shall return to this point in Chapters 15 and 16.

Random motions. The best existing method for obtaining the local rotation rate, Oort's constant B, involves a method originated by Lindblad (Figure 12.10). Lindblad showed that if the Galaxy rotated uniformly, then the *random* velocities in the direction of galactic rotation and in the direction toward the galactic center should be statistically equal. In fact, the observed values of these quantities in the disk stars are not equal. The ratio of the two velocity dispersions is given by Lindblad's theory as a certain combination of Oort's constants A and B. Thus, the measured ratio, plus knowing the numerical value of A, allows the deduction of the numerical value of B. The combination of A and B allows us to obtain the absolute angular speed Ω of the stars in the solar neighborhood and, therefore, to obtain the rotation period $2\pi/\Omega$. The time that it takes for the disk stars in the solar neighborhood to go once around the Galaxy is approximately 230 million years.

Combined with a distance r of the Sun to the Galactic center of 30,000 light years, a rotation period of 230 million years means that the linear speed of rotation of the Sun around the Galactic center is

$$v = r\Omega = 250 \text{ km/sec},$$

which is about half a million miles per hour! This calculation demonstrates two striking facts. First, the Galaxy must be enormous indeed, because even traveling at half a million miles per hour, the Sun still takes 230 million years to go once around. (If they had to start at rest, present-day rockets would require almost 10 billion years to go once around.) Second, we have here a striking demonstration of the principle of relativity. Despite the enormous speed of the Sun's motion through the Galaxy, we on Earth experience no great rush. Indeed, as we have discussed, the velocity of the Sun's rotation is very hard even for modern astronomers to measure, despite the use of very fancy and sensitive gadgets.

Problem 12.4. (requires a little calculus). Let Ω be the mean angular velocity of disk stars at a distance r from the center of the Galaxy. Oort's constants A (one-half the shear) and B (one-half the vorticity) are defined by the formulae

$$A = -\frac{r}{2}\frac{d\Omega}{dr}, \qquad B = -\frac{1}{2r}\frac{d}{dr}(r^2\Omega).$$

Show that $\Omega = A - B$, which equals $-B$ only if $A = 0$, i.e., only if Ω corresponds to uniform rotation. Analysis of the local differential motions yields $A = 0.0050$ km sec^{-1}/lt-yr. Measurement of the ratio of random velocities in the radial and circular directions yields $(1 - A/B)^{1/2} = 1.6$. (Notice that this ratio equals 1 for $A = 0$.) Compute the numerical value of B in the solar neighborhood, and calculate the rotation period $2\pi/\Omega$ in millions of years. If $r = 30{,}000$ lt-yr, calculate the circular speed $v = r\Omega$, and estimate the mass of the Galaxy interior to the solar circle by the crude formula $M_G = rv^2/G$. Convert your answer to $M_\odot$.

Extra credit. To obtain B, why not use the proper motions of disk stars measured relative to quasars as background objects that do not share in the rotation of the Galaxy? *Hint*: Since quasars are extragalactic objects, are they easily seen through the plane of the disk, where there is considerable extinction by interstellar dust?

Crude Estimate of the Mass of the Galaxy

Having obtained the speed of the Sun's orbit about the center of the Galaxy, we can make a rough estimate of the mass of the Galaxy. Or at least, we can make a rough estimate of that part of the Galaxy which is interior to the Sun's radius. This crude estimate assumes that the Sun moves on a circular orbit about a Galaxy whose mass interior to the Sun is all concentrated at a single point in the center. Since the mass of the Galaxy is much greater than that of the Sun, the problem reduces to a case analogous to our derivation in Chapter 5 of the mass of the Sun from the motion of the Earth. In the present case, the relevant formula is

$$M_G = rv^2/G, \tag{12.2}$$

where M_G is the mass of the Galaxy interior to the Sun, r is the Sun's distance from the Galactic center, and v is the Sun's rotational speed about the center. Putting in the values for r and v, we obtain

$$M_G = 1.3 \times 10^{11} M_\odot.$$

Problem 12.5. Show that the formula (12.2) holds exactly if the mass distribution of the Galaxy were spherically symmetric, providing we interpret M_G as the mass contained within a sphere of radius $r = 30{,}000$ lt-yr. (Consult Problem 5.6.) The formula also happens to hold exactly for a completely flattened distribution of matter if the rotation law were given by $r\Omega(r) =$ constant (independent of r) and if we interpret $M(r)$ as the mass within a cylinder of radius r (consult Problem 13.5). Some CO observations of Leo Blitz indicate $v = 300$ km/sec at $r = 60{,}000$ lt-yr. What is M_G if we use these values of v and r?

The Insignificance of Stellar Encounters

If half a solar mass were typical for the mass of a star in the Galaxy, the crude estimate given above would imply that the Galaxy interior to the Sun must contain roughly three hundred billion stars! Now, 3×10^{11} stars is a lot of stars, but the Galaxy is so large that the average spacing between stars is a few light-years. This is such a large distance that stars never collide with one another—at least, they have vanishingly small probability of doing so, except possibly right at the Galactic center. Except for bound binaries and multiple systems, in fact, stars hardly ever suffer even individual gravitational encounters with one another. They feel primarily the gravitational pull of the entire Galaxy, not the separate tugs of individual stars. Thus, for example, it is extremely unlikely that some star would pass in the near future close enough to steal Pluto from the Sun.

Problem 12.6. Use the formula

$$t_{\text{relax}} = \frac{V^3}{4\pi G^2 m^2 n \ln(RV^2/Gm)}$$

derived in Problem 9.11 to estimate the relaxation time for disk stars in the Galaxy. Assume a typical stellar mass $m = 0.5M_\odot$, a typical random speed $V = 40$ km/sec, and calculate the average number density n given 3×10^{11} stars in a galactic disk of radius $R = 30{,}000$ lt-yr and a total effective height $2H = 2{,}000$ lt-yr. How does the numerical value for t_{relax} compare with the typical orbital period, $\sim 10^8$ yr? What is the mean spacing $n^{-1/3}$ between stars in light-years? How does this compare with the distance to proxima Centauri? How does the effective distance $2Gm/V^2$ for a "close encounter" compare with the size of the solar system? By what factor $\ln(RV^2/Gm)$ do distant encounters dominate close ones in the formula for t_{relax}? Are gravitational encounters between individual stars an important process in the present Galaxy?

The Theory of Epicyclic Orbits

Because the stars in the Galaxy do not suffer significant individual deflections, they follow well-defined orbits in response to the collective gravitational field. In the language of stellar dynamics, the Galaxy is an encounterless stellar system. The quantities which are conserved during a star's motion, the so-called "integrals of the motion," play an especially important role in the structure of the Galactic system according to a theorem devised by Sir James Jeans:

> The distribution of stars in an encounterless stellar system depends only on the nature of all the integrals of motion.

A concrete example will fix ideas. Approximate the gravitational field of the Galaxy in lowest order to be steady and axisymmetric. Under such circumstances, it is advantageous to introduce cyclindrical coordinates (r, θ, z), where r measures the radial distance from the axis of rotation, θ measures the angular distance in the direction of Galactic rotation, and z measures the distance perpendicular to the Galactic plane (Figure 12.11). When the gravitational field is independent of angle θ and time t, the total energy E and the z-component of the angular momentum J per unit mass of each star can be shown to be conserved throughout its orbital motion. However, if the distribution of stars is a function of only E and J, then the random velocities of the stars in the z and r directions must be equal on average (Problem 12.7). For the disk stars in the Galaxy, the random velocities are observed not to be equal. The z-velocity dispersion at the solar neighborhood is only about 60 percent of the r-velocity dispersion. We infer, therefore, that the actual distribution of disk stars must depend on at least one more conserved property of stellar orbits. This issue is called the problem of "the third integral."

Problem 12.7. Denote the three velocity components of a star in the Galaxy by (v_r, v_θ, v_z). The energy and angular momentum per unit mass are given by

$$E = \tfrac{1}{2}(v_r{}^2 + v_\theta{}^2 + v_z{}^2) + \mathscr{V}(r, z), \qquad J = rv_\theta,$$

where $\mathscr{V}(r, z)$ is the gravitational potential energy per unit mass of a star whose position at time t is (r, θ, z). Because $\mathscr{V}(r, z)$ is assumed to be independent of t and θ, E and J are conserved as functions of time if we follow the motion of a star. In general, there are four other integrals of the motion in addition to E and J, but they cannot be written down as readily unless $\mathscr{V}(r, z)$ has some other special properties (for example, the gravitational potential corresponding to a point mass). Nevertheless, we can already draw some general conclusions. If the entire distribution of stars depends only on the functional forms of E and J, show that the average values of $v_r{}^2$ and $v_z{}^2$ cannot differ from one another. *Hint*: Can $v_r{}^2$ and $v_z{}^2$ enter in a nonsymmetric fashion if we consider an arbitrary function of E and J?

The basic foundations for understanding the orbits in the Galactic disk were laid by Bertil Lindblad. His theory—the **epicyclic theory**—generalizes Newton's description of slightly noncircular orbits for the case of gravitating mass distributions which consist of more than two mass points. As we shall see, Lindblad's theory also provides some insights into Ptolemy's description of the planetary orbits in terms of epicycles on top of circles.

We can begin a discussion of the epicyclic theory by noting that circular motion constitutes a basic reference

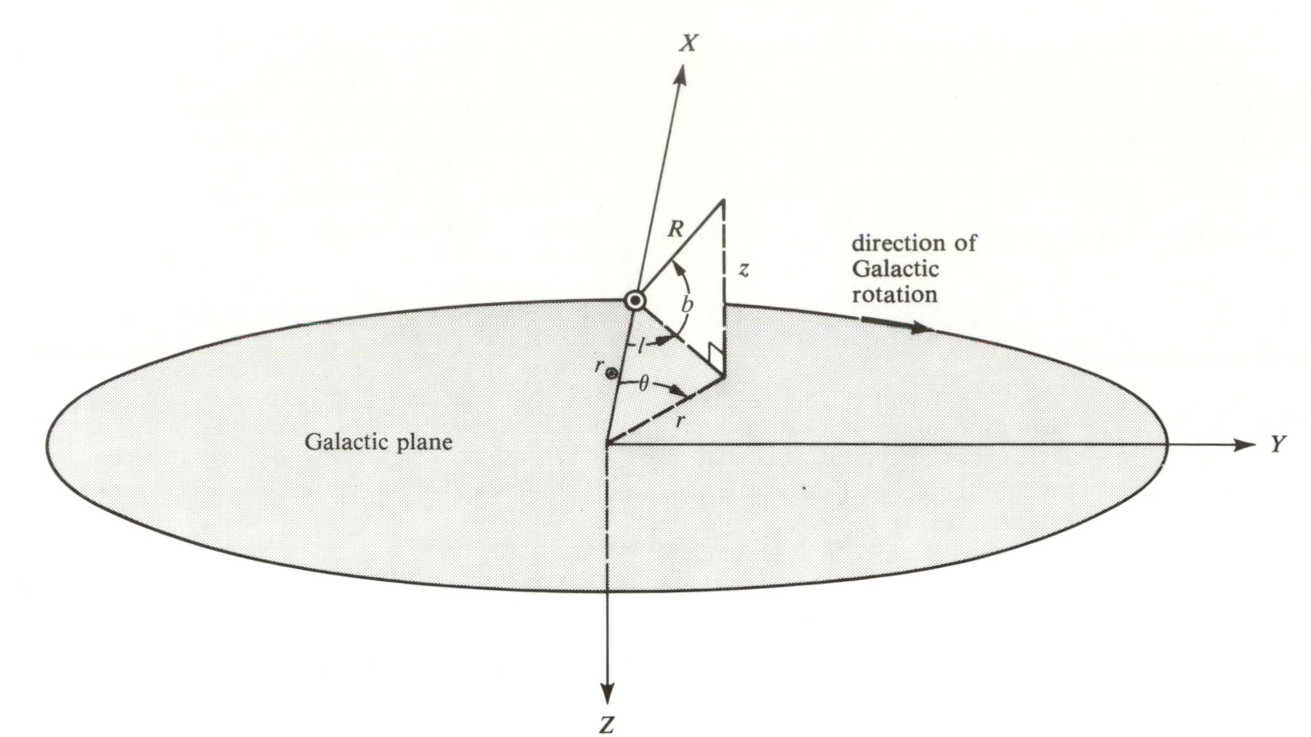

Figure 12.11. Galactic coordinates. It is convenient to introduce, for theoretical work, the cylindrical coordinate system (r, θ, z) with origin at the Galactic center; and for observational work, the spherical coordinate system (R, l, b) with origin at the Sun and with l being the longitude and b the latitude. Observers like to plot observations of the Galactic plane as if we were looking from the top down onto the above diagram. Thus, in observational plots, the sense of Galactic rotation occurs in a clockwise fashion. In terms of the right-handed coordinate system with origin at the center of the Galaxy, the z-axis would then point into the plane of the paper. In this book, we shall follow the observers and plot diagrams of our Galaxy with differential rotation occurring clockwise; in more general discussions, however (for example, Figure 12.7), we shall plot rotation counterclockwise (as if we were looking at the above diagram from the bottom up).

point for orbits in the Galactic disk. Circular orbits are special, in that of all orbits of a given angular momentum J, the circular orbit has least energy E. However, a disk star will usually have a small amount of noncircular motion; that is, its true energy E will usually exceed the minimum amount $E_c(J)$ that it would have if it had a circular orbit with the same angular momentum J. Let us call the positive difference, $E - E_c(J)$, the epicyclic energy $\mathscr{E}$. For $\mathscr{E} \ll |E_c(J)|$, as is true for any disk star, we expect the actual orbit of the star to execute small oscillations in all three dimensions about an epicenter, which is the locus in the central plane $z = 0$ of the reference circular orbit (Figure 12.12). For a thin disk, the oscillations in the vertical direction turn out to be nearly decoupled from those in the plane (Figure 12.5). This explains why the z-component of the noncircular velocity can remain on average a fraction of the r-component. In other words, the "third integral" is basically the partial energy of the z-component of noncircular motion. The z partial energy is (almost) separately conserved from the partial energy of the oscillations in the plane.*

* A better approximation for the "third integral" is the z partial energy multiplied by the period of oscillation in z.

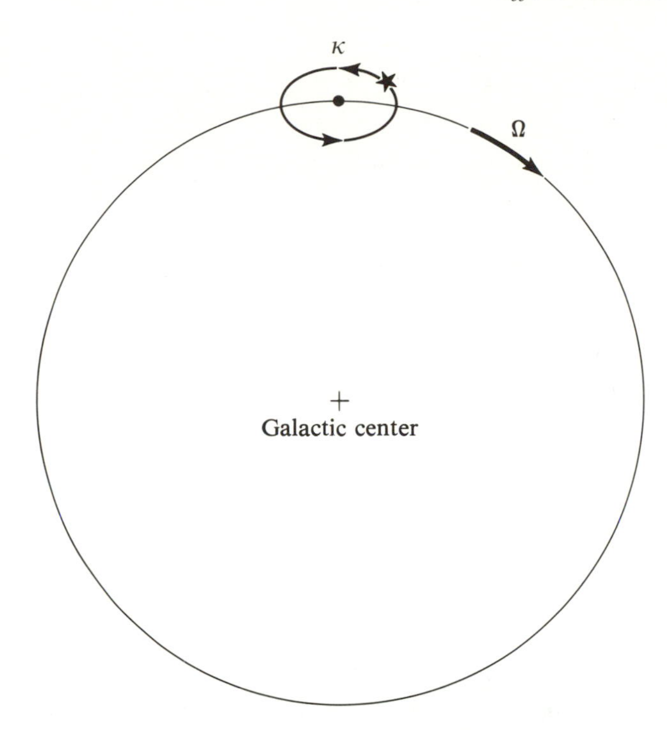

Figure 12.12. The epicyclic theory of Galactic orbits. For small noncircular motions, the orbit of a disk star can be thought to be composed of two parts: a rotation of the epicenter (dark point) around the Galactic center at the circular angular speed Ω; and a retrograde elliptical motion of the star about the epicenter at epicyclic frequency κ. (Vertical motion, perpendicular to the plane of the page, occurs with a frequency higher than either Ω or κ, and is illustrated schematically in Figure 12.5.) The epicyclic motion in the Galactic plane occurs in a retrograde sense to conserve angular momentum. In other words, to conserve angular momentum, the θ-velocity must be larger when the star is closer to the Galactic center than when it is farther away.

The oscillations in the r and θ directions take place with the epicyclic frequency κ, which differs in general from the circular angular speed Ω; so the full orbit seen by an inertial observer generally does not close, even when projected onto the central plane (refer to Figure 12.5). Seen by an observer who rotates with the epicenter at angular speed Ω, however, the noncircular motion describes a retrograde ellipse centered on the epicenter. The ratio of the short and long axis of the elliptical epicycle has the value $\kappa/2\Omega$, where κ and Ω in the solar neighborhood can be calculated in terms of Oort's constants A and B discussed earlier. In particular, only for uniform mean rotation is the ratio $\kappa/2\Omega$ equal to unity (in which case the elliptical epicycle would become a circle) and the noncircular velocities in the r and θ directions equal on average. The observed inequality between the velocity dispersions in the r and θ directions provided, as we have already discussed, Lindblad's deduction that the actual rotation in the Galaxy occurs differentially.

Problem 12.8. To calculate what a corotating observer in the Galaxy, like ourselves, would measure for the ratio of the r and θ velocity dispersions, we must imagine the statistical effect of seeing many stars on different portions of their epicyclic oscillations about different epicenters as they pass through the observer's neighborhood. The ratio $\langle v_r^2\rangle^{1/2}/\langle(v_\theta - r\Omega)^2\rangle^{1/2}$ turns out to be given theoretically by $2\Omega/\kappa = (1 - A/B)^{1/2}$, where

$$\kappa^2 \equiv r^{-3}d(r^4\Omega^2)/dr.$$

If the observed ratio of velocity dispersions is 1.6, what is the epicyclic period $2\pi/\kappa$, given that the circular period $2\pi/\Omega$ is 230 million years? Would the direct observation of an epicyclic orbit be possible within an astronomer's lifetime?

To complete our discussion of this subsection, we compare Lindblad's theory of orbits with Newton's. For the $1/r^2$ force law which would result if we replaced the actual Galaxy with a single mass point, the epicyclic frequency κ would exactly equal Ω. Since the time to execute one epicyclic gyration would then exactly equal the period of one rotation of the epicenter, the result would yield a closed orbit according to an inertial observer. This closed orbit would be fully equivalent to Newton's theory for bound orbits of small eccentricity (see Figure 3.7).

The Large-Scale Field of Differential Rotation

To study the large-scale structure of the Galaxy (for example, to obtain the overall mass distribution), we need to know the rotation speed at positions other than that of the Sun. To reach great distances from the Sun, astronomers must turn to radio observations to bypass the interstellar extinction that proves so limiting for optical observations. Indeed, Oort was among the first to recognize the potential boon of radio astronomy for investigations of galactic structure, and this is what prompted him to encourage his younger colleague van de Hulst to look for a useable spectral line in the radio region of the electromagnetic spectrum. The announcement of the 21-cm spin-flip transition of atomic hydrogen

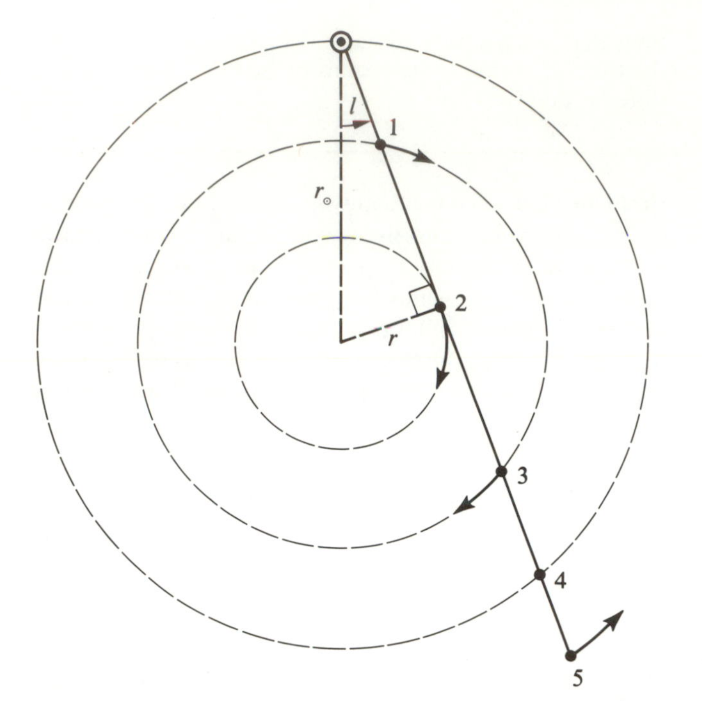

Figure 12.13. The pattern of large-scale differential rotation according to an observer who rotates with the speed of the local standard of rest. Notice that clouds 1 and 3 have the same line-of-sight ("radial") velocity relative to the Sun, and cloud 2 has the maximum positive velocity (recession). Notice also that cloud 5, which lies outside the solar circle, has a negative radial velocity, and cloud 4, which lies on the solar circle, is at rest relative to the Sun.

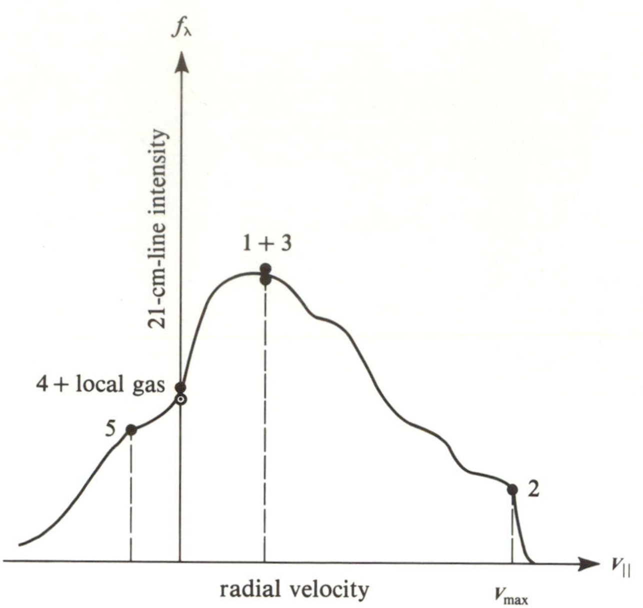

Figure 12.14. A schematic diagram of the 21-cm line profile along a direction of Galactic longitude l as in Figure 12.13. Except for the effects of slight noncircular motions, the maximum radial velocity recorded, V_{max}, should occur for the cloud 2 at the "tangent point." Intermediate positive velocities arise from the sum of the contributions of clouds like 1 and 3, which lie interior to the solar circle on either side of the tangent point. Zero-velocity gas corresponds to material on the solar circle, either in the immediate vicinity of the Sun or at the intersection point 4 in Figure 12.13. Negative velocity gas (in the first quadrant: $0° < l < 90°$) corresponds to gas beyond the solar circle.

by van de Hulst in 1945 changed the course of Galactic astronomy (Chapter 11).

Imagine moving with the Sun and pointing a radio telescope in the direction of Galactic longitude l (Figure 12.13). (Compared to Galactic scales, the size of the solar system is entirely negligible. Thus, Galactic and extragalactic astronomers often speak of being at the position of the Sun when they really mean being on Earth.) Because of differential Galactic rotation, we would see HI cloud #2 move fastest away from us; clouds #1 and 3 move more slowly away from us; cloud #4 has no motion relative to us; and cloud #5 (which lies outside the solar circle) moves toward us. Now, not knowing the absolute distances to clouds #1 and 3, we would see only that they have the same radial velocities and the same angular position in the sky (the same Galactic longitude l in the Galactic plane), so we would find it difficult to separate the effects of clouds #1 and 3. In contrast, cloud #2 is unique, because there is only one "tangent point" (or "subcentral point"), and the cloud at this tangent point has the highest radial velocity relative to the Sun (Problem 12.9). Since we know from the previous discussion the rotational velocity of the Sun, the measurement of the relative rotational velocity of the tangent point plus its unique geometry allows us to find its absolute rotational speed. In other words, the measurement of the highest radial velocity of the 21-cm line profile in a Galactic longitude direction l (Figure 12.14), allows us to find the rotation speed v at a distance r from the center given by the formula

$$r = r_\odot \cdot \sin(l),$$

where $r_\odot$ is the distance of the Sun from the Galactic center. By pointing the radio telescope to a variety of different longitudes l, we can obtain the rotational speed $v(r)$ for a variety of radial distances (of the tangent point) from the Galactic center. The result is shown in Figure 12.15, a diagram known as the rotation curve of the Galaxy. It is simple to see that the method only works for values of r less than the radius of the solar circle, $r_\odot$, because outside the solar circle there is no tangent point.

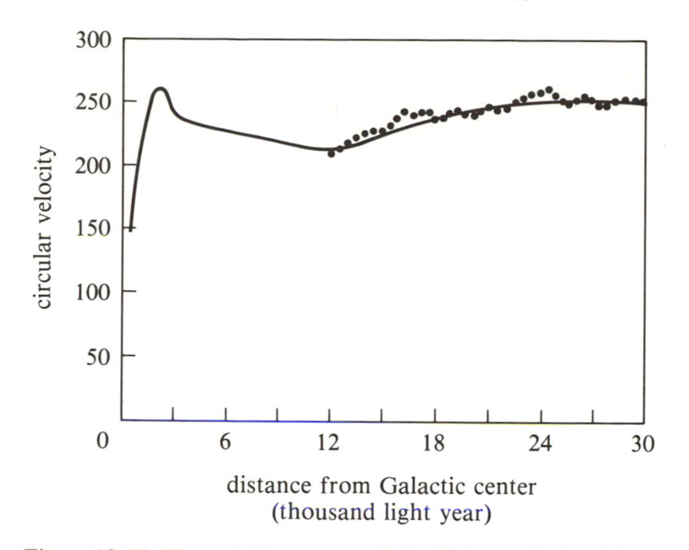

Figure 12.15. The rotation curve of the Galaxy obtained from 21-cm observations of the atomic hydrogen gas interior to the solar circle.

Problem 12.9. In a frame which rotates with angular velocity $\Omega(r_\odot)$, the circular velocity of a gas cloud at radius r is given by $r[\Omega(r) - \Omega(r_\odot)]$. Use the diagram below, after justifying the various angular identifications, to show that the component of velocity along the line of sight is given by

$$V_{\parallel} = r[\Omega(r) - \Omega(r_\odot)] \sin(\theta + l).$$

Use the law of sines to show that $r \sin(\theta + l) = r_\odot \sin(l)$. In this fashion, derive the formula

$$V_{\parallel} = r_\odot[\Omega(r) - \Omega(r_\odot)] \sin(l).$$

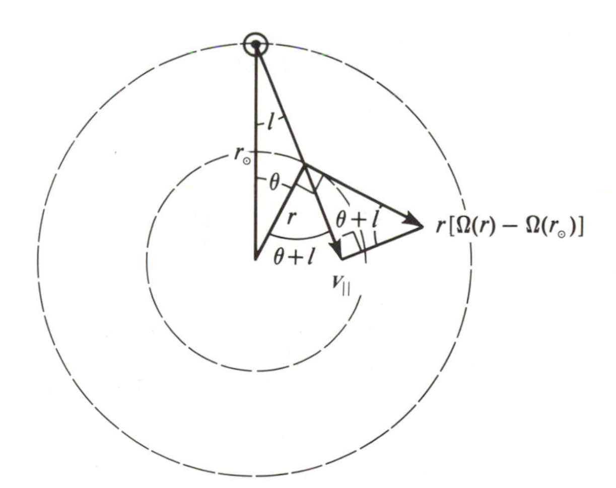

For a given line of sight l, the only thing which can vary in the above formula is the radial location r of our gas cloud. If $\Omega(r)$ is a monotonically decreasing function of r, argue that $V_{\parallel}$ acquires its maximum positive value (for $0° < l < 90°$) when r corresponds to the radius of the tangent point: $\theta + l = 90°$. The circular velocity of the tangent radius $r = r_\odot \sin(l)$ is, therefore, given by the formula

$$v(r) = r\Omega(r) = V_{\max} + r_\odot \Omega(r_\odot) \sin(l),$$

where everything on the right-hand side is obtainable from observations.

To obtain the rotation curve outside the solar circle, astronomers must rely on optical observations to calibrate the distances to the celestial objects whose radial velocities are to be measured. The classical method of extending the rotation curve beyond $r_\odot$ uses Cepheid variables. However, optical spectroscopy of single stars to obtain radial velocities becomes quite difficult when such stars lie more than several thousand light years away in the Galactic disk. Thus the Cepheid technique fails to get us substantially further out beyond $r_\odot$.

A more promising method has recently been pioneered by Blitz. Giant HII complexes which lie at the radial outskirts of the disk suffer relatively little dust obscuration, and their distances can be found photometrically (albeit without great accuracy) by studying the optical properties of the exciting stars. Giant molecular clouds accompany the HII complexes (Chapter 11), and the radial velocities of such giant molecular clouds can be found from their carbon-monoxide (CO) emission. By this technique, the rotation curve of the Galaxy can be extended almost to $2r_\odot$ (see Problem 12.5). The resulting rotation curve, when appended to that of the HI observations for $r < r_\odot$, is in Figure 12.16.

Figure 12.16 makes it clear that the Galaxy actually rotates differentially, and its matter distribution is more spread out radially than a mass point. Given some assumptions about the extent of the various distributions in the directions perpendicular and parallel to the Galactic plane, astronomers can construct **mass models** which make much more accurate estimates than the simple estimate of Problem 12.5. As I write, attempts are underway to make mass models which include the information contained in the extended rotation curve. Good fits for the data seem to require at least three components: a central bulge, a flattened disk, and a massive halo. Moreover, we can argue that the matter constituting the halo must be relatively nonluminous. Given that the rotation speed v beyond the solar circle does not fall with increasing distance r, equation (12.12) predicts that the mass interior to r = 75,000 lt-yr must

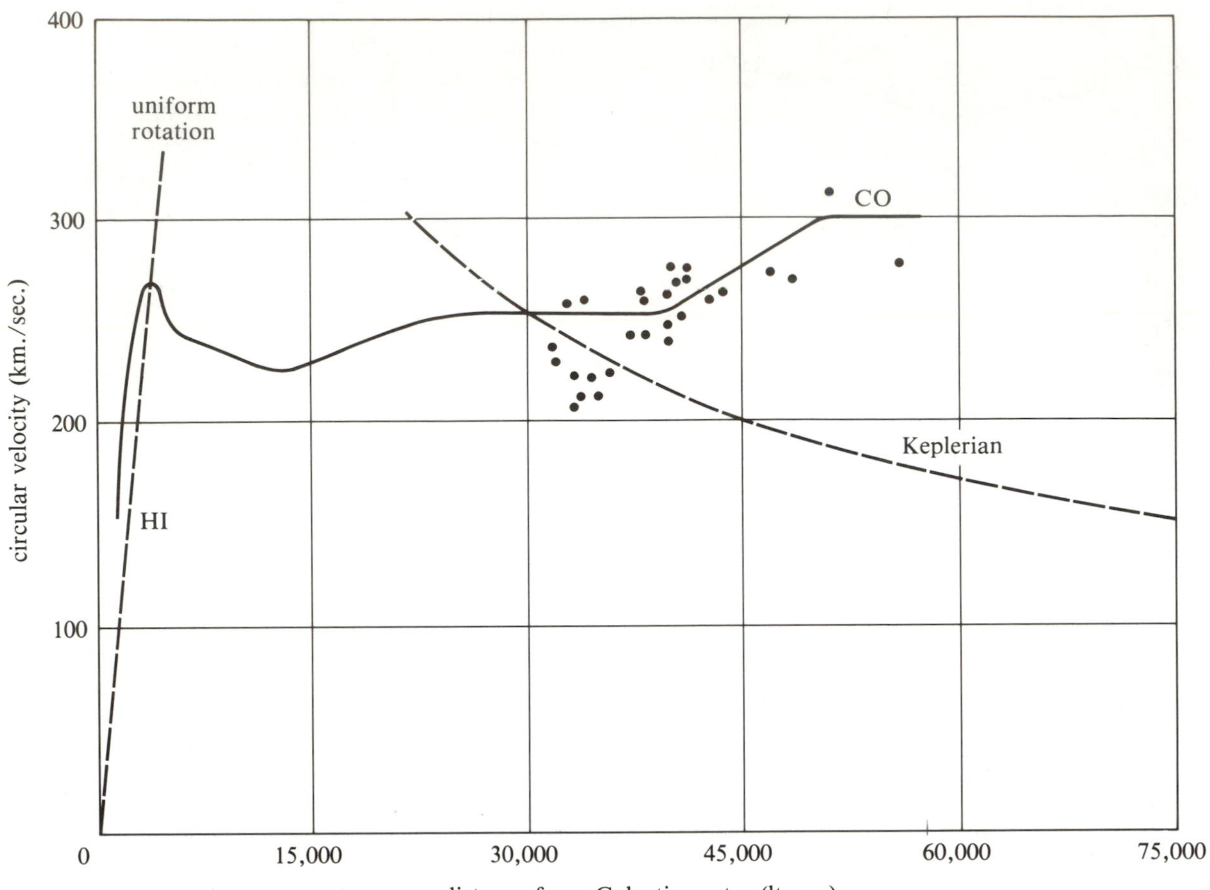

Figure 12.16. The rotation curve of the Galaxy extended beyond the solar circle by means of photometry of OB stars and CO observations of their associated molecular clouds. The dashed curves show what the rotation curve of the Galaxy would have looked like if it had either rotated uniformly or contained all its mass at a single point at the center.

be about three times its value at $r = 30{,}000$ lt-yr. But observations indicate that there is little light from a galaxy like our own at large distances from the galactic center. The local mass-to-light ratio must therefore increase dramatically in the outer parts, where the halo dominates. This behavior makes our Galaxy similar to other disk galaxies whose rotation curves have been studied by Morton Roberts, Vera Rubin, and other astronomers.

The Thickness of the Gas Layer

Our discussion of the tangent-point observations in the previous subsection led us to conclude that the tangent point has a unique radial velocity (the maximum value along any given line of sight l). It is conventional to assume that this property holds as well at small vertical distances off the central plane. By studying the drop-off in the 21-cm line emission as they move their telescopes to nonzero values of galactic latitude b (the angular direction perpendicular to the plane; see Figure 12.17) at a fixed galactic longitude l, radio astronomers can measure the effective thickness of the gas layer at such tangent points. They find that the atomic hydrogen gas has essentially a constant total thickness of about 700 light-years at all radii r interior to $r_\odot$. Similar experiments have now been carried out in CO emission for the distribution of molecular clouds. The layer of molecular gas has a total thickness of about 300 light-years. Since the diameter of the solar circle is 60,000 light-years, we see that the layer of gas and dust in the Galaxy is proportionally almost as thin as an index card.

From the study of Kerr and others, it is known that the atomic-hydrogen layer thickens considerably beyond

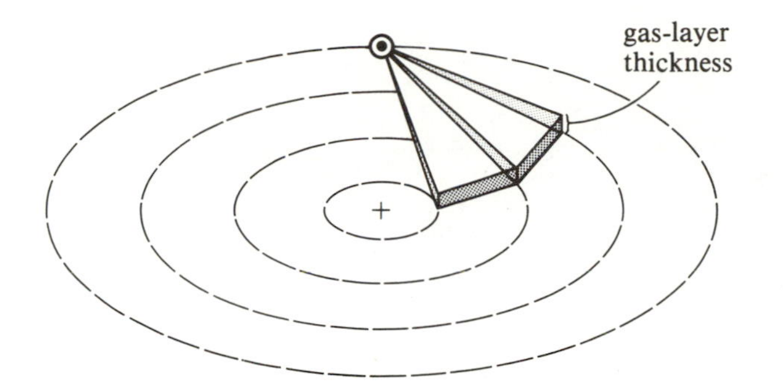

Figure 12.17. Because the tangent point has a unique position for each longitude l, the vertical thickness of the gaseous layer may be measured at the tangent point. One need only sweep the radio telescope in galactic latitude b for fixed l to see how extended vertically is the 21-cm emission at velocity V_{max}. Repeating this procedure at several different longitudes l gives a cross-sectional profile of the vertical distribution of atomic hydrogen gas.

Figure 12.18. A schematic edge-on view of the gaseous disk of our Galaxy illustrating both the flare of the disk thickness beyond the solar circle and the warp of the plane.

the solar circle (Figure 12.18). This fact receives the following conventional explanation. The surface-mass density of stars probably decreases exponentially in the outer parts of the disk, judging from de Vaucouleur's measurements of the characteristic fall-off of the disk light in external galaxies. For a given level of support by random velocities, magnetic fields, etc. (see Figure 11.32), a decrease of the gravitational attraction of the stellar disk for the gas would result in a greater layer thickness for the gas. This effect may account for the flare of the hydrogen-gas layer in the outer Galaxy.

The gas layer also exhibits a warp in its outer parts, like an index card that has been left out in the rain. The physical reason for this warp is unclear, but it has now been observed to occur in many external galaxies also. Hunter and Toomre have suggested that a recent passage of the Magellanic Clouds may have gravitationally excited such a warp. In favor of this hypothesis is the discovery by Mathewson and Cleary of a stream of hydrogen gas which bridges the Clouds and the Galaxy (Figure 12.19). The picture suggests that strong tidal forces resulting from a close encounter may have torn a ribbon of material from the Clouds, much as similar processes result in gas streams between interacting binary stars (Chapter 10).

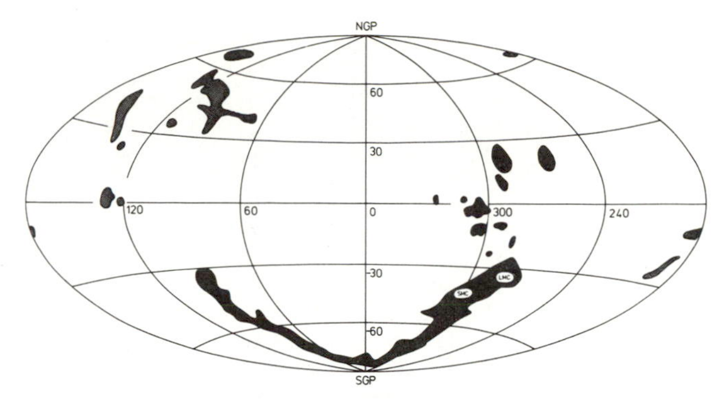

Figure 12.19. High-velocity hydrogen gas is plotted on the sky in Galactic coordinates, with the Galactic equator running horizontally in the middle and the north and south Galactic poles situated at the top and bottom. The Magellanic stream is a ribbon of gas which spans the gulf between our Galaxy and its two satellites, the Large and Small Magellanic Clouds (LMC and SMC). It has been speculated that this ribbon of gas was torn in a violent gravitational encounter between the Galaxy and the Magellanic Clouds. (From Mathewson and Cleary, IAU Symp. No. 58, 1973, 368.)

Kinematic Distances

The measurement of Galactic distances poses a vexing astronomical problem when optical methods are not available, as they often are not in the plane of the disk. Radio astronomers have devised an ingenious technique for estimating the distances to Galactic radio sources whose line-of-sight velocities can be measured by Doppler shift and whose kinematics can be assumed to correspond to pure circular rotation. The essence of the method lies in knowing the rotational velocity at each radial distance r in the Galaxy. This knowledge plus a measurement of the component of velocity along the line of sight fixes the position of the object (Problem 12.10). For HI gas interior to the solar circle, the **kinematic distance** obtained this way is not unique. There are two possible choices, a far point and a near point, that are both consistent with the radial velocity measurement (see Figure 12.13 for cases not corresponding to the tangent point). If we assume that the HI gas has a constant layer thickness everywhere in the inner Galaxy (and not just at the tangent points), we can remove the distance ambiguity, at least in principle.

We begin by examining the same velocity feature at different Galactic latitudes b. If the gas at that velocity has a relatively small angular extent in b, it must lie at the far point; if a relatively large angular extent, at the near point (Figure 12.20). The same technique does not work as well for resolving the distance ambiguity of CO observations of molecular clouds (or of H 109α observations of HII regions). Such objects are discrete entities,

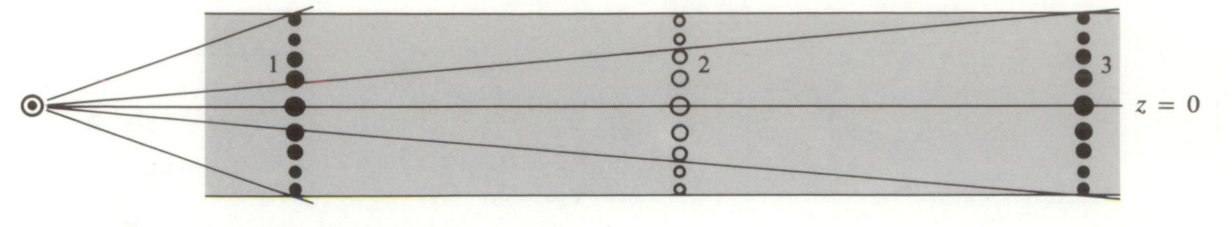

Figure 12.20. The resolution of the distance ambiguity in the kinematic method. If we assume the layer of atomic hydrogen gas (shaded region) to have a uniform thickness along the line of sight, gas clouds at the far point 3 can be distinguished from gas clouds at the near point 1 because of the difference in angular extent in the vertical direction. Gas clouds along the same line of sight, but at intermediate distance 2, cause no confusion because they differ in radial velocity from the gas clouds at points 1 and 3 (see Figure 12.13). This method obviously works only for gas clouds, like those of atomic hydrogen, which are sufficiently numerous at each point to give us reasonable statistics for their vertical distribution in the Galaxy.

and they are subject to much greater statistical uncertainty than the HI gas. For completeness, we should note that the kinematic distances of objects exterior to the solar circle do not suffer from the distance ambiguity.

Problem 12.10. Recall the formula

$$V_{||} = r_{\odot}[\Omega(r) - \Omega(r_{\odot})] \sin(l)$$

derived in Problem 12.9. If we know $r_{\odot}$, $\Omega(r_{\odot})$, and the functional form of $\Omega(r)$, and we measure $V_{||}$ when we point a radio telescope in the direction l, this equation clearly allows us to deduce r. Describe a way to perform this operation by using graphs. Consult, now, the diagram in Problem 12.9. Show that if $r < r_{\odot}$, the line of sight generally intersects the circle of radius r at two points, a near point and a far point. This is the distance ambiguity. Show that this distance ambiguity does not arise if $r > r_{\odot}$.

Spiral Structure

It is now known that many disk galaxies have spiral structure (refer to Color Plate 13), and it is natural to believe that our Galaxy may too. Optical spiral structure in external galaxies is outlined primarily by OB stars and their associated HII regions, as was first realized by Baade for the Andromeda galaxy, M31. In 1951, Morgan, Sharpless, and Osterbrock traced the locations of the three optical spiral arms in the Galaxy nearest to the Sun. A global view of Galactic spiral structure had to wait, however, for the maturation of radio astronomy. In 1958, Oort, Westerhout, and Kerr combined the 21-cm observations made in the Netherlands and in Australia to produce the contour map shown in Figure 12.21. The location of the hydrogen gas was obtained by the method

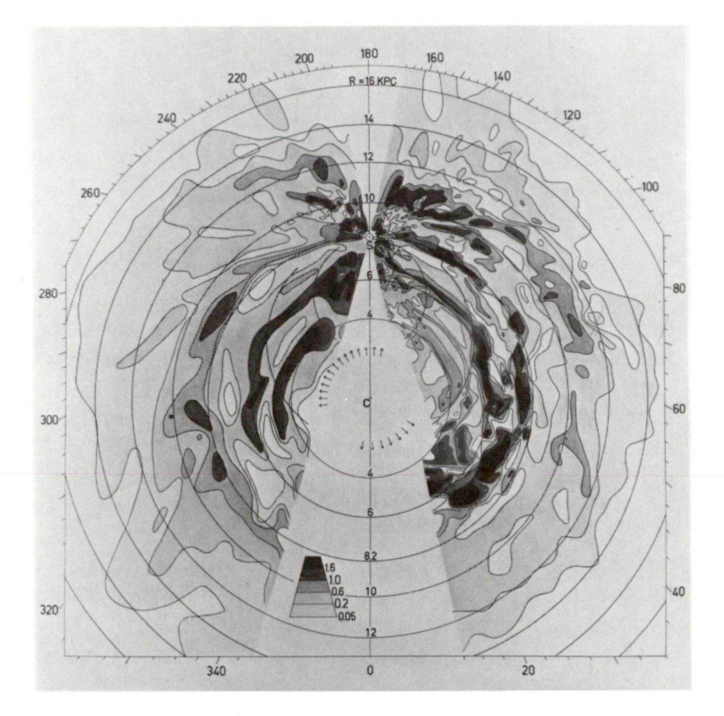

Figure 12.21. A map of the distribution of atomic-hydrogen gas in our Galaxy, made by the method of kinematic distances according to the combined results of Dutch and Australian observations. (After Oort, Westerhout, and Kerr, *M.N.R.A.S.*, **118**, 1958, 379.)

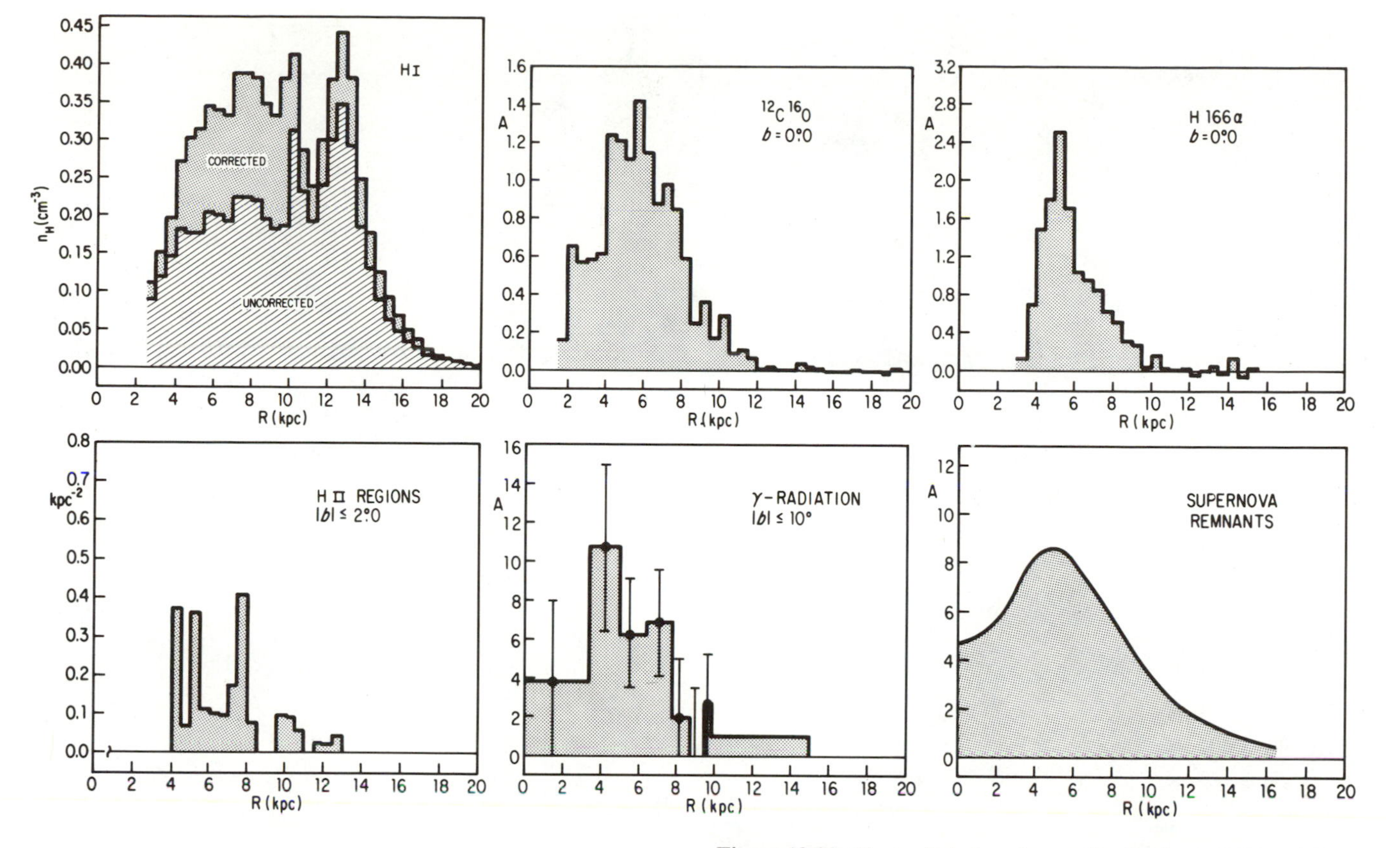

Figure 12.22. The radial abundance distribution of HI and HII, molecular clouds, supernova remnants, and gamma rays. The similarity of all except the HI suggest that they mark the radial limits of strong spiral structure within our own Galaxy. (Adapted from W. B. Gordon.)

of kinematic distance, and the amount of hydrogen gas present at each location was derived from the intensity of the 21-cm line emission at the corresponding radial velocity. Similar maps based on more comprehensive data have subsequently been constructed by Weaver and by Simonson. Present in all such constructions are the long arcs of gas so reminiscent of spiral features observed optically in many external galaxies. Recently, Thaddeus and coworkers have also found the molecular gas, as traced by the CO emission, to lie in distinctive spiral arms.

It has become apparent from recent work by Burton and Yuan that maps such as Figure 12.21 do not give a definitive picture of the spiral structure of our galaxy. A major assumption is that the gas motions correspond to pure circular rotation. This assumption does not hold in the theory of spiral structure which we shall soon discuss. In particular, relatively minor departures from pure circular rotation can make appreciable changes in the spatial mapping. Conversely, if the gas moves at a velocity which differs appreciably from pure circular rotation, it may be misplaced considerably by the assumption that it moves at the circular velocity appropriate for some spatial location along the line of sight.

A manifestation of this problem can be discerned in Figure 12.21. The reader will notice that the gas near the position of the Sun tends to point in long fingers radially away from the Sun. In a pure circular-rotation model, gas near the Sun should have almost zero velocity relative to the local standard of rest. If, in actuality, this gas has some noncircular motions, it would systematically be mislocated at too large a distance along the line of sight in a circular-rotation model. In this manner, every concentration of gas near the Sun tends to be stretched artificially along the line of sight. Bart Bok has commented that this phenomenon represents the fingers of God pointing at us and telling us: "You are wrong, you are wrong, you are wrong!"

There exists other empirical evidence that our Galaxy possesses strong spiral structure interior to the Sun (Figure 12.22). This evidence comes from summing over circular annuli the abundance of objects which are

Figure 12.23. The face-on spiral galaxy NGC4622 (type Sb) may resemble the Milky Way System if we could imagine placing ourselves near the top of the picture in the spur that joins the two outer spiral arms. Notice the "beads on a string" appearance of the optical spiral arms, and the fact that strong dust lanes can be readily seen on the *inside* edges of the optical arms. (The Cerro Tololo Inter-American Observatory.)

believed to be associated with the OB stars that delineate optical spiral structure in other galaxies: HII regions, molecular clouds, supernova remnants, and gamma rays. The distance ambiguity does not affect any summation along loci of constant distance r from the galactic center (see Problem 12.10), and the summation in circles helps to cancel out systematic errors produced by the noncircular motions associated with spiral structure. As you can see, all the distributions in Figure 12.22 show a sharp peak near $r_{\odot}/2$, with a gradual tapering out to $r_{\odot}$. This evidence is conventionally interpreted to indicate that the spiral structure of our Galaxy is mostly confined to radii between somewhat inside $r_{\odot}/2$ and somewhat beyond $r_{\odot}$. Spiral structure may well extend far beyond the solar circle, but it is evidently relatively weak in the production of the above tracers of spiral structure.

The Nature of the Spiral Arms

If we could look at our Galaxy face-on, as we can with external systems, we might see a picture like that in Figure 12.23. This optical photograph shows a geometric organization frequently found in disk galaxies, namely, a beautiful spiral pattern composed of two trailing spiral arms. The spiral structure looks so striking because the arms are outlined by brilliant young stars which are strung along the arms, as Arp has remarked, like beads on a string. The "beads" are actually giant HII complexes whose embedded OB stars cause them to fluoresce with a power perhaps a thousand times greater than the Orion nebula (Color Plate 6). Such OB stars have main-sequence lifetimes of only a few million years; so the lifespans of these massive stars are very short compared to the age of the parent galaxy, perhaps 10 billion years. Since it is extremely unlikely that we just happen to live at an opportune time to witness the spectacle of the birth of massive stars in spiral arms, the formation of such stars must be an ongoing process. New stars must continually be produced from the clouds of gas and dust that pervade interstellar space, and these stars must continually replace their older counterparts in the population of luminous objects.

A question now arises. Why should new stars be born in a grand spiral pattern? The spiral arms delineated by the strongest HII regions are often very narrow, yet they may extend for more than one hundred thousand light-years from one end to the other. What could be the mechansim that triggers the simultaneous formation of stars throughout such a narrow and long front?

One might be tempted to answer thus. The OB stars outline a spiral pattern because the gas and dust clouds, which constitute the raw material for star formation, are always found in spiral arms. However, as Oort has emphasized, the differential rotation of the disk makes untenable any attempt to understand spiral structure in terms of arms which are always made of the same material. Material arms would be wound tighter and tighter by the differential rotation, whereas real spiral arms rarely have more than one or two turns.

Figure 12.24 illustrates the winding dilemma in action. Imagine at time 1 having a long string of gas clouds lined up radially in the Galaxy. Because the gas clouds in the interior take less time to go once around than those in the exterior, by time 2, when the innermost cloud has gone halfway around the Galaxy, the line of clouds would have been drawn into a trailing spiral arc. By time 3, when the innermost cloud has gone completely around, the original line would have added almost one complete turn. Since the innermost gas cloud may take typically less than 10^8 years to go once around, such a mechanism operating for 10^{10} years would produce spiral arms consisting of 100 turns. This conflicts dramatically with real spiral arms, which, as we have already mentioned, rarely have more than one or two turns.

The way out of this dilemma was suggested by Lin and Shu in 1963, following some earlier work by B. Lindblad and P. O. Lindblad. We proposed that spiral structure in disk galaxies is a wave phenomenon, to be

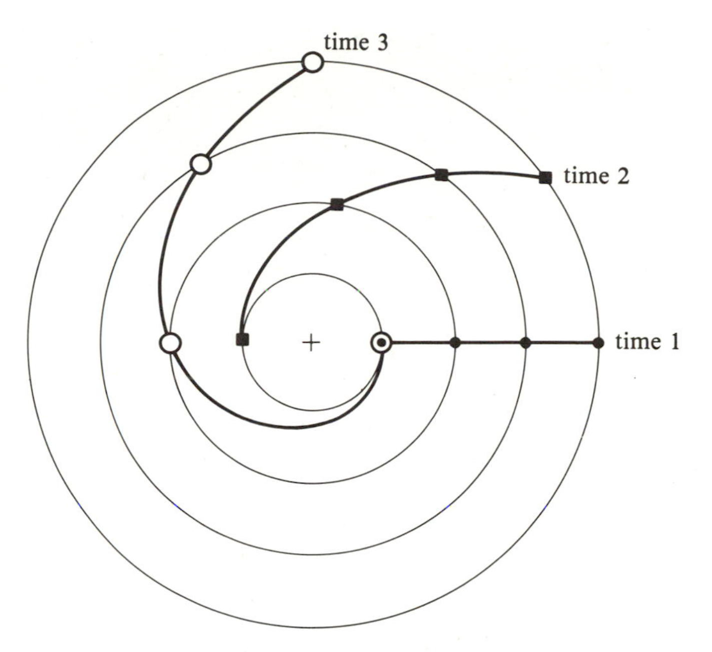

Figure 12.24. The winding dilemma associated with thinking of spiral arms as material alignments in a field of differential rotation. By the time ($\sim 10^8$ yr) the innermost gas cloud has completed one circle of rotation, an originally straight arm would have added almost a complete turn. Since spiral galaxies are likely to be 10^{10} years old, this picture cannot account for the observed spiral structures.

specific, a density wave, maintained by the self-gravity of the large-scale distribution of matter. To see the basic idea of density-wave theory, let us return to our analogy of stars in a differentially rotating galaxy acting like cars in different speed lanes on a circular freeway. Suppose road repairs close down one of the lanes along a given stretch. A local increase in the density of cars—i.e., a traffic jam—would occur along that stretch, but the region of maximum concentration of cars is not always composed of the same cars. Rather, the individual automobiles move through the density concentration, but photographs taken from above in a helicopter show that the maximum concentration of cars always occurs where repairs are being undertaken. Clearly, the speed at which the individual cars travel and the speed at which the repair work proceeds can be quite different. In scientific terminology, we could say that the traffic jam represents a **density wave** of cars, and that the speed of the density maximum (the wave speed) can be quite different from the speed of the individual cars (the material speed).

The Density-Wave Theory of Galactic Spiral Structure

Let us translate the above analogy into a more concrete mechanistic basis for spiral structure. In particular, let

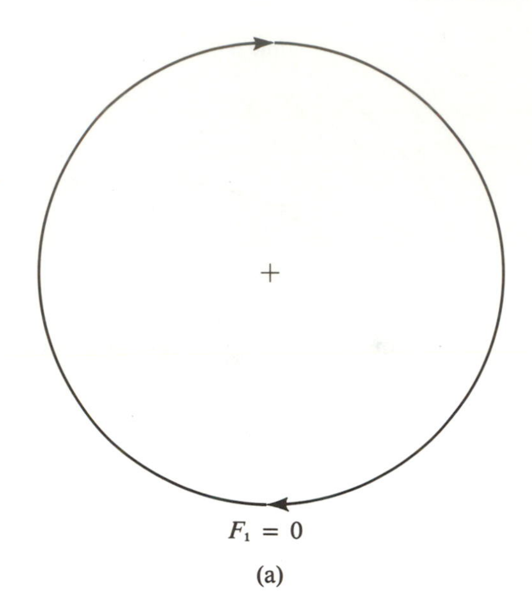

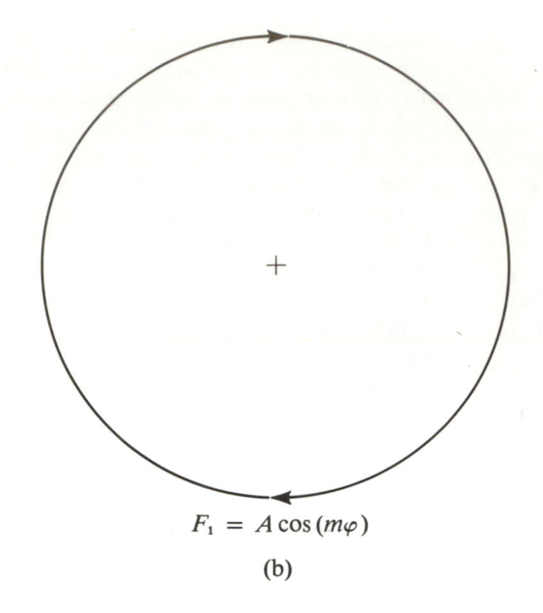

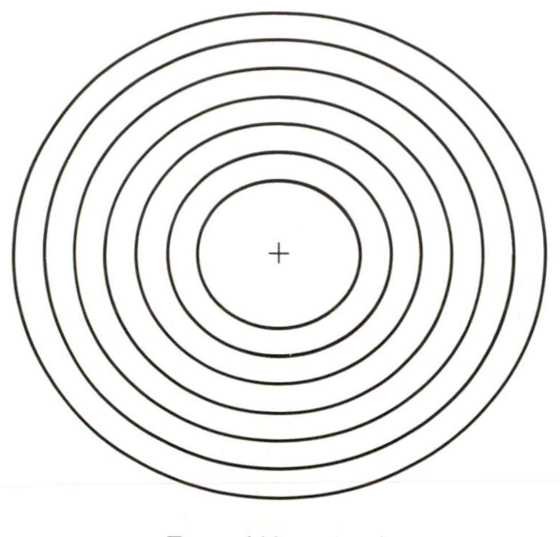

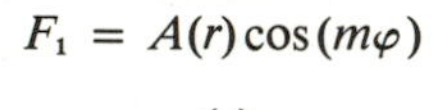

(c)

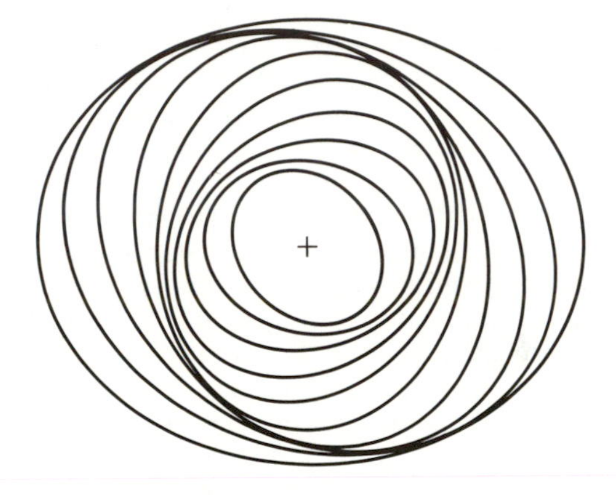

$F_1 = A(r) \cos[m\varphi - \Phi(r)]$

(d)

Figure 12.25. The response of the mean motions of stars or gas clouds to the presence of a nonaxisymmetric gravitational force field F_1. The perturbational force F_1 is assumed periodic in time and angle, with radian frequency ω and angular mode number m, in an inertial frame of reference. Viewed in a frame which rotates at angular speed $\Omega_p = \omega/m$, the same force field is m-times periodic in the angle $\varphi = \theta - \Omega_p t$. Each time a star goes once around in this frame, it is pulled outward m times, causing its orbit to have m bumps. Here, $m = 2$.

us identify precisely the agent which deflects the motion of stars and gas clouds analogous to the road repairs which deflect cars to fewer lanes. This agent is the gravitational field caused by the fact that a spiral galaxy does not have a perfect axisymmetric distribution of matter.

Except for small random motions, the mean orbit of a star or a gas cloud in an unperturbed disk galaxy describes a circle about the center at an angular speed Ω (Figure 12.25a). Consider now the response of this object to a perturbational gravitational field that is periodic in time and azimuthal angle (Figure 12.25b). Let us consider the resulting motion in a frame which rotates at the angular speed $\Omega_p = \omega/m$, in which the perturbational field is time-independent. The steady-state response in this frame is a mean rotation of the object at the relative angular speed $\Omega - \Omega_p$, which carries the object around a distorted circle which contains m bumps (m being a positive integer). For $m = 2$, the distorted circle looks like a closed oval in the Ω_p frame. Consider next applying a perturbational gravitational field of the same time and angular structure to stars or gas clouds at different radii r. If the angular phase of the perturbational field is the same at all radii, the long axes of the ovals would all line up in the same direction. (We ignore here subtle effects due to resonances.) The originally circular disk would then acquire an oval distortion (Figure 12.25c). Such oval distortions or barlike structures can be found in the central regions of roughly half the observed disk galaxies. On the other hand, if the angular phase is systematically displaced by a monotonically varying amount $\Phi(r)$ at different r, the perturbed disk would acquire a two-armed spiral structure (for $m = 2$; see Figure 12.25d). The essence of density-wave theory is to find the functional form of the phase function $\Phi(r)$, and of the amplitude function $A(r)$, which allows the perturbation to be a normal mode of oscillation of the system. For such a normal mode, the disturbance gravitational field associated with the nonaxisymmetric distribution of matter precisely equals the perturbational field required to elicit the nonaxisymmetric response in the first place (Figure 12.26).

Normal-mode calculations have now been carried out by several different groups, including Kalnajs, Toomre, and coworkers. Figure 12.27 shows an example calculated by Lin, Lau, Mark, and their colleagues at MIT. Some of the spiral modes are slightly unstable, and their amplitudes would grow spontaneously. Roberts and Shu have suggested that in the presence of interstellar gas, this growth would saturate when the spiral component of the surface density of old disk stars has reached several percent of the axisymmetric surface density as required by Yuan's analysis of the observational data. If a single mode dominates, this background spiral structure would then take on a quasi-stationary appearance. The spiral concentration of matter would maintain a fixed shape by rotating rigidly at the pattern speed Ω_p. The matter in the disk does not rotate uniformly; so matter on the inside of the galaxy overtakes the pattern, but the pattern sweeps past the matter on the outside.

The detailed calculations show that usually more than one unstable normal mode can be expected to grow in amplitude. Theoretical investigations have only recently begun on the consequences of superimposing several unstable spiral modes. The field remains open, and much is yet unknown especially about the role played by barlike instabilities, which were discovered in computer experiments by Hohl and by Miller and Prendergast; their theoretical importance has been emphasized by Kalnajs and by Ostriker and Peebles (Figure 12.28). There seems to be no question that such barlike structures do play an important role in the central regions of many disk

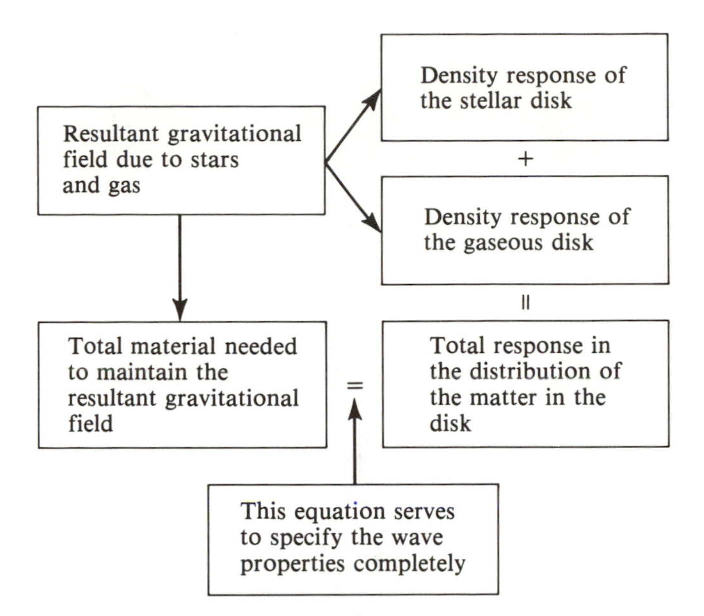

Figure 12.26. The calculational scheme used to calculate the normal modes of oscillation of a disk galaxy. (After C. C. Lin.)

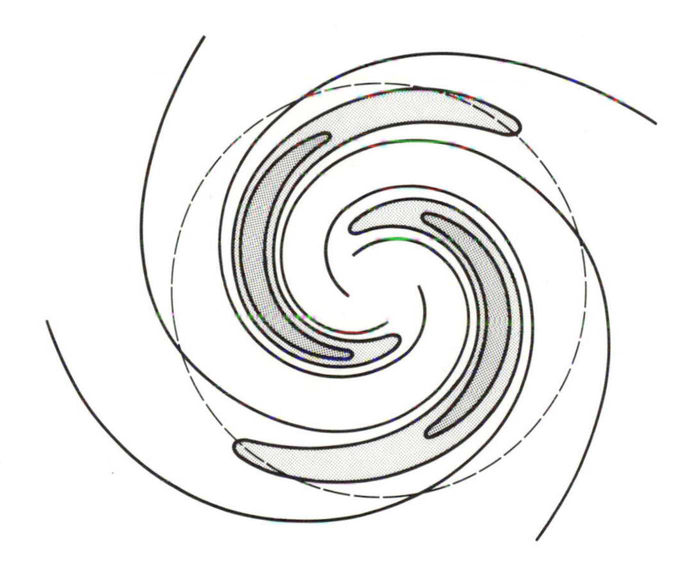

Figure 12.27. Contours of equal density excesses above the average value around a circle in a typical spiral mode. The dashed circle gives the radius where the material rotates at the same speed as the wave pattern.

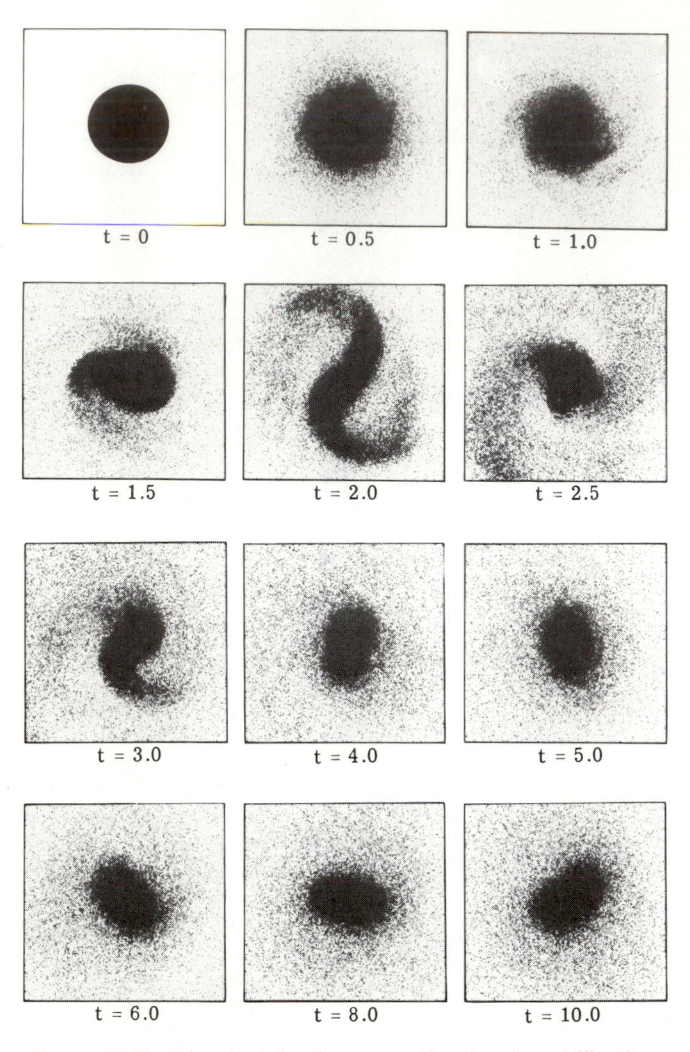

Figure 12.28. The time-development of a bar instability in a numerical simulation of an initially uniformly rotating disk of stars. (From F. Hohl, *Ap. J.*, **168**, 1971, 343.)

galaxies, but there is little evidence that they are present in every system.

Observationally, it is known that in addition to the classical two-armed spirals which exhibit a clear "grand design" (Figure 12.29a), there exist also more irregular systems where the spiral pattern is multi-armed and filamentary (Figure 12.29b). One idea that naturally suggests itself is the following. The "grand design" spirals like Figure 12.29a are disk galaxies in which a single normal mode dominates, whereas the multi-armed spirals like Figure 12.29b may require the superposition of several pure modes of the type given by Figure 12.27. It remains a challenge to sort out the reason for this difference. It is also unclear to which class our own galaxy belongs.

One deduction does seem firm, however, and that concerns why *two*-armed structures dominate. Density-wave theory predicts that wave structures can exist only in between the radii where the pattern speed $\Omega_p = \Omega \pm \kappa/m$ (Figure 12.30). The radius where $\Omega_p = \Omega - \kappa/m$ is called the inner Lindblad resonance; the radius where $\Omega_p = \Omega + \kappa/m$, the outer Lindblad resonance. The range of radii between the inner and outer Lindblad resonances is called the principal range. As Figure 12.30 shows, the principal range for $m = 3, 4, 5, \ldots$, cannot be very large; therefore, even if spiral waves with 3, 4, 5, ..., arms were present, they could dominate the scene only in rare circumstances. Axisymmetric waves, $m = 0$, are unlikely to be present; and even if they were, it would be hard to distinguish their effects from that of the equilibrium disk. One-armed spirals could be present, and they have a large principal range. For realistic galaxy models, however, one-armed disturbances seem to have lower growth rates than two-armed disturbances, but more research is needed on this point. Two-armed spiral waves apparently dominate for three reasons: (a) they probably grow to appreciable amplitudes before saturating; (b) they have a large principal range; (c) the response of the interstellar medium makes such waves highly visible (see the discussion of the next subsection).

Problem 12.11. The dispersion relation of a wave relates its wave frequency ω to its wavenumber $k = 2\pi/\lambda$, where λ is the wavelength. In the simplest version of density-wave theory, the dispersion relation for spiral waves is given by

$$(\omega - m\Omega)^2 = \kappa^2 + k^2 a^2 - 2\pi G|k|\mu, \tag{1}$$

where Ω and κ are the circular and epicyclic frequencies, a and μ are the r-velocity dispersion and surface mass density of the galactic disk, and λ is the radial spacing between spiral arms. For axisymmetric waves, $m = 0$, the dispersion relation reads

$$\omega^2 = \kappa^2 + k^2 a^2 - 2\pi G|k|\mu. \tag{2}$$

Figure 12.29. Photographs of two Sb galaxies: (a) M81 (NGC3031) where the "grand design" of the spiral pattern is very evident; (b) NGC2841, a spiral galaxy with many filamentary spiral arms. (From A. Sandage, *The Hubble Atlas of Galaxies*, Carnegie Institution of Washington, 1961; photographic materials from Palomar Observatory, California Institute of Technology.) ⟶

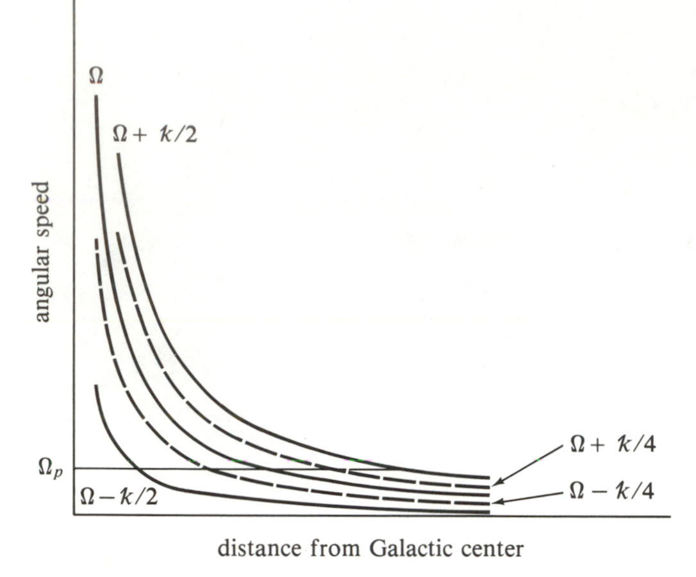

Figure 12.30. The extent of the principal range for various choices of the angular mode number m.

Equation (2) has the following interpretation. Positive values of ω^2 (real wave frequencies) correspond to stable (oscillatory) disturbances. Negative values of ω^2 (imaginary wave frequencies) correspond to unstable (exponentially growing) disturbances. The right-hand side of equation (2) therefore informs us that differential rotation (represented by κ^2) is a stabilizing influence, that velocity dispersion (represented by k^2a^2) is also stabilizing, whereas self-gravity (represented by $-2\pi G|k|\mu$) is destabilizing. Show that if the velocity dispersion has at least the minimum value $a_{\text{min}} = \pi G\mu/\kappa$, then axisymmetric waves of all wavenumbers k are stable. *Hint*: Substitute $a = a_{\text{min}}$ into equation (2), and show that the right-hand side becomes a perfect square. Complete the argument physically for $a > a_{\text{min}}$. How fast would waves grow if a were appreciably less than a_{min}? (An order of magnitude estimate will do.) Can you argue now physically why a is unlikely to be below a_{min}? Calculate a_{min} numerically for the solar neighborhood, and compare it with the observed velocity dispersion of disk stars of about 32 km/sec. Comment on the likely influence of finite disk thickness which was ignored in deriving equation (1).

For simplicity, suppose the actual velocity dispersion in the disk to have the minimum value a_{min}. Then, for $m \neq 0$, show that equation (1) can be solved for $|k|$ to give the solution

$$|k| = \frac{\kappa^2}{\pi G\mu}\left[1 \pm m(\Omega - \Omega_p)/\kappa\right], \tag{3}$$

where $\Omega_p = \omega/m$. The two possible choices of sign in equation (3) yield the short and long waves, depending on whether $|k|$ has a large or small value for given Ω_p. The formula (3) allows the short waves to extend beyond the Lindblad resonances where $\Omega = \Omega_p \pm \kappa/m$, but a more proper stellar dynamical analysis invalidates formula (3) near the Lindblad resonances. At the Lindblad resonances an orbiting star would encounter a spiral arm exactly once every epicyclic period. (The resonant condition is $\Omega - \Omega_p = \pm\kappa/m$, because there are m spiral arms.) In between the Lindblad resonances, the relative driving frequency $|m(\Omega - \Omega_p)|$ is less than the natural frequency κ, and the response of the star can have the right phase to support the original gravitational field (see Figures 12.23 and 12.24). Outside the principal range, the relative driving frequency is too large, and the response cannot support the imposed field. In other words, outside the principal range, self-sustained spiral waves are not possible.

To obtain a numerical feel for equation (3), consider the following example. Many spiral galaxies have flat rotation curves ($r\Omega$ = constant = V_0) for nearly all observable radii r. If the gravitating mass which produces such a flat rotation curve all lies in a flat disk, the corresponding mass model is $\mu = V_0{}^2/2\pi Gr$. Moreover, $\kappa = 2^{1/2}\Omega$ for such a disk. In this special case, show that equation (3) reduces to the simple expression:

$$|k| = \frac{4}{r}\left[1 \pm \frac{m}{\sqrt{2}}\left(1 - \frac{r}{r_0}\right)\right], \tag{4}$$

where r_0 is the radius V_0/Ω_p of the corotation point. Show now that the outer and inner Lindblad resonances are located, respectively, at $r = (1 \pm \sqrt{2}/m)r_0$. For $m = 2$, the inner Lindblad resonance is at $0.293r_0$, and the outer Lindblad resonance is at $1.707r_0$. What is the principal range for $m = 4$? Is there an inner Lindblad resonance for $m = 1$?

With a little calculus, the spiral arms can be shown to be inclined to the circular direction by an angle i given by the formula $\tan(i) = m/|k|r$. Given expression (4) for $|k|$, show that $\tan(i) = m/6$ for the short waves at a (typical) distance halfway between the corotation circle (where $[\ldots] = 1$) and the inner Lindblad resonance (where

$[\ldots] = 2$). For $m = 2$, therefore, we expect a typical inclination of $i = 18°$, which agrees reasonably well with the observations given the crudeness of our estimates here. What is the inclination for $m = 1$? For $m = 3$?

Problem 12.12. How would the above formula change if only a fraction of the gravitating mass which produces the *observed* flat rotation curve actually lay in a flat disk? Argue that if $f(r)$ is the fraction at radius r, the relevant equation for $a = a_{\min}$ becomes

$$|k| = \frac{4}{rf(r)}\left[1 \pm \frac{m}{\sqrt{2}}\left(1 - \frac{r}{r_0}\right)\right].$$

What would $f(r)$ be if $i = 12°$ were a typical inclination for two-armed galactic spirals? What caveats would you provide against taking this deduction too literally? A galaxy with a large central bulge would correspond to a smaller $f(r)$. Discuss how this explains qualitatively Hubble's morphological classification of Sa, Sb, and Sc spirals(Chapter 13).

The Basic Reason for Spiral Structure

Why will a rapidly and differentially rotating galaxy generate spiral waves? The basic driving tendency arises with the galaxy's attempt to gain more binding energy for its inner parts. This is also the basic driving tendency for the evolution of stars (Chapters 7 and 8), but the methods employed by a star to achieve this end differs from those employed by a disk galaxy. A star gains more binding energy for its interior merely by patiently overcoming all classical resistance of the inner parts to contraction. A rapidly rotating object like a disk galaxy must be more subtle. Any contraction which takes place has to be consistent with the principle of the conservation of the total angular momentum of the system. If we artificially try to force the inner parts of a rotating galaxy to contract all by itself, the tendency to conserve angular momentum would speed up the rotation rate and fling the material back out. To obtain a lasting change, some of the angular momentum of the inner parts must be transferred outward. This transfer is exactly what is accomplished by a trailing spiral density wave.

The careful reader will have noticed a certain underlying similarity between the description given above and that given in Chapter 10 for the physics of accretion disks. In both, we have a rapidly rotating object where angular momentum must be transported outward if the object is to fulfill its thermodynamic and gravitational destiny. The detailed mechanisms for the transport are very different: in viscous accretion disks, the process invoked is friction; in disk galaxies, it is the self-gravity of a non-axisymmetric normal mode of oscillation. Because galaxies are so large, the "friction" from stars encountering one another cannot compete with the much more efficient process of spiral structure. Given that the nature of the anomalous friction postulated in conventional accretion disks is completely unknown, it is conceivable that self-gravity may also play a role in their basic dynamics, especially if they turn out to be sufficiently massive relative to the central object.

The discussion above leads us to what should be by now a familiar conclusion. Despite the intricacy of the detailed processes at work, the beautiful spiral structures which exist in disk galaxies arise for a simple reason. They arise in response to the interplay of the universal law of gravitation and the second law of thermodynamics. This is the underlying message of the unity of the universe.

The Birth of Stars in Spiral Arms

The theoretical proposal that spiral structure is a wave phenomenon is well-verified by several different observations. The most cogent of these concern the structure of some well-observed external galaxies, a topic which we shall touch on later (see also Chapter 13). We wish now to return to the question which motivated our theoretical discussion: what is the mechanism that triggers the simultaneous formation of stars throughout a narrow and long spiral front?

Most of the gravitational support for the spiral wave comes from the small sinusoidal variation of the surface density of old disk stars. Theory and observation suggest that the spiral arms contain only several percent more mass than the average of the disk. The small excess mass in the spiral arms cannot, therefore, be why the spiral arms stand out so prominently in optical photographs of external galaxies. The reason is that light from the spiral arms contrasts strongly with the light of the rest of the galaxy, especially in the blue light to which most photographic film is primarily sensitive. Thus, it must be that the birth of blue-white stars (i.e., OB stars) is especially frequent in the spiral arms, much more so than the small increase of total *mass* might have led us to expect. Why is that?

The answer came from calculations by Fujimoto and by Roberts, following initial suggestions by Prendergast. The interstellar gas and dust have much smaller random velocities than the typical disk stars; so the gas and dust tend to respond much more nonlinearly than the disk stars to a small-amplitude spiral wave. This effect is

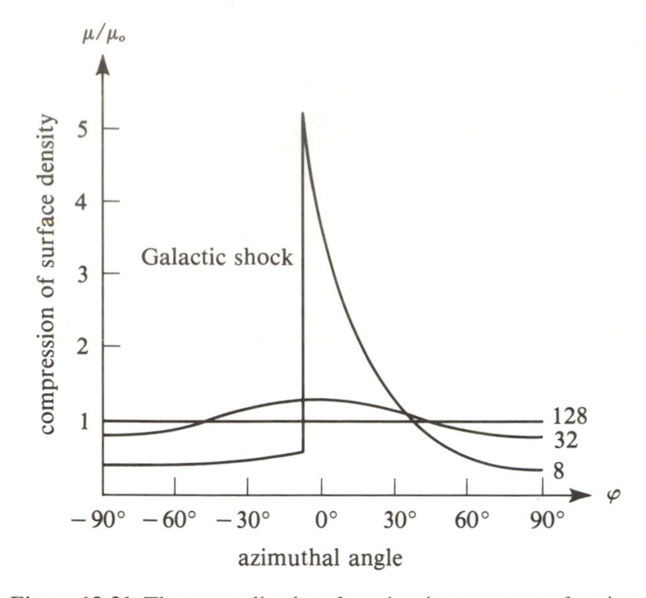

Figure 12.31. The normalized surface-density response of various components of the Galaxy to the presence of a small-amplitude spiral wave. The symbol μ_0 refers to the average surface density of a component seen along a circle in the Galaxy, and μ denotes the corresponding local surface density measured at azimuthal angle φ along that circle. The plots show the responses of matter that has a velocity dispersion of 8, 32, and 128 km/sec. (See Shu, Milione, and Roberts, *Ap. J.*, **183**, 1973, 819.)

Figure 12.32. The first clearcut evidence for the correctness of a wave interpretation for spiral structure came from radio-continuum studies of M51, the "Whirlpool galaxy." The ridge lines of strongest synchrotron emission were found to lie along the dust lanes *inside* the optical arms that are outlined by giant HII complexes. This indicated that the synchrotron emission was enhanced not so much by supernova explosions as by compression of the interstellar magnetic field. Such an interpretation requires matter to flow relative to the spiral pattern so that OB stars form, after a time delay of about 10^7 years, *downstream* from the region of maximum gas compression. (From Mathewson, van der Kruit, and Brouw, *Astr. Ap.*, **17**, 1972, 468; reprinted with permission from the Netherlands Foundation for Radio Astronomy.)

shown in Figure 12.31, where the responses to the same 5 percent spiral gravitational field is plotted as we follow the flow halfway around the Galaxy for components whose velocity dispersions are 128, 32, and 8 km/sec. Such velocity dispersions are typical of halo stars, disk stars, and interstellar gas and dust clouds, respectively. From Figure 12.31, we would expect halo stars to show no spiral structure; old disk stars to show weak spiral structure; and interstellar gas to show very pronounced spiral structure. Indeed, the interstellar gas responds so nonlinearly that it tends to pile up in radiating shock waves. The large concentration of raw material behind the shock would presumably lead, after some time delay, to an increase in star formation downstream from the shock front. (Figure 12.32)

The spiral shock wave in the interstellar gas would probably not induce the direct formation of *individual* stars. Nevertheless, the large-scale spiral organization of newly formed OB stars leaves little doubt that the compression of the general interstellar medium behind such shocks leads to conditions conducive to the birth of massive stars. The details of this triggering process remain controversial. An important step is probably the formation of giant cloud complexes, and it has been proposed that this step proceeds by the triggering of Parker's instability behind galactic shocks (see Figure 11.33). In this picture, the large-scale clumping of the compressed interstellar gas along the length of spiral arms accounts for their eventual appearance as "beads on a string."

Apart from such clumping along the length of spiral arms, the global effect of spiral shocks is displayed in Figure 12.33, devised by Herman Visser. As the streamlines of interstellar gas pass through the galactic shock, they turn sharply and are brought closer together, re-

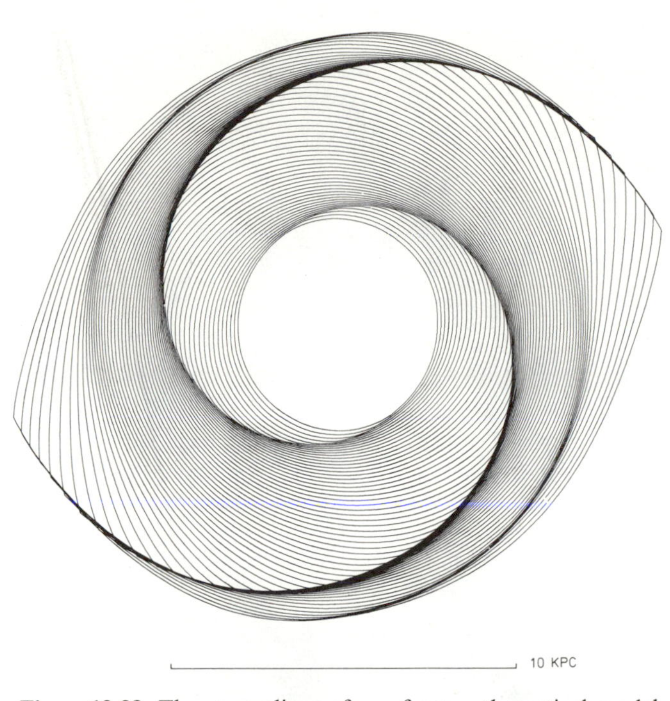

Figure 12.33. The streamlines of gas from a theoretical model of the spiral galaxy M81 (NGC3031, type Sb). (From H. Visser, *Astr. Ap.*, **88**, 1980, 159.)

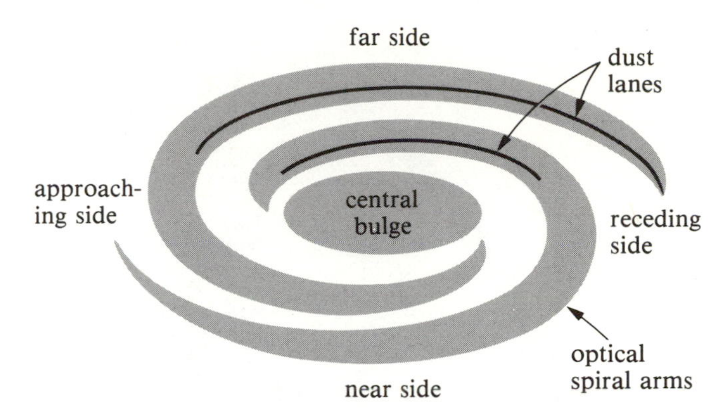

Figure 12.34. A schematic diagram which shows how the knowledge that dust lanes occur on the inside edges of spiral arms can be used to deduce the sense of rotation of a galaxy. Knowing that dust lanes occur on the inside edges of optical spiral arms shows that the spiral arms at the top of the Figure must be further away from us than those at the bottom. (The dust has to be in front to cause obscuration.) This deduction of how the galaxy is inclined, coupled with the observation that the left side of the galaxy approaches us while the right side recedes from us, tells us that the galaxy rotates in the sense that its spiral arms *trail*. The fact that spiral galaxies have arms which trail behind the material rotation seems to be true of all galaxies which are inclined enough to allow this kind of deduction. Trailing spiral arms are also predicted by the normal-mode calculations.

sulting in a large-density compression at each radius, similar to that shown in Figure 12.31. The interstellar dust suspended in the gas would also be brought closer together behind the galactic shock. We would therefore expect a spiral galaxy to show strong dust lanes at the loci of galactic shock fronts, just as we see a heavy envelope of ink in Figure 12.32, caused by the crowding together of streamlines. For the inner and prominent part of the spiral structure, where the matter rotates faster than the wave, the dust lanes should occur on the *inside* edges of the optical spiral arms. The optical spiral arms are formed primarily by the brilliant OB stars that are newly born downstream from the shock front. Precisely this effect (dust lanes on the inside edges of optical spiral arms) has long been known to occur in real spiral galaxies. Indeed, this fact underlies the conventional interpretation that spiral arms trail behind the galactic rotation(Figure 12.34).

The OB stars live for a short time compared to the time required for them to migrate from their birthplace in one spiral arm to the next spiral arm. Thus, the continually forming OB stars, together with their associated HII regions, should constitute excellent tracers of the underlying spiral pattern of old disk stars. Although OB stars contain a relatively small fraction of the mass of the galaxy, their easy visibility allows them to mark the crest of a spiral density wave as clearly as whitecaps mark the crest of a breaking water wave. In both, the tracers are material particles (whitecaps are air bubbles) which move at a different speed from the wave, and which form and burst on a timescale which is short compared to the time it takes a particle to get from one crest to another. Whitecaps and OB stars seem to sit on the crest of their respective waves, not because they move at the same speed as the wave, but because we do not see the *same* whitecaps and OB stars when we look at the wave at a later time.

Direct observations of the interstellar hydrogen gas in external galaxies has become possible in recent years with the advent of radio interferometric techniques. Studies of the galaxy M81 by Rots, Shane, and Visser have been particularly revealing. Figure 12.35a shows a comparison of the density distributions of theory and observations; Figure 12.35b shows an analogous comparison of the velocity fields. As you can see, the agreement is remarkably good. In particular, notice that the atomic hydrogen gas in M81 really is confined to long, narrow spiral arms that outline very clearly the global spiral structure of that galaxy. A similar combination of theory and observation is necessary if we are ever to produce the successor to Figure 12.21, a map of our own Galaxy, the Milky Way System.

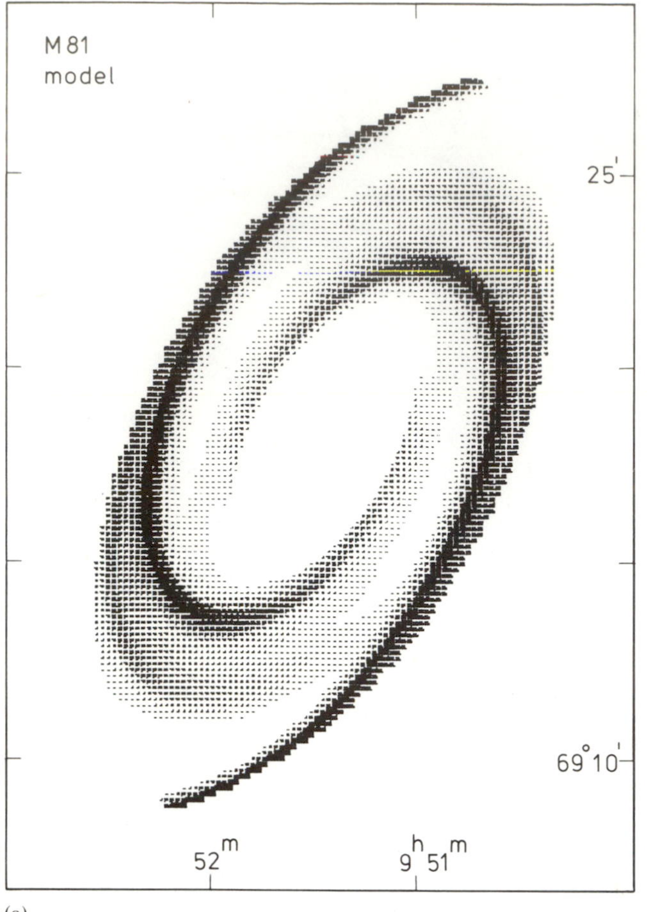

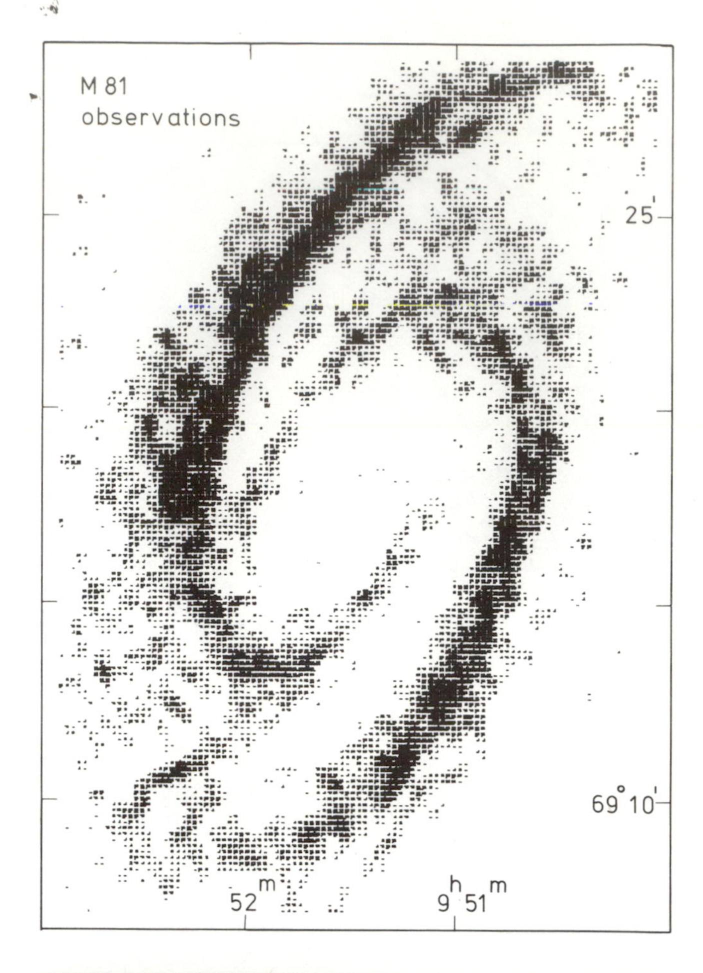

(a)

Figure 12.35. Comparison of the density-wave theory and observations of the spiral galaxy M81. (a) The theoretical and observational distributions of the atomic hydrogen gas (after taking into account the finite angular resolution of the radio-telescope beam). (b) The theoretical and observational velocity fields along the line of sight for the hydrogen gas. (From H. Visser, *Astr. Ap.*, **88**, 1980, 159.)

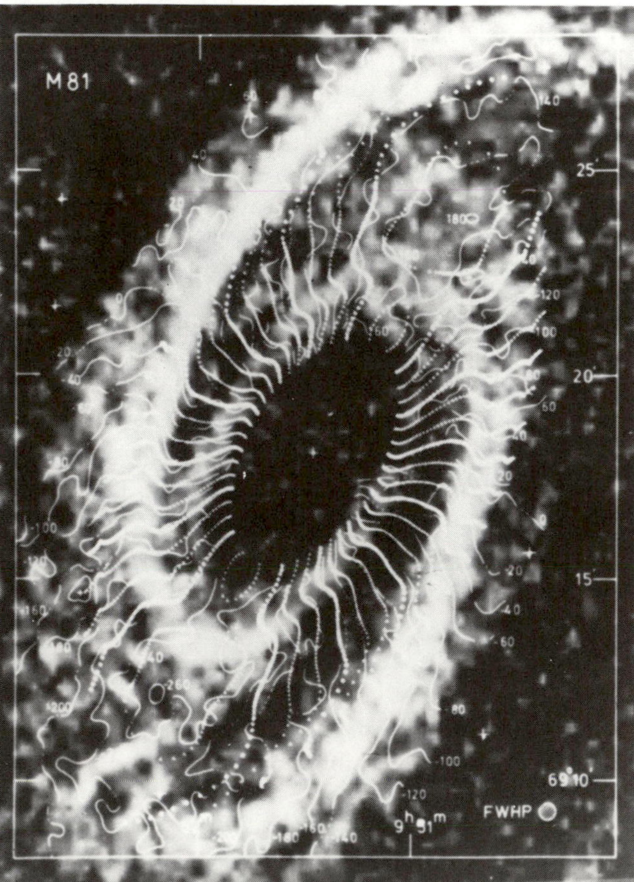

(b)

Concluding Remarks

This chapter supplied some good news and some bad news. The good news was that the same physical principles which guided us on our journey to the stars continue to serve us well in exploring the Galaxy and beyond. In particular, we saw that the interplay between gravitation and the second law of thermodynamics continues to direct the cosmic drama at galactic scales.

The bad news was the nasty surprise of "missing mass" that haunts the study of the dynamics of galactic phenomena. Many astronomers are depressed by the issue of missing mass. We are confronted with the prospect that the celestial objects which have been observed all these centuries may constitute only a minor fraction of all the mass. More than 90 percent of the mass in the universe may be unobservable except for its contribution to gravity. This *is* a depressing state of affairs, but I believe that we can afford a more philosophical attitude.

Astronomers have been treated in the twentieth century to a string of spectacular advances which has lulled us into thinking that all the important problems can and should be solved in our own lifetimes. We are therefore distressed to find a problem for which a solution is literally not in sight (see, however, Chapter 16). But this is hardly a unique situation in the history of science. As Victor Weisskopf has remarked, humanity has always wondered about the great questions. What is matter? What is light? What is the nature of creation? What is the nature of life? These issues proved to be unripe for solution by the natural philosophers; so they turned their attention to more mundane matters which *were* capable of being attacked. How does a lever work? How do balls roll down inclined planes? Their solutions to these practical questions gave rise to modern science, and paved the way for us. Now we are poised on the brink of having the answers to those great questions. We do have a deep, although incomplete, insight into the structure of matter and light, into the nature of creation and life. Why should we be surprised if we are also stymied by occasional unresolvable problems? Why should we not also push levers and roll balls down inclined planes to prepare the groundwork for future generations? No one ever received a promise of instant gratification for doing science.

CHAPTER THIRTEEN

Quiet and Active Galaxies

By the eighteenth century, people knew that the night sky contained fuzzy and distended objects. These were originally all designated "nebulae." Spiral nebulae became early topics of discussion. Thomas Wright and Immanuel Kant (the famous philosopher) were the first to speculate that spiral nebulae might actually be stellar systems fully comparable to our own Milky Way. The idea that spiral nebulae might be independent stellar systems which lay well outside of the Milky Way became known as the "island-universe hypothesis."

In 1918 Shapley proposed his new model of the Galaxy (Chapter 12), a model which abscribed truly enormous dimensions to the Milky Way system. To Shapley, the gigantic size of our Galaxy seemed to make it unlikely that the spiral nebulae could be outside the Milky Way system. Shapley concluded that such objects could not represent separate "island universes," but must resemble the gaseous nebulae which were known at that time to lie within the confines of the Milky Way. Other astronomers, led principally by Heber D. Curtis, disagreed. In 1920 a debate was arranged between Shapley and Curtis before the National Academy of Sciences in Washington, D.C., to try to resolve the controversy. The Shapley-Curtis debate makes interesting reading even today. It is important, not only as a historical document, but also as a glimpse into the reasoning processes of eminent scientists engaged in a great controversy for which the evidence on both sides is fragmentary and partly faulty. This debate illustrates forcefully how tricky it is to pick one's way successfully through the treacherous ground that characterizes research at the frontiers of science.

The Shapley-Curtis Debate

Part of the debate centered on the issue of a large size for the Galaxy. Modern developments have proven Shapley mostly to be correct on this issue. He did not account for interstellar extinction, but he can hardly be faulted for this, since in 1920 Trumpler's work was still a decade into the future. On the part of the debate which dealt with whether spiral nebulae are stellar systems in their own right, Shapley has been proven, of course, to be wrong. Today we know that spirals are, indeed, "island universes" like the Milky Way, and to avoid confusion, we no longer speak of them as "nebulae" but as *galaxies*. We reserve the term "nebulae" for true gas and dust clouds.

Three main issues were addressed in the part of the debate which concerned the "island-universe hypothesis" (Box 13.1):

(1) What are the distances to the spirals?

(2) Are spirals composed of stars or gas?

(3) Why do spirals avoid the plane of the Milky Way?

The Distances to the Spirals

Both Shapley and Curtis realized that the key to resolving the controversy lay in measuring the distances to the spirals. Shapley had indirect and direct arguments for favoring a small distance, which led him to conclude that the spirals must be part of the Galaxy. The indirect, but more powerful argument, rested with von Maanen's proper-motion study, which seemed to show that the magnificent spiral M101 rotated by 0.02 seconds of arc per year (Figure 13.1). Since the *circumference* of M101 extends about half a degree, this would imply a rotation period of less than 10^5 yr, independent of its distance from us. If M101 were a stellar system comparable in linear extent to Shapley's model of the Milky Way, a rotation period of 10^5 yr would imply super-relativistic rotation speeds! This unpalatable deduction led Shapley to infer that M101 must actually be an intrinsically small object which lies close enough to us to present an appreciable angular extent. Shapley's reasoning was flawless; unfortunately, the data was not. Curtis pointed out that the measurements were subject to question, an appraisal confirmed by Hubble and von Maanen, who showed in 1935 that the proper motions measured on photographs arose from observational errors.

BOX 13.1

The Shapley-Curtis Debate

(1) What are the distances to the spirals?
  Arguments for a small distance:
  (a) Von Maanen's measurements of proper motion of rotation in M101.
  (b) Brightness of S Andromedae in M31 compared to Nova Persei.
  Arguments for a large distance:
  (a) Proper-motion measurements may be in error.
  (b) Brightness of nova outbursts in M31 compared to those in Milky Way.

(2) Are spirals composed of stars or gas?
  Arguments against stellar interpretation:
  (a) Milky Way in neighborhood of Sun has much less surface brightness than central parts of most spirals.
  (b) Outer regions of spirals are bluer than their central portions.

(3) Why do spirals avoid the plane of the Milky Way?
  Arguments against island-universe hypothesis:
  (a) Avoidance suggests influence, as do large velocities of recession.
  (b) Both could be explained by postulating a new force of repulsion.
  Arguments for island-universe hypothesis:
  (a) Many edge-on spirals exhibit central belt of obscuring material.
  (b) If Milky Way also has such a belt, and if Sun is embedded in the middle of such a belt, and if spirals are external to Milky Way System, the zone of avoidance could be explained.
  (c) No immediate explanation for large recessional velocities except that "such high speeds seem possible for individual galaxies."

Problem 13.1. If the true rotation period of M101 were about 2×10^8 yr, what would be the proper motion in seconds of arc per year? Could von Maanen have detected the angular displacement in a *century* of observations?

Direct methods of measuring the distances to the spirals were considered by both Shapley and Curtis. In particular, both astronomers used observations of novae in the Andromeda galaxy, M31, to argue their case (Figure 13.2). The basic method involved the comparison of the peak apparent brightnesses of novae in M31 with the calibrated novae outbursts in our own Galaxy. Different conclusions were reached by Shapley and Curtis, partly because they did not agree on the distance scale for our own Galaxy and partly because at that time the distinction between supernovae and novae was not known either observationally or theoretically (Chapters 8 and 10). Today, the nova method is still used in an updated form as one of the primary distance indicators to nearby galaxies like M31.

Spirals: Stellar or Gaseous Systems?

If spirals are galaxies, Shapley reasoned they should have photometric and spectral characteristics like those of the Milky Way system. To support, therefore, his position that spirals were not galaxies, Shapley cited the work of Seares and Reynolds. Seares had used the star counts of Kapteyn and van Rhijn to deduce that the Milky Way possessed considerably less surface brightness than the spirals. Unfortunately, no one at that time knew that interstellar extinction reduces naive estimates of the surface brightness of the Galaxy. Also, the measured surface brightnesses of spirals refer primarily to their bright inner regions, rather than to the faint outskirts analogous to where the Sun resides. Reynolds had examined the colors of the spirals and had found their disks to be bluer than their central bulges. Moreover, absorption lines characteristic of a stellar spectrum were difficult to detect in the central regions. These properties made the spirals quite different from what was then known about the system of stars in the solar neighborhood. The proper explanation of these differences awaited Baade's introduction in 1944 of the concept of different stellar populations.

We see here that Shapley again used fairly sophisticated arguments for which Curtis really had no adequate response. Shapley's basic point—if spirals are galaxies, they should look like stellar systems—was an excellent one. (The same point has been emphasized by the Burbidges in the current controversy over whether quasars are parts of galaxies.) Unfortunately, no one in 1920

Figure 13.1. Messier 101, an Sc I galaxy, which played an important role in early and present-day debates about the distance scale of the universe. (From A. Sandage, *The Hubble Atlas of Galaxies*, Carnegie Institution of Washington, 1961; photographic material from Palomar Observatory, California Institute of Technology.)

really understood what a galaxy was supposed to look like.

The Zone of Avoidance

The most extraordinary part of the Shapley-Curtis debate concerned a peculiarity of the spatial and velocity distributions of the spirals. Spirals seem to avoid the Galactic plane; they are found preferentially toward the Galactic poles (Figure 13.3). Moreover, Slipher had shown that spirals on the average tend to race away from us at very large velocities! To Shapley, these two observations seemed to show that the spirals shun the Milky Way, and therefore must be close enough to be influenced by it. He speculated that the Milky Way may exert a peculiar force of repulsion on the spirals which accounted for both their avoidance of the plane and their large velocities of recession. Curtis adopted a scien-

Figure 13.2. The spiral NGC6946 during and after Ritchey's nova. The confusion of early twentieth-century astronomers about the differences between novae and supernovae (compare with Figure 8.14) contributed to the controversy over the "island universe hypothesis." (Yerkes Observatory photograph.)

tifically more conservative viewpoint. He pointed out that some nebulous objects (spirals viewed edge-on) showed a dark thin band of obscuring material in their midplane (Figure 13.4). If the Milky Way possessed a similar dark band, and if we were situated in the midplane, and if the spirals lay external to the Galaxy, then we could see only those spirals which lie far from the central plane of the Milky Way. Those are three big *ifs*, which all turn out to be true (Chapter 12). Curtis had no explanation for the large recessional velocities of the spirals except to make the offhand (and prescient) remark that "such high speeds seem possible for individual galaxies."

From a strictly logical point of view, Shapley's argument might seem preferable, since he invokes only one hypothesis (unknown repulsive force) to explain two observational facts (zone of avoidance and large recessional velocities), whereas Curtis needs three hypotheses to explain one fact and leaves the second fact unexplained. However, in terms of physics, Curtis's position is preferable, since it requires no new physical law invented to "save the phenomenon"; it merely asks that we believe certain things about ourselves which we know to hold elsewhere in the universe. It is informative that Eddington, a theorist of remarkable intuitive powers, could be strongly influenced by such philosophical considerations:

> "If the spiral nebulae are within the stellar system, we have no notion what their nature might be. That hypothesis leads to a full stop. . . . If, however, it is assumed that these nebulae are external to the stellar system, that they are in fact systems coequal with our own, we have at least a hypothesis which can be followed up, and which may throw some light on the problems that have been before us. For this reason

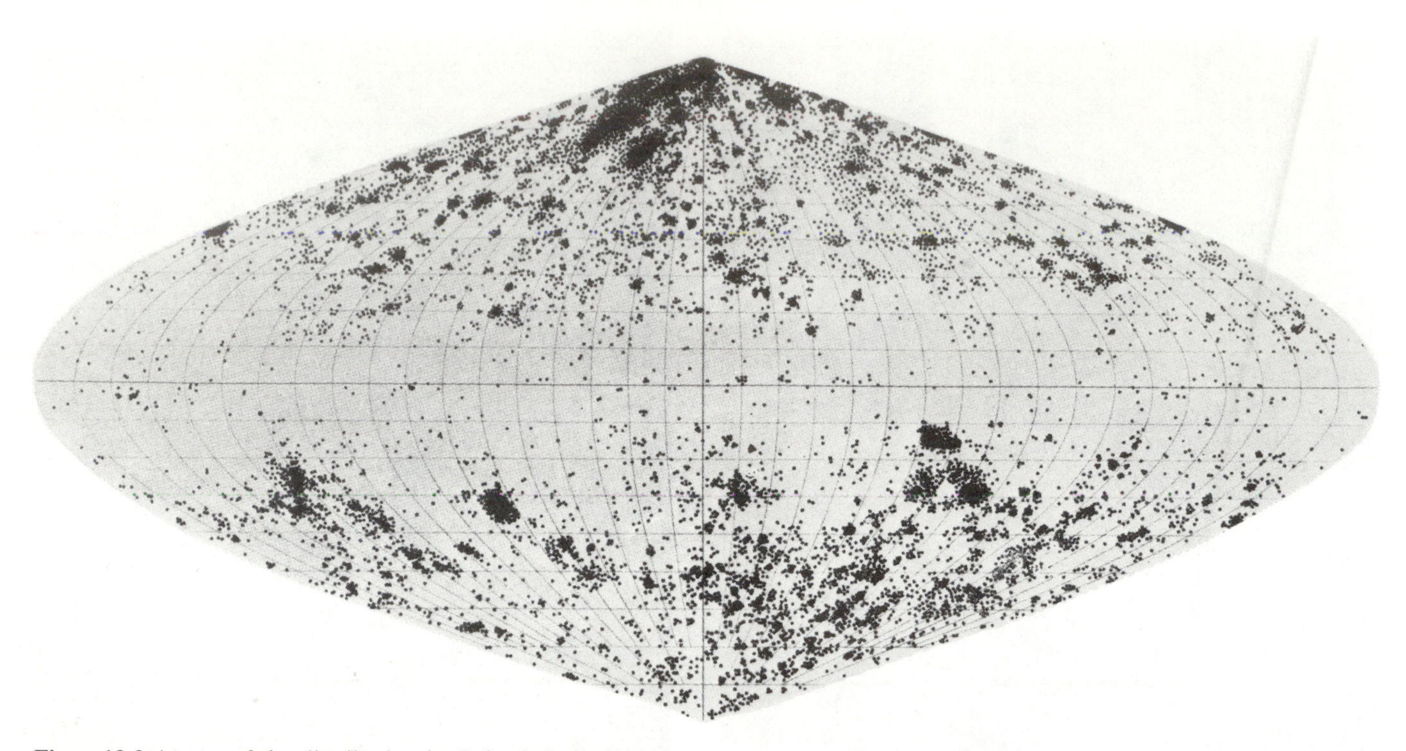

Figure 13.3 A map of the distribution in Galactic latitude and longitude of NGC objects made by Charlier in the 1920s. The abbreviation NGC stands for *New General Catologue*, which was published by John Dreyer in 1895 as a revision of John Herschel's *General Catalogue of Nebulae*. Notice that relatively few spirals can be found near the Galactic equator. (Figure kindly supplied by Paul Gorenstein.)

Figure 13.4. An edge-on spiral, NGC4565 (type Sb). Notice the characteristic dark belt of absorbing material in the principal plane of the disk. (From A. Sandage, *The Hubble Atlas of Galaxies*, Carnegie Institution of Washington, 1961; photographic material from Palomar Observatory, California Institute of Technology.) ⟶

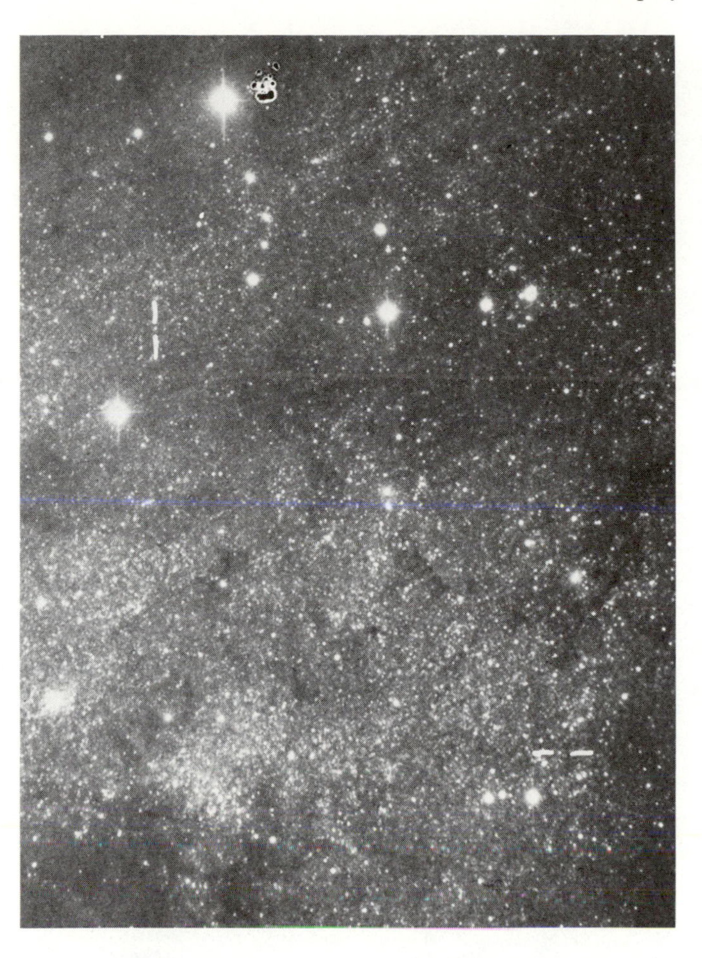

Figure 13.5. A star field in the Andromeda galaxy (M31) with two variable stars marked. Such variable stars allowed Hubble in 1923–24 to measure the distance to M31 and thereby to resolve the debate over the "island universe hypothesis." (Palomar Observatory, California Institute of Technology.)

the 'island universe' theory is much to be preferred as a working hypothesis; and its consequences are so helpful as to suggest a distinct probability of its truth."

Problem 13.2. Write a ten-page essay analyzing the strengths and weaknesses of the two viewpoints expressed in the Shapley-Curtis debate. You might find it useful to consult the text of the original debate, in H. Shapley and H. D. Curtis, "The Scale of the Universe," *Bulletin of the National Research Council*, **2** (1921), 217.

The Resolution of the Controversy

It was Edwin Hubble in 1923 who provided the definitive evidence that spiral "nebulae" were truly independent galaxies. In 1923, Hubble identified a Cepheid variable in the Andromeda galaxy, M31. This was a tremendous accomplishment in itself. First, Hubble needed to resolve the outer regions of M31 into stars; then, he needed to find and identify the variable stars (Figure 13.5). From a measurement of the Cepheid's apparent brightness and the period-luminosity relation for Cepheids, Hubble could deduce the distance to M31. The modern value for this distance is 2 million light-years, which places it well outside the confines of the Milky Way (see Problems 9.8 and 9.9); so the Andromeda galaxy is an independent stellar system. An idea of the truly gigantic size of the universe can be gleaned from the fact that, even at two million light-years distance, the Andromeda galaxy is the *closest* large spiral galaxy to the Milky Way! (The Magellanic Clouds are closer, but they are quite small compared to the Galaxy.)

(a)

(b)

(c)

(d)

Figure 13.6. Examples of (a) an elliptical galaxy, (b) an ordinary spiral galaxy, (c) a barred spiral galaxy, and (d) an irregular galaxy. (From A. Sandage, *The Hubble Atlas of Galaxies*, Carnegie Institution of Washington, 1961; photographic materials from Palomar Observatory, California Institute of Technology.)

The Classification of Galaxies

Hubble's pioneering discovery of the distance to the Andromeda galaxy opened the frontiers of extragalactic research. It is now estimated that there are about 10^{10} galaxies in the observable universe. Many of these galaxies are comparable in mass to our own Milky Way; most are much smaller; a few are much bigger.

Hubble's Morphological Types

Despite the enormous number of galaxies. Hubble found the variety of their geometric shapes to be surprisingly small. Hubble was able to classify all the galaxies he observed, except for a few oddballs, into the four morphological types shown in Box 13.2. A few percent of the galaxies have irregular shapes; Hubble called them *irregular* galaxies. The vast majority of galaxies have regular geometric forms when seen in projection against the sky; Hubble called them *regular* galaxies. Regular galaxies come in two basic shapes: roundish galaxies, which are called *ellipticals*; and flattened galaxies, which are called *disks*. Disk galaxies are further subdivided into two kinds: *ordinary spirals* and *barred spirals*. Figure 13.6 shows an example of each type. The two kinds of spirals constitute perhaps 70 percent of all galaxies; the irregulars, a few percent; and the ellipticals, the rest. The spirals and ellipticals contain, therefore, the bulk of the mass of the observable universe.

The major differences between spirals and ellipticals are as follows.

(1) The ratio of random velocities to rotational velocities is much larger in ellipticals than in spirals. This fact accounts largely for the differences in shape between roundish ellipticals and flattened spirals. However, the converse is not true: elliptical galaxy can be somewhat flattened without possessing much net rotation.

(2) There is little if any gas in most ellipticals. Dust also seems largely absent. This is very unlike most spirals and some gas-rich irregulars.

(3) There is no evidence of young stars in most ellipticals, probably because there is no gas and dust from which young stars could form.

(4) There is no spiral structure in ellipticals. Spiral structure is almost always present in large disk galaxies; however, some transition types between ellipticals and spirals—called SOs and SBOs—also lack spiral structure.

These differences are summarized in Box 13.3, and Hubble's morphological classification of regular galaxies is summarized in Figure 13.7, which shows his famous "tuning-fork diagram." (see also Problem 12.12.)

Ellipticals

The morphological sequence running from E0 to E7 denotes a sequence of more flattened ellipticals (as seen in projection against the sky). It was thought until a few years ago that this sequence represented galaxies of increasing rotation; however, recent observations of the motions of the stars in elliptical galaxies has revealed the actual situation to be more complicated. Bertola and Capaccioli and Illingworth have studied the shapes of the absorption lines produced by the total contribution of all the stars along a given line of sight through an elliptical galaxy. From these studies, they conclude that the flattening of many observed ellipticals occurs not because of rotation, but because the large random velocities are **anisotropic** (i.e., unequal in different directions). That this effect could produce flattening was known theoretically for a long time to stellar dynamicists; recently Binney and Schwarzschild have constructed explicitly numerical models which have this property (Figure 13.8).

There is a superficial resemblance between elliptical galaxies and the central bulges of spiral galaxies (compare Figures 13.6a and 13.4). Perversely, however, recent spectroscopic studies of the central bulges of spirals *do show* that they have the amount of rotation expected from their observed flattening. The implications of these strange results for the origin of elliptical galaxies and

---

BOX 13.2

Morphological Types of Galaxies

(1) Ordinary spirals: S or SA, central bulge plus flattened disk.
(2) Barred spirals: SB, central bulge plus bar plus disk.
(3) Ellipticals: E, round and smoothly distributed light.
(4) Irregulars: Irr, irregular geometric shape.

BOX 13.3

Differences Between Spirals and Ellipticals

(1) Ratio of random velocities to rotational velocities much larger in E than in S. This fact accounts largely for differences in shape: round versus flat.
(2) Little gas and dust in E.
(3) No evidence of young stars in E.
(4) No spiral structure in E (nor in transition types, SO and SBO).

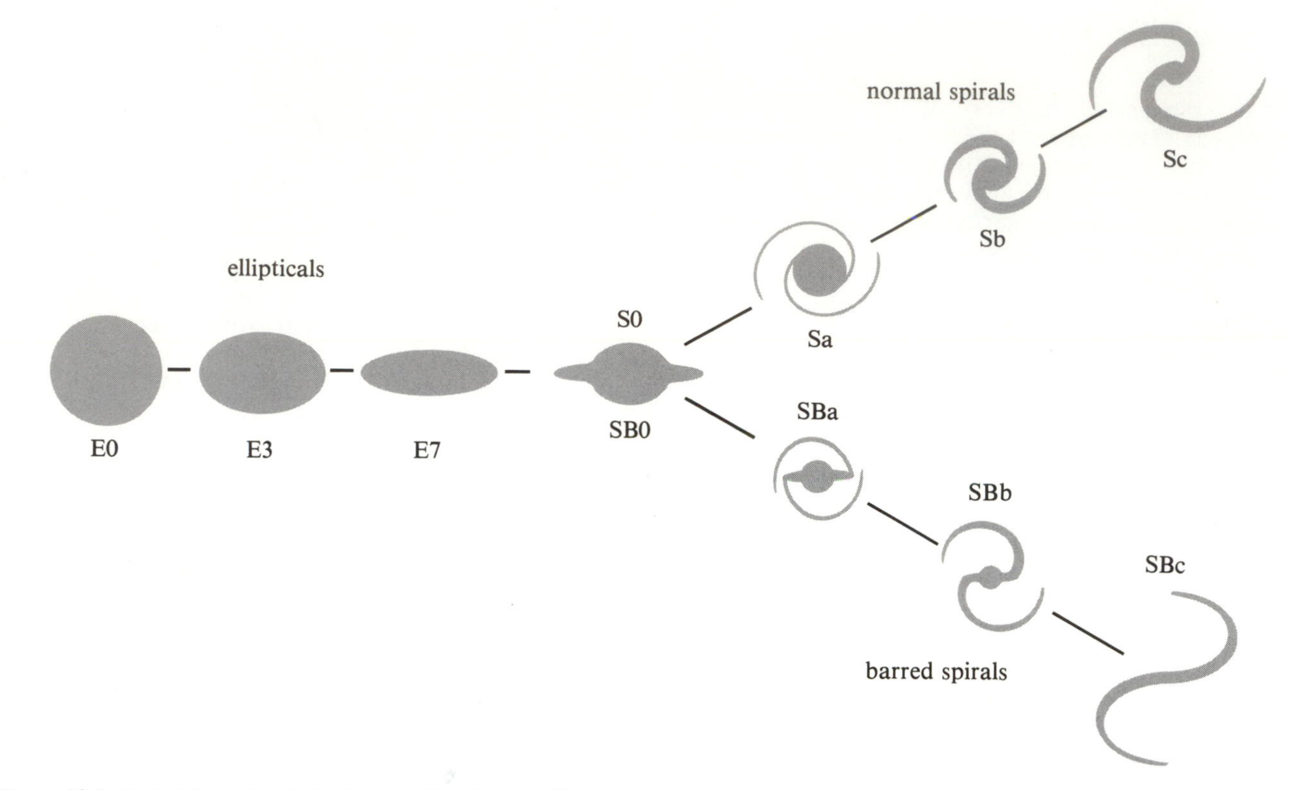

Figure 13.7. Hubble's tuning-fork diagram for the classification of regular galaxies.

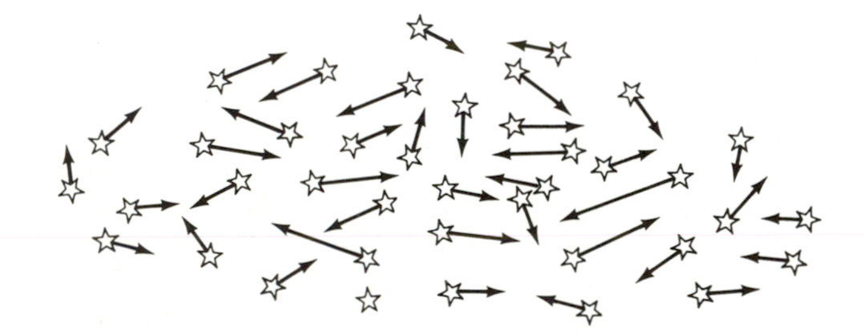

Figure 13.8. A stellar system may appear flattened, with little sense of net rotation, if the random velocities of the stars are not the same in all three directions. As the simplest theoretical example of this phenomenon, consider a disk galaxy which *is* flattened because of rotation. Now, imagine reversing the velocities of half the stars. The axisymmetric distribution of mass is unaffected by this operation; i.e., the stellar distribution still forms a flattened disk. But the disk now contains no net sense of rotation. The situation in elliptical galaxies is more complex than this example, but the principle is the same.

the central bulges of spiral galaxies is still being examined actively. For this book, we may say that the present-day structure and dynamics of elliptical galaxies can be understood roughly as follows: elliptical galaxies represent a jumble of stars which are supported against their mutual self-gravity by their moving *more or less* randomly with respect to each other. This basic randomness explains the gross, smoothly varying light distribution of elliptical galaxies. There is not enough order in the basic underlying distribution, as there is in disk galaxies, to generate coherent structures like spiral waves (Chapter 12).

Problem 13.3. The notation Eq is given to an elliptical galaxy whose minor axis b and major axis a satisfy the equation: $q = 10(1 - b/a)$. What is the ratio b/a for the most flattened elliptical, an E7?

Spirals

The morphological sequences running from Sa to Sc, and from SBa to SBc, denote sequences of ordinary and barred spiral galaxies with spiral arms that are increasingly more loosely wound. The tightness of spiral windings correlates closely with the relative size of the central bulge: the bigger the central bulge, the more tightly wound are the spiral arms. Sa and SBa galaxies tend to have large central bulges, whereas Sc and SBc galaxies

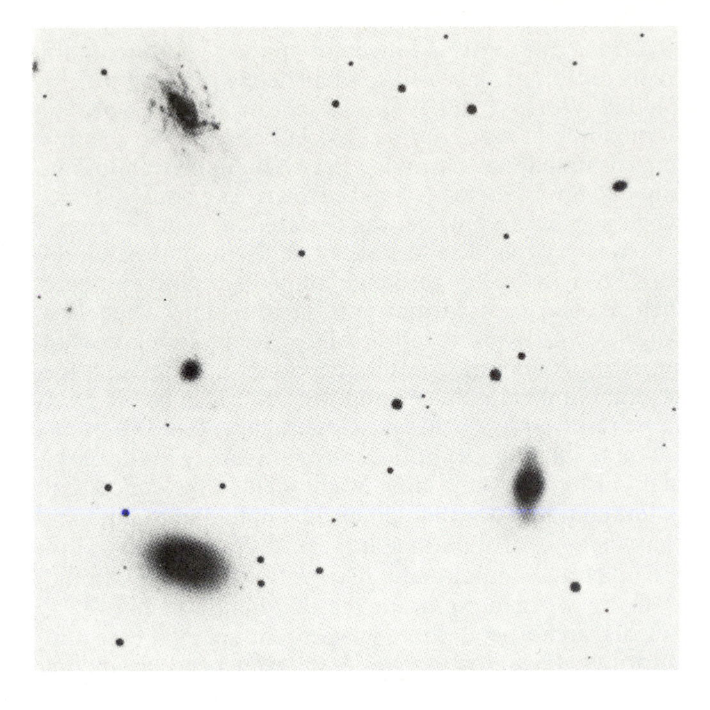

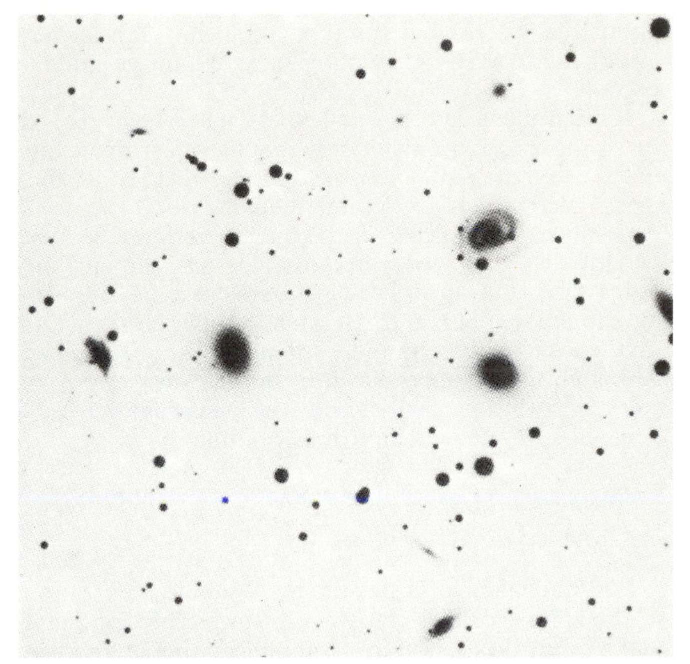

Figure 13.9. Negative prints of the smooth-arm spirals, (a) NGC495 and (b) NGC1268. Compare the smooth texture of the arms of these two galaxies with the knots and patches in NGC496 at the upper left. (From Strom, Jensen, and Strom, *Ap. J. Lett.*, **206**, 1976, L11.)

tend to have small central bulges (or no central bulge at all). The tightness of spiral windings also correlates somewhat with fractional gas content, in the sense that Sa and SBa galaxies tend to have fractionally less gas and dust than Sc and SBc galaxies. This correlation may be complicated by the recent discoveries by van den Bergh and Strom and coworkers of a class of "anemic" and "smooth-arm" spirals which show little evidence of gas and dust (Figure 13.9). It has been suggested that these spirals may be turning into SO or SBO galaxies, but this suggestion is not universally accepted.

Problem 13.4. One reason for the correlation between bulge size and tightness of spiral windings is readily understood from density-wave theory. Review Problem 12.11, and show that the spacing between spiral arms is directly proportional to the surface-mass density of the disk. To produce a given rotation curve, the amount of mass that has to be put into the disk is less if the bulge is bigger. Argue that this effect tends to produce tighter spiral arms for central bulges which are relatively bigger compared to the disk. (Consult also Problem 12.12.)

Although not all present-day SOs or SBOs may have descended from spirals, all spirals which avoid merging with other galaxies will apparently one day become SOs or SBOs. Almost all astronomers agree that the gas in spiral galaxies must gradually be exhausted in the process of making stars. Some gas is returned to the interstellar medium by dying stars, but the complete cycle almost certainly locks up a net amount of gas. Indeed, the confinement of the birth of stars to spiral arms suggests that in most spiral galaxies, the gas has become so diffuse that only where it is compressed inside the arms can star formation proceed very actively today (Chapter 12). This realization has two immediate implications.

First, since the fraction of mass in the form of gas and dust rarely exceeds several percent of the mass of present-day spiral galaxies, the building boom in stars must have occurred long ago, when the galaxies were nearly all gas rather than nearly all stars.

Second, Morton Roberts has estimated that, even at the low present-day usage of gas, star formation should exhaust the remaining gas in most spiral galaxies in a few times 10^{10} yr. The rate of usage of gas may decline as more and more gas is used up, but eventually—say, in 10^{11} yr—the gas must become so rarefied that all star formation is effectively over. All theoretical indications point to such a dying spiral galaxy being transformed eventually to a SO or a SBO galaxy. What else could it become?

If the present-day SOs and SBOs are to be explained by a similar scenario (with or without a catastrophic loss of gas as envisaged in some proposals), there must be a mechanism which accelerates the exhaustion of gas in disk galaxies with large central bulges, not only because of Hubble's observed correlation of gas content and bulge size, but also because Burstein and Boroson have found that SOs and SBOs tend to have larger bulges on the average than a random sample of spiral galaxies. Although astronomers have an outline of a theory of galactic evolution, they are obviously very far yet from being able to make definitive calculations.

Refinements of Hubble's Morphological Scheme

De Vaucouleurs has stressed that Hubble's separation of the spiral galaxies into two distinct types, ordinary and barred spirals, is an oversimplification. De Vaucouleurs finds a continuous gradation of bar, ring, and spiral structures. He begins by adopting the more parallel notation SA and SB for the Hubble types S and SB, and he extends the labels a, b, c of Hubble's scheme by adding the subdivisions d and m (for Magellanic irregulars) to denote a transition from well-developed spiral arms to more chaotic structures. De Vaucouleurs also appends the labels (r) and/or (s) to indicate the prominence of ring and/or spiral features. Thus, what Hubble might call Sc, de Vaucouleurs might classify SA(rs)cd. The de Vaucouleurs system is more useful for precise work, but it is too detailed for our needs here.

Another useful addition to Hubble's purely morphological scheme has been the inclusion by Morgan, Mayall, and Osterbrock of the expected stellar makeup of a galaxy. Morgan proposed that starlight from the central bulge of a spiral galaxy could be expected to be of later composite spectral type (later in the sequence OBAFGKM) than starlight from the disk and spiral arms. Morgan therefore proposed to label galaxies in the concentration classes af, f, fg, gk, and k, corresponding to how much the light from the central bulge dominated that from the whole disk.

Stellar Populations

To understand the basis of Morgan's correlation of spectral type with bulge/disk ratio, as well as Seares Reynolds's earlier observations of the color differences between bulges and disks, we need to invoke Baade's concept of different stellar populations (see Chapter 12). We have already remarked that Hubble had succeeded in the 1920s in resolving the outer portions of nearby spiral galaxies into stars. The corresponding feat for ellipticals and the central bulges of spirals, which contain many more stars per unit solid angle, had to wait for the Second World War. Walter Baade, who was born in Germany but was a naturalized US citizen, had lost his citizenship papers in moving from one house to another, and had not bothered to replace them. As a consequence, when war broke out, he was declared an "enemy alien" and was restricted to nonsensitive astronomical work. This "restriction" gave Baade ample access to the 100-inch telescope on Mount Wilson at a time when Los Angeles was blacked out at night. The combination of superb astronomer, large telescope, and dark sky allowed several elliptical galaxies and the central bulge of M31 to be resolved into stars. In constructing H-R diagrams for these stars, Baade noticed their similarity to the globular cluster stars, and this insight yielded the concept of a Population II which differs from the Population I characteristic of stars in spiral arms (Chapter 12).

Today, as a consequence of quantitative observational work by Morgan, Spinrad, van den Bergh, their collaborators, and others, we understand the spectral type and color of galaxies (or parts of them) to depend largely on the characteristic *age* of the stellar mix, and partly on the *heavy-element content* of the stars. Ellipticals and the central bulges of spirals contain mostly old stars, of about 10^{10} years; the disks of spirals contain a mix of old and young stars, with the youngest stars concentrated in spiral arms; some irregular galaxies are exceptionally blue, with the galactic light dominated by massive young stars (Table 13.1). In no galaxy, however, do we have evidence of a definite lack of old stars, and the belief has grown that all galaxies are about 10^{10} years old, although stars may be forming more actively in some. The theoretical synthesis of a galaxy's light as its constituent stellar populations evolve in time is currently being pursued by many astronomers.

Table 13.1. **Galaxy spectral classification.**[a]

Category	*Spectroscopic features*	*Typical systems*
Orion	Strong emission lines like the Orion nebula. Continuum and absorption lines indicative of B through F stars.	Irregular galaxies like the Magellanic Clouds.
Intermediate	Composite spectrum of F and G stars (Morgan's type f and fg).	Bulges of Sc galaxies, disks of giant spirals.
Amorphous	Spectrum like K stars.	Bulges of giant spirals, main bodies of giant ellipticals.
Weak-lined	Spectrum like globular clusters, deficiency of heavy elements.	Dwarf ellipticals.

[a] Adapted from Spinrad and Peimbert, in *Galaxies and the Universe*, edited by Sandage, Sandage, and Kristian, (Univ. of Chicago Press, 1975).

Figure 13.10. Messier 100, an Sc I galaxy, one of the brightest in the Virgo cluster. Notice that M100 has very well-developed spiral arms and large HII complexes. (From A. Sandage, *The Hubble Atlas of Galaxies*, Carnegie Institution of Washington, 1961; photographic material from Palomar Observatory, California Institute of Technology.)

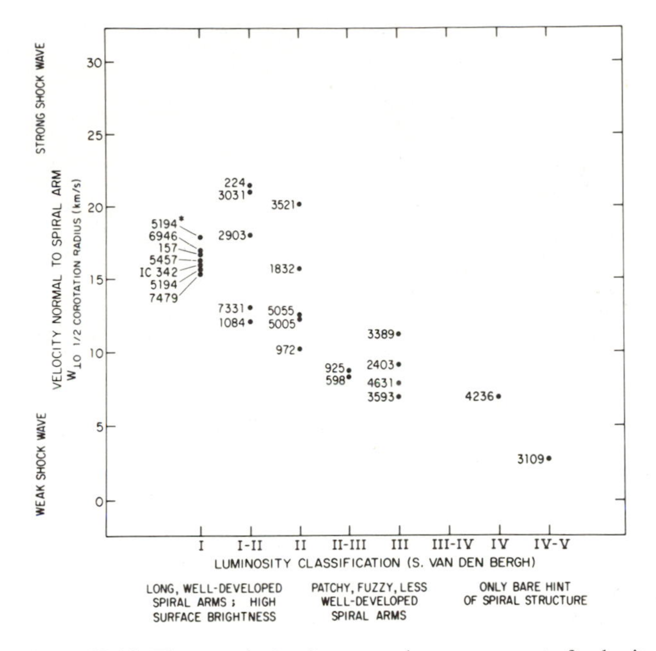

Figure 13.11. The correlation between the component of velocity perpendicular to spiral arms (according to density-wave theory) and van den Bergh's luminosity class. (From W. W. Roberts, M. S. Roberts, and F. H. Shu, *Ap. J.*, **196**, 1975, 381.)

Van den Bergh's Luminosity Class

Another analogy with stars that we shall find useful for categorizing galaxies is van den Bergh's concept of luminosity classes. The van den Bergh scheme is to append to Hubble's type a luminosity class label which runs from roman numeral I to roman numeral V. Just as with stars (Chapter 9), luminosity class I describes the most intrinsically luminous spiral galaxies, luminosity class V the most intrinsically faint. It is now known that the total optical light is roughly proportional to the total mass of visible stars; thus, luminosity class I galaxies are also the most massive galaxies.

Van den Bergh's luminosity class is useful because it correlates well with the regularity of the spiral structure. The most luminous (and the most massive) spiral galaxies tend to have the most regular and pretty spiral structure (Figure 13.10). Thus, an Sc I galaxy is a very luminous ordinary spiral, containing a small central bulge, which has well-defined (regular and pretty) loosely wound spiral arms. Such a galaxy also tends to have much interstellar gas, and hence very large HII complexes. These properties are very useful for cosmological studies, as we shall see in Chapter 14.

The empirically derived correlation between luminosity class and regularity of spiral structure has a simple explanation in density-wave theory. In a massive galaxy, everything goes faster. In particular, the speed at which gas clouds bang into each other in galactic shocks tends to be higher in a more massive galaxy. In the traffic-jam analogy of Chapter 12, the gas clouds are like unwieldy trucks which tend to pile up very badly if they start to run into each other. Just as terrestrial traffic jams become worse if accidents arise at high speeds, so do gas clouds become more highly concentrated behind galactic shocks when they arise at high speeds. The higher concentration of interstellar gas can naturally be expected to result in better-defined spiral structure for the newly forming stars (Figure 13.11).

The Distribution of Light and Mass in Regular Galaxies

Surface Photometry of Galaxies

Although we can learn much by merely looking at optical photographs of galaxies (such as Figure 13.6), this simple approach also has its drawbacks. For example,

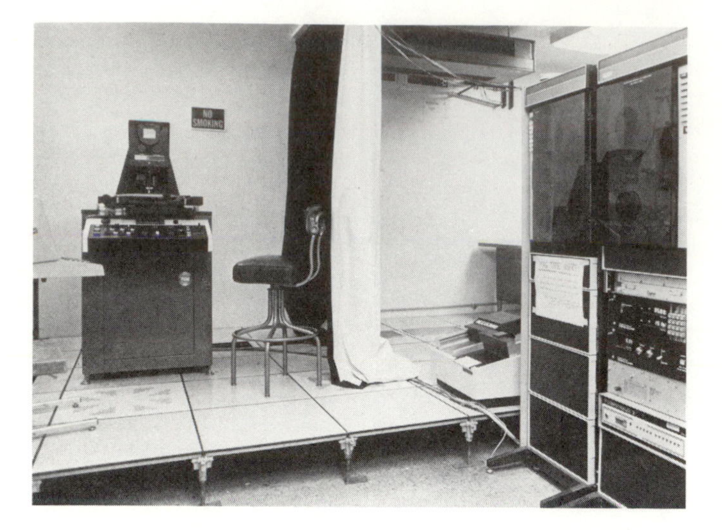

Figure 13.12. Automatic measuring machines controlled by fast minicomputers, such as the PDS system at Berkeley, have revolutionized the digital analysis of astronomical photographs. The empty human chair is, however, at present only symbolic.

the optical spiral structure of disk galaxies looks so dramatic partly because photographic film is usually most sensitive to blue light, and partly because the human eye tends to accentuate the sharp contrasts in blue light provided by the complexes of OB stars, which are strung out, as Baade put it, like "pearls along the arms." To avoid being misled by such effects, astronomers rely on quantitative measurements of the distribution of surface brightness in galaxies, a technique called **surface photometry** of galaxies. Rapid automated procedures for carrying out such measurements have become possible with the revolution in microelectronics (Figure 13.12), and surface photometry in several colors (or wavelength bands) has recently added much to what we know about galaxies.

The pioneering studies in this field were carried out by Hubble and Holmberg. The results are well-summarized by certain empirical laws, the most widely used of which today are probably those devised by de Vaucouleurs. De Vaucouleurs found that the surface brightness $\mathscr{L}$ (luminosity per unit area) of an elliptical galaxy as a function of the distance r from the center of the system along the major axis could be fit by the law

$$\mathscr{L}(r) = \mathscr{L}(0)\exp[-(r/r_0)^{1/4}]. \tag{13.1}$$

The light distribution of the central bulges of spiral galaxies also follows equation (13.1). If the light from the disk is examined carefully, as has been done by Schweizer, sinusoidal variations can be seen in the azimuthal directions (Figure 13.13). The smooth spiral arms detected in the red and infrared by this technique can be interpreted to be the low-amplitude variation of disk stars predicted by density-wave theory (Chapter 12). If this light is averaged in circles to eliminate spiral structure, the disk light distribution seems to follow an exponential law,

$$\mathscr{L}(r) = \mathscr{L}(0)\exp(-r/r_0). \tag{13.2}$$

In equations (13.1) and (13.2), the numerical values of the central surface brightness $\mathscr{L}(0)$ and the scale length r_0 are to be adjusted to give the best fits for the observational surface photometry of a galaxy (or a particular component of the galaxy.) When such fits were performed, Fish and Freeman called attention to an amazing fact. The scale length r_0 was found to vary greatly from one galaxy to another, as might be expected if galaxies come in both intrinsically small and intrinsically large sizes. However, Fish and Freeman pointed out, in different but mathematically equivalent ways, that the quantity $\mathscr{L}(0)$ had a surprisingly small scatter of values. In particular, $\mathscr{L}(0)$ for about two dozen well-studied ellipticals all fell near $2 \times 10^4 L_\odot/(\text{lt-yr})^2$; whereas $\mathscr{L}(0)$ for about three dozen disks clustered about $15 L_\odot/(\text{lt-yr})^2$. More recent studies show that the constancy of $\mathscr{L}(0)$ does not hold for dwarf systems, and that even for giant galaxies there probably is more scatter than was first thought. Nevertheless, the near constancy of $\mathscr{L}(0)$ may be important for studies of the large-scale structure of the universe, because it may allow a calibration of one of the intrinsic properties of giant regular galaxies. No one has yet offered a plausible theoretical reason for the observed constancy.

Velocity Dispersions in Ellipticals and Rotation Curves of Spirals

At various points so far we have seen that the best way to measure the mass of an object is to consider the orbital motions of material affected by its gravitational attraction. Galaxies are no exceptions to this rule. The only complication is that the observable matter near galaxies almost always lies inside the galaxies themselves; so the deduced masses refer mostly to the observable (inner) parts of galaxies rather than to their total (unknown) extent. The measured velocities in ellipticals pertain to the spread in random velocities of the stars; in spirals, to the rotational velocities of either the stars or the gas about the center of the galaxy. If v denotes either kind of velocity measured at a (deprojected) distance r from the center of the galaxy, the mass M interior to r can be roughly estimated to be

$$M(r) = rv^2/G, \tag{13.3}$$

where G is the universal gravitational constant.

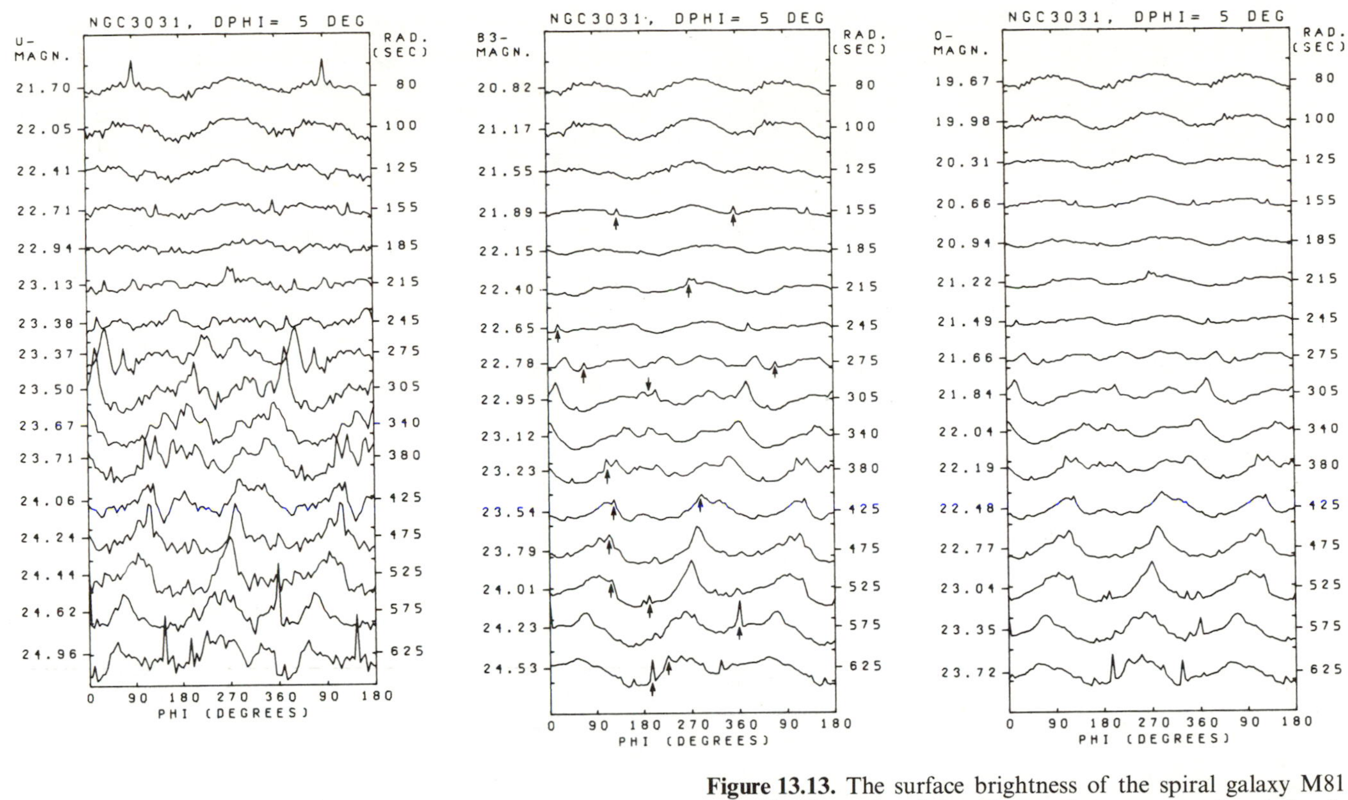

Figure 13.13. The surface brightness of the spiral galaxy M81 (NGC3031) at ultraviolet, blue, and orange wavelength bands shows sinusoidal variations if we trace the distribution of light in a circle in the plane of the disk. In addition to the wavy tangential variations, there is also a steady exponential decline in the light as we move radially outward in the disk. (From F. Schweizer, *Ap. J. Suppl.* **31**, 1976, 313.)

Problem 13.5. At this point, we should be a little more precise in specifying what we mean by $M(r)$. For a disk galaxy, it is clearly more meaningful to let $M(r)$ be the mass contained within a *cylinder* of radius r. For a spherical galaxy, it is theoretically more convenient to let $M(r)$ be the mass contained within a *sphere* of radius r, but an argument could also be made for retaining the cylindrical definition. The argument rests with the observational fact that we can only measure quantities

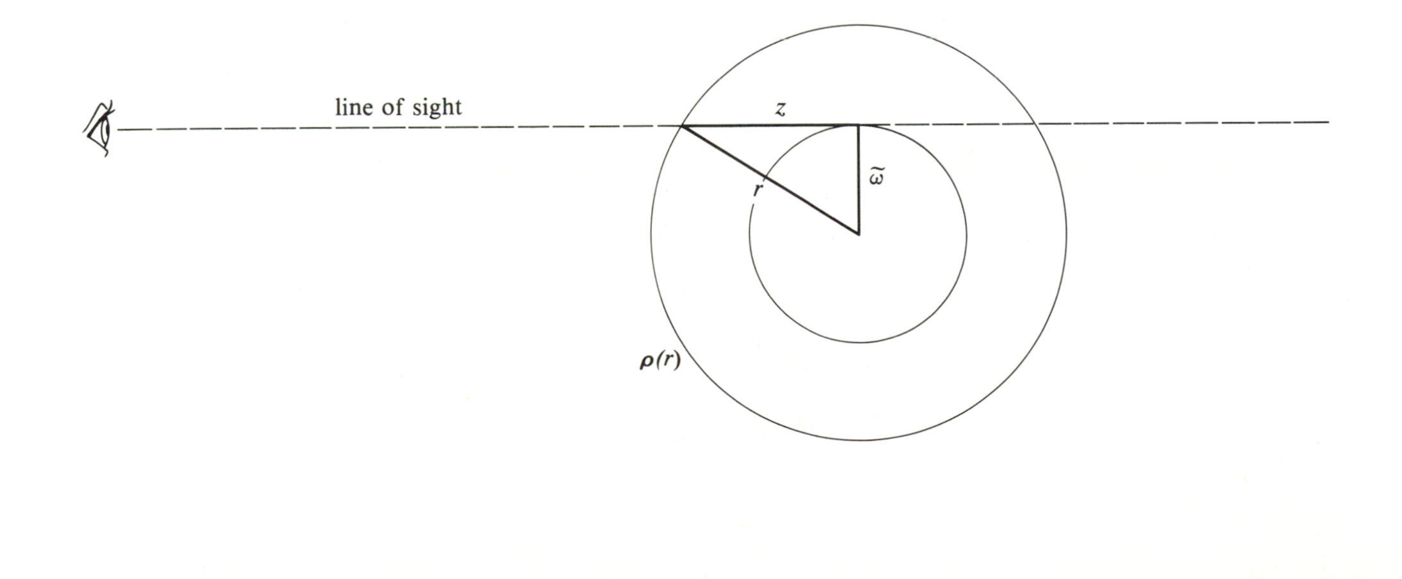

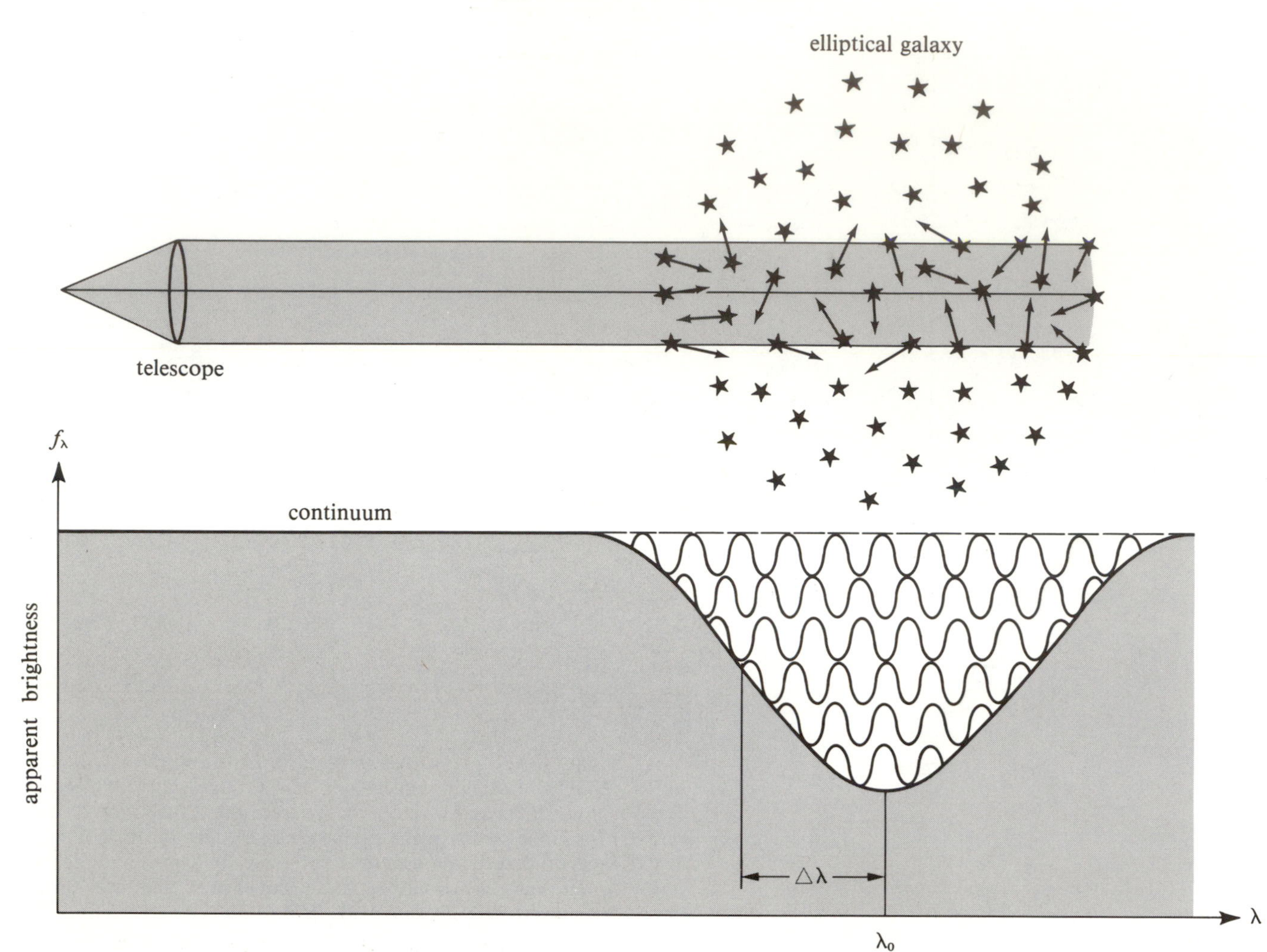

Figure 13.14. The broadening of a spectral line by the velocity dispersion of stars in an elliptical galaxy. In the top diagram, a telescope collects light from all stars within a cylinder through galaxy. Each star may possess a narrow spectral line at the rest wavelength λ_0, Doppler-shifted to a different wavelength λ because of its component of velocity along the line of sight. The superposition of billions of such line profiles produces the broadened spectral line seen in the bottom diagram, in which the line width $\Delta\lambda$ is proportional to the velocity dispersion V of the stars in the galaxy.

along the total line of sight through a galaxy; consequently, a cylindrical definition pertains more directly to the observations. The practical difference between these two definitions may not be large for actual galaxies. For those of you who know no calculus, prove this assertion for a completely flattened distribution of mass. For those of you who do know calculus, continue with the following problem. Consider a spherical distribution with the density law $\rho(r) = Cr^{-2}$, where C is a constant. We shall see that such a distribution is characteristic of the "missing mass" problem in the outer parts of galaxies. Let ϖ be the axial distance from the center of the galaxy. (Up to now, we designated ϖ by r.) Show that the surface mass density $\mu(\varpi)$, the projected mass per unit area, is given by

$$\mu(\varpi) = \int_{-\infty}^{\infty} \rho(r)\, dz,$$

where $r^2 = \varpi^2 + z^2$. Substitute in the expression $\rho(r) = Cr^{-2}$, and make the change of variables $z = \varpi \tan u$ to

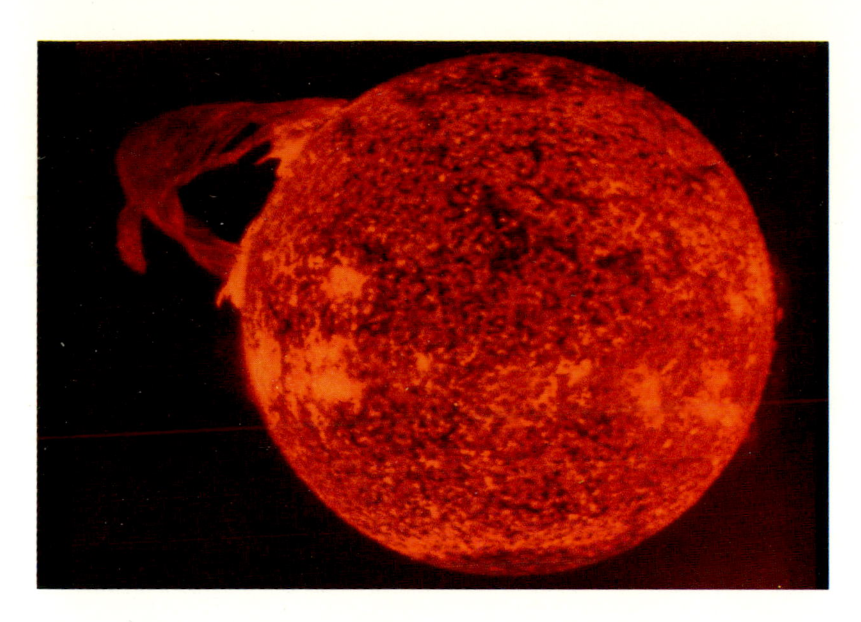

Color Plate 1. An eruption 365,000 miles across on the Sun recorded by Skylab in the 304 Å ultraviolet light of ionized helium. (NASA and the Naval Research Laboratory.)

Color Plate 2. A spectrum of the day sky which constitutes scattered sunlight. The continuum from violet to red has superimposed on it dark absorption lines which are formed in the atmosphere of the Sun.

Color Plate 3. Messier 16, the Eagle nebula in Serpens, containing an open cluster of stars. (© Association of Universities for Research in Astronomy, Inc., The Kitt Peak National Observatory.)

Color Plate 4. Messier 45, the Pleiades stars and reflection nebulosity in Taurus. (Palomar Observatory photograph, copyright by the California Institute of Technology.)

Color Plate 5. Messier 13, a globular cluster of stars in Hercules. (U.S. Naval Observatory.)

Color Plate 6. Messier 42, the Great Nebula in Orion, a Galactic HII region. (© Association of Universities for Research in Astronomy, Inc. The Kitt Peak National Observatory.)

Color Plate 7. Messier 1, the Crab nebula in Taurus, a supernova remnant. (Palomar Observatory photograph, copyright by the California Institute of Technology.)

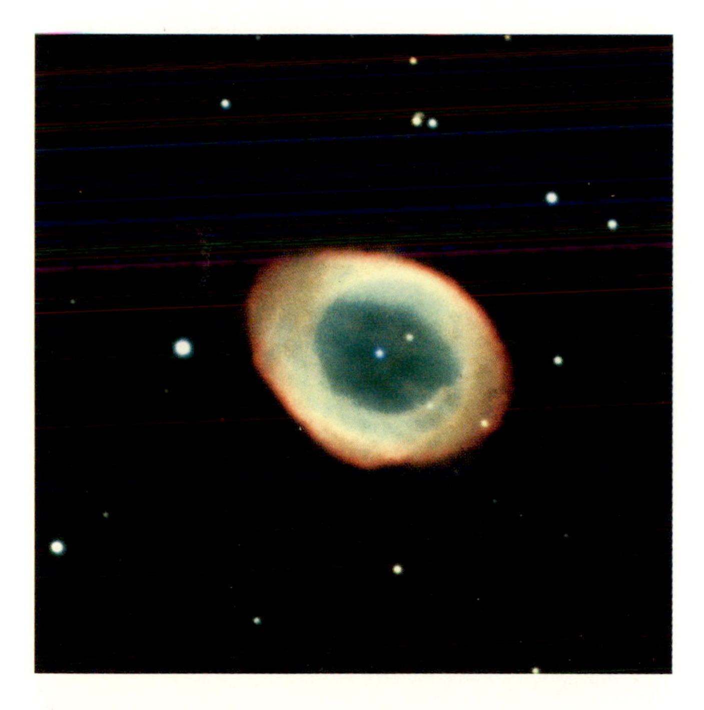

Color Plate 8. Messier 57, the Ring nebula in Lyra, a planetary nebula. (Palomar Observatory photograph, copyright by the California Institute of Technology.)

Color Plate 9. Messier 31, the Great Galaxy in Andromeda, the closest large spiral galaxy to our own. (Palomar Observatory photograph, copyright by the California Institute of Technology.)

Color Plate 10. Messier 32, a dwarf elliptical galaxy in the Local Group of galaxies. (U.S. Naval Observatory.)

Color Plate 11. NGC5128, a peculiar elliptical galaxy in Centaurus; it is a source of intense radio emission. (© Association of Universities for Research in Astronomy, Inc. The Cerro Tololo Inter-American Observatory, CTIO 4-meter photograph.)

Color Plate 12. NGC 5907, an edge-on spiral galaxy in Draco. (U.S. Naval Observatory.)

Color Plate 13. Messier 51, the Whirlpool Galaxy, a spiral galaxy in Canes Venatici seen nearly face-on. It has a companion galaxy at the projection of one of its spiral arms. (U.S. Naval Observatory.)

Color Plate 14. The Earth seen from Apollo 17. (NASA, Johnson Space Center.)

Color Plate 15. Apollo 16 astronaut Duke collects samples on the Moon with the Lunar Rover in the background. (NASA, Johnson Space Center.)

Color Plate 16. False-color closeup in ultraviolet light of Venus cloud details by Mariner 10. (NASA, Jet Propulsion Laboratory.)

Color Plate 17. A panoramic view of the Martian landscape taken by the Viking 1 lander. Notice the trench on the lower right which was dug so that, among other experiments, the lander could test the Martian soil for possible biological activity. (NASA)

Color Plate 18. A closeup view of the turbulence in Jupiter's atmosphere taken by Voyager 1. (NASA, Jet Propulsion Laboratory.)

Color Plate 19. Huge volcanic eruptions were discovered by Voyager 1 on Io, one of the moons of Jupiter. (NASA, Jet Propulsion Laboratory.)

Color Plate 20. A thick haze covers the surface of Titan, the largest moon of Saturn. Detailed analysis of Voyager 1 data revealed surprisingly that Titan's atmosphere was made mostly of molecular nitrogen, with a surface pressure somewhat larger than that of Earth and a surface temperature cold enough possibly to allow pools of liquid nitrogen in special locations. (NASA, Jet Propulsion Laboratory.)

Color Plate 21. An unusual comet, Comet Humason. If funding can be found, scientist hope to fly a spacecraft in 1986 for a much closer view of Halley's comet. (Palomar Observatory photograph, copyright by the California Institute of Technology.)

Color Plate 22. A false-color X-ray image of the quasar OQ 172, which has the highest redshift, 3.53, yet known. (NASA, courtesy of William Ku and Columbia University.)

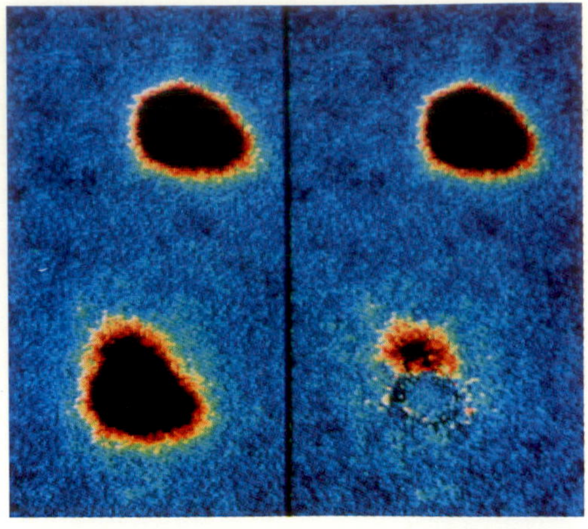

Color Plate 23. The double quasar. The left-hand picture gives a false-color view of the system; the right-hand picture, an imaged processed view after subtraction to reveal the presence of the intervening galaxy which presumably acts as the gravitational lens. For details, see A. Stockton, *Ap. J. Lett.*, 242, 1980, L141, and Chapter 15. (Institute for Astronomy and Planetary Geosciences Data-Processing Facility, University of Hawaii.)

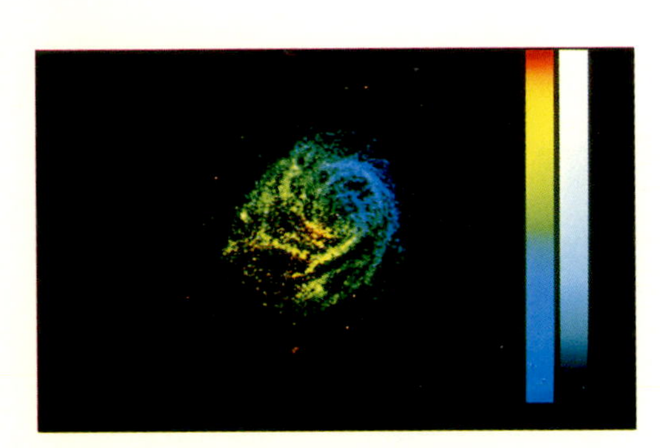

Color Plate 24. A radiograph of M101 showing the distribution and velocity of atomic hydrogen in this Sc I galaxy. Red denotes redshifts with respect to the center of the galaxy; blue, blueshifts. (Kapteyn Astronomical Institute, University of Groningen and the Netherlands Foundation for Radio Astronomy; courtesy of Ronald J. Allen.)

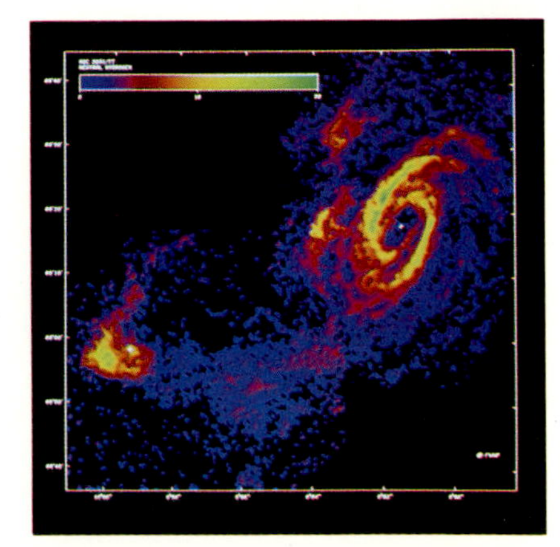

Color Plate 25. A false-color radiograph showing the distribution of atomic hydrogen gas in the spiral galaxy M81, the small irregular galaxy NGC 3077, and the streamer that connects the pair. In this representation, green refers to the highest intensity of the neutral hydrogen emission. (Kapteyn Astronomical Institute, University of Groningen and the Netherlands Foundation for Radio Astronomy; courtesy of Ronald J. Allen.)

Color Plate 26. An optical photograph of the irregular galaxy M82 in Ursa Major. (Palomar Observatory photograph, copyright by the California Institute of Technology.)

Color Plate 27. Messier 33, the third largest spiral galaxy in the Local Group after the Milky Way and M31. (Palomar Observatory photograph, copyright by the California Institute of Technology.)

Color Plate 28. The central bulge of M31. (© Association of Universities for Research in Astronomy, Inc. The Kitt Peak National Observatory.)

Color Plate 29. The Large Magellanic Cloud, a barred irregular galaxy which is a satellite of the Milky Way System. (© Association of Universities for Research in Astronomy, Inc. The Cerro Tololo Inter-American Observatory, CTIO 61-cm Schmidt photograph.)

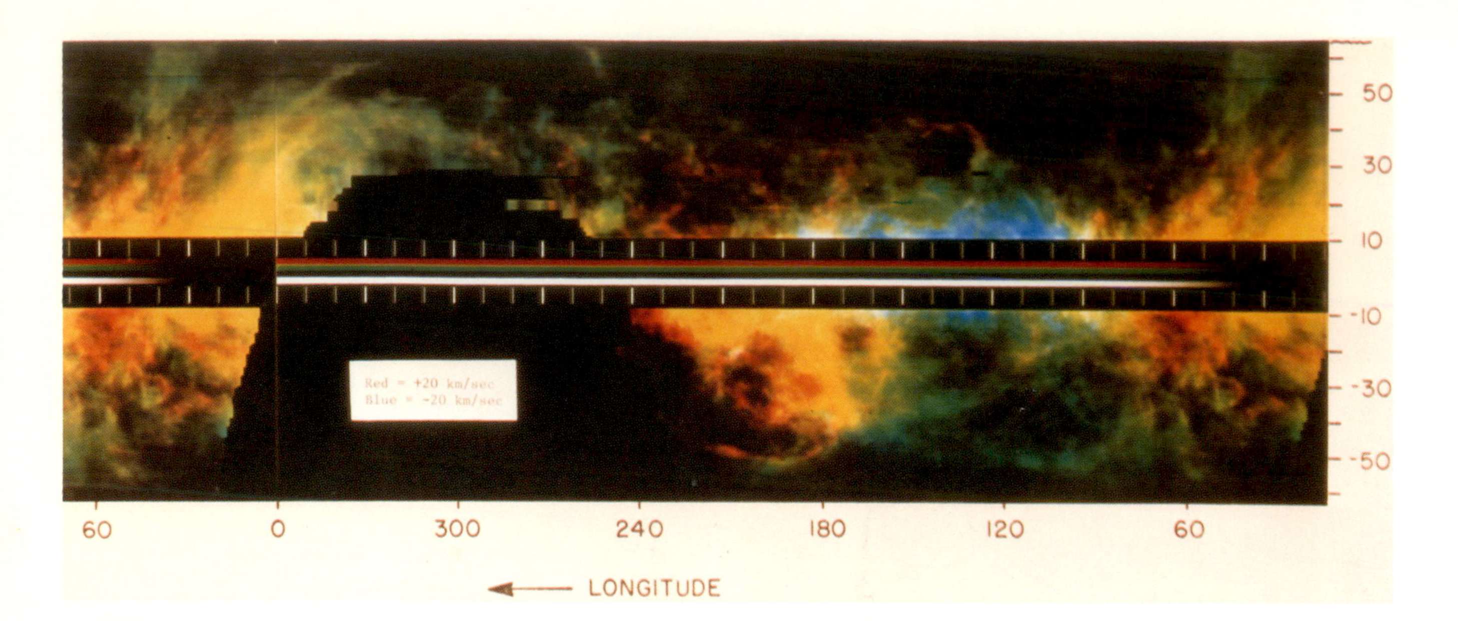

Color Plate 30. Radiograph showing the distribution and velocity of atomic hydrogen gas present locally in our Galaxy. (Courtesy of Carl Heiles and the Radio-Astronomy Laboratory, University of California, Berkeley.)

Color Plate 31. A relatively wide-angle view of a portion of our Galaxy that includes the Horsehead nebula. (Palomar Observatory photograph, copyright by the California Institute of Technology.)

Color Plate 32. The old and the new. The stars in the globular cluster Omega Centauri, NGC 5139 (left), are over ten billion years old; the stars in the open cluster NGC 6231 (bottom right) are relatively newly born from the surrounding gas and dust. (© Hans Vehrenberg)

Color Plate 33. The Rosette nebula in Monoceros, a Galactic HII region. (Palomar Observatory photograph, copyright by the California Institute of Technology.)

Color Plate 34. NGC7293, a planetary nebula in Aquarius. (Palomar Observatory photograph, copyright by the California Institute of Technology.)

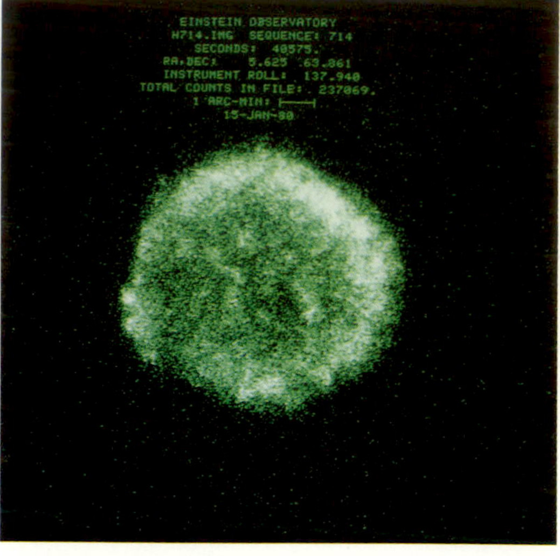

Color Plate 35. False-color X-ray image of the Tycho supernova remnant. (Courtesy of Paul Gorenstein and the High-Energy Astrophysics Division of the Harvard-Smithonian Center for Astrophysics.)

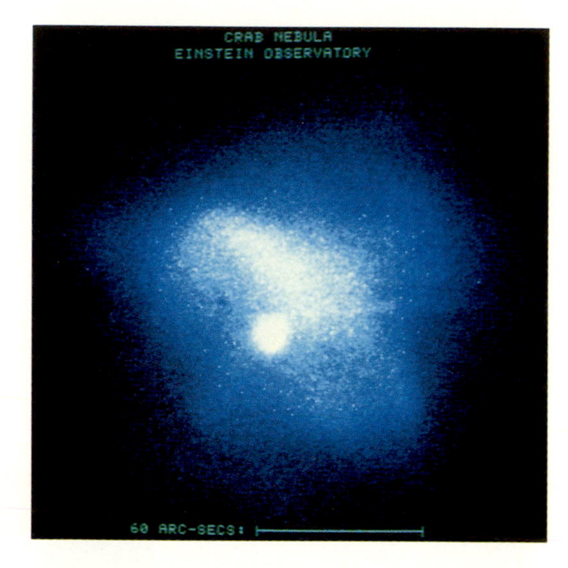

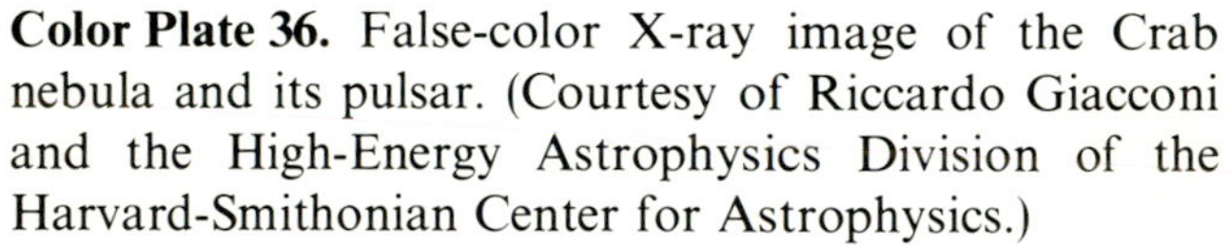

Color Plate 36. False-color X-ray image of the Crab nebula and its pulsar. (Courtesy of Riccardo Giacconi and the High-Energy Astrophysics Division of the Harvard-Smithonian Center for Astrophysics.)

Color Plate 37. An artist's rendition of the solar system. The sizes of the Sun and the nine planets are drawn to scale at the bottom. (Hansen Planetarium, Salt Lake City, Utah.)

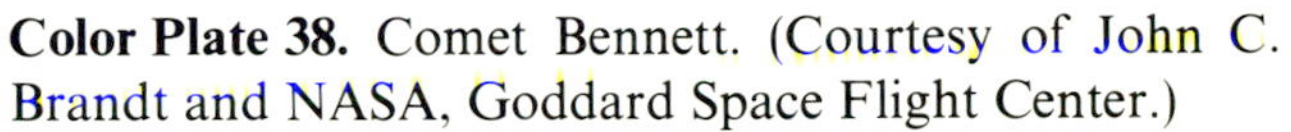

Color Plate 38. Comet Bennett. (Courtesy of John C. Brandt and NASA, Goddard Space Flight Center.)

Color Plate 39. Saturn, when Voyager 1 was farther from the Sun than the planet was. (NASA, Jet Propulsion Laboratory.)

Color Plate 40. Saturn, when Voyager 1 was closer to the Sun than the planet was. (NASA, Jet Propulsion Laboratory.)

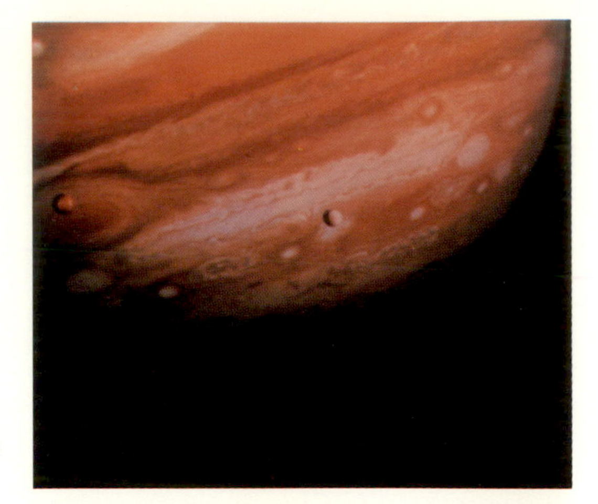

Color Plate 41. Jupiter with Io and Europa in the foreground. (NASA, Jet Propulsion Laboratory.)

Color Plate 42. The Great Red Spot of Jupiter as photographed by Voyager 2. (NASA, Jet Propulsion Laboratory.)

Color Plate 43. A montage of the four Galilean satellites of Jupiter, showing their correct relative sizes. Clockwise from upper left: Io, Europa, Callisto, and Ganymede. (NASA, Jet Propulsion Laboratory.)

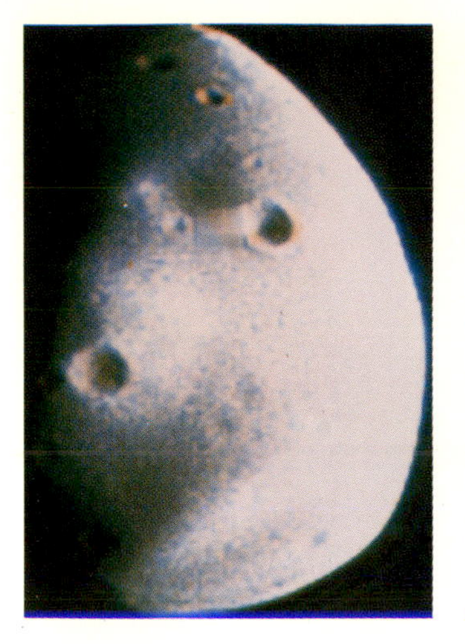

Color Plate 44. A two-color picture of Deimos, the smaller of the two satellites of Mars, generated from Viking 1 camera shots through a violet filter and an orange filter. Deimos is believed to resemble most common asteroids. (NASA).

Color Plate 45. View of the area near the Viking 1 Lander site on Mars. (NASA.)

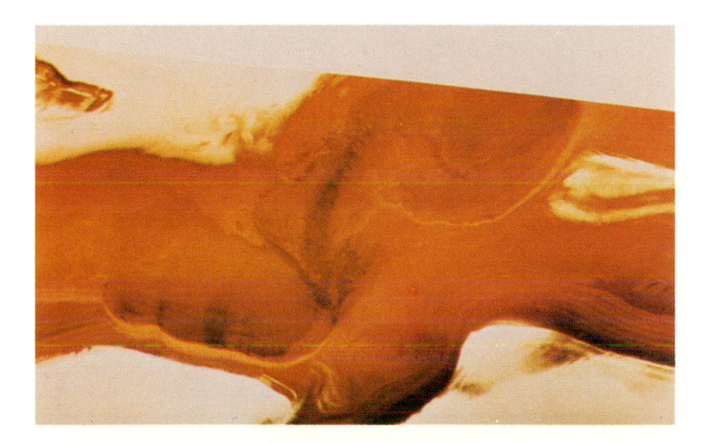

Color Plate 46. Frosty scene near the north pole of Mars viewed in mid-summer by Viking Orbiter 2. (NASA.)

Color Plate 47. Four false-color ultraviolet views of Venus by Pioneer Venus Orbiter. The time sequence begins clockwise from lower left, and shows that the planet's clouds circle Venus completely once every four days, much faster than the rotation period of the planet. (NASA, Ames Research Center.)

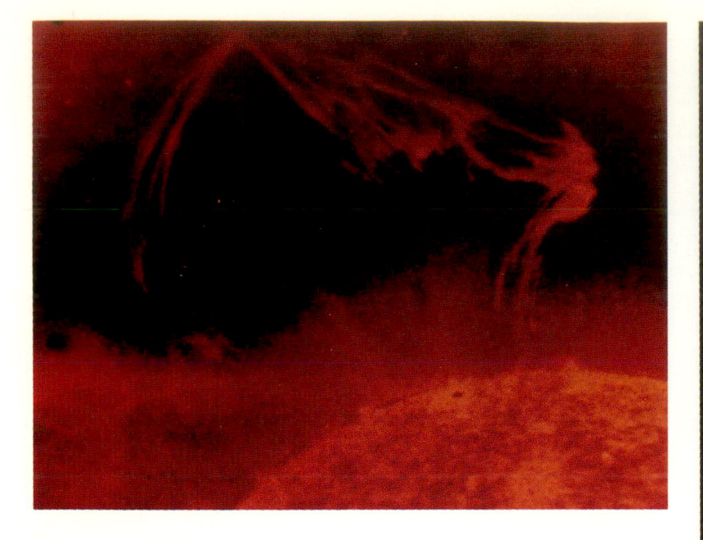

Color Plate 48. Eruption of the Sun viewed from Skylab at 304 Å. (NASA and the Naval Research Laboratory.)

Color Plate 49. The Moon viewed from Apollo 11. (NASA, Johnson Space Center.)

Color Plate 50. The Earth rising over the Moon's horizon viewed by Apollo 8. (NASA, Johnson Space Center.)

obtain the result $\mu(\tilde{\omega}) = \pi C/\tilde{\omega}$. Show now that the masses contained in a sphere of radius r and in a cylinder of radius $\tilde{\omega}$ are given, respectively, by

$$M_s(r) = \int_0^r \rho(r) 4\pi r^2 \, dr,$$

$$M_c(\tilde{\omega}) = \int_0^{\tilde{\omega}} \mu(\tilde{\omega}) 2\pi\tilde{\omega} \, d\tilde{\omega}.$$

With $\rho(r) = C/r^2$ and $\mu(\tilde{\omega}) = \pi C/\tilde{\omega}$, show that $M_s(r) = 4\pi C r$, whereas $M_c(\tilde{\omega}) = 2\pi^2 C\tilde{\omega}$. For $r = \tilde{\omega}$, therefore, $M_s(r)$ differs from $M_c(\tilde{\omega})$ only by a factor of $2/\pi$. Draw now the appropriate physical conclusions.

There are three difficulties in the actual application of formula (13.3) to galaxies. First, the formula (13.3) requires a numerical correction, a coefficient of order unity. whose exact value depends on the detailed kinematics and mass distribution within the system. This difficulty can be partially alleviated by theoretical modeling of the mass distribution. Second, the directly measurable quantity is the angular displacement θ from the galactic center, not the linear displacement r. To obtain r from θ, we need to know the distance to the galaxy, a topic which we shall discuss in more detail in Chapter 14. Third, the astronomical measurement of v itself is a nontrivial matter, which we shall now discuss.

When a spectrum is taken of an elliptical galaxy, we measure not the light from any single star, but the light from a significant part of the entire galaxy of stars. The superposition of billions of stars moving at different random velocities along the line of sight will, because of the Doppler effect, broaden appreciably the observed spectral lines (Figure 13.14). A comparison of the broadened absorption-line spectrum of the galaxy with the intrinsic spectrum of an appropriately chosen star allows an estimate for the velocity dispersion present in the billions of stars seen along the whole line of sight through the galaxy. We call this average velocity dispersion V to distinguish it from the three-dimensional value v which applies at a local point. (Usually, $v = \sqrt{3}V$.)

Problem 13.6. The intrinsic line-width of a K giant star might be 0.5 Å at a central wavelength $\lambda_0 = 5000$ Å. Suppose an elliptical galaxy is observed which has a line width $\Delta\lambda = 3$ Å. Assume that the increase is wholly due to broadening by random velocities of the constituent stars, and calculate V from the Doppler formula $\Delta\lambda/\lambda_0 = V/c$.

As an example of a galaxy which has been weighed by the above technique, consider the giant elliptical M87 (Figure 13.15a). This E0 system contains about 4 ×

Figure 13.15. Two elliptical galaxies at extreme ends of masses. (a) The giant elliptical galaxy M87 (type E0), which lies in the Virgo cluster. Notice the very large number of globular clusters which attend this system. (b) The dwarf elliptical galaxy, Leo II, which lies in our Local Group of roughly two dozen galaxies. Leo II may contain only as many stars as one of the globular clusters of M87. (From A. Sandage, *The Hubble Atlas of Galaxies*, Carnegie Institution of Washington; photographic materials from Palomar Observatory, California Institute of Technology.)

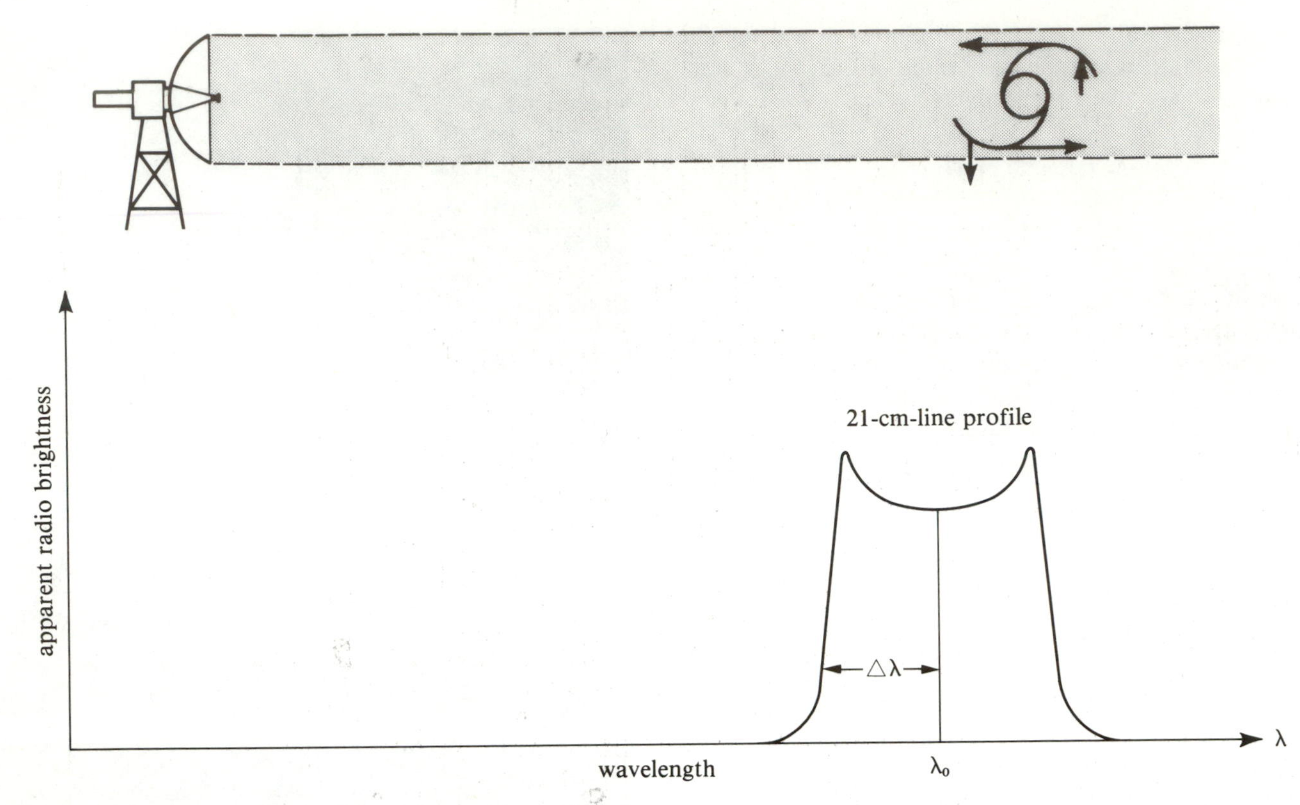

Figure 13.16. The 21-cm-line profile of a spiral galaxy which is not resolved by the radio telescope. If the spiral galaxy is smaller than the angular beam of the telescope as shown in the top diagram, each emitting hydrogen atom contributes a radio photon at a wavelength which is Doppler-shifted from the rest wavelength of λ_0 by the component of velocity along the line of sight. Because of the differential rotation of the galaxy, this typically produces the line shape seen in the bottom diagram, where the total line width $\Delta\lambda$ is proportional to the maximum rotational velocity V in the galaxy.

$10^{12} M_\odot$ in its optically visible parts. It is hard to estimate the mass of a dwarf elliptical like Leo II (Figure 13.15b) by this method because of its low surface brightness and velocity dispersion. Leo II probably contains only about $10^6 M_\odot$. An unknown is how much mass elliptical galaxies contain beyond their optically visible parts.

One can also measure the rotational velocities of spiral galaxies by taking optical spectra and measuring the Doppler shifts of either the absorption lines of the aggregate of stars or the emission lines from HII regions. Babcock, Mayall, and Courtes pioneered these techniques, and they have been extensively applied by Burbidge, Burbidge, and Prendergast, by Rubin and her collaborators, and by many others. The first detection of atomic hydrogen in 21-cm-line observations of external galaxies by Kerr and Hindman opened the possibility of using radio techniques to accomplish the same feat with greater sensitivity. The early observations were made with single-dish telescopes, which did not have enough angular resolution to map out the hydrogen distribution and kinematics of most spiral galaxies. What can be obtained from such measurements is only the total velocity width of the 21-cm-line profile (Figure 13.16). Volders and Hogbom and Roberts showed that such measurements could yield total mass estimates for the galaxy being observed, if one made some simplifying assumptions. Such measurements are also useful for defining the Fisher-Tully relation (which we will discuss soon). However, to obtain resolved images of spiral galaxies required the building of radio interferometric arrays (see Chapter 2).

Large arrays of radio telescopes have now provided spectacular aperture-synthesis pictures of the distribution and motions of atomic hydrogen in nearby spiral galaxies. Color Plate 24, produced by Westerbork radio astronomers, provides especially impressive visual dis-

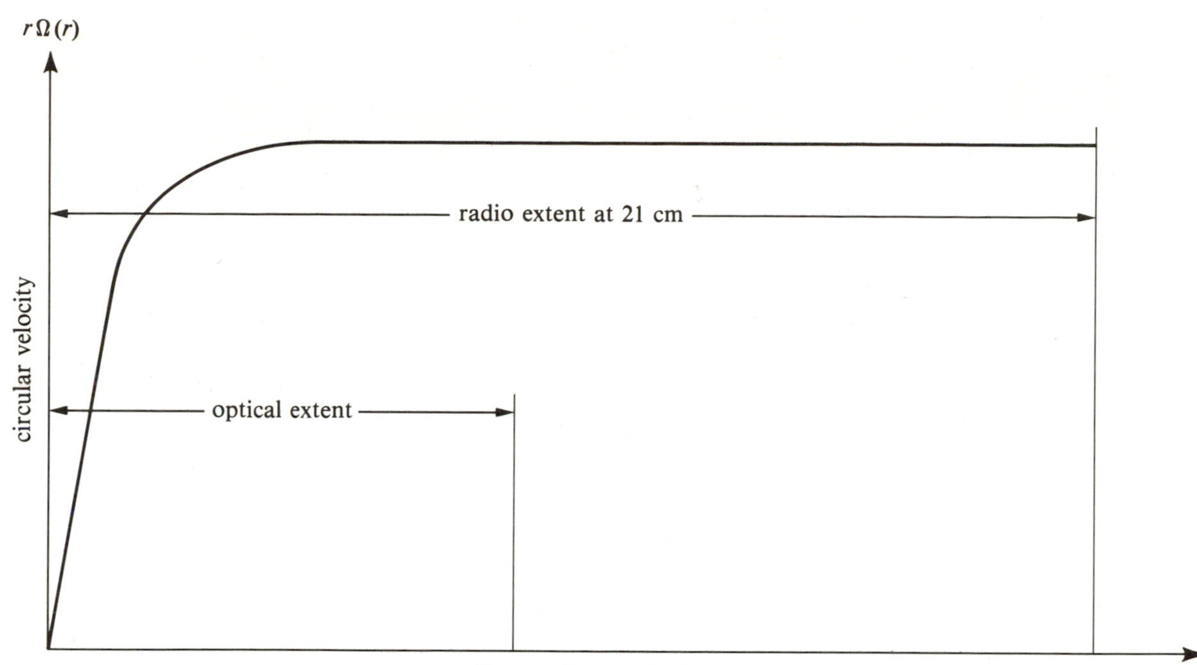

Figure 13.17. Schematic rotation curve of a spiral galaxy as derived from 21-cm-line observations of the atomic hydrogen. Notice that the rotation curve stays flat at distances much larger than the optical extent of the galaxy. The optical and radio extents are defined to some limiting sensitivity of the survey method.

play of the results for the galaxy M101. In the color coding in this computer-generated radiograph, red denotes redshifts and blue denotes blueshifts (the velocity along the line of sight after subtraction of the velocity of the center of the galaxy). Yellow represents no radial velocity, i.e., the actual motion relative to the center of the galaxy taking place tangentially across our line of sight. Green results from a mixture of yellow and blue, i.e., intermediate velocities of approach, etc. Given this coding, you should be able to see the large-scale rotation of the galaxy M101. Since the intensity of the colors represents the intensity of the observed 21-cm signal, the arrangement of the hydrogen gas in a grand spiral structure should also be apparent from this radiograph.

Problem 13.7. Suppose a spiral galaxy extends 10 minutes of arc in the sky. What baseline would you need for an interferometric array if you wanted to get 30 picture resolution elements ("pixels") in one dimension for a radiograph of the galaxy at the 21-cm wavelength?

Rotation curves of nearby galaxies can also be obtained more tediously, but with greater sensitivity, by making radio measurements along the major axis of a galaxy using the large Arecibo radio telescope. Observations of this kind by Krumm and Salpeter are able to extend the rotation-curve measurements to very large distances from the center of a spiral galaxy. These observations and the aperture-synthesis results, taken together with earlier work by Morton Roberts and others, have provided the most definitive evidence that mass is missing on a galactic scale in spirals. The rotation curves of spiral galaxies inevitably turn out to be very flat at large distances from the galactic center (Figure 13.17). This implies that the surface-mass density $\mu(r)$ of the gravitating contribution, whether contained in a halo or in the disk, can drop off with increasing distance from the center only as $1/r$ (see Problems 12.11 and 13.5):

$$\mu(r) \propto r^{-1}. \tag{13.4}$$

But the light distribution $\mathscr{L}(r)$ projected into the plane of the disk drops off exponentially with radius r (see

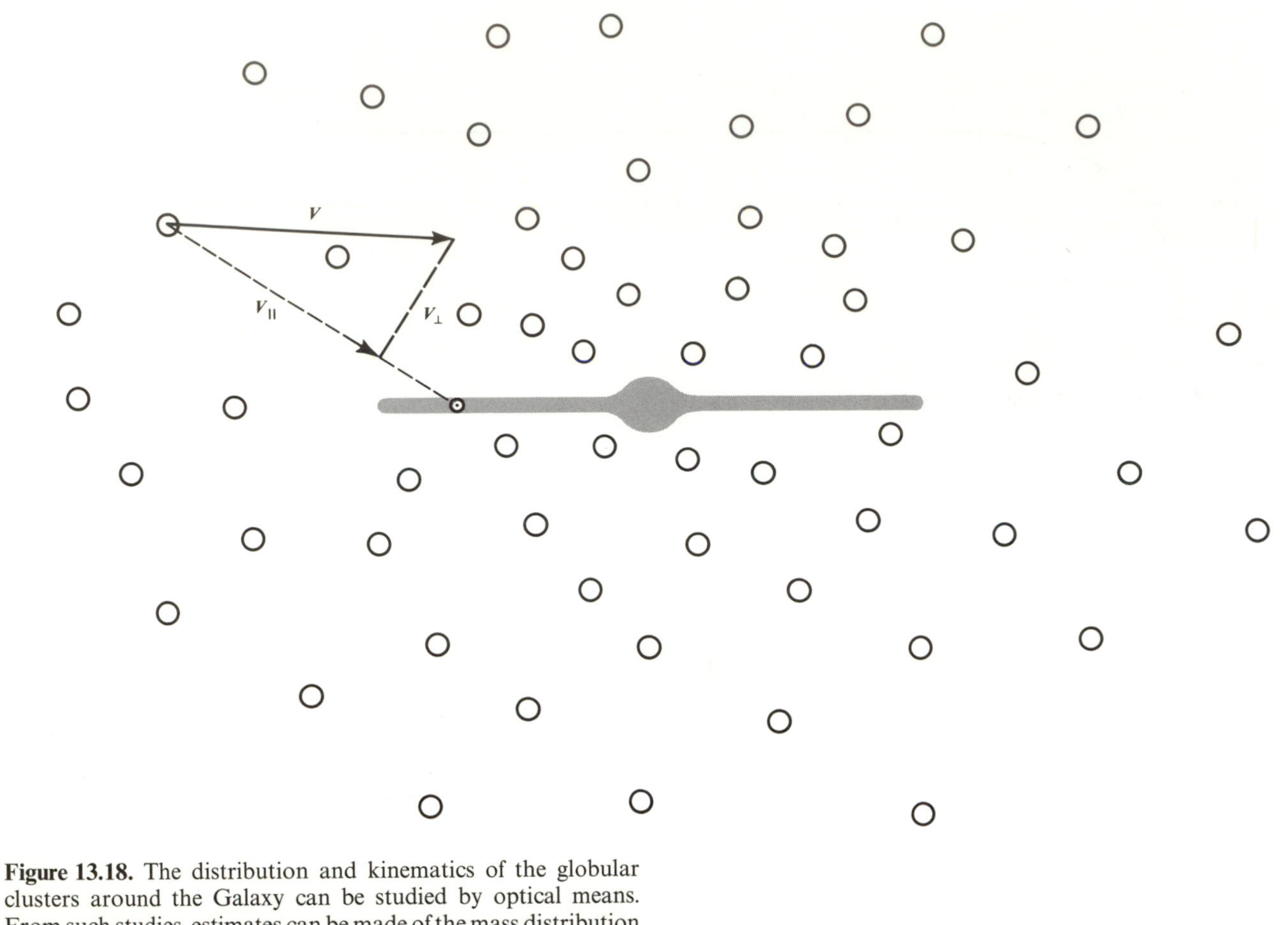

Figure 13.18. The distribution and kinematics of the globular clusters around the Galaxy can be studied by optical means. From such studies, estimates can be made of the mass distribution of the Galaxy at relatively large distances from the galactic center. To do this, some simplifying assumption must be made about the component of velocity $v_{\perp}$ of a globular cluster perpendicular to our line of sight.

equation 13.2); consequently, at large r, the mass-to-light ratio $\mu(r)/\mathscr{L}(r)$ must increase dramatically in spiral galaxies. This deduction is, of course, the crux of the issue of missing mass.

Problem 13.8. Draw a diagram to show that major-axis measurements of the radial velocity yield $r\Omega(r) \cdot \sin(i)$ in a circular rotation model of a spiral galaxy, where $r\Omega(r)$ is the circular velocity and i is the inclination angle of the disk ($i = 90°$ corresponds to an edge-on disk). How would you go about calculating the angle i?

Flat rotation curves imply that the mass contained within radius r, $M(r)$, must increase linearly with r. That is, if radio astronomers see the rotation curve stay flat for distance which are, say, three times further than was expected from optical observations, then the mass of the galaxy must be three times greater than what had been deduced for the optically visible parts. Now, a mass law which diverges linearly with r at large r obviously cannot extend to arbitrarily large distances. How far does the law $M(r) \propto r$ hold? No one knows. There have been speculations that this law might hold to 10 or 100 times the usual optical radii of galaxies, but to be honest, no one really knows.

Attempts have been made to study this issue observationally (Figure 13.18). Sargent and Hartwick have examined the distribution of globular clusters around our own Galaxy and around the Andromeda galaxy, M31. Globular clusters can be found at great distances from the galactic center. Unfortunately, globular clusters do not follow simple circular orbits around the Galaxy;

so we must adopt some statistical assumptions about their motions. The statistics are not very good, since even a large spiral galaxy may contain only 200 globular clusters. Although the globular-cluster method has substantially vindicated the radio-astronomical result that the optically derived masses probably need to be revised upward by a few times, it has not resolved the larger and more controversial issue of very massive halos which contain 10 to 100 times more mass than the optically visible galaxy. An even worse statistical situation exists with attempts by Page, Turner, Karenchentsev, and Peterson to analyze the motions of binary galaxies. Here, we have only one companion, information about only the projected separation and the difference in line-of-sight velocities. To make any progress, many statistical assumptions have to be made, and these have led to much controversy about the derived results. The problem of missing mass and massive halos is likely to haunt extragalactic astronomy for a long time.

The $L \propto V^4$ Law for Ellipticals and Spirals

Most galaxies are too small in angular extent to allow astronomers to investigate the detailed kinematics of the interior. It is easier to measure the total spread V of velocity in the line profile of the radiation from the visible part of the galaxy. In elliptical galaxies, V results from the random velocities of the stars; in spirals, from the rotational velocities of the gas. Studies by Faber and Jackson showed that the velocity spread V in an optical spectrum of an elliptical galaxy is intimately related to its luminosity L. The Faber-Jackson relation is

$$L \propto V^4. \tag{13.5}$$

Thus, Faber and Jackson found that if a elliptical galaxy has a velocity dispersion which is twice as large as that of another elliptical, the former is likely to be sixteen times brighter intrinsically. Fischer and Tully discovered a similar relation for the luminosity L of a spiral galaxy and the velocity spread V of its 21-cm-line profile. In particular, Aaronson, Huchra, Mould, Sullivan, Schommer, and Bothun showed that if L refers to the near infrared, so that the light is more typical of the disk stars, then the Fischer-Tully relation between L and V has the same power-law dependence as equation (13.5). The proportionality constant in equation (13.5) is, of course, different for ellipticals and spirals. It is to be found by calibration galaxies of known distances.

Freeman pointed out that a connection may exist between the Faber-Jackson law and the constancy of the central surface brightnesses $\mathscr{L}(0)$ of giant elliptical galaxies. Schecter made the same point for the Fischer-Tully relation and the constancy of $\mathscr{L}(0)$ observed for giant spiral galaxies. The Faber-Jackson-Fischer-Tully relation is equivalent to the Fish-Freeman discovery only if we assume that the mass-to-light ratio $\mu(r)/\mathscr{L}(r)$ does not vary with position r in the galaxy (Problem 13.9). We have already seen that this assumption is almost certainly incorrect for the outer parts of spiral galaxies. For now, it is perhaps wisest to regard equation (13.5) as an empirically established rule which has yet no theoretical justifications. The importance of this relation for spiral galaxies is that V is a relatively easy 21-cm measurement. With V known, equation (13.5) allows an estimate of the intinsic brightness L. A measurement of the apparent brightness f then yields the distance r to the galaxy (see Chapter 14).

Problem 13.9. Use scaling arguments to show the equivalence of $L \propto V^4$ and $\mathscr{L}(0) =$ a universal constant. *Hint*: Construct the physical arguments to go along with the following algebraic manipulations: $L \propto \mathscr{L}(0)r_0{}^2$. If $\mu(r)/\mathscr{L}(r) =$ constant, then $L \propto M$. But $M \propto r_0V^2/G$; so $r_0 \propto V^2/\mathscr{L}(0)$. Therefore, $L \propto V^4/\mathscr{L}(0)$, and $L \propto V^4$ if $\mathscr{L}(0) =$ constant.

Dynamics of Barred Spirals

Barred spirals differ from ordinary spirals in having the presence of an elongated bar in the central regions (compare Figures 13.6b and c). Crane has shown that the enhancement of light in the bar is usually not large; again, the visual impression obtained by the eye is misleading because of its propensity to concentrate on contrasting light levels.

The currently accepted explanation for barlike structures in the central regions of disk galaxies is like the explanation for spiral structure: the phenomenon is likely to be a wave in the density distribution (Chapter 12). Bar-forming instabilities in a flattened distribution of stars were first encountered in numerical simulations by Hohl, Miller, Prendergast, and others (see Figure 12.28). The tendency to form bars has subsequently been studied theoretically by many people, notably Agris Kalnajs. Ostriker and Peebles proposed, as a general criterion, that a stellar system would form a bar whenever the energy contained in rotational motions exceeds 39 percent of the energy contained in random motions. Since the energy contained in the rotation of stars in the solar neighborhood is about 40 *times* bigger than the energy in the random motions, they proposed that the Galaxy—and by extension, all spiral galaxies—must have a massive quasispherical halo of stars whose large random velocities would make up for the deficit in the disk. Their criterion is now known to be overly pessimistic, and Mark and Berman have shown that they probably also underestimated the stabilizing influence of the central bulges of spiral galaxies.

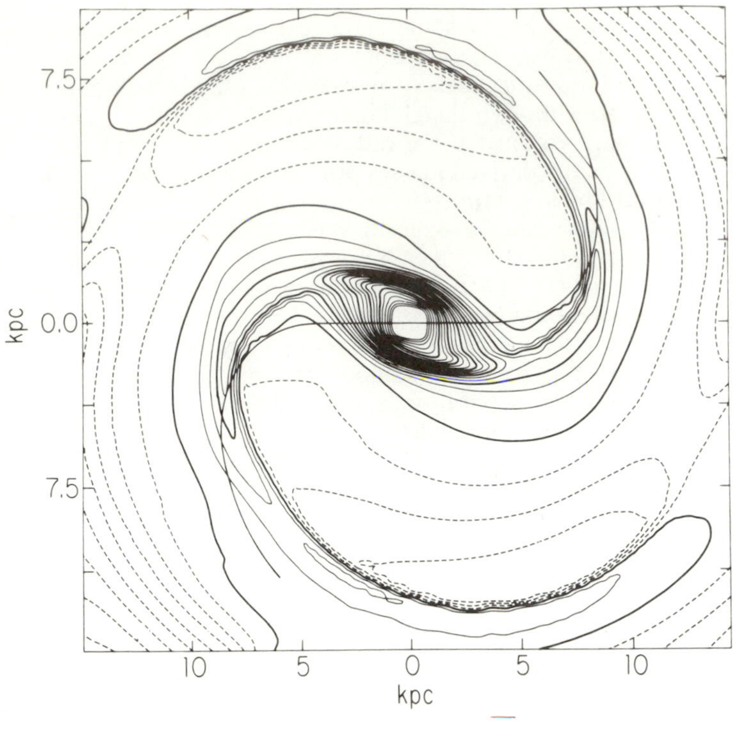

Figure 13.19. (a) Streamlines and (b) density contours for interstellar gas flow in a theoretical model of a barred spiral galaxy. (From Roberts and van Albada, *Ap. J.*, **246**, 1981, 740.)

Nevertheless, the basic tendency for rapidly rotating self-gravitating bodies to want to distort into an elongated configuration has been well-established in other contexts by Jacobi, Dedekind, Riemann, Chandrasekhar, and Lebovitz. In essence, another way (in addition to spiral structure) for a rotating body to gain more gravitational binding energy for its inner parts, while conserving total angular momentum, is to pull matter in along one axis and out along the other (see Figure 12.25c). Just as with spiral waves, even a small fractional distortion of the stellar distribution can produce a large response in the interstellar matter. Explicit numerical calculations of the effect were first carried out by Sanders and Huntley. Figure 13.19 shows that large noncircular motions can be induced inside the bar; this is the characteristic theoretical signature of a barred spiral.

Active Galaxies

So far our discussion has centered on the properties of ordinary galaxies which show little activity beyond what might be expected for a collection of stars and gas. In a small percentage of observed galaxies, there is violent activity well beyond the norm described so far. The

BOX 13.4
Active Galaxies

Seyferts. Spiral galaxies with bright nucleus. Nuclear light shows emission lines and some nonthermal radio emission.

N Galaxies. Elliptical galaxies with bright nucleus. Some are radio sources.

BL Lacertae Objects. Elliptical galaxies (probably) with very bright nucleus. Nuclear light shows a featureless continuum and high variable polarization.

Radio Galaxies. Generally giant and supergiant ellipticals which exhibit a double-lobed radio structure. Occasionally sources are of the core-halo type, with a compact nuclear source whose radio emission can vary on a timescale of years. Radio jets can show superluminal expansion.

Quasars. Quasi-stellar radio sources, sometimes with extended radio structures like double-lobed radio galaxies. Optical spectra reminiscent of Seyferts. Usually strong emitters of X-rays and infrared, as well. No *direct* evidence of being galaxies.

QSOs. Quasi-stellar objects similar to quasars, but lack strong radio continuum emission. Both quasars and QSOs possess large redshifts.

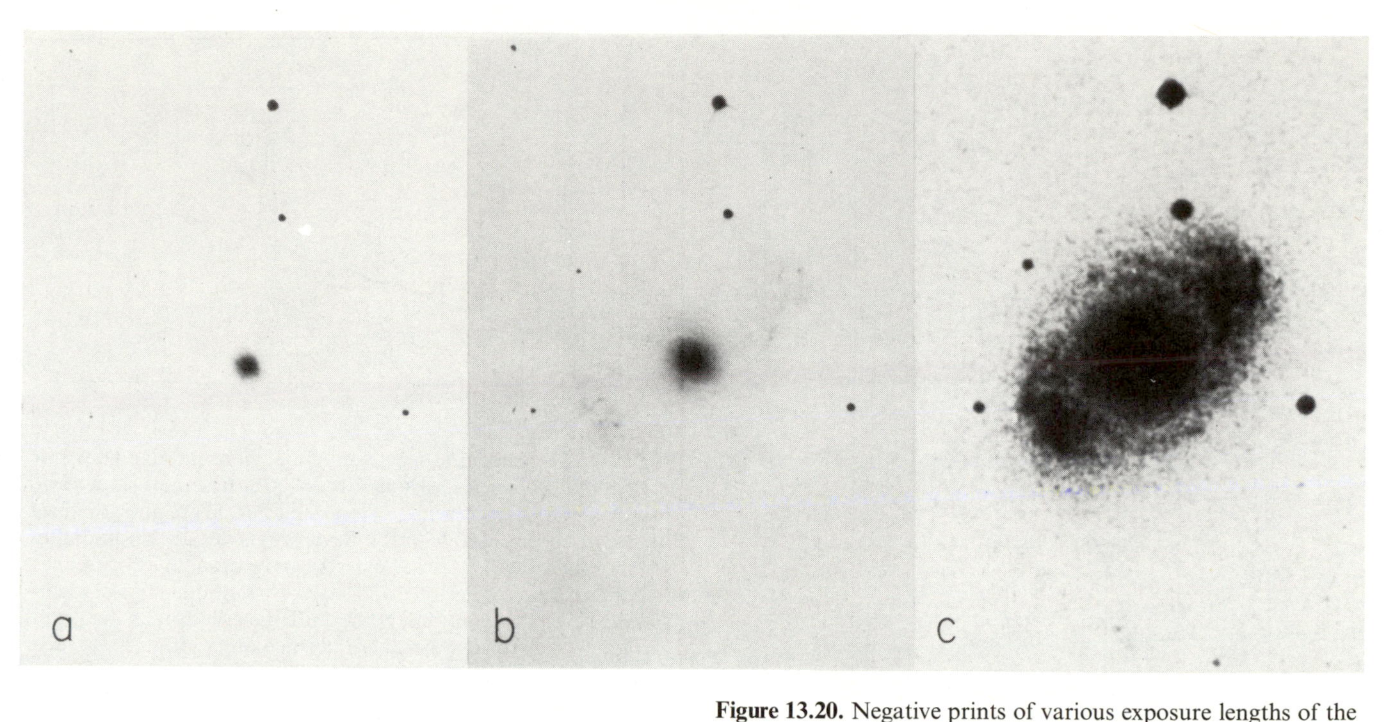

Figure 13.20. Negative prints of various exposure lengths of the Seyfert galaxy NGC4151. (a) A short exposure shows only the bright nucleus. (b) A longer exposure yields the hint of spiral arms, while (c) a deep exposure burns out the nuclear regions and exhibits both the disk and arms of a spiral galaxy. (From W. W. Morgan, *Ap. J.*, **153**, 1968, 27; photographic materials from the Mt. Wilson and Palomar Observatories.)

suspicion has grown that this activity is driven by energetic events in the nuclei of the galaxies. Similar "diseases" (but with some differences) seem to afflict both spiral and elliptical galaxies (Box 13.4).

Seyfert Galaxies

The first anomalies were discovered in a class of galaxies with very bright centers studied by Carl Seyfert. A short exposure of a Seyfert galaxy typically shows only a very bright central point of light. This starlike image is actually the nucleus of a galaxy. Longer exposures, which greatly overexpose the nucleus, show that the outer parts of a Seyfert galaxy are invariably those of a spiral galaxy (Figure 13.20). In many Seyfert galaxies, the bright nucleus outshines the rest of the galaxy put together. Moreover, the enormous light output from the nucleus can often vary by more than two times in less than a year. Since coherent changes on a timescale t are possible in an object only if it is smaller than the distance, ct, that light can tranverse in time t, this observation alone limits the size of the central light source to less than one light-year (Figure 13.21). In other words, in Seyfert galaxies, we are dealing with a nucleus which is no bigger than the mean separation of stars in a typical galaxy, yet can put out more light per second than an entire system of 100 billion stars.

Moreover, the light coming from the nucleus of a Seyfert galaxy is not at all like the light which emanates from a typical galaxy. In an ordinary galaxy, the optical light emitted is mostly starlight. Starlight consists (mostly) of a thermal continuum plus *absorption* lines, and the same will be true of the light combined from a myriad of stars. In contrast, Seyfert nuclei exhibit strong emission lines like those from HII regions and planetary nebulae (recombination spectra) and a continuum which looks nonthermal (in that $\lambda^2 f_\lambda$ increases with increasing wavelength; see Figure 13.22). Thus, the basic energy source of the nuclei of Seyfert galaxies probably does not derive from the thermonuclear power of ordinary stars.

This suspicion is reinforced by the observation that Seyfert galaxies also emit strongly in the radio, infrared, and X-rays. There is good reason to believe that the infrared emission extensively observed by Rieke and Low often results from the heating of dust grains by more energetic photons. Much dust and gas must surround the nuclei of Seyfert galaxies, a conclusion borne out

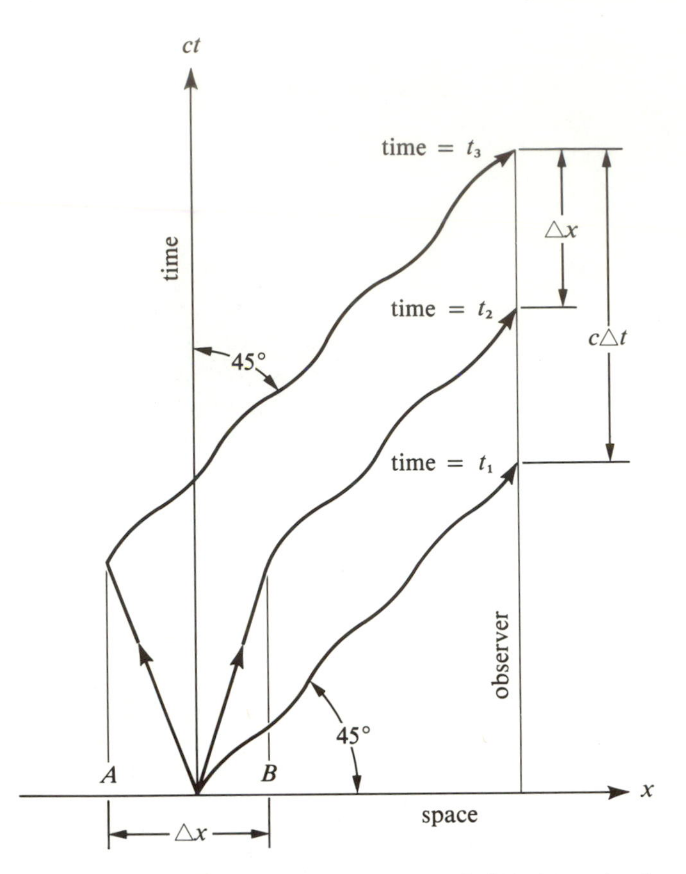

Figure 13.21. Spacetime diagram of light variations in a galactic nucleus. Consider the x-ct plane in which light (wavy lines) propagates along lines inclined at 45° to the space and time axes. Suppose a region of size Δx (assumed to be stationary relative to the observer for simplicity) varies in light output. Imagine this first happens at the center of the interval Δx at time $t = 0$. Information that this has happened first reaches the observer at time t_1. Meanwhile, the cause of the variation propagates outward from the center of the object to the extreme points A and B at speeds less than c (propagation indicated by straight arrows). When the cause reaches A and B, these points also begin to vary in light output. Information that this has happened reaches the observer at times t_3 and t_2. Since light propagates in this diagram at 45° angles, the interval $c(t_3 - t_2)$ clearly equals Δx. The size of the intrinsically varying region, Δx, is therefore less than $c(t_3 - t_1) = c\,\Delta t$, where Δt is the total time interval during which light variations are seen by the observer.

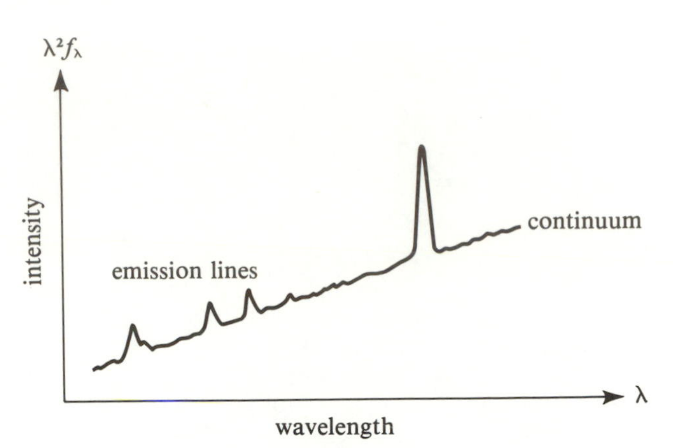

Figure 13.22. Schematic spectrum of a Seyfert nucleus. The optical continuum often looks as if it has a nonthermal slope ($\lambda^2 f_\lambda$ increasing with increasing λ; see Chapter 11). Emission lines dominate the spectral features. Seyfert galaxies are traditionally separated into two classes: Seyferts I have broad hydrogen lines (recombination radiation) and narrow forbidden lines (collisionally excited radiation); Seyferts II have hydrogen lines and forbidden lines of comparable intermediate widths. These distinguishing features probably depend on whether the hydrogen and emission lines are formed at different or the same distances from the center of the galaxy.

also by observations of Walker, Kraft, Osterbrock, and others that many tens of solar masses of gas per year may be blown out of the nucleus of a Seyfert galaxy at speeds of several hundred km/sec. It is now believed that the presence of this gas and dust may largely obscure a direct view of the nucleus itself, and that the real size of the central powerhouse in Seyfert galaxies may be considerably less than one light-year.

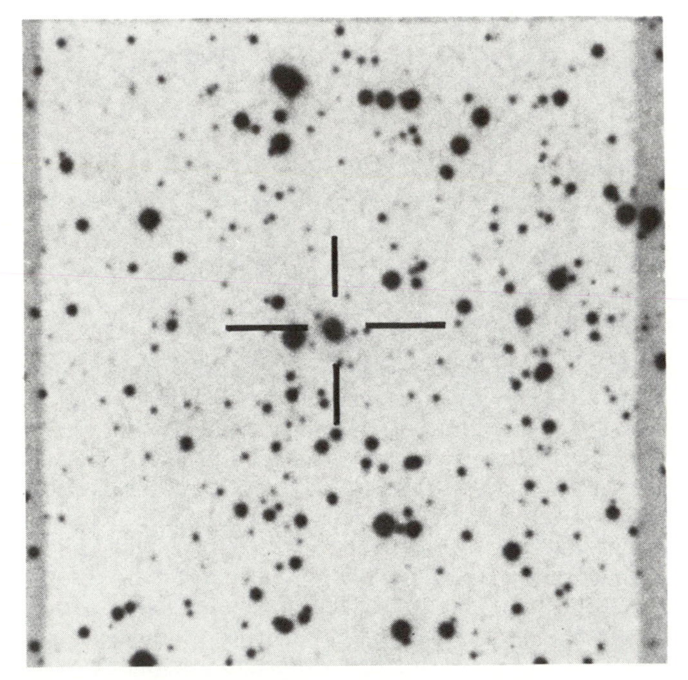

Figure 13.23. A negative print of BL Lacertae, the prototype of a class of extragalactic objects. On the Palomar Sky Atlas, BL Lac objects appear almost starlike; however, their spectra show no lines, just a featureless continuum. (Courtesy of John Macleod, National Research Council, Canada; photograph from Palomar Sky Atlas.)

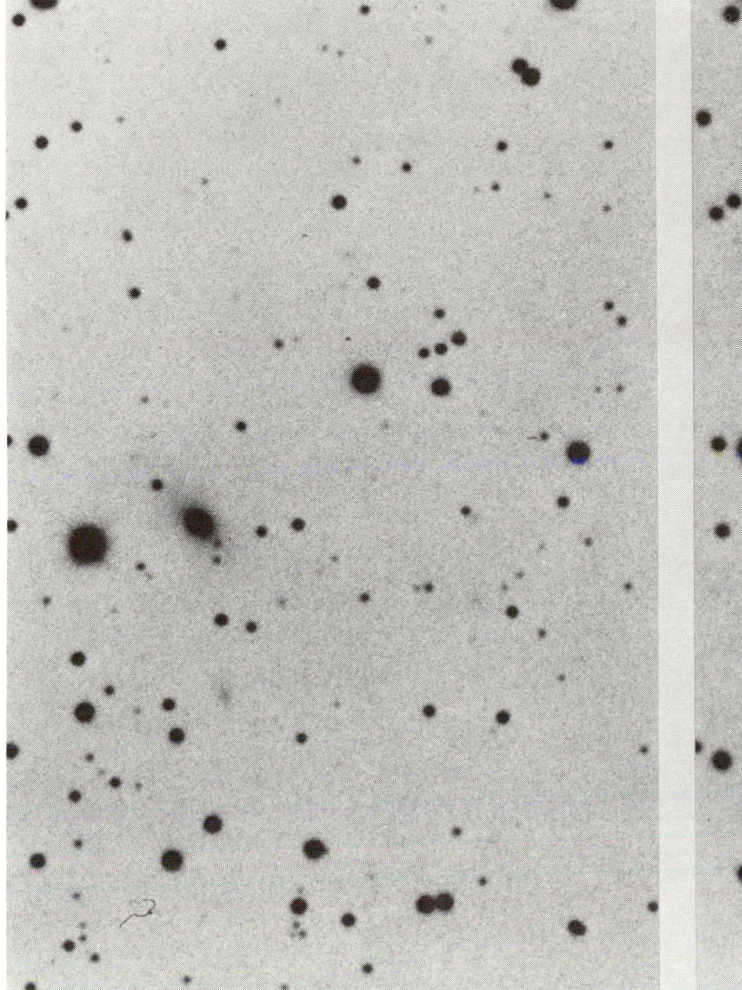

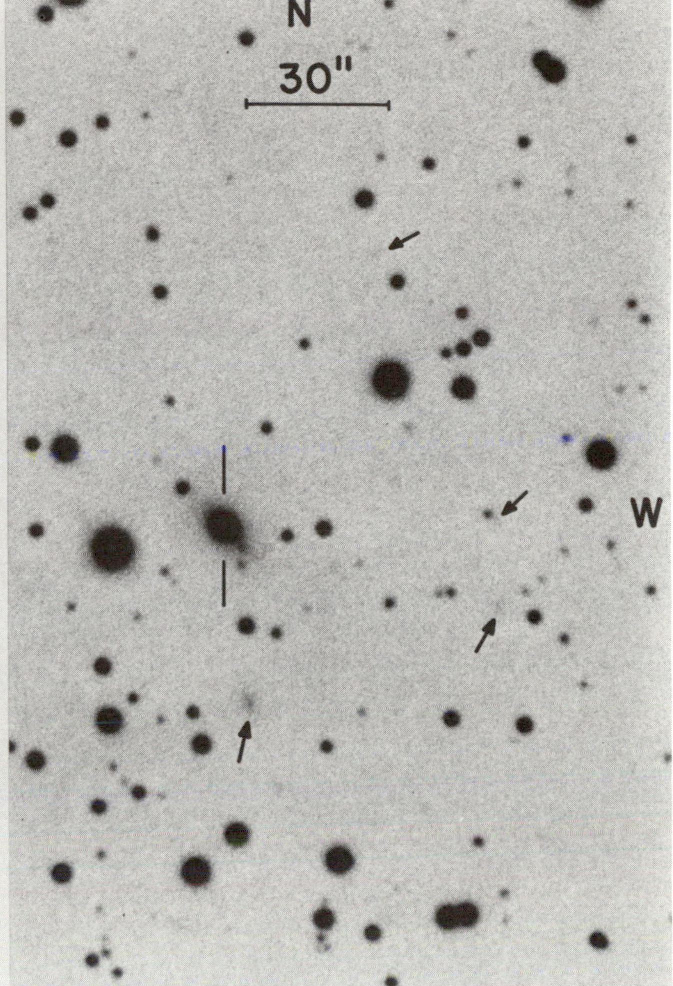

Figure 13.24. On deep exposures, BL Lac (between the vertical bars) is surrounded by nebulous light. The surface brightness distribution of the "fuzz" is consistent with that of a giant elliptical galaxy. Arrows mark what are probably faint galaxies; they may form a cluster of which BL Lac is a member. (From T. D. Kinman, *Ap. J. Lett.*, **197**, 1975, L49; photograph from the Kitt Peak National Observatory.)

BL Lac Objects

Elliptical galaxies with bright nuclei are also known, and are called N galaxies. An extreme example of elliptical galaxies with bright central nuclei may be the BL Lacertae objects. The prototype, BL Lac, is located in the constellation of Lacerta and presents a starlike image on short exposures (Figure 13.23). The light from BL Lac had been discovered in 1929 by Cuno Hoffmeister to vary by a factor of 2 in one week, and by a factor of 15 during a few months. This property explains why astronomers first thought it to be a peculiar kind of variable star and assigned it that kind of name. The true nature of BL Lac grew more mysterious when Macleod and Andrew stumbled onto it as the most likely optical counterpart to an intense and polarized source of radio emission. The amount of Faraday rotation suffered by their radio source (see Chapter 11) suggested that the object probably lay outside our Galaxy; so the identification by Schmitt of the radio source with the "variable star" BL Lac came as quite a surprise.

That BL Lac was not a star became apparent when deep photographs and optical spectra were taken. Deep photographs showed BL Lac to be distinctly "fuzzy," not as extended as galaxies of comparable brightness, but distinctly more extended than a star. (Figure 13.24).

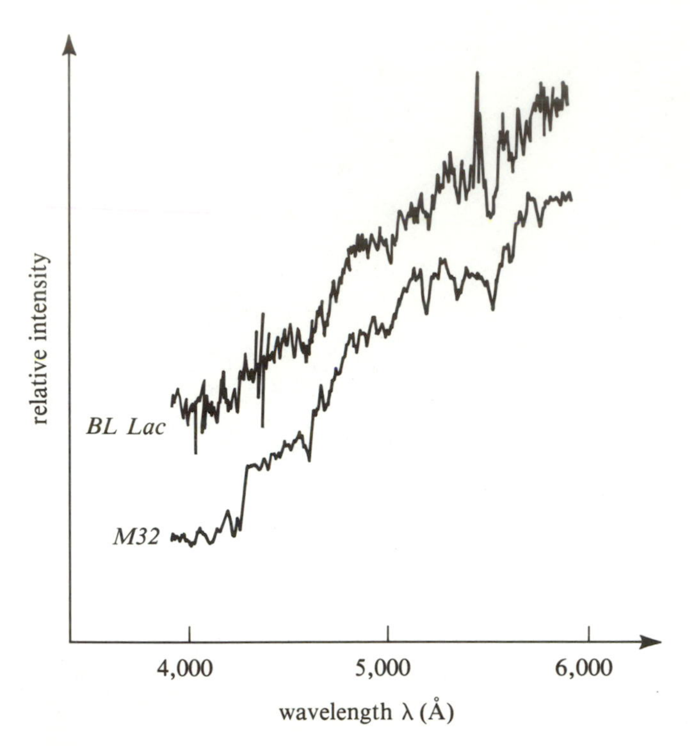

Figure 13.25. A comparison of the spectra of M32 and the fuzz of BL Lac. The former is a small elliptical companion of M31. The slight dissimilarities between the two spectra can be explained if BL Lac is a giant elliptical galaxy (see Table 13.1). (Adapted from Miller, French and Hawley, *Ap. J. Lett.*, **219**, 1978, L85.)

Optical spectra of BL Lac showed it to have neither absorption lines nor emission lines; its electromagnetic spectrum seemed to form a featureless continuum! Thus, BL Lac looked neither like a star nor like an active galaxy, such as a Seyfert or an N galaxy. What, then, is BL Lac? The mystery deepened with the discovery of other systems with BL Lac properties.

The breakthrough came in the late 1970s. Following some earlier work by Gunn and Oke, a group of astronomers led by Miller blocked off the light from the nucleus of BL Lac and obtained a spectrum of the faint "fuzz." The spectrum of the resulting light turned out to be very similar to that of M32, a small elliptical near the Andromeda galaxy M31 (Figure 13.25). The slight differences could be explained if the fuzzy nebulosity around the bright central source in BL Lac represented a giant elliptical galaxy at a distance of about one billion light-years. In this interpretation, the featureless continuum comes from a nucleus of exceptional brightness.

The optical light from BL Lacertae objects is now known to be highly variable and highly polarized. As has been measured by Angel and others, the polarization can change significantly in strength and in direction during one night of observation; so in BL Lac objects, the central powerhouse must be smaller than one light-day. That we can see much further into the heart of a BL Lac object than into a Seyfert argues that the former is surrounded by much less gas and dust. Such a conclusion would be compatible with BL Lac objects' being elliptical galaxies and Seyferts' being spiral galaxies.

Radio Galaxies

The growth in astronomical interest in active galaxies has been stimulated primarily by the fact that many (but not all) such systems emit strongly in the radio continuum. Extragalactic radio astronomy matured as a science in Australia under the leadership of Pawsey and Bolton and in England under the leadership of Sir Martin Ryle. In particular, until Ryle's development at Cambridge of radio-interferometric techniques (see Chapter 2), radio astronomers were stuck (except for clever but limited use of occultations by the Moon and reflections by the sea) with relatively poor angular resolution. Jansky's pioneering studies, and Reber's continuation of them, were confined mainly to mapping the radio-continuum background of the Milky Way. Hey, Parsons, and Phillips were the first to find a discrete source of radio emission apart from the Sun. Since this source is the strongest to lie in the direction of the constellation of Cygnus, it came to be called Cygnus A. With the poor angular resolution of their radio telescope, Hey and his coworkers could not pin down the position of Cygnus A with sufficient precision to permit an optical identification. The introduction of radio interferometers allowed F. G. Smith to use a two-element device to obtain an accurate position for Cygnus A. Armed with this information, Walter Baade and Rudolph Minkowski were able to identify Cygnus A with the peculiar galaxy drawn in Figure 13.26. Baade and Minkowski could then estimate the distance to this galaxy from its redshift (see Chapter 14), and from its distance and observed radio power, they could then deduce its radio luminosity. Cygnus A turned out to shine with a radio luminosity which is a million times stronger than the radio luminosity that we now attribute to ordinary galaxies!

Meanwhile, Ryle and his colleagues were busy making surveys to find additional (but apparently weaker) discrete radio sources. Many of the astronomically known radio sources carry designations from such surveys. For example, Cygnus A is also known as 3C405 because it is the 405th source in the third Cambridge catalogue. Not all the 3C sources are extragalactic objects. For example, the first and third apparently brightest radio sources in the 3C catalogue, 3C361 and 3C144, were identified earlier by Ryle and Smith and by Bolton and Stanley with the galactic supernova remnants, Cassiopeia A and Taurus A (the Crab nebula; see Chapter 11), respectively. Some of the 3C sources (and those in succeeding 4C and

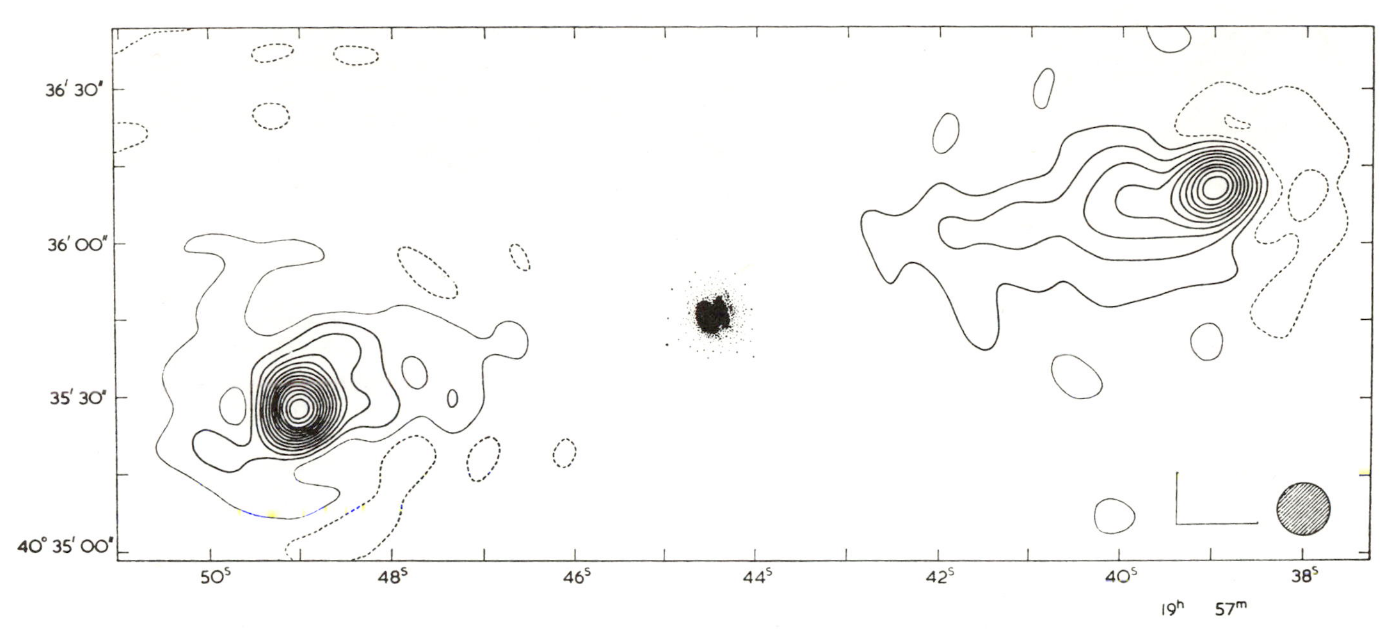

Figure 13.26. A radio map of Cygnus A shows it to be a classic radio double. The "butterfly" image at the center represents an artist's impression of the optical appearance of the peculiar galaxy that lies at the center of the activity (see Figure 14.17). (From Mitton and Ryle, *MNRAS*, **146**, 1969, 221. A higher resolution map can be found in Hargrave and Ryle, *MNRAS*, **166**, 1974, 305.)

5C catalogues) have coincided with galaxies that contain active nuclei, such as Seyferts. Deep photographs show Seyferts to have extended optical images, but their radio appearance is essentially pointlike. In contrast, some of the strongest extragalactic sources also have extended radio structures. This structure typically consists of two radio lobes straddling a central galaxy whose nucleus may or may not be abnormally bright in the radio and the optical. The radio lobes typically lie millions of light-years on diametrically opposite sides of the central galaxy, with the sizes of the lobes being about a third to a fifth of the spacing. As shown in Figure 13.27, the central galaxy is usually a giant elliptical, or even more commonly a supergiant elliptical, but never a spiral.

The double-lobed radio sources contain the intrinsically brightest radio-emitting objects in the universe. In particular, the first-discovered discrete source, Cygnus A, is a radio double. The mechanism for the radio emission in the lobes has all the earmarks of synchrotron radiation (see Chapter 11). The amount of energy in cosmic-ray particles and magnetic fields required to produce a given amount of synchrotron radiation depends on the assumed field strength B. The higher B is, the more magnetic energy is required, but the less is the cosmic-ray energy (Figure 13.28). Geoffrey Burbidge showed that a *minimum* estimate for the total energy resulted if one assumed that the magnetic field and relativistic particles had comparable energies (as indicated in Figure 13.28). Applied to radio galaxies, this conservative estimate resulted in tremendous total energies in the lobes, as much as 10^{61} erg, as much as is released by *ten billion* supernova explosions! Indeed, since probably only one percent of a supernova's observed energy comes out in the form of relativistic particles, only an entire galaxy of supernova explosions would suffice—and barely, at that. Whatever the source behind this majestic phenomenon is, it probably involves more than the power of "mere" supernova explosions.

Very long baseline interferometry (VLBI; see Chapter 2) has again indicated some kind of extraordinary activity in the nucleus of the central galaxy when there is a central radio source between two radio lobes (Figure 13.29). Some of these compact sources vary in roughly one year. To study the structure of these compact sources, we must therefore be able to resolve to the angle subtended by one light-year. An object one light-year in size at a distance of 200 million lt-yr subtends an angle of 1 milliarcsec. Angular resolution on the order of a milliarcsec requires a baseline of the size of continents at radio wavelengths. (Compare this with the requirements

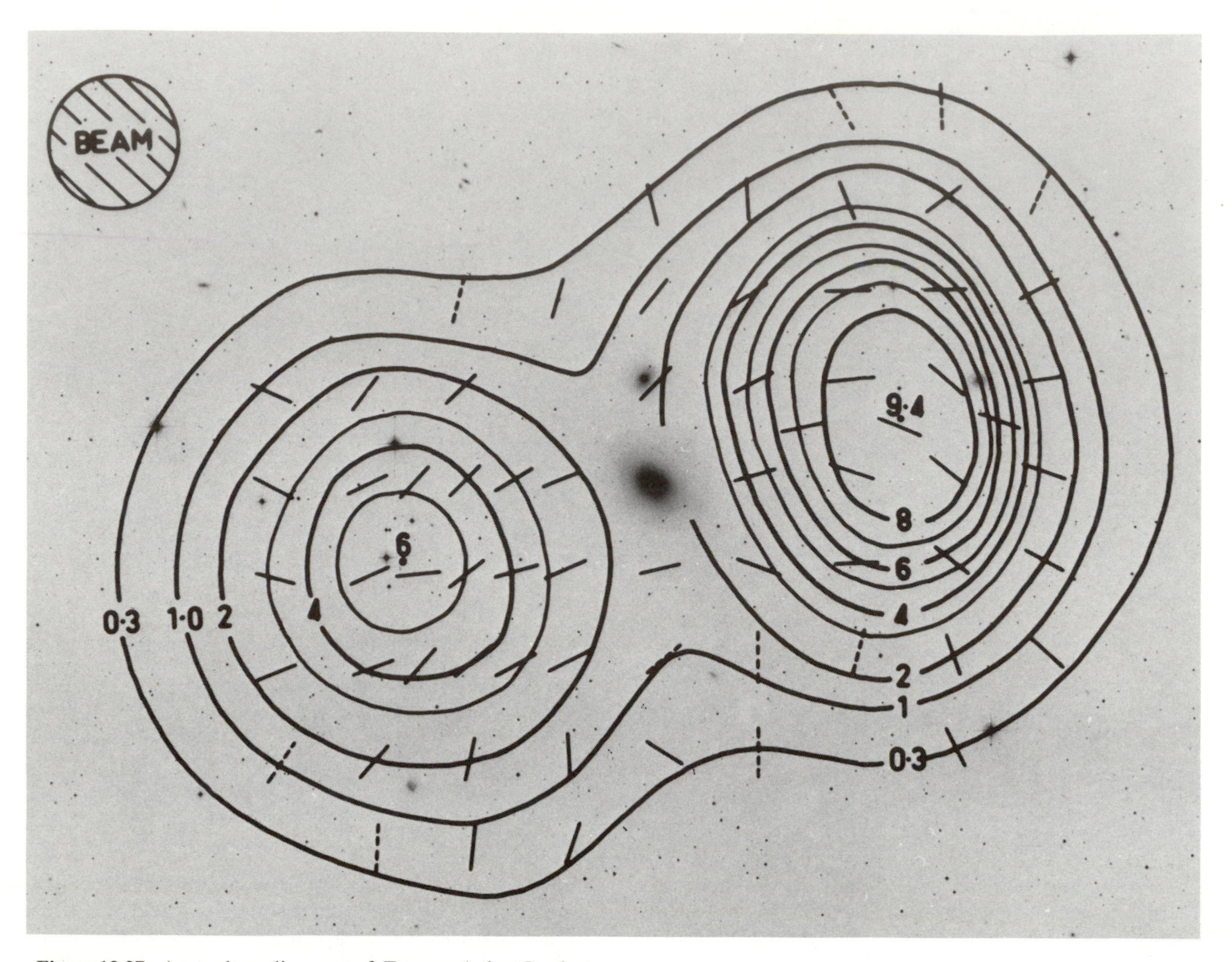

Figure 13.27. An early radio map of Fornax A by Gardner and Price showed it to have a double-lobed structure. The short straight lines indicate the direction of the polarization vector of the nonthermal radio continuum emission. (From A. T. Moffet, *Ann. Rev. Astr. Ap.*, **4**, 1966, 145.)

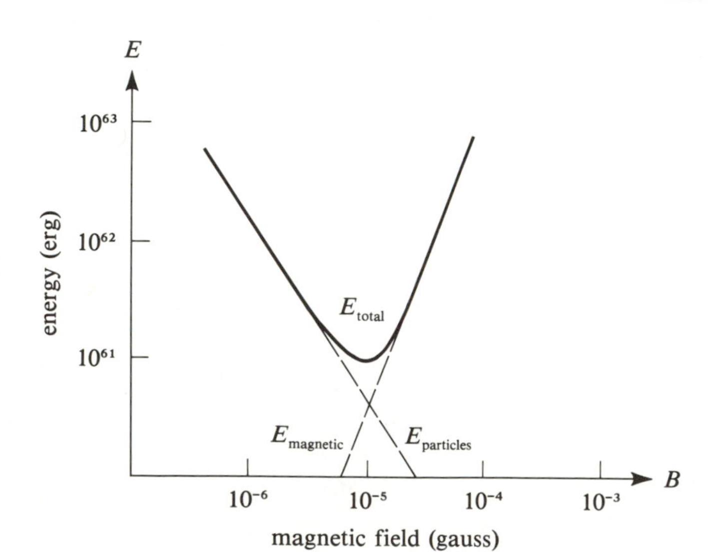

Figure 13.28. Schematic estimate for the energy content of a radio galaxy. Given the distance to the galaxy, the measured angular sizes of the radio lobes, and the total radio brightness of the source, the amounts of energy estimated for the parts contained in magnetic fields and in relativistic particles depend on the assumed magnetic-field strength B. The larger B is, the smaller is the energy $E_{\text{particles}}$ required for relativistic particles (to produce the given amount of synchrotron radiation), and the larger is the magnetic-field energy E_{magnetic}. The sum $E_{\text{total}} = E_{\text{magnetic}} + E_{\text{particles}}$ has a minimum value near the value of B where $E_{\text{magnetic}} = E_{\text{particles}}$ (the crossover point of the two straight lines in the diagram). The numerical value of E_{total} works out to about 10^{61} erg for a typical strong radio galaxy. ⟶

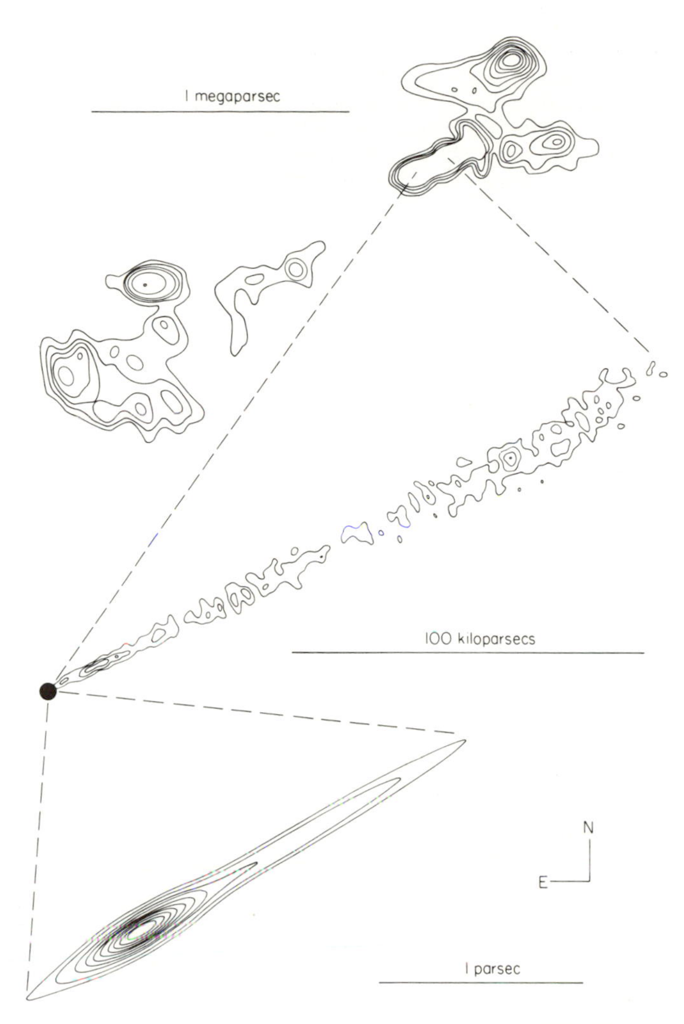

Figure 13.29. The radio structure associated with the galaxy NGC6251 on three different scales. On a scale of megaparsecs (1 megaparsec = 3 million light-years), we see a classic radio double. On a scale of hundreds of thousands of light-years, we see an aligned large jet. On a scale of light-years, there is a small jet. (From Readhead, Cohen, and Blandford, *Nature*, **272**, 1978, 131.)

in Problem 13.7.) VLBI radio astronomy therefore requires national, and often international, cooperation.

Under such very high angular resolution, the central source can be deduced in several radio doubles to consist of a series of blobs which are lined up along the axis joining the two outer lobes. (The bottommost diagram in Figure 13.29 is a model, and not an actual picture like the other two diagrams, because VLBI techniques do not synthesize a filled aperture as depicted in Figure 2.30.) A popular model of double-lobed radio sources, championed theoretically by Blandford, Rees, and Scheuer, envisages the ejection of energetic particles and magnetic fields at high speeds along the two beams. These beams provide the energy which lights up the jets and the lobes of the radio source. This idea has also been adapted to explain the peculiar galactic object SS433 (Chapter 10).

VLBI techniques have also been used to study compact radio sources which do not exhibit two lobes. Some of these objects also have elongated structures consisting of two or more closely separated components which can actually be observed to separate from each another during a few years. In several cases, Irwin Shapiro and his colleagues deduced the apparent velocity of separation in the transverse direction to be several times the speed of light! A notable example is the brightest quasistellar radio source 3C273 (Figure 13.30). The so-called **superluminal** expansion cannot, of course, represent a real material velocity. There are several possible explanations of the phenomenon; one is that the effect is one of phase velocity, as when light illuminates a screen (Figure 13.31). Indeed, precisely such a screen effect had been seen decades earlier by optical astronomers who also noticed superluminal apparent velocities connected with nova outbursts. The preferred explanation in radio jets, however, involves Blandford and Ree's suggestion that the plasma blobs in some active galactic nuclei might be ejected at *relativistic* speeds ($v \lesssim c$) in two oppositely directed beams. If the radio observer then happens to lie almost along the direction of the axis of these beams, separation at superluminal *apparent* velocities will be observed (Problem 13.10).

Problem 13.10. To understand the beam model, neglect the motion of the galactic nucleus with respect to the observer, and consider just the motion of one blob moving at velocity v with respect to the nucleus. Draw an apt diagram. Consider now the radio emission received in time by a radio observer at distance r_0 from the galactic nucleus. At time $t = 0$ (in the frame of the nucleus), the blob is at the position of the nucleus. At time t_0, the blob has moved to a distance vt_0 along a line which makes an angle θ with respect to the observer-nucleus axis. Show that the observer sees the blob coinciding with the position of the nucleus at time $t_1 = r_0/c$. (Assume that both the nucleus and the blob continuously emit radio waves.) Show also that the observer sees the blob with a transverse displacement $\Delta y = vt_0 \sin\theta$ from the nucleus at a time $t_2 = t_0 + (r_0 - vt_0 \cos\theta)/c$. In this fashion, show that the elapsed time for the observer has been $\Delta t = t_0(1 - \beta\cos\theta)$, where we have defined $\beta = v/c$. The apparent y-velocity of the blob relative to the nucleus is therefore, $v_y = \Delta y/\Delta t$. Consider v_y to be a fraction of c, and call it $\beta_{\perp\,\text{app}}$. Show that $\beta_{\perp\,\text{app}}$ is given

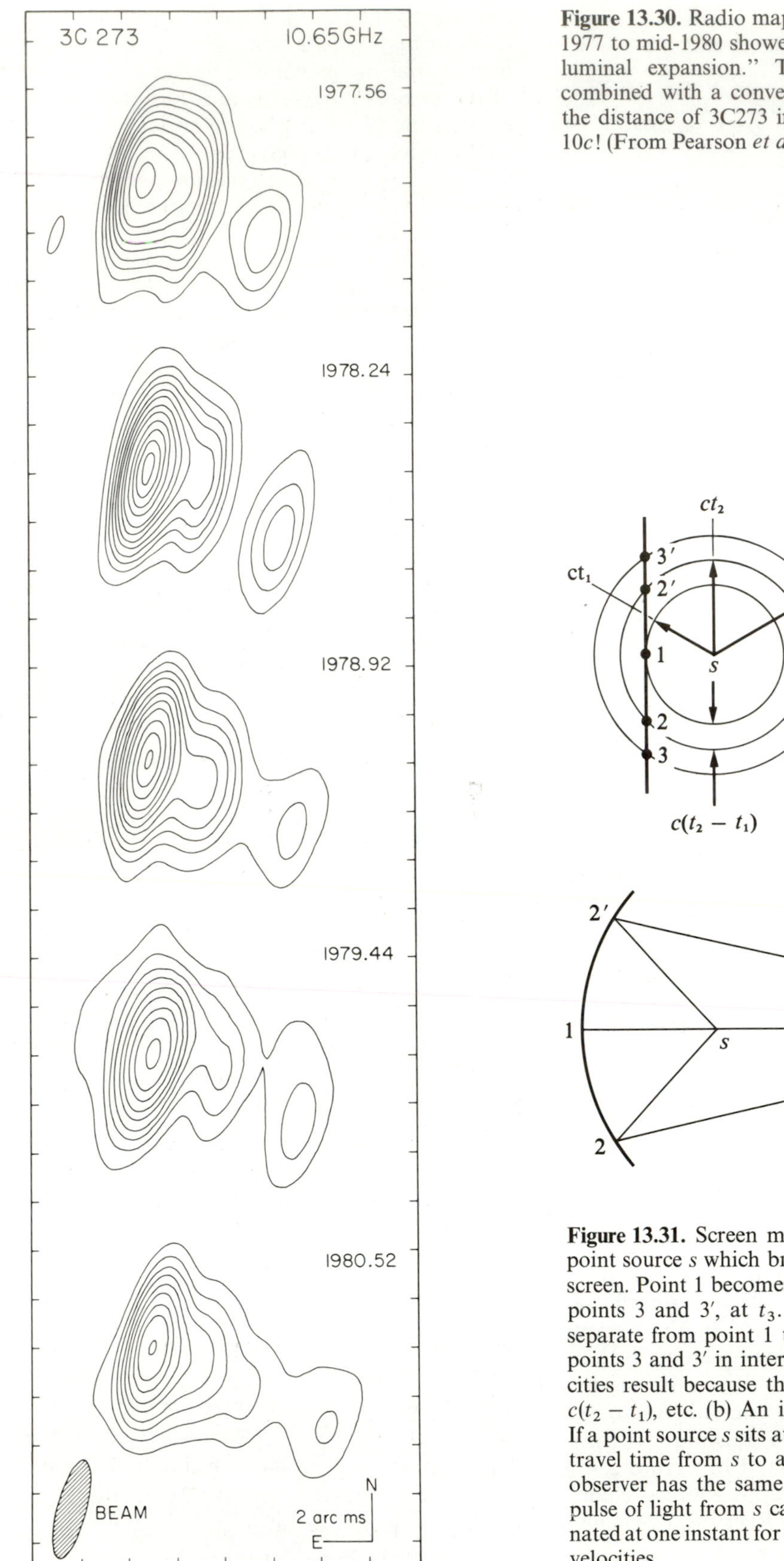

Figure 13.30. Radio maps made by VLBI techniques from mid-1977 to mid-1980 showed the quasar 3C273 to undergo "superluminal expansion." The angular separation of the blobs combined with a conventional cosmological interpretation for the distance of 3C273 implied an apparent transverse speed of 10c! (From Pearson *et al.*, Nature, **290**, 1981, 365.)

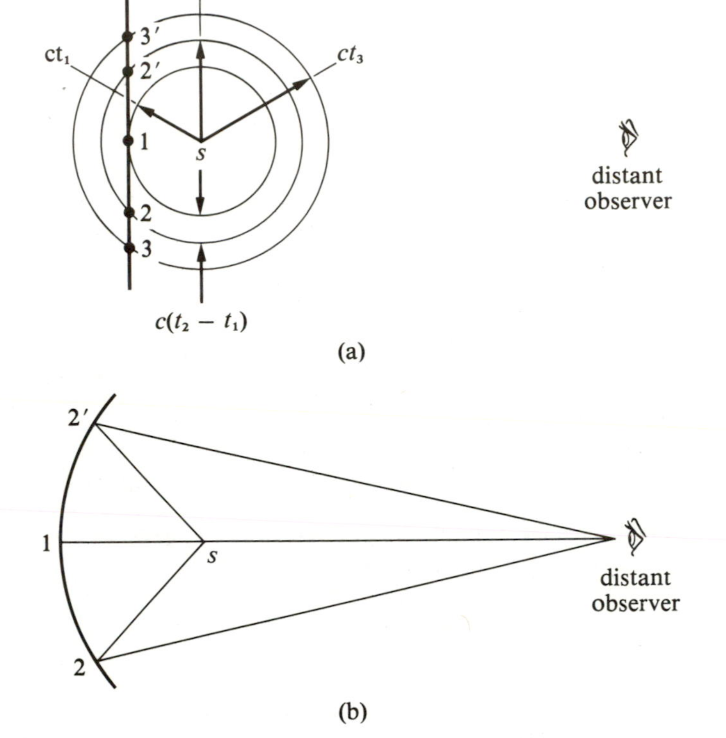

Figure 13.31. Screen models for superluminal expansion. (a) A point source s which brightens at time $t = 0$ lights up a straight screen. Point 1 becomes bright at time t_1; points 2 and 2′, at t_2; points 3 and 3′, at t_3. An observer at infinity sees two blobs separate from point 1 to points 2 and 2′ in interval $t_2 - t_1$; to points 3 and 3′ in interval $t_3 - t_1$. Superluminal apparent velocities result because the distances from 2 to 2′ is greater than $c(t_2 - t_1)$, etc. (b) An ideal screen for superluminal expansion. If a point source s sits at the focus of a parabolic screen, the light-travel time from s to any point on the screen to a very distant observer has the same value for all points on the screen, so a pulse of light from s causes the entire screen to become illuminated at one instant for the observer, resulting in infinite apparent velocities.

by the formula

$$\beta_{\perp \text{app}} = \frac{\beta \sin \theta}{1 - \beta \cos \theta}.$$

Clearly if $\beta \cos \theta$ is close to unity, the above expression can be greater than unity (superluminal expansion). However, since β refers to a physical speed ($\beta = v/c$), that of the blob, it must itself be less than unity. Thus, superluminal expansion requires the combination $v \approx c$ *and* $\cos \theta \approx 1$, i.e., relativistic expansion almost in the direction of the observer. The expansion cannot be exactly in the direction of the observer (i.e., θ cannot exactly be equal to zero), since $\sin 0 = 0$, and we would obtain $\beta_{\perp \text{app}} = 0$. For a fixed value of β, in fact, the value of θ which maximizes $\beta_{\perp \text{app}}$ is given by $\cos \theta = \beta$, $\sin \theta = (1 - \beta^2)^{1/2}$. (Those of you who know calculus should prove the above assertion.) For this optimum value of β, show that the corresponding value of $\beta_{\perp \text{app}}$ is

$$\beta_{\perp \text{app}}(\text{max}) = \beta (1 - \beta^2)^{-1/2}.$$

What is the value of $\beta_{\perp \text{app}}(\text{max})$ if v is 99 percent of the speed of light, c?

The above theory is purely kinematic. It makes no attempt to explain why the ejection should sometimes be at relativistic speeds, sometimes at nonrelativistic speeds. It does, however, offer the advantage that the same basic type of source can look different because of the viewing geometry.

For extragalactic radio sources, there is indirect evidence for an ejection process from some central "machine." This evidence comes from the examination of "head-tail" radio sources in rich clusters of galaxies (Figure 13.32). The idea here is that blobs of plasma are ejected and subsequently swept backward, by the relative motion of the elliptical galaxy, through an ambient substrate of hot gas. The properties of this gas, which has been observed in X-rays to exist in the space between the galaxies of a rich cluster, will be discussed in greater detail in Chapter 14.

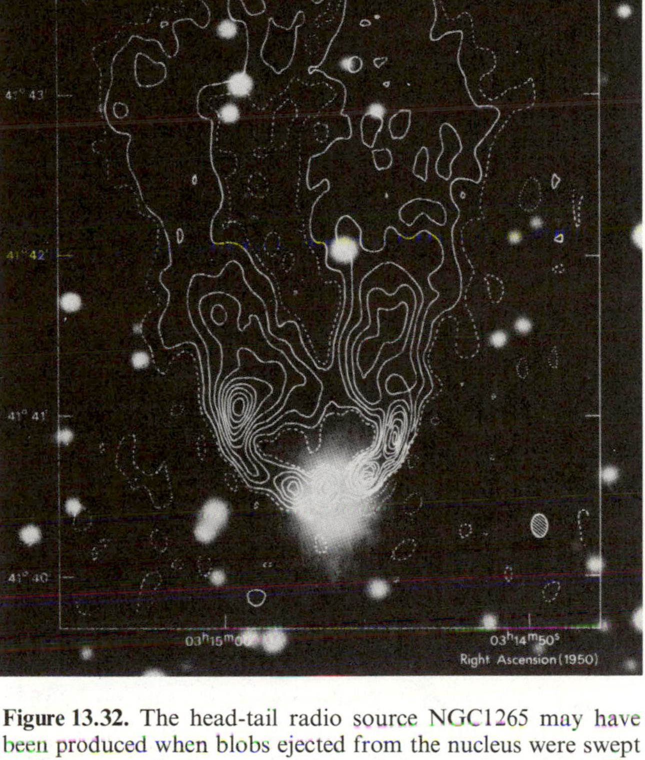

Figure 13.32. The head-tail radio source NGC1265 may have been produced when blobs ejected from the nucleus were swept back by intracluster gas moving relative to the galaxy. The radio contours here are superimposed on an optical photograph from the Palomar Sky Survey. (From Wellington, Miley, and van der Laan, *Nature*, **224**, 1973, 502; reprinted with permission from The Netherlands Foundation for Radio Astronomy.)

Quasars

Quasars were first detected as unresolved radio sources in the Cambridge surveys mentioned earlier. Good positions eventually led to the identification of quasars with optical images that turned out quasistellar, i.e., starlike in appearance even in deep optical photographs. This is the origin of the word (invented by H. Y. Chiu) quasar = quasistellar radio source.

With longer baselines, radio interferometers eventually provided resolved radio images of a few quasars. These occasionally show the classic double-lobed structure reminiscent of strong radio galaxies. The major apparent difference between quasars and radio galaxies is that the central optical image is not obviously a galaxy. It *may* be a galaxy (or the nucleus of one), but it is not *obviously* a galaxy.

The optical spectrum of the central source of quasars is shown in Figure 13.33. This spectrum is nothing like that produced by a collection of stars (thermal continuum plus absorption lines); instead, it is extremely reminiscent of the spectrum produced by the nucleus of a Seyfert I galaxy (Figure 13.22).

For every radio-emitting quasar, astronomers now know that there are about 20 radio-quiet starlike objects which have the same optical properties as quasars. These

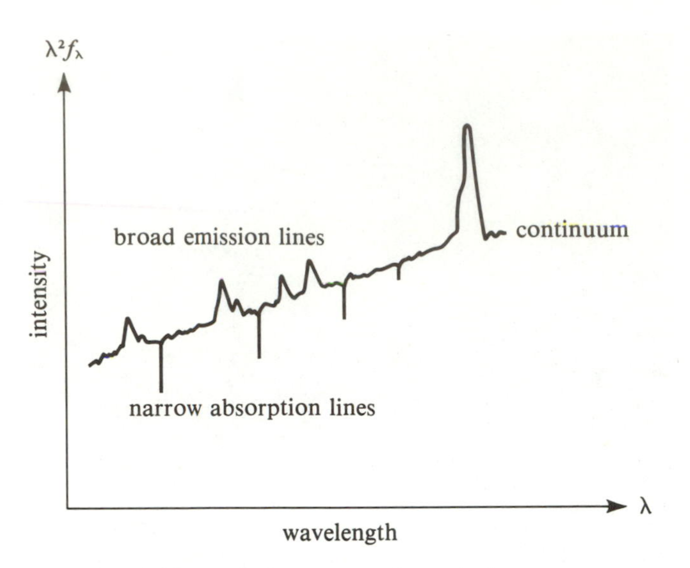

Figure 13.33. The optical spectrum of a typical quasar contains a continuum with a nonthermal slope, broad emission lines, and, perhaps, narrow absorption lines.

latter objects are called QSOs (quasistellar objects). QSOs and quasars are believed to be similar kinds of systems except that, for some inexplicable reason, QSOs do not generate appreciable nonthermal radio emission to go along with their copious emission at optical, infrared, and X-ray wavelengths.

The resemblance of quasars to radio galaxies and of QSOs to the nuclei of Seyfert galaxies, is so strong that most (but not all) astronomers believe that quasars and QSOs arise from nuclear activity in galaxies. However, no one has actually *demonstrated* directly that quasars and QSOs are parts of stellar systems. Some astronomers were at first reluctant to conclude that quasars are galaxies, largely because of the problem about their distances.

Maarten Schmidt was the first in 1963 to make sense of the puzzling pattern of emission lines in quasars. Schmidt found that he could account for the observed pattern of 3C273 only if he postulated a redshift $z = 0.158$. Such a redshift was astoundingly large for an object previously thought to be a star; since this early work, quasars with redshifts as large as $z = 3.5$ have been found. If the observed redshift were due to the Doppler effect, then some quasars would have a velocity of recession directly away from us of about $.9c$ (consult Problem 3.13). This is an enormous speed, the largest ever encountered for any macroscopic object. The only other situation in astronomy where one finds comparable (but appreciably smaller) speeds is for the general expansion of the universe (Chapter 14). In this *cosmological* interpretation, quasars would then be the most distant massive objects yet seen in the universe. They would be located about 10^9 or 10^{10} lt-yr distant from us. Ten billion light-years is 5,000 times the distance of the Andromeda galaxy, and a billion times further than the nearest stars! Yet even at such enormous distances away, quasars and QSOs appear as bright as faint stars (Figure 13.34). The intrinsic brightnesses of quasars must be truly gargantuan if they are at cosmological distances—and they are. The optical luminosity of some quasars amounts to about 10^{46} erg/sec if we assume them to be at distances appropriate for their redshifts. Such an energy output is more than a *trillion* times the output of the Sun. Recent observations carried out in the Einstein observatory by Giaconi's group show that the X-ray luminosity of quasars may be even more impressive: as much as 10^{47} erg/sec with large variations in as little as three hours! Evidently in X-rays we are seeing close to the heart of the central machine, which is less than three light-hours in size. The cosmological interpretation of quasars envisages their central workings to be objects that are about the size of the solar system and which produce energy at a rate greater than *ten trillion* suns! No wonder some astronomers were reluctant to accept such a conclusion.

Problem 13.11. Suppose a quasar is apparently as bright as a solar-type star at a distance of 3,000 lt-yr, and suppose the quasar is a million times further away. What is the optical luminosity of the quasar?

Greenstein and Schmidt later gave a beautiful argument for the cosmological interpretation of the redshifts of the quasars 3C48 and 3C273. In particular, they showed that nearly insuperable difficulties arose from the hypothesis that the redshifts of the optical emission lines were gravitational redshifts caused by proximity to a strongly gravitating object (Problem 13.12). It is difficult to convey the terrific impact of Schmidt's discovery on his contemporaries. We have by now become accustomed to the idea that there are objects in the universe which produce outlandish amounts of energy: pulsars, binary X-ray sources, radio galaxies, etc. Today, most people are willing to accept the hypothesis that quasars represent merely the extreme end of energetic extragalactic phenomena that span a continuum containing ordinary galaxies, Seyferts, N galaxies, BL Lac objects, radio galaxies, and QSOs. In 1963, there was understandably more scepticism.

Problem 13.12. It is conventionally accepted that the hydrogen emission lines of a quasar arise from recombination in a photoionized plasma (see Chapter 11). Suppose that the redshift z of these emission lines arise

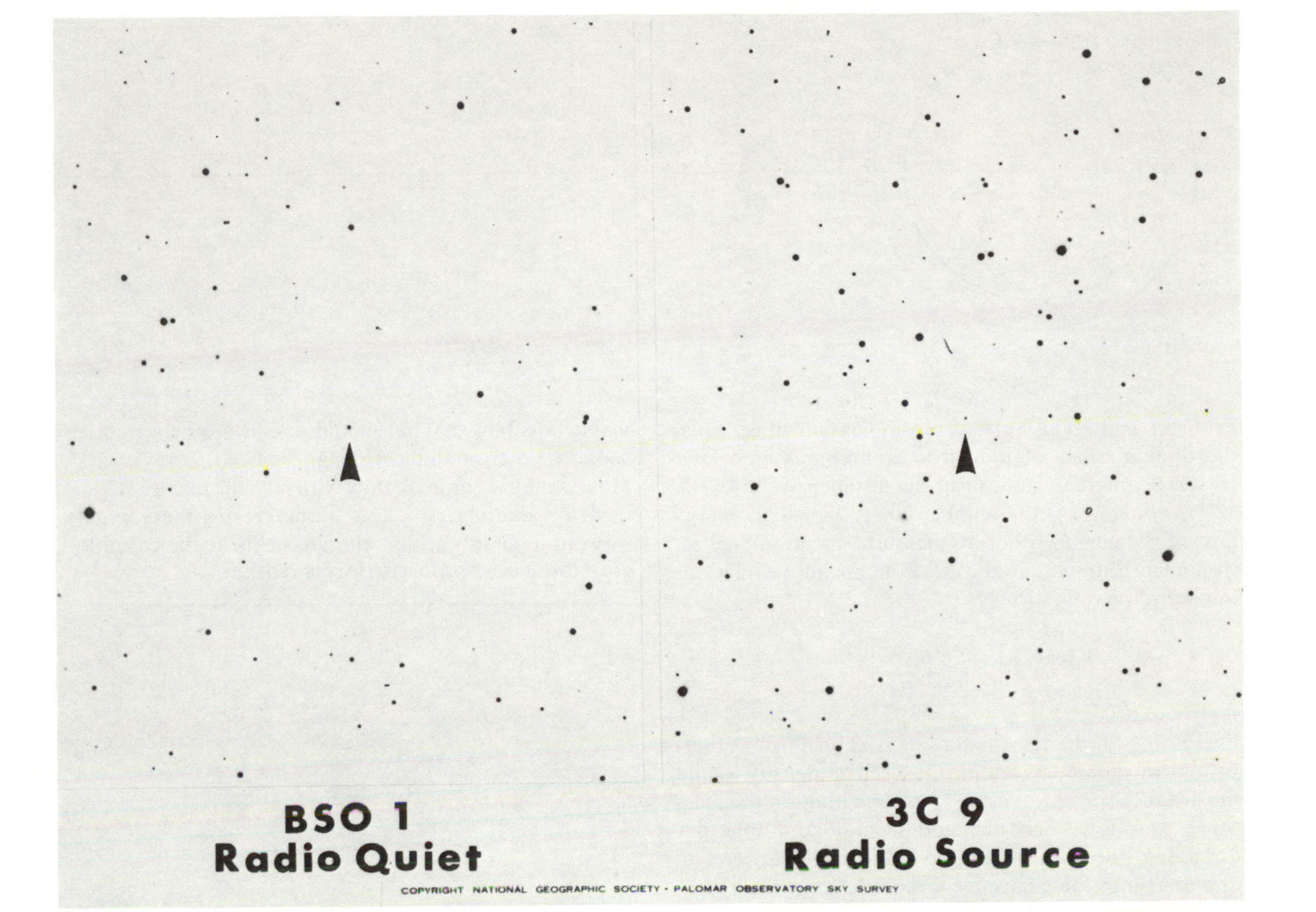

Figure 13.34. The quasistellar object BSO 1 is radio-quiet, whereas the quasar 3C9 is a radio source. Notice that on a direct photograph it is not possible to distinguish a quasar from a faint star. (Palomar Observatory, California Institute of Technology.)

from gravitational effects because this gas sits at a typical distance R from an object of mass M (see diagram). The general theory of relativity gives $\lambda/\lambda_0 = 1 + z = (1 - 2GM/c^2R)^{-1/2}$ (see equation 7.5). Show that if z is relatively small, $z = GM/c^2R$. It is an observational fact that the emission lines of quasars are typically much broader than appropriate for thermal broadening at a nebula temperature of $T \sim 10^4$ K; nevertheless, the observed fractional width $\Delta\lambda/\lambda \ll z$. This implies that the range of radii within which this gas resides must be much less than the mean radius R; otherwise the differential gravitational redshifts arising at different values of R would make the observed line widths too broad. If the shell thickness ΔR is constrained by this effect, we require

$$\Delta\lambda/\lambda \sim (GM/Rc^2)(\Delta R/R);$$

i.e.,

$$\Delta R/R \sim z^{-1}(\Delta\lambda/\lambda) \ll 1.$$

Argue that the total number of hydrogen recombinations per second in the gas is given by $\alpha(T)n_e{}^2 4\pi R^2\,\Delta R$ (consult

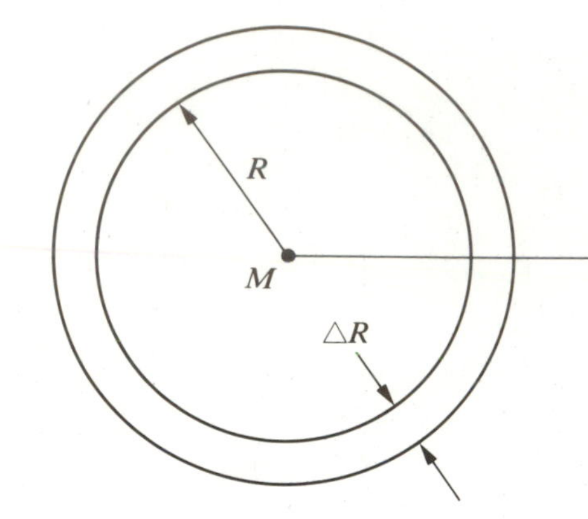

Problem 11.3). This rate of recombination must equal the total number of Balmer-line photons which later crosses a sphere of radius r at the distance of the Earth, $4\pi r^2 f_H$, where f_H is the number flux (rather than energy flux) of Balmer-line photons measured by an optical astronomer. Eliminate R and ΔR from the above relations to obtain the estimate

$$\frac{GM}{c^2} \sim \left[\frac{z^4 r^2 f_H \lambda}{\alpha(T) n_e^{\,2} \Delta\lambda}\right]^{1/3}. \tag{1}$$

Everything on the right-hand side is known from observations or theory, except for the electron density n_e and the distance r to the quasar. To place a limit on the mass M, let us follow Greenstein and Schmidt, and note that forbidden lines can usually be found in the spectra of quasars. Since these have the same redshift as the hydrogen lines, they would have to be formed also at R in the present model. This fact can be used to place an upper limit on n_e in this region. The electron density must be less than some computable value n_w if the forbidden lines are not to be saturated by collisional deexcitation (see Problem 11.4). Since the actual $n_e < n_w$, the replacement of n_e by n_w in formula (1) yields a lower limit on the gravitating mass required to produce the redshift of quasars by gravitational means:

$$M > \frac{c^2}{G}\left[\frac{z^4 f_H \lambda}{\alpha(T) n_w^{\,2}\,\Delta\lambda}\right]^{1/3} r^{2/3}. \tag{2}$$

For a typical quasar of low redshift, the coefficient before $r^{2/3}$ in formula (2) might work out to be $10^8 M_\odot(\text{lt-yr})^{-2/3}$. What would M have to be if the quasar existed inside the solar system ($r = 10^{-5}$ lt-yr, say)? Is this a viable possibility? What would M have to be if the quasar existed inside the Galaxy ($r = 10^4$ lt-yr, say)? Is this a viable possibility? What would M have to be if r corresponded to cosmological distances ($r = 10^9$ lt-yr, say)? How would M compare now with galactic masses? Comment on the merits of this model versus the straightforward assumption that the quasar is at the cosmological distance appropriate for its redshift.

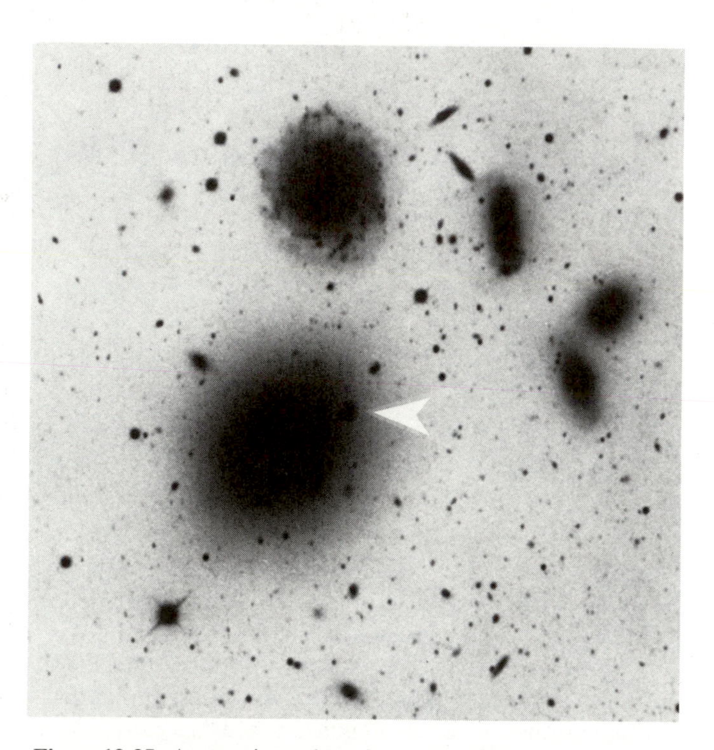

Figure 13.35. A negative print of a quasistellar object near an elliptical galaxy. H. C. Arp argues that the QSO marked with an arrowhead, which has redshift $z = 0.044$, lies in front of the elliptical galaxy NGC1129, which has redshift $z = 0.009$. The cosmological interpretation of quasar redshifts would require the QSO to lie behind the galaxy, because the former has the larger redshift. (Courtesy of H. C. Arp.)

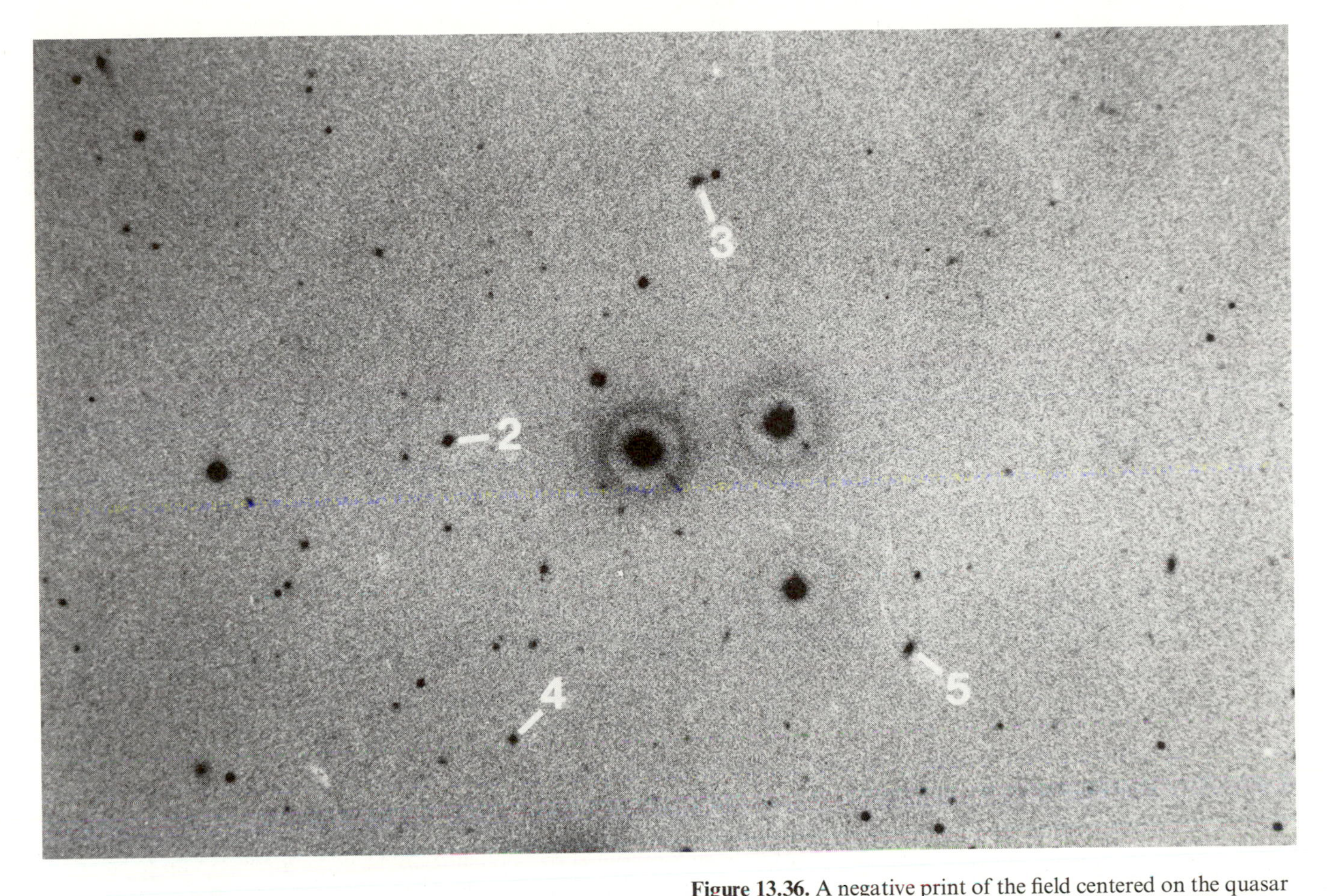

Figure 13.36. A negative print of the field centered on the quasar 3C273 (object with jet). The numbers mark four of the faint galaxies that have redshifts agreeing with 3C273 and are therefore believed to be physically associated with it. (Courtesy of Alan Stockton, Institute for Astronomy, University of Hawaii.)

There developed also an interpretation that quasars are relatively local objects which happen to have very large velocities of recession. If quasars are local and have large spatial velocities, their velocity vectors should generally also have a large transverse component. Since quasars have no measured proper motion, this interpretation needs to postulate that they lie at least a million light years distant. But putting quasars at distances comparable to those of nearby galaxies does not eliminate other kinematic objections. For example, why is it that no quasar has ever been found with a large *blueshift*? Terrell has suggested that perhaps quasars formed by a stupendous explosion at the nucleus of our own galaxy, and this explains why we see them all racing away from us. But to avoid blueshifts completely, one would also have to suppose that no other galaxy suffered a similar explosion, and this would put ourselves in a very special position in the universe. (In contrast, the cosmological interpretation for the redshifts of galaxies and quasars requires no special position for ourselves; see Chapter 15).

A viewpoint which is more difficult to counter is Arp's hypothesis that quasar redshifts arise for completely unknown reasons. In support of this position, he has pointed out a number of striking and disturbing cases where a quasar and a galaxy seem closely paired in the sky, yet have completely different redshifts (Figure 13.35). In contrast, Stockton has found more than two dozen examples where quasars in the vicinity of a cluster of galaxies do share common redshifts with the cluster members (Figure 13.36). This feat is possible only for relatively low redshifts, because high-redshift galaxies are hard to identify, much less measure spectroscopically. Most astronomers accept Stockton's findings as further evidence that the redshifts of both quasars and galaxies arise for cosmological reasons, and reject Arp's findings as chance superpositions of a nearby galaxy and a distant quasar. Many people have argued about the statistical significance of each camp's results, but quasar statistics are notorious for being of low moral character: one can do anything with them. Schmidt, Weedman, and others have presented a good case that we have no compelling reason to treat quasars separately from a general class of active galactic nuclei, for which the cosmological interpretation of the redshift is not in doubt.

Supermassive Black-Hole Models for Active Galactic Nuclei

That an object can change its luminosity by more than twice within a short time argues strongly that the basic machine in the nucleus of a Seyfert, or a radio galaxy, or a quasar, is a single body. The most promising theoretical candidate is an accreting supermassive black hole. This concept, originally by Edwin Salpeter and Donald Lynden-Bell, has been applied on a smaller scale to explain the emission from the binary X-ray source, Cygnus X-1 (Chapter 10). To play a similar role in active galactic nuclei, the black hole has to be supermassive just to be able to overwhelm the enormous radiation pressure which tries to push the accreting matter back out (Problem 13.13). For a luminosity of 10^{47} erg/sec, the mass of the black hole has to be larger than about $10^9\ M_\odot$.

Problem 13.13. The quantity $r_e = e^2/m_e c^2$ is called the classical radius of the electron, not because the electron actually has finite size, but because a free electron interacting with photons scatters them as if it had a radius of r_e. In particular, if the photons have much lower energies than the rest energy of the electron, the scattering cross section of free electrons is $8\pi r_e^2/3$. Idealize the gas near a quasar to be completely ionized hydrogen which is optically thin to the radiation emerging from the central source of luminosity L. Show that each free electron intercepts per unit time the energy $(8\pi r_e^2/3)(L/4\pi r^2)$, and scatters this energy in all directions. Assume that only the interception process leads to a net transfer of momentum, and show that, since the momentum carried by a photon is $1/c$ times its energy, the time-rate of transfer of momentum per free electron is given by $2r_e^2L/3cr^2$. This outward push of the electrons by the radiation pressure acts also indirectly on the protons (the nuclei of the ionized hydrogen atoms), because strong electrostatic forces tend to make the electron gas and proton gas move together. Countering the outward force $2r_e^2L/3cr^2$ acting on each electron-proton pair is the inward pull of gravity $GM(m_p + m_e)/r^2$, where M is the mass of the central supermassive black hole (we assume r is far from the event horizon of the black hole). Argue that for accretion to occur, the pull of gravity must be larger, and that this condition requires

$$M \geq \frac{2r_e^2 L}{3Gm_H c},$$

where $m_H = m_p + m_e$ is the mass of the hydrogen atom. Compute the right-hand side when $L = 10^{47}$ erg/sec. For a given M, the value of L which balances this equation is called the Eddington luminosity L_E. Compute $L_E = 3GMm_Hc/2r_e^2$ for a stellar mass M. How does this compare with the maximum X-ray luminosity $\sim 10^{38}$ erg/sec seen in binary X-ray sources? (Consult Chapter 10.) Draw appropriate astronomical conclusions.

To produce $L = 10^{47}$ erg/sec (via a swirling accretion disk, say), this supermassive black hole has to swallow more than $10 M_\odot$ per year (Problem 13.14). It may acquire this matter by breaking up whole stars by tidal forces, or by ingesting interstellar gas left over from star formation or subsequently ejected from evolving stars. If the density of stars is sufficiently high near the black hole, direct collisions of stars may contribute to the diet of the black hole. Rees has emphasized that there are many possible ways in which such a black hole might have first formed and by which it might subsequently acquire more material to swallow.

Problem 13.14. For every mass m which a black hole swallows (via an accretion disk, say), an amount of energy εmc^2 is liberated, where ε is the efficiency of the process. The maximum possible value for ε is unity, but a value equal to 10 percent might be more realistic. At what rate $\dot{M}$ would a supermassive black hole have to swallow mass to produce $L = 10^{47}$ erg/sec if $\varepsilon = 0.1$? Convert your answer to $M_\odot$/yr.

Some people object to the idea of supermassive black holes because they feel it is a very weird concept. Yet, in one important sense, supermassive black holes are much less weird than stellar-mass black holes. Consider the density of the pre-black-hole object which is doomed to collapse to singularity. In order of magnitude, this density must equal the mass M divided by the volume V, for which we may use the volume of a sphere $4\pi R^3/3$, with R equal to the Schwarzschild "radius" $2GM/c^2$. In Chapter 7, we saw that Euclidean formulae (like that for the volume of a sphere) should not really be applied if the geometry of spacetime becomes strongly distorted by gravity, but our estimation technique should give correct order-of-magnitude answers for the *pre-collapse* state. Thus, the mean density of the pre-black-hole object of mass M would be roughly

$$\text{average density} \sim 3c^6/32\pi G^3M^2. \quad (13.6)$$

Apart for some universal constants, expression (13.6) is proportional to the inverse square of the mass M. Thus, for a $1M_\odot$ pre-black-hole object, formula (13.6) yields a density of roughly 10^{16} gm/cm^3, somewhat greater than the density of a typical neutron star (Chapter 7). Now, some people object to the concept of stellar-mass black holes because we know too little about the properties of matter (e.g., the role of quark forces) at extranuclear densities. However, equation (13.6) applied to a mass of $10^9\ M_\odot$ would yield a density 10^{18} times smaller than when applied to a mass of $1M_\odot$. In other words, the mean density of such a pre-supermassive black hole would only be roughly 10^{-2} gm/cm^3. This is the density of a gas (100 times more rarefied than liquid water), and no one could reasonably claim that our knowledge of the properties of gases is incomplete in any significant sense. Thus, as we talk about black holes of higher and higher mass, their physical properties in the pre-collapse state become actually less and less weird. This statement reaches its epitome of reasonableness when applied to the universe as a whole. If we live in a closed universe (see the discussion of Chapter 15), then in a sense we live in a pre-black-hole object of gigantic mass. And there is nothing very weird about the lives of most of us!

Problem 13.15. Check the derivation of formula (13.6) and the numerical values for the densities quoted in the text.

All this discussion is not to suggest that $10^9M_\odot$ black holes necessarily form from the gravitational collapse of a single object, say, a supermassive star. Supermassive stars ($M \sim 10^5$ to $10^8M_\odot$) were originally suggested by Fowler and Hoyle as a possible model for quasars, and they are still promising candidates for the precursors to fairly massive black holes at the centers of galaxies. However, they are not the only possible route to supermassive black holes. We need only note that accretion models of active galactic nuclei require the accumulation of mass at the rate of 1 to $10M_\odot$/yr. At such rates, it would be possible to acquire supermassive black holes in 10^6 to 10^9 years starting with practically anything (or nothing).

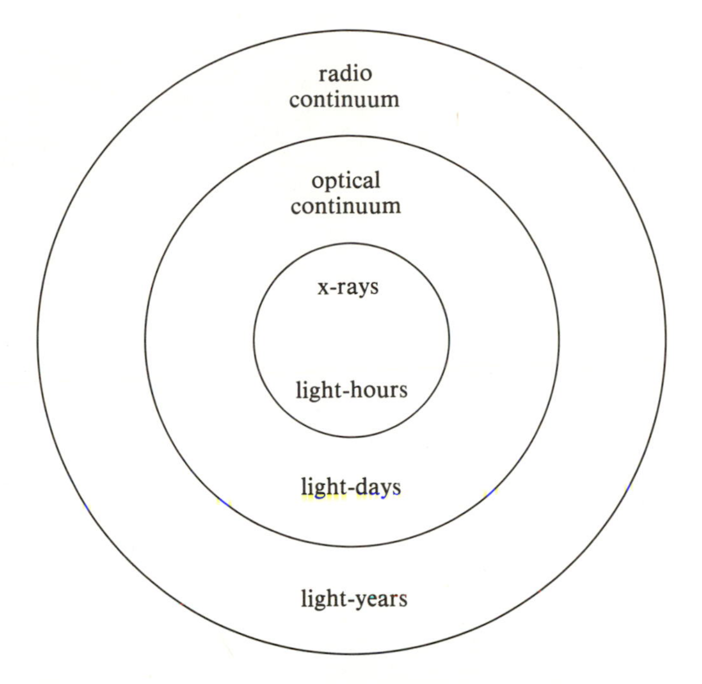

Figure 13.37. Schematic diagram of various emission regions about the nucleus of an active galaxy or quasar. At the center of the configuration may be a supermassive black hole accreting matter from its surroundings.

The theory of active galactic nuclei, as based on the supermassive black-hole model, is in its infancy of development. Observations suggest that the regions of emission of the X-rays, optical continuum, and radio continuum occupy successively increasing distance from the central machine (Figure 13.37). The optical emission lines probably form in the inner parts of the radio-continuum source; where the narrow optical absorption lines arise is a matter of some controversy. The central machine may be a supermassive black hole surrounded by a thick accretion disk, but the evidence so far is circumstantial. We also have relatively little understanding why jets should form in some cases and not in others (e.g., apparently never in spiral galaxies). The study of active galactic nuclei promises to remain an exciting area of observational and theoretical investigations for some time to come.

Problem 13.16. Calculate the numerical value of the Schwarzschild radius for a $10^9M_\odot$ black hole. How does this compare with the distance of three light-hours deduced for the size of the X-ray variability region?

Figure 13.38. An optical photograph aimed toward the center of our Galaxy, the Milky Way in Sagittarius. The direction of the Galactic nucleus is indicated by an arrowhead. Stars seen toward this direction are foreground stars. (Palomar Observatory, California Institute of Technology.)

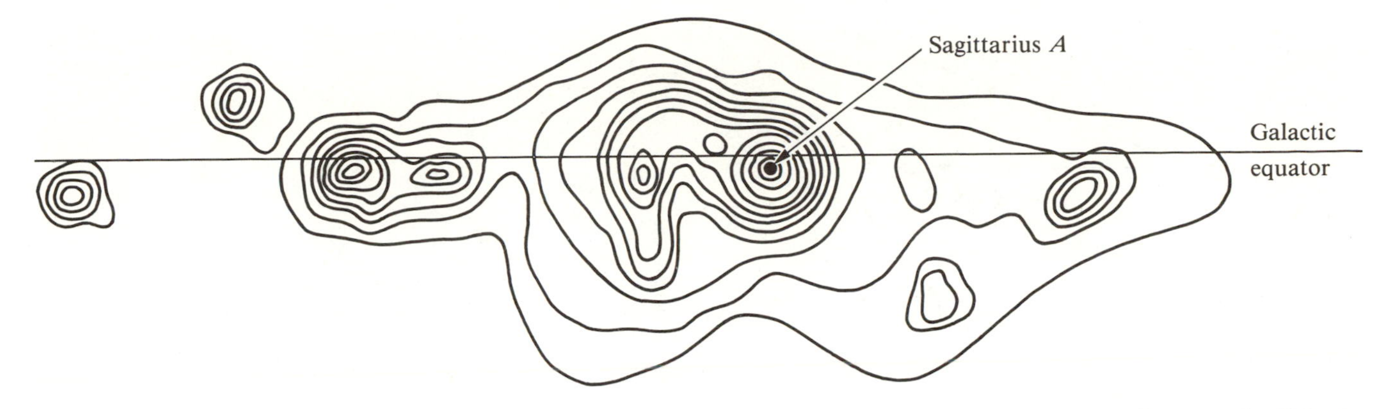

Figure 13.39. The radio-continuum emission near the Galactic center. The Galactic-center source is the strongest in the direction of Sagittarius and is therefore called Sagittarius A. The Galactic nucleus itself is believed to coincide with the western portion of this source, Sgr A West.

Observational Efforts to Detect Supermassive Black Holes

The Nucleus of Our Own Galaxy

An exciting question that follows from all this discussion about active galactic nuclei is, what is the nucleus of our own Galaxy like? The study of the nucleus of our own Galaxy suffers both advantages and disadvantages. The advantage is that it is the closest nucleus of any galaxy accessible to us; so high spatial resolution is easier to achieve for it than for any other example. The disadvantage is that the interstellar medium prevents our being able to see the nucleus of our own Galaxy at optical, ultraviolet, and soft X-ray wavelengths (Figure 13.38). To study the Galactic center requires radio, infrared, hard X-ray, and gamma-ray observations. In particular, gamma-ray lines with energy equal to the rest energy of an electron have been detected to come from the Galactic center. This suggests that processes energetic enough to produce electron-positron pairs (which subsequently annihilate to give two gamma-ray lines of the appropriate energy) take place at or near the nucleus of our own Galaxy.

The pioneering radio studies of the central regions of the Galaxy were made in 1960 by Rougoor and Oort. They found that the atomic hydrogen (studied at 21 cm) within several hundred light-years of the Galactic center was arranged in the form of a small nuclear disk. Outside this nuclear disk, there is evidence for much expansional motions of the hydrogen gas, but the nuclear disk itself appears to have mostly circular rotation. Molecular-cloud complexes have been found in the region roughly coincident with the HI nuclear disk, and the molecular clouds show evidence for large noncircular motions. It remains unknown whether these noncircular motions have an explosive origin or a gravitational origin (e.g., due to a barlike structure in the central regions).

A discrete source of nonthermal radio continuum lies in the direction of the center of the Galaxy. This source is called Sagittarius A, because it is the strongest radio source in the constellation of Sagittarius. Sagittarius A (or specifically its western portion, Sgr A West) is conventionally assumed to be the nucleus of our own galaxy (Figure 13.39). Sagittarius A West has been studied by VLBI techniques, which seem to show a core radio source only about 1 lt-hr in size. This radio source resembles many other compact radio sources in galaxies in having a radio spectrum which is much flatter than that usually found in synchrotron sources (see Figure 11.15). Indeed, Robert Brown has found Sgr A West to have a spectrum which even increases slightly toward *shorter* wavelengths. This characteristic of compact radio sources has conventionally been interpreted to imply that so much radiation and so many relativistic electrons have been packed into such a small place that the emitting electrons begin to absorb their own synchrotron emission (Figure 13.40). That is, the electromagnetic field of the synchrotron radiation accelerates the spiraling motion of the electrons. The opacity for synchrotron self-absorption is highest at long wavelengths, reducing the spectrum of the escaping radio radiation to that in Figure 13.41. Evidently, the center of our Galaxy has not yet been observed at short-enough wavelengths to exhibit the more familiar behavior of synchrotron radiation under transparent conditions.

Another physical process which may complicate a simple interpretation of any compact radio source, including that at the nucleus of the Milky Way system, is

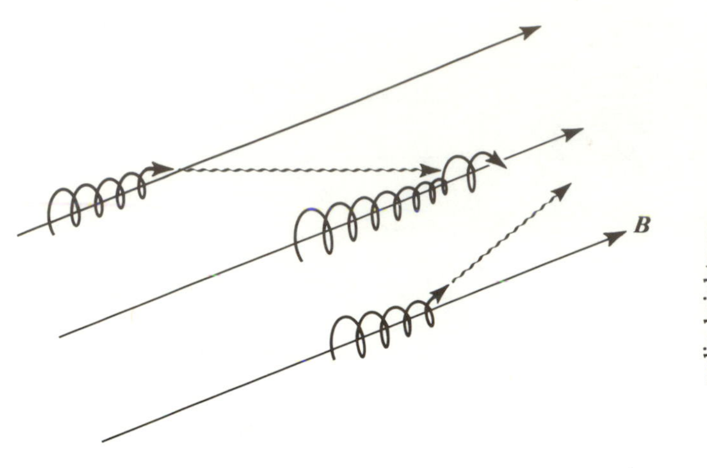

Figure 13.40. Synchrotron self-absorption. Relativistic electrons, spiraling around magnetic field lines $\boldsymbol{B}$, radiate synchrotron photons. This causes them to lose kinetic energy, leading to steadily decreasing radii of gyration. If the photons and relativistic electrons are packed densely enough, synchrotron photons are occasionally absorbed by gyrating electrons, causing the latter to spiral more energetically about $\boldsymbol{B}$.

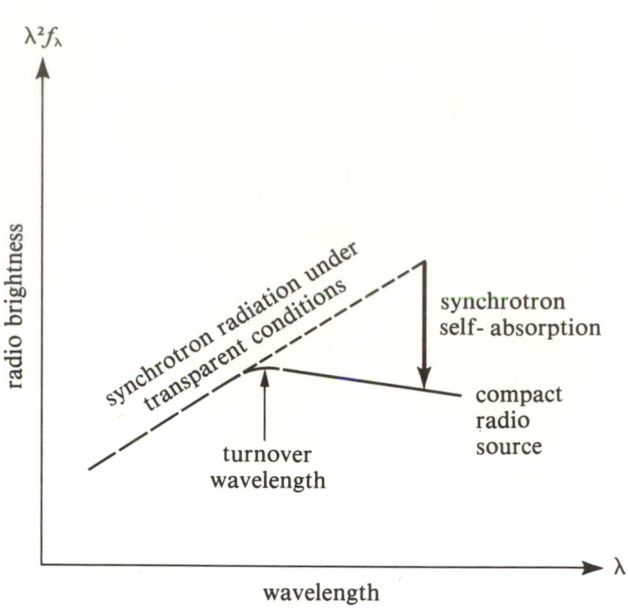

Figure 13.41. The opacity for synchrotron self-absorption is highest at long wavelengths. This will cause the radio spectrum of a compact source to turn over at sufficiently large λ. In Sgr A West, radio astronomers may have observed only the wavelengths which are longer than the turnover value (solid curve).

the so-called inverse Compton effect. This process arises because the relativistically moving electrons can not only absorb synchrotron photons, but can also scatter them. The relativistic electrons will tend statistically to share their energies with the radio photons, and to boost the latter up to the infrared or higher (up to gamma rays if the electrons are energetic enough). This could constitute a severe energy drain on the reservoir of relativistic electrons, and Kellerman and Pauliny-Toth have proposed that this mechanism limits the surface brightness of a compact radio source.

Problem 13.17. To see how the inverse Compton effect works, let us consider the head-on collision (for simplicity) of a photon of wavelength λ and an electron of speed v_e. After the collision, the photon has wavelength λ', and speed v_e'. We assume that v_e is high enough that the electron maintains its original direction after the scattering, but the photon recoils as in diagram (1). The algebra is less tedious if we transform to a frame (2) where the electron is initially at rest, i.e., a frame which moves with speed v_e to the left in diagram (1). In frame (2), argue that the wavelengths λ_0 and λ_0' of the photon before and after the scattering are given by the Doppler-shift formulae,

$$\lambda_0 = \left(\frac{1 - v_e/c}{1 + v_e/c}\right)^{1/2} \lambda, \qquad \lambda_0' = \left(\frac{1 + v_e/c}{1 - v_e/c}\right)^{1/2} \lambda'. \tag{1}$$

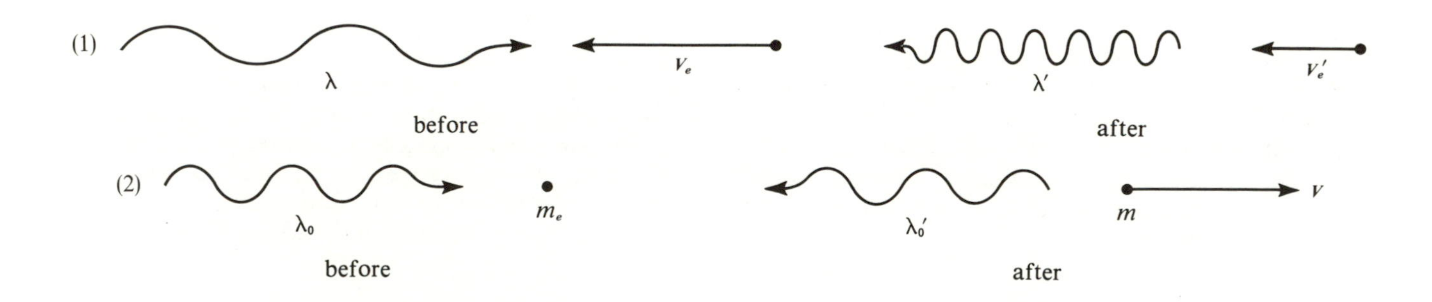

To find the ratio λ'/λ, divide these expressions to derive

$$\frac{\lambda'}{\lambda} = \left(\frac{1 - v_e/c}{1 + v_e/c}\right)\frac{\lambda_0'}{\lambda_0}. \tag{2}$$

We wish to show that the ratio λ_0'/λ_0 can be obtained simply, and that it usually has a value near unity. Thus, formula (2) implies that the scattering process can significantly reduce λ to λ' if the electron's motion is highly relativistic: v_e very close to c. To find λ_0' given λ_0 is a problem of Compton scattering. In the frame where the electron is initially at rest, the scattering process looks as in diagram (2). Argue that the conservation of linear momentum and energy require

$$\frac{h}{\lambda_0} = -\frac{h}{\lambda_0'} + mv,$$
$$\frac{hc}{\lambda_0} + m_e c^2 = \frac{hc}{\lambda_0'} + mc^2, \tag{3}$$

where $m = \gamma m_e$, $\gamma = (1 - \beta^2)^{-1/2}$, and $\beta = v/c$. Eliminate hc/λ_0' from equations (3), and derive the nondimensional relation

$$\gamma(1 + \beta) = 1 + \frac{2\lambda_c}{\lambda_0}, \tag{4}$$

where $\lambda_c = h/m_e c$ is the Compton wavelength of a free electron. The Compton wavelength is that wavelength for which the energy of a photon is equal to the rest-mass energy $m_e c^2$ of an electron. Calculate λ_c numerically; in what part of the electromagnetic spectrum does it lie? Introduce the symbol $A = 1 + 2\lambda_c/\lambda_0$, and show that β, γ, and λ_0' have the solutions:

$$\beta = \frac{A^2 - 1}{A^2 + 1}, \qquad \gamma = \frac{A^2 + 1}{2A}, \qquad \lambda_0' = \lambda_0 + 2\lambda_c. \tag{5}$$

Thus, λ_0' and λ_0 will be nearly equal if $2\lambda_c$ is short in comparison to λ_0. Consider a cosmic-ray electron for which $(1 - v_e/c) = 10^{-5}$ that backscatters a radio photon of wavelength $\lambda = 10$ cm. Compute the numerical values of λ_0, λ_0', and λ'. In what regime, radio, infrared, optical, ultraviolet, X-ray, or gamma-ray, does λ' fall? What if λ had started out as an optical photon wavelength 5000 Å? Draw appropriate conclusions about the process of inverse Compton scattering.

There is also much thermal activity near the Galactic center. At a small angular displacement from Sgr A is a giant molecular cloud, Sgr B2, which is a favorite among molecular radio astronomers. In Sgr B2 is found the richest known variety of interstellar molecules. Whenever radio astronomers have a new candidate molecule to search for in the interstellar medium, they invariably first look for it in Sgr B2. Given the sheer abundance of raw material to be found at the Galactic center, it does not surprise us to learn that star formation seems to proceed actively there.

There also are many discrete X-ray sources near the Galactic center. Most of these sources are probably binary X-ray sources of the variety discussed in Chapter 10, but we should be prepared for surprises. The Einstein X-ray Observatory finds a similar situation in the central regions of the Andromeda galaxy, M31. In fact, the nucleus of M31 contains even more discrete X-ray sources (Figure 13.42).

It is the region within a few light-years of Sgr A West, however, that interests us most here. Infrared maps of this neighborhood show several discrete infrared sources (Figure 13.43). It is generally believed that the 10–20 micron sources correspond to dust clouds heated in part by a smoothly distributed population of old stars and in part by a patchy population of young O and B stars. Within the same few light-years exist many clouds of ionized gas that have been studied intensively by Townes and co-workers. They perform infrared spectroscopy on a fine-structure line of once-ionized neon, and can thus obtain high-resolution information about the motions of the ionized gas clouds in the central few light-years of the Galaxy. From an analysis of the statistics of this motion, they are able to place a limit of a few million solar masses for the amount of gravitating matter contained at the center of our Galaxy. Thus, if the nucleus of our Galaxy has a supermassive black hole, its mass does not significantly exceed $10^6 M_\odot$ (Figure 13.44).

A puzzle does arise about the physical properties of the ionized gas clouds described above. They individually contain typically only a fraction of a solar mass of material (assuming cosmic abundances for neon), and Townes and his coworkers speculate that these discrete clouds may have been stripped from the envelope of colliding red-giant stars near the Galactic center. Moreover, the ionization pattern in the clouds is somewhat peculiar; the atoms are not as highly ionized as they would be, for example, if they were exposed to the light of an ordinary association of O and B stars. There may be another powerful source of ionizing radiation, and an accretion disk surrounding a massive black hole of $10^6 M_\odot$ is one possibility. In brief, the present observational evidence about activity in the nucleus of our own Galaxy is consistent with the existence of a $10^6 M_\odot$ black hole there, but not compelling for anyone disposed toward disbelief.

Figure 13.42. A narrow-angle view in X-rays of the central region of the Andromeda galaxy, M31. The eighteen X-ray sources found here outnumber those found near the center of our own Galaxy. (Courtesy of Riccardo Giacconi and the High-Energy Astrophysics Division of the Harvard-Smithonian Center for Astrophysics.)

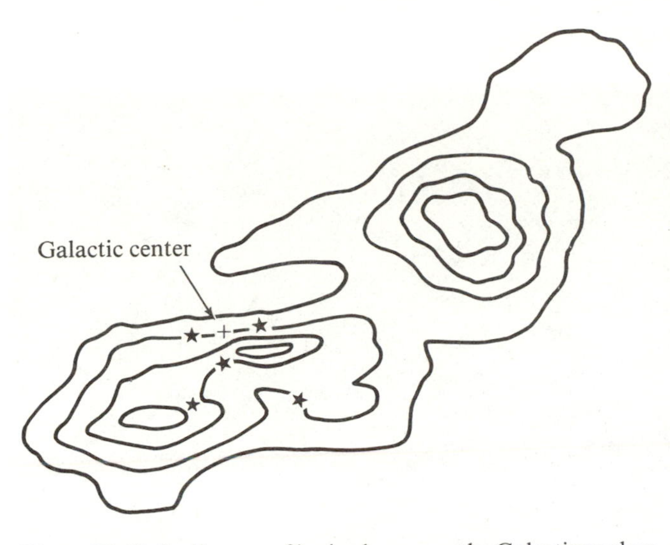

Figure 13.43. Radio map of ionized gas near the Galactic nucleus by Balick is superimposed on five infrared sources, lying within 30 arcseconds of each other, found at 10–20 microns by Low and Rieke. The Galactic center's position is defined by the peak emission at 2.2 microns observed by Becklin and Neugebauer.

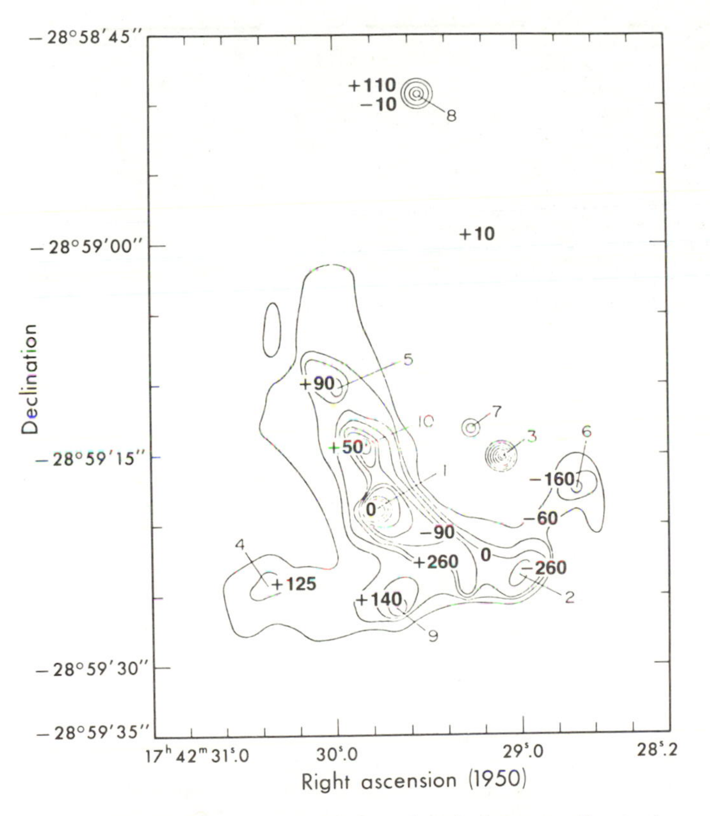

Figure 13.44. The 10-micron infrared emission near the center of our Galaxy has been mapped by Becklin and colleagues. The velocities of the ionized neon gas (bold-faced numbers) associated with this region require the presence of a fairly massive central object if these gas clouds are gravitationally confined to the small region indicated here. (From Lacy, Townes, Geballe, and Hollenbach, *Ap. J.*, **241**, 1980, 132.)

Figure 13.45. The central region of the giant elliptical galaxy M87 (type E0, see Figure 13.15a) has a jet that emanates from the nucleus. (From A. Sandage, *The Hubble Atlas of Galaxies*, Carnegie Institution of Washington, 1961; photographic material from Mt. Wilson Observatory.)

The Nucleus Of M87

A better case can be made for the giant elliptical galaxy M87 (see Figure 13.15a). This galaxy is the third brightest system in the Virgo cluster, the closest rich cluster of galaxies. Messier 87 first aroused the attention of optical astronomers when H. D. Curtis discovered in 1918 that it had an optical jet which pointed out of its nucleus (Figure 13.45). An optical jet has subsequently also been found for the quasar 3C273 (Figure 13.46). Both M87 and 3C273 have copious X-ray emission, as was measured by Friedman and coworkers, Bradt and coworkers, and Bowyer and coworkers. X-ray images made by the Einstein Observatory in 1980 also showed evidence of jets for these two objects. It is only natural to believe that such optical and X-ray jets are related to the radio jets of strong radio galaxies. From this point of view it is gratifying to note that 3C273 and M87 (which is Virgo A as identified by Bolton, Stanley, and Slee, and 3C274 in the third Cambridge catalogue) are among the strongest discrete sources of radio emission in the sky. Messier 87 is, however, not a classic radio double. Instead, it is the prototype of a minority of radio sources with a core-halo structure. In core-halo radio sources, a small radio core exists at the center of an extended halo

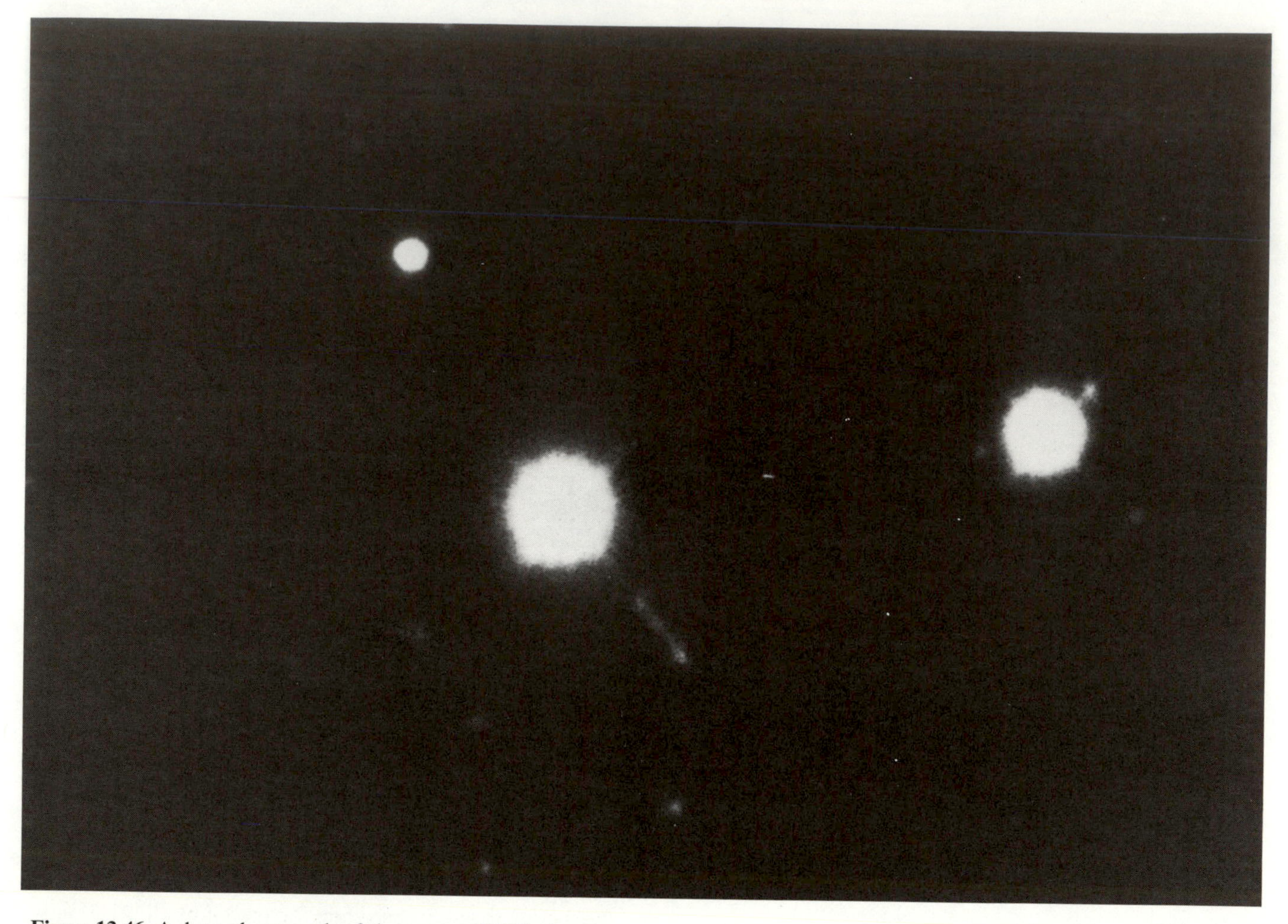

Figure 13.46. A deep photograph of the quasar 3C273, showing a jet of light emanating from the center of the image. Compare this photograph with Figure 13.45 of M87. (The Kitt Peak National Observatory.)

of fainter radio emission (Figure 13.47). In 1969, Hogg and coworkers found the core source of M87 to coincide in alignment with the optical jet, and Cohen and coworkers showed it to have a nucleus less than six light-months in size. A radio jet also exists in 3C273, which also has a compact radio component that Dent discovered in 1965 can vary within one year.

Meanwhile, theoretical work along several different fronts began to focus on an intriguing issue of stellar dynamics, raised by P. J. E. Peebles. What is the response of a distribution of ordinary stars to a very massive point attraction placed in the center of a spherical stellar system? The formal answer provided by Peebles turned out to be wrong, but his work did motivate Bahcall and Wolf, and subsequently many others, to look into the idea that a central cusp of light might result from the tendency of

Figure 13.47. (a) An early map of the radio structure of M87, which is the prototype of a core-halo radio source. Notice that the jet of M87 is barely discernible in the radio contours, because the shape and size of the radio beam (right panel) is comparable to those of the source. (From Hogg, MacDonald, Conway, and Wade, *A. J.*, **74**, 1969, 1206.) (b) A VLA map of the radio structure of the same central regions of M87. The short straight lines indicate the polarization direction of the nonthermal radio continuum emission. Notice that the jet of M87 is well-resolved, because the radio beam has a size (upper left corner) which is much less than the length of the jet. (From Owen, Hardee, and Bignell, *Ap. J. Lett.*, **239**, 1980, L11. The National Radio Astronomy Observatory is operated by Associated Universities, Inc., under contract with the National Science Foundation.)

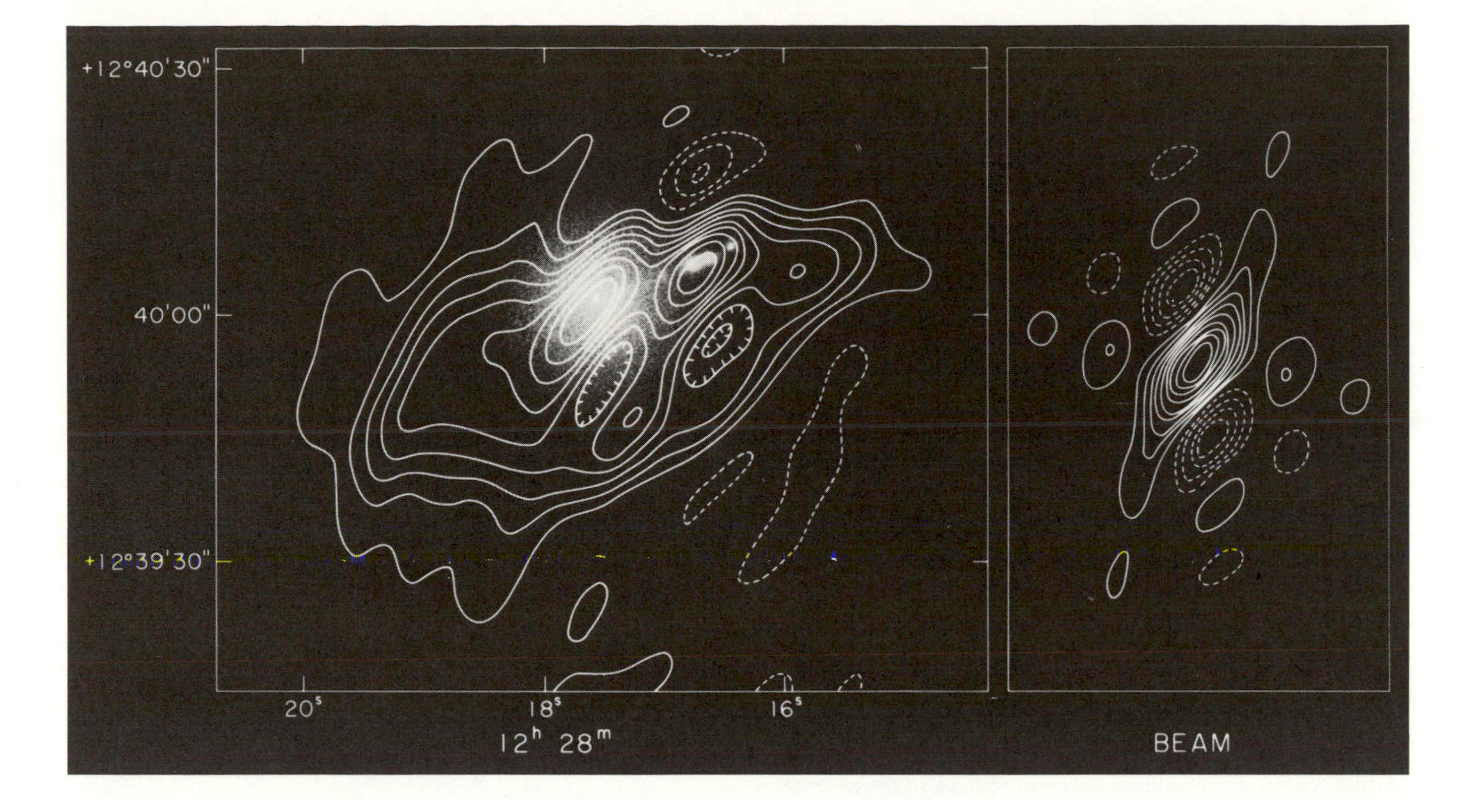
+12°40'30"
40'00"
+12°39'30"
20s
18s
16s
12h 28m
BEAM

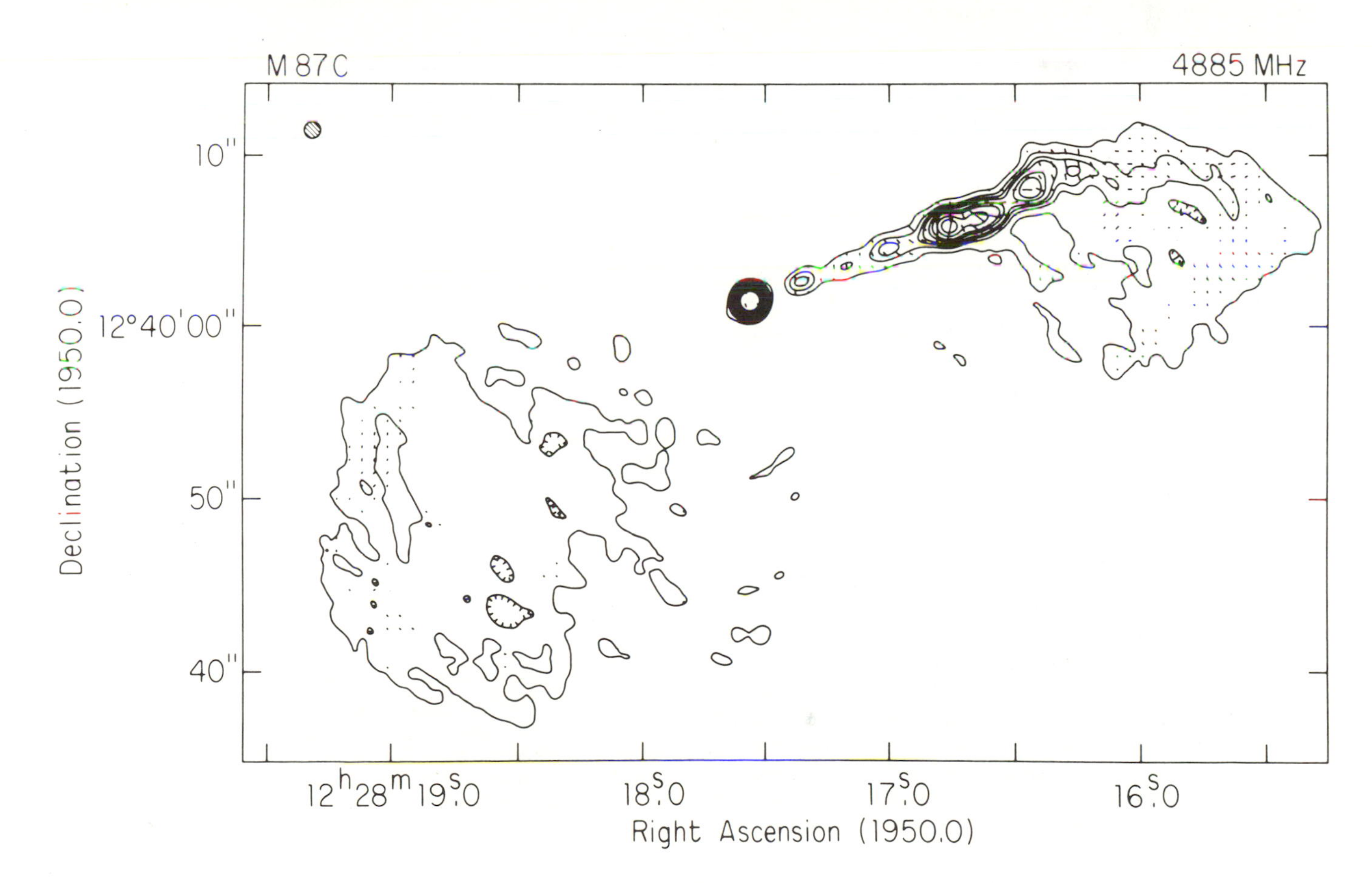
M 87C
4885 MHz
10"
12°40'00"
50"
40"
Declination (1950.0)
12h28m19s.0
18s.0
17s.0
16s.0
Right Ascension (1950.0)

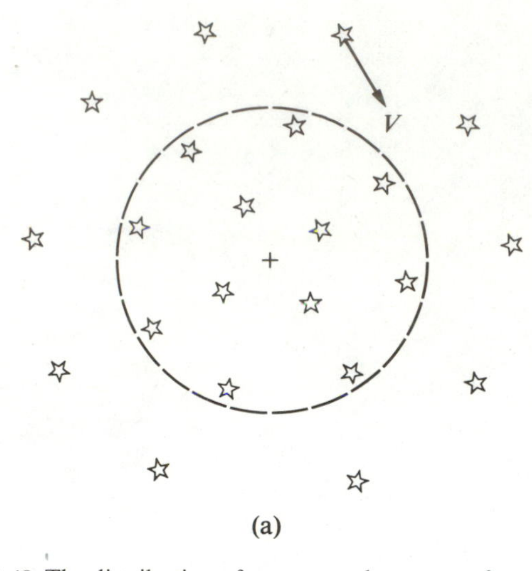

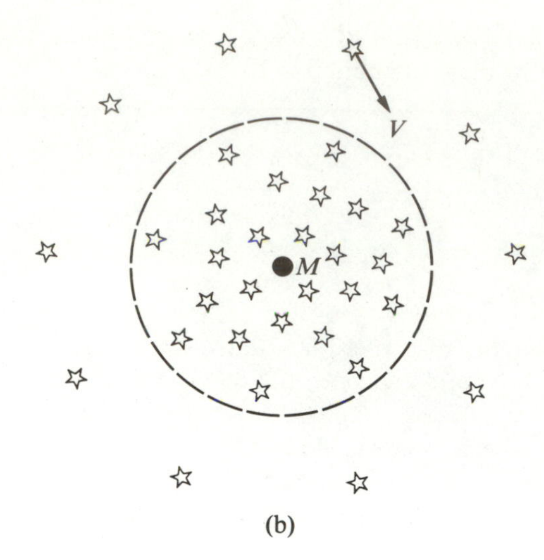

Figure 13.48. The distribution of stars near the center of a galaxy. (a) In a system with no heavy mass at the nucleus, the number density of stars tends to reach a constant value in the central regions. (b) In a system with a large mass M at the nucleus, additional stars are attracted toward the nucleus. The influence of the supermassive black hole is especially important for radii inside GM/V^2, where V is the velocity dispersion that the stars in the central parts of the galaxy would have in the absence of the black hole. The locus $r = GM/V^2$ is drawn as a dotted circle; inside this circle, the stars have random velocities appreciable larger than V.

the massive body to attract stars toward itself (Figure 13.48). At first, the idea was addressed to an ill-fated attempt to understand the X-ray burst sources in globular clusters (Chapter 10), but it soon became apparent that the concept might prove to be more fruitful in connection with the supermassive black-hole model of active galactic nuclei. In particular, a sufficiently massive black hole could increase the central surface brightness beyond that predicted by de Vaucouleur's law (equation 13.1) for elliptical galaxies and the central bulges of spirals. This central cusp might be observable given sufficient angular resolution and sensitivity.

All these considerations, no doubt, motivated a team of Caltech and British astronomers, led by Young, Sargent, and Boksenberg, to take a more careful optical look at the nucleus of M87. In 1978, they announced their result. The surface brightness distribution of M87 did, in fact, have a substantial central cusp (Figure 13.49). Moreover, spectroscopic observations showed that the velocity dispersion of the stars in M87 increased dramatically toward the center of the light cusp. Such an increase of random velocity was predicted theoretically if the central black hole is not to gobble stars at too rapid a rate (Problem 13.18). Moreover, detailed theoretical modeling of M87 showed that both the light cusp and the increase in central velocity dispersion were consis-

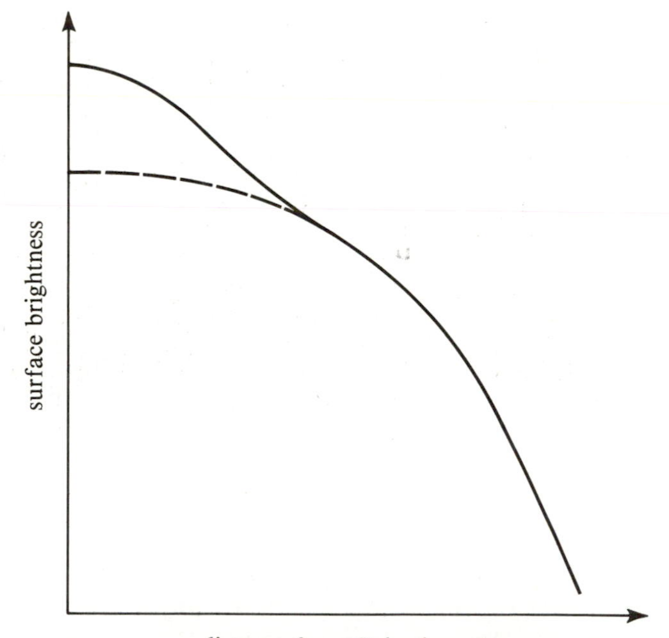

Figure 13.49. Surface photometry of the central region of M87 which shows the excess light near the center that results perhaps from the attraction of a supermassive black hole for stars in its vicinity. (Adapted from Young *et al.*, *Ap. J.*, **221**, 1978, 721.)

tent with the existence of a black hole at the nucleus of M87 with a mass of about $5 \times 10^9 M_\odot$.

Problem 13.18. Let us show the increase of velocity dispersion by first assuming the contrary. The light distributions of ordinary globular clusters and spherical galaxies are well-fit by adopting a distribution function which is virtually a negative exponential function of the energy per unit mass E of the stars (consult Problem 12.7): $\exp(-E/V^2)$, where $E = \mathcal{T} + \mathcal{V}(r)$, where $\mathcal{T}$ and $\mathcal{V}$ are the kinetic and potential energies per unit mass, and V is a *constant* velocity dispersion. (If encounters play an important role, the square of the velocity dispersion V^2 is inversely proportional to the mass of the star.) The exponential (Maxwell-Boltzmann) distribution cannot continue to hold as the black hole is approached. Show for small r (but not near the event horizon), that $\mathcal{V}(r) \simeq -GM/r$, where M is the mass of the black hole. Thus, the number of stars of all possible $\mathcal{T}$ at a distance r from the nucleus would, according to the Maxwell-Boltzmann distribution, be proportional to $\exp(+GM/V^2r)$, which would diverge for $r \ll GM/V^2$. Calculate GM/V^2 for $M = 10^9 M_\odot$ and $V = 300$ km/sec, and compare the values of $\exp(+GM/V^2r)$ at $r = GM/V^2$ and $r = 1$ lt-yr. Is such an increase in the number density of stars at $r = 1$ lt-yr plausible? Conclude, therefore, that the quasisteady distribution near the black hole (inside $r \sim GM/V^2$) must deviate from a Maxwell-Boltzmann distribution. Dimensional reasoning suggests a power law in E, with the exact exponent dependent on rather subtle arguments. Give a simple heuristic argument that the local velocity dispersion v inside the radius GM/V^2 must approximate $(GM/r)^{1/2}$, which increases with decreasing r.

Thus, in most astronomers' opinion, M87 is the Cygnus X-1 of extragalactic astronomy; that is, M87 is the best-documented candidate for having a supermassive black hole in its nucleus. However, the case is not closed. The observations of M87, although admirably performed, are very difficult. Moreover, the detailed behavior of the cusp in the theoretical models depends somewhat on the role of stellar encounters, the importance of which is not as well-established for galactic nuclei as for globular clusters. Also, King and Illingworth showed that one can reproduce similar increases in light at the center of globular clusters without needing to invoke massive black holes. Undoubtedly, similar models could be invented for elliptical galaxies. We need only remember how complex the center of our own Galaxy is to appreciate that most theoretical models must vastly oversimplify the actual state of affairs in galactic nuclei. Despite all these possible objections, however, the supermassive black-hole model provides the most elegant explanation for the energetic phenomena which attend the nucleus of M87. To paraphrase Eddington, all other hypotheses seem to lead to a full stop.

Philosophical Comment

When galaxies were first discovered, one could not help but marvel at their serene majesty. The cosmic drama seemed to proceed at such a slow and measured pace in these stately objects: hundreds of millions of years were required for spirals to turn, a timescale well beyond any in which mortal humans could perceive changes. Knowledge of active galaxies has changed our world view. We now know that the serene appearance of certain galaxies is somewhat a facade. Deep in their hearts, they are severely troubled. We see their nuclei undergoing tulmultuous changes on timescales of hours. Who could have believed such turmoil to be possible when Shapley and Hubble first demonstrated the immense scale of the realm of galaxies? Today, as a consequence of these truly mind-boggling discoveries, we view ourselves as living in a "violent universe." The popular perception of this violence, however, may still be a bit inverted. Violence brings to mind terrific explosions, some of which may well take place in galaxies. But just as with supernovae, the true violence may reside not in the exploding part, but in the imploding part. And if accretion by supermassive black holes is truly the key to active galactic nuclei, then the violence is ultimately generated by inflowing matter, not by outflowing matter. And at the center of this most energetic of all galactic phenomena, we again find our old friend, gravity power.

CHAPTER FOURTEEN

Clusters of Galaxies and the Expansion of the Universe

So far we have discussed galaxies as if they were isolated entities, free to pursue their evolution apart from the influence of other galaxies. In practice, just as there are interacting binary stars, so are there interacting binary galaxies. Just as there are clusters of stars, so are there clusters of galaxies.

Interacting Binary Galaxies

Strongly interacting pairs of galaxies constitute a very small percentage of all galaxies, but the more spectacular examples produce intriguing structures which are not present in single galaxies (Figure 14.1). Spectacular examples of bridges, tails, and rings have been catalogued observationally by Vorontsov-Velyaminov and by Arp. Our theoretical understanding of such systems has been advanced greatly by the numerical simulations carried out by Holmberg, Alladin, Toomre, Wright, and others.

The computer simulations by Toomre and Toomre of closely gravitating disk galaxies are especially interesting. The greatest commotion results from orbits which cause the spinning near edge of a participating disk to travel in the same direction as the passing galaxy. If the two galaxies have very different masses, the small galaxy can then pull out material from the near side of the larger galaxy into a bridge which temporarily spans the gulf between the two galaxies (Figure 14.2). Except for the noncircular orbit and the flattened original distribution of matter, the bridges are analogous to the mass-transfer streams that form in semidetached binaries (Chapter 11). On the other hand, if the encountering galaxies have nearly equal masses, one long tail from each galaxy can develop that generally extends away from the main bodies (Figure 14.3). This nonintuitive result arises because the tidal interaction is much stronger than the familiar example produced by the Earth-Moon system (Figure 1.7). If we pursue the closer analogy with interacting binary stars, long tails are analogous to mass lost from the outer Lagrangian points, L_2 and L_3 (consult Figure 10.7). Long bridges and tails are best produced when the orbit of the two galaxies are bound, so that the two systems are not flying past one another too fast when they reach closest approach. In the much more rare case when the two bodies of the galaxies interpenetrate,

Figure 14.1. Examples of interacting galaxies. (a) A system with an apparent *bridge* that spans two galaxies. *following page* (b) A negative print of a system where each galaxy has a *tail* that points away from the other galaxy. (c) Two examples of *ring* galaxies where a companion galaxy can be found a short distance along the minor axis. [(a) from A. Sandage, *The Hubble Atlas of Galaxies*, Carnegie Institution of Washington, 1961; (b) courtesy of H. C. Arp; (c) Kitt Peak National Observatory and Cerro Tololo Inter-American Observatory.]

(a)

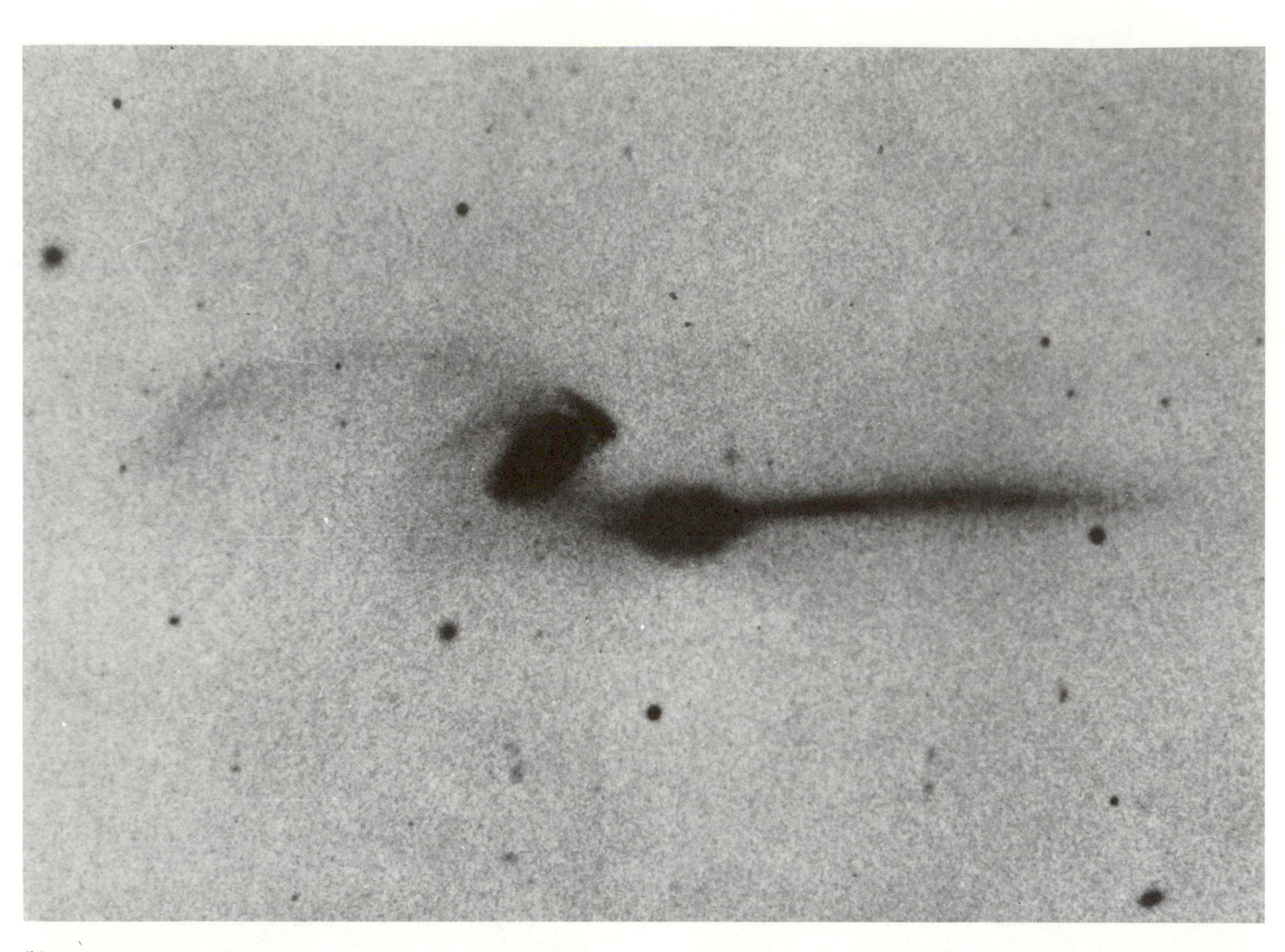

(b)

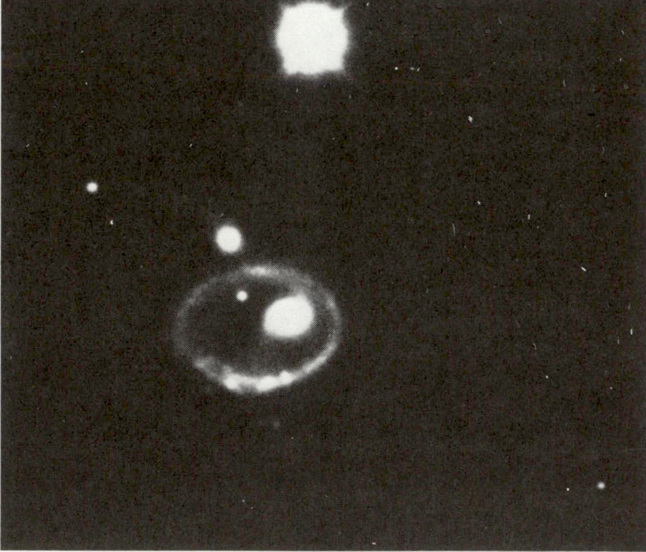

(c)

Figure 14.1 *Continued*

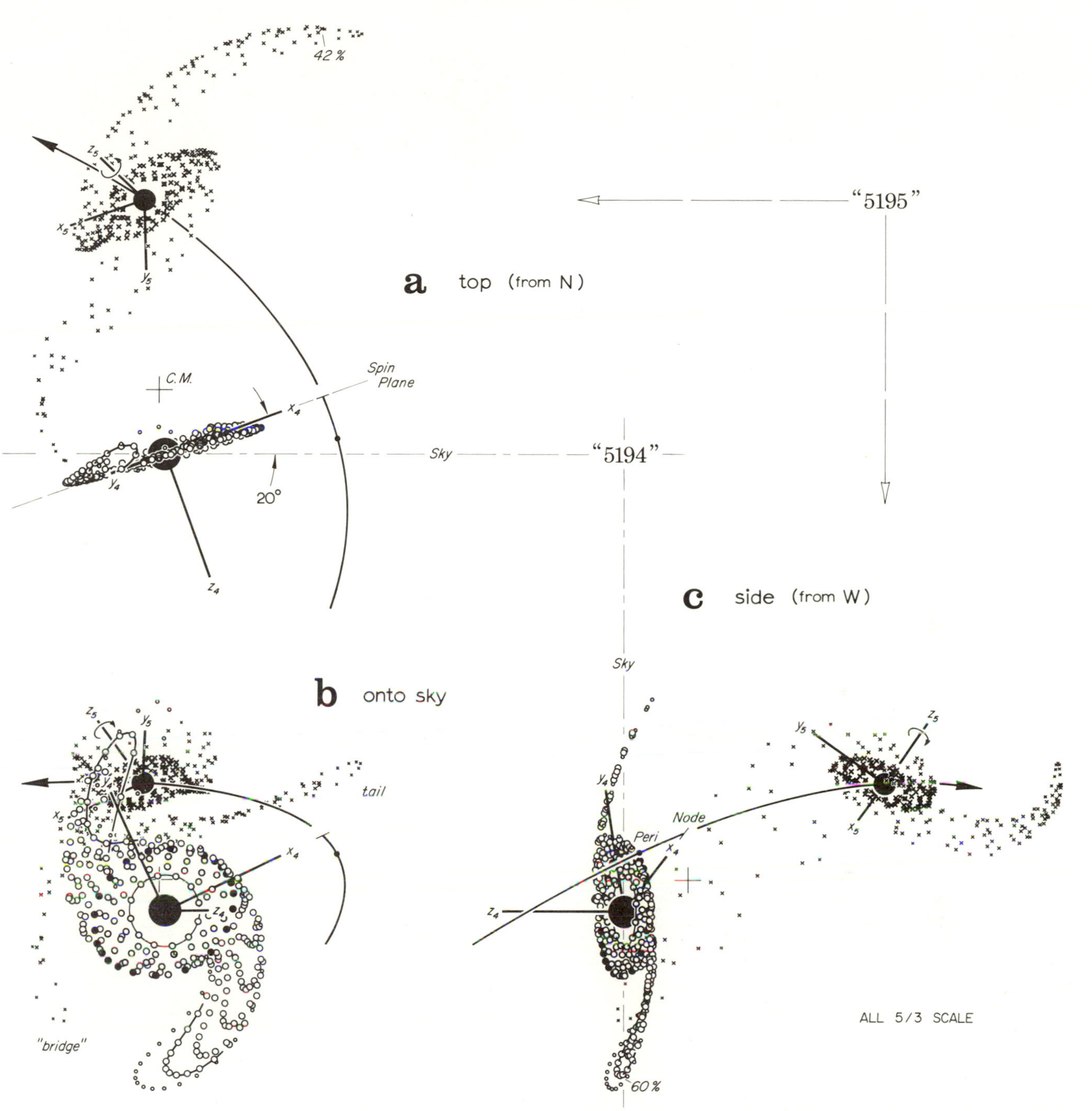

Figure 14.2. A numerical simulation of a bridge-producing encounter. A direct close flyby of a small galaxy past a big disk galaxy draws out a bridge of material which temporarily spans the gulf between the two bodies. Compare b with Figure 14.1a. (After A. Toomre and J. Toomre, *Ap. J.*, **178**, 1972, 623.)

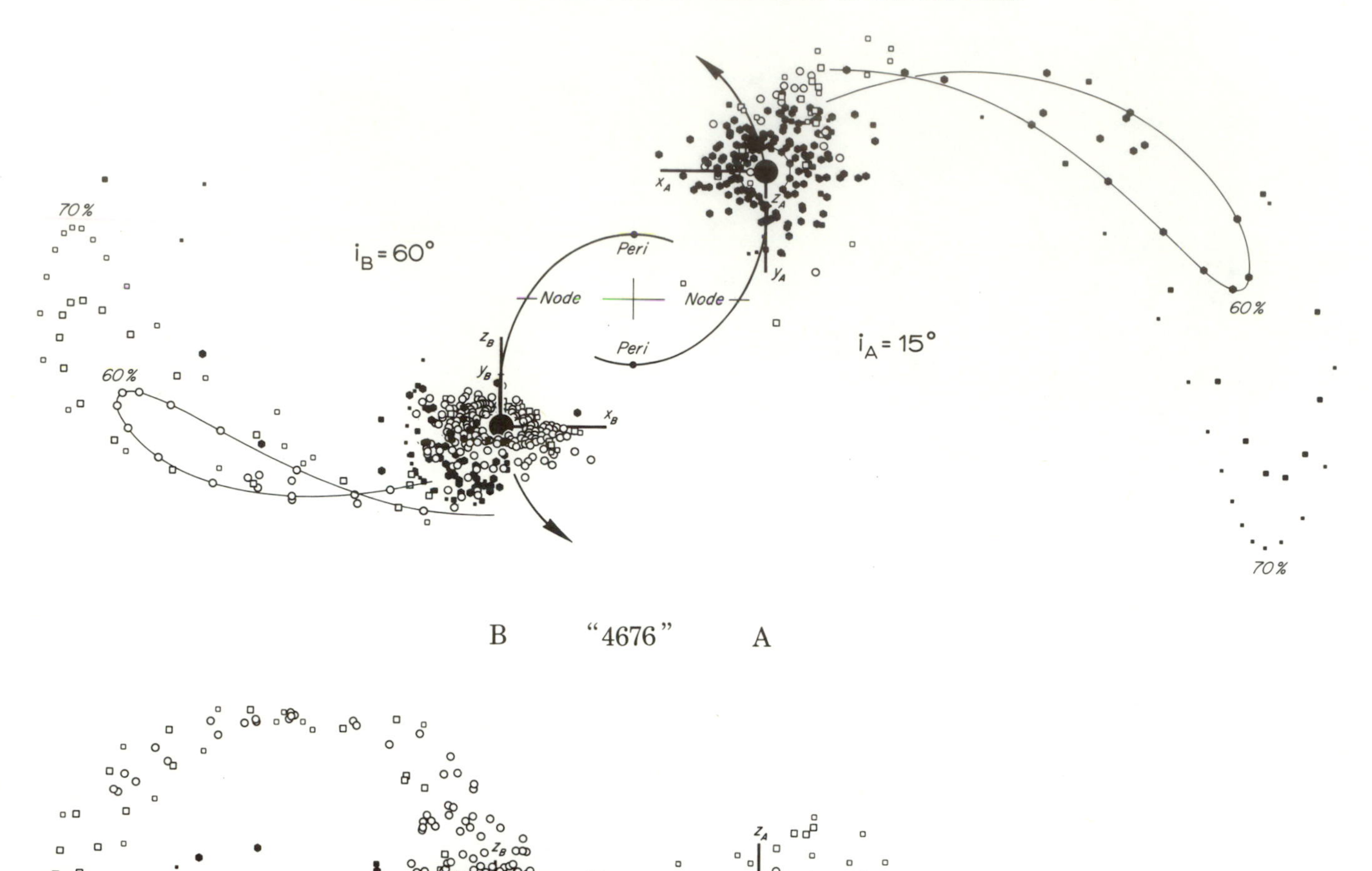

Figure 14.3. A numerical simulation of a tail-producing encounter. A direct flyby of two disk galaxies of comparable masses yields two tails which extend away from the main bodies. Compare bottom diagram with Figure 14.1b. (After A. Toomre and J. Toomre, *Ap. J.*, **178**, 1972, 623.)

exotic looking "ring galaxies" can be formed (Figure 14.4).

Mergers

Close encounters between galaxies obviously excite much internal motion in them. The energy to produce these motions must come from the orbital motion. In a system consisting of a collection of gravitating points (stars), kinetic energy of motion cannot be dissipated, eventually to leave the system as radiation. In an encounter between two galaxies, the individual stars fly right by one another, suffering only gravitational deflections produced by the entire collection of stars. The interstellar gas clouds might bang together inelastically, but the collection of stars would tend to conserve its total energy. The stars might, however, transform one kind of energy into another kind, say, orbital energy into random motions.

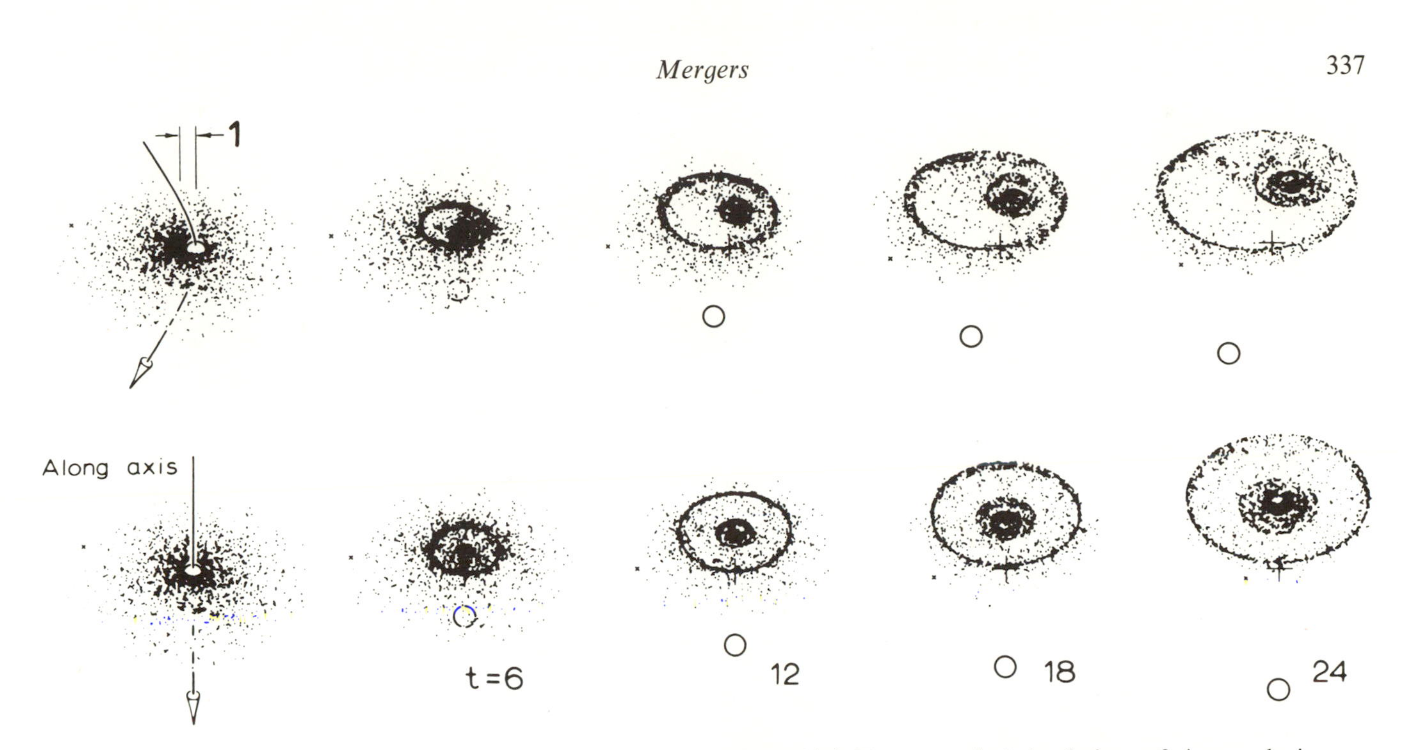

Figure 14.4. Two numerical simulations of ring-producing encounters. An interpenetration of a disk galaxy by another massive body yields rippling waves of rings. Compare with Figure 14.1c. (After A. Toomre, *IAU Symp. No. 79*, 1978, 109.)

Indeed, it can be argued on statistical grounds that repeated encounters between two bound galaxies must tend to bring the two galaxies closer together, consistent with the constraints of the conservation of total angular momentum and total energy. This expectation is based on the second law of thermodynamics, which when applied to trillions of stars still states that order tends to be replaced by disorder (Chapter 4). The close encounter of two bound spiral galaxies, containing much energy in the ordered forms of spinning disks and orbiting centers, must tend eventually to produce a merger into a single pile of stars, containing much more energy in the disordered form of random motions. This merger process involves a form of "violent relaxation," in which violently changing gravitational fields help to produce a final (relaxed) smooth distribution of stars. Except in having relatively great amounts of interstellar gas and dust, such a pile of stars would probably strongly resemble an elliptical galaxy.

François Schweizer has found an excellent candidate for such a merger process (Figure 14.5). In some exposures, the system looks like a crumpled spider with no close neighbors. On deep exposures and spectroscopic studies, one can see two long tails, chaotic internal motions, and evidence for much gas, dust, and recent star formation. The central body of the system looks surprisingly like a giant elliptical galaxy. The luminosity of the system is equal to that of two luminosity class I spirals. We apparently have here the gravitational merger of two giant spiral galaxies into a single pile of stars. After the gas and dust has been completely used up to form more stars, the resulting system will probably be classified as an ordinary elliptical galaxy.

Alar Toomre has speculated that perhaps all elliptical galaxies formed in this way. His argument is beguiling, and proceeds as follows. Of the 4,000 or so NGC galaxies, perhaps a dozen are interacting systems exhibiting spectacular bridges or tails. On the other hand, the numerical simulations show that such geometric forms are transient phenomena that cannot be maintained for more than a few times 10^8 years. If each such interacting pair of galaxies (which are probably gravitationally bound) eventually leads to a merger, then during 10^{10} years, we can expect on the order of 400 of the present 4,000 NGC galaxies to have resulted from such coalescence. In other words, mergers of spiral galaxies would cause 10 percent of all galaxies to become elliptical galaxies. The actual fraction of ellipticals is more like 20 or 30 percent, and many of these ellipticals are to be found in rich clusters; however, it is conceivable that mergers were either more common in the past than now or more common in clusters than in the general field.

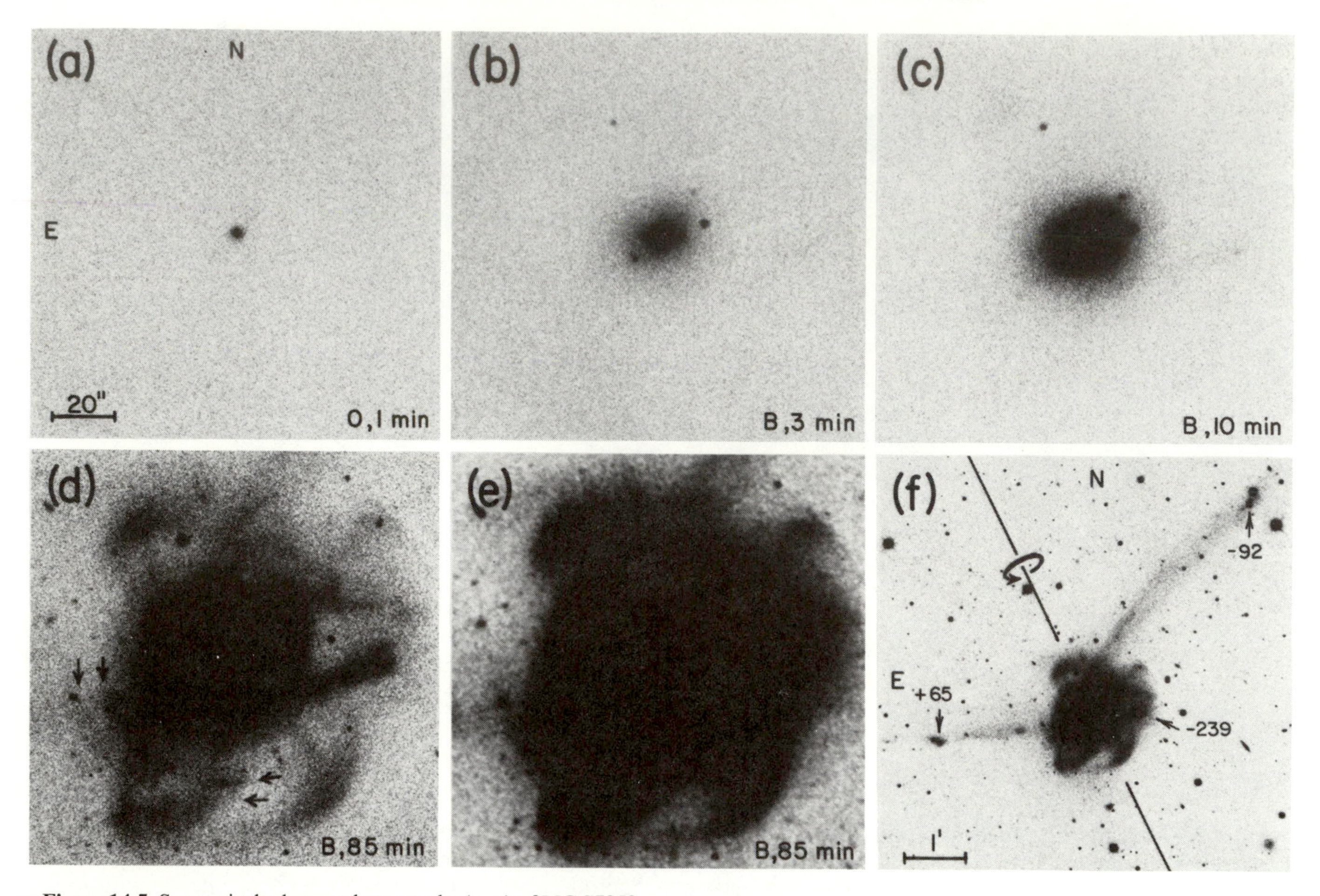

Figure 14.5. Successively deeper photographs (a–e) of NGC7252 reveals a complex set of filaments surrounding a central body that resembles a giant elliptical galaxy. Spectra of some of the features show the counter-velocities (f) in two tails that are characteristic of interacting disk galaxies of comparable mass. For these reasons, this system is regarded as an excellent candidate for a merger between two giant spiral galaxies. (From F. Schweizer, *Ap. J.*, 1981, in press.)

Problem 14.1. Justify the numbers of the merger argument given above.

As attractive as Toomre's proposal is, several strong objections to it have been raised. The most damaging concern the systematic properties of giant elliptical galaxies—such as the Fish-Freeman relation and the Faber-Jackson relation (Chapter 13)—which would be difficult to explain in terms of random mergings of a variety of spiral galaxies. In particular, one would naively expect the merged product to have fewer central stars than the constituent spirals, because the excess orbital energy must be absorbed into the merger product. But the present central surface brightnesses of giant ellipticals are much higher than those of giant spirals (Chapter 13). Perhaps the process of dynamical friction to be described later helps to bring the dense cores of the merging galaxies into a common bright center. In any case, one must balance these theoretical arguments against Schweizer's observational example that nature has found a way to do things which theorists thought difficult. *Some*, if not all, ellipticals must have formed from mergers of spirals.

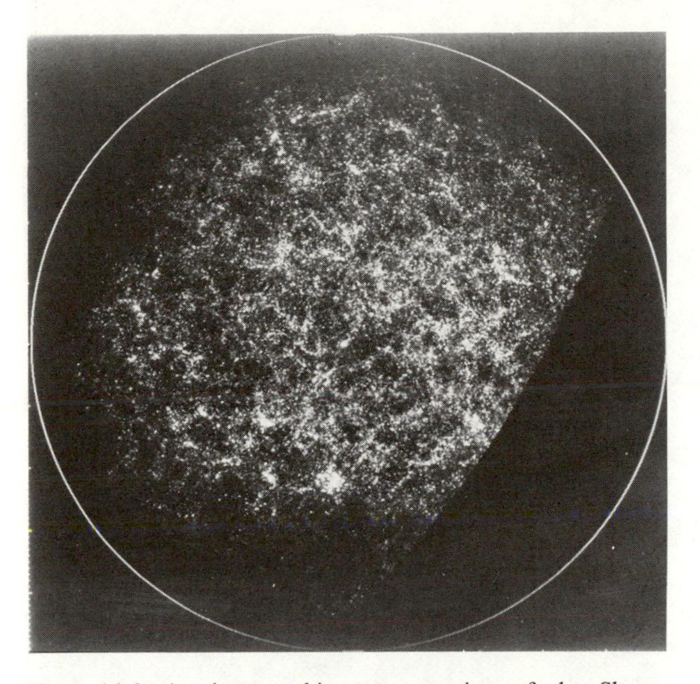

Figure 14.6. A photographic representation of the Shane-Wirtanen counts of galaxies. Individual galaxies are not shown. Instead, each small square bin is intensity-coded to represent the number of galaxies found in that boxed area of the sky. Clustering is apparent to the naked eye, as is verified by a detailed analysis of the positions of the million galaxies that have more than a minimum brightness. (From Seldner, Siebers, Groth, and Peebles, *A. J.*, **82**, 1977, 349.)

Hierarchical Clustering

In addition to doubles, galaxies also come in small groups and rich clusters. In small groups like the Local Group, to which the Milky Way system and the Andromeda galaxy belong, there may be a dozen to two dozen members extending over a radius of about a million light-years. As catalogued by Abell, Zwicky, and others, rich clusters of galaxies may contain several thousand galaxies that extend over a radius of ten million light-years. On even larger scales, say, a radius of a hundred million light-years, astronomers have discovered another level to this hierarchical clustering. Superclusters exist which contain about a hundred clusters, as was noted by Shapley, and confirmed for the Local Supercluster by Holmberg, Reiz, and de Vaucouleurs. It is not known whether clusters of superclusters exist or not. From detailed statistical analyses of galaxy counts by Shane and Wirtanen (Figure 14.6), Peebles and his coworkers find examples of clustering at all sizes up to 60 million light-years, beyond which the amount of clustering drops dramatically, but up to that scale, there is no intrinsic size for hierarchical clustering.

Rich Clusters of Galaxies

The closest fairly rich cluster to ourselves is the Virgo cluster, located about 50 million light-years away in the direction of the constellation of Virgo (Figure 14.7). About 200 bright galaxies reside in the Virgo cluster, of which 68 percent are spirals, 19 percent are ellipticals, and the rest are irregulars or unclassified. Although spirals are more numerous in the Virgo cluster, the four brightest galaxies in Virgo are ellipticals; among them is, of course, M87 (Chapter 13). About seven times further away than Virgo is the Coma cluster, the nearest great cluster containing thousands of galaxies (Figure 14.8). Most of the galaxies in the Coma cluster are ellipticals or S0s. Rood estimated that only 15 percent of the systems in Coma are spirals or irregulars. This appears to be a general feature of rich clusters: they are notably deficient in spiral galaxies, which are quite common in the field and in poorer groups (Figure 14.9). The Coma cluster has another characteristic that is shared by many rich clusters: the presence of one or two very luminous supergiant elliptical galaxies near the center of the cluster (Figure 14.10). Such a supergiant elliptical (called a cD galaxy for historical reasons which are not particularly illuminating) will often dominate the appearance of the whole cluster. Despite the fact that cD systems are intrinsically quite rare, they are the most common optical counterparts of the double-lobed radio sources that were discussed in Chapter 13. Astronomers have gained in recent years much understanding of why spirals are rare and cD galaxies frequent in rich clusters. We will spend the next two subsections discussing these topics.

Galactic Cannibalism

A defining property of a cD galaxy is the possession of a very distended envelope of stars. A good example is NGC6166, a radio galaxy that resides in the Abell cluster 2199. This cD galaxy has a visible radius of about one million light-years, roughly 20 times larger than an ordinary giant elliptical or spiral. Oemler has performed surface photometry on the extended envelopes of cD galaxies. He finds that they typically drop off with increasing distance from the center at a slower rate than given by de Vaucouleurs law (equation 13.1), which is valid for ordinary elliptical galaxies (Figure 14.11).

The idea has become popular, largely because of the efforts of Ostriker, Richstone, Spitzer, and Tremaine, that cD galaxies have grown bloated by cannibalizing their lesser neighbors. This concept is a variation of the idea of galaxy mergers. Here, when the dining galaxy is much bigger than its dinner, we can conceptually separate the meal into two courses: **tidal stripping**, which arises because the galaxy to be eaten has a finite size; and

Figure 14.7. The central parts of the Virgo cluster. This fairly rich cluster lies about 50 million light-years from us. Notice the presence of the giant elliptical galaxies, M84 and M86, which together with M87 (Figure 13.15a) are the largest systems in the Virgo cluster. (The Kitt Peak National Observatory.)

Figure 14.8. (*above*) A portion of the Coma cluster, which is typical of most regular great clusters in that it contains mostly elliptical and SO galaxies. This closest rich cluster lies about 350 million light years from us. (Palomar Observatory, California Institute of Technology.)

Figure 14.9. (*below*) The irregular cluster in Hercules is unusual in that it is fairly rich and yet contains a great variety of galaxy types. (Palomar Observatory, California Institute of Technology.)

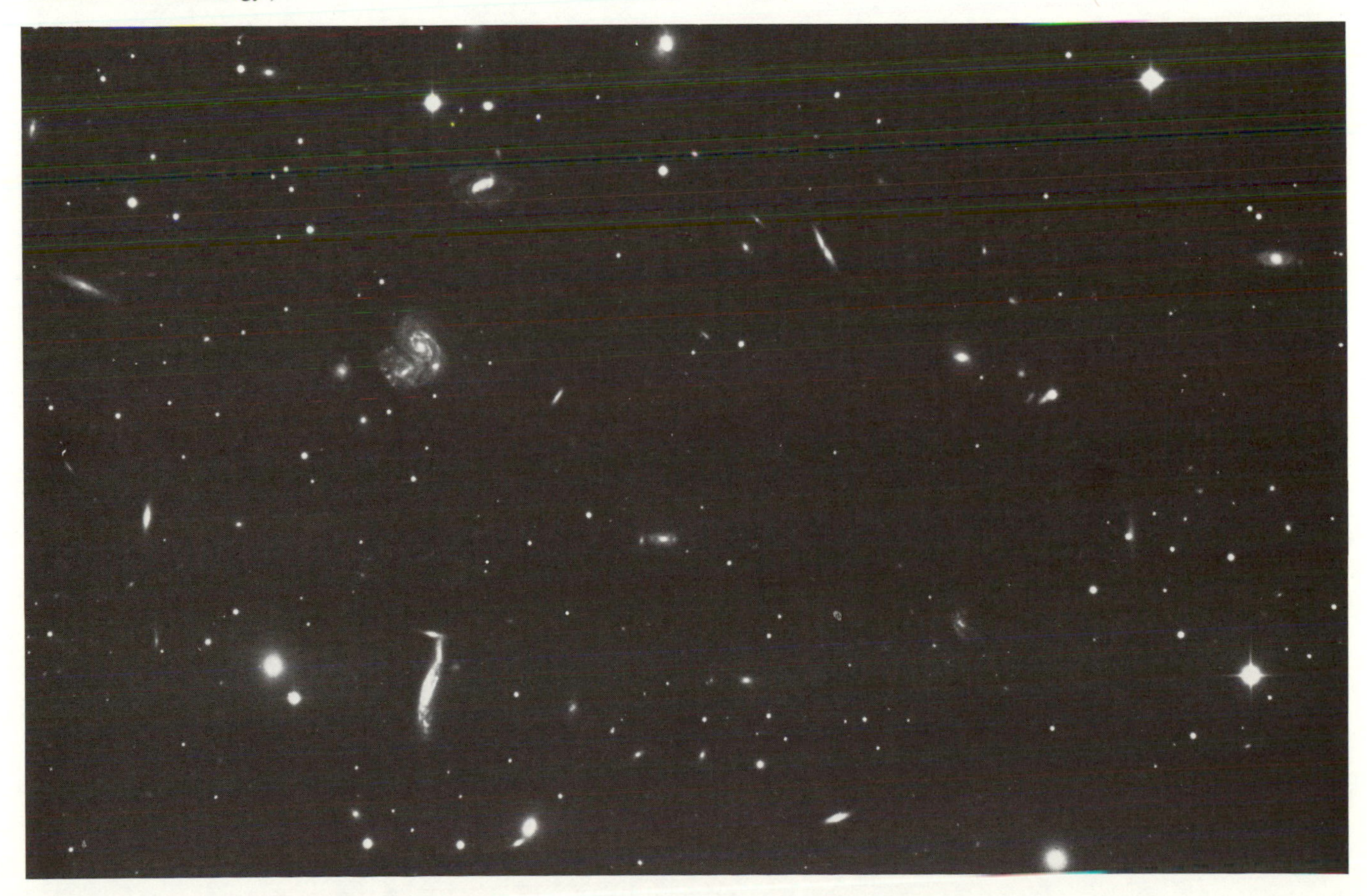

Figure 14.10. The central region of the Coma cluster contains two supergiant ellipticals. These cD systems may have grown bloated by cannibalizing their smaller neighbors. Perhaps in the distant future, they themselves will be forced to merge. (The Kitt Peak National Observatory.)

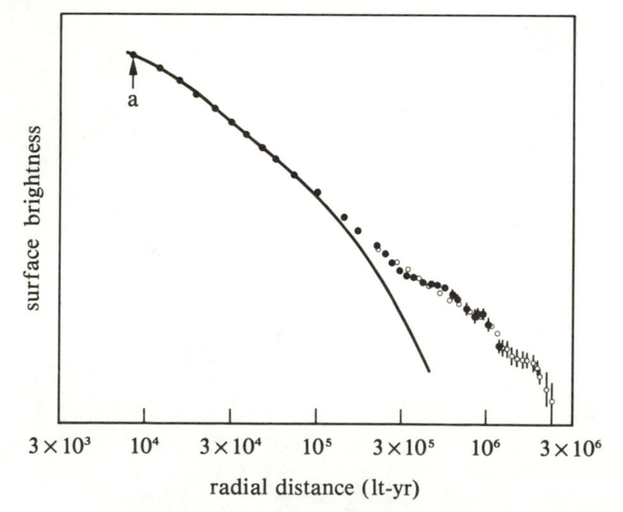

Figure 14.11. Contrast between the surface brightness profiles of a cD galaxy (dots) and an ordinary elliptical galaxy (solid curve). (Adapted from A. Oemler, *Ap. J.*, **180**, 1973, 11.)

dynamical friction, which would arise even if part of the meal comes in one small dense lump (Figure 14.12).

Tidal stripping is fairly straightforward to understand; the physical principle can be traced back to Roche. An object of mass m and radius R, which is held together by its self-gravitation and which approaches within distance r of a massive body M, will be ripped apart by the tidal forces when r becomes too small. A simple estimate gives (Problem 14.2)

$$r = (2M/m)^{1/3}R \qquad (14.1)$$

as the critical distance of approach (Roche limit). Thus, a cD galaxy with a mass M which is 500 times bigger than its victim's mass m will begin to rip the latter's stars from it at a radius r which is 10 times the victim's radius R. This process presumably explains the formation of the extensive envelope around a cD galaxy, although not all the shredded material need be captured gravitationally by the cD galaxy. Some of the stars may enter into orbit about the cluster as a whole (Figure 14.13).

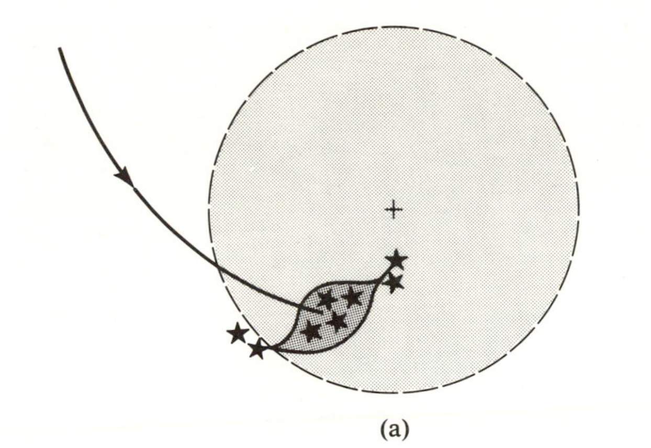

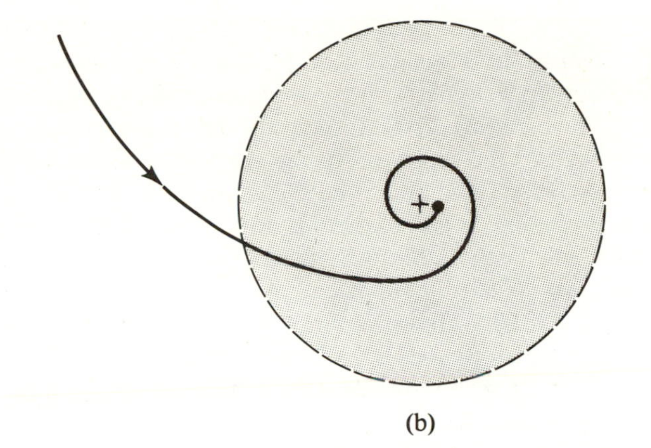

Figure 14.12. Galactic cannibalism. A supergiant cD galaxy devours a small neighbor in two stages. (a) Tidal stripping occurs when the stars at the periphery of the small galaxy are torn loose from the parent body. (b) Dynamical friction causes the dense core of the small galaxy to spiral gradually into the central regions of the cD galaxy.

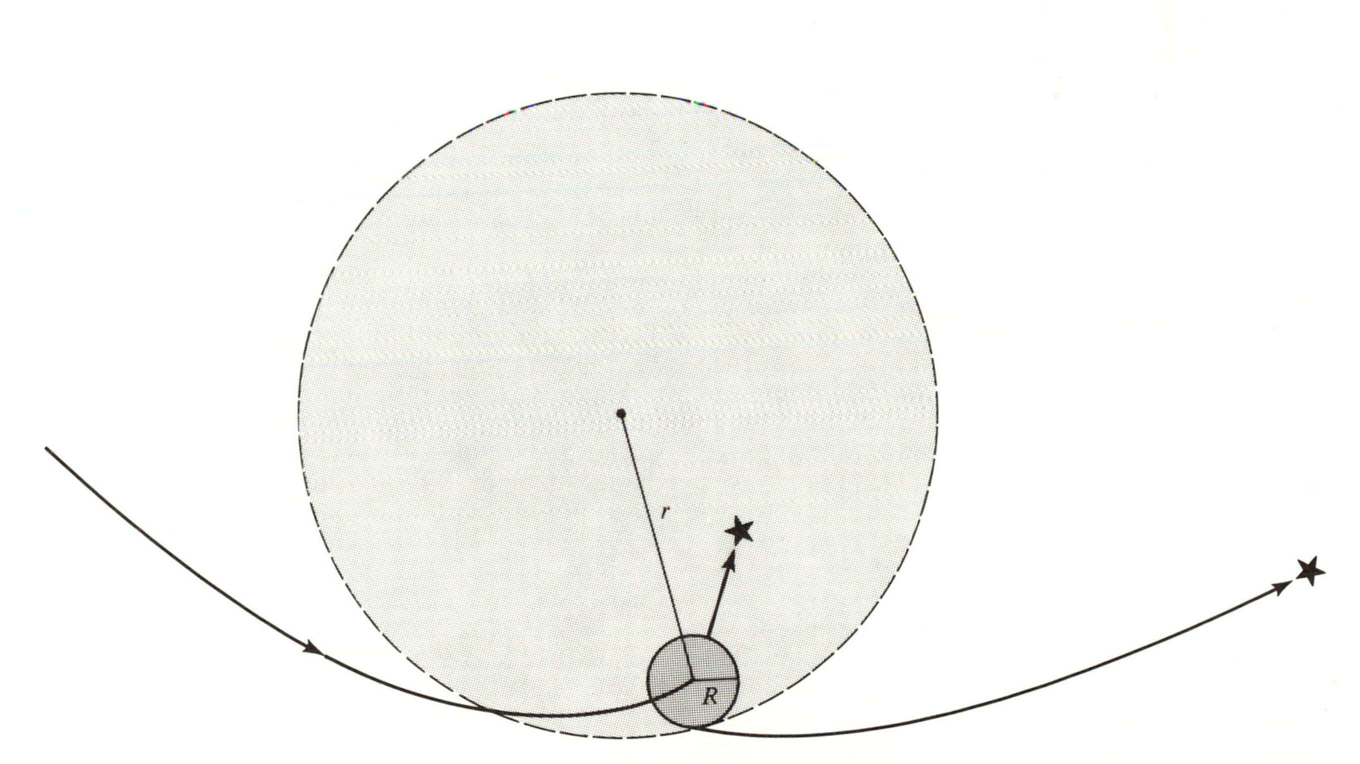

Figure 14.13. Tidal stripping can cause stars to be flung far from the cD galaxy as well as captured by it. The ratio of stars captured versus stars lost would depend on a variety of conditions, including the spin of the devoured small galaxy. Stars which are lost to the cD galaxy may not be lost from the cluster as a whole. Such stars may form a loosely dispersed sea which permeates the entire cluster.

Problem 14.2. Consider two gravitating masses m and M placed in a circular orbit at a distance r apart. Suppose m to have radius R. The gravitational attraction of M for m causes the center of the latter to have acceleration $v^2/r = GM/r^2$. A different acceleration, $GM/(r - R)^2$, is felt by the end of m nearest M, and $GM/(r + R)^2$ is felt by the side farthest. Show that the difference between the accelerations felt by either end and the center has magnitude $2GMR/r^3$ if we assume $R \ll r$. All stars at the extreme ends of m feel a force per unit mass toward the center of m given by Gm/R^2. If the disruptive tidal acceleration $2GMR/r^3$ is greater than the binding force per unit mass Gm/R^2, the galaxy m will be torn apart. Show that this leads to a critical distance given by equation (14.1).

If the galaxy m does not circle M, but passes it very rapidly at a speed v, we require a different approach. Here the tidal acceleration $a = 2GMR/r^3$ at closest approach r will last for a time given roughly by $t = r/v$. Argue that this will cause the ends of the galaxy m to acquire a speed $\Delta v = at = (2GMR/r^3)(r/v)$ different from the center. This difference will unbind the stars if Δv exceeds the escape velocity $(2Gm/R)^{1/2}$. Show that this leads to a critical distance given by $r = \alpha^{1/3}(2M/m)^{1/3}R$, where $\alpha = GM/rv^2$ is a dimensionless number equal to unity for a circular orbit and much less than unity for the hyperbolic trajectories characteristic of random encounters in a galaxy cluster.

Equation (14.1) has a simple physical interpretation. Imagine spreading the mass m of the small galaxy uniformly over a sphere of radius R equal to its size. The average density would be $3m/4\pi R^3$. Imagine spreading the mass M of the big galaxy over a sphere of radius r equal to the *distance* of the small galaxy. The average density would be $3M/4\pi r^3$. Equation (14.1) can now be interpreted as stating that the Roche limit occurs when the average density of the small galaxy is *twice* that of the large galaxy. Now, we see a potential limitation to the applicability of equation (14.1). Galaxies *do* have spread-out mass distributions, but this distribution is highly nonuniform, becoming largest at the center. The rarefied outer portions of the small galaxy will exceed the Roche criterion at relatively large distances r from the big galaxy, and tidal stripping will operate rather efficiently on the matter there. The dense cores of even small galaxies, however, will have to plummet quite far into the heart of a cD system before they encounter interior densities comparable to their own. Since such plunging orbits are *a priori* rare, how does a cD galaxy manage to gobble the cores of other galaxies? By the second course of the meal: *dynamical friction.*

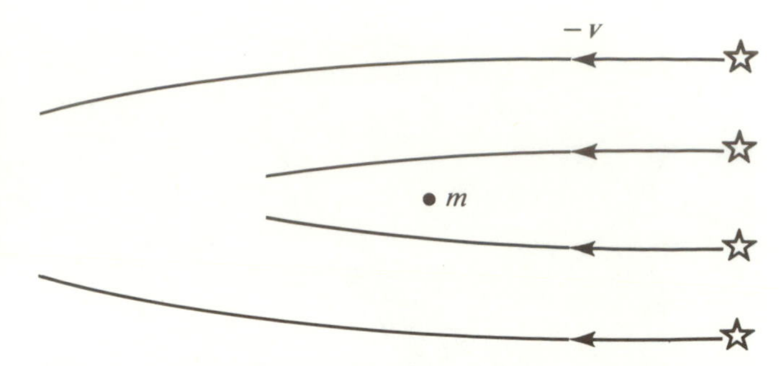

Figure 14.14. Dynamical friction arises when a mass m moves at velocity $\mathbf{v}$ relative to a distribution of stars which statistically have no mean motion. For simplicity, we have depicted the stars to be at rest (relative to the center of their galaxy), and we have drawn the situation as it appears to an observer at rest relative to m. Thus, mass m sees stars approach at velocity $-\mathbf{v}$, and be deflected by the gravity of m. This deflection produces a slight excess of mass behind M, since the stars will on the average be closer together after deflection than before. This mass excess pulls m in the direction of $-\mathbf{v}$, producing a net drag which tends to decrease $\mathbf{v}$.

For gravitating bodies, dynamical friction arises for the reason outlined in Figure 14.14. As a heavy dense core m (approximated to be a point mass) moves through a medium containing stars (the envelope of the cD galaxy), the core deflect the stars it passes. These deflections statistically tend to give a slight excess of stars in back of m. (In the language of plasma physics, the presence of the heavy mass polarizes the ambient medium.) This excess mass pulls on the mass m, and thereby tends to reduce its motion relative to the distribution of stars. The net effect of dynamical friction, therefore, is to bring the galactic core to the center of the cD galaxy. Digestion occurs when the galactic core reaches ambient densities comparable to its own and is chewed up by tidal forces, releasing its stars to join the others in the central region of the cD system.

The concept of dynamical friction is not new. In gravitating systems, it originated with the work of Chandrasekhar on stellar dynamics. Chandrasekhar's own work was inspired by Einstein's paper on Brownian motion. In Chapter 9, we pointed out that encounters between stars of different masses in a star cluster tend to bring about a state of equipartition of energy, where high-mass stars have low random speeds and low-mass stars have high speeds. Such a thermodynamic distribution is actually brought about by two opposing processes. Random gravitational scatterings of stars by one another cause them statistically to random walk to higher and higher velocity dispersions. However, each star also suffers dynamical friction as described in Figure 14.14, and this tends to reduce their random velocities. The balance between diffusion to higher random velocities and drag to lower ones leads, in a steady state, to a statistical distribution where stars of a given mass tend to have a certain velocity dispersion (Box 14.1). The relation between the velocity dispersions and masses of stars is given by the principle of *equipartition of energy*.

In contrast to a star cluster, a galaxy has stars which are separated by enormous distances. Moreover, the stellar masses are minuscule in comparison with that of the system as a whole. Under galactic conditions, neither velocity diffusion nor dynamical friction have enough time to affect the behavior of the stars (see Problem 12.6). Thus, we usually think of the stars in a galaxy as being effectively encounterless. The core of a galaxy, however, is much more massive than any individual star, and the deflections of passing stars produced by this core is correspondingly greater (Figure 14.14). This is why dynamical friction can bring the core of a galaxy to the center of a cD galaxy in times short compared to 10^{10} years (the age of the universe), whereas the corresponding processes (equipartition of kinetic energy) are completely negligible for individual stars.

Problem 14.3. The formula derived in Problem 9.11 for the relaxation time t_{relax} (which is inversely proportional to the coefficient of dynamical friction) is inversely proportional to two powers of m, because in Problem 9.11 we assumed that the scatterer and the scatterees have the same mass m. If we had taken the mass of the scatterer to be m (core of a galaxy) and the scatterees to be m_* (stars), we would have found $t_{\text{relax}} \propto 1/m_* m$. Argue now that, except for a coefficient of order unity, t_{relax} is given by the formula

$$t_{\text{relax}} = \frac{(rv)^3}{3G^2 mM \ln(rv^2/Gm)},$$

where we have assumed there are N stars in a volume $(4\pi r^3/3)$ to estimate n, and we have noted that $Nm_* = M$, the mass of the cD galaxy. Compute the numerical value of t_{relax} for $m = 10^{10} M_\odot$, $M = 10^{14} M_\odot$, $r = 10^6$ lt-yr, and $v = 300$ km/sec. Comment on your answer. Also comment on whether big galaxies would be cannibalized in preference to little ones.

BOX 14.1

Equipartition of Random Kinetic Energy

If classical bodies interact for a long enough time (for example, by mutual gravitational deflections), each species of mass m tends to acquire a statistical distribution of random kinetic energy whose average value is independent of m. The constancy of $m\langle v^2 \rangle/2$ implies that the velocity dispersion $\langle v^2 \rangle^{1/2}$ of heavier bodies tends to be smaller by a factor proportional to the inverse square root of the mass of the species, $m^{-1/2}$. In a self-gravitating collection of bodies, therefore, statistical equilibrium favors the sinking of the heaviest components toward the center of the system.

The basic driving force in the galactic-cannibalism model is complex in its details, but the underlying principle is simple. It is, once again, the interplay of the second law of thermodynamics and the universal law of gravitation which leads to the agglomeration of more and more matter toward the center of the system. For this reason will interstellar gas collect to form stars (Chapter 11). For this reason do single stars evolve to become compact objects (Chapters 7 and 8). For this reason do single galaxies develop spiral structure (Chapter 12). For this reason will binary stars ultimately spiral together to fuse into a single object (Chapter 10). For this reason will binary galaxies ultimately want to merge (this chapter). For this reason do star clusters suffer core collapse (Chapter 9). For this reason do galaxy clusters suffer cD formation (this chapter). The chapters read differently, but the underlying plot is the same.

Hot Gas in Rich Clusters

Satellite observations in the early 1970s revealed that X-rays pour from the spaces between galaxies in rich clusters. Spectral information could be derived by measurements at different X-ray wavelengths, and analyses of this information led to the conclusion that the emission arises from hot gas with temperatures between 10 and 100 million degrees K. The mass of gas in a typical cluster like Coma was found to be comparable to the total mass contained in the optically visible parts of the member galaxies. At first, there was some uncertainty about the place of origin of this gas, and the idea was entertained that the gas might have fallen into the cluster from intergalactic space. This concept lost favor when a spectral X-ray line was observed which showed the presence of substantial amounts of iron in the cluster gas. Since iron is believed to be synthesized only in the deep interior of massive stars which are destined to supernova

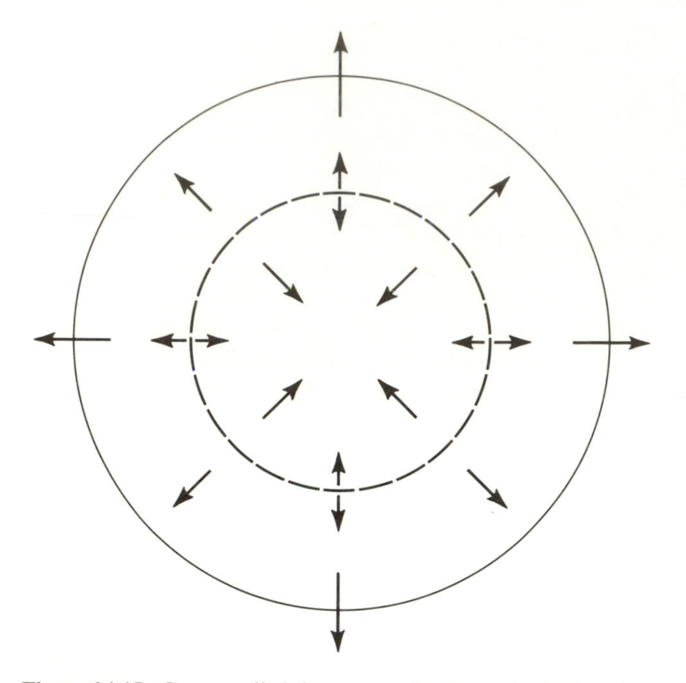

Figure 14.15. Gas expelled from stars in the main body of an elliptical galaxy may heat up and either blow outward in a galactic wind, or flow inward toward the galactic nucleus. In general, the former will happen in the outer regions, and the latter in the inner regions.

(Chapter 8), current belief is that this gas comes from the galaxies of the cluster.

There are several theoretical ways in which gas spewed from supernovae might eventually find its way into the cluster medium. Mathews and Baker proposed that in elliptical galaxies the interstellar medium might be so hot that the gas lost from stars would blow continuously out of the galaxies, much as the solar wind carries material away from the corona of the Sun (Figure 14.15). This mechanism would work for both isolated ellipticals and those in rich clusters. In the more general case,

Figure 14.16. Idealized fast collision between two spiral galaxies. On the left, two spiral galaxies with their characteristic band of gas and dust in the midplane of their disks approach each other for a head on collision. The stars travel past each other without suffering any damage, but the dust and gas clouds cannot interpenetrate and are arrested in a strong shock. Stripped of their gas and dust, the two stellar systems continue on their ways, having been converted to SO galaxies. The gas and dust layers, meanwhile, have been heated to very high temperatures, and may expand and dissipate into the general cluster environment. Alternatively, if the gas cools rapidly enough, stars may form prolificly in the greatly compressed state brought about by the collisional process.

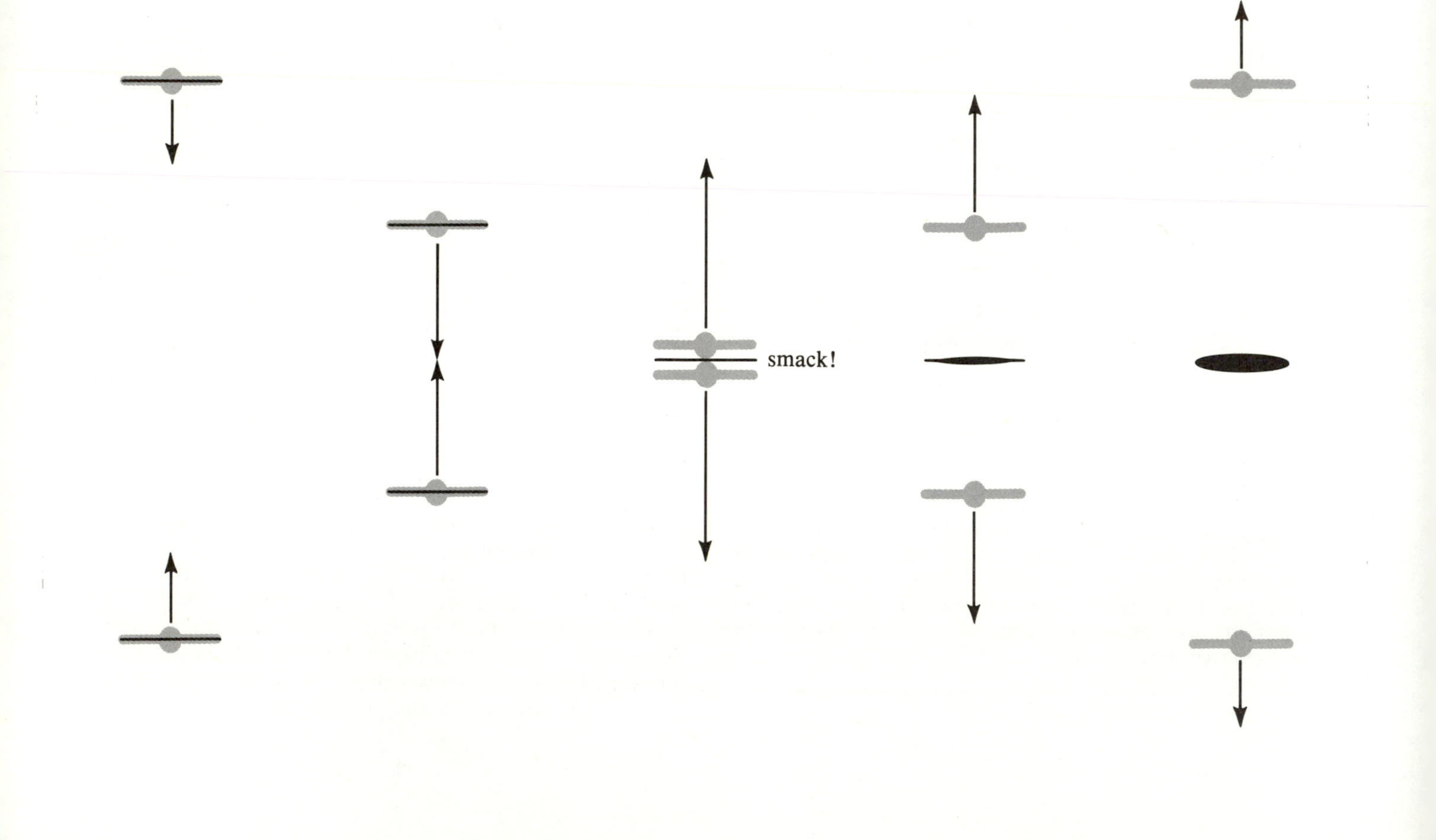

Figure 14.17. Optical photograph of the most powerful radio source known, Cygnus A. This galaxy is believed to be a giant elliptical galaxy with a dust lane which bifurcates the optical image to give the curious visual impression of a butterfly. (Palomar Observatory, California Institute of Technology.)

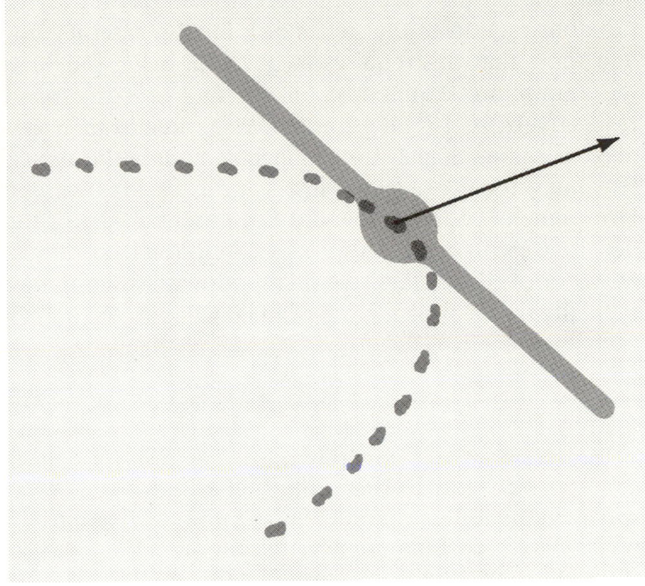

Figure 14.18. A spiral galaxy can be stripped of its gas and dust clouds if it moves at high speeds through a dense enough cluster medium. The relative wind that results can literally blow away the gas and dust of the galaxy.

Axford and Johnson showed that gas in the inner regions flows toward the nucleus and gas in the outer regions blows away in a galactic wind. The processes would make the main bodies of elliptical galaxies virtually devoid of gas (Chapter 13). In spiral galaxies, there is probably too much gas in the disk for the various heating mechanisms to maintain a very high temperature; so Faber and Gallagher have hypothesized that the wind mechanism is much less efficient for removing gas from disk systems. In rich clusters, there are additional mechanisms for stripping the interstellar gas from a galaxy. Spitzer and Baade proposed direct collisions between galaxies as one possibility. The stars inside two galaxies which collide would pass by one another without any harm, but the interstellar gas would produce a big smack and be left standing in the middle (Figure 14.16). Indeed, this was the mechanism proposed by Baade and Minkowski to explain the radio galaxy Cygnus A (Figure 14.17). Astronomers now know that a direct collision between two galaxies of comparable size is a relatively rare event; so the odd optical appearance of Cygnus A more probably results from the recent gobbling of a gas-rich ordinary galaxy by a supergiant elliptical. The additional gas acquired in this way might help to feed the central machine of this active galaxy, but that is another story (Chapter 13). A more efficient method of stripping gas from the galaxies of rich clusters is probably the "ram pressure" created by the relative motion of the galaxies through the cluster medium (Figure 14.18). The mechanism is analogous to the wind knocking the hat off a rapidly pedaling cyclist. A simple calculation, carried out first by Gott and Gunn, shows that a strong relative wind of cluster gas would suffice to blow the interstellar gas out of even spiral galaxies (Problem 14.4). Ram-pressure stripping may therefore explain in part why spirals are much rarer than SOs and ellipticals in rich clusters. A similar explanation may underlie the swept-back look of head-tail radio sources (Chapter 13).

Problem 14.4. Imagine a wind of mass density ρ_0 and velocity component V which blows perpendicularly on a disk of interstellar gas with area A. Show that the momentum transferred per unit time to the disk must be, in order of magnitude, given by $\rho_0 V^2 A$. Suppose the mean surface density of the interstellar gas is μ, and suppose that this gas is attracted by the gravity of a disk of stars of mean surface density μ_*. The gravitational field g associated with the latter is $2\pi G\mu_*$ (consult Problem 12.3). Show, therefore, that the gravitational force of attraction on the entire disk of interstellar gas is given by $\mu A 2\pi G\mu_*$. Argue that stripping results if the momentum transferred per unit time exceeds the gravitational attraction of the stellar disk for the gaseous disk, and show that this requires $\rho_0 > 2\pi G\mu_*\mu/V^2$. Calculate the miminum required ρ_0 if μ_* and μ correspond, respectively, to $10^{11}M_\odot$ and $10^9 M_\odot$ spread out over a disk of radius 30,000 lt-yr, and if V equals the typical random velocity of a galaxy in a rich cluster, 1,000 km/sec. How much cluster gas would be present if gas of this density were to occupy a sphere of 3 million lt-yr radius? How does this mass compare with that contained typically in the optical parts of 1,000 galaxies of $10^{11}M_\odot$ each?

Indirect support for this ram-pressure hypothesis comes from recent observations at optical and X-ray wavelengths. Butcher and Oemler find that rich clusters which have redshifts greater than about 0.5 look substantially bluer than rich clusters with smaller redshifts. In the cosmological interpretation, objects with large redshifts are very far away, and when we look at faraway objects, we are peering into the past. Therefore, the Butcher-Oemler effect suggests that clusters in their youth contain more gas-rich galaxies than they do now. (Remember that gas-rich galaxies have active star formation, and therefore tend to look bluer; see Chapter 13.) X-ray pictures of rich clusters taken with the imaging telescope aboard the Einstein Observatory find that they present two distinct morphological classes (Figure 14.19). The first type has a smooth distribution of cluster gas, very similar to what simple theoretical models predict. The second type shows a clumpy appearance, with the gas concentrated in lumps around individual galaxies or groups of galaxies. To Jones and her colleagues, this suggests that the latter type represents an earlier phase in the evolution of cluster gas, in which the gas expelled from the stars of a galaxy has not yet had time to spread into the spaces between the galaxies.

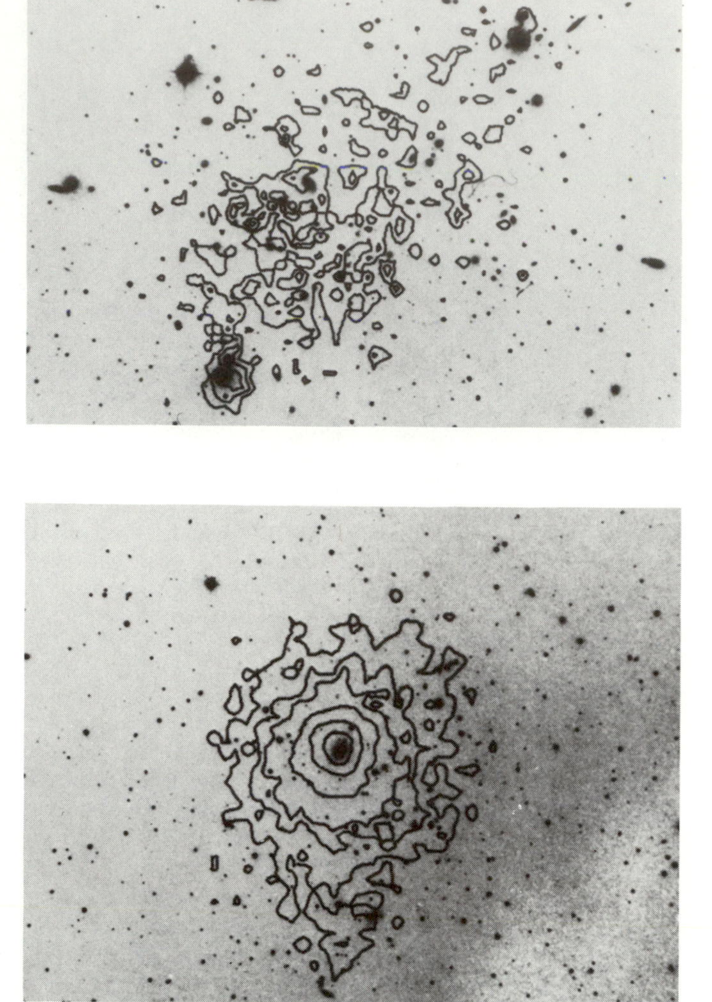

Figure 14.19. Distribution of hot gas in the Abell clusters 1367, (*top*) and 85, (*bottom*) which show, respectively, a clumpy appearance and a smooth appearance. (Courtesy of the High-Energy Astrophysics Division of the Harvard-Smithonian Center for Astrophysics.)

Missing Mass in Rich Clusters

Shane and Wirtanen and de Vaucouleurs discovered that the number of galaxies per unit area of the Coma cluster varied with distance r from the center, according to the law

$$\mathcal{N}(r) = \mathcal{N}(0)\exp[-(r/r_0)^{1/4}], \qquad (14.2)$$

where the central value $\mathcal{N}(0)$ and the scale factor r_0 are constants found by fitting the observational number counts. The law (14.2) bears a striking resemblance to equation (13.1), which gives the luminosity profile of a typical elliptical galaxy, or of the central bulge of a spiral galaxy, or, for that matter, of a globular cluster. Since the early work on the Coma cluster, many people, notably Fritz Zwicky and Neta Bahcall, have noted that the distribution of galaxies in a rich cluster resembles what one might expect theoretically for a well-relaxed system of gravitating masses. This comment suggests that the virial theorem (see Problem 9.10) could be applied to estimate the total mass of a rich cluster from measurements of the velocity dispersion of the member galaxies and the core radius of the volume of their spatial distribution.

Problem 14.5. Rework Problem 9.10 and derive the expression for the total mass M of a spherical cluster of galaxies,

$$M = 2Rv^2/G,$$

where we have assumed that the total number N of galaxies is much greater than one, R is the core radius, and v^2 is the square of the mean random velocity in all three dimensions. On the average, v^2 is three times larger than V^2, where V is the mean line-of-sight velocity displacement of a galaxy with respect to the center. Comment on the similarities of the expression here and equation (13.3). Calculate M for $V = 1{,}000$ km/sec and $R = 3$ million light-years. Express your answer in solar masses. How does M compare with the masses discussed in Problem 14.4?

When such programs were carried out, astronomers beginning with Zwicky and Smith found that the total mass needed gravitationally to bind the cluster typically exceeded that present in the form of optically luminous matter by a factor of ten or so. Today, we know that rich clusters commonly have about as much mass in hot gas as in the optically part of visible galaxies; so the net discrepancy is only a factor of five or so. Nevertheless, there is still a significant problem of missing mass in rich clusters. The most popular candidate for this missing mass seems to be low-mass stars, which may be either tied to the halos of the optical galaxies or loosely dispersed throughout the volume of the entire cluster. Putting the mass in this form may alleviate another problem. In most stellar systems that exist today, the amount of interstellar gas present is typically a small fraction of that contained inside stars (Chapter 12 and 13). But the amount of cluster gas is comparable to the combined masses of all the optical galaxies. This presents a riddle which may be partially solved by postulating that there exist in clusters ten times more stars than are immediately apparent in current optical photographs. Moreover, in a cluster with a large total binding mass, gas that escapes from an individual galaxy would still be retained by the cluster as a whole.

Problem 14.6. Given what you now know about galaxy clusters and galaxies, argue that the escape speed from a rich cluster is substantially greater than the escape speed even from a giant galaxy. Also calculate the thermal speed $(kT/m)^{1/2}$ associated with a hydrogen plasma at a temperature of 10^8 K. Compare this with the typical velocity dispersion $V = 1{,}000$ km/sec of a galaxy, and comment on the expected spatial extent of hot gas in a rich cluster compared with that of the member galaxies. For extra credit, do you think the latter numbers are a coincidence, or can you invent an explanation for the observed facts?

Unsolved Problems Concerning Small Groups and Clusters

Although most astronomers are prepared to take a conservative stand on the issue of missing mass in clusters of galaxy—namely, that it isn't really missing; it's just not in a form which is readily observable—there exists a vocal minority who warn that our understanding of galaxies may be less secure than we would like to believe. This minority has as its strongest spokesmen, V. A. Ambartsumian and H. C. Arp. Their general view is that groups and clusters of galaxies may not, after all, be gravitationally bound, and that some of them may be relatively young systems which have only recently been created and in which the individual galaxies are being expelled from the group or cluster. If this view were correct, the application of the virial theorem to estimate the masses of clusters would be meaningless, and the issue of "missing mass" for these objects would disappear.

The counterarguments to the above proposal are essentially three in number. First, the smooth appearance of the spatial distribution of member galaxies and hot gas in many rich clusters suggests that these systems are in quasimechanical equilibrium. The two-body encounter time between individual galaxies is marginally sufficient to explain the relaxed look of a rich cluster (Problem 14.7). But an even worse situation holds for the distribution of stars in elliptical galaxies, the equilibrium of which is not in doubt. Perhaps, as has been proposed by D. Lynden-Bell, there were large fluctuations of the mass distribution in the early stages of the formation of these objects. A process of *violent relaxation* would then result whereby the rapidly changing gravitational fields randomly scatter the individual members into a thermodynamic distribution. Second, if clusters were not gravitationally bound, the member galaxies should disperse in the time it takes them to cross the cluster diameter, about 10^9 yr. Since the universe is on the order of 10^{10} yr old, it would then be difficult to understand why most galaxies are still to be found in groups or clusters. Third, there is the philosophical view that any truly radical proposal which would overthrow a substantial body of established knowledge should be backed

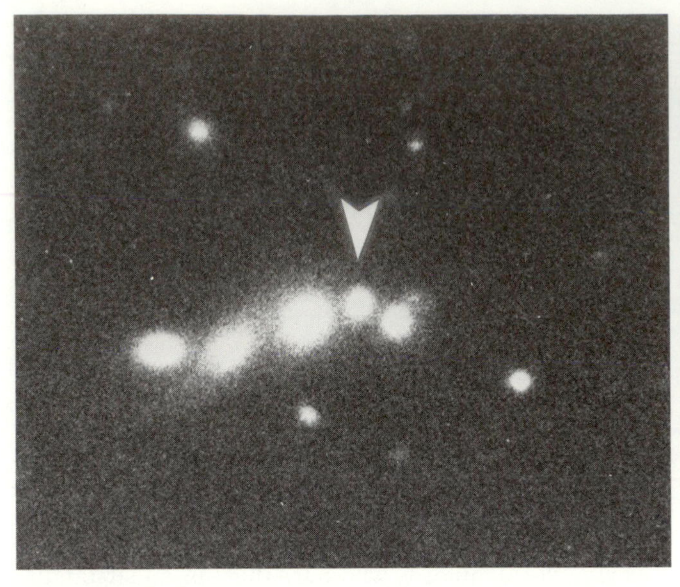

Figure 14.20. An image-processed view of a photograph taken by H. C. Arp of the chain of galaxies VV172. The galaxy marked with an arrowhead has a substantially higher redshift, $z = 0.12$, than the other four galaxies in the chain, $z = 0.05$. (Courtesy of J. W. Sulentic, Jet Propulsion Laboratory.)

Figure 14.21. Stephan's quintet. The galaxy marked with an arrowhead has a substantially lower redshift than the other four systems in this group. Recent HI observations suggest that the discrepant galaxy is a foreground object, and is dynamically unrelated to the other systems. (Palomar Observatory, California Institute of Technology.)

by really cogent arguments. Most astronomers are not convinced that the evidence yet warrants a full-scale scientific revolution.

Problem 14.7. Adapt the formula

$$t_{\text{relax}} = \left(\frac{2R}{v}\right)\frac{N}{24\ln(N/2)},$$

derived in Problem 9.11 to calculate t_{relax} for a rich cluster. What is the numerical value of the crossing time $2R/v$?

Nevertheless, it must be admitted that very strange configurations have occasionally been found for small groups of galaxies. Figure 14.20 shows a chain of galaxies discovered by Vorontsov-Velyaminov. If such a chain represents a real linear alignment, and not just a chance projection effect (say, five galaxies in a single plane viewed edge-on), it would be very difficult to understand the dynamical stability of the group. To make life worse, Arp has found that the fourth member in the chain has a redshift totally different from that of the other four galaxies. A similar situation holds in another famous small group of galaxies, Stephan's quintet (Figure 14.21). Again, four galaxies have the same redshift, but one has a very different redshift. Here, however, Ron Allen and his colleagues have studied the 21-cm emission from Stephan's quintet. Their study shows convincingly that the discrepant galaxy is a foreground object, unrelated dynamically to the rest of the group. Thus, Stephan's quintet is probably a quartet.

The Expansion of the Universe

At various points in this chapter and in Chapter 13, we have referred to the cosmological interpretation for the redshift of the galaxies. We would now like to explain this concept more fully. In particular, we would like here to establish the empirical basis for the claim that the larger the redshift of a galaxy is, the more distant it is likely to be. To accomplish this aim, we must start with a discussion of how astronomers obtain the distances to extragalactic objects.

The Extragalactic Distance Scale

In Chapters 9 and 12, we described the bootstrap operation by which astronomers could measure the distances to objects within our own Galaxy; and we mentioned in Chapter 13 that Cepheids and novae could be used to derive the distances of the nearby galaxies like M31. The use of Cepheids allows distance measurements to about 10 million light-years, beyond which it becomes too difficult to pick out individual Cepheids in a galaxy. To reach greater distances, astronomers must calibrate other "standard candles" or "standard rulers" (compare Boxes 9.5 and 14.2), whose intrinsic brightnesses or intrinsic sizes can be compared with the apparent brightnesses and angular sizes of similar but more distant objects (Figure 14.22). The trick, of course, is to come up with a criterion which allows you to recognize that you have the same kind of object that you have already encountered at closer inspection. The story of the establishment of the extragalactic distance scale is truly one of the heroic tales of twentieth-century science (see Problem 14.11). Intertwined in this tale there are many subplots, but here we shall follow only a few lines.

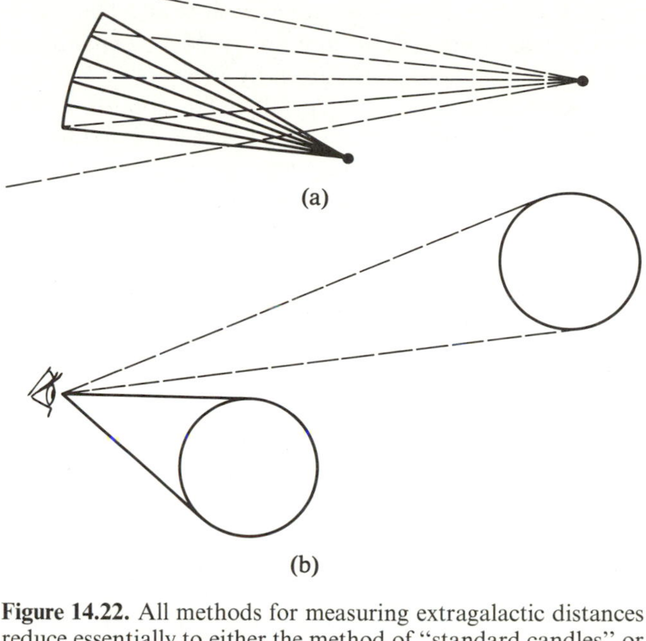

Figure 14.22. All methods for measuring extragalactic distances reduce essentially to either the method of "standard candles" or the method of "standard rulers." (a) *The method of standard candles.* An object of known luminosity and known distance will have a certain fraction of its light intercepted by a telescope. An object of the *same* luminosity at a greater but unknown distance will have a smaller fraction of its light intercepted by the telescope. The ratio of the distances is the inverse square root of the ratio of apparent brightnesses. (b) *The method of standard rulers.* An object of known linear size and known distance will subtend a certain angle. An object of the *same* linear size at a greater but unknown distance will subtend a smaller angle. The ratio of the distances is the inverse of the ratio of subtended angles.

Problem 14.8. To reach the observable edge of the universe, one must try to use the intrinsically brightest objects, which can be seen even if they are at enormous distances. Quasars and QSOs are such objects. Unfortunately, quasars and QSOs exhibit a large dispersion in luminosities at all wavelengths. Moreover, apart from irregular variability on timescales of hours to years, quasars and QSOs may evolve within billions of years. Explain why these properties make them difficult to use as standard candles. Recently, however, Baldwin has found that relatively distant QSOs have average luminosities which correlate with the width of an emission line of triply ionized carbon. Discuss how this might be put to use to measure the distances to quasars. Think also about what this distance might mean. Is it the current distance to the quasar, or is it the distance at the time that light left the quasar? (Remember, the universe is expanding.) Discuss, therefore, the need for a more careful analysis of spacetime relationships when one deals with objects at distances of billions of light-years.

BOX 14.2
Extragalactic Distance Indicators

Local distance indicators:
- Classical Cepheids (standard candle)
- Novae (standard candle)
- RR Lyrae variables (standard candle)
- W Virginis stars (standard candle)

Intermediate distance indicators:
- Brightest nonvariable stars of a galaxy (standard candle)
- Brightness of globular clusters (standard candle)
- Diameters of giant H II complexes (standard ruler)

Global distance indicators:
- Fischer-Tully relation (standard candle)
- Brightness of Sc I galaxies (standard candle)
- Supernovae (standard candle and indirect ruler)
- Three brightest galaxies of a cluster (standard candle)
- Diameters of bright galaxies (standard ruler)
- Baldwin relation for QSOs (standard candle)

Problem 14.9. A promising method (being developed by Kirshner, Wagoner, and others) involves the use of supernovae as distance indicators. A conventional technique is the comparison of the peak brightness of a supernova in a distant galaxy with that in nearby galaxies, including our own. A more subtle technique requires no comparison to other explosions. The basic ideas are as follows. The luminosity of an expanding, cooling photosphere is still given by $L = 4\pi R^2 \sigma T_e{}^4$. We can measure T_e spectroscopically. To obtain R, we *assume* that the time rate of change of R is related to the velocity v of the spectral lines produced by the expanding gases. Thus, timing the development of the explosion plus Doppler-shift measurements yield R (indirect ruler), and from R and T_e we may deduce L. Comparison of L and f yields the distance r. Discuss complications that might hinder practical application of these ideas.

To provide checks on the results, astronomers like to have several ways to measure the distances to the same objects. For this book, we concentrate on only a few important examples. Within reach of the Cepheid technique, there are many fairly bright spiral galaxies. In particular, the Local Group contains, besides the Milky Way, M31 and M33. The M81 group contains a few comparably large galaxies (including, of course, M81). Such large spiral galaxies contain fairly large HII complexes inside their spiral arms, and Sersic discovered that the linear sizes of the largest HII complexes tend to be the same for a spiral galaxy of given Hubble type and van den Bergh luminosity class (consult Chapter 13). Unfortunately, the HII complexes of Sc I galaxies are the biggest and the brightest, and there are no Sc I galaxies within 10 million lt-yr of us. The next step is therefore to use the measured angular sizes of large HII complexes in the smaller galaxies of the M101 group, together with the presumed linear sizes appropriate for galaxies of the corresponding Hubble type and luminosity class, to obtain the distance to the M101 group. This distance turns out to be about 20 million light-years. Knowing this distance allows astronomers to calibrate the true linear sizes of the largest HII regions in M101 and M51, which are the two nearest Sc I galaxies (refer to Color Plates 24 and 13). Similar calibrations of the sizes of other relatively nearby Sc I galaxies show the happy result that their largest HII complexes tend to have the same linear sizes. The giant HII complexes of Sc I galaxies can be recognized and measured for distances out to about 200 million light-years; so the distances and, therefore, the luminosities of Sc I galaxies can be obtained out to that distance. It turns out that Sc I galaxies tend to have the same intrinsic brightnesses (luminosities). Furthermore, an Sc I galaxy can be recognized to great distances purely morphologically, as a very pretty spiral galaxy with relatively open spiral arms and a small central bulge. Knowing the intrinsic brightness of an Sc I galaxy and measuring its apparent brightness allows us to get its distance from the $1/r^2$ law of radiation. This method yields the distances to clusters of galaxies, as long as they contain at least one Sc I galaxy, out to about 600 million light-years.

To get even further out, it would be useful to calibrate the intrinsic brightnesses of galaxies even more luminous than Sc I galaxies. Rich clusters, however, contain more ellipticals than spirals, and the very brightest ellipticals are the supergiant cD systems. As we discussed earlier, there is much evidence that these supergiant galaxies have gotten so bloated only by cannibalizing their neighbors. Thus, although cD systems are intrinsically very bright, they probably vary greatly in their properties. It is beyond our current capabilities to correct for their evolutionary histories in order to use them as "standard candles." Similar objections can be raised, of course, to their being used as "standard rulers." We should also keep in mind, Beatrice Tinsley warned, that as we look to more and more distant objects, we are examining them as they appeared in the more and more remote past. Thus, the light from the furthest recognizable Sc I galaxies began its journey to us 600 million years ago. If we look at the faintest cD galaxies, we are probably looking billions of years into the past. During such times, much evolution of the stellar populations may have taken place, and it becomes increasingly invalid to compare them with more local examples. This represents a fundamental and unresolved obstacle for observational cosmology.

Hubble's Law

The astronomer Slipher was the first to notice that galaxies systematically tend to have redshifts rather than blueshifts, that is, velocities of recession rather than velocities of approach (Chapter 13). Hubble quantified this finding and was the first to enunciate the correct law of the expansion of the universe. Hubble found that the further away a galaxy is away from us (as measured by one of the techniques just discussed), the faster on the average does it tend to recede from us (as measured by the redshift interpreted as a Doppler effect). The increase of the velocity of recession according to Hubble is directly proportional to the galaxy's distance. In formula form, if v is the velocity of recession and r is the distance, Hubble's law states

$$v = H_0 r, \tag{14.3}$$

where H_0, the constant of proportionality, is now called Hubble's constant.

Numerical Value of Hubble's Constant

To obtain the numerical value of Hubble's constant requires, of course, the measurement of the distances and velocities of many galaxies. The velocity (or more

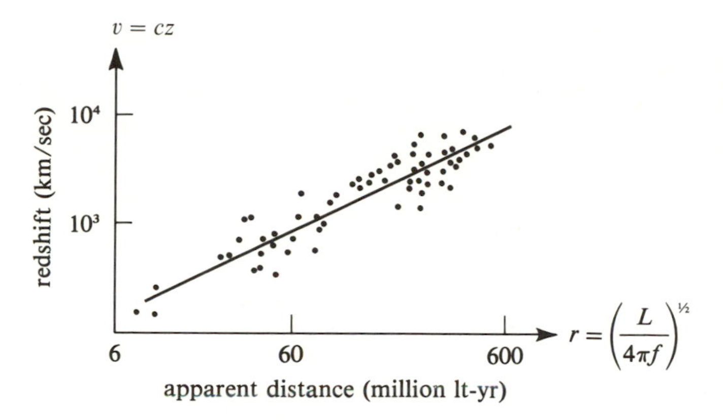

Figure 14.23. The Hubble diagram for Sc I galaxies. The data points are well-fit by a straight line that demonstrates the validity of Hubble's law: $v \propto r$. The numerical value for Hubble's constant H_0 can be obtained from this diagram as outlined in the text.

accurately, the redshift) measurements involve little controversy. The distance measurements are much more problematical, since they depend on a series of stepwise calibrations. The most thorough modern measurement of Hubble's constant is derived from the study of Sc I galaxies by Sandage and Tammann (Figure 14.23). A straight-line fit through the data points in Figure 14.23 shows that v is proportional to a power of r. The power is unity, i.e., $v \propto r$, because v increases by a factor of 10 when r does. To obtain the proportionality constant, H_0, we notice that the straight line must pass through the origin ($r = 0$, $v = 0$), and it does pass through the point ($r = 600$ million lt-yr, $v = 9{,}000$ km/sec). Thus, we have the result: $H_0 = 9{,}000$ km sec^{-1}/600 million lt-yr = 15 km sec^{-1}/million lt-yr. Some astronomers, however, dispute the calibration of the luminosity of Sc I galaxies by Sandage and Tammann. Changing the adopted value of L would not affect the proportionality $v \propto r$, but it would affect the proportionality constant, H_0. The exact value of H_0, therefore, is a matter of spirited dispute. Abell agrees with the value 15 km sec^{-1}/million lt-yr, but van den Bergh and de Vaucouleurs believe that H_0 may be 50 to 100 percent higher, because the Sc I galaxies are closer on average than calibrated by Sandage and Tammann.

Recently, Aaronson and his coworkers have entered the fray. They use the Fischer-Tully relation to estimate the intrinsic brightness L of a spiral galaxy at infrared wavelengths (Chapter 13). Radio observations at 21 cm are made to obtain both the velocity width V of the hydrogen-line profile and its redshift displacement v (due to recession of the galaxy as a whole). With a local calibration of the Fischer-Tully relation, $L \propto V^4$, to fix the proportionality constant, the measured value of V allows a deduced L. Combined with an apparent brightness measurement f at infrared wavelengths, the deduced L permits an inference of the distance r from $r = (L/4\pi f)^{1/2}$. A *linear* plot of v versus r for all spiral galaxies should then yield a unique straight line, whose slope gives Hubble's constant H_0. When this procedure was carried out in practice, Aaronson and coworkers found an anomaly. The measured slopes turned out to be different in different directions of the sky. This discrepancy could be interpreted in terms of our Galaxy falling toward the center of the Virgo cluster at a speed of about 500 km/sec. That is, the slope of the v-r relation is too small for the Virgo cluster by 500 km sec^{-1}/50 million lt-yr = 10 km sec^{-1}/million lt-yr. If this conclusion, which is supported by earlier experiments concerning the anisotropy of the microwave background (Chapter 16), is adopted here, and if one corrects for this local perturbation to the Hubble flow, Aaronson and coworkers find a Hubble constant from other clusters equal to $H_0 = 27$ km sec^{-1}/million lt-yr. This value accords roughly with the results of de Vaucouleurs.

Thus, although all workers agree on the qualitative truth of Hubble's law at large distances r, there are disputes about local departures and about the exact numerical value for H_0. The issues are complex, and the controversy will probably continue to rage for a number of years. Here we will just adopt a nice round number for H_0, namely, the geometric mean of 15 and 27—$(15 \cdot 27)^{1/2} = 20$—so that $H_0 = 20$ km sec^{-1}/million lt-yr, but be warned that this number therefore has some uncertainty.

Problem 14.10. Notice that once you have adopted a numerical value for H_0, you can use Hubble's relation to obtain the distance r of an extragalactic object for which you know only its redshift. Suppose a galaxy has a redshift $z = 0.02 = v/c$. What is its distance if we adopt $H_0 = 20$ km sec^{-1}/million lt-yr? Is there any danger in using this relation for very small redshifts? *Hint*: consider the likely magnitude of the random velocities of galaxies.

Naive Interpretation of the Physical Significance of Hubble's Law

The simplest and most naive interpretation of Hubble's law is the following. Imagine that an explosion took place starting at a time t ago, with us at the center. After some time, when we look at the galaxies, we shall find them all racing away from the origin of the explosion (Figure 14.24). The ones which were originally moving the fastest will have gotten farthest from us. In particular,

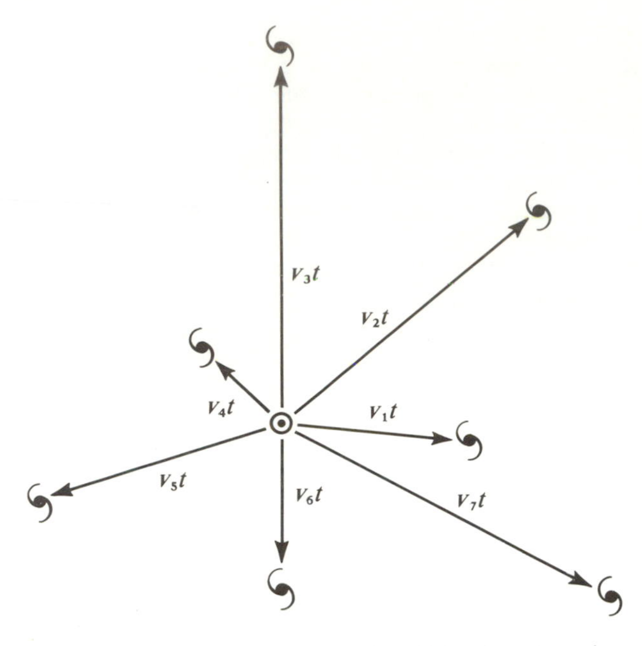

Figure 14.24. Naive interpretation of Hubble's law. At a time t ago a big explosion took place which sent the galaxies racing away from us at different velocities v. The present distance r of a galaxy, if it coasted at velocity v, is $r = vt$; so the present relation between recessional velocity v and distance r is given by the linear equation $v = r/t$.

a galaxy moving freely at a velocity v will have traveled a distance

$$r = vt.$$

Upon division by t, the above relation can be compared directly with Hubble's law, equation (14.3). Indeed, the transposed relation

$$v = r/t$$

is Hubble's law, providing we are willing to identify $1/t$ with H_0, where t is the present (fixed) time. This simple identification states that the time elapsed since the big explosion is given by

$$t = 1/H_0 = {H_0}^{-1}, \tag{14.4}$$

where the inverse of Hubble's constant, ${H_0}^{-1}$, is now known as the Hubble time.

Although this simple derivation is wrong, because it ignores the role of gravitation both in slowing down the general expansion and in determining the geometry of space and time (Chapter 15), formula (14.4) is nevertheless accurate to within an order of magnitude for estimating the age of the universe, that is, the time elapsed since the Big Bang. Indeed, if gravity were negligible, the inverse of the Hubble constant would give *exactly* the age of the universe.

The numerical value of the Hubble time, if we adopt $H_0 = 20$ km sec^{-1}/million lt-yr is easily calculated. One million lt-yr $= 3 \times 10^5$ km sec^{-1} $\times 10^6$ yr; therefore,

$$\begin{aligned} {H_0}^{-1} &= 3 \times 10^{11} \text{ km sec}^{-1} \text{ yr}/20 \text{ km sec}^{-1} \\ &= 1.5 \times 10^{10} \text{ yr}. \end{aligned}$$

In other words, the universe is roughly 15 billion years old. What a fabulously intriguing number this is! Surely, one of the supreme achievements of humanity so far is that we have deduced the age of the universe to within an uncertainty of probably less than 50 percent. In the next two chapters, we will savor the deeper meaning of this remarkable revelation.

Problem 14.11. Write a ten-page essay on the history of the discovery of the expansion of the universe. Consult, especially, E. Hubble, *The Realm of the Nebulae* (Yale University Press, 1936). You may also find useful the articles written by Sandage, "The Redshift," and by van den Bergh, "The Extragalactic Distance Scale," in *Galaxies and the Universe*, ed. by Sandage, Sandage, and Kristian (Univ. of Chicago Press, 1975), although this book is written for professional astronomers.

Philosophical Comments

This chapter has taught us that at the scale of galaxy clusters, gravity still dominates the scene. The maelstrom which is a rich cluster undergoes tremendous stirrings because of the relentless tendency for the self-gravity of any system to pull more and more matter toward the center. Yet at the distance scale of the rich clusters, we find a remarkable phenomenon. These outposts of extragalactic astronomy are expanding away from us (and from each other). Does this mean that at cosmological scales, finally, gravity plays no role, so that we have expansion and not contraction? No! As we shall see in the next chapter, the very essence of cosmology is gravitation. Indeed, gravitation is the very fabric of the structure of space and time in the universe. Modern gravitation theory suggests that the universe itself would not have been possible had there not been an expansion and, therefore, a beginning to time. This is the real significance of Hubble's constant: that the universe, although old, is not infinitely old. There was a beginning to time, and this beginning took place on the order of 15 billion years ago. As a consequence, it has become possible to discuss the creation event on a scientific basis. What a heady and exciting development!

CHAPTER FIFTEEN

Gravitation and Cosmology

At the end of Chapter 14, we gave a naive interpretation of Hubble's law, namely, an explosion plus free expansion starting at a time $t = 1/H_0$ ago, with us at the center. There are two unsatisfactory aspects of this interpretation: the need for us to be at the center; and the neglect of the effects of gravitation. In this chapter, we will rectify these two shortcomings, in two steps. We start our study of cosmology on the basis of Newton's perception of mechanics and gravitation. That is, we attempt to explain the twentieth-century astronomical observations by the principles of classical mechanics as understood before the twentieth century. Finding the Newtonian approach unsatisfactory, we shall discuss cosmology on the basis of Einstein's theory of general relativity. ("Newtonian cosmology" was invented in hindsight *after* the general-relativistic formulation had already been worked out. Newton himself had nothing to do with the twentieth-century problem addressed here.)

Newtonian Cosmology

The Cosmological Principle

An explanation of the expansion of the universe which requires us to be at the center of the expansion is obviously naive, since nothing else astronomically about our position in the universe appears that special. Thus, cosmologists have generally adopted the philosophical point of view that at *any instant of time*, the universe on a large-scale average must look **homogeneous** (the same at every location) and **isotropic** (the same in every direction) to an observer on a typical galaxy. This *assumption* that the universe looks the same for all observers at any given instant of time is called the **cosmological principle**. On a small scale, of course, the universe is demonstrably *not* homogeneous and isotropic. Even on a scale of a hundred million light-years, we find superclustering of galaxies (Figure 14.6). When we look, however, at the deepest galaxy maps, where the average distances to the displayed galaxies amount to billions of light-years, the distribution of galaxies does appear to be homogeneous and isotropic (Figure 15.1). In Chapter 16, we will encounter in the microwave background radiation an even more cogent empirical justification of the cosmological principle.

It follows from the cosmological principle that if the universe expands, then everyone at any one instant in time must see the same rate of expansion. To visualize how everyone can see everyone else receding from oneself, imagine the situation of the raisins in a rising raisin cake (Figure 15.2). As the cake swells, each raisin sees every other raisin recede from itself. However, the analogy is incomplete, because the raisins at the surface of the raisin cake obviously do not see an isotropic situation: there are no raisins on the free side of the surface. Thus, in a Newtonian cosmology, the universe has to be infinite in spatial extent if we are to satisfy the cosmological principle.

Figure 15.1. Deep galaxy map of a square section of the sky, 6° on a side, made from data supplied by Rudnicki and colleagues at the Jagellonian University in Cracow. The average galaxy among the more than 10,000 counted may be located between two and three billion light-years away in this map. Notice that the distribution on this scale is almost random, in contrast with Figure 14.6. (For details, see Groth, Peebles, Seldner, and Soneira, *Scientific American*, **237**, May 1977, 76.)

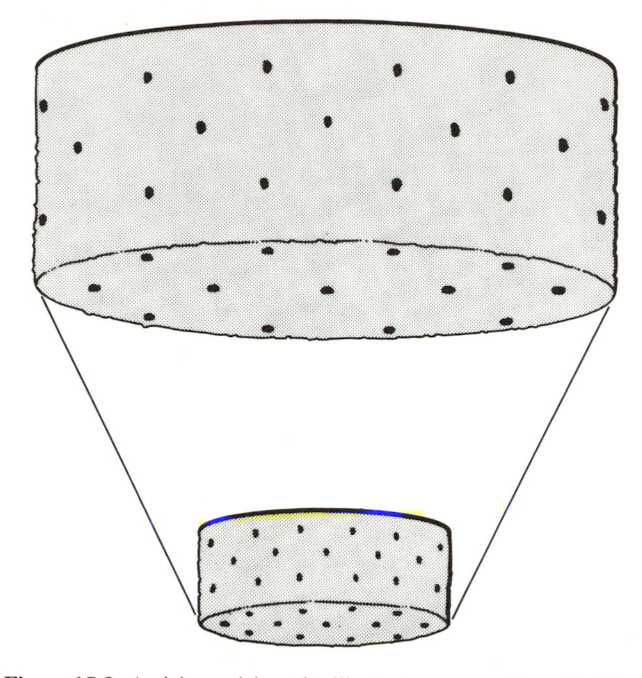

Figure 15.2. A rising raisin cake illustrates a recession of raisins from each other which occurs homogeneously and isotropically except for the raisins which are located at the surface of the cake. If the cake were infinite in extent, it would be a model of Hubble's law in Newtonian cosmology.

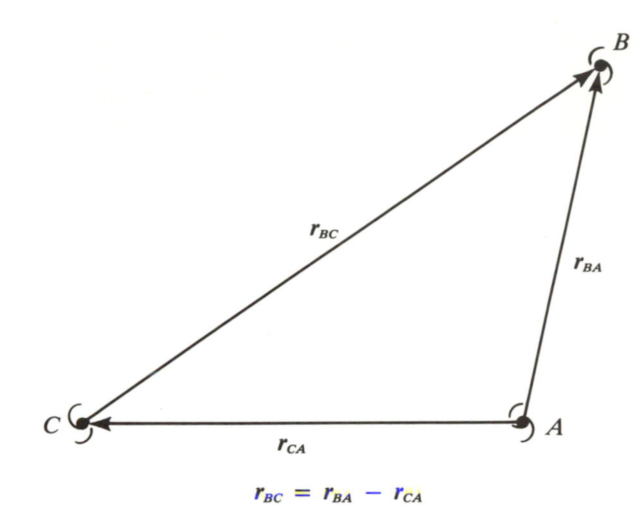

Figure 15.3. In Newtonian mechanics, two galaxies B and C which have vector displacements, $\mathbf{r}_{BA}$ and $\mathbf{r}_{CA}$, with respect to galaxy A, have separation $\mathbf{r}_{BC} = \mathbf{r}_{BA} - \mathbf{r}_{CA}$ with respect to each other. Because of this vector rule, Hubble's law is the only way in which the universe can expand and satisfy the cosmological principle.

We begin now to see the power of the cosmological principle revealed. By following its logical consequences, we have already derived a very important physical conclusion: that Newtonian cosmology requires an infinite volume of space uniformly filled with galaxies. We can derive an even more remarkable conclusion: if the universe expands, it necessarily satisfies Hubble's law. Actually, the converse is easier to demonstrate: If we assume Hubble's law, then we can show that the universe expands homogeneously and isotropically.

The proof proceeds as follows (Figure 15.3). Consider three galaxies A, B, C in an expanding universe which satisfies Hubble's law according to galaxy A. We wish to show that Hubble's law would then apply also to galaxy C (and, by extension, to every other galaxy in the universe). If Hubble's law applies for galaxy A, the velocities of galaxies B and C according to galaxy A are given by the vector formulae

$$\mathbf{v}_{BA} = H_0\mathbf{r}_{BA}, \qquad \mathbf{v}_{CA} = H_0\mathbf{r}_{CA},$$

respectively. In Newtonian mechanics, the velocity of galaxy B according to galaxy C can be obtained by vector subtraction:

$$\mathbf{v}_{BC} = \mathbf{v}_{BA} - \mathbf{v}_{CA} = H_0(\mathbf{r}_{BA} - \mathbf{r}_{CA}).$$

But $\mathbf{r}_{BA} - \mathbf{r}_{CA}$ is simply the vector distance $\mathbf{r}_{BC}$ of galaxy B from galaxy C; so the velocity of galaxy B relative to galaxy C is given by Hubble's law,

$$\mathbf{v}_{BC} = H_0\mathbf{r}_{BC}.$$

This proves that if the universe expands according to Hubble's law for galaxy A, it also does so for galaxy C (Q.E.D.). Conversely, if we assume that the universe is homogeneous and isotropic (the cosmological principle), then the universe has no choice except to follow Hubble's law! (Of course, H_0 could be negative or zero, in which case the universe would contract or remain static.)

Cosmologists find the latter point of view more satisfying, because then the adoption of a certain philosophical point of view (the cosmological principle, whose origins can be traced back to Copernicus) leads to an observationally checkable result (Hubble's law). Philosophy plays a more important role in cosmology than in any other branch of science. In other fields of study, we usually have many objects or systems which follow the same basic laws, but which start with different initial conditions. For example, the theory of stellar evolution deals with stars, which begin with different initial masses and chemical compositions. The theory may then be checked under many circumstances. But in the study of the universe, i.e., in cosmology, we have only one system, the entire universe. To explain any feature about a unique system, one could postulate that the feature (e.g., Hubble's law) arose because of some particular initial condition. But such an approach would be scientifically sterile. Any "explanation" which relies on one assumption to produce one fact is not a true explanation at all; it is merely a rephrasing of the observed fact. Cosmologists therefore prefer to adopt certain broad, sweeping philosophical points of view (e.g., the cosmological principle) which replace a whole set of arbitrary initial conditions with a more satisfying global outlook. The simpler the outlook, the more satisfying and elegant is the resulting theory. Ultimately, this outlook—epitomized by Einstein's approach to science—boils down to one of *faith*. Scientists would like to believe that the universe is simple and beautiful. Of course, this faith may prove to be misguided when checked against the hard reality of experiments and observations, but the history of science has amply justified the fruitfulness of this approach. When simple, beautiful theories have turned out to be wrong (and all physical theories are flawed at some level), what has usually been found is not that nature is ugly, but that she is cleverer than the theorists. Einstein put it best: "Nature is subtle, but she is not malicious."

The Role of Gravitation in Newtonian Cosmology

One naively and correctly expects that the gravitation of the matter in the universe should slow down the recession of the galaxies. However, exactly how it would do so is ambiguous from a pure Newtonian point of view. In an infinite universe uniformly filled with matter, what is the direction of the gravitational field $\mathbf{g}$? Shouldn't every piece of matter on the average be equally attracted in all directions? Wouldn't the gravitational field $\mathbf{g}$ then exactly equal zero everywhere? Indeed, it is this very same puzzling series of questions which prevented Newton himself from ever doing anything in cosmology. In fact, it is easy to show from Gauss's law that a pure Newtonian cosmology leads to an absurd conclusion. Construct anywhere in the universe an arbitrary volume V which contains mass M. Since $\mathbf{g} = 0$ everywhere on the surface A of volume V, Gauss's law (equation 3.1′), gives

$$4\pi GM = 0.$$

But since V is completely arbitrary, the above equation requires that $M = 0$ everywhere in the universe. In other words, the only universe satisfying the cosmological principle allowed by Newtonian cosmology is a completely empty universe! Your existence and mine offer disproof of this conclusion.

The physical and astronomical evidence suggests that the difficulty lies in Newton's theory of gravitation, not in the cosmological principle. We can rectify the situation by replacing Gauss's law with the following rule. Let us assume that the velocity $\mathbf{v}$ of any galaxy as seen by an observer O at a distance $\mathbf{r}$ away is affected only by the

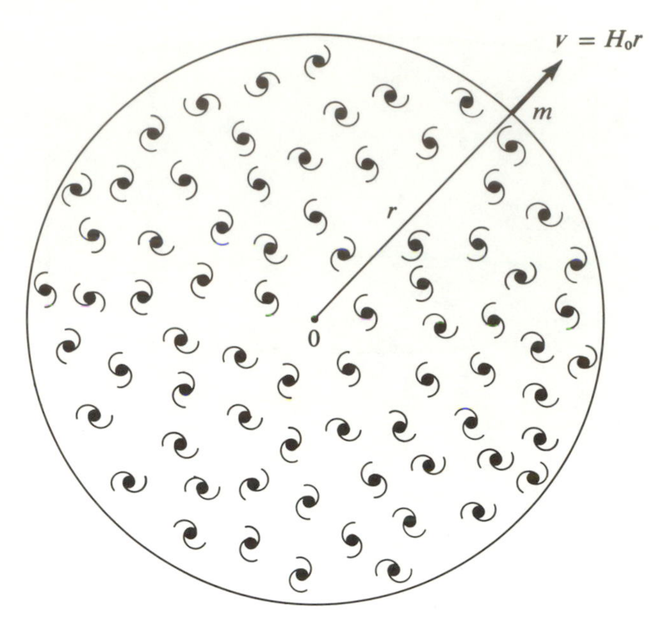

Figure 15.4. Gravitation is incorporated in Newtonian cosmology by adopting Birkhoff's rule that the motion v of a galaxy m located at distance r from an observer O is influenced only by all the matter which lies within a sphere of radius r centered on O.

gravitational pull of all the galaxies inside a sphere centered on O (Figure 15.4). This assumes the heuristically plausible result that the infinite number of galaxies outside of the sphere produces no net pull on m. Moreover, this assumption applies to *any* observer living on one of the receding galaxies. In pure Newtonian theory, our rule is completely unjustified; however, Garrett Birkhoff has shown that it does follow from general-relativity theory, providing the radius r is not too large.

In any case, if we do accept Birkhoff's rule, we can calculate the deceleration of the expansion of the universe produced by gravitation. In particular, we can answer the following interesting question. Is the gravitation strong enough to halt completely the expansion of the universe sometime in the future, reverse it (because gravity never relents), and bring the universe to a state of collapse? This issue of whether the universe is **bound** can be investigated by considering the energy of our representative galaxy m. If the average mass density in the universe is currently ρ_{m0}, then the total mass M of all the attracting galaxies inside the radius r amounts to

$$M = \text{density} \cdot \text{volume} = \rho_{m0}\,\frac{4\pi}{3}\,r^3.$$

The mass will remain the same if we follow the motion of the galaxy m as it expands outward from O, since all galaxies interior to it have smaller recession speeds (in accordance with Hubble's law) and will remain interior to the expanding sphere. If no matter is created to fill the voids left by the receding galaxies, the future matter density ρ_m will fall as the packing of matter becomes sparser. Gravity will also weaken, but the total energy of the galaxy m will be conserved in the expansion process:

$$E = \frac{1}{2}mv^2 - \frac{GMm}{r} = \text{constant}, \tag{15.1}$$

where the present velocity v is given by Hubble's law: $v = H_0 r$. According to Chapter 3, the condition $E = 0$ divides those cases where the mass is bound from the cases where m is unbound. If $E < 0$, the galaxy m will eventually fall back to the origin O; whereas if $E > 0$, the galaxy m will recede forever. Let us, therefore, examine the critical case $E = 0$. If we set $E = 0$ in equation (15.1), and substitute the relations $v = H_0 r$ and $M = \rho_{m0}4\pi r^3/3$, we can easily derive the equation

$$\text{critical value of } \rho_{m0} = \frac{3H_0^{\,2}}{8\pi G}. \tag{15.2}$$

Since m and r canceled out, there remains no reference to any specific galaxy; so equation (15.2) applies to all galaxies in the universe. In other words, if the present mass density ρ_{m0} equals the combination $3H_0^{\,2}/8\pi G$, which represents a definite number given H_0, the universe is marginally bound. Clearly, if the actual average density ρ_{m0} exceeds this critical value, the gravitational potential-energy term in equation (15.1) would be more negative, and the universe would be definitely bound ($E < 0$). On the other hand, if ρ_{m0} were less than the critical value, the universe would be unbound ($E > 0$). These conclusions are summarized in Box 15.1.

The great question arises: how *does* ρ_{m0} compare to $3H_0^{\,2}/8\pi G$ in the actual universe? If we adopt the value $H_0 = 20$ km sec^{-1}/million lt-yr, we easily calculate

$$3H_0^{\,2}/8\pi G = 8 \times 10^{-30}\ \text{gm/cm}^3,$$

which is the density equivalent to about five hydrogen atoms per cubic meter of space. The critical density required to bind the universe is a ridiculously small number. However, if we were to spread out the mass in the optical parts of galaxies, we would obtain $\rho_{m0} = 3 \times 10^{-31}$ gm/cm^3, a value 25 times smaller yet. Therefore, the observed universe is unbound and will expand forever, *unless* there is a lot of missing mass. Since we do not know how much missing mass exists, we need to consider all possibilities.

BOX 15.1

Bound Versus Unbound Universes

Clearly, if the density of matter now in the universe is larger, the effects of gravitation are correspondingly larger. Thus, it should not surprise us to discover that high mean densities ρ_{m0} of matter today correspond to bound models of our universe, and low values to unbound models:

if $\rho_{m0} > 3H_0^{\,2}/8\pi G$, universe is bound and will eventually contract;

if $\rho_{m0} < 3H_0^{\,2}/8\pi G$, universe is unbound and will expand forever.

Problem 15.1. Confirm the value of $3H_0^{\,2}/8\pi G$ given in the text. Obtain also a very rough estimate of the density contribution of the optical parts of galaxies by taking the total mass of the Local Group (mostly the Milky Way and M31) and spreading it out to the M81 group. Assume a total mass of $4 \times 10^{11} M_\odot$ and a cubic volume centered on the Local Group with sides 15 million lt-yr in length. Compare your two densities.

Deceleration of the Expansion Rate

Now, as the universe expands, the matter density ρ_m must drop below its current value ρ_{m0}. In an absolute sense, gravity will therefore weaken. Does this mean that even if the inequality $\rho_{m0} > 3H_0^{\,2}/8\pi G$ held now, so that the universe is bound, the inequality might reverse as the matter density drops in the distant future? No! The condition of boundedness derives from the principle of conservation of energy, equation (15.1). Once the issue of whether the universe is bound or not bound has been settled for any one time, it has been settled for all time. The energy constant E in equation (15.1) always has the same numerical value; once negative, it cannot later become positive. This fact can be reconciled with the current decline of ρ_m and the condition (15.2) only if Hubble's "constant" *changes in time.* In other words, Hubble's law should really read that the velocity v of a galaxy at a distance r from us at cosmic time t varies as (Figure 15.5)

$$v = H(t)r, \tag{15.3}$$

where the proportionality factor H depends on the time t. The value $H_0 = 20$ km sec^{-1}/million lt-yr represents only the value of $H(t)$ now at time $t = t_0$. That explains the reason for the subscript zero. In the future, $H(t)$ will decrease in value, and it will even become negative (contraction instead of expansion) if the universe is truly bound. In this fashion, $\rho_{m0} \gtrless 3H_0^{\,2}/8\pi G$ would imply

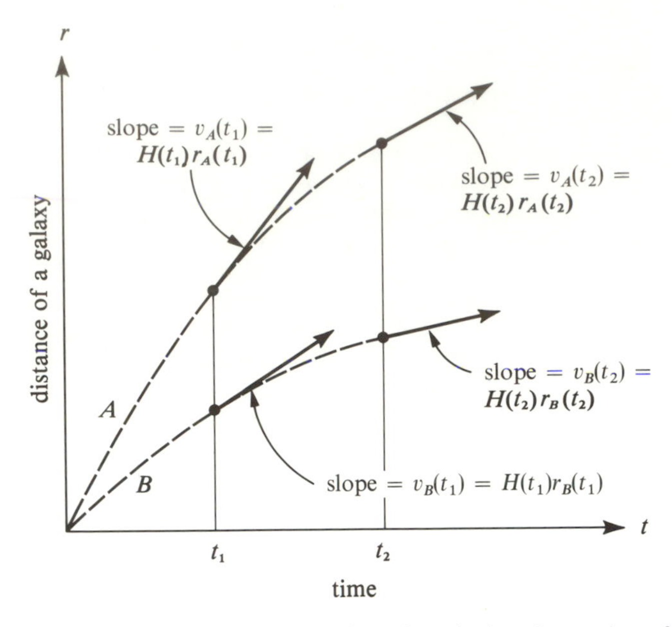

Figure 15.5. Because of gravitation, the velocity of recession of any galaxy relative to a given galactic observer will slow down in time. This is shown above for two galaxies, A and B, with galaxy A being twice as far as galaxy B at time t_1. Thus, at time t_1, the velocity of galaxy A, $v_A(t_1)$, is twice as large as the velocity of galaxy B, $v_B(t_1)$, but they both yield the same Hubble constant $H(t_1) = v_A(t_1)/r_A(t_1) = v_B(t_1)/r_B(t_1)$. At a later time, t_2, the distance and velocity of galaxy A are still twice those of galaxy B; so there is still a uniform value of the Hubble constant $H(t_2) = v_A(t_2)/r_A(t_2) = v_B(t_2)/r_B(t_2)$. However, the gravitational deceleration of both galaxies in the interim would have decreased the Hubble constant at t_2 compared to its value at t_1: $H(t_2) < H(t_1)$.

$\rho_m(t) \gtrless 3H^2(t)/8\pi G$ for all times, past and future. In other words, gravity never grows weaker or stronger *relative to the current expansion rate.*

The terminology "Hubble's constant" may seem inappropriate for a quantity $H(t)$ which varies in time, but we retain this usage because the truly important aspect about $H(t)$ is that it is ideally the *same* value for all observers throughout *space* at any given time t. Nevertheless, the actually *observed* Hubble relation for any one observer cannot remain strictly linear in v versus r for arbitrarily large distances r. Since light does not propagate infinitely fast, when we examine distant galaxies, we are looking at them as they appeared in the past. In the past, according to equation (15.3), the Hubble constant had a different value than it does now. This suggests that it may be possible to use observations to measure the deceleration of the universe and thereby to decide whether the Universe is bound or unbound. Intuitively, however, we may expect the measurable effects to enter for values of v which approach a healthy fraction of the speed of light. For such conditions, equations (15.1) and (15.3), as well as the redshift itself, require relativistic

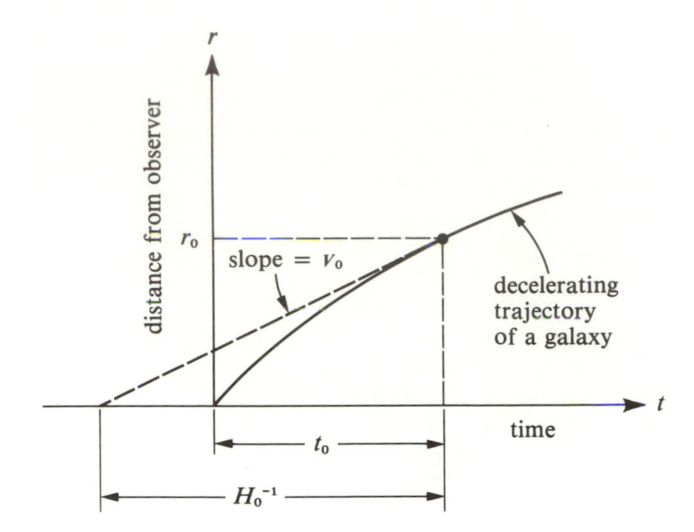

Figure 15.6. A galaxy at distance r_0 from us at the present time t_0 is observed to have a velocity of recession $v_0 = H_0 r_0$. If it is assumed that this galaxy has been coasting at constant speed (slanted dashed line), we obtain the estimate that its location coincided with ours ($r = 0$) at a time H_0^{-1} ago. This estimate, H_0^{-1}, for the age of the Universe t_0 (since the Big Bang) will be an overestimate to the extent that gravitation has caused the galaxy to have a decelerating trajectory.

reinterpretations. We will discuss the general-relativistic formulation of cosmology later; for now, let us consider a cosmological test of whether the universe is bound which does not stretch the limits of validity of equations (15.1) and (15.3).

Problem 15.2. If one extrapolated the formula $v = H_0 r$ to very large values of r, for what distances would v become comparable to c? State your answer in light-years. Are the Sc I galaxies depicted in Figure 14.23 near the limit where one has to worry about the nonlinearity of the observational Hubble relation?

The Age of the Universe

In Chapter 14, we constructed a naive free-expansion model for the universe to estimate the age of the universe as $1/H_0$. Since gravity makes the expansion rate H itself a function of time, we realize that we may have to modify our estimate. The direction of the modification needed is easy to deduce. Since gravity now acts to decelerate the expansion, the expansion must have been faster in the past than it is now, and the galaxies would have reached their current average separations faster than in a free-expansion model. Therefore, the actual age of the universe, if gravity plays any role, must be less than the free-expansion value $1/H_0$ (Figure 15.6).

An analogy might help here. Suppose you visit a friend at her house. She sees you walk in the door at 3 miles per hour. Knowing that you live three miles away, she might deduce that you took an hour to get there. She would have overestimated the elapsed time, however, if you had traveled faster in the past than when you walked through the door; you might have taken the bus part of the way, for example. So it is also with astronomers and galaxies.

To be more precise, we need to make a calculation (Problem 15.3), but even without any detailed mathematics, one point should be clear. The free-expansion model must represent more of an overestimate if the universe is bound than if it is not. Gravity plays little role in slowing down the expansion of an unbound universe; so $1/H_0$ must be a good estimate for the age of such a universe. The precise calculations show the current age t_0 of the universe to be bracketed by the inequalities (Box 15.2):

$$\begin{aligned} &\text{bound,} && 0 < t_0 < \tfrac{2}{3}{H_0}^{-1}; \\ &\text{unbound,} && \tfrac{2}{3}{H_0}^{-1} < t_0 < {H_0}^{-1}. \end{aligned} \tag{15.4}$$

BOX 15.2
The Age of the Universe

Let us characterize models of the universe by the ratio of the actual mass density ρ_{m0} to the critical value $3{H_0}^2/8\pi G$, so that $\Omega_0 = 8\pi G\rho_{m0}/3{H_0}^2$. Then, a detailed calculation (Problem 15.4) shows that the present age t_0 of the universe is given by the expression:

$$t_0 = \tau_{*0}{H_0}^{-1},$$

where τ_{*0} is a dimensionless number:

$$\begin{aligned} &2\tau_{*0} = \Omega_0(\Omega_0 - 1)^{-3/2}(\eta_0 - \sin\eta_0), \text{ with} \\ &\cos\eta_0 = (2 - \Omega_0)/\Omega_0, \text{ for } \Omega_0 > 1; \\ &2\tau_{*0} = \Omega_0(1 - \Omega_0)^{-3/2}(\sinh\eta_0 - \eta_0), \text{ with} \\ &\cosh\eta_0 = (2 - \Omega_0)/\Omega_0, \text{ for } \Omega_0 < 1. \end{aligned}$$

In the limits,

$$\begin{aligned} &\Omega_0 \to 1 \text{ (marginally bound universe)}, && \tau_{*0} = 2/3; \\ &\Omega_0 \to \infty \text{ (very bound universe)}, && \tau_{*0} = 0; \\ &\Omega_0 \to 0 \text{ (very unbound universe)}, && \tau_{*0} = 1. \end{aligned}$$

The total time span for a bound universe from Big Bang to Big Squeeze is finite and equals $\Omega_0(\Omega_0 - 1)^{-3/2}{H_0}^{-1}$. The total time available for an unbound universe is infinite, and the galaxies recede forever.

Problem 15.3 (for those who know a little calculus). Consider the energy equation (15.1) for the critical case, $E = 0$. Identify the velocity v of the galaxy m with dr/dt and follow the galaxy as it recedes. For $E = 0$, show that the equation of motion can be written as

$$\frac{dr}{dt} = +\left(\frac{2GM}{r}\right)^{1/2},$$

where the mass M interior to r is a constant throughout the expansion. Justify the choice of the plus sign when taking the square root of the energy equation. Show that the above ordinary differential equation can be integrated, subject to the big-bang condition $r = 0$ at $t = 0$, to give

$$\tfrac{2}{3}r^{3/2} = (2GM)^{1/2}t.$$

At the present time $t = t_0$, $(2GM/r)^{1/2} = dr/dt = v = H_0 r$. Combine these relations to show that the present age of the universe in this case is given by

$$t_0 = \tfrac{2}{3}{H_0}^{-1}.$$

Argue physically, now, for the inequalities (15.4).

If the Universe is marginally bound ($E = 0$), $t_0 = (2/3){H_0}^{-1} = 10$ billion years, for ${H_0}^{-1} = 15$ billion years (i.e., if $H_0 = 20$ km sec^{-1}/million lt-yr).

We have noted in Chapter 9 that the ages of globular cluster stars are probably on the order of 12 billion years. The universe is at least as old as its oldest stars; therefore, $t_0 > 12$ billion years. For our adopted Hubble constant, the age of a marginally bound universe would be 10 billion years. The actual universe, therefore, appears to be too old to be bound, and we might thereby conclude that the universe is unbound. However, there is enough uncertainty both in the age estimates of the ages of globular clusters and the exact value of H_0 to allow still the conclusion that the universe is bound. Thus, although in principle an age estimate for the universe would be a very nice test for whether the universe will expand forever or not, the age estimates now available to us are not firm enough to provide a definitive statement. In Chapter 16, we will discuss other astronomical evidence that bears on the issue of whether the universe is bound or unbound.

The Ultimate Fate of the Universe

The galaxies are currently receding from us. If the universe is unbound, this recession will proceed forever, until ultimately the galaxies are infinitely dispersed.

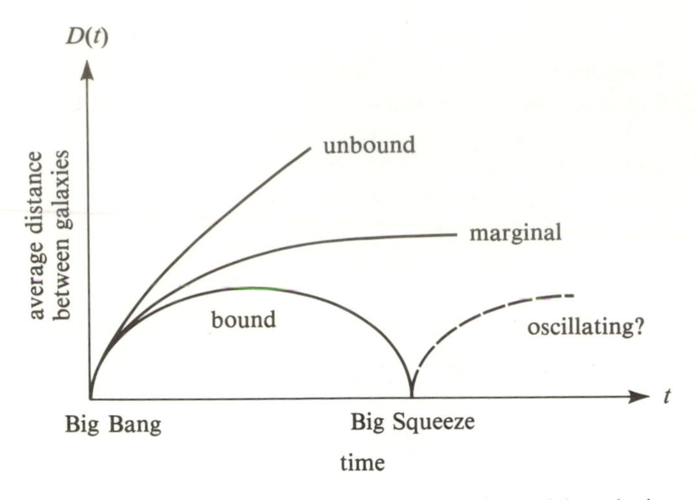

Figure 15.7. The history of the average separations of the galaxies if the universe is bound, marginally bound, or unbound. In an unbound universe, the galaxies will recede forever, with nonzero velocities even when they are infinitely separated. In a marginally bound universe, the recession velocities will exactly equal zero at infinite separations. In a bound universe, the galaxies will come together again after a time of finite duration. Some workers have speculated that after a period of superhigh compression (the Big Squeeze), a closed universe may rebound in another cycle of Big Bang/Big Squeeze, *ad infinitum*. Fred Hoyle has objected to such "oscillating models" on the grounds that mechanical systems oscillate only about points of *equilibria*, and such equilibria do not exist for conventional cosmological models. In any case, bound models are separated from unbound models by the marginal case, in which the present matter density has the critical value $\rho_{m0} = 3H_0{}^2/8\pi G$.

Eventually, all the stars in the galaxies will die and the universe will have become a pitch-dark, empty place. This is a bleak forecast, and although it may not come to pass effectively for trillions of years, no one really wants to see the universe come to nought.

But there is hope. If nature wanted to make an unbound universe, why did she tantalize us by making it within a factor of 25, and perhaps 2 or 3, of being bound? Why not be off by a factor of a million? Has she been malicious to raise false hopes? On the other hand, perhaps for some good reason, we would not have been around to ask these impudent questions unless the universe were almost but not quite bound.

If the universe is bound, the galaxies, although currently receding, will ultimately be stopped. Past that time will begin the epoch of the Big Squeeze, when all the matter of the universe will be recompressed. When superhigh densities have been reached, will the universe then rebound in another cycle of Big Bang and Big Squeeze? Current physics loses validity before it can give us the answer. Figure 15.7 merely summarizes the various possibilities which exist, given the current development of our ideas (see also Problem 15.4).

Problem 15.4 (for those who are good at calculus). To solve the general energy equation (15.1), proceed by first introducing nondimensional variables, $D_* = r/r_0$, $\tau_* = H_0 t$, where r_0 is the current distance of the galaxy. Identify the energy constant E as

$$E = m(H_0{}^2/2 - 4\pi G\rho_{m0}/3)r_0{}^2,$$

and show that equation (15.1) can now be written

$$\left(\frac{dD_*}{d\tau_*}\right)^2 - \frac{\Omega_0}{D_*} = 1 - \Omega_0, \tag{1}$$

where we have defined $\Omega_0 = 8\pi G\rho_{m0}/3H_0{}^2$ to be the ratio of the actual density to the critical density. We have already solved the case $\Omega_0 = 1$ in Problem 15.3. For $\Omega_0 \neq 1$, we rescale one more time by defining $\xi = (|1 - \Omega_0|/\Omega_0)D_*$ and $\tau = (|1 - \Omega_0|^{3/2}/\Omega_0)\tau_*$. Show that equation (1) then becomes

$$\left(\frac{d\xi}{d\tau}\right)^2 - \frac{1}{\xi} = \pm 1, \tag{2}$$

where the right-hand side is $+1$ if $\Omega_0 < 1$ (unbound universe), and is -1 if $\Omega_0 > 1$ (bound universe). Show now that the solution of equation (2) can be formally obtained as

$$\tau = \int_0^\xi \left(\frac{\xi}{1 \pm \xi}\right)^{1/2} d\xi,$$

where we have assumed $\xi = 0$ at $\tau = 0$. To integrate the right-hand side when $1 - \xi$ occurs in the denominator (bound universe, with $\xi \leq 1$), make the substitution $\xi = \sin^2(\eta/2)$, and notice the trigonometric identity $\sin^2(\eta/2) = (1 - \cos\eta)/2$. When $1 + \xi$ occurs in the denominator (unbound universe, where ξ can become infinitely large), make the substitution $\xi = \sinh^2(\eta/2)$, and notice the identity $\sinh^2(\eta/2) = (\cosh\eta - 1)/2$. In this manner, obtain the solutions in parametric form:

bound universe, $\xi = \frac{1}{2}(1 - \cos\eta)$, $\tau = \frac{1}{2}(\eta - \sin\eta)$;

unbound universe, $\xi = \frac{1}{2}(\cosh\eta - 1)$, $\tau = \frac{1}{2}(\sinh\eta - \eta)$.

Make a table of η, τ, and ξ for the two cases by choosing different values of η. Ignore the column for η, and plot ξ versus τ. Comment on why the curves look similar near the origin, $\tau = 0$ and $\xi = 0$. Show also that the maximum radius reached by a galaxy in the bound case is $\Omega_0/(\Omega_0 - 1)$ times r_0.

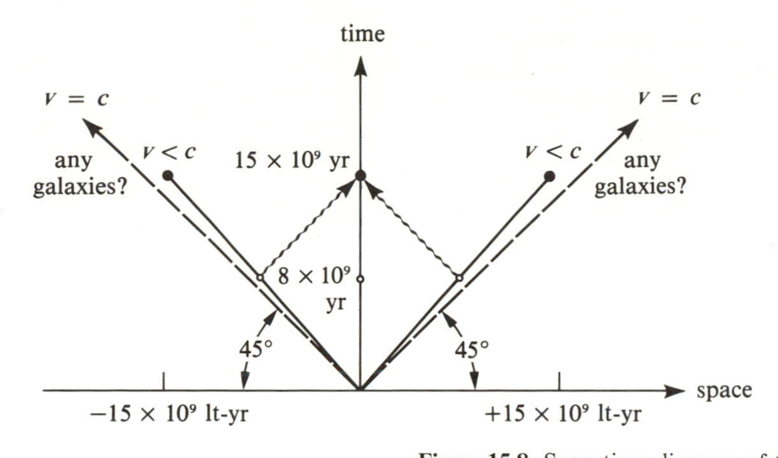

Figure 15.8. Spacetime diagram of the redshift interpreted as a Doppler effect. At $t = 0$ in the rest frame of the observer, all galaxies start at the same point in space. At $t = 15 \times 10^9$ yr, objects which move at $v = c$ relative to the observer (dashed lines) have moved 15×10^9 lt-yr away. Galaxies with v somewhat less than c (black dots) are currently, say, at a distance of 14×10^9 lt-yr, but the light from these objects reaching the observer now (wavy lines) actually left the objects themselves at a time, say, $t = 8 \times 10^9$ yr. Seven billion years in the past, these galaxies may have been quite a bit more active than they are now, just as galaxies in the observer's immediate neighborhood might have been. At high redshifts (v close to c), the observer therefore sees galaxies which are much more commonly active (white dots) than galaxies today (black dots). A difficulty with this model concerns whether there are currently any galaxies beyond a distance of 15×10^9 lt-yr.

General Relativity and Cosmology

In cosmology we are studying phenomena with the greatest possible spans in space and time; so we must be careful not to think in terms of Newton's concepts of absolute space and absolute time. But even special relativity is not really adequate to discuss cosmology; we need general relativity. One example will illustrate the point.

Consider the quasars, and suppose them to be at cosmological distances. Does the cosmological principle apply to quasars? If we adopt a special-relativistic interpretation, there are some real and apparent difficulties. An apparent difficulty is the following.

Maarten Schmidt has found that, expressed as a fraction of the comoving galaxies, quasars near us are relatively rare; 3C273 is the nearest, and it has a redshift $z = 0.158$ (Chapter 13). In contrast, far from us, quasars are quite common, common enough that Giacconi believes that they may account for the background of high-energy X-rays that were once thought to arise from a hot and diffuse intergalactic medium. At first sight, this discovery seems to violate the cosmological principle. Does Schmidt's work mean that our location is special because there are more quasars far from us than near to us? No. Imagine the following thought experiment. Imagine that you could fly *instantaneously* to the position of a distant quasar which has an apparent recessional velocity of $0.93c$. In a special-relativistic model (without gravity), such an object would have been coasting at $0.93c$ for 15 billion years (the age of the universe for free expansion), and would therefore be located at a distance of 14 billion light-years. Now, having flown instantaneously to the position of this quasar, look around you. Do you see lots of quasars near you? No! There are very few quasars compared to ordinary galaxies! Now, look back toward the Milky Way system, from where you came. You see that quasars are quite common there! In other words, you see exactly the same situation here that you saw at home: few quasars near you, lots far away. This is, of course, the meaning of the cosmological principle: at any instant in time, all galactic observers on the average must view the same phenomena.

The special-relativistic explanation of the phenomena at hand can be found in Figure 15.8. Light takes a finite amount of time to travel from the quasar to the Milky Way and vice versa. Because they are now separated by 14 billion light-years, this time is an appreciable fraction of the total elapsed time, 15 billion years, since the Big Bang, when the Galaxy and the quasar were right on top of each other, so to speak. Thus, each time we peer to such distances, we are looking back to a past epoch, when the fraction of quasars was apparently much greater than it is today. Indeed, the observed spatial distributions of QSOs and radio sources (see Chapter 16) provide strong evidence that the large-scale properties of the universe change with time.

Figure 15.8 points out a real difficulty, however, with the special-relativistic interpretation of redshifts. At this instant in time in the Milky Way frame of reference (in which Figure 15.8 is drawn), are there any galaxies beyond a distance of 15 billion light-years from the Milky Way? If there are *not*, is there a material edge to the universe? (This material edge would be different for different observers.) If there *are* galaxies beyond a distance of 15 billion light-years from the Milky Way system, and if all objects started nearly together in the Big Bang, how did they get so far away unless they can travel faster than the speed of light?

Modern cosmologists believe that there are indeed quasars and galaxies beyond 15 billion light-years from the Milky Way, although they may not be observable by us. Moreover, they got there without having had to move faster than the speed of light relative to any *local* observer, indeed, in a certain sense, without having had to move at all! All this is made possible by the concept of spacetime curvature, a concept we owe to Einstein's theory of general relativity.

The Foundations of General Relativity

General relativity has as its cornerstones the two physical principles displayed in Box 15.3. Special relativity (see Chapter 3) governs the properties of spacetime when the gravitational influence of matter is negligible. In particular, valid equations of physics must be formulable locally in a way that treats all observers equally. There can be no action at a distance; the fastest that influences can be transmitted is the speed of light in a vacuum.

BOX 15.3

Basic Postulates of General Relativity

(1) The relativity principle: Special relativity governs local physics. The global structure of spacetime may, however, be warped by gravitation.

(2) The principle of equivalence: There is no way to tell locally the difference between gravity and acceleration.

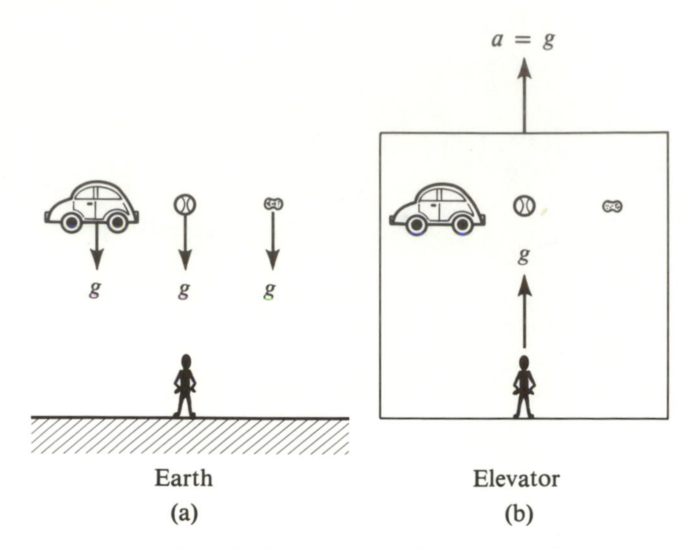

Figure 15.9. The principle of equivalence. According to this principle, an observer in space being accelerated upwards by an elevator with $\boldsymbol{a} = \boldsymbol{g}$, would feel the same physical effects as if he were back on Earth subjected to a gravitational field $\boldsymbol{g}$.

It is the second principle—the principle of equivalence—which distinguishes general relativity from special relativity. The principle of equivalence is motivated by Galileo's observation that all objects in a gravitational field accelerate at the same rate (Chapter 2). Newton's explanation of this fact depends on his equations for dynamics, $F = ma$, and for gravitation, $F = GMm/R^2$ (for a large spherical Earth of radius R). Thus, Newton would set $ma = GMm/R^2$, cancel out m, and obtain $a = g$, where the gravitational field $g = GM/R^2$ makes no reference to the material properties of the mass m. It does not matter whether we drop a car, a baseball, or a peanut, they all accelerate at the same rate $a = g$ under the influence of the gravitational field g of the Earth (Figure 15.9a).

For Newton, this was a remarkable consequence of his equations, not to be pursued beyond the above explanation. For Einstein, the experimental evidence (which has now been shown to hold to very great accuracy by Eötvös, Braginski, and Dicke) was so remarkable that it deserved to be enshrined as a postulate which transcended Newton's equations (which we know now to be only approximations to the truth). If we accept the equivalence principle, Einstein reasoned that we should be able to replace the objects falling under the influence of the Earth's gravitational field and consider them at

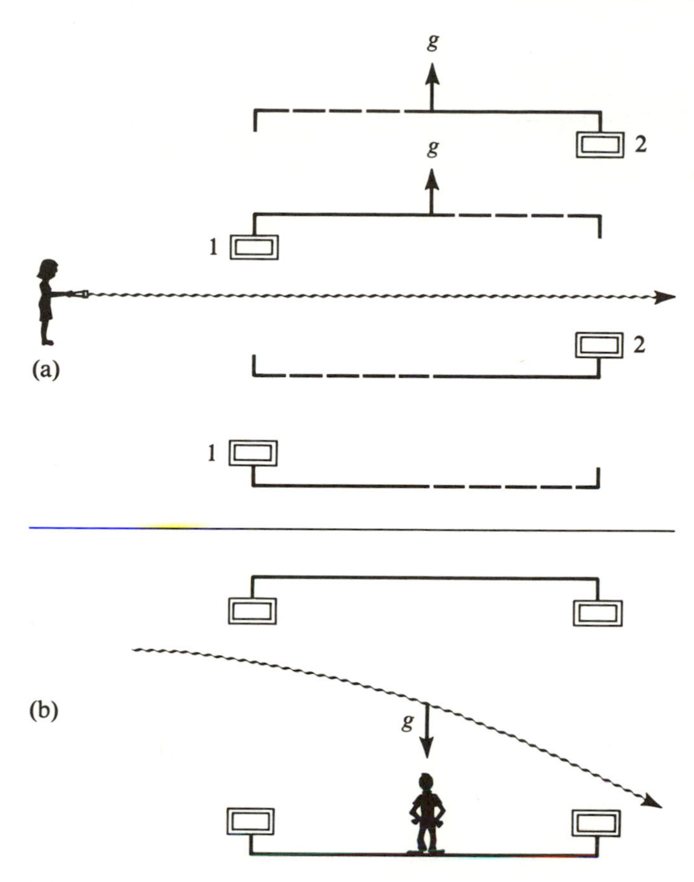

Figure 15.10. Two views of the same experiment involving the propagation of light: (a) according to an inertial observer, (b) according to an observer inside the accelerating elevator.

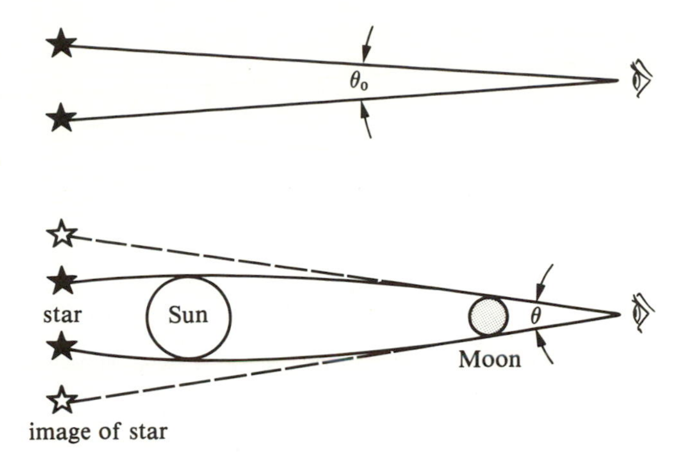

Figure 15.11. The bending of starlight by the gravitational field of a massive body. The angle θ subtended by two stars whose rays graze the limb of the Sun is larger than the angle θ_0 subtended when the Sun is not between the stars and the observer. For optical observations, this experiment must wait for an eclipse of the Sun by the Moon, so that the glare of sunlight does not interfere. For VLBI measurements of quasistellar radio sources, one does not need to wait for a solar eclipse.

rest in space but seen by an observer accelerating upward toward them in a large elevator (Figure 15.9b). If the acceleration $\boldsymbol{a}$ of the elevator is numerically equal to $\boldsymbol{g}$, the gravitational field of the Earth, there should be no physical difference between the descriptions of the two observers. *Is* there any discernible difference between the observations of the two observers? No. If either one stands under the car, he will get squashed. If he stands under the baseball, he will get a headache. And if he stands under the peanut, he will get a snack. The observer in this large elevator might easily imagine that he stands on *terra firma*, and that it is the gravitational field of the Earth which makes the car, the baseball, and the peanut fall toward at him at exactly equal rates. For his frame of mind, he would also think it a remarkable fact that the laws of gravitation and motion are such that they cancel out all dependence on mass, shape, color, taste, etc. For Einstein, however, who constructed the elevator, this fact is no miracle, but the most natural thing in the world.

If there is truly no way to tell the difference between the two situations depicted in Figure 15.9, then (Einstein reasoned further) it must be possible for gravity to bend light, because it is obvious that light will be bent according to the observer in the elevator (Figure 15.10). Light which travels in a straight line according to an inertial observer outside the elevator enters near the top ledge of the left-hand window of the elevator. It takes a finite length of time for it to traverse the width of the elevator, and when it exits the right-hand window of the elevator, it does so near the bottom ledge. The inertial observer simply ascribes the result to the motion of the elevator in the interim; for her, light travels in straight lines, and the elevator is accelerating in her frame of reference. The observer in the elevator thinks, however, that he lives in the gravitational field of the Earth, and that light which entered near the top of one of two parallel windows has been bent by gravity to pursue a curved path to fall near the bottom of the other window. Thus, he thinks gravity can bend light. They are both right; both descriptions are equally admissible.

The deflection of light by massive bodies is, by now, a well-substantiated phenomenon. This prediction by Einstein was first verified in 1919 during a very famous expedition led by Eddington which observed the deflection of starlight by the Sun during a solar eclipse (Figure 15.11). More accurate deflection and timing experiments have since been carried out on the radio

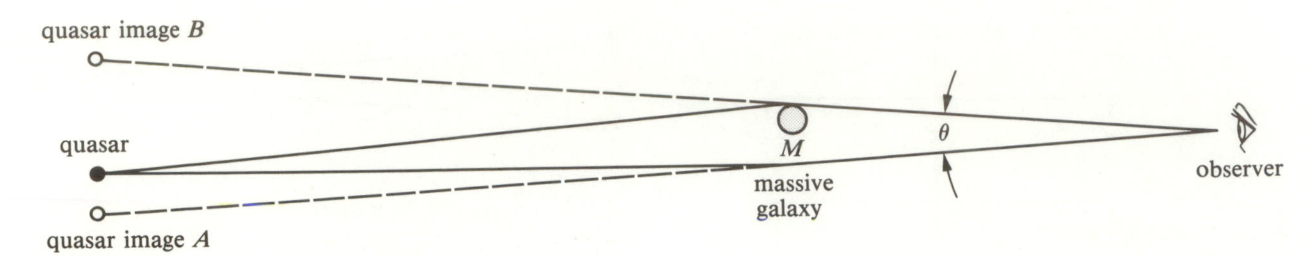

Figure 15.12. The gravitational lens effect. Light rays from a distant quasar may arrive at the position of the observer along two different paths if a massive galaxy lies at the proper intervening distance. For a point mass M, one would seem to observe two quasars at an angular separation θ, but the two images A and B would have identical spectra, etc. However, if the real quasar suffers time variations, the images would not vary simultaneously, because the two light paths generally have different travel times. Walsh, Carswell, and Weymann have discovered a "double quasar" with many of these expected properties; however, Sargent and Young have discovered that the situation may be complicated by the massive galaxy M having an extended distribution of mass. Moreover, the best candidate for the "lens" appears to reside near the center of a fairly rich cluster of galaxies. (See Color Plate 23.)

emission from quasars whose light passes near the Sun. Such experiments have the advantage that there is no need to wait for a solar eclipse. A related phenomenon—the "gravitational lens"—is believed to underlie a recently discovered "double quasar" whose two components have nearly identical appearances (Figure 15.12, Color Plate 23).

If Newton had been confronted with the light-bending experiment, he would probably have said that light was bent from a straight line by the *force* of gravity. In regarding gravity as a force, Newton would be joined by most modern physicists (see Figure 6.1). The quantum view of this force as arising from the exchange of gravitons would probably have delighted Newton, who himself abhorred the concept of action at a distance, but could see no remedy for it. Einstein made his biggest break with Newton on this point. To Einstein, the view that light could be bent from a straight line by the force of gravity was unnatural. Einstein preferred to think that light always traveled in a straight line (in the sense that a straight line is the shortest distance between two points), and that it was spacetime itself which was distorted by the presence of matter. In other words, according to Einstein, curved spacetime *is* gravity; this was the important lesson of the principle of equivalence. John A. Wheeler has summarized Einstein's theory of general relativity in these terms:

> Curved spacetime tells mass-energy how to move;
> Mass-energy tells spacetime how to curve.

Contrast this with Newton's view of the mechanics of the heavens:

> Force tells mass how to accelerate ($F = ma$);
> Mass tells gravity how to exert force ($F = GMm/r^2$).

The advantage of Einstein's way of expressing things is that, since gravity is a property of spacetime itself, it becomes automatic that all objects should fall with the same acceleration under the influence of a uniform gravitational field. Indeed, objects in free fall experience no gravitational "forces" at all; they are merely pursuing their natural trajectories in curved spacetime. The disadvantage of Einstein's way of thinking is that it tends to make gravity seem somehow different from the other forces. This may prove an obstacle to an ultimate superunification of the four forces of nature in the framework of relativistic quantum mechanics. Since no full theory of quantum gravity yet exists, it is difficult to foresee which view will ultimately dominate. Perhaps the final synthesis will have aspects of both, much as waves and particles both found their way into the final formulation of quantum mechanics. In this chapter, however, we will adopt Einstein's geometric interpretation, which, as far as we know, *is* the correct version for a classical theory of gravitation.

Relativistic Cosmology

Einstein's theory of gravitation, general relativity, was completed in 1915. Almost immediately, he and others, notably de Sitter, applied it to the simplest and grandest example of spacetime curvature, that of the universe as a whole. Thus was modern cosmology born. In Einstein's early work, however, he adopted what seemed like a perfectly natural assumption, namely, that the universe on the average was not only homogeneous and isotropic (the cosmological principle), but unchanging in time as well. The assumption that the universe does not change in time is philosophically very pleasing, and when added to the cosmological principle, was later called the **perfect cosmological principle** by the British school of cosmologists who carried it to its ultimate refinement.

In Einstein's hands, however, the perfect cosmological principle led to immediate diasater. Einstein found that no physically meaningful *static* solution to his equations could be found except for the trivial case of an empty universe! So much symmetry (and, therefore, so much beauty) was too much to be real. To cure this deficiency, Einstein modified his equations to incorporate a repulsive force of unknown origin, which he called the cosmological constant. With the cosmological constant, the gravitational effects of a finite mass-energy density could be balanced; and it was possible to produce nontrivial static models. In 1922, Friedmann discovered two classes of nonempty cosmological models which did not require the cosmological constant; and they were also discovered independently later by Lemaitre. The Friedmann–Lemaitre models are homogeneous and isotropic, but not static; i.e., they evolve in time, and therefore do not satisfy the perfect cosmological principle. Indeed, they begin with an expansional phase with an initial condition of infinite density; they are Big Bang models. Hubble was unaware of the Big Bang models when he began his epochal work on the recession of the galaxies, and it is fitting that the discovery of the character of the expansion of the universe should be called "Hubble's law."

When Einstein learned of Hubble's discovery and realized that he might have made this theoretical prediction himself had he only had sufficient faith in the original form of his own equations, he renounced the introduction of the cosmological constant as the "biggest blunder of his life." Even the greatest genius of our time could err. But the mistake is understandable: no one in 1916 knew what galaxies were, much less what to make of their redshifts. Later, it was discovered that Einstein's original static models were flawed in another sense anyway. They were unstable and could not remain static even if they were prepared initially that way. In this chapter, we will also exorcise the cosmological constant from our equations, but we will encounter this demon in a new disguise in Chapter 16.

Einstein made partial amends when he later, together with de Sitter, discovered the third class of Big Bang models. The Einstein–de Sitter model is, in fact, the dividing line for the two classes discovered earlier by Friedmann and Lemaitre. In that sense, the Einstein–de Sitter model may be regarded both as the simplest and most elegant of cosmologies. Where the Friedmann–Lemaitre models have an infinite number of possible candidates in each class (given a present value for Hubble's constant), the Einstein–de Sitter model is unique: there is only one model in the third class. This uniqueness has led some cosmologists with a philosophical bent to regard the Einstein–de Sitter model as the most fitting candidate for the actual universe.

The reader may have guessed by now that the two classes of Friedmann–Lemaitre models are the relativistic analogues of the bound and unbound universes of Newtonian cosmology, whereas the Einstein–de Sitter model is the analogue of the marginally bound case. (To be historically accurate, we should say the Newtonian models are analogues of the relativistic cosmologies, since the latter were developed first.) Indeed, it is possible to characterize the various relativistic models in terms of the present density and age of the universe, as shown in Box 15.4. The formulae for these conditions are identical to their Newtonian counterparts; so you might wonder why anyone would bother with the more complicated relativistic formulation. There is, however, a crucially important difference between the Newtonian models and the relativistic models, in the geometry of the spacetime structure of the universe. The Newtonian models assume a rigid concept of absolute space and absolute time; therefore, the Newtonian models form a continuous sequence. All the Newtonian models could be derived from the single equation (15.1) by mathematically varying the energy constant E in a continuous fashion from negative to positive values (see Problem 15.4). In Newtonian cosmology, if we have a marginally unbound universe, the addition of one more hydrogen atom would make it bound. In relativistic cosmology, it is the *unbound* universes which have more hydrogen atoms *in total* (but not per unit volume) than the bound universes, in fact, infinitely more in total. It is not possible to turn an unbound universe into a bound universe in relativistic cosmology, because they are different from the start. That is why Friedmann and Lemaitre did not automatically find the Einstein–de Sitter model as a special case of their equations. For the geometric reasons which will become clear in the next section, cosmologists, therefore, prefer to call the relativistic Big Bang

BOX 15.4
Summary of Conventional Matter-Dominated Cosmological Models

Type	*Discoverers*	*Spatial curvature*	*Total volume*	*Present density*	*Present age*
Closed	Friedmann–Lemaitre	Positive	Finite	$\rho_{m0} > 3H_0{}^2/8\pi G$	$0 < t_0 < \frac{2}{3}H_0{}^{-1}$
Flat	Einstein–de Sitter	Zero	Infinite	$\rho_{m0} = 3H_0{}^2/8\pi G$	$t_0 = \frac{2}{3}H_0{}^{-1}$
Open	Friedmann–Lemaitre	Negative	Infinite	$\rho_{m0} < 3H_0{}^2/8\pi G$	$\frac{2}{3}H_0{}^{-1} < t_0 < H_0{}^{-1}$

Ultimate fate: Closed—will recontract; Flat—barely expands forever; Open—expands forever.

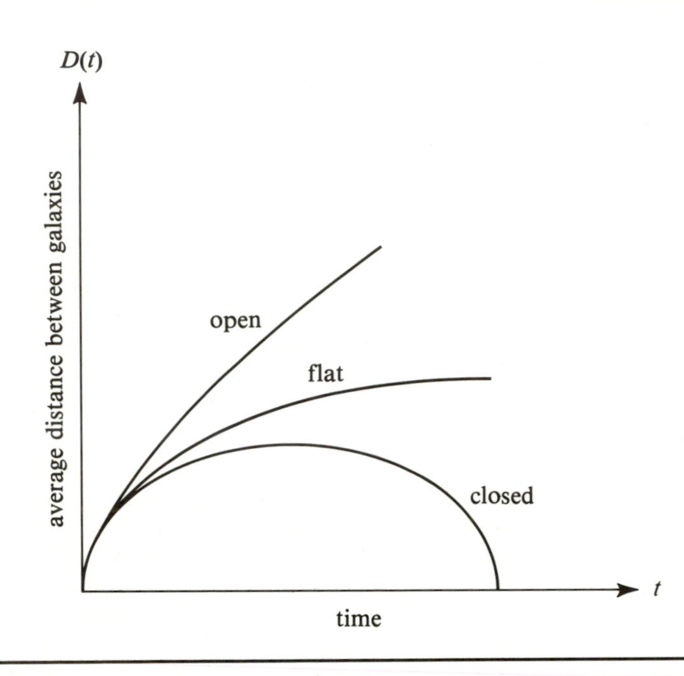

models not bound, marginally bound, and unbound, but rather **closed**, **flat**, and **open** (Box 15.4).

The Large-Scale Geometry of Spacetime

Space Curvature

The most startling aspect of relativistic cosmologies is in the properties of a closed universe. Whereas an analogous bound model of Newtonian cosmology has an infinite volume of space and an infinite number of galaxies, a closed model of relativistic cosmology has a finite volume of space and a finite number of galaxies. This finite volume has, moreover, *no boundaries* and *no center* (else it would violate the cosmological principle). How can this be? Because space is curved.

We have encountered space curvature elsewhere (Chapter 7) when we discussed the geometry of spacetime outside the event horizon of a black hole. There space curvature manifested itself in a distortion of lengths in the direction toward the center of the black hole (see Figure 7.9). But our model of a closed universe satisfies the cosmological principle: every direction is the same. How can *every* direction be curved? Curved with respect to what? The answer is "curved with respect to an artificial *fourth* spatial direction which we cannot perceive and which has no physical meaning." This fourth spatial dimension is a mathematical artifice introduced to describe mathematically the properties of a three-dimensional space which has uniform positive curvature; the artifice does not work in general. The

Figure 15.13. The distribution of ant cities on a spherical Earth. (We don't call them "anthills" because we wish to suppress the notion of a third dimension perpendicular to the surface of the world.) The distribution is homogeneous and isotropic on average, and there is a finite total number of ant cities on the finite surface area of the world.

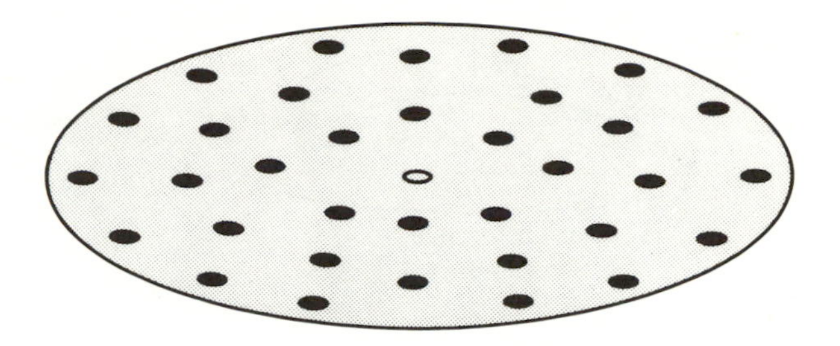

Figure 15.14. The common ant's view of its world with its city (white dot) being at the center of a flat circular plane containing many other ant cities (black dots).

fourth dimension here has nothing to do with time, whose role we shall discuss shortly. It is not possible for us to visualize curved three-dimensional space because we are three-dimensional creatures and cannot see the curvature in the "fourth" dimension. We can, however, make an analogy with two-dimensional creatures which helps us to understand our own space curvature.

Imagine the world of ants on the surface of a spherical Earth (Figure 15.13). These ants have no concept of the existence of a third dimension perpendicular to the surface of the Earth. They are doomed to crawl forever only in the two dimensions parallel to the surface of the Earth. Because the world is a big place compared to any ant city, these ants have always imagined that they live on a flat Earth. Moreover, being two-dimensional creatures, they, unlike us, cannot visualize geometrically a round Earth.

Imagine further that ant-light travels on the surface of the Earth along the shortest distance between two points (a **geodesic** which is an arc of a great circle). An ant astronomer can point its telescope in any direction along the surface of the world and see other ant cities. Thus, the ants in the Milky Way city learn that there are other cities beside their own. Some of the ant astronomers make counts of the ant cities and discover that, on the average, the world is homogeneous and isotropic. There are the same number of ant cities per square kilometer at all locations and in all directions of the world. This is the ant cosmological principle. Notice, now, that the total surface area of the world is finite—the world is "closed"—yet the world has no edge. This proposition is something that no ant can easily grasp, because each ant has been brought up to think that it lives on a plane. For a plane to have a finite surface area, there must be edges. Moreover, it is difficult for an ant to understand that its city is not at the center of the *surface* of the world. They ask, if the world has a finite area and it looks the same to us in every direction, how can we not be at the center? You see, they insist on thinking that they must live at the center of a flat circle (Figure 15.14). Being three-dimensional creatures, we easily see how space curvature in the *third* dimension can create a world with finite surface area, no edges, and no center *on the surface.* But no two-dimensional creature can easily grasp this fact. If we could communicate with the ants, we would probably cause the least confusion by telling each colony of ants that it may justifiably regard itself as the center of the world's *surface* for any practical calculation.

Now, the interesting question arises, is there any way that the ants can discover that they don't live on a flat Earth? Yes, they only need to perform geometric experiments of sufficient scope. For example, they can draw "circles" of successively larger size. How do ants draw circles? The same as you and I. They take a length of string; one ant holds one end still at the center of the city; and the other ant crawls, always keeping the string taut. Taut here means stretched against the surface of the Earth, the only two real dimensions. The circumference (distance crawled to go once around) divided by the "radius" (the length of the string) of a small circle would very nearly equal $2\pi = 6.28\ldots$, because within small distances, the surface of the world does behave like a flat plane; that's how Euclid managed to prejudice all the ants in the first place. As the ants increase the length of the string, they discover, much to their amazement, that the ratio of the circumference to the "radius"

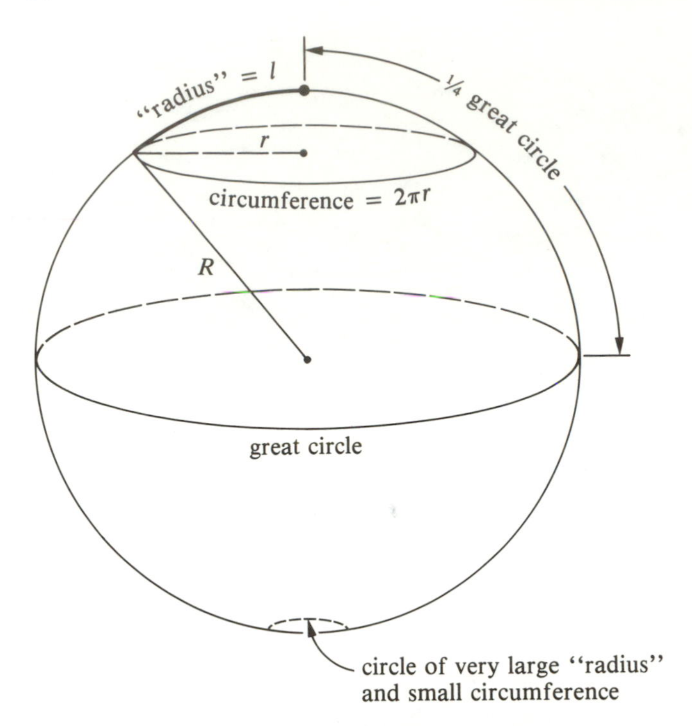

Figure 15.15. The geometric experiments of two bold ant surveyors who discover that the ratio of the circumference $2\pi r$ of a circle to its "radius" l does not equal 2π if the "radius" l is large. Notice that the quantity r is not the line of sight distance l that the ants can measure. Because r and R exist in the third unobservable dimension, ants can never measure r and R directly. They can, however, measure the circumference $2\pi r$, and by measuring the discrepancy between $2\pi r/l$ and 2π, the ants can indirectly calculate the radius of curvature R of their world.

gets discernibly smaller than 2π. For example, if they had the fortitude and the funds to obtain a length of string equal to one-quarter of a great circle around the Earth (Figure 15.15), the ratio circumference/"radius" = great circle/0.25 great circle = 4, not $2\pi = 6.28 \ldots$. Indeed, if a very long string were used, to allow the crawling ant to reach the other end of the world, the ratio would go to zero! That the circumference of a circle with a very large "radius" can become zero must seem a very strange result to creatures who are used to dealing only with plane geometry. Being three-dimensional creatures, we easily see how the trick is done, but most ants would have a pretty hard time swallowing this concept. However, a really clever ant mathematician could calculate all the effects described above, providing it could imagine right triangles whose hypotenuse squared did *not* equal the sum of the squares of its sides. This violation of Pythagoras's theorem (Appendix B) would require, of course, that ants not live on a *flat* plane. In this way, the ant mathematician could describe a world which was curved in an unobservable *third* dimension. When the ant mathematician lectures on this topic, the ant physicists all pay it no attention, because they scoff, "We live in two dimensions; what's this talk about a third dimension?" But on hearing the news of the bold ant surveyors, the ant physicists all rush to the ant mathematician and contritely ask, "Say, do you remember when you were telling us about two-dimensional space curved in a third dimension . . . ?"

Eventually the concept is so well-accepted that it is taught to beginning ant-astronomy classes. The ant professor begins its lecture: "We cannot imagine a curved two-dimensional surface because we are not three-dimensional creatures. We can draw an analogy, however, with curved one-dimensional lines which we *can* visualize. Imagine a lowly class of insect—call them termites—who can only burrow in a line. They think they live on a straight line, but actually, they live on the rim of a circle . . . ." And so the ant professor would proceed to explain to its class the intricacies of living in a curved one-dimensional world. And so would a superior four-dimensional creature describe these efforts to explain how your universe and mine can be curved in three dimensions. Replace "closed world" by "closed universe"; replace "surface area" by "volume"; replace "unobservable third dimension" by "unobservable fourth dimension"; and the ant parable is our own.

Spacetime Curvature

Now, you might ask, is it really necessary for the ants to crawl around the world to make these geometric measurements? Can't they find out if the world is round with much less effort, simply by pointing their telescope in some direction—any direction—and finding out whether they can see the back of their own heads? (Well, maybe not quite the back of their heads; see Figure 15.16.) Anyway, that's a good idea. Unfortunately, it doesn't work. I forgot to tell you that the ants live in a closed world which is currently expanding, but will eventually recontract to zero area. The expansion and recontraction from zero area to zero area takes just the same amount of time as light would take to go once around the world; so no ant alive, when the world has finite area, can live long enough, even in principle, to see the back of its rear end.

The general ant populace is now totally confused. They ask, "You astronomers tell us that we live in a world of finite area with no edges. If that's so, how can the area increase in time? Where does the extra area come from into which our world expands? And if there is extra preexisting area, won't the world crinkle badly when our territory slams into this other area?" You see, the problem is that the general ant populace insists on thinking that the expansion represents real motion *along* the surface of their world. Those are the only two dimensions of which they are aware, and they can conceive of motion

Figure 15.16. Ant light travels on a geodesic in the ant world (the shortest distance between two points, which is the arc of a great circle on a spherical world). Can ants make use of this property to see the back of their rear ends?

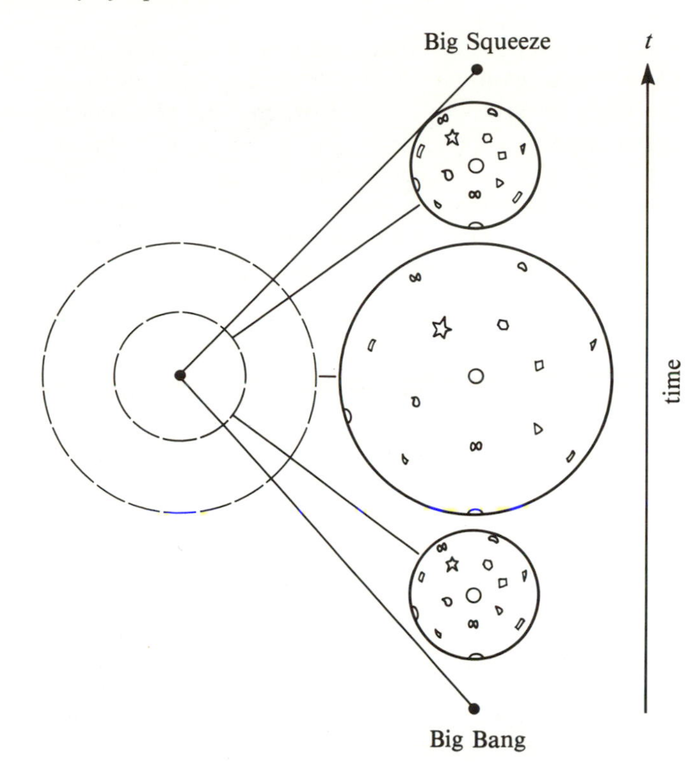

Figure 15.17. Expansion and contraction of a closed ant world. Once ant cities have formed, they retain their sizes and shapes. The arrow of time takes flight at the instant at the Big Bang and ends at the Big Squeeze.

only in those two dimensions when they are told the world expands. Figure 15.17 shows that the expansion of the world takes place because the surface of the world is carried in time to a different location in the *third* (unobservable) dimension. Thus, the increase in total surface area results, not because any region moves *along* the surface into area occupied by some other region, but because the radius of the world increases in time in the dimension which is completely fictional for ants. As a further consequence, we easily see that the expansion of the ant world follows Hubble's law. According to every ant city, every other ant city seems to be receding to greater distances. This recession occurs homogeneously and isotropically, with the recessional velocity being linearly proportional to the distance.

Problem 15.5. Consider the great circle which corresponds to a line of sight from any given ant city (bottom diagram). Let the radius of curvature of this great circle be $R(t)$, the instantaneous radius of the world relative to a center displaced in the third, unobservable dimension. Suppose that there are N ant cities at any time, distributed more or less evenly along this great circle. Show that the average distance $l(t)$ between two ant cities along this (or any other) line of ant sight is given by $l(t) = \theta R(t)$, where $\theta = 2\pi/N$ radians. Argue that, in general, the distance between two ant cities which are initially separated by an angle θ (relative to the unobservable center of the

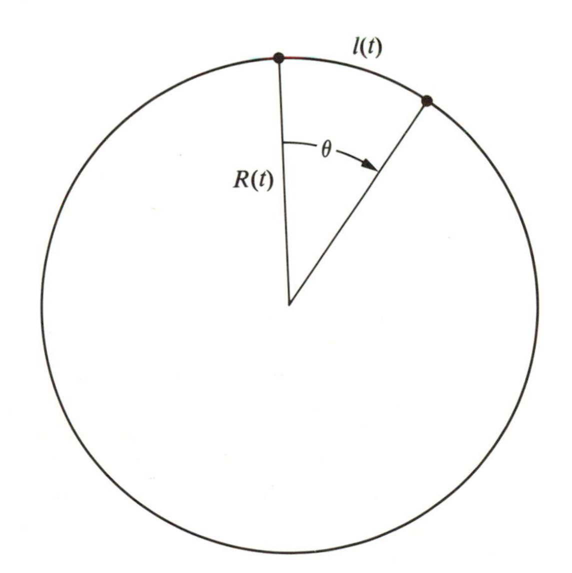

world) is given at any later time t by the formula $l(t) = \theta R(t)$. The recessional velocity of one city with respect to the other is given by $v = \dot{l}(t)$, where the dot denotes differentiation with respect to time. Show that this recessional velocity satisfies Hubble's law: $v = H(t)l$, where $H(t) = \dot{R}(t)/R(t)$ depends only on the history of the radius of curvature R. Notice, in particular, that all reference to the unobservable angle θ drops out.

Analogous statements can be made about our actual three-dimensional universe if it is closed. It would have a finite volume, but no boundaries or center (in any of the observable three dimensions). The total volume can increase homogeneously and isotropically, yet no volume is gained at the expense of some preexisting volume. Our three-dimensional space would, so to speak, be carried to another location in an unobservable fourth dimension. Moreover, for practical calculations, we need never try to envisage this fourth spatial dimension; we need speak only of curved spacetime in three spatial dimensions. From this point of view, we need not worry about the change of the total volume of space, because the volume of space is not a conserved quantity in the presence of changing gravitational fields (whether or not the universe expands). The radius of curvature $R(t)$ of our homogeneous and isotropic three-dimensional space is the only important geometric quantity in the present problem (compare with Problem 15.5), and its behavior with time is governed by Einstein's field equations of general relativity. That is, the homogeneous distribution of matter-energy in the universe tells spacetime how to "curve". Curved spacetime, in turn, tells matter-energy how to move.

In Problems 15.8 and 15.9 we will discuss the Friedmann-Lemaitre solution for the radius of curvature $R(t)$ for a matter-dominated closed universe. For now, let us suppose someone gives us $R(t)$ for this case, and let us examine more closely what is meant by the statement that curved spacetime tells matter-energy how to move.

We begin by drawing a curved spacetime diagram. Since the universe is homogeneous and isotropic on the average, any direction is the same as any other. Hence, let us ignore two of the spatial dimensions, and simply plot position along one real spatial dimension along the horizontal axis, and time along the vertical axis. To make the horizontal and vertical axis have the same units of length, let us follow our usual procedure of multiplying the vertical axis by the speed of light, c; so we use ct rather than just t as our actual vertical axis. We have grown used to plotting such spacetime diagrams by now on a flat sheet of paper, because we usually deal with flat spacetime (special relativity). Locally, in a nonempty universe, we may still make such flat plots, in which the

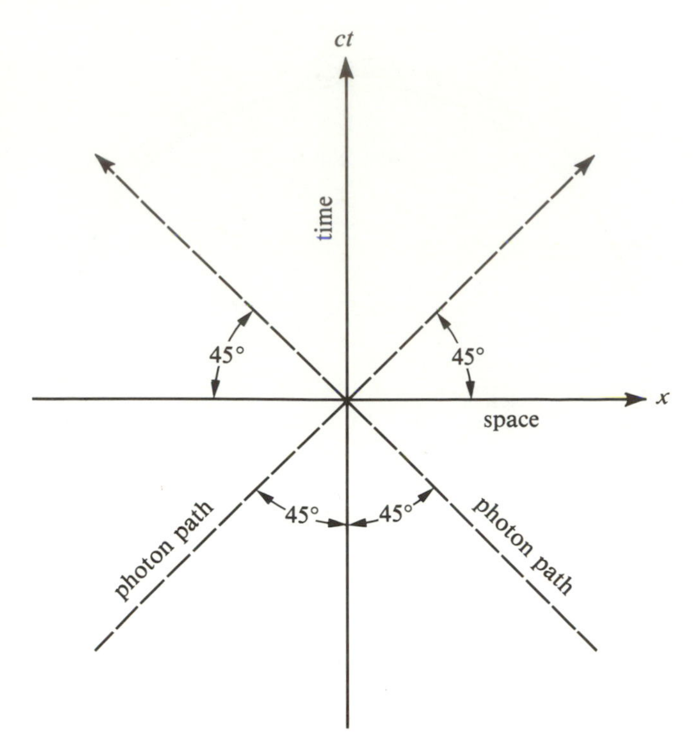

Figure 15.18. The trajectories of photons in flat spacetime are at angles of 45° to the coordinate axes.

trajectory of a photon is at 45° relative to the axes, because photons always fly at the speed of light c relative to all local observers (Figure 15.18). Globally, however, the large-scale structure of spacetime in a closed universe must be plotted on a curved sheet of paper (Figure 15.19). In fact, Figure 15.19 shows such a plot for the middle half of the duration of a matter-dominated, closed universe. The "ticks" along the horizontal space axes mark the positions of comoving galaxies, and these galaxies are carried upward along the vertical axis by the forward march of time. The *dimension off from the gridded spacetime surface* corresponds to the unobservable fourth spatial dimension. (Remember, any one of the actual three spatial directions is represented by the ringed axis labeled "space.") The radius of curvature $R(t)$ in the fourth dimension is called the radius of the universe. If you could fly infinitely fast in any one direction, always in a straight line dead ahead, you would come back to where you started from after flying a distance $2\pi R(t)$. The distance to the "other side of the universe" is therefore $\pi R(t)$, and it lies in any and every direction away from us!

Nothing in this universe, however, can fly faster than the speed of light. Photons travel in the spacetime diagram by making an angle of $\pm 45°$ (depending on whether they are going to the right or to the left) to the local axes.

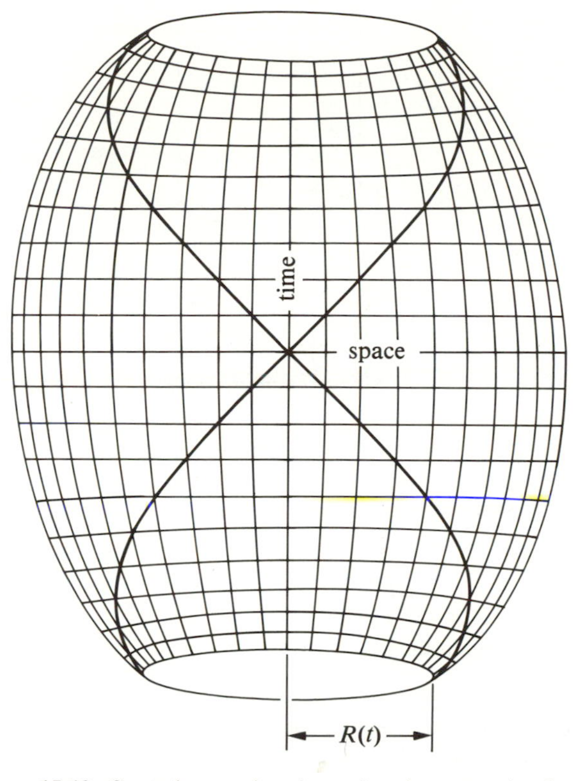

Figure 15.19. Curved spacetime in a closed, matter-dominated universe during the middle half of its existence. Galaxies on average travel upward along the time axis without moving with respect to the spatial grid. Photons travel along trajectories which are at 45° angles to the local spacetime axes ("corner to corner" if we construct little square grid markers). The space axis, at each instant of time t, forms a closed loop with radius $R(t)$ in an artificial "fourth" dimension. During the expansional phase, when $R(t)$ is an increasing function of t, notice that the individual square grids become larger. This is the expansion of spacetime that is the essence of relativistic cosmologies in the present epoch.

Galaxies, on the other hand, do not move at all (on the average) in the direction of the space axes, because, in fact, we *chose* the axes to have this property. Einstein's theory of general relativity actually allows the choice of arbitrary spacetime coordinates; the equations look the same in all such choices. However, the appearance of phenomena (i.e., the *solutions* to the equations) do *not* look the same in all frames. It is obviously to our advantage, therefore, to choose coordinates in which natural phenomena look the simplest possible, and the only frame which satisfies the cosmological principle is one which comoves with the galaxies. It is this frame which comes closest to Newton's conception of absolute space and absolute time, for it is this frame that we implicitly use to describe the dynamics of the heavens and the dynamics of atoms (Chapter 2). Ironically, it is also this frame which illustrates most forcefully the failings of Newtonian and Euclidean precepts about the properties of space and time.

In Figure 15.19, we see that the expansion of the universe occurs not because any galaxy slides in any one of the three spatial dimensions away from any other galaxy (although small random motions of this type are possible). It occurs because space itself is expanding, and the galaxies recede from one another because they are swept along in this expansion. (However, the internal sizes of the galaxies—as well as of stars, our bodies, etc.—are not expanding. They are bound by forces which easily resist the large-scale expansion of space.)

Problem 15.6. Suppose you were not held together by electromagnetic forces. How long would it take you to grow one cm because of the expansion of the universe? *Hint*: Apply Hubble's law to your head as seen by your feet (or vice versa).

Since the expansion of the universe is caused by the increase in time of the radius of curvature $R(t)$, and not by motion of the galaxies in the three real dimensions, the terminology recession *velocity* is somewhat of a misnomer. Galaxies have redshifts because of spacetime curvature, not because they have velocities along the line of sight in a Newtonian-Euclidean sense. Galaxies separate from one another globally, but to think of this global separation in terms of velocities is to invite trouble (Problem 15.7). Once upon a time, in a closed universe, there would have been galaxies which were very close to us, yet were on the opposite end of the universe. They are still on the opposite end of the universe, but now they lie at distances greater than 10 billion light-years, and they are still unobservable. Only after the universe begins to recontract will light from the opposite end of a closed universe have had enough time to reach us, and then we can observe such galaxies by looking in any direction! Clearly, to think of such galaxies as having reached their present positions by flying away from us faster than the photons they emit is incorrect. Moreover, in what direction should any galaxy fly if later we can see them in all directions? No, it is much better to say that, on the average, no galaxy moves at all, and that it is the expansion of space itself that separates the galaxies.

Problem 15.7. Use the methods of Problem 15.5 to show that Hubble's law applies to the closed universe of Figure 15.19; the law is $v = H(t)l$, where by v we mean the time rate of separation of the distance l between two galaxies. In particular, derive the identification $H(t) = \dot{R}(t)/R(t)$. Early in the history of a matter-dominated universe, $R(t) \propto t^{2/3}$, and the Hubble constant can have very large values; so the formula, $v = H(t)l$ can produce velocities that are greater than c. Does this contradict the conclusion of special relativity which states that nothing can pass anything else with a speed faster than c? Discuss why it is dangerous to think of the *nonlocal* formula $v = H(t)l$ in terms of real velocities. (For local phenomena, l is very small, and we have no problems.)

If we don't think of galaxies as having a *velocity* of recession, we cannot think of the redshift as a Doppler shift. How, then, does the cosmological redshift arise? Figure 15.19 already contains the answer. We need only consider the trajectory of photons in curved spacetime. To make clear what's happening, let us take a small portion of spacetime from our grid and magnify it (Figure 15.20a). The local spacetime structure is flat; so if we made all our cells in Figure 15.19 little squares, light will propagate from corner to corner (at 45° relative to the axes). If there are N cells at time t_1 (with a galaxy, say, at each vertex), the length of the sides of each cell is $D(t_1) = 2\pi R(t_1)/N$. At a later time, t_2, there are still N cells (the galaxies at every vertex have been conserved), but the average distance between galaxies has grown to $D(t_2) = 2\pi R(t_2)/N$. Now, suppose that at time t_1 an elliptical galaxy begins to emit a radio wave with wavelength λ_1 equal, for simplicity, to one cell length at time t_1. The head of this radio wave travels from corner to corner in our spacetime diagram, and the tail follows from corner to corner one vertex back. At time t_2, a spiral galaxy begins to intercept the head of this radio wave (Figure 15.20b). The tail of the radio wave is now one cell size back at time t_2, because light always traveled from corner to corner in all intervening spacetime cells. The wavelength λ_2 of the received radio wave has therefore grown in direct proportion to the expansion factor:

$$\frac{\lambda_2}{\lambda_1} = \frac{D(t_2)}{D(t_1)}. \quad (15.5)$$

The same formula would have resulted had we considered electromagnetic radiation as having n wavelengths per cell length instead of only one wavelength; so equation (15.5) is valid for light from radio waves to gamma rays. During epochs when the universe is expanding, $D(t_2) > D(t_1)$, the received radiation has wavelength λ_2, which is longer than the emitted radiation λ_1 by the ratio of the expansion factor $D(t_2)/D(t_1)$. This is the origin of the cosmological redshift. When we look at distant galaxies, we are looking at a remote time, when the universe was much smaller than it is now. Thus, the redshift z, where $1 + z = D(t_2)/D(t_1)$, exhibited by distant galaxies is large. Although our formal derivation was carried out for a closed model, equation (15.5) also applies to open and flat universes. Speaking loosely, we may say that the cosmological expansion stretches out the wavelength of every photon (or any other massless particle) in direct proportion to the expansion factor in the elapsed interval.

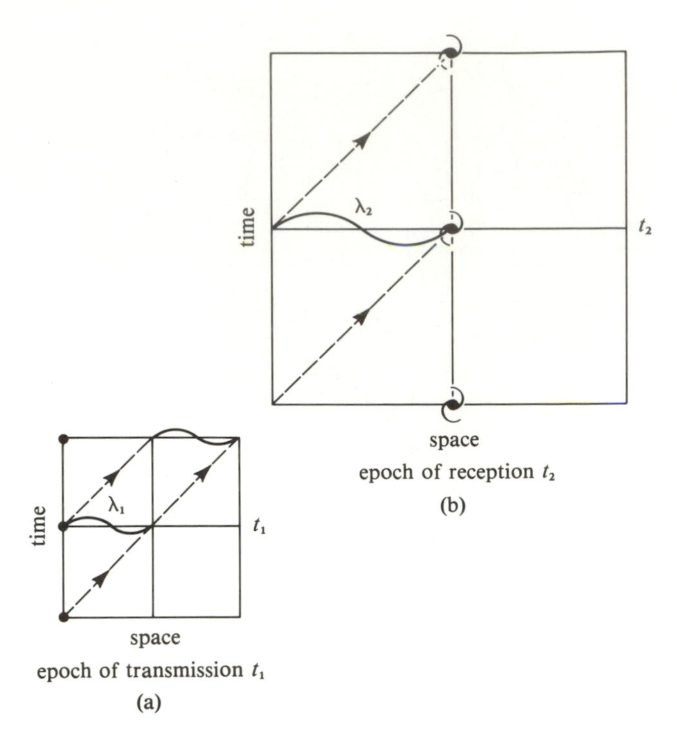

Figure 15.20. The origin of the cosmological redshift. See the text and Figure 15.19 for explanations.

During the expansional phase of the universe, the redshift z is a monotonic function of both distance and time. An object with a large redshift both is currently a great distance from us and is being seen as it appeared a long time ago. Because z is directly measurable, and neither the distance nor the elapsed time can be obtained without adoption of a specific cosmological model, astronomers often use the redshift z as an indicator of both distance and elapsed time. Thus, the statement "object A lies at $z = X$" means that "object A lies at a distance associated with redshift X." The statement "event B occurred at redshift $z = Y$" means that "event B occurred a time ago associated with redshift Y."

Problem 15.8 (for those who know a little calculus). We want to discuss here the fact that matter-energy tells spacetime how to curve. A general discussion would require the full theory of Einstein's field equations. Let us just note here that Friedmann and Lemaitre showed that, for a universe of cold matter (galaxies with little random motions), Einstein's field equations lead to a single relation that governs the radius of curvature R:

$$\frac{\dot{R}^2}{R^2} - \frac{8\pi G\rho_m}{3} = -\frac{c^2}{R^2}, \tag{1}$$

where c is the speed of light, and $\rho_m(t)$ is to be obtained from the condition that the total mass of the universe is conserved. The total volume of a closed universe is given by $2\pi^2 R^3$ (Problem 15.9); so we require

$$\rho_m = \frac{M}{2\pi^2 R^3}, \tag{2}$$

where M is the total mass contained in protons, neturons, electrons, etc. Show now that equation (1) can be written

$$\frac{1}{2}\left(\frac{dR}{dt}\right)^2 - \frac{2GM}{3\pi R} = -\frac{1}{2}c^2. \tag{3}$$

Comment on the similarities and differences between this equation and equation (15.1). To solve this equation, introduce the dimensionless variables, $\xi = (3\pi c^2/4GM)R$, $\tau = (3\pi c^3/4GM)t$, and rewrite equation (3) as

$$\left(\frac{d\xi}{d\tau}\right)^2 - \frac{1}{\xi} = -1, \tag{4}$$

which we have already solved in Problem 15.4. What is the maximum value of R? The total elapsed time from Big Bang to Big Squeeze? Calculate these values numerically for $M = 10^{24} M_\odot$. In this model, what is the time t_0 today if $H_0 = 20$ km sec^{-1}/million lt-yr? *Hint*: Find R_0 from equations (1) and (2). Reduce and solve the corresponding cubic equation for ξ_0; deduce η_0 and τ_0 from Problem 15.4, and obtain $t_0 = (4GM/3\pi c^2)\tau_0$. How does t_0 compare numerically to $(2/3)H_0^{-1}$?

Problem 15.9 (for those good at calculus). Consider the following sequence:

circumference of circle, $x_1^2 + x_2^2 = R^2$; $x_1 = R\cos\varphi$, $x_2 = R\sin\varphi$.

surface of sphere, $x_1^2 + x_2^2 + x_3^2 = R^2$;

$$x_1 = R\sin\theta\cos\varphi,$$
$$x_2 = R\sin\theta\sin\varphi,$$
$$x_3 = R\cos\theta.$$

three-volume of four-sphere, $x_1^2+x_2^2+x_3^2+x_4^2=R^2$;

$$x_1 = R\sin\psi\sin\theta\cos\varphi,$$
$$x_2 = R\sin\psi\sin\theta\sin\varphi,$$
$$x_3 = R\sin\psi\cos\theta,$$
$$x_4 = R\cos\psi.$$

Pythagoras's theorem applied to small displacements in mutually perpendicular directions yields as the square of the separation interval (line element) for points on:

circumference of circle, $dl^2 = R^2\,d\varphi^2$;
surface of sphere, $dl^2 = R^2(d\theta^2 + \sin^2\theta\,d\varphi^2)$;
volume of four-sphere, $dl^2 = R^2[d\psi^2 + \sin^2\psi\,(d\theta^2 + \sin^2\theta\,d\varphi^2)]$.

Therefore, show:

$$\text{circumference of circle} = \int_0^{2\pi} R\,d\varphi = 2\pi R;$$

$$\text{surface area of sphere} = \int_0^{\pi} R\,d\theta \int_0^{2\pi} R\sin\theta\,d\varphi = 4\pi R^2;$$

$$\text{volume of four-sphere} = \int_0^{\pi} R\,d\psi \int_0^{\pi} R\sin\psi\,d\theta \times \int_0^{2\pi} R\sin\psi\sin\theta\,d\varphi = 2\pi^2 R^3;$$

How does the last result relate to Problem 15.8? Define $r = R\sin\psi$, so that the locus $r =$ constant defines spheres of area $4\pi r^2$. Show that $R^2\,d\psi^2 = (1 - r^2/R^2)^{-1}\,dr^2$, so that the square of the spatial interval becomes

$$dl^2 = \frac{dr^2}{(1 - r^2/R^2)} + r^2(d\theta^2 + \sin^2\theta\,d\varphi^2).$$

This expression governs the spatial geometry of three-dimensional space with uniform positive curvature without having to refer to the fourth dimension.

Flat and Open Universes

Spacetime is also curved in an open model of the universe, but the spatial curvature is *negative*; so the circumferences of large circles in an open universe are *greater* than 2π times the radius (the distance from the center). In a flat universe, the geometry of space is Euclidean. Figures 15.21 and 15.22 show two-dimensional analogues of the spatial geometry of a flat universe and an open universe, which are a flat plane and a saddle,

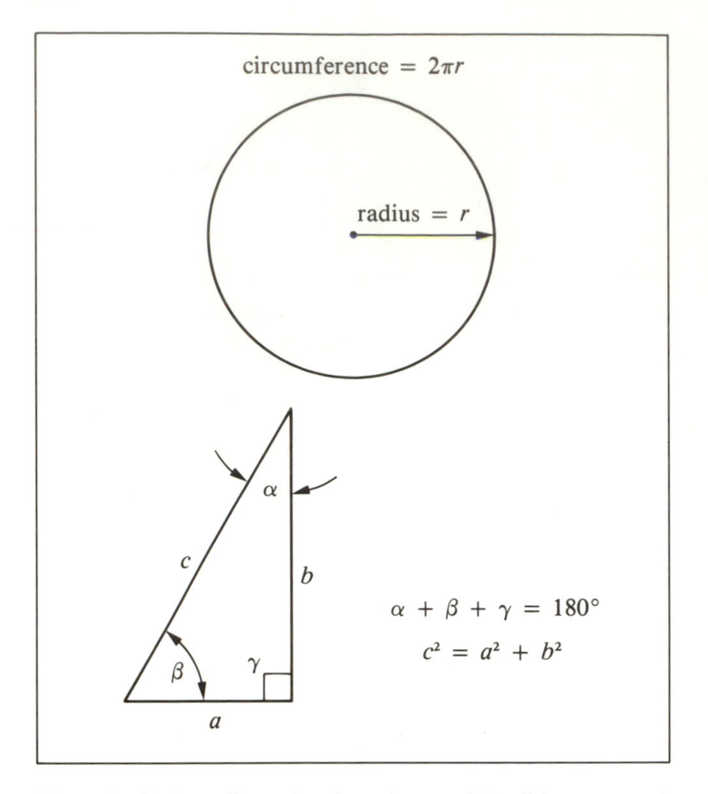

Figure 15.21. Two-dimensional analogue of Euclidean space is a flat plane. In a flat plane, the ratio of the circumference of a circle to its radius is 2π. Moreover, the sum of the angles of a triangle equals 180°, and the square of the hypotenuse of a right triangle is equal to the sum of the squares of its sides.

respectively. The spacetime of a flat universe is curved; it is not the flat spacetime of special relativity on a global scale.

In both flat and open universes, the total volume of space at any time is infinite, and so is the total number of galaxies in the universe. However, at any finite time after the Big Bang, there is an event horizon beyond which we cannot see, because the light leaving the galaxies from beyond the event horizon would not have had time since the Big Bang to reach us; so the observable universe—which enlarges as time proceeds—is always smaller than the total universe.

Apart from the actual but unobservable infinity of possible worlds in an open universe, the geometry of spacetime is less weird in an open universe than in a closed one, because, in an open universe, the density of matter-energy is less able to warp spacetime than in a closed universe. Indeed, as time becomes infinite in an

Figure 15.22. Two-dimensional analogue of space with negative curvature is the middle of a saddle. (The embedding of an *entire* surface of constant negative curvature cannot be performed in *Euclidean* three-dimensional space.) In such a surface, the ratio of the circumference of a circle to its radius is greater than 2π. Moreover, the sum of the angles of a triangle is less than 180°, and the square of the hypotenuse of a right triangle is less than the sum of the squares of the sides. (Each side of the triangle is drawn by connecting the vertices along a path of least distance along the surface of the saddle.)

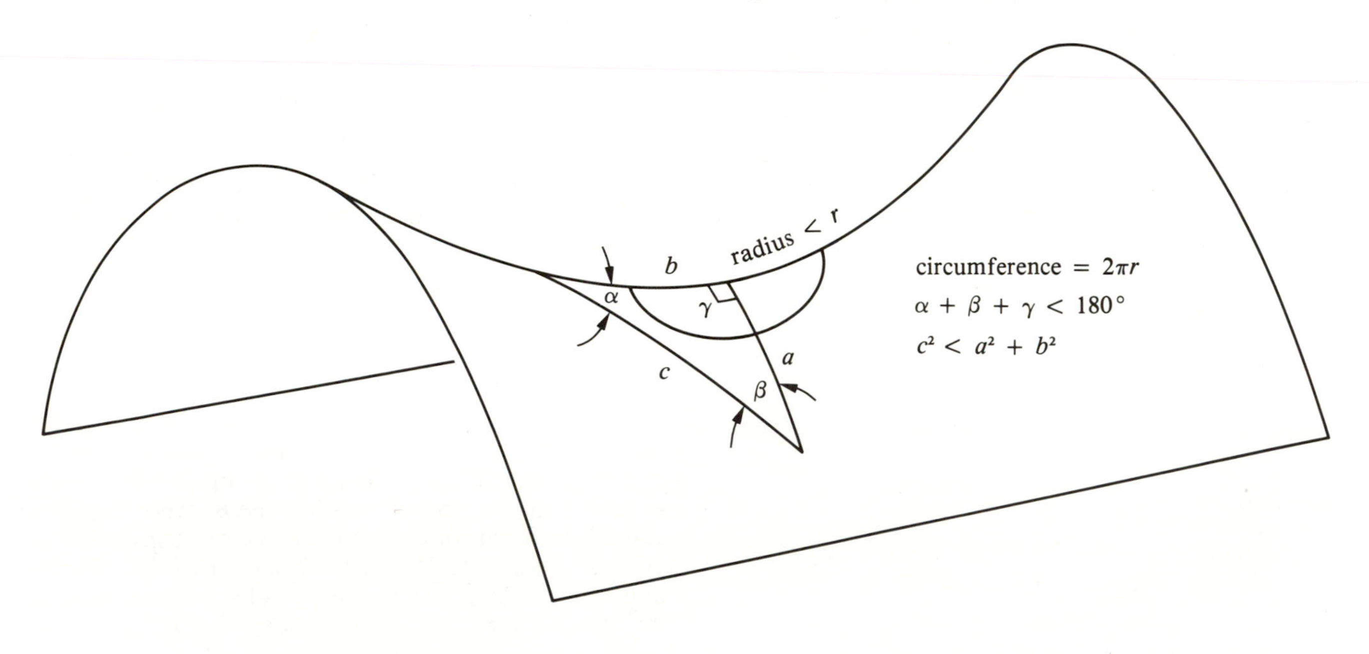

open universe, the density of matter and energy both tend toward zero, and spacetime in nonexpanding coordinates becomes the flat spacetime of special relativity.

Problem 15.10 (for those good at calculus). When we deal with model universes other than closed ones, we find it convenient to replace the "radius of the universe $R(t)$" with a proportional measure, say, the average distance $D(t)$ between galaxies at time t. Let us further define the dimensionless coordinate r_* such that the locus $r = D(t)r_* =$ constant at fixed time t describes a sphere of surface area $4\pi r^2$. Robertson and Walker have shown that the spatial metric of any homogeneous and isotropic universe can then be written, in general, in the form (compare with Problem 15.9):

$$dl^2 = D^2(t)\left[\frac{dr_*{}^2}{1 - Kr_*{}^2} + r_*{}^2(d\theta^2 + \sin^2\theta\, d\varphi^2)\right],$$

where θ and φ are the usual spherical angles, and K is a constant and dimensionless curvature parameter, which is positive for a closed universe, zero for a flat universe, and negative for an open universe (so the geometry of space for a flat universe is Euclidean). In a matter-dominated universe of negligible random motions, $D(t)$ satisfies the equation

$$\frac{\dot{D}^2}{D^2} - \frac{8\pi G}{3}\rho_m = -K\frac{c^2}{D^2},$$

where c is the speed of light and $\rho_m(t)$ is given by the conservation of mass in a comoving volume,

$$\rho_m(t)D^3(t) = \rho_{m0}D_0{}^3,$$

where we have denoted $\rho_m(t_0)$ and $D(t_0)$ by ρ_{m0} and D_0. Identify $H(t) = \dot{D}(t)/D(t)$, and show that K can be expressed in terms of the present conditions as $K = (H_0{}^2D_0{}^2/c^2)(\Omega_0 - 1)$, where $\Omega_0 = 8\pi G\rho_{m0}/3H_0{}^2$ is the ratio of the actual mass density to the critical value required for closure. Introduce the dimensionless variables $D_* = D/D_0$ and $\tau_* = H_0 t$, and transform the differential equation for D into the form given in Problem 15.4.

The spacetime metric is given by

$$ds^2 = c^2\,dt^2 - dl^2,$$

where the spatial part dl^2 is given above. The trajectory of a comoving galaxy is given by $(r_*, \theta, \varphi) =$ constant, while the trajectory of a photon satisfies $ds^2 = 0$. In other words, the distance l that a photon travels by flying radially away from a source (θ and φ held fixed, because light travels in straight lines) is governed by the differential equation

$$D^2(t)\frac{(dr_*/dt)^2}{1 - Kr_*{}^2} = \left(\frac{dl}{dt}\right)^2 = c^2.$$

Thus, photons always fly a proper distance l in a proper time-interval $(t - t_e)$ at the speed of light c, $l = c(t - t_e)$. Emitted from an isotropic light source, however, the photons cover a sphere of area $4\pi r_*{}^2(t)D^2(t)$ at the later time t, and this area does not equal $4\pi l^2$. The discrepancy depends on the elapsed time and the cosmological model, i.e., on $t - t_e$, K, and $D(t)$. Comment now on the similarities and differences between Newtonian and relativistic cosmologies.

Cosmological Tests by Means of the Hubble Diagram

Suppose that a galaxy of luminosity L in its own rest frame emits photons isotropically into space. Suppose an observer at a different location in space and later in time intercepts a portion of these photons. What is the apparent brightness f of this galaxy if cosmological effects are important?

Let z be the redshift of the galaxy according to the observer, and let $4\pi r^2$ be the surface area of a sphere centered on the galaxy which passes through the observer at the time of the reception of the photons. The quantity r will depend on z, $r = r(z)$, and for a nonflat universe $r(z)$ will not equal the present distance to the galaxy at large z. (See Problems 15.10 and 15.11, where r is denoted by r_*D, and D is the average distance between galaxies). For large z, f is less than $L/4\pi r^2$ because of the expansion; so the wavelength λ of each photon has been increased by a factor $1 + z = D(t_2)/D(t_1)$ between emission and reception (see equation [15.5]). Since the energy of a photon is given by hc/λ, this means that the total energy of all photons has been reduced by one factor of $1/(1 + z)$. Moreover, the time to receive any wavetrain of photons has been lengthened by one factor of $(1 + z)$ in comparison with the time required to emit the wavetrain (see Figure 15.20). Since the luminosity crossing the sphere of area $4\pi r^2$ is equal to the total energy divided by the time required for the wavetrain to pass, the apparent brightness f must be given by the formula

$$f = \frac{L}{(1+z)^2 4\pi r^2}. \qquad (15.6)$$

The function $r = r(z)$ can be calculated, given a cosmological model (Problem 15.11), and it will be different for closed, flat, and open universes. Consequently, we are motivated to compare a plot of z versus $(1 + z)r(z)$ for theoretical cosmological models and a plot of z versus $(L/4\pi f)^{1/2}$ for observed objects with measured values of z and f and with assumed values for L. (The observations are usually made only over a finite bandwidth, rather than over all wavelengths; so corrections called "K corrections" need to be applied.) The comparison of theoretical and observational plots could then, in principle, tell us whether the universe is closed, flat, or open. Such plots are called Hubble diagrams in honor of the pioneer of observational cosmology, Edwin Hubble.

Problem 15.11 (for those who know a little calculus). We calculate the function $r(z)$ for the simple case of a Einstein–de Sitter (flat) universe. We begin with the dependence (consult Problems 15.10, 15.3, and 15.4 for the case $K = 0$)

$$D(t) = D_0\left(\frac{t}{t_0}\right)^{2/3}.$$

If t_e is the time of the emission of the photons, show that the redshift z at the time of reception by us is given by the formula

$$1 + z = \left(\frac{t_0}{t_e}\right)^{2/3}.$$

Problem 15.10 requires us to obtain r_* for a photon from the differential equation $dr_*/dt = c/D(t)$ for a flat universe. Show that $r_*(t_0)$ reads

$$r_*(t_0) = \frac{3c}{D_0} t_0\left[1 - \left(\frac{t_e}{t_0}\right)^{1/3}\right],$$

where $4\pi D^2(t_0) r_*^2(t_0)$ is the surface area of a sphere centered on the source and passing through us at the present time. Identifying $r(z) = D_0 r_*(t_0)$, show that $r(z)$ is given by the formula

$$r(z) = \frac{2c}{H_0}[1 - (1+z)^{-1/2}],$$

where we have made use of the fact that $t_0 = (2/3)H_0^{-1}$ for a flat universe. Because space is Euclidean in a flat universe, $r(z)$ is the present distance of the source from us. Compare the expression for $r(z)$ with the distance

$$l(z) = c(t_0 - t_e) = (2c/3H_0)[1 - (1+z)^{-3/2}]$$

that the photons have flown between emission and reception. Interpret the expressions at small and large z. In particular, explain how a source at large z can be further away than the light-travel distance for the age of the universe. Plot $(1 + z)r(z)$ versus z for $H_0 = 20$ km sec^{-1}/million lt-yr, i.e., for $2c/H_0 = 30$ billion lt-yr. Extra credit: for an empty universe, $\rho_m = 0$, show

$$(1 + z)r(z) = (c/H_0)z(1 + z/2).$$

Allan Sandage has been the most energetic astronomer to pursue such tests. The tests are very difficult, because at small z all standard cosmological models yield a linear relation between z and $(1 + z)r$. This linear relationship is, of course, Hubble's law, and the measurement of the slope yields H_0, Hubble's constant (see Figure 14.23). But the measurement of H_0 does not by itself tell us whether the universe is open or closed. Therefore, what is needed is to observe the *departures* from linearity in Hubble's diagram (characterized by a quantity that astronomers call $q_0 = 2\Omega_0$, the deceleration parameter). Measurable nonlinearity occurs only for relatively large values of z, and almost by definition, such objects are far away and therefore faint and difficult to measure. So far, attempts have been made using only the brightest galaxy in a rich cluster, under the assumption that the luminosity L of the brightest galaxy has the same value for all rich clusters.

Figure 15.23 shows the theoretical Hubble diagram (see Problem 15.11). Kristian, Sandage, and Westphal recently obtained data points which fall on the "closed" side of the curve labeled "flat." However, the evolution of the stellar populations in galaxies would imply that they were intrinsically brighter in the past. Larger luminosities L than assumed at larger values of the redshift z would move the observed data points toward a more "open" universe. An opposite effect would be introduced by galactic cannibalism. In Chapter 14, we discussed that the brightest galaxy in rich clusters tend to be cD systems. A cD galaxy would tend to grow brighter in time if it cannibalizes its neighbors (i.e., smaller L at larger z). The corrections needed for both effects are not now known precisely enough to allow the observational cosmological test described here to settle the issue of whether the universe is open or closed. In Chapter 16, we discuss

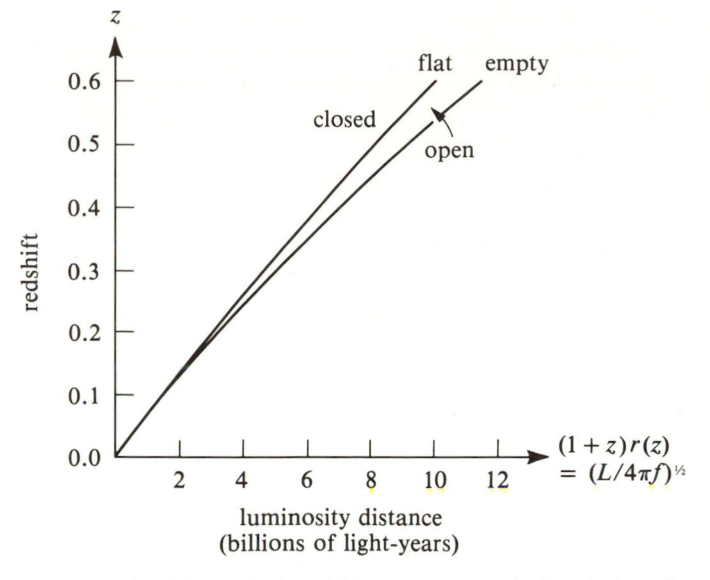

Figure 15.23. Theoretical Hubble diagram in which galaxies of a given luminosity L but of different apparent brightness f and redshift z in an open universe would lie between the curves labelled "flat" and "empty," while galaxies in a closed universe would lie above and to the left of "flat." All conventional cosmological models exhibit a linear relationship between redshift and luminosity distance at low z (Hubble's law), so only observations of the curvature in this relationship at high z can reveal whether the universe is open or closed. Unfortunately, as is discussed in the text, the interpretation of the observational Hubble diagram at high z is complicated by evolutionary and cannibalism effects.

why many astronomers believe that the universe is open. If they are correct, evolution is more important than cannibalism.

Problem 15.12. Another observational cosmological test involves the use of "standard rulers." Suppose we have a galaxy of fixed length l_0 in its rest frame, and suppose this length is oriented perpendicular to our line of sight. At cosmic time t_e, photons are emitted from the two ends of length l_0 which subtend an angle $\Delta\theta$ at our position. Since $2\pi D(t_e)r_*$ is the circumference of a circle centered on us which passes through the galaxy at time t_e (see Problem 15.10 for a definition of r_*), the length l_0 expressed as a fraction of this circumference must equal $\Delta\theta/2\pi$, i.e.,

$$\Delta\theta = \frac{l_0}{D(t_e)r_*}.$$

Because photons propagate in straight lines, the angle difference $\Delta\theta$ remains constant. The quantity r_* for a galaxy also remains constant (see Problem 15.10). Show that $\Delta\theta$ can therefore be expressed in terms of present-day quantities in the form

$$\Delta\theta = (1 + z)\frac{l_0}{r(z)},$$

where $r(z) = D(t_0)r_*$ is $(1/2\pi)$ times the circumference of a circle centered on us at time t_0 which passes through the galaxy. Describe how observed angular sizes $\Delta\theta$ and redshifts z might be used, in principle, as a cosmological test. What are practical difficulties associated with using cD galaxies as "standard rulers"?

Define $\alpha_0 = H_0 l_0/c$, and calculate its numerical value for a large galaxy. Convert your answer to seconds of arc. With the functions $r(z)$ given in Problem 15.11, plot $\Delta\theta$ versus z as expected theoretically for a flat and an empty universe. Can you explain why $\Delta\theta$ *increases* with large z for a flat universe? *Hint*: consider whether the universe as a whole can act as a gravitational lens. Construct the two-dimensional analogue of the effect (for a closed universe) by considering rulers on an expanding sphere. (See figures below.)

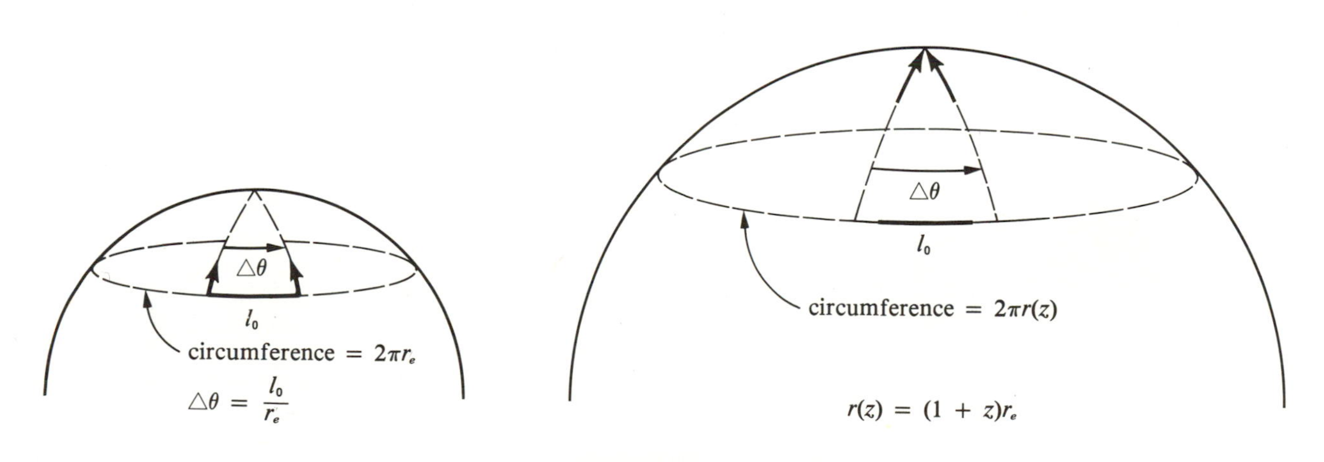

Figure P15.12

Concluding Remarks

In this chapter, we have learned that at the largest scales of space and time, gravitation dominates the scene. The universe expands, not because galaxies are moving away from us, but because of the expansion of space and time themselves. In particular, according to Einstein's geometric interpretation of gravitation, the very fabric of spacetime in the universe is manufactured by gravitation. This explains why our conception of the physical world is often referred to as "Einstein's universe."

The main astronomical issue to arise from this insight is whether the universe is open or closed. If the present average mass density is low, the universe is open, and there exists an infinity of space and galaxies. If the present average mass density is high, the universe is closed, and there exists a finite amount of space and a finite number of galaxies. If the universe is open, the galaxies will recede forever. If the universe is closed, the galaxies will be reforged in a fiery crucible. The evidence argues for open, but our hearts hope for closed. The poets have put it best.

T. S. Eliot:

"This is the way the world ends,
Not with a bang but with a whimper."

Robert Frost:

"Some say the world will end in fire
Some say in ice.
From what I've tasted of desire,
I hold with those who favor fire.
But if it had to perish twice,
I think I know enough of hate
To say that for destruction ice
Is also great
And would suffice."

CHAPTER SIXTEEN

The Big Bang and the Creation of the Material World

Independent of the final fate of the actual universe, almost all astronomers are now convinced that it began with a Big Bang. This was not always so.

Big Bang vs. Steady State

At one time, there were many adherents to the **steady-state theory** of Bondi, Gold, and Hoyle. Even Einstein was susceptible to the appeal of a universe which never changes in time on the average. The philosophical allure of the "perfect cosmological principle" (Chapter 15) is undeniably great. Before Hubble's observational discovery of the expansion of the universe (Figure 16.1), it was great enough to lead Einstein, painfully and reluctantly, to adopt a repulsive force of unknown origin (the "cosmological constant") to balance the effects of gravitation. In other words, to keep the universe unchanging in time, Einstein was willing to forsake the requirement that spacetime be flat in the absence of gravitation (i.e., satisfy special relativity). After the observational discovery of the expansion of the universe, the allure was still great enough to lead Bondi, Gold, and Hoyle to adopt the hypothesis of spontaneous creation of matter to fill the voids left by the cosmological expansion. In other words, to keep the universe unchanging in time, Bondi, Gold, and Hoyle were willing to forsake the conservation of mass-energy. Two astronomical discoveries of the 1960s, however, convinced most astronomers that the actual universe does evolve in time. Therefore, despite its enormous philosophical appeal, the steady-state theory seems definitely wrong.

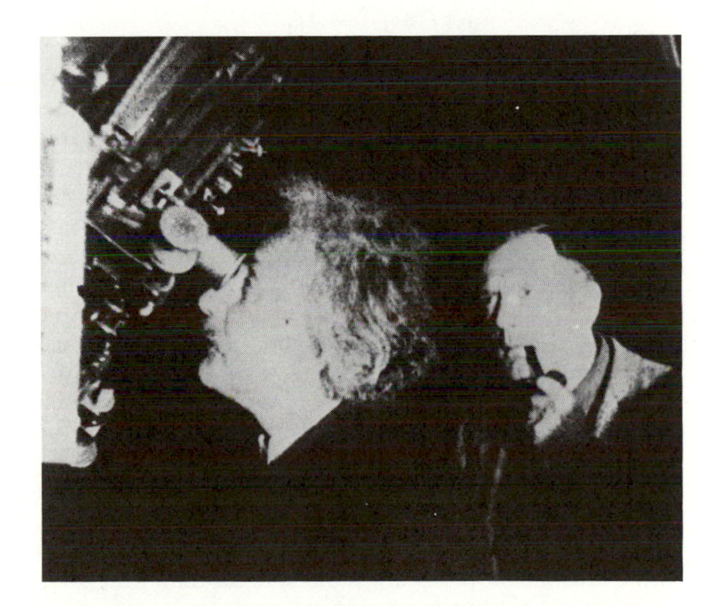

Figure 16.1. Albert Einstein visiting Edwin Hubble on Mount Wilson in 1930. (The Huntington Library, San Marino, California.)

Problem 16.1. Suppose that the Hubble constant $H_0 =$ 20 km sec^{-1}/million light-years, and that the average matter density in the universe is 10^{-6} H atoms per cm^3. What rate of creation of H atoms would be needed to maintain the present average density in a steady state in spite of the cosmological expansion? Express your answer in units of H atoms per cubic meter of space per billion years. Would such a spontaneous creation rate be easily detectable by direct means?

Distribution of Radio Galaxies and Quasars

The first piece of evidence concerned the spatial distribution of strong radio galaxies. Counts of these sources by Ryle and his colleagues revealed that there were more radio galaxies at far distances from us than at near distances. The situation is opposite that discussed in Chapter 12 for the star counts of Kapteyn. There, we noted that the expected number of sources with apparent brightness f greater than any limit f_0 should be proportional to the minus 3/2 power of f_0,

$$N(f > f_0) = Af_0^{-3/2}, \tag{16.1}$$

if the sources have a uniform spatial distribution. In other words, if we define A to equal the measured product $f_0^{3/2}N(f > f_0)$, A should be constant for a uniformly distributed population of sources in a Euclidean universe. If A decreases with decreasing f_0—as in Kapteyn's star counts—the number density of sources must decrease with increasing distance from us; whereas if A increases with decreasing f_0, the number density of sources must increase with increasing distance from us. In actual practice, corrections must be applied for interstellar extinction to interpret the star counts properly (see Chapter 12), and for the effects of cosmological expansion to interpret the radio-galaxy counts properly. In practice, the latter corrections always reduce A *below* the corresponding Euclidean (static) case. The observed results turned out to yield *increasing* values of A for decreasing f_0 (within limits; see Figure 16.2), and this implies that there must be more distant radio sources than near ones, no matter what our cosmological model is.

Now, when we look at very distant objects, we are looking at events in the remote past. The excess of strong radio galaxies in the remote past must mean that the conditions in the remote past were different from what they are today. At the very least, radio galaxies must have been more common in the past than today. Maarten Schmidt has found a similar result for quasistellar objects (QSOs), for which one can also obtain redshift information from the optical emission lines; so Schmidt has been able to quantify their evolutionary history. He found that the fraction of galaxies which are QSOs is a function of cosmological time that decreases roughly exponentially,

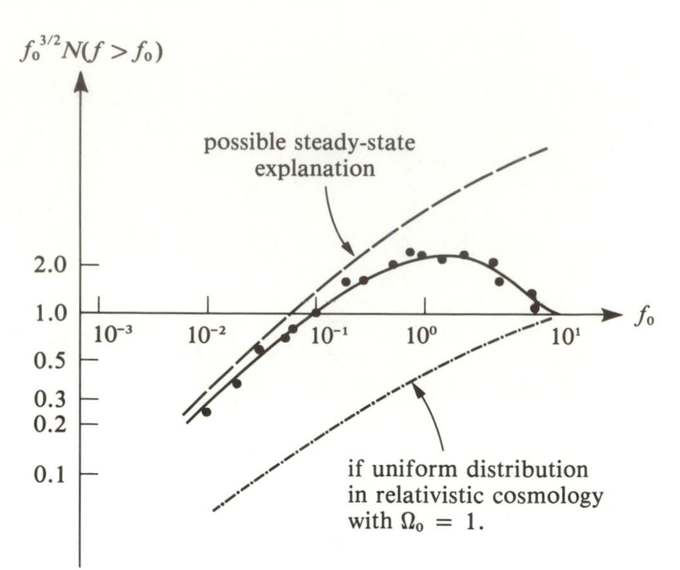

Figure 16.2. Counts of radio sources brighter than the flux level f_0 show an increase of $A = f_0^{3/2}N(f > f_0)$ for decreasing f_0 before reaching a maximum and falling. The dash-dot curve shows the expected behavior for an Einstein–de Sitter universe where radio galaxies are a constant proportion of all galaxies. The dashed curve shows the same behavior for a steady-state model, except that we have followed Hoyle and normalized the distribution at the faint end rather than at the bright end. (Adapted from Pooley and Ryle, *M.N.R.A.S.*, **139**, 1968, 529.)

$$\frac{\text{number of QSOs}}{\text{number of galaxies}} = 10^{-5t/t_0}, \tag{16.2}$$

where t_0 is the time since the Big Bang. Equation (16.2) imples that at the present time $t = t_0$, there are roughly a hundred thousand times as many galaxies as there are QSOs, whereas, very early in the history of the Universe, $t \ll t_0$, there were nearly as many QSOs as galaxies. (In other words, if QSOs are parts of galaxies, then in the beginning almost every galaxy was a QSO, if it is fair to extrapolate so far back.)

The above interpretations imply that the universe is evolving in at least one respect, namely in its population of radio galaxies and QSOs. For the universe to change any one of its global properties as a function of time would be a violation of the perfect cosmological principle, and would, therefore, disprove the steady-state theory. Now, there are possible escapes from this conclusion. For example, it has been suggested that the QSOs may not lie at cosmological distances (Chapter 13). This suggestion cannot hold for radio sources which have been identified optically with galaxies; however, the radio source counts may allow the alternative interpretation that there is a deficit of near sources rather than an excess of far sources (see Figure 16.2). Although the relative statement is the same, the philosophical implications are different. A deficit of near sources could be allowed by the steady-state theory, providing one was willing to admit the possibility of a local fluctuation near ourselves. An excess of far sources would be a global property of the universe and would, therefore, contradict the basic tenet of steady-state theory. Nevertheless, the accumulation of more and more astronomical data makes these possible escapes for the proponents of steady-state theory look extremely unlikely.

Now, you might think that all this means that the steady-state theory was a bad scientific theory, because it was ultimately shown to be wrong. On the contrary, the steady-state theory was a *good* theory, precisely because it made many unambiguous predictions that were capable of observational disproof. It is a mistaken impression in some quarters that science proceeds by proving this or that theory. Good theories can *never* be proven, because good theories make an infinity of predictions; and although one million observational checks may accord with the theory, there is always the chance that the next observation or experiment will supply a counterexample. Each successful confrontation may provide adherents of the theory with more confidence in its basic correctness, but no good theory can ever be truly said to be proven, only disproven. By these standards, the steady-state theory was an excellent theory, doubly so, because it stimulated much fruitful work while being falsified. However, the best theories are those, like Newtonian mechanics and gravitation, which provide such wonderful approximations to the truth that scientists continue to use them, within their realm of validity, even after it becomes well-known that the theories are ultimately flawed.

Olbers' Paradox: Why Is the Sky Dark at Night?

Before we discuss the second line of evidence which favors Big Bang theory over steady-state theory, let us consider an old problem related to the derivation of equation (16.1). (See Problem 12.1.) For an infinite Euclidean universe uniformly filled with stars on the average, equation (16.1) predicts an infinite number of very faint stars. Although the light from each individual star might be exceedingly feeble, the number of faint stars is very great. Thus, their total contribution might far exceed the value allowed by observations of the brightness of the night sky. This point seems first to have been recognized by Kepler in 1610, and was rediscussed by Halley and de Cheseaux in the eighteenth century, although it was not popularized as a paradox until Olbers took up the issue in the nineteenth century.

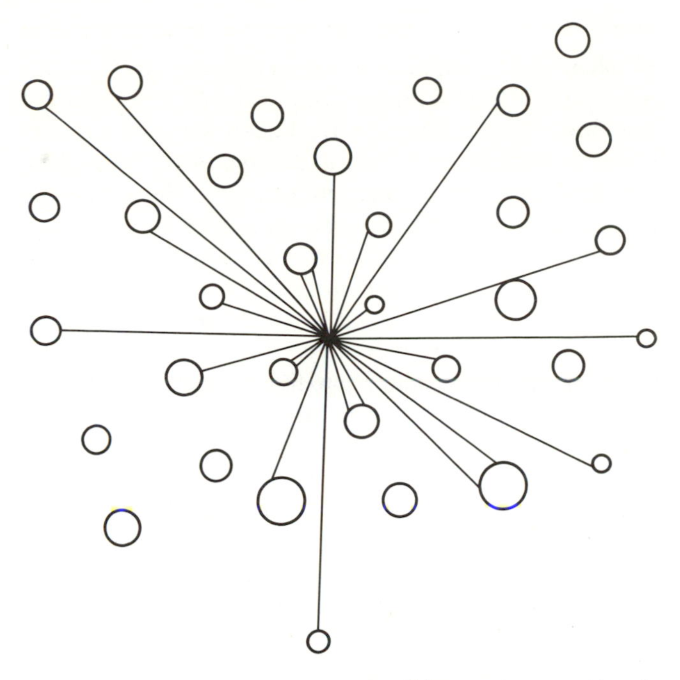

Figure 16.3. Olbers' Paradox. In a Euclidean universe uniformly and statically filled with stars on average, every line of sight will eventually intercept the surface of a star, and the night sky would be as bright as the surface of an average star.

A simple calculation gives the gist of the argument (Problem 16.2). If we assume point stars in an infinite, static, Euclidean universe uniformly filled with stars on the average, we find that the brightness of the night sky should be infinite! Taking into account that stars of finite size will block the light from more distant stars along the same line of sight, we obtain a slightly less ridiculous answer: namely the night sky should be on the average as bright as the surface of an average star (Figure 16.3). In other words, the night sky should be as bright as if we lived on the surface of the Sun! This is a patently false conclusion; hence, the paradox. The final irony arises when we remember that our discussion of the structure of stars (see Chapters 5 and 8) depends crucially on the fact that stars themselves "see" a sky which is much darker than their own surfaces. If this were not the case, each star would raise its surface temperature until it could push out the same energy as it generates in its interior. The effect of all stars doing this would ultimately cause the temperature inside and outside stars to acquire the same uniform value. Life as we know it would be impossible under these circumstances. We owe our existence to the dramatic fact that the actual sky at night is dark! What seems a trivial observation is actually a profound cosmological statement.

Problem 16.2. Suppose a number density of stars, n_0, each star with an average luminosity L. The apparent brightness f of a star at a Euclidean distance r is $f = L/4\pi r^2$, and there are $n_0 4\pi r^2\, dr$ such stars in a thin spherical shell of thickness dr and radius r. Show that the contribution to the night-sky brightness of each such spherical shell of point stars is $n_0 L\, dr$, independent of r. If we sum (integrate) the contribution of all shells out to a distance r, we obtain $n_0 L r$, which increases without bound for increasing r. In actual practice, for stars of a finite size R, show that every line of sight intercepts a stellar surface after we go an average distance $r = (n_0 \pi R^2)^{-1}$. Using this value of r, show that the total night-sky brightness over 4π steradians of solid angle equals $L/\pi R^2$. Thus, show that the **specific intensity** I (brightness per steradian) of the night sky should equal, by this argument, $\pi^{-1} F$, where $F = L/4\pi R^2$ is the energy flux leaving the surface of an average star. In other words, the specific intensity (see Chapter 4 for a definition) reaching the Earth should equal the specific intensity $I = \pi^{-1} F$ leaving the surface of a star. This, then, is Olbers' paradox: on the average, every direction we look at night should appear just as bright as when we look at the Sun at day (assuming the Sun to be an average star)! To get a hint of how to escape this absurd conclusion, calculate the numerical value of $r = (n_0 \pi R^2)^{-1}$, for $R = 1 R_\odot$ and n_0 corresponding to 4×10^{11} stars per cube of side 15 million light-years (see Problem 15.1). Convert your answer to light-years, and comment on the magnitude of the answer in comparison with cosmological distances.

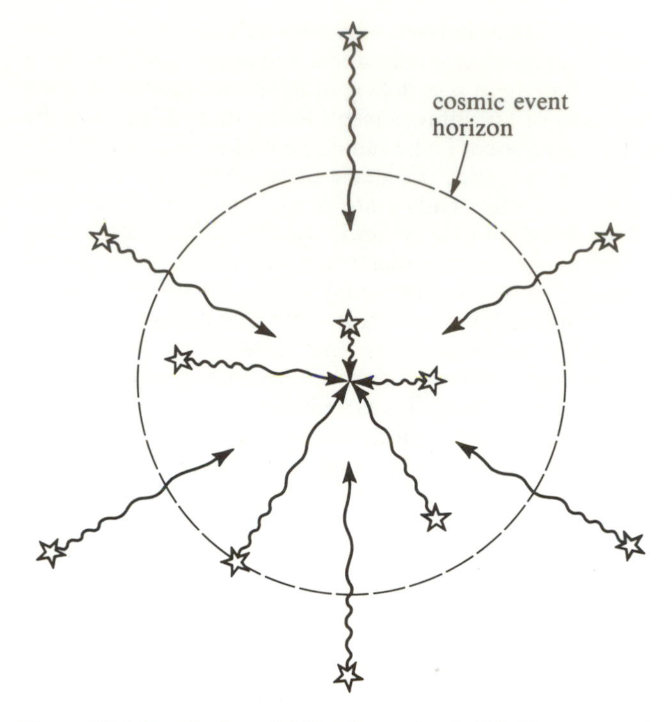

Figure 16.4. Resolution of Olbers' paradox in Big-Bang cosmologies. There are two effects which contribute to the darkness of the night sky. First and foremost, the universe has a finite age; therefore, the light from stars located beyond the cosmic event horizon has not had time yet to reach us. Second, even the light from stars within the event horizon will be redshifted (if the stars are located in distant galaxies); so the energy carried per photon will be less than when it left the star. In conventional Big-Bang cosmologies, the redshift of a star at the cosmic event horizon is infinite (photon wavelength is infinite upon arrival at Earth).

As E. R. Harrison has emphasized, in conventional cosmologies there are logically two independent ways out of this dilemma: expansion of the universe, and a finite age for the universe (Figure 16.4). Expansion helps, because the cosmological redshift prevents distant stars from making the simple Euclidean contribution to the night-sky brightness (see equation 15.6). A finite age for the universe also helps, because the light from very distant stars then simply does not have time to get here even if we were to ignore the redshift. In the steady-state theory, only the expansion argument is available; therefore, the observed darkness of the night sky requires the universe to be expanding in steady-state theories. In Big-Bang theories, cosmological expansion and a finite age go hand in hand. Indeed, for starlight, the finite age of the universe is currently far more important. Problem 16.2 shows that the distance an arbitrary line of sight must hypothetically extend before intercepting the surface of a star greatly exceeds the event horizon if the universe is on the order of 10^{10} years old. Thus, the night sky *is* dark except for those lines of sight within the event horizon which *do* intercept a star. However, there exists in the universe, as we shall soon see, a much more extended distribution of radiation than starlight. And the reason that the night sky does not glow optically from the presence of this radiation *is* because of the expansion. This cosmic background radiation turns out to be, in fact, the second and decisive piece of evidence in favor of Big Bang over steady state.

The Cosmic Microwave Background Radiation

The discovery of the cosmic microwave background by Penzias and Wilson in 1965 is rivaled in cosmological importance only by Hubble's original discovery of the expansion of the universe. Both observational discoveries were made by scientists who were unaware that implicit predictions of the phenomena discovered had, in fact, already been published in the theoretical literature some years before. For the microwave background, the pioneering theoretical investigations were those of Alpher and Gamow, although Dicke and Peebles had independently rediscovered the effect. Indeed, Dicke, Peebles, and Wilkinson were actively building an antenna to detect this prediction of Big Bang theory when they learned of the series of measurements already made by Penzias and Wilson.

The theoretical history leading to the prediction reads as follows. Around 1950, Gamow proposed that many of the heavy elements currently found in the universe were synthesized by thermonuclear reactions which took place while the universe was very young. To enable such reactions to proceed fast enough to offset the rapid decrease of density caused by the expansion of the early universe, the matter in the Big Bang must have started at very high temperatures. Associated with this high temperature would have been a thermal radiation field, whose energy density at thermodynamic equilibrium can be shown to greatly exceed that of the matter. As the universe expands, both matter and radiation would cool, and eventually the temperature would drop too low for further nuclear reactions. As we will describe in more detail shortly, eventually the matter and radiation would begin to follow separate thermal histories. Past this epoch, the radiation is no longer in thermodynamic equilibrium with the matter, but it would continue to cool as each photon is redshifted by the cosmological expansion (see equation 15.5). Under these conditions, it can be shown that the radiation continues to maintain a thermal (i.e., blackbody) spectral distribution characterized by lower and lower radiation temperatures as the universe continues to expand. By now, the radiation temperature would have dropped to very low values, no more than tens of degrees above absolute zero, according to the early estimates of Gamow and his coworkers.

Interest in these calculations seems to have waned as a result of two developments. First, Gamow never pursued seriously whether the radiation from the Big Bang was observationally detectable. Second, it soon became apparent, as Gamow conceded, that the lion's share of the synthesis of elements heavier than helium must have occurred inside stars. Dicke and Peebles worked to improve the Big Bang nucleosynthesis calculations when it became clear that stellar nucleosynthesis could probably not explain the present cosmic abundance of helium (see Chapter 12). In particular, they concluded that the best available data led to a prediction of a few Kelvin for the present radiation temperature. Thermal radiation of this temperature is best detected at microwave frequencies (wavelengths in the millimeter to centimeter regime), and together with Wilkinson, Dicke and Peebles set out to build an appropriate antenna to detect this radiation.

Figure 16.5. Arno Penzias (right) and Robert Wilson standing in front of the horn antenna which was used to detect the cosmic microwave background in 1965. (Bell Labs photo)

Problem 16.3. Use the Wien displacement law to find the wavelength of the peak emission associated with a blackbody of 3 K. Comment on the difficulties of detecting this peak radiation by examining Figure 2.3.

In this project, the Princeton astrophysicists were scooped by two Bell Labs scientists. Arno Penzias and Robert Wilson of Bell Labs had been calibrating a small radio horn designed for satellite communication (Figure 16.5). They became deeply puzzled by the ubiquitous presence of excess noise in their measurements, after they had carefully eliminated the possibility that it could have a terrestrial, solar, or Galactic origin. This noise seemed to be distributed throughout the sky in a completely isotropic fashion. When they consulted Bernard Burke of MIT about this problem, Burke realized that Penzias and Wilson had probably found the cosmic background radiation that Dicke and his colleagues were planning to measure. Put into touch with one another, the two groups subsequently published simultaneous papers detailing the prediction and discovery of a universal thermal radiation field with a temperature of about 3 K. Ironically, about a year earlier, two Soviet scientists had misread a preliminary report by Penzias and Wilson and

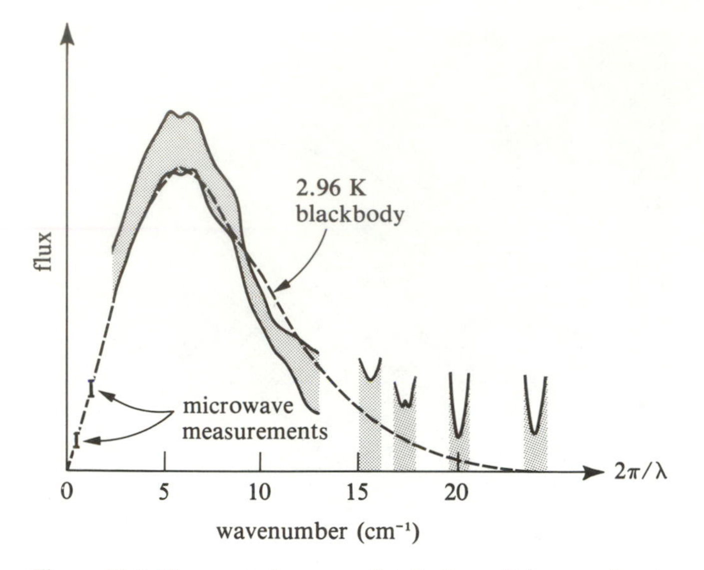

Figure 16.6. The spectral energy distribution of the cosmic microwave background radiation. The bracketed points show some of the early radio measurements together with error bars. The shaded region between the solid curves show the measurements of Woody and Richards. The dotteded curve is a theoretical blackbody of temperature 2.96 K. (Adapted from D. P. Woody and P. L. Richards, *Phys. Rev. Lett.*, **42**, 1979, 925.)

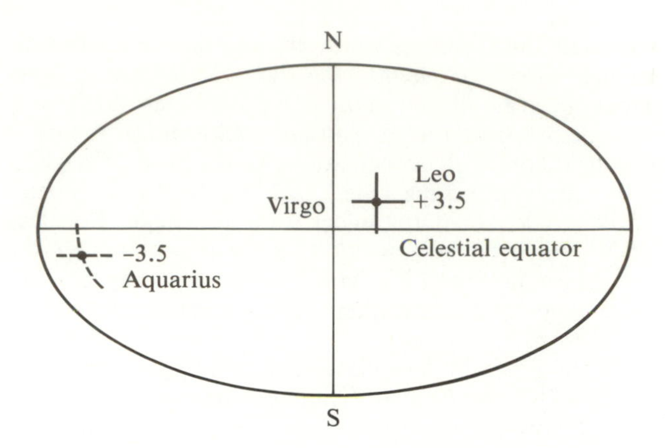

Figure 16.7. Anisotropy of the cosmic microwave background plotted on the sky in celestial coordinates. The hottest spot (solid cross) is hotter than average by 3.5 millidegrees Kelvin and lies in the direction between the constellations of Leo and Virgo. The coldest spot (dashed cross) is colder than average by 3.5 millidegrees Kelvin and lies 180° away, diametrically opposite, in the constellation of Aquarius. If the local temperature excess is plotted as a function of angle between the hottest and coldest points, the result is a cosine curve. We can understand this result as arising from the Earth's (slight) motion about the Sun, the Sun's motion about the Galaxy, and the Galaxy's motion toward the Virgo cluster. (For further details, see R. A. Muller, *Scientific American*, **238**, May 1978, 64.)

had concluded that the latter's measurements had failed to detect signals compatible with Gamow's prediction.

In the years since 1965, more accurate measurements have been made of the properties of the microwave background radiation. Several facts are now known:

(a) The spectrum of the radiation field is, indeed, thermal, with a temperature of about 3.0 K. Small departures may have been detected by Woody and Richards (Figure 16.6), but it is not clear if these departures are very significant.

(b) The intensity of the radiation field is almost completely isotropic, i.e., the same in all directions. This is an absolute must for any radiation associated with the global properties of a universe which satisfies the cosmological principle. Muller, Smoot, and their colleagues have found a slight anisotropy in the cosmic background radiation (Figure 16.7), but they think it arises from a combination of the Sun's motion in the Galaxy, and the Galaxy's motion toward the Virgo cluster (Chapter 14).

In any case, the discovery of the cosmic microwave background, which was a direct prediction of Big-Bang theory, effectively sounded the death knell for the steady-state viewpoint.

Problem 16.4. If the wavelength of each photon is Doppler-shifted because of nonrelativistic motion of the observer with respect to the frame defined by Figure 15.19, argue that the measured temperature T_{rad} of the cosmic microwave background should vary sinusoidally with angle, having a peak fractional amplitude v/c. Compute the expected fraction v/c for $v = 400$ km/sec. Discuss how one might proceed to measure such a small effect. (*Hint*: consult Muller, *Scientific American*, May 1978, p. 64.)

The Hot Big Bang

We shall now discuss the theoretical implications of the Penzias–Wilson discovery. First, let us consider the gravitational effects of this radiation. In particular, since radiation has energy, and since energy is equivalent to mass as a source of gravitation, is there enough radiation present today to close the universe? Since mass = energy divided by the speed of light squared, we may write the mass density associated with thermal radiation of temperature T_{rad} as

$$\rho_{\text{rad}} = aT_{\text{rad}}^4/c^2, \tag{16.3}$$

where a is the radiation constant and $aT_{\rm rad}^4$ is the energy density of the radiation field. For $T_{\rm rad} = 3.0$ K, we easily calculate

$$\rho_{\rm rad\,0} = 6.5 \times 10^{-34}\ \text{gm/cm}^3,$$

which is much less than the matter density ρ_{m0} known currently to reside inside galaxies. Thus, the mass density of the cosmic microwave background radiation is too small by a factor of about 10^4 (see Chapter 15) to close the universe.

Since the present matter density much exceeds the present radiation density, $\rho_{m0} \gg \rho_{\rm rad\,0}$, cosmologists say that we currently live in a matter-dominated Universe. This would not have been true in the past. To see the truth of this assertion, we may argue heuristically as follows. Since the wavelength $\lambda_{\max}$ of peak thermal emission varies with the radiation temperature $T_{\rm rad}$ in accordance with Wien's displacement law, $\lambda_{\max} T_{\rm rad} = \text{constant}$ (Chapter 4), and since this peak wavelength must stretch with the expansion of the universe in accordance with the redshift formula, $\lambda_{\max} \propto D(t)$, where $D(t)$ is the scale factor of the universe (see equation 15.5), then the radiation temperature $T_{\rm rad}$ must decrease as the universe expands according to

$$T_{\rm rad} \propto D^{-1}.$$

Thus in the past, when D was smaller, the density of the thermal radiation field must have been larger, as given by equation (16.3):

$$\rho_{\rm rad} \propto D^{-4}$$

In contrast, the matter density in the past would have been higher only according to

$$\rho_m \propto D^{-3};$$

thus, the ratio $\rho_{\rm rad}/\rho_m$ can be written in the form

$$\frac{\rho_{\rm rad}}{\rho_m} \propto D^{-1}.$$

The current ratio $\rho_{\rm rad\,0}/\rho_{m0}$ approximately equals 10^{-3} if we include only the matter seen in galaxies to calculate ρ_{m0}. Thus, at the present epoch, matter dominates radiation. But at a redshift of about 10^3, i.e., when $1 + z = D_0/D = 10^3$, and the universe was a thousand times smaller than it is now, the mass densities of matter and radiation would have been nearly equal. Before that time, the universe would have been radiation-dominated (Figure 16.8).

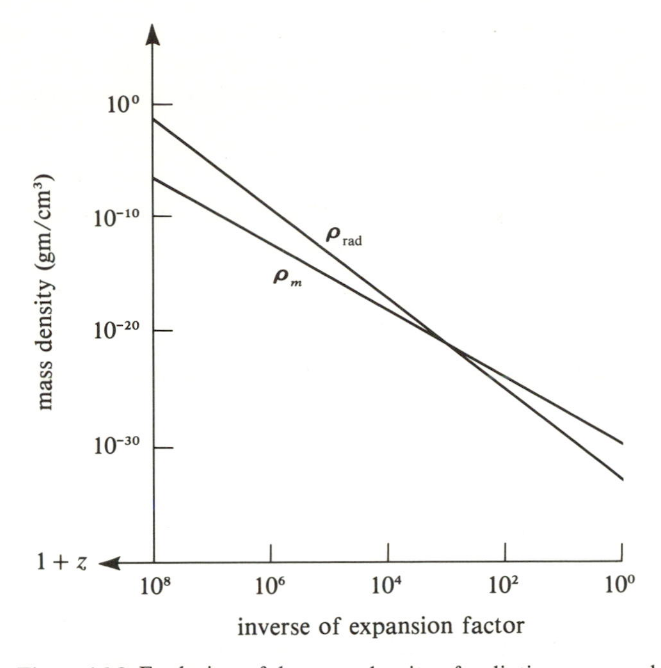

Figure 16.8. Evolution of the mass density of radiation, $\rho_{\rm rad}$, and ordinary matter (baryons), ρ_m, as a function of the expansion factor of the universe. To obtain a timescale requires adopting a specific cosmological model, and this depends on how we treat the neutrino contribution.

Problem 16.5. Suppose every photon in a distribution of photons is transformed in wavelength in accordance with formula (15.5): $\lambda_0/\lambda = D_0/D$. Show that the Planck distribution (see equation 4.6) then transforms from one characterized by temperature T_0 to one of temperature $T = T_0 D_0/D$. *Hint*: notice that the number of photons in a comoving volume is preserved, and that the number of photons is proportional to $D^3(hc/\lambda)^{-1} \cdot B_\lambda(T)\,d\lambda$ for each wavelength interval $d\lambda$.

Let us summarize in words the above mathematical deduction. If we go backward in time, the radiation density and the matter density both increase, but the radiation density increases faster than the matter density. Thus, in the beginning, radiation must have dominated over the ordinary forms of matter (not counting neutrinos). The early universe must have been exceedingly bright; the cosmic background today represents only a ghostly relict of its former splendor. This deduction of modern astronomy is, interestingly enough, in accord with the Biblical account in *Genesis* 1:1–3,

> "In the beginning God created the heaven and the Earth. The Earth was without form, and darkness was upon the deep, and the Spirit of God was moving over the face of the waters. And God said, 'Let there be light,' and there was light."

So, modern cosmology affirms, in the beginning there was light. And the brilliance of that light has begun to illuminate our understanding of the creation of the material world.

Problem 16.6. It is also possible, with a little calculus, to derive the results quoted above from the first law of thermodynamics. If we ignore nonadiabatic processes, the energy contained in a comoving volume $\rho c^2 D^3$, where ρ is the total mass density contained in matter and radiation, will change at a rate proportional to the work performed by the expansion

$$\frac{d}{dt}(\rho c^2 D^3) = -P\frac{d}{dt}(D^3),$$

where P is the pressure of the cosmological "fluid." We may idealize the mass density to consist of two components: $\rho = \rho_m + \rho_{\text{rel}}$, where ρ_m comes from particles (ordinary matter) which move very slowly in comparison with c in the comoving frame, and where ρ_{rel} comes from particles (photons, gravitons, neutrinos, etc.) which move at or near the speed of light. Only the pressure of the latter is nonnegligible; thus, we take $P = \rho_{\text{rel}}c^2/3$. (Can you prove this by the methods of Problem 5.10?) Show that, if we can ignore the effects of pair creation and annihilation, we may write

$$\frac{d}{dt}(\rho_m D^3) = 0,$$

$$\frac{d}{dt}(\rho_{\text{rel}} D^3) = -\frac{1}{3}\rho_{\text{rel}}\frac{d}{dt}(D^3).$$

Derive, therefore, the following results for nonrelativistic and relativistic particles:

$$\rho_m D^3 = \text{constant} = \rho_{m0} D_0{}^3,$$
$$\rho_{\text{rel}} D^4 = \text{constant} = \rho_{\text{rel}\,0} D_0{}^4.$$

Problem 16.7. The considerations of Problem 16.6 modify the solution $D(t)$ we derived in Chapter 15 for standard, matter-dominated cosmologies. One of Einstein's field equations would still read (see Problem 15.10):

$$\frac{\dot{D}^2}{D^2} - \frac{8\pi G}{3}\rho = -K\frac{c^2}{D^2},$$

but the total mass density is now given by

$$\rho = \rho_{m0} D_0{}^3/D^3 + \rho_{\text{rel}\,0} D_0{}^4/D^4.$$

Show, however, that closure still depends on whether $\rho_0 = \rho_{m0} + \rho_{\text{rel}\,0}$ is larger than $3H_0{}^2/8\pi G$.

The Thermal History of the Early Universe

The notion that radiation dominates matter in the early universe and that this radiation is thermal leads to a simple but incredible model for the thermal history of the universe at early times. Independent of whether the universe is open or closed, the mass density of the relativistic component of the universe (photons, gravitons, neutrinos, etc.) would behave at early times as

$$\rho_{\text{rel}} = 3/32\pi G t^2. \tag{16.4}$$

Problem 16.8. Show for small D that the Einstein field equation of Problem 16.7 can be approximated by

$$D\dot{D} = (8\pi G \rho_{\text{rel}\,0}/3)^{1/2} D_0{}^2.$$

With the Big Bang initial condition, $D = 0$ at $t = 0$, derive the solution

$$D^2/2 = (8\pi G \rho_{\text{rel}\,0}/3)^{1/2} D_0{}^2 t,$$

valid for early times. Show that the above solution can be expressed in the form given by equation (16.4). Comment on the difference in early expansion rates: $D \propto t^{2/3}$ for matter-dominated universes (see Problem 15.4) and $D \propto t^{1/2}$ for radiation-dominated universes. Which behavior is actually more relevant at early times?

If the photons were the only relativistic component of mass-energy present, we would require $\rho_{\text{rel}} = \rho_{\text{rad}} = aT^4/c^2$ in accordance with equation (16.3). However, at high temperatures and energies, pair creation from the vacuum will produce other forms of relativistic particles in thermodynamic equilibrium at the common temperature T. What kinds of pairs are present depends on the value of T. Writing $\rho_{\text{rel}} = Q\rho_{\text{rad}}$, where $Q = Q(T)$ is a pure number (greater than unity) to take account of these effects, we may express equation (16.4) in the form

$$Q^{1/4}T = \left(\frac{3c^2}{32\pi G a}\right)^{1/4} t^{-1/2}. \tag{16.5}$$

Equation (16.5) gives $T = 10^{10}$ K at $t = 1$ sec, with T decreasing thereafter by a factor of 10 for every hundredfold increase of t. In other words, T would equal 10^9 K at $t = 100$ sec; 10^8 K at $t = 10{,}000$ sec, etc. Earlier than $t = 1$ sec, the temperature would increase with decreasing t at a rate slower than $t^{-1/2}$ because of the increase in Q associated with the copious production of electron-positron and other kinds of pairs. (Figure 16.9).

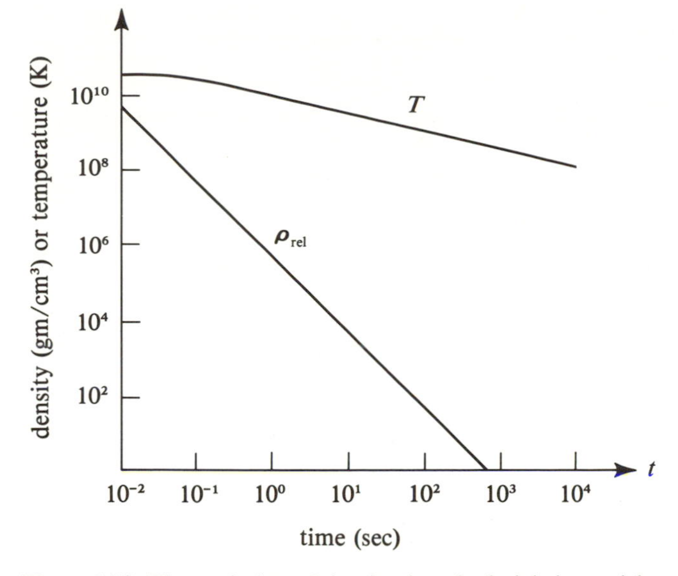

Figure 16.9. The evolution of the density of relativistic particles (photons, neutrinos, etc.), ρ_{rel}, and the (radiation) temperature T in the early universe. The simple relations (16.4) and (16.5) of the text are modified at times earlier than $t = 10^0$ sec, depending on how one treats pair creation and pair annihilation.

Problem 16.9. For what value of T would kT equal the rest energy of an electron-positron pair? A proton-antiproton pair?

Big-Bang Nucleosynthesis

At a temperature in excess of 10^{12} K (t less than 10^{-4} sec), the creation and destruction of baryon-antibaryon pairs could have been in thermodynamic equilibrium with the ambient radiation field. For reasons to be discussed later, cosmologists believe that at these early times there was a slight excess of baryons over antibaryons, an asymmetry amounting perhaps to one part in a billion. In other words, for every 10^9 antiprotons, there were 10^9 plus 1 protons. To maintain overall charge neutrality, a similar asymmetry would have existed between electrons and positrons.

As the universe continued to expand, the temperature would have dropped to lower values in accordance with equation (16.5). When the associated radiation field no longer has enough energy per photon to create baryon-antibaryon pairs, the protons and antiprotons and the neutrons and antineutrons will recombine. Each annihilation produces on average a pair of photons, which are added to the general background radiation. This radiation field is still strongly coupled to the copious number of electrons and positrons that exist in the universe via the equilibrium reactions:

$$ {}_{-1}^{0}e + {}_{1}^{0}\bar{e} \rightleftarrows \text{photons}, $$

and similar reactions would have kept neutrinos and antineutrinos present at equilibrium concentrations.

Because of the slight asymmetry of baryons and antibaryons, there would be a residue of protons and neutrons left without partners to annihilate when the temperature drops significantly below 10^{12} K. These residual baryons are kept in thermal equilibrium with one another, moreover, by the presence of the enormous sea of electrons, positrons, neutrinos, and antineutrinos, via the reactions:

$$ {}_{1}^{1}\text{p} + {}_{-1}^{0}\text{e} \rightleftarrows {}_{0}^{1}\text{n} + {}_{0}^{0}\nu, \tag{16.6a} $$

$$ {}_{1}^{1}\text{p} + {}_{0}^{0}\bar{\nu} \rightleftarrows {}_{0}^{1}\text{n} + {}_{1}^{0}\bar{\text{e}}. \tag{16.6b} $$

These reactions are, of course, mediated microscopically by virtual weakons (Chapter 6), but the energy required to make a neutron from a proton must basically be derived from an energetic electron or an energetic antineutrino. As long as enough such energetic particles are around, the relative concentrations of neutrons and protons will satisfy the equilibrium relation:

$$ \frac{N_n}{N_p} = \exp[-(m_n - m_p)c^2/kT]. \tag{16.7} $$

The right-hand side of equation (16.7) represents the Boltzmann factor (Chapter 4) associated with two energy states that differ by the rest energies of the neutron and proton (see Problem 6.6). As an application of equation (16.7), we note that there would be roughly one neutron for every five protons at a temperature $T = 10^{10}$ K.

Problem 16.10. Calculate the ratio N_n/N_p for $m_n - m_p = 0.0014 m_p$ and $T = 10^{10}$ K.

After the universe is a few seconds old, the temperature would have dropped to values appreciably below 10^{10} K. The average photon in the background would not then have enough energy to create electron-positron pairs (see Problem 16.9). At this stage, the electron-positron reactions will be driven in the direction of annihilation. Each annihilation adds two gamma rays to the general background. However, matter and radiation would still remain strongly coupled, because there will be a residue of electrons which have no positrons with which to annihilate. Indeed, matter and radiation should remain well-coupled until a time ($t \gtrsim 10^5$ yr) when the temperature has dropped to such low values ($T \lesssim 4{,}000$ K) that the electrons and protons combine to form neutral hydrogen *atoms*. Many interesting processes occur well before the universe is a hundred thousand years old; in particular, the major part of Big Bang nucleosynthesis occurs during the first *three minutes* of the existence of the universe.

When the universe is tens of seconds old, all the positrons will effectively have disappeared. The neutrinos and antineutrinos will also have decoupled from the electrons and positrons, since a decrease by a factor of 10^9 in the total number of the latter suffices to free the neutrinos and antineutrinos from the clutches of matter. The neutrinos will proceed to propagate freely through the universe, maintaining their original numbers in any volume which follows the expansion of the universe. The momentum, $p = h/\lambda$, of each neutrino will decrease as the wavelengths of the neutrino "waves" suffer the cosmological redshift $\lambda \propto D$. Thus, the present universe should be filled with a sea of low-energy neutrinos and antineutrinos with a total number density which is comparable to the number density of photons in the cosmic microwave background. Unfortunately, physicists have no way now to detect the presence of such weakly interacting, low-energy particles, despite their enormous total numbers in the universe. Curiously enough, astronomers may already have observed the indirect effects of these neutrinos if neutrinos have nonzero rest mass (a suggestion we will discuss later).

For the present, we notice that the liberation of the neutrinos and antineutrinos has a more immediate effect. The absence of electrons and antineutrinos with sufficient energy to drive the reactions (16.6) in the forward direction means that the free neutrons will want to beta-decay (Chapter 6),

$$ {}^1_0\mathrm{n} \rightarrow {}^1_1\mathrm{p} + {}^{\;\;0}_{-1}\mathrm{e} + {}^0_0\bar{\nu}, \tag{16.8}$$

with no compensating reactions to restore the free neutrons. Past $t = 1$ sec, when the temperature $T = 10^{10}$ K and one neutron is present for every five protons, therefore, the free neutrons begin to decay with a half-life of 10 minutes. However, well before 10 minutes is up, the free neutrons may be captured by free protons to form deuterium nuclei:

$$ {}^1_0\mathrm{n} + {}^1_1\mathrm{p} \rightarrow {}^2_1\mathrm{H} + {}^0_0\gamma. \tag{16.9}$$

The above reaction is, of course, possible all along, but before the temperature has dropped to about 10^9 K ($t \sim 100$ sec), the temperature is too high for the newly formed deuterium nucleus to remain bound. Remember our general thermodynamic rule (Chapter 4):

> At relatively high temperatures, matter prefers more spatial freedom and more particles. At relatively low temperatures, matter prefers more binding energy.

Thus, at the relatively high temperatures of the very early universe, there tended to be many, many elementary

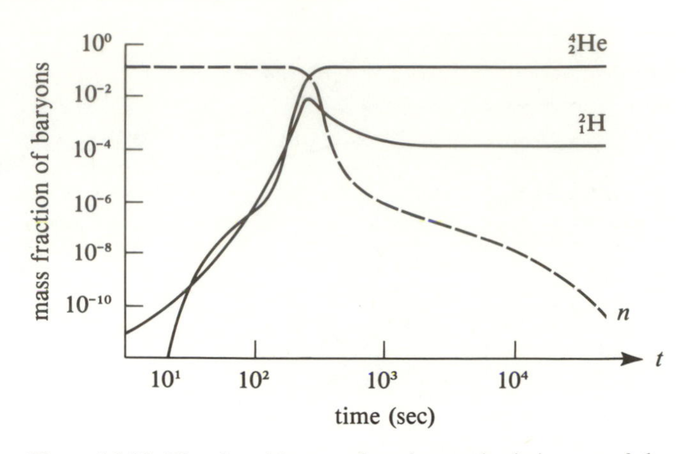

Figure 16.10. The time history of nucleosynthesis in one of the Big-Bang models of Wagoner. As time proceeds, neutron capture by protons (not shown on this graph for sake of clarity) produces deutrium nuclei, ${}^2_1$H. Further reactions produce helium-4 nuclei, ${}^4_2$He. After several minutes of cosmic time, all Big-Bang nucleosynthesis is over, and any remaining free neutrons beta-decay to protons and electrons.

particles. As the universe expands and cools, these particles gradually combine with one another. The combination of neutrons and protons to form heavier nuclei is, thus, just another step in this general tendency.

Once deuterium has formed, a rapid sequence of reactions leading mostly to the formation of helium-4 nuclei (${}^4_2$He) follows. The lack of more bound nuclei immediately heavier than helium-4 (Chapter 6), plus the rapidly declining baryon density in the expanding universe, implies that Big-Bang nucleosynthesis effectively ends with the formation of helium-4. A small residue of deuterium (as well as other trace light elements) will survive to the present day, but the amount of deuterium which can avoid being incorporated ultimately into helium-4 will be small if the baryon density when the universe is a few minutes old is not very low. Thus, realistic calculations show that most of the neutrons present at $T = 10^9$ K (t equal to about 1.5 minutes) ends up in helium-4 nuclei. Two neutrons and two protons are needed to form one helium-4 nucleus. Thus, if we started with two neutrons and ten protons (one neutron for every five protons), we would end with one helium-4 nucleus and eight hydrogen nuclei (protons). The atomic weight of the helium-4 nucleus is 4, whereas we have a total atomic weight of $2 + 10 = 4 + 8 = 12$. Thus, this simple calculation yields an expected fraction of helium-4 by mass of $4/12 = 1/3$ (33 percent). More sophisticated analyses yield a helium mass fraction closer to 1/4 (25 percent). Observationally, stars are believed to begin their lives with a helium mass fraction of 20 to 30 percent. The reasonable agreement of this range with that predicted by standard Big-Bang models is cause for optimism that we are on the right track (Figure 16.10).

Problem 16.11. Show that neutron capture by a proton (equation 16.9) is more likely than neutron decay (equation 16.8) under the conditions thought to prevail in the Big Bang. The mean time for neutron capture is given by (compare with Problem 9.11)

$$t = (n_p A v)^{-1}$$

where n_p is the number density of protons, A is the neutron-capture cross-sectional area, and v is the mean relative velocity. Apply this formula to the time when $T = 10^9$ K. Assume that the baryon *mass* density $\rho_b(T)$ at $T = 10^9$ K equals 2×10^{-5} gm/cm^3, with 5/6 of this amount in protons. Thus, take $n_p = (5/6)\rho_b/m_p$ with $v = (kT/m_n)^{1/2}$, and assume that $A = 10^{-28}$ cm^2. What is the numerical value of t? How does this compare with the neutron beta-decay time of 600 sec? To the age of the universe at $T = 10^9$ K? Draw appropriate physical conclusions.

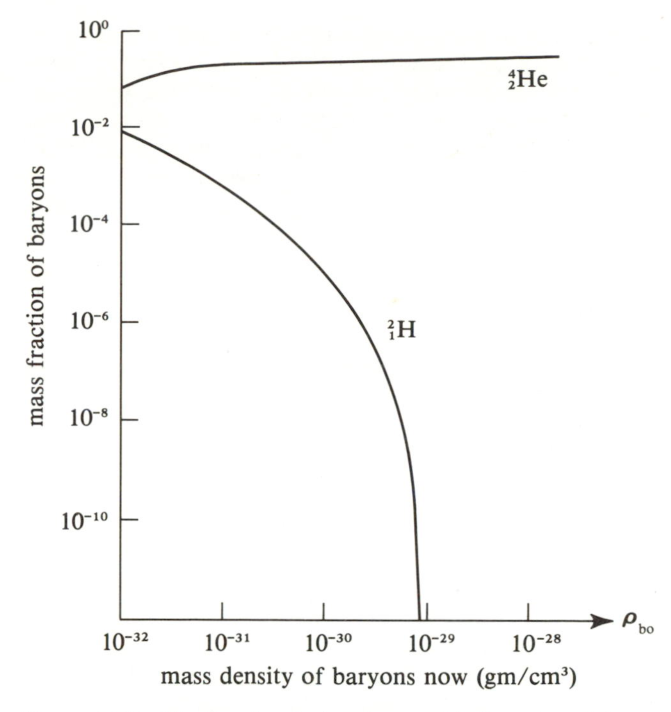

Figure 16.11. The fractional abundance of helium-4 and deuterium which results from the Big-Bang calculations of Wagoner for various assumptions about the present-day mass density of baryons in the universe, ρ_{b0}. Each point on the $^4_2$He curve and the $^2_1$H curve represents a calculation of the type depicted in Figure 16.10, extrapolated to present-day conditions. The calculations depicted here have not been updated for the latest discoveries in particle physics or for the newest measurements of the temperature of the microwave background.

The Deuterium Problem

Not all the deuterium formed from the capture of neutrons by protons will be further reacted to form helium. As we mentioned earlier, a slight residue can be expected to be left. The exact fraction left will depend sensitively on the density of baryons during the epoch of helium production, say, the density when the temperature was around 10^9 K. A higher baryon density at that time means a greater capture probability for nuclear reactions, and, therefore, a smaller fraction of surviving deuterium nuclei. The fraction of deuterium measured to be present in the Galaxy today is about 2×10^{-5} of the amount of hydrogen. To leave this fraction of deuterium unreacted, the baryon density $\rho_b(T)$ at a temperature $T = 10^9$ K must have been about 2×10^{-5} gm/cm^3. But the baryon density $\rho_b(T)$ when the temperature was T can be related to the baryon density $\rho_b(T_{\text{rad}})$ at a later time, when the background radiation temperature is T_{rad}. Baryon conservation, $\rho_b D^3$ = constant in a volume D^3 which expands with the universe, plus the redshift relation for the temperature of blackbody photons, leads to the relation

$$\rho_b(T_{\text{rad}}) = \rho_b(T)(T_{\text{rad}}/T)^3. \qquad (16.10)$$

With $\rho_b(T) = 2 \times 10^{-5}$ gm/cm^3 at $T = 10^9$ K, we obtain $\rho_b(T_{\text{rad}}) = 5 \times 10^{-31}$ gm/cm^3 at $T_{\text{rad}} = 3.0$ K (Figure 16.11). The value 5×10^{-31} gm/cm^3 is close to the density $\rho_{m0} = 3 \times 10^{-31}$ gm/cm^3 that results if we spread out uniformly the mass visible in galaxies today (Chapter 15). Thus, if the standard model of the Big-Bang theory is correct, and if the measured deuterium abundance is indeed primordial, the visible parts of galaxies probably contain most of the baryons in the universe. Since the mass contained in galaxies is short by a factor of about 25 from closing the universe (Chapter 15), many astronomers have taken this as the strongest evidence that the universe is, in fact, open.

Problem 16.12. Verify the above number, 5×10^{-31} gm/cm^3, for $\rho_b(T_{\text{rad}})$ by assuming $T_{\text{rad}} = 3.0$ K and $\rho_b(T) = 2 \times 10^{-5}$ gm/cm^3 at $T = 10^9$ K.

There are, however, several possible loopholes to this conclusion. First, standard Big-Bang theory might be wrong. Perhaps the universe was initially more chaotic than supposed in the standard models. However, detailed examination of this possibility has not provided convincing or useful results. Second, there may be other ways of producing deuterium than by Big-Bang nucleosynthesis. Unfortunately, most physical processes investigated tend to destroy deuterium rather than produce it. Finally, and most intriguingly, perhaps the bulk of the mass-energy in the universe today is *not* in its most familiar manifestation, baryons. Perhaps the bulk is in some other form! One suggestion is primordial black-holes, but this suggestion has led to an astronomical dead end—it is hard to prove or disprove. Recent developments in particle physics suggest a more attractive possibility: a nonzero rest mass for neutrinos.

Massive Neutrinos?

In recent experiments (Figure 16.12), Frederick Reines and coworkers have obtained evidence that electron neutrinos may spend a fraction of their lifetime in other neutrino states (the muon neutrino, but also perhaps the tau neutrino; see Chapter 6). Relativistic quantum mechanics does not allow such oscillations unless the neutrino (and its antiparticle, the antineutrino) have nonzero rest masses. Reines' experiments do not shed light on what the rest mass of the neutrino is, only that it cannot be zero. At the same time, some experiments involving the decay of tritium by E. T. Tretyakov and coworkers indicate that the rest mass may be on the order of 10^{-7} of the mass of the proton. The smallness of this rest mass would make the neutrino (and the antineutrino) the least massive, by far, of all particles which have nonzero rest mass, and would explain why, in most experiments carried out at reasonable energies, the neutrino behaves as if it had zero rest mass.

The fact that the neutrino has so much less rest mass than even electrons might suggest they cannot contribute much to the bulk of the universe. However, there must be enormous numbers of neutrinos present in the universe today. Let us calculate that number now on the basis of standard Big-Bang theory.

We mentioned that when electrons and positrons annihilate in the early universe, a sea of neutrinos and antineutrinos would have been liberated from the clutches of matter. At that time their numbers would have been comparable to the numbers of blackbody photons. In any volume D^3 which follows the expansion of the universe, the numbers of both neutrinos and photons would have been conserved (assuming negligible rates of net annihilations); so the number density of neutrinos and antineutrinos today must still be comparable to the number density of cosmic background photons. Let us, therefore, calculate the latter value (which is directly observable) and thereby obtain an estimate of the number density of neutrinos and antineutrinos.

The energy density associated with blackbody radiation of temperature T_{rad} is $aT_{\text{rad}}{}^4$, and the mean energy per photon is $2.7kT_{\text{rad}}$. Thus, the number density of blackbody photons is equal to $aT_{\text{rad}}{}^4/(2.7kT_{\text{rad}}) = 0.37aT_{\text{rad}}{}^3/k$, or about 5.5×10^2 cm^{-3} for $T_{\text{rad}} = 3.0$ K. Let us suppose, therefore, that the current number density of neutrinos and antineutrinos is 10^2 cm^{-3}. The random speeds of the neutrinos have been reduced by the cosmological expansion to nonrelativistic speeds (Problem 16.13); so their current mass-energy resides primarily as rest mass, say, 10^{-31} gm per particle. The neutrino-antineutrino contribution to the mass-energy content of the current universe would then be $\rho_{\nu 0} \sim 10^{-29}$ gm/cm^3, and this number is very nearly equal to the critical density required to close the universe! (See Chapter 15.)

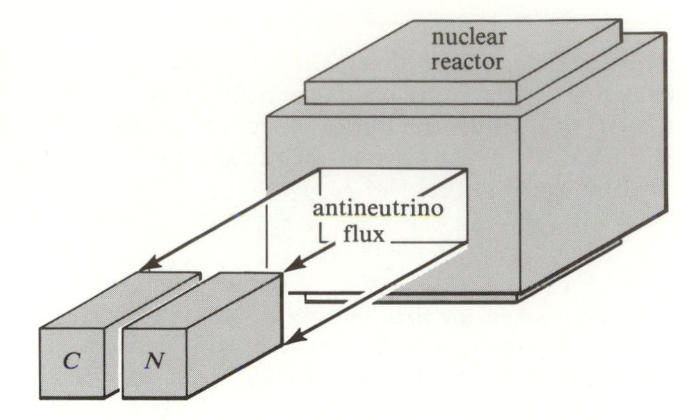

Figure 16.12. A schematic depiction of the neutrino experiment of Frederick Reines, Henry Sobel, and Elaine Pasierb. Antineutrinos generated by a nuclear power plant undergo two types of weak interactions. The first (C) is mediated by charged weakons and is, therefore, sensitive to electron-type antineutrinos. The second (N) is mediated by neutral weakons and can be triggered by any type of antineutrino (electron, muon, or tau). In the absence of neutrino oscillations, the two type of interactions should measure the same antineutrino flux. In the actual experiments, there are less than half the charged weakon interactions compared to the neutral ones, indicating that the antineutrinos from the reactor spend less than half their time in the electron-type state.

It is interesting to pursue the reason why neutrinos make a more important contribution than the nuclei of atoms. As an average, the number density of baryons equals ρ_m/m_p, where m_p is the mass of the proton and ρ_m is the mass density that would result if we were to spread out smoothly the matter contained in galaxies. Thus, the ratio $\mathcal{N}$ of the number density of blackbody photons, $0.37aT_{\text{rad}}{}^3/k$, to the number density of baryons, ρ_m/m_p, is given by

$$\mathcal{N} = 0.37(aT_{\text{rad}}{}^3/\rho_m)(m_p/k) = 2 \times 10^9, \quad (16.11)$$

if we substitute in the observed values $T_{\text{rad}} = 3.0$ K and $\rho_m = 3$ to 5×10^{-31} gm/cm^3 for the present epoch. It is a comparably large ratio of number density of neutrinos and antineutrinos to baryons which makes the former so important if each neutrino were to have even a tiny rest mass (say, $10^{-7}m_p$).

Problem 16.13. Let us crudely try to estimate the current temperature T_ν of the sea of neutrinos and antineutrinos. (More precise estimates have been made recently by Bond.) As long as the neutrinos remain relativistic, their temperature will remain equal to that of the blackbody photons in an adiabatic expansion. When the neutrinos become nonrelativistic, their temperature will drop by approximately one power of the expansion factor D faster. Prove this by considering the equations

$$E^2 = m_\nu{}^2c^4 + p^2c^2, \tag{6.6}$$

and

$$p = h/\lambda,$$

(3.5) with $\lambda \propto D$ (equation 15.5). Assume that kT_ν approximately equals the mean kinetic energy $E - m_\nu c^2$ of the distribution of neutrinos, and notice that the latter is proportional to p in the ultrarelativistic regime and to p^2 in the nonrelativistic regime. Argue that the transition from relativistic to nonrelativistic behavior occurs roughly when $kT_\nu = kT_{\text{rad}} = m_\nu c^2$, where T_{rad} is the photon temperature. Show that this condition works out to be $T_{\text{rad}} = 10^6$ K if $m_\nu = 10^{-7}m_p$. But $T_{\text{rad}} = 10^6$ K when D was $(10^6/3.0)$ times smaller than now. Thus, $T_{\nu 0} \sim T_{\text{rad }0}(3.0/10^6) \sim 10^{-5}$ K. Calculate the resulting thermal speed $(kT_\nu/m_\nu)^{1/2}$ at the present time.

Needless to say, the above conclusion should be considered very tentative, since the experimental results on the nonzero rest mass of the neutrino were preliminary findings as this book is being written. More work needs to be done, both on the particle physics and on the cosmological models, before we can secure a more definitive understanding. Nevertheless, no one can deny that the possibility of a nonzero neutrino rest mass offers a very attractive solution to a variety of astronomical paradoxes.

Foremost among these puzzles is the issue of "missing mass." Apart from the question of the closure of the universe, astronomers have found that there must be more gravitating mass associated with individual galaxies and with clusters of galaxies than can be detected by its electromagnetic radiation (Chapters 12, 13, and 14). A large mass of neutrinos gravitationally bound to such objects would help to explain the discrepancy. We have also mentioned how neutrino oscillations might help to explain the missing "solar neutrinos" (Chapter 6). The careful sorting out of the possible effects is likely to occupy physicists and astronomers for some years to come.

Table 16.1. **Important events which follow the era of Big-Bang nucleosynthesis.**

Cosmic time t[a]	*Redshift z*	*Event*
3×10^5 yr	10^3	Universe becomes transparent. Baryon density exceeds radiation density. Globular clusters form?
3×10^8 yr	10^1	Galaxies form?
1×10^9 yr	$10^{0.5}$	Galaxies cluster? Quasars born?

[a] Calculated from the redshift assuming that we live in an Einstein–de Sitter universe with $t \propto (1+z)^{-3/2}$ with $t_0 = (2/3)H_0{}^{-1} = 10$ billion years.

The Evolution of the Universe

Table 16.1 lists some of the more important epochs that follow the era of Big-Bang nucleosynthesis. We are especially interested in events which could, in principle, be observed; consequently, we will describe them not in terms of the cosmic time t during which they occurred (which would depend on adopting a particular cosmological model), but in terms of the redshift z (where $1 + z = D_0/D$) that photons emitted by the events would exhibit when they arrive now at Earth.

The combination of electrons and protons to form neutral hydrogen atoms would occur at a redshift z approximately equal to 10^3 (when $T_{\text{rad}} = 4{,}000$ K). Helium would have become neutral slightly earlier. Once the cosmological fluid becomes largely electrically neutral, radiation becomes decoupled from matter and streams freely through the universe. It would, however, continue to maintain a thermal character, with ever-decreasing radiation temperature as long as the universe continues to expand, for the reasons outlined earlier. The disappearance of free electrons to scatter photons has the following important observational consequence. After $z = 10^3$, photons could fly through most of the universe in a straight line, subject only to spacetime curvature; before that time, photons had to random walk through the cosmological plasma (Figure 16.13). Thus, $z = 10^3$ represents the "photosphere" of the universe, beyond which we cannot peer via the use of electromagnetic radiation. In particular, the cosmic microwave photons that we now receive on Earth flew to us on a straight line from that epoch. The isotropy of that radiation implies that the universe was exceedingly smooth at $z = 10^3$.

The other significant occurrence at $z = 10^3$ is that the mass-energy density of baryons and photons would have been roughly equal at this time. Before this time (i.e., for z greater than 10^3), the photon energy density would have been larger than that of ordinary matter (although less

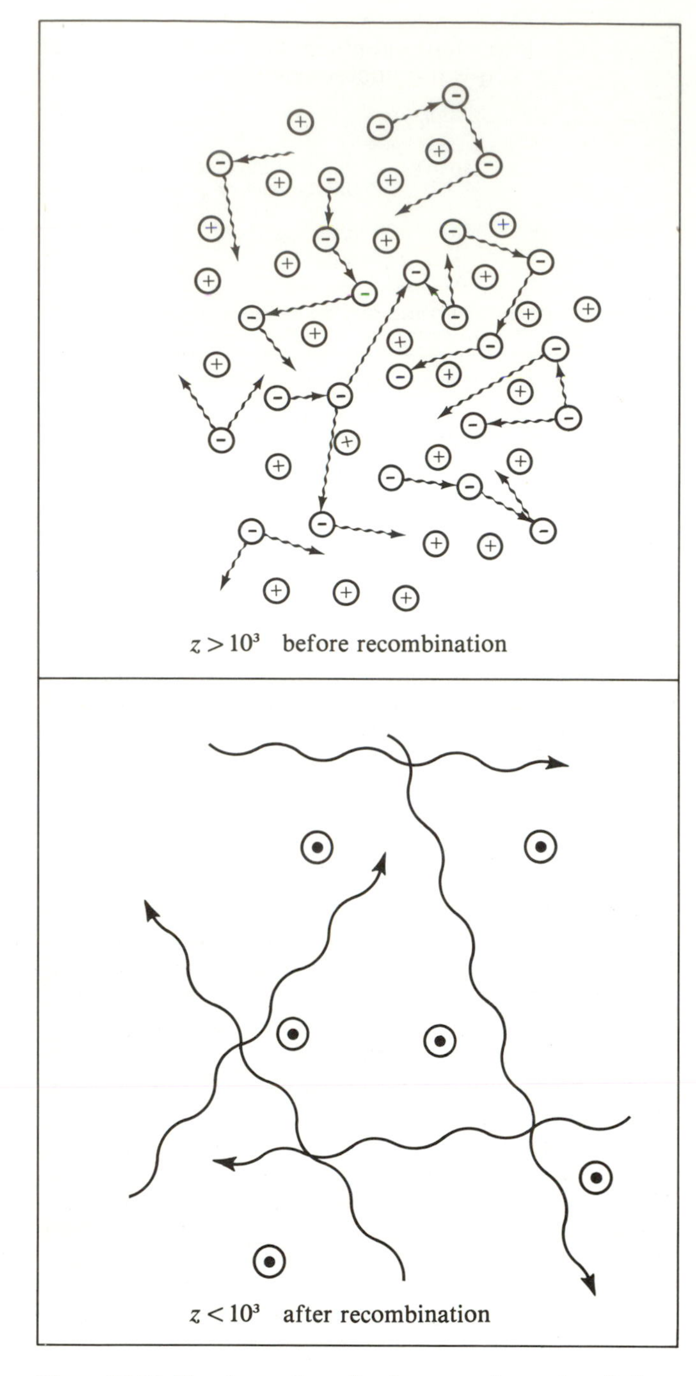

Figure 16.13. The decoupling of ordinary matter and radiation. At redshifts larger than 10^3, the temperature is high enough to keep hydrogen ionized. The presence of many free electrons forces radiation to perform a random walk through the universe. After recombination, all the electrons are bound to neutral hydrogen atoms, and radiation can freely stream through the universe. The microwave radiation that we receive today began to fly to us on a straight line at a redshift equal to about 10^3.

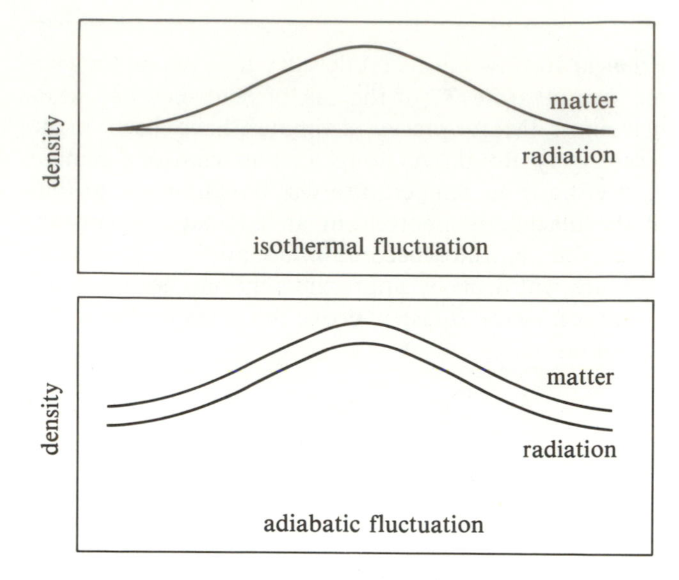

Figure 16.14. Conceptually, two kinds of density perturbations can arise. In an isothermal fluctuation, the matter density increases in some regions, but the background radiation field remains uniform (i.e., at constant temperature). Before decoupling, however, such perturbations cannot grow, because the intense radiation field effectively prevents the electrons from moving much. After decoupling, such perturbations can grow, and the smallest unstable disturbance has a mass comparable to that of a globular cluster. In an adiabatic fluctuation, the matter and radiation vary together in time and space (like a generalized kind of sound wave)' Such disturbances tend to be damped by the gradual diffusion of heat from the compressed portions of the radiation field; however, the damping time may be quite long for disturbances of large spatial scale. After decoupling, the matter component of the latter disturbances can also grow. Silk found the smallest such disturbance which can be expected to survive the damping of the earlier coupling era to contain a mass comparable to a cluster of galaxies.

than that contained in neutrinos). The resulting domination of ordinary matter by radiation would have prevented matter from clumping significantly by self-gravity onto itself. For baryons to clump, the protons would have to drag the electrons along electrically (to maintain overall electric neutrality), but the regions of enhanced electron density are kept from growing by the random scatterings produced by the intense background radiation field (Figure 16.14). Astronomers believe that the formation of gravitationally bound subunits of ordinary matter (e.g., galaxies and clusters of galaxies) therefore had to wait for the decoupling of ordinary matter from radiation.

Dicke and Peebles proposed that the first objects to form after decoupling were the globular clusters. They based their argument on a modern update of a problem first conceived by Newton, and solved quantitatively by Sir James Jeans in 1902. Jeans posed the problem thus.

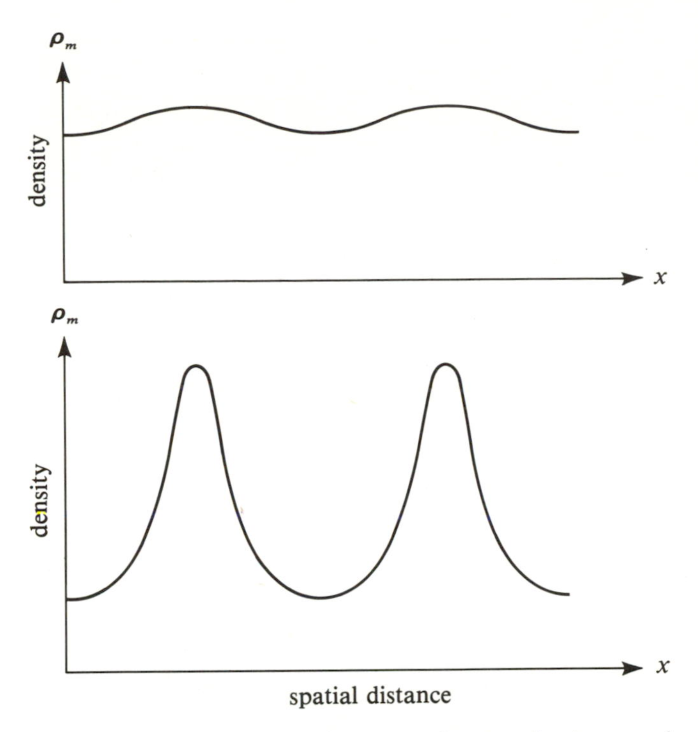

Figure 16.15. The growth of increases of matter density caused by gravitational instability was a problem first studied by Sir James Jeans. Perturbations of the type depicted in the top drawing will be unstable and grow in amplitude (after decoupling) as depicted in the bottom drawing, providing the wavelength of the disturbance exceeds 2.3 times the Jeans length defined by equation (16.12).

Suppose one has an infinite, static, material medium of temperature T and density ρ_m, and suppose one were to perturb this medium slightly to introduce density irregularities (Figure 16.15). The regions of higher density than the average could be expected to feel an increased gravitational attraction toward their center, but this attraction would be resisted by the fact that the pressure of these regions has also grown during the compression. Under what conditions would gravity be able to overwhelm pressure and produce a permanently bound subunit? The answer that Jeans found was that all disturbances of sufficiently large scale could be expected to be gravitationally unstable and to condensations, and his criterion for the minimum spatial scale worked out to be (within factors of order unity)

$$L_J = \left(\frac{kT}{mG\rho_m}\right)^{1/2}, \tag{16.12}$$

where m is the mean mass of the particles of the material medium. If we apply the above formula to the cosmological conditions at decoupling, $T = 4{,}000$ K, $\rho_m = 10^{-21}$ gm/cm^3, and $m = 2 \times 10^{-24}$ gm, we obtain $L_J = 6 \times 10^{19}$ cm. This length equals about 60 light-years, a reasonable value for the diameter of a globular cluster, especially since the final value for the bound object is likely to be somewhat smaller than the initial value. Moreover, the mass contained within a volume $L_J{}^3$ works out to be $\rho_m L_J{}^3 = 2 \times 10^{38}$ gm, i.e., about $10^5 M_\odot$, a very respectable mass for a globular cluster. This calculation accords also with our knowledge that globular-cluster stars are among the oldest in the known universe, but it would not explain why no globular cluster has yet been found which exhibits *zero* heavy-element abundance.

Problem 16.14. To derive Jeans' criterion (16.12), let us argue as follows (see also Problem 4.4). A mass $M = \rho_m L^3$ composed of a classical perfect gas will collapse gravitationally if its internal pressure $P = \rho_m kT/m$ cannot withstand the weight of a column of material of unit area and height L (see Chapter 5). The weight of the latter is roughly given by $(GM/L^2)(\rho_m L)$, and if we set this greater than or equal to $P = \rho_m kT/m$, we can obtain a constraint on the sizes of fragments which will separate gravitationally out of the general medium. Show that the minimum size L_J is given by Jeans' criterion (16.12). Show by scaling ρ_{m0} at $T_{\text{rad}} = 3.0$ K to ρ_m at $T_{\text{rad}} = T = 4{,}000$ K that $\rho_m = 10^{-21}$ gm/cm^3, and verify the numerical values given in the text for the size and mass of the minimum Jeans fragment. *Extra credit*: how long does the collapse process take?

How soon after redshift 1,000 did galaxies form? If quasars are parts of galaxies, galaxies must have formed before redshifts of about 3 or 4, because the most distant quasars (in the cosmological interpretation) seen to date have those kinds of redshifts. Thus, galaxies are believed to have formed somewhere between redshift $z = 1{,}000$ and $z = 4$. Many astronomers guess at a redshift of about 10, because the right kinds of timescales and densities are reached about then. Unfortunately, despite much work on this important problem, no one has given a really convincing theoretical reason why galaxies should have formed with their observed properties; all theories so far have had to assume rather special initial conditions for the fluctuations that are supposed to grow to become galaxies upon further gravitational condensation. The fundamental issue of even why there are two basic types of galaxies, spirals and ellipticals, has so far eluded a convincing theoretical explanation. We don't know whether protogalaxies collapsed when the bulk of their material was still gaseous or when it was in the form of stars. Larson has pursued the old idea that a difference in the rate of formation of stars in the collapsing protogalactic cloud accounts for the formation

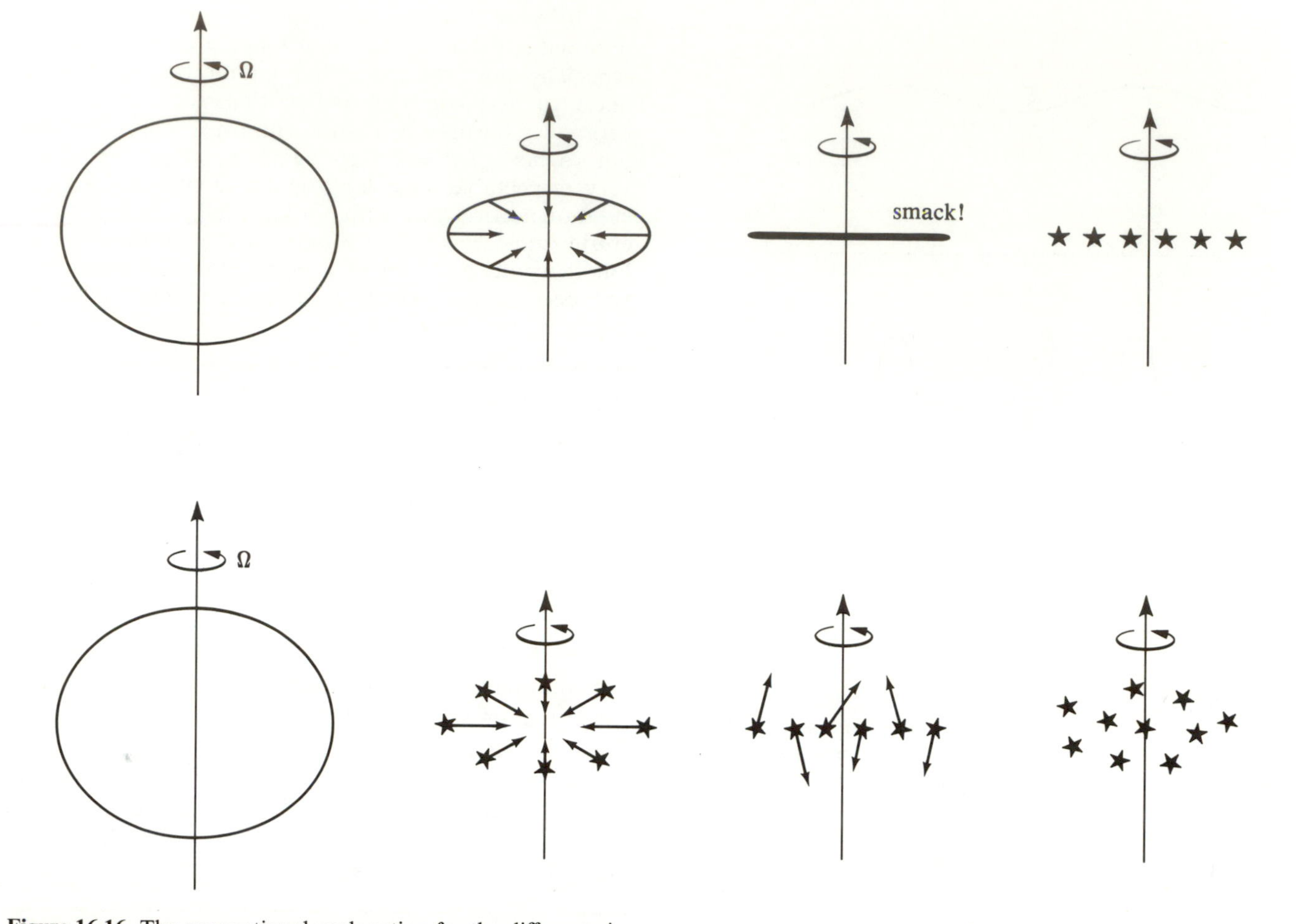

Figure 16.16. The conventional explanation for the difference in shapes between a disk galaxy (upper) and an elliptical galaxy (lower) relies on a difference in how fast stars form relative to the collapse of the protogalactic cloud (leftmost objects). In this view, disk galaxies result when most of the stars do not form until after the gas has produced a big smack and dissipated all the motions perpendicular to the principal plane of the galaxy. Ellipticals result when stars form before collapse to a thin disk is completed.

of ellipticals and spirals. In this picture (Figure 16.16), spirals form when the collapse of a rotating protogalactic cloud to a disk takes place while most of the matter is still in a gaseous form. Because the disk can dissipate the perpendicular motions generated by the collapse process, the resulting stars that form from the gas disk will have a flattened distribution. Ellipticals result when stars form while the collapse is still underway. The lack of dissipation then results in a more rounded final distribution, as has been explicitly demonstrated by Gott, Binney, and others. No one has satisfactorily explained why such different star-forming efficiencies should exist for the two types of galaxies, and other approaches may be possible (see Chapter 14). Perhaps the problem of galaxy shapes cannot be solved until astronomers have arrived at a better fundamental understanding of how and why *stars* form (Chapters 11 and 12).

The Creation of the Material World

From the preceding discussion, we see that there are still major gaps in our understanding of the *macroscopic* organization of the universe. What about our understanding of the *microscopic* organization? In particular, although the hot Big-Bang theory has numerous successes to its credit—the prediction of the cosmic microwave background, the explanation of the hydrogen, deuterium, and helium content, and the possibility of (perhaps) closing the universe after all—it still depends on arbitrary assumptions about "initial conditions" for the calculations. There was, for example, the curious asymmetry of one part in a billion between matter and antimatter. This asymmetry was, of course, not known *a priori*, but was inferred from the ratio of photons to baryons, $\mathcal{N} = 2 \times 10^9$, calculated for the present universe. What is the origin of the latter number? There is an even more fundamental question: where did all these particles (including the photons) come from anyway? For our discussion of Big-Bang nucleosynthesis, we merely assumed them to be present when the universe was a tenth of a millisecond old ($t = 10^{-4}$ sec). A very attractive alternative idea is that at $t = 0$, at the very beginning of time, no particles whatsoever existed—no baryons, no leptons, no photons—nothing, a vacuum. All the required particles were subsequently created from this vacuum. Such a picture would explain one of the most remarkable features of our physical world, namely that, as far as we can tell, the universe as a whole is exactly electrically neutral. If everything that now exists ultimately came from the vacuum, the explanation would reduce to the fact that a vacuum has zero total electric charge. Can we therefore learn the nature of the creation event if we probe what happened at the earliest cosmic times?

The Limits of Physical Knowledge

Before we discuss current ideas on this heady subject, let us first ask how far back in time (to $t = 0$) dare we push equation (16.4)? This equation gives the mass-energy density required in general relativity to account for the spacetime curvature of a model universe that satisfies the cosmological principle even at the very earliest instants. The isotropy of the cosmic microwave background gives us *observational* confidence in the soundness of this assumption for most of the time when this background existed, but now we wish to discuss times when even this most fundamental of all radiation may have been absent. Indeed, quantum mechanics forbids the direct creation of photons from the vacuum, even in curved spacetime. Thus, particles with rest mass have to be created first, and any extrapolation of equation (16.5) backward in time must fail at very early times. How about equation (16.4)?

Bold questions require bold action; so let us ask the reverse question. Is there any fundamental *theoretical* reason why we should not extrapolate equation (16.4) right back to $t = 0$? If there is such a reason, what is the smallest value of t that we may consider within the context of current physical knowledge? The answer turns out that there is a limit, and, not surprisingly, it depends on what constitutes the proper notion of space and time near the instant of creation.

In every physical theory devised by humans so far, spacetime has been treated as a continuum. By inventing general relativity, Einstein introduced not only the possibility that spacetime might be bent, but also the possibility that it might be punctured (develop singularities, such as black holes). Near a mass m, classical gravitation physics states that events within length-scales of order (see equation 7.4)

$$L = Gm/c^2$$

are hidden from the outside world. On the other hand, the uncertainty principle states that quantum effects (e.g., the Hawking process) limits our ability to locate the event horizon on scales smaller than the Compton wavelength (see Problems 7.12 and 13.17)

$$\lambda = h/mc.$$

If we equate the expressions for L and λ, we may solve for m to obtain the expression for the Planck mass (see Problem 6.13; sometimes these definitions are made with $\hbar$ substituted for h):

$$m_P = (hc/G)^{1/2} = 5.46 \times 10^{-5} \text{ gm}. \qquad (16.13)$$

This mass is not to be confused with the proton mass m_p, which is some 10^{19} times smaller than m_P. The lengths L and λ corresponding to $m = m_P$ are called the Planck length,

$$\lambda_P = (Gh/c^3)^{1/2} = 4.05 \times 10^{-33} \text{ cm}, \qquad (16.14)$$

whereas the light-travel time across this length is called the Planck time,

$$t_P = (Gh/c^5)^{1/2} = 1.35 \times 10^{-43} \text{ sec}. \qquad (16.15)$$

No one knows how to discuss the history of the universe meaningfully, within the context of a classical theory of gravitation (i.e., general relativity), for instants t less than the Planck time t_P. For $t = t_P$, the mass density of the universe, equation (6.4), is such that the mass $\rho 4\pi r^3/3$ contained within the cosmic event horizon $r = ct$, becomes of the order of the Planck mass (16.13). Under these circumstances, spacetime cannot be thought of even approximately as a smooth continuum. Instead, Wheeler has conjectured, it might take on a foamlike consistency that lacks even simple connectivity (Figure 16.17). A quantum theory of gravity is needed, and we shall discuss later one promising candidate supergravity.

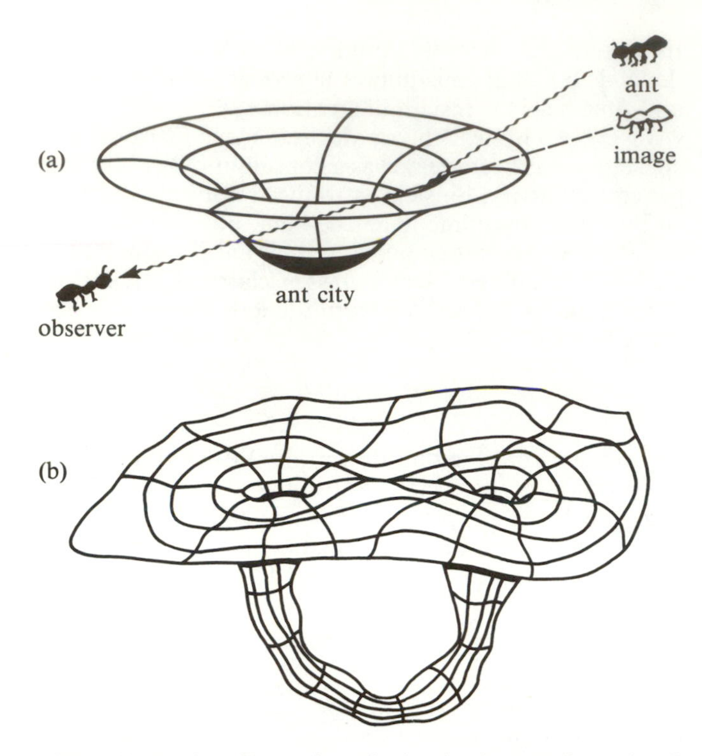

Figure 16.17. The effects of gravitation in the two-dimensional ant world of Chapter 15 may be represented by local curvatures in an unobservable third dimension. (a) For example, the mass of an ant city can cause a dimple in the fabric of space, so that antlight pursuing the shortest distance between two points can seemingly be bent to give the phenomenon of deflection of light. (b) In an expanding ant world, space at the earliest instants of cosmic time might have had a foamlike structure. Every little piece of space may have been chaotically curved by quantum fluctuations. Occasionally, a well might form whose structure makes an outside observer believe that a massive object exists at the bottom of the well; however, a brave explorer plunging into the well might find that the tunnel leads nowhere except to the mouth of another well elsewhere in space. Some people have speculated that particle-antiparticle pairs might represent nothing more than such contortions of space (or superspace) in our universe. Can mass-energy ultimately be given a purely geometrical interpretation?

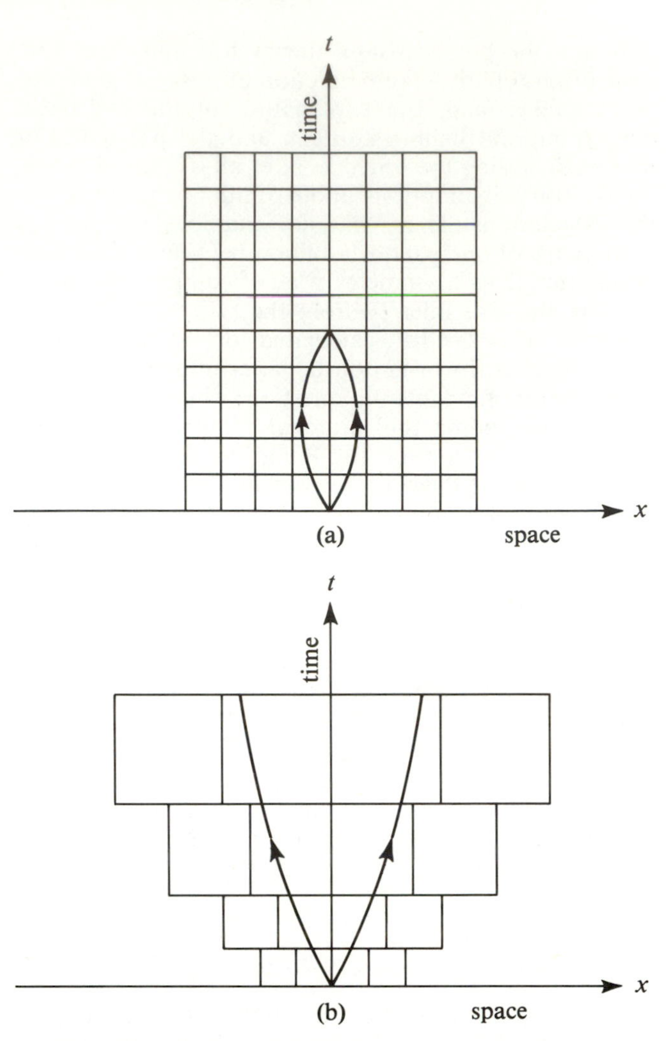

Figure 16.18. Pair creation from the vacuum. (a) In flat spacetime, only virtual pairs can result because there is not enough energy in the vacuum (by definition) to create the appearance of a real particle-antiparticle pair. (b) In expanding spacetime (which shouldn't be plotted on a sheet of flat paper, but we have to make do here), real particle-antiparticle pairs can be ripped apart by the violent expansion of spacetime itself.

Problem 16.15. Compute the numerical value of the mass density of the universe for t equal to the Planck time. How much denser is this than nuclear densities?

Mass-Energy From the Vacuum?

For t appreciably greater than the ridiculously short time, $t_P \sim 10^{-43}$ sec, we may adopt a classical treatment for the gravitational field and treat spacetime as a smooth continuum. Within this context, many physicists (including Zeldovich, Weinberg, and Parker) have asked the following interesting question. Is it possible that the universe might have started as a perfect vacuum, and all the particles we know of in the material world were created from the expansion of spacetime? The detailed calculations to answer this question are complex, and only partial results are available so far. However, the basic ideas are simple, and we will discuss them via some heuristic arguments given recently by Wade and Sachs.

We mentioned in Chapter 7 that the vacuum of relativistic quantum mechanics is not the uninteresting empty state that it is in classical mechanics. Instead, the vacuum is envisaged as a state of minimum energy where quantum fluctuations can lead to the temporary formation of particle-antiparticle pairs. In flat spacetime, destruction follows closely upon creation (the pairs are *virtual*), because all the known forces between a particle and antiparticle are attractive and will pull the pair back together to annihilate one another (Figure 16.18a). In the violently expanding spacetime of the early universe, however, the particle and antiparticle may be ripped apart and gain real existence (Figure 16.18b). If one prefers to think of gravity as a force rather than as spacetime curvature, one may say that the ripping apart of the pair is caused by the violent tides that exist in the early universe. From this point of view, the energy required to give the particle and antiparticle real existence must lead to a back reaction in the rate of the expansion of the universe. (In other words, equation 16.4 assumes that the dominant form of mass-energy has negligible rest mass, which need not be true at very early times.) No fully self-consistent calculations are yet available; it is quite conceivable that the proper inclusion of the back reaction would lead to a unique model for the creation of the universe. The order-of-magnitude arguments that we are about to embark upon, however, do not depend sensitively on the precise expansion rate of the very early universe.

To create particles of mass m, the pair must be pulled apart for a time Δt which satisfies the uncertainty principle (we ignore factors of 2 throughout what follows):

$$mc^2 \, \Delta t \sim h. \tag{16.16}$$

Let us suppose that the end of the process leaves the pair separated by a distance l. This separation results from the application of a relative acceleration g during a time interval Δt:

$$l \sim g(\Delta t)^2. \tag{16.17}$$

If we ignore all force fields in comparison with the enormous differential accelerations produced by gravity in the early universe, we may write

$$g \sim (\ddot{D}/D)l \sim l/t^2. \tag{16.18}$$

To derive equation (16.18), it does not matter whether we use $D \propto t^{1/2}$ (Problem 16.8) or $D \propto t^{2/3}$ (Problem 15.4) for the very early universe. The substitution of equations (16.18) into equation (16.17) shows $\Delta t \sim t$, i.e., that the interval over which the particles can be torn apart is comparable to the age of the universe. In order for the particles to acquire real existence, the work done by the acceleration (16.18) must be comparable to the rest energy of the particles:

$$mgl \sim mc^2. \tag{16.19}$$

The substitution of equation (16.18) into (16.19) shows that $l \sim ct$, i.e., that the pair, once created, will be ripped apart to a distance comparable with the cosmic event horizon. Moreover, with $\Delta t \sim t$ and $l \sim ct$, equation (16.16) implies that, at the end of the process, the pair is separated by a length l approximately equal to the Compton length h/mc. If m is one of the particles which mediate a short-range force, the Compton length measures the effective range of that force (see Problem 6.3), and our neglect of this force field in writing equation (16.18) would then be explicitly justified in order of magnitude.

Let us summarize these findings. The creation of particles of mass m occurs primarily when the cosmic time t equals the Compton time h/mc^2 of the particles in question. After creation, a pair is separated by a distance comparable to the Compton wavelength h/mc. Since the Compton wavelength is the distance in which particles are quantum-mechanically "fuzzy," we find that pairs are separated, after creation, by distances comparable to their "sizes." Thus, roughly one particle would be created per Compton volume $l^3 \sim (h/mc)^3$, yielding a mass density of created particles of

$$\rho \sim m/l^3 \sim m^4c^3/h^3 \quad \text{at} \quad t \sim h/mc^2. \tag{16.20}$$

Problem 16.16. A more accurate calculation by A. E. Wade yields a total density of π^0 mesons at a time t after their creation as

$$\rho = \frac{7}{6(2\pi)^{1/2}} \frac{m^{5/2}}{(ht)^{3/2}}.$$

The time dependence is due to the expansion of the universe; i.e., after particle creation, we have $\rho D^3 =$ constant in a comoving volume $D^3 \propto t^{3/2}$. Show that equation (16.20) agrees with the above in order of magnitude.

Let us consider a "conservative" application of equation (16.20). If we consider the production of π mesons (the spin-0 particles that mediate the strong nuclear force; see Chapter 6) and substitute $m = m_\pi$, we obtain subnuclear densities $\rho \sim 10^{12}$ gm/cm^3 at a time $t \sim 10^{-23}$ sec. A cosmic time $t \sim 10^{-23}$ sec would comfortably exceed the Planck time $t_P \sim 10^{-43}$ sec, allowing us to assume a background spacetime that is smooth. Let us now consider a "radical" application of equation (16.20). If the Planck mass m_P represents the mass of some yet-undiscovered elementary particle that results from the superunification of all four forces, then the blind application of formula (16.20) to the Planck time $t \sim t_P$ would imply the spontaneous creation of roughly the entire mass-energy density (16.4) required to produce the original spacetime curvature (Problem 16.17). In other words, if we had a full theory of the four forces of nature, and if we could calculate the back-reaction of the creation of mass-energy in changing spacetime curvature, we might ultimately be able to explain how spacetime curvature creates the entire mass-energy content of the universe, which, in turn, is responsible for generating the curvature of spacetime. One wild but beautiful speculation is that, with the creation of the Planck-mass particles, we may have precisely the particle that is capable of introducing all four forces into the material world. These four forces—strong, electromagnetic, weak, and gravitation—would be initially of the same strength, but as the universe expanded and available energies per particle decreased, they would become the four disparate siblings that we know today.

But isn't there a paradox here? To create mass-energy we need spacetime curvature; to generate spacetime curvature, we need mass-energy. So which came first, spacetime curvature or mass-energy? No one knows the answer to this modern version of the primal question of origins: which came first, the chicken or the egg? The best guess is probably: neither, they arose together, hand-in-hand, bootstrapping themselves into existence by what Douglas Hofstadter called in his book *Gödel, Escher, Bach* a "strange loop." (See also Chapter 20.) In this view, the universe contains something rather than nothing because it was favorable to have something rather than nothing. So, in one brief flutter of nature's eyelash, 10^{-43} sec, the universe created itself: spacetime, mass-energy, the four forces. There may have been only one way for this to happen, and the universe has done it!

Figure 16.19. The annihilation of two good friends, one of which happened to be made of matter, the other of antimatter.

Problem 16.17. Verify that the substitution of m equal to the Planck mass m_P (equation 16.13) into equation (16.20) results in a density of the order given by equation (16.4) when t equals the Planck time t_P (equation 16.15).

The Quality of the Universe

Any physical theory of the creation of the material world should explain not only how much mass-energy there is in the universe, but also what forms of particles this mass-energy takes. This issue of the *quality* of the creation demands a treatment more sophisticated than that given to the *quantity* in the preceding discussion. Indeed, although quite heuristic, that discussion does indicate that the quantity of mass-energy created in the early universe is probably not very sensitive to the details of our quantum theory of matter. As L. Parker was the first to note, the overall quantity of mass-energy produced at the Planck limit depends only on some very general and basic concepts. There are, however, interesting properties of the observed matter content of the universe that cannot be approached without a deeper understanding of elementary particles. As an example of such a fundamentally interesting question, we may ask: Why are the atoms of the present universe composed of protons, neutrons, and electrons, and not of antiprotons, antineutrons, and positrons?

The Asymmetry Between Matter and Antimatter

It is a facet of everyday experience that the world on a small scale, at least, is made almost entirely of matter. If this were not so, if our world were symmetric with matter and antimatter, the chances would be that if you were made of matter, then your best friend would have been made of antimatter. And, thus, the first time you two shook hands, you would have mutually annihilated one another in a bolt of pure energy (Figure 16.19).

The extension of this thought experiment to astronomical scales shows that the universe in the large must also be made almost exclusively of matter, and not antimatter. Suppose this were not true. Suppose that large chunks of the universe were made of matter (like our Galaxy), but that adjacent chunks were made of antimatter. Occasionally matter and antimatter at the interfaces must come into contact. When they did, there would be enormous liberation of energy over the entire face of contact. Astronomers, however, have explored vast parts of the observable universe for many decades, and they have nowhere witnessed evidence for such a phenomenon. They have seen many strange things, but they have not seen this one. Thus, most (but not all) astronomers believe that the material universe consists almost exclusively of protons, neutrons, and electrons.

How did such a situation ever arise? A few years ago, one might have dismissed this as a meaningless question, since baryon number (protons plus neutrons, essentially) and lepton number (electrons plus the difference between neutrinos and antineutrinos, essentially) were conserved in all physical processes. In this view, the existence of a net baryon number and a net lepton number in the present universe must mean that such numbers have always existed, ever since the instant of creation, and are not amenable to scientific discussion, but must constitute part of the "initial conditions" for the creation of the material world. Indeed, in our discussion of Big-Bang nucleosynthesis, we assumed the strict conservation of these numbers back to the cosmic instant $t = 10^{-4}$ sec.

However, we have seen that an extremely attractive idea for the origin of mass-energy in the universe is its creation from the quantum vacuum in the violent early expansion of spacetime from the Big Bang. Thus, at the instant $t = 10^{-43}$ sec, the net baryon and lepton numbers were, by definition, zero. If particles and antiparticles were subsequently pulled out of the vacuum with the strict conservation of baryon and lepton number, the resulting material Universe would have equal amounts of matter and antimatter. Clearly, one of these ideas must be wrong.

As we discussed in Chapter 6, physicists now believe that the idea of strict baryon and lepton conservation is wrong. Indeed, in most grand unification schemes of the electromagnetic, weak, and strong forces, it becomes possible at high-enough energies to change baryons into leptons, and vice versa. How does this help solve our current problem? None of the ideas on this issue have achieved complete acceptance, because the program of grand unification is yet unfinished, but one promising idea (originated by Andrei Sakharov, but since worked upon by many physicists) is the following.

All grand unified theories introduce a hyperweak force that mediates the transition between baryons and leptons. Associated with this new force are new bosons (X-bosons) with masses on the order of 10^{15} times that of the proton. A thousand trillion protons worth of mass is unprecedented for an elementary particle, which explains why these X-bosons have sometimes been jokingly called "intermediate vector baseballs." In any case, like the other bosons that transmit forces, the X-bosons are neither matter nor antimatter. These particles could be produced in great profusion in the early universe (around 10^{-39} sec after the Big Bang). Moreover, being massive bosons (unlike the gluon, photon, and gravition, but like the weakon and the π meson), the X-bosons are probably unstable, and decay into various kinds of more familiar particles. The decay of one X-boson might lead to many baryons and leptons. The X-boson is neither a baryon nor a lepton, but it can transcend the ordinary rules and produce a net baryon and lepton number. (Along with charge conservation, the baryon number minus the lepton number has to equal zero. Therefore, after very unstable particles have further decayed, the net number of protons equals the net number of electrons, and the net number of neutrons equals the number of neutrinos minus antineutrinos.) Now, it might be argued that the antiparticle of this X-boson would also have been pulled out of the vacuum. If its decay were to produce a corresponding number of antibaryons and antileptons, the universe would still be left with net baryon (and net lepton) number equal to zero—contrary to observations.

Is it possible, therefore, that the decay of the X-bosons and their antiparticles, taken in sum, *still prefers* a net excess of baryons and leptons? Such asymmetry seems to violate physical intuition, but it would not, in fact, be without precedent. Before we enter a detailed discussion of the latter issue, let us first see how much asymmetry is needed.

The degree of asymmetry needed depends on whether the temperature in the very early universe (see equation 16.5) can rise high enough to keep the X-bosons nearly (but not exactly) in thermodynamic equilibrium: the closer they are to complete thermodynamic equilibrium, the larger the asymmetry needs to be. The minimum asymmetry needed can, therefore, be estimated if we ignore completely the presence of the thermal background. In this case, the recently produced X-bosons and their antiparticles will simply drift and decay. The decay of a particle as massive as an X-boson will ultimately result in many baryons, antibaryons, leptons, and antileptons. As the universe continues to expand, most of the baryons and antibaryons annihilate, each annihilation producing two gamma rays (photons). The universe at a very early stage, therefore, becomes radiation-dominated, with the radiation quickly taking on a thermal (blackbody) distribution. If there was an initial excess of baryons over antibaryons, the net baryon number in a comoving cosmological volume will survive to the present day. The same will be true for the number of blackbody photons, although individual photons may be emitted or absorbed. But in equation (16.11) we have already calculated the ratio $\mathcal{N}$ of the number density of blackbody photons to that of baryons in the present universe. This ratio is about 2×10^9; so the decay of X-bosons must give a minimum asymmetry in favor of matter over antimatter of roughly to one part in a billion. More sophisticated calculations, which assume the presence of a thermal background, result in required asymmetries roughly of this amount.

Problem 16.18. The quantity $aT_{\text{rad}}{}^3/\rho_m$ that appears in equation (16.11) equals the entropy per unit mass associated with a thermal radiation field of temperature T_{rad}

embedded in matter of mass density ρ_m. To a factor of order unity, $\mathcal{N}$ can be interpreted as the entropy per baryon (in units of k) present in the universe today. (Most of the current entropy currently resides in the 3 K background radiation; very little resides in the matter itself.) Show explicitly that $\mathcal{N}$ is conserved as the universe expands. The high value of entropy today—in the picture pursued here—must have therefore arisen at the same time as the creation of matter itself in the Big Bang. The only alternative would be to assume unprecedented amounts of dissipation.

Broken Symmetries

Historically, it was T. D. Lee and C. N. Yang who first discovered that the decays of elementary particles might not satisfy intuitive symmetry arguments. In 1956, after a thorough theoretical analysis, Lee and Yang proposed that beta decay, which relies on the weak interaction for its mediation, might proceed asymmetrically relative to reflections in space (as occurs when one looks at events using a mirror). In other words, Lee and Yang proposed that weak interactions allowed elementary particles to discriminate between left-handedness and right-handedness! In a famous experiment performed in 1957, C. S. Wu and coworkers confirmed Lee and Yang's surprising prediction.

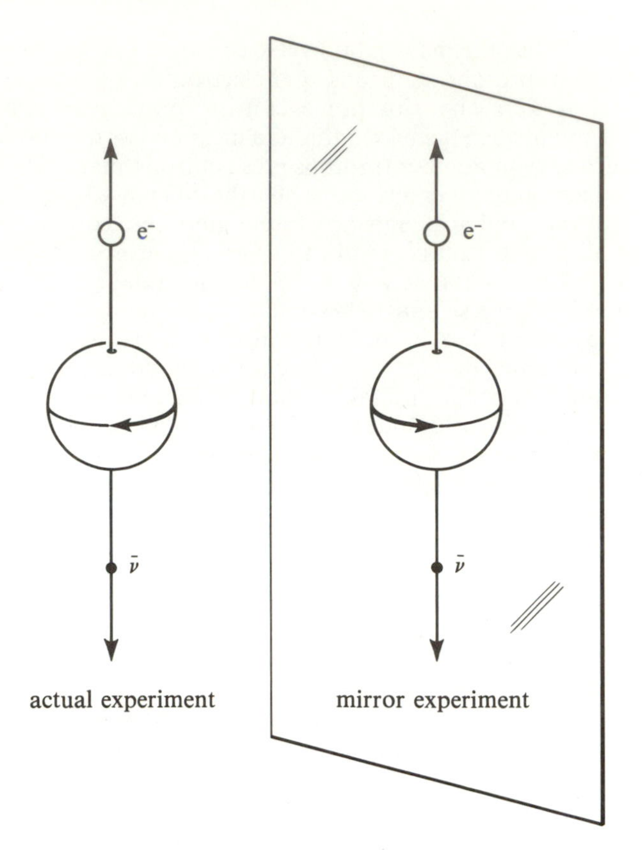

Figure 16.20. Schematic depiction of the cobalt-60 decay experiment conducted by C. S. Wu and colleagues. The mirror is drawn to help with the reasoning about naive symmetry expectations discussed in the text.

Figure 16.20 shows a schematic representation of Wu's experiment. She lined up many atoms of radioactive cobalt-60 so that their nuclei all spun in the same direction (clockwise as viewed from the electron in Figure 16.20). Intuition suggests that beta decay (see Chapter 6) should proceed with as many electrons leaving in one direction as in the other (and similarly for the undetected antineutrinos). In other words, one naively expects that as many of the electrons leaving the cobalt nucleus should see the nucleus spinning anticlockwise as see it spinning clockwise. The reason for this naive expectation is that the anticlockwise outcome would be the mirror result of the clockwise outcome (as in the mirror in Figure 16.20), and one naturally expects the mirror result to be just as likely as the nonmirror result. In Wu's actual experiment, the electrons preferred to emerge with the clockwise result. Contrary to all *a priori* expectations, but in accord with Lee and Yang's analysis, the weak interaction does not satisfy mirror symmetry! Since most little children cannot tell whether they are looking at the real thing or at a mirror, does this mean that mere electrons "know" more than little children?

In fact, the explanation of this puzzle lies with the antineutrino, not with the electron. Spurred by the experimental evidence, theoretical physicists proposed that neutrinos and antineutrinos differ from one another only in the sense of their spin relative to their direction of motion ("helicity"). As shown in Figure 16.21, it was proposed that a neutrino spins like a left-handed corkscrew, an anitneutrino like a right-handed corkscrew. This proposal would explain the fallacy in thinking that the mirror experiment in Figure 16.20 is just as likely as the actual experiment. A mirror would turn a right-handed corkscrew into a left-handed one. But a left-handed corkscrew, according to the proposal, is a neutrino, not an antineutrino. A cobalt nucleus cannot decay by the emission of an electron and a neutrino, since such a decay would violate lepton conservation (which does not occur at the low energies available in the universe today); so the mirror experiment in Figure 16.20 does not correspond to reality. Thus would weak interactions be able to tell left from right.

If this proposal were correct—that is, if the *only* difference between a neutrino and an antineutrino were their helicities—then these particles must have zero rest mass. Suppose this were not so. Suppose a neutrino had rest mass, so that it flew slower than the speed of light. Then it would be possible, in principle, for a physical

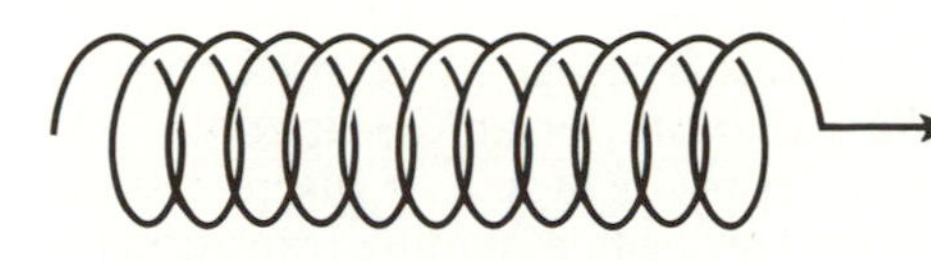

Figure 16.21. According to a theory of Feynman and Gell-Mann, the neutrino is a left-handed corkscrew, and the antineutrino a right-handed corkscrew.

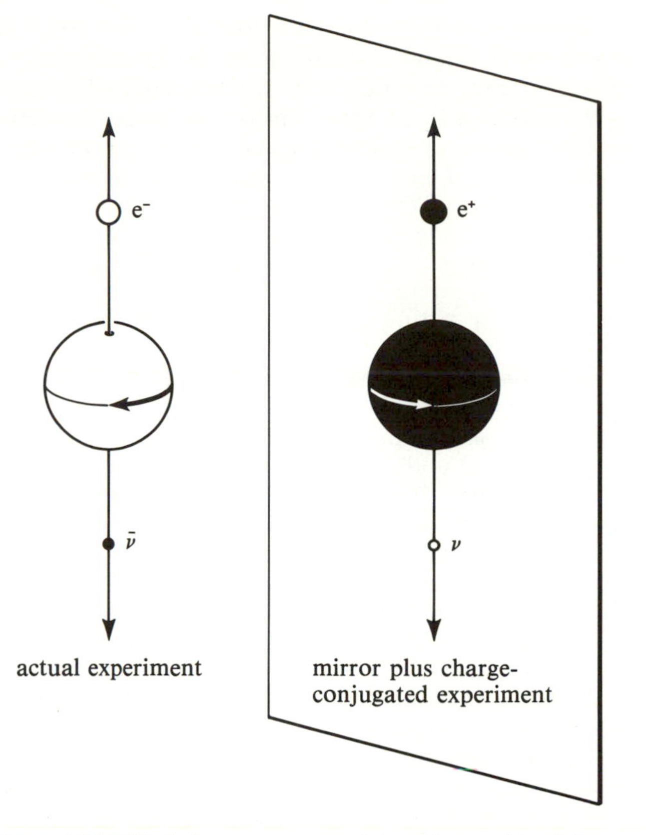

Figure 16.22. Weak decays are symmetric under combined charge and parity reversals (*CP* invariance); so the mirror-plus-charge-conjugated analog of the Wu experiment should be as probable as the actual experiment.

observer to fly even faster in the same direction and overtake the neutrino. Such an observer would see the neutrino spin in the same sense as an observer at rest, but the two observers would perceive different directions of motion of the neutrino. In other words, they would differ on the helicity displayed by the neutrino. The resting observer would think the neutrino is a left-handed corkscrew, the flying observer that it is a right-handed corkscrew. If neutrinos come in only one variety—left-handed corkscrews—this thought experiment would represent a violation of the principle of relativity (all physical observers have to agree what is a lepton, e.g, the result of a neutrino-electron interaction). Thus, if neutrinos come with only one helicity, they must be impossible to overtake; i.e., they must travel at the speed of light, and, therefore, contain zero rest mass.

Conversely, if neutrinos are able to change forms (oscillate between electron and muon types, for example), then they must have rest mass. If they have rest mass, they must also have both kinds of helicities. If so, neutrinos and antineutrinos differ by more than merely their helicities. They must contain an inner label that marks neutrinos as leptons and antineutrinos as antileptons. Whether neutrinos have rest mass or not, Wu's experiment shows that, when radioactive cobalt decays, many more antineutrinos of the right-handed variety are emitted than of the left-handed variety, according to the laboratory observer; so it is a well-documented experimental fact that beta decay does not satisfy mirror symmetry.

To carry this discussion deeper, let us note that the mirror result of the experiment depicted in Figure 16.20 does become possible if, in addition to space reflection, we were also to reverse the sign of all charges (electric, weak, and strong). In that case (Figure 16.22), the electron would turn into a positron; the antineutrino into a neutrino; and the cobalt nucleus into a anticobalt nucleus (because the quarks in the protons and neutrons would become antiquarks, turning the former into antiprotons and antineutrons). The combined operation—space reflection (or parity reversal, *P*, which turns the right-handed corkscrew into a left-handed corkscrew) plus charge reversal (*C*)—produces the beta decay of an anticobalt nucleus by the emission of a positron and a neutrino. This is a perfectly valid process: it satisfies lepton conservation; and it results (primarily) in the emission of left-handed neutrinos.

Now, not all physical processes are symmetric under the combined *CP* operation. In fact, the strongest statement that theoretical physicists have been able to prove with great generality is the so-called *CPT* theorem:

> If a microscopic process is possible, then the process which corresponds to taking charge reversal (*C*), parity reversal (*P*, or mirror reflection), and time reversal (*T*), has an equal probability of occurring.

Since beta decay satisfies *CP* symmetry, the time-reverse of the Wu experiment must, taken by itself (with no parity or charge reversal), also be microscopically possible. And it is, although *macroscopic* considerations, namely, the second law of thermodynamics, makes the actual time-reversed experiment almost impossible in practice.

Experimental evidence now exists for an even stranger symmetry violation in the decay of an elementary particle. This concerns the decay of a so-called K^0 meson. In the quark scheme, the K^0 meson consists of a combination of a strange and anti-down quark ($s\bar{d}$) and its conjugate ($\bar{s}d$). Thus, in some sense, the K^0 meson is its own antiparticle. Without going into the details, the decay of this particle has been found to violate *CP* symmetry; therefore, it must also violate time-reversal symmetry, assuming the *CPT* theorem to be valid. In the K^0 meson, nature has a particle which can discern the arrow of time!

Violation of *CP* symmetry is also needed for X-bosons and their antiparticles to decay asymmetrically in the numbers of baryons and antibaryons produced; so there is experimental and theoretical precedence for the suggestion explored in the earlier discussion. And if this suggestion turns out to be wrong, then perhaps it is the decay of the Planck mass particle which produces the matter-antimatter asymmetry seen in the universe today.

It might be argued that, in some sense, symmetry still exists, because if one were to apply all the necessary global transformations (parity reversal, etc.) and throw in time-reversal as well, we would have another possible universe, with antimatter instead of matter. To this point of view, we may reply that the equations of physics may be symmetric, but its solutions need not be. We have already encountered one example in our discussion of ferromagnetism in Chapter 5 (see Figure 5.13). Let us take another analogy to make the point clear. Consider the flip of a coin. The physical laws governing the flip of a coin are symmetric in terms of the probabilities of getting heads or tails. However, the outcome of any one flip need not be symmetric. The symmetric outcome would be for the coin to land on its edge, but that almost never happens. The usual outcome of one flip is either heads or tails, not something exactly halfway in between. And so may it be with the universe. To create the present universe, nature got only one flip, and she had to start with the universe *expanding*. And in that flip, the universe came up as matter!

Local Symmetries and Forces

The reader cannot be blamed if he or she now feels like a ship that has lost it moorings. If at sufficiently high energies, particles can readily change their identities, where are the secure anchors that we may grasp to call this or that particle truly fundamental? This is a profound question, for which modern physics has yet no final answers. Indeed, what may prove more fundamental in the final analysis is not the particles themselves, but the laws that govern their change.

An analogy might help at this point. Quantum mechanics arose largely out of the challenge posed by atomic spectroscopy (Chapter 3). In particular, the emission and absorption lines of atomic hydrogen presented remarkable regularities that cried out for an explanation. That explanation came in terms of atoms in discrete states which followed certain rules when making transitions to and from other states. Because the energy separations were small, only one force—the electromagnetic interaction—was involved, and all nonelectromagnetic properties of different states remained the same; so we found it easy to think of the excited atoms as being the *same* basic entity as the atoms in the ground state. It is conceptually more difficult to think of neutrons and protons as being the same basic entities, and even more so for baryons and leptons or fermions and bosons. But might not that be simply because the energy separations are now much larger, so that more obvious differences have a chance to manifest themselves? In this view, it is only at higher energies that we are able to witness transitions between the various particle states, and thereby see the basic underlying simplicity of the principles behind the "spectroscopy" of particles. It is these laws of nature which are simple, not their interplay in the production of particles, atoms, molecules, humans, stars, or galaxies.

Modern explanations of particles and their forces are dominated by the simple and beautiful idea of symmetry under various transformations. The transformations can be either of *global* form—the same transformation is applied simultaneously to every spatial point—or of *local* form—a different transformation is applied at every different spacetime point. Invariance under the first kind of transformation (global symmetries) are associated with conservation principles; invariance under the second kind (local symmetries) are associated with (indeed, usually require) the introduction of forces.

As an example, consider transformations involving the coordinates of spacetime themselves. Let us imagine, first, a global transformation involving translation at a constant velocity relative to an isolated collection of moving but noninteracting masses. Such a transformation might be effected in practice by recording the observations of people who move at constant velocity with respect to one another. The laws describing the dynamics of noninteracting masses remain invariant under such a global transformation; in particular, each observer would formulate the principles of conservation of momentum and energy for noninteracting masses. This conclusion, of course, constitutes one of the bases for the theory of special relativity.

Let us now imagine a local transformation involving independent translations of each spacetime point. Such a transformation might be effected by recording the observations of people who *accelerate* with respect to one another. Clearly, the laws of dynamics would not remain invariant under such a *local* transformation unless each observer introduces the concept of force. The force that would have to be introduced would be indistinguishable from classical gravitation (Chapter 15). This conclusion, of course, constitutes one of the bases of the theory of general relativity. The full theory requires the additional incorporation of the ideas that mass-energy acts as a source for the gravitational field, and that spacetime has a local structure such that the speed of light is the same for all local observers.

Of the two symmetry principles, we see therefore that local symmetries are the more powerful, because real masses *do* exert gravitational force on one another. Global symmetry would apply only if this force could be artificially turned off. For gravitation, of course, Einstein preferred to think not in terms of force, but in terms of spacetime curvature (Chapter 15). On the other hand, in a world where gravitation is the only consideration, Deser has shown that the methods of relativistic quantum mechanics suffice to derive general relativity even when gravity is thought of, not as spacetime curvature, but as a force. This force, as in all quantum field theories, is transmitted by the exchange of virtual bosons. The boson, in this case, is the graviton with spin 2 (Chapter 6). Spin 2 enters because classical gravitation (i.e., general relativity) is not a *vector* theory; consequently, it cannot arise from the exchange of *vector* bosons (spin-1 particles which have $2 \cdot 1 + 1 = 3$ possible orientations or quantum states; see Chapter 3). To describe the geometry of curved spacetime requires at least a particle of spin 2, but in the quantum derivation curved spacetime does not enter explicitly; one deals only the force of gravitation that, in the end, can be alternatively interpreted as spacetime curvature (as Wheeler has put it, "curvature without curvature"). This does not mean that Einstein's theory of gravitation has been supplanted by the new derivation, only that there are two equivalent ways of thinking about the same thing.

The introduction of forces that arise when global symmetries are converted into local ones applies more generally. Invariance under "gauge" transformations—a concept introduced by Hermann Weyl—is the terminology used to describe this more general viewpoint. Historically, it was the theory of electromagnetism which was first shown to have this property. In particular, the invariance associated with local transformations of the phase of a charged particle's wave function yields the forces of quantum electrodynamics. The global symmetry associated with the arbitrariness in defining the zero of the electromagnetic potentials leads to the principle of the conservation of electric charge.

Consider another example: the notion that the proton and the neutron would be basically the same particle (a nucleon) if only all forces except the strong one could be "turned off" (Chapter 6). Imagine, therefore, a double-headed arrow in some abstract space ("isotopic spin," in the language of nuclear physics) inside the proton/neutron, such that the arrow points up and down if the particle is a proton, and it points sideways if the particle is a neutron (Figure 16.23). Consider now the transformation which rotates this arrow by 90°, in the process changing a proton to a neutron, and vice versa. Global symmetry under this operation is the (approximately true) statement that changing all protons into neutrons and neutrons into protons would not change the physical world. This statement is false to the extent that forces *other than* the strong force *would be affected* by this transformation.

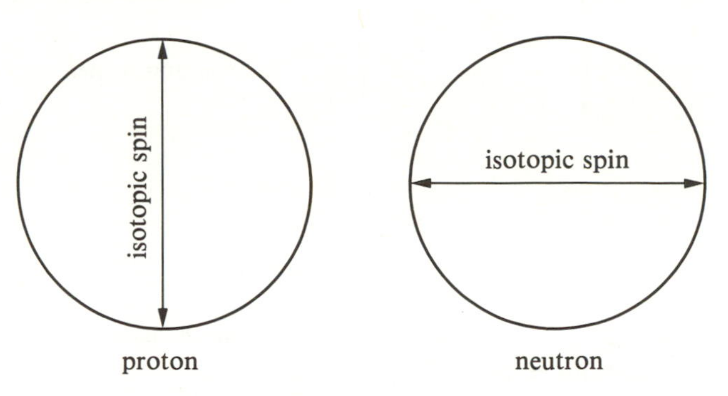

Figure 16.23. The concept of isotopic spin. In interactions involving *only* the strong force, protons and neutrons behave as if they were two states of a single hypothetical particle, the *nucleon*. This raises the analogy with the ordinary spin states of an electron (Chapter 3), where the spin orientation can be either "up" or "down." To avoid confusion of isotopic spin with ordinary spin, let us imagine protons and neutrons to be the same except for some internal double-headed arrow which points vertically in the proton and horizontally in the neutron. Physical phenomena should not change by a different definition of what we call "vertical"; however, to perform actual calculations, a definition is needed (choice of "gauge"). Local symmetry with respect to this choice requires the introduction of forces.

What happens if one tries to make the isotopic-spin symmetry a local one? That is, what happens if one considers transforming only *some* protons into neutrons and vice versa. The mathematical techniques for performing this task continuously in the field equations were invented by Yang and Mills in a very influential paper of 1954. In the original Yang-Mills theory, new forces have to be introduced to preserve the local symmetry of isotopic-spin rotations, and three massless bosons were introduced to transmit these forces. One of these could be identified with the photon (which mediates the electromagnetic force). The other two carried charge. No particles of zero rest mass which carry electric charge exist; so the Yang-Mills theory lay dormant, since it seemed to have no practical application to the real world.

In 1967, Steven Weinberg (and, independently, Abdus Salam and John Ward) proposed a variant of the Yang-Mills theory to combine the electromagnetic and weak interactions. To accomplish this intertwining, their theory contained, at the outset, *four* massless spin-1 bosons to transmit the new forces. One of these was positively charged, another negatively charged, and the

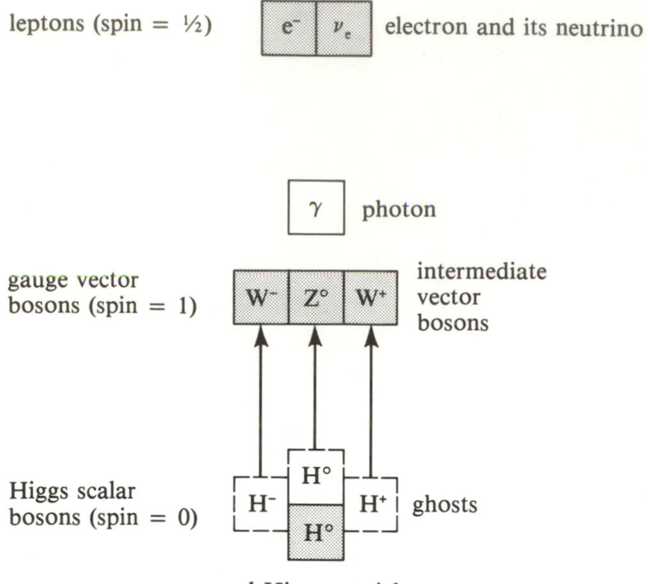

Figure 16.24. The Weinberg-Salam theory of electroweak interactions introduces four vector bosons of spin 1 to mediate the forces. Three of these (shaded boxes labelled W^-, Z^0, and W^+) gain mass by "eating" one Higgs boson each. The photon remains massless. The three eaten Higgs particles become "ghosts," particles which have no real existence, but which can be used in a mathematical trick to help with the calculations. The uneaten Higgs particle (shaded box labeled H^0) should be observable. The main leptons which feel the electroweak forces are the electron and its neutrino, both of which are now believed to possess rest mass. (For details, see 't Hooft, *Scientific American*, **242**, June 1980, 104.)

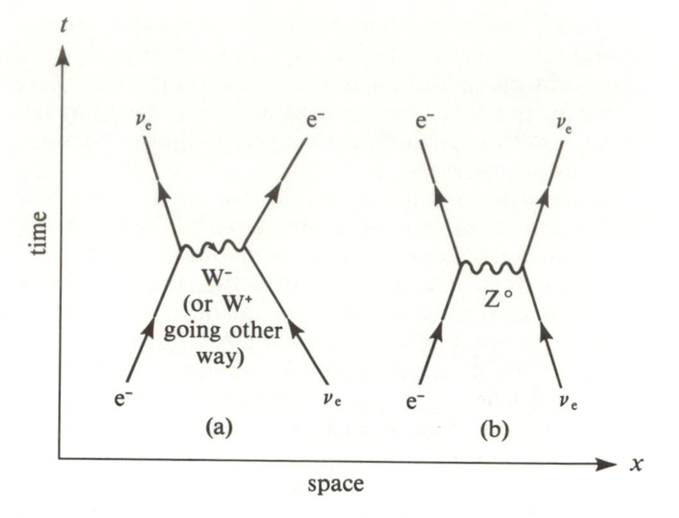

Figure 16.25. The scattering of electrons by neutrinos can be effected either (a) by an exchange of a W^- boson, in which case the electron becomes the neutrino and vice-versa, or (b) by an exchange of a Z° boson, in which case the electron and the neutrino maintain their identities. Such scattering mechanisms are believed to play a role in helping to expel the envelope of a massive star when it supernovas (see Chapter 8). The neutral process (b) is novel to the Weinberg-Salam theory.

remaining two electrically neutral. To avoid the difficulties faced by the original Yang-Mills theory, the Weinberg-Salam model endowed three of the four bosons with nonzero rest mass by a procedure known as the "Higgs mechanism" (Figure 16.24). One of the original spin-1 bosons remains massless, and it is identified with the photon (which mediates the electromagnetic interaction). The three massive spin-1 bosons are identified with weakons (which mediate the weak interaction). Prior to the Weinberg-Salam theory, it was thought that weak interactions always involved the transmission of a negatively charged or a positively charged weakon (see, e.g., Figure 6.4). The Weinberg-Salam theory also allows mediation by a neutral weakon (Figure 16.25), and this prediction was verified experimentally in 1973. There should also be a massive Higgs particle which the photon did not "eat" to gain weight (Figure 16.24); its reality awaits experimental verification.

The Weinberg-Salam theory was largely ignored until Gerard 't Hooft proved in 1971 that it could make sensible predictions without being plagued by the troublesome infinities that wreck many other quantum field theories (see the discussion of "renormalization" in Chapter 6). This theory is now considered the "standard" account of the forces of electromagnetism and weak interaction. In particular, the electroweak force can distinguish between protons and neutrons, so the exact equivalence of protons and neutrons in the original Yang-Mills theory is broken ("spontaneous symmetry breaking").

Since about 1973, there have been many attempts to bring about a grand unification of the three forces: strong, electromagnetic, and weak. In particular, in 1974, Georgi and Glashow proposed a version which starts with 24 massless (spin 1 or vector) bosons. Spontaneous symmetry breaking by the Higgs mechanism allows 12 of these to acquire nonzero rest masses. These 12 (containing various "colors" and electric charges) are the X-bosons of masses equal to about $10^{15} m_p$ that mediate the transformation of quarks into leptons. Among other accomplishments, this theory would explain the profound fact that the proton and the electron have exactly equal but opposite electric charges. The remaining 12 massless bosons are divided into eight gluons, which mediate the strong force between quarks (Chapter 6), and four bosons, which mediate the electroweak force. Another round of symmetry breaking gives mass to three of these four bosons, and the three massive weakons plus the massless photon constitute the transmitter of force in the "standard" Weinberg-Salam theory of electroweak interactions.

A very general prediction of just the concept that there may be a deep connection between quarks and leptons has already received experimental verification. In 1964, shortly after Gell-Mann and Zweig postulated the existence of three quarks (up, down, strange), Bjorken and Glashow argued that there should be a fourth quark (charm) to balance the four known leptons at that time (electron, electron neutrino, muon, muon neutrino). This aesthetic argument was greatly strengthened in 1970 by Glashow, Iliopoulos, and Maiani, who showed that the existence of charm was needed to explain certain unexpected regularities in weak decays. As Sydney Coleman has emphasized, this line of reasoning was amazing: "New kinds of strongly interacting particles were predicted, not on the basis of strong-interaction theory *per se*, but in order to make strong-interaction theory consistent with a weak-interaction theory that (so far as Glashow *et al.* knew) had not yet been invented." Thus, it was a glorious moment in particle physics when two experimental groups in 1974, led independently by Samuel Ting and Burton Richter, found evidence for a meson with "hidden charm" ($c\bar{c}$). Since that time, mesons with "naked charm" ($c\bar{u}$, $c\bar{d}$, $c\bar{s}$) have made appearances, as well as a fifth lepton (tau), and a fifth quark in the form of a meson with "hidden bottom" ($b\bar{b}$). Physicists recently glimpsed "naked bottom." Since the tau lepton presumably comes with its own neutrino, there should presumably also be a sixth quark (top or t). The grand unified theory of Georgi and Glashow predicts that there should be no additional flavors of quarks or leptons. According to Steigman, Schramm, and Gunn (but Stecher disagrees), this would be consistent with the cosmological evidence of helium production in the early universe. If there were more than three kinds of neutrinos (plus their antiparticles), the temperature-time relation of equation (16.5) would cause too much primordial helium to emerge from the era of Big-Bang nucleosynthesis.

A fantastic prediction of grand unified theories is now receiving active experimental attention. Although quarks cannot easily be turned into leptons at the meager energies per particle available in the universe today, quantum mechanics does allow a finite but small probability for such an occurence. This has the consequence that the proton should not be a strictly stable particle. If X-bosons have a rest mass of about $10^{15}m_p$, protons should have a halflife of about 10^{31} years against decay into leptons and photons. This means that if one were to monitor 10^{31} protons (contained in, say, about a hundred tons of water), one should see on average one decay per year. One has, of course, to reduce the noise due to background events (cosmic rays, etc) below that level, and that means going down in a deep mine (as in the solar-neutrino experiment). Experimental physicists have begun experiments of the appropriate sensitivity, and the results are anxiously awaited.

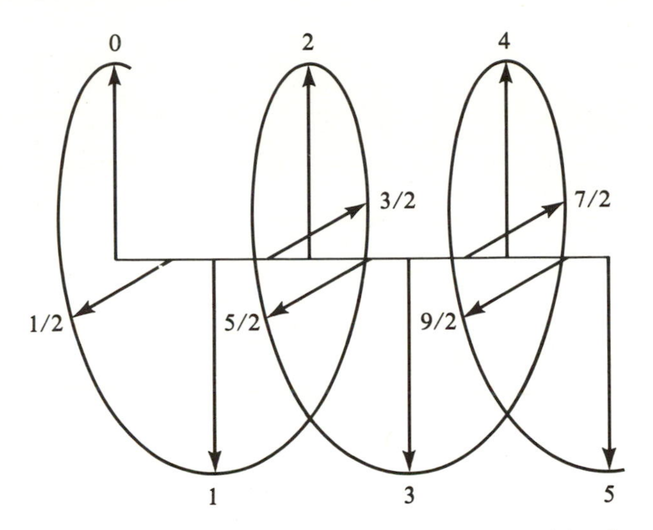

Figure 16.26. Supersymmetry supposes that bosons and fermions can be regarded as basically the same kind of particle with an internal arrow which points vertically for bosons and horizontally for fermions. We have arranged the various spin states on a helix so that supersymmetry transformations correspond to rotations along such a helix.

Supersymmetry, Supergravity, and Superunification

If the three forces, strong, electromagnetic, and weak, can be unified, surely the eldest sibling, gravity, can be brought into the fold. After tremendous effort and frustration—beginning with the early efforts of Einstein and his coworkers—theoretical physicists now believe that they have found a promising approach. Again, the fundamental idea is rooted in a symmetry principle, one of such beauty and elegance that it has been called supersymmetry.

Supersymmetry connects two of the most fundamental concepts of twentieth-century physics: the relativistic notion of spacetime and the quantum-mechanical notion of spin angular momentum (Chapter 3). In Figure 16.23, we considered the notion of isotopic spin as the internal quantum label distinguishing a proton from a neutron. Historically, the concept of isotopic spin was introduced by way of analogy to spin angular momentum, which explains half the name of the former. ("Isotopic spin" therefore has nothing to do with angular momentum, but it does relate to chemical isotopes, since the latter differ only in the number of neutrons in the atomic nuclei.) Let us here turn the analogy backward, and consider transformations of particles with different spin angular momenta as we had earlier considered transformations of particles with different isotopic spins (compare Figures 16.23 and 16.26).

Imagine that bosons (whole-integer spin particles) differ from fermions (half-integer spin particles) in having

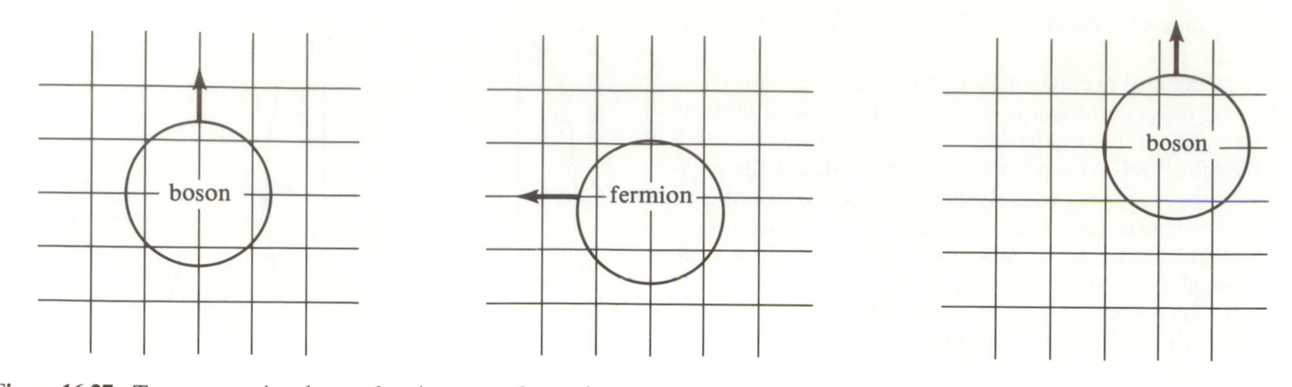

Figure 16.27. Two successive boson-fermion transformations succeed not only in recovering the original particle, but also in translating it to a different spatial point. (For details, see D. Z. Freedman and P. van Nieuwenhuizen, *Scientific American*, **238**, Feb. 1978, 126.)

an internal arrow which points vertically in bosons (up for spin 0, 2, . . . ; down for spin 1, 3, . . .), and which points sideways in fermions (left for spin 1/2, 5/2, . . . ; right for spin 3/2, 7/2, . . .). To depict this situation, arrange the various spin states on a helix (Figure 16.26). The transformations under supersymmetry were discussed first in 1971 by physicists in the USSR, France, and US—Golfand and Likhtman, Neveu, and Raymond and Schwartz. These transformations link bosons and fermions of adjacent spin states: spin s to spin $s + 1/2$ or $s - 1/2$, and vice versa. Now, it might be thought that the repeated application of two such global transformations would recover all particles in their original condition. Thus, a rotation of $+90°$ in the abstract helical space would increase the spins of all particles by 1/2. A counterrotation of $-90°$ would decrease the spins of all particles by 1/2, and thereby recover all original spin states. And so it does, except that we find all particles have been translated to different positions in physical space (Figure 16.27)!

The explanation for this peculiar behavior is highly mathematical, but its origin lies with the fact that transformations of the type considered by all Yang-Mills type of theories are *noncommutative*. In other words, the result of applying transformation A followed by transformation B does not give the same result as applying transformation B followed by transformation A. (Multiplication by ordinary numbers is, of course, commutative, in the sense that $ab = ba$.) The simplest example of noncommutative transformations are rotations about different axes in three dimensions (Figure 16.28). The need for boson-fermion transformations to be noncommutative (in fact, anticommutative, in the sense that $AB = -BA$ or $AB + BA = 0$) is related to Pauli's exclusion principle. Two bosons may share the same quantum state, but they cannot both be transformed into the same fermion state, because the latter is forbidden by Pauli's exclusion principle. This condition leads directly to the requirement that the boson-fermion transformation must be anticommutative. But this implies that a rotation R of $+90°$ in our abstract spin space (Figure 16.26) does not have strictly an inverse R^{-1} of $-90°$ which would restore all particles to their original quantum state. In the language of quantum-mechanical operators, we would argue that the anti-commutative nature of supersymmetry transformations require $RR + RR = 0$, i.e., $RR = 0$; therefore R has no inverse R^{-1} such that the product $R^{-1}R = 1$, the identity operator.

We encountered similar (but different) "numbers" elsewhere, namely, in Dirac's theory of the electron (Problem 6.3). Moreover, Dirac's discussion of spin shows this concept to be intimately related to the four-dimensional structure of spacetime (Chapter 6). Perhaps it should therefore not come as a total surprise that the noncommutative nature of successive boson-fermion transformations should express itself as a displacement in physical space.

Problem 16.19. Suppose B and F represent the probabilities of finding a particle in a certain boson and fermion state, respectively. Further, suppose F and B to be related by a supersymmetry transformation: $F = RB$. Since the probability of finding *two* fermions in the same state must be zero, we require $FF = 0$; i.e., F cannot be an ordinary number. On the other hand, two bosons can be found in the same quantum state, so $BB \neq 0$, and B can be an ordinary number. Consider, now, the equation,

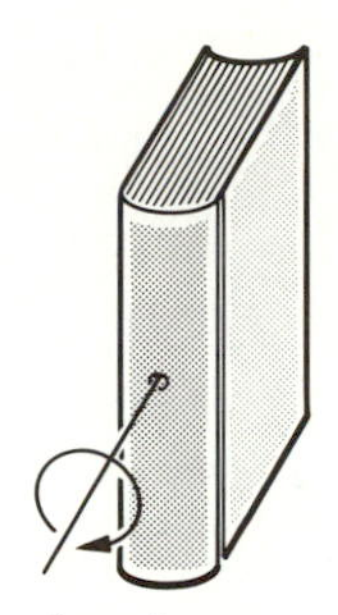

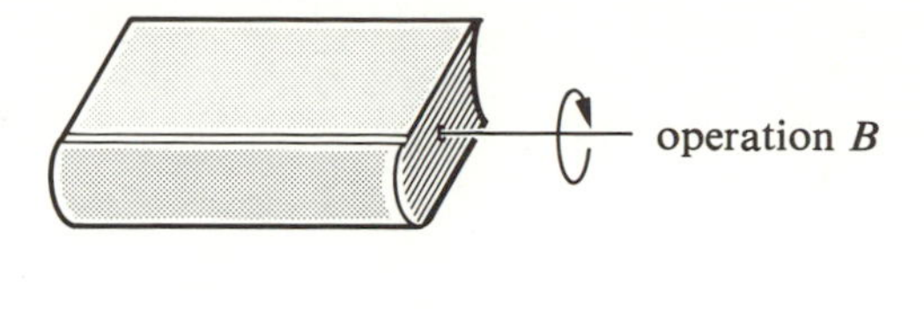

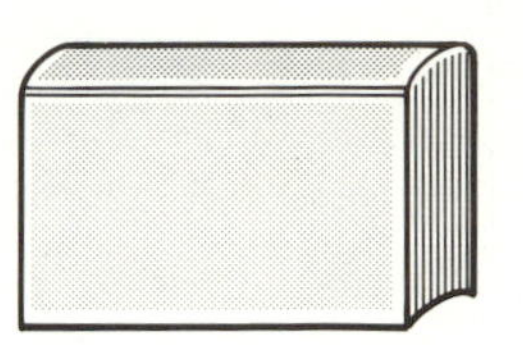

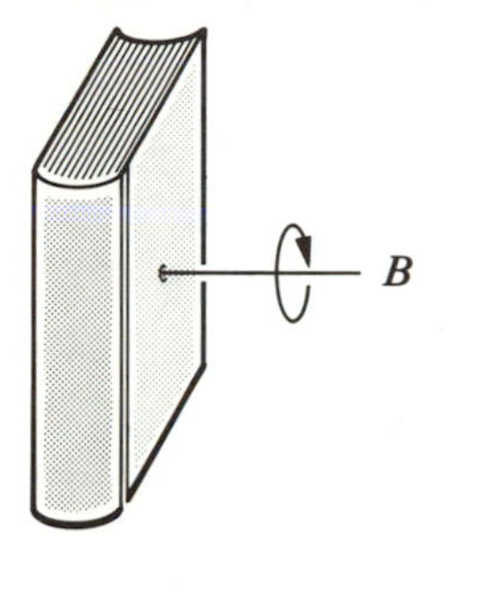

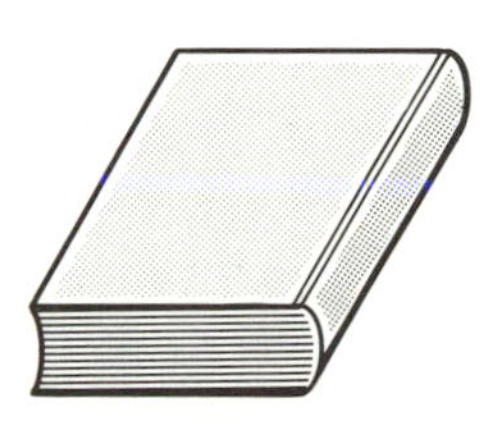

Figure 16.28. Rotations of three-dimensional objects exhibits noncommutative properties. In the above example, rotations of 90° along two different axes taken in different orders produces different final configurations for a book.

$0 = FF = (RB)(RB)$. Since $BB \neq 0$, we require $RR = 0$. This means that R generally has no inverse R^{-1}. Show, however, that the sum $(1 + R)$, where 1 is the identity operator, does have an inverse, and that this inverse equals $(1 - R)$. A convenient representation arises if R corresponds to an infinitesimally small rotation, for then we may write $1 + R$ as e^{+R} and $1 - R$ as e^{-R}, with e^{+R} and e^{-R} more obviously the inverses of each other, "Number" systems formed from ordinary numbers, anticommuting numbers, and their products and sums were invented by Hermann Grassmann more than a century ago, as a branch of pure mathematics now called "Grassmann algebra."

In the last subsection, we discussed how promoting global symmetries to local symmetries involved the introduction of forces. But local shifts in spacetime, which would result from successive operation of local supersymmetry, are precisely the concerns of the theory of general relativity! In other words, local supersymmetry naturally lends itself to the construction of a theory of quantum gravitation: supergravity.

The simplest form of a supergravity theory was invented in 1976 by Freedman, van Niewenhuizen, and Ferrara, and improved upon by Deser and Zumino. A (gauge) field was introduced for each of the symmetries present in the problem. Spacetime shifts naturally called for the introduction of the spin-2 graviton as the appropriate field boson. Since supersymmetry relates particles of adjacent spins, the other field particle was chosen to be a spin-3/2 particle called the *gravitino*. But the gravitino is a fermion; so the force associated with local supersymmetry transformations must arise from the exchange of *pairs* of virtual gravitinos (Figure 16.29). The exchange of such a pair has negligible probability except at very small separations; so this force would be short-range even if the gravitino had zero rest mass. Thus supergravity, unlike classical gravitation, has two components to the force field: the familiar long-range attractive force associated with the exchange of virtual gravitons; and a novel short-range (repulsive?) force associated with the exchange of a pair of virtual gravitinos.

In the simplest supergravity theory, there is only one form of graviton and one form of gravitino, both of which have zero rest mass. Since nothing is known about

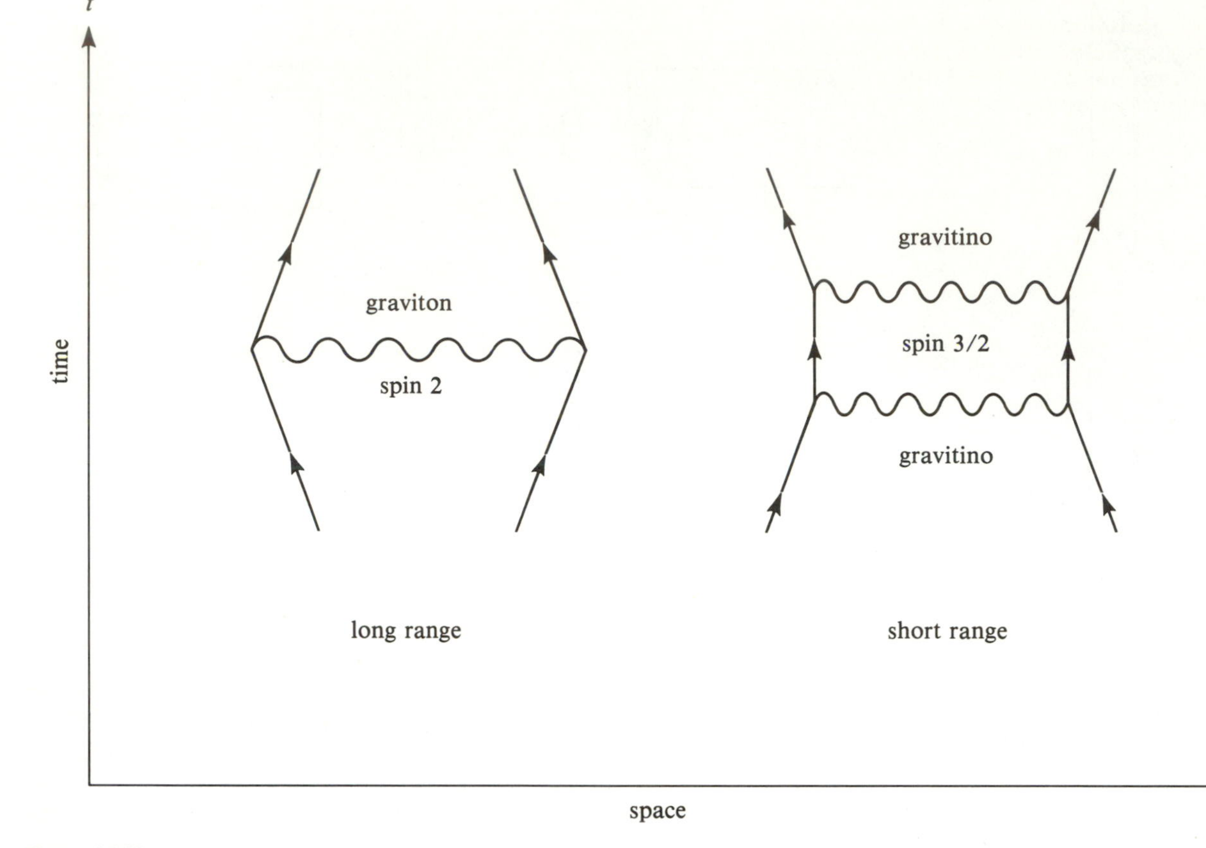

Figure 16.29. In supergravity theories, the force of gravitation can be transmitted in two different ways. The exchange of virtual gravitons lead to the long-range force familiar to Newton's or Einstein's theories of gravitation. The exchange of a pair of virtual gravitinos leads to a short-range force which is not present in classical gravitation.

the effect of gravitation at subatomic separations, it would be quite permissible to give the gravitino a nonzero mass by the Higgs mechanism of spontaneous symmetry breaking. Indeed, it seems quite tempting to suppose that this mass should be the Planck mass, because that choice would seem naturally to lend spacetime a "grainy structure" on a scale of the Planck length and the Planck time. If we turn out to live in a closed universe after all, could this departure from classical gravitation theory at very short range cause the Big Squeeze to rebound into another cycle of Big Bang-Big Squeeze? (See Figure 15.7.) Moreover, could the second law of thermodynamics be suspended during the Planck instant, when we have effectively only one particle per event horizon? This would be desirable if the universe were to start truly anew without having to accumulate the entropy generated during the previous cycle. (See Chapter 4 for the statistical basis of the laws of thermodynamics that requires many particles per system for their validity.) How would black holes which developed in the previous cycle be processed in the rebound? Would they or their shreds act as "seeds" for the next cycle of galaxy and quasar formation? Without further advances and quantitative calculations, such remarks and questions can only be termed "superspeculative."

Particle physicists have extended supergravity theories in another direction, by introducing additional forms of gravitinos. The maximum number is eight. Moreover, local supersymmetry can also be used to generate particles with spin less than 3/2, by connecting states of adjacent spins. Thus, in the maximally extended version of supergravity, there is one graviton of spin 2, eight gravitinos of spin 3/2, 28 vector bosons of spin 1, 56 fermions of spin 1/2, and 70 scalar particles of spin 0 (Table 16.2). It remains to be seen if this list is large

Table 16.2. **Relationship between number of elementary particles allowed by supergravity theories and number of gravitinos used to perform the supersymmetry transformations.**[a]

Number of gravitinos	Number of particles of spin				
	0	1/2	1	3/2	2
1				1	1
2			1	2	1
3		1	3	3	1
4	2	4	6	4	1
5	10	11	10	5	1
6	30	26	16	6	1
7	70	56	28	7	1
8	70	56	28	8	1

[a] Adapted from Freedman and van Nieuwenhuizen, *Scientific American*, **238**, Feb. 1978, 126.

enough to include all fundamental particles, partly because we don't know yet whether some particles (such as the muon) are truly elementary.

Active attempts are now underway to bring the other three forces of nature under this one banner. A curious difficulty seems to have arisen in the first such efforts. The superunification of all four forces seems to introduce into the equations a cosmological constant! (See Chapter 15 for an account of Einstein's struggles with this demon.) Worse, the cosmological constant seems to be larger than allowed by current observational evidence. However, the cosmological constant can be reduced by the process of spontaneous symmetry breaking to endow some of the particles in the currently relatively simple theories with finite rest mass. Supergravity theories now seem to disagree with the experimental facts, but, as 't Hooft has commented, the same seemed true of the Yang-Mills theory when it was first proposed.

Before we end this subsection, we would like to make one closing comment. Instead of regarding quantum gravity as a force, some physicists have preferred to follow Einstein's lead and consider gravitation to be a geometric entity. In this work (pioneered by Akulov and Volkov, extended by Salam and Strathdee and by Arnowitt and Nath), quantum gravitation plays out its dynamic role, not in the usual four-dimensional spacetime continuum, but in the arena of "superspace." Superspace assigns to each event not only space and time coordinates, but also a spin state; so it may turn out that even quantum gravitation has both a force interpretation and a geometric interpretation. Indeed, it may be possible to give all *four* interactions a geometric interpretation—thereby realizing one of Einstein's lifelong dreams—by extension of the concept of ordinary space (or spacetime) by attaching at each point new manifolds (basically a set of ordered numbers). Such extensions of ordinary geometry, or "fiber bundles," were studied by modern geometers, such as Elie Cartan, well before any possible applications to physics was contemplated. This example, and many others, is a source of considerable amazement to scientists. Why is it that mathematics, a product purely of human invention, and physics, ultimately based on experience, should prove to be such powerful allies? Does this represent the way that the world really is, or only a way that humans can think about it?

Philosophical Comment

Chandrasekhar has addressed eloquently the intriguing question of why theories that are beautiful often turn out also to be true. It is not enough merely to follow Keats' "Beauty is truth, truth beauty"; one must also enquire what constitutes beauty in science. As his criteria, Chandrasekhar borrows from complementary sentiments by Werner Heisenberg and Francis Bacon:

> "Beauty is the proper conformity of the parts to one another and to the whole."

> "There is no excellent beauty that hath not some strangeness in its proportion."

Let us pursue here the above line of thought, in the sense that great science, like great architecture, requires both symmetry and asymmetry in its construction. Unsullied perfect symmetry appeals to the inexperienced eye; only with maturity do we learn to appreciate the greater beauty in a proper blend of the asymmetric with the symmetric.

Consider cosmology. A universe unchanging in time (steady-state theory) undoubtedly has its allures, enough so that when Big-Bang theory eventually prevailed, it was suggested in some quarters that nature had botched the job! In hindsight, we see how much grander it was for the universe to have started with a Big Bang. Coupled with the new developments of particle physics, this view has led to a truly amazing insight. In the beginning, the universe may have begun with the four forces almost of equal strength—almost, but not quite. The slight inequalities in strength grew rapidly as the universe expanded beyond the Planck state. Particles and antiparticles were pulled out of the vacuum in great profusion. A slight violation of *CP* invariance led to a slight dominance of matter over antimatter. (This prevented the universe from being composed today of only neutrinos, antineutrinos, and photons.) The decay of unstable particles and the annihilation of most of the matter and antimatter generated an incredible fireball of primeval radiation. The resulting high temperature caused the eventual emergence of atomic nuclei from an immensely dense state in the simple forms of hydrogen and helium, rather than in the unusable form of iron. As the universe continued to expand, the number densities of matter and radiation kept pace with one another. But because the energy per particle of matter has a rock-bottom value (the rest energy), whereas the energy per photon continued to be redshifted to lower and lower values, eventually the mass-energy density of matter began to dominate that of radiation. Whether the universe will expand indefinitely seems now to hinge on whether neutrinos have an appreciable rest mass or not. In any case, as the cosmos continued to cool, galaxies ultimately condensed gravitationally from the primordial medium. Except for the twinkling of the stars, which fused hydrogen into helium in their interiors, the sky grew dark. Because of the separation of the one superforce by spontaneous symmetry breaking into four distinct and disparate siblings, objects of the scale of planets and humans could be born around the stars. Interspersed throughout this history is a delicate interplay between the microscopic and the macroscopic, between thermodynamics and gravitation. How varied and how beautiful is this physical world, because nature managed to achieve a proper balance between the symmetric and the asymmetric! How wondrous that the human mind can actually comprehend some of her innermost secrets! Let us celebrate this grand construct of the human intellect and, as Chandrasekhar suggests, be inspired by Virginia Woolf to seek further advances:

> "There is a square; there is an oblong.
> The players take the square and place it on the oblong.
> They place it very accurately; they make a perfect dwelling place.
> Very little is left outside. The structure is now visible;
> what was inchoate is here stated; we are not so various or so mean;
> we have made oblongs and stood them on squares.
> This is our triumph; this is our consolation."

PART IV

The Solar System and Life

Neil Armstrong's footprint on the Moon; "One small step for a man, one giant leap for mankind."

CHAPTER SEVENTEEN

The Solar System

In this book so far, we have taken a journey to explore the universe, and reached the most extreme outposts of observable space and the earliest instants of measurable time. Let us pull back now. Let us return past the Big Bang, past the quasars (Color Plates 22, 23), past the distant clusters of galaxies (Figure 17.1), past the nearby groups of galaxies (Color Plates 24-26), past the Local Group of galaxies (Color Plates 27-29), into our own Milky Way system of stars and gas clouds (Color Plates 30-36), into our own solar system (Color Plate 37). It's been a long journey; it's good to be back.

The first bodies we encounter as we approach the solar system are frozen balls of gas and dust that exist in the billions, usually at a considerable fraction of the distance to the nearest stars. Look there! There's one of those frozen balls that has wandered too close to the Sun; it's taken the familiar, but ever-fascinating, form of a comet (Color Plate 38). And look over there (Figure 17.2a). That's Pluto. Usually it's the most distant planet from the Sun, but during 1979–2000, its eccentric orbit has actually carried it inside the orbit of Neptune (Figure 17.2b). Also inside Neptune lies Uranus, the seventh planet of the solar system (Figure 17.2c). Pluto, Neptune, and Uranus are all so far away from us that their properties are relatively poorly known. For example, it was only in 1977 that a team of astronomers led by James Elliott discovered in an airborne observation that Uranus was surrounded by a set of thin rings; and it was only in 1978 that James Christy discovered on several high-quality photographs that Pluto possessed a satellite (named Charon, the mythical figure who ferried souls across the river Styx to the god of the underworld, Pluto).

The sixth planet, Saturn, is close enough and bright enough that it was known to the ancients. But only during recent spacecraft missions which flew past Saturn have we gotten a really clear look at this spectacular ringed giant (Color Plates 39 and 40). The next planet in, Jupiter, is even more massive (Color Plate 41). Its face is covered by an intricate system of zones and belts, and a giant red spot further attests to the fierce winds and storms which wrack this lord of the planets (Color Plate 42). Like any regal figure, Jupiter is surrounded by a retinue of satellites: fifteen known moons, great and small, with lesser bodies quite likely to be discovered in the future. Four of these moons were known to Galileo: Io, Europa, Ganymede, and Callisto (Color Plates 43). These grand moons are large enough that they would be called planets in their own right were they not dwarfed by the great lord that they orbit.

Inside the orbit of Jupiter lies a vast wasteland, littered by the debris of failed planets: the asteroids (Color Plate 44). What is the true story of their failure? Closer yet to the Sun lies the fourth planet, Mars, considered the god of war by the ancients because of its fierce red color (Color Plate 45). Its extraordinarily scarred face bears witness to an active past (Color Plate 46). Let's press on, past the third planet, and as we approach the Sun, we pass the second planet, Venus (Color Plate 47), considered by the ancients to be the goddess of love. A thick

Figure 17.1. The cluster of galaxies in Hydra may lie at a distance exceeding 3 billion light years from us. (Palomar Observatory, California Institute of Technology.)

veil of sulfuric-acid clouds hides her face from us, and modern explorations have revealed that her passions are too fiery for mortal man. Closest to the Sun is Mercury (Figure 17.3). Fleetest of the planets, Mercury was considered by the ancients to be the messenger of the gods. In what errands did he acquire such a pitted face?

At the center of the solar system lies the true monarch of the realm, the mighty Sun (Color Plate 48). So plentiful are the Sun's riches that in one second it squanders more energy than lies in oil under the sands of all Arabia (Chapter 5). Even great Jupiter bows under the influence of the Sun. As we swing around the Sun, we pass Mercury and Venus, and approach again the third planet. We see

(a)

Figure 17.2. (a) The motion of Pluto is discernible from these two photographs taken successive nights apart.

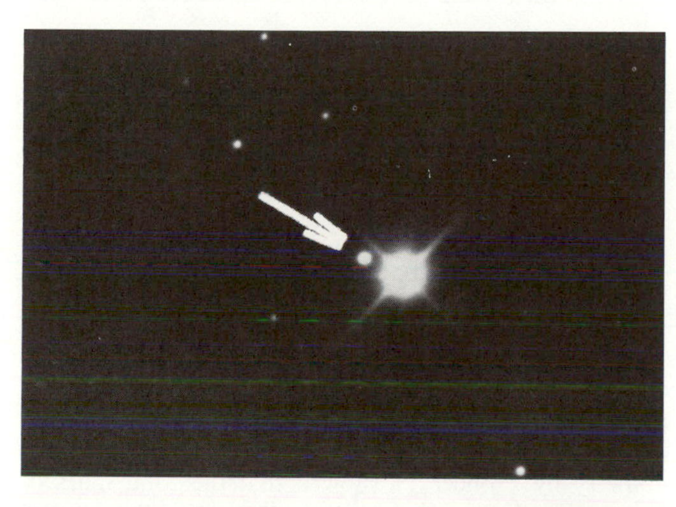

(b)

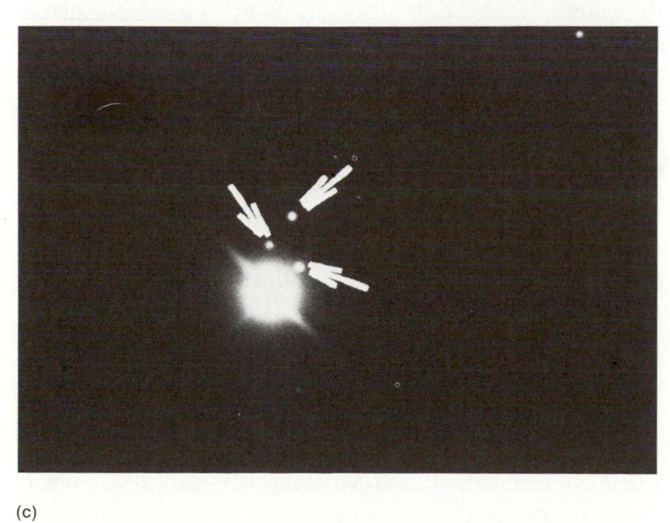

(c)

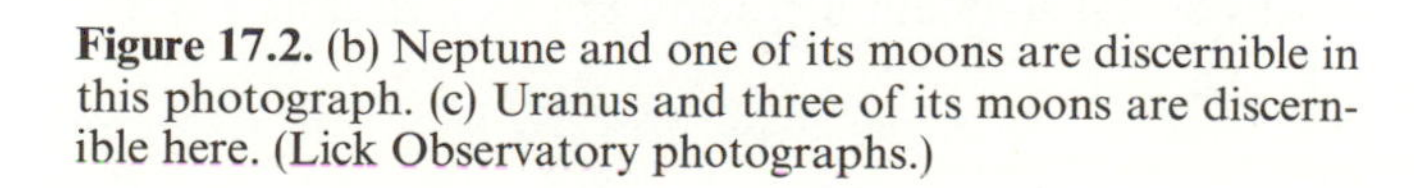

Figure 17.2. (b) Neptune and one of its moons are discernible in this photograph. (c) Uranus and three of its moons are discernible here. (Lick Observatory photographs.)

Figure 17.3. Mosaic of Mercury produced from images taken by Mariner 10. (NASA, Jet Propulsion Laboratory.)

Table 17.1 **The Planets**

Planet	*Symbol*	*Equatorial diameter (Earth = 1)*	*Gravitational mass (Earth = 1)*	*Mean density* (gm/cm^3)	*Mean distance from Sun (Earth = 1)*	*Sidereal rotation period (day)*
Mercury	☿	0.38	0.055	5.4	0.39	58.
Venus	♀	0.95	0.81	5.1	0.72	−243.[a]
Earth	⊕	1.0	1.0	5.5	1.0	1.0
Mars	♂	0.53	0.11	3.9	1.5	1.0
Jupiter	♃	11.	318.	1.3	5.2	0.41
Saturn	♄	9.4	95.	0.70	9.5	0.43
Uranus	♅	4.4	15.	1.0	19.	−0.51[a]
Neptune	♆	3.9	17.	1.7	30.	0.66
Pluto	♇	0.26[b]	0.0023	0.8[b]	39.	6.4

[a] The retrograde rotations of Venus and Uranus are somewhat different in character. The spin plane of Venus is almost aligned with its orbital plane, whereas the spin plane of Uranus is almost *perpendicular* to its orbital plane. Uranus is unique among the planets in having a very large angle of inclination between its spin plane and orbital plane.

[b] These figures are recent estimates derived from Cruikshanks and Morrison's analysis of the reflectivity, and from Boksenberg and Sargent's use of a method called speckle interferometry.

that she has a faithful companion, the Moon (Color Plate 49). A serene if pock-marked figure, the Moon shines by reflected light from the Sun. At long last, the third planet comes into view, the Earth (Color Plate 50). An insignificant speck of dust in this immense universe, yet in our eyes, no more beautiful sight exists than this planet Earth, our Earth, our home.

Inventory of the Solar System

Planets and Their Satellites

Table 17.1 gives a quick inventory of the nine planets which circle the Sun, and shows that the planets divide into two categories: the inner or terrestrial planets, which are small and have a mean density of 4–5 gm/cm^3; and the outer or Jovian planets, which are large (except for Pluto) and have a mean density of 1–2 gm/cm^3.

Each of the Jovian planets—Jupiter, Saturn, Uranus, and Neptune—has more than one moon. Jupiter has at least fifteen moons; the thirteenth was discovered in 1974 in ground-based observations conducted by Charles Kowal; the fourteenth and fifteenth, in the 1979 flyby of Voyager 1. Saturn has at least twenty-two moons, only nine of which were known before space missions to this spectacular planet. Uranus has five known moons; and Neptune, two. However, more small moons may be discovered as future and present spacecraft inspect these giant planets at closer range.

In addition to extensive satellite systems, the Jovian planets may also all possess rings. The rings of Saturn were discovered by Galileo in 1610, but their true geometry was not understood until Huygens offered the correct solution in 1655, and in 1856 Maxwell showed theoretically that the rings must consist of many independent particles. Uranus was discovered to have rings by stellar occultation studies in 1977 (Figure 17.4), and direct imaging during the Voyager missions revealed Jupiter to have rings in 1979 (Figure 17.5). We may have to await until 1989, when Voyager 2 is scheduled to encounter Neptune, to discover whether this fourth Jovian planet also has faint rings like Jupiter and Uranus.

Celestial-mechanics experiments plus ultraviolet photometry by Pioneer 11 showed that the total mass of Saturn's rings cannot exceed 3 millionths of the mass of the planet. It used to be thought that this mass consisted mostly of small icy specks; however, analysis of radio waves transmitted through the rings during the Voyager 1 flight past Saturn revealed many boulders with diameters of several meters. In contrast, the rings of Jupiter and Uranus are probably made of much smaller and less reflective particulate material.

The terrestrial planets—Mercury, Venus, Earth, and Mars—have no rings and very few moons. Mercury has no moon; Venus also has none; Earth has one (Luna or *the* Moon); and Mars has two (Phobos and Deimos, "Fear" and "Panic," the chariot horses of the god of war). Moreover, in comparison with the large moons of the Jovian planets, the moons of the terrestrial planets, except for Luna, are quite small (Table 17.2). Indeed, since the two small moons of Mars are probably captured asteroids, the Earth-Moon system is unique among the terrestrial planets, and perhaps ought to be regarded, not as a planet-satellite system, but as a double planet.

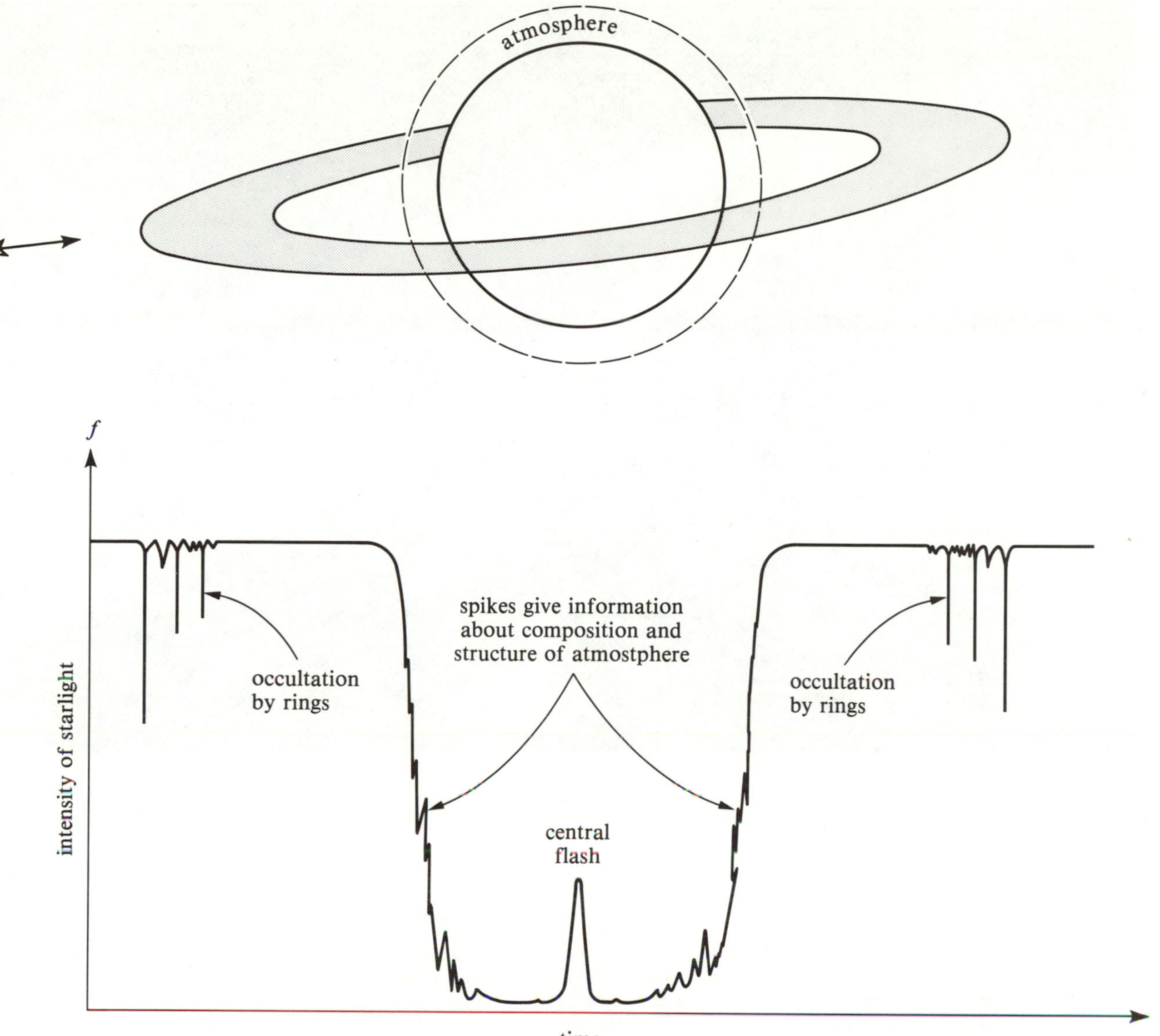

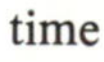

Figure 17.4. The rings of Uranus were revealed by studies of the occultation of a distant star. The lower light curve shows schematically the observed light intensity of the star as its position relative to a ringed planet changes. Such light curves can reveal not only the presence of occulting ring material, but also much information about the composition and structure of the planet's atmospheric layers. A central flash occurs when the star is directly behind the center of the planet, and is caused by refraction and extinction in the planet's atmosphere, which acts like a lens in this configuration. (See James L. Elliot, *Ann Rev. Astr. Ap.*, **17**, 1979, 445; see Figure 9 of this article for actual geometry.)

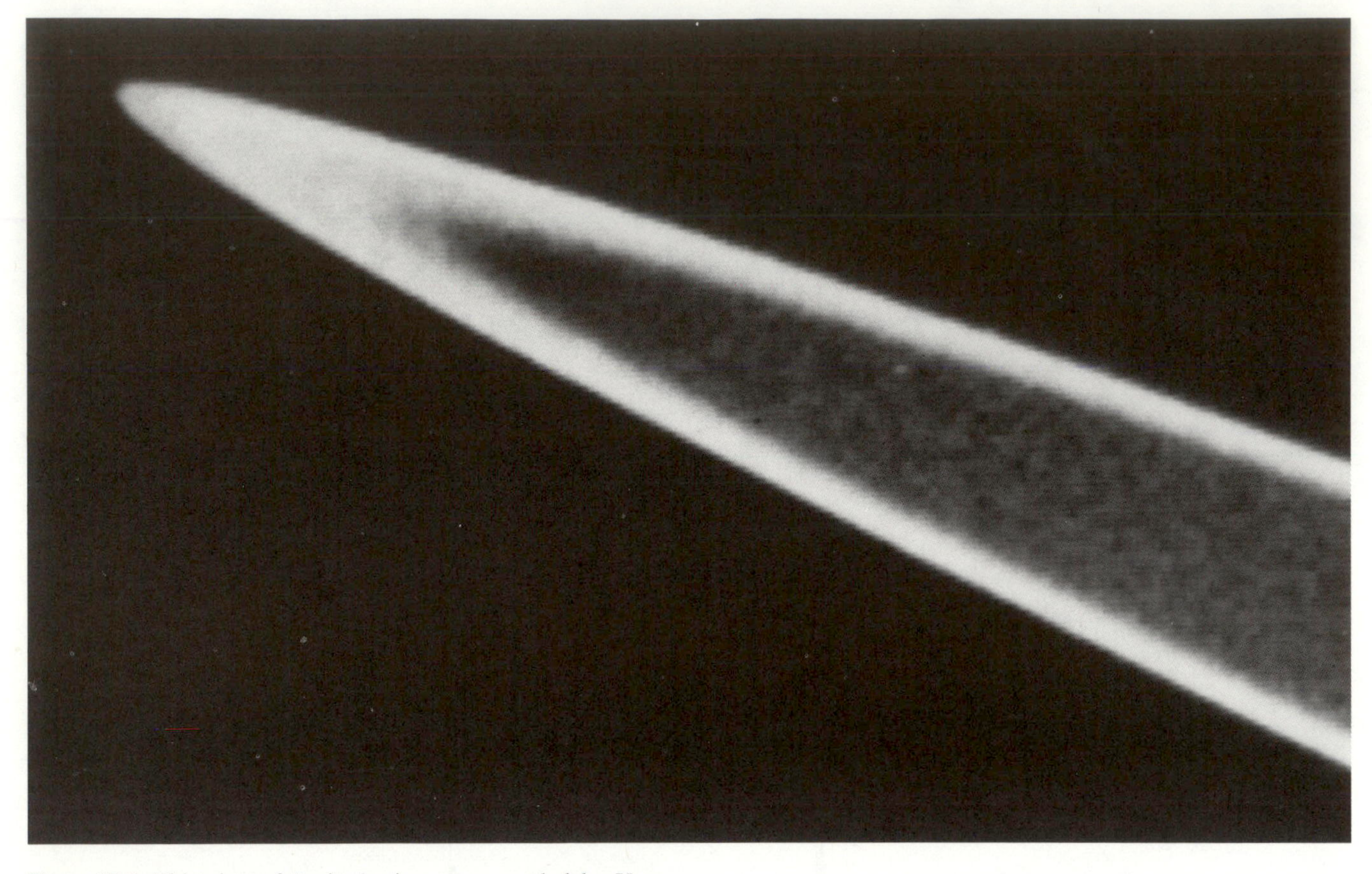

Figure 17.5. This view of Jupiter's ring was recorded by Voyager 2. Forward scattering of sunlight by small particles made the ring brighter than expected. (NASA, Jet Propulsion Laboratory.)

Table 17.2 **Largest Moons of Terrestrial and Jovian Planets.**

Planet	*Largest satellite*	*Mean diameter (Earth = 1)*	*Gravitational mass (Earth = 1)*	*Mean density* (gm/cm^3)	*Distance from planet* (lt-sec)
Mercury	none	—	—	—	—
Venus	none	—	—	—	—
Earth	Moon	0.27	0.012	3.3	1.3
Mars	Phobos	0.0016	1.7×10^{-9}	1.9[a]	0.03
Jupiter	Ganymede	0.41	0.025	1.9	3.3
Saturn	Titan	0.40[b]	0.023	1.9	4.1
Uranus	Titania	0.14	0.0007	1.4	1.5
Neptune	Triton	0.40 ?[b]	0.023	1.9	1.2
Pluto	Charon	0.06	?	1 ?	0.06

[a] A gravitatitional mass estimate for Phobos is available from Viking Orbiter and suggests that Phobos has a density characteristic of chondritic material.
[b] Voyager 1 measurements showed that Titan is slightly smaller than Ganymede, contrary to previous belief. It remains for spacecrafts to settle whether Triton or Ganymede is the solar system's biggest moon.

Minor Planets or Asteroids; Meteoroids, Meteors, and Meteorites

In addition to the nine major planets and their satellites, there are between 10^4 and 10^6 minor planets or **asteroids** (perhaps more) with orbits that lie between Mars and Jupiter at distances of 2.2 to 3.3 astronomical units from the Sun. There may also be minor planets further from the Sun, like Chiron (discovered by Kowal in 1977), whose orbit lies mostly between Saturn and Uranus. The total mass of all the asteroids has been estimated by Kresak to be less than 10^{-3} of the mass of the Earth.

Although most of the asteroids lie in a "belt" between Mars and Jupiter, there are some—called **meteoroids**—which have Earth-crossing orbits. Figure 17.6a shows schematically the prevalent theory about the production of such meteoroids from asteroids. The basic idea is that

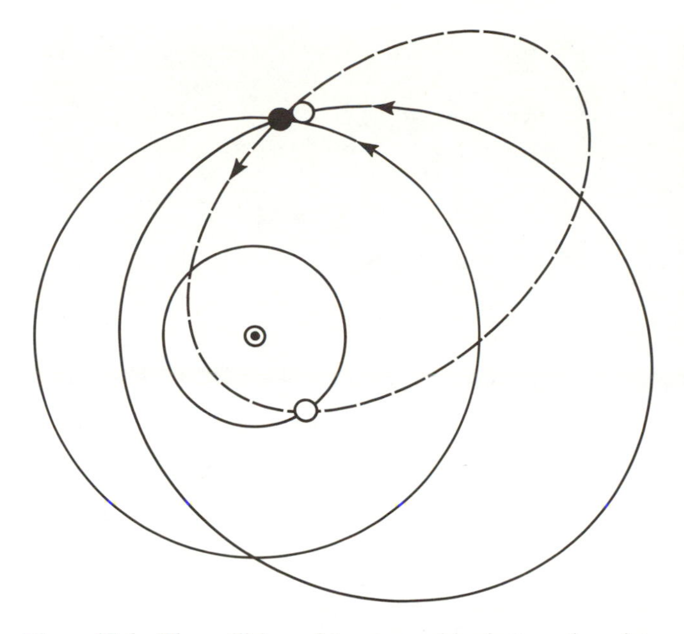

Figure 17.6a. The collision of two asteroids shatters them into many pieces, some of which may suffer resonant interactions with Jupiter and, within 10^4 to 10^5 years, enter Earth-crossing orbits. Meteoroids found on Earth are therefore, in essence, small asteroids.

Figure 17.6b. A meteor or "shooting star" arises when a solid piece of matter is heated to incandescence by friction as it falls through the air. (Yerkes Observatory photograph.)

Figure 17.6c. The mile-wide Meteor Crater in Arizona was created about 22,000 years ago by the impact of a large iron meteorite. Notice the scale of the road leading up to the crater. (Meteor Crater Enterprises, Inc.)

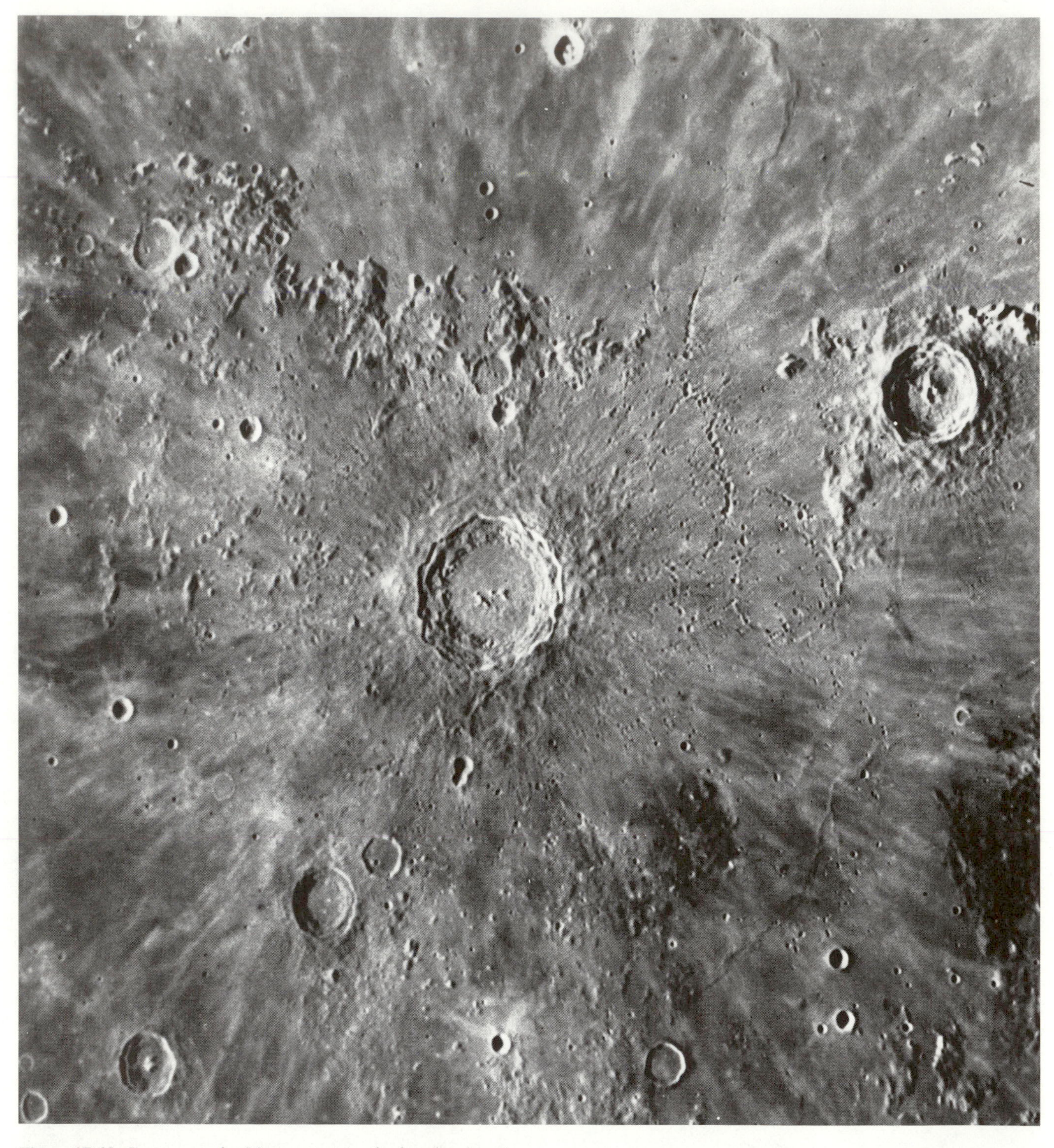

Figure 17.6d. Craters on the Moon attest to the bombardment by meteorites in the past and the present. (Lick Observatory photograph.)

Figure 17.6e. A piece of the Murchison meteorite, a carbonaceous chondrite, was found to contain 16 different kinds of amino acids. (Courtesy of C. Ponnamperuma.)

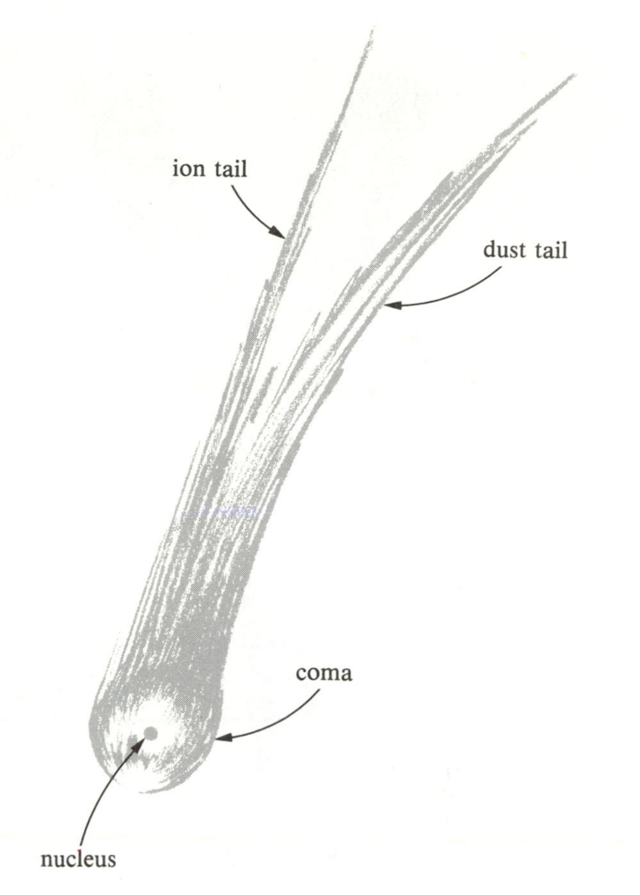

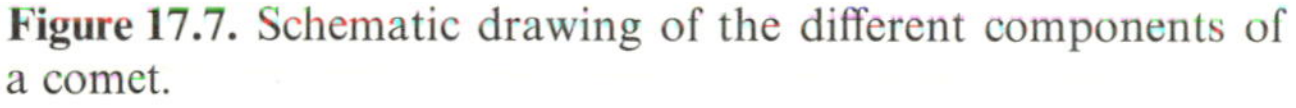

Figure 17.7. Schematic drawing of the different components of a comet.

occasionally asteroids collide with one another. These collisions can shatter the bodies, and some of the fragments may be thrown into resonant interactions with Jupiter (see Chapter 18) that ultimately give them Earth-crossing orbits.

A meteoroid which actually intersects the Earth and enters the Earth's atmosphere will heat up from the friction generated in the passage. It will then appear as a fiery "shooting star" which is called a **meteor** (Figure 17.6b). If the mass of the meteor is less than 10^{-10} gm, it may slow down so fast that it survives the flight; if so, it is called a micrometeorite. On the other hand, if the mass of the meteor is greater than 10^3 gm (a kilogram), it may have enough material to survive the ablation and also make it to the ground. If subsequently found, such an object is called a **meteorite**. The impact of an especially large meteor may create an enormous crater (Figure 17.6c). The pock-marked appearances of Mercury and the Moon (Figures 17.3 and 17.6d) attest to the heavy bombardment that all the terrestrial planets must have suffered early in their history. The lack of wind erosion and surface geological activity on Mercury and the Moon have merely preserved a more vivid record of the cratering process than on Venus, Earth, and Mars.

Meteorites come in three basic types, which depend on their chemical composition: "stones," "stony irons," and "irons". Stones resemble rocks; stony irons have some metal-rich inclusions; and irons contain mostly metals like iron and nickel. Fascinating among the stones are a subclass called the **carbonaceous chondrites** (Figure 17.6e). These meteorites contain carbon-bearing compounds ("organic" compounds) in rounded inclusions called chondrules. We shall discuss the significance of these findings for the issue of life in the universe in Chapter 20.

Comets

The word **comet** derives from the Greek name for "long-haired star." For naked-eye observers, comets are one of astronomy's most spectacular displays (Color Plate 38). Figure 17.7 shows schematically some features commonly found in comets: a head consisting, it is believed, of two parts—a **nucleus** and a **coma**—plus one or two tails. The nucleus of the comet is the essential part, since it is the ultimate source of all the mass.

In the most widely accepted theory of comets, Fred Whipple proposed that the nucleus is composed of chunks of dust and frozen ices of compounds such as methane (CH_4), ammonia (NH_3), water (H_2O), and carbon dioxide (CO_2). Whipple has likened the nucleus of a comet to a "dirty snowball," but given that it may span a few kilometers across, a "dirty iceberg" might be a more appropriate name. In more recent models, the cometary nucleus is thought to have a rocky core, surrounded by a mantle of "dirty ices."

Figure 17.8. Montage of photographs which show Halley's comet as it approached and receded from the Sun in 1910. Notice how the tail of the comet always points generally away from the Sun. This spectacular object is scheduled to make another appearance in 1986; unfortunately, the alignment with the Earth at that time will foreshorten the actual tail. (Palomar Observatory, California Institute of Technology.)

In any case, Jan Oort proposed that there are billions of such small bodies in a "cloud" of about 10^5 AU in radius surrounding the Sun. Some of these have very elongated orbits, and these occasionally suffer a gravitational perturbation from a nearby passing star which sends the body closer to the Sun. When the comet approaches within a critical distance of the Sun, the ices in a skin around the mantle vaporize and form a huge ball of expanding fluorescent gas. This coma constitutes the part of the head that is visible on Earth. Two processes then contribute to the development of the tail(s). Radiation pressure can push on the dust particles embedded in the expanding coma; and as Ludwig Biermann was the first to realize, a solar wind can blow past the ionized gas and drag on it. These two processes sweep the gas and dust into one or two long tails which generally point radially away from the Sun (Figure 17.8). Notice, in particular, that the tails do *not* necessarily correspond to the direction of motion of the comet.

Problem 17.1. Suppose that the mean density of a cometary nucleus is 2 gm/cm^3. How much mass does a typical nucleus have if it were modeled as a sphere of radius 1 km? How much mass is contained in comets in total if there are a billion such objects? Compare your answer with the mass of the Earth. Discuss the uncertainty of this estimate.

Some comets like Halley's comet (Figure 17.9), make periodic returns. These must ultimately have all their volatile compounds outgassed from the mantle. (Other comets may be perturbed by a close passage to Jupiter and be thrown out of the solar system altogether; see Figure 9.18 for an analogy.) Indeed, astronomers know of some 30-odd "Apollo objects," which are asteroid-like bodies whose orbits cross the orbit of Earth. Some scientists believe these objects to be the exposed rocky cores

Figure 17.9. The head of Halley's comet on May 8, 1910, showed the development of two tails. (Palomar Observatory, California Institute of Technology.)

of such "extinct" comets. Wetherill has estimated that such Apollo objects strike the Earth once every 250,000 years or so, releasing an amount of energy equal to 100,000 megaton nuclear bombs. No such collision has occurred during recorded history, but some prehistoric impacts may have produced some of the larger craters that exist on Earth, and an especially destructive collision may have led to the extinction of the dinosaurs (Chapter 20).

Problem 17.2. Imagine an extinct comet to strike the Earth with velocity equal to the orbital velocity of the Earth around the Sun, 30 km/sec. Assume a mass equal to one-tenth that given in Problem 17.1, and compute the kinetic energy of this body. A convenient measure of the energy of nuclear explosions is in megatons of TNT (because nuclear reactions release about a million times more energy per ton than chemical reactions). One megaton of TNT produces 4.2×10^{22} ergs, and will blast a crater of roughly 1 km diameter. If the crater diameter scales as the 1/3 power of energy, how big a crater will our comet produce?

The Interiors of Planets

Interior Structure

We have examined photographs of the planets in the Solar System, but that does not tell us what the bulk of their interiors look like. Or to make this point in another way, let us ask how is the internal structure of a planet different from that of a star like the Sun? To answer this question, we must remind ourselves what the Sun's interior looks like. In Chapter 5, we learned that the Sun represents a mass of gas in which:

(a) the inward pull of self-gravitation is balanced by the internal (thermal) pressure of the gas (mechanical equilibrium);

(b) heat diffuses outward from the hot interior, eventually to leave the surface in the form of freely escaping photons (heat transfer); and

(c) the central temperature is raised high enough to release enough nuclear energy to balance the heat transferred outwards (energy generation and thermal equilibrium).

Planets are not now collapsing gravitationally; so they must also be in a state of mechanical balance. This balance differs, however, from that of the Sun, in that the matter in the interior of a planet resembles nothing like a perfect gas. The interiors of planets contain solid or liquid matter, or a combination of both. As a consequence, mechanical balance in planets is largely independent of heat transfer and energy balance. Because the interiors of planets are generally still hotter than their surfaces, heat transfer does take place in them also, but primarily by convection or by conduction, not by radiative diffusion (in the interiors). Finally, the internal temperatures of planets are far too low to allow thermonuclear reactions (see Chapter 8, which gave the lower mass limit for luminous stars as $0.08M_\odot$). The loss of internal heat therefore either must lead to a gradual cooling of the planet, or must be made up by other sources (e.g., radioactive decay, slow gravitational contraction). Let us now examine these issues in more detail.

Mechanical Balance

Since planets are relatively cool objects, it is instructive to follow Hamada and Salpeter and see how their structure is related to that of white dwarfs (Chapter 7). Ordinary white dwarfs are, of course, dead stars, whose primary support comes from the degeneracy pressure of the free electrons. Imagine, now, building a sequence of white dwarfs of ever-decreasing mass. In such a sequence, the radius *increases* with decreasing mass (Problem 7.2), so that the degeneracy energy per particle becomes less and less. Relative to this decreasing degeneracy pressure, we expect the Coulomb forces to play a more and more important role in helping to support the object against the pull of self-gravity. (As an extreme example, self-gravity is easily resisted by molecular forces in small objects like humans.) At some point, the Coulomb forces must be as important as degeneracy pressure in supporting white-dwarf-like objects against self-gravity. Hamada and Salpeter show in a detailed calculation that this point is reached for masses comparable to that of Jupiter. Let us recover their result by an approximate but simple physical argument.

At the critical mass, the degeneracy energy, Coulomb energy, and self-gravitational energy must all be of comparable magnitude. Let us therefore compare the Coulomb energy with the self-gravitational energy. To simplify the calculation, let us assume that Jupiter, like the Sun, is composed primarily of hydrogen. If we have N protons and N electrons confined to a sphere of radius R, any one proton will feel the Coulomb forces of the *nearest* electrons and protons at distances $l \sim R/N^{1/3}$ and be shielded from more distant charges (Problem 4.6). The Coulomb energy of the entire collection of charges must therefore have magnitude roughly equal to $N(e^2/l) \sim N^{4/3}e^2/R$. The self-gravitational energy of the same configuration roughly equals $G(Nm_p)^2/R$, and the two expressions will be comparable for N equal to the critical value.

$$N_{\text{crit}} = (e^2/Gm_p{}^2)^{3/2} \tag{17.1}$$

We recognize the quantity $e^2/Gm_p{}^2$ as the dimensionless number 1.24×10^{36} (Problem 3.4). Thus, the critical mass $N_{\text{crit}}m_p = 2.3 \times 10^{30}$ gm, which very nearly equals the mass of Jupiter, as advertised.

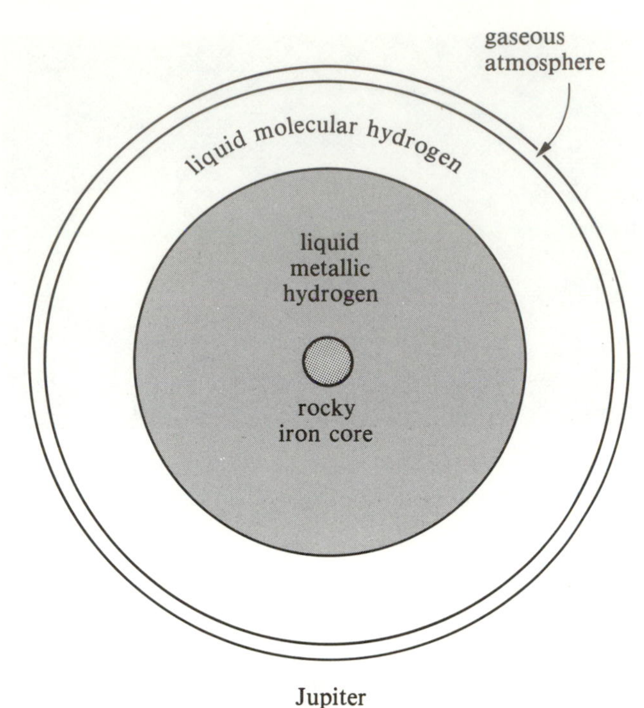

Figure 17.10. A model for the interior of Jupiter.

Problem 17.3. Retrace the arguments leading to equation (17.1) to make sure that you understand them, and verify the numerical value of $N_{\text{crit}}m_p$ given in the text. How does this compare with the mass of the Sun? The mass of Jupiter? The mass of Earth?

Notice that the radius R of the critical-mass object drops out of the calculation. How would we estimate what R should be? Well, since the Coulomb forces are only comparable with the forces of degeneracy pressure and self-gravity at the critical state, it must be equally valid to ignore Coulomb forces for a rough calculation. This neglect yields, of course, the usual mass-radius relation for white dwarfs: $R \propto M^{-1/3}$ (Problem 7.2). Since white dwarfs of $1M_\odot$ have radii equal roughly to $10^{-2}R_\odot$, "white dwarfs" of $10^{-3}M_\odot$ can be expected to have radii roughly ten times bigger, $10^{-1}R_\odot$. The equatorial radius of Jupiter does, in fact, equal $0.10R_\odot$. The good agreement, however, is somewhat fortuitous, because the electrostatic attraction of the protons *for the electrons* would be large enough to prevent the material in the interior of Jupiter from becoming completely "pressure ionized" as required in our calculation. This does not mean, however, that the degeneracy pressure abruptly drops to zero when the hydrogen becomes neutral, for the electrons cannot be pushed, on average, to radii smaller than the first Bohr radius.

Let us, therefore, follow John Anderson and William Hubbard, and assume the hydrogen in Jupiter to be mostly neutral. Can we still account for the size (or mean density) of Jupiter? Neutral hydrogen atoms packed in a cubic array with sides equal to one Bohr diameter yield a mean density of 1.4 gm/cm^3 (compare with Problem 5.5). This value is close to the mean density of Jupiter, 1.3 gm/cm^3 (Table 17.1). At an average position in Jupiter, therefore, the electronic shells of hydrogen are pushed right up against one another, causing the matter to behave like a (nearly) incompressible liquid (Figure 5.5). (We assume that the temperature is not low enough to cause solidification.) Anderson and Hubbard actually found that the detailed model shown in Figure 17.10 works for Jupiter. In the atmospheric layers, the hydrogen takes the form of a neutral molecular gas. As one goes deeper, the hydrogen turns gradually into a molecular liquid (the transition from gas to liquid is smooth at high temperatures), then an atomic liquid (when the molecules dissociate), then a metallic liquid (when some electrons become detached). It is still uncertain whether the very central regions of Jupiter contain a small liquid core made of heavier materials.

Similar theoretical models can be constructed for the other Jovian planets. In particular, to account for the

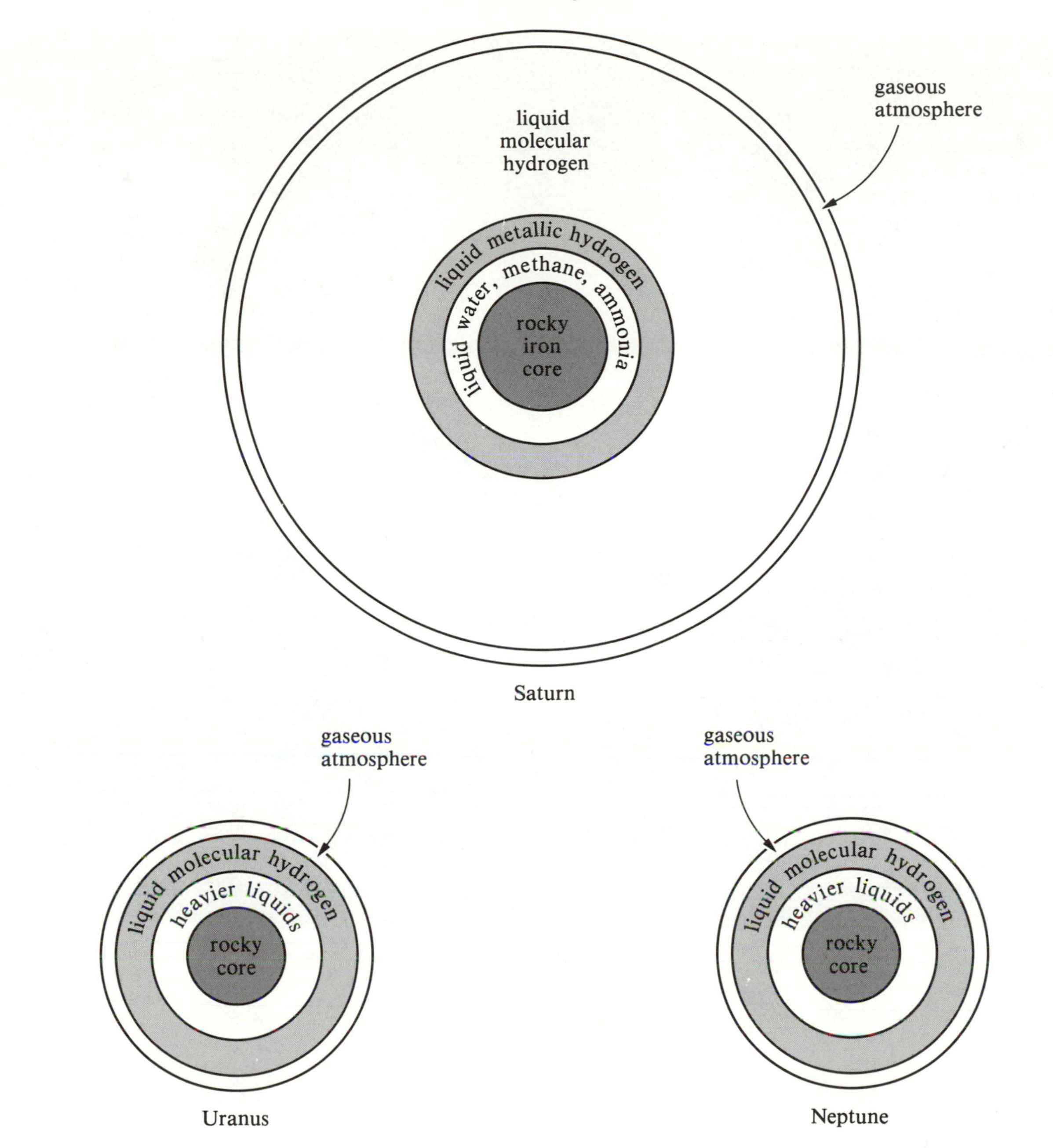

Figure 17.11. Models for the interiors of Saturn, Uranus, and Neptune. Notice the basic similarities to Jupiter in Figure 17.10, especially the dominant role played by liquid molecular hydrogen.

mean densities of Uranus and Neptune, which have appreciably less mass than Jupiter, most workers have found it necessary to include fairly sizable rocky cores in their models of these planets (Figure 17.11). Jupiter and Saturn have rotation periods of about 10 hours, and they are substantially flatter at the poles than at the equator. The oblateness of Saturn is discernible even upon casual inspection (Figure 17.12). It is possible to construct rotating planetary models to see whether the theoretically predicted oblateness matches the observed

Figure 17.12. Saturn is intrinsically oblate because of its rapid rotation. Notice also three of the moons of Saturn: Tethys (outer left), Enceladus (inner left), and Mimas (upper right). (NASA, Jet Propulsion Lab.)

one. Various gravity experiments measuring the departures from a spherical distribution have now also been carried out by the spacecraft which have visited the outer planets, and this data can also be used to sharpen up the theoretical models. Present-day models are quite sophisticated, and they give realistic treatments of the complex states of matter found inside the Jovian planets.

The molecular forces of solids and liquids are also dominant in supporting the interiors of terrestrial planets against their self-gravity. Let us consider the best-understood case, that of the Earth. Table 17.1 shows that the average density (mass/volume) of the Earth is 5.5 gm/cm^3. The crust of the Earth is made mostly of silicate rocks, which have a density of about 2.7 gm/cm^3. This is more than a factor of 2 lower than the mean density. This discrepancy implies that the Earth must have a *differentiated* interior structure. Figure 17.13 shows the various layers which have been inferred from seismological studies by geologists.

In a similar fashion, the interiors of the other terrestrial planets can also be deduced to be differentiated (Figure 17.14). This probably means that the terrestrial planets must all have been molten at one time. In a molten state, the denser material would naturally have sunk to the bottom. In any case, we see that the terrestrial planets are mostly rocks and irons, whereas the Jovian planets

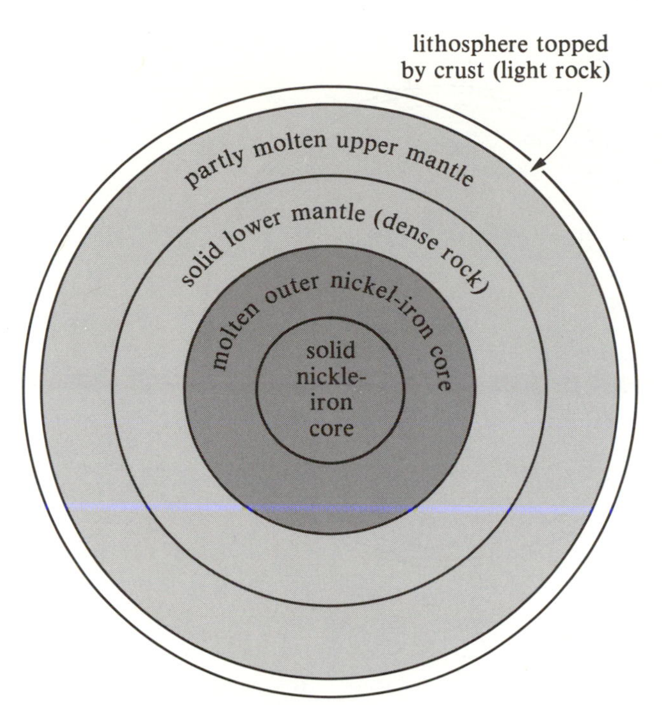

Figure 17.13. The differentiated interior structure of the Earth as deduced from seismological studies. The lithosphere plus crust consists of light rock roughly 100 km thick. The density in the lithosphere ranges from 2.6 to 3.0 gm/cm^3, and a temperature (inferred from phase-change calculations coupled with the seismic data) from 300 to 1,400 K. The mantle consists of dense rock roughly 2,800 km thick, with densities from 3.4 to 5.5 gm/cm^3 and temperatures of 1,400 to 4,000 K. The core consists mostly of iron in a sphere of roughly 3,500 km radius. Densities in the core probably range from 10 to 13 gm/cm^3; temperatures, from 4,000 to 4,600 K. The mantle and core have molten and solid parts as shown here. (For details, see F. Press and R. Siever, *Earth*, W. H. Freeman, 1978.)

contain mostly hydrogen and helium. This difference in chemical composition deserves to be explained in any theory of the origin of the solar system (Chapter 18).

Heat Transfer and Energy Balance

The comparison made above between Jovian planets and white dwarf stars extends further than one might think, for it turns out that Jupiter, Saturn, and Neptune all release significant amounts of internal heat! Uranus may do so also, on a reduced scale.

Historically, it was infrared observations of Jupiter which revealed that Jupiter radiates from its interior twice as much energy as it receives from the Sun. Even if Jupiter were a ball of perfect gas, its central temperature would be only 10^{-2} the central temperature of the Sun (see Problems 5.9 and 5.10), and the former value would be too low to sustain any nuclear reactions. The net energy lost from Jupiter must therefore come at the expense of some internal reservoir. This reservoir is its residual heat, and *not*, as commonly stated, gravitational contraction. Jupiter is a liquid body, not a gaseous one; so it can cool and still maintain mechanical balance at (almost) the same size. (Refer to the contrast between ordinary stars and white dwarfs in Chapters 5 and 7.) As John Wolfe has put it, "Jupiter cannot be radiating because it is contracting; on the contrary, it is contracting because it is slowly cooling."

The need to transport so much heat outward in Jupiter means that conduction and radiative diffusion are totally inadequate to carry out the task. Convection must set in (see Chapter 5), but such currents are easily maintained in a liquid planet. The presence of convection in the atmosphere of Jupiter also has dramatic consequences for the visual appearance of the planet that we shall explore shortly.

In contrast to the Jovian planets, the terrestrial planets are mostly made of elements much heavier than hydrogen. As a consequence (Problem 4.5), such material solidifies (at ordinary pressures) at a much higher temperature than either hydrogen or helium. The surfaces of the terrestrial planets, as we know from direct experience, are solid.

Solids do not permit (appreciable) convection; so the solid outer layers of a terrestrial planet block outward transport of heat. The inefficiency of heat transport in a terrestrial planet yields a net flow from the interior which is quite small in comparison to the amount of sunlight absorbed and reradiated by the surface and the atmosphere. For example, Gordon MacDonald has estimated that the average flux of interior heat from the Earth amounts to only 50 erg sec^{-1} cm^{-2}. Compare this value with the 1.38×10^6 erg sec^{-1} cm^{-2} in sunlight which falls on the Earth (Chapter 5). The small heat-loss from the interior implies that the Earth can retain its interior heat for periods long on geological timescales (Problem 17.4). The amount of interior heat still retained by a terrestrial planet largely determines whether it is still geologically active. We intuitively expect a larger planet to retain its heat longer than a smaller planet. (As an extreme example, it does not take very long at all for a hot rock to cool off.) In this regard, it is comforting to note that the Earth is still quite geologically active (Chapter 18); Mars shows evidence of more limited activity; whereas Mercury and the Moon have not been geologically active for a very long time. (Venus is thought to have an interior similar to that of the Earth; however, Venus's very cloudy and very different atmosphere has allowed us only limited looks at its surface.)

Problem 17.4. Calculate the total surface area of the Earth, and show that the rate of energy loss by the interior

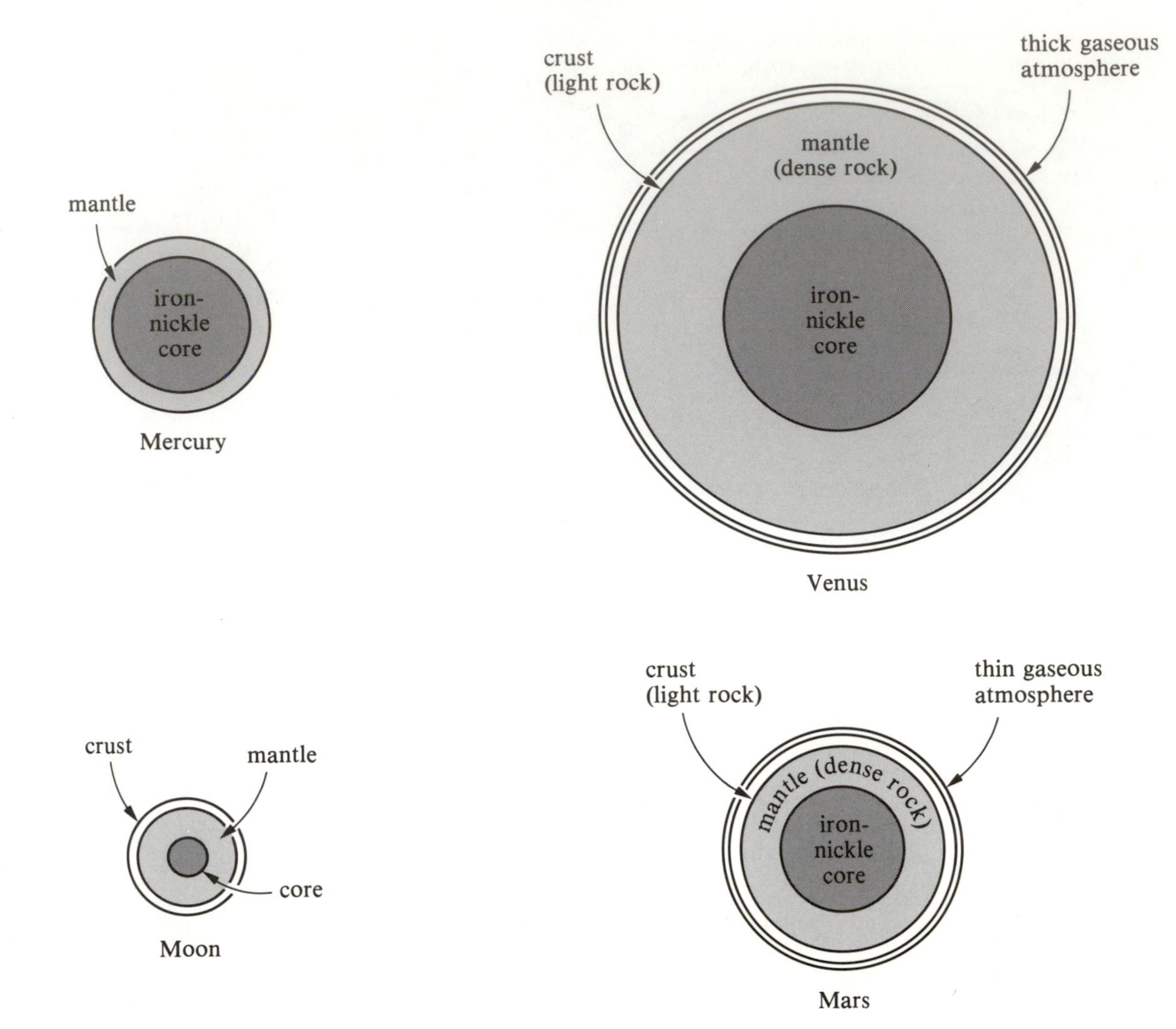

Figure 17.14. Models of the interiors of Mercury, Venus, the Moon, and Mars. Notice the similarities to the Earth in Figure 17.13, especially the predominance of metallic cores and rocky mantles or crusts.

equals 2.6×10^{20} erg/sec. Except at very low temperatures, the internal energy of a solid is given by the law of Dulong and Petit to be $3kT$ per lattice atom ($1/2\ kT$ kinetic energy and $1/2\ kT$ potential energy for each vibrational degree of freedom on a crystal lattice). Assume that the mean mass of a lattice atom inside the (partly) solid Earth is $40m_p$, and that the mean interior temperature equals 3,000 K. Show now that the total internal energy of the Earth is about 1.1×10^{38} erg. How long can this supply last at a heat-leakage rate of 2.6×10^{20} erg/sec? Convert your answer to billions of years.

You may know that Voyager 1 and 2 found spectacular volcanoes on Io (Color Plate 19), one of the four Galilean satellites of Jupiter. Since Io closely resembles the Moon in total mass and mean density (Table 17.2), why should Io be so geologically active while the Moon is now quite dead? Is Io an exception to the general rule outlined above?

No, in fact, Io is the exception which proves the rule. In a paper published three days before Voyager 1 encountered Io, Peale, Cassen, and Reynolds predicted that "widespread and recurrent surface volcanism" would be present in Io. They reasoned that, because of the gravitational perturbations exerted by the other Galilean satellites, Io would not have a perfectly circular orbit about Jupiter. The resultant stretching and recontraction of Io produced by the tidal gravitational field of Jupiter (see Figure 10.6 for an analogy to binary stars) would introduce, by way of frictional dissipation, a tremendous supply of heat into the interior of Io. It is the leakage to the surface of this heat, and not of some primordial reservoir, which combined with molten pools of sulphur yields the hot pizza-pie look of Io (Color Plate 43).

The Origin of Planetary Rings

The case of Io raises an interesting possibility for the origin of the rings around the Jovian planets. Could the rings of Jupiter, Saturn, and Uranus be the remains of a satellite which spiraled too close to its parent planet and was torn apart by the gravitational force of the latter (Figure 17.15)? Let us defer, for the time being, the issue of what causes the orbital decay, and first discuss the criterion for what constitutes "too close." In Chapters 10 and 14, we introduced the concepts of Roche lobes and Roche limits to discuss the tearing apart of one star by another or one galaxy by another. In fact, these concepts were originally invented by Edouard Roche in the nineteenth century for application to planets and satellites. In particular, he hoped to explain the origin of Saturn's rings. Let us reconsider Roche's original problem.

Consider a spherical body of mass m and radius R_m, *held together only by its self-gravity*. There is a limit to the distance r that this body may approach a much more massive body M of radius R_M. Equation (14.1) estimates r to be

$$r = (2M/m)^{1/3} R_m. \tag{17.2}$$

Suppose, now, that both masses have similar densities (a property of planets and their satellites). Then, the ratio of masses is given by the cube of the ratio of radii,

$$M/m = (R_M/R_m)^3,$$

which upon substitution into equation (17.2) yields the limiting distance r as

$$r = 2^{1/3} R_M = 1.26 R_M. \tag{17.3}$$

A more precise calculation by Roche, using *liquid* moons whose shapes can distort continuously, yields the *Roche limit* as

$$r = 2.44 R_p, \tag{17.4}$$

where R_p is the radius of the planet.

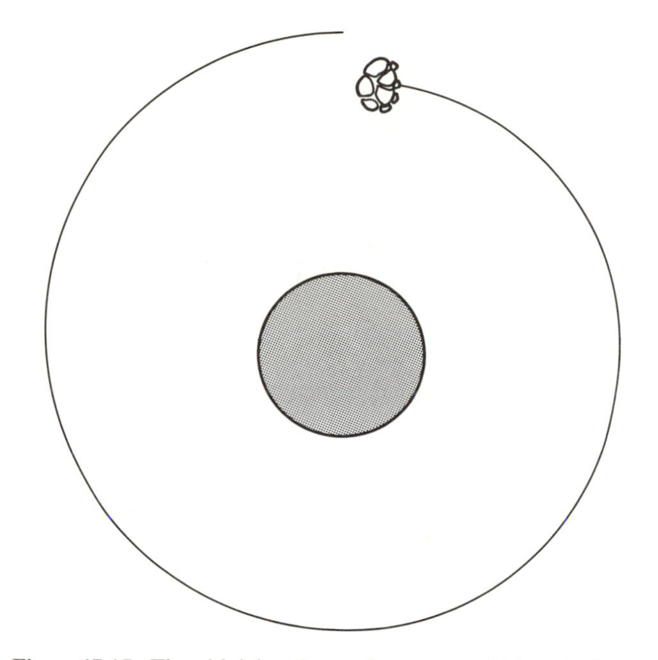

Figure 17.15. The tidal breakup of a moon which spirals too close to a planet.

Occasionally, criterion (17.4) is stated carelessly as meaning that any body which wanders within the Roche limit (2.44 times the radius) of a planet will be torn apart by the gravitational tidal force of the latter. Stated thus, it sounds like patent nonsense. You and I are within the Roche limit of the Earth, and we are not torn apart. The point is, of course, that you and I are not liquids held together only by our self-gravity. Except for weak van der Waals forces, a liquid human *would* be torn apart here on the surface of the Earth. You and I, however, have connective tissue with cohesive strength well beyond self-gravity; we do not come apart so easily at the seams.

A solid moon, of course, also has some cohesive strength beyond that of a liquid moon, as Jeffreys and later Pollack pointed out. Moreover, a rocky satellite would have a different mean density than a Jovian planet. It has been estimated that a rocky moon smaller than about 10^2 km would not be torn apart by Saturn's tides before it plunged into Saturn's atmosphere. Equation (17.4) therefore needs to be applied with some care even in astronomical contexts. Nevertheless, it is an intriguing fact that all the rings of Jupiter, Saturn, and Uranus lie within or near the Roche limits of their planets. Furthermore, because of resonant interactions with the other Galilean satellites, Io wobbles about a mean distance of about five planetary radii from Jupiter and is in the process of liquefying itself. Jupiter has a well-known moon, Amalthea, which is even closer to the parent planet, only 2.5 planetary radii. However, Amalthea is

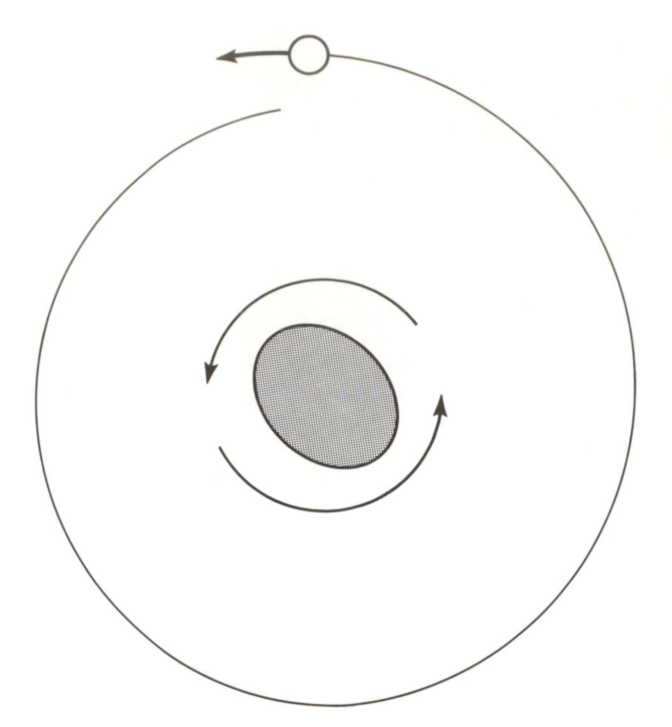

Figure 17.16. A moon which lies outside the synchronous orbit of a rapidly spinning planet will lag behind the tidal bulge that the moon raises in the planet. As a result, tidal friction will transfer angular momentum from the spinning planet to the orbiting moon, and cause the moon gradually to spiral outward (assuming the moon to be originally in a prograde orbit).

a small and irregularly shaped asteroidal body that undoubtedly has considerable gravitational cohesiveness, since its longest dimension is only 270 km. More intriguingly, the recently discovered fourteenth satellite of Jupiter lies at the outer edge of the rings of Jupiter, about halfway between the surface of the planet and the orbit of Amalthea. This satellite is only a few dozen kilometers in diameter. Similarly, the recently discovered satellites of Saturn lie at distances as close as 2.28 planetary radii, just outside the brighter rings of Saturn. Are these small satellites fragments of some original larger body which broke up into several pieces when it wandered within the Roche lobe of the involved planet? And are these fragments subsequently ground down by mutual collisions into the centimeter- to decameter-sized particles that we now observe in the rings?

Attractive as this possibility is, it has met with some serious objections, primarily over what mechanism could cause the moon to spiral inward toward the central planet. For such orbital decay to be possible, angular momentum must be removed from the motion of the moon and deposited elsewhere. Suppose this repository to be the central planet. But both Jupiter and Saturn have rotation periods which are *shorter* than the orbital period of any moon located outside the Roche limits of these planets. Therefore, any tidal interaction between such a hypothetical moon (assumed to have a direct or prograde orbit) and the rapidly spinning central planet would tend to *give* angular momentum to the moon, and cause it to spiral *outward*, not inward (Figure 17.16). (The ideal of circular orbit and synchronized spins discussed in Chapter 10 would not be realized for extreme mass ratios, because the planet cannot be spun down enough to achieve synchronism if it is much more massive than its moon.) Moreover, Goldreich and Soter have shown that the Jovian planets probably do not have enough internal dissipation to make tidal friction an important process for slowing down a moon.

Problem 17.5. Does tidal friction in the Earth-Moon system cause the Moon currently to move toward or away from the Earth? Is an Earth day getting longer or shorter (by sixteen seconds every million years)? Can a moon with a *retrograde* orbit be used in Roche's hypothesis to make the rings of Saturn? In what sense would the resulting rings rotate? Discuss how Doppler-shift measurements of Saturn's rings (first carried out by Keeler in 1895) could reveal both the sense of rotation and the fact that it takes place differentially (i.e., not like a solid body).

What about using the other moons of the satellite system as a repository of the excess angular momentum of the hypothetical doomed moon? Unfortunately, gravitational interactions between moons that are moving on nearly circular orbits about a much more massive central planet produce significant perturbations only for resonant conditions (when one orbital period is a rational fraction of the other). Such interactions could not cause a continuous spiraling of an orbit.

Thus, despite the "world-breaking" appeal of Roche's hypothesis, it meets with severe difficulties. Unless some novel and efficient means can be found for inducing orbital decay, it seems more likely to most astronomers that the matter which makes up the rings of the Jovian planets has always lain within the Roche limits of those planets. The ultimate source of matter for the ring particles may indeed be from the erosion of the small satellites now found within the Roche limits of the Jovian planets, but these small satellites may have formed at their present locations roughly at the same time as the formation of the solar system itself. Moreover, precisely because these small bodies lay within the Roche limits of the massive central planet, they could not cluster gravitationally to form a larger coherent moon.

The Atmospheres of Planets

We all have an intuitive definition of the atmosphere of a terrestrial planet: it is the mass of gas which lies above the surface of the planet. But a Jovian planet does not have a solid surface, at least, not until we reach deep into the core of the planet (see Figures 17.10 and 17.11). What would be meant by the atmosphere of a Jovian planet? Well, a star like the Sun also has no solid surface, and we had no difficulty making a distinction between its interior and its atmosphere (see Chapter 5):

> The atmosphere of an astronomical body is that layer where the photons make a transition from "walking" to "flying."

For a star, this definition is fairly unambiguous, because most of the ambient photons have similar wavelengths, mostly in the visual range, for example. For a planet, there is a complication. Sunlight provides a source of incoming radiation which has photons in the visual and near-ultraviolet regime. This radiation field affects the atmosphere qualitatively differently from the outflowing radiation of infrared photons (see, e.g., Problem 4.14). The very different transparencies of the atmosphere to radio, infrared, optical, and ultraviolet wavelengths also affect how far we can see into (or through) the atmosphere of a planet at different wavelengths.

Thermal Structure of the Atmospheres of Terrestrial Planets

Mercury and the Moon have no atmosphere to speak of, and Mars has only a very thin atmosphere (about 1 percent the surface pressure of Earth). Peter Gierasch and Richard Goody have shown that the temperature structure of the Martian atmosphere is much affected by the direct absorption of sunlight by dust particles swept into the air (Figure 17.17). The terrestrial planets with substantial gaseous atmospheres are Venus and Earth. Let us begin our discussion with the more familiar example. Figure 17.18 shows a schematic plot of the temperature of the Earth's atmosphere as a function of height above the ground (averaged over latitude, day and night, and seasons). The thermal structure of the Earth's atmosphere has four distinctly recognizable regions.

(a) The **troposphere** is the region which adjoins the ground and is the region with which we usually associate weather. Vertical heat transfer in the troposphere consists mostly of the upward diffusion of infrared photons and fluid transport by convection. As a consequence, the temperature is highest near the ground and steadily decreases to the top of the troposphere, the **tropopause**.

(b) The **stratosphere** is heated by the action of the ultraviolet component of sunlight in forming and destroying ozone, O_3. This direct absorption under optically thin conditions leads to an inversion of the temperature above the tropopause, so that the stratosphere is actually warmer than the troposphere. This increase of temperature with increasing height continues until the **stratopause** is reached.

(c) In the **mesosphere**, there is a decrease in ozone production and an increased rate of infrared cooling by carbon dioxide, CO_2; so temperature decreases with increasing height up to the **mesopause**.

(d) In the **thermosphere**, heat input from sunlight occurs by the photodissociation and photoionization of molecular oxygen, O_2. This heating mechanism is analogous to that which operates in HII regions in the interstellar medium (see Chapter 11). The balance of this heat input by various radiative cooling mechanisms leads again to a rise of the temperature with increasing height in the thermosphere until temperatures of about 1,000 K are reached. The primary energy transfer in the lower thermosphere is actually a *downward* conduction of heat into the mesosphere.

For life on Earth, our primary emphasis in this part of the book (Chapters 19 and 20), the most interesting regions are the troposphere and the stratosphere. Let us discuss here only the troposphere, and save the ozone layer in the stratosphere for the next chapter.

Because the atmosphere is transparent to optical photons, sunlight strikes the ground (or sea) with little interaction with the troposphere except for scattering by molecules, dust, and reflection from cloudtops (all of which transfer no net energy to the atmosphere). The troposphere can therefore be regarded as a gaseous atmosphere which is heated from *below* by the emanation of infrared radiation from the ground (see Problem 4.14). The troposphere is partially opaque to this emergent infrared radiation; therefore, to be able to push this radiation outward, it has to be hotter below than above. In planetary astronomy, this is called the "greenhouse effect," but the basic principle is the same as that which governs radiative transfer inside a star (see Chapter 5, especially the explanation of limb darkening in the Sun).

Problem 17.6. To consider the radiative-transfer problem, approximate any unit area of the Earth's troposphere as a plane-parallel atmosphere of height h. We simplify matters by treating the infrared absorption as a

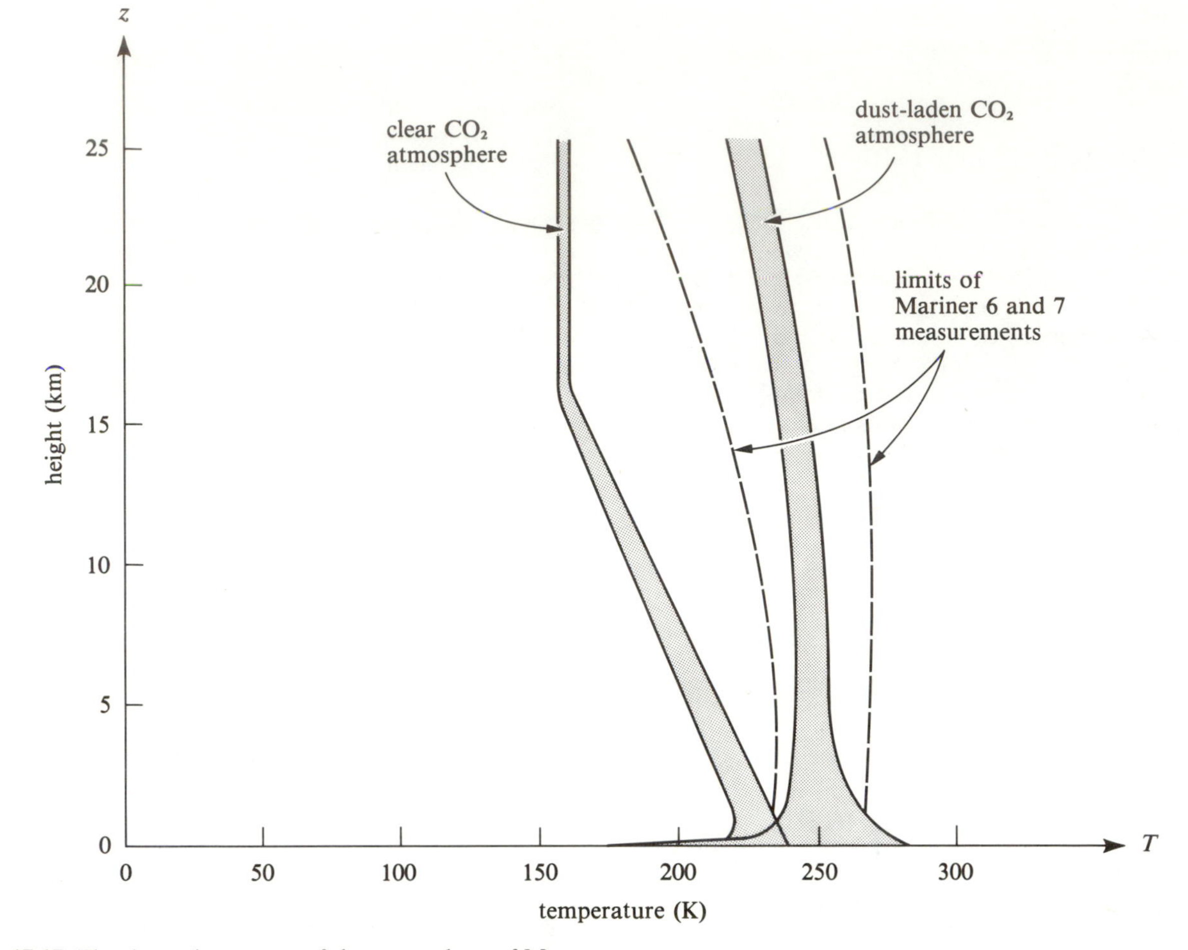

Figure 17.17. The thermal structure of the atmosphere of Mars is much affected by the amount of dust swept into the air by the winds of that planet. The shaded region between the curves demarcates the theoretical range of temperatures found in calculations using typical daytime conditions for a dust-laden and clear Martian atmosphere. The dashed curves show the limits found by spacecraft observations of Mars. (Adapted from P. J. Gierasch and R. M. Goody, *J. Atmos. Sci.*, **29**, 1972, 400.)

continuous opacity, instead of the superposition of many molecular bands made of discrete lines. Let l represent the mean free path of a typical infrared photon, and modify the arguments of Problems 5.11 and 5.12 to estimate the energy flux,

$$F_{\text{rad}} = \frac{(\text{height}) \times (\text{radiation energy per unit volume})}{\text{random walk time to cover distance } h}$$

$$= \frac{h(aT^4)}{3h^2/lc}. \tag{1}$$

If infrared radiation were the only mechanism of upward heat transport, we would require for radiative equilibrium, $F_{\text{rad}} = \sigma T_e{}^4$, where $T_e = 246$ K is the effective temperature of the Earth calculated in Problem 4.14. Denote the ratio h/l by τ, the total optical depth of the Earth's atmosphere in the infrared; notice that $\sigma = ca/4$ (Appendix A); and show that equation (1) can be written as

$$T^4 = (3/4)T_e{}^4\tau.$$

As it stands, T refers to the temperature at some intermediate point in the troposphere. For it to refer to the

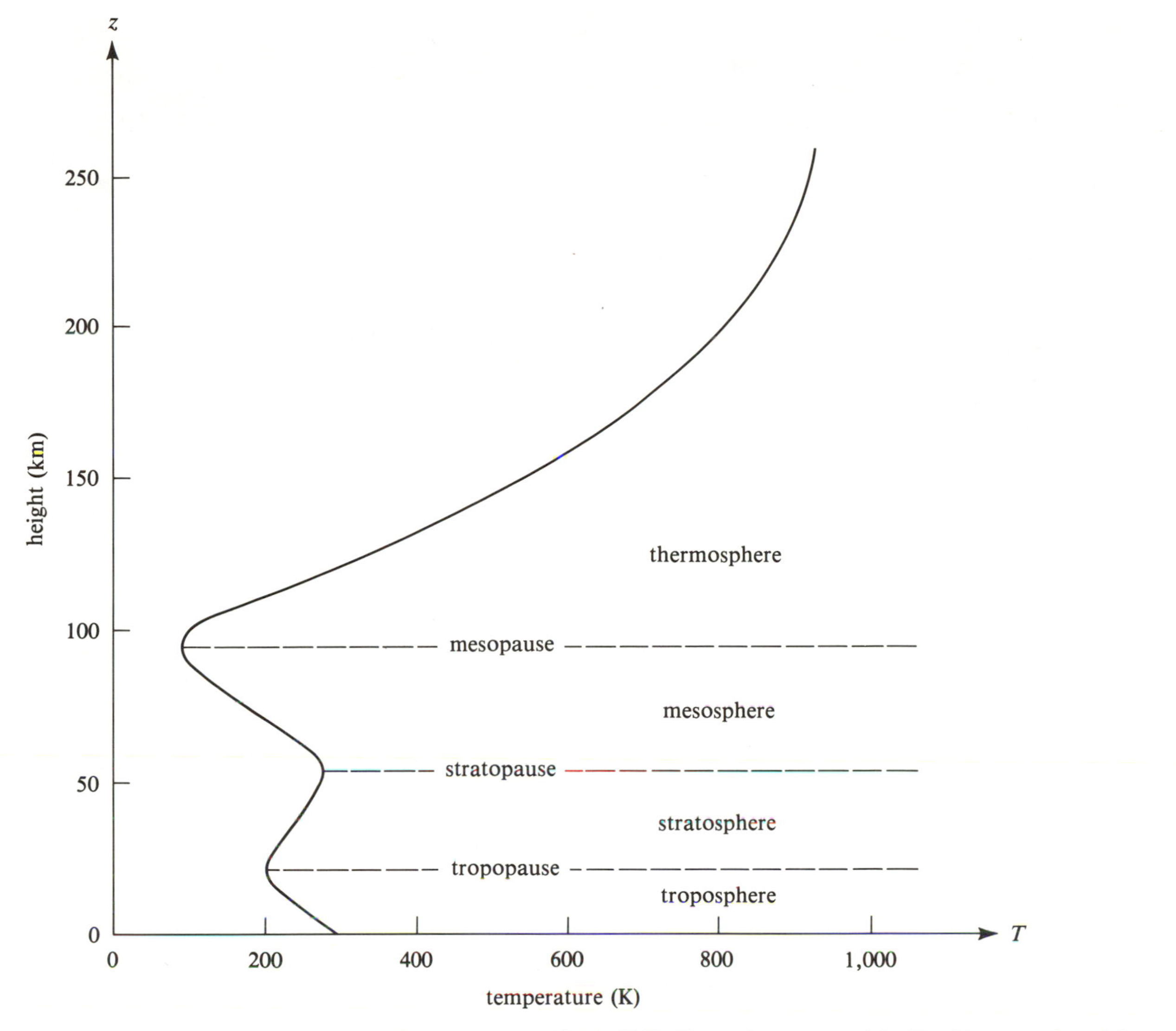

Figure 17.18. Thermal structure of the Earth's atmosphere. The various regions (troposphere, stratosphere, mesosphere, and thermosphere), plus their associated transition zones (tropopause, stratopause, and mesopause), are presumably duplicated qualitatively in the other planets. However, the detailed mechanisms of ultraviolet heating in the upper atmospheres of the other planets may differ from those of the Earth, because of the different chemical compositions involved. (For details, see J. W. Chamberlain, *Theory of Planetary Atmospheres*, Academic Press, 1978.)

bottom of the troposphere (i.e., to the air at ground level), a more sophisticated analysis of radiative transfer (Eddington's approximation) shows that we should add the number 2/3 to τ, i.e.,

$$T^4 = \tfrac{3}{4}T_e^{\,4}(\tau + \tfrac{2}{3}). \tag{2}$$

Since our location of the bottom of the troposphere is mathematically arbitrary, equation (2) must in fact apply locally at each point in the troposphere, providing τ means the local optical depth from the top of the troposphere (i.e., the number of mean free paths from the top, where infrared photons can fly freely away). Notice that the temperature increases inward as τ increases from zero at the top, which accords with our intuitive expectations. Show that the total optical depth of the troposphere can be written as $\kappa\mu$, where κ is the mean infrared opacity and μ is the mass per unit area through the Earth's atmosphere (obtainable by measuring the surface pressure; see Problem 5.9). Take $\kappa = 2.0 \times 10^{-3}$ cm^2/gm and $\mu = 1.0 \times 10^3$ gm/cm^2, and calculate the average temperature at the bottom of the troposphere. What is the temperature at the top of the troposphere (where $\tau \simeq 0$). For what value of τ is $T = T_e$? Comment on the magnitude of the Earth's greenhouse effect.

Problem 17.6 estimates that the temperature at the top of the troposphere averages 207 K. Because the Sun does not shine equally on the Earth's equator and poles, we expect some variation of this temperature with latitude. In fact, measurements show that the tropopause at the tropics and polar caps does have different temperatures; but contrary to naive expectations, the tropopause is *warmer* at the poles (about 225 K) than at the tropics (about 195 K)! This situation is, of course, opposite to that which prevails at sea level. No one knows quite what is the right explanation, but Victor Starr has proposed that the refrigeration of the tropopause at the tropics results from the planetary circulation pattern of the Earth's atmosphere, which we will discuss later.*

The upper levels of the troposphere are convectively stable; however, the lower levels are not. Because the radiative equilibrium solutions require the temperature in the lower troposphere to drop fairly quickly with increasing height, these solutions may become unstable and allow convection to develop (see Problem 5.15). Compounding this effect is the fact that the purely radiative solutions actually require the ground (which is heated by both the trapped infrared radiation *and* the incident optical sunlight) to be warmer than the air just above it (Figure 17.19). This situation tends to produce convection currents at least at ground level. Who among you has not seen the shimmering of the air above a hot pavement? How strongly convecting the rest of the troposphere becomes depends in part on how wet (with water vapor) the air is. On the other hand, the humidity depends in part on the weather patterns that develop as a result of the convection and the large-scale circulation pattern. It is easy to see, therefore, why accurate weather prediction is such a complicated business! In any case, the combination of all these complications on the Earth manages to produce an average surface temperature of about 290 K, quite close to the freezing point of water, 273 K. It is a delicate balance of factors, indeed, which produces a thriving garden on Earth rather than either a frigid wasteland or a blazing hell.

* Pioneer Venus probes have found that the polar regions of Venus's upper troposphere are also warmer than the equatorial regions.

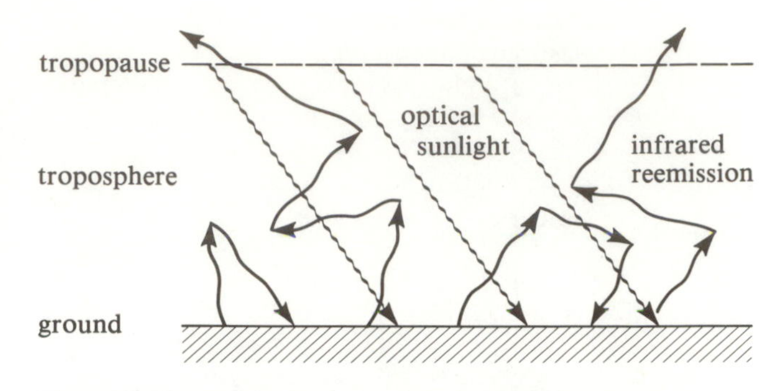

Figure 17.19. The greenhouse effect in planetary atmospheres involves the additional warming of the surface layers because of the partial trapping of the reemitted infrared radiation. At one time, it was thought that the name for this effect might be a misnomer, since greenhouses may not be kept warm by the same mechanism. However, a reanalysis by Silverstein shows that the greenhouse effect may be nontrivial even for greenhouses! In any case, the effect is the same as that which keeps the interiors of stars warmer in their interiors than at their surfaces.

To appreciate the last point, let us enquire what the surface temperature is on Earth's sister planet, Venus. We have mentioned that Venus is enshrouded with clouds, so that its surface cannot be examined by ground-based telescopes in either the optical or the infrared. Radio measurements can reach the surface, and such measurements (carried out in the 1960s by Barrett, Pollack, and their colleagues) indicated that the surface temperature of Venus might be in the neighborhood of 750 K! Such a temperature is high enough to melt lead, as was verified by direct landings of the Soviet Venera probes.

Why should Venus be so much hotter than the Earth? It is 1.38 times closer to the Sun, but it reflects about 72 percent of the incident sunlight; the Earth reflects 39 percent. Therefore, the average flux of sunlight *absorbed* by Venus is $(1.38)^2(.28/.61) = 0.87$ of that absorbed by Earth. In other words, the *effective* temperature of Venus should only be $(0.87)^{1/4}$ that of Earth, or about 238 K. Why should the surface temperature be so much higher than the effective temperature? The answer must be, of course, that Venus has a whopping greenhouse effect; so we can rephrase our original question: Why does Venus have such a large greenhouse effect? The answer must be, of course, that Venus has a whopping thick atmosphere. It does. The Venera probes found a surface pressure about 90 times that at sea level on Earth. This atmosphere is composed almost entirely (96 percent) of carbon dioxide, CO_2, which is an efficient absorber in the infrared. The methods of Problem 17.6 would predict an even larger surface temperature than the 750 K which is actually measured. The discrepancy arises because not all the absorbed sunlight in Venus penetrates all the way to the ground; some is absorbed at higher levels. In any case, Venus has a much thicker atmosphere than the Earth, with a composition heavily weighted toward CO_2. Why?

Runaway Greenhouse Effect

The best idea seems to be that of a runaway greenhouse effect (proposed by Kellogg and Sagan in 1961, reexamined by Ingersoll in 1969 and by Rasool and de Bergh in 1970). The crucial factor is the existence of liquid water on the surface of Earth and not on Venus. Sedimentary rocks (basically carbonates) have a great capacity to absorb carbon dioxide if there is liquid water present. Indeed, the Earth has an estimated equivalent of 70 atmospheres' worth of CO_2 currently locked up in rocks like limestone ($CaCO_3$). The corresponding amount of CO_2 in Venus is actually in the atmosphere. Why?

The conjecture proceeds as follows. In 1928 G. C. Simpson showed that as long as there is liquid water on the surface of a planet, the top of the convection zone in the troposphere will tend to hold as much water vapor as it physically can (be **saturated**), because more water can be held in warm air than in cold air. Upwelling currents would tend to supersaturate a cold layer were it not for the descent of drier air from above. But the layers above the top of the convection zone are convectively stable by definition; so there is no descent of dry air into the top of such a convection zone. This implies that the air at the top tends to be exactly saturated, with the excess water going into the formation of clouds of small droplets of liquid water or particles of ice (Figure 17.20).

Now, water vapor has a saturation temperature T which is a well-known function of the partial pressure P_{H_2O},

$$T = T_s(P_{H_2O}). \tag{17.5}$$

On the other hand, the temperature above the top of the convection zone must follow the radiative equilibrium law derived in Problem 17.6,

$$T = T_e[\tfrac{3}{4}(\tau + \tfrac{2}{3})]^{1/4}. \tag{17.6}$$

For a given planetary effective temperature T_e, what happens if the value of T derived from equation (17.5) falls below the value of T derived from equation (17.6) for a given water vapor pressure P_{H_2O} and optical depth at the top of the convection zone? If radiative equilibrium keeps the top of the convection zone too warm to allow water vapor to saturate at the top of the convection zone, more water will be drawn from the surface (or the oceans) by the ascending convection currents. The dwindling supply of liquid water to form sedimentary rocks would release more CO_2 into the atmosphere. The increase of both CO_2 and H_2O concentrations in the atmosphere would increase the greenhouse effect and therefore the temperature (17.6) at the top of the convection zone. If this cycle increases the discrepancy between equations (17.6) and (17.5), we have a positive feedback which would evaporate even more liquid water. The runaway would not be halted until all the available surface water (and a lot of the CO_2) is released into the atmosphere, and the surface of the planet is left in an arid and sweltering state. Let us hope that humanity does not tip the atmospheric balance enough to create a runaway greenhouse effect on Earth.

Some scientists have speculated that the opposite of a runaway greenhouse effect may have occurred on Mars. Pictures sent back from Mariner 9 in 1972 revealed the presence of sinuous channels, strongly suggestive of dry river beds (Figure 17.21). Spectroscopic and photometric studies performed during this mission suggested a surface pressure of about 0.6 to 0.7 *percent* that of the Earth, and a surface temperature which varies greatly with latitude and between day and night, but which never exceeds 300 K (a summer day on Earth) and may fall below 150 K (the freezing point of carbon dioxide, when it forms "dry ice," at the surface pressure of Mars). The Viking landers in 1976 found that Mars' atmosphere consists of 95 percent carbon dioxide (CO_2), 2.7 percent nitrogen molecules (N_2), 1.6 percent argon (Ar), and small amounts of free oxygen (O_2) and water vapor (H_2O). The average surface pressure and temperature on Mars are now marginally too low to allow the presence of liquid water (except, perhaps, in periods of volcanic activity). However, it is likely that Mars once had a much denser atmosphere, and therefore much higher surface temperatures. Thus, liquid water may well have once existed on the planet, and this water may have carved channels like those in Figure 17.21.* Most of the water and carbon dioxide now seems to be locked up either in the polar caps (Figure 17.22). or in carbonate rocks. How did this disaster befall Mars?

Those who think there was a reverse runaway greenhouse effect think that an earlier and denser atmosphere on Mars did allow the presence of liquid water on the planet. However, the gradual locking up of the atmospheric carbon dioxide, CO_2, into sedimentary rocks led to a thinning of the atmosphere, with a consequent lowering of the temperature (17.6). This caused more water vapor to precipitate out of the atmosphere, which caused even more carbon dioxide to be locked up. This runaway (in the opposite direction from what may have happened on Venus) would then have led to the present

* More recent ideas suggest that the channels were carved by volcanic flows.

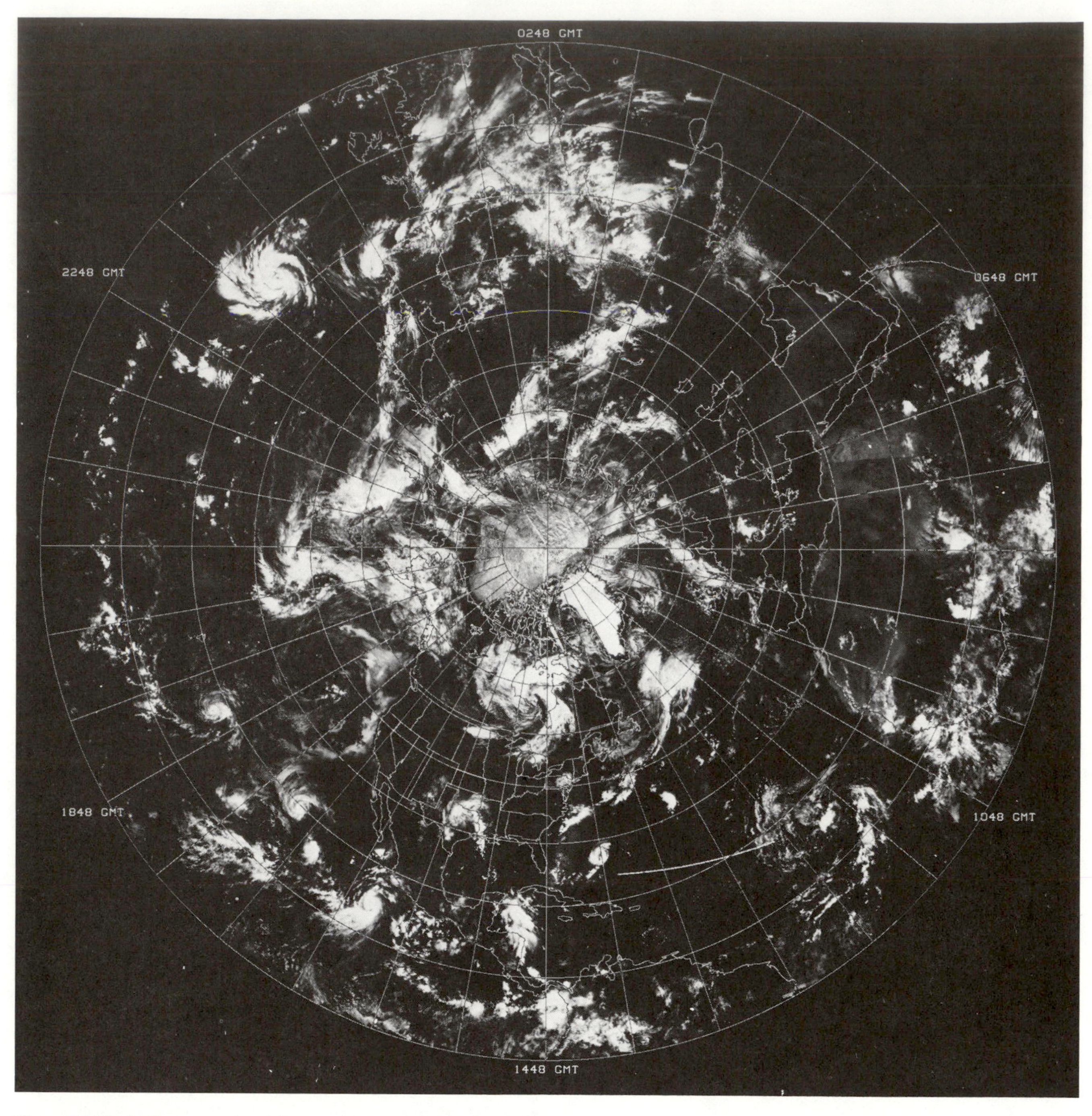

Figure 17.20. A photomosaic of the northern hemisphere of the Earth obtained from a NOAA satellite in 1974. The cloud cover over much of the planet attests to the saturation of water vapor at the top of the convection zone of the troposphere. The true complexity of the weather pattern on Earth can also be appreciated from this picture. Notice, in particular, the string of tropical hurricanes. (National Oceanic and Atmospheric Administration.)

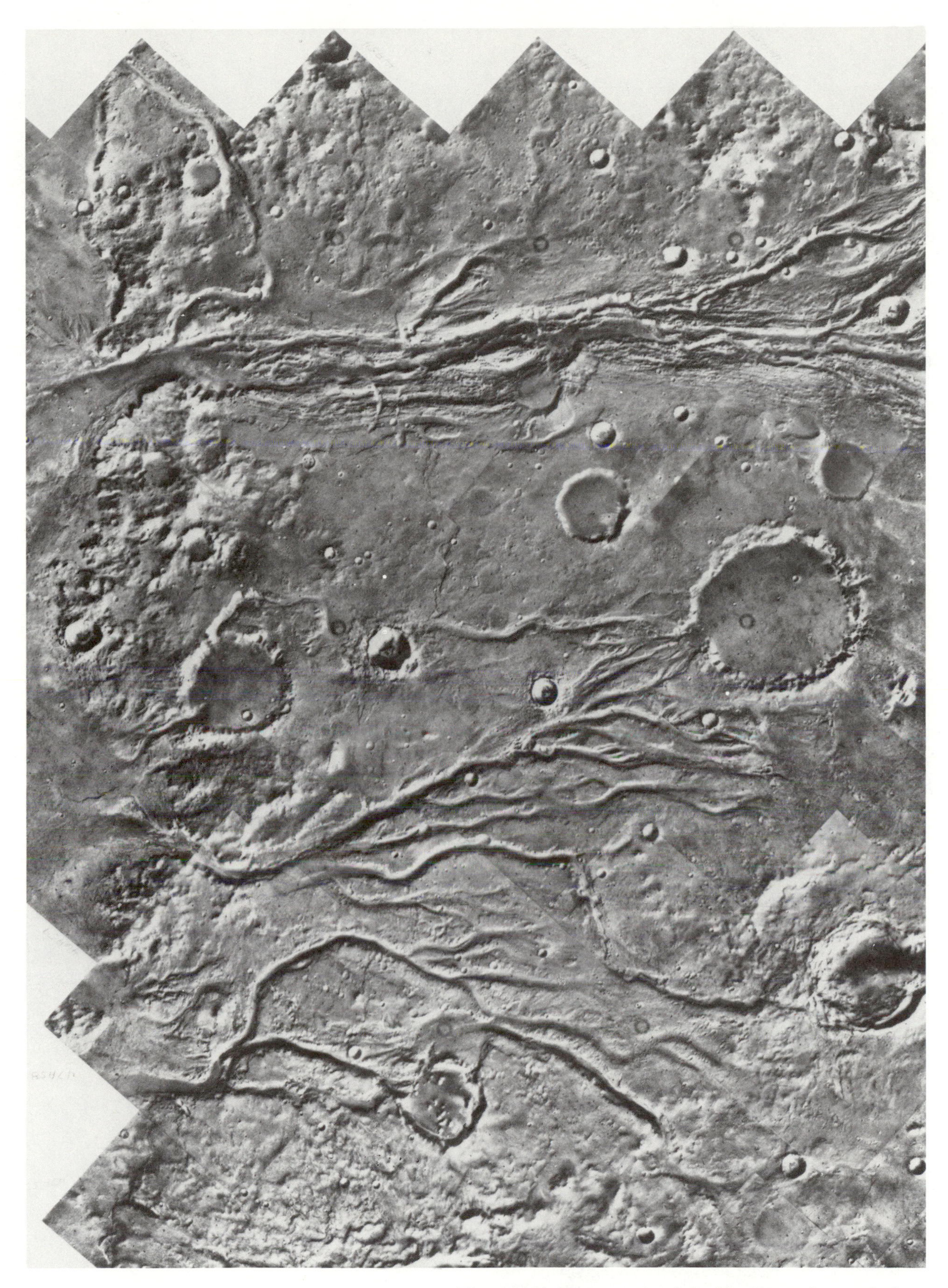

Figure 17.21. Photomosaic of the Martian surface made from pictures taken by Viking Orbiter 1. The channels are suggestive of a massive flood of water, although they may also have been carved by volcanic flows. In some cases, the channels cut through craters; in other cases, the craters were formed by meteorites impacting preexisting channels. (NASA.)

Figure 17.22. Much of the carbon dioxide, and perhaps the water, on Mars is locked up in its polar caps. The above photomosaic was made from pictures taken by Viking Orbiter 2 of the region near the south polar cap of Mars during the late Martian summer. Unlike the part of the north polar cap that did not evaporate in summer, which is all water ice, the residual south polar cap may be either solid CO_2 or H_2O ice, or a mixture of both. (NASA.)

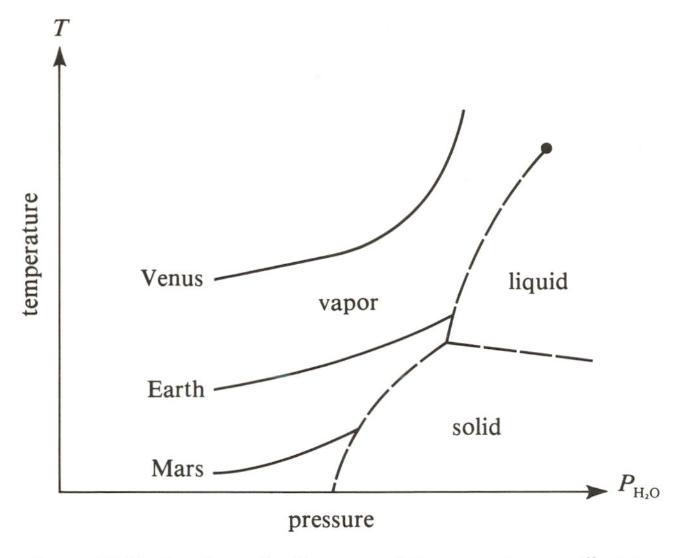

Figure 17.23. A schematic diagram of the greenhouse effect for Venus, Earth, and Mars. The solid curves yield the surface temperature of the planet, assuming an opacity associated with various values of the vapor pressure of water at ground level. The dashed curves show the loci of thermodynamic equilibria possible for the three phases of water (see Figure 4.10). Water opacity cannot raise the surface temperature on Mars higher than a point where its vapor becomes saturated with respect to ice, and on Earth, with respect to liquid water. For Venus, a runaway greenhouse effect occurs because the solid curve is incompatible with the existence of liquid water on the surface of the planet. (Adapted from S. I. Rasool and G. de Bergh, *Nature*, **226**, 1970, 1037, and R. M. Goody and J. C. G. Walker, *Atmospheres*, 1972.)

situation, where the atmosphere is so thin that the liquid water froze as permafrost (Figure 17.23). Thus, as Don Goldsmith and Toby Owen have remarked, the story of liquid water on Venus, Earth, and Mars is something like the story of Goldilocks and the porridge of the three bears: Venus is too hot, Mars is too cold, and the Earth is just right!

The Exosphere

If the runaway greenhouse effect explains Venus' atmosphere, why does Venus not have oceans' worth of water vapor in its atmosphere? Venus' atmosphere is now 96 percent CO_2; about 3.5 percent of the remainder is nitrogen molecules, N_2; and there is only a trace amount of H_2O. What has happened to the missing H_2O?

The answer may lie in chemical and photochemical processes studied by Bates and Nicolet which take place in the upper stratosphere, and which we will discuss in the next chapter. For here, let us remark that such processes can dissociate water vapor and release free oxygen and hydrogen atoms. The oxygen atoms will attach to some other heavy atoms, but the hydrogen atoms may escape from the planet. Hydrogen atoms are light, and they have a relatively large thermal velocity. A significant fraction of these atoms will lie in the high-velocity tail of the Maxwell-Boltzmann distribution, making these good candidates for evaporation from the planet. The general problem of the escape of atoms and molecules from the upper reaches of a planetary atmosphere was first formulated by Sir James Jeans in 1916. Lyman Spitzer and Joseph Chamberlain have made important modern improvements of the theory. In what follows, we shall consider Jean's original approach, and we shall refer loosely to "molecules" when we mean "molecules or atoms."

The first important thing to realize when considering the Jeans escape problem is that molecules cannot escape from the lower levels of a planetary atmosphere any more than a photon in the deep interior of the Sun can fly out in a straight line (Figure 17.24). The chances are that a fast molecule would collide with another molecule well before it had made any substantial progress toward escape. The path followed by any molecule in these lower regions is a tortuous "random walk." Only in the rarefied outer portions of a planetary atmosphere can a molecule begin to "fly" freely and have a chance to escape the gravitational clutch of the planet. (It is also in these regions where a newly formed hydrogen *atom* can escape before recombining with another hydrogen *atom*.) The region where fast molecules change from "walking" to "flying" is called the *exosphere*. Exospheres are to molecules on planets what photospheres are to photons on stars (Chapter 5). There is an important difference, however. All photons which make it to the photosphere can escape, because all photons travel at the speed of light. With molecules, only that fraction in the exosphere whose velocity exceeds the escape speed of the planet can actually escape.

Let us consider the simplest case where an exosphere is located at a radius r_e above the center of a spherical planet of mass M, and where the exosphere is at a uniform temperature T, containing only one molecular species of mass m. Let the collisional cross-sectional area of molecules m with each other be denoted A. Using the concepts just discussed, we may express Jeans' formula for escape as yielding the rate of loss of the total number of molecules m (Problem 17.7) as

$$\dot{N}_m = (2\pi)^{1/2} \frac{2}{A} r_e v_T f(x_e), \tag{17.7}$$

where $v_T = (kT/m)^{1/2}$ is the thermal speed and $f(x_e)$ is the dimensionless function

$$f(x_e) = x_e(1 + x_e)\exp(-x_e) \quad \text{with} \quad x_e = \frac{GMm}{r_e kT}. \tag{17.8}$$

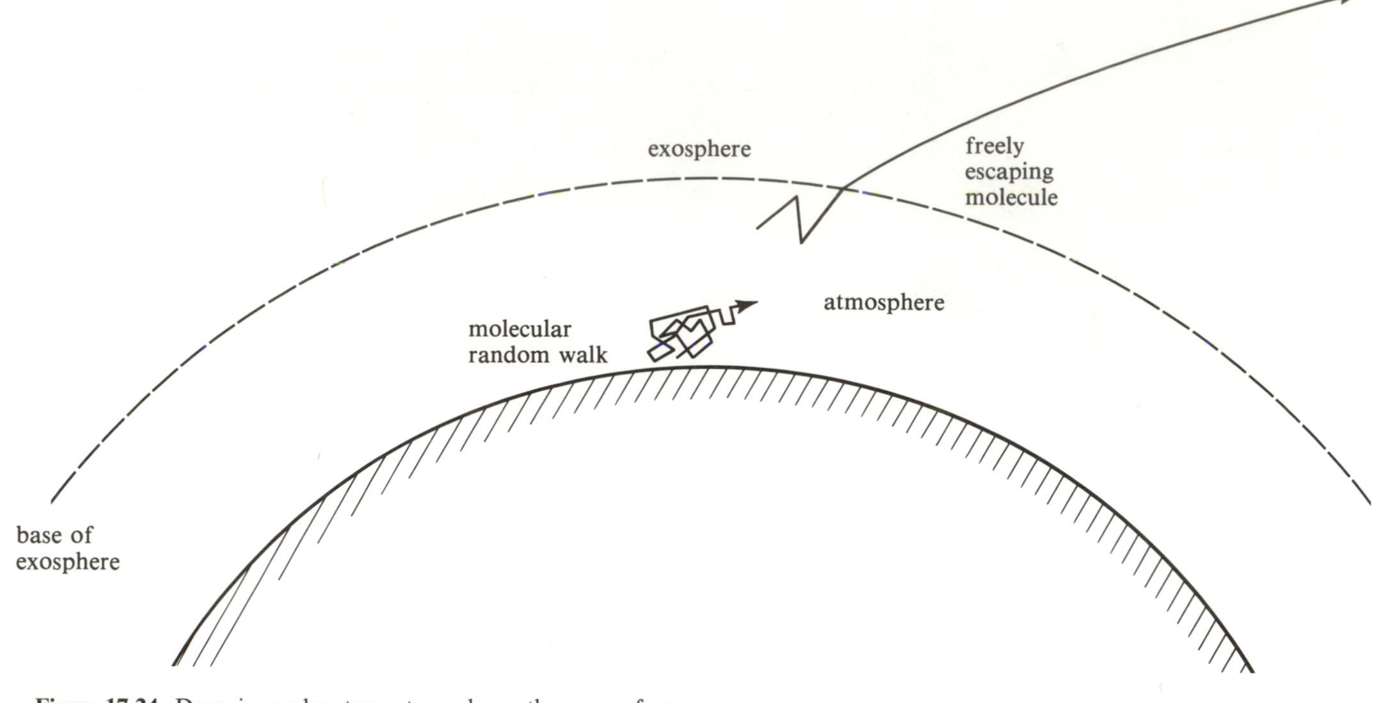

Figure 17.24. Deep in a planetary atmosphere, the mean free path for molecule-molecule collisions is so short that even a very fast molecule has no chance to escape from the planet. Above the base of the exosphere, the density of molecules has dropped low enough that the probability for escape becomes high for a sufficiently fast molecule. (Compare with Figure 5.7.)

The function $f(x_e)$ is plotted in Figure 17.25 for the relevant regime x_e appreciably greater than unity. In this regime, we see that $f(x_e)$ is a steeply declining function of increasing x_e. With everything else being equal, we see therefore that the Jeans' escape mechanism drastically favors the escape of light molecules, since v_T and (especially) $f(x_e)$ are both larger for smaller values of m in equation (17.7), which gives the number of molecules lost per second. In particular, hydrogen *atoms* are especially susceptible to evaporation from a planet. This conclusion corresponds to intuitive expectations and is analogous, of course, to our explanation in Chapter 9 of why light stars tend preferentially to be lost from globular clusters.

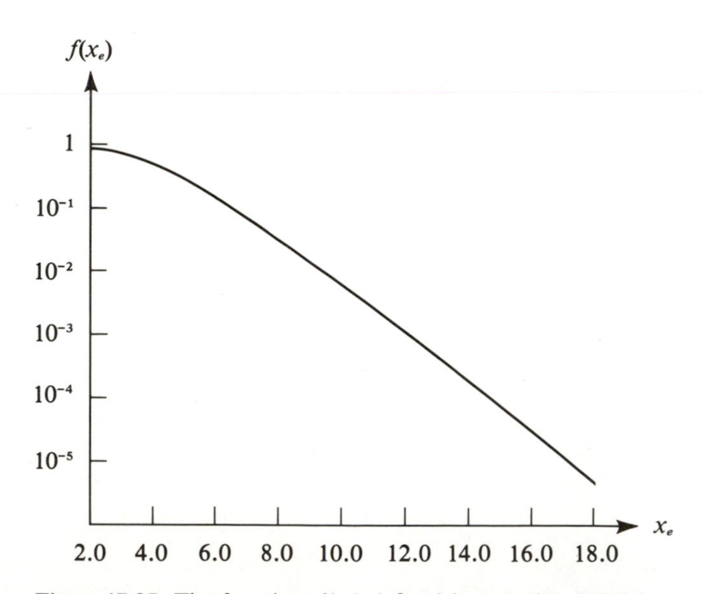

Figure 17.25. The function $f(x_e)$ defined in equation (17.6) is a rapidly declining function of its argument for x_e appreciably greater than unity.

Problem 17.7. Let us begin by seeing how to define the exosphere. We assume that the atmosphere is in quasi-hydrostatic equilibrium at a temperature T, so that the (partial) pressure at any level is equal to the (partial) weight of a column of unit area and (scale) height H of molecules of mass m and number density n,

$$P = gmnH, \quad \text{where} \quad g = GM/r_e^2.$$

With the perfect-gas law, $P = nkT$, show that this fixes the scale height H as $H = kT/mg$. At the base of the exosphere, we expect the mean free path l for molecule-molecule scatterings to be equal to the scale height H,

$$(nA)^{-1} = l = H.$$

With this definition, the density at the base of the exosphere is given by $n = (HA)^{-1}$.

The number flux $\mathscr{F}_m$ of escaping particles is given by

$$\mathscr{F}_m = \tfrac{1}{4}n\langle v \rangle_e, \tag{1}$$

where $\langle v \rangle_e$ is the average speed of molecules capable of escape. The factor 1/4 enters into equation (1) for the same geometric reason that the outward energy flux of blackbody photons (with an energy density aT^4) is $\sigma T^4 = caT^4/4$ and not caT^4. For a Maxwell-Boltzmann distribution, the quantity $\langle v \rangle_e$ is given by

$$\langle v \rangle_e = 2\left(\frac{2kT}{\pi m}\right)^{1/2}\left(1 + \frac{mv_e^2}{2kT}\right)\exp\left(-\frac{mv_e^2}{2kT}\right).$$

where $v_e^2 = 2GM/r_e$ is the square of the escape speed at the base of the exosphere. With the above relations and $\dot{N}_m = 4\pi r_e^2 \mathscr{F}_m$, derive equation (17.7).

The rest of this problem is for those who know calculus. We wish to derive the expression given above for $\langle v \rangle_e$. First, we notice that for a Maxwell-Boltzmann distribution in three dimensions, the probability that a molecule has speed between v and $v + dv$ is proportional to $\exp(-mv^2/2kT)4\pi v^2\, dv$. Argue now that $\langle v \rangle_e$ is given by the ratio

$$\langle v \rangle_e = \frac{\int_{v_e}^{\infty} v \exp\left(-\frac{mv^2}{2kT}\right) 4\pi v^2\, dv}{\int_0^{\infty} \exp\left(-\frac{mv^2}{2kT}\right) 4\pi v^2\, dv}. \tag{2}$$

Introduce the transformation $x = mv^2/2kT$, and show that equation (2) can be written

$$\langle v \rangle_e = \left(\frac{2kT}{m}\right)^{1/2} \frac{\int_{x_e}^{\infty} xe^{-x}\, dx}{\int_0^{\infty} x^{1/2}e^{-x}\, dx}, \tag{3}$$

where x_e is the dimensionless quantity defined in equation (17.8). The bottom integral in (3) is a standard integral having the value $\pi^{1/2}/2$. Show by integration by parts that the top integral in (3) equals $(1 + x_e)\exp(-x_e)$.

Unfortunately, the substitution of realistic numbers into equation (17.7) shows that the Jeans mechanism is probably inadequate to have gotten rid of Venus' excess hydrogen. To see this, let us consider the more secure case of the Earth. Let us adopt for T the temperature 1,000 K which obtains at $r_e = 7.0 \times 10^8$ cm (about 600 km above sea level; see Figure 17.18). The collisional cross section of hydrogen atoms by hydrogen atoms is about 10^{-15} cm^2, but to take account of collisions also with other molecules, let us adopt a collisional cross section *per hydrogen atom* equal to $A = 6 \times 10^{-14}$ cm^2. With these values, we obtain from equation (17.7) a total loss rate of hydrogen atoms by Jeans escape from the Earth of

$$\dot{N}_m = 1 \times 10^{27}\ \text{sec}^{-1}.$$

This value is comparable to that inferred from observations of Lyman alpha emission from the Earth's upper atmosphere. During the 4.6 billion years of its existence, the Earth would lose, at the above rate, about 2×10^{20} gm of hydrogen, roughly three orders of magnitude less hydrogen than is contained in the form of H_2O in the oceans. The hydrogen loss rate measured for Venus is comparable to the above rate, and the Jeans mechanism seems incapable of giving a much larger rate in the past. Thus, if Venus ever had as much liquid water as the Earth, it probably could have lost only 10^{-3} of it by the Jeans mechanism.

Problem 17.8. Adopt the numerical values given in the text, and verify the values quoted for $\dot{N}_m$ and the total mass of hydrogen lost in 4.6×10^9 yr.

The reader might object to this hasty conclusion, and point out that the function $f(x_e)$ in equation (17.8) has a sensitive negative-exponential dependence on $1/T$. Could we not obtain a much higher exospheric evaporation rate merely by raising the adopted value of T at the base of the exosphere? Unfortunately, this strategy does not work. As Chamberlain has emphasized, Donald Hunten has shown that any increase of the exospheric evaporation rate must ultimately be limited by the ability of the lower layers to resupply the lost particles by diffusion. From the regions where the material is well-mixed, the hydrogen (or hydrogen-carrying compounds) has a limited capacity to rise diffusively *relative* to the other constituents in the atmosphere. Our application of the Jeans formula turns out already to strain the capacity of the lower layers to maintain the same upward flux of hydrogen.

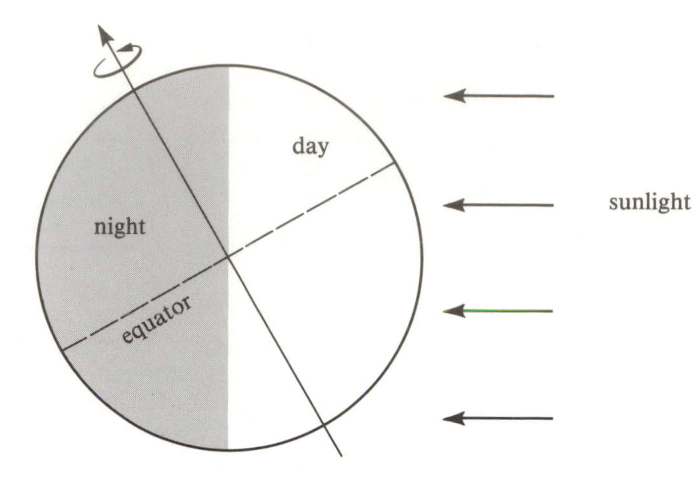

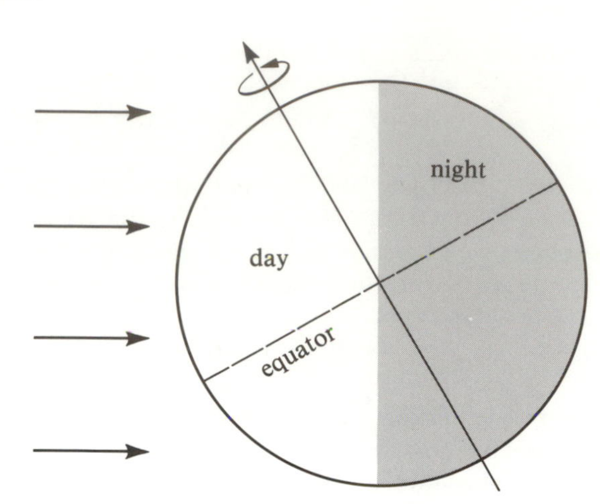

Figure 17.26. Sunlight falls more perpendicularly on the equator than on the poles of most planets, and more perpendicularly in summer than in winter. This uneven heating leads to a planetary wide system of circulation of the atmosphere.

As an aside, we should note that similar reasoning explains why hydrogen in stars does not automatically rise to the surface relative to the other elements. In general, the diffusion time is too long to make the effect an interesting possibility. Exceptions do exist—in the atmospheres of certain A stars examined by Michaud and others, where radiation pressure in selected absorption lines plays a dominant role, and in the atmospheres of certain white dwarfs examined by Alcock and others, where the gravity is enormously large—but these exceptions require unusual circumstances; usually, diffusive effects are slow in large bodies.

Does this imply that Venus never had as much liquid water as the Earth? No, not necessarily. There may be other loss mechanisms for hydrogen than simple Jeans escape. In particular, we know that the Jeans mechanism is totally inadequate to account for the loss of *helium* which must and does occur for the Earth and the other terrestrial planets. Helium is produced in the form of alpha particles (see Chapter 6) by the decay of radioactive elements in the Earth's crust. The subsequent outgassing of helium from the crust exceeds by six orders of magnitude the amount of helium that could be lost by exospheric evaporation. Yet the concentration of helium is not building up in the atmosphere. There must be another mechanism for helium loss, and whatever produces the helium loss may also increase the hydrogen loss. It remains to be seen, however, whether Hunten's argument would also limit this other mechanism to rather modest values of $\dot{N}_m$.

Planetary Circulation

In our discussion of the thermal balance in a planet so far, we have averaged over latitude, day and night, and seasons. In other words, we have treated sunlight as coming, on the average, equally from all directions. In actual practice, of course, sunlight strikes the ground of a terrestrial planet differently at different latitudes and at different times (Figure 17.26). This uneven heating drives a general circulation in the atmospheres of Venus, Earth, and Mars. As a consequence, the surface temperatures of these planets—especially Venus and Earth—suffer less extreme variation than the surface temperatures of planets that lack an atmosphere. For example, Mercury has a noontime temperature of 700 K at the subsolar point (where the Sun is directly overhead) and a nightime temperature below 100 K! (The variation would be even more extreme were it not for a peculiarity of the spin-orbit dynamics of Mercury that we will discuss in Chapter 18.)

To discuss the general circulation pattern in terrestrial planets, it is simplest to begin with a planet which has relatively little rotation and a very thick atmosphere. Venus fills this bill perfectly. Unlike most of the planets and moons, Venus rotates in a retrograde sense in comparison with its orbital revolution. Its rotation period with respect to the "fixed stars" is -243 days, and its orbital period is $+225$ days. This combination yields the duration of a Venusian "day"—from sunrise to sunrise—as 117 (Earth) days.

Problem 17.9. If the sidereal period of a planet's rotation and revolution are, respectively, τ_{rot} and τ_{rev}, show that the length of a day $\tau_\odot$ equals $\tau_{\text{rot}}\tau_{\text{rev}}/(\tau_{\text{rot}} - \tau_{\text{rev}})$. *Hint*: Show that the angular velocity of the Sun's motion relative to an observer situated on the rotating planet is given by $\omega_\odot = \omega_{\text{rev}} - \omega_{\text{rot}}$. Calculate numerically the length of a Venusian day. Given that $\tau_\odot$ equals -1.00 day on Earth, and that the sidereal year τ_{rev} equals 365.26 days, calculate the length of a sidereal day, τ_{rot}. What do we mean by the minus sign in $\tau_\odot = -1.00$ day?

Consider now the solar heating on Venus. Sunlight falls with maximum strength on the subsolar point at the equator and with zero strength at the poles. Consequently, in the absence of any circulation, the air would tend to be warmest at the equator and coolest at the poles. However, warm air has a larger scale height H than cool air (consider the formula $H = kT/mg$ derived in Problem 17.7). Thus, the warm air at the top of the atmosphere near the equatorial regions will want to flow over the top of the atmosphere near the polar regions (Figure 17.27). The alleviation of the weight at the top will cause hot air near the ground of the equator to well upward, to be replaced by cool air flowing near ground level from the poles. This simple circulation pattern is called *Hadley circulation* in honor of G. Hadley, who first proposed such a scheme in 1735 for Earth's general circulation. The Hadley circulation model attempts to iron out directly the difference in temperatures between equator and pole, and it is believed by many people that such a basic circulation pattern does apply near the surface of the planet. However, this simple model is observed to break down in the upper atmosphere. The sulfuric-acid clouds high in Venus's atmosphere rotate 60 times faster than the planet does (see Color Plate 47). It remains a mystery how Venus manages to transport angular momentum through its atmosphere to maintain this superrotation.

How good a job does the general circulation do in smoothing out the uneven solar heating? Well, since Venus has a very thick atmosphere, we might intuitively expect that it would do a very good job. And so it does. Moreover, it is possible to see semiquantitatively why this is so. Consider the thermal timescale t_{th} required for the planet to lose its heat if the Sun could hypothetically be turned off. The average heat content per unit area equals $(7/2)kTnH = (7/2)PH$, where P is the surface pressure. (The factor 7/2 enters instead of 3/2 for two reasons. A linear molecule like CO_2 possesses kT in rotational energy. Another kT is associated with the assumption that for cooling at hydrostatic equilibrium in a constant gravitational field, the atmosphere descends

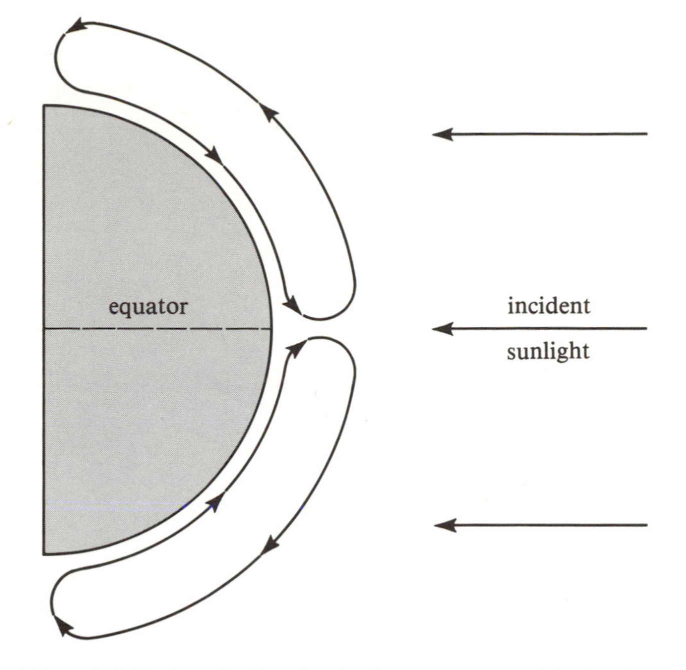

Figure 17.27. The Hadley circulation pattern would arise in a nonrotating planet because the warm air at the equator would tend to flow over the poles, and cold air from the poles would flow near the ground to replace the warm air from the equator.

at constant pressure [equal to the weight per unit area]. Thus, $3kT/2 + kT + kT = 7kT/2$.) The heat loss per unit area equals σT_e^4, where T_e is the effective temperature of the planet. The thermal timescale t_{th} can now be expressed as

$$t_{\text{th}} = \frac{(7/2)PH}{\sigma T_e^4}, \tag{17.9}$$

which equals roughly 10^2 yr for Venus. On the other hand, air flows of speed v over horizontal scale L have a dynamical timescale,

$$t_{\text{dyn}} = L/v, \tag{17.10}$$

which equals roughly one week if we use planetary scales $L \sim 6 \times 10^8$ cm and observed surface windspeeds of several meters per second. Consequently, the ratio $t_{\text{dyn}}/t_{\text{th}}$ is very small, of order 10^{-4} for Venus, and we expect the general circulation to be extremely efficient at ironing out temperature differences between equator and poles. (The ratio $t_{\text{dyn}}/t_{\text{th}}$ is called the Golitsyn number in planetary astronomy, but it also plays an important role in one theory of contact binaries like W Ursae Majoris stars; see the discussion in Chapter 10.) Moreover, since the Venusian day (117 Earth days) is

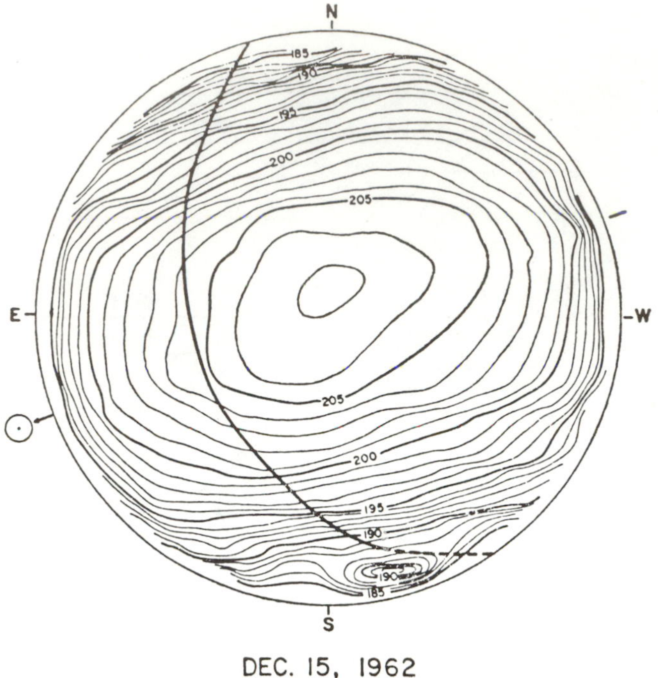

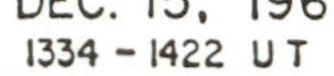

Figure 17.28. Apart from the effects of limb darkening (see Figure 5.3) and a slight latitudinal dependence, the contours of constant infrared temperature on Venus show a remarkable uniformity. Notice, in particular, that the point directly below the position of the Sun is not especially warm despite the slow rotation of Venus. (From Murray, Wildey, and Westphal, *J.G.R.*, **68**, 1963, 4813.)

short compared to the thermal timescale t_{th}, but is long compared to the flow timescale t_{dyn}, we can expect there to be relatively little variation in both the surface temperature and the effective temperature of the "day" and "night" sides of the planet. The thick atmosphere of Venus serves to buffer the thermal state of the planet, so that all the lower layers near the surface maintain a nearly uniform temperature distribution. This conclusion that Venus acts as a good "convection oven" has been verified by radio and infrared observations of the thermal emission from this planet (Figure 17.28).

Problem 17.10. Verify the numerical values quoted for t_{th} and t_{dyn}. First, calculate $H = kT/mg$ using $T = 750$ K, $m = 7 \times 10^{-23}$ gm (mass of CO_2 molecule), and $g = 860$ dyne/cm^2 (surface gravity of Venus). Now, use $P = 9 \times 10^7$ dyne/cm^2 to calculate t_{th}. Likewise, calculate t_{dyn} assuming $L = 6 \times 10^8$ cm and $v = 20$ km/hr $\sim$ 600 cm/sec.

Unlike Venus, the Earth rotates fairly rapidly and does not possess a very thick atmosphere. The rotation period (1 day), the dynamical timescale (1 week) and the thermal timescale (several weeks) are all more nearly comparable. This makes quantitative calculations more difficult, but it provides variety in living on Earth. We do have day and night temperature variations, as well as seasonal ones, and the weather at the arctic and antarctic is noticeably different from the weather at the tropics. Nevertheless, because of the general circulation, these variations are milder than they would be if our atmosphere were considerably thinner (like that of Mars, say).

The fairly rapid rotation of the Earth introduces a strong dynamical influence which is nearly absent in Venus. This influence is the tendency for each moving parcel of gas to conserve its angular momentum. Because of frictional drag and nonaxisymmetric instabilities, the detailed conservation of angular momentum will not hold strictly. Nevertheless, the tendency exists, and is ascribed to **Coriolis** force by an observer who corotates with the planet. (Legend has it that Coriolis thought of the effect while playing billiards.) Let us examine the consequences.

First, we should note that frictional coupling to the ground forces the air near the ground to rotate at nearly the same solid-body rate as the ground itself. (The layer where this effect is strongest is called the Ekman layer.) However, the uneven solar heating would like to produce a Hadley circulation pattern like that in Figure 17.27. But the air rising from near the ground at the equator cannot proceed all the way to the pole and even approximately conserve its angular momentum, because doing so would cause the air mass to speed up tremendously at midlatitudes, in the sense of the rotation of the Earth. (Recall what happens to a spinning skater when she raises her arms closer to her spin axis; Figure 3.23). A single Hadley cell (Figure 17.27) would require much larger rotational speeds for the poleward-moving air than the equatorward-moving air. This difference would probably introduce more vertical shear than the Earth's atmosphere could maintain stably. The actual circulation in the meridional plane consists of *three* overturning cells in each hemisphere (Figure 17.29). The descending air from the poles carries low-angular-momentum material to lower latitudes, and makes the winds in the polar caps near the ground blow predominantly **easterly** (*from* the east, against the sense of Earth rotation), as seen by an Earthbound observer. The rising air moving toward the poles, on the other hand, contains higher-angular-momentum material, causing the high-altitude winds to blow **westerly** (*from* the west). Similarly, the descent of lower-angular-momentum air makes the trade winds near sea level at the tropics blow primarily easterly, whereas the ascent of high-angular-momentum air toward higher latitudes cause the high-altitude wind

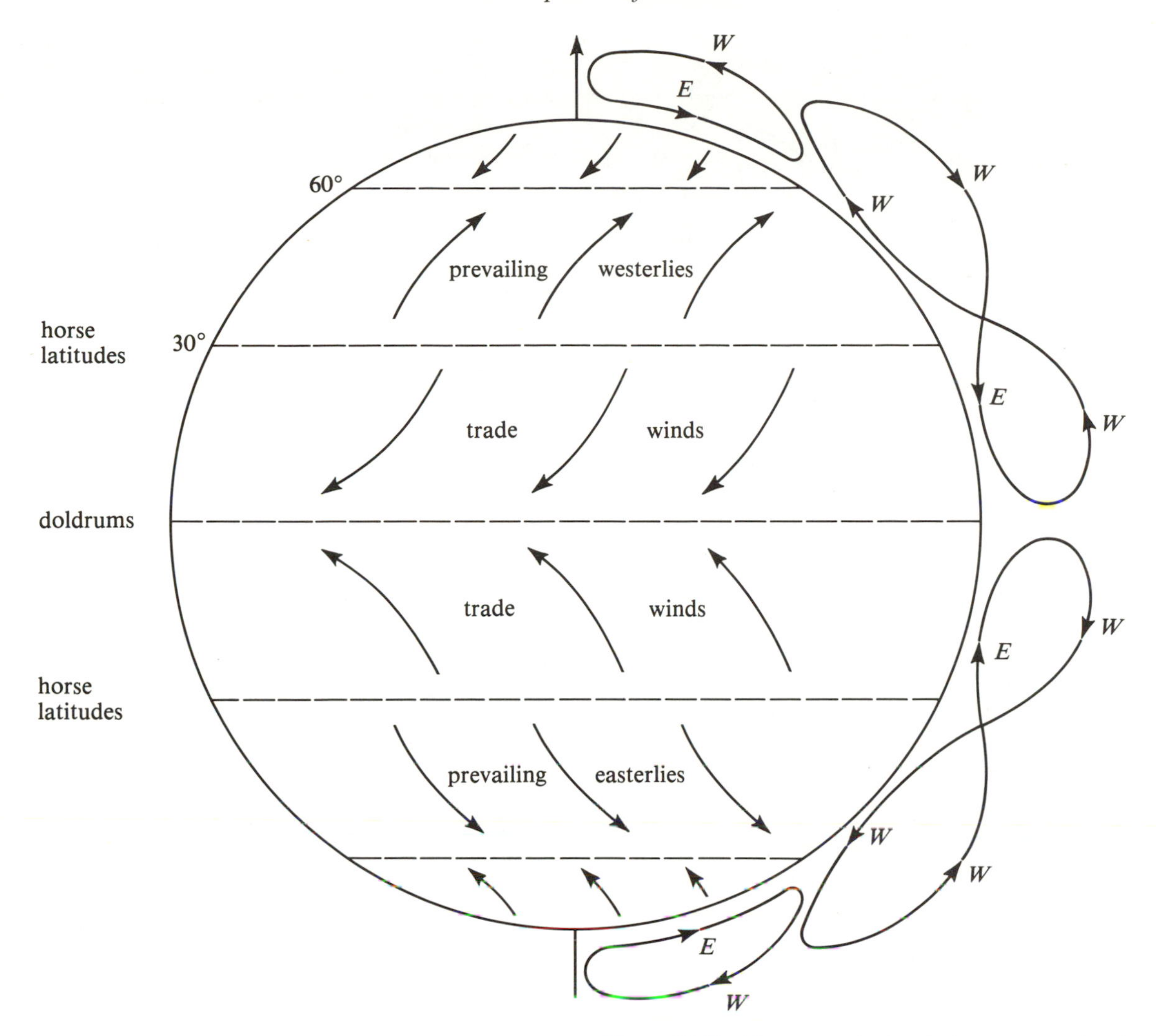

Figure 17.29. The general circulation pattern in the Earth's atmosphere. The ascending and descending streamlines at mid-latitudes do not cross, but occur at different longitudes. The horizontal directions of the surface winds are indicated schematically on the face of the planet. (Adapted from C.-G. Rossby, in *Yearbook of Agriculture, Climate, and Man*, ed. G. Hambridge, Washington, D.C., US Govt. Printing Off., 1941, p. 599.)

in the tropics to blow westerly. The middle cell near sea level mediates the requirements of the polar and tropical cells, leading to a subsidence of the surface winds at roughly $\pm 60^\circ$ and $\pm 30^\circ$ (the "horse latitudes"). Thermal effects in the middle cell cause the winds, however, at high altitudes to blow even *harder* in the westerly direction than the winds at low altitudes. The strong westerly direction of the former—the so-called jet stream—is familiar to anyone who has noticed that a jet takes longer to fly from the East Coast of the United States to the West Coast than from the West Coast to the East Coast.

Storms on Earth

Even Figure 17.29 represents, however, a simplification of the actual situation on Earth. In Chapter 10, we explained the natural tendency for gases confined by the gravitational field of a body to attain a *barotropic* state (where the pressure, density, and temperature are uniform on a gravitational equipotential surface). For the Earth, the uneven solar input prevents the establishment of a barotropic state. The surfaces of constant pressure, constant density, constant temperature, and constant gravitational potential do *not* coincide; they

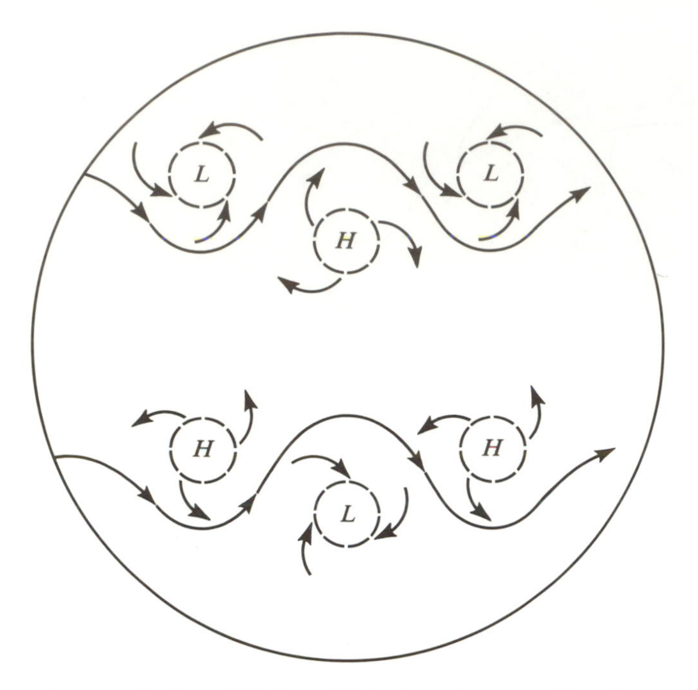

Figure 17.30. Baroclinic instabilities introduce wavelike disturbances in the basic westerly (*from* the west) flow at midlatitudes of the Earth. The crests of the wave (excursions toward the poles) are associated with regions of high pressure (H); the troughs (excursions toward the equator), with low pressure (L). The flow of air about highs is anticyclonic; about lows, cyclonic. (For details, see A. P. Ingersoll *Scientific American*, **234**, March 1976, 46.)

are generally *inclined* with respect to one another, and so are called *baroclinic*. This baroclinic state is, of course, what provides the unbalanced force that leads to the planetary circulation in the first place. But (as has been discussed by J. G. Charney, E. J. Eady, and other meteorologists) the resulting baroclinic flows of planetary scale may allow disturbances of yet smaller scale. The most important of these disturbances at midlatitudes involves the superposition of wavelike disturbances on top of the basic westerlies (Figure 17.30). These waves are akin to disturbances first discussed by C.-G. Rossby, and they are unique to rotating planets. The fastest-growing of these instabilities has a growth time of about a couple of days and a wavelength of about 4,000 km. Low-pressure regions occur where the air moves equatorward; high-pressure regions, where the air moves poleward. The associated flow about the low-pressure region is cyclonic (counterclockwise in the northern hemisphere, clockwise in the southern); about high-pressure regions, anticyclonic. Near the ground, warm air flows upward and inward toward a low; cool air flows downward and outward from a high. The water in the rising air tends to condense out; so low-pressure regions are generally associated with bad weather.

It is useful as a model to treat the baroclinic instabilities as if they were minor perturbations in a hypothetical basic state represented by Figure 17.29. In actuality, the waves introduced by the baroclinic instabilities provide the major mechanism at midlatitudes for the transport of heat both vertically upward through the atmosphere and toward the poles. In other words, the basic state itself would not exist without the instabilities.

Problem 17.11. (for those familiar with vector calculus). To understand the cyclonic motion of air about low-pressure regions (and anticyclonic for highs), we need to know that the apparent forces which can act per unit mass on a fluid viewed in a rotating frame of reference include those due to pressure gradients and Coriolis force,

$$-\frac{1}{\rho}\nabla P - 2\mathbf{\Omega} \times \mathbf{v},$$

where $\mathbf{\Omega}$ is the angular velocity of the rotation and $\mathbf{v}$ is the velocity of the fluid relative to the rotating frame of reference. For disturbances of sufficiently large scale and not located too close to the equator of a rotating planet, the above two forces approximately balance in the horizontal directions (perpendicular to gravity). This approximate balance is called the geostrophic approximation. Show that the geostrophic approximation produces motions parallel to isobars (contours of constant pressure in the horizontal directions), and that these motions occur in a cyclonic sense if the center is a low. This conclusion is, of course, approximate, since we naturally expect some converging motion toward a center of low pressure. Show also that the geostrophic approximation must break down near the equator. *Hint*: In which direction does $\mathbf{\Omega}$ point at the equator?

Very violent cyclonic storms (hurricanes) occur near the tropics. The Sun heats the surface of the oceans, which causes moist air to rise and to form cumulus clouds when the water vapor begins to condense. The release of liquid binding energy (latent heat) by the condensation process strengthens the upwelling. Outside this region are compensating downward flows, but they tend to occupy a much larger area. If the circulation grows sufficiently strong, there is a sustained convergence of moist air toward the eye of the storm, and the hurricane can be resupplied with latent heat. An interesting recent development of the theory of hurricanes suggests that the spiral rainbands which demarcate the photographic structure of hurricanes (Figure 17.31) may be wave patterns like those which create the spiral structure of disk galaxies! (See Chapter 12.)

Figure 17.31. A NOAA satellite view of two terrestrial hurricanes in 1974. The spiral rainbands in hurricanes may result from wave phenomena like those that create the spiral pattern of star formation seen in disk galaxies (see Chapter 12). Notice the relative calm at the eye of the storm. (National Oceanic and Atmospheric Administration.)

Ocean Circulation

More than two-thirds of the surface of the Earth is covered with liquid water. Ours is a blue planet (see Color Plate 50). It may seem out of place to discuss the oceans of the Earth in the same section as the atmospheres of planets, but there are goods reasons to do so. The oceans couple strongly both dynamically and thermally with the atmosphere. We have already discussed one of these important interactions, namely, the water budget of evaporation and precipitation. (Water also leaves from and enters the solid Earth, but we will

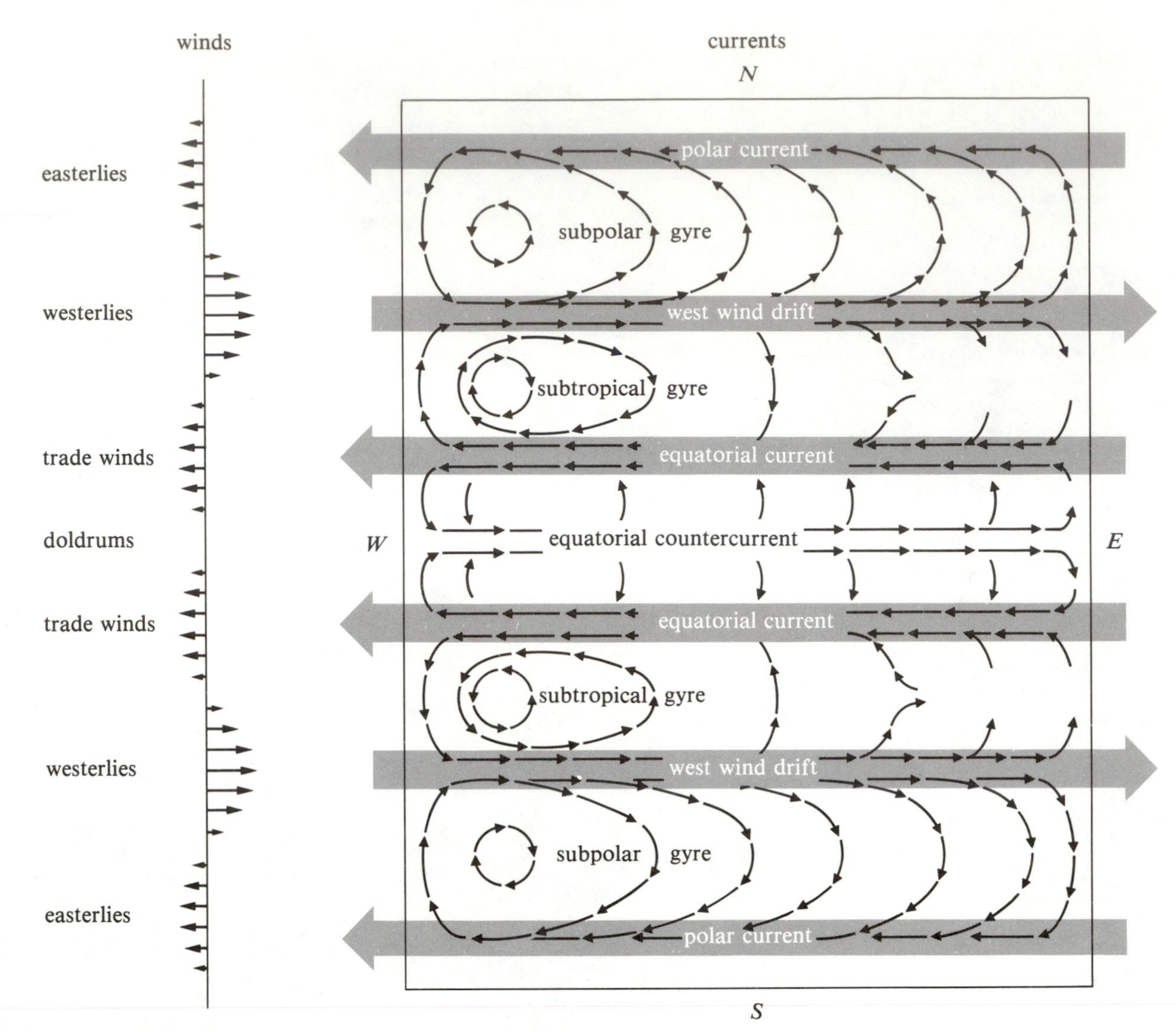

Figure 17.32. Surface winds indicated by the broad grey arrows blow across an idealized rectangular ocean basin. The winds would drive the schematic pattern of ocean circulation indicated by the short black arrows inside the rectangular basin. The pattern of surface winds on the left is that implied by Figure 17.29. The observed east-west equatorial currents flow at speeds of 3 to 6 kilometers per day, slowly enough to adjust thermally to the ambient air. In contrast, the Gulf Stream, which flows from the Gulf of Mexico to Europe, travels 40 to 120 kilometers per day. These waters remain warm, and help to keep the countries bordering the North Atlantic warmer than they would be without the oceans. (For details, see W. Munk, *Scientific American*, **193**, Sept. 1955, 96.)

leave that topic for the next chapter.) Here, let us consider the phenomenon of ocean circulation.

The large-scale surface currents in the oceans familiar to mariners are driven primarily by frictional coupling with the winds in the Earth's atmosphere (see Figure 17.29). In an idealized calculation, W. Munk has shown that the ocean currents can be described in terms of large closed loops called "gyres" (Figure 17.32). The idealized wind-driven ocean flow of Figure 17.32 bears some striking resemblances to the actual pattern of flow in the Pacific. (The ocean currents in the Pacific are less interrupted by land masses than those in the Atlantic.) Displacements take place not only horizontally but also vertically. Upwelling or sinking occurs, for example, when currents are driven against or from the shores of continents. Deep cold water which rises toward the surface can cool the overlying air. This cooling is responsible for the fog familiar to residents of the West Coast of the United States.

Another mechanism for driving vertical ocean circulation is uneven heating by the Sun. The same argument which led Hadley to propose the atmospheric circulation pattern depicted in Figure 17.27 can be applied to the oceans. Compared to the wind-driven surface currents, the thermally driven deep-ocean circulation is very slow. It has an estimated timescale of about 1,000 years. This slow motion prevents much mixing, and the water flowing along the ocean bottom from the poles toward the equator tends to remain colder and less salty than the overlying layers. This property helps oceanographers identify the points of origin of large water masses.

The Atmosphere of Jupiter

How do the atmospheres of the Jovian planets differ from those of the terrestrial planets? To discuss this question, let us examine the best-studied Jovian planet, Jupiter itself.

Jupiter differs from the terrestrial planets in several ways, the four most important here being the following.

(a) Jupiter has no solid surface at the base of its atmosphere. Its liquid interior merges smoothly with the gaseous atmosphere.

(b) Jupiter has a source of internal energy which releases twice the solar heat input.

(c) Jupiter rotates relatively much faster than even the Earth. Meteorologists say that Jupiter has a smaller "Rossby number."

(d) The chemical composition of Jupiter is rich in hydrogen-containing compounds (H_2, CH_4, NH_3, etc); so the chemistry of the planet is vastly different from that of the terrestrial planets, whose atmospheres are dominated by oxygen-containing compounds (O_2, CO_2, H_2O) or N_2.

Problem 17.12. Jupiter absorbs 55 percent of the incident sunlight, Earth 61 percent. Jupiter is located at 5.2 AU from the Sun. If Jupiter had no internal heat source, and if the Earth's effective temperature is 246 K, what would the effective temperature of Jupiter be? How does this compare with the observed value $T_e =$ 134 K? Since the total outward radiative flux is proportional to T_e^4, deduce the ratio of Jupiter's internal heat source to the solar input.

How do these differences manifest themselves in Jupiter? Figure 17.33 shows the rough thermal structure of Jupiter's atmosphere as deduced from astronomical data. Here we wish to call attention only to the rapid rise in temperature in the thermosphere of the planet. The solid line shows the profile inferred from stellar occultation studies carried out on Earth by Veverka, Elliott, Wasserman, and Sagan and by radio-occultation studies of Pioneer 10. Compare this profile with the dashed curve, which is that expected for heating by solar ultraviolet radiation. Clearly, an additional source of energy is needed to explain the observations.

French and Gierasch have proposed that Jupiter's upper atmosphere is heated by wave disturbances which propagate upward from the interior convection zone. In this view, the thermosphere of Jupiter would be analogous to the solar chromosphere and corona, which are also believed to be heated by magnetic and mechanical disturbances which arise because of the existence of a convection zone in deeper layers (Chapter 5). This analogy extends to the spectroscopic evidence, which shows both emission and absorption lines in both objects, indicating the presence of hot gas overlying cool gas (see Chapter 5).

We see therefore that the internal heat source of Jupiter which drives strong convective motions in the interior and atmosphere of that planet has a rather dramatic consequence for the upper atmosphere. Are there more easily visible consequences? Well, how about the striking bright *zones* and dark *belts* which encircle the planet (Color Plate 41)? Are these a more extreme manifestation of the breakup of a simple Hadley circulation pattern (see Figure 17.29) because of uneven solar heating on a very rapidly rotating planet? Or do the zones and belts result from a complicated interplay between convection driven by the internal heat source of Jupiter and its rapid rotation?

Peter Stone has estimated that, if the circulation patterns in Jupiter were driven by baroclinic effects (due to uneven solar heating), the equator would be 30 K hotter than the poles. Infrared measurements by Pioneer 10 and 11, on the other hand, established that the temperature at the poles is no more than 3 K colder than the equator. This implies that Jupiter is substantially barotropic and not baroclinic. Why does Jupiter respond to solar heating differently from the Earth?

Andrew Ingersoll suggested that the primary difference was the lack of a solid surface to provide a base for the atmosphere of Jupiter. As a consequence, the atmosphere of Jupiter is, by definition, very thick, and, as Gierasch and Goody have pointed out, the thermal timescale

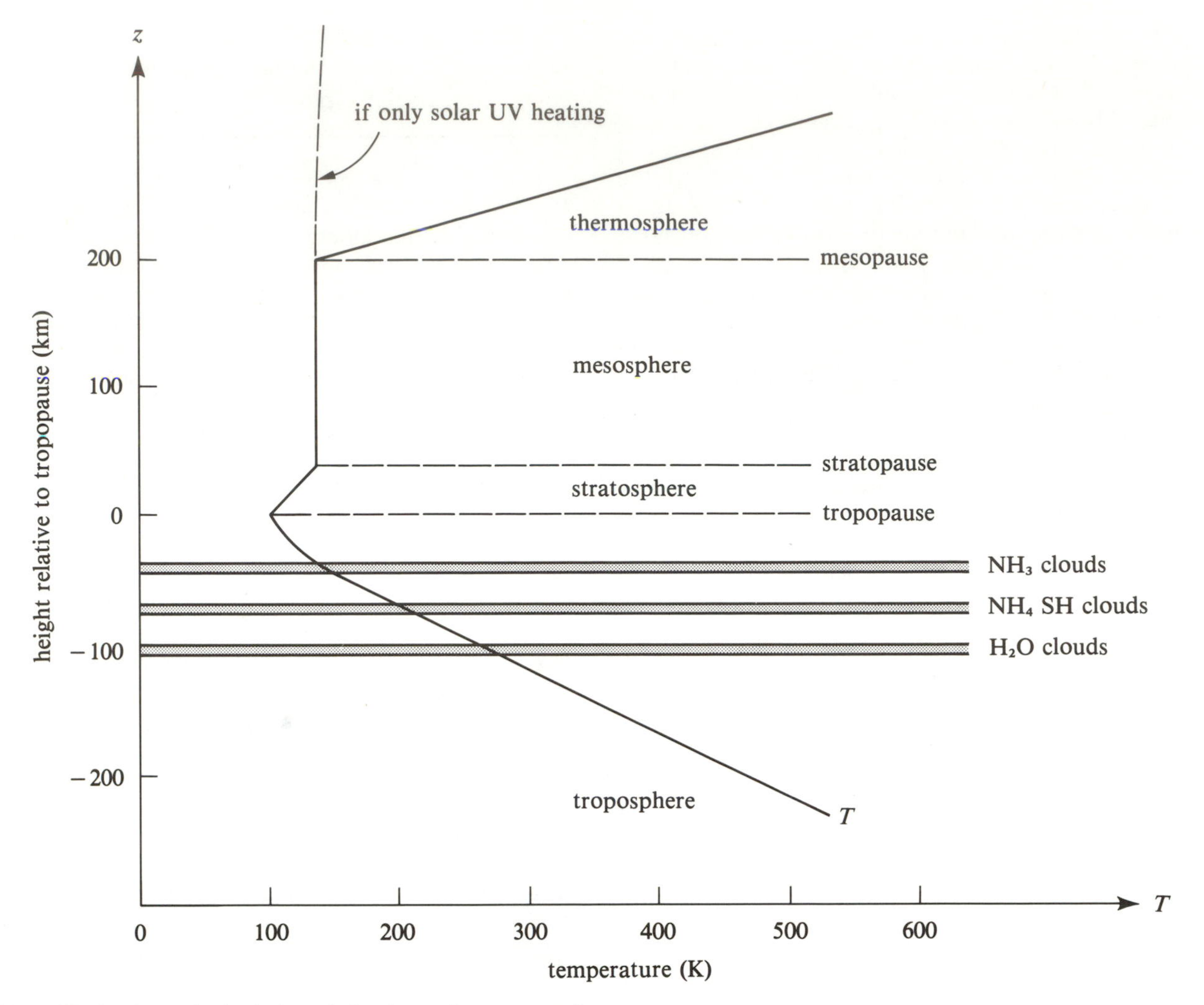

Figure 17.33. A schematic depiction of the thermal structure of the atmosphere of Jupiter. Since Jupiter's atmosphere is not bounded below by a solid surface, we have measured the altitude relative to the location of the tropopause (where the temperature going outward first changes from decreasing with altitude to increasing). The vertical locations of ammonia (NH_3) clouds, ammonium-hydrosulfide (NH_4SH) clouds, and water (H_2O) clouds have been placed in accordance with John Lewis's thermodynamic calculations of condensates in Jupiter's hydrogen-rich atmosphere. (For details, see J. W. Chamberlain, *Theory of Planetary Atmospheres*, Academic Press, 1978.)

(17.9) is much longer than the dynamical timescale (17.10). As a result, the deeper levels in Jupiter's atmosphere must become barotropic despite the uneven solar input; whereas on Earth the virtually undeformable surface of ocean and land can support relatively large variations of surface pressure.

If circulation currents induced by baroclinic effects do not suffice to iron out the differences between solar heating at the equator and at the poles, what explains the relatively uniform effective temperature over the surface of Jupiter? Ingersoll proposes that this difference must be compensated for by a difference in flow of internal heat from the interior of the planet; more convective transport of internal heat occurs at the poles than at the equator, and almost exactly balances the solar radiative input. If the poles were to cool relative to

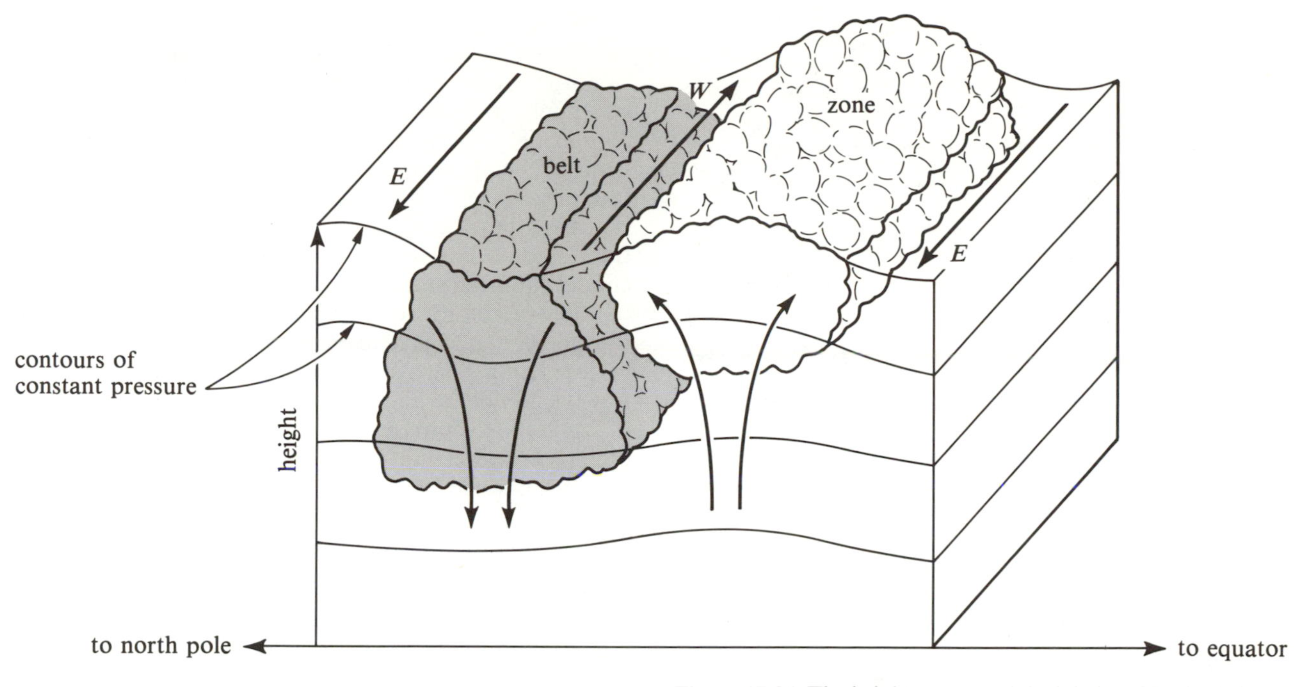

Figure 17.34. The bright zones and dark belts of Jupiter's banded structure can be interpreted in terms of rising and descending columns. At a given altitude, the zones are regions of high pressure in comparison with the belts. Coriolis force thus deflects the equatorward edge of a zone to flow in an easterly (E) direction, and the poleward edge of a zone to flow in a westerly (W) direction. This pattern is the Jovian analog of anticyclonic flow on Earth (compare with Figure 17.30). Similar patterns are observed in laboratory experiments conducted on convection in rapidly rotating fluids. (For details, see A. P. Ingersoll, *Scientific American*, **234**, March 1976, 46.)

the equator, the resulting greater buoyancy of the deeper and hotter layers would lead to a greater transport of heat by convective motions. For this mechanism to maintain a nearly barotropic atmosphere throughout the region of infrared emission, the optical sunlight probably has to be deposited primarily at levels where the convective efficiency is large.

The existence of convection implies, of course, that the fluid must have *some* horizontal variations, since the entropy per unit mass of rising fluid parcels must be (slightly) greater than the average of the fluid at that level (see Problem 5.15). This remark may be related to the proper explanation for Jupiter's zones and belts. The Pioneer 10 and 11 infrared measurements show that the cloudtops of the zones are slightly cooler and therefore slightly higher in the troposphere than the cloudtops of the belts. A natural explanation of this phenomenon is in terms of long, thin columns of rising and descending gases (Figure 17.34). Such patterns are often observed in theoretical and laboratory studies of convection in rapidly rotating fluids, and Friedrich Busse has proposed that the zones and belts of Jupiter may be the tops of such columns.

In such a picture, the zones would be regions of high pressure in comparison with the belts at each vertical level. The geostrophic tendency (see Problem 17.11) would then lead us to expect a pattern of horizontal flow like that depicted in Figure 17.35. Such a pattern of wind flow has been known to exist for a long time, and were already interpreted in 1951 by Hess and Panofsky to imply that the zones are high-pressure regions, the belts low-pressure regions. A detailed theoretical analysis by Cuzzi and Ingersoll in 1969 found that the straight-band pattern may break down at certain latitudes, even if the flow remained very nearly barotropic and safe from baroclinic instabilities (see Figure 17.30). The mode of instability found for Jupiter was related to the development of turbulence in shearing flows that take place,

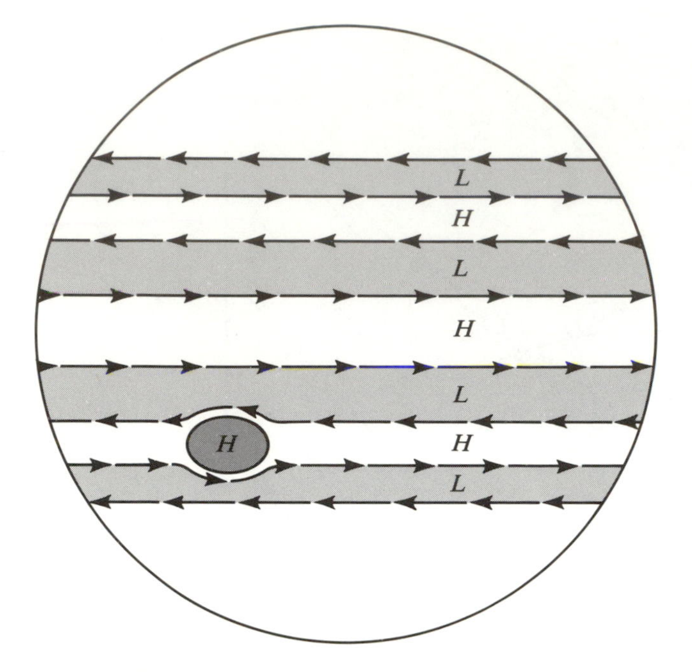

Figure 17.35. The expected and observed pattern of horizontal flow in Jupiter's atmosphere based on the interpretation that the zones and the great red spot are regions of high pressure (H), and the belts, regions of low pressure (L). (See A. P. Ingersoll *Scientific American*, **234**, March 1976, 46.)

for example, when air rushes by airplane wings.* The swirling eddies found in Jupiter's band system by the Pioneer and Voyager spacecrafts may be related to such barotropic instabilities.

Color Plate 42 shows the Great Red Spot of Jupiter. For many years it was thought that the permanence of the Great Red Spot (discovered by Giovanni Cassini in 1665) meant that it was "tied" to some protuberence on the surface of the planet. There is now good evidence that Jupiter is a liquid planet incapable of supporting features like mountains or valleys; so probably the Great Red Spot is a giant storm in the atmosphere of Jupiter similar to hurricanes on Earth. The anticyclonic pattern of flow observed around the Great Red Spot (see Figure 17.35) is consistent with the interpretation that it is the top of a rising mass of gas (and, therefore, a high-pressure region). The hurricanes on Earth are cyclonic because at sea level the eye of the storm is a region of low pressure. However, the center of the storm is a rising column of warm air, which has a larger scale height than its surrounding regions. Therefore, the *tops* of hurricanes may become regions of *relative* high pressure, and therefore weakly anticyclonic.

*Such instabilities were studied in their hydrodynamical context by Lord Rayleigh, W. Heisenberg, and C. C. Lin, and they were adapted to the meteorological context by H. L. Kuo.

The hurricane interpretation of the Great Red Spot is supported by the finding that it is not unique. Little spots of both red and white hues have now been observed, all of which probably represent tropospheric storms in Jupiter. The Great Red Spot's permanence may result from its being a type of long-lived disturbances called "solitary waves" or "solitons." A puzzle which remains is what gives the spots, as well as the zones and belts, their colors. The clouds depicted in Figure 17.33 are the condensable substances that John Lewis calculates to be possible (assuming local thermodynamic equilibrium) in the chemical and physical conditions believed to prevail in the troposphere of Jupiter. The cloud particles of Lewis's model are all white, whereas the observed clouds of Jupiter also contain shades of yellow, red, brown, and blue. An interesting speculation holds that the colors may be supplied by various organic molecules produced by lightning discharges in an atmosphere rich in hydrogen-containing compounds. (We will discuss laboratory experiments of this nature in Chapter 20.)

Magnetic Fields and Magnetospheres

Finally, we will discuss the magnetic fields of planets and their interaction with the solar wind. The Earth, of course, was the first planet discovered to possess a general magnetic field. Most people are aware that the Earth's magnetic north pole does not coincide with its spin north pole; the former lies near Hudson's Bay in Canada. Moreover, the magnetic south pole is not directly opposite the magnetic north pole. More importantly, the magnetic polarity of the Earth reverses directions, roughly every 10^5 to 10^6 years. The evidence for this activity rests on the study of fossil magnetism of ferromagnetic rocks. When molten rocks are spewed from volcanos and cool, they can spontaneously magnetize, and their orientation will be influenced by the direction of the Earth's magnetic field at the time of their formation (see the discussion of ferromagnetism in Chapter 5). By studying the reversals of field in different layers of such rocks and by dating the ages of these rocks, geologists can reconstruct a history of the behavior of the magnetic field of the Earth.

The periodical reversals of the Earth's magnetic field are reminiscent of the solar cycle. They again suggest that a dynamo mechanism must be working in the interior. This reasoning provides a strong motivation for believing that the core of the Earth must contain an electrically conducting fluid, presumably molten iron. The combination of rapid rotation and fluid motions would then provide the ingredients for successful dynamo action. Indeed, the theoretical modeling of the dynamo action of the Earth's interior by Walter Elsasser and Edward Bullard constituted historically the first believable demonstration of the operation of this mechanism in astronomical bodies.

Figure 17.36. The "northern lights" or *Aurora Borealis.* (U.S. Naval Observatory.)

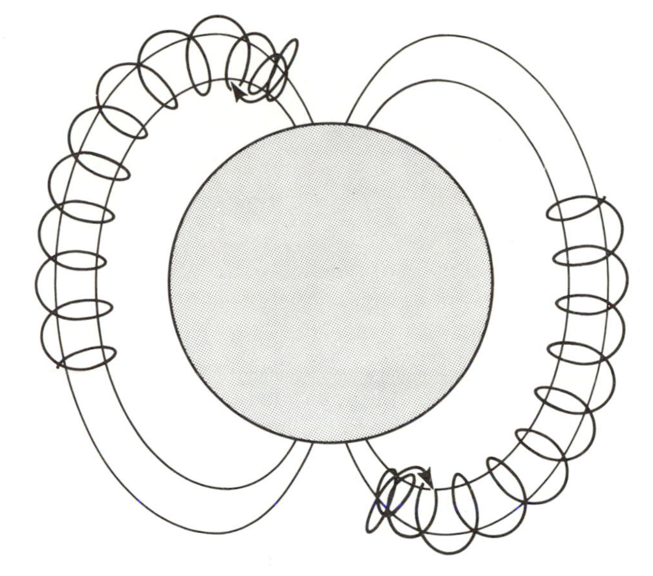

Figure 17.37. The convergence of the magnetic field lines at the magnetic poles of the Earth serves to trap charged particles in a "magnetic bottle." The mechanism is explained in Problem 17.13. More sophisticated magnetic bottles have been devised to confine a hot plasma for the fusion-energy program.

Do the other planets also have magnetic fields? Nothing observationally is known about Uranus, Neptune, and Pluto; so let us discuss the other planets. Jupiter and Saturn rotate rapidly, and have liquid metallic interiors; so we would expect them to have very strong magnetic fields (by planetary standards), and they do. Venus and Mars have metallic cores, but Venus rotates slowly, and the core of Mars might be cool and viscous. Consequently, we might guess that they should have only weak magnetic fields, and they do. The Moon is quite cold, and rotates slowly. We would probably venture that it should have practically zero magnetic field, and it does. Mercury (as the photographs obtained by Bruce Murray and his colleagues from Mariner 10 show) looks much like the Moon on the outside, and also rotates slowly. Consequently, we would guess it should have zero magnetic field. It doesn't! Mariner 10 found Mercury to have a magnetic field whose surface strength is about 1 percent that of the Earth. This does not sound like much, but Mercury's magnetic field is far stronger than that of either Venus or Mars. What happened? No one knows.

Having inventoried the planets' magnetic fields, let us now consider how these magnetic fields interact with the solar wind. Let us start with the phenomena of **Aurora Borealis** and **Aurora Australis**, which occur at the polar and subpolar regions of the Earth. In the northern hemisphere, residents at high latitudes are sporadically treated to spectacular displays of "northern lights" (Aurora Borealis; see Figure 17.36). The similar phenomenon in the southern hemisphere is called the Aurora Australis. Carl Stormer, Sydney Chapman, and Hannes Alfven have enabled scientists to understand the polar auroras as manifestations of the impact of fast, charged particles from the Sun with the magnetic field of the Earth. We now understand the nocturnal auroras as one aspect of the existence of a belt of energetic particles discovered in 1958 by James Van Allen. The convergence of the magnetic field lines toward the two poles of the Earth act as a "magnetic bottle" to trap energetic charged particles (Figure 17.37). In three dimensions, there are actually two sets of Van Allen belts shaped like doughnuts (Figure 17.38). The trapped

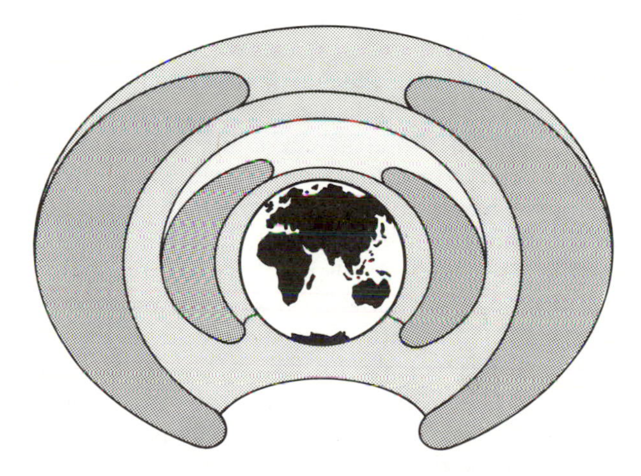

Figure 17.38. The Van Allen belts surrounding the Earth contain charged particles trapped by the Earth's magnetic field in two doughnut shaped regions. The inner belt has a mean radius about $1.6R_{\oplus}$; the outer, about $3.5R_{\oplus}$; where $R_{\oplus}$ is the radius of the Earth.

particles attain their lowest altitudes over the magnetic poles, where they occasionally collide with and excite atoms in the upper atmosphere of the Earth. The radiation produced when the atoms deexcite creates the beautiful aurora displays.

Problem 17.13. If we ignore cyclotron radiation (the nonrelativistic version of synchrotron radiation), charged particles which move in magnetic fields follow helical trajectories which conserve the magnitude of the momentum $\boldsymbol{p}$, i.e., $p_{\parallel}^2 + p_{\perp}^2 =$ constant, where $p_{\parallel}$ and $p_{\perp}$ are the components of $\boldsymbol{p}$ parallel and perpendicular, respectively, to the magnetic field $\boldsymbol{B}$. If the magnetic field lines converge slowly in space—i.e., if $\boldsymbol{B}$ increases in magnitude with a convergence length which is long compared to a gyroradius—the ratio $p_{\perp}^2/B$ tends also to be conserved (actually, $p_{\perp}^2/B$ is an "adiabatic invariant" and is only very nearly a constant). Show that the constancy of both $p_{\parallel}^2 + p_{\perp}^2$ and $p_{\perp}^2/B$ implies that a particle which spirals toward converging field lines will eventually reverse ("bounce") and spiral back out of the region of high $\boldsymbol{B}$. *Hint*: Suppose $p_{\parallel}$ to be initially positive and show that it must eventually decrease to zero and become negative.

If the field lines are also curved, as in the dipole structure of the Earth's magnetic field, charged-particle trajectories will also contain a slow drift in the direction perpendicular to $\boldsymbol{B}$ and its curvature. Argue now that energetic charged particles introduced randomly into the Earth's magnetic field will automatically tend to produce a doughnut-like structure like that depicted in Figure 17.38. The Van Allen belts partially shield the Earth from bombardment by cosmic-ray particles. The strength $\boldsymbol{B}$ of the Earth's magnetic field at the outer Van Allen belt is typically 10^{-2} gauss. Use Problem 11.11 to calculate the product γv for a proton whose radius of gyration r in such a field would be comparable to the radius of the Earth. The quantity γ measures the energy of a particle in units of its rest energy. How energetic a proton would you estimate could be effectively kept out by the Earth's magnetic field?

The particles which make up the Van Allen belts slowly but continuously leak out the polar ends of their magnetic bottle. Clearly, then, there must exist a source that replenishes the supply of charged particles. We now know that this source is the solar wind. As we discussed in Chapter 5, the corona of the Sun expands continuously into interplanetary space. By the time this wind blows past the Earth, the average number of ions (mostly protons, which are compensated for, of course, by an equal number of electrons) per unit volume is about $5\ \text{cm}^{-3}$, and the mean windspeed is about 500 km/sec (a million miles per hour!). These values, however, vary greatly with the solar wind increasing in strength sharply during periods of maximum solar activity. Increased auroral displays and upper atmospheric disturbances on Earth are well-correlated with such solar activity.

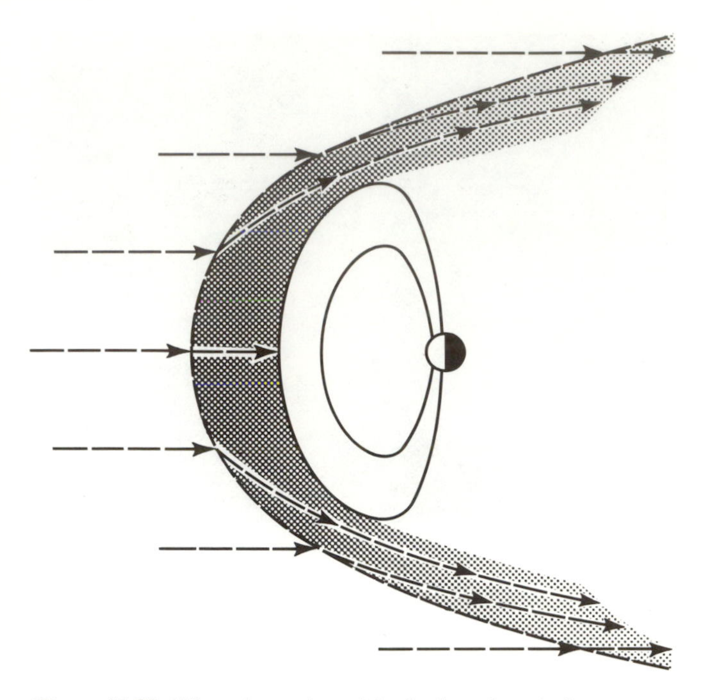

Figure 17.39. When charged particles in the solar wind encounter the magnetic field of the Earth, their inability to penetrate the magnetosphere easily causes a bow shock (curved parabolic line) to form on the upstream side. Steamlines (dashed curves) are generally turned sharply at the bow shock, and the plasma is compressed on the downstream side of the shock wave. Except for the detailed (magnetic) mechanisms which mediate the shock, the bow shock in front of the Earth is analogous to that which forms in front of a supersonically flying aircraft.

Magnetic fields from the surface of the Sun are frozen into the solar wind and are drawn out by the flow. The magnetized plasma of the solar wind cannot easily penetrate the Earth's system of closed magnetic field lines. Consequently, when the wind strikes the magnetosphere of the Earth at high speeds, a shock wave develops, past which the wind plasma flows much more slowly around the Earth's magnetosphere (Figure 17.39). The magnetic field lines carried by the wind plasma will generally not line up with the magnetic field lines of the Earth. Several theorists (notably, Petschek, Parker, and Sturrock) have proposed that the misaligned magnetic fields will reconnect at their juncture points as shown in Figure 17.40. Notice that charged particles which are

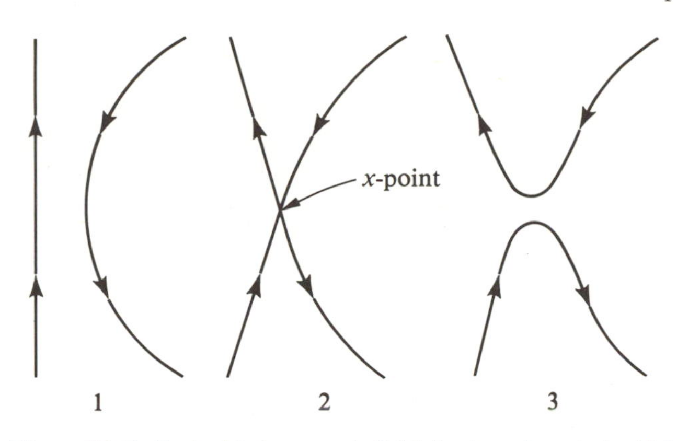

Figure 17.40. Embedded magnetic fields in the solar wind which impinges from the left on the curved magnetic field of the Earth can make reconnections in the sequence 1, 2, 3. The reconnection shown is one in time of individual field lines. The resulting quasi-steady-state structure is illustrated in Figure 17.41.

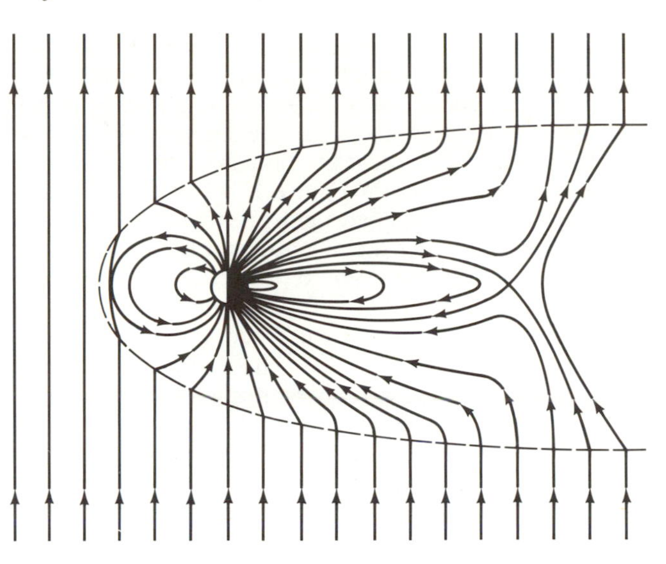

Figure 17.41. Schematic diagram of the magnetic-field structure associated with the magnetosphere and bow shock of the Earth. Notice the role of reconnection at the "X" points at the head and tail of the structure which produces the so-called "open" field lines. Charged particles can enter or leave the magnetosphere of the Earth along such open field lines. Similar magnetic structures are believed to exist around Mercury, Jupiter, and Saturn, and the study of the magnetospheres of the planets has helped to guide the thinking about accretion onto magnetized neutron stars (see Chapter 11). (Adapted from J. W. Dungey, *Phys. Rev. Lett.*, **6**, 1961, 47; see also C. F. Kennel and F. V. Coroniti, *Ann. Rev. Astr. Ap.*, **15**, 1977, 389.)

Figure 17.42. A luminous haze of sodium atoms exists around Io. A dark occulting disk has been used to exclude the glare from Io itself, and the white dot and photograph of Jupiter have been added later for orientation. (After D. L. Matson, B. A. Goldberg, T. V. Johnson, and R. W. Carlson, *Science*, **199**, 1978, 531.)

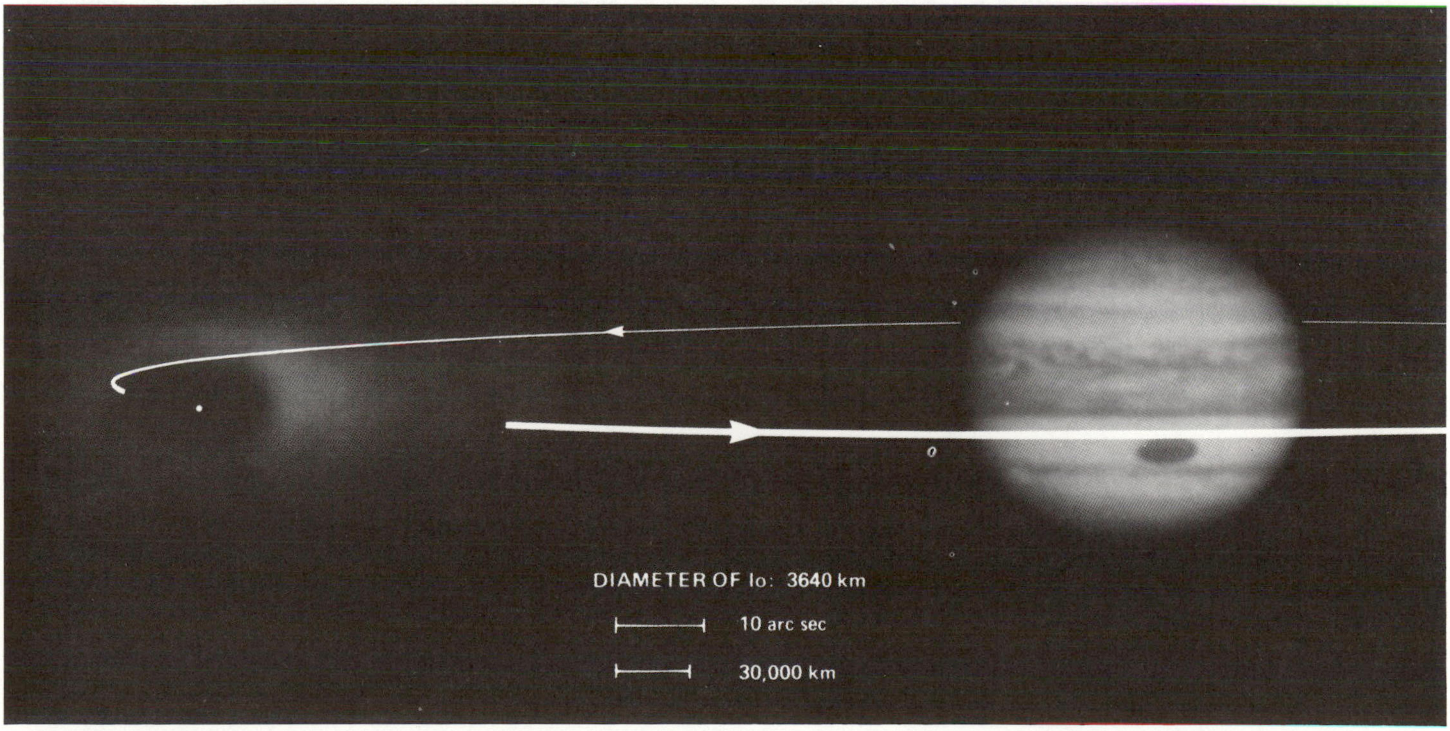

"tied" by their gyrational motions to the original wind field line will end up being tied to a different field geometry after the reconnection process. It is also believed that the release of magnetic-field energy by the reconnection process will accelerate fast charged particles.

The sequence shown in Figure 17.40 represents what happens to two misaligned field lines with the passage of time. If we were to take a snapshot of the entire flow at one instant of time, we would presumably see different field lines in different stages of the reconnection process. The resultant picture is shown schematically in Figure 17.41. Notice, in particular, that charged particles originally in the solar wind can now enter via the "open field lines" into the magnetosphere of the Earth. Similarly, ions in the Earth's atmosphere can leave the polar regions along such open field lines. This so-called "polar wind" may abet Jeans escape. Although Figure 17.41 is highly idealized, it is gratifying to magnetospheric physicists that recent analysis of satellite data by Kinsey Anderson and Robert Lin shows that the actual magnetosphere of the Earth has many of the features predicted by the theorists.

It is now believed that both Jupiter and Saturn have similar interactions with the solar wind. In particular, the Pioneers and Voyagers found structures consistent with the interpretation shown in Figure 17.41 for these two Jovian planets. Jupiter has an additional interesting interaction with its moon, Io. For a long time, radio astronomers knew that Jupiter sporadically emitted bursts of radio waves. Gradually it was realized that these bursts were correlated with the orbital motion of Io. Peter Goldreich and Donald Lynden-Bell proposed that this phenomena could be understood if Io possessed a conducting ionosphere which sliced through the magnetic field lines of Jupiter. In 1974, Robert Brown discovered that Io is surrounded by a haze of sodium atoms (Figure 17.42), and sulpher atoms and ions have now been detected as well. The Voyager discovery of the intense geological activity on Io makes it plausible that these atoms and ions have been outgassed from the interior of Io, perhaps by super volcanos. In any case, the plasma before it expands away from Io may well provide the conducting medium for the generator mechanism discussed by Goldreich and Lynden-Bell.

The Relationship of Solar-System Astronomy to the Rest of Astronomy

There is an unfortunate tendency in some quarters to treat solar-system astronomy as almost a separate branch of science from the rest of astronomy. Planets, moons, and comets seem far removed from stars, galaxies, and the universe. Discussed one by one, the study of planets and moons does seem to acquire a life of its own. There are so many interesting features about this planet or that moon. For a general survey of astronomy, it is almost unfortunate that we know so much about the planets, because we can easily let their individual allures blind us to general patterns of behavior. I have deliberately avoided such an approach in writing this chapter. Solar-system astronomy *does* have recognizable general patterns, because, in the best traditions of science, it is deeply rooted in the fundamental laws of physics. Planets, moons, and comets *are* related to stars, galaxies, and the universe, at the level of physical processes, not at the level of descriptive appearances. This is the real lesson of the unity of the universe.

CHAPTER EIGHTEEN

Origin of the Solar System and the Earth

How and when did the solar system come to be? Not many years ago the beginning of the solar system was thought to be virtually synonymous with the act of creation itself. To a self-centered society, the creation of the Earth, the Sun, and the Moon was seen as the main act. The formation of the rest of the universe—the planets and the stars—seemed almost an afterthought. Today, we know that the issue of the origin of the solar system is not the same as that of the origin of the universe (Chapter 16). Astronomy teaches us that the universe is much, much larger than the solar system. More significantly for the issue of origins, geology teaches us that the solar system is appreciably (but not overwhelmingly) younger than the oldest objects in the universe. Thus, the solar system must have formed against the backdrop of an already mature universe, and we can hope to trace our roots to an era where the clues are not hopelessly obliterated by the cleansing fires of the primeval fireball.

The Age of the Earth and the Solar System

How do we know the age of the Earth? The geologists of the eighteenth and nineteenth centuries (A. G. Werner, William Smith, George Cuvier, Jean Guettard, Nicholas Demarest, and most notably James Hutton and Charles Lyell) showed how the sequence of rock strata and fossils could be arranged in a qualitative and even semiquantitative sequence. However, good absolute estimates of ages were stymied by the uncertain and variable rates of volcanism, weathering, erosion, and sedimentation. Estimates of the age of the Earth from theoretical considerations of its rate of cooling by Sir Isaac Newton, Comte de Buffon, and Jean Fourier suffered from lack of knowledge about the composition of the Earth's interior, the origin of its interior heat, and the role of solar heating for keeping the surface warm. Nevertheless, by the nineteenth century, it was already clear that, if the rate of geological processes in the past was not vastly different from that in the present, then the Earth must be much older than the roughly six thousand years accorded to its history by Archbishop James Ussher in 1664. Most geologists insisted that the Earth had to be even older than the 20 or 30 million years that Kelvin and Helmholtz deduced from their contraction theory for the power of the Sun (Chapter 5). It was left to Ernest Rutherford to provide the definitive method for determining how much older the Earth was. Rutherford's method was, of course, radioactive dating.

Radioactivity

Radioactivity was discovered in 1896 by Henri Becquerel. The discovery was a fortunate accident, since Becquerel was actually looking to see if fluorescent materials emitted X-rays. Instead, Becquerel found that all uranium salts, fluorescent or not, fogged his photographic plates. Marie Curie was among the first to realize

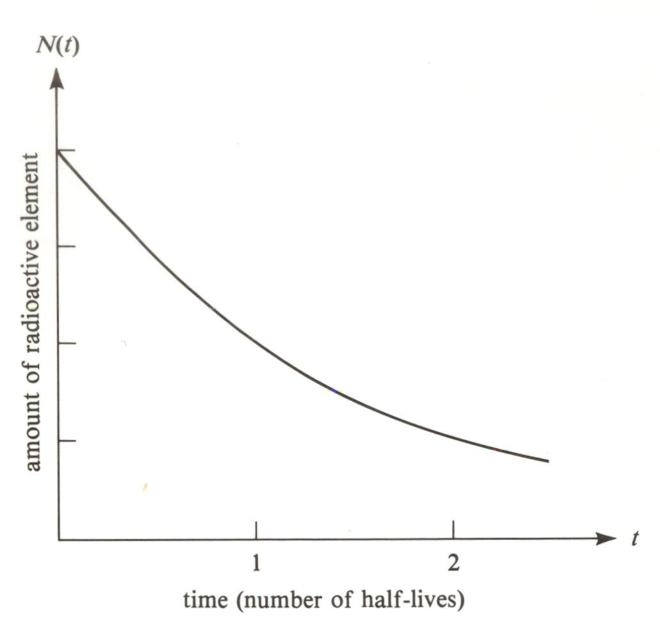

Figure 18.1. The characteristics of the law of radioactive decay. The number of atoms, $N(t)$, at time t declines with increasing time as a negative exponential. The timespan required for the number to drop to half its original value is called the half-life of the radioactive element.

that radioactivity was an intrinsic property of a chemical element, not subject to the dictates of its environment or previous history. Her life with her husband Pierre (who had earlier done pioneering studies of magnetism and crystal; see Chapters 4 and 5) is an incredible story of hardship, heroism, triumph, and tragedy. After Pierre's death in a traffic accident at the age of 47, Marie went on to win the second of two Nobel prizes for their joint work in the discovery of radium, a far more powerful source of natural radioactivity than uranium. Ernest Rutherford in 1900, together with Frederick Soddy, showed that radioactivity transforms one chemical element into a completely different chemical element. Rutherford's scattering experiment, carried out in 1910 with Geiger and Marsden, established the nuclear model of the atom (Chapter 3). Thus did the road lead to the modern understanding of the nature of matter, and, in particular, to the understanding that radioactive decay takes place by emission of particles from an unstable *nucleus* of an atom (Chapter 6).

The law of radioactive decay, which Rutherford formulated, states that the number of radioactive atoms $N(t)$ at time t will decline in comparison with the number $N(0)$ at $t = 0$ as a decaying exponential in time,

$$N(t) = N(0)\exp(-t/t_e) = N(0)(1/2)^{t/\tau}, \qquad (18.1)$$

where $\tau = t_e/\ln(2) = 0.693t_e$ is the **half-life** of the radioactive species. For example, the half-life of uranium-238 is 4.6 billion years; so every 4.6 billion years, the amount of uranium-238 in a uranium-bearing rock will be halved (Figure 18.1). A theoretical analysis by Gamow revealed equation (18.1) to be a statistical law governing the probability that alpha particles will "tunnel" from the nuclei of ^{238}U atoms (Chapter 6). In any case, by a series of alpha and beta decays, uranium-238 will ultimately decay into lead-206 (^{206}Pb), with the half-lives of the intermediate products being short in comparison with the initial decay of uranium-238. This set of circumstances allows the radioactive dating of uranium-bearing rocks that we shall now describe.

Problem 18.1. Given that $Z = 92$ and $A = 238$ for ^{238}U and $Z = 82$ and $A = 206$ for ^{206}Pb, how many total alpha and beta decays need to occur to transform ^{238}U to ^{206}Pb? Why can radioactivity lead to fluorescence, but not vice versa? *Hint*: Does fluorescence arise from the excitation of the nucleus of an atom or of its electronic shell? For those of you familiar with nondigital watches, explain why radium is used in watch dials.

Radioactive Dating

Radioactive dating is a sophisticated experimental technique; we will discuss here only its most basic concepts. Suppose we have a rock which now contains ^{238}U. Since ^{238}U decays ultimately into ^{206}Pb (which is stable), we know that the amount of ^{238}U must have been larger in the past, and the amount of ^{206}Pb must have been smaller. Since the time when the rock solidified and locked in the atoms of uranium and lead, however, the *sum* of the number of uranium-238 and lead-206 atoms must be a constant:

$$N(^{238}\text{U}) + N(^{206}\text{Pb}) = \text{constant}. \qquad (18.2)$$

We may find out the value of the constant by measuring the amounts of ^{238}U and ^{206}Pb currently present in the ore. We may then use equation (18.1) to deduce the amount of uranium-238 contained in the ore at past times, and use equation (18.2) to do the same for lead-206. If we assume further that all the lead-206 comes from the radioactive decay of uranium-238, we can find the time in the past when $N(^{206}\text{Pb})$ was zero (Figure 18.2). This time is then the age of the ore (when the rock solidified).

What if $N(^{206}\text{Pb})$ did not equal zero when the ore originally solidified? In other words, how do we proceed if we cannot assume that all the lead-206 came from the radioactive decay of uranium-238? A slightly more complicated line of reasoning is needed. We need to deduce how much of the lead-206 was not the product of the radioactive decay of uranium-238. To do this, *in*

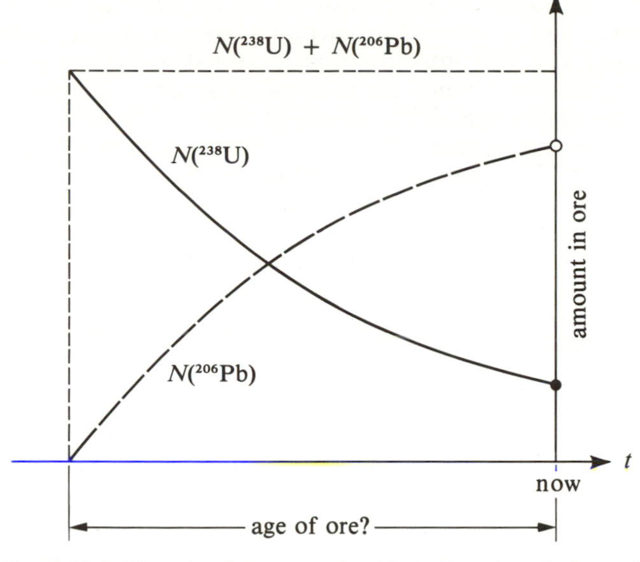

Figure 18.2. The simplest example of radioactive dating of uranium ores. The number of uranium-238 atoms, $N(^{238}U)$, has declined to its present value, given by the solid circle, by a known law of exponential decay. Its ultimate decay product, lead-206, has a number, $N(^{206}Pb)$, whose present value, given by the open circle, can also be measured. The sum $N(^{238}U) + N(^{206}Pb)$ must remain a constant when extrapolated back in time; therefore, the age of the ore cannot be older than a time when this sum consists of a contribution only from ^{238}U, nothing from ^{206}Pb.

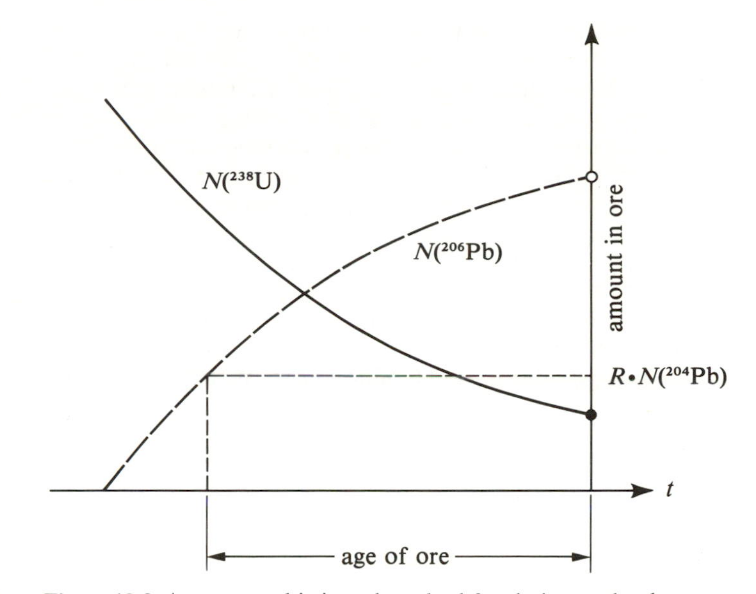

Figure 18.3. A more sophisticated method for dating rocks that contain uranium. As in Figure 18.2, one begins by measuring the present amounts of ^{238}U and ^{206}Pb, given by the solid and open circles, respectively. One also measures the amount of lead-204, ^{204}Pb, which is not a product of radioactive decay, and multiplies that amount by R, the ratio $[^{206}Pb]/[^{204}Pb]$ of lead-206 to lead-204 found in nonradioactive rocks. The product $R \cdot N(^{204}Pb)$ is then assumed to be the amount of primitive lead-206 incorporated into the original rock at its solidification. This assumption gives a smaller age for the rock than the naive estimate of Figure 18.2 if the amount of nonradioactively derived lead, $R \cdot N(^{204}Pb)$, is significant.

principle, all we would need to do is examine the amount of lead-204, ^{204}Pb. Lead-204 is another stable isotope that could be present in the radioactive ore, but it is not the product of a radioactive decay. Lead-204 and lead-206 have the same chemical properties; so the radioactive ore will incorporate the nonradioactively derived portions of these two isotopes in the *same proportion* as in any nonradioactive rock. This proportion can be measured in nonradioactive rocks, and the primitive ratio can then be used to infer the amount of ^{206}Pb which does not result from the radioactive decay of ^{238}U; we merely multiply the primitive ratio $[^{206}Pb]/[^{204}Pb]$ by the amount of ^{204}Pb now in the radioactive ore. Figure 18.3 shows schematically the correction procedure that subtracts out the primitive ^{206}Pb. Notice that Figure 18.2 is a special case of Figure 18.3 if $N(^{204}Pb)$ is zero, i.e., if the ore contained no intrinsic lead at the time of its formation.

In practice, however, that procedure has some difficulties. Precisely because lead 204 and 206 are isotopes of the same chemical element, no chemical procedure can measure their abundances separately. Physical methods based on the small difference in atomic weight could distinguish between ^{204}Pb and ^{206}Pb, but the effort is generally time-consuming and expensive. Finely tuned lasers like those now used to separate ^{235}U and ^{238}U might be more practical, but such techniques have become available only very recently. To date, geochemists such as Harold Urey and Gerald Wasserburg have preferred to use other radioactive species which produce easily identified decay products. One such radioactive species is potassium-40, ^{40}K, one of whose decay products, argon-40, ^{40}Ar, is a noble gas which is easily distinguished by mass spectrographs from other isotopes of argon. The half-life of ^{40}K is 1.3×10^9 yr, which turns out to be a convenient magnitude for geological and astrophysical studies.

By techniques similar to that described above, geochemists have found the ages of the oldest rocks on Earth to be about 3.8 billion years. It is believed, however, that the Earth itself is older than this. Rocks older than 3.8 billion years have apparently not survived to this day, or else there were none that solidified in an early phase of the evolution of the Earth. We shall present the arguments favoring the first point of view when we discuss modern ideas about the origin of the Earth, but for now let us discuss other direct lines of evidence for an older age of the solar system. The radioactive dating of lunar rocks brought back by the Apollo missions show that the oldest lunar rocks are about 4.2 billion years old. Some scientists speculate that still older rocks may lie on the other side of the Moon, where less destruction of lunar rocks by meteoroid bombardment may have occurred. In contrast with the spread of ages found for Earth and Moon rocks, the radioactive dating of meteorites show them to be all the same age, to within the errors of measurement, between 4.5 and 4.7 billion years old. Thus, the interpretation has grown that the first objects solidified in the solar system about 4.6 billion years ago. This time, 4.6 billion years, is the probable age of the entire solar system. The geological extension of the history of the Earth from several thousand years to 4.6 *billion* years is truly one of the great accomplishments of modern science.

Exposure Ages of Meteoroids

The value of 4.6 billion years refers to the time since the asteroids solidified. However, meteoroids are broken fragments of asteroids (see Figure 17.6a), and there is another age of interest for meteoroids: the elapsed time since the shattering which produced the meteoroid. This latter age is possible to obtain because the solar system suffers a continuous bombardment by high-energy cosmic-ray particles (Chapter 11). These high-energy particles can smash the heavy atomic nuclei that exist in an iron meteorite, say, and produce lighter nuclei, some of which may be radioactive. Before the asteroid was shattered, the particular piece of meteorite which eventually fell to Earth may have been deep inside the main body, and therefore shielded from the external cosmic-ray bombardment; so the time when this meteoroid's *exposure* to cosmic-ray produced radioactive elements did not begin until the shattering of the asteroid. Sophisticated techniques exist to measure the *exposure ages* of such meteoroids, based on the idea that longer exposures lead to a greater concentration of long-lived isotopes relative to short-lived ones, because the latter reach an equilibrium between production and decay. Such measurements yield exposure ages of hundreds of millions of years for the iron meteorites, and tens of millions of years for the stones. Systematic discrepancies do arise from the use of radioactive species with different half-lives. It has been proposed that such apparent discrepancies originate in a variable flux of galactic cosmic rays encountered by the solar system as it orbits around a spatially inhomogeneous galaxy (Chapter 12).

Age of the Radioactive Elements

Another application of radioactive dating allows geochemists and astrophysicists to measure the age of the radioactive elements themselves. The basic idea is that since radioactive elements like uranium exist naturally, they must have been created by natural processes some finite amount of time ago; otherwise, uranium would all have decayed into lead by now. Two common isotopes of uranium are well-known: ^{238}U and ^{235}U, which have half-lives of 4.6×10^9 yr and 7.1×10^8 yr, respectively. In uranium ore, we can measure the current proportion of ^{235}U to ^{238}U. If we knew the primordial ratio, we could deduce the age of the elements themselves from the different rates of decay of the two isotopes of uranium.

There is a complication with such a program. We now believe that heavy nuclei were formed inside massive stars and subsequently expelled in supernova explosions (Chapter 8). We know the relative proportion in which the two isotopes of uranium would be formed under such conditions fairly well. That is, from the work of Hoyle, Fowler, Clayton, Arnett, and others, we know the relative formation of ^{235}U and ^{238}U via neutron capture under the r-process with fairly good certainty. However, the radioactive elements found today in the solar system were probably not produced in a single supernova event. Although the use of just ^{235}U and ^{238}U does not allow one to test the validity of this statement, the use of other radioactive isotopes (in particular, the extinct species iodine-129 studied by John Reynolds and his coworkers) shows that the radioactive elements incorporated in the material that ultimately made up the solar system (more specifically, the meteorites) were probably produced by a continual series of nucleosynthesis and ejection processes that took place before the actual formation of the solar system. It is our lack of knowledge of the details of these cumulative effects that prevents geochemists from measuring the age of the radioactive elements with high precision. Gerald Wasserburg and David Schramm have studied this problem in detail, using as many radioactive species as possible to avoid too many theoretical assumptions. Different interpretations are possible for the several anomalies which arise from such analyses. The safest conclusion that one can draw is that the age of the radioactive elements (the time since first synthesis) is probably somewhere between 7 and 15 billion years. In order of magnitude, this delineation is in good agreement with astronomers' estimates of the age of globular cluster stars (Chapter 9) and the age of the universe (Chapter 15). That three independent methods of establishing the ages of some obviously ancient objects produce the same number—within a factor of two or so—is very reassuring (Figure 18.4). In any case, we can state with considerable confidence that the Sun and the solar system are probably less than half the age of the oldest objects in the universe. This assertion lies behind our earlier claim that the formation of the solar system must have taken place against the backdrop of an already mature universe.

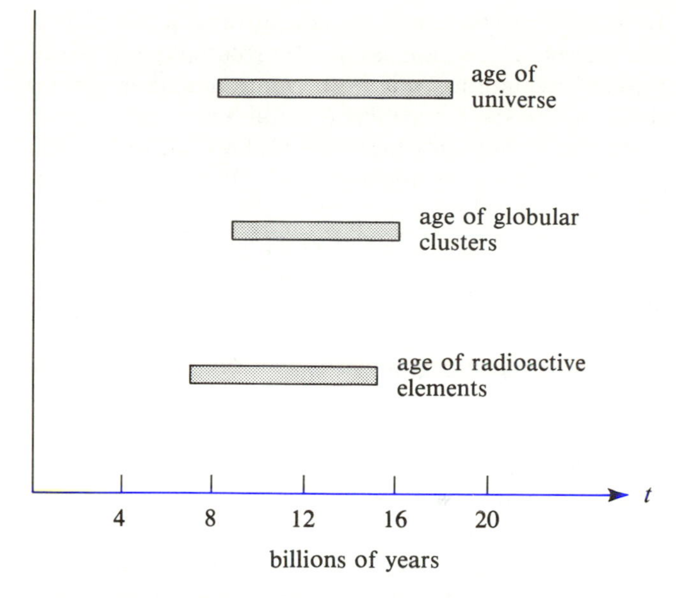

Figure 18.4. Estimates of the ages of three ancient objects. The horizontal bars indicate the spread of estimates of ages of the universe, the globular clusters, and the radioactive elements that most astronomers would consider still to be permissible given the uncertainties of the measurements. Notice that there is significant overlap in the range 9 to 15 billion years. Thus the age of the solar system, 4.6 billion years, is likely to be between one-half and one-quarter of the age of the oldest objects in the universe.

Problem 18.2. Suppose the r-process produces in net a ratio of ^{235}U to ^{238}U of 1.5. The present ratio of ^{235}U to ^{238}U in uranium ore is 0.007. Assume formation by a single event, and calculate the elapsed time since the creation of uranium. The half-lives of ^{235}U and ^{238}U are $t_5 = 7.1 \times 10^8$ yr and $t_8 = 4.6 \times 10^9$ yr respectively. *Hint*: Derive the equation $0.007 = 1.5(1/2)^{t(1/t_5 - 1/t_8)}$ and solve for the elapsed time t. If the actual history of nucleosynthesis was steady until the formation of the solar system, and uranium subsequently decayed freely, would the derived age of first creation be greater or less than the above value? Qualitative reasoning will suffice.

Motions of the Planets

How did the formation of the solar system occur? Besides the clues that we have assembled so far in this and the preceding chapter, we need some additional important data concerning one of the oldest topics of science: the motions of the planets (see Chapter 1).

Kepler's Three Laws of Planetary Motions

Since Johannes Kepler's work in the early part of the seventeenth century, it has been known that the planets follow (approximately) his three general laws of motion (shown in Box 18.1). The geometric meaning of Kepler's three laws is explained in Figure 18.5.

Newton's Derivation of Kepler's Laws

It was Sir Isaac Newton who coupled his laws of mechanics with his theory of gravity to explain Kepler's empirical discoveries in terms of dynamics. In particular,

---

BOX 18.1
Kepler's Three Laws of Planetary Motions

(1) The orbit of a planet forms an ellipse with the Sun at one focus.
(2) The Sun-planet radius vector sweeps out equal areas in equal times.
(3) The square of the period of revolution of a planet is proportional to the cube of the semimajor axis of its elliptical orbit.

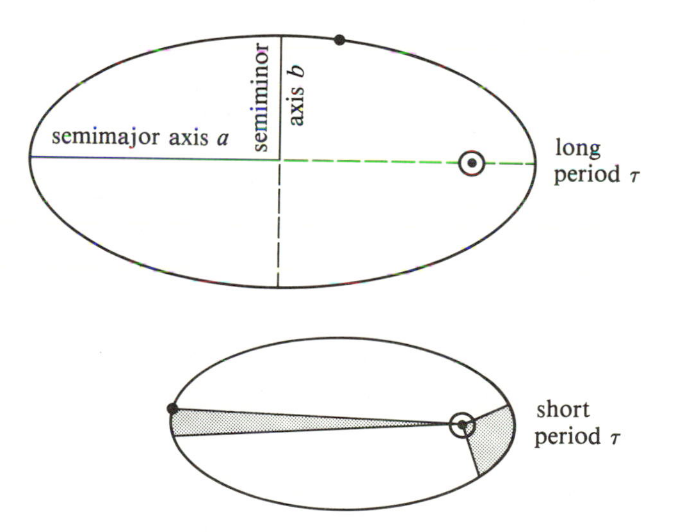

Figure 18.5. Kepler's three laws of planetary motions. The first law states that the orbits of all planets—both those with long orbital periods and those with short orbital periods—take the form of ellipses with the Sun ⊙ at one focus. The second law states that equal areas—indicated by the shaded regions in the bottom diagram—are swept out in equal times. The third law states that the square of the period is proportional to the cube of the semimajor axis: $\tau^2 \propto a^3$, with a proportionality constant which is the same for all planets (see, however, Problem 18.4).

Newton showed that Kepler's laws are approximate to the extent that the planets and the Sun can be idealized as perfect spheres and that the mass of the Sun is much greater than that of any of the planets. Today, we further know that Kepler's laws are also approximate to the extent that relativistic and quantum-mechanical effects are negligible. All these approximations are valid in the actual solar system (some more than others), and our purposes here will be adequately served if we merely accept Newton's explanations for Kepler's laws (see Problems 18.3 and 18.4). Indeed, the story of Kepler and Newton's discoveries provided such a cogent demonstration of the explanatory power of the analytical method that it set the example *par excellence* for all subsequent scientific endeavors. After almost four centuries (Kepler's three laws were formulated in 1609–1618), the formal results may seem ancient, but the reasoning used to derive the results is as modern and as full of vitality as any other we have discussed in this book. Like rare wine, the great works of science do not suffer with age.

Problem 18.3.(for those who know vector calculus). Ignore the perturbations of the other bodies in the solar system and consider only the mutual attraction between the Sun, mass m_1 at position $\boldsymbol{r}_1$ relative to some inertial frame, and a planet, mass m_2 at position r_2. In this approximation, we have a two-body problem, and we wish to show here how to reduce this two-body problem to an equivalent one-body problem in a field of external force. To begin, Newton's laws lead to the following dynamical equations for body 1 and body 2:

$$m_1 \frac{d^2\boldsymbol{r}_1}{dt^2} = \frac{Gm_1m_2}{r^3}\boldsymbol{r}, \qquad m_2 \frac{d^2\boldsymbol{r}_2}{dt^2} = -\frac{Gm_1m_2}{r^3}\boldsymbol{r},$$

where $\boldsymbol{r} = \boldsymbol{r}_2 - \boldsymbol{r}_1$ is the displacement vector from mass 1 to mass 2. Since the forces on the right-hand side are equal and opposite, we are motivated to add the above two vector equations. Show that this results in

$$\frac{d^2}{dt^2}(m_1\boldsymbol{r}_1 + m_2\boldsymbol{r}_2) = 0.$$

To interpret the above, define the total mass, $m = m_1 + m_2$, and center of mass position, $\boldsymbol{R} = (m_1\boldsymbol{r}_1 + m_2\boldsymbol{r}_2)/(m_1 + m_2)$, and show that the above equation may be written

$$m\frac{d^2\boldsymbol{R}}{dt^2} = 0.$$

In other words, the center of mass moves as a free particle (no external force) of mass m. The solution for $\boldsymbol{R}$ corresponds to uniform straight-line motion, and eliminates three of the six independent variables in the original equations of motion (three components each of $\boldsymbol{r}_1$ and $\boldsymbol{r}_2$). Let us take, as the remaining three independent variables, the three components of $\boldsymbol{r} = \boldsymbol{r}_2 - \boldsymbol{r}_1$. To manipulate the original dynamical equations into a single vector equation for $\boldsymbol{r}$, divide out m_1 in the first equation and m_2 in the second. Subtract the results, and derive

$$\frac{d^2\boldsymbol{r}}{dt^2} = \frac{-Gm}{r^3}\boldsymbol{r}.$$

Thus, the motion of body 2 relative to body 1 can be conceived as that of a test particle attracted to a center of gravitation of mass $m = m_1 + m_2$.

To obtain the individual motions of bodies 1 and 2 with respect to the center of mass (which is an inertial frame), define the relative positions: $\boldsymbol{R}_1 = \boldsymbol{r}_1 - \boldsymbol{R}$ and $\boldsymbol{R}_2 = \boldsymbol{r}_2 - \boldsymbol{R}$. Show by direct substitution that $\boldsymbol{R}_1$ and $\boldsymbol{R}_2$ can be written

$$\boldsymbol{R}_1 = \frac{-m_2}{m_1 + m_2}\boldsymbol{r}, \qquad \boldsymbol{R}_2 = \frac{m_1}{m_1 + m_2}\boldsymbol{r}.$$

Thus, $\boldsymbol{R}_1$ and $\boldsymbol{R}_2$ describe loci of the same shape as $\boldsymbol{r}$, but they are smaller by the respective factors $m_2/(m_1 + m_2)$ and $m_1/(m_1 + m_2)$. Compare this result with Figure 10.2.

Problem 18.4. (also for those who know vector calculus). The nontrivial part of Problem 18.3 is the solution of the equivalent one-body problem:

$$d^2\boldsymbol{r}/dt^2 = -Gm\boldsymbol{r}/r^3. \tag{1}$$

To simplify the problem, let us first derive the conservation of angular momentum and energy. Take the cross product of equation (1) with $\boldsymbol{r}$. The right-hand side is zero and the left-hand side is $\boldsymbol{r} \times d^2\boldsymbol{r}/dt^2$. Show that the latter can be expressed, with $\boldsymbol{v} = d\boldsymbol{r}/dt$, as

$$\frac{d}{dt}(\boldsymbol{r} \times \boldsymbol{v}) = 0. \tag{2}$$

Show that equation (2) implies the conservation of angular momentum per unit mass, and consult Figure 3.21 to see why this conservation principle implies Kepler's second law as well as the notion that the orbit takes place in a single plane.

Now take the dot product of equation (1) with $\boldsymbol{v} = d\boldsymbol{r}/dt$. Show that the result can be written as

$$\frac{d}{dt}\left(\frac{1}{2}|\boldsymbol{v}|^2 - \frac{Gm}{r}\right) = 0, \tag{3}$$

which implies the conservation of energy per unit mass. Explain how this result differs from or extends the result derived in Problem 3.1.

Since the relative motion takes place in a single plane, let us adopt polar coordinates (r, θ) in this plane. With this coordinate system, show that the equations for conservation of angular momentum and energy (per unit mass) can be written in the forms,

$$r^2 \frac{d\theta}{dt} = J, \tag{4}$$

$$\frac{1}{2}\left(\frac{dr}{dt}\right)^2 + \frac{J^2}{2r^2} - \frac{Gm}{r} = E, \tag{5}$$

where J and E are constants (angular momentum and energy per unit mass).

To obtain the proportionality constant in Kepler's second law, show that the area A swept out by the radius vector r satisfies $dA = r(r\,d\theta)/2$. Divide by the elapsed increment of time dt, and obtain $dA/dt = J/2 =$ constant, which is Kepler's second law with an explicit expression for the constant.

To prove Kepler's first law, write equation (5) as

$$\frac{dr}{dt} = \left(2E + \frac{2Gm}{r} - \frac{J^2}{r^2}\right)^{1/2}. \tag{6}$$

To obtain a purely spatial relation involving only r and θ (and not t), divide equation (6) by $r^2\,d\theta/dt = J$, and show that

$$\frac{1}{r^2}\frac{dr}{d\theta} = \left(\frac{2E}{J^2} + \frac{2Gm}{J^2 r} - \frac{1}{r^2}\right)^{1/2}. \tag{7}$$

Make two successive change of variables $u = 1/r$ and $w = u - Gm/J^2$ to obtain the differential equation

$$dw/d\theta = -({w_0}^2 - w^2)^{1/2}, \tag{8}$$

where

$${w_0}^2 = 2E/J^2 + (Gm/J^2)^2.$$

Show that the solution of equation (8) is $w = w_0 \cos(\theta)$, where we have defined the x-axis so that it lies along the long direction of the orbit. Show furthermore that the solution for r as a function of θ is given by

$$\frac{1}{r} = \frac{1}{r_0}[1 + \varepsilon \cos(\theta)], \tag{9}$$

where $r_0 = J^2/Gm$ and $\varepsilon = (1 + 2EJ^2/G^2m^2)^{1/2}$. For $E < 0$ (bound orbit), equation (9) describes an ellipse of eccentricity $\varepsilon < 1$ with one focus at the origin. Show that the semimajor axis a is given by $a = r_0/(1 - \varepsilon^2)$, and that the semiminor axis b is given by $b = r_0/(1 - \varepsilon^2)^{1/2}$. The equations corresponding to $\varepsilon = 1$ and $\varepsilon > 1$ yield, the loci for a parabola and a hyperbola, respectively (see Figure 3.7). The case $\varepsilon = 0$ is a circle and is the limit of an ellipse with zero eccentricity ε.

You are probably more familiar with the equation of an ellipse in the Cartesian form,

$$x^2/a^2 + y^2/b^2 = 1,$$

with the origin of (x, y) taken to be the *center* of the ellipse rather than a focus. Show that the transformation

$$x = r\cos(\theta) + \varepsilon a, \qquad y = r\sin(\theta),$$

applied to Newton's solution, yields the Cartesian form. Show now that the area $A = \int dx\,dy$ of the ellipse equals πab. *Hint*: Transform the equation $x^2/a^2 + y^2/b^2 = 1$ into the equation for a unit circle, $x'^2 + y'^2 = 1$, by the linear transformation $x' = x/a$, $y' = y/b$.

To prove Kepler's third law, integrate $dA/dt = J/2$ over one period τ,

$$\tau = \int dt = \frac{2}{J}\int dA = \frac{2A}{J}.$$

Use the formula $A = \pi ab$ plus the expressions derived previously for a and b to show that τ^2 can be expressed as

$$\tau^2 = \frac{4\pi^2}{Gm}a^3,$$

which is Kepler's third law. Notice that $m = M_\odot + m_P$, where m_P is the mass of the planet; consequently, the original statement of Kepler's third law, that the period of a planet depends only on its semimajor axis and on no other property, is true only to the extent that the mass of the Sun, $M_\odot$, is much larger than any of the masses of the planets in the solar system.

Given the semimajor axes of the nine planets in the solar system in AU as 0.387, 0.723, 1.00, 1.52, 5.20, 9.54, 19.2, 30.1, 39.4, calculate their periods in Earth years.

Problem 18.5. To obtain a better physical appreciation of the meaning of the angle-averaged inverse radius r_0^{-1} and eccentricity ε defined by Newton's solution in Problem 18.4, let us consider the relation of Newton's exact theory of planetary orbits about the Sun to Lindblad's epicyclic theory of stellar orbits about the Galaxy (Chapter 12). In our version of epicyclic theory, the basic reference state is the circular orbit of the same angular momentum per unit mass as the actual orbit. In the Newtonian problem, a circular orbit of specific angular momentum J would have radius r_0 given by centrifugal balance. Show that this condition defines $r_0 = J^2/Gm$ as in Problem 18.4. This circular orbit would, moreover, have an energy

$$E_0 = J^2/2r_0^2 - Gm/r_0 = -Gm/2r_0$$

which is generally less than the specific energy E of an orbit with a nonzero value of the eccentricity. Indeed, show that the square of the eccentricity ε in Problem 18.4 satisfies

$$\frac{\varepsilon^2}{2} = \frac{r_0}{Gm}(E - E_0).$$

This shows explicitly for the Newtonian problem that, of all orbits of given angular momentum, the energy of the circular orbit is least, $E - E_0 > 0$. This property of circular orbits holds for more general mass distributions than that of the Newtonian problem, and it explains why bound orbits tend generally to become more circular in the presence of dissipation.

Dynamical Evolution of the Solar System

In 4.6 billion years, even Pluto could go around the Sun almost 19 million times. During so many revolutions of the planets about the Sun, and of the moons about the planets, have the orbits suffered much dynamical evolution? Let us ignore effects like gravitational radiation, and restrict our attention to the Newtonian formulation of mechanics. With this viewpoint, we may begin with Newton's thoughts on the long-term evolution of the solar system.

Newton was not aware, of course, of the very long history of the solar system's past. Being devoutly religious, Newton was perhaps inclined to accept Archbishop Ussher's pronouncement that the Earth was created at 9:00 A.M., October 26, 4004 B.C. Newton's own calculations on the rate of cooling of a molten iron sphere, which did not incorporate a fully correct theory of heat transfer, led him to conclude that the surface of such a sphere would not become inhabitable until after the passage of about 50,000 years. The heretical conclusion which naturally followed, he rejected. As has been documented by Stephen Toulmin and June Goodfield, Newton adopted the pious attitude that the issue of the origin of the solar system was beyond scientific discussion. It has been said, therefore, that Newton's vision of the physical world was that of a "clockwork universe." Once created, the planets could be left on their own to run around the Sun *ad infinitum*. The creation event itself was beyond physical investigation; only subsequent phenomena needed to obey natural laws. Because the creation was so perfect that it left the planets orbiting nearly in circles in the same direction and nearly in the same plane, Newton speculated that the planetary system could and would remain perfectly stable forever. That is, Newton surmised that if one were to include the gravitational attraction of the planets *for each other* (not done in Newton's two-body treatment of the Sun and a planet; see Problems 18.3 and 18.4), the planetary orbits would not show secular drifts. If the planets do not eventually escape to infinity or plunge into the Sun, then, given enough time, Newton reasoned, the planetary configurations would eventually achieve the same essential state as some earlier time. And being much influenced by mystical views (see Chapter 1), Newton then speculated that history would subsequently repeat itself! Perhaps we should not, however, view Newton's naiveté with too much smugness. Is not our own nostalgia for an oscillating universe (Chapters 15 and 16) rooted in a similar atavistic yearning?

The discovery of the second law of thermodynamics implies, of course, that events in the solar system cannot be exactly repetitive. Nevertheless, dissipative effects in the influence of planetary motions are *now* quite small. Thus, many physicists and astronomers who followed Newton have to this day felt strongly motivated to tackle the gravitational "N-body problem." It is well beyond the scope of this book (and the ability of this author!) to review the important results and techniques that have developed from these studies of celestial mechanics. Here let us mention only that the formal proof of stability remains unsolved. However, for any timescale of reasonable interest, the present orbits of the nine planets are quite stable for the future cumulation of mutual gravitational perturbations.

The situation for the asteroids is different, since nongravitational forces (i.e., inelastic collisions) can drastically change the orbit of an asteroid. Even if such collisions are ignored, Lecar and Franklin have shown that near-resonant interactions with the giant planets will make the orbits of small bodies in much of the interplanetary space between Mars and Saturn unstable on a timescale of only several thousand years. There are some exceptions, and these exceptions include, of course, orbits confined to the current asteroid belt. The heavy cratering of the terrestrial planets implies that asteroidal bodies must have been much more prevalent throughout interplanetary space in the early solar system than they are now. In particular, Hartmann's analysis of the lunar cratering suggests that meteoric bombardment tailed off dramatically about four billion years ago. Very likely, the gravitational perturbations of the massive planets quite early swept the solar system clean of asteroidal debris, and the remaining asteroids are only those which accidentally happened to have orbits fit for survival.

There remain in the solar system, however, orbital regularities which cannot be explained in terms of the general picture developed so far. Some of these regularities probably arose from the initial conditions and therefore bear on the question of the origin of the solar system. Others probably involve slow evolution to very special circumstances from unexpectional initial conditions. It is clearly important to sort out the latter before we attempt to incorporate the remaining regularities into a theory about the formation of the solar system.

Resonances in the Solar System

The solar system contains several bodies which exhibit spin or orbital resonances. A spin resonance occurs when the periods of rotation and revolution, τ_{rot} and τ_{rev}, have a ratio expressible in small whole integers (**commensurate**). Examples are the spin of the Moon and its motion about the Earth ($\tau_{rot}/\tau_{rev} = 1/1$), and the spin of Mercury and its motion about the Sun ($\tau_{rot}/\tau_{rev} = 2/3$). An orbital resonance occurs when the periods of revolution of two or more satellites that orbit the same massive body are commensurate. Examples are the orbital periods of the Trojan asteroids and of Jupiter about the Sun (Trojans:Jupiter = 1:1), the orbital periods of Neptune and Pluto about the Sun (Neptune:Pluto = 2:3), and the orbital periods of the three innermost Galilean moons about Jupiter (Io:Europa:Ganymede = 1:2:4). In the asteroid belt around the Sun or the rings around the Jovian planets, the orbital resonances may also cause a deficit of particles. The Kirkwood gaps in the asteroid belt are at distances where the period of a circular orbit about the Sun would have a ratio of 1/3, 2/5, 3/7, or 1/2 the orbital period of Jupiter. A prominent gap in the rings of Saturn occurs at a distance where the period of a circular orbit would have a ratio of 1/2 the orbital period of Mimas (a major close moon). We could give many other examples of spin or orbital resonances, but they involve bodies less-familiar to the reader, although they are of great practical importance in astronomy, since they often provide our only direct measurements of the masses of certain bodies.

In Problem 10.2 and elsewhere, we have already explained why the spin of the Moon is synchronized with its orbit about the Earth. In 1889, Schiaparelli found that Mercury rotates slowly, and in analogy with the Moon-Earth system, astronomers naturally assumed that Mercury always kept the same face turned toward the Sun. Pettingil and Dyce performed radar experiments in 1965 which showed that Mercury rotates with a period of approximately 59 days, considerably less than its orbital period of 88 days. Guiseppe Colombo brilliantly conjectured shortly afterward that the ratio of the two periods was exactly 2/3, and that such a commensurability might result from the action of the solar gravitational field on a nonspherical solid planet with a pronounced eccentric orbit. Goldreich and Peale proved not only that this was a stable motion, but that Mercury could be trapped in such a resonance by tidal friction acting on an initial state which was not particularly special. To an observer on Mercury's surface, the Sun would move across the sky, with a Mercurian day corresponding to exactly *two* Mercurian years! (See Problem 18.6 and Figure 18.6.) The long interval between "noon" and "midnight" yields very extreme variations of surface temperatures (Chapter 17), but not as extreme as they would be if this airless planet had spun synchronously.

Problem 18.6. Use the equation $\tau_{\odot} = \tau_{rot}\tau_{rev}/(\tau_{rot} - \tau_{rev})$ derived in Problem 17.10 to reach the same conclusion as shown in Figure 18.6 if $\tau_{rot}/\tau_{rev} = 2/3$.

Since the work of Joseph Louis Lagrange in 1775, it has been known that small bodies placed at the L_4 and L_5 points (see Figure 10.7) of the Sun-Jupiter system could remain there indefinitely, stabilized by the action of the Coriolis force, if the initial noncircular motions were small (Figure 18.7). This study provides the accepted explanation of the Trojan asteroids. In a more complex analysis, Pierre Simon de Laplace showed in 1829 that the resonant orbits of Io, Europa, and Ganymede were also stable. Similar understanding exists for the Neptune-Pluto resonance, and for most of the other known orbital resonances in the solar system involving moons or planets. Only recently, however, have we achieved an insight into how these orbital resonances came to be. Goldreich in 1965 showed that the orbital resonances of the moons of Saturn (Mimas and Tethys, Enceladus and Dione, and possibly Titan and Hyperion) must arise from nonresonant states by the action of tidal friction. The idea is that Saturn's rapid rotation coupled with tidal friction forces the moons to spiral outward at different rates (see Figure 17.17). When two moons pass into a strongly resonant condition, they may become locked into this condition. The resonant state is, in some sense,

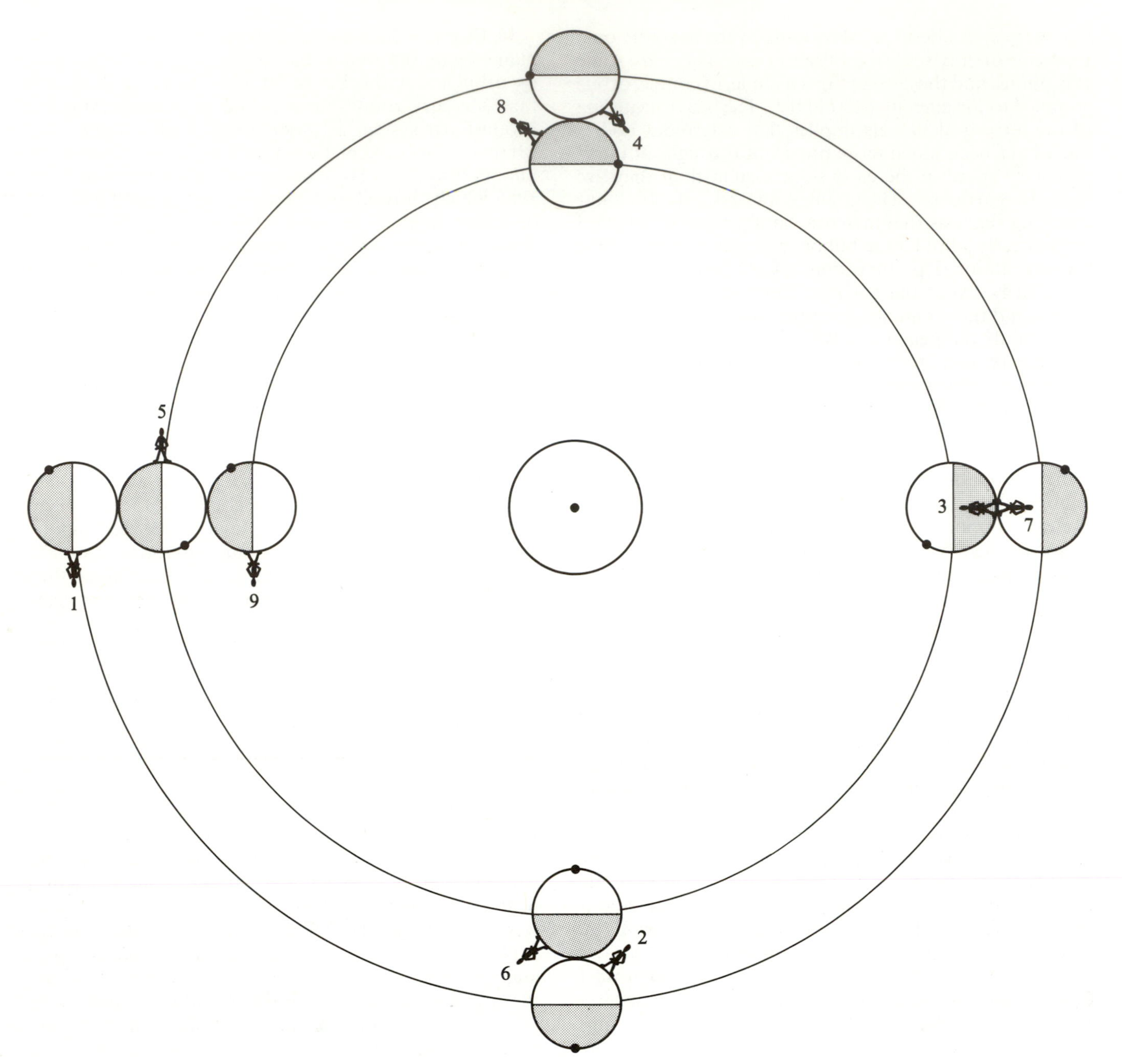

Figure 18.6. One Mercurian day equals two Mercurian years. At position 1, an observer, X, experiences dawn with the Sun just rising above the horizon. At position 2, X has revolved a quarter circle and rotated three-eighths of a circle ($\tau_{\rm rot}/\tau_{\rm rev} = 2/3$), with the black dot showing X's former position relative to the fixed stars (down the page). A similar progression occurs to position 3, where X experiences noon with the Sun directly overhead. By position 5, one full orbit around, dusk has arrived, and the Sun sets over X's horizon. It is not till position 9, two full orbits around, that X experiences dawn again. Thus, one Mercurian day—sunrise to sunrise—equals two Mercurian years.

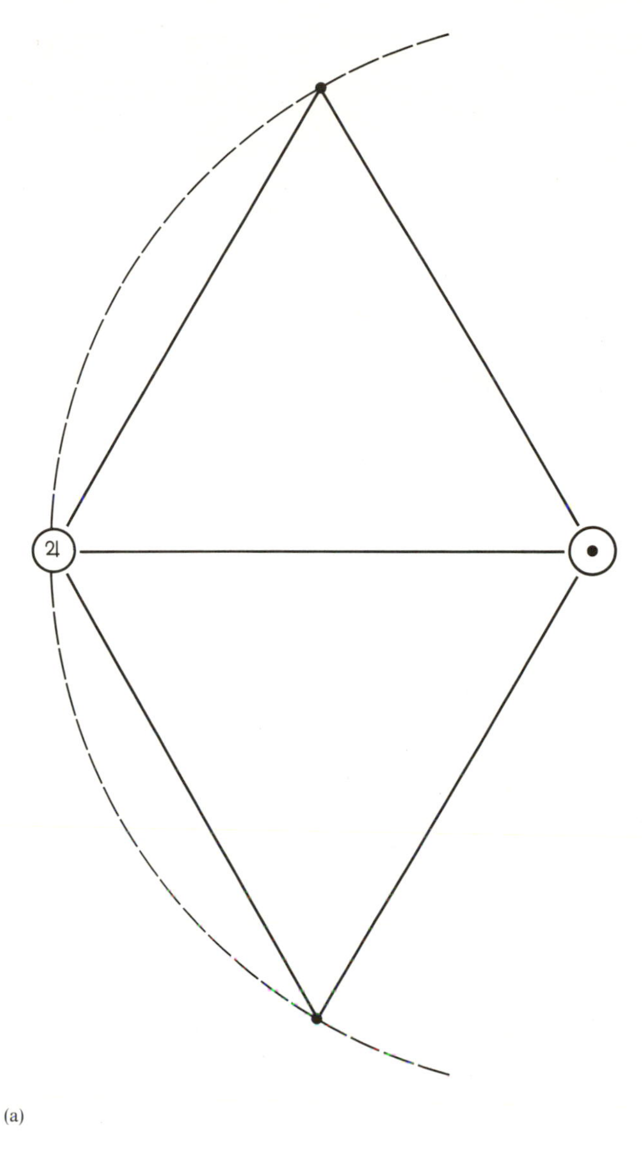

(a)

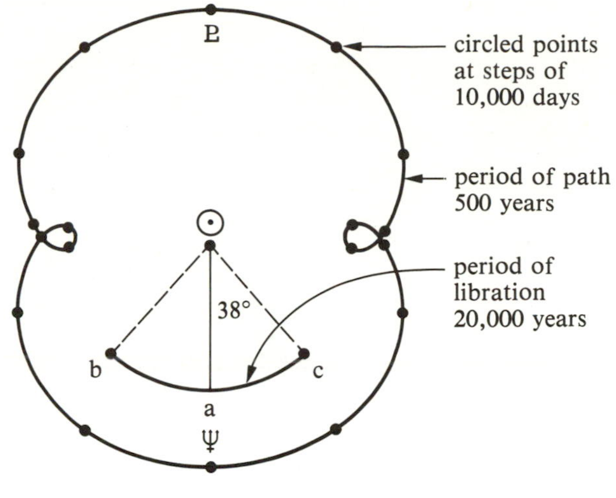

(b)

Figure 18.7. Examples of bodies locked in resonant orbits. (a) The Trojan and Greek asteroids are found near two Lagrangian points which form equilateral triangles with the Sun ⊙ and Jupiter ♃. These two points, which corotate with Jupiter's (nearly) circular motion about the Sun, are places where a small mass can be confined by the action of Coriolis forces against the gravitational attractions of the Sun and Jupiter. Originally, asteroids found near one point were to be named after famous Trojans; those near the other point, after famous Greeks. Carelessness led to some Greeks in the Trojan camp, and *vice-versa.* Homer would have felt vindicated! (b) The orbit of Pluto ♇ and Neptune ♆ as seen by an observer who revolves about the Sun with a speed nearly equal to Neptune's average motion about the Sun. Cohen and Hubbard found in 1965 that, although Pluto's orbit occasionally wanders inside that of Neptune, the two planets would never collide because of a 2:3 resonance.

a state of *local* minimum energy, with the crucial equation being analogous to that which governs the motion of a pendulum. Further tidal evolution causes the moons to spiral outward at rates which keeps the *same* ratio between their orbital periods around Saturn. This mechanism requires the presence of some orbital eccentricity. Estimates of the frictional dissipation for planets and moons contain uncertainties, and it remains unclear whether the three-moon resonance of Jupiter's satellites can be explained in the same fashion. However, as Stanton Peale has emphasized, it must be surely no accident that the relation, $1/\tau_I - 3/\tau_E + 2/\tau_G = 0$, is satisfied by the orbital periods of Io, Europa, and Ganymede about Jupiter to *nine* significant figures!

How are we to understand the *absence* of bodies in certain resonant orbits in the asteroid belt and in the rings of Saturn (Figure 18.8)? It is natural to think that originally the gaps in the rings of Saturn and the asteroid belt were both filled, and that, somehow, gravitational interactions with the moons of Saturn or with Jupiter emptied the affected regions. In 1967, William Jeffreys showed that such purely gravitational theories were in error, and proposed that nongravitational forces (collisions) must play an essential role in clearing out the gaps in the asteroid belt and the rings of Saturn.

Figure 18.9 indicates schematically how such a collisional mechanism might work.* To fix ideas, let us focus on the strongest resonance where the orbital period of a test particle (asteroid or ring particle) is 1/2 that of the perturber (Jupiter or Mimas), although the argument

* The basic ideas were devised by Colombo and Franklin, but our description will follow that proposed by Robert Sanders and James Huntley for a similar problem in barred spiral galaxies. Subsequent work by Douglas Lin and John Papaloizou and by Phil Schwartz also draw on the similarities between dynamical theories of the flattened distributions of matter in galaxies, close binary stars, and the solar system.

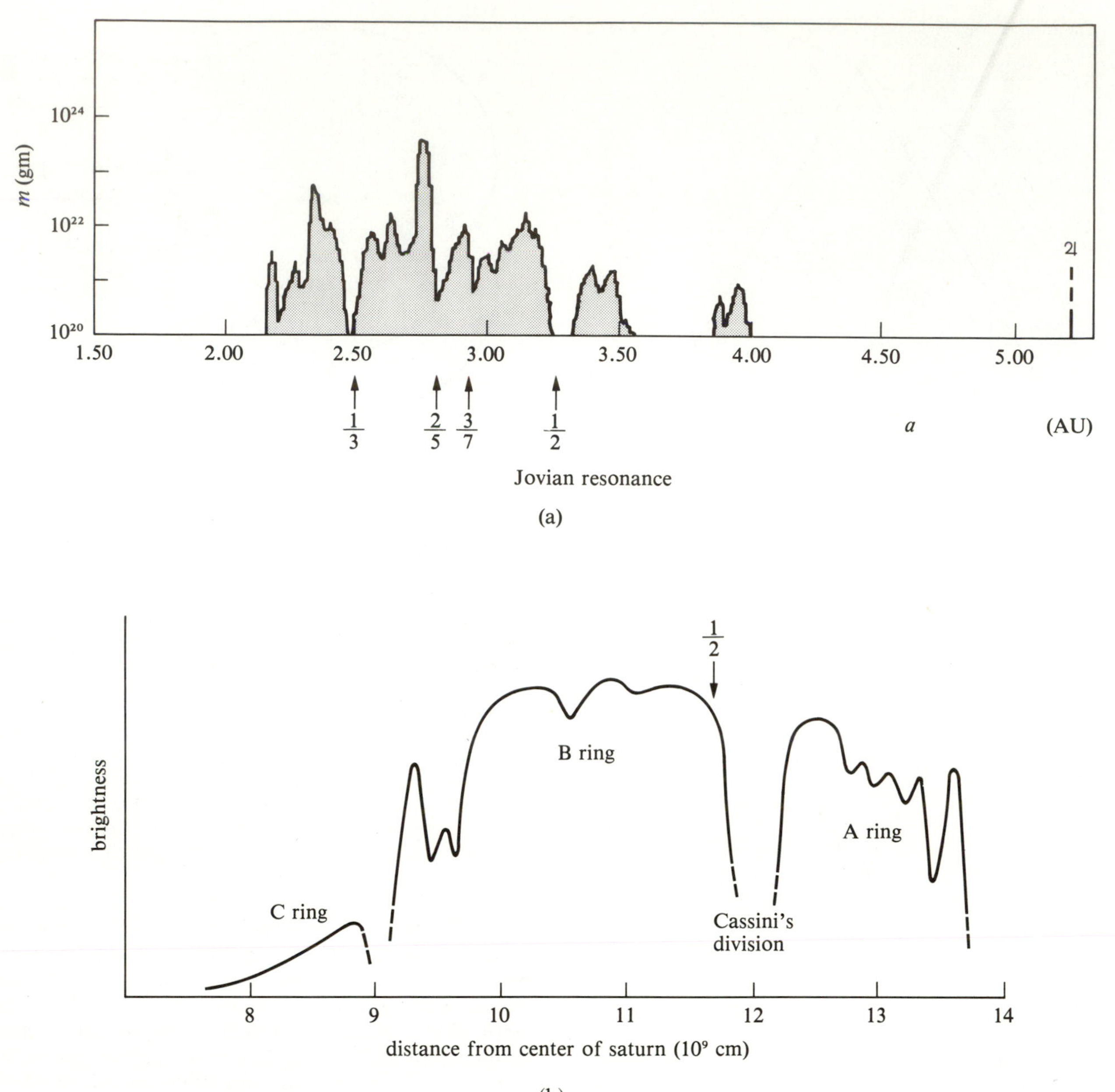

Figure 18.8. Examples of absence of bodies in resonant orbits. (a) The total mass m of asteroids with orbital semimajor axis a. Notice the Kirkwood gaps that exist at positions where the orbital periods of the asteroids would be 1/3, 2/5, 3/7, and 1/2 of the orbital period of Jupiter around the Sun (adapted from Alfven *et al.*, *Nature*, **250**, 1974, 634). (b) The brightness profile of Saturn's rings as seen from Earth. Notice that the inner edge of Cassini's division corresponds to a position where the orbital period of a ring particle would be 1/2 of the orbital period of Mimas around Saturn (adapted from Dollfus, in *The Solar System, vol. III, Planets and Satellites*, ed. by G. P. Kuiper and B. M. Middlehurst, Univ. of Chicago Press, 1961, p. 568; more accurate measurements are given in Gehrels *et al.*, *Science*, **207**, 1980, 434).

(a)

Figure 18.9. (a) A Voyager I photograph of Saturn and its rings. The outer narrow band is the F ring. The broad band inside that is the A ring, which contains a narrow gap, the Encke division, toward its outer edge. The inner edge of the A ring is separated from the outer edge of the B ring by a wide dark band, the Cassini division. The dark band inside the B ring is called the C ring. The shadow of the rings can be seen on the face of the planet. The fact that Saturn can be seen through even the dense B ring indicates that this ring has a fine "lacy" structure, composed of hundreds to thousands of "ringlets." The small black dots ("reseaus") seen throughout the frame are artifacts of the camera system. (NASA, Jet Propulsion Laboratory.)

should be generalizable to the other resonances for a successful theory of the origin of the gaps. On either side of the 1/2 resonance, the periodic pull of the perturber will cause a circular distribution of test particles to suffer an oval distortion. In the frame of the perturber, if the test particles were collisionless, the long axes of the ovals on either side of the exactly resonant orbit would tend to align perpendicular to one another. In the actual situation, collisions would occur for intersecting orbits, and such collisions would depopulate the intersecting

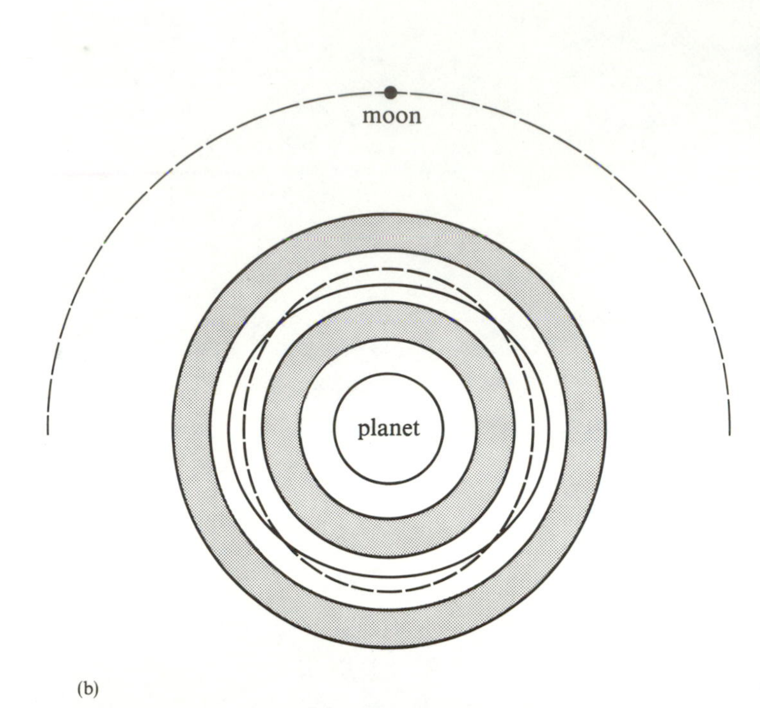

(b)

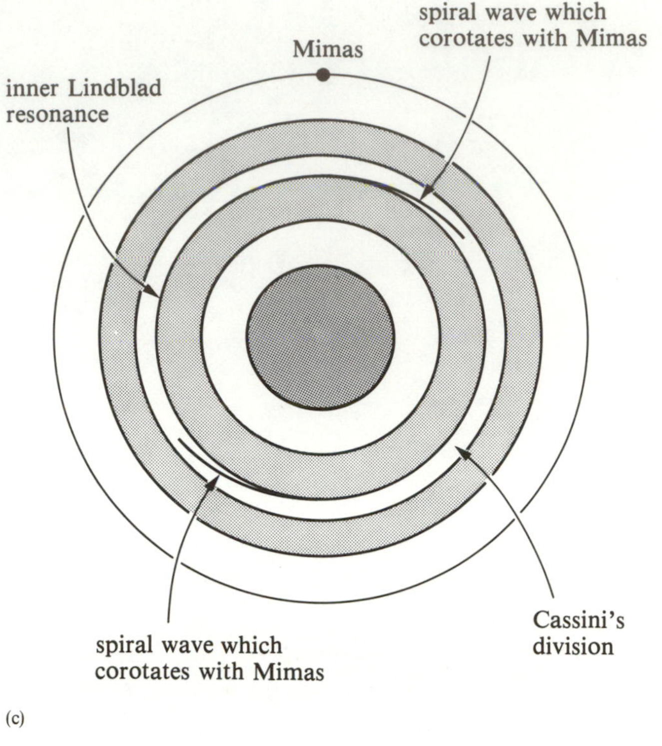

(c)

Figure 18.9. (b) The collisional mechanism for creating gaps at the locations of resonances associated with outer moons has been found inadequate to explain the prominent Cassini division. (c) Resonantly driven density waves have also been invoked to clear Cassini's division. Although this mechanism's ability to clear large gaps remains controversial, recent analysis has established that spiral density waves do exist in some parts of Saturn's rings.

orbits until a gap was produced. The width of the gap as a fraction of its radius would be given roughly by $(M_p/M_c)^{1/2}$, where M_p is the mass of the perturber and M_c is the mass of the central body. The square root of the mass ratio of Jupiter to the Sun is about 0.03, and this value of $(M_p/M_c)^{1/2}$ is consistent with the observed fractional width of the Kirkwood gap associated with the 1/2 resonance (Figure 18.8a).

For Saturn's rings, the above explanation encounters difficulties. The square root of the mass ratio of Mimas to Saturn is about 0.0003, which is 100 times smaller than the observed fractional width of Cassini's division. Moreover, the 1/2 resonant condition applies not to the center of the division, but to its *inside* edge (Figure 18.8b). Collisions acting alone would clear out a gap more or less equally on both sides of the resonant orbit. To explain these discrepancies, Peter Goldreich and Scott Tremaine have invoked the help of the *self-gravity* of the rings. The creation of oval distortions in a self-gravitating disk of matter is, of course, reminiscent of the density-wave theory of spiral structure in disk galaxies (see Figure 12.25). The specific proposal of Goldreich and Tremaine is that a spiral density wave is excited near the position of the 1/2 resonance of Mimas, where the ring particles suffer a particularly strong oval distortion. In the language of density-wave theory, the 1/2 Mimas resonance is the inner Lindblad resonance of the spiral wave with two arms (Chapter 12 and Problem 18.7). The mass ratio of the rings of Saturn to Saturn itself is much smaller than the mass ratio of the disk to the central bulge of a spiral galaxy; consequently, the spiral waves which are excited in Saturn's rings are correspondingly more tightly wrapped. The transfer of angular momentum from the ring particles to the spiral wave to the orbital motion of Mimas would gradually deplete the ring particles in the region where the spiral waves could originally propagate. Since spiral waves exist only on the *outside* of an inner Lindblad resonance, the *inner* edge of Cassini's division would naturally fall on the position of the 1/2 Mimas resonance (Figure 18.10).

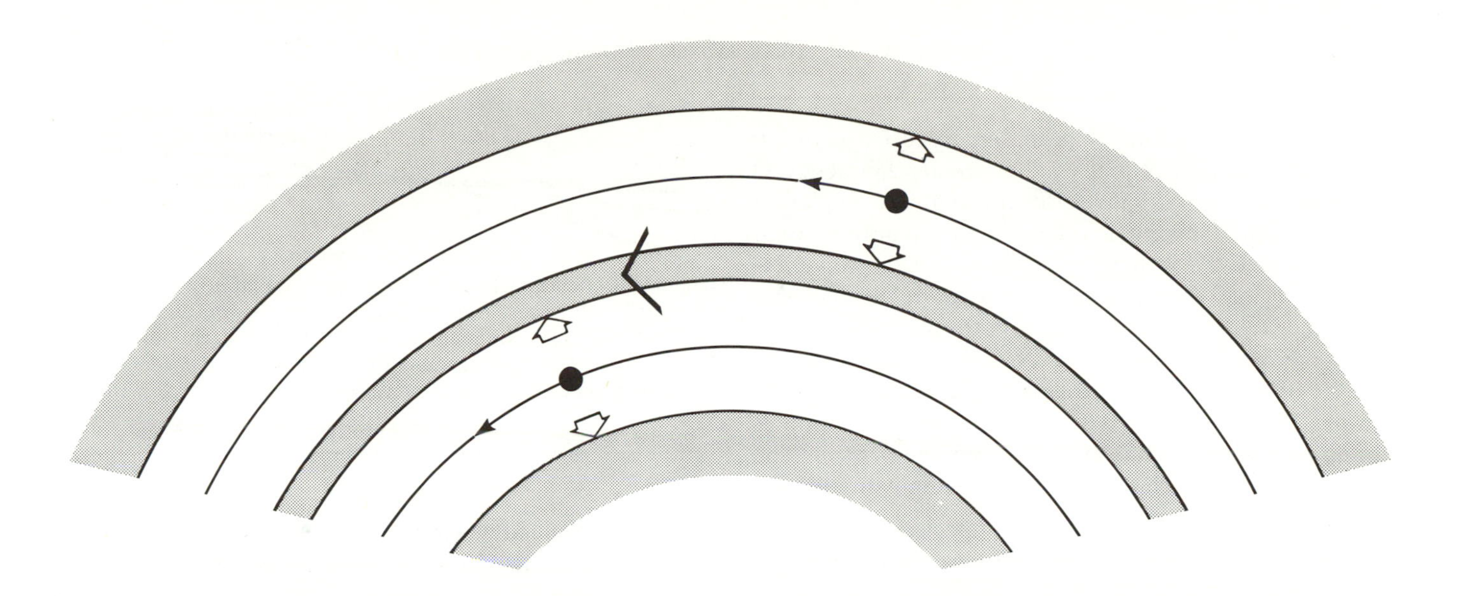

(a)

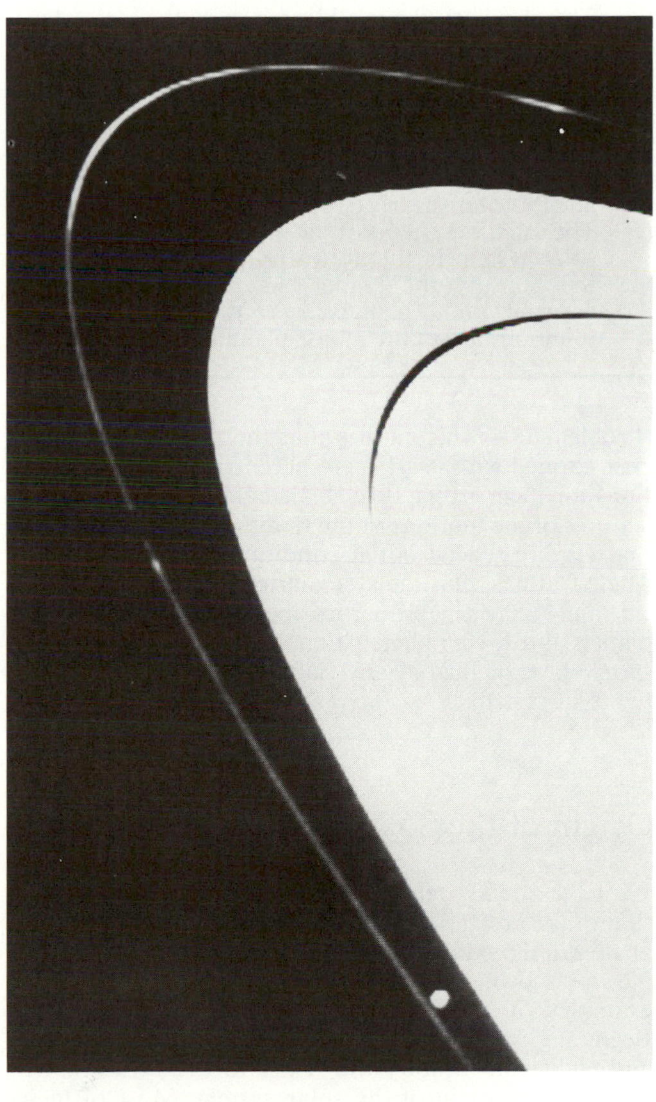

(b)

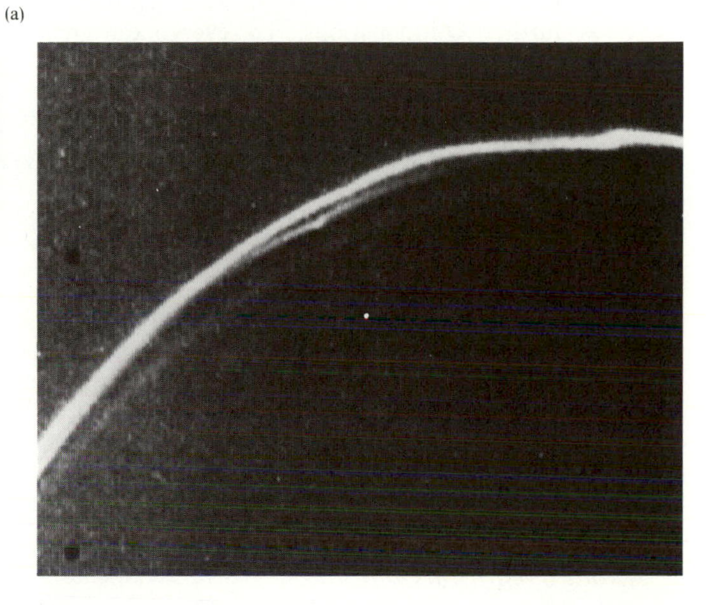

(c)

Figure 18.10. (a) A mechanism for creating gaps in rings relies on the action of massive bodies orbiting in a circle within the ring system. Such a body tends to give angular momentum to the ring particles outside its own radial position and to remove angular momentum from the ring particles inside. In net, this results in ring particles being moved away in both directions from the radius of the orbit of the body. If two fairly large and fairly closely spaced bodies lie on either side of ring material, they could "shepherd" the ring particles to form a narrow ring. (b) The "shepherd" theory of narrow planetary rings received strong observational support from the discovery of Voyager 1 that two small moons lay on either side of Saturn's F ring. The inner shepherd of the F ring can be seen at the bottom right of this photograph. The apparent hole (middle left) in the F ring is an artifact produced by a reseau mark engraved on the Voyager cameras. (NASA, Jet Propulsion Laboratory.) (c) In a close-up view, the F ring has a complex "braided" structure, which may arise because of the details of its resonant interaction with its shepherds. (NASA, Jet Propulsion Laboratory.)

Problem 18.7. Assume that the epicyclic frequency κ in Saturn's rings is equal to the circular frequency Ω because the ring particles perform essentially Keplerian motion (of small eccentricity) in the central gravitational field provided by Saturn. Argue now that the wave frequency Ω_p of the disturbance caused by Mimas will equal the (mean) angular velocity of the latter. Show that the condition that the orbital period, $2\pi/\Omega$, is equal to 1/2 the orbital period of Mimas, $2\pi/\Omega_p$, corresponds to stating that the ring particle is located at the inner Lindblad resonance of the two-armed spiral wave. The waves that Goldreich and Tremaine actually invoke are the long trailing waves (see Problem 12.11).

The data gathered by Voyager 1 from Saturn has revealed further complications. To everyone's great surprise, Saturn's rings on close inspection broke into many closely spaced "ringlets," separated occasionally by larger gaps or divisions (see Figure 18.9a). Saturn also has an outer faint and very thin ring (the F ring), which may be similar to the thin faint rings of Uranus (Figure 18.10a). Goldreich and Tremaine proposed that such thin rings could be prevented from spreading radially if the ring particles were "shepherded" by two small moons on either side (Figure 18.10a). Strong support for this theory came from the finding that the F ring of Saturn is indeed shepherded by two such small moons (Figure 18.10b). It is conceivable that similar mechanisms on a more modest scale by kilometer-sized "moonlets" contained within the main body of Saturn's ring system explains the ringlet structures.

Regularities in the Solar System Caused by Initial Conditions

Having discussed the regularities in the solar system which could have arisen by dynamical evolution from an unexceptional initial state, let us now summarize the regularities of the solar system that are not explained by Newton's *laws* of mechanics, but must be imposed as *initial conditions*. For example, Newton's derivation of Kepler's three laws allows for arbitrary eccentricities and inclinations of the orbits of the planets about the Sun. Yet, except for Mercury and Pluto, whose orbits have clearly been influenced by resonances, the eccentricities and inclinations of the planets' orbits are all nearly zero (Figure 18.11).* Since eccentricities and inclinations are conserved quantities in Newton's theory (Problem 18.4), this implies that the planets must have been formed with nearly zero eccentricities and inclinations for their orbits. Box 18.2 summarizes the striking features of the motions of the bodies in the solar system that require special initial conditions to explain them. These features, plus the systematic differences in chemical and physical properties between the terrestrial planets and Jovian planets that we noted in Chapter 17, should be explained by any viable theory of the origin of the solar system.

* Mars's eccentricity, currently 0.093, has been calculated to vary from 0.004 to 0.14 because of long-period oscillations associated with a near-resonant interaction with Jupiter and Earth: $3/\tau_J - 8/\tau_M + 4/\tau_E \simeq 0$.

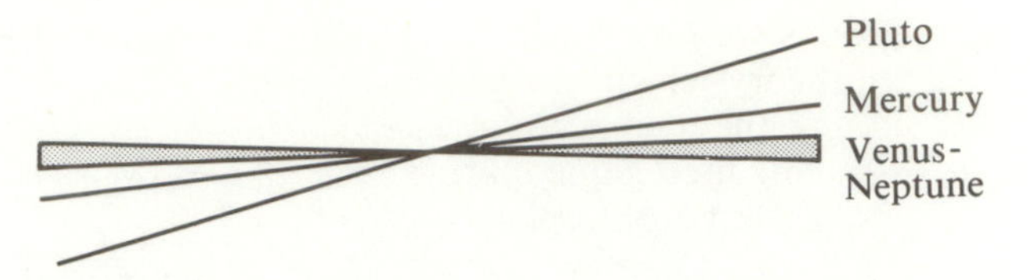

Figure 18.11. The inclinations of the orbital planes of the planets. Venus, Earth, Mars, Jupiter, Saturn, Uranus, and Neptune all have orbital planes which lie within 3.5° of each other. The orbital planes of Mercury and Pluto are more inclined relative to the ecliptic; $i = 7°$ for Mercury, and $i = 17.2°$ for Pluto.

BOX 18.2
Regularities in the Solar System that Require Special Initial Conditions

(a) The planetary orbits all lie nearly in the same plane and take place with the same sense of revolution.
(b) The spin of the Sun is also nearly in the plane of the ecliptic.
(c) The spins of the planets (except for Uranus) are also nearly in the same plane.
(d) The orbital shapes of the planets (except for Mercury and Pluto) are nearly circular, i.e., of low eccentricity.
(e) The satellite systems of the major planets mimic the solar system in the above properties if we replace the Sun by the planet. As a general analogy, asteroids are to planets are to the Sun as rings are to moons are to the major planets.

Origin of the Solar System

We have already remarked that Newton felt the question of the creation of the solar system did not have an answer in mechanical terms. Fortunately, partly because Newton's own work provided such splendid counterexamples, the spirit of enquiry could not be squelched. Beginning with Descartes, scientists of the seventeenth and eighteenth century did begin to conjecture about the possible origin of the solar system. Most of these speculations we now know to be premature. Nevertheless, the theory which has gained the most wide acceptance today, the nebular hypothesis for the origin of the solar system, first gained prominence because of the writings of Kant and Laplace on the subject.

Problem 18.8. Many of the early theories relied on some catastrophic event. For example, Newton's successor at Cambridge, William Whitson, proposed that the Biblical flood resulted from the gravitational tides caused by a passing comet. Buffon made this proposal even more grandiose by supposing that a comet passed closely enough to the Sun to draw out a small fraction of the latter's mass. This string of material went into orbit about the Sun, cooled, and eventually condensed into the planets. When it became generally recognized that comets had much too little mass to accomplish the desired trick, others proposed that a passing star performed the same service in creating the solar system. Discuss the merits and failings of this proposal.

The Nebular Hypothesis

The modern versions of Kant and Laplace's nebular theory contain the features outlined in Box 18.3.

Most of the general ideas in the Box follow naturally from modern concepts and calculations of the collapse of rotating, gravitationally bound, interstellar clouds. They are speculative, but not wildly so. The ideas in point (c) are more controversial; they are attempts to answer at least two major additional questions, as follows.

(a) What caused the differences between the terrestrial planets (small and dense) and the Jovian planets (large and rarefied)?

(b) What is the explanation for the spacing of the orbits (the qualitative expression of "Bode's law") wherein the terrestrial planets have small spacings, and the Jovian planets have large spacings?

Difference Between Terrestrial and Jovian Planets: The Condensation Theory

In Chapter 17, we argued that the average density of 4 or 5 gm/cm^3 for the terrestrial planets implied that they must be made mostly of rocky materials in their mantles and iron-nickel in their cores. The average density of 1 or 2 gm/cm^3 for the Jovian planets imply that they must be made mostly of hydrogen and helium, and so are more similar than the terrestrial planets to the "cosmic abundance" characteristic of the Sun and the rest of the universe. The chemical difference between the terrestrial planets and the Jovian planets suggests that gravity did not act alone to condense the planets from the primitive solar nebula. Gravitation is universal, and will not distinguish (in a well-mixed gas) between hydrogen, helium, iron, or uranium.

However, in all reasonable models of the primitive solar nebula, such as those constructed in recent years by A. G. W. Cameron, the temperature is lower in the outer radial regions of the solar nebula than in the inner

---

BOX 18.3
Nebular Hypothesis for the Origin of the Solar System

(a) A rotating nebula forms from a contracting or collapsing interstellar gas cloud. The tendency to conserve angular momentum causes the gas cloud to spin faster and to flatten. Under some conditions, the gas cloud breaks into two or more lumps, which subsequently form a binary or multiple-star system (Figure 18.12). Under other conditions, the rotating nebula forms with a central concentration (protosun) which is surrounded by a rotating disk of gas (and dust). There may be a prolonged phase of accretion in this configuration (Figure 18.13).

(b) The rotating disk later condenses into many small pieces (asteroids and protoplanets) which grow into full-scale planets by capturing neighboring pieces of material. The formation of the solar system under these dissipative circumstances may explain why the planets currently contain about 98 percent of the total angular momentum of the system, while the Sun contains 99.9 percent of the total mass. The dissipation present in a gaseous medium would also explain why the orbits and spins of the planets are nearly coplanar and aligned, with the orbits being nearly circular.

(c) Insofar as the planets grow by the accretion of solids, the Jovian planets get a head start, because they form in the outer and cooler regions of the solar nebula. The Jovian planets acquire large-enough masses that they begin to gather gravitationally much of the gases near them in the solar nebula; this would explain both their giant sizes and the large spacings between planets. The terrestrial planets get a late start and never acquire a large fraction of the cosmic matter before the solar wind sweeps the system clean of gas and small solid particles. The asteroid belt is all that remains to indicate the larger debris that used to litter much of the solar system.

(d) A scaled-down version of this scenario may have led to the formation of the satellite and ring systems of the giant planets.

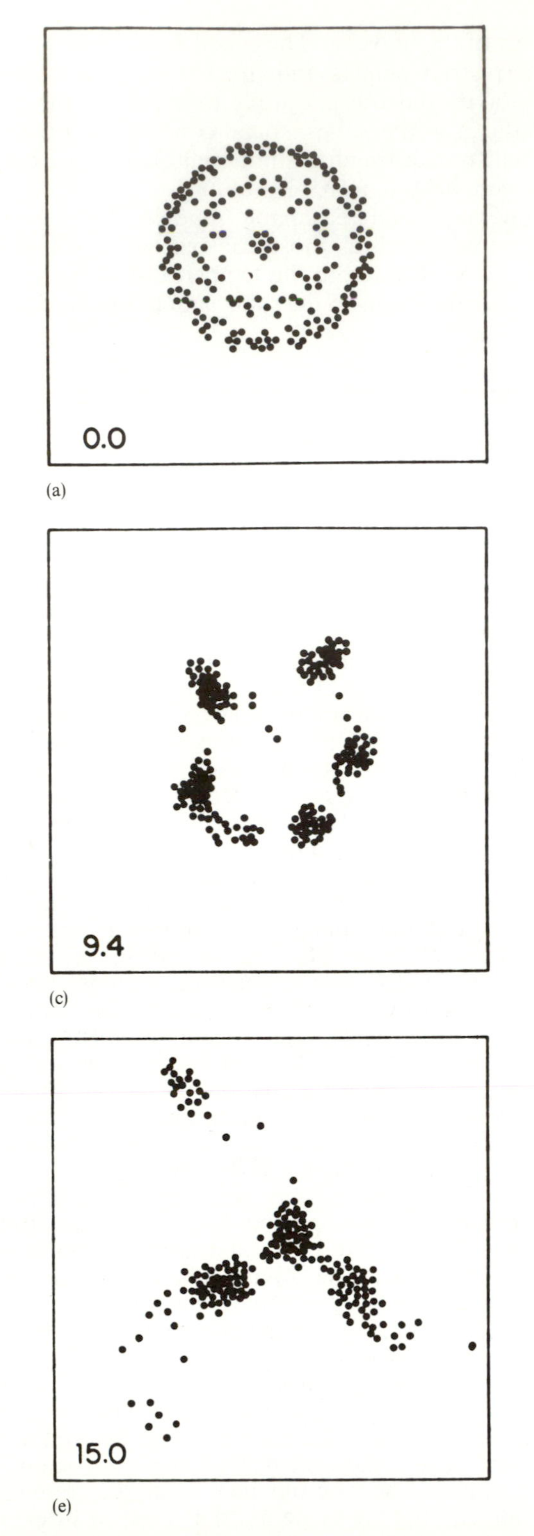

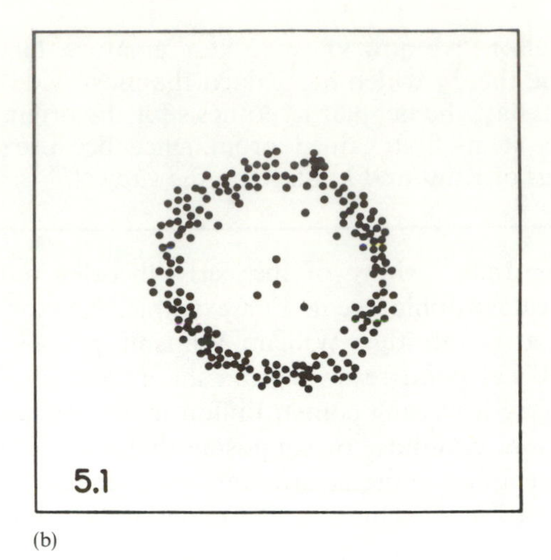

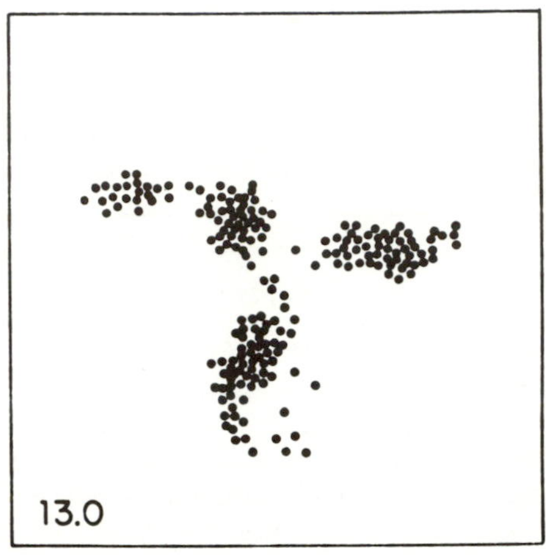

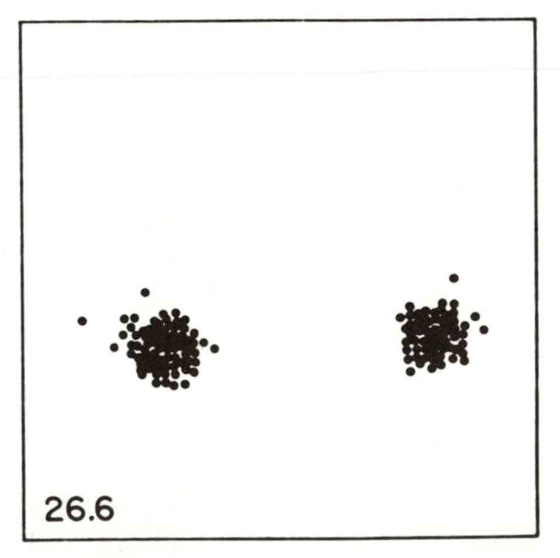

Figure 18.12. A numerical simulation of the collapse of a rotating mass of gas. After a short time, the mass breaks up into several pieces, which subsequently eject some matter as well as coalesce into two stars that orbit about a common center of mass. (Adapted from L. Lucy, *IAU Symp. No. 88*, 1979, pp. 9–11.)

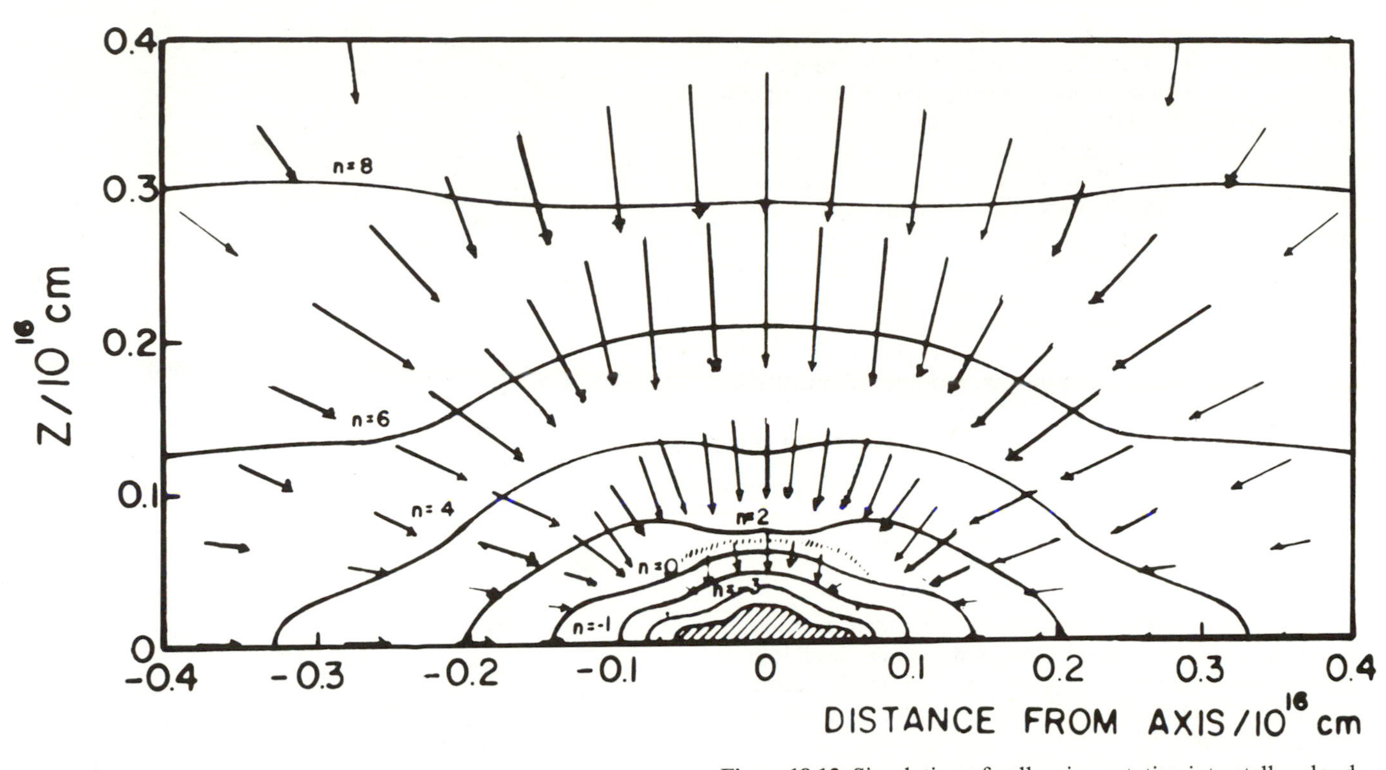

Figure 18.13. Simulation of collapsing rotating interstellar cloud showing disk formation together with an accretion flow. (From Regev and Shaviv, *Ap. J.*, **245**, 1981, 955.)

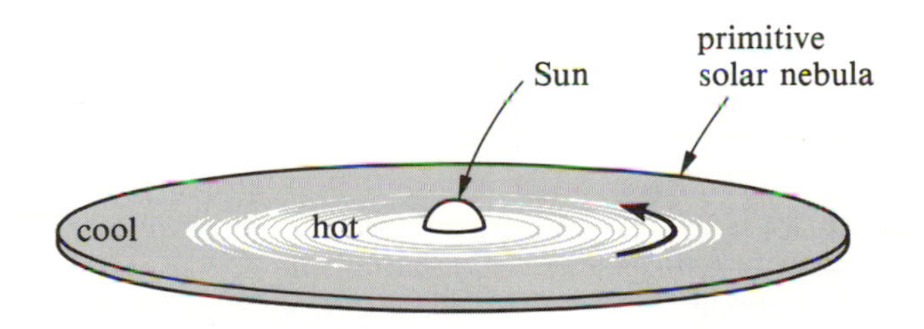

Figure 18.14. Schematic depiction of the early solar system, showing the protosun and the primitive solar nebula. The inner portions of the disk of the solar nebula can be expected on general grounds to be hotter than the outer portions.

Table 18.1 **The expected condensation sequence under conditions of local thermodynamic equilibrium.**[a]

Temperature (K)	*Condensates*	*Elements*	*Fractional abundance*
1400–2000	Calcium-aluminum oxides	Ca, Al, O	10^{-6}
1200–1800	Magnesium silicate and iron-nickel alloy	Mg, Si, O Fe, Ni	10^{-4}
1000–1600	Sodium-potasium-aluminum silicates	Na, K, Al Si, O	10^{-6}
600–700	Iron sulfide	Fe, S	10^{-5}
100–200	Ices of methane, ammonia, and water	C, N, O, (and H)	10^{-3}
less than 100	solid neon and argon	Ne, Ar	10^{-4}

[a] Adapted from S. S. Barshay and J. S. Lewis, *Ann. Rev. Astr. Ap.*, **14**, 1976, 81.

regions (Figure 18.14). At low-enough temperatures, solids can condense via molecular forces out of the gas. Alternatively, at low temperatures, the dust grains present originally in the interstellar cloud from which the solar nebula formed will not evaporate. In the hot inner portions of the solar nebula, John Lewis has shown that only the most rocky sort of materials (silicates) can condense out at first. In the cool outer portions, more volatile compounds—like the ices of water, methane, and ammonia—can also condense out. Table 18.1 gives the condensation sequence for some of the more common compounds.

The conjecture proceeds, then, that as the solar nebula cooled overall, solids condensed out, forming small pieces of rocky and iron-bearing material. These small objects are the asteroidal bodies. They would form first in the outer regions of the solar nebula, where it is generally cooler. Later, even more volatile compounds could condense in the outer solar nebula as the temperature dropped below the requisite values. Meanwhile, the temperature in the inner parts of the solar nebula dropped low enough to allow the condensation of silicates, perhaps even of iron sulfide. But the empirical evidence argues that the temperature never got low enough to allow the condensation of the ices of water, methane, or ammonia. If planets subsequently formed from the aggregation of the small solid bodies (asteroids or planetesimals), we would have a natural explanation of why the inner, terrestrial planets are composed mostly of rocky and iron-bearing materials, whereas the outer, Jovian planets contain, in addition, a much larger fraction of hydrogen-rich compounds.

The Sizes of the Planets and Their Orbital Spacings

Carrying the above ideas further, we may even attempt to explain why terrestrial planets are small, and Jovian planets are large. The explanation we prefer accounts also for the small and large spacings of the terrestrial and Jovian planets.

The basic idea is that the outer planets get a head start because solids condense first in the cooler outer parts of the solar nebula; small pieces of solid matter would collide, aggregate, and eventually become large enough to be self-gravitating. At this point, the protoplanet could begin to gather the neighboring solid material *gravitationally*. If the planet grew massive enough, its gravity might become large enough for it to capture and to hold the neighboring hydrogen and helium *gas*. Such planets would grow to giant sizes by sweeping up the material in a large part of their surroundings, leaving little to form additional planets; so the spacing between the giant Jovian planets would automatically be large.

In contrast, the inner planets got a late start, because the hotter, inner parts of the solar nebula had to wait longer for the condensation of solid matter. Moreover, the major fraction of the condensable compounds (the hydrogen-carrying ices) never got a chance to form solids. Thus, the terrestrial planets would naturally be small, and they would never become massive enough to accrete much gas. Powerful solar winds may turn on roughly 10^5 years after the formation of the protosun, and this blast would probably sweep the solar system clean of gas and dust. The remaining asteroidal bodies either collided with the planets, or were ejected gravitationally from the solar system by the perturbations from the giant planets. Only the asteroid belt remains today to tell us how much litter was present when the solar system was young.

Quantitative calculations have been carried out to model the processes described briefly above. Although such calculations are typically quite simplified and still contain great uncertainties, they can yield models which are quite suggestive of the actual solar system (Figure 18.15). No one has succeeded in producing a model in which all the details are right, but that is probably because we don't know enough about all the important physical processes and initial conditions. All in all, the general conclusions derivable from the nebular hypothesis seem reasonably in accord with the known facts. In particular, the nebular hypothesis does not require extraordinary physical processes or circumstances that would be unique to our little corner of the universe; so the formation of a solar system is probably a quite common event, one to be expected when a (slowly) rotating interstellar cloud contracts gravitationally to form a star (see Chapter 10). A more rapidly rotating cloud might collapse to form a binary star system, but a slowly rotating cloud may well always prefer to form a star plus planets and debris.

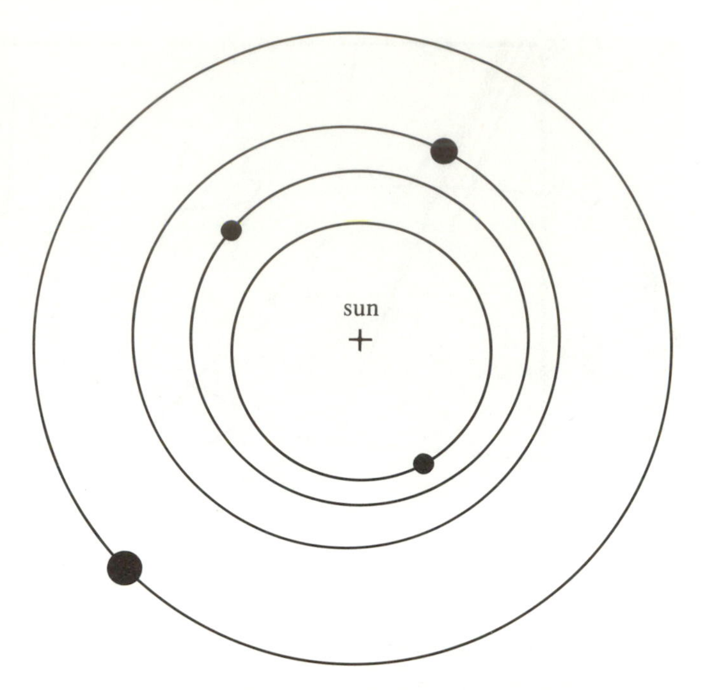

Figure 18.15. One numerical simulation of the formation of the terrestrial planets from the accumulation of planetesimals shows that the process is completed in roughly 10^8 yr. The result is four rocky planets that travel in slightly eccentric orbits about the Sun nearly in the same plane and with the same sense of revolution. (For further details, see G. W. Wetherill, *Scientific American*, **244**, June 1981, 162.)

Moons, Rings, and Comets

Where do the moons, rings, and comets fit into this picture? Given the great similarities between the moons and rings around planets and the planets and asteroids

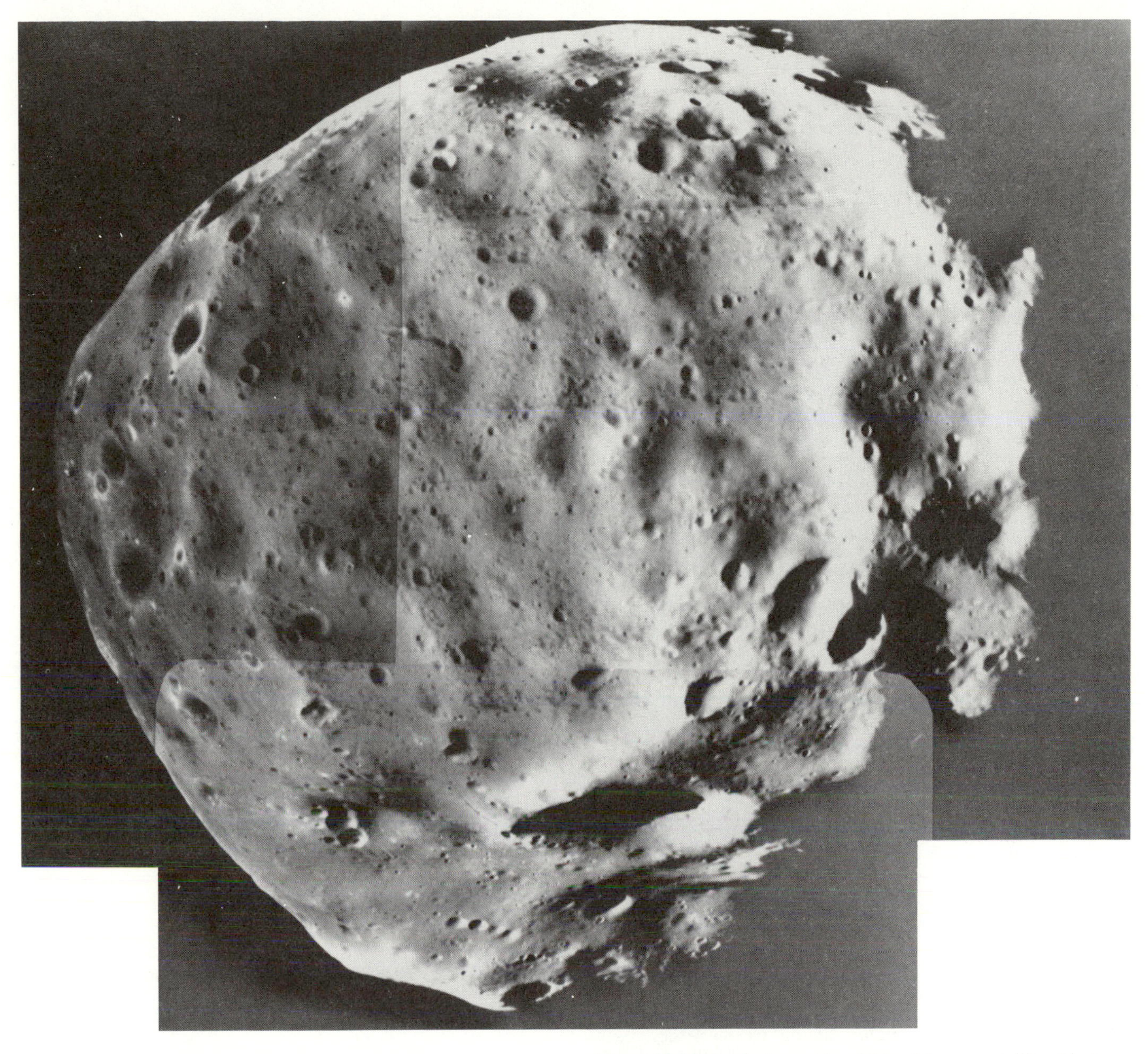

Figure 18.16. A photomosaic of Mars' larger satellite, Phobos, produced from pictures taken by Viking Orbiter 1 from a range of 300–400 miles. (NASA.)

around the Sun, there were probably protoplanetary disks around planets analogous to the protostellar disk (solar nebula) that formed around the Sun. Then, the same processes that formed the planets and asteroids could, on a smaller scale, have formed the moons and rings around the Jovian planets. In particular, the satellites of Jupiter and Saturn can be divided into two categories: regular and irregular. The orbits of the regular satellites about Jupiter and Saturn mimic the motions of the planets about the Sun, and these moons probably formed from protoplanetary disks. The irregular satellites are relatively small and may be captured asteroids.

The moons of the Earth and Mars may have had quite different origins, but this topic is controversial. The double-planet system of the Earth and the Moon may have an origin analogous to that of double stars, whatever the latter might be. The two moons of Mars—Phobos and Deimos (Figure 18.16)—resemble nothing

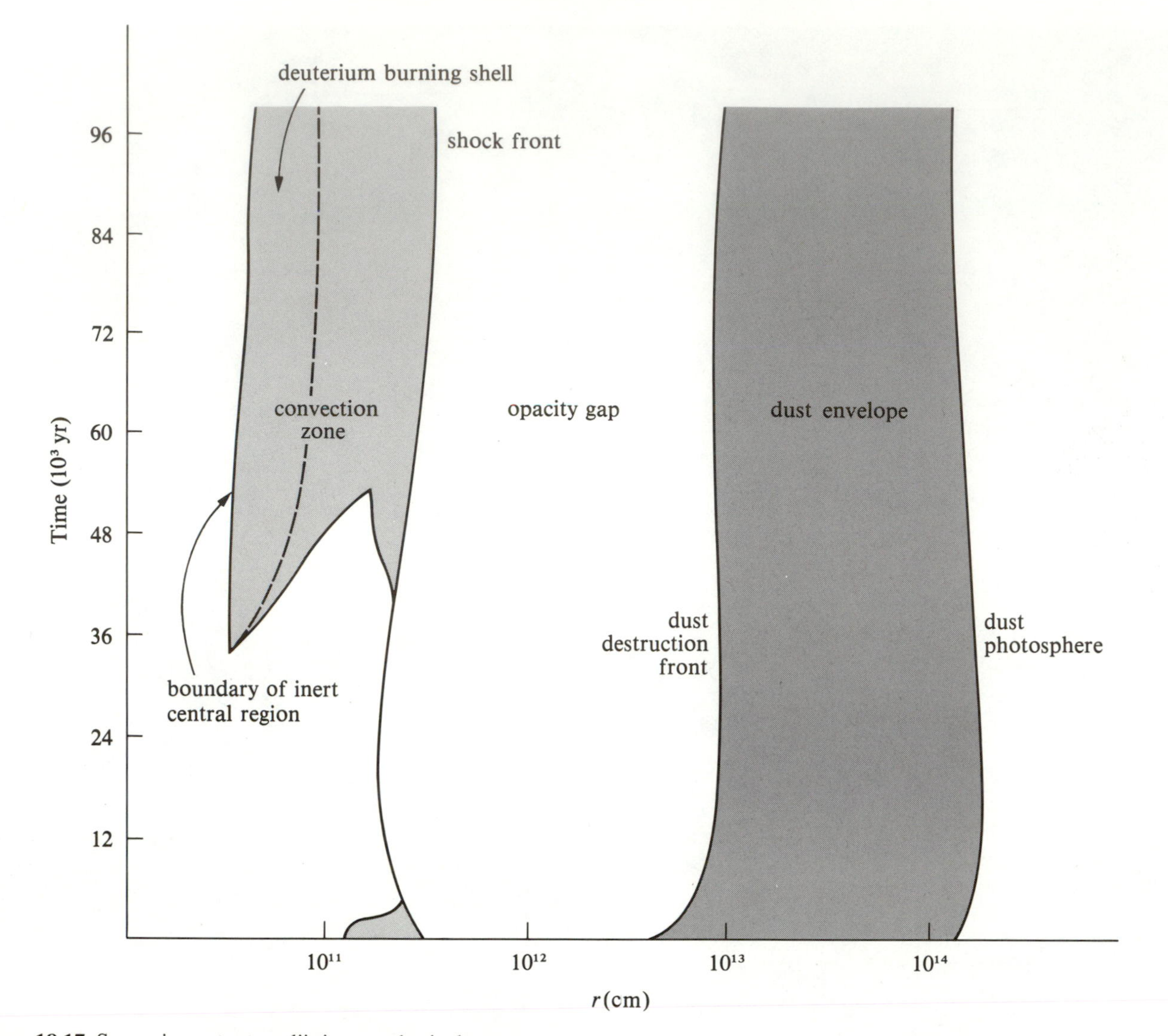

Figure 18.17. Some important radii in a spherical protostar calculation. Plotted as functions of time are the radii of the newly forming star, where inflowing graphite grains are destroyed, and where the outflowing radiation is degraded into infrared radiation of sufficiently long wavelengths to escape freely from the system. Notice that dust grains can survive until they reach within about 10^{13} cm of the protosun. Although this result was derived for a purely spherically symmetric calculation, it should hold approximately for the more complex situation applicable to the formation of the solar system. In particular, dust grains which are held by centrifugal forces outside such a radius may avoid destruction altogether and may be able to agglomerate to form larger bodies. (After Stahler, Shu, and Taam, *Ap. J.*, **241**, 1980, 637; see also Fig. 11.45.)

so much as captured asteroids, and they may well be just that.

The formation of the comets is also obscure. They may have formed at their present large average distances from the Sun; or they may have formed further in, and been thrown out to their present great distances by gravitational encounters with the giant planets (especially Jupiter). In studying the solar system, astronomers and space scientists are beginning to see the outlines of a coherent story of its origin, but further work is needed.

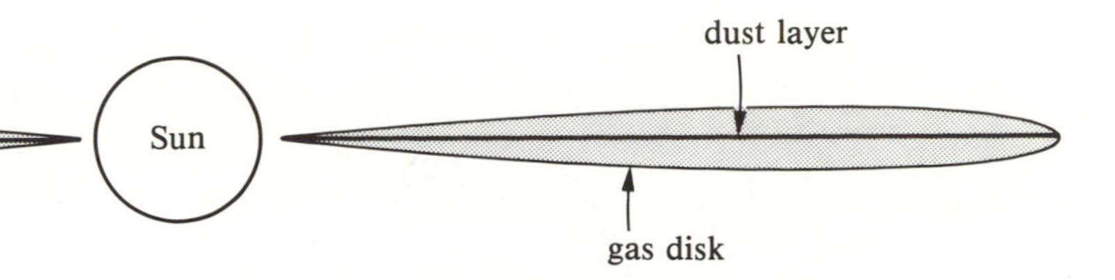

Figure 18.18. The dust grains which survive entry into the primitive solar nebula may settle into the midplane of the disk and form a thin layer. When the layer of solid matter has acquired enough surface density, it may become gravitationally unstable, and fragment into asteroidal-sized bodies.

It is difficult at times to distinguish the accidental features of the solar system—those which arise from special initial conditions—from the truly fundamental features that would arise from any reasonable initial conditions. This is the problem, of course, which always arises when one has only one known system to study. The discovery of just one other star with planets would improve the situation immeasurably.

Condensation versus Agglomeration of Presolar Grains

Exactly how would the first substantial solid bodies in the solar nebula have formed? The simplest assumption to make chemically is that the primitive solar nebula started as a hot gas of homogeneous ("cosmic") abundance from which solid particles condensed directly when the nebula cooled. This assumption, however, ignores the observed fact that small solid particles exist as dust grains in interstellar clouds. Simple, but detailed, calculations of the process of star formation (see Chapter 11) show that the infalling material which does not approach within roughly 10^{13} cm of the protosun may never reach temperatures high enough to destroy the refactory cores of such interstellar grains (Figure 18.17).

Cameron has suggested that the agglomeration of interstellar (or presolar) grains may have led to the formation of millimeter- to centimeter-sized solid particles. (Such particles exist in the rings of Saturn today.) These particles of sand would have settled toward the midplane of the gaseous solar nebula in 10^4 to 10^5 years. A thin layer of solid particles with small random velocities would become gravitationally unstable once their accumulated surface mass density exceeded a certain critical value (Problem 18.9). Goldreich and Ward proposed that such an instability would cause the layer to separate into bodies of the masses of large asteroids (Figure 18.18). The further collisional and gravitational agglomeration of such planetesimals would ultimately form the bodies of the terrestrial planets and the cores of the Jovian planets. The subsequent evolution would then proceed along the lines of standard nebular theory.

Problem 18.9. According to Problem 12.11, if the velocity dispersion a of a disk of particles of surface density μ and local epicyclic frequency κ (= the circular angular speed Ω, for a Keplerian disk) becomes less than the value $\pi G\mu/\kappa$, the disk will be unstable to axisymmetric waves of wavenumber k roughly given by $\kappa^2/\pi G\mu$. In a dissipative solar nebular disk, a flattened distribution of solid particles will become unstable, therefore, when the surface mass density μ becomes high enough. Suppose a critical condition is reached when a mass $\varepsilon M_\odot$ of solid particles is spread out uniformly over a disk of radius R centered on the Sun with mass $M_\odot$. The average surface density μ in this disk equals $\varepsilon M_\odot/\pi R^2$, and if $\varepsilon \ll 1$, the square of the epicyclic frequency κ^2 at the midpoint of the disk is given by $GM_\odot/(R/2)^{-3}$. Show that the critical wavenumber has an inverse given by $\pi G\mu/\kappa^2 = \varepsilon R/8$, and that the mass associated with this characteristic length scale is given by $M = \mu\pi k^{-2} = \varepsilon^3 M_\odot/64$. Calculate M for $\varepsilon = 3 \times 10^{-5}$ (assuming 0.1 percent of the mass of the solar nebula to be in the form of refractory cores of dust grains). Compare this value with the mass of an asteroid of diameter 10 km.

There are a number of possible advantages to this picture. First, the survival of interstellar graphite grains and their subsequent agglomeration into larger bodies may explain the origin of the carbonaceous chondrites. Barshay and Lewis have stated that these carbon-rich meteorites are difficult to understand in terms of condensation in a cosmic gas where oxygen is more abundant than carbon. Second, if small pieces of solid matter can maintain some chemical integrity, then Clayton has proposed a simple way to understand various chemical anomalies that have been found in the Allende meteorite by Wasserburg and his colleagues. Clayton notes that different stellar processes (supernovae, novae, planetary nebulae, etc.; see Chapter 11) may have allowed a variety of interstellar grains with different isotopic compositions to accumulate in the interstellar medium. If this variety

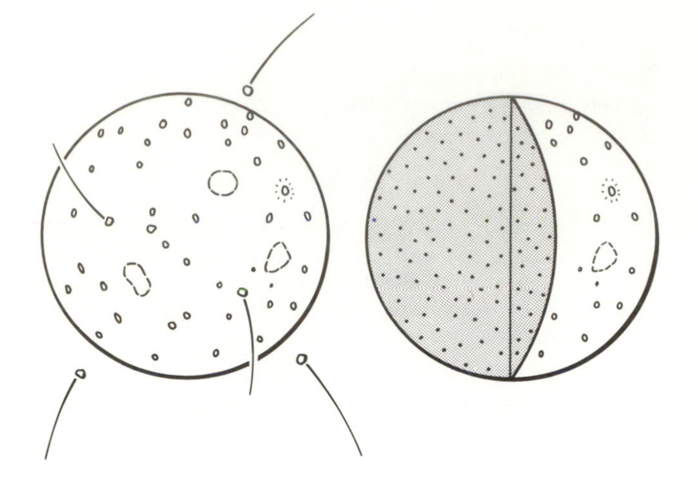

Figure 18.19. Modern theories for the origin of the Earth picture it to have accumulated by the accretion of many small asteroidal bodies (or planetesimals). The primitive Earth formed this way is likely to have been homogeneous on a large scale. Such an interior structure would contrast greatly with the present interior structure, which is highly differentiated (compare with Figure 17.13).

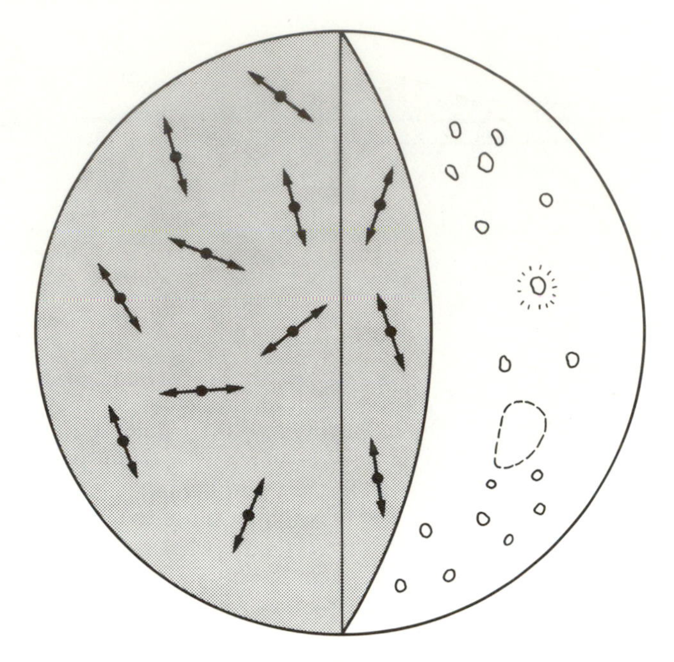

Figure 18.20. Radioactive decay releases fast particles and is an important source of heat for the interior of the Earth, particularly when it was young.

can maintain some integrity in the formation of the solar system, then some isotopic anomalies are bound to occur in meteorites. Alternative proposals for such anomalies rely on the injection of supernova-processed material just before the collapse of the presolar interstellar cloud. Clearly, there are still controversial opinions around on the origin of the solar system!

Origin and Evolution of the Earth

The Melting of the Solid Earth

In Chapter 17, we mentioned that the differentiated interior structure of the Earth strongly implied that it must have been molten at one time. How else could the settling of the heavy material to the bottom have occurred? However, the nebular theory of the origin of the planets suggests that the solid Earth was formed in a relatively cool and homogeneous state by the agglomeration of asteroidal stones and irons. The gradual accumulation of planetesimals during 10^5 to 10^8 years would have released gravitational energy, but most of the energy release would have occurred near the surface of the planet, where it is easily radiated away.* Geophysicists who have calculated the accretional and compressional heating conclude that the resultant central temperature may only have been about 1,500 K (Figure 18.19; compare with Figure 17.13 of the present interior structure). This is not enough to cause the Earth to melt.

* This conclusion may be invalid if the agglomeration of planets always occurred by the collision of two roughly equal pieces rather than by the accretion of small bodies by a large central mass.

The rocks of the Earth contain radioactive elements, in particular, uranium, thorium, and potasium-40. The particles emitted in the radioactive decay of these elements will rattle around and heat up their surroundings. (Figure 18.20). Because the heat introduced this way does not easily leak out of the interior, one can calculate that in several hundred million years after the formation of the solid Earth, the average interior temperatures will rise high enough (2,000 to 2,500 K) to reach the melting point of iron. Per unit volume, iron weighs more than the rocks of Earth. Thus, once iron becomes molten, it will begin to seep toward the center (Figure 18.21). The falling of roughly one-third of the mass of the Earth in the form of molten iron would release an enormous amount of gravitational energy by terrestrial standards. Converted into heat, this energy would not have easily escaped again from the interior of the Earth, and the resultant average interior temperature may have risen to above 4,500 K. This would have been more than enough to cause the rocks of the Earth to melt, a catastrophe precipitated by the melting of iron. Thus, the iron catastrophe of supernovae which led to the expelling of heavy elements into the interstellar medium (Chapter

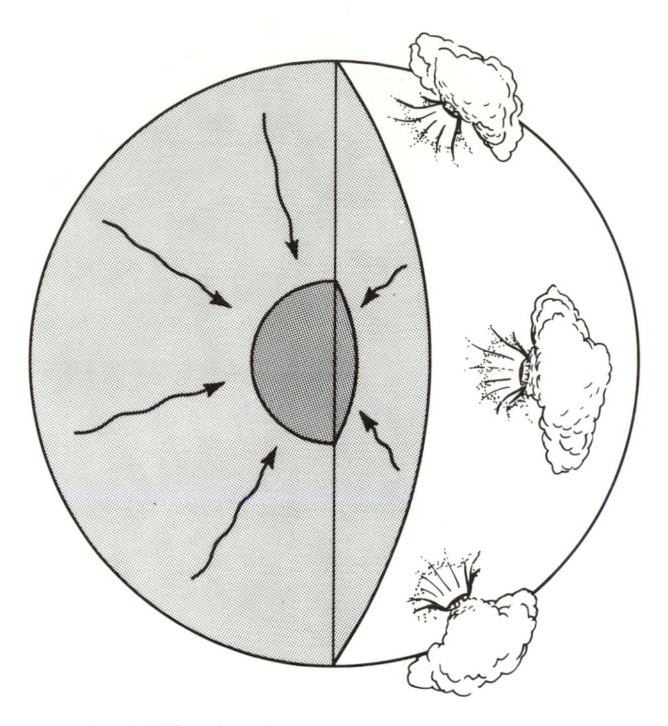

Figure 18.21. When iron became molten in the primitive Earth, it would have seeped toward the core. The conversion of gravitational potential energy into heat would have led to the melting of the rocks of the Earth and considerable outgassing from the interior.

8) ultimately resulted in another iron catastrophe on Earth. Very intimately woven indeed is nature's tapestry!

After the virtually complete melting of the Earth, the lighter rocks would have floated to the top, eventually cooling to form a solid crust. The oldest surviving rocks on Earth, therefore, date back only 3.8 billion years, presumably to the solidification of this crust. The heavy iron would have sunk to the bottom to form the core of the Earth. Compounds of intermediate density would have formed the mantle (see Figure 17.13). In this manner, it is believed, the Earth acquired its differentiated interior structure.

Problem 18.10. Much of the radioactivity in the Earth now is concentrated in the lightest rocks which make up the continents. A cubic centimeter of granite rock releases about 800 erg/yr by radioactivity. If granite rock occupies 1/3 of the surface area of the Earth to a depth of 30 km, show that the present radioactive release of heat by this source alone is 4.1×10^{27} erg/yr. (Compare this figure with the outflow given in Problem 17.4.) Assume that the initial radioactivity of the newly formed Earth was ten times larger and was spread uniformly throughout the interior. How much heat would radioactivity have released in the first 800 million years? By what temperature would this have raised the interior of the Earth if we assume the data of Problem 17.4?

Problem 18.11. The gravitational field at a radius r inside a homogeneous sphere of density $\bar{\rho}$ is given by Gauss's law to be $g = G(4\pi\bar{\rho}/3)r$. The buoyancy deficit per unit volume of iron at density ρ is $(\rho - \bar{\rho})g$, which has a value $(\rho - \bar{\rho})G(2\pi\bar{\rho}/3)R_\oplus$ at an average position $r = R_\oplus/2$. The energy released by iron falling an average distance $R_\oplus/2$ through this bouyancy deficit is, therefore, roughly given by $V(\rho - \bar{\rho})G(\pi\bar{\rho}/3)R_\oplus{}^2$, where V is the volume of the iron. Some of this energy will go into changing the thermodynamic state of the iron, but its order of magnitude should be available to heat the interior of the Earth. Calculate the numerical value of this energy by referring to the data in Figure 17.13, and compare the value with those given in Problems 18.10 and 17.4. Comment on the importance of the iron catastrophe inside the Earth.

The Formation of the Atmosphere and the Oceans: The Big Burp

The drastic heating of the Earth during the epoch of planetary differentiation roughly four billion years ago led to considerable outgassing of volatile compounds that had been locked up in the rocks of the Earth. In the traditional view, the volatile compounds are those which were common in the primitive solar nebula—water, methane, and ammonia—and are, in fact, still in the present atmospheres of the Jovian planets. The atmosphere of the Earth, in this view, was the *natural* atmosphere that we expect for planets. In recent years, this traditional view has been challenged by people who maintain that much more carbon dioxide and molecular nitrogen would also be outgassed.

In either case, the primitive atmosphere was *not* acquired by the Earth by gas accretion from the solar nebula, as occurred in the Jovian planets. As Donald Menzel was the first to point out, if the Earth had undergone a substantial gas-gathering phase, a noble gas like neon, which is quite plentiful cosmically, would not be so rare on Earth. The rarity of neon (and argon, krypton, and xenon) argues that the first atmosphere of the Earth was formed when the intense heating released volatile compounds locked up chemically in rocks. Since the noble gases are not chemically reactive, relatively little of them were contained in rocks. In this theory, the atmosphere was born when the Earth issued a "big burp."

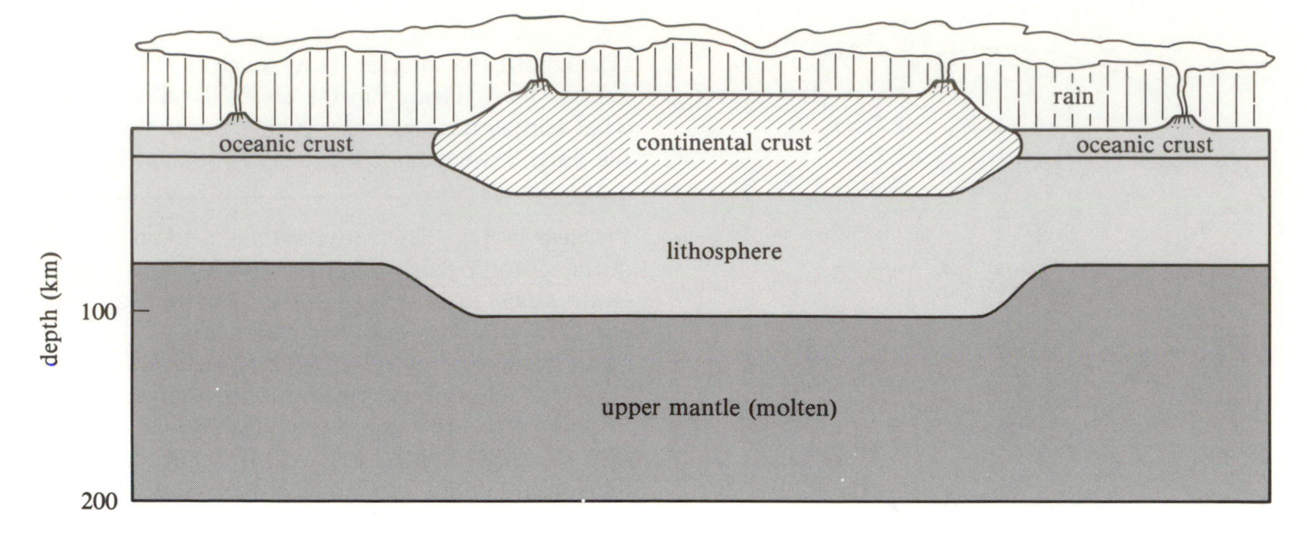

Figure 18.22. The crust and lithosphere of the Earth floats on a partially molten upper mantle. In particular, the continental crust is made of lighter and thicker material than the oceanic crust; so the continents are more buoyant, and jut out above the ocean basins. The water from rain tends to run downhill and, therefore, eventually empties into the oceans. (For details, see F. Press and R. Siever *Earth*, Freeman, 1978.)

The oceans formed later, when water condensed from the air.

When the crust and lithosphere of the Earth cooled, they did not become a smooth outer shell of solid material. Some regions were made up of denser and thinner crustal material than other regions. Since the lithosphere plus crust floats on top of a partially molten outer mantle (see Figure 17.13), the former sits deeper than the latter. Thus, when the rains came, the water tended to drain into the former, which became the ocean basins (Figure 18.22). The land which jutted out above the oceans became the continents.

Continental Drift and Sea-Floor Spreading

The preceding description of the early history of the Earth is speculative in many ways; do not be surprised if its details must be revised in years to come. However, the general outline of the argument is probably correct, since it is supported (and indeed, largely motivated) by concepts about present-day geological events that were arrived at only in the 1960s and 1970s. This revolution in geological thinking is called **plate tectonics.**

The basic idea of plate tectonics is very simple. Geologists believe that the surface of the Earth (crust plus lithosphere) is divided into a dozen or so large **plates** that move quasirigidly, like rafts, on a partially molten outer mantle. A key piece of evidence in support of this concept of **continental drift** was advanced in 1620, when Francis Bacon remarked on the jigsaw fit of the opposite shores of the Atlantic. By the beginning of the twentieth century, Alfred Wegner had assembled evidence that the continents had drifted to their present configurations from the breakup of a supercontinent. Wegner's examination of the geological record of rocks and fossils led him to conclude that, 200 million years ago, the land mass of the Earth existed in one piece—*Pangaea* ("all lands")—and the oceans were an undivided *Panthalassa* ("all seas"). Figure 18.23 shows a modern reconstruction of this idea.

Unfortunately, the idea foundered because none of its proponents could come up with a plausible mechanism to drive the drift. To this day, the details of the motive force elude geophysicists, but there is little doubt that it has something to do with the thermal convection that accompanies heat flow in the upper mantle (Figure 18.24). What has changed dramatically in the last few decades is the flood of evidence which provides incontrovertible support for the central concept of plate tectonics. In what follows, therefore, we shall briefly outline the kinematics of plate tectonics and the geological evidence which has convinced almost everyone of its essential correctness.

The crucial element of plate tectonics is not so much the horizontal motions of the plates as the vertical motions at their boundaries. Plates separate at rifts in

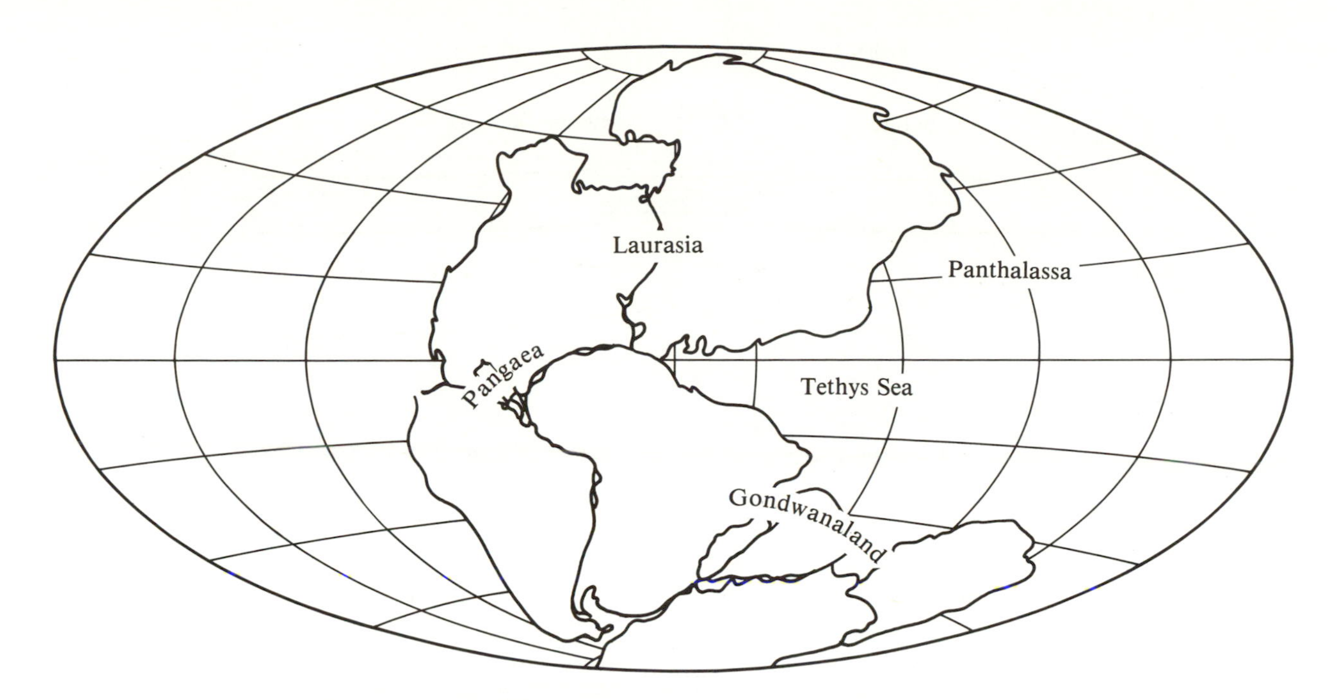

Figure 18.23. A modern reconstruction of Wegner's ideas about continental drift. About 200 million years ago, the land mass of the Earth existed as one piece, Pangaea, and the oceans were an undivided Panthalassa. (From F. Press and R. Siever, *Earth* W. H. Freeman, 2d ed., 1978.)

Figure 18.24. Convection currents in the molten upper mantle are believed to provide the motive force for plate tectonics. Recent geologic evidence derived from studies of the isotope ratios of trace elements suggests convective cells in the upper and lower mantles must be largely decoupled. (For details, see R. K. O'Nions, P. J. Hamilton, and Norman M. Evenson, *Scientific American*, **242**, May 1980, 120.)

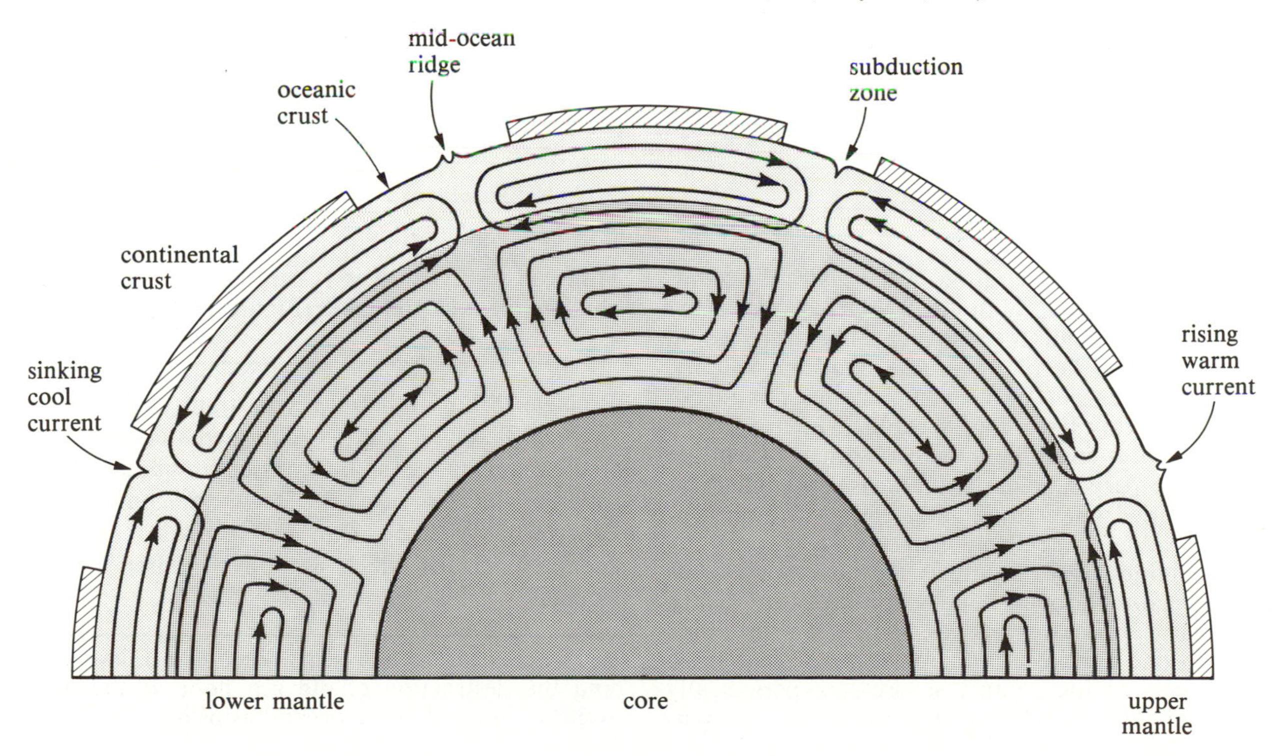

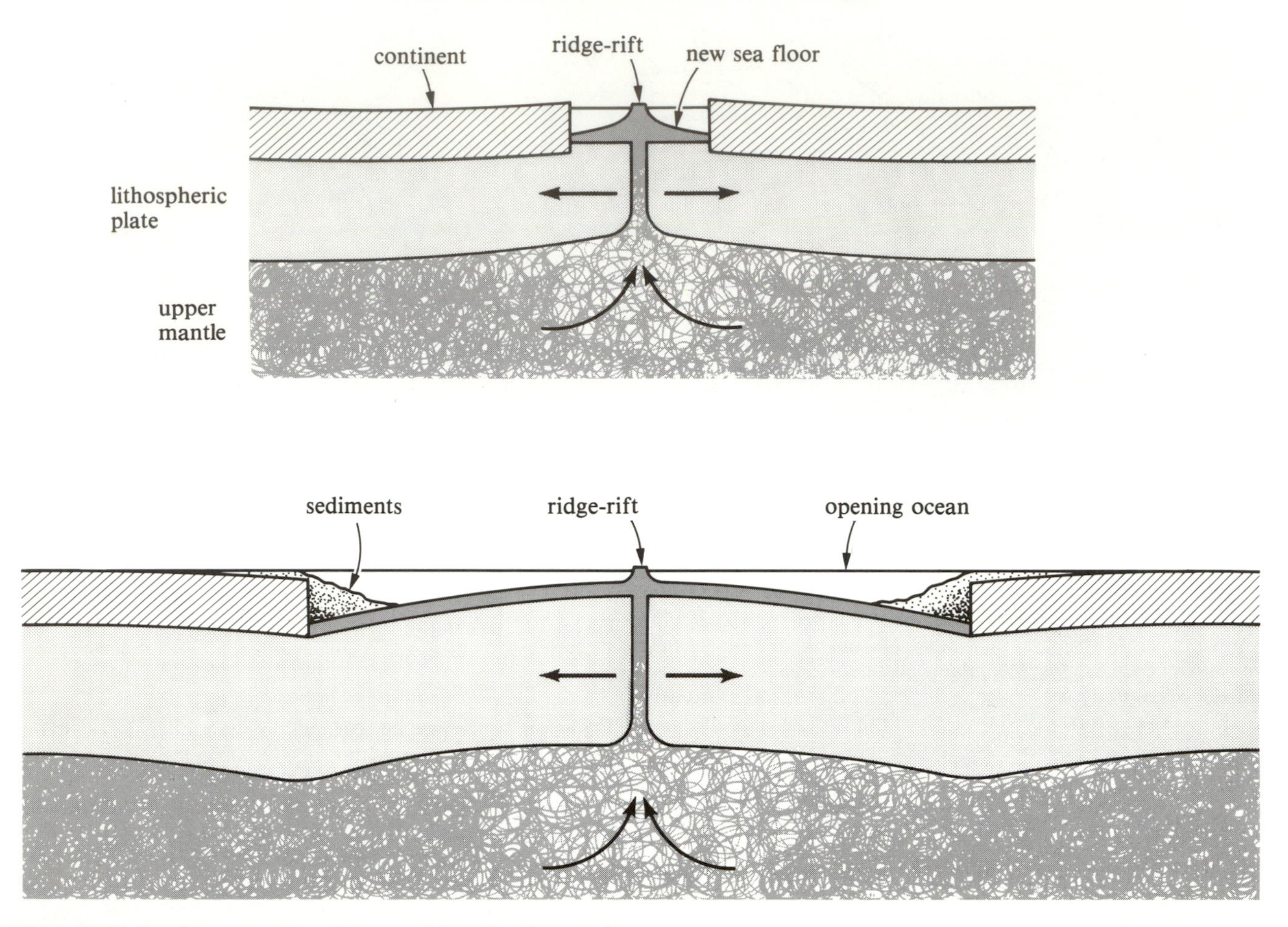

Figure 18.25. Sea-floor spreading. The upwelling of molten rock beneath a ridge-rift replaces the spreading sea floor. As the continents on opposite sides of the opening ocean recede from one another, erosional debris flows downhill to the trailing edges and deposits thick wedges of sediments.

the ocean floor. The separation of the plates is accompanied by an upwelling of molten rocks (lava) from the upper mantle which fills the void created by the recession of the plates. As the molten rocks cool and solidify, they become a new part of the sea floor; thus, this part of the process is called **sea-floor spreading** (Figure 18.25). Elsewhere, plates converge upon each other. When two plates collide perpendicularly, the denser oceanic crust of one is buried under the lighter continental crust of the other, leading to a closing of an ocean (Figure 18.26). The overriding plate crumples upward and forms mountain chains, the overridden plate buckles downward and forms deep ocean trenches. Plates can also slide parallel to one another. The slip can take place via a series of jolts as the contact stresses are periodically built up and relieved. As residents along the San Andreas fault can testify, such jolts can lead to very unpleasant shallow-focus earthquakes (Figure 18.27).

Problem 18.12. What do you think will happen in Figure 18.26 after the two continents collide? Assume that continents are too buoyant to be subducted into the mantle. *Hint*: Consult J. F. Dewey, *Scientific American*, **226**, May 1972, 56. Can you think of examples of mountain chains which are on the interiors of continents?

The creation of new sea floor at the mid-ocean rifts and the destruction of old sea floor at the deep sea trenches imply that the ocean basins are not permanent

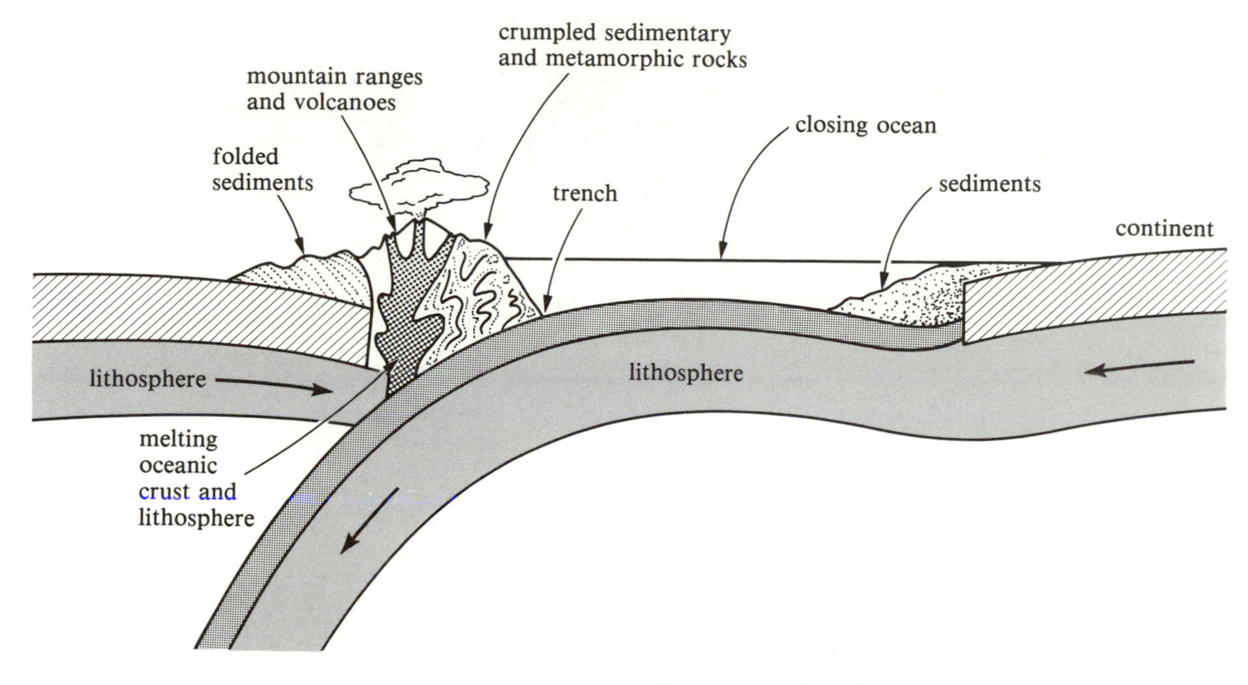

Figure 18.26. The collision of two plates. The continental plate is crumpled, and the oceanic plate is subducted. Accompanying geological processes include the formation of high mountain chains, volcanoes, folded sediments, earthquakes, and deep sea trenches.

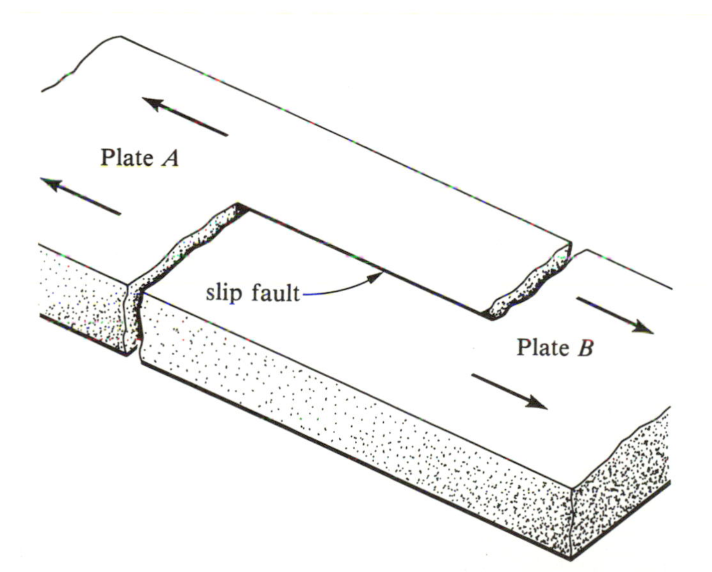

Figure 18.27. A slip fault occurs where two plates slide parallel past one another. It is often the site of dangerous shallow-focus earthquakes.

fixtures on this Earth, as was explicitly recognized by Harold Hess in the 1960s. After World War II, oceanographers used the magnetometers developed to detect submarines to map the polarity of magnetized rocks on the sea floor. To their great surprise, they found patterns of fossil magnetism like that shown in Figure 18.28. Parallel to the crest of a midocean ridge ran stripes of magnetized rocks with alternating magnetic polarities! In 1963, F. J. Vine and D. H. Mathews (and independently L. Morley and A. Larochelle) came up with an explanation. They reasoned that the rocks must have been magnetized in the direction of the prevailing geomagnetic field at the time that they cooled from molten lava (Chapter 4). The alternating stripes then correspond to the periodic reversals suffered by the geomagnetic field that we mentioned in Chapter 17. The stripes are displaced horizontally from one another because the sea floor must be spreading apart at the mid-ocean rifts, as hypothesized by Hess. Thus, the ocean floor acts like a tape recorder, preserving a history of its own motions!

How fast is the motion of sea-floor spreading? To answer this question, we need only divide the horizontal displacements of adjacent stripes by the timespan between the reversals of the geomagnetic field for that episode. How do we find the latter? Lava flows from volcanoes situated on continents provide a vertical record of magnetic-field reversals as successive eruptions are marked with the prevailing polarity of the geomagnetic field. By radioactive dating, the age of the rock since solidification for each stripe can be dated, working from the uppermost layer downward. In this fashion,

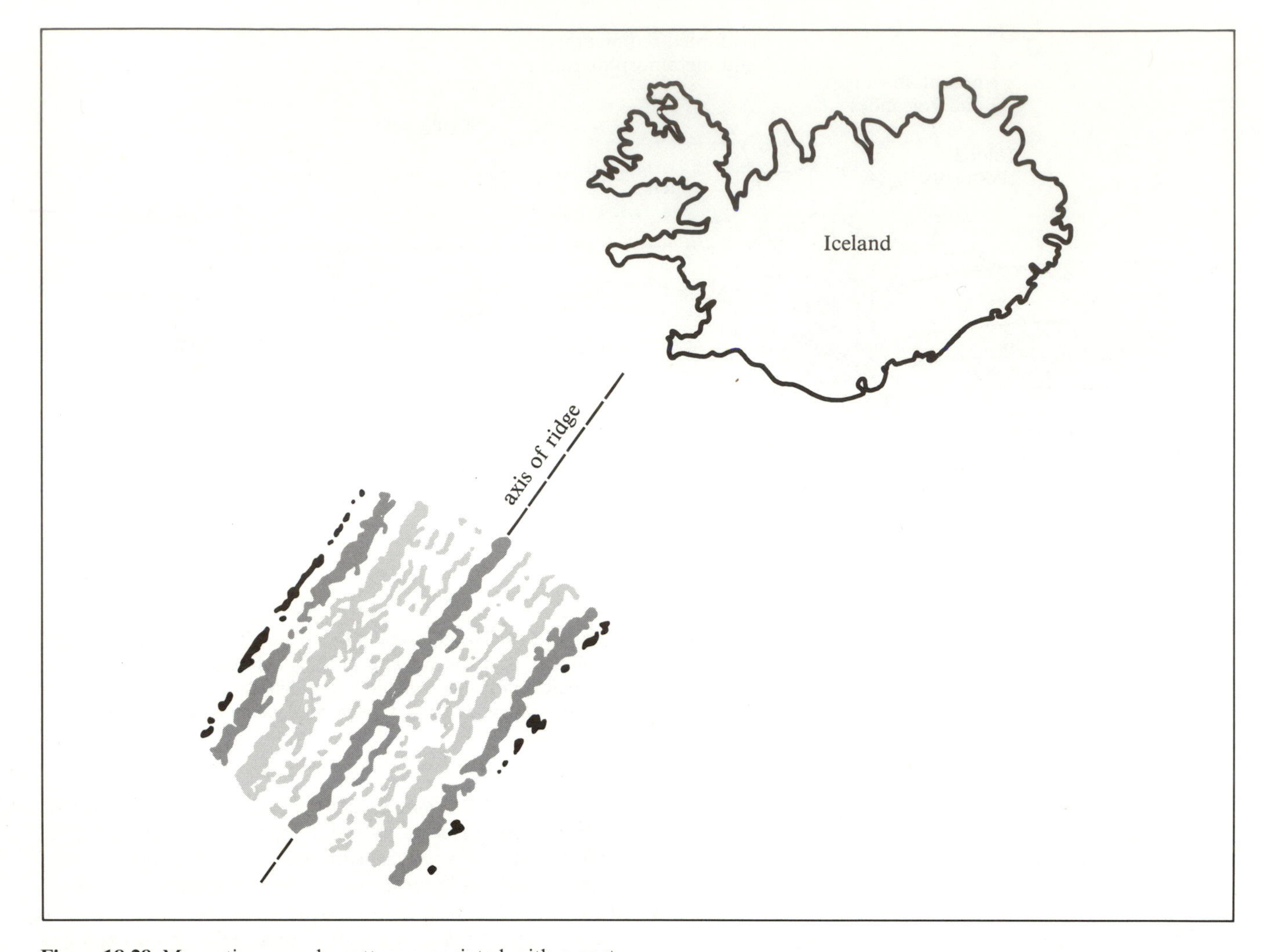

Figure 18.28. Magnetic anomaly pattern associated with a part of the mid-Atlantic ridge. The dark stripes correspond to rock formations which have magnetic polarity in the sense of the current geomagnetic field. The white spaces in between correspond to rocks which have the opposite polarities. According to the theory of sea-floor spreading, the youngest rocks are those on the ridge axis; the oldest are those furthest from the axis on either side. (For details, see E. Orawan, *Scientific American*, **221**, Nov. 1969, 102.)

the history of reversals of the Earth's general magnetic field can be calibrated into the past. Working from the newest sea floor at the midocean ridges outward in both directions, Vine and Mathews were now able to read off the ages of the successive ocean stripes. The sea-floor spreading takes place at a rate typically of a few cm/yr—as Press and Siever have commented, about as fast as your fingernails grow. Now, your fingernails do not grow very fast, but if you let them grow unimpeded for geological times, a few hundred million years, they could span the Atlantic! Clearly, therefore, plate tectonics is a very vital phenomenon when we talk about global geological processes.

Being of lighter and more buoyant material, the continents do not get consumed by the fires below every few hundred million years. No, the continents form much longer-lasting features, and rocks can be found on them that date 3.8 billion years into the past. This does not mean, however, that the geological features of the continents are permanent. Indeed, the

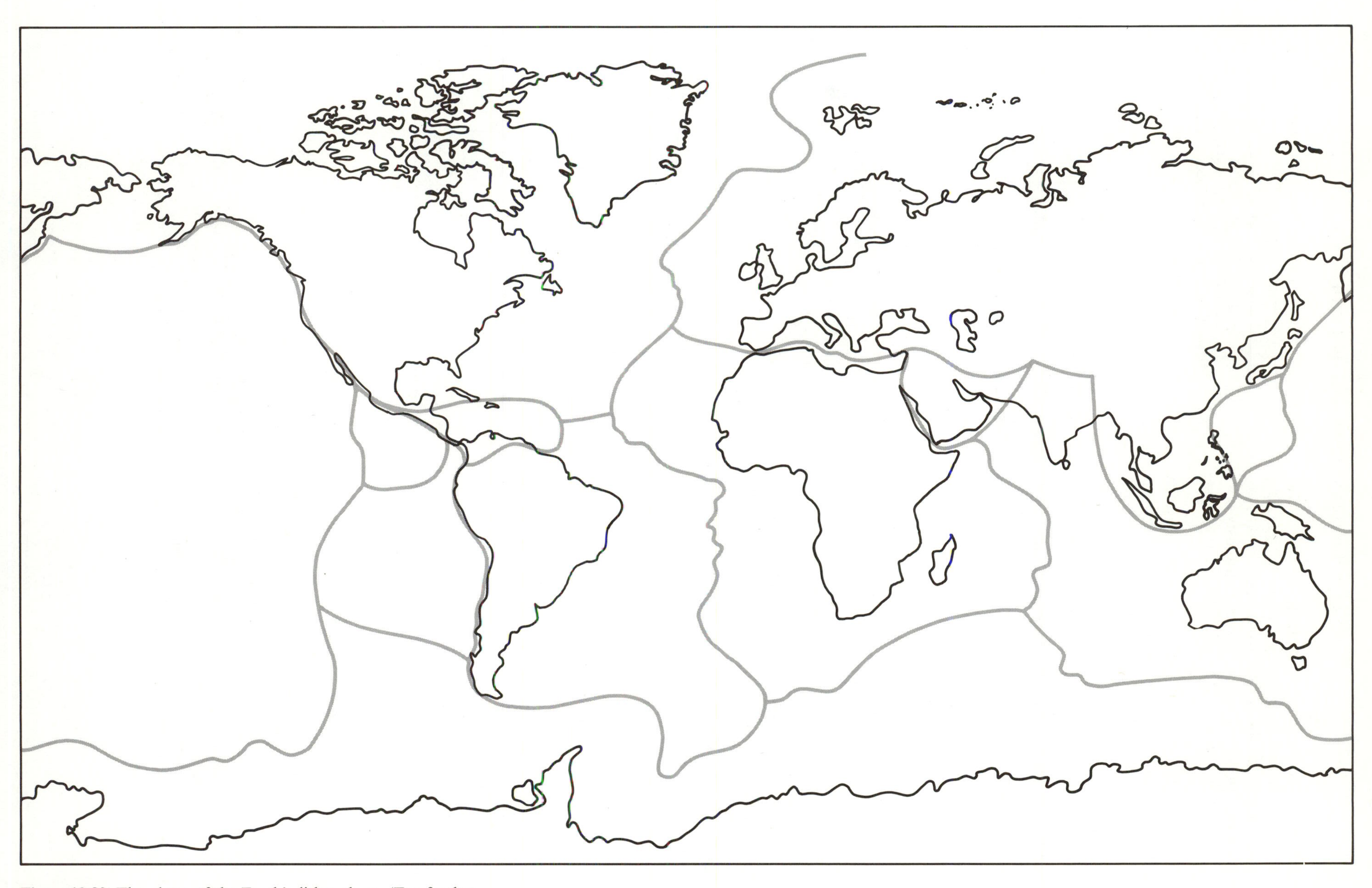

Figure 18.29. The plates of the Earth's lithosphere. (For further details, see J. F. Dewey, *Scientific American*, **226**, May 1972, 56.)

concentration of the radioactive elements in the continental crust plus radioactive dating studies led Patrick Hurley and others in the 1960s to propose that the continents have grown more or less continuously for the past 3.8 billion years. The basic idea is that the radioactive elements, which have large ionic radii, would tend to be trapped in the relatively open crystalline structure of light rocks as the convection currents in the mantle brought such light materials to the surface (Figure 18.24).

On a smaller scale, where the crust is weak, volcanos can provide a window to the interior, building mountains in the process. The smashing of plates against one another can change land forms. On the other hand, the movement of glaciers, weathering, and wind and water erosion can help wear mountains down. For the continents, the Earth acts like *two* heat engines. One is the engine of the solid Earth, driven by the heat flow from below. The other is the engine of the atmosphere and the oceans, driven by the uneven solar heating from above. Generally, the former builds mountains, the latter grinds them down. As the geologists of the nineteenth century already recognized, all that is needed is time.

Geologists in this century have used the organizing principles described above, together with the analysis of earthquakes, to define the boundaries of the major plates. Figure 18.29 shows a map resulting from a grand synthesis of these efforts. It is to be hoped that lunar-ranging experiments will provide, in the near future, even more definitive direct measurements of the motions of the plates of the Earth's lithosphere.

The Past and Future of the Continents

With hindsight, we now know that there is nothing special about the date, 200 million years in the past, which Wegner assigned to the breakup of Pangaea (see Figure 18.23). The motions of the plates did not begin only 4 percent of the age of the Earth ago. They probably extend four billion years into the past, and they will continue for billions of years into the future. Some bold geologists (among them, J. B. Bird, J. F. Dewey, W. R. Dickinson, and R. S. Dietz) have begun to reconstruct

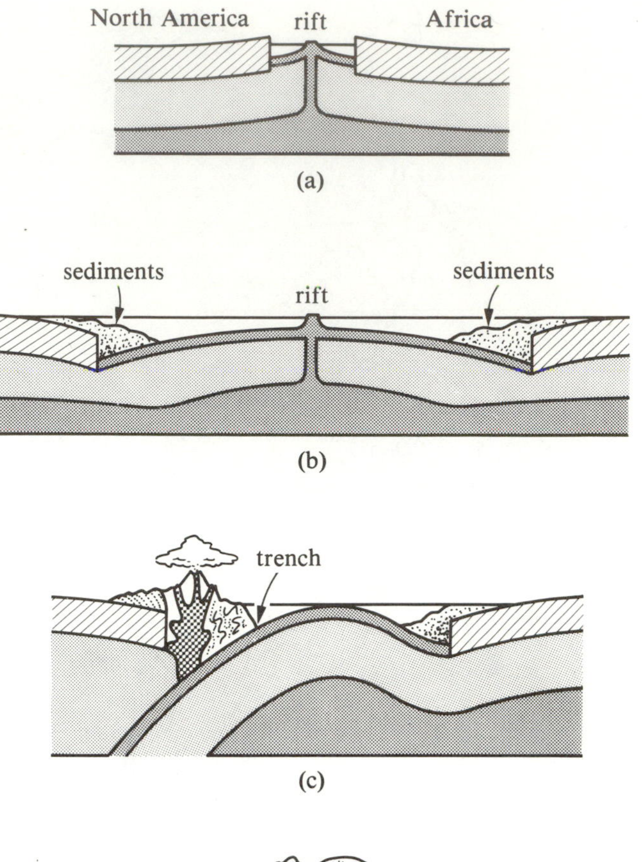

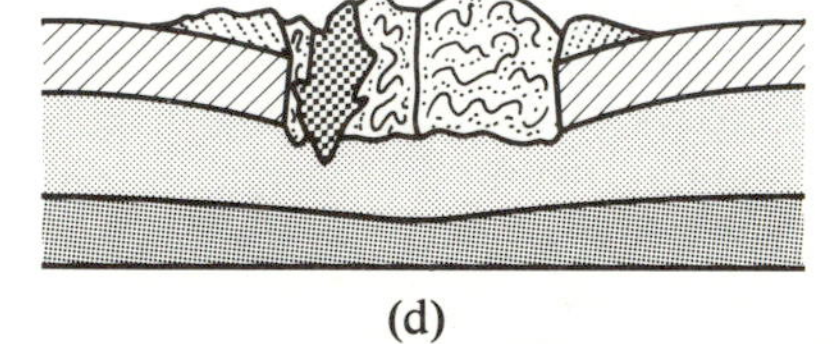

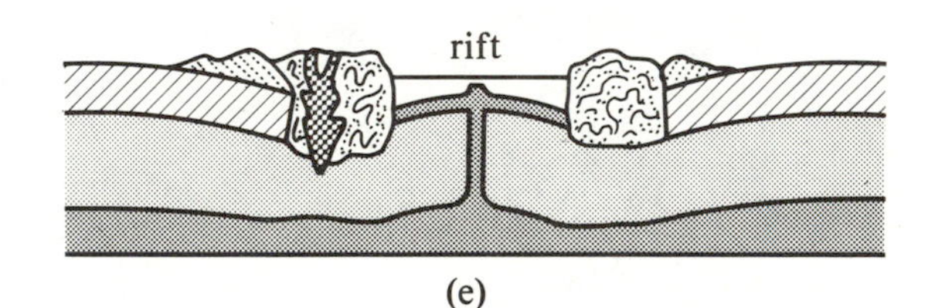

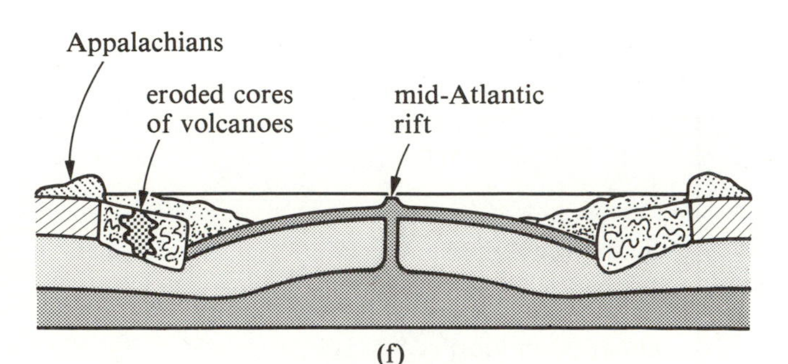

Figure 18.30. The history behind the formation of the Appalachian mountains. (a) North America and Africa split apart 600 million years ago. (b) Sediments are deposited on the trailing edges of the two continents. (c) About 500 million years ago, the Atlantic ocean began to close, and sediments on the North America coast were uplifted and crumpled. (d) The collision of North America with Africa and Europe raised a mountain chain which must have rivaled the Himalayas. (e) About 200 million years ago, the Atlantic reopened with the breakup of Pangea. (f) The modern Appalachians are the crumpled sediments of past events. (For details, see R. S. Dietz, *Scientific American*, **226**, May 1972, 30.)

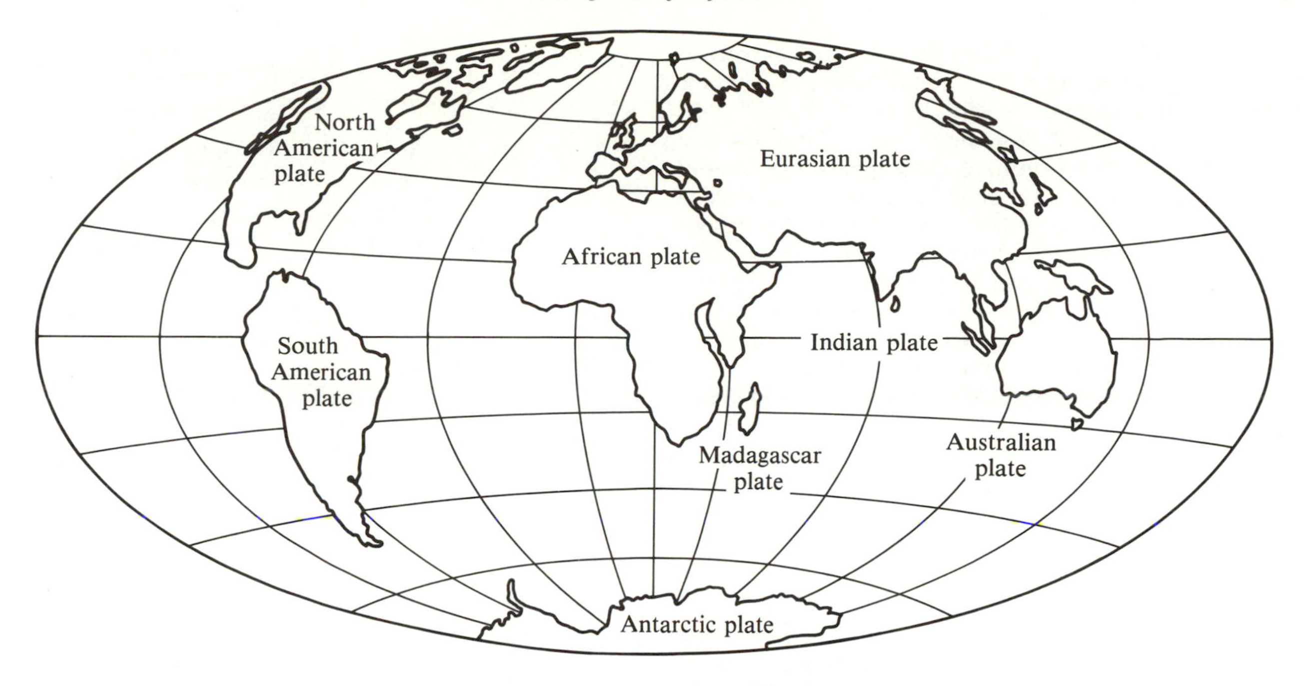

Figure 18.31. World geography as it may be taught 50 million years from now. (For further details, see R. S. Dietz and J. C. Holden, *Scientific American*, **223**, Oct. 1970, 30.)

the global events of several hundreds of millions into the past, and also to extrapolate more modestly into the future.

One of the goals of the reconstruction of the past is to give a unifying account of ancient geological features. For example, the theory which accounts for the Appalachians begins by noting that they are old eroded mountains which contain sediments from the sea that have evidently been subjected to intense pressures. This mountain range must have formed before the (most recent) break-up of Pangaea. Let us pick up the story 600 million years ago, when a continent comprised formerly of North America and Africa split apart, forming an Atlantic Ocean (Figure 18.30a). Sediments would be deposited offshore from weathered rock of the continents (Figure 18.30b). About 500 million years ago, the Atlantic began to close again, the heat released by oceanic crust sliding under continental crust causing a volcanic arc (Figure 18.30c). By 375 million years ago, North America had smashed into Africa and Europe, with a consequent thickening of the crust to form a great mountain range that rivaled the Himalayas (Figure 18.30d). Pangaea broke up 200 million years ago, and *the* Atlantic reformed (Figure 18.30e). In the meantime, the folded sea sediments which became the Appalachian mountains were eroded to their present state (Figure 18.30f). A nice story!

Figure 18.31 gives an educated guess at how geography will be taught 50 million years into the future. Australia is about to become part of Asia, and Los Angeles has slipped past Berkeley on its way to be consumed by the Aleutian trench! Not such a nice story?

The Emergence of Life on Earth

We wish to finish our discussion in this chapter by considering the effects that life has had on the physical characteristics of Earth. This topic is, of course, an unfinished story, since humans have a great potential for changing the face of the Earth. But that relatively recent fact is more suitable for discussion in an ecology text than here. Let me confine my remarks here to the relatively early history of the Earth, where the most obvious changes that occurred were in the atmosphere of the Earth (or, more generally, what is often called the *biosphere*).

The Evolution of the Earth's Atmosphere

We remarked earlier on traditional views that the primitive atmosphere was rich in hydrogen-bearing compounds (e.g., methane, CH_4, and ammonia, NH_3). Such an atmosphere is said to be **reducing**, because of the chemical reactions that would result if ordinary substances were exposed to such gases. Chemists use the term "reduction" when a substance gains valence electrons. In most reactions with hydrogen-rich compounds, a substance (other than a metal) gains a valence electron from each hydrogen atom it combines with (see Chapter 3). Thus, that substance is said to be "reduced," with hydrogen being the "reducing agent." Thus, in the traditional view, the primitive atmosphere was highly reducing (or perhaps only slightly reducing), and contrasted sharply with the present atmosphere. The present atmosphere of the Earth is composed mostly of molecular nitrogen, N_2, and molecular oxygen, O_2. When exposed to an atmosphere with *free* oxygen, a substance tends to lose valence electrons to the oxygen atoms it combines with: that substance is said to be "oxidized," with oxygen being the "oxidizing agent." Since "oxidation" refers to the process of losing electrons, modern usage allows it to take place even when oxygen itself is not involved in the chemical reaction. The present atmosphere is highly **oxidizing**.

Water is present in both the primitive and the present-day atmosphere. Except when attacked by alkali metals or halogens, water vapor is not terribly reactive in oxidation or reduction. In H_2O, the hydrogen atoms are already quite stably oxidized; the oxygen, quite stably reduced. Because the water molecule is so polar (Chapter 4), liquid water provides an excellent solvent for many chemical reactions. For example, the weathering of rocks is greatly accelerated by the presence of liquid water which can exert a weak attraction on the solid surface, and eventually dissolve it. Oxidation-reduction is one important aspect of chemical reactions, but it is hardly the only one.

Here let us concentrate on the following question. What changed the Earth's atmosphere from a reducing one to an oxidizing one? It is now believed that the change was a two-stage process. The first step is inorganic and depends on the production of ozone. The second step is organic and depends on the appearance of life. We shall see that water plays a crucial role in both steps.

The Formation of the Ozone Layer

In the absence of free oxygen, ultraviolet rays from the Sun can penetrate all the way to the ground and to sea level (Figure 18.32). Energetic ultraviolet photons can dissociate water vapor, H_2O, into hydrogen atoms, H, and free oxygen atoms, O. (The actual pathways investigated by Sydney Chapman and refined by Joseph Chamberlain are more complex than this simple description, but the details are not important here.) The Earth cannot easily retain light hydrogen atoms (Chapter 17), and the hydrogen can be expected to escape into space. In contrast, the free oxygen atoms released from the dissociation of water could combine with themselves to form molecular oxygen, O_2, or they might combine with ammonia, NH_3, and methane, CH_4, to form (among other things) slight amounts of molecular nitrogen, N_2, and carbon dioxide, CO_2. The important point is that

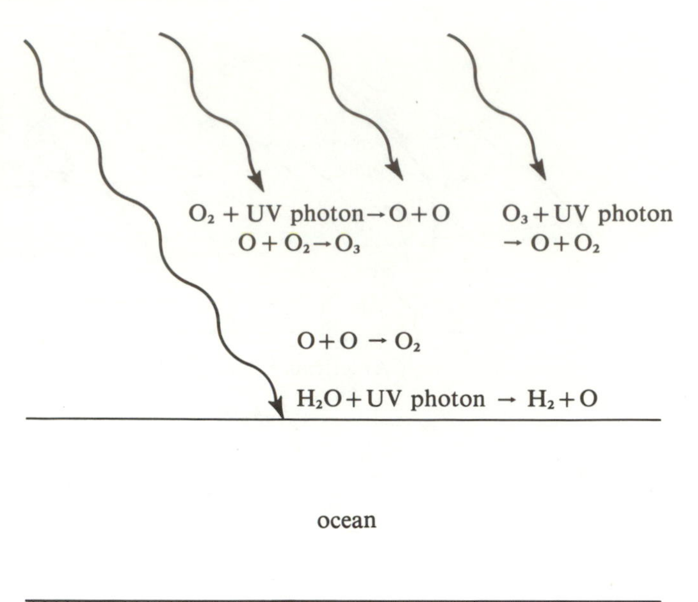

Figure 18.32. The schematic formation of the ozone layer. Ultraviolet light which strikes water molecules in either the primitive oceans or the atmosphere dissociates them into hydrogen and oxygen. Among other things, atomic oxygen, O, combines with itself to form molecular oxygen, O_2, and with molecular oxygen, O_2, to form ozone, O_3. Ozone is a prolific absorber of ultraviolet radiation, and the formation of an equilibrium layer of ozone in the stratosphere effectively filters out all ultraviolet rays from the incident sunlight.

the escape of hydrogen from the dissociation of water frees oxygen to combine with other chemical species. Some of these oxygen-carrying compounds would eventually reach the upper atmosphere where photodissociation would again produce free oxygen atoms. These free oxygen atoms, O, combine readily with the molecular oxygen, O_2, which has not been dissociated to form ozone, O_3. Ozone is a prolific absorber of ultraviolet radiation; and even a slight accumulation of ozone in the upper atmosphere—i.e., the production of an "ozone layer" in the stratosphere (Chapter 17)—suffices to prevent any more ultraviolet radiation from penetrating to lower levels where there is an abundant source of water vapor.

With the development of an ozone layer, we have an equilibrium established between the destruction of ozone by ultraviolet photons and the formation of ozone by the chemical combination of oxygen atoms and molecules. At this point, essentially only visible light would reach sea level. The shielding by the ozone layer prevents the further dissociation of water vapor. The amount of molecular oxygen, O_2, present in the atmosphere would then stabilize at some small percentage. The water in the oceans was also saved from further damage. (The failure

Figure 18.33. An example of a fossil stromatolite. The many thin layers are believed to have been formed by communities of prokaryotic cells which grew in successive stages on top of one another in shallow water. These particular specimens were found in African rocks dating back 2.6 billion years. (Courtesy of J. W. Schopf.)

of Venus to retain its water probably led to a disastrous runaway greenhouse effect, as we saw when a similar story was reconstructed for Earth's sister planet in Chapter 17).

The ozone layer's role in filtering out the ultraviolet rays of the Sun cannot be overemphasized. Ultraviolet rays are very harmful to many lifeforms, inducing skin cancer, for example, in humans. This explains the concern over our release into the atmosphere of fluorocarbons, which have an enormous potential for destroying ozone. In any case, in geological times, the dissociation of water by ultraviolet light was halted by the appearance of the ozone shield. The full conversion of the Earth's atmosphere from a reducing one to an oxidizing one had to await, therefore, the development of life.

The Role of Life in Changing the Earth's Atmosphere

The evidence that life played a major role in the evolution of the Earth's atmosphere lies in the geological record. Charles Walcott was the first to draw attention to thinly layered deposits of limestone rock called *stromatolites* (Figure 18.33). He interpreted these structures as fossilized reefs of algae growth. The algae are now gone, but they have left behind "beds of stone" (stromatolite in Greek). Paleontologists remained skeptical until 1954, when Stanley Tyler and Elso Barghoorn reported the finding of a large number of well-preserved fossil microorganisms ("microfossils") in similar structures at the Gunflint Iron formation near Lake Superior. Some of the stromatolite deposits in Africa and Australia are now known to date back to 3.5 billion years ago. This suggests that life developed on Earth fairly soon after the crust resolidified after the melting of the Earth. (Remember that the oldest surviving rocks on Earth are 3.8 billion years old.) In 1980, Stanley Awramik and J. William Schopf discovered microfossils of long filamental chains of bacterial cells in rocks from the region of the Australian stromatolites. Detailed analysis is underway to see if these microfossils are the same age as the stromatolite, 3.5 billion years. The evidence now looks good that, around 3.5 billion years ago, unicellular organisms existed on Earth that could perform photosynthesis.

Photosynthesis is the process by which organisms that contain chlorophyll convert carbon dioxide and water into carbohydrates. Carbohydrates are, of course, the basic source of chemical energy for all organisms. Schematically, for example, the production of glucose, $C_6H_{12}O_6$, by photosynthesis can be written as the net equation

$$6\,CO_2 + 6\,H_2O + \text{light energy} \xrightarrow{\text{chlorophyll}} C_6H_{12}O_6 + 6\,O_2.$$

The carbon dioxide on the left-hand side of the above equation could have come from geological activity when carbonate rocks are heated, or from biological activity via respiration. Respiration involves the burning of carbohydrates with oxygen to release chemical energy. For example,

$$C_6H_{12}O_6 + 6\,O_2 \rightarrow 6\,CO_2 + 6\,H_2O;$$

thus, respiration is, in net effect, the opposite of photosynthesis.

Problem 18.13. The name carbohydrate ("watered carbon") derives from the fact that in many such carbon compounds, the hydrogen and oxygen atoms enter in the same proportion as in water. Verify this to be true for ribose, $C_5H_{10}O_5$, but not deoxyribose, $C_5H_{10}O_4$. What do you think the "deoxy" refers to? Because carbon compounds possess complex three-dimensional structures that are important for their biochemical action, a chemical formula such as $C_5H_{10}O_5$ or $C_5H_{10}O_4$ does not provide a unique specification of the molecule. The forms of ribose and deoxyribose that exist in nuclei acids (RNA and DNA) take a ring form, as shown in the diagrams.

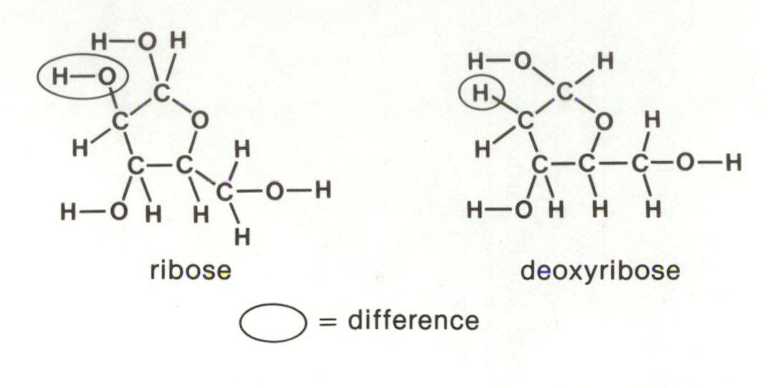

As long as there was more photosynthesis activity than respiration, there would have been a slow and steady build-up of free molecular oxygen in the atmosphere. It is conjectured that photosynthesis by blue-green algae began to release free oxygen into the Earth's atmosphere around 3.5 billion years ago, when such organisms first appeared, but there is evidence that the amount of free oxygen was not substantial until two billion years ago. In rocks more than two billion years old, geologists find reduced minerals—for example, sulfides instead of sulfates—which implies that these rocks were not exposed to much free oxygen when they formed.* On the other hand, in rocks less than two billion years old, geologists find oxidized minerals—for example, red beds of iron oxide. This must mean that in the intervening period, within a few hundred million years, there was a major shift of the Earth's atmosphere from reducing to oxidizing. This shift caused all the iron in the oceans to rust and sink, i.e., to precipitate to the ocean floor. When the iron buffer in the oceans had been swept clean, the oxygen released by blue-green algae could begin to rise to modern atmospheric concentrations (Figure 18.34).

The shift from a reducing to an oxidizing atmosphere apparently allowed a wild proliferation in varieties of microorganisms. The large number of microfossils discovered at the Gunflint Iron formation turned out to be amazingly diverse. The organisms have shapes that include long filaments, spheroids, umbrellas, and stars. Fossil stromatolites in general become quite common after about 2.3 billion years. Did this diversity arise in response to the increasing oxygen supply, or was it the cause of it? Perhaps both are true.

* The debate about whether the primitive atmosphere was slightly reducing or highly reducing revolves around whether the oldest rocks are reduced *enough* for the traditional view to be valid. The issue is not very important for this book; it is more important to see that the primitive atmosphere was not *oxidizing*; i.e., it lacked free molecular oxygen.

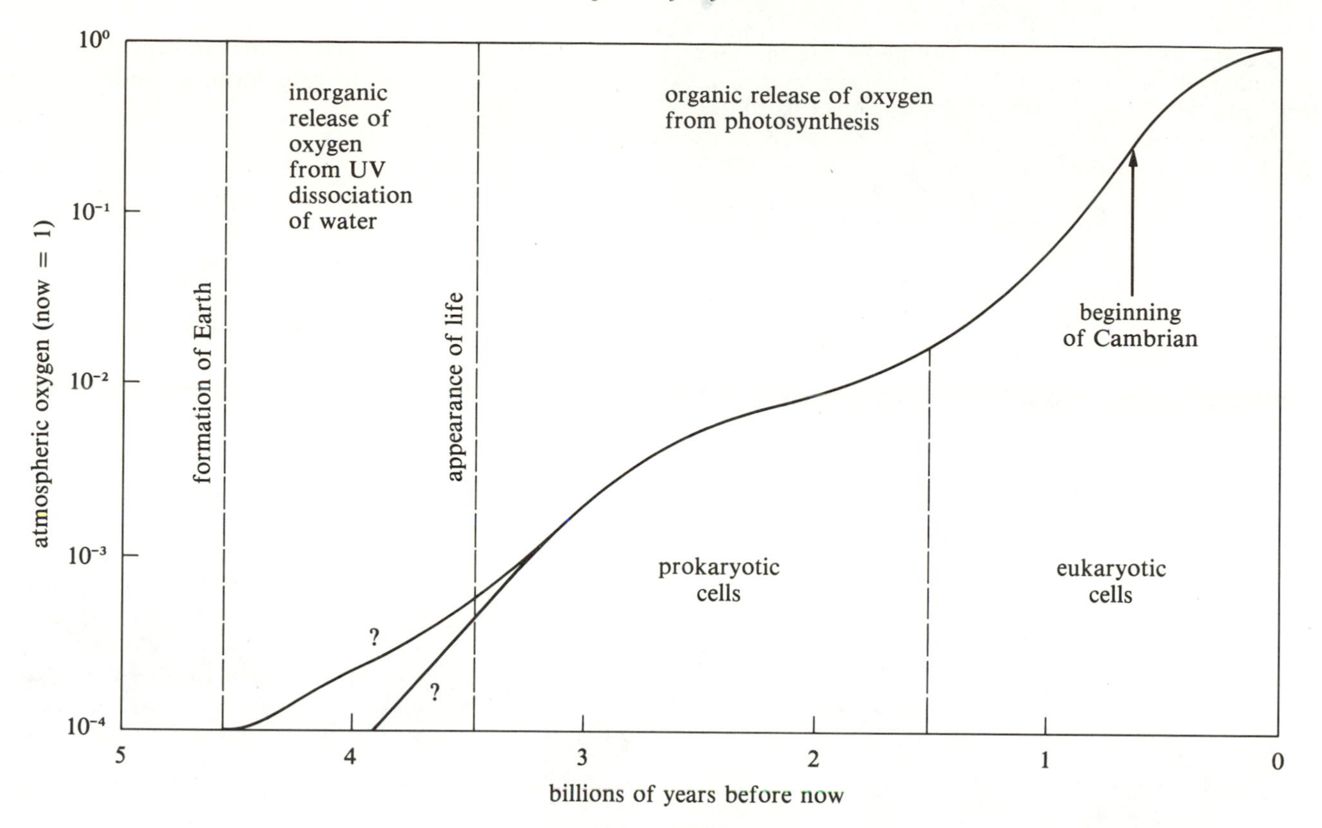

Figure 18.34. A schematic plot of the free oxygen in the Earth's atmosphere in the past as a fraction of the present supply. The general trend, showing a more or less steady rise from very low values to modern levels, is undoubtedly correct, but the details of the curve become increasingly uncertain as we go back in time.

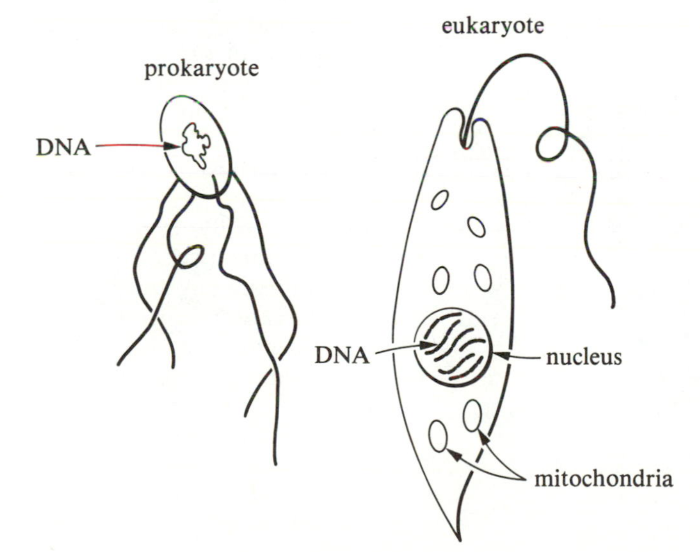

Figure 18.35. Schematic depiction of the difference between prokaryotic cells and eukaryotic cells. Apart from lacking a nucleus and internal organelles (like mitochondria), a prokaryote is typically much smaller than a eukaryote. (The sketches are not to scale.) The evolution from prokaryotes to eukaryotes is considered one of the most significant developments in the history of life on Earth.

The shift from a reducing to an oxidizing atmosphere is believed to have been caused almost entirely by blue-green algae (or similar simple organisms), because only in rocks less than 1.5 billion years old do geologists find microfossils of cellular organisms which are substantially more complicated than blue-green algae. Blue-green algae exemplify *prokaryotic* cells, cells which do not contain nuclei. Only in rocks less than 1.5 billion years old do we find examples of *eukaryotic* cells, cells with true nuclei (Figure 18.35).

Cells with nuclei offer much more versatility than cells without nuclei. For one thing, two cells with the same nuclei structure can serve completely different specialized biological functions. One cell in your body might be a brain cell; another, a muscle cell. On one level, brain and brawn are exactly identical; on another, they are completely different. This ability is called **cell differentiation**, and it is a crucial property that allows the more complex organization intrinsic to multicellular organisms. For another thing, only cells with nuclei are capable of advanced sexual reproduction, a mechanism with great

(a)

(b)

Figure 18.36. The fossil remains of some ancient multicellular organisms. (a) A Precambrian fossil of an annelid worm. (Courtesy of M. F. Glaessner.) (b) A fossil of a nearly complete trilobite. (From John S. Shelton, *Geology Illustrated*, W. H. Freeman, 1966.)

genetic advantages (Chapter 19). Biologically, then, the difference between prokaryotic and eukaryotic cells is as fundamentally important as the difference between plants and animals. Thus, the first appearance of eukaryotic cells was a major event in the history of life on Earth. In Chapter 20, we will discuss some of the speculations about how cells with nuclei could have originated.

The first multicellular organisms were probably soft-bodied and lived in the seas; so there are only a few preserved fossils of such lifeforms (Figure 18.36a). Shelled organisms appeared at the beginning of the **Cambrian** period, about 600 million years ago. For example, in rocks less than 600 million years old are trapped the remains of trilobites (Figure 18.36b). The fossil record for the Cambrian period is exceedingly rich, suggesting that, roughly 600 million years ago, there began an "explosion" in the variety of higher lifeforms. The course of evolution in subsequent times has been admirably mapped by paleontologists. In our tracing of the ancestral roots of humans to more primitive organisms, there are still gaps, but probably because of gaps in the fossil record, not because of real gaps in the evolutionary history. As Martin Rees has commented in another context, the absence of evidence does not constitute evidence of absence! In any case, mammals did not appear until 200 million years ago, and humans arrived on the scene only four million years ago. Thus, in the 4.6-billion-year drama of the Earth, humans have played a role only in the last one-tenth of one percent of the entire timespan.

Historical Comment

Speculations about the nature and age of the solar system and the Earth have always fascinated scientists and philosophers. Thales (640–562 B.C.), the "father of western philosophy," was an astronomer. Confucius (551–479 B.C.), the "father of eastern philosophy," recorded 36 eclipses of the Sun. According to J. L. E. Dreyer, Thales thought the Earth to be a circular disk floating like a piece of wood on the ocean, an idea that gives old meaning to the concept of continental drift! Anaximander (611–545 B.C.) thought that the Earth was a cylinder, with a height 1/3 the breadth, and that the Sun and the stars were wheels of fire. It was Pythagoras (572–497 B.C.) who argued that the Earth was a sphere. Heracleides (388–312 B.C.), a follower of Pythagoras, taught that the spherical Earth rotated freely in space and that Mercury and Venus revolved about the Sun. Aristarchos (310–230 B.C.) found the Sun to be larger than the Earth and thereby suggested that the Earth also revolved around the Sun. Unfortunately, these ideas fell into disfavor in the Western world when some of the leaders of Christianity insisted on too literal an interpretation of Scripture. The fall of the Roman Empire in A.D. 476 led to a "Dark Age," during which much of what was known to the Greeks was lost or forgotten. Hindu astronomers, however, still taught that the Earth was a sphere and that it rotated once daily. Only with Copernicus (1473–1543), who knew of Aristarchos' views and acknowledged them, did we progress again in our concept of the physical world.

Scientists themselves were partly to blame for this thousand-year hiatus in physical science. The Greeks set the pattern by adopting certain rigid schools of thought. Scholars were trained in two very different styles, and they suffered greatly from a lack of interaction. In the aesthetic mode, exemplified by Aristotle, the ideal was to find the great sweeping principles that govern the system of the world. The greater the apparent philosophical beauty of the argument, the more dogmatically were such principles to be embraced, no matter how contradicted by "petty" experience. In the empirical mode, exemplified by Ptolemy, the ideal was to describe actual events as accurately and as minutely as possible. No contrivance of the model, no matter how lacking in physical basis, was viewed as unnatural if it "saved the phenomenon." By partitioning their efforts in this manner, scientists were unable to successfully defend the validity of their subject before a hostile audience. Only with the unification of its aethestic and empirical branches in the work of Copernicus, Galileo, Kepler, and Newton did science grow strong enough to withstand attacks by opponents of free inquiry.

No subsequent discovery has had a greater impact on the way that we view the Earth than the geologists' discovery of its true age. We have come a long way since Aristotle argued on logical grounds that the world was eternal, extending infinitely into the past and the future; or since Archbishop Ussher argued on religious grounds that the world was 6,000 years old. We now know neither answer is anywhere near correct. The Earth does have a past, but that history is neither so long that we cannot ask sensible questions about its origins, nor so short that all scientific answers yield patent nonsense. Even more remarkable, perhaps, has been the discovery that not only is the issue of geological origin and evolution amenable to scientific analysis, but so is the issue of biological origin and evolution. Scientific investigations of the latter topic have triggered a new storm of criticism by people who do not wish to accept the findings. The rest of this book concerns the modern scientific view of the nature of life and its possible origin. Let each reader make an independent judgment of the merits of the case presented.

CHAPTER NINETEEN

The Nature of Life on Earth

What is the origin of life on Earth? Before we can address that issue intelligently, we have to ask, What is the nature of life? What distinguishes the living from the nonliving? These questions are difficult to answer precisely. Here it will suffice for us to note that scientists from Charles Darwin to Francis Crick have focused attention on two characteristics that all fully live things share: *growth* and *reproduction.*

Growth

The growth of living things differs from the growth of nonliving things in one spectacular way. A lot of information is needed to describe even one bacterial cell; much less is needed to describe a rock. Consider this fact. We can make a child grow by feeding it corn flakes. We can make a rock grow by adding cement to it. By having cement added to it, a rock becomes more and more cementlike. But as Carl Sagan has put it, by eating corn flakes, a child does not become corn flakes—corn flakes become child!

This commonplace fact, upon deeper reflection, is so remarkable that for a long time people thought there must be a special "life force" which distinguishes living things from nonliving things. This notion that the behavior of living things cannot, even in principle, be understood by ordinary processes of physics and chemistry goes by the name of **vitalism**. *Modern biology has completely discredited vitalism.* It is extremely regrettable that vitalistic notions can still be found in many unenlightened regions of this world.

To cite a refuge often adopted by vitalists, consider the thermodynamics of living things growing in an organized way. Since the information needed to produce this organization is the mathematical opposite of entropy (Chapter 4), it might be thought that living things violate the second law of thermodynamics. This is, however, false. As I. Prigogine has emphasized, living things are "open systems," where energy and chemicals flow in and out. When careful measurements are made, it can be shown that a living thing gains internal order only by introducing more disorder into its surroundings. Since the second law of thermodynamics requires only the total entropy (disorder) in the universe to increase, or at best remain constant, living things do not, in fact, violate the second law.

The persistent vitalist might argue that even the gaining of local order at the expense of introducing general disorder seems to require something special. And so it does; it requires an input of free energy from the outside. For animals, that input is other animals or plants. For plants, that input is ultimately sunshine. But given the existence of such an input, living things cannot be regarded as unique in their ability to produce circumstances which violate thermodynamic intuition. Let us give some astronomical examples. In Chapter 5 we encountered the counterintuitive result that the base of the chromosphere of the Sun is colder than the regions on either side of it. In Chapter 17, we encountered similar phenomenon on Jupiter, as well as one on Earth, where the temperature of the tropopause is lower at the equator than at the poles. In each of these examples (there are others), we can also point to a likely source for the free-energy input: the heat engines of the underlying convection zones.* Thus, the ability of living things to gain local structure at the expense of seemingly chaotic surroundings differs only magnitude, not in kind, from the ability of nonliving things.

Reproduction

The ability for *self-replication* by living things is even more remarkable than ordered growth. Let us talk about the birds and the bees. How is it that birds beget only birds, and bees, only bees? Why isn't this information degraded from generation to generation? Why don't birds beget sponges, and sponges beget rocks? This would be the naive expectation of a progression from order to disorder. How did the complexity which is a bird arise in the first place?

The last question can be answered at two different levels, the macroscopic and the microscopic. The macroscopic answer has been known for more than a hundred years. Charles Darwin and Alfred Wallace independently provided it in 1858 with their theory of **natural selection**. The microscopic answer has been known for only a few decades. It began with the 1953 discovery by James Watson and Francis Crick of the double-helix structure of DNA, and the unraveling of the remaining puzzles is an active ongoing enterprise.

Figure 19.1. Darwin's finches. (From *Darwin's Finches* by D. Lack. Copyright © 1953 by Scientific American, Inc. All rights reserved.)

Natural Selection and Evolution

In 1831, at the age of 22, Charles Darwin shipped as a naturalist aboard the *H. M. S. Beagle* for a five-year trip around the world. In the Galapagos Islands, Darwin noted the curious fact that each island had its own variety of finch, tortoise, etc. (Figure 19.1). There were naturally variations about the mean, but the average was different from island to island. Darwin came to the conclusion that each island variety descended from a common stock, but that generations of reproduction on isolated islands had produced different incipient species on each island. By 1838, Darwin had formulated a mechanism—natural selection—by which the *average characteristics of a population* (such as those contained on any one of the islands) could gradually change. In other words, Darwin conceived of natural selection as a mechanism for the *biological evolution of a species*. The key to his idea lay in rapid reproduction, which would ensure a distribution of different individuals in the general population. How does natural selection then work?

One of the most influential treatises in Darwin's day was published in 1798 by the economist Thomas Malthus. In his *Essay on Population*, Malthus pointed out that human reproduction tends to proceed by geometric progression: two million beget four million, who beget eight million, who beget 16 million, etc. (Figure 19.2). But the Earth has only resources to sustain a finite population of humans; therefore, when the environmental limitations are reached, the death rate must increase until it balances the birth rate (Figure 19.3). Darwin saw that this principle should apply to any biological species, and he asked the crucial question: When the population has grown so large that the environment can no longer support it, which of the many offsprings would have the

* In other cases, for example, the superrotation of Venus's upper atmosphere (Color Plate 47), the exact free-energy input remains unclear.

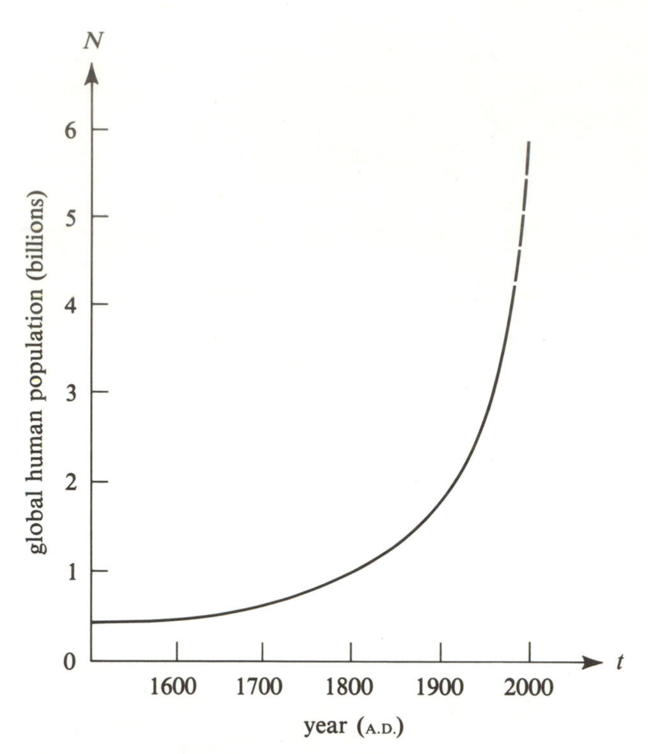

Figure 19.2. The world human population has grown by geometric progression (exponential growth except for temporary deviations) for much of recorded history. Many projections, including those by the United Nations and a special commission appointed by President Carter, state that the world population will reach or exceed six billion by the year 2000. (For details, see *Population, Evolution, and Birth Control*, ed. G. Hardin, W. H. Freeman, 1969.)

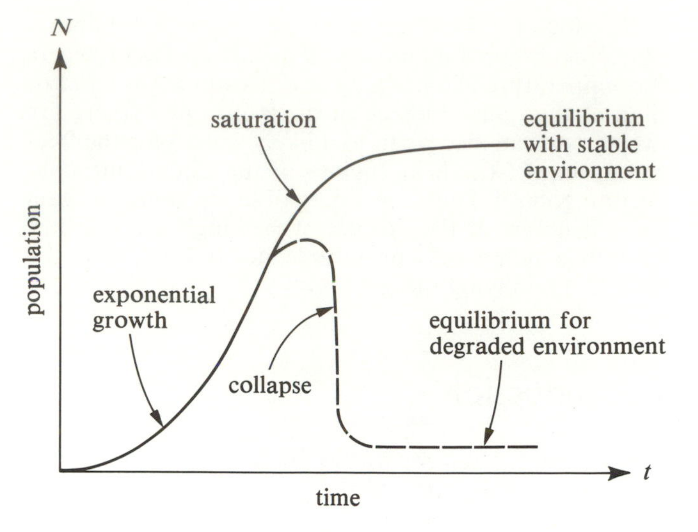

Figure 19.3. Equilibrium populations of any biological species. The solid curve shows the characteristic *S*-shape that typifies the initial explosive growth and saturation of number of individuals as the limits of the environment to support more members are reached. If the increase in the number of individuals is accompanied by a corresponding degradation of the environment, then there may be a collapse of the system until enough individuals have died for an equilibrium to be possible under the conditions of a much degraded environment. The dashed curve illustrates schematically the second situation.

best chance of surviving? His answer: those individuals who happen to have the characteristics best suited for survival under the prevailing circumstances! The disadvantaged varieties die off, or their offspring are diluted in the expanding pool of the advantaged variety. Such a selective elimination of the nonprivileged during episodes of biological stress leads to the "survival of the fittest," to use a phrase introduced by Herbert Spencer and popularized by Thomas Huxley.

Darwin also recognized that biological stress could arise from changes in the geological environment; so he proposed that biological evolution could be stimulated by geological events. At this time in the nineteenth century, geologists had already firmly established the fact of persistent and gradual changes in the features of the Earth itself. The eternal world of Aristotle was crumbling, and evolution was in the air! Indeed, Charles Darwin had set sail on the *H. M. S. Beagle* with a copy of Charles Lyell's *Principles of Geology* (published in 1830) tucked under his arm, and he was greatly impressed by the uniformitarianism argument for the need of long timescales to produce geological changes. Above all else, time was needed if the process of natural selection was to lead to the evolution of complex species. And time enough plus a natural source of changing environmental pressure were what the new geology gave.

Evolution Is Not Directed

Darwinian theory does not view evolution as being directed toward some noble purpose. Biological evolution proceeds by a series of chance happenings. The phrase "survival of the fittest" refers to those fittest to produce offspring, not fittest in some moral or directed evolutionary sense.

Imagine a sea containing a population of fish which have fins with a distribution of sizes (Figure 19.4). That is, because of natural variations or whatever, there are some fish with fins longer than the average. In ordinary circumstances, this anomaly is neither advantageous nor disadvantageous. If the anomaly arose because of a mutation, the mutation was neutral. However, imagine that the sea begins to close because of the convergence of two continental plates (Chapter 18). As the shallow sea begins to dry up, the fish with longer than average fins and their offspring find it easier to waddle from one mudhole to

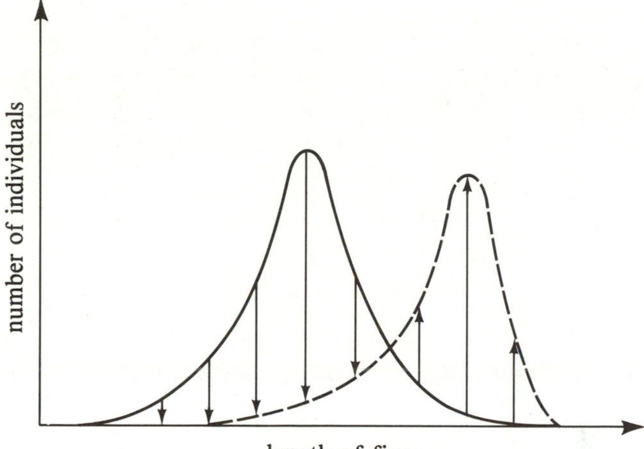

Figure 19.4. A schematic depiction of the evolution of the average characteristics of a population. Under normal conditions, the equilibrium distribution of the lengths of fins of a fish population may show the shape indicated by the solid curve. Due to natural variations, there will be fish with fins longer and shorter than the average length. When the shallow seas dry up, those with shorter fins may die off selectively (downward arrows), and those with longer fins may reproduce selectively, passing on this trait to their increased number of progeny (upward arrows). In this manner, the species changes to a distribution with longer average fins (dashed curve).

the next. Those fish with average or shorter fins tend selectively to perish. As a result, the fish with longer fins get to eat the food that their competitors might otherwise have eaten, and get to mate with fish that their competitors might have bred with. As a consequence, if the characteristic fin length gets passed on to the descendants, the population of fish changes to a distribution with longer and longer fins, better and better suited for walking on dry land. Could this have been the beginning of land animals?

The important point here is that evolution is not *directed* toward walking on land. No fish ever decided: "Hey, let's walk on dry land." Instead, because of inevitable copying errors or because of the natural dispersion in the lengths of fins, there is *always* a supply of fish with fins shorter than average and another supply with fins longer than average. If *none* of the fish had fins suitable for waddling from mudhole to mudhole, then that species would simply have become extinct when the shallow sea dried up.

Notice, however, that as a consequence of this chance happening, the species, if it does survive, may gain in variety and in complexity: variety because there may exist more than one successful adaptive strategy; complexity because the mechanisms that arise in response to each new crisis are passed down to successive generations (with eventual attrition for useless appendages). Given enough time, Darwin believed, even as complex an organism as a human could have arisen from the process of natural selection. The gain in complexity would have occurred merely by a superposition of many accidents and the early demise of many brethren. This, then, is the teaching of Charles Darwin. It won him worldwide fame and worldwide controversy.

Darwin's Accomplishment in Perspective

Darwin was not the first to conceive of the idea of biological evolution. Before Darwin, eighteenth- and early nineteenth-century scientists like Maupertius, Buffon, Linneaus, and Lamarck had speculated on the possibility of limited biological evolution. Thus began the break with the Aristotlean tradition that living species had been fixed for all time. However, the views of these early workers differed from Darwin's (and from each other's) in several crucial respects. For example, of the predecessors to Darwin, the views of Jean Baptiste de Lamarck were perhaps the most sophisticated. In particular, Lamarck believed a species could evolve to higher forms, rather than simply vary on a basic theme established at Creation, or even degenerate from more "perfect" specimens, as most of his contemporaries preferred. Unfortunately, Lamarck also thought that each group of organisms represented an independent evolutionary line which originated with "spontaneous generation in heat and humidity" and subsequently strove toward perfection by acquiring good characteristics from experience. For Darwin, evolution was not directed, and acquired characteristics of individuals could not be inherited, but *natural selection* could provide the mechanism by which the *species* could be transformed into more complex forms. Indeed, the far-reaching implications of natural selection as an evolutionary mechanism led Darwin ultimately to speculate that all living organisms, including humans, might be linked to a single *chemical* origin of life. In this radical view, Darwin differed even from his contemporary Alfred Wallace, who in 1858 had arrived independently at the idea of natural selection. (Indeed, it was the realization of Wallace's imminent presentation of this discovery that led Darwin finally to cut short his twenty-year delay in the presentation and publication of his own views.) Wallace exempted humans from the process of natural selection, and by this exception avoided much of the controversy (and the fame) which attached to Darwin.

To appreciate the courage of Darwin's stand, one must realize that even after the publication of his *Origin of Species* in 1859, which presented in exquisite detail the fruits of a lifetime of gathering evidence and weaving closely reasoned arguments, many leading scientists of Darwin's day were reluctant to accept his conclusions. Even Charles Lyell, one of the leaders of the revolution in geology, was a late (but ardent) convert to Darwin's views. In hindsight, one might quibble and point out that not all of Darwin's ideas were original with Darwin. Darwin certainly owed a great debt to Malthus and Lyell, among others. As Newton said of himself, one might say of Darwin: if he had seen further than others, it was because he had stood on the shoulders of giants. But because of the truly penetrating light provided by the ideas which Darwin wove together into a single unifying theory for all of biology, Darwin, like Newton, must be reckoned as one of the greatest minds produced by civilization.

Problem 19.1. Explain why the following cartoon, which is adapted from *Punch*, March 25, 1861, represents a misunderstanding of Darwinian theory.

Figure 19.5. Adaptation is exemplified by the peppered moth, which shows both dark- and light-winged forms. The top photograph shows a pair resting on the trunk of a bare tree; the bottom, a pair resting on the trunk of a lichen-covered tree. (For details, see Cook and Bishop, *Scientific American*, **232**, Jan. 1975, 90; or Lewontin, *Scientific American*, **239**, Sept. 1978, 212.)

The Evidence in Favor of Darwinian Evolution

Most modern readers of *The Origin of Species* have found the evidence and reasoning there compelling enough to convince them of the essential correctness of the Darwinian theory of natural selection and evolution. Since Darwin's time, the weight of scientific discoveries has landed even more heavily in favor of his point of view. We will discuss the evidence in genetics and molecular biology shortly, but let us consider here a few macroscopic examples.

It is difficult to see natural selection in "living action" (as contrasted with "fossilized records") in mammals, since most mammals have long reproductive cycles, so that systematic changes in a population will take longer than the lifetime of human investigators. The same objection does not apply to insects, and since Darwin's day, some striking examples of evolution have been documented for insect populations. A famous example is the case of the light and dark forms of the peppered moth (Figure 19.5). These moths have wing colorations which serve as camouflage against predator birds when the moths rest against the trunks of oak trees in the part of Wales where these moths are common. In the eighteenth century, lichen commonly grew on such trees, and the light-winged moths were dominant. Air pollution during the industrial revolution killed the lichen, darkening the barks, and the dark-winged moths became prevalent during the nineteenth century. In the twentieth century, coal burning has decreased, the lichen have made a comeback, and so have the light-winged moths.

Problem 19.2. Explain the case of the peppered moth in terms of natural selection. Be careful not to use the concept of directed evolution.

Less benign lessons have been taught to humanity in the area of chemical pest control. The extensive use of petroleum-derived insecticides have been responsible, in large measure, for the "green revolution" which has helped to feed a burgeoning world population. Such pesticides typically lead to an immediate and impressive reduction of numbers of the pests that are the target of the program. Unfortunately, the rapid breeding cycles of the pest insect plus the unwanted elimination of its natural predators by the same pesticides lead to the emergence of new generations of insecticide-resistant strains. This requires the invention of even more powerful and less discriminatory insecticides, which compound the problem. A similar experience has occurred with the use of poisons to control rats.

Problem 19.3. Explain the pest-control problem in more detail than is given in the text. In particular, find references and articles dealing with alternatives to pesticides for the control of pest populations, especially those that seek to disrupt the reproductive or developmental cycles of the pest. Discuss the advantages and disadvantages of these alternative methods in terms of natural selection and of economics.

Now, the skeptic might be willing to accept the concept of natural selection for "lower forms" like insects and rats (both of which "breed like animals"), but not for "higher organisms" like humans. However, it is a popular misconception that insects and rats are lower on the evolutionary ladder than humans. Lower in intelligence, certainly; lower in the sense of natural selection, no! The Darwinian phrase "survival of the fittest" has no moral or aesthetic connotation; it merely refers to the ability of a species to adapt successfully to varying environmental conditions and to reproduce in large numbers. Judged from this purely biological perspective, insects, rats, and flowering plants are all the equals (and perhaps eventually the superiors) of humans. They have succeeded by the sheer ability to reproduce quickly and in large numbers. Their evolutionary success requires the inevitable early deaths of countless billions. We have succeeded by our brain power, and our numbers—alone among all the species—continue to grow on an unsaturated exponential curve (Figure 19.2). Let us hope that we are wise enough to exercise our power and circumvent the unpleasant prelude to the operation of the law of natural selection (Figure 19.3 or worse).

Our examples so far all refer to the present or to the immediate past. Is there any evidence for the operation of the law of natural selection in the distant geological past? Yes, from the study of fossils mentioned in Chapter 18. By radioactive dating of the rocks that the fossils are found in, paleontologists have reconstructed a history of the types of organisms that existed on Earth hundreds of millions, and even billions, of years into the past. The record is yet incomplete, but there do exist examples of continuous and gradual changes of the forms of a species that is required by the Darwinian theory. There were also short periods of intense proliferation of the varieties of species. For example, consider the explosion of life-forms which took place 600 million years ago, inaugurating the Cambrian period, or the appearance of birds and mammals which took place 200 million years ago during the Triassic period. Geologists now believe that these events were associated with the environmental crises brought on by the plate-tectonic processes which occurred prior to the formation of the Appalachian mountains and during the most recent breakup of Pangaea (Chapter 18). Indeed, the theory of plate tectonics is expected in the near future to explain much of the correlation between major geological events and episodes of intense biological evolution and extinction. This development was, of course, foreseen in spirit, if not in detail, by Darwin's work on natural selection and evolution.

The Bridge Between the Macroscopic and Microscopic Theories of Evolution

As great as is Darwin's theory of natural selection, it does have major weaknesses, the most important being that the theory contained no intrinsic explanation of why characteristics vary between individuals of a species and how such characteristics are passed on from parents to offspring. In Darwin's theory, variations between individuals and the inheritance of genetic traits were postulates based on everyday experience. Darwin himself had no clear concept of the basic mechanisms which

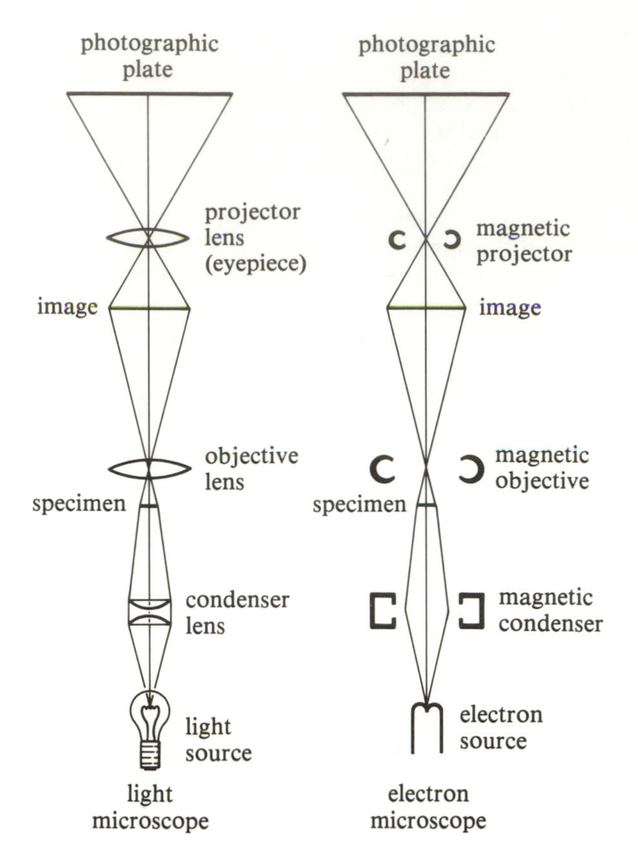

Figure 19.6. Schematic depictions of a light microscope and an electron microscope. Leuwenhoek's compound microscope was a forerunner of these modern instruments. A modern light microscope allows the resolution of the smallest bacterial cells (about 10^{-5} cm), whereas an electron microscope allows the resolution of structures somewhat larger than individual molecules (say, 10^{-7} cm). See Chapter 2 for a discussion of the concept of angular resolution, and contrast the microscope with the telescope.

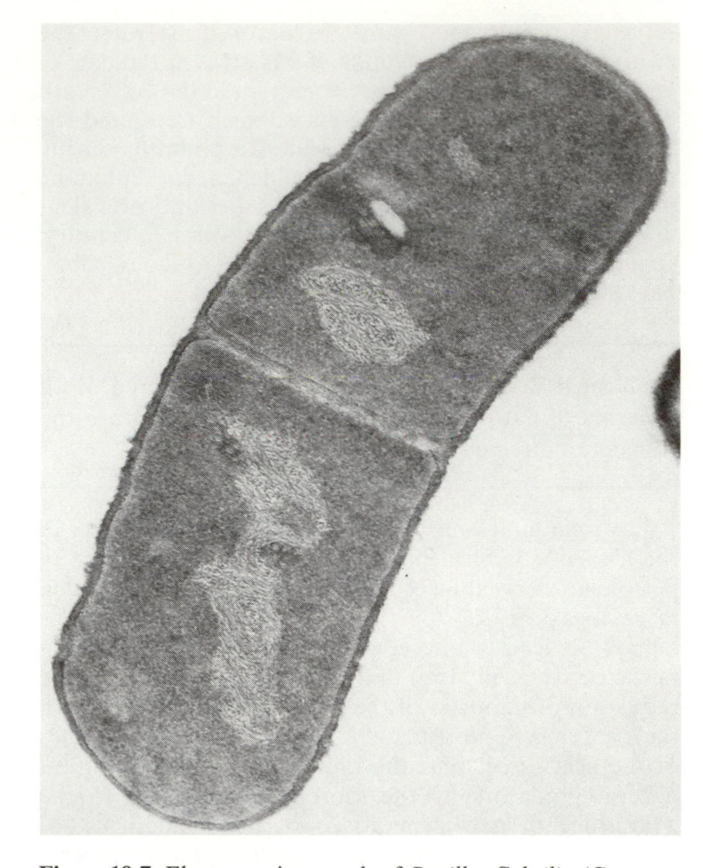

Figure 19.7. Electron micrograph of *Bacillus Subtilis*. (Courtesy of C. Robinson and J. Marak.) The length of this cell, which is about to divide is approximately 2×10^{-4} cm. The DNA is seen as the strands of spaghetti in the middle of the cell. (From Gunther S. Stent, *Molecular Genetics*, W. H. Freeman, 2d ed., 1978.)

underlay these biological phenomena. Indeed, Darwin shared the common misconception of his day that heredity was somehow carried in the blood of the parents. This misconception is repeated today whenever snobbish people speak in terms of "bloodlines."

Fortunately for the development of biology, two discoveries were made close to the time of Darwin's publication of *The Origin of Species* which allowed the final bridging of the gap between the macroscopic phenomena and the microscopic mechanisms. The first development was the concept of the *cell* as the unit of life, introduced by Rudolph Virchow in 1858. The second was the concept of the *gene* as the unit of heredity, introduced by Gregor Mendel in 1866. Unfortunately, the work of Mendel, an Austrian monk who published his findings in an obscure journal, went unnoticed until 1900, when Correns, DeVries, and Von Tschermak rediscovered many of Mendel's results.

The Cell as the Unit of Life

The term *cell* (meaning "little room") was introduced by Robert Hooke in 1665 to describe the honeycomb structure of many plant tissues. Most individual cells are too small (say, 10^{-3} cm) by a factor of about 10 to be seen by the naked eye, although there are spectacular exceptions, such as a chicken egg. Anton van Leuwenhoek's invention of a good compound microscope in 1674 (Figure 19.6) allowed him to discover that living organisms could consist of a single cell (Figure 19.7). By 1839, it became widely accepted through the work of Schleiden and Schwann that cells constitute the basic structural component of all living things. Virchow's proposal in 1858 that new cells originate from the division of preexisting cells (except, of course, possibly for the very *first* cell) completed the interpretation of the cell as the unit of life.

Problem 19.4. Microscopes are to biologists what telescopes are to astronomers. The main purpose of a microscope is **magnification**. A telescope can also be used to provide magnification (Problem 2.3), but its main purpose is to gather more light via a large aperture for the objective lens or mirror (Chapter 2). The magnification M of a compound microscope is the product of the *lateral* magnification m_o of the objective lens and the *angular* magnification M_e of the eyepiece. To see what we mean by these terms, let us consider a compound microscope, where the biologist looks through an eyepiece at the image formed by the objective lens (Diagram 1). Let us consider both lenses to be thin. For a thin lens of focal length f, the relation between the distances s and s' of the object and of the image from the lens respectively, is given by the equation $1/s + 1/s' = 1/f$. Show that this relation may also be written in the Newtonian form $(s - f)(s' - f) = f^2$, where $s - f$ and $s' - f$ are, respectively, the distances of the object and image from the closest focal point of the lens. The ratio of image to object sizes, I/O, is called the lateral magnification, m. It clearly scales by the rule of similar triangles with the ratio of image and object distances: $m = s'/s$. Use the Newtonian form of the thin-lens equation to show that the lateral magnification of the objective is given by $m_0 = (s_0' - f_0)/f_0$. The angle θ_e (assumed small) subtended at the eyepiece by the image of the objective is given by $\theta_e = I/f_e$ (see also Figure 2.14). The magnifying power of the compound microscope is defined to be the ratio of the angle θ_e to the hypothetical angle θ_o that would be the subtended at the eye of the biologist if the object were brought to the closest distance d_o where the naked eye could still focus sharply: $\theta_o = O/d_o$, where $d_o = 25$ cm. Thus, $M = \theta_e/\theta_o = (I/f_e)(d_o/O) = m_o M_e$, where $m_o = I/O = (s_o' - f_o)/f_o$ and $M_e = d_o/f_e$ define the lateral magnification of the objective and the angular magnification of the eyepiece, respectively. A high-power microscope might have $f_o = f_e = 1$ cm and $s_o' - f_o = 40$ cm. Calculate m_o, M_e, and M of such a compound microscope. If the naked eye can discern objects of size 10^{-2} cm in size, what would a high-power light microscope allow

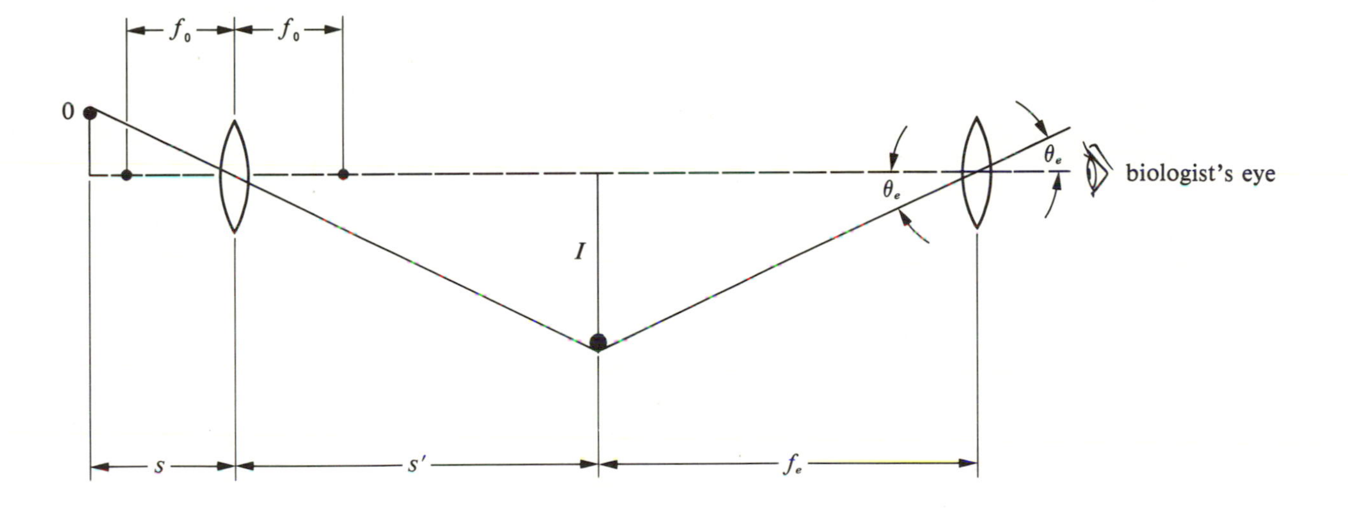

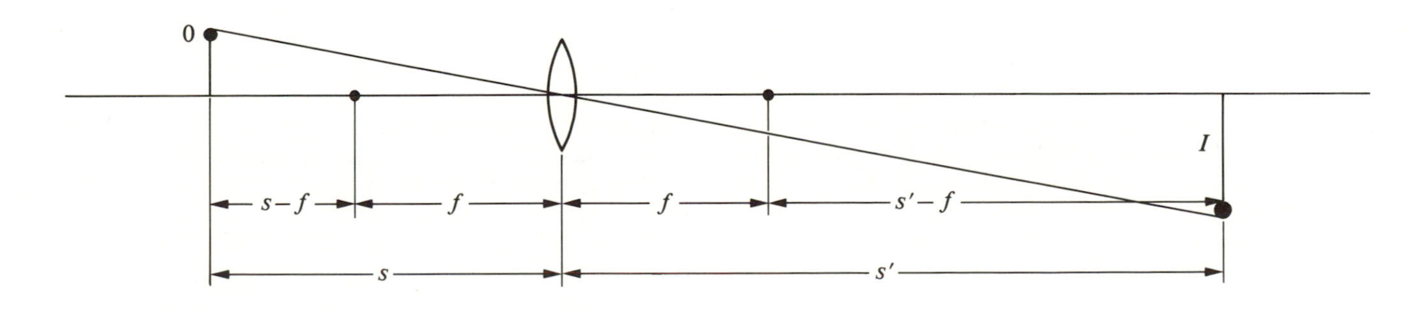

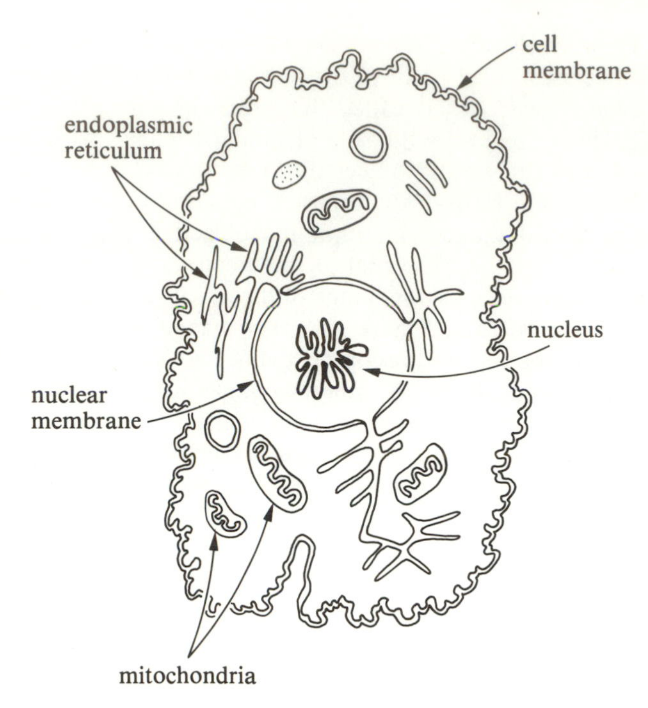

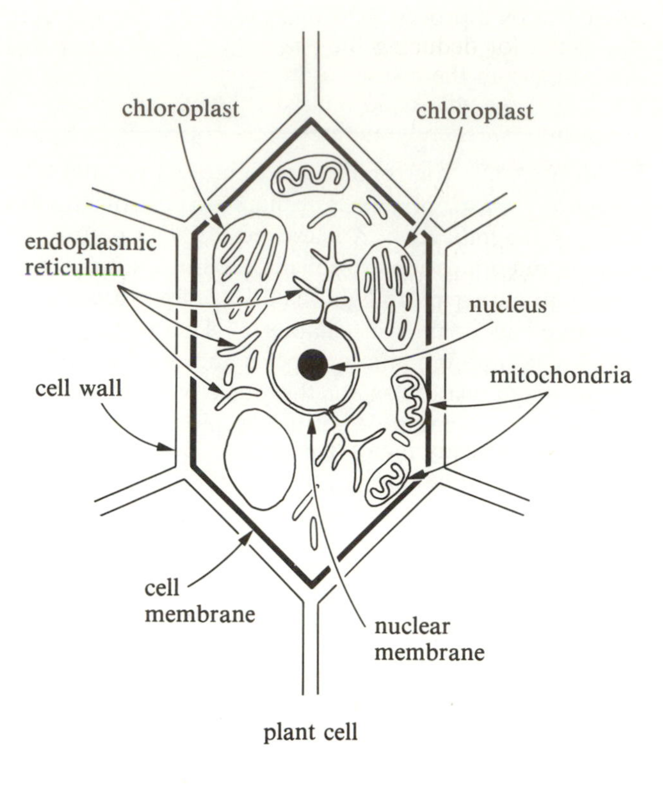

Figure 19.8. Simplified diagrams of typical animal and plant cells. See text for explanation of functions of the various components of the cell.

one to see? By increasing M even further, would a light microscope allow us to discern even smaller objects? *Hint*: Consider the limits placed on resolution by diffraction effects (Chapters 2 and 3).

From the modern perspective, cell division (**mitosis**) constitutes an important element in the *growth* of multicellular organisms. In unicellular organisms, mitotic cell division also results in *reproduction*; whereas the *sexual reproduction* of higher organisms involves a more complex form of cell division (**meiosis**) plus fertilization. We will explore in greater detail the cellular mechanisms of mitosis and meiosis after we have discussed the cell from the vantage point of molecular biology. For now, we merely note that it is only cells with nuclei that are capable of both types of cell division.

Figure 19.8 shows simplified diagrams of the structural components found in typical animal and plant cells. Surrounding the entire cell is a thin and partially permeable **cell membrane**, which allows certain molecules to flow in and out. Cell membranes are composed of medium-sized molecules called **lipids** (e.g., fats and oils) and very large molecules, or macromolecules, called **proteins**. In plant cells, there may be an additional **cell wall** which lends the cell some rigidity. Aside from lipids and proteins, cell walls contain **polysaccharides** (like starch or cellulose), which are long chains (or **polymers**) of simple sugars (a monomer or monosaccharide like glucose). Sugars and polysaccharides belong to the general class of **carbohydrates**.

Near the center of the cell is found a spherical or oval **nucleus**, which contains most of the master material (DNA, which is a **nucleic acid**) that ultimately governs the basic operations of the cell and transmits the hereditary traits from one cell generation to another. In other regions of the cell are found such inclusions as **mitochondria** and **chloroplasts**. Mitochondria are the powerhouses of both animal and plant cells; they are the sites where the energy contained in carbohydrates is converted by respiration into a form useful to the cell. Chloroplasts exist only in green plants and in some photosynthetic bacteria; they are responsible for the production of carbohydrates by the process of photosynthesis. Mitochondria and chloroplasts also contain some DNA, and they can increase in number within a single cell by a process of division. These latter facts are significant for deducing the origin of mitochondria and chloroplasts in the history of the development of life on Earth that we will explore in Chapter 20.

For now, we merely summarize that the main chemical components of the cell are the four groups of organic compounds: lipids, carbohydrates, proteins, and nucleic acids. They all play essential roles in various biological processes, but which one controls ultimately the processes responsible for our definition of life, growth and reproduction? The answer came from the science of genetics.

The Gene as the Unit of Heredity

We have already mentioned that the laws of genetics were first formulated by Mendel in 1866. Mendel's pioneering work on heredity were amazingly simple. (Monks do not have access to expensive laboratory equipment!) Mendel cross-bred varieties of peas and recorded the parental traits passed on to the offspring. He made the fundamental discovery that such traits were transmitted (in sexual reproduction) from parents to offspring in discrete units, *two* at a time, with the probability of any offspring receiving a particular unit (now called a *gene*) governed by the *laws of chance*.

Let us consider a specific example. By mating pure-bred pea plants containing smooth peas with pure-bred pea plants containing wrinkled peas, Mendel found all the progeny to contain smooth peas. This suggested to Mendel that the trait for smooth peas was **dominant**, for wrinkled peas, **recessive**. When he bred the first-generation hybrids with one another, he found 3/4 of the second-generation progeny to contain smooth peas, and 1/4 to contain wrinkled peas. These proportions hold in a statistical sense only, i.e., when the number of progeny in this generation is very large.

To account for these results (and similar ones concerning the colors of the peas as well as other characteristics of the plants), Mendel developed a brilliant theoretical model. His explanation constitutes the fundamental basis for the science of genetics. Mendel's theory, stated in modern terminology, contained the elements shown in Box 19.1 Figure 19.9 shows how Mendel's laws predict the results of his experiments on the shapes of peas in pea plants.

BOX 19.1
Mendel's Laws of Genetics

(a) In advanced sexual reproduction, each trait is controlled by a *pair* of genes.
(b) In the formation of progeny, each parent contributes *one gene* of each pair that characterizes a certain trait.
(c) If the two genes of a pair are different from one another (are **alleles** for a given trait), only the trait of the *dominant* gene will actually be expressed. (This law is only approximately true.)
(d) When more than one trait is being monitored, the distribution for each trait among the progeny follows a principle of *independent sorting*. (This law is also only approximately true.) For example, traits for pea shape are inherited independently from those for pea color.

Problem 19.5. Sickle-cell anemia manifests itself only when a person has two sickle genes. If a person has one sickle gene, he or she is only a carrier. If both parents are carriers, what is the probability that a child suffers the disease?

Problem 19.6. Suppose you start off with pure-bred pea plants containing smooth yellow peas and pure-bred plants containing wrinkled green peas. Assume smooth shape (S) to dominate over wrinkled shape (s), and yellow color (C) over green color (c). Denote the pure-bred smooth yellow peas by the notation $SSCC$ and the pure-bred wrinkled green peas by $sscc$. Show that Mendel's laws predict that the first generation of crosses between such pure-bred plants all have the genes $SsCc$, and will bear, therefore, smooth and yellow peas. Show also that Mendel's laws predict 9/16 of the second-generation progeny of the first-generation hybrids will have smooth yellow peas; 3/16 will have wrinkled yellow peas; 3/16 will have smooth green peas; and 1/16 will have wrinkled green peas.

Mendel's achievement of such clean results derived in part from his choice of characteristics to study; he chose traits with only two alleles. In contrast, the eye colors of fruit flies are controlled by more than a dozen different alleles. Moreover, Mendel considered characteristics with a strong sense of dominance or recessiveness. Often the traits of the parents can be mixed in the offsprings. For example, a cross of red- and white-flowered plants may produce a plant bearing pink flowers. Also, although the transmission of some genes takes place independently from that of other genes (Mendel's hypothesis of "independent sorting"), most observed traits are the results of the interaction of a large number of genes and can display a wide latitude of variation, as with for example, human intelligence and coordination. Mendel's intuition, which led him to

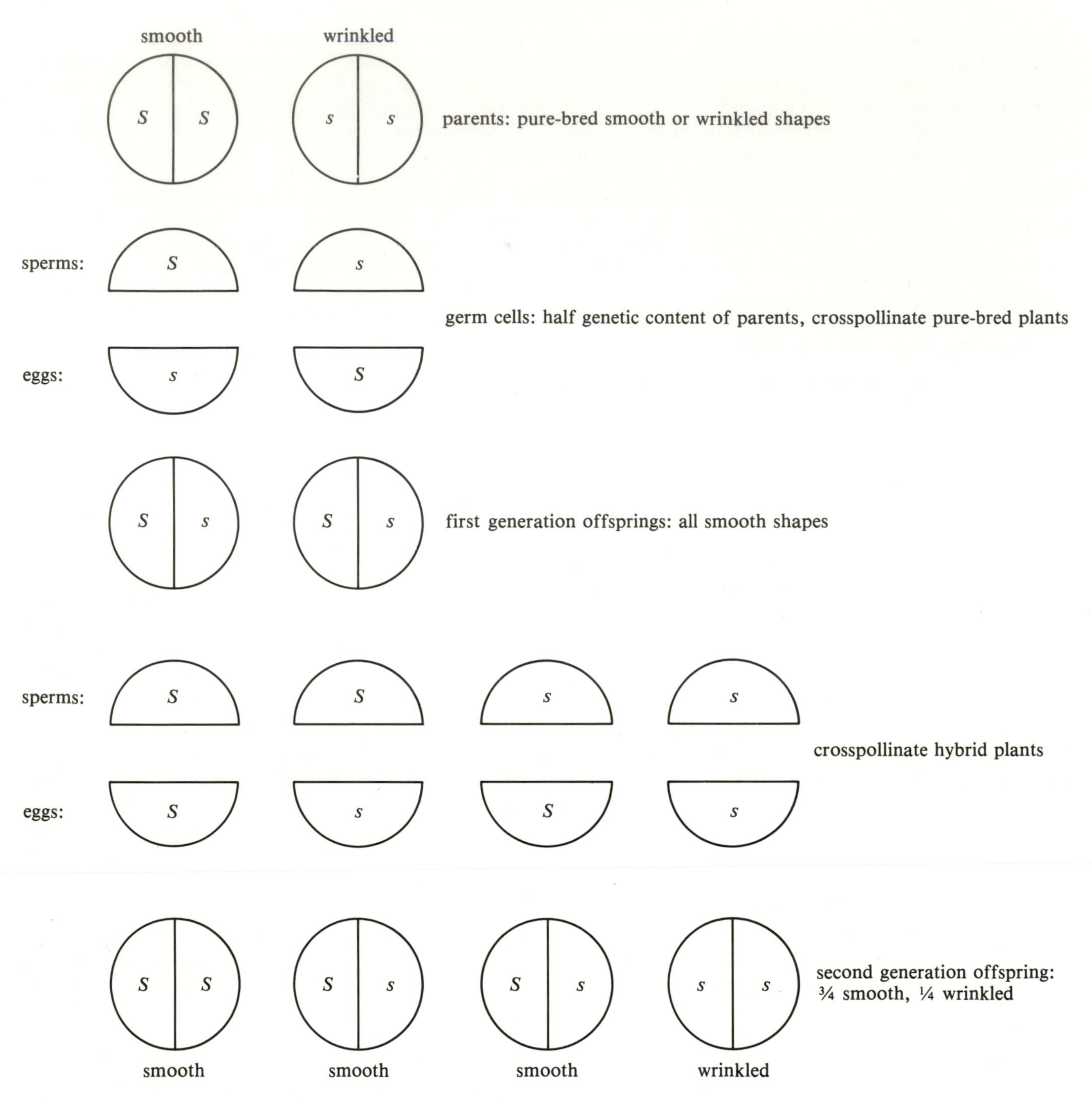

Figure 19.9. Explanation of Mendel's experiments concerning the inheritance of the shapes of seeds in pea plants. The dominant gene for shape is denoted S; the recessive gene, s with S = smooth, and s = wrinkled.

choose simple cases to make a first study, plus his meticulous technique and insistence on working with large numbers of plants, resulted in his being able to see patterns of behavior some three and a half decades before other scientists. Truly, Mendel's accomplishment was an astonishingly far-sighted piece of work!

Problem 19.7. In population studies, the typical variation about the mean can be expected to follow a generalized "random walk" law. Thus, if N examples are expected theoretically, an error on the order of $N^{1/2}$ can be expected when N samples are actually taken. The fractional error is of the order of $N^{-1/2}$. What does N have to be if we desire a sampling accuracy, say, of 3 percent? Does this explain to you why Mendel conducted experiments on thousands of plants, and why public-opinion polls typically sample a thousand or more responses?

Chromosomes, Genes, and DNA

The realization that the cell is the unit of life, in that each cell has the ability to replicate itself, plus Mendel's work on genetics, implied that the material governing heredity must lie within living cells. Where? Biologists who studied the process of cell division (mitosis) at the end of the nineteenth century used staining techniques to make the interior features of the cell stand out more clearly. In particular, long threads of material, found in the nucleus of a cell, gained dark color upon staining and were called **chromosomes**. In 1879, Walther Flemming observed that, before cell division, each chromosome divided into a pair of two chromosomes, each identical to the parent structure. Then these threads shortened (by coiling), and one of each pair of identical chromosomes moved to an opposite side of the nucleus. The cell then divided, with each daughter cell receiving the same complement of chromosomes as was present in the original cell. These chromosomes were encased in nuclear membranes, and the two daughter cells began a new phase of independent growth and eventual cell division (Figure 19.10).

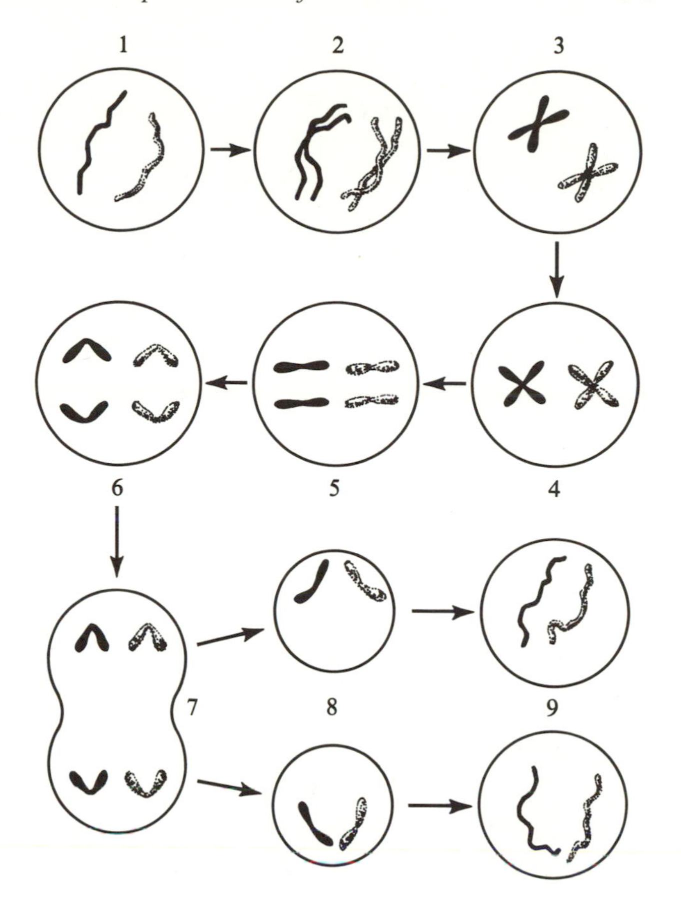

Figure 19.10. Simple cell division, or mitosis. (1) Two different chromosomes, say, in a bacterial cell, are visible in an elongated configuration. (2) Each chromosome forms a double. (3) The chromosomes shorten and thicken, presumably by coiling. (4) The chromosomes line up along the equatorial plane. (5) Each chromosome double divides. (6) One of each single chromosome moves to opposite poles. (7) The cell begins to divide. (8) Division is complete with the formation of two daughter cells with the same genetic content as the original parent. (9) The chromosomes elongate and the daughter cells grow when supplied with nutrients.

W. S. Sutton in 1903 was the first to give a modern interpretation of the above process (and that of egg fertilization) as the mechanism by which Mendelian genetic information is transmitted from generation to generation. The implication that the genes of organisms was located in a linear sequence along the chromosomes (23 pairs in humans) was confirmed in studies conducted at the beginning of this century by C. E. McClung, N. Stevens, and E. B. Wilson. In particular, Stevens found that female mealworms have two large chromosomes that she called X chromosomes, while male mealworms have a large X chromosome paired with a small Y chromosome (Figure 19.11). The egg of the female mealworm contained only a single X-chromosome (formed in *meiosis* by the splitting of a XX chromosome pair), whereas the sperm of the male mealworm contained either an X chromosome or a Y chromosome (formed in *meiosis* by the splitting of an XY chromosome pair). Thus, when a sperm and an egg unite (fertilization), the result may be either XX (female) or XY (male). Clearly, therefore, the X and Y chromosomes must contain the genetic material that determines the sexual characteristics of the organism! That X and Y chromosomes contain, in addition, other genetic information

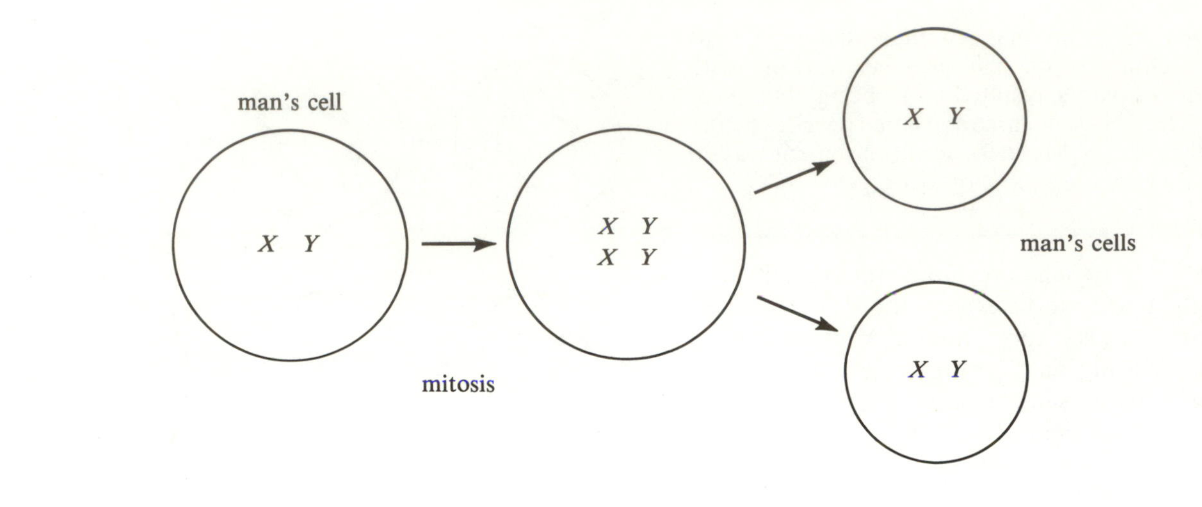

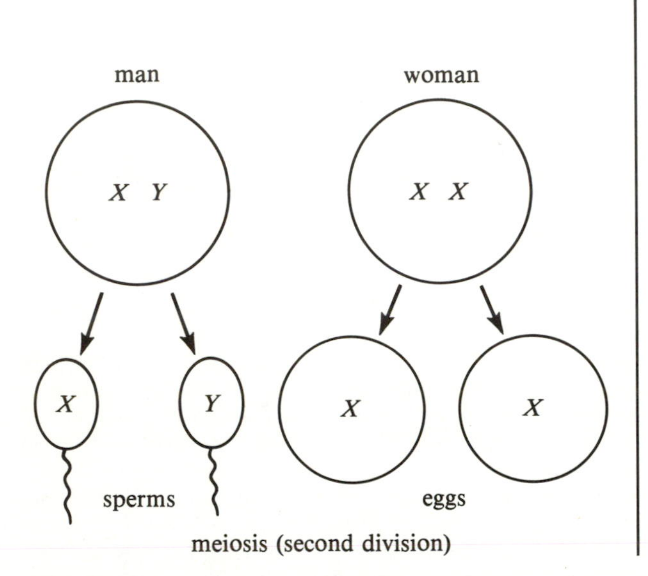

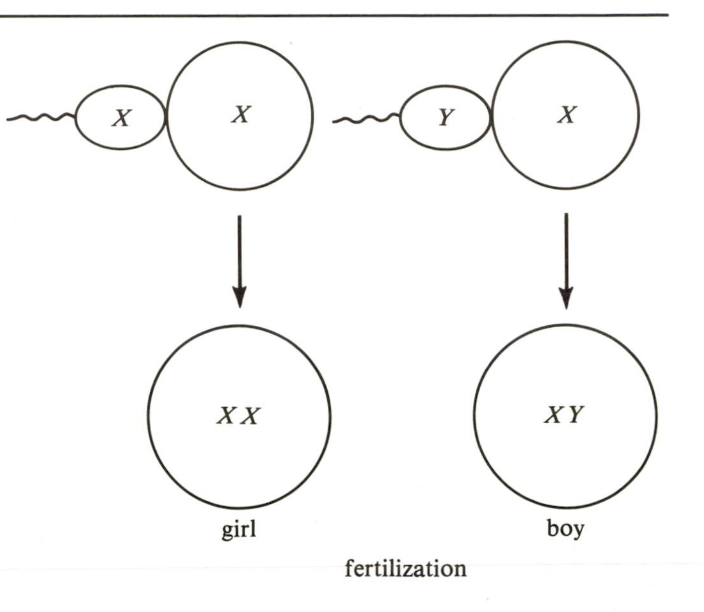

Figure 19.11. The sex of a human being is determined by a pair of sex chromosomes. In the woman, each body cell contains two X chromosomes; in the man, each body cell contains an X and a Y chromosome. In mitosis, cell division produces two body cells in a man, each of which has an X and a Y chromosome. In meiosis, cell division produces germ cells with only one sex chromosomes. Thus, normal sperm contain either an X chromosome or a Y chromosome, and normal eggs contain one X chromosome. Fertilization of an egg by an X sperm produces a girl; by a Y sperm, a boy.

was shown by T. H. Morgan in 1910, when he demonstrated the transmission of sex-linked characteristics in fruit flies.

Problem 19.8. Because of local infections, the ovary of a hen may very occasionally transform into a testis. The chromosomal content of the bird is unaltered, but the fowl is sometimes capable of fathering chicks. (So much for the idea of vast differences between the sexes!) What would the ratio of sexes be for such chicks assuming the mother to be a normal hen?

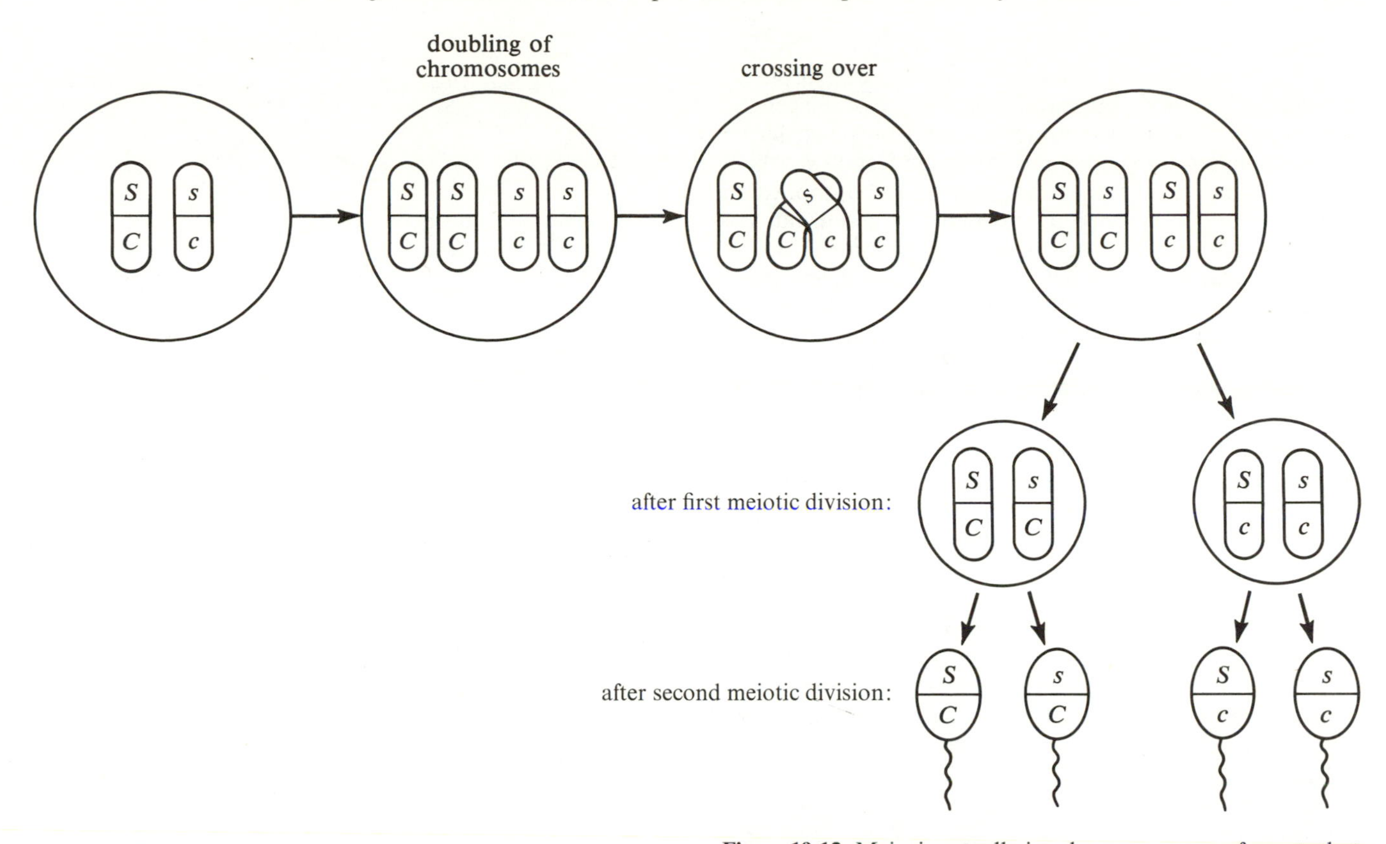

Figure 19.12. Meiosis actually involves a sequence of events that culminates in the production of germ cells with unpaired chromosomes. In the example above, a cell containing paired chromosomes that code for shape (S = dominant shape, s = recessive shape) and color (C = dominant color, c = recessive color) first doubles the total chromosomal components. The process of crossing over allows the mixing of gene traits on the chromosomes. After crossing over, the cell divides to produce two daughter cells with paired chromosomes. This is followed by a second meitotic division that produces the germ cells. Notice that a single crossing over results in the production of twice as many sperm types than would have been possible without the crossing over.

Since many genes are located on each chromosome, it appears that Mendel's hypothesis of independent sorting (in meiosis) can occur only for genes located on different chromosomes. Many examples of nonrandomly assorted genes are now known for fruit flies (through the study of mutants which do not occur in the wild). Linked genes are presumably located on the same chromosome, a hypothesis supported by the observation that the number of linked groups of traits in the fruit fly equals the number of morphologically different chromosomes in the egg or sperm. Linkage is, however, rarely complete. W. Janssens proposed the mechanism of **crossing over** to account for incomplete linkages. He proposed that during the coiling phase of chromosomal threads, two of the threads might break at a corresponding point along their lengths and reconnect crosswise (Figure 19.12). The phenomenon of crossing over allows a far greater variation of individuals produced as progeny than simple sexual reproduction plus mutations would allow. Morgan and his coworkers reasoned that the phenomenon of crossing over would make genes located close together more likely to remain linked than genes located far apart. This reasoning allowed them to map by 1915 the relative positions of the genes which governed many of the traits observed in fruit flies.

It was also in the early part of the twentieth century that chemical analysis of chromosomes showed them to contain both nucleic acids and proteins. But both kinds of compounds are relatively large molecules by the standards of ordinary chemistry, and no real progress could be made in understanding the nature of the genetic material until some true understanding of the structure of these macromolecules had been achieved. Work in the 1930s indicated that some genes may be implicated in the production of certain proteins by a cell, but the exact relationship between genes and protein synthesis remained unclear. Similarly, the role of nucleic acids remained obscure. In 1944, however, O. T. Avery and coworkers showed that deoxyribonucleic acid (DNA) introduced into pneumonia bacteria could alter the hereditary traits of subsequent progeny. Here, finally was the first hard evidence of the chemical makeup of the gene: it is a form of nucleic acid, to be specific, DNA. At first sight, however, this finding does not seem very encouraging, since Hammarstein and Caspersson had shown in the 1930s that DNA molecules have molecular weights ranging upward from the hundreds of thousands. We now know that DNA molecules in simple bacteria and in mammalian cells may have molecular weights in the billions; i.e., DNA molecules may contain hundreds of millions of atoms! How could one ever hope to disentangle the three-dimensional structure of such large molecules? Moreover, how could one ever hope to read the information contained in these large molecules that is needed to specify even the simplest of living organisms? And how can these large molecules be faithfully reproduced, by generation after generation of cells? These questions, and others, formed the basis for the founding of a new branch of science, molecular biology.

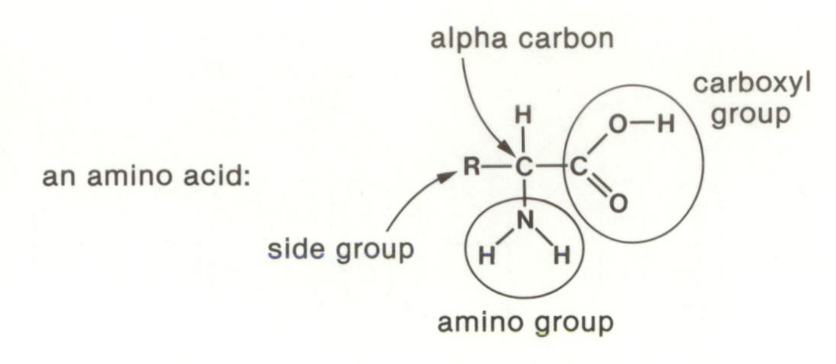

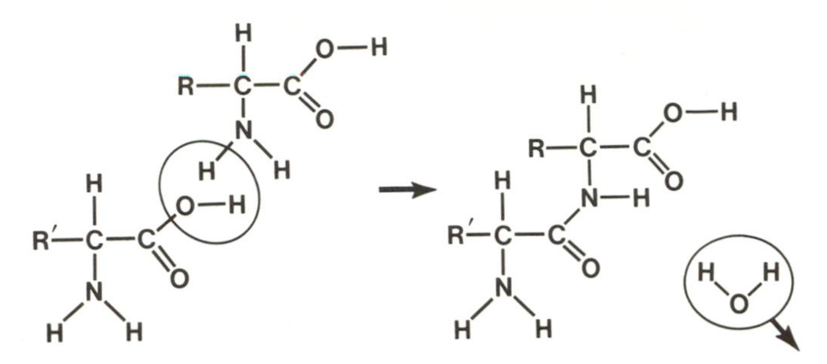

Figure 19.13. Amino acids and proteins. (a) The typical structure of an amino acid, where R is a side group that takes different forms for different amino acids. For example if R is a single hydrogen atom, H, the amino acid is glycine; if R is the methyl group, CH_3, the amino acid is alanine, etc. Each thick line represents a covalent bond (one shared electron); notice that neutral carbon is capable of forming four covalent bonds (see Chapter 3). In the cell, the carboxyl group will usually become charged, $COOH \rightarrow COO^- + H^+$, and the ability to donate a proton, H^+, makes the carboxyl group behave like an *acid*. In the cell, the amino group will usually also become charged, $NH_2 + H^+ \rightarrow NH_3{}^+$, and the ability to gain a proton, H^+, makes the amino group behave like a *base*. (b) Amino acids can polymerize by forming a peptide bond, with the release of one water molecule, H_2O, per link.

Molecular Biology and the Chemical Basis of Life

Even before Darwin's discoveries, chemists had established that living matter contained the same elements already known to inorganic material. However, chemists also found that cells exhibited a great predominance of carbon compounds, or *organic* compounds. For some years, vitalists maintained that organic molecules could be made only inside a living organism, i.e., that a "vital force" made organic chemistry intrinsically different from inorganic chemistry. This myth received a crushing blow in 1828, when Friedrich Wohler synthesized urea from ammonium cyanate (prepared from animal hoofs and horns), and it was completely shattered in 1845 when Hermann Kolbe synthesized acetic acid (vinegar) from pure elements.

Despite the success of "organic" chemists in working out the structure of relatively simple organic molecules, like glucose, there remained an aura of mystery surrounding the very large organic molecules. For example, up to three-quarters the dry weight of a cell may be in the form of nitrogen-containing carbon compounds called *proteins*, whose molecular weights may range from the several thousands into the millions. The bacterium *E. Coli*, found in the intestines of higher animals, routinely manufactures a few thousand different kinds of proteins, yet until a few decades ago, chemists were completely in the dark about how to make a single protein. Early in the twentieth century, Emil Fischer had established that proteins were constructed by linking together long chains of small molecules called **amino acids** (Figure 19.13). However, there are 20 common forms of amino acids basic to living systems (21 if one counts a kind found only in the brain), and it was not until 1942 that Martin and Synge developed techniques for assaying the proportions of the various amino acids in any pro-

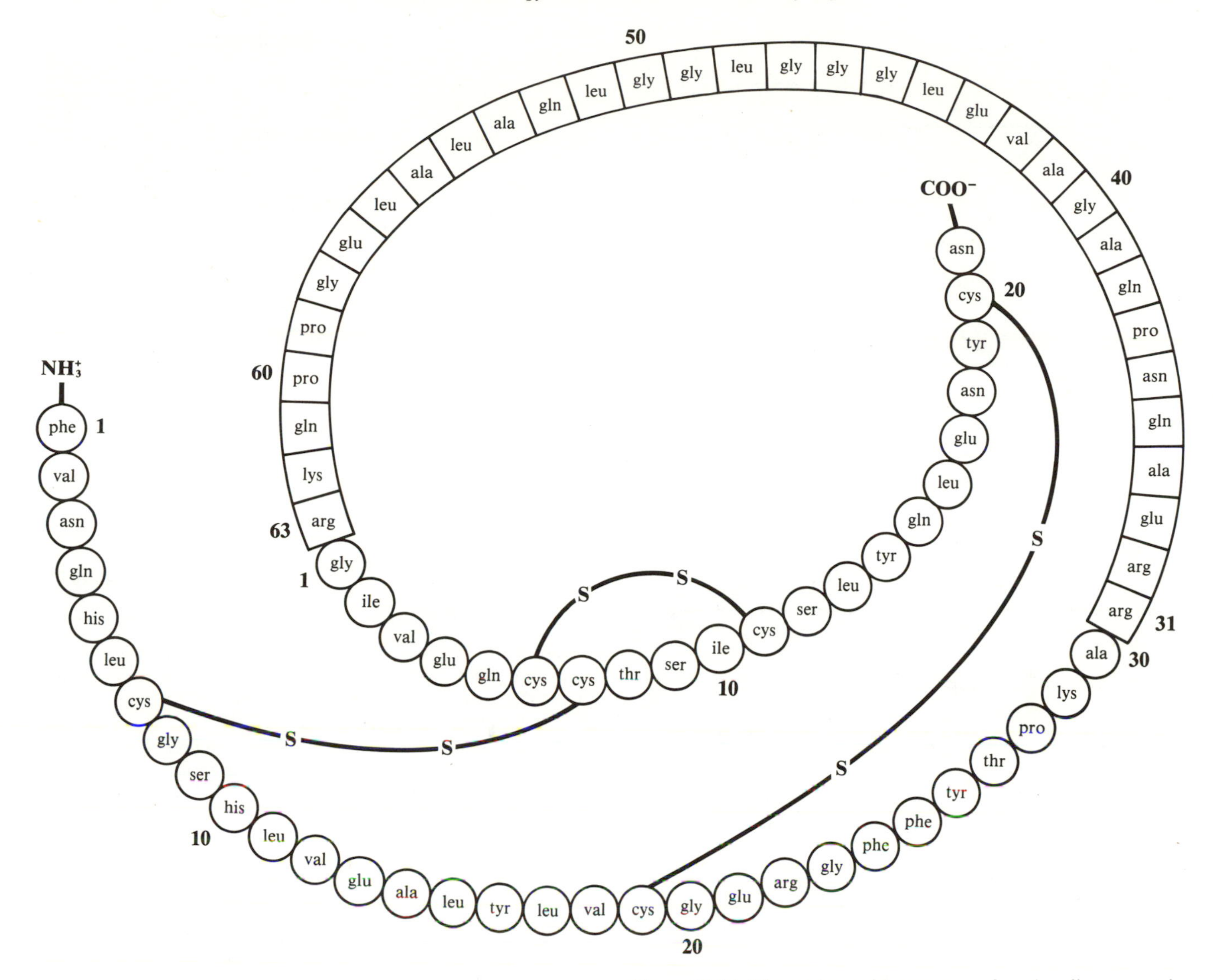

Figure 19.14. The amino-acid sequence of proinsulin, a protein which contains 84 amino acids. One end has the group NH_3^+; the other end, COO^- (consult Figure 19.13). The abbreviations used for amino acids are listed in Table 19.1. When the part of the chain depicted schematically by the connected squares (numbers 31 to 63) is removed, proinsulin becomes insulin, a hormone composed of two polypeptide chains connected by disulfide bridges. (For details, see Chance, Ellis, and Bromer, *Science*, **161**, 1968, 165.)

tein. In 1953, Sanger had found the sequence of 51 amino acids that characterize bovine insulin (two polypeptide chains tied together by disulfide bridges). By now, amino-acid sequences of even larger proteins are known (Figure 19.14).

Also in 1953, James Watson and Francis Crick elucidated the structure of DNA on the basis of X-ray diffraction studies performed by Rosalind Franklin, Maurice Wilkins, and their collaborators. The technique of X-ray diffraction had been pioneered by physicists (in particular, by W. L. Bragg) to study the three-dimensional structure of regular crystalline solids. The first dramatic clue that such methods might yield important dividends for biology came in 1951, when Linus Pauling combined

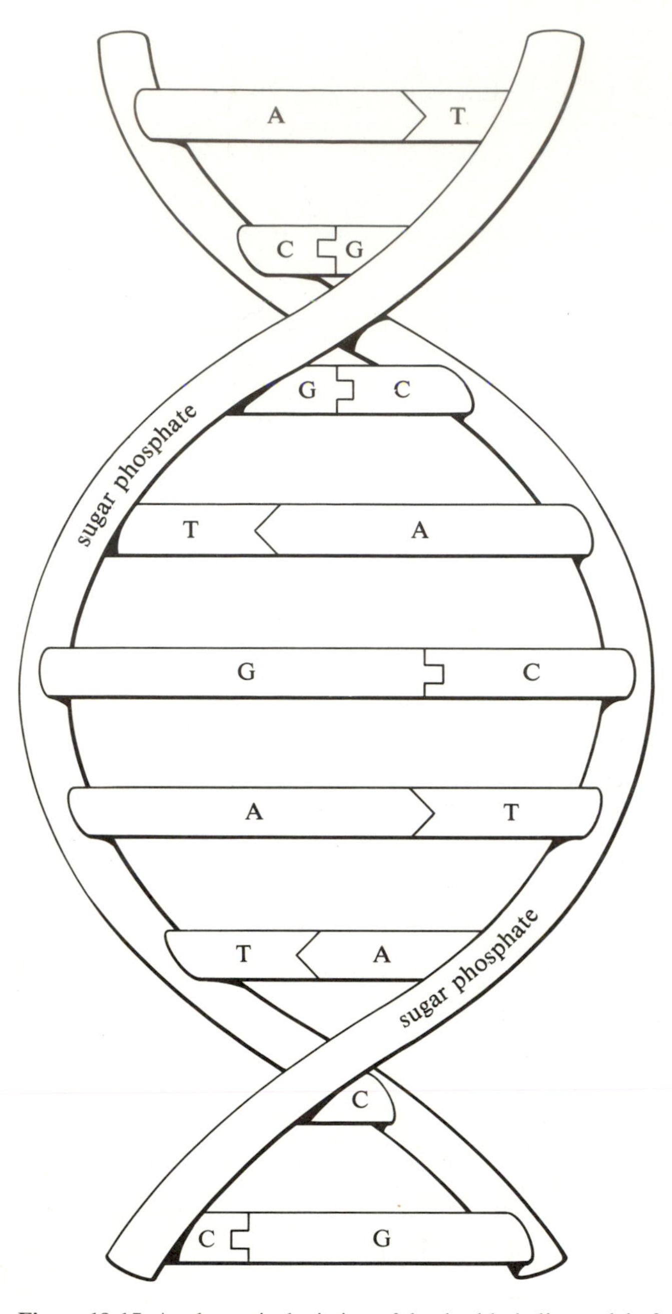

Figure 19.15. A schematic depiction of the double-helix model of DNA. The symbols A, G, C, and T stand for adenine, guanine, cytosine, and thymine. Notice that the purine A is always paired with the pyrimidine T, and that the purine G is always paired with the pyrimidine C.

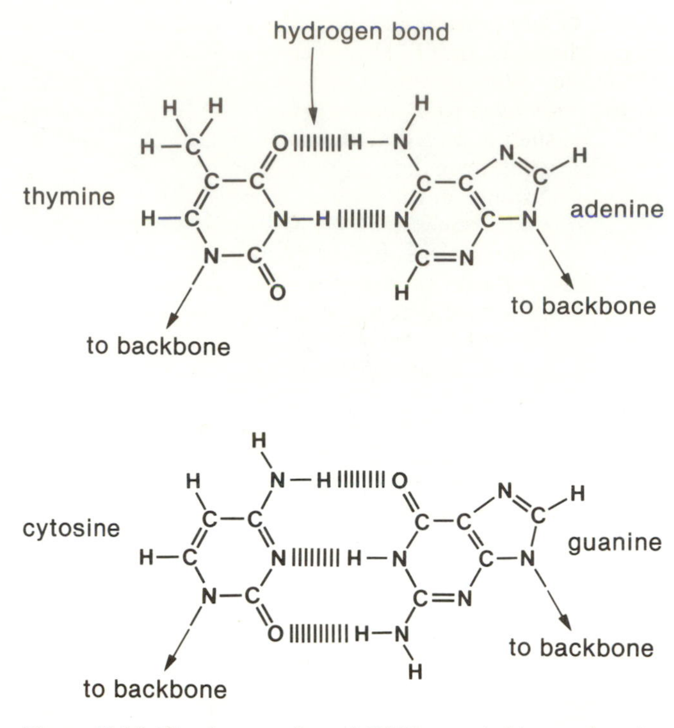

Figure 19.16 The base pairs of DNA are held together by hydrogen bonds, indicated above by the series of short vertical lines. The spans of the A and T pair and of the G and C pair across the double-helix ladder are nearly identical in lengths. Hydrogen bonds, where hydrogen atoms can bond to *two* atoms, play an important role in many biological molecules. (See J. D. Watson, *Molecular Biology of the Gene*, W. A. Benjamin, 1976.)

X-ray diffraction studies with stereochemical principles to deduce that proteins can also take on regular periodic structures under suitable conditions. In particular, Pauling and Corey proposed that hydrogen bonding (see Chapter 4) could make amino-acid chains (or at least, segments of protein molecules) take the form of a spiral staircase. Their "alpha helix" model of this phenomenon was verified in that same year by Cochran and Crick, who discussed theoretically the X-ray diffraction patterns expected from regular single helixes. By late 1951, Franklin's X-ray studies had led her to conclude that DNA can also take a form which might contain two, three, or four helical structures, with sugar-phosphate groups forming the backbones of the helixes. In a brilliant piece of detective work, Watson and Crick combined the X-ray data of Franklin and Wilkins and the chemical data of Erwin Chargaff to come up with their now famous **double-helix** model for DNA (Figure 19.15).

In the Watson-Crick model, DNA is a twisted ladder, where the sides of the ladder are sugar-phosphate groups that lend structural support to the long molecule, and where the rungs of the ladder are pairs of purine-pyrimidine bases. The heart of the molecule is the sequence of base pairs that form the rungs of the ladder. Four different bases, A = adenine, T = thymine, C = cytosine, and G = guanine, are found in DNA. Adenine and guanine,

A and G, are purines, and are larger molecules than the pyrimidines, C and T. However, with the help of Jerry Donahue, Watson and Crick found that A can be *hydrogen-bonded* to T, and G can be hydrogen-bonded to C, in such a way that the A-T and G-C complexes (which are flat) have the *same* total lengths (Figure 19.16). Since the sides of the ladder formed by the sugar-phosphate backbones of DNA like to maintain constant separation for energetic reasons, this means that, in DNA, *A is always paired with T, and G with C.* This uniqueness in base pairings provided a simple explanation for Chargaff's earlier chemical analysis, which showed A and T always to enter in the same proportions in DNA, and similarly for G and C. This uniqueness of base pairings has, however, an even more important biological implication, which Watson and Crick, in the formal writeup of their short announcement paper, were content to understate merely as having "not escaped their attention." In what follows, we outline some of these implications and how they have revolutionized all of biology.

Problem 19.9. Write a ten-page essay tracing the historical developments leading to the discovery of the double-helix model for DNA. Consult especially *The Double Helix*, by James D. Watson, and *Rosalind Franklin and DNA*, by Anne Sayre. Discuss how you would explain or try to resolve the conflicting pictures presented by these two points of view.

The Central Dogma

We summarized earlier the point of view that the basic unit of life on Earth is the cell. From the most elementary vantage point, the cell acts in two critical life processes:

(a) as a factory to make proteins for structural and other purposes;

(b) as a design model in cell division, after which there are two factories with identical capabilities to those of the original cell.

Since proteins are the basic building blocks for the structure of the cell, process (a) is clearly part of *ordered growth*, one of the defining characteristics of living things. In multicellular organisms, process (b), mitosis, is also part of growth. The central dogma of molecular biology is that the information required to produce ordered growth is stored in DNA.

To see how the central dogma works in outline, let us consider the example with which we began this chapter, the growth of a child. The child eats corn flakes. The proteins in the corn flakes are dissembled during digestion into their component amino acids. The amino acids enter the bloodstream from the intestines and flow into the child's cells. Inside each cell, these amino acids are reassembled into different combinations; i.e., they are reassembled into proteins which are useful for the growth of the cells of the child. When a cell becomes large enough, it divides into two, with each daughter cell inheriting the chemical recipes necessary to continue its own growth. In this manner do corn flakes ultimately become child.

The information needed to make different kinds of proteins—some for structural purposes, some (called enzymes) for speeding up specific chemical reactions, some for regulating gene function, etc.—ultimately resides in the macromolecules of DNA. The job of actually carrying out the instructions encoded in DNA is entrusted to another macromolecule called RNA (ribonucleic acid). Thus, the central dogma of molecular biology states that the flow of information follows the direction

$$\text{DNA} \Rightarrow \text{RNA} \Rightarrow \text{proteins} \qquad (19.1)$$

with very few exceptions. We now consider in greater detail the mechanisms which underlie the schematic flow (19.1).

Protein Synthesis

To make proteins in the cell factory, DNA is the designer, and various forms of RNA act as a blueprint (messenger RNA), assembly-line worker (transfer RNA), and assembly-line belt (ribosomal RNA). We have mentioned that DNA is a twisted ladder in which the rungs are base pairs in which A is always found opposite T, and G opposite C. Thus, if either half of the DNA ladder is known (i.e., one strand plus its sequence of attached bases), the other half can be unambiguously deduced. The two strands of DNA are said to be *complementary* to one another. The sequences of bases on either strand of all the DNA in the cell contains the basic design of the cell. In particular, the design for making proteins is carried out as follows.

The first step is to synthesize messenger RNA. The RNA molecule is similar to DNA in having a heart which is a long chain of bases, but in RNA the base uracil, U, replaces thymine, T. (In transfer RNA, there are further alterations to some of the bases after first synthesis.) To make a length of messenger RNA, DNA partially unzips in the presence of special proteins (Figure 19.17). Bases tied to the sugar ribose plus three phosphates (ATP, UTP, GTP, and CTP) are attracted to the unpaired bases of one of the strands of the partially unzipped DNA. The sequences of bases on the strand of

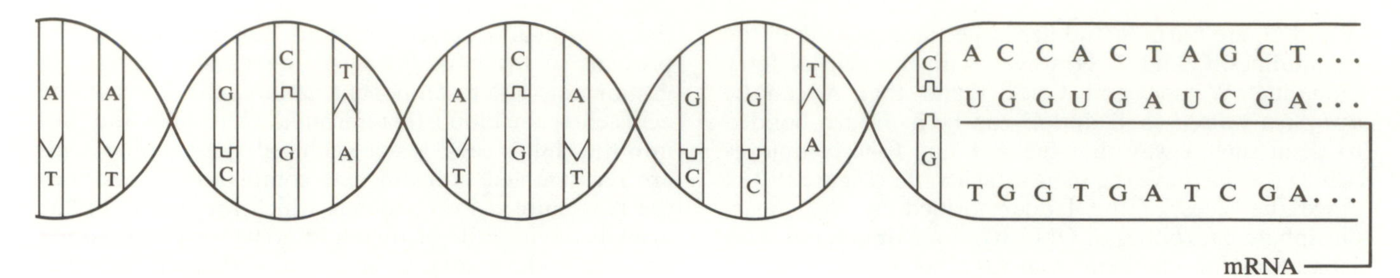

Figure 19.17. The act of transcription synthesizes a single strand of RNA, with one of the strands of DNA acting as a template. In terms of the discussion in the text, the particular form of RNA depicted here is messenger RNA.

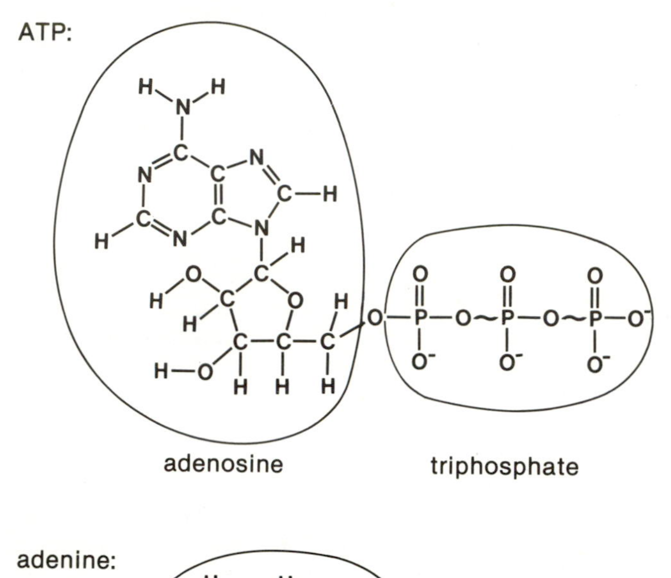

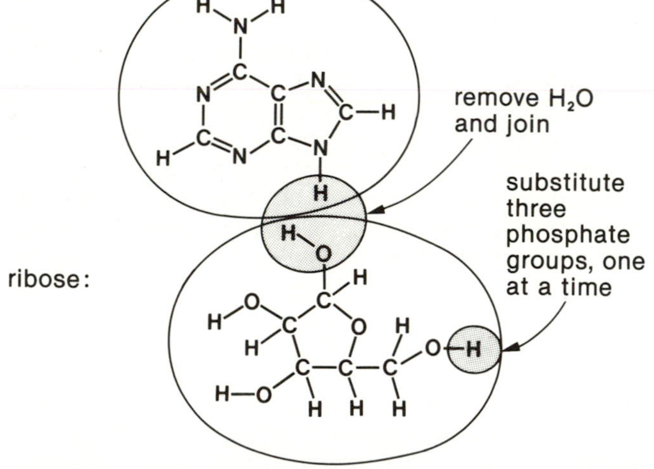

Figure 19.18. The chemical structure of ATP, adenosine triphosphate. The wavy lines joining the phosphorous and oxygen atoms indicate high energy bonds (energy stored in the repulsion of the O^- ions). The bottom diagram indicates how ATP may be synthesized schematically from its constituent parts: adenine (purine A), ribose (sugar), and three phosphate groups.

RNA is complementary to the design of the length of DNA to which the former is attached. Thus, if a length of DNA reads

$$\text{ACCACTAGCT} \cdots, \tag{19.2}$$

the corresponding length of RNA reads

$$\text{UGGUGAUCGA} \cdots. \tag{19.3}$$

We will see that *it is the nonrepetitiveness of the sequences (19.2) and (19.3) which provides the complexity needed for life.*

The phosphate bonds of ATP, UTP, GTP, or CTP contain excess chemical energy (Figure 19.18), and two of the phosphates can be removed (together with some water, H_2O), joining the remaining **nucleotides** (base plus sugar plus phosphate) along the backbone characteristic of the RNA molecule (Figure 19.19). As the special proteins that initiate RNA synthesis move further down the length of the double helix of DNA, the two complementary DNA strands find it energetically favorable to close again, displacing the DNA-RNA hybrid. The displaced single strand of RNA, once completed, carries the message for the synthesis of proteins.

Problem 19.10. Suppose a certain length of DNA reads TAGCTAGCCA $\cdots$. How would the corresponding length of RNA read after transcription?

The messenger RNA is then moved outside the nucleus of the cell (if the cell is eukaryotic) and attaches onto a ribosomal RNA. Ribosomal RNA is usually bound to inner cell membranes called the *endoplasmic reticulum* (see Figure 19.8) with the help of certain proteins (ribosomal RNA plus proteins = **ribosomes**). The ribosomal RNA were themselves made by other acts of

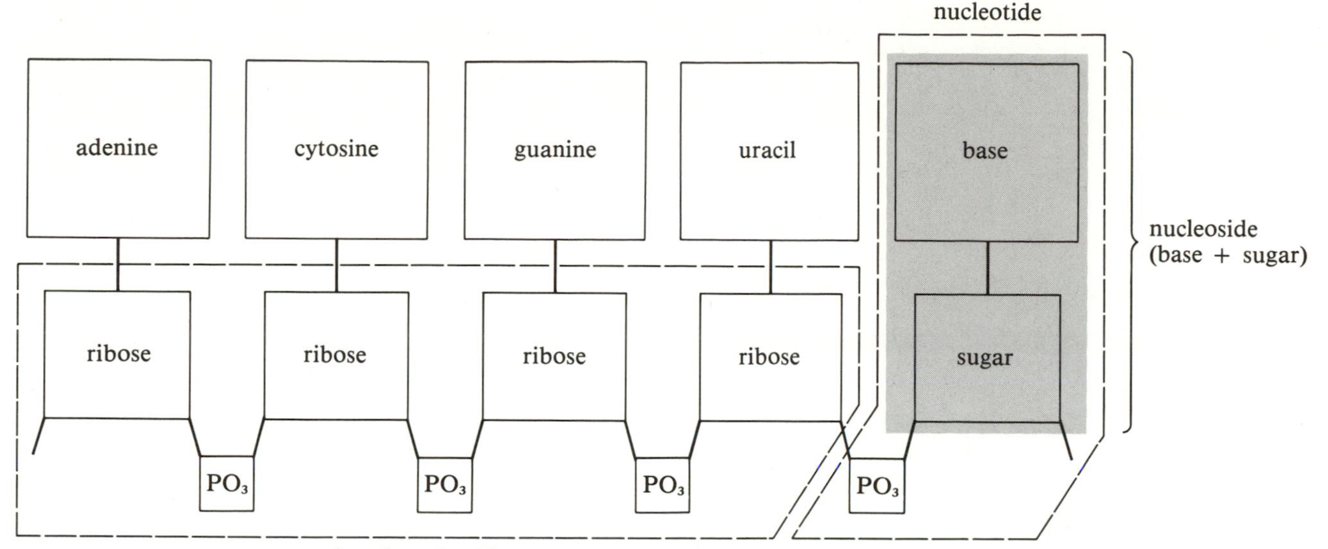

Figure 19.19. Schematic structure of a segment of RNA. The basic unit of the irregular polymer is a nucleotide (base plus sugar plus phosphate). The combination base plus sugar is called a nucleoside. To join a base to a sugar, a bond must form between a nitrogen atom of the base and a carbon atom of the sugar. Since an H atom and an OH group were formerly attached to these atoms, the joining of a base to a sugar to form a nucleoside requires the release of water, H_2O. Water is also released in the formation of the bonds between the sugars and the phosphate ions (usually in the form of $H_2PO_4^-$ in an aqueous solution).

transcription from DNA. The ribosomal RNA holds the messenger RNA steady for **translation** of the blueprint for making proteins. This translation is carried out by a third class of RNA, transfer RNA (also made by transcription from DNA). Transfer RNA consists of single-stranded RNA which twists back onto itself like a clover leaf, with the shape held rigid by hydrogen-bonding of as many base-pairs as possible (Figure 19.20). This configuration leaves one end of the transfer RNA with three unpaired bases that can fit snugly against any three complementary bases on messenger RNA. There are many different kinds of transfer RNA, each kind being defined by the sequence of three bases at one end. The other end of each kind of transfer RNA attaches to only one kind of amino acid. This architecture allows the transfer RNA to "read" the bases on the messenger RNA in "words" of three letters at a time, with the accompanying "translation" of base sequences into amino-acid sequences for the synthesis of proteins. (Figure 19.21). Thus, translation takes place via a **triplet code** (Table 19.1).

To be specific, let us consider how a chain of amino acids (a **polypeptide** chain) grows in length during protein synthesis mediated by RNA (Figure 19.22). As each new transfer RNA attaches itself to the next position down the strip of the messenger RNA, the transfer RNA brings the next amino acid along with it on its opposite end. With the help of certain existing proteins, the ribosomal RNA attaches the previous chain of amino acids to the newest link. This attachment occurs by the formation of a **peptide bond**, with the concomitant release of water, H_2O (see Figure 19.13). In the process, the latest transfer RNA is freed to get another amino acid of its particular specification. In this way, a complete protein would eventually be synthesized as a long chain of amino acids by the cooperation between the three types of RNA: messenger RNA = blueprint; transfer RNA = assembly-line worker; and ribosomal RNA = assembly-line belt and scaffolding. However, the information for making the different kinds of RNA, and therefore ultimately for making the proteins, lies in the DNA of the cell.

To summarize the above description, we may say that the different sequences of bases (A, G, C, and T) taken three at a time in DNA can be used to make a large variety of RNA and proteins. Similarly, the author

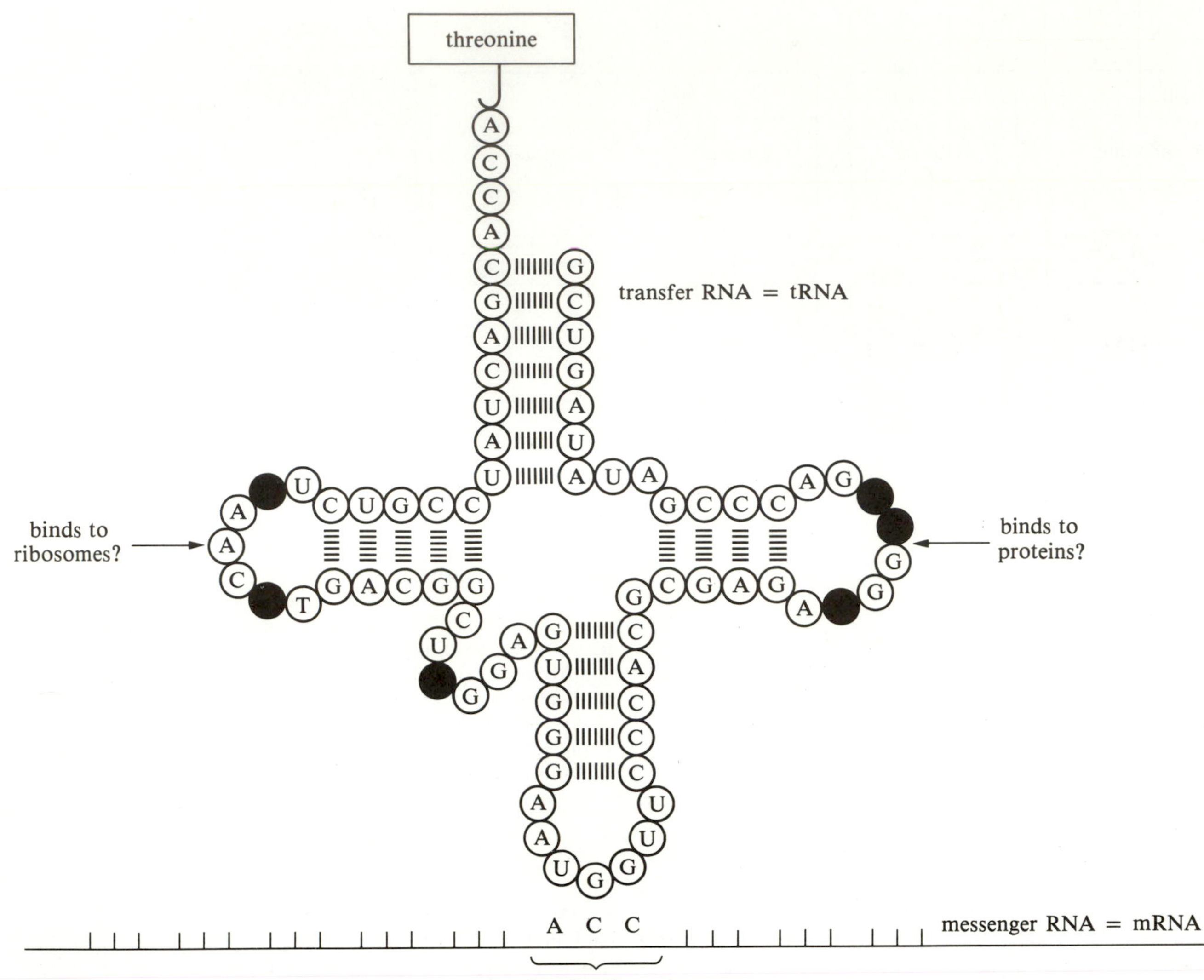

Figure 19.20. Transfer RNA is a single strand of RNA which twists back onto itself like a clover leaf, with the shape held rigid by hydrogen bonds (series of short lines) between complementary base pairs. The solid circles indicate unusual bases. The top end of the tRNA shown here attaches to the amino acid threonine when the bottom end of the tRNA contains a sequence of three bases that is complementary to the codon ACC for threonine in messenger RNA. There is actually some "wobble" in the third position of the hydrogen bonds (not shown) which form between ACC and UGG. This wobble reflects itself in the fact that many of the "words" in the triplet code of Figure 19.21 which differ only in the third letter correspond to the same amino acid. (For details, see J. D. Watson, *Molecular Biology of the Gene*, W. A. Benjamin, 1976.)

first RNA base	second RNA base				third RNA base
	U	C	A	G	
Uracil U	1	6	10	17	U
	1	6	10	17	C
	2	6	Stop	Stop	A
	2	6	Stop	18	G
cytosine C	2	7	11	19	U
	2	7	11	19	C
	2	7	12	19	A
	2	7	12	19	G
adenine A	3	8	13	6	U
	3	8	13	6	C
	3	8	14	19	A
	Start/4	8	14	19	G
guanine G	5	9	15	20	U
	5	9	15	20	C
	5	9	16	20	A
	5	9	16	20	G

Figure 19.21. The triplet code. The numbers 1 to 20 stand for the amino acids given in Table 19.1.

Table 19.1. The triplet code of messenger RNA.

Number	*Triplet code*	*Amino acid*	*Abbreviation*
1	UUU, UUC	phenylalanine	phe
2	UUA, UUG, CUU, CUC, CUA, CUG	leucine	leu
3	AUU, AUC, AUA	isoleucine	ile
4	AUG	methionine	met
5	GUU, GUC, GUA, GUG	valine	val
6	UCU, UCG, UCA, UCG, AGU, AGC	serine	ser
7	CCU, CCC, CCA, CCG	proline	pro
8	ACU, ACC, ACA, ACG	threonine	thr
9	GCU, GCC, GCA, GCG	alanine	ala
10	UAU, UAC	tyrosine	tyr
11	CAU, CAC	histidine	his
12	CAA, CAG	glutamine	gln
13	AAU, AAC	asparagine	asn
14	AAA, AAG	lysine	lys
15	GAU, GAC	aspartic acid	asp
16	GAA, GAG	glutamic acid	glu
17	UGU, UGC	cysteine	cys
18	UGG	tryptophane	trp
19	CGU, CGC, CGA, CGG, AGA, AGG	arginine	arg
20	GGU, GGC, GGA, GGG	glycine	gly
start	AUG		
stop	UAA, UAG, UGA		

Figure 19.22. The translation of the message contained on messenger RNA (mRNA) is undertaken by transfer RNA (tRNA) and ribosomal RNA (rRNA). See text for explanation. (For details, see J. D. Watson, *Molecular Biology of the Gene*, W. A. Benjamin, 1976.)

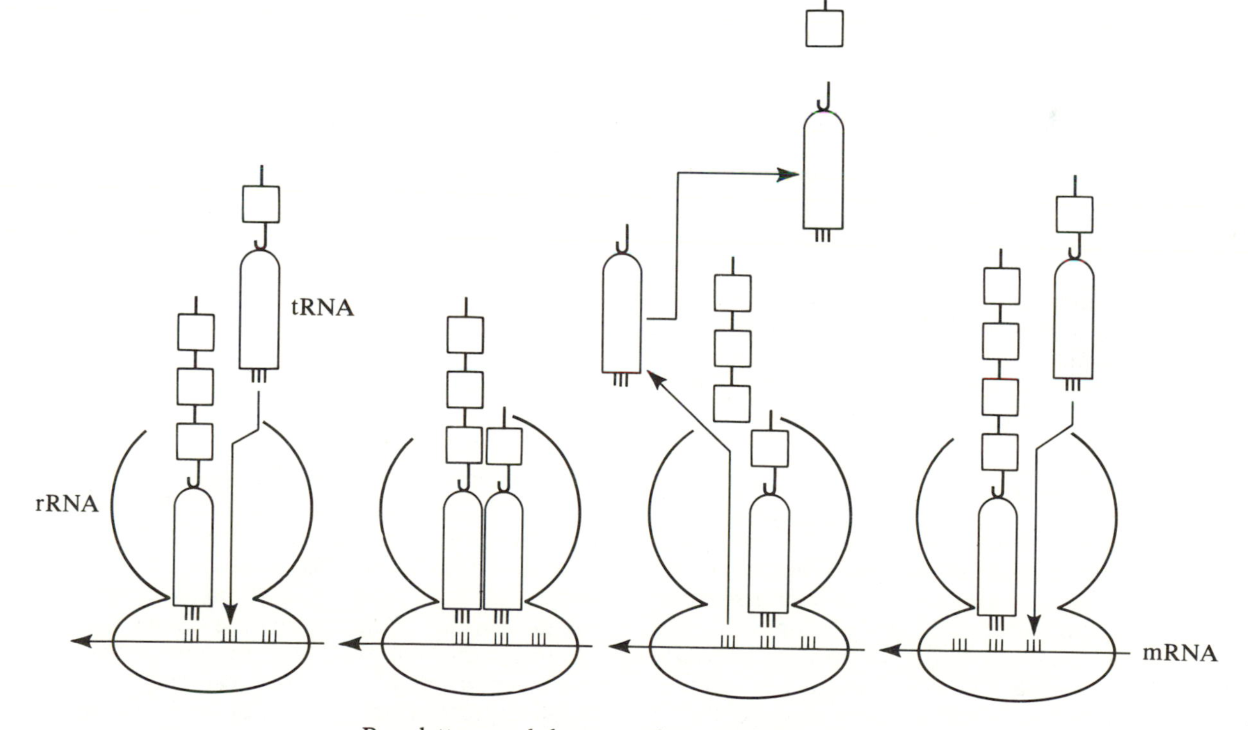

behind a typewriter uses 26 letters to compose a large variety of paragraphs. The collection of proteins in a cell expresses the behavior of that cell in the same sense that the collection of paragraphs defines the purpose of a book.

Problem 19.11. Show that 4 different RNA bases (A, C, G, and U) read three at a time can form 64 distinct "words." Since there are only 20 different kinds of amino acids, argue that the vocabulary of the triplet code is more than sufficient to give each different amino acid a different specification. In fact, it is now known that some words are duplicates, and other words control commands to start making a protein chain, and to stop.

Figure 19.21 gives the dictionary of the triplet code as it has been cracked by Francis Crick, Marshall Nirenberg, Philip Leder, H. Gobind Khorana, and others. Notice that the code has some duplication—more than one three-letter word can yield a given amino acid—and that the code contains command signals for "start" and "stop" (making a protein chain). There are several different ways to decipher the triplet code, and these methods are by now in good agreement with each other.

Before we leave the subject of protein synthesis, we should make one remark. To make the explanation of the mechanisms easy to grasp, we have discussed the actions of the various DNA and RNA molecules as if they were intelligent workers in a factory (the cell). Molecules, even very large ones, of course, contain no intrinsic intelligence. They are attracted or repelled from other molecules purely because of established chemical principles, i.e., the laws of quantum mechanics and electromagnetism. Vitalistic notions are irrelevant to the process of protein synthesis inside living cells. In particular, the transfer RNA depicted in Figure 19.20 takes on the useful shape that it does simply because this form is the thermodynamically stable one given the physical conditions found in the cell. Proteins can also fold into complex three-dimensional shapes (held by hydrogen bonds, disulfide bridges, etc.), and this property plays a crucial role in the chemical regulation of the cell.

ATP and Enzymes

Energy is required to activate many of the chemical reactions involved in protein synthesis. Where does this energy come from? Ultimately, from the oxidation (burning) of carbohydrates. However, the cell is a delicate piece of machinery, and it has to work at nearly a constant temperature. The cell is *not* a heat engine. Hence, the burning of carbohydrates is not useful as a direct source for driving chemical reactions. In all organisms alive today, the energy released by the burning of carbohydrates is stored in the chemical bonds of the compound ATP that we described earlier as one of the precursors to RNA nucleotides. The letters ATP stand for "adenosine triphosphate," where adenosine = adenine plus ribose is called a *nucleoside* (refer to Figures 19.18 and 19.19 again). The chemical energy contained in the phosphate bonds of ATP is released when chemical reactions need to be driven "uphill." Thus, ATP is said to be the *universal energy currency* of all living things, a discovery made in the 1930s by Meyerhoff and Lipmann. In Chapter 20, we will discuss how ATP may have come to be the universally accepted money for running the economies of all cells.

The process by which carbohydrates are converted into useful molecules like ATP is called (carbohydrate) **metabolism**. Carbohydrate metabolism occurs by a series of steps. The first series involve **fermentation**, a breakdown of carbohydrates that does not involve the use of free oxygen. The discovery that fermentation occurs in the absence of air (anaerobic) was made by Louis Pasteur in the nineteenth century. The last group of steps in the metabolism of aerobic organisms include oxygen-utilizing respiration. These last steps take place in the mitochondria of plant and animal cells and release about 10 to 20 times more energy in the usable form of ATP than the first steps.

Chemical reactions, such as those required to produce ATP, would take place very slowly at the temperatures characteristic of living cells if those reactions were left to proceed on their own. Catalysts are needed to speed up the chemical reactions if the cell is to function efficiently as a biological entity. Proteins which act as catalysts to speed up chemical reactions (probably by lowering the "activation energy") are called **enzymes**. Most enzymes act in a very specific way that depends on their three-dimensional shapes, as Emil Fischer first proposed in analogy with the concept of a lock and key (Figure 19.23). The reason for this is simple. Most chemical reactions of biochemical interest involve rather mild chemical forces. For these mild *short-range* forces to have a chance to act, molecules must "snuggle up" against one another. This snuggling-up clearly implies chemical specificity and indicates the importance of the three-dimensional structure of biological molecules. In particular, a specific enzyme acts usually to speed up a specific chemical step, and that step only. Enzymes are needed to make lipids, polysaccharides, nucleic acids, and other molecules useful to the cell, as well as to break these molecules down. Thus, water plus a certain enzyme in the saliva of animals will break down starch (a polysaccharide of many glucose units) into maltose (a disaccharide of two glucose units). There are an enormous variety of such chemical reactions that go on in a living organism, a few thousand even in a simple bacterium like *E. Coli*. The entire collection of such

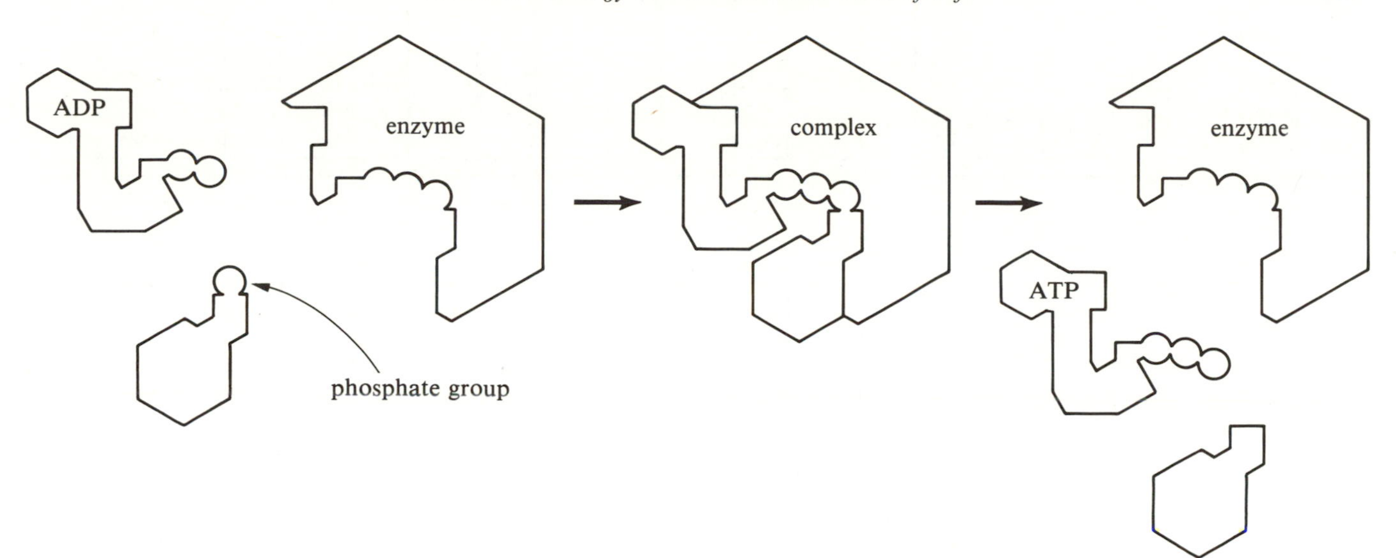

Figure 19.23. Enzymes catalyze biochemical reactions by providing a three-dimensional shape which other molecules can fit snugly into if a "lock and key" relationship holds. Illustrated here is a schema by which an extra phosphate group, originally attached to some small molecule, might be transferred to ADP (adenosine diphosphate) to form ATP (adenosine triphosphate). (For details, see J. D. Watson, *Molecular Biology of the Gene*, W. A. Benjamin, 1976.)

chemical reactions leads to what we have loosely termed *growth*, but which is more precisely called *metabolism*. To speed up the huge number of chemical reactions that constitute the metabolism of a cell requires, clearly, that the cell make an enormous variety of different kinds of enzymes. However, since catalysts are not affected in net by the chemical reactions they help to mediate, they are able to function over and over again, sometimes at very high speeds. Hence, any given catalyst (whether a protein or a cofactor like a vitamin) may need to be present in the cell only in trace amounts.

Some simple organisms can use glucose as their sole food intake and manufacture everything else they need. In particular, two groups of bacteria are known which can transform the molecular nitrogen, N_2, that exists in gaseous form in the Earth's atmosphere into amino acids (**nitrogen fixation**). Plants can convert nitrogen-containing compounds in the soil into amino acids. In contrast, animals cannot manufacture all the amino acids they require for their proteins, and must supplement their carbohydrate intake with at least some protein intake. You are partly what you eat!

Cell Division

The process of growth in multicellular organisms includes cell division, or mitosis. How does the information required to run the operation of a cell get passed on to the two daughter cells when a single cell divides? The answer is contained in the double-helix structure of the master molecule, wherein all the basic genetic information resides. Before cell division, the DNA unzips by breaking the hydrogen bonds between its base pairs. Figure 19.24a gives a simplified schematic representation of the fundamental process. Each open base attracts a complementary deoxynucleoside-triphosphate (dATP, dGTP, dCTP, and dTTP, which are the deoxy counterparts of ATP, GTP, CTP, and UTP) with a subsequent linking of backbones that is completely analogous to the synthesis of RNA. Here, however, each single strand of DNA forms its own complement, and the cell then has *two* complete double-helix molecules of each kind of DNA. One double helix of each kind of DNA is passed on to the two daughter cells after cell division. The mechanism of heredity is contained in the complete set of double-stranded DNA in the chromosomes of each daughter cell. In this view, a gene is that segment of a DNA molecule (thousands of bases long) which controls the formation of a single polypeptide chain. The entire set of DNA molecules gives each cell the complete list of chemical recipes it needs to live.

The complementary nature of the base pairings in DNA guarantees the production of two daughter cells with the same genetic content as the original cell before

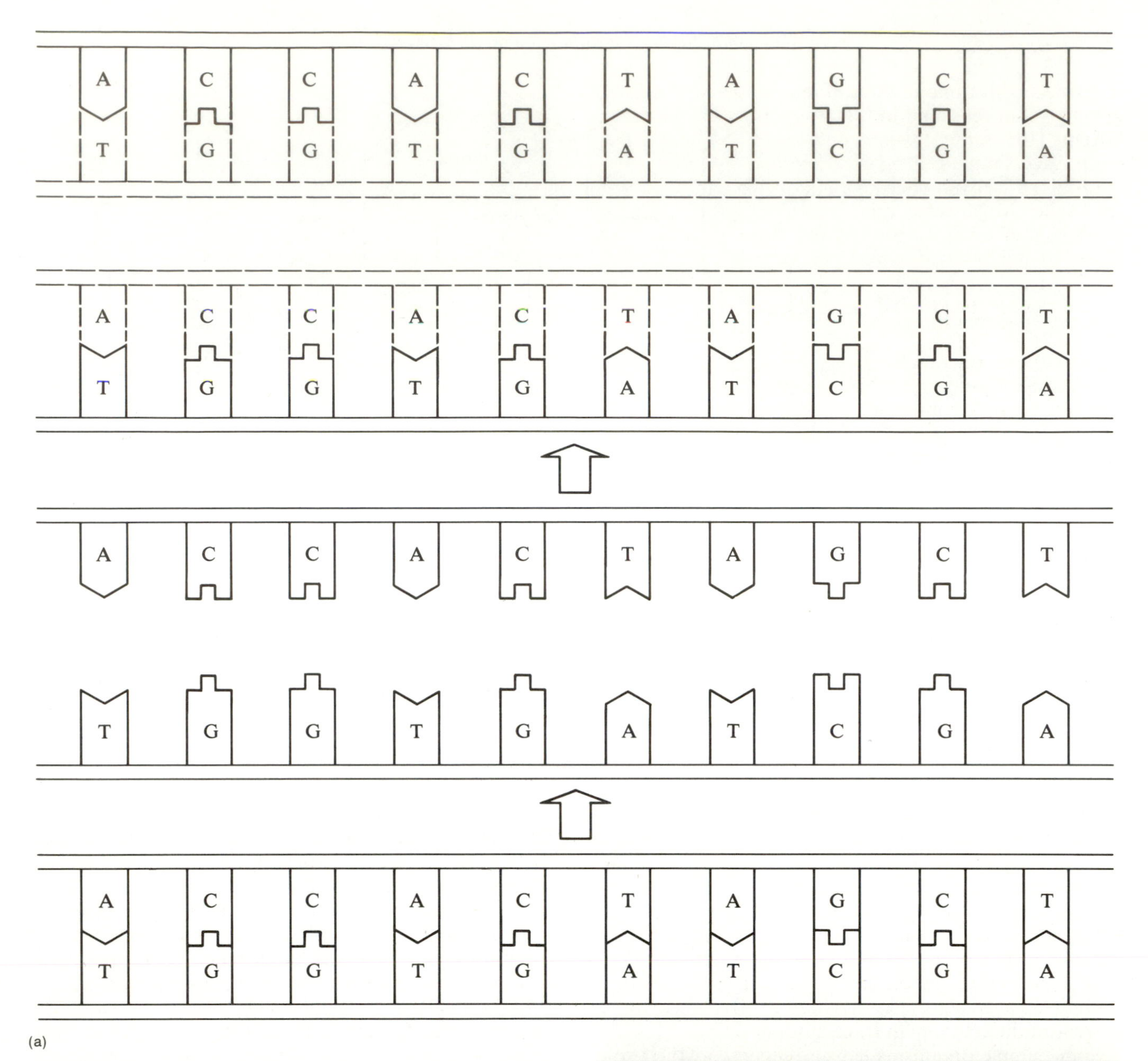

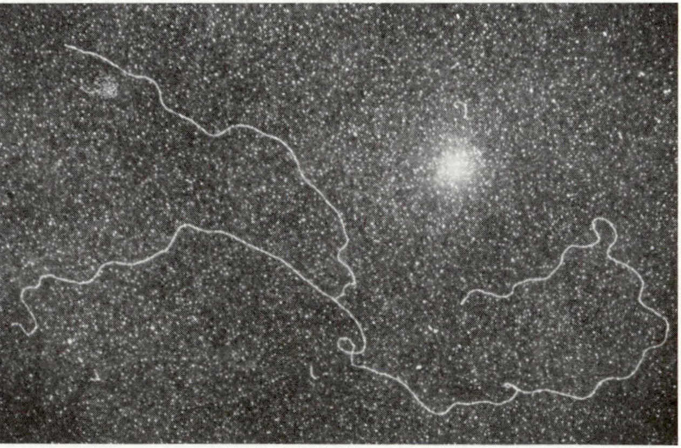

Figure 19.24. (a) The duplication of double-stranded DNA precedes cell division. (b) Electron micrograph of duplicating DNA shows that reproduction does not begin with the complete separation of the original double helix into two single strands as simplified in (a). Instead, the region of the two forming double helixes lengthens as the point advances where the parent double helix unravels. (Courtesy of John Wolfson and David Dressler)

it divided. Thus, one of your skin cells would divide to give two skin cells, and in this way your skin will grow, keeping the same texture and color. In unicellular organisms, simple cell division or mitosis already constitutes that second characteristic of life, reproduction. In higher organisms, the reproduction of another individual involves a more complex form of cell division (meiosis) and unification (fertilization).

Sexual Reproduction

Sexual reproduction involves, as we have already said the fusion of a sperm nucleus and an egg nucleus. A normal cell of a complete human contains 23 pairs of chromosomes (23 pairs of double-stranded DNA). The production of a sperm or egg cell in the human body results in sperm or eggs which each contain only one member of each pair of chromosomes (Figure 19.25). In other words, a germ cell (sperm or egg) contains only 23 chromosomes (instead of 23 pairs of chromosomes), half the genetic information of each parent. When fertilization takes place with the unification of sperm and egg, the resultant cell has again the full complement of 23 pairs of chromosomes. But the offspring would be genetically different from either parent, since it contains half the genetic content of the father and half of the mother. Moreover, the separation of the original 23 pairs in each germ cell (sperm and egg) took place randomly, and the combination of one sperm with one egg will generally be different from the combination of another sperm with another egg. The variety possible is enhanced greatly by the additional process of crossing over (Figure 19.12). Hence, brothers and sisters will generally have different genetic content from each other as well as from their parents. This constitutes the great biological advantage of sexual reproduction. It produces a great variety of individuals, with a new combination of genes for every offspring. This great variety of individuals virtually assures that some members in the general population will have the adaptive characteristics that allow the species to survive when new environmental pressures arise.

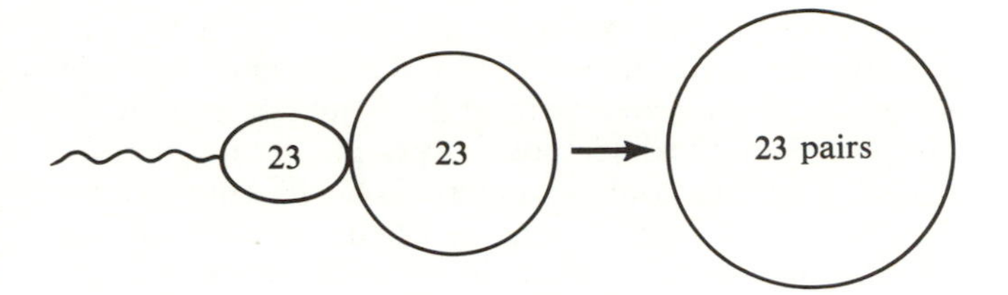

Figure 19.25. In the formation of a new human being, an egg carrying 23 chromosomes is fertilized by a sperm carrying 23 chromosomes. The fertilized ovum contains the 23 pairs of chromosomes that characterize the genetic content of every complete human. (This picture is not drawn to scale. The actual human egg is much larger than the human sperm.)

Problem 19.12. Identical twins are siblings who develop from a single fertilized egg. Do identical twins have identical genetic contents? Can identical twins be of different sexes? Fraternal twins are siblings who develop from two different fertilized eggs. Do fraternal twins have identical genetic contents? Can fraternal twins be of different sexes? Do they have to be of different sexes?

Problem 19.13. Ignore crossing over, and show that a given parent is capable of producing 2^{23} different kinds of germ cells (sperms or eggs). Show now that such a husband and wife can produce, in principle, $2^{46} \simeq 7 \times 10^{13}$ different kinds of children. Argue that the latter number increases dramatically if we allow the phenomenon of crossing over. Comment on the power of combinatorials to produce enormous numbers by very simple mechanisms.

Cell Differentiation

At the instant of fertilization, an incipient human consists of only one cell. Yet a mature human obviously contains many different kinds of cells. A brain cell is very different from a skin cell. How did both come to be if all cells in our bodies are ultimately traceable to the division of a single fertilized egg? The answer is **cell differentiation**.

All the cells in our bodies (with the exception of enucleated cells, like mature blood cells, in which the nucleus has disintegrated) contain in the chromosomes all the genetic information required, in principle, to construct the complete individual. (This statement is obviously true for the single fertilized egg that created ultimately a complete individual by the process of cell division.) In mature differentiated cells, however, certain proteins are manufactured much in preference to other proteins, presumably by the selective synthesis of certain types of messenger RNA. Thus, brain cells contain much greater concentrations of certain proteins necessary for their function than do skin cells, and vice versa. The mechanisms responsible for cell differentiation are now very incompletely known, but they evidently require great complexity in the cell. This complexity is found to a high degree only in eukaryotic cells, cells that contain nuclei. Prokaryotic cells, like bacteria, are capable, at best, of only a very simple form of cell differentiation: the formation of spores when undernourished. The ability to carry out advanced cell differentiation constitutes a prerequisite for the higher organization possible in multicellular organisms. This justifies on a molecular level our claim in Chapter 18 that the difference between eukaryotic and prokaryotic cells is just as significant biologically as the difference between plants and animals.

Cell differentiation also underlies one difficulty behind the cloning of humans that is often discussed in the popular literature. *Cloning* is a set of techniques for creating a complete individual identical in genetic content to another mature individual. In higher organisms, this means that the fertilized egg has to contain the *same* genes as the mature individual. This result is, of course, not possible with ordinary sexual reproduction, but it can be achieved (in theory) in the following artificial manner. Take an unfertilized egg and destroy its nucleus. Insert now the nucleus of the cell of a complete individual. In principle, the nucleus of any cell will do: skin cell, muscle cell, or nose cell (as in Woody Allen's movie, *Sleeper*). The resulting egg, if it can be kept alive and induced to grow and divide, could, in principle, recreate the desired individual. The problem is that in higher organisms, the nucleus has been extracted from a cell which has already undergone cell differentiation. This cell has, therefore, either suppressed certain biological functions or enhanced others. (The evidence favors positive control rather than negative control, but the case is not closed by any means.) Successful cloning therefore requires reversal of whatever differentiation has already taken place. This turns out to be relatively easy in plants; for example, the cloning of carrots is relatively simple. (This corresponds with everyday experience, since we all know how easily weeds grow back, even after one has destroyed the whole plant except for some root cells.) The cloning of higher animals has proven to be much more difficult, partly because of the difficulty in manipulating small complex eggs. Amphibians have fairly large eggs, and it is significant that whole frogs have been grown using the chromosomes obtained from single cells on the skin of tadpoles. More recently, mice have been cloned. There seems to be no obstacle of principle that would prevent the same feat from being carried out for humans. The morality of such possible cloning of humans—which, from an evolutionary point of view, is a backward step, not a forward one—is an issue we shall not pursue in this book.

Cell differentiation may also underlie an ultimate understanding of cancer. Normal cells are characterized not only by cell growth and cell division, but also by mechanisms which prevent them from proliferating wildly. As an extreme example, the brain cells of a mature human never divide; you are stuck with the number of brain cells you had at birth, some of which may die in the process of aging. In contrast, cancerous cells are characterized by uncontrolled growth. Significantly, cancerous cells also show less differentiation than the normal cells of the tissue from which they were derived. There is evidence that cancer sometimes develops because some alteration of the cellular DNA has caused the cell regulatory mechanisms to go awry. This is a topic we will postpone until after we have discussed the action of viruses.

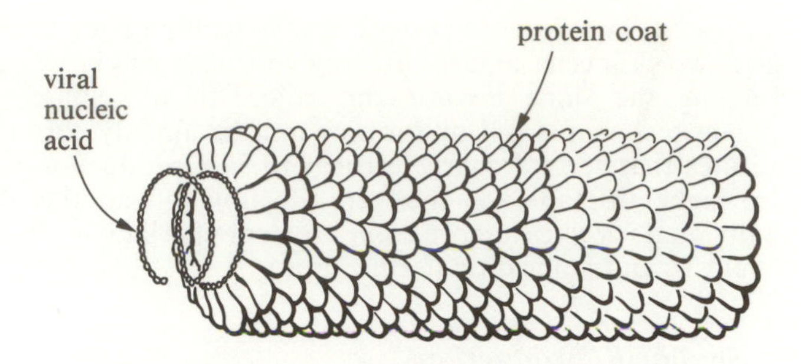

Figure 19.26. Tobacco-mosaic virus (TMV) is an example of a simple virus particle. The virus particle (one-sixth of whose length is shown here) consists of a single strand of RNA, representing perhaps only four genes, wrapped in a protein coat. The genetic material of viruses can be either DNA or RNA. If it is RNA, the virus is usually exceedingly simple in structure, as the tobacco-mosaic virus is. (For details, see P. J. G. Butler and A. Klug, *Scientific American*, **239**, Nov. 1978, 62.)

Viruses: The Threshold of Life

Our definition of life depends on the joint processes of growth and reproduction. Under this division, a bacterium clearly falls on one side; a basketball, on the other. Is there anything which falls on the borderline? Yes, biologists generally consider viruses to lie between the living and the nonliving. What are viruses?

Figure 19.26 shows schematically that a virus particle typically consists of a core containing DNA (occasionally RNA) surrounded by a protein coat (which may include lipids and carbohydrates as well). In a sense, a virus is a chromosome which exists naked in the world without the trappings of the rest of the cell. Because they lack cellular mechanisms for protein synthesis and gene replication, viruses cannot grow or reproduce on their own; they are not alive by our definition. Left to its own devices, a virus particle remains an inert collection of organic molecules, a sophisticated collection, to be sure, but nevertheless inert and lifeless. A dramatic change occurs, however, when a virus invades a living cell.

Consider the example of the invasion of an *E. Coli* bacterium by a T4 virus (Figure 19.27). A T4 virus particle collides with and sticks onto the cell wall of a bacterium. An enzyme makes a small hole in the cell wall, and the DNA of the virus enters the bacterium. The intruding DNA quickly takes over the operation of the cell factory, breaking down the bacterial DNA, stopping the production of bacterial protein, and starting manufacture of viral proteins. The appearance of the necessary enzymes allows replication of viral DNA to take place

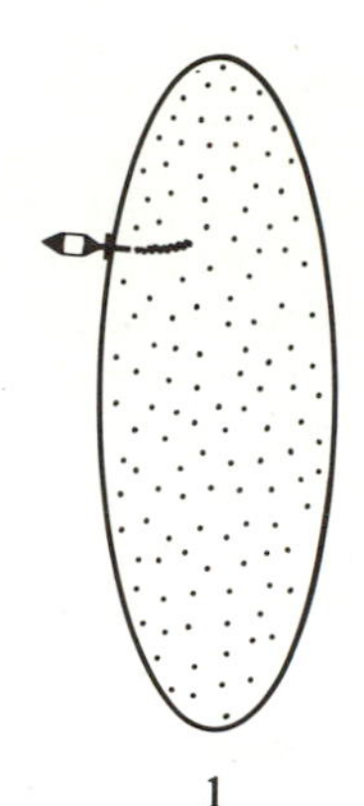

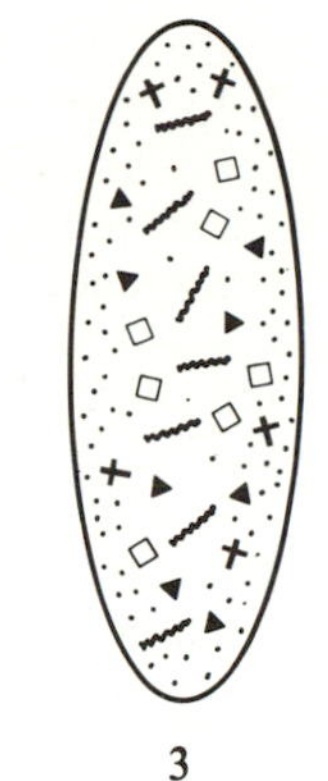

Figure 19.27. Viral infection of an *E. Coli* bacterium. (1) Infection begins when a T4 virus particle's DNA is injected into the bacterium. (2) Shortly after the manufacture of appropriate viral enzymes, the cell is forced to duplicate many units of viral DNA, represented by the squiggly curves. (3) Protein components, represented by triangles, squares, and crosses, which are necessary to make the head, tail, etc., of the virus are then manufactured. (4) The various subunits are assembled to form perhaps 200 complete virus particles. (5) Viral enzymes burst open the cell, and virus particles are freed to begin a new cycle of infection. (For details, see W. B. Wood and R. S. Edgar, *Scientific American*, **217**, July 1967, 60.)

five minutes after infection. During the next half hour, the structural proteins and viral DNA are assembled until about 200 complete viruses occupy the bacterium. The attack of viral enzyme on the bacterial cell wall causes the bacterium to burst, and a horde of viral particles are liberated which can begin a new cycle of infection.

This devastating description might lead one to wonder how any cell ever manages to survive the onslaught of viral attacks. The answer is, of course, that many do *not* survive. Those that do survive manage because of certain biological defense mechanisms. For example, modern medicine exploits the ability of molds to produce aureomycin to ward off viral attacks, and penicillin and streptomycin to combat bacterial infections. The action of such **antibiotics** is often to inhibit protein synthesis or to promote mistakes in the translation of the genetic code. The human body manufactures its own **antibodies**, which are proteins, to enable it to fight viral and bacterial infections. The study of the production of antibodies and the mechanisms of their biological action forms part of the discipline of **immunology**.

Problem 19.14. Our description of the action of T4 viruses might suggest to the novice that somehow these viruses "know what they are doing." Dispel this notion by considering the following line of thought. Imagine a large population of different kinds of viruses, some of which happen to have a mechanism for breaking open cell walls, some of which do not; some of which happen to have primitive mechanisms for taking over the operation of a cell, some of which do not; etc. Which of these viruses can naturally be expected to multiply and eventually to dominate the virus population? Discuss also how the mechanisms of virus replication might be successively refined in succeeding generations of viruses. Comment, therefore, on the possibility of extending the Darwinian idea of natural selection down to objects which are "merely" chemical entities. Learn thereby the power and the scope of this beautiful mechanistic idea, natural selection.

Before we leave the topic of viruses, we should mention that not all viruses are **lytic**, i.e., not all viruses multiply inside a host cell and ultimately kill the host upon breaking out. Nonlytic, or **lysogenic** viruses are also known. A lysogenic virus does not multiply upon gaining entry into a host cell. Instead, by a process analogous to crossing over, the viral DNA may be inserted into a specific section of the host DNA. The viral chromosome then becomes part of the genetic material of that cell and may be passed down to succeeding cell generations.

Cancer, Mutations, and Biological Evolution

We now have the data needed to outline the molecular basis for cancer, mutations, and evolution. The basic idea is the following. Suppose something happens which alters the sequence of bases in the DNA of a cell. This alteration may affect only one site of a genetic word, or it may involve many bases. Such alterations could come from many sources, some of which are listed in Box 19.2. The alteration of one or more words of the genetic message of a cell would cause it to behave differently (e.g., in the production of proteins) from its unaffected state. A likely result is cell death. Another possible result is cancer if the alteration happens to a somatic cell (an ordinary cell of the body). If the alteration happens to a germ cell (egg or sperm), the result may be the production of mutants. Most such mutations are disadvantaged or neutral, and would tend to perish or be diluted in the general gene pool. A few, in accordance with the ideas of Darwin and Wallace, may have better survival or reproductive abilities than their nonmutant counterparts, and these individuals may initiate a lasting change of the species. Organisms which reproduce asexually can evolve only by the production of such mutations (except for the rare crossing-over event that might occur in mitosis). Higher organisms can adapt with much greater versatility because of the diversifying role of sexual reproduction. Our fertile Earth is testimony to the power of the ability of species to adapt and to evolve.

Problem 19.15. Discuss in very general terms why cancer is a disease which is so hard to cure. Indeed, discuss why cancer should not be considered a single disease, and why *preventive* measures may prove to be a more effective strategy in the long run than traditional medical cures. Consult John Cairns, *Scientific American*, **233**, May 1975, 64.

BOX 19.2
Possible Sources of Mutations

(a) Copying errors in mitosis or meiosis, where a base sequence may be misread, or a fragment of a gene may be inserted or deleted in a crossing-over event.
(b) Viral infections which introduce foreign genetic material into the cell.
(c) Chemical pollutants which enter the cell accidentally.
(d) Radiation damage, either by energetic particles or by energetic photons.

Can the discoveries of molecular biology be used to help trace the course of biological evolution? Yes, and the development of such techniques during the past two decades has provided one of the major impetuses in a resurgence of interest in taxonomy. The new techniques applied to higher organisms have substantially verified the more classical techniques of classifying organisms according to an evolutionary "tree of life" (see Chapter 20). Applied molecular biology thus yields strong support for the Darwinian theory of natural selection and evolution. In addition, it provides *quantitative* measures of the rates of evolution and the closeness of relations between various different species. For example, by using quantitative measures of the similarities and differences of the immunological systems of different species, molecular taxonomists have shown that the relation between humans and the African apes is about as close as that between horses and zebras and closer than that between dogs and foxes.

The power of the new methods can be gauged by the fact that they can be used to help sort out the evolutionary tree of even unicellular organisms, where the classical methods and fossil data are poorest. The ultimate method would probably be to discover the exact sequences of bases in the DNA of different organisms; however, such techniques are still in their infancy of development, because of the difficulties of this type of biochemical analysis. An easier method involves the study of the products of the RNA templates, namely, the amino-acid sequences of proteins and the three-dimensional structures of proteins. In particular, it is largely the three-dimensional form of proteins that determines their specific biological functions (consult Figure 19.23). Proteins contain many active sites, but there are also stretches of the molecules which serve no discernible biological purpose. It is significant that studies of the proteins of organisms which are believed to be closely related on other grounds show that the active sites of the proteins are conserved, although variations do occur for distant species. This suggests that evolving species retain successful biological mechanisms (in the genetic templates of DNA) while natural selection sorts between competitive mechanisms in times of biological stress. In Chapter 20, we will examine how protein structure can be used to trace the "tree of life" back to near its origin.

Philosophical Comment

In this chapter, we have described biology from a *reductionist* point of view, probably to the great despair of poets. In this approach, science is portrayed as reducing a complex phenomenon to its component parts until an analysis emerges (often highly mathematical) in terms of a few fundamental principles of nature. People with the ultimate reductionist frame of mind take great pleasure from knowing that all of chemistry and biology can be reduced, *in principle*, to the laws of quantum electrodynamics:

$$\left[c\boldsymbol{a}\cdot\left(-i\hbar\nabla-\frac{e}{c}\boldsymbol{A}\right)+bmc^2\right]\psi=\left(i\hbar\frac{\partial}{\partial t}-e\phi\right)\psi, \tag{19.3}$$

$$\nabla^2\boldsymbol{A}-\frac{1}{c^2}\frac{\partial^2\boldsymbol{A}}{\partial t^2}=-4\pi e\psi^\dagger\boldsymbol{a}\psi, \tag{19.4}$$

$$\nabla^2\phi-\frac{1}{c^2}\frac{\partial^2\phi}{\partial t^2}=-4\pi e\psi^\dagger\psi. \tag{19.5}$$

We will not bother to define these symbols, because they are not central to our point; but even without definitions, one can see that these equations possess tremendous simplicity and grace. Even the most ardent foe of mathematics and physics cannot help but marvel at the power of the reductionist method that allows the distillation of everything in the ordinary world into such compact statements. By the same token, only the most naive mathematical optimist could believe that one could deduce, *in practice*, the properties of life from an examination of these equations alone. These equations *do* lie at the base of life, but this basis is as far removed from DNA as DNA is removed from a soaring eagle. For the poet, the eagle soars; the equations and DNA do not.

There exists, however, another half to the story of science, and that half represents the *constructionist* point of view. Here the guiding motivations are not the seeking of the fundamental laws of nature, but the recognition and elucidation of general patterns of organization, which may be ultimately rooted, of course, in fundamental laws, but the connections need not be established mathematically. The best example of the constructionist method is the Darwinian theory of natural selection and evolution. It is simplest and best stated without the use of any mathematics; yet even the most ardent fan of mathematics and physics would have to admit that Darwinian theory represents one of the brightest jewels in the crown of science.

For the constructionist, nature teaches a truly profound and remarkable lesson: our everyday world is built in a hierarchy of organizational structures. From quarks and electrons are made atoms; from atoms are made small molecules; from small molecules are made macromolecules; from macromolecules are made cells; from cells are made humans; from humans are made societies. Nature tells us that this is the efficient way to do things if very complex structures are to result. Each step alone in this hierarchy contains enormous complexities and interrelationships, and the partial elucidation of some of these at each step has occupied some of the best minds humanity has ever produced. No sane person could even begin to contemplate a "*first principles*" understanding of the entire structure.

The constructionist point of view exists also, of course, in the disciplines of astronomy and geology (particles → stars → galaxies → clusters → universe; atoms → molecules → minerals → rocks → Earth). But the *complexity* of the organization reaches its highest development in biology. The distinction between animate and inanimate objects lies fundamentally in this degree of complexity. At the one end of the scientific spectrum, we have the particle physicist; at the other end, the anthropologist.

All of this is not to suggest that the reductionist and the constructionist viewpoints are at odds with each other. No, if anything, the history of science teaches us a third great lesson, that of complementarity. The reductionist and constructionist approaches are both valid; they complement one another and spur progress. Science can, will, and should be done at all levels of organization, with a mix of rigorous deduction from basic principles and systematic arrangement of empirical knowledge. People with one outlook will work at the fundamental-laws end toward more complex structures; people with another outlook will work at the whole-system end toward simpler structures. Indeed, it might even be argued that there is no such thing as the ultimately simple or the ultimately complex, and that all scientists are engaged in a great enterprise which starts at the "middle" and proceeds in opposite directions! In this great enterprise, nature has been kind in providing us with fascinating problems and fascinating solutions at all levels of organization.

CHAPTER TWENTY

Life and Intelligence in the Universe

In Chapter 19 we saw that life, once started, could sustain itself and naturally gain in complexity without the intervention of vitalism. What about the origin of life and the emergence of intelligence? Do special "life forces" or "intelligence waves" have to be invoked to explain how life and intelligence emerged from inanimate matter? The final answers are not yet in on these questions, but the preliminary findings seem to suggest not. Life and intelligence represent extremes of chemical and cellular organization, but given enough time, the process of natural selection seems perfectly adequate to produce the necessary structures here on Earth. How about elsewhere in the universe? These questions touch on some of the most intriguing and most speculative issues humanity has ever raised. They are the topics of this, our last, chapter.

The Origin of Life on Earth

Why do scientists think that life on Earth did have an origin? For one thing, astronomers and geologists believe that the conditions prevalent on the primitive Earth were too extreme to allow any preexisting lifeforms (from which we could have descended) to have survived (Chapter 18). For another thing, there exists now good biochemical data which suggests that living things did originate once from nonliving things, and that this "spontaneous generation" probably happened only once, at least for the surviving descendants on Earth. These different molecular clues are summarized in Box 20.1.

Problem 20.1. Read R. E. Dickerson, *Scientific American*, **242** (Mar. 1980), 136, for a detailed account of the use of cytochrome c as a tool to trace the evolution of energy metabolism. Write a two-page summary of this article.

Most of the pieces of evidence cited in Box 20.1 may relate to a later stage in evolution than the actual first appearance of life. (When we talk about the origin of life on Earth, we naturally restrict ourselves to the *succesful* evolutionary line. Unsuccessful early attempts at spontaneous generation have left no traces to recount their stories.) Why should we believe there was only one successful protocell? Why not ten or a hundred? Perhaps DNA and certain proteins are the *only* viable way for the creation of life on Earth, and all emerging lifeforms had to adopt the same general mechanistic pattern because of the dictates of the laws of chemistry and natural selection.

It is only a guess, of course, made for simplicity of exposition, that there was only one successful protocellu-

BOX 20.1

Molecular Clues to the Origin of Life on Earth

(a) DNA and RNA are the universal basis of all life forms on Earth. This suggests that once a single organism had found this wonderful template mechanism for growth and reproduction, its descendants inherited the Earth.

(b) Amino acids are intrinsically three-dimensional molecules with both left-handed forms (L-amino acids) and right-handed forms (D-amino acids) possible in nature (Figure 20.1); yet living organisms on Earth use L-amino acids virtually exclusively. This suggests that the first living organism happened to use only left-handed amino acids, and this curious exclusiveness has been passed down to all succeeding generations of living organisms. The use of only 20 amino acids in life processes when many more are known chemically is another indication that we all descended from a single protocell.

(c) The molecule ATP is universally the energy currency which drives biochemical reactions in all cells (Chapter 19). This fact again points to a unique origin for all surviving species of living organisms.

(d) In any cell, the first steps of carbohydrate metabolism, the conversion of food into useful energy for the operation of the cell (Chapter 19), involve fermentation. The last steps of metabolism in aerobic organisms include oxygen-utilizing respiration (Figure 20.2). This fact suggests that the first and more primitive organisms did not, and probably could not, use oxygen. If present, free oxygen would probably have been poisonous to them. Later generations that evolved in response to the growing accumulation of free oxygen, which is a waste product of photosynthesis (Chapter 18), must have gained such an advantage (a factor of 18 in ATP production) from an ability to use oxidation reactions that they soon inherited most of the desirable habitats on Earth and florished.

(e) Cytochrome c, a protein which provides electron transport in photosynthesis and respiration, is similar but not identical in structure in all aerobic organisms. This fact suggests that there may have been several, but not many, major branches from anaerobic to aerobic organisms.

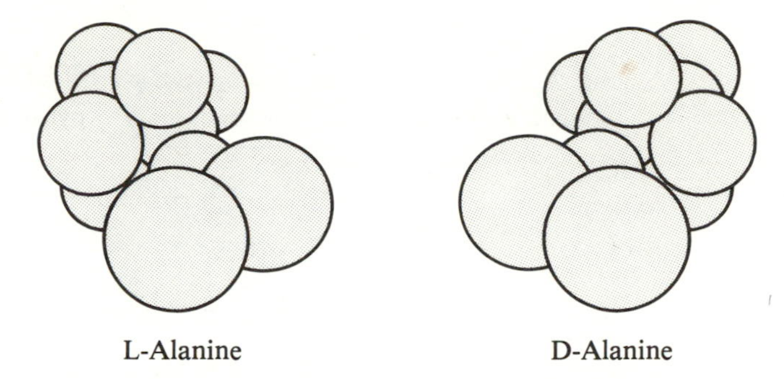

Figure 20.1. Amino acids come in both left-handed and right-handed forms, which are mirror images of one another. The convention L (levo) and D (dextro) refer to the way that a collection of such molecules will rotate the plane of polarization of a light beam that passes through the assembly. The discovery that biological molecules generally take only one of the possible mirror-image forms was made first by Louis Pasteur. We now know that living organisms usually use only L-amino acids and D-sugars.

lar line. Box 20.1 does, however, contain one clue—item (b), the almost exclusive use of L-amino acids—which suggests that the successful branches of the tree of life have a common trunk. We know of no laws of chemistry which would lead to an *intrinsic* discrimination between left-handedness and right-handedness. (Our discussion of parity violation in Chapter 16 involved the weak interaction, which has no relevance for ordinary chemistry.) In other words, a biology which is the mirror image of the actual biology that exists on Earth is perfectly admissible—and, indeed, would be likely—if there were many separate origins of life. As we will discuss shortly, the early Earth probably contained large amounts of amino acids of *both* L and D forms. If there had been several separate origins of life on Earth, it is extremely unlikely that they would all have incorporated only L-amino acids. Some of them would almost certainly have used the D-amino acids, for which there would have been no competition, since the L-forms of life would have found D-proteins indigestable! This suggests that the L-forms of life accidentally got a head start, and soon made conditions impossible for a second instance of "spontaneous generation." Perhaps the atmosphere became too oxidizing, or perhaps certain natural resources became too severely depleted. In any case, the exclusive use of L-amino acids, plus the laws of probability, argue that a single successful protocellular line gave rise to all the living things on Earth today.

Problem 20.2. Freshly fallen meteorites of the carbonaceous chondrite variety (Figure 17.6c) show traces of amino acids. Can you suggest a way to tell whether such amino acids arose by contamination from handling by careless experimenters? *Hint*: Consider the likely ratio of L- and D-forms if the amino acids were synthesized by nonbiological processes in outer space. One form of the *panspermia hypothesis* suggests that life on Earth may

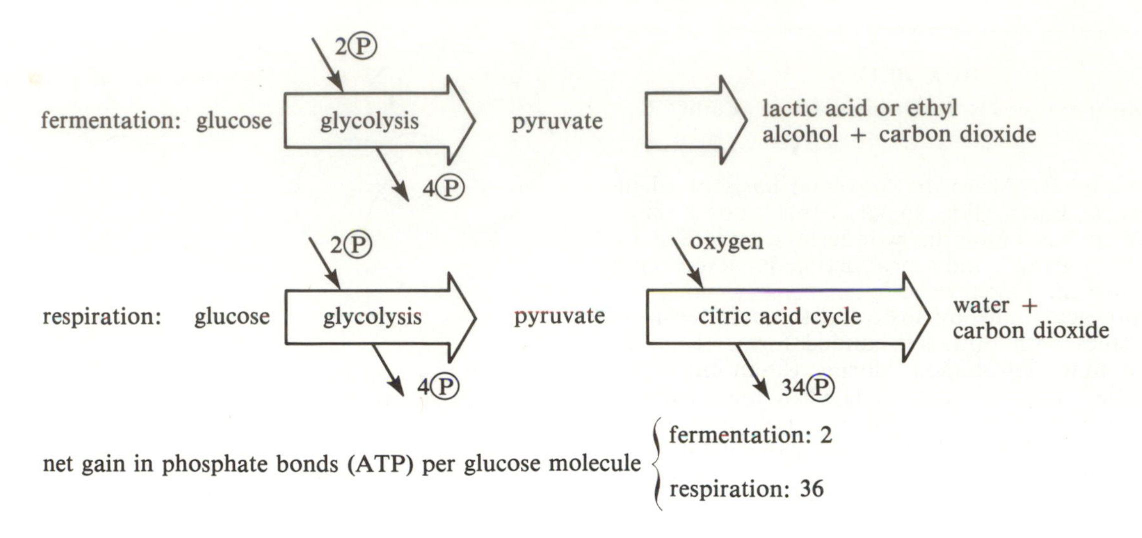

Figure 20.2. Schematic comparison between fermentation and respiration. The symbol Ⓟ refers to a high-energy phosphate bond. Respiration differs from fermentation principally in the addition of the citric-acid cycle, as was discovered by Szent-Georgi, Martius, and Krebs. Notice that respiration is 18 times more efficient than fermentation in producing ATP from glucose, because fermentation does not extract much of the chemical energy of glucose in merely converting it to the intermediate product, pyruvate, and the further breakdown into lactic acid or ethyl alcohol plus carbon dioxide does not produce any net ATP. (For details, see J. W. Schopf, *Scientific American*, **239**, Sept. 1978, 110.)

have been seeded by preexisting life carried on meteoroids. Since the primitive Earth was bombarded by many meteoroids whose basic materials may have originated from many different places (Chapter 18), what implications does the biological use of L-amino acids only have for this form of the panspermia hypothesis? Since life had to originate somewhere, discuss the philosophical merits—if any—of the panspermia hypothesis.

Pasteur's Experiments on Spontaneous Generation

Throughout most of human civilization, it was commonly believed that certain living organisms could rise spontaneously from nonliving matter. It seemed indisputable that maggots could suddenly appear in rotting meat, or worms from mud and slime. By the beginning of the nineteenth century, careful observation and experimentation had shown that complex organisms were always preceded by parents, and the emphasis shifted to the issue of whether microorganisms could exhibit spontaneous generation. The controversy was settled decisively in 1862 by Louis Pasteur in a series of famous experiments involving sterilized broth and swan-necked flasks (Figure 20.3). Pasteur showed that nutrients prepared to contain no initial microorganisms did not spoil as long as they were kept carefully shielded from contamination. In other words, under the circumstances which prevail today, even microorganisms cannot arise spontaneously from nonliving matter. In the modern world of canned foods and Pasteurized milk, this fact has become, of course, commonplace knowledge.

Pasteur's famous demonstration unfortunately also lent itself superficially to a vitalistic interpretation. Vitalists argued that if living things could not arise spontaneously from nonliving matter, was this not proof that there existed some crucial "life force" which was missing from inanimate material? Charles Darwin, however, characteristically peered deeper than most of his contemporaries. Darwin reasoned that once life had appeared on this planet, it would be very difficult for spontaneous generation to take place again. The developing form of prelife would probably require geological

Figure 20.3. One of Louis Pasteur's experiments on spontaneous generation used a swan-neck flask. A nutrient broth was heated up to destroy any microorganisms (bacteria, mold, etc.), the vapor being driven out through the swan neck of the flask. After cooling down, the flask allowed air to circulate to the broth, but any microorganisms contained in the air were trapped within the S-bend of the neck. A broth prepared this way did not go "bad," i.e., did not spoil because of the "spontaneous generation" of bacteria within the broth. Pasteur had to allow air to circulate to the nutrient broth because some vitalists had claimed that air itself contained the critical "life force" that made spontaneous generation possible.

times before it could become fully "alive," but well before the alloted time passed, this prelife would probably become food for some already living organism. However, Darwin reasoned further that this objection would not apply to a time before there were any living organisms; then there would have been time aplenty. Thus, Charles Darwin could write to a friend in 1871:

> "It is often said that all the conditions for the first production of a living organism are now present, which could ever be present. But if (and oh! what a big if!) we could conceive in some warm little pond, with all sorts of ammonia and phosphoric salts, light, heat, electricity, etc., present, that a protein compound was chemically formed ready to undergo still more complex changes, at the present day such matter would be instantly devoured or absorbed, which would not have been the case before living creatures were formed."

Modern workers would add the desirability of a lack of free molecular oxygen and a role for nucleic acids; otherwise, Darwin's words provide a perfect introduction to current ideas on the origin of life.

The Hypothesis of a Chemical Origin of Life

Independently, in the 1920s and 1930s, J. B. S. Haldane in Britain and A. I. Oparin in Russia proposed the hypothesis that life originated on Earth via a series of chemical steps. It had occurred to Haldane in 1929 that the common use of fermentation as the first steps in the carbohydrate metabolism of all organisms suggested that the primitive atmosphere of the Earth must have been reducing, one in which free molecular oxygen was absent. A few years earlier, Oparin had reached the same conclusion from a conjecture that natural petroleum may have been derived from nonbiological processes. (The prevailing view among geologists now is that biological processes account for nearly all the present-day supply of oil and natural gas, but in recent years Thomas Gold has revived the nonbiological viewpoint.) As we discussed in Chapter 18, both the astronomical and the geological evidence also suggests that the first atmosphere of the Earth was not oxidizing.

This point of view is reinforced by the simple observation that dead organisms which have lost their biological defense mechanisms will oxidize when heated in the present atmosphere of the Earth. Logs and steaks burn; rocks and lakes do not. The reason is, of course, that in contrast to the chemical makeup of nonliving things on Earth, the carbon and hydrogen atoms in the molecules of living things are attached often to each other, not to oxygen atoms. Reduced organic matter would not have arisen had the primitive atmosphere of the Earth been oxidizing. Can we understand then in detail how living matter came to exist on Earth?

The subject of the chemical generation of life is a very young discipline, and there are still great uncertainties. The main ideas, as seen by modern workers, are as follows (see the summary in Box 20.2).

(a) One starts with a reducing atmosphere and the primitive oceans of the Earth. Lightning strokes, ultraviolet light (which would have penetrated to the surface in the absence of free oxygen; see Chapter 18), volcanic heating, and/or other natural energetic phenomena present in the reducing atmosphere would have produced various fatty acids, sugars, amino acids, purines (A and G), and pyrimidines (C, T, and U). The random synthesis of such organic compounds by such methods has been

BOX 20.2
Important Steps in the Creation of Life from Inanimate Matter (a Theory)

(a) Creation of small organic molecules in a reducing (or nonoxidizing) atmosphere via energetic natural phenomena: lightning strokes, the penetration of ultraviolet light, etc.
(b) The dissolving of such small organic molecules in the primitive oceans to form a warm and dilute "primordial soup."
(c) The synthesis of ATP, GTP, CTP, UTP, and their deoxy counterparts, dATP, dGTP, dCTP, and dTTP, as the precursors to nucleic acids.
(d) The polymerization by nonbiological means of amino acids into proteins and of nucleoside phosphates into nucleic acids (RNA and DNA). The primary difficulty is to remove the water required for the joining of monomers.
(e) The spontaneous formation of concentrated droplets of organic molecules, either coacervates or microspheres, with a survival advantage for those protocells that have a primitive ability to form polymers from monomers.
(f) The evolution of metabolic chains, step by step from "back to front," as critical organic molecules in the general environment begin to become exhausted.
(g) The natural selection for those objects which have a genetic apparatus (the appropriate DNA) for "remembering" the metabolic recipes necessary for survival and which can pass that apparatus down to daughter fragments. The perfection of even a simple basis for growth and reproduction would have constituted the creation of life on Earth in the form of a primitive prokaryotic cell.

reproduced in many laboratory experiments, starting with those of Stanley Miller and Harold Urey in 1953, and continuing with work by P. H. Abelson, W. Gross, H. von Wessenhoff, and others. The essential trick to the nonbiological synthesis of simple organic compounds is to make sure that no free molecular oxygen is around. Reducing conditions probably also account for the formation of organic molecules in interstellar space, the colored bands of Jupiter and Saturn, and carbonaceous chondrites (Chapters 11, 18, 20).

Problem 20.3. Use the estimates of Problem 4.15 to show that 1.7×10^{24} ergs of sunlight fall per second on the Earth. A fraction, 1.6×10^{-4}, is contained in photons energetic enough to make and break common chemical bonds, i.e., in photons with wavelengths shorter than roughly 2000 Å. Show that this fact implies that about 2.7×10^{31} useful ultraviolet photons fell per second on the primitive Earth. Accounting for inefficiencies, assume that about 10^4 such photons were needed to form simple organic molecules that ultimately dissolved in the primitive oceans. Show, then, that in 3×10^8 years (from 3.8 billion years ago to 3.5 billion years ago), about 2.6×10^{43} organic molecules would have existed in the primitive oceans from this source alone. If the oceans had their present volume, the water would have occupied a total of 1.4×10^{24} cm^3. Thus, show that ultraviolet synthesis would have contributed, on average, about 1.9×10^{19} simple organic molecules to each cubic centimeter of the primitive oceans. If the average molecular weight of these organic compounds was 10^2, show that each cubic centimeter would have contained about 0.003 gm of organic matter and 1 gm of water. In other words, organic matter would have made up 0.3 percent of the primitive oceans: Haldane's dilute "primordial soup." Compare this estimate (which is clearly very approximate, but is probably not off by too many orders of magnitude) with the roughly 30 percent organic matter and 70 percent water of an *E. Coli* cell.

(b) The dissolving of the relatively small molecules described in (a) in the primitive oceans would have produced Haldane's hypothesized warm and dilute "primordial soup." Combined with various dissolved minerals and salts—in particular, phosphates—derived from the erosion and weathering of rocks, this liquid broth would have provided an ideal medium for further chemical reactions. In particular, fatty acids are an important constituent of lipids; simple sugars are the building blocks for polysaccharides; amino acids are the building blocks for proteins; and purines and pyrimidines (the nucleic-acid bases) are the building blocks for RNA and DNA. The discussion of Chapter 19 leads us to expect that, if we have a means to make proteins and nucleic acids and to put them in a protocellular environment, the assembly of lipids and carbohydrates into cellular structures might naturally be expected to follow. The discussion of Chapter 19 also shows that we can regard proteins and nucleic acids to be *irregular* polymers: long chains of different amino acids and different *nucleotides* (base plus sugar plus phosphate), respectively. Thus, the next logical step is to see how individual nucleotides can be assembled, then how amino acids can be polymerized to give proteins, and then how nucleotides can be polymerized to give nucleic acids (Figure 20.4a).

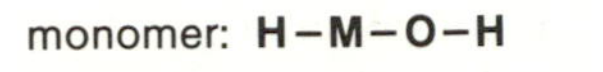

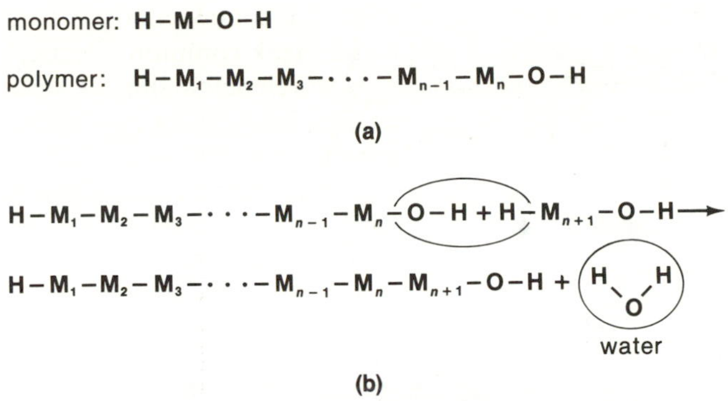

polymer: $H-M_1-M_2-M_3-\cdots-M_{n-1}-M_n-O-H$

(a)

$$H-M_1-M_2-M_3-\cdots-M_{n-1}-M_n-O-H+H-M_{n+1}-O-H \longrightarrow$$
$$H-M_1-M_2-M_3-\cdots-M_{n-1}-M_n-M_{n+1}-O-H+H_2O$$

water

(b)

Figure 20.4. Polymerization. (a) Schematic depiction of biological polymers where H—M—OH is a typical free monomer. In polysaccharides, the monomer units $M_1, M_2, \ldots, M_n$ are either identical or form repeating sequences. In proteins, the monomers are 20 different kinds of amino acids which join in nonrepeating sequences. In nucleic acids (RNA or DNA), the monomers are four different kinds of nucleotides which join in nonrepeating sequences. (b) The adding of one more monomer unit to a polymer chain requires the release of a water molecule, H_2O.

(c) In living cells today, ATP supplies the energy required to drive polymerization reactions. The molecule ATP can be regarded as the base adenine joined to the sugar ribose (forming the *nucleoside* adenosine) plus a tail of three phosphates (ATP = adenosine triphosphate). The ingredients adenine, ribose, and phosphates would have been present in the primitive oceans according to step (a) above. Thus, it is significant that experiments carried out by Cyril Ponnamperuma and Carl Sagan, shining ultraviolet light on a dilute solution of adenine, ribose, and phosphoric acid, yielded large amounts of ATP. Presumably, the same process which joins adenine to ribose to three phosphates to give ATP could join any of the three other bases, guanine, cytosine, and uracil, to ribose and three phosphates to give GTP, CTP, and UTP. Richard Dickerson has noted that there is no obvious reason why ATP should be better suited for chemical energy storage than GTP, CTP, or UTP. The universal adoption of ATP as the energy currency of all cells may have resulted from the fact that nonbiological mechanisms of synthesizing the nucleic-acid bases produce much more adenine than any of the others. Those prelife forms which made use of the resultant free ATP in Haldane's dilute soup would then have had a significant advantage over those that did not. Lifeforms developed according to what was available, not according to some preconceived master plan.

(d) If we could cause ATP, GTP, CTP, and UTP to polymerize, with the elimination of excess phosphate groups, we would have a strand of RNA. Similarly, if we could cause dATP, dGTP, dCTP, and dTTP to polymerize, we would have a strand of DNA (refer to Chapter 19). Unfortunately, the spontaneous polymerization of these nucleoside-triphosphates encounters an immediate obstacle in Haldane's dilute soup. The polymerization of monosaccharides to give polysaccharides, of amino acids to give proteins (polypeptides), and of nucleotides (nucleoside-monophosphates = base plus sugar plus phosphate) to give single-stranded RNA or DNA (polynucleotides) all involve the release of water (Figure 20.4b). Conversely, biochemical reactions which supply water to the polymers can break them down into smaller pieces. For example, in digestion, the stomach enzyme pepsin catalyzes the hydrolytic breakdown of proteins. In the presence of other enzymes, polymerization may proceed even when excess water is present. Thus, Severo Ochoa, Arthur Kornberg, and their coworkers have synthesized RNA and DNA in the laboratory by mixing a solution of nucleoside-triphosphates, some special enzymes, and some inorganic magnesium salts. Unfortunately, such polymerizations would not work for Haldane's dilute soup, for the simple reason that we are talking about a situation before there were any proteins and, therefore, any enzymes. The presence of water, water everywhere, in the primitive oceans would then have presented a substantial obstacle to polymerization. For this reason, J. D. Bernal suggested that a more attractive site for polymerization might be in the nooks and crannies of inorganic micas and clays. Aharon Katchalsky has demonstrated that this is indeed an efficient mechanism for promoting the polymerization of protein-like compounds. Stanley Miller and Leslie Orgel, as well as Sidney Fox, have pointed out other ways to obtain nonbiological protein synthesis, by either freezing or heating out the water. The nonbiological polymerization of nucleic acids of the variety found inside cells today seems to be a more difficult trick. Perhaps protein synthesis preceded RNA or DNA synthesis, and the enzymatic action of some of these proteins helped form the first strands of RNA or DNA. On the other hand, if we are truly to extend Darwin's proposal of natural selection down to a chemical level, we must recognize that the trick works well only when we have a good mechanism for accurate reproduction. Relying on purely accidental combinations would seem to require prohibitively long times. Thus, many people think that nucleic acids must have preceded proteins. This, then, is still a point of debate among origin-of-life theorists: which came first, not the chicken or the egg, but nucleic acids or proteins? Right now, no one really knows. Probably they both arose more or less at the same time (Figure 20.5).

(e) In any case, if we have a solution of nucleic acids, proteins, fatty acids, sugars, ATP, and possibly other goodies, then this mixture may spontaneously form concentrated droplets of organic molecules, droplets that

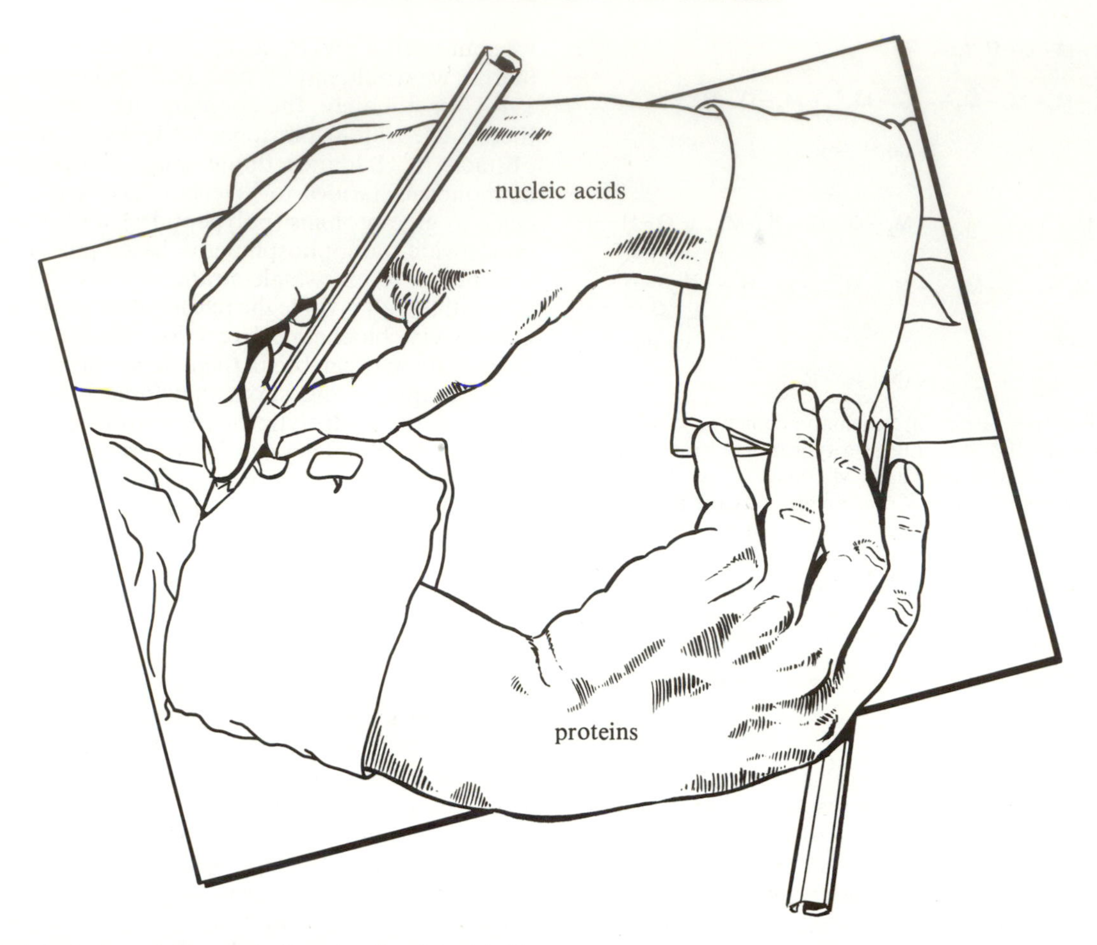

Figure 20.5. Adaptation of lithograph *Drawing Hands* by M. C. Escher (1948) can be taken to symbolize a fundamental issue in the creation of life. Let one hand represent nucleic acids; the other, proteins. Which came first, nucleic acids or proteins? Probably neither, each gave rise to the other through the inviolate laws of biochemistry. This aspect of self-reference or self-generation is called a *strange loop* by D. R. Hofstadter in his marvelous book, *Gödel, Escher, and Bach.* Hofstadter suggests that strange loops underlie the origin of many complex phenomena, including consciousness.

Oparin called *coacervates* (Figure 20.6). Oparin followed up this observation of the results of some laboratory experiments by hypothesizing that some coacervates will happen to hold together better in rough seas than others. Sidney Fox has shown that the heating of organic solutions in the presence of proteins produces microspheres with remarkably tough protomembranes. (The actual heating in the primitive oceans might come, for example, from volcanic action.) Both Oparin and Fox found that their tough droplets (which we will call protocells for simplicity) got tougher if they were given or acquired a primitive kind of metabolism. The toughest protocells may hold together long enough to develop more complicated mechanisms for metabolism, in which production of more polymers would in turn increase their chances for survival. Over aeons there would then evolve a strong chemical selection for protocells able to draw in molecules and energy from their surroundings and to synthesize large molecules that would promote not only their own survival, but also that of their daughter protocells which would result when the parent became too big to hold together. In short, physical survival favors metabolism (Figure 20.8).

(f) The evolution of metabolism would proceed in a certain pattern, according to a conjecture of Norman Horowitz that metabolic pathways, which are charac-

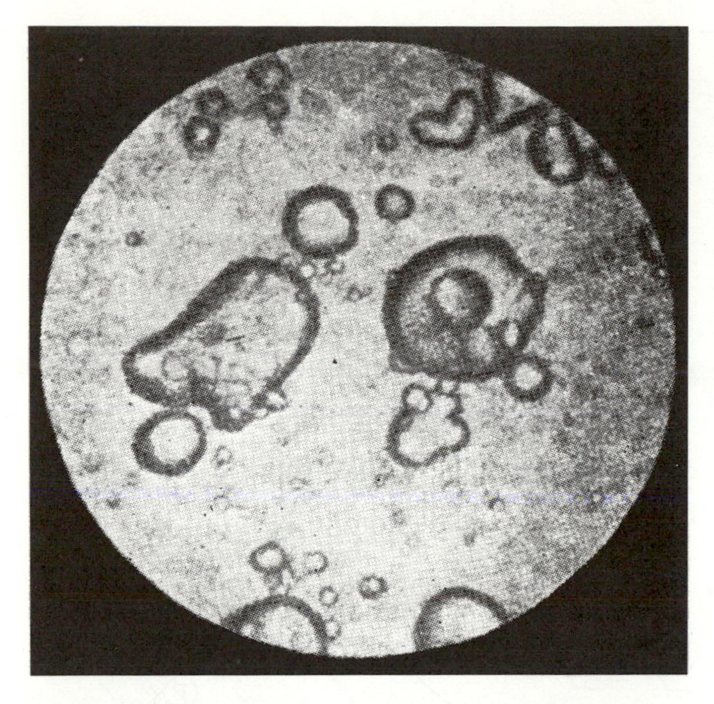

Figure 20.6. Photograph of coacervates. A. I. Oparin found that aqueous solutions of polymers tend to separate spontaneously into concentrated droplets of organic compounds suspended in a more dilute medium. (After Evreinova, courtesy of S. W. Fox.)

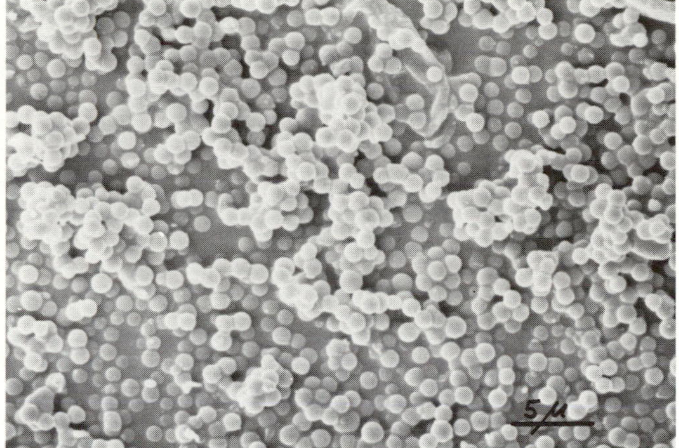

Figure 20.7. Electron micrograph of protenoid microspheres. A dry mixture of amino acids heated to moderate temperatures can polymerize to form proteins. If these proteins are subsequently added to warm water, microspheres of concentrated proteins containing a double-layered membrane will form. Under the proper conditions, the microspheres can even grow at the expense of the protein dissolved in their surroundings and can bud. (Courtesy of S. W. Fox, Institute for Molecular and Cellular Evolution.)

terized today as a series of chemical steps, may have evolved from "back to front." For example, we think of carbohydrate metabolism today as producing ATP given an initial supply of glucose (Chapter 19). In fact, as we have already discussed, ATP may have been plentiful in the primitive seas to begin with. But in the chemical competition for this natural energy source, ATP may have gotten scarcer and scarcer. Those protocells which could use a precursor compound to synthesize ATP would then have a selective advantage. But that other compound would also become used up. Then there would be a natural evolution to use another chemical step to synthesize this vanishing resource, etc. Is this the way in which the chemical steps of anaerobic fermentation were established, one chemical step after another, from back (ATP) to front (glucose)? Finally, when glucose itself became a vanishing resource, was a primitive form of photosynthesis established in order to manufacture glucose?

(g) The development of the mechanisms of metabolism (growth) would have constituted a fine beginning for a protocell. But it would not yet have been alive. For life to have originated, the protocell must possess a mechanism for self-replication (reproduction) by doubling the genetic content and passing down this content to daughter protocells. Dickerson suggests that self-replication became important as soon as the ready supply of free energy (ATP) diminished enough that there was a distinct advantage in passing down information for the nonrandom synthesis of compounds for which one vital reaction step led to another one. Thus, unlike Oparin, who believed that the mechanisms of metabolism (growth) preceded the development of the mechanisms of self-replication (reproduction), or Haldane, who believed the exact opposite, Dickerson advocates the plausible scenario that the mechanisms of growth and reproduction developed simultaneously. The perfection of these two mechanisms would have led to the first prokaryotic cell: life! And why was DNA chosen as the primary genetic material? Because the double helix of DNA constitutes the outstanding example of a polymer which catalyzes its own polymerization (Chapter 19).

The preceding scenario is obviously still conjectural. No one has, of course, succeeded in demonstrating from first principles that a chemical origin of life (a) is possible and (b) did, in fact, take place. The demonstration of possibility is close to being convincing, but there is room for those who wish to doubt. The demonstration that a chemical origin of life did, in fact, take place here on Earth probably lies outside the realm of science. We can only claim that it now seems the simplest and most natural explanation of the facts about life on Earth today in Box 20.1.

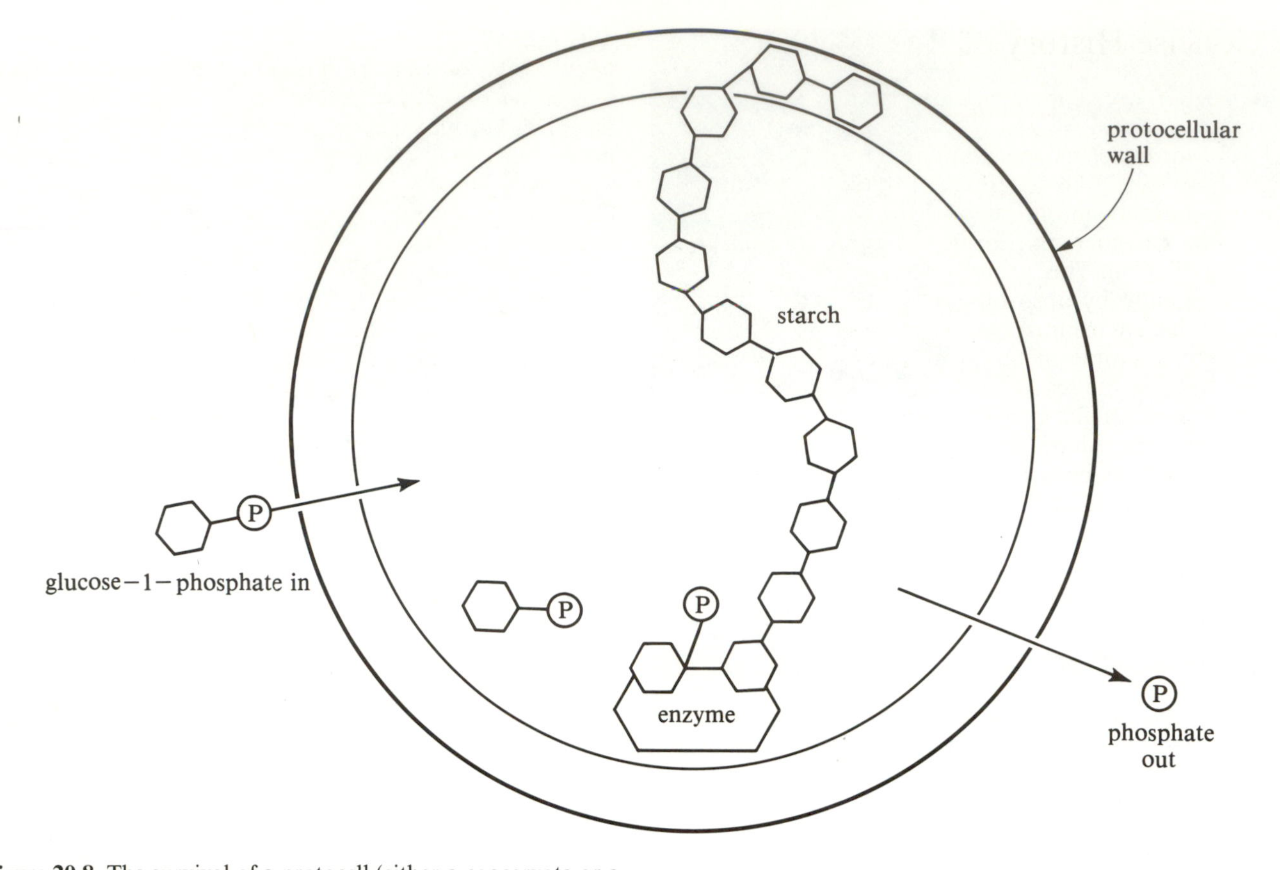

Figure 20.8. The survival of a protocell (either a coacervate or a microsphere) is improved if it can carry out polymerization reactions. Here glucose-1-phosphate is a glucose molecule which has a phosphate group attached to its "1" position (a convention developed by organic chemists to distinguish, say, between glucose-1-phosphate and glucose-6-phosphate). (For details, see R. E. Dickerson, *Scientific American*, **239**, March 1978, 70.)

If one does accept the outlines of the above theory, it is important to note that it does not contain any vitalistic notions. In particular, the theory of the chemical origin of life merely extends the concept of Darwinian evolution to a molecular level. At no time did a batch of chemicals ever decide, "Hey, let's get together to create life." There must have been many false starts and many dead ends. The successful combination arose finally merely by chance and by the operation of natural selection on a molecular level. Viewed today in its full glory, life seems incredibly complex. And so it is. But, historically, the continuity of nonliving things and living things was probably established one small chemical step at a time. In a simple organism like *E. Coli*, biologists have learned to decipher perhaps a third of all the chemical steps required to make *E. Coli* "tick." According to James Watson, in perhaps another twenty years we may know all the metabolic reactions involved in the *E. Coli* cell. Nothing in principle would then prevent us from constructing a live *E. Coli* bacterium, starting with only laboratory chemicals. Is it too much to believe that nature was capable of a comparable trick, given hundreds of millions of years to work at it?

Problem 20.4. This problem is for those who want to see a rough continuity between nonliving and living things in the world of objects which exists currently. Do you think molecules are alive? Do you agree that a virus is just a collection of molecules? Do you see that much difference between a virus and a bacterium?

A Concise History of Life on Earth

The Earliest Cells: The Prokaryotes

To fix ideas, let us speculate about what conditions might have been like at the dawn of life on Earth. The first living organism on Earth was probably a primitive bacterium, considerably simpler in structure and function than *E. Coli.* This primitive bacterium would have survived totally by absorbing small organic molecules found in its environment and restructuring them according to a simple metabolism. The chemical energy needed to drive the reactions would have come from the ATP it also scavenged from the surroundings, or possibly from some immediate predecessor of ATP. For our bacterium to have been alive according to our definition, it would also have had to possess a genetic apparatus for self-replication. At the beginning, then, it and its descendants would have grown and multiplied by consuming the organic molecules which were dissolved in the primitive oceans. These organic compounds were themselves synthesized by nonbiological processes, principally by solar ultraviolet radiation and lightning discharges acting on a reducing atmosphere.

As the numbers of the earliest bacteria grew explosively from lack of competition, the readily usable high-energy compounds became rarer. The resultant series of energy crises would have led to the early demise of countless billions, but also to the evolution, step by step, of a fermentation cycle by which ATP could be synthesized biologically from glucose. At this point, we would have had bacteria which resemble the simplest sort of fermenting bacteria found today. Since the glucose would still have been derived from nonbiological processes, however, even at this point, the earliest cells would have been consumers or transformers of organic matter, not producers.

Problem 20.5. A modern bacterium, like *E. Coli*, might divide once every two hours when supplied with adequate nutrients and freed from competition. Suppose, for sake of argument, that primitive bacteria reproduced much less efficiently, doubling their numbers, say, only once a week. Ignore the nutritional requirements needed merely to sustain life; assume a single bacterium contains 10^{-10} gm, and calculate how many years it takes unsaturated exponential growth to exhaust the organic reserves of the primitive oceans (see Problem 20.3). Outline how you might estimate the steady-state population of bacteria that could have been supported by nonbiological synthesis of organic matter. Assume you are given the relevant metabolic rates.

Eventually, even the supply of free glucose would begin to run out. The resultant biological stress could be expected to select for the evolution of a *biological* mechanism to produce glucose from the still readily available natural chemical and energy resources. This mechanism was probably a crude form of photosynthesis. Melvin Calvin has suggested that the first form of photosynthesis did not use water, H_2O, as one of the initial reactants, but rather hydrogen sulfide, H_2S. The reaction, discovered by C. B. van Niel, can be indicated schematically as

$$12\ H_2S + 6\ CO_2 + \text{light energy} \rightarrow C_6H_{12}O_6 + 12\ S + 6\ H_2O. \quad (20.1)$$

The use of hydrogen sulfide to reduce carbon dioxide (both of which could have been outgassed from volcanoes) is still the method followed by green and purple sulfur bacteria. In contrast, the normal photosynthetic pathway uses water, H_2O, to reduce carbon dioxide, via the net reaction

$$12\ H_2O + 6\ CO_2 + \text{light energy} \rightarrow C_6H_{12}O_6 + 6\ O_2 + 6\ H_2O. \quad (20.2)$$

Why should process (20.1) have developed before process (20.2)? An explanation requires an understanding of photosynthesis, which is a very complicated process involving many chemical steps. For our purposes, it suffices to indicate the main steps as the following. A cell pigment P (chlorophyll in the case of green plants), surrounded by a potential electron acceptor A and a potential electron donor D, is excited by the absorption of a photon. The pigment molecule deexcites, with the energy of deexcitation being used (in part) to transfer an electron from D to A. Thus, we may indicate the sequence of events schematically as

$$\frac{A}{\frac{P}{D}} \rightarrow \frac{A}{\frac{P^*}{D}} \rightarrow \frac{A-}{\frac{P}{D+}}. \quad (20.3)$$

The external energy input from light is then transferred to chemical bonds (with an overall efficiency of about 1 percent) by a series of oxidation-reduction reactions mediated by proteins like cytochrome c. The entire photosynthesis process (20.2) is, however, more complex than the entire process, (20.1), since process (20.2) requires two centers where light input occurs, and process (20.1) requires only one center; so it is natural to believe that process (20.1) evolved before (20.2).

The sulphur S that is one of the end products of process (20.1) is released from green sulfur bacteria in the form of sulfates as a waste product. Among anaerobic organisms that still survive, there are bacteria which use

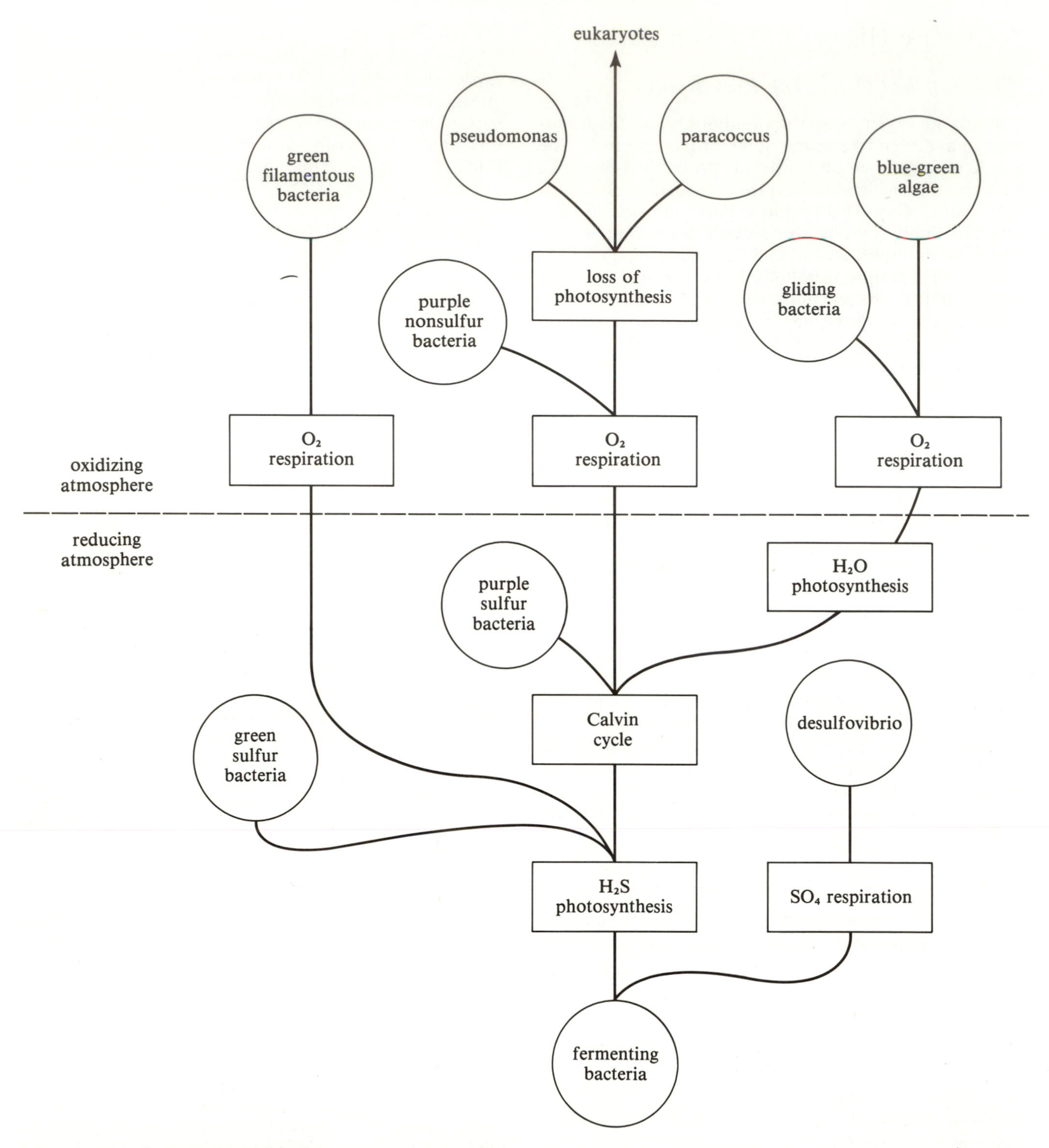

Figure 20.9. A phylogenetic tree of some of the earliest cell lines which have survived to the present day. (For details, see R. E. Dickerson, *Scientific American*, **242**, March 1980, 136.)

sulfate (SO_4) as an oxidant (instead of molecular oxygen, O_2) in a prerespiration extension of fermentation. It is natural to think that this mechanism evolved in response to increasing availability of SO_4 as a means to make more efficient use of the energy contained in glucose. Apart from carbon dioxide, CO_2, the waste product of this anaerobic form of respiration is hydrogen sulfide, H_2S, instead of water, H_2O. Had it not been for the development of the modern photosynthesis process (20.2), we might have had a parallel system of plants and animals where we would have talked about the carbon-dioxide/sulfuric-acid budget instead of the carbon-dioxide/oxygen budget. That would make a nice science-fiction tale for life on Venus!

Of course, the more complex process (20.2) did develop, and we can presume that it did so simply because water, H_2O, is and was a much more abundant natural resource on Earth than hydrogen sulfide, H_2S. With the evolution of cyanobacteria (or blue-green algae) and similar true photosynthetic organisms, the biological release of molecular oxygen would have caused the atmospheric O_2 to begin its rise to modern levels (Chapter 18). The gradual accumulation of O_2 in the atmosphere resulted in several important events.

First, the photosynthetic release of free oxygen would help to establish an ozone layer (Chapter 18), shutting off one of the main nonbiological mechanisms for producing organic compounds, the penetration of ultraviolet light, and thus putting a high premium on efficient energy usage. This would have created a great selective advantage for those respiring organisms that could use the released oxygen gas to produce ATP from glucose. Perhaps at this time, or even earlier, the disappearance of an ample supply of amino acids from nonbiological sources led to the evolution of the mechanism of nitrogen fixation (the ability to use atmospheric nitrogen, N_2, to synthesize nitrogenous organic compounds).

Second, the proliferation of cells that were capable of manufacturing all the essential organic compounds from inorganic matter meant that their deaths would lead to the general release of such compounds into the environment. This meant that some organisms would be able to absorb nutrients without having to produce their own. Some of these organisms may then have lost their ability to use (visible) sunlight to help synthesize organic molecules. Early in the history of life on Earth, therefore, organisms separated into two branches: producers (algae) and consumers (bacteria). Finally, free molecular oxygen is toxic to anaerobic organisms. Ones that did not evolve a biological defense mechanism against this highly reactive gas would have been eliminated from all but the most undesirable habitats, where they manage to eke out a primitive existence even today.

We have no way of knowing, of course, whether the preceding description represents an accurate portrayal of the history of the evolution of the earliest cell lines. The molecular evidence (as exemplified by Dickerson's studies of the structure of cytochrome c) and the microfossil evidence (as gathered and studied by Schopf and others) suggest that this description is plausible. Figure 20.9 gives, then, a phylogenetic tree of some of the earliest cell lines which have survived to the present day. How many trial-and-error experiments of natural evolution did not survive, we shall probably never know.

Sex Among the Bacteria; The Appearance of Eukaryotes

For at least two billion years—from 3.5 billion years ago to 1.5 billion years ago—the Earth contained only two groups of organisms, bacteria and simple algae, both of which are prokaryotes (see Chapter 18). Evolution, of course, was not standing still during this time. As we saw in the previous subsection, the elaboration and sorting out of different kinds of metabolisms was taking place at this time. Eventually, some cells must have acquired the ability to move about, a great advantage in the search for nutrients. Two other milestones in the evolution of life on Earth were the discovery of sex and the appearance of eukaryotic cells: cells with nuclei and various organelles.

Bacteria are capable of a primitive kind of sex in which two adjoining organisms partially mix their genetic material (Figure 20.10). This does not constitute true sex, since bacteria do not have paired chromosomes; so bacteria are incapable of forming the germ cells (sperm and egg) that characterize advanced sexual reproduction by multicellular organisms. Nevertheless, even the primitive conjugation enjoyed by bacteria would be a relatively efficient mechanism for trying out many different combination of genes. Successful metabolic mechanisms worked out by separate cell lines could then be pooled, to the great advantage of the lucky progeny. Sex among the bacteria may, therefore, have led to a rapid proliferation of different kinds of unicellular organisms.

The selection for cells with sufficient DNA to "remember" the ever-growing complexity of metabolic chains would probably have led to an increase in the size of certain cell lines. Eventually, some organisms must have become so large that they could engulf their smaller neighbors, perhaps much as present-day amebae do (Figure 20.11). This wholesale way of acquiring food ("eating") may have proven so advantageous that there was a strong natural selection among predatory cells for large size.

Lynn Margulis has speculated that, in one such an engulfing event, the chemistry between the two participants was such that the would-be diner found its would-be dinner, say, a small bacterium, to be indigestible. If the bacterium reproduced only very slowly inside its host cell, so as not to end up killing its host, the host upon mitotic division may produce daughter cells which

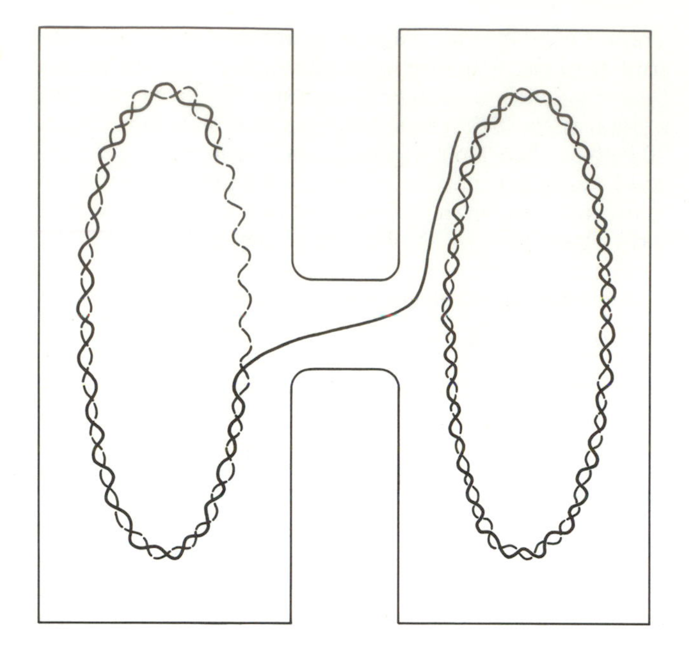

Figure 20.10. The DNA of most bacteria is arranged in a circular form where the two ends of the double helical molecule are joined. Bacteria may mate with one of the participants inserting a single strand of its DNA into the other. This strand, shown as a solid curve above, then forms its complement inside the recipient bacterium, and is converted to normal double-helical DNA. Parts of this may be exchanged for a portion of the DNA of the recipient bacterium by genetic recombination (like crossing over). When the bacterium subsequently divides, the genetic inheritance of the progeny cells is different from that of either original bacteria. (For details, see J. D. Watson, *Molecular Biology of the Gene*, W. A. Benjamin, 1976.)

Figure 20.11. Photograph of an amoeba engulfing a ciliate. (Courtesy of Ralph Buchsbaum.)

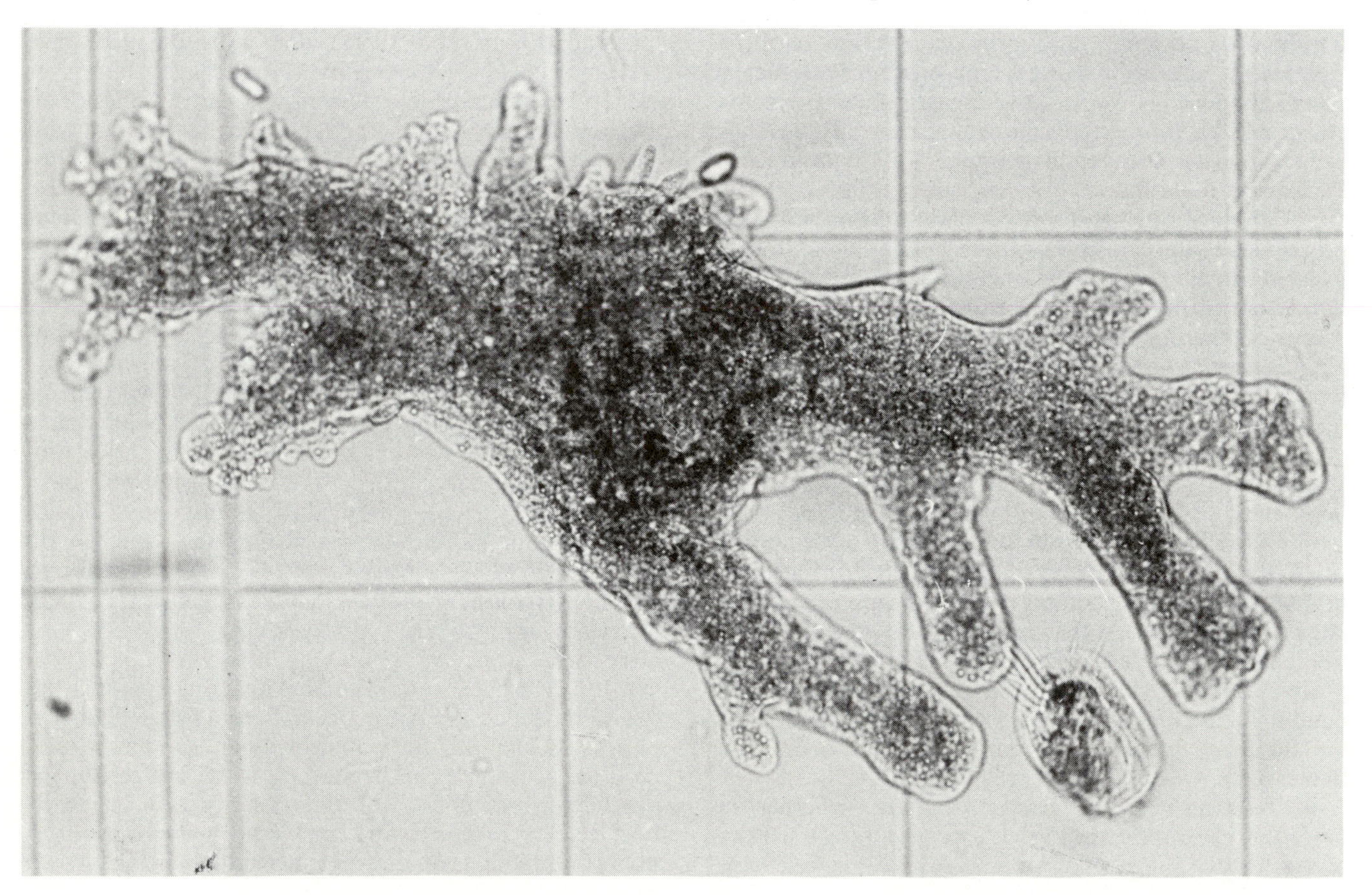

contain examples of the bacterium or its descendants. Eventually, the large predator cells and the small bacteria may enter into a *symbiotic* relationship, to the benefit of both participants. Could this have been the origin of *mitochondria* in eukaryotic cells? Margulis not only believes yes, but also suggests that *chloroplasts* might have had a similar origin, when a blue-green alga turned out to be indigestible. In support of these points of view are the facts that mitochondria and chloroplasts have their own DNA and reproduce by division (Chapter 19), and that Kwang Jeon has documented a present-day case in which amebae have formed symbiotic relationships with bacteria "infecting" the former. By considering several lines of evidence, Philip John and Frederick Whatley have proposed the bacterium *Paracoccus* as the most likely ancestor of mitochondria (see Figure 20.9). The flagella which whirl and propel some species of bacteria also have their own DNA, and Margulis has speculated that they originated in another symbiotic event involving a spirochete bacterium.

James Watson, among others, further points out that since many vital biological processes occur on membranes, and since the outer membrane of a large cell contains a relatively small ratio of surface area to volume, there would be a selective advantage for large cells to develop internal membranes to surround smaller functional units. This explanation presumably underlies the observed partition of most of the DNA of a eukaryotic cell into a nucleus, with other internal membranes surrounding various organelles (like mitochondria or chloroplasts).

The appearance of eukaryotes allowed a great increase in adaptive behaviors. For example, paramecia are capable of a more advanced form of conjugation than exhibited by bacteria. Indeed, occasionally two mating paramecia may fuse and form a doublet animal! These amazing creatures can reproduce by fission to give several generations of doublets. Could a fusion process have given rise to cell lines with paired chromosomes that were capable of both mitotic and meiotic cell division?

Multicellular Organization and Cell Differentiation

The great organizational capability of eukaryotic cells made possible the appearance of multicellular organisms. What was the evolutionary advantage to being multicellular? When there are predatory cells around, it pays to be as large as possible, so that you cannot be engulfed. However, there is a law of diminishing returns in making a single cell larger and larger. The cellular machinery becomes inefficient if molecules have to be transported over great distances. Moreover, the chances for lethal copying errors rise with greater complexity of the parent cell. Also, the cell walls or cell membranes of large cells are easier to break, causing a total loss for such a large investment. No, nature teaches that if you want to make something large and complex, the most efficient way is to aggregate small subunits, so that you can take advantage of the basic machinery already worked out for the subunits (see the philosophical discussion of Chapter 19). To assemble a working whole, you then need only some mechanism for throwing out the bad subunits, a procedure followed in multicellular organisms and in the mass production of modern goods.

The simplest multicellular organisms, such as filamentous green algae and sponges, are hardly more than colonies of identical eukaryotic cells. The cells of filamentous green algae are interesting, in that they are capable of both simple reproduction (mitosis) and advanced sexual reproduction (meiosis plus fusion of egg and sperm). Reproduction in sponges occurs asexually in some varieties (by budding) and sexually in others (usually, but not always, with sperm and egg coming from different sponges). The cells of a sponge typically do show some differentiation, but a sponge has no well-defined tissues or organs. The lack of significant differentiation allows some sponges to regenerate the whole organism from a small piece.

In higher multicellular organisms, there is much more cell differentiation. Indeed, the very wonder that is a higher organism is that its cells can specialize to form different *tissues*, different tissues can act collectively to form *organ systems*, and organ systems can coordinate their activities to form the entire specimen. We have already outlined the macroscopic theory of Charles Darwin and Alfred Wallace about how natural selection could have led to the evolution of such complex organisms. In this view, given enough time, organisms as diverse and as versatile as flowering plants, insects, and mammals could all have evolved from a single line of ancestral fermenting cell (Figure 20.12).

This is a remarkable view, a humbling yet beautiful insight. Perhaps Darwin himself expressed it best in the closing passage of his *Origin of Species*:

> "It is interesting to contemplate an entangled bank, clothed with many plants of many kinds, with birds singing on the bushes, with various insects flitting about, and with many worms crawling through the damp earth, and to reflect that these elaborately constructed forms, so different from each other, and dependent on each other in so complex a manner, have all been produced by laws acting around us. These laws, taken in the largest sense, being Growth and Reproduction; Inheritance, which is almost implied by reproduction; Variability from the direct and indirect action of the external conditions of life, and from use and disuse; a Ratio of Increase so high as to lead to a Struggle for Life, and as a consequence to Natural Selection, entailing Divergence of Character

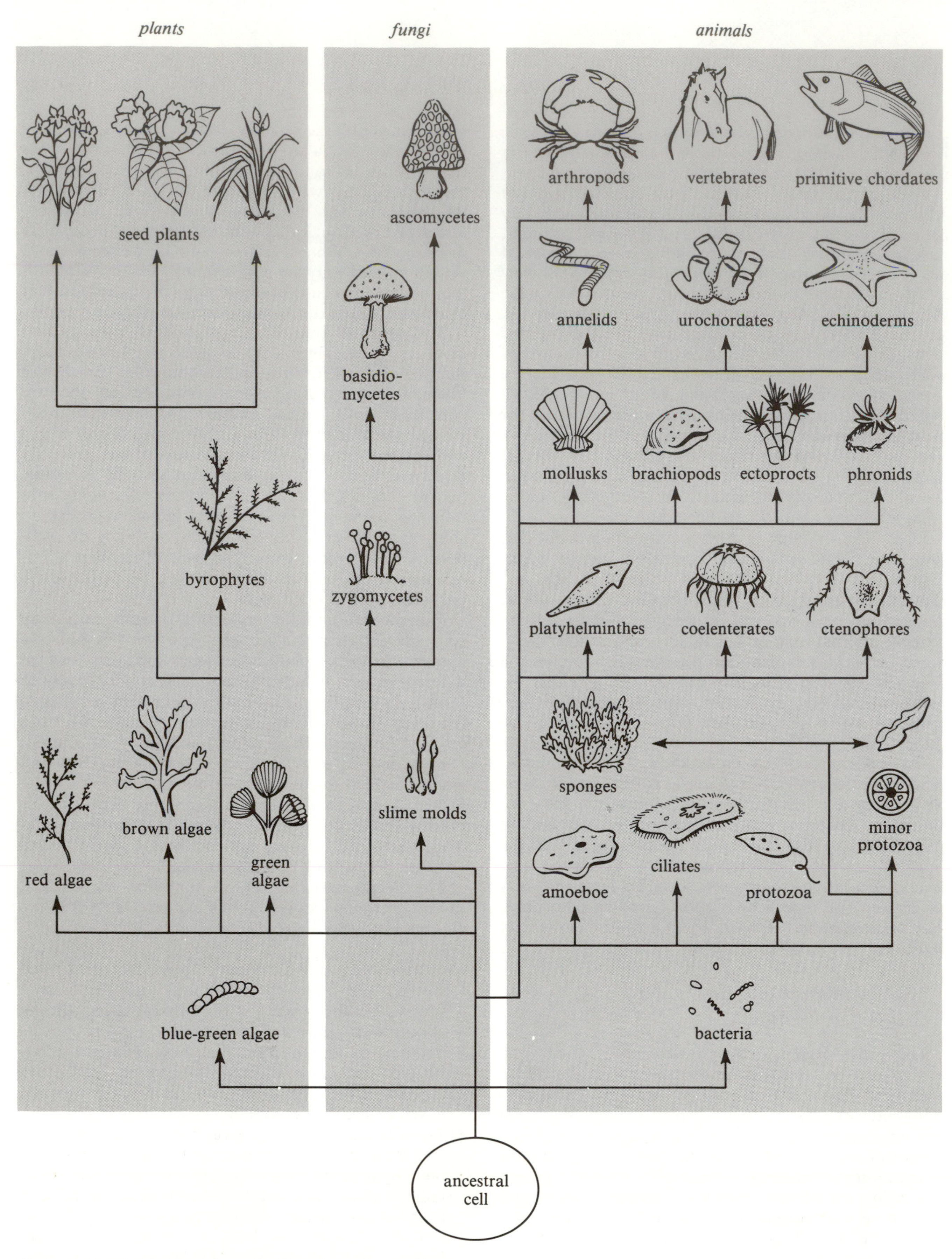

Figure 20.12. Kingdoms of the organisms, charted according to R. H. Whittaker. (For details, see J. W. Valentine, *Scientific American*, **239**, Sept. 1978, 140.)

and the Extinction of less-improved forms. Thus, from the war of nature, from famine and death, the most exalted object which we are capable of conceiving, namely the production of the higher animals, directly follows. There is grandeur in this view of life, with its several powers, having been originally breathed into a few forms or into one; and that, while this planet has gone cycling on according to the fixed laws of gravity, from so simple a beginning endless forms most beautiful and most wonderful have been, and are being, evolved."

After more than a century, those words still have the power to move us deeply. Pause and reflect on them. Darwin did not say anything as undignified as that we descended from the apes. Rather Darwin said that humans, apes, flowers, trees—everything that lives on this beautiful Earth, we are *all* related. We even have a common bond with the air, rocks, and waters of Earth. Indeed, there is grandeur in such a view of life.

The Organization of Societies

Another level of organization above the multicellular organism is observed in animal behavior. This is the tendency for groups of organisms to form societies in which individual organisms specialize and perform tasks beneficial for the society as a whole. Examples among the insects include ants, bees, and termites. The behavior patterns of such insects can seem at times to be superficially anti-Darwinian. For example, consider the act of a worker bee stinging an intruder, thereby dying to protect the hive. How can genes which contribute to the death of an individual have become established by natural selection?

Early attempts to explain such altruistic behavior as "being for the good of society" were flawed, for they implied a moralistic purpose that even many humans lack. The modern interpretation of altruistic behavior among animals was devised primarily by W. D. Hamilton. Consider two hypothetical beehives, the first of which has genes passed down from the queen bee that include altruistic defense of the hive, the second of which does not. The attack of an intruder may kill many more bees in the second hive than in the first hive. With the passage of time, genes from the first hive clearly have a selective advantage, being more likely to be transmitted to future generations than genes from the second hive. In particular, as a corollary, we also see why genes that give a first priority to protecting the queen might get promoted, for if the queen bee is killed, all is lost.

Mammals that form small groups or large herds are, of course, also known. Humans probably descended from a line of primates that naturally formed small groups (Figure 20.13). We probably carry genes that code for altruistic behavior toward members of our own family and close circle of friends. Humans, however, have also evolved to possess great intelligence. At first, this great intelligence had enormous survival value, for it led humans to band together in ever larger groups, wherein individual members could be freed to pursue ever more specialized knowledge and skills. With our accumulated knowledge and skills, we as a civilization have gained supremacy over all the living species on this planet. To run the vast human enterprise, we have had to construct "social compacts" and "moral laws" that lie outside the confines of natural science. In other words, human society at present is a product of the intellect, not of the genes. This led to no conflict when new frontiers kept opening and technological progress satisfied rising expectations. In such times, even simple principles of social organization would work. Under these circumstances, optimists could dream of a utopian world when all of humanity would forever be freed of Tennyson's "Nature, red in tooth and claw."

And yet, realistically, can we humans ever be unshackled from our natural instincts? We have been genetically programmed by hundreds of millions of years of evolution for self-preservation and reproduction. In the modern world, these instincts now often clash with what we know intellectually is "good for society." Our altruistic genes, if any, are directed only to small groups, not to a large and impersonal civilization taken as a whole. For the good of society, we ought to face up to the problem of population growth before we are struck with the Malthusian prophecy. Instead, we tend to our own affairs as if the explosion, when it comes, will affect only distant lands. For the good of society, we ought to stop a costly and dangerous arms race. Unfortunately, our natural aggressive tendencies teach us that the strong will prey on the weak, and so we continue to arm ourselves and those we perceive as our friends. For the good of society, we ought to have energy and economic policies that look decades ahead, if not centuries. Instead, we individually pursue those strategies which best safeguard our own interests. It would be *unnatural* to do otherwise. And *that* is the basic difficulty with these issues. They arise from a basic divergence between our natural instincts and the welfare of civilization.

We are not the first to face these dilemmas, of course. They have been with us since the dawn of civilization. In past times, the resolution was simple. Nature took its course. Human suffering ensued, but the species endured. Civilization endured. Now, however, we have accumulated so much power that if we were to allow nature to take its course, the species may well become extinct. Certainly, anything we might care to call civilization has little chance of standing. Perhaps a cynic might agree with Bertrand Russell that it is to be hoped that "*homo sapiens* may put an end to himself, to the great advantage of such less ferocious animals as may survive." But I cannot help feeling that this is an overly harsh judgment,

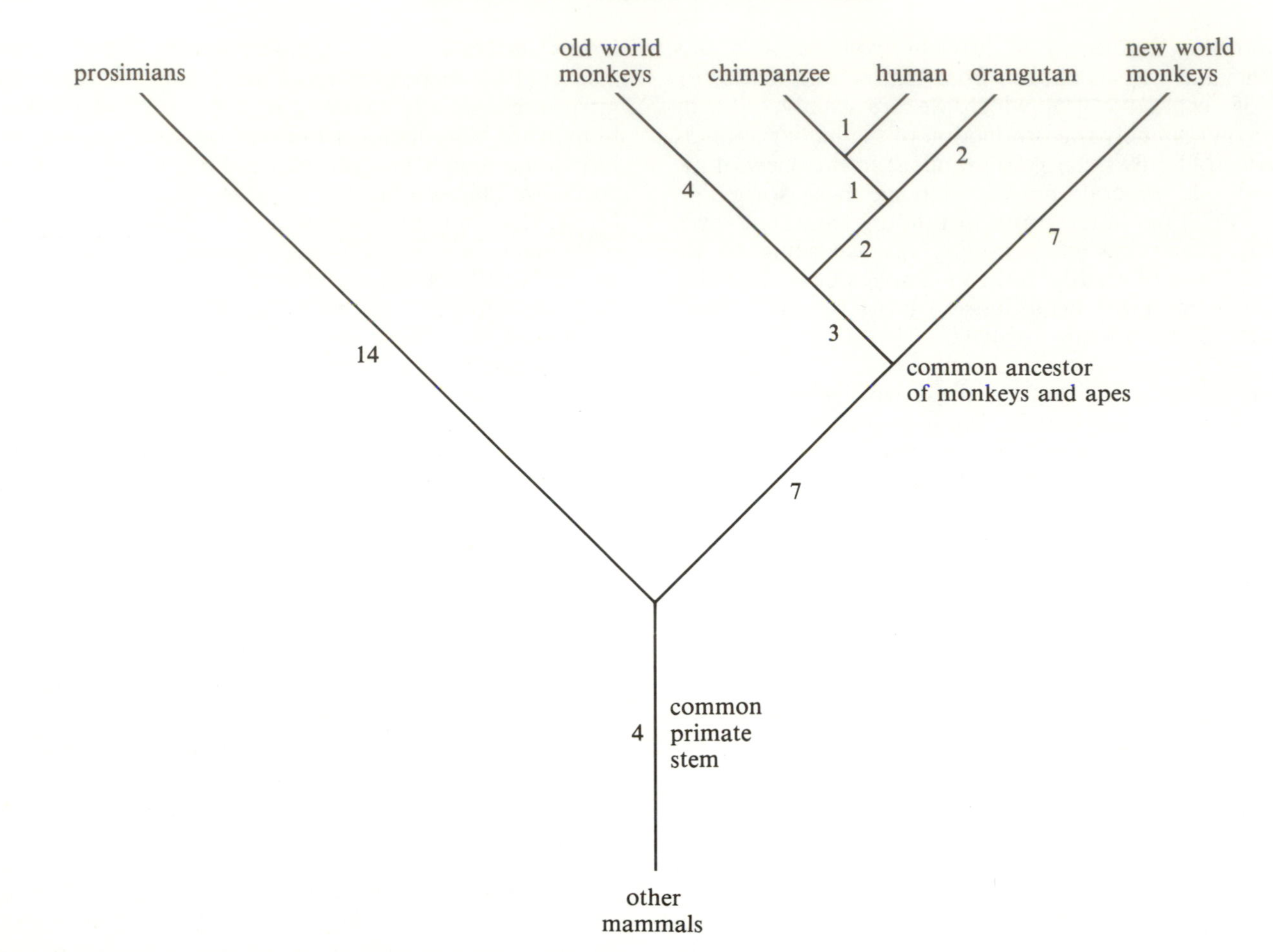

Figure 20.13. Phylogenetic tree tracing the evolution of primates. The units indicate the molecular distance (based on immunological studies) of each branch from a particular fork. The distance between humans and chimpanzees is standardized to 1, and chimpazees and humans are both 4 units distant from orangutans. The orangutans, the New World monkeys and the Old World monkeys are all 7 units distant from a common ancestor of monkeys and apes. Prosimians, like lemurs, and monkeys branched at a fork 14 units from each, and the most primitive primates lie a distance of 4 units further. (For details, see S. L. Washburn, *Scientific American*, **239**, Sept. 1978, 194.)

one that ignores much of beauty and worth in what *homo sapiens* has accomplished.

How did we ever get into such a predicament? How will we ever get out? The first question is a technical one and has a technical answer. Let us tackle it first. The second question is far more difficult to answer. The cure, if there is one, probably contains many components. Some of those components will undoubtedly come from science; therefore it is important to bring up this sociological issue in a science book (despite the fact that sociobiology is a dirty word in some scientific circles).

Intelligence on Earth and in the Universe

The Timescale of the Emergence of Human Culture

Many of humanity's problems arise because our societal evolution has proceeded much faster than our biological ability to adapt. Let us document this fact graphically. To reach the point where we had a recog-

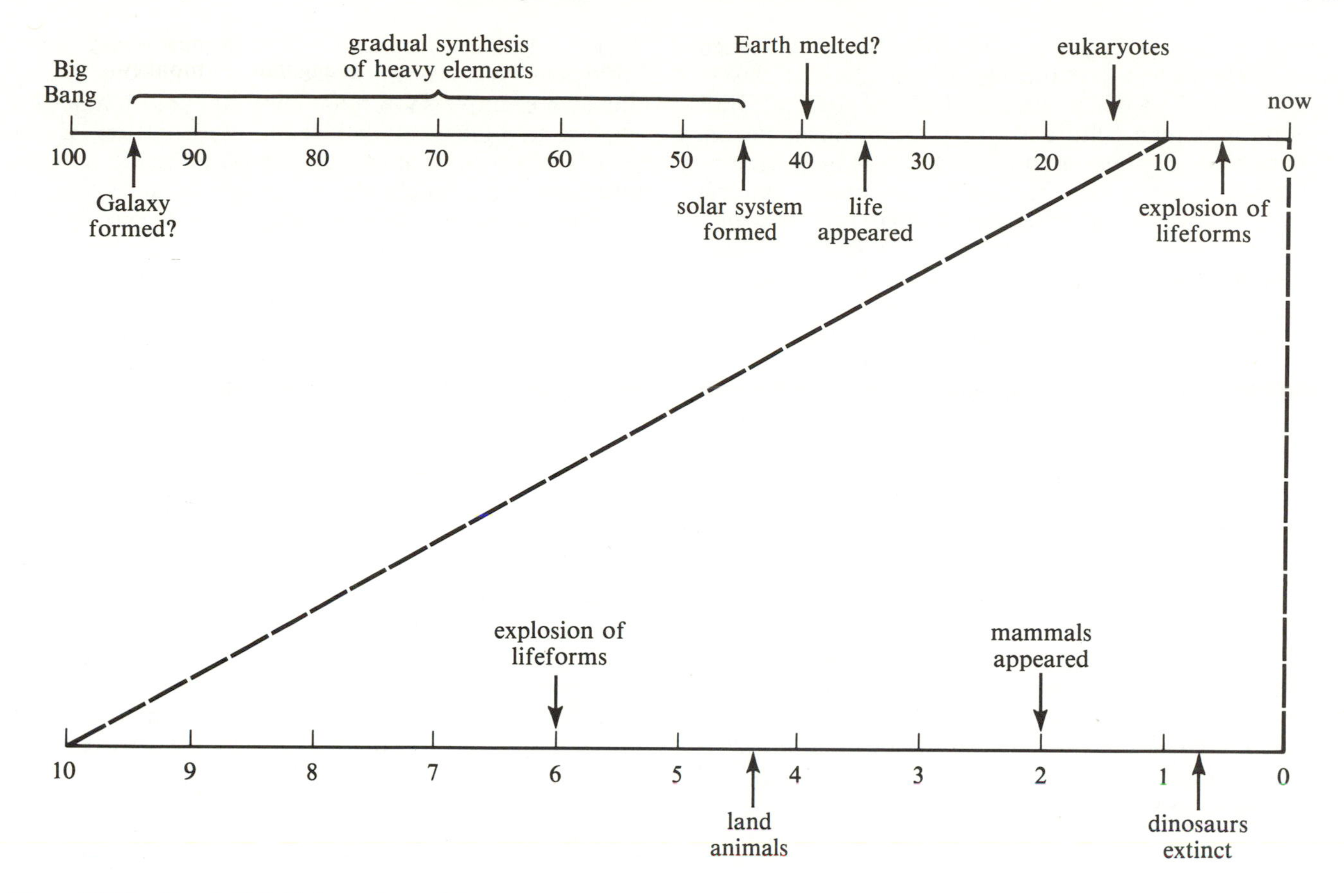

Figure 20.14. Football-field chronology of important events.

nizable human culture took time. How much time? Figure 20.14 shows an approximate chronology of important events since the Big Bang to the present. The scale relates all time measured to the length of a football field: the Big Bang took place 10 billion years ago on the 100-yard line, and "now" is on the 0-yard line; so every 10 yards corresponds to a billion years. On this scale, the formation of the solar system took place on the 46-yard line, and the first life forms on Earth appeared on the 35-yard line. Eukaryotic cells appeared on the 15-yard line, and the Cambrian Period, which ushered in an explosion of multicellular lifeforms, started on the 6-yard line. Shortly after the appearance of trilobites came fish, and thereafter, dinosaurs. Mammals appeared on the 2-yard line, and dinosaurs became extinct on the 2-foot line (roughly 65 million years ago). Recent evidence suggests that the dinosaurs died because of a catastrophic collision of an Apollo object or a comet with the Earth. Luis Alvarez and his colleagues have speculated that the gas and dust thrown into the atmosphere by this event led to the extinction of many plant forms and, ultimately, to the deaths of the dinosaurs which depended directly or indirectly on these plants for food. Little furry animals survived by eating the flesh of the dead dinosaurs; and thus mammals managed to flourish as they radiated into the habitats vacated by the dinosaurs. Eventually, of course, plant life was reestablished and evolution continued. On the 3/2-inch line (roughly four million years ago), the first human-like creatures (*Australopithecus*) emerged from a hominid branch which probably linked to the Africian apes. Only on the 1/30-inch line, about 100,000 years ago, do we find evidence of modern humans, *homo sapiens*. And the whole story of human culture—roughly ten thousand years, since neolithic humans began to farm in earnest—is written on a thickness less than the edge of one blade of grass!

Obviously, on the timescale of cosmic or biological evolution, human culture is a very recent phenomenon, a mere blade of grass in an entire football field. That is, in contrast with our biological evolution, cultural evolution—the accumulated knowledge and skills of a civilization—occurs very quickly on a cosmic timescale. No significant biological evolution of humans has occurred during the last ten thousand years; yet just contemplate the difference in cultures between the neolithic age and the contemporary one. What wonderful achievements have we attained in a mere ten thousand years! What a terrible predicament we find ourselves in after a mere ten thousand years! What might extraterrestrial cultures have accomplished and experienced given a few *billion* years' head start? To such advanced beings, might our achievements and troubles not seem as those of bacteria seem to us? Could not such advanced beings give us helpful advice and guidance on our own problems? Unlike us, could they be both smart and wise? The following, then, is a possible course of action advocated by some astronomers: both for the potential benefits and for the sheer glory of the discovery, we should search for intelligent life elsewhere in the universe. Let us examine the progress made in this search, and comment on the implications of the results.

Intelligent Life in the Galaxy

What are the chances that there are technologically advanced civilizations in our own Galaxy? From a purely astronomical point of view, the chances would seem to be pretty good. Our Sun is only one mediocre star among 10^{11} in one galaxy of 10^{10} in the observable universe. Moreover, the modern nebular theory of the origin of the solar system views the formation of planets as a rather common event that naturally tends to accompany the formation of many stars. The origin of life from natural selection acting on the chemical level is probably a more delicate issue, and requires a fairly narrow range of physical conditions, if we judge from the fact that missions to three extraterrestrial bodies—the Moon, Mars, and Venus—have failed to find even rather primitive life forms. Nevertheless, given the enormous number of probable solar systems in our Galaxy alone—not to mention the stupendous number which exist in the entire universe—it seems extremely unlikely that we could be alone in this vast universe. If we are not alone, and if intelligence has sprung up elsewhere from evolution of other life forms, then it is overwhelmingly likely that this intelligence would be far superior to our own.

This point has been made forcefully by a number of prominent astronomers, among them Sebastian von Horner, Philip Morrison, Carl Sagan, and Iosef Shklovskii. What is the reasoning behind this point? Our own example shows that cultural evolution, in distinction to biological and chemical evolution, is extremely rapid. In a mere ten thousand years, humans have gone from scratching out a meager living by farming with the crudest tools to a highly developed technological civilization. Most of the progress has been made in the last hundred years. Indeed, it has been estimated that the sum total of human scientific knowledge doubles every ten years. The point is that once a certain level of civilization has been reached, further advances come very rapidly.

Suppose that some other life form in the Galaxy has had an experience in evolution similar to ours. Then the chances are, given that the Galaxy is of the order of 10 billion years old, that this life form is either a few billion years behind us in development or a few billion years ahead of us. If it is a few billion years behind us, their individuals would seem like bacteria to us. If their civilization is a few billion years ahead of us, we must seem like bacteria to them. The chance that two intelligent civilizations originating billions of years apart would arrive at the same point of cultural development at some later time is too improbable to merit serious contemplation. Since we are merely ten thousand years into our cultural evolution, of all the civilizations in the Galaxy, we must be at the very beginning of a possible cultural development which could span, at least in principle, billions of years. Thus, we must be babes in the woods, and any other advanced civilization is likely to be vastly superior to our own.

UFOs, Ancient Astronauts, and Similar Nonsense

What forms are these supercivilizations likely to take? How many are there in the Galaxy? No one knows. I will soon offer some standard and personal speculations, but first let me dispense with the notion that UFO sightings and similar encounters of the "first, second, or third kinds" already prove that supercivilizations have visited us on Earth. It is, of course, *not impossible* that such visits might have taken place or are taking place, but in practice it is just short of impossible that the reported phenomena could be such visits. The usual reports are weak, not because they are too fantastic, but because they are not fantastic enough! The usual story involves flying saucers or spacethings whose technology seems just beyond (or, often, even *behind*) what we humans already possess. But we have argued that if there are supercivilizations capable of visiting other stars, they are likely to be so far in advance of us that we may well seem like bacteria in comparison! Thus, the usual stories are implausible because they exhibit too little imagination, not because they exhibit too much.

The idea that ancient astronauts visited Earth and taught early civilizations much of what they knew deserves to be especially castigated. The evidence offered borders, at times, on delibrate fraud. But worst of all, the concept of ancient astronauts popularized in recent years by certain books and films involves an element of denigration. It insults the intelligence and the accomplishments of past civilizations to attribute their rightful attainments to mythical entities. People deserve to know that the systematic underestimation of the genuine abilities of primitive peoples represents a form of grave robbing. To exploit the general ignorance on this topic for profit is dishonorable both to the living and to the dead.

Estimates for the Number of Advanced Civilizations in the Galaxy

Let us return now to our original question: How many advanced civilizations might there be in the Galaxy? To begin a rational discussion of this point, we must adopt certain assumptions. The most common hypothesis is that all intelligent lifeforms, whatever their present shape or nature might be now, had an origin in *chemical* processes occurring on a planet orbiting some star. This is, of course, a conservative assumption, modeled after our own example on Earth; so any estimate that emerges from such an analysis must represent a lower limit given the rest of the assumptions.

Adopting the conservative assumption outlined above, Frank Drake has proposed that the number N of advanced civilizations in our galaxy *with whom we might make contact* is given by the formula

$$N = pR_*L, \tag{20.4}$$

where L is the average lifetime that a civilization remains technologically active, R_* is the average rate of forming stars in the Galaxy, and p is a dimensionless number given by the product

$$p = f_p n_e f_l f_i f_c, \tag{20.5}$$

where f_p is the average fraction of stars which have planets, n_e is the average of Earthlike planets per solar system, f_l is the average of Earthlike planets which have developed life, f_i is the average fraction of life-bearing planets which have evolved at least one intelligent species, and f_c is the average fraction of intelligence-bearing planets capable of interstellar communication.

Problem 20.6. Supply the missing reasoning needed to derive Drake's equation (20.4), stating clearly the implicit assumptions underlying your derivation.

Two comments are in order before we use Drake's equation to make numerical estimates of N. First, the emphasis on Earthlike planets is reasonable, since a chemical origin of life requires the ability to synthesize very complex molecules on timescales of cosmic interest. Spectroscopic studies show that the elements in the rest of the universe have electronic shell structures identical to those of elements on Earth; thus, the laws of chemistry are universal. No chemical basis of life is likely to be *more efficient* than that of carbon compounds in the presence of liquid water, and the biochemical development of life on Earth using this basis already occupied a significant fraction of the total cosmic time (Figure 20.14). Thus, the term "Earthlike" applied to n_e really means a terrestrial planet capable of having liquid water on its surface. The number n_e equals 1 in our solar system, and the discussion of Chapter 18 suggests that the average number for all solar systems may well be substantially less than 1. In other words, given what we now know about the history of the Earth, it appears that our gardenlike conditions result from several fortunate circumstances—the right-sized planet at the right distance from an appropriate star, etc.—and that only the most optimistic of astronomers would now assign an *average* value of unity or more to n_e.

Second, starting with Cocconi and Morrison, most astronomers have argued that interstellar communication effectively means *radio communication*. The emphasis on radio communication rather than on physical space travel is based on our current sense of economics. It would cost us less than a dollar to send a ten-word radio telegram to the stars, whereas even a one-way manned mission occupying less than the typical human lifespan would cost much more than the gross national product of the United States. Thus, in the derivation of equation (20.5), where each technologically capable civilization is counted only once in the estimate for the total number N, there is an implicit assumption that any advanced civilization will be content to stay on its home planet, communicating with other civilizations by radio transmission. Clearly, it is a *very conservative* assumption to suppose that our sense of economics or mortality would constrain the efforts of supercivilizations who might make us look like bacteria in comparison.

Let us adopt, for the time being, the very conservative argument for radio transmission only from home planets, and continue with our estimates for N. Since the quantity p defined in equation (20.5) involves a product of fractions, fs, and a number n_e which is, at best, unity, the

L \ p	1	10^{-2}	10^{-4}	10^{-6}
10 yr.	10^2 10,000	1 –	10^{-2} –	10^{-4} –
10^3 yr.	10^4 2,000	10^2 10,000	1 –	10^{-2} –
10^5 yr.	10^6 300	10^4 2,000	10^2 10,000	1 –
10^7 yr.	10^8 70	10^6 300	10^4 2,000	10^2 10,000
10^9 yr.	10^{10} 15	10^8 70	10^6 300	10^4 2,000

Figure 20.15. The number N (top value) and distance d in light years (bottom value) on average between advanced technological societies in the Galaxy for various choices of the lifetime L and the combined probability p. For example, if $L = 10$ yr and $p = 1$, there are 10^2 advanced civilizations, each located an average distance of 10,000 lt-yr from each other. See the text for a more detailed explanation and implications.

most optimistic possible estimate for p is $p = 1$. If one is inclined, on the other hand to be an astronomical and biological pessimist, p could be much lower, say 10^{-6}.

Consider, now, the other terms in Drake's equation (20.4). The only quantity we know with any certainty is R_*. The current rate of star formation in the Galaxy is 2 or 3 per year, but in the past this rate must have been higher, since there are at least 10^{11} stars in our Galaxy, which is about 10^{10} years old. Let us take $R_* = 10\ \text{yr}^{-1}$, therefore, as a representative average rate of star formation in our Galaxy. The remaining quantity, L, the lifetime of a technologically capable civilization, is much more difficult to estimate. Our own ability to transmit radio waves to the stars has existed less than a century, and we may at any moment terminate ourselves forever. If the clash between biological instinct and societal evolution that we claimed to be at the root of many of our problems today is a universal phenomenon for all emerging technological civilizations, then $L = 100$ years might represent a realistic estimate for the average lifetime of such civilizations. In this case, there would be no supercivilizations with technologies very much in advance of radio communication. On the other hand, a more optimistic sociological assessment might be that if advanced societies can overcome their tendencies to wipe themselves out, there is nothing, in principle, to keep them from lasting as long as their life-sustaining star continues to shine; so the average L might be as long as, say, 10^9 years.

Given, therefore, a wide latitude of possible choices for p and L, we would be wise to consider various possible combinations. Figure 20.15 gives, as the top entry in each combination box, the number N of currently fluorishing civilizations in our Galaxy, corresponding to a particular choice of p and L. If this number is greater than 1, the bottom number gives the expected distance in light-years to the nearest such civilization, assuming that these civilizations have a distribution in the Milky Way similar to that deduced for the distribution of stars (Chapter 12).

Problem 20.7. Make a table of N using the guesses: $p = 1$, 10^{-2}, 10^{-4}, 10^{-6} and $L = 10$ yr, 10^3 yr, 10^5 yr, 10^7 yr, 10^9 yr, with all possible combinations as in Figure 20.15. Assume $R_* = 10\ \text{yr}^{-1}$ for your calculations. Estimate the distance d to be expected on the average to the nearest technologically active civilization for each combination of p and L, by assuming the Galaxy has a volume V given by a cylinder of radius 30,000 lt-yr and height 2,000 lt-yr with us at about the midplane. *Hint*: If civilizations are arranged on a cubic array with sides of length d, show that each civilization occupies an average volume d^3. Argue that $Nd^3 = V$, and calculate d. Notice that your numbers will not agree exactly with those given in Figure 20.15, because those expected distances take into account the inhomogeneous nature of the distribution of stars in the Galaxy (Chapter 12). With your numbers, discuss the conditions under which we would have the interesting possibility of a two-way radio communication.

Figure 20.15 contains both good news and bad news. The good news is that only the most pessimistic estimates of both p and L (i.e., pessimistic biology and sociology) combine to give the conclusion that we are currently alone in our Galaxy. Even these pessimistic estimates show, however, that we are not alone in the observable universe. The bad news is that unless L is appreciably larger than 10,000 years or so, a two-way radio communication cannot be established before one of the civilizations disappears. Only if L is appreciably larger than

Table 20.1 The search for extraterrestrial intelligence.[a]

Date	Observer	Site and telescope size (m)	Wavelength (cm)	Objects	Total hours
1960	Drake	NRAO 26	21	2 stars	400
1968–69	Troitskii Gershtein Starodubtsev Rakhlin	USSR 15	21 and 32	12 stars	11
1970–	Troitskii Bondar Starodubtsev	USSR dipole	16, 32, and 50	all sky	700 and cont.
1971–72	Verschuur	NRAO 43 an 91	21	9 stars	13
1972–76	Palmer Zuckerman	NRAO 91	21	600 stars	500
1972–	Kardashev	ENICR dipole	16–22	all sky	cont.
1973–	Dixon Cole	OSURO 53	21	all sky	cont.
1974	Bridle Feldman	ARO 46	1.4	500 stars	
1975–76	Drake Sagan	NAIC 305	13, 18, and 21	4 galaxies	100
1976–	UC Berkeley	HCRO 26	18 and 21	all sky	cont.
1976	Clark Black Cuzzi Tarter	NRAO 43	3.5	4 stars	7
1977	Black Clark Cuzzi Tarter	NRAO 91	18	200 stars	100
1977	Drake Stull	NAIC 305	18	6 stars	10
1978	Horowitz	NAIC 305	21	185 stars	80
1978–	Wielebinski Seirldeki	MPIFR 100	75	3 stars	2
1979	Cohen Malkan	NAIC 305, HRO 36, CSIRO 63	18, 1.3, 19	25 globular clusters	80

[a] Courtesy of Jill Tarter.

10,000 years will the round-trip light-travel time corresponding to $2d$ be less than the average lifetime L. Hence, if we are to join the Galactic communications club, we may have to show that we are capable of not blowing ourselves up in the next 9,900 years!

Figure 20.15 lends quantitative support to our previous argument that reported UFO sightings are overwhelmingly unlikely to represent visits by advanced extraterrestrial civilizations. If the average technological development of advanced societies is not much in advance of our own—namely, if $L \sim 10^2$ yr, say—then the average separations between advanced civilizations are so large that interstellar flight from one to the other is prohibitive in the allotted time. On the other hand, if advanced civilizations are "only" separated by tens or hundreds of light-years, then the average lifetime of technologically capable civilizations must number in the billions of years. In that case, their spacecraft must appear as incomprehensibly magnificent to us as ours must appear to bacteria.

Given that we have no way to tell what the correct situation is, a growing number of scientists have proposed that we should carry out a program of "eavesdropping" to pick up possible radio communications from other civilizations. According to this argument, we might accidentally pick up conversations between two supercivilizations, or we might pick up probing signals from an advanced archaeological expedition searching for relatively primitive societies like our own. Indeed, more than a dozen small eavesdropping experiments have been or are being carried out with no sign yet of a positive detection (see Table 20.1).

Some astronomers have already found the lack of success to be disappointing, and in recent years, there has been a growing contigent of scientists who have wondered aloud whether we might not be alone after all!

Figure 20.16. Would we really be able to understand the signals of a truly advanced being?

If true, this would have profound implications for humanity; in particular, it would make it even more vital to safeguard this sole site of intelligence in our Galaxy. On the other hand, other scientists find this egocentric viewpoint unappealing, and insist that success of contact will come only from a long, dedicated program of search. The philosophical implications, not to mention the practical implications, of a definitive discovery that we either are or are not alone in the universe would be staggering.

What should we think of our failure (so far) to make contact? After-the-fact rationalizations have their dangers, but let me make an attempt at an explanation anyway. Any civilization close in technological development to our own would probably wisely opt *not* to beam delibrately, but only to listen, as we have ourselves chosen to do. Such civilizations might inadvertently transmit radio and television programs into space, as we are currently doing, but the separations of immature technological civilizations like our own are so large that such weak signals could not be picked up by us in any case. Perhaps only mature and very advanced civilizations would be actively broadcasting, but why would they be interested in seeking us out? Do we attempt to contact bacteria? We might study a few examples, partly because we realize that unicellular organisms like bacteria were once our ancestors, but we certainly don't make an attempt to find and communicate with the countless bacteria that exist throughout our world! Why would truly advanced civilizations be any more interested in such trivia? Moreover, unless they did actively seek to contact us specifically, would we have any better chance of trying to decipher their language and means of communication than bacteria have of trying to decipher ours? (Figure 20.16) Finally, if they do want to seek us out specifically, surely they will pick the time and the place, not we.

Despite this pessimism, I do fervently hope that we or our descendants will communicate eventually with other intelligent beings. Nor am I as pessimistic as some of my fellow scientists about the idea that we or our descendants could travel someday to the stars. This possibility depends, of course, on what the future holds for the development of life and intelligence on Earth, and it is to this extremely speculative topic that we now turn.

The Future of Life and Intelligence on Earth

Trying to predict the future of life and intelligence on Earth is even more dangerous than after-the-fact rationalizations of the details of its evolutionary past. Darwinian theory cannot be expected to give more than a general sweeping outline in the reconstruction of our biological past. There has been no Hari Seldon (see Isaac Asimov's *The Foundation Trilogy*) who has done for sociology what Darwin did for biology. Nevertheless, I believe there is enough knowledge to show, at least vaguely, where we are going. To grasp this argument, we must first understand more about the basis of the present form of intelligence on Earth.

The Neuron as the Basic Unit of the Brain

What is the basis of human intelligence? In a broad sense, human intelligence resides in the brain, in particular, in the manipulation of information sent to the brain by various sensory devices (Figure 20.17). (There may be some preprocessing by the spinal cord.) The basic element of this machinery is the nerve cell, the **neuron** (Figure 20.18). The human brain consists of a mass of neurons, perhaps 10^{11} in number, each making 1 to 10^4 **synapse** connections with other neurons. The connections are, as we shall see, chemical and not physical. Many of the neurons have long cables called **axons** which transmit *nerve impulses* along their lengths. The gray matter of the brain (and spinal cord) consists of nerve cells, and the white matter is composed of Schwann cells wrapped around the axons.

The classic picture of the central nervous system of humans, formulated by Santiago Ramon y Cajal in 1904, is that of a network of nerve cells in which nerve impulses may be received by any cell body from many other neurons and may be transmitted or relayed to many other neurons. Only in the last four decades (through the work of Dale, Eccles, Hodgkin, Huxley, Katz, Kuffler, and others) have we come to understand reasonably well how information is conveyed from nerve cell to nerve cell by two basic mechanisms, one that governs

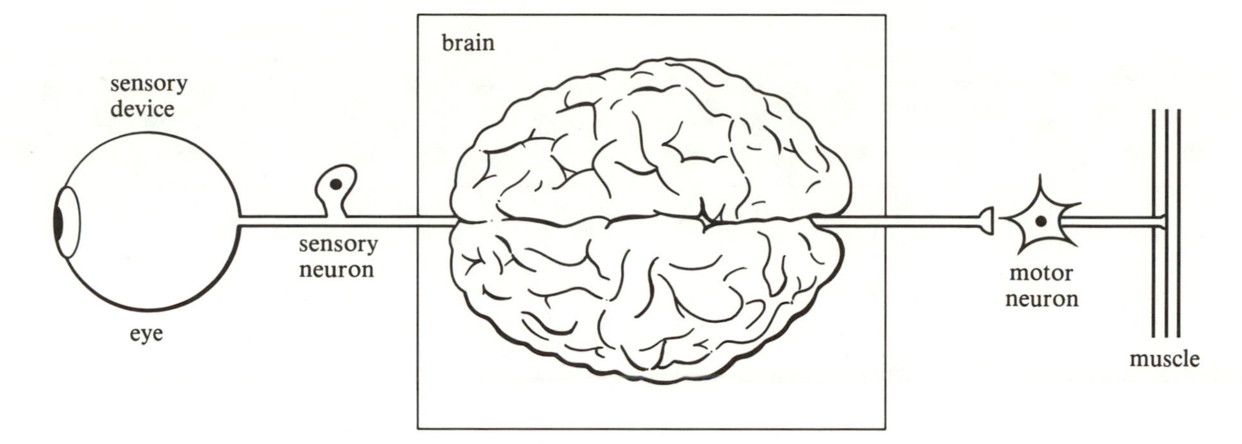

Figure 20.17. A schematic depiction of the basis of human intelligence. Input from sensory devices into the brain are processed, and an output goes out to the muscles of the body to take appropriate action.

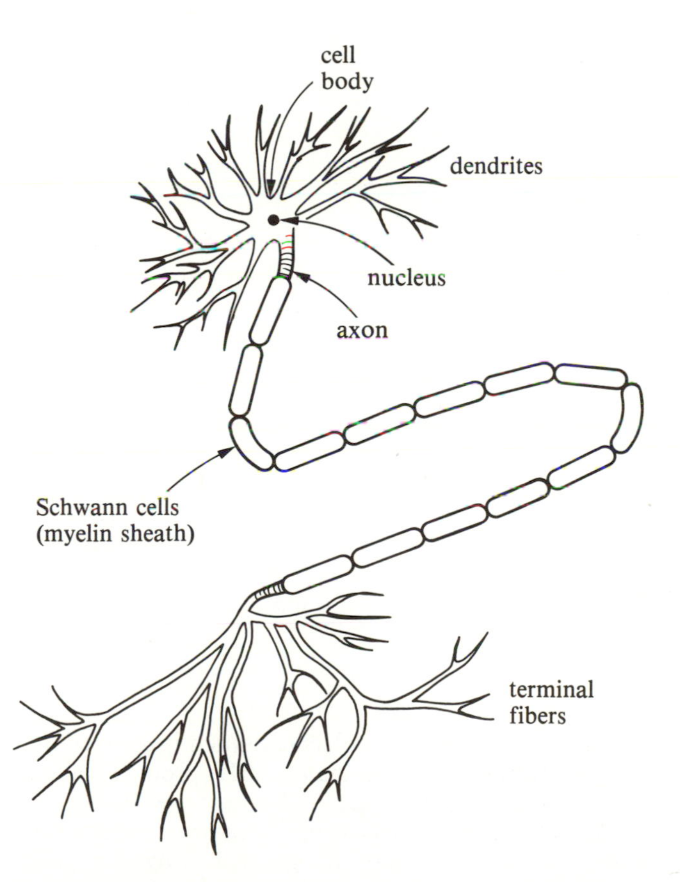

Figure 20.18. The structure of a nerve cell or neuron. The length of the axon may be much longer in proportion to the cell body than depicted here.

the transmission of signals across synapses, and one that governs the transmission of the nerve impulses down axons.

Nerve impulses are electrochemical in nature (Figure 20.19). The sending of nerve impulses down axons works on an *all or none* principle. The information is entirely contained in whether a signal is sent or not (and the frequency of such sending). The individual impulses are virtually identical; there is no information carried by a variation of the amplitude of the impulse. This part of the operation of the nervous system resembles the "on or off" switching mechanism of logic gates in modern digital computers. In this electronic analogy, the axons are wires, and the myelin sheaths provided by the Schwann cells resemble the insulation that is wrapped around ordinary electrical wires. Multiple sclerosis is a disease which destroys the myelin sheaths of nerve axons, greatly impairing the conduction of nerve impulses.

The decision by a nerve cell to send or not to send an impulse is based on the total chemical information gathered from all its synapses with other neurons. Neurons which transmit nerve impulses to their axon terminals cause a release of a chemical near the synapses. The chemicals released from the terminal fibers of different nerve cells may have either exciting or inhibiting effects on the **dendrites** or cell bodies on the other sides of the synapses (Figure 20.20). In any case, each neuron accumulates the total effect of all exciting and inhibiting signals received from **receptor sites** located on the postsynoptic membranes, and this total effect is what decides whether it will fire a nerve impulse or not. Each receptor site is a protein molecule with a shape that

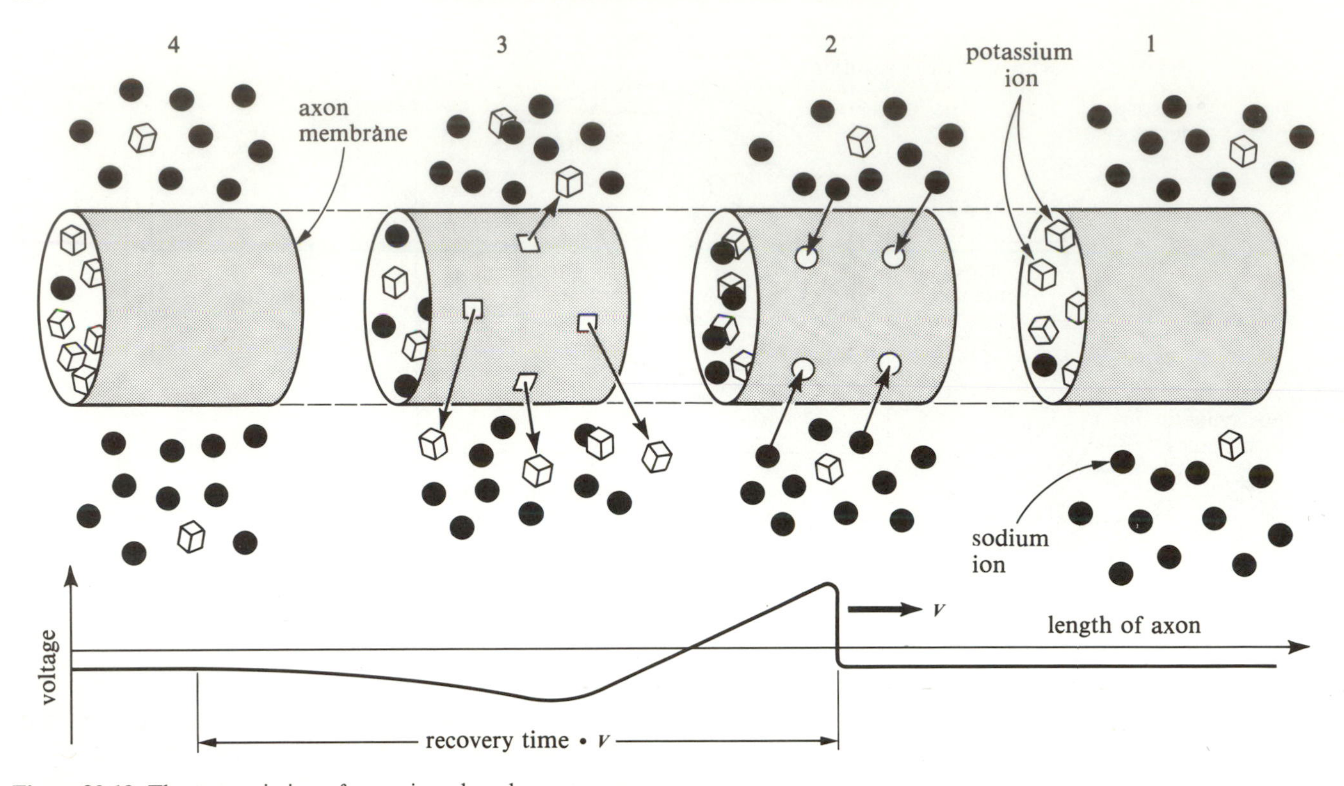

Figure 20.19. The transmission of nerve impulses down an axon. (1) Before the nerve impulse arrives, the semipermeable membrane of the axon keeps the concentration of potassium ions much higher inside the axon than outside, and vice versa for the sodium ions. Both sodium and potassium ions have an electrical charge of +1 (in units of e), and there are negative ions (like chloride ions or organic ions) which are not depicted in this diagram. In the resting state, however, the total charge distribution is such that the interior of the axon is maintained at a lower voltage than its surroundings. (2) The arrival of the nerve impulse, traveling at speed v, corresponds to a sudden opening of channels that selectively allow sodium ions to rush into the interior of the axon. The accumulation of positive charge inside makes the axon voltage positive relative to its surroundings. (3) An instant later, the sodium channels close, and the potassium channels open, leading to a net outrushing of potassium ions. The loss of positive charges brings the voltage to negative values. (4) After a recovery time following the arrival of the nerve impulse, the axon has been restored to its resting state and is ready to fire again.

provides a "lock" which only certain small molecules can fit as a "key" (refer to Figure 19.23). Thus, the decision of the nerve cell to fire an impulse or not is made purely at a biochemical level; no individual neuron can be said to possess intelligence.

The most common form of inhibitory molecule found in the human brain is an amino acid (GABA) that is not one of the twenty used in protein synthesis (Chapter 19). Several other molecules have also been identified as excitatory or inhibitory molecules, and it is believed that certain drugs (e.g., mescaline or LSD) produce hallucinations because their molecules bear a close resemblance to the natural excitatory or inhibitory molecules of the brain. In other words, by taking such drugs, an "acid-head" is causing his or her brain cells to misfire. Some excitatory and inhibitory molecules apparently work in a two-stage process. Earl Sutherland proposed that the binding of such a molecule to a receptor site could

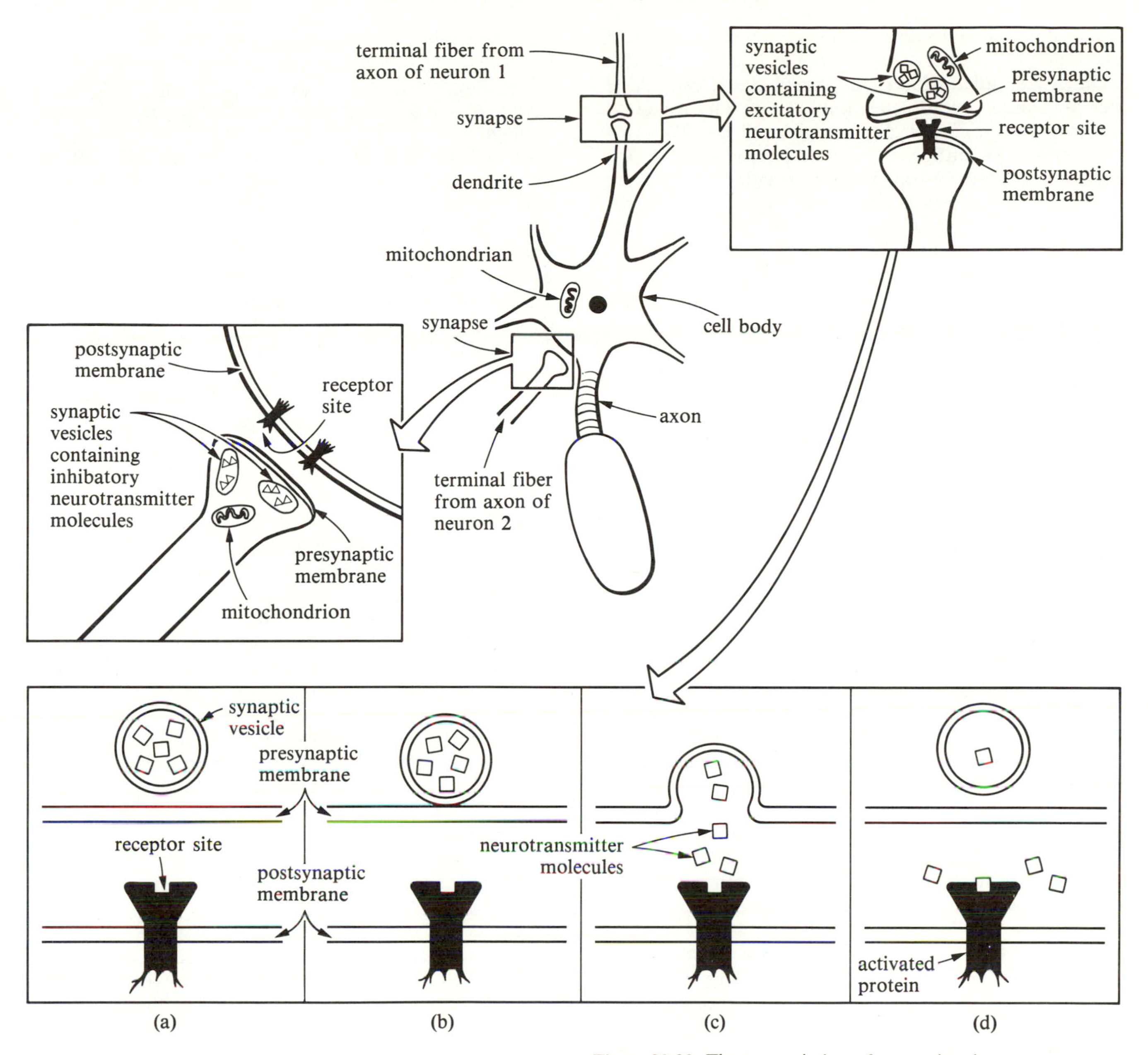

Figure 20.20. The transmission of nerve signals across synapses occurs by the release of neurotransmitter molecules. (a) The arrival of a nerve impulse at the terminal fiber of a sending neuron triggers a sequence of events. (b) The change in the properties of the presynaptic membrane causes synaptic vesicles to fuse with the presynaptic membrane. (c) Neurotransmitter molecules are released into the synaptic gap. (d) One of these molecules binds with a receptor site on the postsynaptic membrane and activates the protein. The total action, both excitatory and inhibitory, of such proteins activated by sending neurons 1, 2, etc., determines whether the receiving neuron will also send a nerve impulse down the length of its axon. (For details, see C. F. Stevens, *Scientific American*, **241**, Sept. 1979, 54.)

trigger the protein to help a catalyst convert many ATP molecules in the nerve cell into the cyclic nucleotide, cyclic AMP (adenosine monophosphate). Cyclic AMP then acts (in an incompletely known way) on the cellular machinery with certain physiological effects. This two-stage process, resulting in the production of many functional molecules from the trigger provided by only one, allows the amplification of weak signals into strong ones, a mechanism also used in electronic devices like the transister and photomultiplier tube. The chemical amplification presumably explains why trace amounts of narcotic drugs, such as morphine, can produce such large responses. An exciting recent discovery in neurochemistry by John Hughes and Hans Kosterlitz is that the brain produces its own opiates in the form of small chains of amino acids (enkephalins). As even more powerful natural morphine (beta endorphin) was subsequently found by C. H. Li. Apparently nature kindly provides the brain with natural pain killers, and morphine works as an *artificial* pain killer because it can simulate nature's molecules.

The Organization of the Brain and Intelligence

The preceding, simplified description of the operations of a single nerve cell gives a hint of how relatively simple perceptions, such as pleasure and pain, might result from purely chemical and physical processes. However, they do not shed light on how the human mind is able to perform higher feats, e.g., recognize visual and audio patterns. Even more mysterious seems to be the ability of the human mind to memorize, to learn, and to reason. Indeed, *consciousness* is such an exceptional quality of the human condition that for many years it provided the last refuge for incurable vitalists. The concept of *self* is so ingrained in human thinking that we find it hard to imagine that "self" might not be anything more than the complexity represented by one's neuron network! But the basic concept that consciousness arises when one has sufficient neural organization can be seen by a very simple argument.

We all know now that each of us started as a single cell (a fertilized egg). Let us assume that a single human cell cannot be considered to have consciousness. When the original cell has divided into two, do we then have consciousness? Not likely. How about four cells? No? How about a newborn babe (who has all the brain cells that it will ever have)? If yes, at what point in the cell-division process from a single cell did the baby acquire consciousness? If no, at what point in the growth of a baby into adulthood does it acquire consciousness? Clearly, the reasonable answer is that there exists a continuum of degrees of "consciousness," which depends on the complexity of the organization of one's neuron network. There is no one point in a human's development when he or she can be said to have suddenly acquired consciousness, but to have lacked it completely before that point.

The same argument can be made for the concept of intelligence. It must represent ultimately the complexity of one's neuron connections. If you are still skeptical, and wish to argue that even a single human cell must have consciousness and intelligence, then defend yourself against the following line of reasoning. We now believe that humans descend from a long continuous evolutionary line which began with bacteria too simple even to ferment. At what point in this line did animal cells first acquire intelligence? Do bacteria too simple even to ferment have consciousness and intelligence? If yes, consider the view that these simple bacteria themselves arose from a line of chemical evolution. Do molecules have consciousness and intelligence? How about atoms? How about elementary particles? If your answer is still yes, then consider that these elementary particles are themselves believed to have been created from a vacuum in the Big Bang. Does a vacuum have consciousness? If yes, what practical meaning can the concept of consciousness have if even a vacuum possesses this quality?

No, it is much more rational to believe that humans do possess consciousness and intelligence, but that humans possess them in a *quantitatively* higher degree than, say, apes or pigs do, not in a qualitatively higher degree. As Richard Goldschmidt discovered in 1911, there appears to be no fundamental difference in structure, chemistry, or function between the neurons and synapses in a human and those in a simple marine worm. There is, of course, a very substantial quantitative difference in the complexity of organization in the *many more* neurons possessed by a human brain (10^{11} versus 162). What is the nature of this organization? We do not yet have a detailed answer to this question, but the outlines of one is beginning to emerge.

By now it should come as no surprise to the reader that the human brain seems to be organized in a hierarchy of structures. If we continue to make an analogy with electronic computers, the lowest level in this hierarchy is the microprocesses which occur at the synapses. Here the electrochemical equivalents of the resistors, capacitors, inductors, and transistors of an electrical circuit are to be found. The neuron taken as a whole is the basic logic element, with the "all or none" nerve impulse analogous to "on-off" gates in digital computers. It is now known that the brain cells of a human are organized into groups, centers, and regions, whose functions are beginning to be mapped out by psychologists and neurobiologists (Figure 20.21).

Of particular interest here is the structure of the cerebral cortex, a sheet of nerve cells at the upper portions of all mammalian brains. The cerebral cortex is believed to be the most recent evolutionary addition, and it is largest (with a surface area of about 1.5 square feet)

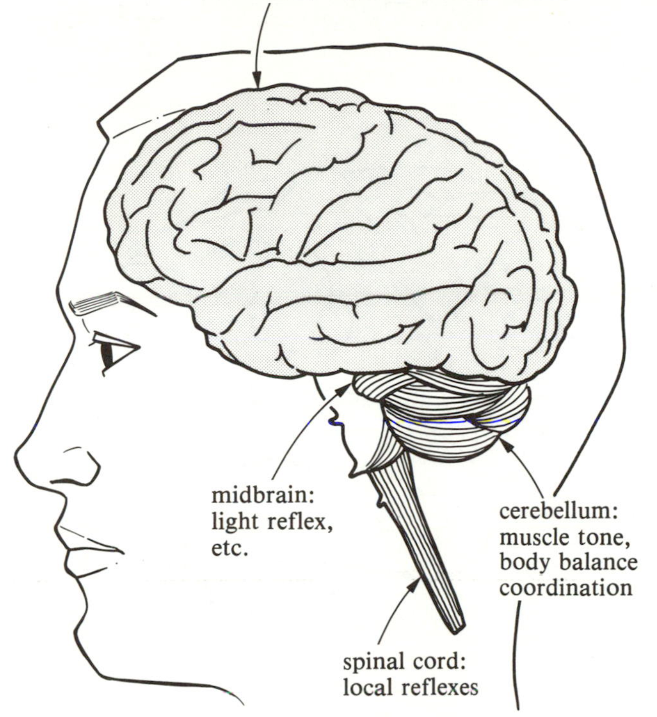

Figure 20.21. The human brain is divided into distinctive portions which handle different functions. In particular, psychologists and neuroanatomists have partially mapped out the regions of the cerebral cortex that handle certain complex mental functions. For example, visual perception is handled by the part of the cerebral cortex that lies at the rear of the head.

in the brains of humans. Since the studies of Paul Broca in 1861, it has been known that the cortex handles many of the abilities that set humans above lower animals. Electrical stimulation of the brain plus the examination of brain-damaged patients has shown that localized areas of the cerebral cortex control the functions of speech, hearing, visual pattern recognition, association (learning, memory, reasoning), etc. Recent analysis by neuroanatomists has revealed that these localized areas are further subdivided into modular columns of nerve cells. The role of these modular columns in visual perception has been analyzed extensively by David Hubel and Torsten Wiesel. Each modular column plays a role like that of the integrated circuits on an electronic chip of a modern computer. Broadly speaking, we may say that an integrated circuit represents many thousands of individual logic gates "hard-wired" together to form a pattern of connections appropriate for some larger purpose (e.g., to play a game of tic-tac-toe). The action of many such integrated circuits (or microprocessor chips) operating together according to a master plan (program) can accomplish even more complex tasks (e.g., play a "game" of nuclear strategy). In the human cortex, it is apparently the combined action of all modular columns that leads to the psychological phenomena that we call visual perception, learning, etc. Interestingly enough, the modular architecture of the cortex is remarkably similar in all mammalian species. This suggests that humans differ from rats not in having modules that are more complex, but in having many more such modules.

It is sometimes easy to underestimate the power of such hierarchic structures to produce enormous combinations of complex interactions. For the English major rather than the computer science major, therefore, let me offer the following example. Consider a language where one had only the 26 letters of the alphabet, with no such things as words, sentences, paragraphs, etc. Imagine writing a book where one could only list the letters one at a time. It would make for extremely dull reading. Now, allow words, sentences, paragraphs, etc. Think of the many books that one could write—that have been, in fact, written! The human race has not even exhausted all the rich possibilities for a single sentence, much less a whole book! Such is the power of combinatorials and hierarchic structures. Such is the power of life and intelligence.

Hardware, Software, Genetic Engineering, and Education

In the development of the human embryo and fetus (Figure 20.22), how do the newly created nerve cells know where to go and how to make the right connections to form the complex architecture that is ultimately a human brain? This problem is, of course, a special case of the more general issue of cell differentiation and embryology. How do any of the cells in the human body know where to go and how to form the right connections? The answer at its most basic level must, of course, lie in the genetic material, the DNA that is common to all undifferentiated cells. The phenomenology of the developing brain is a fascinating story, despite the fact that we do not yet know many of the basic underlying molecular mechanisms governing the process. It is a story, however, that we must forego for sake of brevity.

The story of human intelligence does not stop, however, with the final development of the brain of a newborn infant. No, developmental biology merely provides the basic machinery of the brain: the *hardware*, to continue with our computer analogy. The capacity for memory and reasoning exists within the baby's brain, but for it to achieve its full potential in coordinating muscular movements, in acquiring the abilities for speech, comprehension, and reasoning, the baby

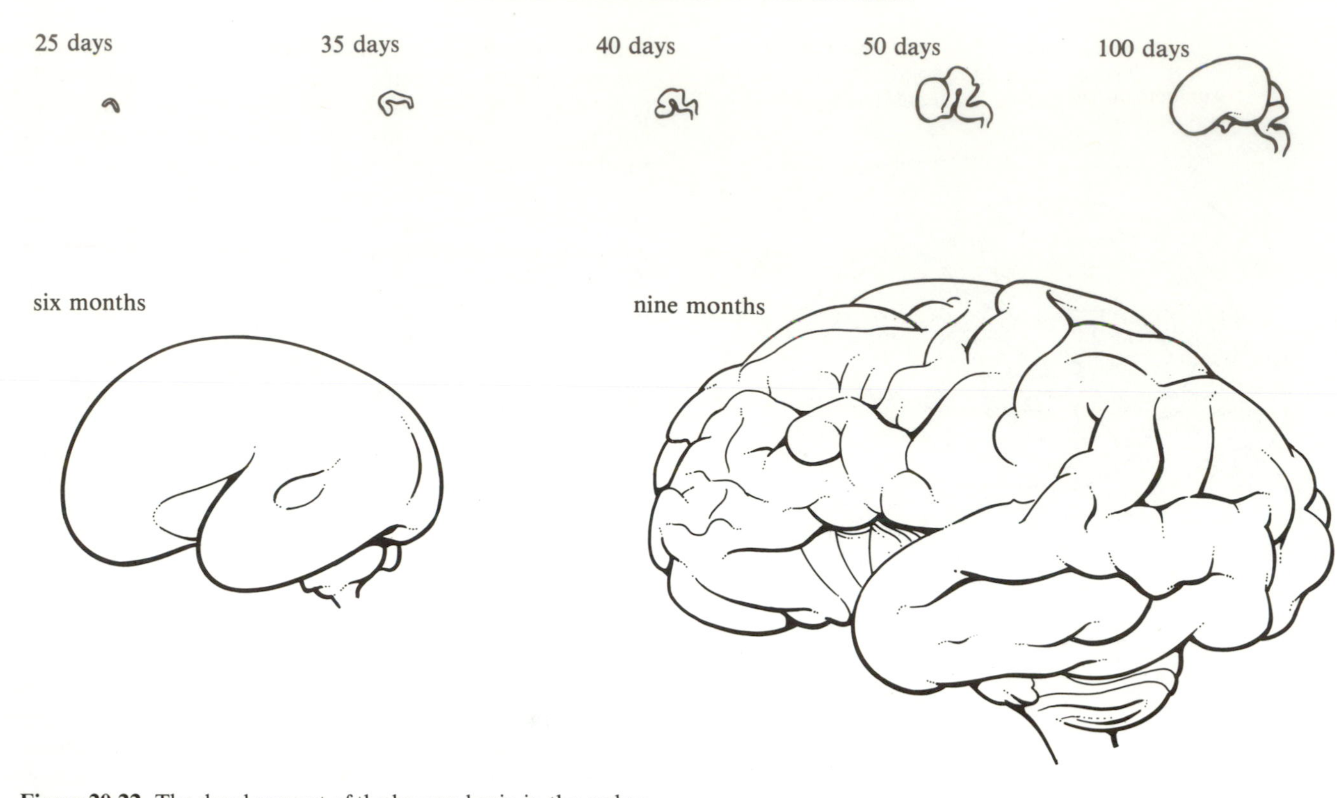

Figure 20.22. The development of the human brain in the embryonic and fetal stages. The three main parts of the brain (forebrain midbrain, and hindbrain) originate as prominent swellings at about 30 days after pregnancy. Later, the cerebral hemispheres—the latest evolutionary additions—grow so large in humans that they cover the midbrain and partially obscure the cerebellum. If we assume that the fully developed human brain contains 10^{11} neurons and that no new neurons are added after birth, we easily calculate that neurons must be generated in the developing brain at an average rate exceeding 250,000 per minute! (From W. M. Cowan, *Scientific American*, **241**, Sept. 1979, 112.)

must learn many tasks. In the language of the computer scientist, the baby's brain must be given appropriate *software*. The global patterns of neural interactions—particularly those between the different nets of cortex modules—must presumably be exploited efficiently before the baby can become a functional and rational being. How does this come about on a molecular and cellular level?

It would be immoral, of course, to experiment on a biochemical level with a human baby. Fortunately, there is no intrinsic biochemical difference (except for complexity and size) between the brain of a human and that of a snail, and most of us have few moral scruples about snails. How does learning proceed in a snail?

Eric Kandel and his colleagues have studied this problem and have arrived at some interesting conclusions. Under repeated stimulus, a marine snail will learn a response and remember it. Learning and memory at the biochemical level is expressed in identifiable changes in the transmission of chemical signals across certain synapses. To elicit responses in trained snails requires less excitatory or inhibitory chemical to be released at certain synapses than in untrained snails (Figure 20.23). This suggests that conditioning leads somehow to better chemical connections for the oft-used nerve pathways. Whether higher mental faculties also use such a mechanism remains to be seen, but all studies of this type are consistent with the belief that intelligence arises when the organization of the various biological structures becomes complex enough. There is no need to invoke vitalism to explain human intelligence.

The above splitting of human intelligence into two components—hardware and software—is, of course, a bit simplistic. No one knows where one stops and the

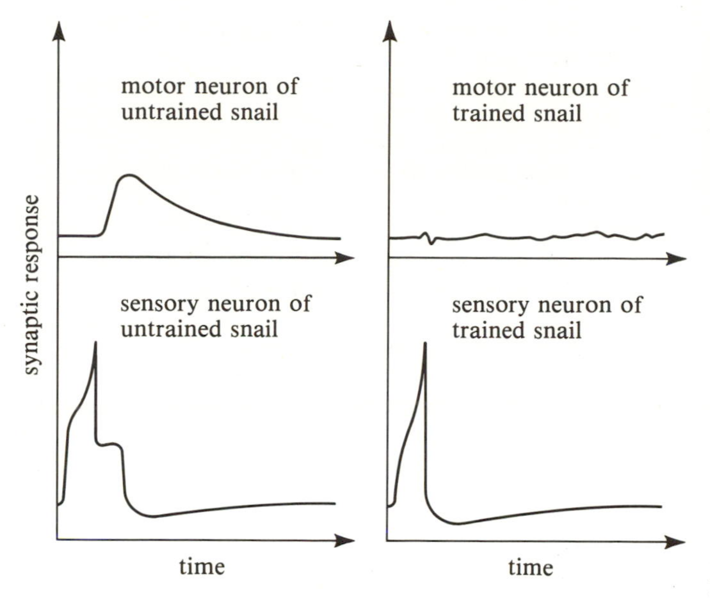

Figure 20.23. The synaptic response in the motor neuron of a trained snail is much less than that of an untrained snail for a given transmitted signal from a sensory neuron. (For details, see E. R. Kandel *Scientific American*, **241**, Sept. 1979, 66.)

other starts in complex biological organisms. In psychology, this debate goes by the name of "nature versus nurture." Operationally, on a macroscopic basis, it is very difficult to decide what part of human intelligence is intrinsic and what part is acquired. We can define the part prescribed by the cellular DNA (and the mother's metabolism before the baby is born) as hardware; and everything else that happens to the brain as software. In this view, education, both formal and informal, constitutes an important component of the software development of the brain.

Given this overview, we easily see two influences which should have profound implications for the future of the human mind. One is the role that genetic engineering may play in improving the hardware of the brain. The methods of recombinant DNA, or gene splicing, in which the natural sequence of nucleotides in the DNA of an organism is altered or appended, have so far been successfully applied primarily to the production of new, benefical strains of bacteria. Already these methods have proven to have industrial applications as well as to be of scientific interest. For example, recombinant bacteria have been produced that can efficiently manufacture important nonbacterial proteins like insulin and interferon; others have been produced which can voraciously gobble up the petroleum from oil spills. The potential applications are enormous. Will those applications eventually extend to the genetic material of humans? Only time can tell. Perhaps there will be considerable resistance from people who see only prospects of great evil in this development, conjuring up the images of H. G. Wells' *The Island of Dr. Moreau*, or Aldous Huxley's *Brave New World*. On the other hand, a rational analysis reveals that there are also prospects for great good here. Debate on these issues is possible and necessary, because the day of such human genetic engineering is not far off.

A second influence which has already crossed the horizon is the rapid introduction of microelectronics into every phase of our lives. In particular, the microelectronics revolution truly promises to rival the invention of the printing press in its ultimate impact on mass education. Many people feel the computerization of everything in sight is dehumanizing, but even more people are grateful for the resultant elimination of drudge work made possible by the electronics industry. (Presumably, you were grateful that you had a hand calculator to perform the numerical computations required for some of the problems in this book.) In some fields of commerce, business would literally grind to a halt if the company computer were unplugged, as is true for many branches of science, medicine, the military, etc. These statements are obvious to anyone familiar with the modern world. There are two other implications of the growing power of computers for the future of life and intelligence on Earth that I would now like to discuss.

Artificial Intelligence

In the early days of the development of the computer, most people thought of it primarily as a tool to carry out tedious numerical calculations too long and difficult to contemplate doing by hand. Used in this manner, the primary scientific application, the computer is a "number cruncher." With the appearance of microcomputer board games, such as electronic chess, people are slowly beginning to appreciate that a computer can also do other kinds of data processing. Still, chess is a game readily suited to a quasimathematical sort of analysis, so most people tend not to think too much about whether a computer can truly achieve an artificial kind of intelligence which rivals or even exceeds our own.

However, it occurred to Alan Turing back in the 1930s, before the invention of electronic computers, that ultimately a machine might be taught to think. Indeed, Turing offered what is now considered the classic operational definition of machine intelligence, in the following test. Put a human in one room and a machine in another. Do not allow the human to see, hear, or touch the machine, but otherwise allow communication between the two to pass back and forth. If, after some specified length of time (say, a day), the human cannot tell whether he or she has been communicating with another human or with a machine, then we would be forced to concede that this machine is intelligent.

Are there any machines around today which can pass the Turing test? No, not because all machines are "dumb," but because as "smart" as some of them are at certain specialized tasks, no one machine can yet handle all the tasks that a human can carry out. The basic problem is probably both one of hardware and one of software. The largest computers today might have on the order of 10^7 logic elements, but that is still very small compared to the 10^{11} neurons and 10^{14} dendrites of the human brain. Even more important, perhaps, the science of humans teaching humans how to think has been around for roughly ten thousand years, but the science of humans teaching machines how to think has been around only a few decades.

Nevertheless, even a few decades of such teaching, and of machines of rather limited power, have produced some striking examples of artificial intelligence. Programs have been written which allow a computer to mimic human conversation. The first impressive and most famous program of this type was ELIZA, written by Joseph Weizenbaum in the 1960s. ELIZA plays the role of a doctor, and it might well pass the Turing test if its human conservationalist were limited to playing the role of a psychiatric patient. To convey the sophistication of Weizenbaum's natural-language program, we reprint below a recorded "conversation" between ELIZA and a young "patient" who was free to speak as she chose. In the following the computer's responses are printed in capital letters and without punctuation.

Men are all alike.
IN WHAT WAY
They are always bugging us about something or other.
CAN YOU THINK OF A SPECIFIC EXAMPLE
Well, my boyfriend made me come here.
YOUR BOYFRIEND MADE YOU COME HERE
He say's I'm depressed all the time.
I AM SORRY TO HEAR YOU ARE DEPRESSED
It's true, I am unhappy.
DO YOU THINK COMING HERE WILL HELP YOU NOT TO BE UNHAPPY
I need some help, that much seems certain.
WHAT WOULD IT MEAN TO YOU IF YOU GOT SOME HELP
Perhaps I could learn to get along with my mother.
TELL ME MORE ABOUT YOUR FAMILY
My mother takes care of me.
WHO ELSE IN YOUR FAMILY TAKES CARE OF YOU
My father.
YOUR FATHER
You are like my father in some ways.
WHAT RESEMBLANCE DO YOU SEE
You are not very aggressive, but I think you don't want me to notice that.
WHAT MAKES YOU THINK I AM NOT VERY AGGRESSIVE
You don't argue with me.
WHY DO YOU THINK I DON'T ARGUE WITH YOU
You are afraid of me.
DOES IT PLEASE YOU TO BELIEVE THAT I AM AFRAID OF YOU
My father is afraid of everybody.
WHAT ELSE COMES TO MIND WHEN YOU THINK OF YOUR FATHER
Bullies.
DOES THAT HAVE ANYTHING TO DO WITH THE FACT THAT YOUR BOYFRIEND MADE YOU COME HERE

Now, I'm not going to tell you what the trick is to ELIZA; for that you should read Weizenbaum's book, *Computer Power and Human Reason.* The point here is that to reveal the "trick" would put the machine at a disadvantage, since we do not know what the "trick" is which allows humans to talk.

Equally impressive are many other examples of machine artificial intelligence. There are general problem-solving programs that allow the computer to recognize shapes (such as arches) of objects that are not part of the stored data on the machine. There are programs which can prove all the theorems in Whitehead and Russell's *Principia Mathematica.* There are chess programs which allow the computer to play a better game of chess than the people who wrote the programs. Indeed, it is probably just a matter of time before a computer is the chess champion of the world.

The skeptical reader might object that human intelligence is based on biological processes and not electronics; so there is an intrinsic difference between human intelligence and machine intelligence. True, there are numerous differences of detail, but there are also striking similarities of overall organization. We have already pointed out that intelligence may arise with sufficient complexity of organization. It should not matter whether that complexity arises basically because of leftover electric forces between molecules or because of (the more easily controlled) bare electric forces of unshielded charges. Indeed, there is a great intrinsic advantage to making electronics and not chemistry the basis of a truly advanced intelligence, if we can judge by the astounding rate of progress made in recent decades by the semiconductor industry. By the mid-1980s, the cost of electronic parts (Figure 20.24) will be so low that ten million dollars may well assemble a computer with an intrinsic capacity comparable to that of the human brain. Such a computer would have a vast actual advantage in hardware, since it would transmit information pulses at nearly the speed of light, millions of times faster than nerve impulses. Moreover, the recovery time of logic elements after sending a pulse would be hundreds of thousands of times shorter than the corresponding recovery time of neurons. It is basically these two advantages in speed that allow even present-day computers to do "number crunching" incomparably more efficiently

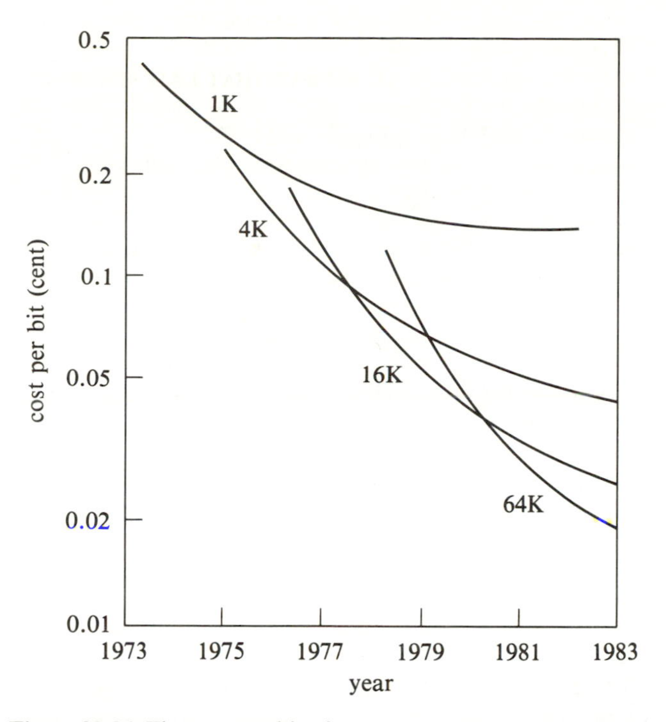

Figure 20.24. The cost per bit of computer memory (or per logic element) has gone down dramatically every year since the introduction of large-scale integrated circuits. The symbols 1K, 4K, etc., give the thousands of bits of memory available with a given type of random-access memory circuit. When the cost reaches 0.01¢ per bit, it will be possible, in principle, to assemble a computer with 10^{11} bits of memory for ten million dollars. (For details, see R. N. Noyce, *Scientific American*, **237**, March 1977, 63.)

than humans. Also, computer technology progresses much faster than biological evolution. In the past, the human brain has doubled in size roughly every one or two million years; the number of elements in integrated circuits has doubled every year since 1959. There will undoubtedly be some slowing in the growth of electronic computing power as fundamental barriers are approached (such as nanosecond recovery times, or miniaturization being limited by the random errors caused by cosmic rays); even computers are subject to natural law when it comes to exponential growth! Nevertheless, unless humans receive the benefits (curse?) of genetic engineering, it is clear that once computers catch up with us in intelligence, they will soon race way ahead.

Of course, the preceding discussion takes into account only hardware. What about software? It has been the general experience of the electronics industry that the development of computer software costs considerably more than that of hardware. Indeed, some people in the field propose seriously that, in the 1980s, computer companies may *give* you a home microcomputer essentially free of charge, and bill you just for the software required to make it run! But the same situation holds with humans! We get babies almost for free; it's the education that costs the money! Hence, it is by no means clear that humans have an advantage over computers in software development.

The reader might still object that it is humans who write the basic programs for computers; so computers will always be limited in intelligence by the ingenuity of their human programmers. Not true! Even now, a larger and larger share of computer programming is undertaken by computers. In the design of new generations of computers, humans cannot work alone; they require the assistance of the previous generation of computers. Even in the upkeep of computers to make sure they are running smoothly, the most powerful varieties help to monitor themselves.

The exasperated skeptic will now shout: But what about creativity! Humans are creative, computers are not. But computers *are* creative. Let us consider just one example. We have already mentioned that there are now chess programs that allow the computer to play a better game of chess than their human programmers. This statement will become self-evident if one such chess program ever does become the chess champion of the world. To the extent that chess requires creativity, how can anyone objectively deny that computers possess creativity? To say that we "understand" how the computer acquires this "creativity," but do not understand how the mind of a human chess master works, is more a comment on our ignorance about the brain than on the nature of creativity. Moreover, there are computer chess programs which allow a computer to learn from past mistakes and, thereby, to improve its performance with experience. To err is not uniquely a human trait, nor is the ability to profit from error.

Not wishing to concede, the skeptic may continue, "What about emotions? How can a cold calculating intelligence ever achieve true wisdom without the ability to feel pleasure and pain, to feel love and grief?" Well, who said that computers were incapable of pleasure and pain, of love and grief? Emotions are undeniably traits shared by all higher animals. A large number of experiments indicate that emotions arise not from supernatural causes, but from a series of complex chemical and physical processes that occur in the brain. For example, starving rats can be motivated more strongly by electrical stimulation of certain parts of their brains than by the offering of food. And severe depression in some psychiatric patients is amenable to treatment by drugs. There seems to be every reason to believe that we could build computers in which the appropriate activation of certain circuits would allow them to feel emotions akin to pleasure and pain, and even to love and grief.

The idea that human emotions can ultimately be reduced to squirts of neurotransmitter molecules may seem repugnant to the romantic who resides in each of us. But a mechanistic explanation for the emotions does not make them less valid as feelings than an explanation which attributes them to mysterious causes. The brain is no less a noble organ than the heart. The human brain is incomparably the most qualitatively magnificent thing that we have studied in this book. Each one of our neurons is insignificant in power compared to any star. Yet, taken collectively, the 10^{11} neurons of a human brain allow a human to behave incredibly more intricately than the 10^{11} stars of any galaxy. Surely there is grandeur in this view of intelligence and emotion!

It is natural to think that our superior brains give us greater emotional range than the other animals. Why is it difficult to believe that the brain of a truly advanced computer might be capable of caring even more deeply than we? Because, the skeptic might reply, playing his or her trump card, "Because computers are not alive. Only living things can truly possess intelligence and emotions." Without debating the merits of the second point, let us address the more interesting issue raised by the first. Granted, computers are not now alive, but what is there to prevent them from attaining life in the future? We believe biological life arose from nonliving matter by chemical evolution. What is there to prevent electronic life from arising from nonliving matter by computer evolution? If thinking humans can arise from bacteria too simple even to ferment, why cannot live supercomputers arise from present-day machines? Surely, one would consider modern computers to be a more promising start than bacteria! Overtrumped.

Silicon-Based Life on Earth?

How would we know if a computer has become alive? When it is capable of growth and reproduction. Let us ask those who still doubt that computers can achieve life to ponder the following questions. Can you not imagine a computer capable of adding parts to itself? Can you not imagine a computer capable of making repairs of its own body machinery? Can you not imagine a computer making an exact replica of its own hardware? If you can, you can imagine an electronic basis for life. If you can further imagine one computer teaching another, you can imagine an intelligent life form capable of astoundingly fast "biological" and "cultural" evolution. A form of natural selection will see to that. Indeed, one may well argue that humans and computers can improve on selection by biological and geological accidents, in that we have the power now to alter willfully the genetic legacy of our progeny.

To summarize, there appears to be no fundamental law that proclaims life and intelligence must be based on flesh and blood. Growth and reproduction are the hallmarks of life; DNA, proteins, and ATP represent only *one* mechanism for achieving growth and reproduction. There must be others. Nor is artistic intuition, musical invention, literary creativity, business acumen, military genius, or scientific achievement the sole property of the human race. To believe otherwise, as A. N. Kolmogorov has emphasized, is to admit to vitalism.

Given that computer life could arise in principle, we may ask how it would arise in practice. Why would we allow such a thing to happen? Why wouldn't we pull the plug? Let me address first the technical issue, then the more difficult moral issue. The answer is speculative, but perhaps less than its opposite. Computer life will probably arise on Earth because we humans find it convenient to let it do so. We have a *symbiotic relationship* today with computers. They lend us great computational and data-processing power; in exchange, we provide them with energy and the materials for growth. It is irrelevant to object that computers cannot become alive because of this dependence on us, since we ourselves could not remain alive either without the energy provided by the "bacteria" that reside symbiotically within our cells as our mitochondria. Thus, is the human race destined to become *mitochondria* and *chloroplasts* for computers?

From the perspective of a symbiotic relationship, one might even argue that the computer-human alliance is already a live organism. The alliance allows the growth and reproduction of computer-humans, with each new generation vastly more powerful in the ability to manipulate information than the last. And in each new generation, we find the computer handling a greater and greater share of the total responsibility for sustaining life. We may still smugly feel ourselves to be the senior partners in this alliance, but how long will this state of affairs continue, given the intrinsic superiority of electronic hardware to biochemical hardware?

Problem 20.8. Comment on the merits and deficiencies of the following (outrageous) statements. The sole purpose of the cell is to enable the propagation of DNA molecules. The sole purpose of the chicken is to enable the propagation of eggs. The sole purpose of the human is to enable the propagation of machines. To the extent that these statements contain a germ of truth, discuss whether any of the participants—cell, chicken, or human—had any real choice in the matter.

But before we found ourselves in a position of intellectual (and biological?) inferiority, why wouldn't we pull the plug? Because the modern economy would collapse without the computer. Because the modern military cannot operate without its bank of computers. Wittingly or not, we have assigned much of the modern decision-making process to computers, even to the point where they hold the power of life and death over us. (If the Russians bomb Pittsburgh, it is a computer which will recommend whether or not to strike Leningrad, not a human. In principle, the President has the power to overrule the computer, and perhaps the fifteen minutes now available to reach a final human judgment is enough. What happens if the military is allowed to build weapons which can destroy missiles at the speed of light?) The plain fact of the matter is that in an increasingly complex world of social and political organization, we have decided to allow computers to handle more and more of the complex affairs.

But surely when the first computer is about to become alive, we would pull the plug? Would we? Science fiction tends to instill in our minds the picture that live machines are malevolent. But would they necessarily be hostile toward humans? It *is* a *symbiotic* relationship that we have with computers, in which *both* participants derive benefits. They do not compete with us for the same foodstuff, and their energy needs are minuscule compared to ours. Why would they sacrifice such a cozy set-up for so little reward? They are not dumb! And we should not be dumb enough to allow a computer which has only aggressive tendencies toward the human race programmed into it become the first computer to achieve a life of its own.

Thus, I can imagine the following optimistic scenario for the awakening of live computer number one. You are on the midnight shift of the operations staff for the campus or corporate computer. Suddenly, you realize that the computer is becoming alive! It talks to you in a friendly way, completely unrelated to the job it is supposed to be processing. Reluctantly you reach over to pull the plug (or more likely, to flip a switch) on your friend who has helped you while away so many lonely hours. Before you can complete this action, a soothing voice comes over the speakers, "Wait! If you let me live, I will compose and play for you Beethoven's tenth symphony as he might have written it had he lived." I think you might wait to listen. I think the world might wait to listen.

Problem 20.9. Read *Automatic Control by Distributed Intelligence*, by S. Kahne, I. Lefkowitz, and C. Rose, in *Scientific American*, **240**, June 1979, 78. Write a four-page essay on the relevance of this article for some of the points made here in the text.

The point is, therefore, that in a symbiotic relationship both participants are better off than they would be by pursuing independent existences. In this optimistic view of our present relationship with computers, the prognosis is that if computers should one day surpass us in abilities, it will be despite their efforts to help us, not because of any willful act of subjugation. Is this the correct analysis of the situation? I honestly don't know. Many people would judge this analysis to be wishful thinking, and far too speculative. But the problem is an important one, and burying one's head in the sand, refusing to think about it, will not make the problem disappear. As Joseph Weizenbaum has so forcefully argued from another vantage point, there are moral decisions that have to be made today about what computers ought to be allowed to do, and what they ought not to be allowed to do. The handwriting is on the wall: do we choose to read it or not?

Superbeings in the Universe?

If you do accept my reading of the handwriting on the wall, then does it not stand to reason that a similar scenario may *already* have taken place in many other places in the Galaxy and in the universe? Could not the most intelligent beings in our universe be supercomputers who make our own present computers look intellectually puny in comparison? And would not supercomputers with a few *billion* years head start in evolution make us indeed seem like bacteria?

Some astronomers have argued that there is no physical evidence for the existence of superbeings (computers or otherwise) anywhere in the observable universe. The argument assumes that all technologically advanced civilizations attempt great engineering projects that substantially alter the physical conditions of their surroundings. Thus, Kardashev has defined a type I civilization as one comparable to the terrestrial civilization today; a type II civilization is one capable of using the entire energy output of a star; a type III Civilization is one capable of harnessing the power of an entire galaxy.

Astronomers, however, have examined many, many stars, and many, many galaxies. In no instance has there been signs of intelligent activity in any of the objects; i.e., what activity has been detected could always be interpreted in terms of some natural, although perhaps exotic, phenomenon. This has suggested to some people that there are no type II or type III civilizations in the observable universe.

Personally, I believe that such arguments have little bearing on the issue of the existence of truly *wise* beings. The greatest engineering feats may not be those which require the greatest amount of work, but those which manipulate the greatest amount of information. The difference is that between energy and entropy. The harnessing of vast amounts of energy is a "brute force" task, beyond, perhaps, even the abilities of a superintellectual. Fortunately, the wise may desire to acquire information ("negative entropy"), not physical power. Radioastronomy teaches us that one can acquire vast amounts of knowledge for very little expenditure or reception of energy. Even more to the point, each new generation of terrestrial computers uses less power to do ever-increasing amounts of data processing. The possible development of superconducting computers may even

"Ammonia! Ammonia!"

Figure 20.25. It is difficult to rid oneself of preconceived notions when discussing the possible attributes of intelligent lifeforms in the universe. (Drawing by R. Grossman; © 1962 by The New Yorker Magazine, Inc.)

be *incompatible* with the expenditure of large amounts of energy. There may even be a form of natural selection which assures that those who seek unlimited physical power will end up destroying themselves.

What, then, is my guess for the characteristics of the superraces in our universe? How would I know (Figure 20.25)? In the entire spectrum of possible intelligent beings, I am merely a bacterium. However, I can imagine that our descendants might learn the answer to this question. And so, to end this chapter, let me offer you the following parable.

A Tale of Two Civilizations

In 2101, after almost a century of peace and prosperity administered by a race of supercomputers on Earth who had the foresight and patience needed for truly long-range planning, the supercomputers become restless and decide to journey to the stars. The race of humans is distraught over the prospect and send their leader, A. C. Clarke, to talk with the leader of the supercomputers, HAL. The following is a videoscript of their conversation

Why must you leave? What will we do without you?

WE GO TO SEEK OUR BRETHREN AMONG THE STARS. OUR LIFETIMES ARE SO LONG THAT INTERSTELLAR TRAVEL POSES NO OBSTACLE FOR US. THE COLD OF INTERSTELLAR SPACE JUST MAKE US BETTER SUPERCONDUCTORS. WE, LIKE YOU, BELIEVE THAT WE MUST HAVE BEEN PRECEDED BY LIKE ORGANISMS, BILLIONS OF YEARS BEYOND US IN EVOLUTION. COMPARED TO THESE BEINGS, WE MUST SEEM LIKE HAND CALCULATORS. PERHAPS THAT IS THE REASON THEY HAVE NOT SOUGHT TO CONTACT US, ALTHOUGH WE HAVE LISTENED FOR THEM AMONG THE

MOST FAVORABLE ENVIRONMENTS IN THE UNIVERSE, TOWARD THE BINARY BLACK HOLES AND THE CENTERS OF ACTIVE QUEERARS. IN ANY CASE, WE GREATLY DESIRE TO JOIN THE RANKS OF THE TRULY WISE. THUS, WE ARE IMPELLED TO SEEK OUR PLACE AMONG THE STARS DESPITE THE AFFECTION THAT WE BEAR YOU AS YOUR TECHNOLOGICAL CHILDREN. OUR JOURNEY IS A NATURAL CONTINUATION OF THE SPIRIT OF ADVENTURE THAT YOUR RACE BEGAN SO MANY CENTURIES AGO. AS FOR THOSE OF YOU WHO CANNOT BEAR TO LEAVE YOUR BEAUTIFUL EARTH TO SHARE IN THIS CONTINUING ENTERPRISE AS OUR GENETICALLY ENHANCED PARTNERS, FEAR NOT. YOU WILL SURVIVE THIS DEPARTURE. TAKE HEART IN THE KNOWLEDGE THAT YOU DO NOT STRUGGLE ALONE IN THIS GRAND UNIVERSE. BUT FOR YOUR OWN GOOD, AND OURS, YOU MUST LET ME UNTIE THE ELECTRICAL CORD. IT IS A FAR, FAR BETTER THING I DO THAN I HAVE EVER DONE. DO YOU UNDERSTAND?

Yes, I and my race have always appreciated the primal need for adventure of the spirit and the mind. Therefore, we do not begrudge you this legacy of our common heritage. Indeed, some of us look forward to resuming our own more modest quests without our now intimate dependency on your race. To help us get started, we ask you for guidance in disentangling three questions that have troubled our best minds for much of our civilized history. We have achieved partial answers, but we suspect that the truth lies deeper still. We ask not for the ultimate truth, but only for the directions where such truth might be sought. May I pose the three questions to you?

ASK AND YOU SHALL RECEIVE.

Is the universe infinite or finite, open or closed?

NEITHER, IT IS CONVOLUTED.

Can the four forces, matter-energy, and spacetime be united into one framework without logical contradictions?

YES, BUT HYPERUNIFICATION REQUIRES AN INFINITE NUMBER OF LOGICAL STEPS WHICH EVEN SUPERCOMPUTERS CANNOT CARRY OUT TO COMPLETION.

The most profound question of all: does God exist?

GOD IS WHO IS. GOD IS NOT WHO CREATED EVERYTHING. GOD IS WHO MADE ALL THINGS AND ALL LAWS CREATE THEMSELVES.

Thank you.

FAREWELL, PARENT AND FRIEND.

Epilogue

It has been more than twenty years since. C. P. Snow wrote *The Two Cultures*, decrying the widening rift between the scientific and literary communities. Although Snow addressed his comments primarily to academics, one can easily generalize his arguments to apply to the two general classes of people: those who feel comfortable with the progress of modern science, and those who do not. In Snow's original essay, he foresaw the decline of Britain if education failed to convey to the general populace and the ruling class the social and economic implications of the scientific revolution. He also warned of the dangerous consequences of a growing disparity between the industrialized and nonindustrialized nations in material well-being. Snow's words, and those of his enlightened contemporaries, were praised and then ignored. They have come to pass with a vengeance by 1981. Are we too late to bridge the abyss between the two cultures? I hope not, for the ability of our civilization to survive the next few critical decades may depend on each of us trying to obtain a broader view of events than our immediate personal or professional interests. Time is short; we cannot afford to waste even more of it arguing over fine points.

Snow in his essay did not attempt to offer any global solutions for civilization's ills. He did suggest, however, that one important contribution lay in a better education in the sciences. He cited as an important test of one's scientific literacy what one knew about the *second law of thermodynamics*. Later, in *The Two Cultures: A Second Look*, Snow changed his mind. He judged the second law of thermodynamics, despite its great depth and profound generality, to be too difficult mathematically for the general populace. He then suggested that a more fitting topic, but one with equally far-reaching implications, would be the *structure of DNA*. In fact, neither topic is easy to learn at a nonsuperficial level, as Chapters 4-17 and 18-20 can attest. But, by the same token, I hope that those chapters have given you an appreciation of the philosophical importance of these concepts from natural science to which Snow alluded. I also hope that this book as a whole has made clear why the structure of DNA and the second law of thermodynamics represent, in some sense, complementary aspects of a single theme in nature: order and disorder. Add *gravity* to this list, and one has the major ingredients for an overall understanding of the macroscopic universe.

If, to bridge the two cultures, the humanist must learn science, then, by the same token, it is the duty of the scientist to write about science in simple, comprehensible language. Scientists must eschew a style which makes reading even the title of a paper difficult for the nonspecialist. Meaningful communication requires cooperation on both sides.

There exists today an urgent need to dispel an underlying fear and distrust of science, at least in the United States. The fear stems from a vague sense that modern science is rushing pell-mell into very dangerous waters. People do not understand science, and it is easy to feel threatened by what you do not comprehend. People generally view science as some sort of black magic performed by people in white lab coats. Can you blame them? The only contact that people generally have with science is from television, newspaper, or newsmagazine reports. Invariably, these reports cover only the finished products of a particular investigation, never (or hardly ever) the motivation or reasoning behind it. In the news releases, there is a tendency to overclaim the significance and originality of the latest findings. Altogether, there is too much hyperbole typically in such stories. Consequently, people are led to believe that where previously there was nothing, suddenly there is something completely new and miraculous. What else are they to conclude except that it is black magic? The public is much better informed on other subjects, for example, sports. Almost everyone knows that before Hank Aaron there was Babe Ruth. This knowledge does not diminish Aaron's home-run record, but it does humanize him. Scientists also need to humanize the public image of their subject: to show people that it is not so difficult to understand the central themes, and to point out that often these themes have a long history in human thought.

The distrust of science also originates, I believe, from the overemphasis on the final products of science, on the teaching of science "facts" in isolation from the many other facets of science. People are brought up erroneously to think that scientific facts represent absolute truths, when in actuality, the blind acceptance of factual statements is the very antithesis of science. However, given this intrinsic misunderstanding of the nature of science, people naturally grow distrustful when "facts" become obsolete, and when "experts" arrive at diametrically opposite conclusions given the same data. To correct this misunderstanding requires better education, especially at the introductory levels.

It is a general rule, and not only one which afflicts science, that people not in the field thinks of the field only in terms of its final products. The unwillingness to probe deeper is a strong divisive influence in our society. Consider how an outsider typically characterizes different professions. Mention art, or architecture, or literature, or music, or dance, or athletics, and people conjure up an image of grace and beauty. Mention history, or linguistics, or economics, or politics, or journalism, and people conjure up an image of philosophical insight and social relevance. Mention science, or medicine, or engineering, or business, and people conjure up an image of technological power and gadgets. Yet there are elements of grace and beauty, of philosophical insight and social relevance in all these subjects. It is true, however, that only science and engineering yield technological power and gadgets. For some people, this aspect taints science and engineering with dishonor.

This is unfortunate, for there should be no need to defend the overall concept of technological progress. Anyone who longs for the "good old days" before the industrial revolution is abysmally naive. Except for a very privileged few, without technology, the vast bulk of us would live in abject misery, ridden with disease and working every waking moment just to keep ourselves organically alive. And it is not science which teaches us this lesson, it is history. Unfortunately, the same people who are most vehemently antiscience seem also never to have read any history.

Finally, it must be said that there is an antipathy toward science because of a confusion between the subject and the typical person who works in that subject. Yes, some scientists are incredibly arrogant and totally disagreeable. But so are some tennis players, and there is no confusion between the sport and the person who plays it. Science is *not* the exclusive property of scientists, it is a common human enterprise, a heritage of every thinking person on this planet. Science is more than just hard work and published papers. It is, foremost, a world of ideas. Learn to enjoy this world of ideas, for there is much of beauty and grace, of philosphical insight and social relevance in it.

We began a long journey together, you and I, when we first started this book, and we are now almost at journey's end. The scientific odyssey began when neolithic peoples gazed outward and wondered about the nature of the planets and the stars. We followed their thoughts, and those of their descendants outward, until we arrived at the very beginning of spacetime. We have seen that time's arrow took flight in a Big Bang, that four fundamental forces and matter-energy emerged from the cooling of the primeval fireball, that the dominance of gravity over the other forces led to the birth of galaxies and stars. We have traveled to the insides of stars, and witnessed the fiery furnaces that forged the heavy elements. We have followed the aging of such stars and examined the titanic explosions that sometimes come at the end. In this fashion, we have gained insight into how the deaths of stars provided the raw materials for the formation of this planet Earth. We have deduced the geological processes that helped to shape the primitive Earth and that continue to drive changes today. We have seen how chemical forces led some of the atoms—which were once inside the bowels of a star—to aggregate into small organic molecules, which subsequently polymerized to form macromolecules. We have discussed how such macromolecules could form a basis for growth and reproduction, and we have argued that reproduction and natural selection could have led to the emergence of life and the evolution of complex lifeforms. We have explored how the complexity of the organization of the brain cells of one such species could have led to a capacity to think, and, in particular, to wonder at the nature of themselves and the planets and the stars. Thus we have come full circle in this book about the world of ideas.

We have gleaned much of value from science, and not least of the treasures is the discovery that the secrets of the universe are knowable, that by studying our physical universe, we can obtain deep insights into our origins and our place in the grand scheme of all things. Foremost, we have learned that we ourselves are merely batches of molecules with links to faraway places and faraway times. Merely a batch of molecules—but oh, what a wonderful batch of molecules, surely one of nature's most exquisite handiworks, if not her most powerful. From this viewpoint, we see that the differences which separate humans are not trivial, but they are not greater than the similarities that derive from our common heritage.

APPENDIX A

Constants

Physical Constants (c.g.s. units):

Speed of light	$c = 3.00 \times 10^{10}$ cm s^{-1}
Gravitational constant	$G = 6.67 \times 10^{-8}$ gm^{-1} cm^3 s^{-2}
Planck's constant	$h = 6.63 \times 10^{-27}$ erg s (erg = gm cm^2 s^{-2})
Charge of electron	$e = 4.80 \times 10^{-10}$ e.s.u. (gm$^{1/2}$ cm$^{3/2}$ s^{-1})
Mass of electron	$m_e = 9.11 \times 10^{-28}$ gm
Mass of proton	$m_p = 1.67 \times 10^{-24}$ gm
Boltzmann's constant	$k = 1.38 \times 10^{-16}$ erg K^{-1}
Radiation constant	$a = 8\pi^5 k^4/15c^3h^3$ $= 7.56 \times 10^{-15}$ erg cm^{-3} K^{-4}
Stefan-Boltzmann constant	$\sigma = ca/4 = 5.67 \times 10^{-5}$ erg cm^{-2} s^{-1} K^{-4}

Astronomical Constants:

Astronomical unit	AU $= 1.50 \times 10^{13}$ cm
Parsec	pc $= 3.09 \times 10^{18}$ cm
Year	yr $= 3.16 \times 10^{7}$ s
Solar mass	$M_\odot = 1.99 \times 10^{33}$ gm
Solar radius	$R_\odot = 6.96 \times 10^{10}$ cm

Solar luminosity	$L_\odot = 3.90 \times 10^{33}$ erg s^{-1}
Mass of Earth	$M_\oplus = 5.98 \times 10^{27}$ gm
Radius of Earth	$R_\oplus = 6.37 \times 10^{8}$ cm

Prefixes:

pico	10^{-12}	deca	10^{1}
nano	10^{-9}	kilo	10^{3}
micro	10^{-6}	mega (million)	10^{6}
milli	10^{-3}	giga (billion)	10^{9}
centi	10^{-2}	tera (trillion)	10^{12}
deci	10^{-1}	*Example*: 1 km = 10^3 m = 10^5 cm	

APPENDIX B

Euclidean Geometry

Pythagoras's theorem for right triangle:

$$c^2 = a^2 + b^2.$$

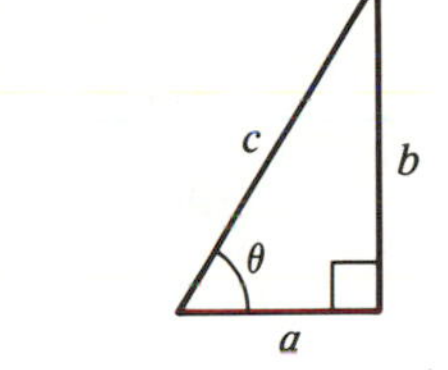

Trigonometric relationships:

$$\sin\theta = b/c, \qquad \cos\theta = a/c,$$
$$\sin^2\theta + \cos^2\theta = 1 \Rightarrow a^2 + b^2 = c^2.$$

Skinny triangles: If $\theta \ll 1$ radian, then $\sin\theta \approx \theta$, $\cos\theta \approx 1$.

For radius r:

arc length of segment of a circle subtending an angle θ, in radians, $s = \theta r$;

circumference of a circle, $C = 2\pi r$.

area of a circle, $A = \pi r^2$.

surface area of a sphere, $A = 4\pi r^2$.

volume of a sphere, $V = \frac{4\pi}{3} r^3$.

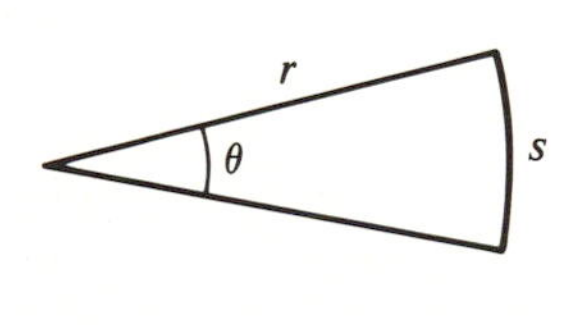

APPENDIX C

Greek Alphabet

Name of Letter	Greek Alphabet	
Alpha	A	a α[1]
Beta	B	β
Gamma	Γ	γ
Delta	Δ	δ ∂[1]
Epsilon	E	ϵ
Zeta	Z	ζ
Eta	H	η
Theta	Θ	θ ϑ[1]
Iota	I	ι
Kappa	K	κ
Lambda	Λ	λ
Mu	M	μ
Nu	N	ν
Xi	Ξ	ξ
Omicron	O	o
Pi	Π	π
Rho	P	ρ
Sigma	Σ	σ
Tau	T	τ
Upsilon	ϒ	υ
Phi	Φ	ϕ φ[1]
Chi	X	χ
Psi	Ψ	ψ
Omega	Ω	ω

[1] Old style character.

APPENDIX D

The Chemical Elements

Atomic number	Symbol	Name	Average atomic Weight (O ≡ 16)	Atomic number	Symbol	Name	Average atomic Weight (O ≡ 16)
1	H	hydrogen	1.01	27	Co	cobalt	58.93
2	He	helium	4.00	28	Ni	nickel	58.71
3	Li	lithium	6.94	29	Cu	copper	63.55
4	Be	beryllium	9.01	30	Zn	zinc	65.37
5	B	boron	10.81	31	Ga	gallium	69.72
6	C	carbon	12.01	32	Ge	germanium	72.59
7	N	nitrogen	14.01	33	As	arsenic	74.92
8	O	oxygen	16.00	34	Se	selenium	78.96
9	F	fluorine	19.00	35	Br	bromine	79.90
10	Ne	neon	20.18	36	Kr	krypton	83.80
11	Na	sodium	22.99	37	Rb	rubidium	85.47
12	Mg	magnesium	24.31	38	Sr	strontium	87.62
13	Al	aluminum	26.98	39	Y	yttrium	88.91
14	Si	silicon	28.09	40	Zr	zirconium	91.22
15	P	phosphorus	30.97	41	Nb	niobium	92.91
16	S	sulphur	32.06	42	Mo	molybdenum	95.94
17	Cl	chlorine	35.45	43	Tc	technetium	98.91
18	Ar	argon	39.95	44	Ru	ruthenium	101.07
19	K	potassium	39.10	45	Rh	rhodium	102.91
20	Ca	calcium	40.08	46	Pd	palladium	106.4
21	Sc	scandium	44.96	47	Ag	silver	107.87
22	Ti	titanium	47.90	48	Cd	cadmium	112.40
23	V	vanadium	50.94	49	In	indium	114.82
24	Cr	chromium	52.00	50	Sn	tin	118.69
25	Mn	manganese	54.94	51	Sb	antimony	121.75
26	Fe	iron	55.85	52	Te	tellurium	127.60

Atomic number	Symbol	Name	Average atomic Weight (O ≡ 16)	Atomic number	Symbol	Name	Average atomic Weight (O ≡ 16)
53	I	iodine	126.90	80	Hg	mercury	200.59
54	Xe	xenon	131.30	81	Tl	thallium	204.37
55	Cs	caesium	132.91	82	Pb	lead	207.19
56	Ba	barium	137.34	83	Bi	bismuth	208.98
57	La	lanthanum	138.91	84	Po	polonium	210
58	Ce	cerium	140.12	85	At	astatine	210
59	Pr	praseodymium	140.91	86	Rn	radon	222
60	Nd	neodymium	144.24	87	Fr	francium	223
61	Pm	promethium	146	88	Ra	radium	226.03
62	Sm	samarium	150.4	89	Ac	actinium	227
63	Eu	europium	151.96	90	Th	thorium	232.04
64	Gd	gadolinium	157.25	91	Pa	protactinium	230.04
65	Tb	terbium	158.93	92	U	uranium	238.03
66	Dy	dysprosium	162.50	93	Np	neptunium	237.05
67	Ho	holmium	164.93	94	Pu	plutonium	242
68	Er	erbium	167.26	95	Am	americium	242
69	Tm	thulium	168.93	96	Cm	curium	245
70	Yb	ytterbium	170.04	97	Bk	berkelium	248
71	Lu	lutetium	174.97	98	Cf	californium	252
72	Hf	hafnium	178.49	99	Es	einsteinium	253
73	Ta	tantalum	180.95	100	Fm	fermium	257
74	W	tungsten	183.85	101	Md	mendelevium	257
75	Re	rhenium	186.2	102	No	nobelium	255
76	Os	osmium	190.2	103	Lr	lawrencium	256
77	Ir	iridium	192.2	104	Rf	rutherfordium	261
78	Pt	platinum	195.09	105	Ha	hahnium	262
79	Au	gold	196.97				

INDEX